Study Guide for

Organic Chemistry
Structure and Reactivity

Study Guide for

Organic Chemistry

Structure and Reactivity

Fifth Edition

Seyhan N. Eğe
University of Michigan

Roberta W. Kleinman
Lock Haven University

Peggy Zitek
University of Michigan

BROOKS/COLE
CENGAGE Learning

Australia • Brazil • Japan • Korea • Mexico • Singapore • Spain • United Kingdom • United States

BROOKS/COLE
CENGAGE Learning™

Study Guide for Organic Chemistry: Structure and Reactivity, Fifth Edition
Seyhan N. Eğe, Roberta W. Kleinman, Peggy Zitek

Vice President and Publisher: Charles Hartford

Executive Editor: Richard Stratton

Development Editor: Rita Lombard

Editorial Associate: Rosemary Mack

Editorial Assistant: Lisa C. Sullivan

Senior Manufacturing Coordinator: Priscilla Bailey

Senior Marketing Manager: Katherine Greig

Marketing Associate: Alexandra Shaw

Cover image: Cover image of a single-walled carbon nanotube wrapped in an amylose coil, based on research by Professor J. Fraser Stoddart (A. Star, D. W. Steurman, J. R. Heath, and J. F. Stoddart, *Anglewandte Chemie, International Edition 2002*, 41, 2508–2512).

About the cover: Single-walled carbon nanotubes consist of small test-tube like structure about 1 nm in diameter and some micrometers in length made up essentially of rolled up sheets of graphene, the carbon structure of graphite (Figure 10.7, p. 365 in the text). They have high strength and elasticity and thermal and electric conductivities. Much research is being done on applications for these new materials.

One area that has been difficult to examine is possible biological applications of these nanotubes because they are not soluble in water. Professor J. Fraser Stoddart of the Department of Chemistry and Biochemistry at the University of California, Los Angeles, and his coworkers discovered that the rod-shaped nanotubes could be made soluble in water if they were wrapped in the coiled structures that a fraction of starch known as amylose assumes when it complexes with molecules of iodine to give the familiar starch-iodine blue color. Figure 23.7, page 1004 in the text, shows how the carbon nanotube displaces iodine molecules from the coiled amylose. The interior of the amylose coil is hydrophobic enough to provide a hospitable environment for the carbon rod. The exterior of the coil hydrogen bonds to the surrounding water molecules to bring the whole complex into solution. The image on the cover of this book shows a single-walled carbon nanotube (the blue rod showing the six membered rings characteristic of graphite) wrapped in a coil of amylose (shown in red, violet and silver).

Once the nanotubes are made water soluble by wrapping in amylose, they dissolve to the extent of 3 grams per liter. The nanotubes in water solution can be separated from other carbon materials that are the side products of the production of the nanotubes, but that do not have the shapes that allow them also to complex with amylose. The water solutions of the nanotubes are quite stable as long as nothing is done to destroy the amylose coil. Our saliva contains *x*-amylose, an enzyme that digests starch. Professor Stoddart reports that the nanotube solutions "are stable for weeks provided you do not spit on them!" Addition of saliva (or solutions of starch-splitting enzymes) precipitate the nanotubes as the enzyme breaks amylose down into smaller and smaller carbohydrate fragments, finally resulting in the formation of glucose. Once the three-dimensional structure of amylose is destroyed, it can no longer surround and solubilize the nanotubes.

Mass spectra adapted from *Registry of Mass Spectral Data*, Vol. 1, by E. Stenhagen, S. Abrahamsson, and F. W. McLafferty. Copyright © 1974 by John Wiley & Sons, Inc. Reprinted by permission of John Wiley & Sons, Inc.

For product information and technology assistance, contact us at **Cengage Learning Customer & Sales Support, 1-800-354-9706**.

For permission to use material from this text or product, submit all requests online at **www.cengage.com/permissions**. Further permissions questions can be emailed to **permissionrequest@cengage.com**.

ISBN-13: 978-0-618-31810-0
ISBN-10: 0-618-31810-0

Brooks/Cole
20 Davis Drive
Belmont, CA 94002-3098
USA

Cengage Learning is a leading provider of customized learning solutions with office locations around the globe, including Singapore, the United Kingdom, Australia, Mexico, Brazil, and Japan. Locate your local office at: **www.cengage.com/global**.

Cengage Learning products are represented in Canada by Nelson Education, Ltd.

To learn more about Brooks/Cole, visit **www.cengage.com/brookscole**.

Purchase any of our products at your local college store or at our preferred online store **www.cengagebrain.com**.

Printed in the United States of America
8 9 10 11 12 13 14 13 12 11

Contents

To the Student xi

Learning Tips for Students of Organic Chemistry xiii

From the Outside Looking In xiii
Learning Skills. Specific Strategies and Tactics xiv

*A few things we know (and a few we do not) about learning
organic chemistry* xiv
Learning skills xv

Learning skills include memorization, but memorization alone
is not enough
Specific Strategies
• **Restatement**
• **Connections**
• **Review and reconnect**
• **Self-constructed summaries and aids**
• **Self-constructed assessments**
• **Information and meaning**
• **Diagnosis and treatment**

The role of teaching in learning xix

Learning to be a critical listener
Working with others is more than a social occasion
Learning to use vocabulary actively and accurately

Examinations xxi
Getting the "A" grade xxii

Chapter 1   An Introduction to Structure and Bonding in Organic Compounds   1
Map 1.1   Ionic compounds and ionic bonding 1
Map 1.2   Covalent compounds 1
            Workbook Exercises 4
Map 1.3   Covalent bonding 12
Map 1.4   Isomers 13
Map 1.5   Polarity of covalent molecules 16
Map 1.6   Nonbonding interactions between molecules 19
            Supplemental Problems 24

Chapter 2   Covalent Bonding and Chemical Reactivity   27
            Workbook Exercises 27
Map 2.1   Molecular orbitals and covalent bonds 29
Map 2.2   Hybrid orbitals 32
Map 2.3   Functional groups 33
Map 2.4   Bond lengths and bond strengths 34
            Supplemental Problems 43

Chapter 3   Reactions of Organic Compounds as Acids and Bases   45
            Workbook Exercises 45
Map 3.1   The Brønsted-Lowry theory of acids and bases 47
Map 3.2   The Lewis theory of acids and bases 48

Map   3.3   Relationship of acidity to position of the central element in the periodic table 53
Map   3.4   The relationship of pK_a to energy and entropy factors 56
Map   3.5   The effects of structural changes on acidity and basicity 61
                 Supplemental Problems 68

Chapter 4   **Reaction Pathways**   **71**
                 Workbook Exercises 71
Map   4.1   The factors that determine whether a chemical reaction between a given
                 set of reagents is likely 72
Map   4.2   A nucleophilic substitution reaction 73
Map   4.3   The factors that determine the equilibrium constant for a reaction 73
Map   4.4   The factors that influence the rate of a reaction as they appear in the rate equation 74
Map   4.5   Some of the factors other than concentration that influence the rate of a reaction 74
Map   4.6   Factors that influence the rate of a reaction and the effect of temperature 74
Map   4.7   An electrophilic addition reaction to an alkene 75
Map   4.8   Markovnikov's rule 76
Map   4.9   Carbocations 77
Map   4.10   Energy diagrams 78
                 Supplemental Problems 86

Chapter 5   **Alkanes and Cycloalkanes**   **88**
                 Workbook Exercises 88
Map   5.1   Determining connectivity 91
Map   5.2   Conformation 93
Map   5.3   Representations of organic structures 94
Map   5.4   Conformation in cyclic compounds 98
                 Supplemental Problems 107

Chapter 6   **Stereochemistry**   **111**
                 Workbook Exercises 111
Map   6.1   Stereochemical relationships 113
Map   6.2   Chirality 113
Map   6.3   Definition of enantiomers 113
Map   6.4   Optical activity 116
Map   6.5   Formation of racemic mixtures in chemical reactions 117
Map   6.6   Configurational isomers 119
Map   6.7   Diastereomers 122
Map   6.8   The process of resolution 123
                 Supplemental Problems 128

Chapter 7   **Nucleophilic Substitution and Elimination Reactions**   **132**
                 Workbook Exercises 132
Map   7.1   A typical S_N2 reaction 137
Map   7.2   The S_N1 reaction 138
Map   7.3   The factors that are important in determining nucleophilicity 140
Map   7.4   The factors that determine whether a substituent is a good leaving group 143
Map   7.5   A comparison of E1 and E2 reactions 145
Map   7.6   Overall view of nucleophilic substitution and elimination reactions 146
                 Supplemental Problems 158

Chapter 8   **Alkenes**   **162**
           Workbook Exercises 162
Map 8.1   Reactions that form carbocations 165
Map 8.2   Reactions of carbocations 166
Map 8.3   The hydroboration-oxidation reaction 168
Map 8.4   Catalytic hydrogenation reactions 169
Map 8.5   The addition of bromine to alkenes 174
Map 8.6   The oxidation reactions of alkenes 176
           Supplemental Problems 189

Chapter 9   **Alkynes**   **195**
           Workbook Exercises 195
Map 9.1   Outline of the synthesis of a disubstituted alkyne from a terminal alkyne 195
Map 9.2   Electrophilic addition of acids to alkynes 196
Map 9.3   Reduction reactions of alkynes 198
           Supplemental Problems 206

Chapter 10   **The Chemistry of Aromatic Compounds. Electrophilic Aromatic Substitution**   210
           Workbook Exercises 210
Map 10.1 Aromaticity 212
Map 10.2 Electrophilic aromatic substitution 214
Map 10.3 Essential steps of an electrophilic aromatic substitution 214
Map 10.4 Reactivity and orientation in electrophilic aromatic substitution 215
Map 10.5 Electrophiles in aromatic substitution reactions 218
           Supplemental Problems 228

Chapter 11   **Nuclear Magnetic Resonance Spectroscopy**   230
           Workbook Exercises 230

Chapter 12   **Ultraviolet-Visible and Infrared Spectroscopy. Mass Spectrometry**   **240**
Map 12.1 Visible and ultraviolet spectroscopy 240
Map 12.2 Infrared spectroscopy 243
Map 12.3 Mass spectrometry 246
           Supplemental Problems 254

Chapter 13   **Alcohols, Diols, and Ethers**   **257**
Map 13.1 Conversion of alkenes to alcohols 260
Map 13.2 Conversion of alcohols to alkyl halides 262
Map 13.3 Preparation of ethers by nucleophilic substitution reactions 264
Map 13.4 Ring-opening reactions of oxiranes 267
Map 13.5 Oxidation and reduction at carbon atoms 269
Map 13.6 Reactions of alcohols with oxidizing agents 273
Map 13.7 Summary of the preparation of and reactions of alcohols and ethers 275
           Supplemental Problems 292

Chapter 14   **Aldehydes and Ketones. Addition Reactions at Electrophilic Carbon Atoms**   **298**
           Workbook Exercises 298
Map 14.1 Some ways to prepare aldehydes and ketones 301
Map 14.2 The relationship between carbonyl compounds, alcohols and alkyl halides 303
Map 14.3 Organometallic reagents and their reactions with compounds
           containing electrophilic carbon atoms 305
Map 14.4 Some ways to prepare alcohols 306

Map 14.5   Hydrates, acetals, ketals 309
Map 14.6   Reactions of carbonyl compounds with compounds related to ammonia 312
Map 14.7   Reduction of carbonyl groups to methylene groups 315
                   Supplemental Problems 334

Chapter 15   **Carboxylic Acids and Their Derivatives. Acyl Transfer Reactions.**   **337**
                   Workbook Exercises 337
Map 15.1   Relative reactivities in nucleophilic substitutions 340
Map 15.2   Preparation of carboxylic acids 344
Map 15.3   Hydrolysis reactions of acid derivatives 346
Map 15.4   Mechanism of acyl transfer reactions 347
Map 15.5   Reactions of acids and acid derivatives with alcohols 352
Map 15.6   Reactions of acids and acid derivatives with ammonia or amines 354
Map 15.7   Protection of functional groups in peptide synthesis 355
Map 15.8   Activation of the carboxyl group in peptide synthesis 357
                   Supplemental Problems 371

Chapter 16   **Structural Effects in Acidity and Basicity Revisited. Enolization**   **374**
Map 16.1   Enolization 380
                   Supplemental Problems 392

Chapter 17   **Enols and Enolate Anions as Nucleophiles.**
                   Alkylation and Condensation Reactions   **394**
                   Workbook Exercises 394
Map 17.1   Reactions of enols and enolates with electrophiles 395
Map 17.2   Alkylation reactions 397
Map 17.3   The aldol condensation 399
Map 17.4   Acylation reactions of enolates 401
Map 17.5   Electrophilic alkenes 402
Map 17.6   Reactions of electrophilic alkenes 405
                   Supplemental Problems 422

Chapter 18   **Polyenes**   **425**
Map 18.1   Different relationships between multiple bonds 425
Map 18.2   A conjugated diene 425
Map 18.3   Addition to dienes 426
Map 18.4   The Diels-Alder reaction 427
                   Supplemental Problems 446

Chapter 19   **Free Radicals**   **448**
Map 19.1   Chain reactions 448
Map 19.2   Halogenation of alkanes 449
Map 19.3   Selective free radical halogenations 451
Map 19.4   Free radical addition reactions of alkenes 453
Map 19.5   Oxidation reactions as free radical reactions 457
                   Supplemental Problems 466

Chapter 20   **The Chemistry of Amines**   **468**
Map 20.1   Preparation of amines 473
Map 20.2   Nitrosation reactions 475
Map 20.3   Reactions of diazonium ions 476
                   Supplemental Problems 487

Chapter 21   **Synthesis**   **489**
Map 21.1   Protecting Groups 494
Map 21.2   Reduction of acids and acid derivatives 497
Map 21.3   Thermodynamic and kinetic enolates 499
Map 21.4   Ylides 500
Map 21.5   The Wittig reaction 501
Map 21.6   Dithiane anions 502
Map 21.7   Reactions of organometallic reagents with acids and acid derivatives 504
Map 21.8   The Diels-Alder reaction 510
Map 21.9   Diazonium ions in synthesis 514
Map 21.10   Nucleophilic aromatic substitution 515
                Supplemental Problems 545

Chapter 22   **The Chemistry of Heterocyclic Compounds**   **550**
Map 22.1   Classification of cyclic compounds 550
Map 22.2   Synthesis of heterocycles from carbonyl compounds 554
Map 22.3   Electrophilic aromatic substitution reactions of heterocycles 556
                Supplemental Problems 573

Chapter 23   **Structure and Reactivity in Biological Macromolecules**   **576**
Map 23.1   Classification of carbohydrates 576
Map 23.2   Amino acids, polypeptides, proteins 578
Map 23.3   Acid-base properties of amino acids 578
Map 23.4   Proof of structure of peptides and proteins 587
Map 23.5   Conformation and structue in proteins 589
                Supplemental Problems 611

Chapter 24   **Macromolecular Chemistry**   **614**
Map 24.1   Polymers 614
Map 24.2   Properties of polymers 614
Map 24.3   Stereochemistry of polymers 619
Map 24.4   Types of polymerization reactions 622
                Supplemental Problems 633

Chapter 25   **Concerted Reactions**   **635**
Map 25.1   Cycloaddition reactions 639
Map 25.2   Electrocyclic reactions 642
Map 25.3   Woodward-Hoffmann rules 643
Map 25.4   Sigmatropic rearrangements 647
Map 25.5   Carbenes 648
                Supplemental Problems 657

                Chapter 2 triangles to construct tetrahedra .... endpage

To the Student

The *Study Guide* that accompanies your text has been prepared to help you study organic chemistry. The textbook contains many problems designed to assist you in reviewing the chemistry that you need to know. The *Study Guide* contains the answers to these problems worked out in great detail to help you to develop the patterns of thought and work that will enable you to complete a course in organic chemistry successfully. In addition, notes that clarify points that may give you difficulty are provided in many answers.

The *Study Guide* also contains *Workbook Exercises,* created by Professor Brian Coppola of the University of Michigan. These exercises are designed to help you review previous material and to introduce you to the problem-solving skills you will need for new material. They are found only in the *Study Guide,* and no answers are given for them. Many of the exercises can be explored with other students in your class.

Suggestions for the best way to study organic chemistry are given on pages ~ - ~ of the textbook and in the essay, "Learning Tips for Students of Organic Chemistry," immediately following this introduction. This essay was written by Professor Coppola as a way of sharing with you his long experience with helping students to learn. Before you work with the *Study Guide,* please review the essay as well as those pages in the text.

This *Study Guide* contains two features to help you to study more effectively. The sections on the Art of Problem Solving in the textbook show you how to analyze problems in a systematic way by asking yourself questions about the structural changes in and the reactivity of the reagents shown in the problem. These same questions are used in arriving at the answers shown in the *Study Guide* for some of the problems. If you follow the reasoning shown in these answers, you will review the thinking patterns that are useful in solving problems.

The *Study Guide* also contains concept maps, which are summaries of important ideas or patterns of reactivity presented in a two-dimensional outline form. The textbook has notes in the margins telling you when a concept map appears in the *Study Guide.* The concept maps are located among the answers to the problems. The Table of Contents of the *Study Guide* on pp. iii - vii will tell you where each concept map is. The concept maps will be the most useful to you if you use them as a guide to making your own. For example, when you review your lecture notes, you will learn the essential points much more easily if you attempt to summarize the contents of the lecture in the form of a concept map. At a later time you may want to combine the contents of several lectures into a different concept map. Your maps need not look like the ones in the *Study Guide.* What is important is that you use the format to try to see relationships among ideas, reactions, and functional groups in a variety of ways.

The *Study Guide* will be most helpful to you if you make every attempt to solve each problem completely before you look at the answer. Recognition of a correct answer is much easier than being able to produce one yourself, so if you simply look up answers in the book to see whether you "know how to do the problem" or "understand" a principle, you will probably decide that you do. In truth, however, you will not have gained the practice in writing structural formulas and making the step-by-step decisions about reactivity that you will need when faced by similar questions on examinations. Work out all answers in detail, writing correct, complete formulas for all reagents and products. Build molecular models to help you draw correct three-dimensional representations of molecules. Consult the models whenever you are puzzled by questions of stereochemistry.

If you do not understand the answer to a problem, study the relevant sections of the text again, and then try to do the problem once more. The problems will tell you what you need to spend most of your time studying. As you solve the many review problems that bring together material from different chapters, your knowledge of the important concepts of organic chemistry will solidify.

In addition to the answers to the problems in the textbook, the *Study Guide* also contains Supplemental Problems for most chapters. These are additional drill and thought questions for which answers will be available to you only through your instructor. These problems are excellent ways to review the material for a test.

We hope that the *Study Guide* will serve you as a model for the kind of disciplined care that you must take with your answers if you wish to train yourself to arrive at correct solutions to problems with consistency. We hope also that it will help you to develop confidence in your ability to master organic chemistry so that you enjoy your study of a subject that we find challenging and exciting.

Seyhan N. Ege
Roberta W. Kleinman
Peggy Zitek

Learning Tips for Students of Organic Chemistry

by Brian P. Coppola
University of Michigan

From the Outside Looking In...

Every year, I am more and more on the outside looking in when it comes to learning the subject of organic chemistry. The reason is simple: As a practicing organic chemist who has been an instructor of this subject for over fifteen years, I cannot see organic chemistry the way that a new student sees it. Students see this subject with the eyes of a fresh learner, with one new idea following another with few previous reference points. One of the things I value in my interactions with students is that they bring their unique perspectives as new learners to my course. The fresh eyes of my students are the greatest tool I have to improve my understanding of "learning organic chemistry", which greatly impacts my ability to help others learn it, too.

As a student, I was a chemistry major thinking about a career in science, so I was predisposed to take my chemistry courses seriously. Although most classmates in my own undergraduate courses were not prospective majors, I was still like many of them, as well as my own students today, in some other respects. One purpose (perhaps a motivation) for learning a subject was to get a good grade on exams. I wanted good grades because I took great pride in doing well in my academic studies. I also knew I needed good grades to get into graduate school. But there was something else. Only in retrospect did I realize that some of my college instructors were trying to get me to see learning from the broader perspective of improving myself through higher education. I think that understanding this lesson was difficult for two reasons. First, I did not have any reference points or experiences for this advice to make sense until much later in life (in fact, in some cases, not until I became an instructor myself). Second, as far as I can recall, these broad lessons in improvement never seemed to show up in my science classes, except maybe as a spoken line or two on the first day of class. These ideas never seemed to appear anywhere else. The book, the homework, the class time and the exams were all "just chemistry problems." Once I became responsible for organizing courses as a faculty member, I found myself wanting to address these two problems. As an instructor, I cannot do anything about the first difficulty. I cannot provide students with 10 years of experience in four months (although the students in my Honors course might disagree with that statement). Experience being what it is, generally, you have to get it in order to have it. One of the things that motivates me as an instructor, though, is the thought that I (and all instructors) can help out with the second difficulty, that of bringing evidence of a broader perspective to multiple aspects of a subject.

Although I may be on the outside looking in when it comes to learning organic chemistry for the first time, my knowledge continues to increase in two other areas. First, I understand better every year how the nuances of this subject fit together, often because of questions my students ask. Second, I continue to learn how students learn organic chemistry, which answers one of the most common questions students ask their instructors: How can teaching the same old thing year after year be interesting? For me, that is easy: I never do it the same way twice. There is always something new I learn about how students learn that makes me improve what I do the next time.

I wrote the phrase "bringing evidence of a broader perspective to multiple aspects of a subject" to describe an instructional goal. What does this mean?

In order to answer this, I have to start with a summary of all of my goals for students in my courses. Many times, when faced with the question of goals, faculty will drag out a copy of the syllabus and say "Here are my goals: On the first week of class we will cover chapter one, then chapter two...." If such statements are examples of goals, I find them unsatisfactory. Over the years, I have found it useful to categorize the goals that I have for student learning in my courses. I think there is an important hierarchy to goals that has been lost in higher education. At the most immediately obvious level are what I call "professional technical goals." These are the goals most directly related to the subject matter of the course: The factual understanding and operational skills you are supposed to develop in your studies, and on which examinations are generally based. In calculus this might be learning how to take a derivative; in French this might be learning how to construct the past participles of some regular verbs. In an organic chemistry course, one early goal is for students to be able to translate the drawings used to represent chemical structures into an inventory of the atoms involved and how they are connected to one other. The technical (subject matter) goals usually become more sophisticated as a course proceeds. The kind of skills you are supposed to develop are gauged by the type of problems that you are supposed to solve. An increasing number of individual skills are combined and balanced into ways for solving new problems. Later in a course, enough specific examples should have been assembled to allow students to make sense of broader categories and concepts. These larger categories and concepts are what define a discipline ("calculus," "French," "chemistry") and identify what I call "professional intellectual goals." These concepts and generalizations also allow me to understand new and unfamiliar information both by applying the larger ideas to any specific new situation and by creating analogies based on other factual information that I know. Indeed, I want my students to develop the skills that are used by a practicing chemist.

Courses and subjects are filled with professional technical goals. The professional intellectual goals are what keep a subject from becoming just an endless list of things that have to be remembered. There are professional intellectual goals that relate to chemistry, such as explaining and predicting everything from bonding to bonding changes (chemical reactions) on the basis of electrostatic interactions (the attraction between positive, or electron-poor atoms, and negative, or electron-rich atoms). There are also other professional intellectual goals that relate to science and scientific practice, such as understanding the role of reproducibility in experimental science or the significance of the Uncertainty Principle in understanding observations. It is my obligation to demonstrate consistently how and why the specifics of chemistry interact with larger ideas of both chemistry and science. It is my students' obligation to appreciate the validity and operational importance of these relationships. Finally, there are "general intellectual goals" that are, to a degree, the overriding purpose of an education. These are the skills acquired from the study of a subject that transcend the subject itself, especially new strategies, insights and experiences about the process of learning and understanding new things.

Learning Skills. Specific Strategies and Tactics

A few things we know (and a few we do not) about learning organic chemistry

You should expect that learning organic chemistry, for the reasons outlined above, may be different from other learning experiences that you have had. The myths that surround the subject of chemistry, and especially organic chemistry, do not help at all.

"Organic chemistry is the most difficult course at the University."

"Organic chemistry is the 'weeder' course for medical schools."

"Memorizing tons of information is the only way to get through."

"Look to your left in class, then look to your right. One of those people will not be there at the end of the term."

"Only students with previous college chemistry, a good AP background, and an organic chemistry prep course can do well."

"I just can't do science classes."

Is it any wonder that it is difficult to concentrate on the course with all of these anxieties lurking around? These statements are simply not true.

Structure and Reactivity is the large introductory course based on organic chemistry and taken by first-year students at the University of Michigan. Since 1989, the University of Michigan faculty have presumed that the precollege chemistry background of our students is adequate to the task of learning organic chemistry. One of the most gratifying aspects of teaching this course has been feedback that we are fulfilling one of the unstated expectations of new university students, that their academic program will be different and challenging and not a repeat of their high school experience (something that is certainly true in the non-academic portion of a student's experience in attending college).

We have looked carefully at what characterizes students who are successful in our *Structure and Reactivity* courses. Here is what we know:

(1) The amount and type of previous chemistry does not make a difference, but learning skills do.

• The fact that students did not take an AP Chemistry course does not matter. On average, the 35-40% of 1000-1200 students in our first-term Structure and Reactivity course who took AP Chemistry perform at the same level as the non-AP students do. We have strong indications that it is not the background in chemistry content that matters, but rather the learning skills of the student.

• A second thing that points to learning skills is the fact that the Math SAT score is the only other background factor that predicts anything significant about student performance in the first-term *Structure and Reactivity* course, even though the course is 100% narrative, or descriptive, and primarily non-mathematical in nature. Historically, the Math SAT score is thought to be representative of general learning skills.

• Students in *Structure and Reactivity* courses tend to develop their deeper learning skills more than their counterparts in a General Chemistry course, therefore the willingness to make these sorts of changes is an important characteristic of those who succeed in the course.

(2) Psychological motivation plays an important role.

• We also have observed that students' beliefs in their own abilities play as large a role in predicting success as the Math SAT score. A person who believes that he or she has developed a degree of control over learning (or over any task), tends to develop better understanding. Part of this is a feedback cycle, where those who do well to begin with get the message that what they were doing was the right thing. On the other hand, we also know from our course that the first exam does a relatively poor job of predicting course outcome. This means that many students who end up doing well develop their successful strategies later, after some less satisfactory experiences have motivated them to make a change. It is important for students to be patient and persistent, and not to let the first discouragement drag them down.

• Student responsibility is also significant. If students find themselves thinking "I did not learn because the instructor did not teach me well enough," then they are requiring far too much of the instructor and not enough of themselves. Similarly, if students conclude that "This course just did not match the way I learn," then they are missing the point about building new skills on the foundation of old ones. I think that becoming a more flexible learner has no down side. Why just reward the same old skills?

Learning skills

Learning skills include memorization, but memorization alone is not enough.

The subject matter of organic chemistry is particularly well-suited to encouraging students to develop deeper learning skills. This is because you spend an entire year with one specific subject that builds upon itself in a meaningful way. Many times, introductory courses are called surveys, where one topic follows another without much linkage. There can be advantages to this approach. For instance, each new topic becomes a fresh start without the immediate need to master a previous topic. On the other hand, you never spend long enough with any one topic to make deep connections that truly challenge your learning skills. Organic chemistry begins with a relatively few general principles for which you can develop a ever deepening understanding as the year goes on. At least, that's the plan! On the other hand, no one can force you to do anything, including learning differently. All an instructor can do is to create a situation where you will come to realize that your old skills are inadequate.

My instructional goal for students is to use chemistry as a way to encourage them to develop new learning skills. To accomplish this goal, students must be faced with learning situations where their old skills are inadequate but not abandoned. The skills with which students begin a course are their strengths, their point of reference. For the most part, students begin a course with what are called "surface skills." Surface skills include the ability to memorize, to organize, to recall and connect one set of symbols or representations with another. A concrete example of such skills is the multiplication tables. You can connect the symbols "2 × 2" with "4" without ever understanding the multiplication relationship. This is also your level of understanding when you learn to do multiplication with an electronic calculator. The multiplication tables or a calculator are just starting points. Your current understanding of multiplication has not replaced the times table or a calculator. Rather it has become broadened and deepened with alternative ways to think about multiplication. Notice that you have not abandoned your fluency with the multiplication tables or calculators because you now have a mastery of multiplication. Rather this fluency is inadequate when faced with a problem that is not on the table you have memorized. Without using a calculator to solve "345.8 × 45.5," a problem that you probably have not seen before but nonetheless can solve easily, you use your more general knowledge of multiplication as well as your specific recollection of the multiplication tables. Even when you use a calculator, your general understanding of multiplication combined with estimation skills would allow you to reject an answer such as "157.339" if it showed up on the display. The additional skills you need to combine with surface skills in order to solve this problem are called "deep processing skills." To solve this unfamiliar problem, the deeper skills interact with your surface skills in ways that allow you to judge whether you are adequately performing the task at hand.

Specific Strategies

Why should students develop new learning skills? I hope that the answer is self-evident: Such development is one of the objectives of higher education. In order to end up with an intellectually rewarding career, you have to be able to walk into a new and unfamiliar situation with the confidence that your skills will see you through. All people who are truly successful at what they do bring these kinds of skills to new problems, and new problems are the interesting ones! Experiences in (and out) of college classes are meant to model these situations. The behaviors and habits students develop during these years define their character for the future, long after the details of specific courses have faded away. I am deeply committed to the idea that we are all life-long learners, and that a necessary goal in education is to encourage the habits of the life-long learner. What does this mean? Mainly, it means that you become more and more responsible for your own education. Rather than having your interests defined by a course or curriculum, you begin to identify what you want to learn, including how to learn it, because it serves some greater, self-defined goal.

What does a deeper or higher order learning skill mean? The skills that more experienced learners bring to a task are complex, and vary from challenge to challenge. The process of making appropriate selections from a menu of existing strategies, or knowing when to invent new ones, is a skill unto itself, analogous to matching the right tool to a mechanical job. For an introductory course, I encourage students to master the following skills:

- **Restatement**
  Restatement is more than just putting it in your own words. It is the process of making new ideas make sense in terms of what you know. As you encounter a new concept, try to imagine having to give a short lesson to another beginning student to get them to understand it. Do not just rehearse the words of the text over and over, and do not just say them to yourself, in your head. If you have to, say your lesson out loud to yourself. If you can, find another classmate to talk to. More will be said on this in the section on "The role of teaching in learning."

- **Connections**
  One hallmark of the best learners I have known is their belief that everything is connected. What you learn in one place can help you understand something else. When I face new and unfamiliar information, one of my first reactions is to find an appropriate analogy. Rather than answering the question "What is this like...?" I start with the certainty of "This *is* like..."

• Review and reconnect

Connections are not enough. As you develop the map that is your understanding, it is also important to review what you once knew in terms of what you now know. New information should give you a new perspective on old information. Another version of this is the idea that your understanding should be sketched out rather than defined too specifically too early. When dealing with a new chapter in your text, for example, you can elect to move very linearly and deliberately through the book, one page at a time, digesting each adverb before you permit yourself to turn the page. Unfortunately, this approach minimizes opportunities to make connections. Another approach is to think of your understanding as a painting. First you start by making a sketch, which is a process filled with erasure and correction, a time when what once seemed right is now out of place, and a time to get a look at the whole canvas and try to see the big picture, even if it is a bit blurry. After this comes a time of refinement and elaboration, where self-consistency across the canvas allows the newly defined parts to complement one another.

• Self-constructed summaries and aids

As you build towards self-reliance, you must begin to solve problems with no information other than what will be available at your exams. Any amount of time you spend "getting little hints" or using anything other than the information in the problem to help solve it is wasted. If you tie your skills to an answer key, your notes, or where you are in the text, then you will be practicing skills that are useless for the exam. At some point, you must be willing to look at an unfamiliar problem and say "I don't know how to do that, yet." and move on to the things you can do with the knowledge you have. If you do construct aids, such as mnemonics, lists, or other associations, make sure they are the kinds of things you have actually used to solve problems.

• Self-constructed assessments

Whatever your course of study, the object of your study will be ideas and how people deal with information. One way to test your own proficiency is to create your own problems. This can be done many ways for many different subjects. In my chemistry courses, I usually recommend two things for everyone. First, take any general subject heading in the course ("resonance forms," "Brønsted acid-base reactions," and so on) and write it on a blank piece of paper. Now create (*do not* look up or recall) 10-20 examples of that phenomenon based on the general principle. One of the best uses students can make of their instructors is to share these creations. Other versions of this exercise might be to see if two or more of the general ideas can be combined, or to get together with others and use these problems as the basis for testing one another. The other advice I have is related to creating exam questions. Instead of creating examples under the topic heading, students can do what the faculty do: Go to chemistry journals. In my course, nearly all of the exam questions have a citation because it is very convenient to thumb through the journals and use simple sorting skills to look for specific examples of general phenomena. You can do this, too.

• Information and meaning

A theme that links the five skills listed above is the distinction between "information" and "meaning." When I write "cat" or "table," these words are just collections of symbols that are meant to represent the idea of a cat or a table. Without prior knowledge about these symbols, it is not possible to extract the meaning of "cat" from the letters c-a-t. The word "cat" is not a cat! Similarly, the symbolic representation "H_2O" is not water, but it is meant to represent all that water is and how it behaves and interacts. One of the things that make organic chemistry so interesting is that once you learn the basics of the structure/reactivity relationships, you will be able to predict the behavior of substances the structures of which you have never seen before, much the way a very complete knowledge of Greek, Latin, and word origins might allow you to understand words you had never seen. Information collects all of the surface features, while meaning gathers all of the inferences. One of the common mistakes made by instructors is to advise students that learning organic chemistry is like learning a foreign language; not so. When you learn any second (or third, etc.) language, you do so with an idea of what the objects that need to be described are. In other words, there is a great deal of translation. If you already know what a cat is, and you have a word for it in your first language ("cat"), then learning that "chat" is how this idea is represented in French benefits from your preconception of what a cat is. Now imagine that some other animal (or maybe you are not even sure it is an animal; it is like nothing you have ever seen before, actually) is not only named in a language with which you are unfamiliar, but that the descriptions of this thing are also only available in this unfamiliar language. Learning organic chemistry is not like learning a second language at all; it is more like learning your first language.

• Diagnosis and treatment

Diagnosis. Solving problems follows a medical metaphor quite well. There are two parts to the problem-solving process: Diagnosis and treatment. Diagnosis is the part where general classifications are made, and perhaps a general strategy is developed. On a chemistry exam, it simply means deciding which of 6 or 7 major ideas is represented by the problem. If you have created such a list before the exam, and practiced using it, then you can use it as a guide while taking the exam. The exam problems must represent the ideas from the chapters in question. This raises an interesting idea to keep in mind about textbooks. Textbooks themselves can allow you to underdevelop or avoid using your skills in diagnosis. For example:

(1) *Problems within the chapter are diagnosed for you before you get there!* Not surprisingly, the problems relate to the preceding section. One way to demonstrate that diagnosis is a real skill is to take photocopies of the in-chapter problems after you have done them, cut them apart from identifying markers, randomize them, and then try to answer the question, "What kind of problem is this?" The same problems that were *so easy* before are now difficult. Ready to learn diagnosis?

(2) *Problems at the end of chapters are still mostly associated with the chapter,* and are sometimes drill-like (Problem 23 had parts a, b, c, d...w). Once you struggle with 23a and 23b, all of a sudden 23c is easy. You are not actually getting any better at diagnosis because you can do 23c; you are just anticipating what the problem is about. It is being done for you. Any time you know what a problem is about before you have read any part of the problem, you probably should not do it. Skip it and come back later and see if you can still tell what it is about.

(3) *Keep book-reading and note-reading time separate from problem-solving time.* Try the problems in a new chapter before you read the chapter, just to see that you cannot do any of them. Even by reviewing the problems, you may begin to get a sense of the ideas that you will need to pay attention to. If something is unclear after a respectable effort, move on. Try to treat chapters as whole entities, as stories where all the parts are interwoven. As you make your initial fast pass through the text, see if any of the problems make more sense. If so, try them out. If you can't solve them, you will come back to them again. If you recognize that you do not know how to do a problem with the understanding that you have at that point, that is an important thing to know. Concentrate your efforts on learning what you *can* do with what you know, and work from there as you reread (and reread and reread). If you do not spend overly long with parts of the chapter that are not clicking, you will free up time for future readings. No knowledge can be presented so linearly that you can't learn from page 54 without getting page 53. And many times, what you learn on page 54 can help you understand page 53. Give yourself permission to turn the page!

The bottom line in learning to do diagnosis correctly is quite compelling: If you don't get this part right, it doesn't matter how well you do the next part, because it will be wrong. The correct answer to the wrong question never gets any points. After all, a physician may know how to treat two different diseases perfectly well, so the most important thing is first to make the diagnosis correctly! A physician does not get "partial credit" for prescribing the right medication for the wrong disease.

Treatments The following suggestions are, by definition, incomplete. These ideas are meant to inspire you to think about learning in ways you might not have before.

(1) *Practice useful skills.* Always ask yourself, "Am I doing this work honestly? Am I just rationalizing someone else's answer in the *Study Guide*? Am I using a resource that I will not have at an exam? Did I know what this problem was about before I did it?" You can learn how to do the wrong thing very well. It feels as though you are making progress, but it is in the wrong direction, or simply allowing you to generate incorrect answers more efficiently. It is fine to get the advice that you must spend a little time studying the subject every day, but this is the beginning of the story, not the end. How you spend your time

matters too. Many high school and beginning university students equate time spent with effort when what is needed is productive effort. The only way to know whether what you are doing is productive is to examine it honestly. Can you do problems when they are out of context, can you explain your ideas in writing, and can you explain them out loud?

(2) *Concentrate on your strengths.* Build on what you know and learn to identify the problems you can do with what you know. Learn to admit when you do not know something as well as when you do. Sketch out your understanding by trying to keep the broad arguments in mind when you are concentrating on the details. Work back and forth as you master new ideas, asking, "How does this fit into the overall picture?"

(3) *On an exam, you are the teacher.* Like it or not, instructors demand performance of one kind or another. If you always keep in mind that you need to express your ideas as well as learn them, you will be ahead of the game. You do not necessarily need to work with another person, but it is generally easier to develop such skills if you do. Self-examination and quizzing your study partners is a chance to practice those skills before the exam. During the examination, your role is that of the instructor, and your instructor is the student to whom you are explaining the ideas. If you have practiced this skill before the exam, you will not be forced to learn it there.

(4) *Constant, daily building of ideas.* If you play catch-up, you will be caught. Listen and think in class. Respond to questions. Create your own tools for solving problems, and do not wait until just before the exam. If you are allowed an index card of information, it should be created and refined throughout your study of the chapter. Even if you are not allowed to bring it to the exam, you can still think about developing a card's worth of information that is useful for solving problems. Look at the general statements and topic headings and conclusions from the lecture and ask yourself, "Do I believe these? Do I believe that the examples support the ideas?" Even if you wait until the last minute, at least give yourself a few days for the longer-term connections to begin to form. If your exam is Tuesday night, then pretend it is really on Saturday or Sunday and use the intervening days to review and allow the ideas to percolate. Whatever your time frame for study, push it back a few days, even if all you intend to do is cram for the exam.

(5) *Exams transmit expectations.* More than anything, the exam is where you really learn what the course is about. You must pick up your graded exam and analyze why you made errors. The "correct answer" simply does not count for that much compared with correcting the process by which you made the error. If you think an exam question was written poorly, then one thing to do is try to rewrite it yourself. Write out in words the thought process you used to create an answer and look for where you went wrong. Having this process written out also is a good way to engage your instructor. Avoid avoidance; when the exam is taken and graded, pick it up and look at it. If you do not pick it up, you are only making things worse, not better. The old exam is a place where you can inspect your real errors, the ones pertaining to how you were learning.

The role of teaching in learning

Learning to be a critical listener

I started with a discussion of how teaching impacts my learning. A phrase familiar to all instructors is based on their first teaching experiences: "I never really learned this subject until I had to teach it." Most instructors understand that the most important advice to give is that students should work together in their learning. The reason for this is that you develop teaching skills when you work with others. Developing teaching skills is relevant to all students who take exams, write papers and give presentations, which includes everyone. All of these events are fundamentally teaching events, that is, situations that call for explanations to be given. When a learner is consciously aware as a goal of the need to explain things to others (in other words, teaching), then learning is improved.

One useful teaching skill is to become a critical listener. When you work with others, don't decide only whether what they are saying is right or wrong according to your rules and ideas. Try to understand the rules and ideas being used by the other person in what they say or do. Let me use a specific example outside of chemistry. If you were helping a grade school student to learn multiplication, asking him or her to *create* 10 or 20 new problems, and their solutions, would be a good idea. Your expertise at multiplication would allow you to scan all 20 problems in a very short amount of time, and give that student some valuable feedback. The interesting thing is that understanding is hardly ever completely correct or completely incorrect. Usually, understanding is incomplete, adequate in some places, inadequate in others. The challenge in monitoring your own learning is to put yourself in situations where you can distinguish between adequate and inadequate understanding. For example, this young learner might come to you with 20 examples, and the first few you see are:

(a) $1.1 \times 11 = 12.1$ (b) $2 \times 2 = 4$ (c) $3.5 \times 1.4 = 4.9$ (d) $2 \times 4 = 6$

As an instructor, you can react in many ways to these examples. The worst thing to do is to say: "Letter (d) is wrong, you need to study more." In my experience, I have always noticed that I can learn about the way students understand something by assuming that what I see is the result of a consistent application of some set of rules. This is an example of critical listening. I am less concerned about only getting across my perspective and more concerned about understanding the perspective of the person I am with. The reason that I like the multiplication example is that it demonstrates something I see often; a student's inadequate rules and my generally more adequate rules can overlap. This means that we can both come to the same factual conclusion for different reasons. If I want to probe the deeper understanding of my students, so that I can better know that they are using the correct *process* to obtain their answers, then I must try to push the edge of understanding. By learning only how to produce (and evaluate) answers with surface strategies, students can end up learning how to do the wrong thing very well; that is, they master inadequate rules that just happen to produce the same answer as the better rules do. It is easy to make this mistake in teaching: Just because another person's interpretation or answer looks correct does not mean it was obtained by the same pathway as yours or that it means the same thing as it does to you. I do not mean to imply that multiple interpretations are not possible; I mean that better communication depends on double checking that I understand the connection between the process and the product of a student's effort before I build what I do on incorrect assumptions. For example, the cases of "multiplication" presented by your student were in fact created by the consistent application of the rules of addition, instead of those of multiplication. To tell the student that he or she was doing something inconsistently would have been very bad advice.

There are a number of different ways to practice your teaching skills as a way to improve your learning. The one prerequisite is that you learn how to open yourself up for interactions with other students: Good communication (speaking and listening) skills, mutual trust, and a willingness to be publicly incorrect and to be corrected are all necessary. You must examine the conditions under which an answer is provided in addition to the answer itself, and you need problems that are both difficult and for which you cannot easily obtain solutions. Better learning through teaching is a fact. It has worked for me, as I described at the beginning of this essay, and it can work for you, too.

Working with others is more than a social occasion

There are additional good reasons for having conversations about things you are trying to learn. Sitting by yourself alone somewhere, you can convince yourself of just about anything. Time on your own is a good beginning, but sooner or later you need to see if you can share what you know. Certainly, as discussed earlier, this is what happens at an exam! When you have the opportunity to say what you are learning out loud, you must consider organizing the ideas for someone else. In fact, when you know you are going to be in the situation of describing what you understand to someone else, you actually learn it differently. If you naturally learn by having discussions, that is fortunate. It probably means you are thinking appropriately about the exam situation. Anticipating the need to make explanations is at the core of this advice. A person does not necessarily have to work with someone else to achieve these benefits. On the other hand, in my experience, students do not seem to take this need enough into account. Editing your own ideas is a difficult task. An external editor, or proofreader, for your ideas, makes sense. Whether you like it or not, the exam will put you in the position of explaining ideas. If you wait until then to develop and practice that skill, you are overburdening the exam time with things that you could have practiced ahead of time.

Learning to use vocabulary actively and accurately

Ideas are represented by words and other symbols. In order to work with ideas, you must also work with words and symbols. As you test your ideas, speak out loud without the safety net of your books nearby. While you are walking across campus, talk chemistry with a friend. While you are out at dinner, get out a napkin and draw out chemical ideas. These are the only ways to build the proper confidence that you can actually communicate using chemistry. When I work with students, I am intolerant (in a nice way) about imprecise language. I will stop students who use phrases such as "that thing over there" or "you know, the one from class" and encourage them to think about the proper terminology and phraseology for communicating ideas in chemistry. These are important skills to practice before examinations. During your exams, you have no choice but to represent your ideas correctly. Your answers will be incorrect if the wrong symbols are used, or if a structure is drawn the wrong way, if the wrong words are used...even if you "knew it." Incorrect representation is an error. "I know I didn't write it that way, but I meant..." never ever works. Does the importance of vocabulary also apply to courses where multiple choice problems are used on examinations? Yes and, unfortunately, no. There are many strategies that rely on recognition and recall that can be used in preparing for these kinds of exams (which I have never given, by the way). This does not mean that students cannot develop a good idea about chemistry in courses with multiple choice exams, but I do think that there is less reason to do so. Learning strategies based only on memorization are familiar and well-practiced by students, so they feel quite comfortable with them. As you can probably tell by now, I think that this degree of comfort is exactly why moving away from those strategies is a good idea!

Examinations

Exams are the real curriculum, not the syllabus. Think about that one more time. Nothing I say about learning in this course matters if a student does not see clearly how it relates to the examinations. Like it or not, the structure and expectations of higher education include grading, and grading results, for the most part, from examinations. Students learn about my expectations at examinations, not from what I say in class. It is therefore quite important to ensure that there is *congruence* between (a) the stated goals in a course, (b) the instructional method, (c) the instructional tasks, and (d) the examinations and how they are evaluated.

I believe that nothing I say will matter, and that nothing I do in class will matter if the examinations do not fulfill the expectations created by the classwork. If I do not want memorization to be the only tool students develop, then my exams must ensure that this strategy alone will not work. In addition, I must think about instruction in a way that encourages the development of new learning strategies. To that end, nearly all of the examination questions in my course are derived from examples taken from current chemistry journals. There is no better way to demonstrate two important ideas. First, simply becoming familiar with the textbook examples cannot lead to success. Students must develop the skill to identify major ideas and themes and then use these concepts as their basis for drawing analogies. Second, we demonstrate that the subject is vital. The major ideas still appear and reappear in current research month after month. All the learning strategies previously discussed can apply to courses that use multiple choice exams. In my view, however, this style of exam does not obviously require this way of learning and can cause students to default to more familiar and comfortable strategies. Many educators debate whether a student's choice to use lower level learning skills should have any bearing on the decisions made by an instructor in choosing a testing strategy. After all, almost any exam will create a distribution of students in the class, and "good students" will probably learn well in any situation. This last sentence highlights my reason for giving the kind of course that I do: I do not see it as my professional responsibility to find the "good students" who are already sitting in my course on the first day of class. My responsibility is to provide an opportunity for improvement by all the students in my class (including me, by the way).

The course packs for the *Structure and Reactivity* courses at the University of Michigan have two parts: Essays such as this one, which constitute one way of transmitting the course goals to students, and actual pages from the last 4-5 years worth of old examinations, with no solutions provided. Sometimes no matter what I say or do, only an inspection of these exams will convince students that my words mean what they say.

Interestingly enough, every year all students think that their exam is more difficult than the one given the year before. This is just not true. There is also an aspect to taking examinations that is characteristic of any situation where performance is called for known as "performance anxiety" (or, in this case, "test anxiety"). Test anxiety is not attached to the subject, even though many students would like to think so. Test anxiety is like stage fright for acting or a musical recital, or like the tension at a sporting event when you are at the starting line for the race that counts. It does not matter how smoothly things have gone before. You are human, and human beings become anxious at the event that matters. There are many different strategies to use to combat this kind of anxiety, including what your teacher, mentor, or coach has to say to you prior to the event. What you need is confidence from two sources: Yourself and the people you respect. Clearly, the more practice you have that allows you to develop skills you can use during the event, the better off you are. An examination, at its core, is an event that requires you to make an explanation about things you have not necessarily thought about before. The more you have practiced this, the better off you will be. The more you have avoided it, the less prepared you will be.

We do not provide solutions to the old exams because we want students to work together. Although it is frustrating not to have an answer key, there are plenty of problems with solutions in the *Study Guide* associated with the textbook. In addition to encouraging students to work together, the old exams can help individuals to regulate their learning. Every day, after class or after studying, a student can go to this set of problems and use them to answer two questions:

(1) What can I do with the knowledge I have now?

(2) Can I identify the kind of knowledge I need to solve the problems I cannot now do?

One of the more sophisticated skills of the expert problem-solver is learning how to develop a sense about whether their solutions "seem reasonable" or "make sense." This is an intuition that only comes through solving problems in a way where problem-solvers are honest with themselves about the confidence they have in their abilities.

Getting the "A" grade

The techniques outlined in the section on specific learning strategies are meant to give you an alternative to simply "doing problems" and constantly re-working them. These techniques should become second nature to you. They will serve you in all courses, including organic chemistry. Working on these skills is like taking an art class. You must take some time to sketch out your ideas and practice your skills nearly every day. You need to show your creations to other people so that errors in your technique can be corrected. Learn how to share your chemistry ideas – especially your incomplete ones – with your peers and with your instructor. Remember, you cannot simply persist in old study practices if they are not working for you and expect to see different results no matter how much time you invest.

You want to get a good grade in your courses, and I want you to learn something about how to learn along with your mastery of chemistry. I want you to do well on your exams because I believe that if you do, there is a good chance you also will have done the following:

1.   *You will have learned how to be successful at something very new to you.*

2.   *You will understand that science operates as a narrative, where sophisticated stories are told by people just like you by using their common sense and reasoning skills.*

3.   *You will realize that information or facts, alone, are not terribly interesting, but they can point to a fascinating understanding or meaning of the world.*

4.   *Best of all, you will develop confidence that your new learning skills will be something you can carry into other parts of your academic life.*

For your part, I would like you to begin to attach a more meaningful value to getting good grades. Your introductory classes can be a valuable learning experience in being with a group of students who have been as successful in their previous work as you have, and in developing the kinds of skills you will need for more challenging courses in the future. "Getting an 'A'" is really not a goal; making sure you have learned how to do new things, including how to double-check yourself in those new abilities, is a goal. I am perfectly comfortable knowing that the majority of students who take chemistry classes will not become chemists who use the information from this course on a routine basis; therefore there must be some other value that goes beyond a grade for your transcript. I do hope students will exit a course like this understanding why some people find a career in chemistry an interesting place to spend their professional lives.

1 An Introduction to Structure and Bonding in Organic Compounds

Concept Map 1.1   Ionic compounds and ionic bonding.

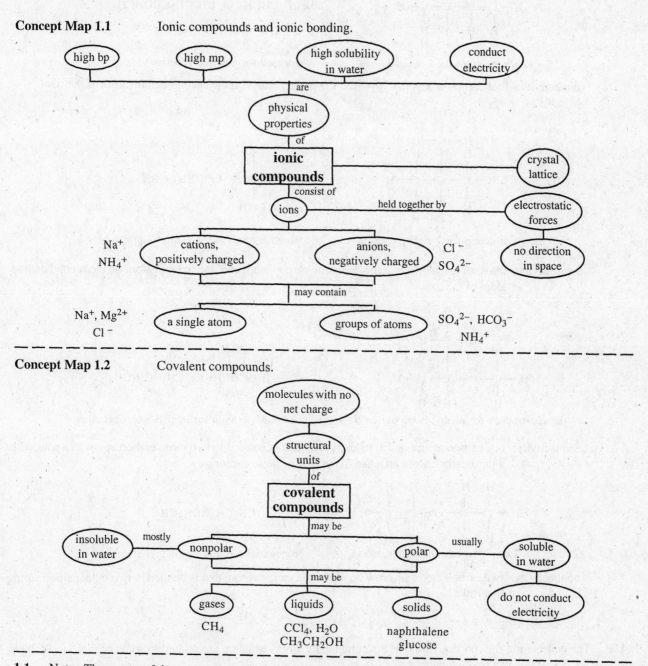

Concept Map 1.2   Covalent compounds.

1.1 Note: The names of the compounds shown on the next page are given for information. You are not yet expected to know how to name the compounds, but an examination of the names to see if you recognize an emerging pattern is fun and will be valuable when you do learn nomenclature.

1.1 (cont)

$$H-\overset{\overset{\displaystyle H}{|}}{\underset{\underset{\displaystyle H}{|}}{C}}-\overset{\overset{\displaystyle H}{|}}{\underset{\underset{\displaystyle H}{|}}{C}}-\overset{\overset{\displaystyle H}{|}}{\underset{\underset{\displaystyle H}{|}}{C}}-\overset{\overset{\displaystyle H}{|}}{\underset{\underset{\displaystyle H}{|}}{C}}-\overset{\cdot\cdot}{\underset{\cdot\cdot}{O}}-H$$

$CH_3CH_2CH_2CH_2OH$

Lewis structure for 1-butanol   condensed formula for 1-butanol

connectivity: four carbon atoms in a row with the oxygen atom at the end of the row

$$H-\overset{\overset{\displaystyle H}{|}}{\underset{\underset{\displaystyle H}{|}}{C}}-\overset{\overset{\displaystyle H}{|}}{\underset{\underset{\displaystyle H}{|}}{C}}-\overset{\overset{\displaystyle H}{|}}{\underset{\underset{\displaystyle \cdot\overset{\cdot}{O}-H}{|}}{C}}-\overset{\overset{\displaystyle H}{|}}{\underset{\underset{\displaystyle H}{|}}{C}}-H$$

$$CH_3CH_2CHCH_3 \text{ or } CH_3CH_2CHOHCH_3$$
$$\underset{OH}{|}$$

Lewis structure for 2-butanol   condensed formula for 2-butanol

connectivity: four carbon atoms in a row with the oxygen atom on the second carbon atom of the row

$$H-\overset{\displaystyle H}{\underset{\displaystyle H}{C}}-H$$

$$H-\overset{\overset{\displaystyle H}{|}}{\underset{\underset{\displaystyle H}{|}}{C}}-\overset{\overset{\displaystyle |}{|}}{\underset{\underset{\displaystyle :\overset{\cdot}{O}-H}{|}}{C}}-\overset{\overset{\displaystyle H}{|}}{\underset{\underset{\displaystyle H}{|}}{C}}-H$$

$$\underset{CH_3CCH_3}{\overset{CH_3}{|}} \text{ or } (CH_3)_3COH$$
$$\underset{OH}{|}$$

Lewis structure for 2-methyl-2-propanol   condensed formula for 2-methyl-2-propanol

connectivity: three carbon atoms in a row with one carbon atom and one oxygen atom on the second carbon atom of the row

$$H-\overset{\displaystyle H}{\underset{\displaystyle H}{C}}-H$$

$$H-\overset{\overset{\displaystyle H}{|}}{\underset{\underset{\displaystyle H}{|}}{C}}-\overset{\cdot\cdot}{\underset{\cdot\cdot}{O}}-\overset{\overset{\displaystyle |}{|}}{\underset{\underset{\displaystyle H}{|}}{C}}-\overset{\overset{\displaystyle H}{|}}{\underset{\underset{\displaystyle H}{|}}{C}}-H$$

$$\underset{CH_3OCHCH_3}{\overset{CH_3}{|}} \text{ or } CH_3OCH(CH_3)_2$$

Lewis structure for methyl isopropyl ether   condensed formula for methyl isopropyl ether

connectivity: one carbon atom bonded to an oxygen that is bonded to two more carbon atoms in a row, with a third carbon atom attached to the first of these carbons

$$H-\overset{\overset{\displaystyle H}{|}}{\underset{\underset{\displaystyle H}{|}}{C}}-\overset{\overset{\displaystyle H}{|}}{\underset{\underset{\displaystyle H}{|}}{C}}-\overset{\cdot\cdot}{\underset{\cdot\cdot}{O}}-\overset{\overset{\displaystyle H}{|}}{\underset{\underset{\displaystyle H}{|}}{C}}-\overset{\overset{\displaystyle H}{|}}{\underset{\underset{\displaystyle H}{|}}{C}}-H$$

$CH_3CH_2OCH_2CH_3$

Lewis structure for diethyl ether   condensed formula for diethyl ether

connectivity: two carbon atoms in a row bonded to an oxygen atom that is bonded to two other carbon atoms in a row

1.2   The names (and the connectivities) of the compounds in this problem are related to some of those in Problem 1.1. See if you can find the pattern.

1.2 (cont)

$$H-\underset{\underset{H}{|}}{\overset{\overset{H}{|}}{C}}-\underset{\underset{H}{|}}{\overset{\overset{H}{|}}{C}}-\underset{\underset{H}{|}}{\overset{\overset{H}{|}}{C}}-\underset{\underset{H}{|}}{\overset{\overset{H}{|}}{C}}-\overset{\cdot\cdot}{\underset{\cdot\cdot}{Br}}:$$

$$CH_3CH_2CH_2CH_2Br$$

Lewis structure for 1-bromobutane   condensed formula for 1-bromobutane

connectivity: four carbon atoms in a row with the bromine atom at the end of the row

$$H-\underset{\underset{H}{|}}{\overset{\overset{H}{|}}{C}}-\underset{\underset{H}{|}}{\overset{\overset{H}{|}}{C}}-\underset{\underset{:Br:}{|}}{\overset{\overset{H}{|}}{C}}-\underset{\underset{H}{|}}{\overset{\overset{H}{|}}{C}}-H$$

$$CH_3CH_2CHCH_3 \text{ or } CH_3CH_2CHBrCH_3$$
$$|$$
$$Br$$

Lewis structure for 2-bromobutane   condensed formula for 2-bromobutane

connectivity: four carbon atoms in a row with the bromine atom on the second carbon atom of the row

$$H-\underset{\underset{H}{|}}{\overset{\overset{H}{|}}{C}}-\underset{\underset{:Br:}{|}}{\overset{\overset{\overset{\overset{H}{|}}{C}}{|}}{C}}-\underset{\underset{H}{|}}{\overset{\overset{H}{|}}{C}}-H$$

$$\overset{\overset{CH_3}{|}}{CH_3}\underset{\underset{Br}{|}}{C}CH_3 \text{ or } (CH_3)_3CBr$$

Lewis structure for 2-bromo-2-methylpropane   condensed formula for 2-bromo-2-methylpropane

connectivity: three carbon atoms in a row with the bromine atom and a carbon atom on the second carbon atom of the row

$$H-\underset{\underset{H}{|}}{\overset{\overset{H}{|}}{C}}-\underset{\underset{H}{|}}{\overset{\overset{\overset{\overset{H}{|}}{C}}{|}}{C}}-\underset{\underset{H}{|}}{\overset{\overset{H}{|}}{C}}-\overset{\cdot\cdot}{\underset{\cdot\cdot}{Br}}:$$

$$\overset{\overset{CH_3}{|}}{CH_3}CHCH_2Br \text{ or } (CH_3)_2CHCH_2Br$$

Lewis structure for 1-bromo-2-methylpropane   condensed formula for 1-bromo-2-methylpropane

connectivity: three carbon atoms in a row with the bromine atom at the end of the row and a carbon atom on the second carbon atom of the row

1.3

$$H-\underset{\underset{H}{|}}{\overset{\overset{H}{|}}{C}}-\underset{\underset{H}{|}}{\overset{\overset{H}{|}}{C}}-\underset{\underset{H}{|}}{\overset{\overset{H}{|}}{C}}-\overset{\cdot\cdot}{N}-H$$

$$CH_3CH_2CH_2NH_2$$

Lewis structure for propylamine   condensed formula for propylamine

connectivity: three carbon atoms in a row with the nitrogen atom at the end of the row

$$H-\underset{\underset{H}{|}}{\overset{\overset{H}{|}}{C}}-\underset{\underset{:N-H}{|}}{\overset{\overset{H}{|}}{C}}-\underset{\underset{H}{|}}{\overset{\overset{H}{|}}{C}}-H$$

$$CH_3CHCH_3 \text{ or } (CH_3)_2CHNH_2$$
$$|$$
$$NH_2$$

Lewis structure for isopropylamine   condensed formula for isopropylamine

connectivity: three carbon atoms in a row with the nitrogen atom on the second carbon atom

1.3 (cont)

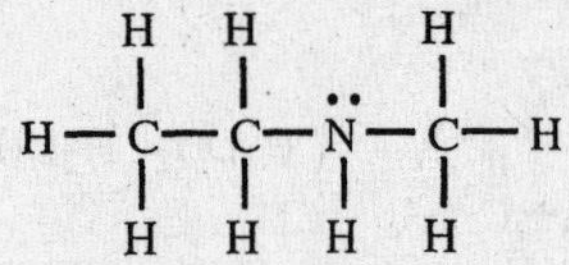

Lewis structure for methylethylamine

$CH_3CH_2NHCH_3$

condensed formula for methylethylamine

connectivity: two carbon atoms in a row attached to a nitrogen atom that is bonded to one other carbon atom

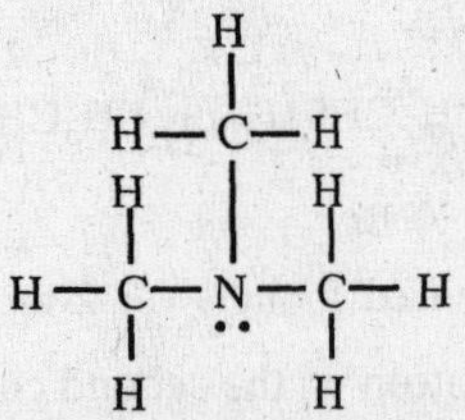

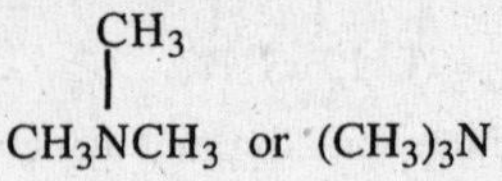

CH_3NCH_3 or $(CH_3)_3N$

Lewis structure for trimethylamine

condensed formula for trimethylamine

connectivity: a nitrogen atom with three single carbon atoms bonded to it

_ _

Workbook Exercises

In most chemical reactions, molecules undergo changes in the connectivity, that is, the bonding, of their atoms. Learning how to identify these changes rapidly is a skill you should master.

EXERCISE I.   Identify these features in the following chemical reactions:

   (1) bonds broken
   (2) bonds formed
   (3) redistribution of bonding and nonbonding electrons

EXAMPLE

SOLUTION

While you are becoming comfortable with molecular structures, it is a good idea to use complete Lewis structures to follow the changes in connectivity. You should also examine the formal charges and electron configurations of the atoms involved in those changes.

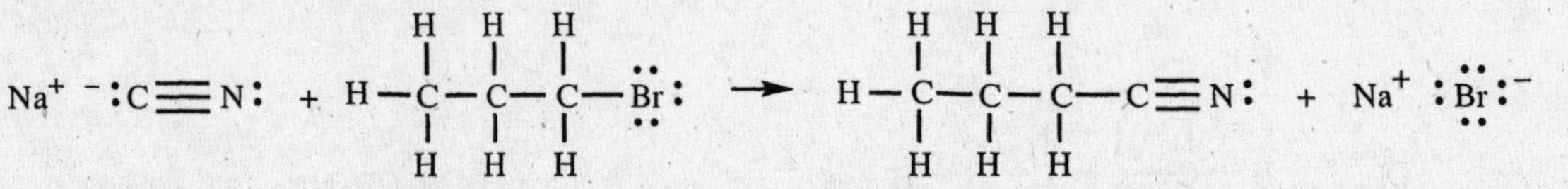

One bond is broken: the two-electron C—Br bond breaks, and this electron pair becomes the fourth nonbonding electron pair on the bromide ion that forms. One bond is formed: the nonbonding electron pair on the carbon atom of cyanide ion (CN^-) is used to form the new C—C bond.

EXERCISE I.   Identify the bonds broken, the bonds formed, and the way electrons have been redistributed in the processes of chemical change for the following reactions.

1. $(CH_3)_2CHCH_2\overset{..}{\underset{..}{O}}H$ + $Na^+ H\overset{..}{\underset{..}{:}}{}^-$ $\longrightarrow$ $(CH_3)_2CHCH_2\overset{..}{\underset{..}{O}}\overset{..}{:}{}^- Na^+$ + H—H

2.

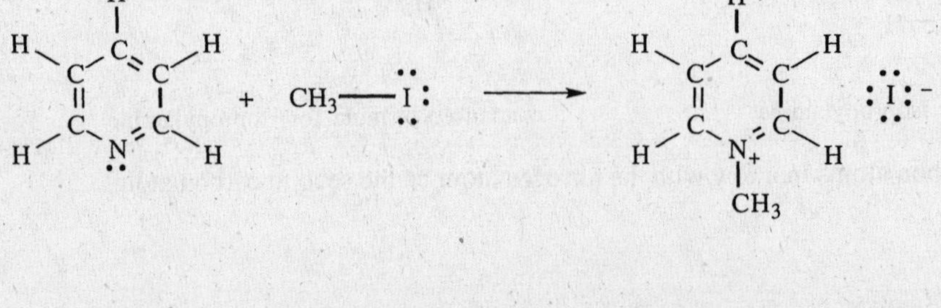

Workbook Exercises (cont)

3. 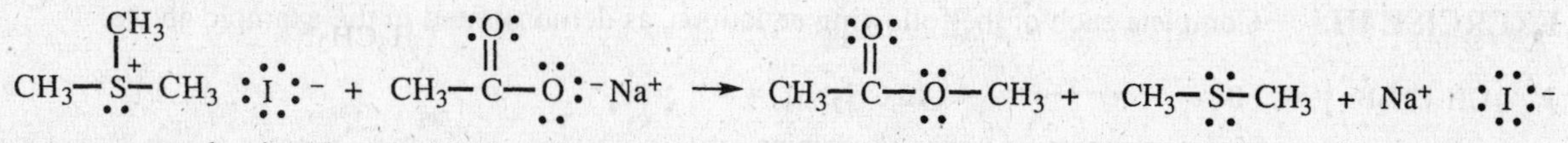 $CH_3-\overset{+}{\underset{CH_3}{S}}-CH_3$ $:\ddot{I}:^-$ + $CH_3-\overset{:\overset{..}{O}:}{\underset{}{C}}-\overset{..}{\underset{..}{O}}:^- Na^+$ $\longrightarrow$ $CH_3-\overset{:\overset{..}{O}:}{\underset{}{C}}-\overset{..}{\underset{..}{O}}-CH_3$ + $CH_3-\overset{..}{\underset{CH_3}{S}}-CH_3$ + Na^+ $:\ddot{I}:^-$

The next two examples involve double bonds, introduced in Section 1.5 of the text. Can you also describe these changes?

4. 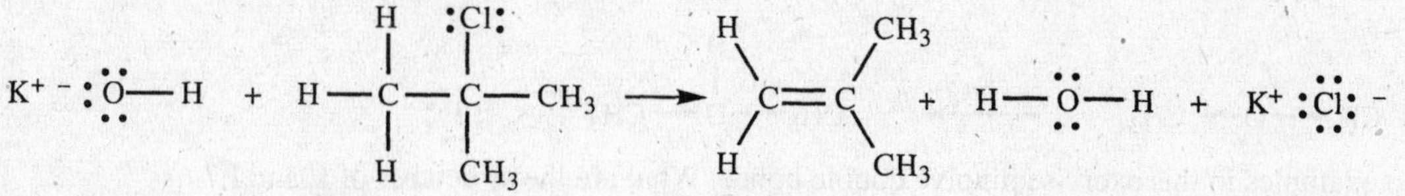 $K^+\,^-:\overset{..}{\underset{..}{O}}-H$ + $H-\overset{H}{\underset{H}{C}}-\overset{:\overset{..}{\underset{..}{Cl}}:}{\underset{CH_3}{C}}-CH_3$ $\longrightarrow$ $\overset{H}{\underset{H}{C}}=\overset{CH_3}{\underset{CH_3}{C}}$ + $H-\overset{..}{\underset{..}{O}}-H$ + $K^+\,:\overset{..}{\underset{..}{Cl}}:^-$

5. 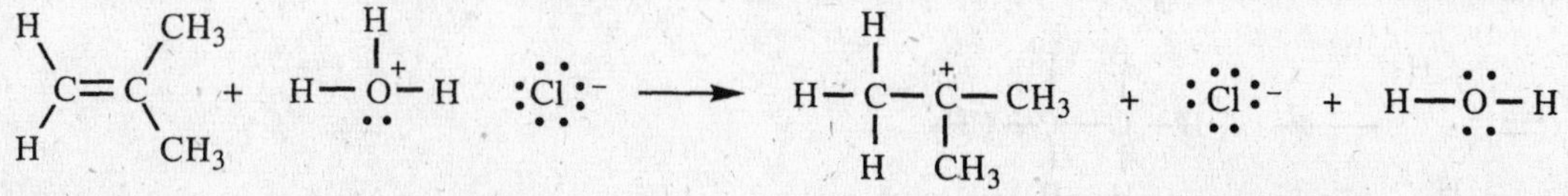 $\overset{H}{\underset{H}{C}}=\overset{CH_3}{\underset{CH_3}{C}}$ + $H-\overset{+}{\underset{H}{O}}-H$ $:\overset{..}{\underset{..}{Cl}}:^-$ $\longrightarrow$ $H-\overset{H}{\underset{H}{C}}-\overset{+}{\underset{CH_3}{C}}-CH_3$ + $:\overset{..}{\underset{..}{Cl}}:^-$ + $H-\overset{..}{\underset{..}{O}}-H$

EXERCISE II.   Complete each of the equations by providing the structure of the molecule that will balance the chemical equation. In a balanced equation, there must be an equal number of each kind of atom on both sides of the equation. The overall charge must also be the same on both sides of a chemical reaction equation. All of the atoms in the compounds in these exercises have closed shell configurations.

EXAMPLE

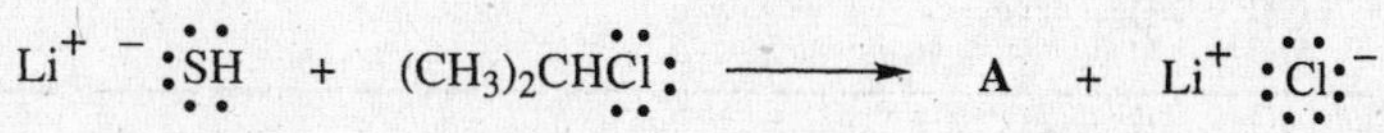

$$Li^+\,^-:\overset{..}{\underset{..}{S}}H + (CH_3)_2CH\overset{..}{\underset{..}{Cl}}: \longrightarrow A + Li^+ :\overset{..}{\underset{..}{Cl}}:^-$$

SOLUTION

As in Exercise I, we can identify some of the bonding changes. From this information, and the fact that the equation must balance, we determine the structure of the unknown substance **A**.

What we see: the C—Cl bond breaks, and the electron pair from the single bond becomes the fourth nonbonding electron pair of chloride ion on the right side of the equation. The unknown compound A must (a) incorporate the atoms from the left hand side of the equation that are not shown on the right, (b) have an overall neutral charge in order to keep the charges balanced, and (c) consist of atoms with closed shell configurations.

So, compound **A** needs to include (a) the SH atoms derived from the ionic compound LiSH, and (b) the atoms from $(CH_3)_2CHCl$, except for chloride ion which appears as a product. The product ion, Cl^-, must come from the uncharged molecule $(CH_3)_2CHCl$ because there is no other source of chlorine atoms on the left hand side of the equation. Although we cannot identify the structure of **A** with certainty at this point, we can account for the atoms from $(CH_3)_2CHCl$ that remain when the chloride ion, Cl^-, is removed. Perhaps only temporarily, we can imagine the presence of a positively-charged fragment, $CH_3-\overset{H}{\underset{+}{C}}-CH_3$, that comes from removing Cl^- from the uncharged starting compound.

In our imagination, the fragments we can use to make **A** are the cation, $CH_3-\overset{H}{\underset{+}{C}}-CH_3$, and the anion, $^-:\overset{..}{\underset{..}{S}}H$. There are two possible ways to create a compound from these fragments:

the first way gives us an ionic compound:

$$CH_3-\overset{H}{\underset{+}{C}}-CH_3$$
$$^-:\overset{..}{\underset{..}{S}}H$$

the second way gives us a covalent compound:

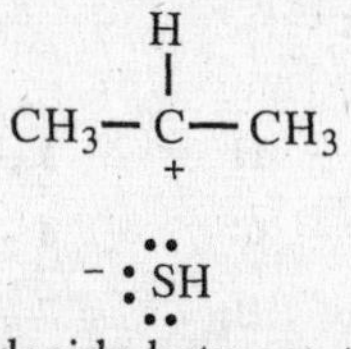

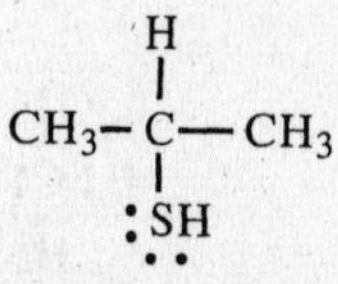

$$CH_3-\overset{H}{\underset{\overset{..}{\underset{..}{S}}H}{C}}-CH_3$$

Is there a way to decide between these two possible structures for **A**? Or are they both acceptable answers? When solving an open-ended problem such as this one, either in the homework or on an examination, it is important to check the assumptions and information given in the problem. In this case, rereading the question tells us that "All of the atoms in these compounds have closed shell configurations." Therefore, the covalent structure is the only one that satisfies this criterion. In the ionic structure, the positively charged carbon has an open shell.

Workbook Exercises (cont)

EXERCISE III.   Complete each of the following equations, as demonstrated in the example above.

1. $CH_3CH_2\overset{..}{\underset{..}{Br}}: + \textbf{B} \longrightarrow CH_3CH_2\overset{+}{N}H_3 + :\overset{..}{\underset{..}{Br}}:^-$

2. $CH_3-\overset{\overset{\textstyle :\overset{..}{O}:}{\|}}{C}-\overset{..}{\underset{..}{Cl}}: + Li^+ \; ^-:\overset{..}{\underset{..}{O}}-CH_3 \longrightarrow \textbf{C} + Li^+ \; :\overset{..}{\underset{..}{Cl}}:^-$

3. $\textbf{D} + CH_3-\overset{..}{\underset{..}{O}}-CH_3 \longrightarrow CH_3-\overset{\overset{\textstyle H}{|}}{\underset{..}{O}}\!\!^+\!\!-CH_3 + :\overset{..}{\underset{..}{F}}:^-$

The next two examples in this exercise involve double bonds. What are the structures of E and F?

4. $\textbf{E} + \overset{H}{\underset{H}{\diagdown}}C=C\overset{CH_3}{\underset{H}{\diagup}} \longrightarrow H-\overset{\overset{\textstyle Br}{|}}{\underset{\underset{\textstyle H}{|}}{C}}-\overset{\overset{\textstyle Br}{|}}{\underset{\underset{\textstyle H}{|}}{C}}-CH_3$

5. $Na^+ \; ^-:\overset{..}{\underset{..}{O}}-H + H-\overset{\overset{\textstyle :\overset{..}{Cl}:}{|}}{\underset{\underset{\textstyle :\overset{..}{Cl}:}{|}}{C}}-\overset{\overset{\textstyle :\overset{..}{Br}:}{|}}{\underset{\underset{\textstyle :\overset{..}{Br}:}{|}}{C}}-\overset{..}{\underset{..}{Br}}: \longrightarrow H-\overset{..}{\underset{..}{O}}-H + \textbf{F} + Na^+ \; :\overset{..}{\underset{..}{Br}}:^-$

-- --

1.4 The molecular formulas for the following compounds match the C_nH_{2n+2} rule, either right away or after hydrogen atoms are mentally substituted for halogen atoms.

    (c) $C_2HBrClF_3$ has 5 halogen atoms, therefore it is the equivalent of C_2H_6.

    (e) $C_{20}H_{42}$

    (g) $CHCl_3$ is the equivalent of CH_4

The structural formulas of these three compounds will contain only singly-bonded atoms. None of them can have a ring structure.

1.5 Formal charge = number of valence electrons − number of nonbonding electrons − $^1/_2$ number of bonding electrons.

(a) CCl_4

Cl $7 - 6 - \frac{1}{2}(2) = 0$

C $4 - 0 - \frac{1}{2}(8) = 0$

Each chlorine atom has 6 nonbonding electrons and 1 electron from a pair of bonding electrons. Therefore each chlorine atom has 7 valence electrons, the number it needs, and no formal charge. Carbon shares 8 bonding electrons. It effectively has 4 electrons around it, the number it needs to have no formal charge.

(b) CH_3Br

H $1 - 0 - \frac{1}{2}(2) = 0$

C $4 - 0 - \frac{1}{2}(8) = 0$

Br $7 - 6 - \frac{1}{2}(2) = 0$

(c) $CH_3\overset{+}{O}H_2$

H $1 - 0 - \frac{1}{2}(2) = 0$

C $4 - 0 - \frac{1}{2}(8) = 0$

O $6 - 2 - \frac{1}{2}(6) = +1$

(deficiency of one electron)

1.5

(d) $^-NH_2$ H:N:H   H—N—H

H	$1-0-\frac{1}{2}(2)=0$
N	$5-4-\frac{1}{2}(4)=-1$

(excess of one electron)

Each hydrogen atom shares 2 bonding electrons and therefore effectively has 1 electron and no formal charge. The nitrogen atom has 4 nonbonding electrons, and shares 4 bonding electrons. It effectively has 6 electrons around it, one more than the 5 electrons it needs to be uncharged. Nitrogen therefore has a formal charge of –1.

(e) $CH_3NH_3^+$

H	$1-0-\frac{1}{2}(2)=0$
C	$4-0-\frac{1}{2}(8)=0$
N	$5-0-\frac{1}{2}(8)=+1$

(deficiency of one electron)

(f) H_2NNH_2

H	$1-0-\frac{1}{2}(2)=0$
N	$5-2-\frac{1}{2}(6)=0$

(g) PH_3

P	$5-2-\frac{1}{2}(6)=0$
H	$1-0-\frac{1}{2}(2)=0$

(h) H_2S

H	$1-0-\frac{1}{2}(2)=0$
S	$6-4-\frac{1}{2}(4)=0$

(i) CH_3CH_2OH

H	$1-0-\frac{1}{2}(2)=0$
C	$4-0-\frac{1}{2}(8)=0$
O	$6-4-\frac{1}{2}(4)=0$

(j) $HOCH_2CH_2OH$

H	$1-0-\frac{1}{2}(2)=0$
C	$4-0-\frac{1}{2}(8)=0$
O	$6-4-\frac{1}{2}(4)=0$

(k) $CH_3OCH_3^+$ CH_3

H	$1-0-\frac{1}{2}(2)=0$
C	$4-0-\frac{1}{2}(8)=0$
O	$6-2-\frac{1}{2}(6)=+1$

(deficiency of one electron)

Each hydrogen atom shares 2 bonding electrons and therefore effectively has 1 electron and no formal charge. Each carbon atom shares 8 bonding electrons and thus effectively has 4 electrons and no formal charge. The oxygen atom has 2 nonbonding electrons and shares 6 bonding electrons. It effectively has 5 electrons, 1 fewer than the 6 electrons it needs to be uncharged, and therefore has a formal charge of +1.

1.6 (a)

$$CH_3-\overset{\underset{\displaystyle CH_3}{|}}{\overset{\displaystyle \overset{CH_3}{|}\overset{+}{}}{N}}-CH_3$$

N $5-0-\frac{1}{2}(8)=+1$

Each hydrogen atom shares 2 bonding electrons, and therefore effectively has 1 electron and no formal charge. Each carbon atom shares 8 bonding electrons, and thus effectively has 4 electrons and no formal charge. The nitrogen atom shares 8 bonding electrons. It effectively has 4 electrons, 1 fewer than the 5 electrons it needs to be uncharged, and, therefore, has a formal charge of +1.

(b) $:\!\ddot{Br}\!-\!\ddot{C}\!-\!\ddot{Br}\!:$

C $4-2-\frac{1}{2}(4)=0$

Each bromine atom has 6 nonbonding electrons and shares 2 bonding electrons, and, therefore, effectively has 7 electrons and no formal charge. The carbon atom has 2 nonbonding electrons and shares 4 bonding electrons, and thus effectively has 4 electrons and no formal charge.

(c)

$$CH_3-\overset{\displaystyle \overset{H}{|}\overset{+}{}}{\underset{\displaystyle ..}{O}}-CH_3$$

O $6-2-\frac{1}{2}(6)=+1$

(e) $:\!\ddot{Cl}\!-\!\overset{\displaystyle \overset{:\ddot{Cl}:}{|}\overset{-}{}}{\underset{..}{C}}\!-\!\ddot{Cl}\!:$

C $4-2-\frac{1}{2}(6)=-1$

Each chlorine atom has 6 nonbonding electrons and shares 2 bonding electrons, and, therefore, effectively has 7 electrons and no formal charge. The carbon atom has 2 nonbonding electrons and shares 6 bonding electrons. It effectively has 5 electrons around it, one more than the 4 electrons it needs to be uncharged, and, therefore, has a formal charge of –1.

(d) $CH_3-\overset{..}{\underset{..}{N}}-H$ (with − charge)

N $5-4-\frac{1}{2}(4)=-1$

(f)

$$CH_3-\overset{\displaystyle \overset{CH_3}{|}}{\underset{}{C}}\overset{+}{-}CH_3$$

C $4-0-\frac{1}{2}(6)=+1$

1.7

$$:\!\ddot{F}\!-\!\overset{\displaystyle \overset{:\ddot{F}:}{|}\overset{-}{}}{\underset{\displaystyle \overset{:\ddot{F}:}{|}}{B}}\!-\!\overset{\displaystyle \overset{H}{|}\overset{+}{}}{\underset{\displaystyle \overset{H}{|}}{N}}\!-\!H$$

F $7-6-\frac{1}{2}(2)=0$

H $1-0-\frac{1}{2}(2)=0$

B $3-0-\frac{1}{2}(8)=-1$ (excess of one electron)

N $5-0-\frac{1}{2}(8)=+1$ (deficiency of one electron)

Each fluorine atom has 6 nonbonding electrons and shares 2 bonding electrons. Each fluorine atom therefore effectively has 7 electrons and no formal charge. Each hydrogen atom shares 2 bonding electrons and therefore effectively has 1 electron and no formal charge. The nitrogen atom shares 8 bonding electrons. It effectively has 4 electrons, 1 fewer than the 5 electrons it needs to be uncharged, and therefore has a formal charge of +1. The boron atom shares 8 bonding electrons. It effectively has 4 electrons around it, one more than the 3 electrons it needs to be uncharged, and, therefore, has a formal charge of –1.

1.8 BF$_3$ has room in its orbitals to accept a pair of electrons. We expect it to react with any uncharged or negatively charged species that has nonbonding electrons.

$$:\!\ddot{F}\!-\!\overset{\displaystyle \overset{:\ddot{F}:}{|}}{\underset{\displaystyle \overset{:\ddot{F}:}{|}}{B}} \quad + \quad H-\overset{..}{\underset{..}{O}}-H \quad \longrightarrow \quad :\!\ddot{F}\!-\!\overset{\displaystyle \overset{:\ddot{F}:}{|}\overset{-}{}}{\underset{\displaystyle \overset{:\ddot{F}:}{|}}{B}}\!-\!\overset{\displaystyle \overset{H}{}\overset{+}{}}{\underset{..}{O}}\!-\!H$$

1.8 (cont)

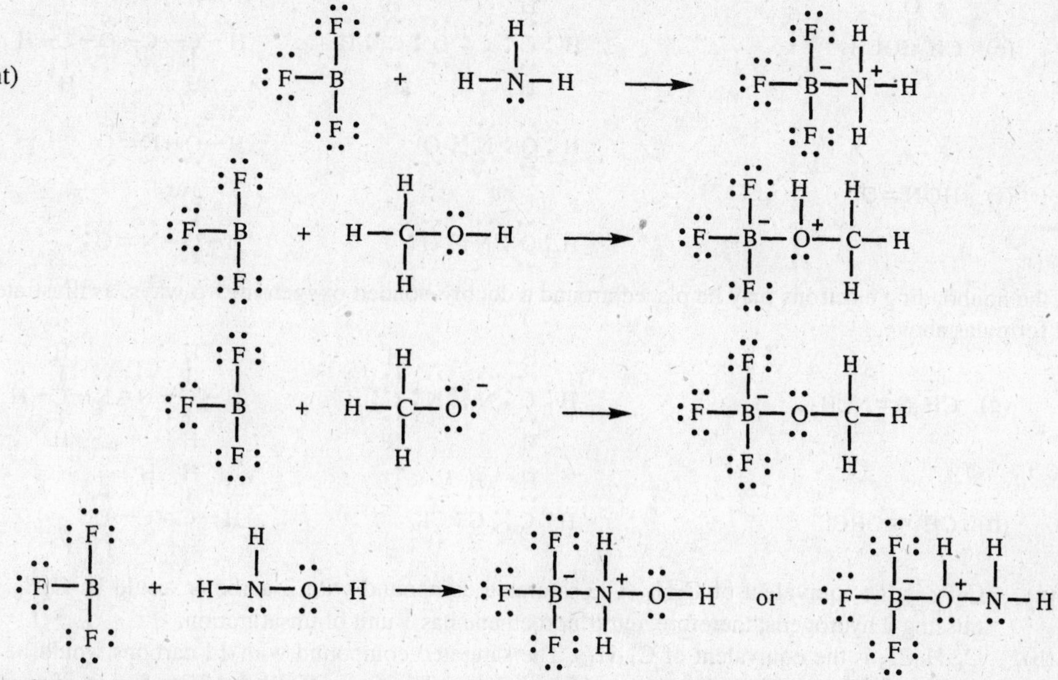

Even though the hydronium ion has a pair of nonbonding electrons, the positive charge on the ion makes them unlikely to participate in further bonding. The electrons on any positively charged species are tightly held, and not easily donated to another atom. Exploring the result of the reaction of BF_3 with H_3O^+ is useful. The resulting species is quite unstable because of the double positive charge on the oxygen atom next to the negative charge on the boron.

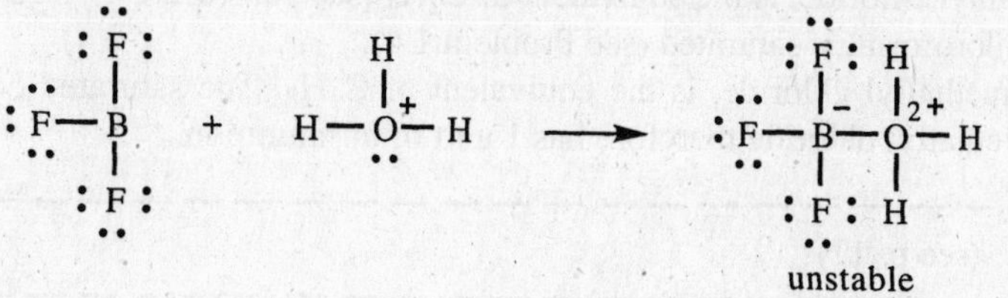

unstable

1.9   Condensed Formula   Lewis Structures

(a) CH_3CCH_3 (with O double-bonded to central C)

(b) $CH_3C{\equiv}CH$

(c) $HCNH_2$ (with O double-bonded to C)

(d) $HCOH$ (with O double-bonded to C)

1.9   (e) CH_3COCH_3

(f)   HON$=$O

Note that the nonbonding electrons may be placed around a doubly-bonded oxygen in two ways, as illustrated in the structural formulas above.

(g) $CH_3N=NCH_3$

(h) $CH_2=CHCl$

1.10   (a) C_2F_4 is the equivalent of C_2H_4. The saturated compound with 2 carbons would be C_2H_6. C_2H_4 is missing 2 hydrogens; therefore, tetrafluoroethene has 1 unit of unsaturation.

   (b) $C_{14}H_9Cl_5$ is the equivalent of $C_{14}H_{14}$. The saturated compound with 14 carbons would be $C_{14}H_{30}$. $C_{14}H_{14}$, DDT, is missing $(30-14)=16$ hydrogens; therefore, DDT has $16/2=8$ units of unsaturation.

   (c) $C_2HBrClF_3$, halothane, has 0 units of unsaturation (see Problem 1.4).

   (d) $C_{10}H_{16}$, adamantane, has 3 units of unsaturation ($C_{10}H_{22}$ is saturated).

   (e) $C_{20}H_{42}$, icosane, is saturated (see Problem 1.4).

   (f) C_2H_3Cl, vinyl chloride, is the equivalent of C_2H_4 [see part (a)].

   (g) $CHCl_3$, chloroform, is saturated (see Problem 1.4).

   (h) C_4H_7Cl, methallyl chloride, is the equivalent of C_4H_8. The saturated compound with 4 carbons is C_4H_{10}. Methallyl chloride therefore has 1 unit of unsaturation.

Concept Map 1.3   (see p. 12)

1.11   (a)

①
major and equivalent;
no separation of charge;
complete octets

②
minor;
separation of charge;
complete octets

③
minor;
separation of charge;
carbon has sextet

   (b)

①
major;
no separation of charge;
complete octets

②
minor;
separation of charge with
negative charge on more
electronegative element;
carbon has sextet

③
very minor;
separation of charge with
negative charge on less
electronegative element;
carbon has sextet

Note that resonance contributor 2 may nevertheless be significant in the chemical reactivity of a species (see Problems 1.12 and 1.13 for examples). Resonance contributor 3 does not explain the known chemistry of this compound and is, therefore, not considered a resonance contributor. It is shown only to complete the set. See (e) for another example of a very minor resonance contributor that is not useful in explaining the properties of the known species.

1.11 (c)

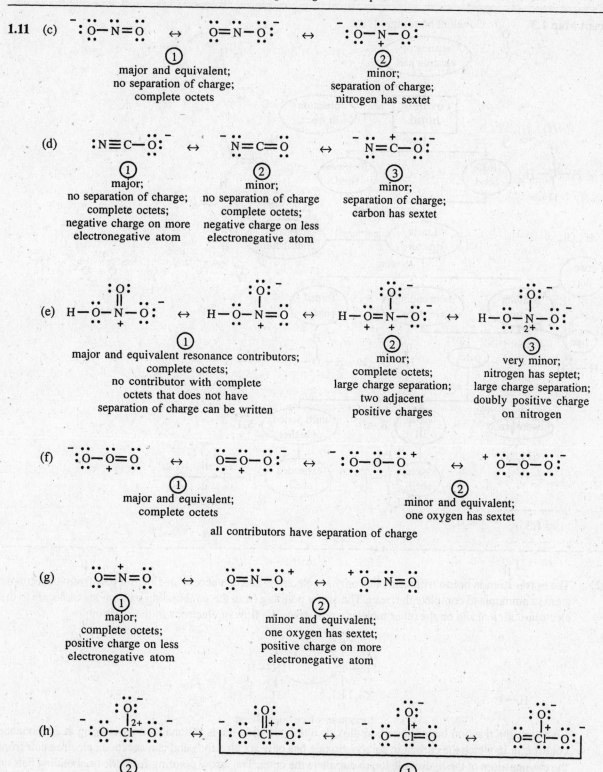

major and equivalent;
no separation of charge;
complete octets

minor;
separation of charge;
nitrogen has sextet

(d)

① major;
no separation of charge;
complete octets;
negative charge on more
electronegative atom

② minor;
no separation of charge
complete octets;
negative charge on less
electronegative atom

③ minor;
separation of charge;
carbon has sextet

(e)

① major and equivalent resonance contributors;
complete octets;
no contributor with complete
octets that does not have
separation of charge can be written

② minor;
complete octets;
large charge separation;
two adjacent
positive charges

③ very minor;
nitrogen has septet;
large charge separation;
doubly positive charge
on nitrogen

(f)

① major and equivalent;
complete octets

② minor and equivalent;
one oxygen has sextet

all contributors have separation of charge

(g)

① major;
complete octets;
positive charge on less
electronegative atom

② minor and equivalent;
one oxygen has sextet;
positive charge on more
electronegative atom

(h)

② minor;
large separation of charge

① major and equivalent;
less separation of charge;
more covalent bonds

Note that chlorine is in the third period of the periodic table and may, therefore, have more than eight electrons around it.

Concept Map 1.3   Covalent bonding,

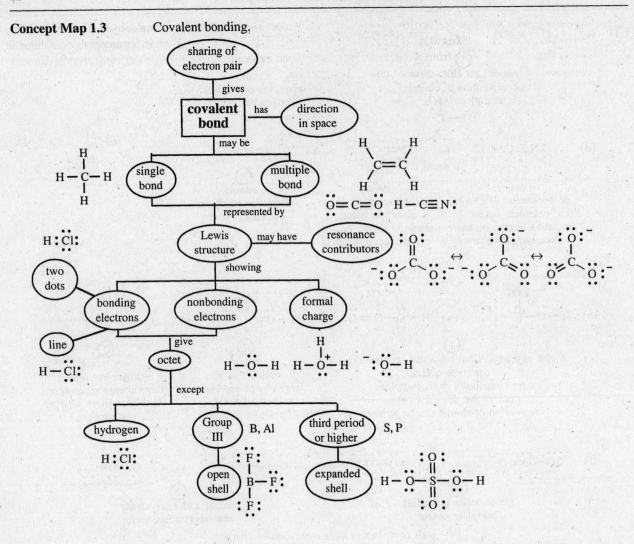

- -

1.12   The boron atom in boron trifluoride has only six electrons and can accept an electron pair from the nitrogen atom in ammonia to complete the octet. The arrow pointing from the nonbonding pair on one molecule to the electron-deficient site on the other molecule represents the flow of electrons in the reaction.

formation of covalent bond

Likewise, the reaction between carbon dioxide and hydroxide ion is rationalized by looking at a resonance contributor in which the carbon in carbon dioxide has only six electrons and can accept an electron pair from the oxygen atom of the hydroxide ion to complete the octet. The arrow pointing from the nonbonding pair on one molecule to the electron-deficient site on the other molecule again represents the flow of electrons in the reaction.

formation of covalent bond

1.13 Ammonia will react with formaldehyde. In the minor resonance contributor of formaldehyde (Problem 1.11b), the carbon atom has a sextet and can accept an electron pair from the nitrogen atom in ammonia to complete the octet. The arrow pointing from the nonbonding pair on one molecule to the electron-deficient site on the other molecule represents the flow of electrons in this reaction.

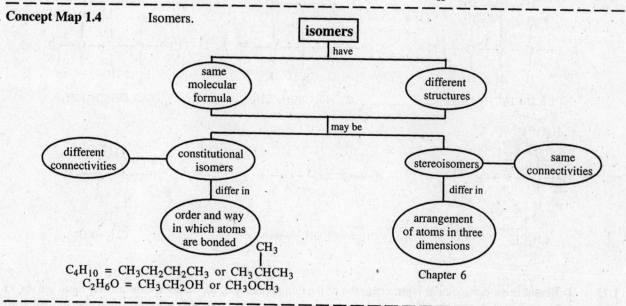

formation of covalent bond

Concept Map 1.4   Isomers.

isomers

have

same molecular formula

different structures

may be

different connectivities — constitutional isomers

stereoisomers — same connectivities

differ in

order and way in which atoms are bonded

$C_4H_{10} = CH_3CH_2CH_2CH_3$ or $CH_3\overset{\overset{\displaystyle CH_3}{|}}{C}HCH_3$

$C_2H_6O = CH_3CH_2OH$ or CH_3OCH_3

differ in

arrangement of atoms in three dimensions

Chapter 6

1.14 C_3H_7Cl

$CH_3CH_2CH_2Cl$

$CH_3CHClCH_3$

C_3H_8O

$CH_3CH_2CH_2OH$

$CH_3CHOHCH_3$

$CH_3OCH_2CH_3$

$C_4H_8Cl_2$

$ClCH_2CH_2CH_2CH_2Cl$

$ClCH_2CH_2CHClCH_3$

$ClCH_2CHClCH_2CH_3$

1.14 (cont)

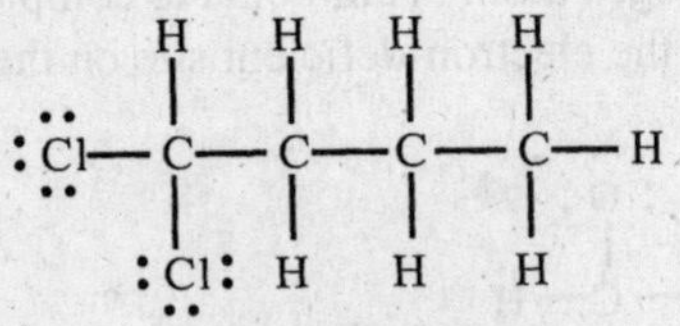

CHCl₂CH₂CH₂CH₃   CH₃CH₂CCl₂CH₃   CH₃CHClCHClCH₃

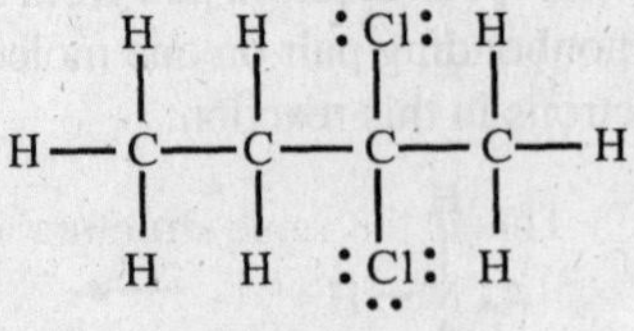

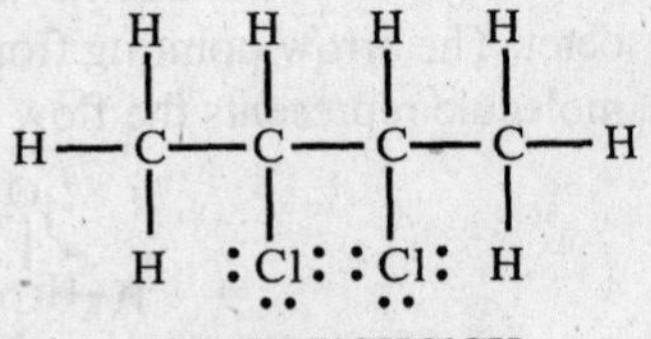

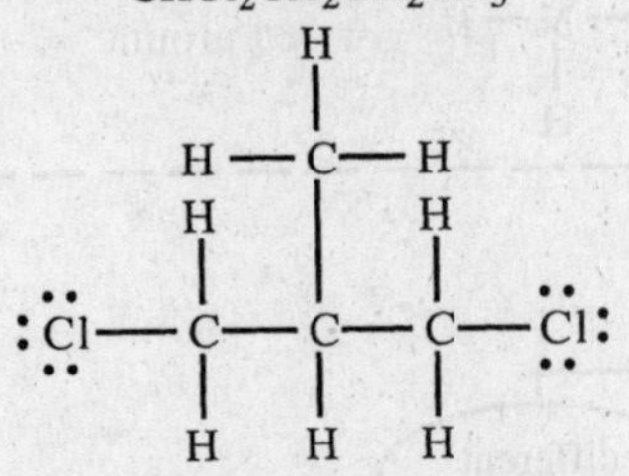

ClCH₂CH(CH₃)CH₂Cl   ClCH₂CCl(CH₃)CH₃   CHCl₂CH(CH₃)CH₃

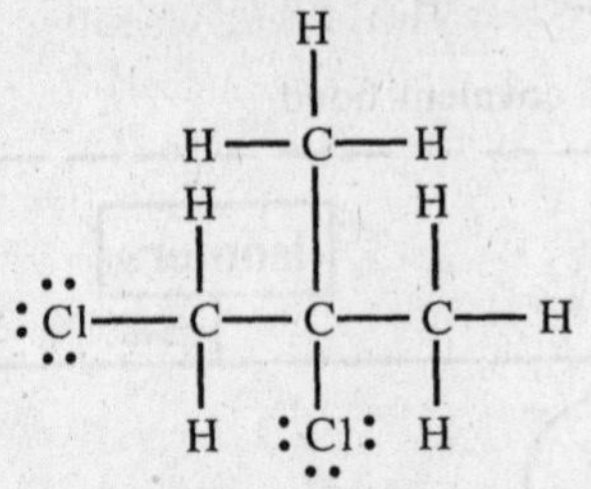

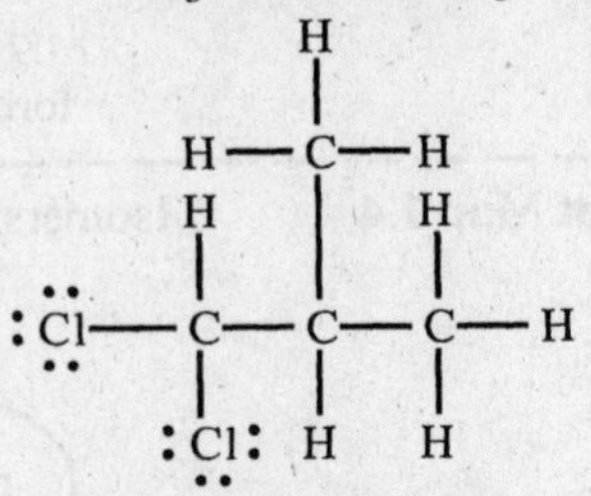

C₂H₄O

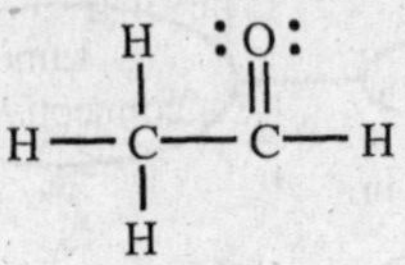

CH₃CH   CH₂=CHOH   (enol)   CH₂−CH₂
(with O)

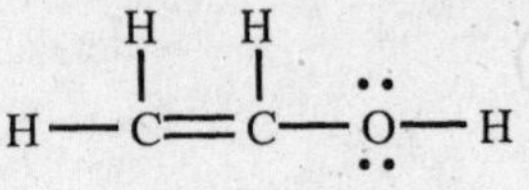

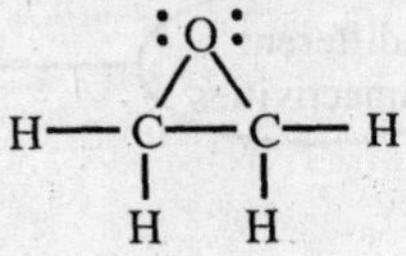

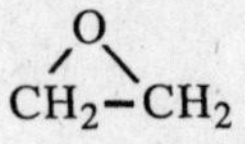

1.15 Different three-dimensional representations of the compounds and the relationships among them are shown below and on the next page. Note that there are many more possibilities for each compound than those that are shown. It will be helpful to work with molecular models to see the different orientations in space that molecules and parts of molecules may have.

(a) 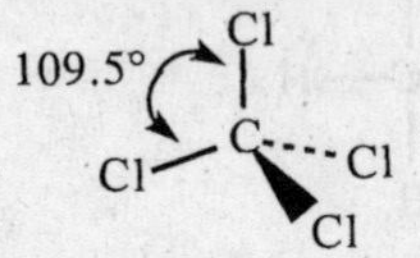 109.5°

This is the same structure as

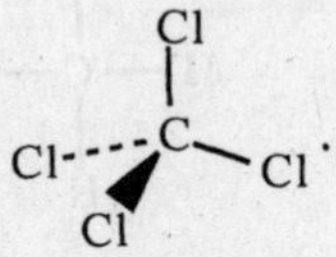

The formula representing the molecule has been rotated around an axis through the top chlorine atom and the carbon atom.

(b) 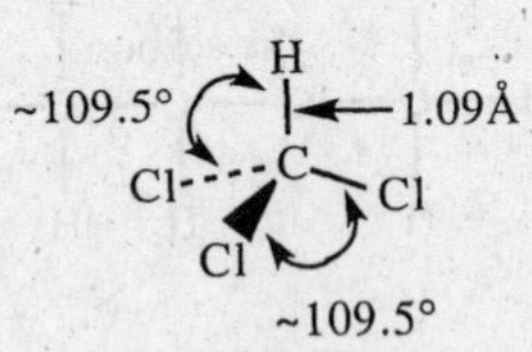 ~109.5°   1.09Å   ~109.5°

This is the same structure as

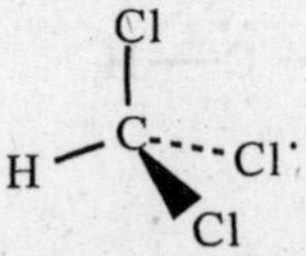

The formula representing the molecule has been rotated in the plane of the paper.

(c) 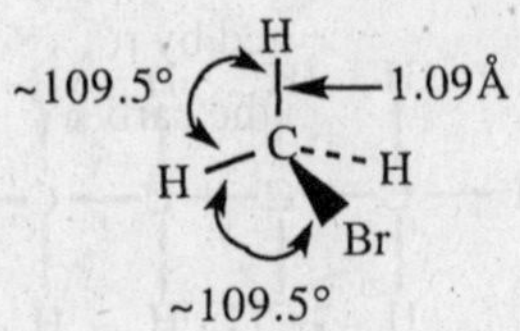 ~109.5°   1.09Å   ~109.5°

This is the same structure as

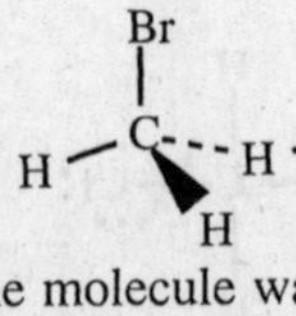

The formula representing the molecule was first rotated around an axis going through the top hydrogen atom and the carbon atom and then rotated in the plane of the paper.

1.15

(d)

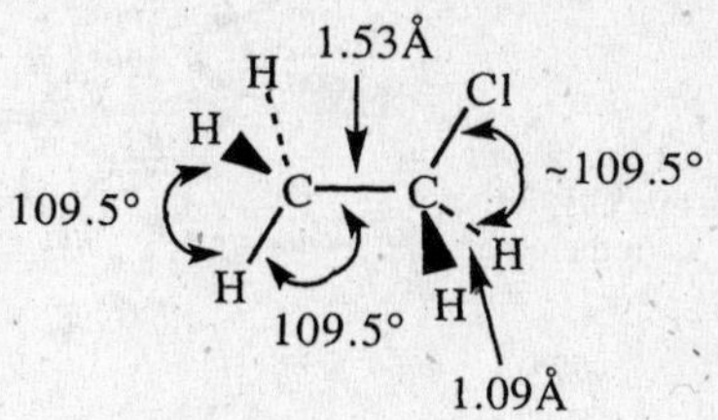

This is the same structure as

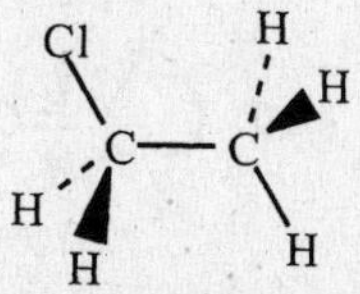

The formula representing the molecule was rotated around an axis perpendicular to the carbon-carbon bond.

(e)

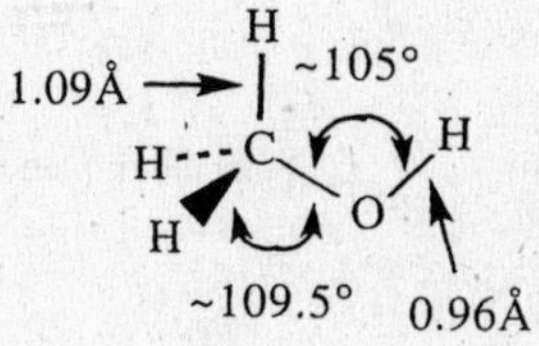

This is the same structure as

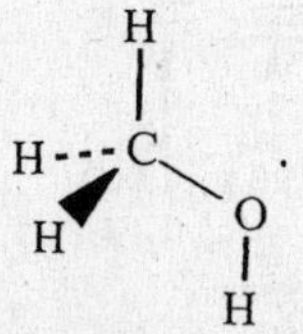

The formula representing the molecule was changed by rotating the O—H bond around an axis going through the carbon atom and the oxygen atom without rotating the carbon atom. This type of transformation will be discussed in greater detail in Chapter 5.

(f)

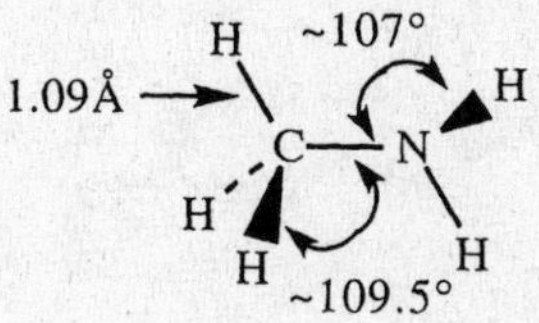

This is the same structure as

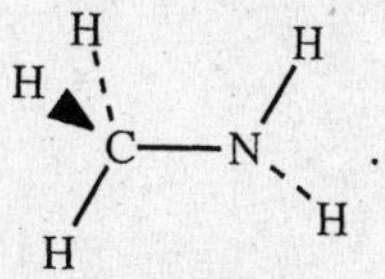

The formula representing the molecule was rotated around an axis going through the carbon atom and the nitrogen atom.

(g)

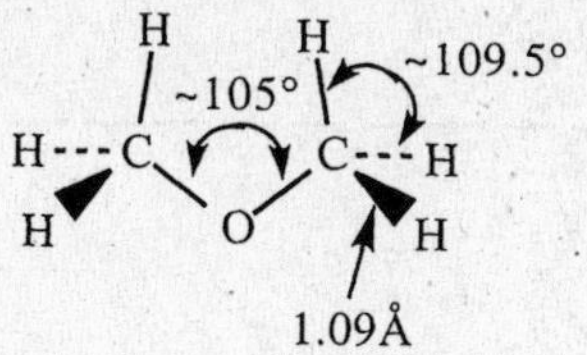

This is the same structure as

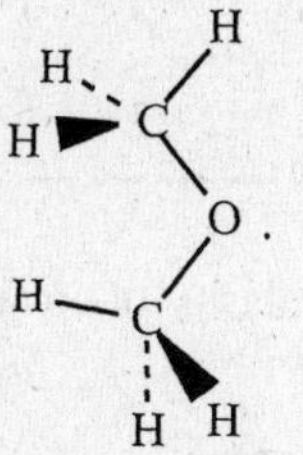

The formula representing the molecule was changed by rotating the CH_3 (methyl) group on the right about an axis through the carbon-oxygen bond on the left.

(h)

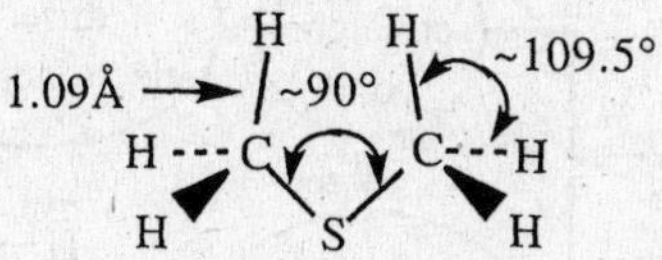

This is the same structure as

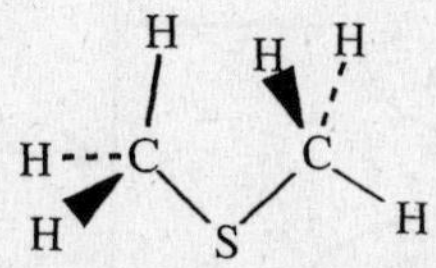

The formula representing the molecule was changed by rotating the CH_3 (methyl) group on the right about an axis through the carbon-sulfur bond on the left.

1.16

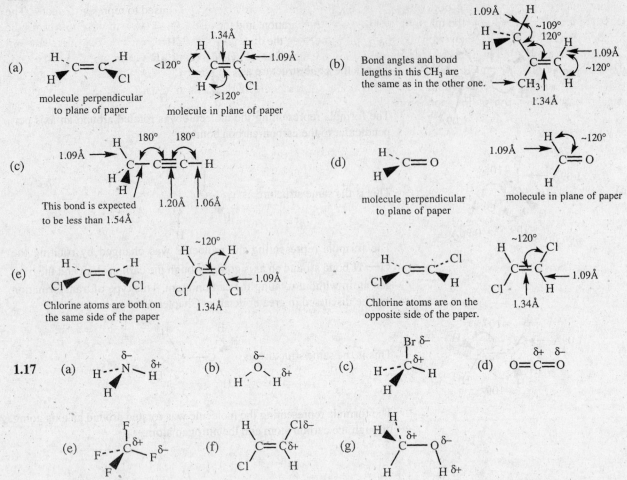

1.17

- -

Concept Map 1.5   Polarity of covalent molecules.

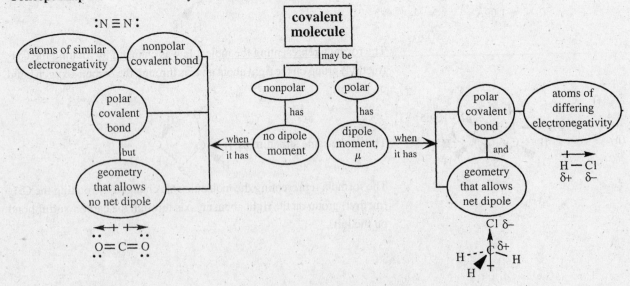

1.18   The molecular dipole is the **vector sum** of the individual bond dipoles in a molecule. A bond dipole is a **vector** quantity which by definition has both **magnitude** and **direction.** An arrow generally is used to represent a vector. The length of the arrow represents the magnitude of the charge separation, and the direction of the arrow points toward the negative end of the dipole. When two vectors are added together, the direction of the dipole must be considered as well as the magnitude. Thus, two vectors of equal magnitude and direction will yield a **resultant** vector pointing in the same direction with twice the magnitude as the original vectors, but two vectors of equal magnitude and opposite direction will exactly cancel out. For example, carbon dioxide is a linear molecule that has no dipole moment because it has equal bond dipoles that point in opposite directions.

$$O = C = O$$

vectors of individual bond dipoles exactly
cancel because the angle between them is 180°

Water is a bent molecule that has two equal bond dipoles pointing toward the oxygen atom away from the hydrogen atoms. The angle between the two bond dipoles is 104°. The resultant vector, the dipole moment, points toward the oxygen atom and bisects the angle between the two hydrogen atoms. The magnitude of the dipole moment is more than that of each individual bond dipole but less than the sum.

vectors of individual bond dipoles at
an angle result in a vector sum

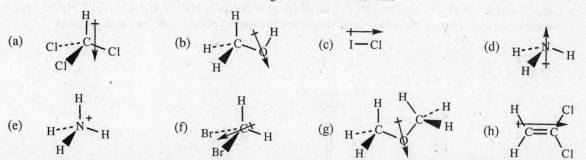

No dipole moment is shown for (e) because ions have full charges, either positive or negative, and thus do not have dipole moments, which are partial positive and negative charges separated within the same molecule.

1.19   (a) $CH_3CH_2CH_2CH_2OH$ has the highest boiling point because of hydrogen bonding.

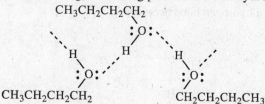

(b) CH_3OH has the highest boiling point because of hydrogen bonding.

(c) CH_3OH has the highest boiling point [see(b)]. The hydrogen bonding in CH_3SH is not significant (see Section 1.10C in the text).

1.20 (a) CH_3CH_2OH

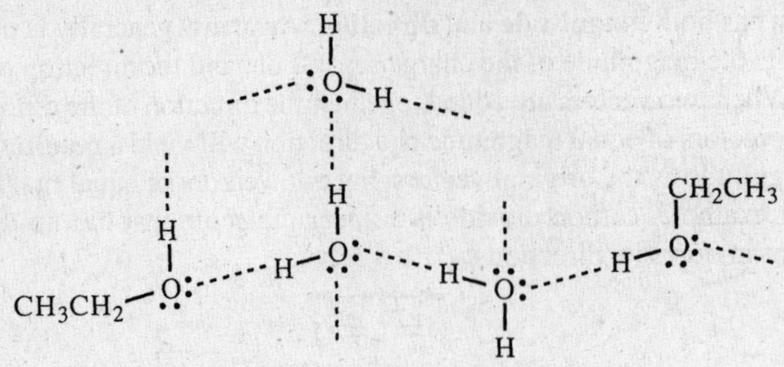

(b) $CH_3CH_2CH_2OH$
see (a) above and substitute $CH_3CH_2CH_2—$ for $CH_3CH_2—$

(c) $HOCH_2CH_2CH_2CH_2CH_2OH$

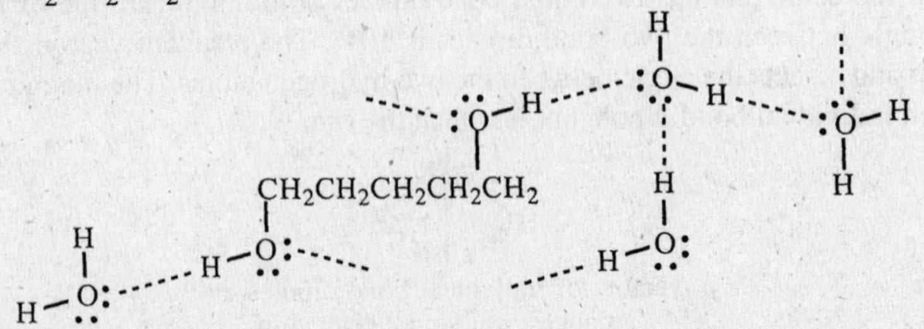

$CH_3CH_2CH_2CH_2CH_2OH$ has one half as many hydrogen bonds as $HOCH_2CH_2CH_2CH_2CH_2OH$ since it has only one —OH group. Note that the hydrocarbon (nonpolar) portions are the same length.

(d) $CH_3CH_2\overset{\displaystyle O}{\overset{\displaystyle \|}{C}}OH$

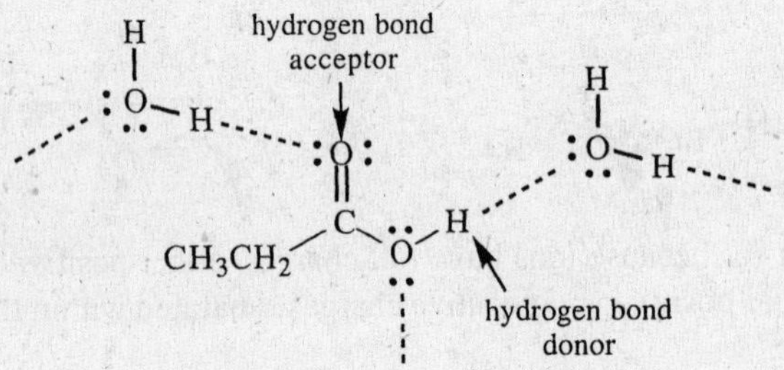

hydrogen bond acceptor

hydrogen bond donor

$CH_3\overset{\displaystyle O}{\overset{\displaystyle \|}{C}}OCH_3$ is only a hydrogen bond acceptor.

(e) $CH_3CH_2CH_2NH_2$

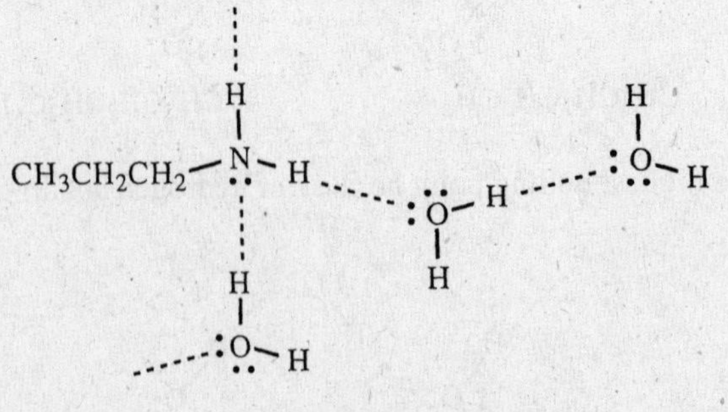

$CH_3CH_2CH_2Cl$ does not participate in hydrogen bonding to any significant extent and has little solubility in water.

Concept Map 1.6   Nonbonding interactions among molecules.

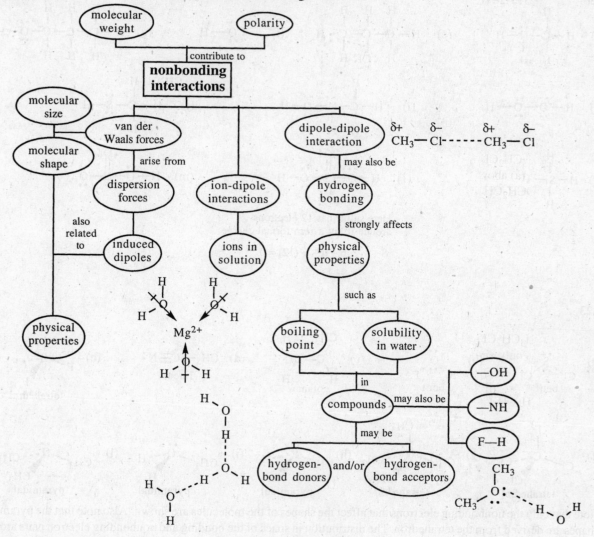

1.21 (a) MgF_2, ionic

(b) SiF_4, covalent
Si—F, polar bond

(c) NaH, ionic
(the anion is H$:^-$)

(d) ClF, covalent
polar bond

(e) SCl_2, covalent
S—Cl, polar bond

(f) OF_2, covalent
O—F, polar bond

(g) SiH_4, covalent
Si—H, polar bond

(h) PH_3, covalent

(i) $NaOCH_3$, ionic
(the anion, CH_3—$\overset{\cdot\cdot}{\underset{\cdot\cdot}{O}}:^-$,
contains covalent bonds)

(j) CH_3Na, ionic
(the anion, $^-:CH_3$,
contains covalent bonds)

(k) Na_2CO_3, ionic
(the anion, CO_3^{-2},
contains covalent bonds)

(l) BrCN, covalent
Br—C and C≡N,
polar bonds)

1.22

(a) $:\overset{\cdot\cdot}{F}—N—\overset{\cdot\cdot}{F}:$
       $\overset{|}{:\underset{\cdot\cdot}{F}:}$

(b) $:\overset{\cdot\cdot}{\underset{\cdot\cdot}{Cl}}—Al—\overset{\cdot\cdot}{\underset{\cdot\cdot}{Cl}}:$
       $\overset{|}{:\underset{\cdot\cdot}{Cl}:}$

(c) $H—\overset{H}{\underset{H}{C}}—\overset{\cdot\cdot}{\underset{\cdot\cdot}{S}}—\overset{H}{\underset{H}{C}}—H$

1.22

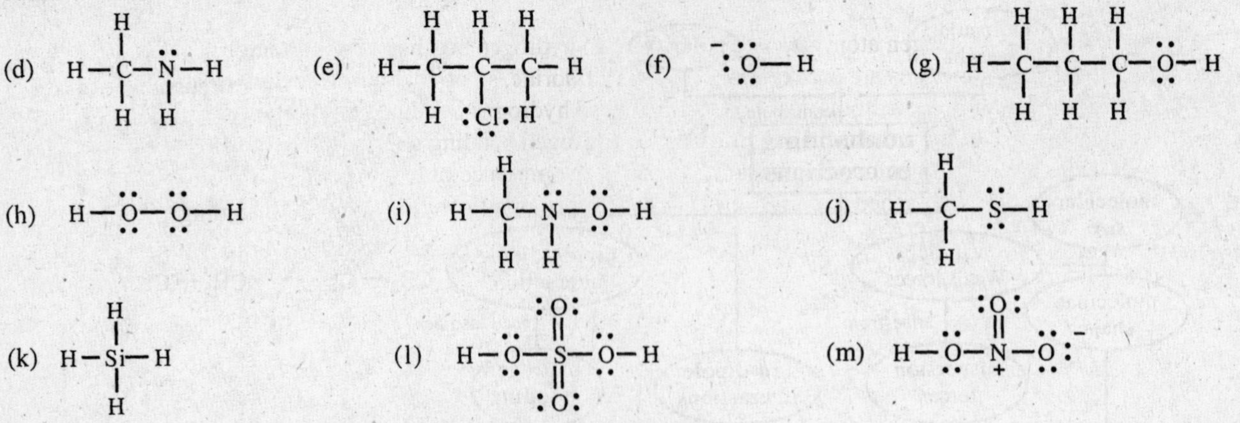

(d) H—C—N—H structure

(e) H—C—C—C—H structure

(f) :O—H

(g) H—C—C—C—O—H structure

(h) H—O—O—H

(i) H—C—N—O—H structure

(j) H—C—S—H structure

(k) H—Si—H structure

(l) H—O—S—O—H structure

Here sulfur has 12 electrons
around it but a zero formal charge

$$S \qquad 6 - 0 - \frac{1}{2}(12) = 0$$

(m) H—O—N—O⁻ structure

1.23

(a) Cl—S—Cl
bent

(b) F—O—F
bent

(c) CH₃, H / C=C / H, H
planar

(d) CH₃—C≡N:
linear

(e) B with F's
tetrahedral

(f) C with F, Cl, Cl, Cl
tetrahedral

(g) CH₃—N⁺(CH₃)(CH₃)—CH₃
tetrahedral

(h) H—P—H, H
pyramidal

(i) CH₃—O⁺—H, H
pyramidal

(j) CH₃—N—CH₃, CH₃
pyramidal

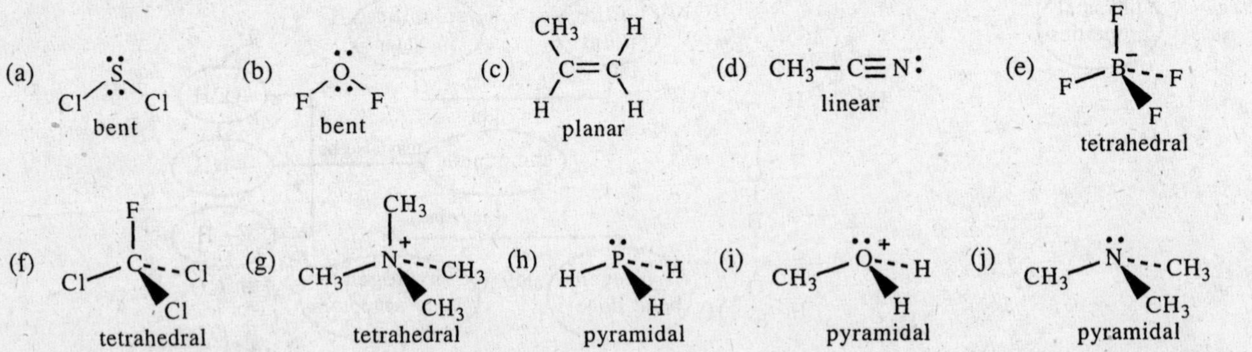

Note that only the nonbonding electrons that affect the shapes of the molecules are shown. Also note that the pyramidal shapes are derived from the tetrahedron. The distribution in space of the bonding and nonbonding electron pairs around these atoms is tetrahedral. The geometry around these atoms appears pyramidal because our experimental methods detect only the positions of the atoms and not the positions of the electrons.

1.24

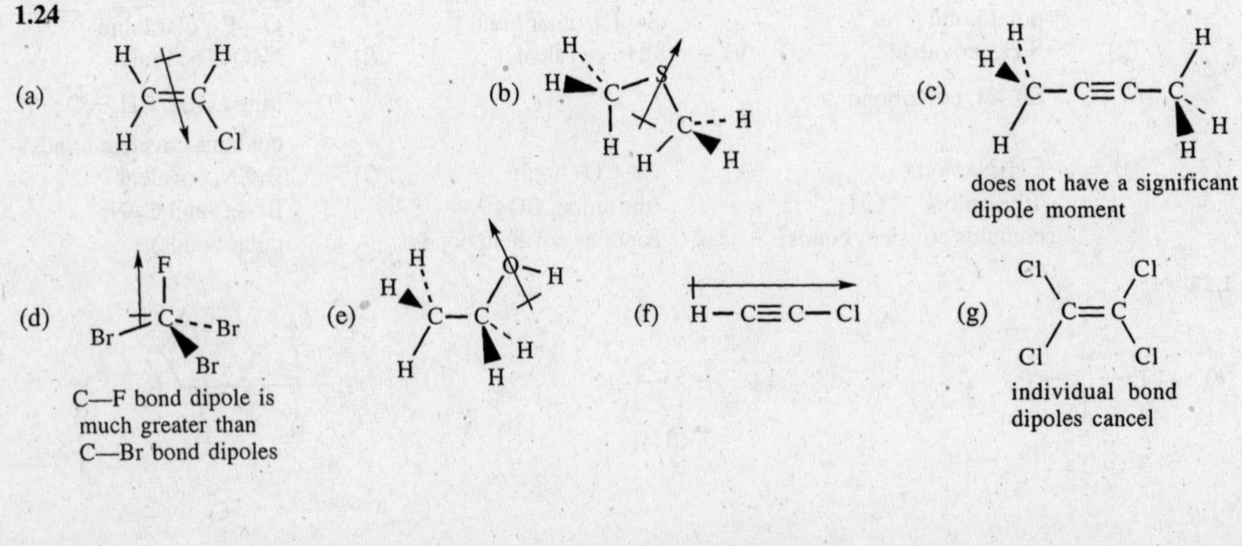

(a) C=C with H, H, H, Cl

(b) C—S—C structure

(c) C—C≡C—C structure
does not have a significant
dipole moment

(d) Br—C(F)—Br, Br
C—F bond dipole is
much greater than
C—Br bond dipoles

(e) C—O—H structure

(f) H—C≡C—Cl

(g) Cl, Cl / C=C / Cl, Cl
individual bond
dipoles cancel

1.25 (a) Hydrogen atoms are attached to carbon, not oxygen, nitrogen or fluorine. Hydrogen bonding is not important.

(b) There is a hydrogen atom attached to oxygen, so hydrogen bonding is important.

(c) No hydrogen atoms are on oxygen, nitrogen or fluorine, so no hydrogen bonding occurs.

(d) There are hydrogen atoms attached to oxygen, so hydrogen bonding is important.

(e) Hydrogen atoms are attached to nitrogen, so hydrogen bonding does occur.

(f) Here, too, hydrogen bonding occurs because of the presence of N—H bonds.

(g) The hydrogen attached to sulfur does not participate significantly in hydrogen bonding.

1.26

Sulfur is not a significant hydrogen-bond acceptor

1.28   (a) The important intermolecular interactions between the host molecule and creatinine are hydrogen bonds.

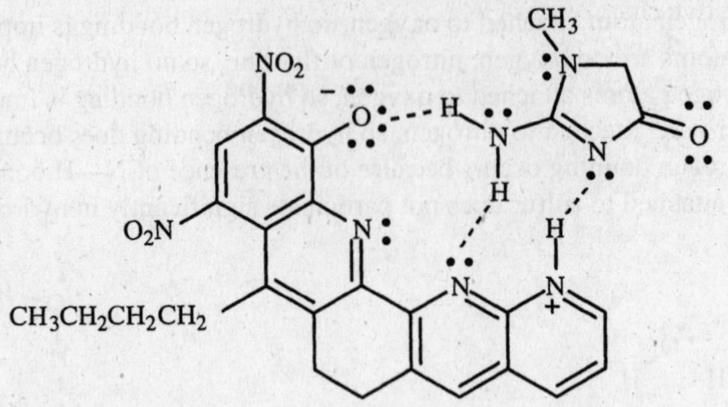

(b) Structure A is a constitutional isomer of creatinine in which a hydrogen atom has moved from one nitrogen atom to another nitrogen atom. Structure B is a resonance contributor of creatinine in which only electron pairs have moved.

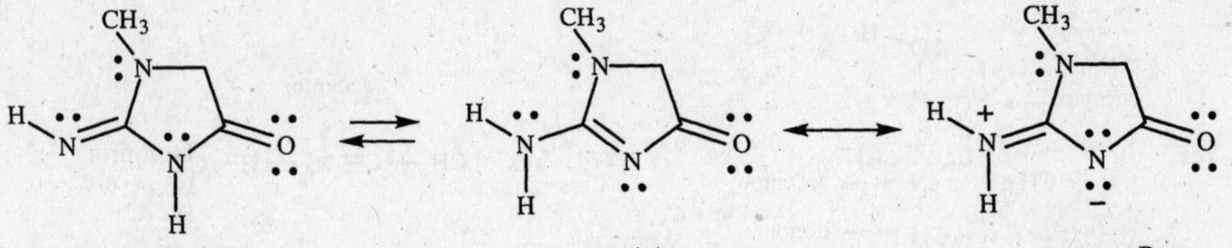

structure A
a constitutional isomer
of creatinine

creatinine

structure B
a resonance contributor
of creatinine

1.29

The shorter than expected B—F bond lengths suggest the compound has resonance contributors in which there is a double bond between the boron and each of the fluorine atoms. Such resonance contributors have the advantages of having an octet around each atom and additional bonding, even though they also put a positive charge on the electronegative fluorine atom.

1.30   (a) 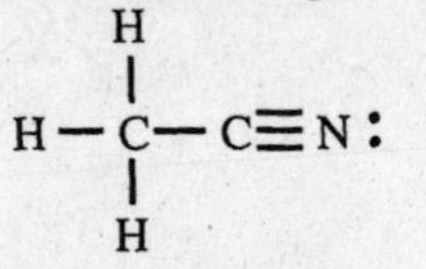   Lewis structure

(b) 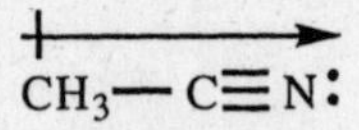 $CH_3 — C \equiv N$:   dipole moment

(c) 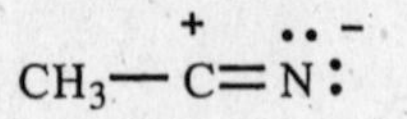 $CH_3 — C = N$:   Resonance contributor with separation of charge and negative formal charge on the more electronegative element. A separation of charge increases the dipole moment.

1.31   (a) NO has eleven electrons for bonding (5 from nitrogen and 6 from oxygen). It must, therefore, have one unpaired electron.

A   B   C   D

1.31 (b) A is the major resonance contributor. It has the maximum number of covalent bonds, one atom (O) has an octet, and the contributor has no separation of charge. B is the next best. Even though it has separation of charge and a positive charge on the more electronegative atom, one atom (N) has an octet, and the contributor has more bonds than C or D.

(c) NO is highly reactive because it has an open shell on one of its atoms.

1.32 (a)

| carbon does not have octet | both atoms have octets but negative charge is on less electronegative atom | carbon does not have an octet but negative charge is on more electronegative atom |

(b) Fe^{2+} is an electron pair **acceptor.** The carbon atom of carbon monoxide bonds to iron; it must be an electron pair **donor.** Therefore, the resonance contributor in which carbon has a negative charge and each atom has a complete octet is more important than the other contributors. The resonance contributor with separation of charge and a positive charge on the carbon atom would lead to electrostatic repulsion between Fe^{2+} and the carbon atom and must, therefore, not make a significant contribution to the structure.

1.33 (a) Dipole-dipole interactions must be the most important here. Acetone and 1,1,1-trifluoropropanone are hydrogen bond acceptors, but not donors. If van der Waals forces were the important factor, we would expect 1,1,1-trifluoropropanone to have the higher boiling point.

(b) The carbonyl group is highly polarized by resonance in acetone.

The corresponding resonance contributor in 1,1,1-trifluoropropanone is not as important because of the electron-withdrawing effects of the fluorine atoms. Therefore, the molecule as a whole has less of an overall dipole.

1.34

| ① | ② | ③ |
| major complete octets; no separation of charge | minor but important complete octets; separation of charge with negative charge on more electronegative atom | minor carbon has sextet; separation of charge |

1.35 SO$_2$ is an 18-electron molecule.

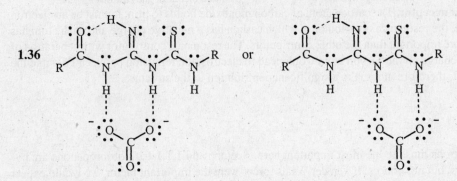

> Note that this structure is possible because S is in the third period of the periodic table.

These resonance contributors would lead us to expect the bond lengths in SO$_2$ to be somewhere between 1.49 Å and 1.70 Å. The experimental bond length (1.43 Å) is shorter than expected. Chemists point to the polarity of the bonds (the strong attraction between the positive charge on sulfur and the negative charge on oxygen, for example) as a further factor in rationalizing such observations.

1.36

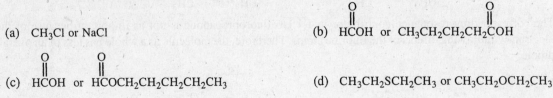

Supplemental Problems

S1.1 Which compound in each pair do you expect to be more soluble in water?

(a) CH$_3$Cl or NaCl

(b) HCOH or CH$_3$CH$_2$CH$_2$CH$_2$COH   (each with C=O)

(c) HCOH or HCOCH$_2$CH$_2$CH$_2$CH$_2$CH$_3$ (each with C=O)

(d) CH$_3$CH$_2$SCH$_2$CH$_3$ or CH$_3$CH$_2$OCH$_2$CH$_3$

(e) HOCCH$_2$CH$_2$CH$_3$ or HOCCH$_2$CH$_2$COH   (each with C=O)
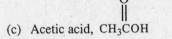

S1.2 The boiling point of water (100 °C) is higher than the boiling point of hydrogen fluoride (19 °C) or ethanol (78 °C). Use drawings to explain this experimental observation.

S1.3 Draw resonance contributors for each of the following species. Be sure to show any formal charges. Decide which resonance contributors are major and which are minor.
(a) Nitrous oxide, N$_2$O (a linear molecule in which the two nitrogen atoms are bonded to each other)
(b) Sulfate anion, SO$_4^{2-}$ (an ion in which all of the oxygens are equivalent)

(c) Acetic acid, CH$_3$COH   (with C=O)

S1.4 The azide anion, N$_3^-$, is linear. The nitrogen-nitrogen bond lengths in the ion are all 1.15Å. Write resonance contributors for the azide anion that account for these experimental observations. (The typical nitrogen-nitrogen double bond length is 1.20Å; the typical nitrogen-nitrogen triple bond length is 1.10Å.)

S1.5 Methyl isothiocyanate has the following connectivity: CH_3NCS.

(a) Draw a Lewis structure for the molecule in which all atoms have closed shells and no separation of charge.

(b) Methyl isothiocyanate is highly reactive toward reagents with nonbonding electrons to share such as hydroxide ion (OH^-). Draw structural formulas other than the structure in part (a) for two resonance contributors for methyl isothiocyanate that may be used to rationalize this reactivity.

(c) Methyl thiocyanate is a constitutional isomer of methyl isothiocyanate. It can be prepared from the thiocyanate ion, SCN^-, and methyl bromide, CH_3—Br, by a process called displacement. Draw a Lewis structure for this isomer, in which all atoms also have closed shells and no separation of charge.

S1.6 The anion derived from nitromethane is important in synthesizing new compounds. Chemists postulate that such an anion is easy to form because it is stabilized by resonance. The connectivity of the anion is $H_2CNO_2^-$.

(a) There are three important resonance contributors in which all atoms have closed shells. Draw these three resonance structures, being sure to show all nonbonding electrons and formal charges. Identify the major resonance contributors and give the reasons for your choice.

(b) The nitromethane anion reacts with positively charged species (such as the proton from an acid) to form a bond to the carbon atom. Which resonance contributor best rationalizes the reactivity described here?

S1.7 The concept of resonance is used to rationalize experimental properties of molecules or ions. Structural formulas for some species and experimental evidence about them are shown below. Draw resonance contributors for the structures that support the experimental evidence given.

the electrons on this nitrogen atom are not available for bonding as are those on $CH_3-\ddot{N}-CH_3$ (with CH_3 on N), for example

(a)

this carbon atom is reactive toward reagents that are deficient in electrons such as protons in acids

(b)

this molecule reacts easily with electron-deficient reagents at the carbon atoms indicated by arrows

(c)

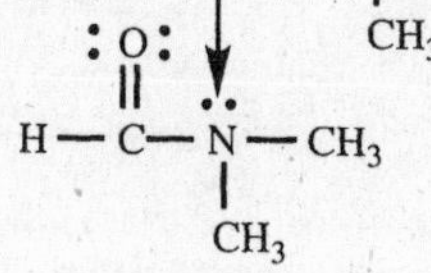

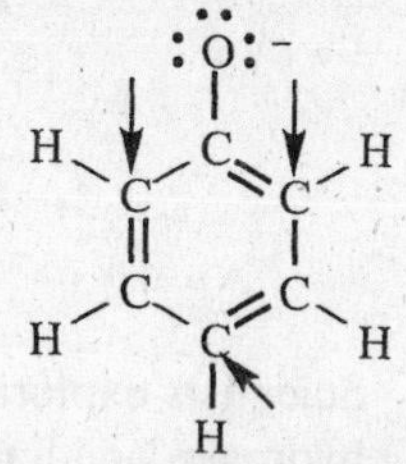

S1.8 Acetone, $(CH_3)_2CO$, is the solvent used in the laboratory to clean and dry glassware. Acetone is protonated on the oxygen atom to give a cation, $(CH_3)_2COH^+$.

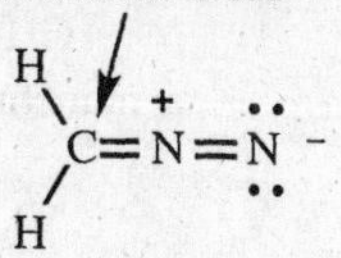

(a) Draw a correct Lewis structure for the protonated form of acetone and for its other resonance contributor.

(b) The carbon–oxygen bond in the protonated form of acetone is about 1.27 Å long. A typical carbon–oxygen double bond length is 1.20 Å; a carbon–oxygen single bond length is 1.43 Å. What do these data tell you about the relative importance of the two resonance contributors you wrote for protonated acetone? Why is the one you picked the major one?

S1.9 There are many compounds with the molecular formula $C_5H_{12}O$. They fall into two categories, those with boiling points above 100 °C and those with boiling points below 70 °C. Propose structural formulas for four of the constitutional isomers of $C_5H_{12}O$, and indicate whether you expect the boiling point of the compound you have drawn to be higher than 100 °C or lower than 70 °C. Your answer should include at least one compound of each category.

S1.10 Urea, $H_2N\overset{\overset{\large O}{\|}}{C}NH_2$, is the compound used by mammals to excrete nitrogen.

(a) What is the geometry of urea at the central carbon atom? Draw a structural formula that clearly illustrates this.

(b) Even though urea has a low molecular weight, it is a solid that melts at 135 °C. Use structural formulas to propose a rationalization for its high melting point relative to its molecular weight and size.

S1.10 (c) Urea has high solubility in water. Draw structural formulas showing the factors responsible for its water solubility.

(d) Urea easily loses a proton to a base to give an anion, H_2NCONH^-. This anion has two important resonance contributors, both of them closed shell species. What are they, and which one is the major one?

S1.11 Compounds with the molecular formula $C_5H_{10}O$ have one unit of unsaturation. Draw structural formulas for four possible constitutional isomers having this molecular formula, one containing a ring, one an alcohol function, one an ether, and one a carbonyl group. Other functional groups may also be present, depending on the structures you choose.

S1.12 About 20 years ago a new class of compounds was discovered in Munich (München in German), Germany, and named "munchnones." These compounds have a permanent separation of charge in a ring containing nitrogen and oxygen atoms and a carbonyl group. A munchnone has a much smaller dipole moment than would be expected for a compound with such a separation of charge. Chemists use the idea of delocalization of charge over the entire molecule by resonance to rationalize this experimental observation.

There are six resonance contributors to a munchnone (besides the one show below) that have closed shells around each of the atoms. Draw the four best (most important) ones of these six.

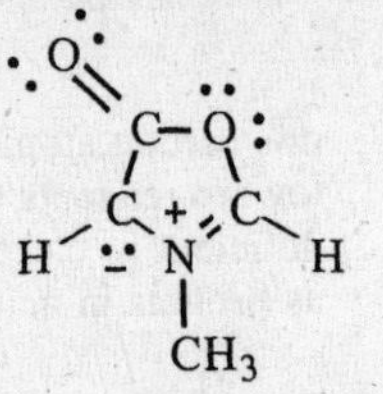

a munchnone

S1.13 Scientists exploring ways to create materials that can be easily taken apart and reassembled have turned to hydrogen bonding as a sort of molecular "Velcro." An array of hydrogen bonds gives rise to strong interactions between different polymer chains. The planar molecular units used to create this array of hydrogen bonds are shown below. Both intramolecular and intermolecular hydrogen bonds are possible within this set of molecules. Draw in the hydrogen bonds that will hold the two sets of molecules together, as well as any intramolecular hydrogen bonds that are possible. Label clearly one intermolecular and one intramolecular hydrogen bond, and use arrows to point to atoms that are hydrogen bond donors.

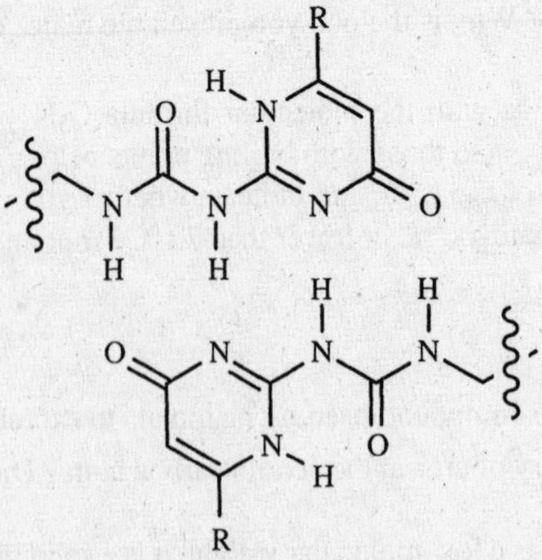

2 Covalent Bonding and Chemical Reactivity

Workbook Exercises

Simple geometrical relationships are an important part of chemistry. Some of these can be demonstrated with the two tetrahedral blocks you can construct using the two large cardboard triangles at the back of the *Study Guide*. Fold each along the three dotted lines to bring the three darkened corners together, tuck the flaps inside, and then tape the sides together to make a tetrahedral block.

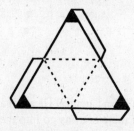 

Hold the two tetrahedral blocks as indicated in the pictures below. Make special note of the spatial relationships between the vertices that are *not* touching.

(1) Sharing a single vertex
(touch A to A)

(2) Sharing an edge
(A to A and B to B)

(3) Sharing a face
(A to A, B to B, C to C)

 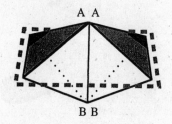 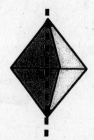

the remaining 6 vertices
can have many orientations,
one of which is shown

the remaining 4 vertices
are coplanar
(What happens if you connect
A to B and B to A instead?)

the remaining 2 vertices
are colinear

Many good molecular model sets are available commercially. You should develop the habit of using molecular models to help visualize the three-dimensional relationships of atoms in molecules. Using your set of molecular models, assemble 2 tetrahedral atoms, and then join them together in a singly bonded connection. You will find that a vertex or bond of one tetrahedral atom will need to be connected directly to the other tetrahedral atom, rather than to a vertex of the other atom. Learn how to construct double and triple bonds with your model set, and compare the geometry of the atoms in those constructions with those of the tetrahedral blocks. Keep the cardboard tetrahedra you have constructed. You will need them for the exercises in later chapters.

		1s	2s	2p			3s	3p		
2.1	sodium	↑↓	↑↓	↑↓	↑↓	↑↓	↑			
	magnesium	↑↓	↑↓	↑↓	↑↓	↑↓	↑↓			
	aluminum	↑↓	↑↓	↑↓	↑↓	↑↓	↑↓	↑		
	silicon	↑↓	↑↓	↑↓	↑↓	↑↓	↑↓	↑	↑	
	phosphorus	↑↓	↑↓	↑↓	↑↓	↑↓	↑↓	↑	↑	↑
	sulfur	↑↓	↑↓	↑↓	↑↓	↑↓	↑↓	↑↓	↑	↑
	chlorine	↑↓	↑↓	↑↓	↑↓	↑↓	↑↓	↑↓	↑↓	↑
	argon	↑↓	↑↓	↑↓	↑↓	↑↓	↑↓	↑↓	↑↓	↑↓

2.2

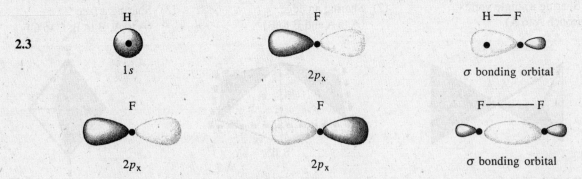

In the He_2^+ molecule-ion, there are only three electrons to be distributed, two from a neutral He atom and one from a positively charged He ion with only one electron and two protons. Two of the electrons go into the low energy bonding molecular orbital, but only one goes into the high energy antibonding molecular orbital. The net result is a lowering of energy of the system, resulting in a bond. The He_2^+ molecule-ion does exist.

2.3

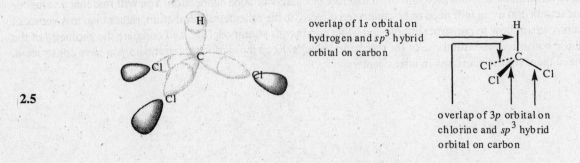

2.4   (a)   No overlap possible; wrong orientation of *p* orbital in space. The *s* orbital is approaching the nodal plane of the *p* orbital where there is no electron density.

         (b)   No overlap possible; wrong orientation of one *p* orbital in space.

         (c)   No bonding will occur. The orbitals are out of phase, and only an antibonding interaction will occur.

         (d)   Bonding will occur. The orbitals are in phase and are approaching each other so that overlap is possible.

2.5

Concept Map 2.1   Molecular orbitals and covalent bonds.

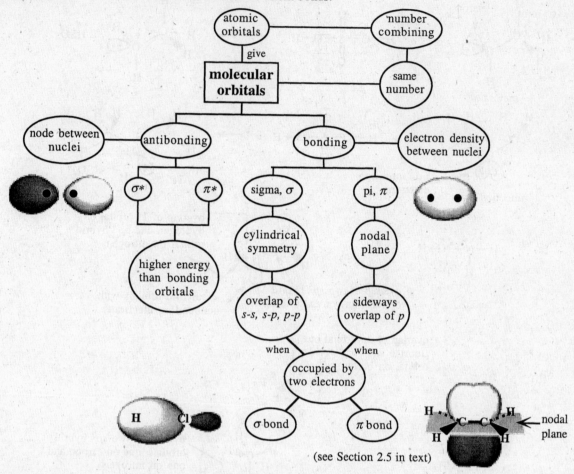

- -

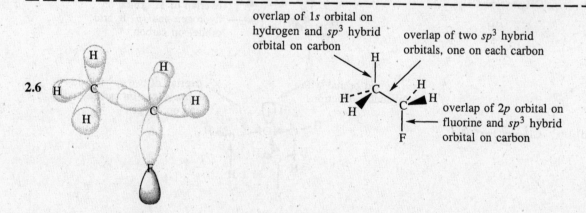

overlap of 1*s* orbital on
hydrogen and *sp*^{3} hybrid
orbital on carbon

overlap of two *sp*^{3} hybrid
orbitals, one on each carbon

overlap of 2*p* orbital on
fluorine and *sp*^{3} hybrid
orbital on carbon

2.7 The three 2*p* orbitals on nitrogen are 90° apart. Bonds formed by the overlap of these orbitals would have bond angles close to 90°. The actual bond angles in ammonia have been experimentally determined to be 107°.

2.8

nonbonding
electrons

2.9

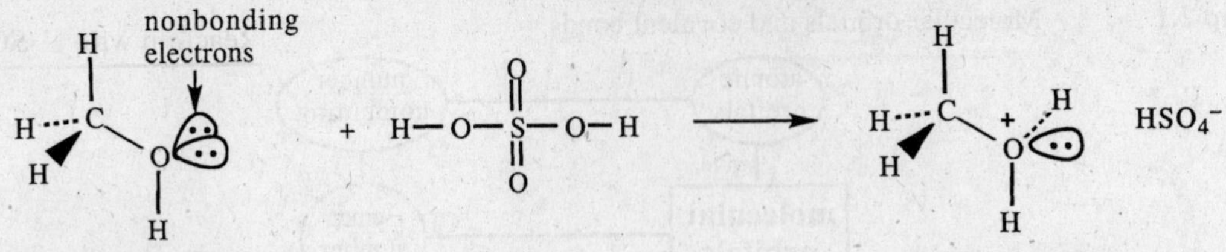

methanol

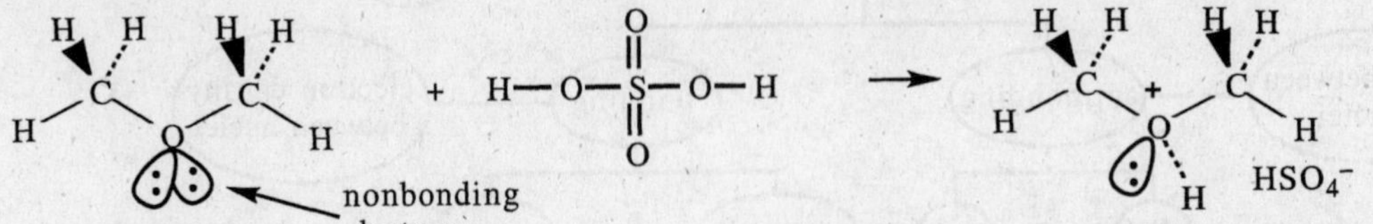

dimethyl ether

2.10   (a)

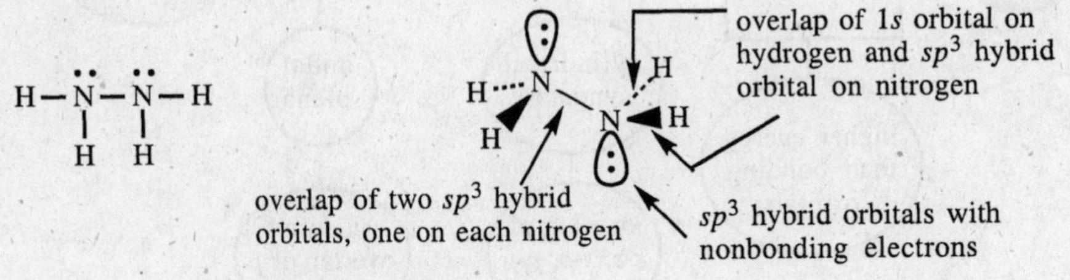

overlap of 1s orbital on hydrogen and sp³ hybrid orbital on nitrogen

overlap of two sp³ hybrid orbitals, one on each nitrogen

sp³ hybrid orbitals with nonbonding electrons

(b)   BF₄⁻

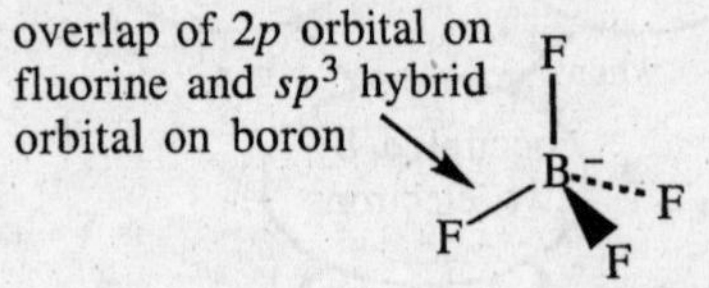

overlap of 2p orbital on fluorine and sp³ hybrid orbital on boron

(c)

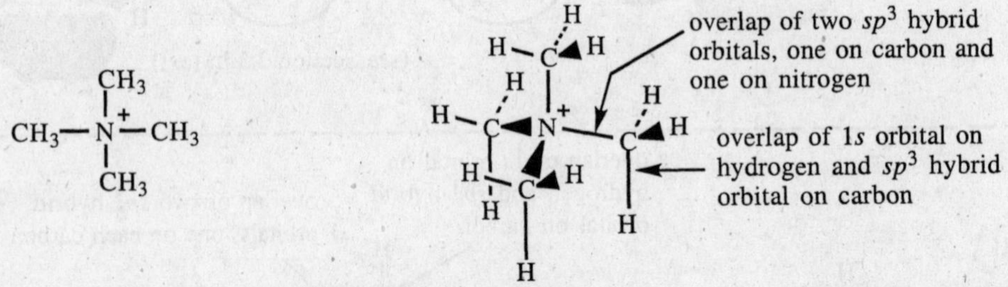

overlap of two sp³ hybrid orbitals, one on carbon and one on nitrogen

overlap of 1s orbital on hydrogen and sp³ hybrid orbital on carbon

sp³ hybrid orbital with nonbonding electrons

overlap of two sp³ hybrid orbitals, one on carbon and one on oxygen

(d)

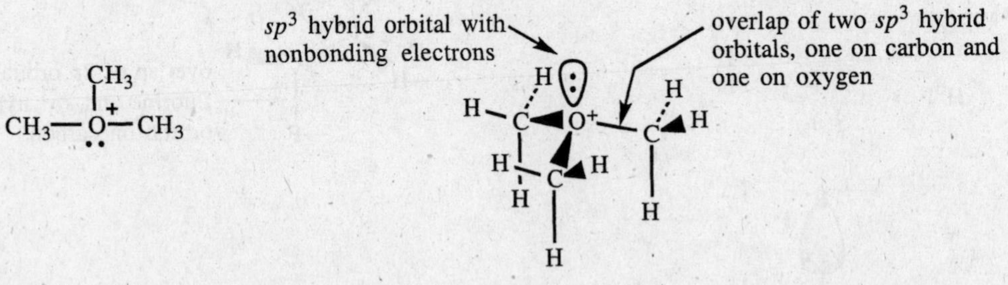

2.11   (a) alcohol   (b) ether   (c) alkane   (d) amine   (e) alcohol   (f) amine

2.12

Reaction with H_2SO_4?   Reaction with H_2SO_4?

(a) $CH_3CH_2CH_2CH_2CH_2$—O—H   Yes   (b) $CH_3CH_2CH_2$—O—CH_3 Yes

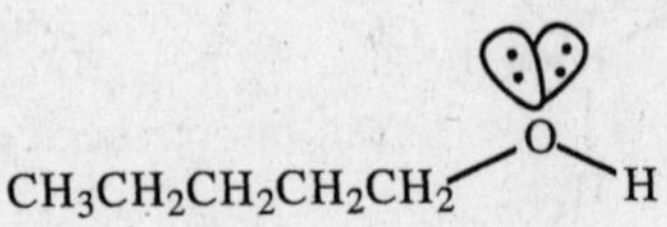

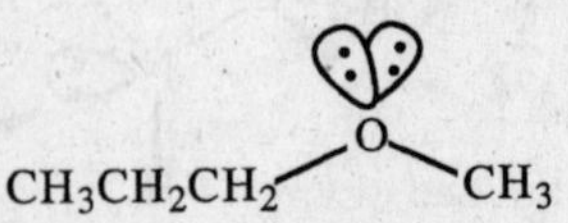

2.12   Reaction with H_2SO_4?   Reaction with H_2SO_4?

(c)   No   (d)   Yes

(e)   Yes   (f)   Yes

2.13

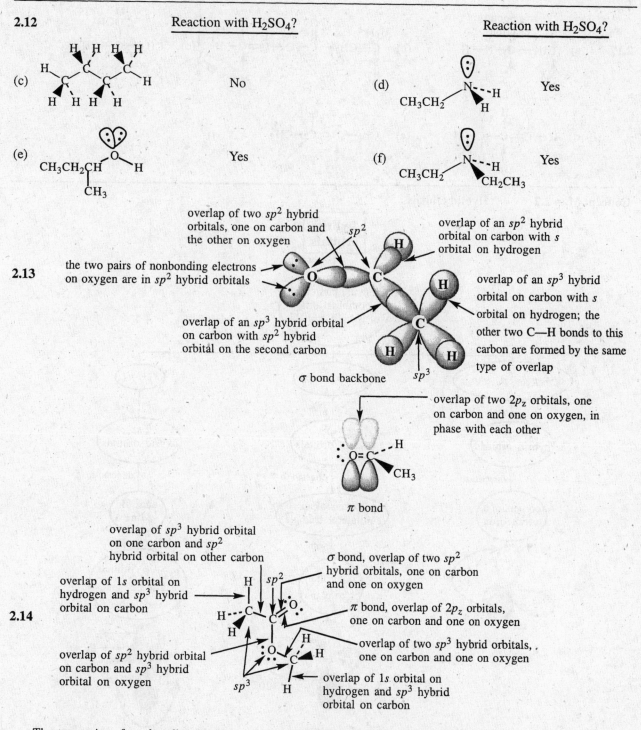

the two pairs of nonbonding electrons on oxygen are in sp^2 hybrid orbitals

overlap of two sp^2 hybrid orbitals, one on carbon and the other on oxygen

overlap of an sp^2 hybrid orbital on carbon with s orbital on hydrogen

overlap of an sp^3 hybrid orbital on carbon with s orbital on hydrogen; the other two C—H bonds to this carbon are formed by the same type of overlap

overlap of an sp^3 hybrid orbital on carbon with sp^2 hybrid orbital on the second carbon

σ bond backbone

overlap of two $2p_z$ orbitals, one on carbon and one on oxygen, in phase with each other

π bond

2.14

overlap of sp^3 hybrid orbital on one carbon and sp^2 hybrid orbital on other carbon

overlap of $1s$ orbital on hydrogen and sp^3 hybrid orbital on carbon

σ bond, overlap of two sp^2 hybrid orbitals, one on carbon and one on oxygen

π bond, overlap of $2p_z$ orbitals, one on carbon and one on oxygen

overlap of two sp^3 hybrid orbitals, one on carbon and one on oxygen

overlap of sp^2 hybrid orbital on carbon and sp^3 hybrid orbital on oxygen

overlap of $1s$ orbital on hydrogen and sp^3 hybrid orbital on carbon

The two pairs of nonbonding electrons on the tetrahedral oxygen are in sp^3 hybrid orbitals. The two pairs of nonbonding electrons on the oxygen of the carbonyl group are in sp^2 hybrid orbitals.

2.15   (a) alcohol   (b) alkene   (c) aldehyde   (d) ketone   (e) amine
           (f) alkane   (g) carboxylic acid   (h) ether   (i) ester

2.16   (a) alcohol and carboxylic acid   (b) amine and carboxylic acid   (c) ketone and carboxylic acid
           (d) alcohol and aldehyde   (e) alkene and carboxylic acid

2.17 (a) $CH_3 \overset{\overset{\displaystyle :\!\overset{..}{O}\!:}{\|}}{\underset{\underset{\displaystyle sp^2}{\uparrow}}{C}} \overset{..}{\underset{\underset{\displaystyle H}{|}}{N}} - H$ (b) $CH_3CH_2 - \overset{\overset{\displaystyle :\!\overset{..}{O}\!:}{\|}}{C} - \overset{..}{\underset{..}{O}} - \overset{\overset{\displaystyle H}{|}}{C} = \overset{\overset{\displaystyle H}{|}}{C} - H$ (c) $CH_3 - \overset{\overset{\displaystyle :\!\overset{..}{O}CH_3}{|}}{\underset{\underset{\displaystyle :\!\overset{..}{O}CH_3}{|}}{C}} - CH_3$

(d) $\overset{\overset{\displaystyle H}{|}}{H - C} = \overset{\overset{\displaystyle H}{|}}{C} - \overset{\overset{\displaystyle H}{|}}{\underset{\underset{\displaystyle H}{|}}{C}} - H$ (e) $CH_3 - \overset{..}{N} = \overset{\overset{\displaystyle CH_3}{|}}{C} - CH_3$

_ _

Concept Map 2.2   Hybrid orbitals.

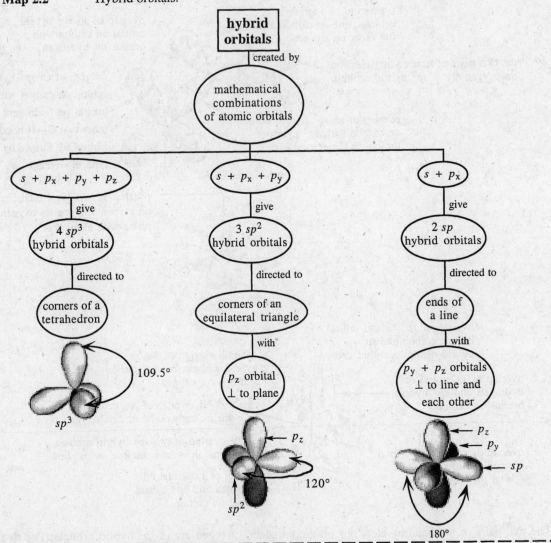

_ _

2.18 (a) alkyne   (b) alkene   (c) ketone   (d) ester
    (e) nitrile   (f) aldehyde   (g) carboxylic acid   (h) amine

2.19 (a) $\overset{\overset{\displaystyle }{}}{H - \underset{\underset{\displaystyle H}{|}}{C} = O}$ (b) $H - \overset{\overset{\displaystyle H}{|}}{\underset{\underset{\displaystyle H}{|}}{C}} - \overset{\overset{\displaystyle H}{|}}{\underset{\underset{\displaystyle H}{|}}{C}} - O - H$ (c) $H - C \equiv C - C \equiv C - H$

$3\sigma, 1\pi$ 8σ $5\sigma, 4\pi$

2.19 (d)   H—C=C—C≡N   (e)   (f)

6σ, 3π   10σ, 1π   10σ

Concept Map 2.3   Functional Groups.

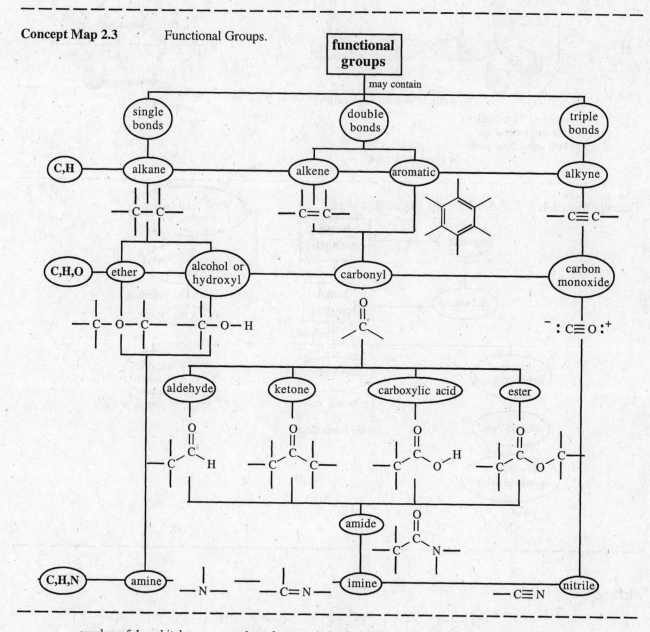

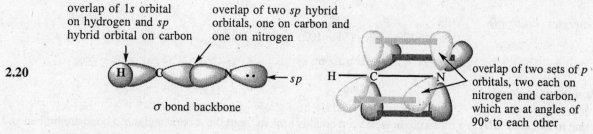

overlap of 1*s* orbital on hydrogen and *sp* hybrid orbital on carbon

overlap of two *sp* hybrid orbitals, one on carbon and one on nitrogen

2.20

σ bond backbone

overlap of two sets of *p* orbitals, two each on nitrogen and carbon, which are at angles of 90° to each other

π bonds

2.21

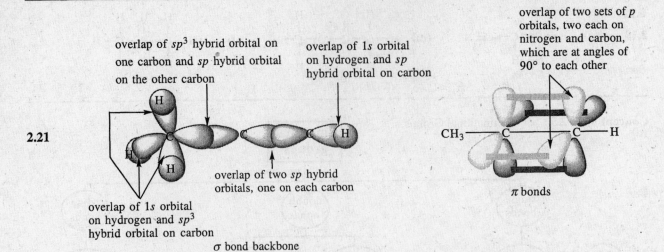

overlap of sp^3 hybrid orbital on one carbon and sp hybrid orbital on the other carbon

overlap of $1s$ orbital on hydrogen and sp hybrid orbital on carbon

overlap of two sets of p orbitals, two each on nitrogen and carbon, which are at angles of 90° to each other

overlap of two sp hybrid orbitals, one on each carbon

overlap of $1s$ orbital on hydrogen and sp^3 hybrid orbital on carbon

σ bond backbone

π bonds

Concept Map 2.4   Bond lengths and bond strengths.

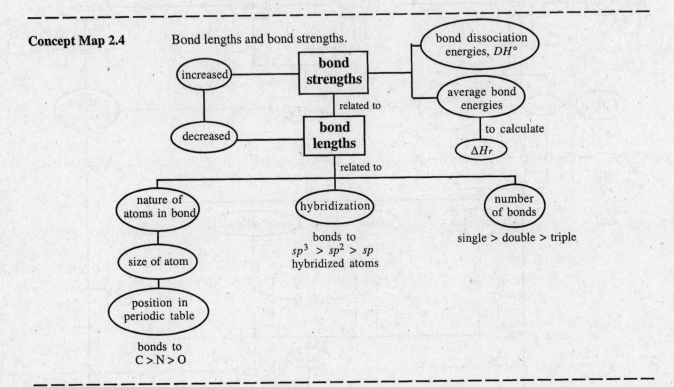

bond dissociation energies, $DH°$

increased

bond strengths

related to

average bond energies

decreased

bond lengths

to calculate

ΔHr

related to

nature of atoms in bond

hybridization

number of bonds

size of atom

bonds to $sp^3 > sp^2 > sp$ hybridized atoms

single > double > triple

position in periodic table

bonds to C > N > O

2.22

(a)

$$CH_3-\xi-H \quad + \quad Cl-\xi-Cl \quad \longrightarrow \quad CH_3-\xi-Cl \quad + \quad H-\xi-Cl$$

Bond energies (kcal/mol)   99   58   81   103

$$\Delta H_r = (99 + 58) - (81 + 103) = -27 \text{ kcal/mol}$$

(b)

$$H_2C=\xi=O \quad + \quad H-\xi-C\equiv N \quad \longrightarrow \quad H-\xi-O-\xi-CH_2-\xi-C\equiv N$$

Bond energies (kcal/mol)   176   99   111   86   83

$$\Delta H_r = (176 + 99) - (111 + 86 + 83) = -5 \text{ kcal/mol}$$

(Note that to calculate ΔH_r in a molecule in which a π bond is broken, both the π bond and the σ bond are broken and the σ bond is reformed in the product.)

2.22 (cont)

(c) $H_2C = \{ = CH_2$ + $H - \{ - Br$ $\longrightarrow$ $H - \{ - CH_2 - \{ - CH_2 - \{ - Br$

Bond energies (kcal/mol) 146   87   99   83   68

$$\Delta H_r = (146 + 87) - (99 + 83 + 68) = -17 \text{ kcal/mol}$$

(d)

$CH_3 - CH - \{ - CH_2$ $\longrightarrow$ $CH_3 - CH = \{ = CH_2$ + $H - \{ - O - H$

86 ~~~ HO   99 ~~~ H

Bond energies (kcal/mol) $\Delta H_r = (86 + 83 + 99) - (146 + 111) = +11 \text{ kcal/mol}$

2.23   (a)   H — C — C — Cl
                    | |
                    H   H   H   H

chloroethane   chloroethene

(b)

A resonance contributor of chloroethene has a double bond between carbon and chlorine, making the bond shorter (1.72 Å) than expected for a carbon-chlorine single bond (1.78 Å). A bond with partial double bond character is also stronger than a single bond, and therefore has a higher bond dissociation energy.

(c)

(d)   The carbon atom to which the chlorine is bonded in chloroethene is an sp^2-hybridized carbon atom. An sp^2-hybridized carbon atom forms a shorter and stronger bond to chlorine than the sp^3-hybridized carbon atom of chloroethane (Section 2.8 in the text).

2.24   (a) $CH_3CH_2CH = CHCH_3$ + Br_2 $\xrightarrow[\text{dark, 25 °C}]{\text{carbon tetrachloride}}$ $CH_3CH_2CH - CHCH_3$
                                                                                    | |
                                                                                   Br   Br

(b) $CH_3C \equiv CCH_3$ + $2 Br_2$ $\xrightarrow[\text{dark, 25 °C}]{\text{carbon tetrachloride}}$ $CH_3C - CCH_3$
                                                                            | |
                                                                           Br   Br

Br_2 ↘ Br_2 ↗

CH_3 Br
   \   /
    C = C
   /   \
  Br CH_3

2.24 (c) $HC\equiv CCH_2C\equiv CH$ + 4 Br$_2$ $\xrightarrow[\text{dark, 25 °C}]{\text{carbon}\atop\text{tetrachloride}}$

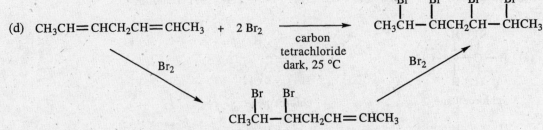

2.25

Alkanes: Tetrahedral carbon atoms bonded only to other tetrahedral carbon or hydrogen atoms

Alkenes: $-\overset{|}{C}=\overset{|}{C}-$ **Alkynes:** $-C\equiv C-$ **Alkyl halides:** $-\overset{|}{\underset{|}{C}}-X$

Alcohols: $-\overset{|}{\underset{|}{C}}-OH$ **Ethers:** $-\overset{|}{\underset{|}{C}}-O-\overset{|}{\underset{|}{C}}-$ **Aldehydes:** $-\overset{\overset{O}{\|}}{C}H$

Ketones: $-\overset{\overset{O}{\|}}{C}-$ **Carboxylic acids:** $-\overset{\overset{O}{\|}}{C}-OH$ **Esters:** $-\overset{\overset{O}{\|}}{C}-O-\overset{|}{\underset{|}{C}}-$

Amines: $-\overset{|}{\underset{|}{C}}-\overset{|}{N}-$ **Nitriles:** $-C\equiv N$ **Aromatic hydrocarbons:**

2.26

No delocalization of electrons in this structure.
We would expect the compound to behave like an alkene.

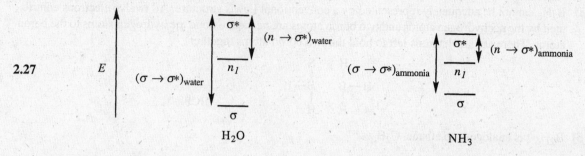

2.27

The $\sigma \rightarrow \sigma^*$ and the $n \rightarrow \sigma^*$ transitions are higher energies (shorter wavelengths) for water than for ammonia.

2.28 The halogen with the highest electronegativity will have the highest energy (shortest wavelength) $n \rightarrow \sigma^*$ transition. Compound A (173 nm) is CH_3Cl, Compound B (204 nm) is CH_3Br, and Compound C (258 nm) is CH_3I.

2.29

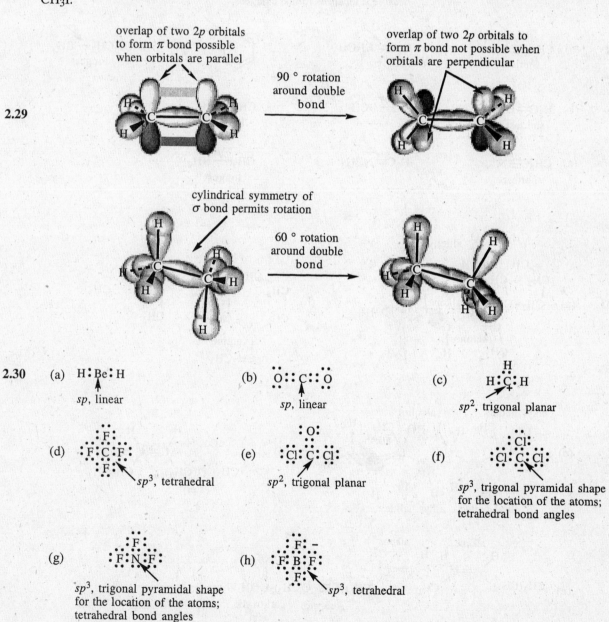

2.30 (a) H:Be:H

sp, linear

(b) Ö::C::Ö

sp, linear

(c) H:C:H   H

sp^2, trigonal planar

(d) :F:C:F: :F:

sp^3, tetrahedral

(e) :O: :Cl:C:Cl:

sp^2, trigonal planar

(f) :Cl: :Cl:C:Cl:

sp^3, trigonal pyramidal shape for the location of the atoms; tetrahedral bond angles

(g) :F: :F:N:F:

sp^3, trigonal pyramidal shape for the location of the atoms; tetrahedral bond angles

(h) :F: –   :F:B:F: :F:

sp^3, tetrahedral

2.31 (a) B_2H_6 cannot be adequately represented by a conventional Lewis structure. All twelve electrons contributed by the six hydrogen atoms and two boron atoms are needed to bond the hydrogen atoms to the boron atoms. There are no electrons left to hold the two boron atoms together.

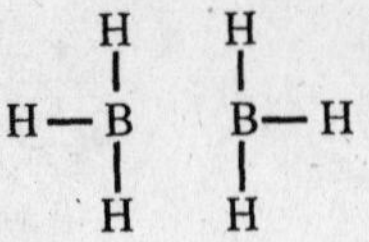

(b) $B_2H_6{}^{2-}$ is analogous to ethane, C_2H_6.

(c) 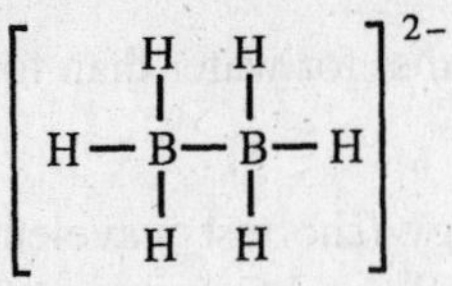 or 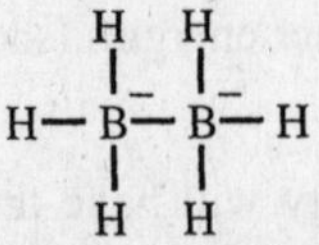

This structure is more correct because it localizes formal charges.

(d) 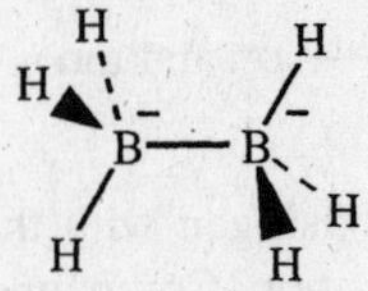

2.32 (a) $CH_3C{\equiv}CH$ shortest; $H_2C{=}CHCH_3$; 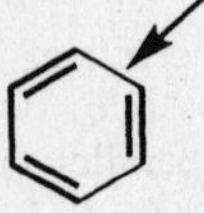; $CH_3CH_2{-}CH_3$ longest

(b) $H_2C{=}O$ shortest; $HC({=}O){-}OCH_3$; $CH_3{-}OH$ longest

(c) $CH_3C{\equiv}N$ shortest; $H_2C{=}NCH_3$; $CH_3{-}NH_2$ longest

2.33 (a) 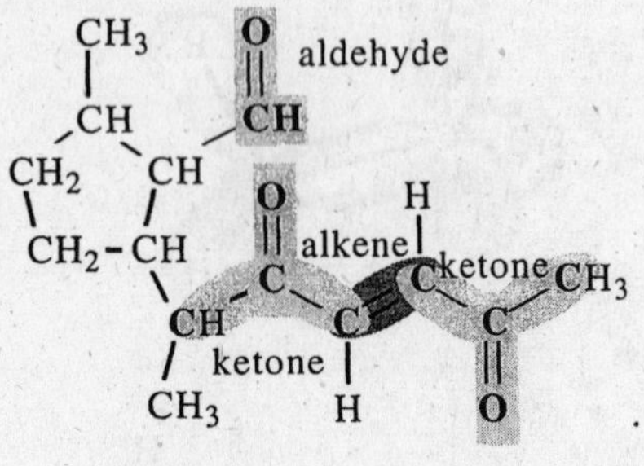

(b) 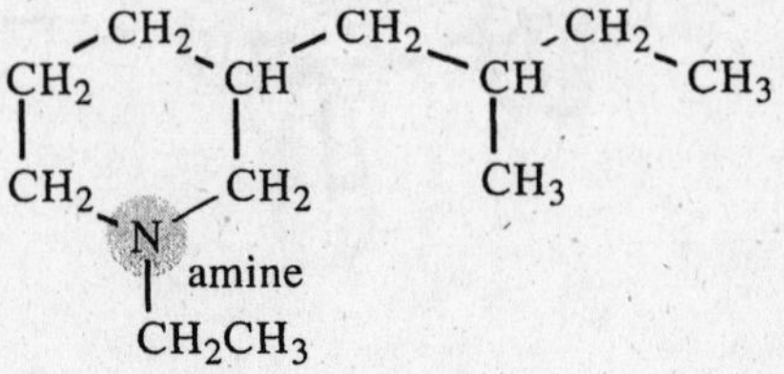

(c) 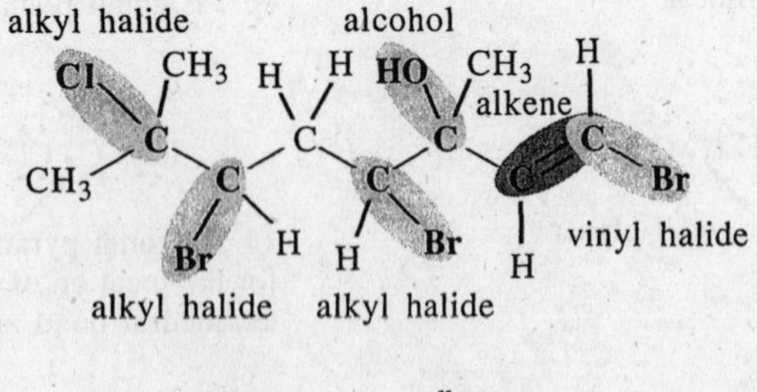

(d) 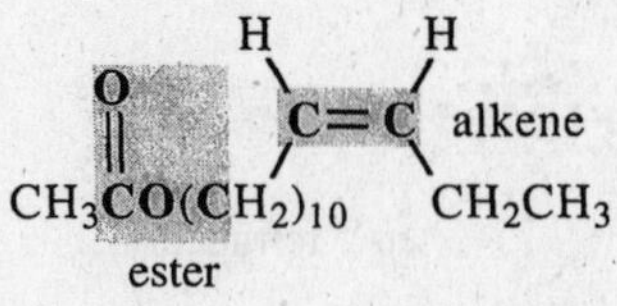

(e) 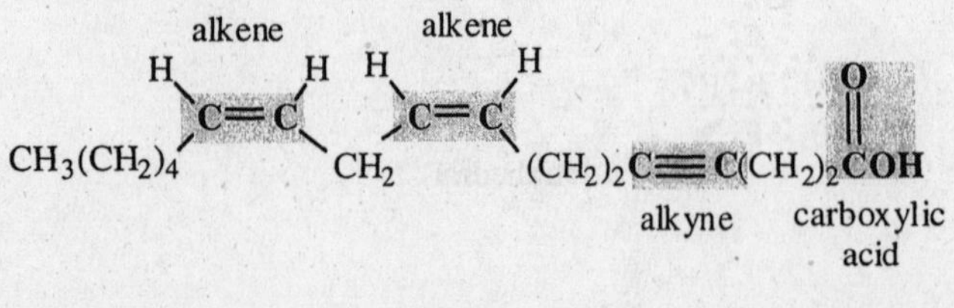

2.34

(a) $H_2C=O$

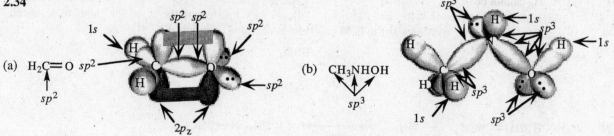

(b) CH_3NHOH

(c) $CH_3CH=CHCH_3$

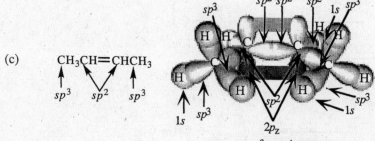

one of two isomers

(d) $CH_3C\equiv CCH_3$

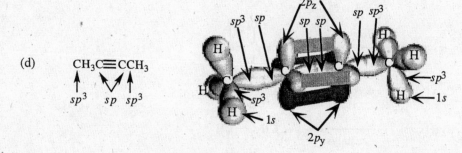

2.35 (a) The carbon and nitrogen atoms are sp^2-hybridized. The oxygen atom is sp^3 hybridized.

(b)

2.36 (a) $CH_3C\equiv N: \ + \ H-OSOH \longrightarrow CH_3C\equiv \overset{+}{N}-H \quad HSO_4^-$

(b) $CH_3\overset{:O:}{\overset{||}{C}}CH_3 \ + \ H-OSOH \longrightarrow CH_3\overset{:\overset{+}{O}-H}{\overset{||}{C}}CH_3 \quad HSO_4^-$

(c) $CH_3\overset{CH_3}{\underset{\cdot\cdot}{C}}=NCH_3 \ + \ H-OSOH \longrightarrow CH_3\overset{CH_3}{C}=\overset{+}{\underset{H}{N}}CH_3 \quad HSO_4^-$

2.36 (d)

$$CH_3\overset{:O:}{\underset{\cdot\cdot}{\overset{||}{C}}}OCH_3 \; + \; H{-}O\overset{O}{\underset{O}{\overset{||}{S}}}OH \longrightarrow CH_3\overset{:\overset{+}{O}{-}H}{\underset{\cdot\cdot}{\overset{||}{C}}}OCH_3 \;\; \text{or} \;\; CH_3\overset{:O:}{\underset{\overset{|}{\underset{H}{}}}{\overset{||}{C}}}\overset{+}{O}{-}CH_3$$

$$HSO_4^- \qquad\qquad HSO_4^-$$

2.37 (a) $CH_3CH{=}CHCH_3 \; + \; Br_2 \xrightarrow[\substack{\text{carbon}\\\text{tetrachloride}\\\text{dark, 25 °C}}]{} \underset{\underset{Br}{|}}{CH_3CH}{-}\underset{\underset{Br}{|}}{CHCH_3}$

(b) $CH_3CH_2CH{=}CH_2 \; + \; H_2 \xrightarrow[\text{catalyst}]{} \underset{\underset{H}{|}}{CH_3CH_2CH}{-}\underset{\underset{H}{|}}{CH_2}$

(c) $CH_3CH_2CH_2CH_3 \; + \; Br_2 \xrightarrow[\substack{\text{carbon}\\\text{tetrachloride}\\\text{dark, 25 °C}}]{} \text{no reaction}$

(d) $\langle\!\!\bigcirc\!\!\rangle{-}CH{=}CH_2 \; + \; Br_2 \xrightarrow[\substack{\text{carbon}\\\text{tetrachloride}\\\text{dark, 25 °C}}]{} \langle\!\!\bigcirc\!\!\rangle{-}\underset{\underset{Br}{|}}{CH}{-}\underset{\underset{Br}{|}}{CH_2}$

(e) $CH_3C{\equiv}CH \; + \; Br_2 \text{ (excess)} \xrightarrow[\substack{\text{carbon}\\\text{tetrachloride}\\\text{dark, 25 °C}}]{} \underset{\underset{Br}{|}}{\overset{\overset{Br}{|}}{CH_3C}}{-}\underset{\underset{Br}{|}}{\overset{\overset{Br}{|}}{CH}}$

(f) $CH_3C{\equiv}CH \; + \; H_2 \text{ (excess)} \xrightarrow[\text{catalyst}]{} \underset{\underset{H}{|}}{\overset{\overset{H}{|}}{CH_3C}}{-}\underset{\underset{H}{|}}{\overset{\overset{H}{|}}{CH}}$

2.38 (a)

$$H_2C{=}{\lbrace}{=}CH_2 \; + \; Br{-}{\lbrace}{-}Br \longrightarrow Br{-}{\lbrace}{-}CH_2{-}{\lbrace}{-}CH_2{-}{\lbrace}{-}Br$$

Bond energies (kcal/mol)   146   46   68   83   68

$$\Delta H_r = (146 + 46) - (68 + 83 + 68) = -27 \text{ kcal/mol}$$

(b)

$$\underset{\underset{CH_3}{|}}{\overset{\overset{CH_3}{|}}{CH_3{-}C}}{-}\underset{86}{{\lbrace}}{-}OH \; + \; H{-}\underset{103}{{\lbrace}}{-}Cl \longrightarrow \underset{\underset{CH_3}{|}}{\overset{\overset{CH_3}{|}}{CH_3{-}C}}{-}\underset{81}{{\lbrace}}{-}Cl \; + \; H{-}\underset{111}{{\lbrace}}{-}OH$$

$$\Delta H_r = (86 + 103) - (81 + 111) = -3 \text{ kcal/mol}$$

(c)

$$\overset{179}{CH_3{-}\overset{\overset{O}{||}}{C}{-}CH_3} \; + \; H{\cdot}\overset{111}{{\lbrace}}{\cdot}OH \longrightarrow \underset{\underset{O{-}H}{86}}{\overset{\overset{O{-}{\lbrace}{-}H}{86 \quad 111}}{CH_3{-}C{-}CH_3}}$$

$$\Delta H_{\bar{r}} = (179 + 111) - (111 + 86 + 86) = +7 \text{ kcal/mol}$$

2.39

$$CH_2 = C = CH_2$$

$$sp^2 \quad sp \quad sp^2$$

In order to determine the sigma bond skeleton, we must first set up the pi bonds. When we do so, we can see that, in order for overlap to occur at the central carbon atom, the *p* orbitals on the two end carbon atoms must be perpendicular to each other. This causes the *sp*² hybrid orbitals that are bonded to the hydrogen atoms on the end carbon atoms also to be perpendicular to each other.

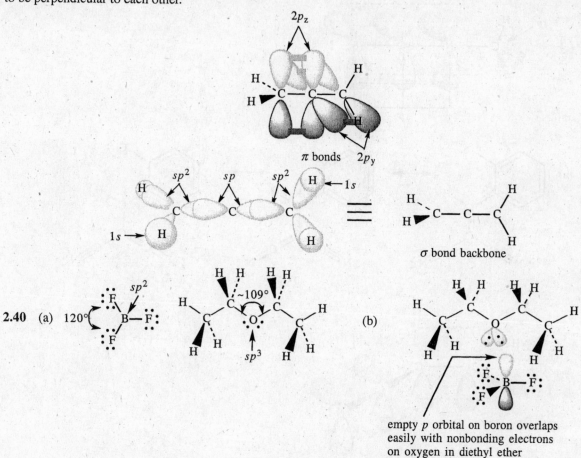

σ bond backbone

2.40 (a) 120°

(b)

empty *p* orbital on boron overlaps
easily with nonbonding electrons
on oxygen in diethyl ether

(c) Boron trifluoride etherate has a higher boiling point than either of the starting materials because the molecular weight of the etherate is higher and because the etherate is a more polar compound than either of the starting materials.

$$F - B^{-} - O^{+} \begin{array}{c} F \\ | \\ | \\ F \end{array} \begin{array}{c} CH_2CH_3 \\ \\ CH_2CH_3 \end{array}$$

separation of charge in boron trifluoride etherate makes it polar;
dipole-dipole interactions between molecules leads to higher boiling point

2.41 (a) H—C—N̈=N⁺=N̈⁻ ⟷ H—C—N⁻—N⁺≡N: ⟷ H—C—N⁺≡N⁺—N:²⁻

This resonance contributor has the
greatest separation of charge and
makes the smallest contribution.

(b) The N—N—N bond angle is 180° with *sp* hybridization for the central nitrogen atom.

2.42 (a) $CH_3 - \ddot{N} = C = \ddot{S}\!:$ $\longleftrightarrow$ $CH_3 - \overset{+}{N} \equiv C - \ddot{\underset{..}{S}}\!:^-$ $\longleftrightarrow$ $CH_3 - \overset{-}{\ddot{N}} - C \equiv \overset{+}{S}\!:$

  A   B   C

(b) Form A is most important because there is no separation of charge.

(c)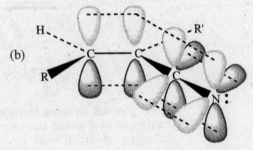

2.43 (a)

(b)

2.44 (a)

(b) The one in which the oxygen atom is sp^2-hybridized. The electron pair in the p orbital should have a different energy than the electron pair in the sp^2 hybrid orbital. The model of water in which the oxygen atom is sp^3-hybridized has both pairs of nonbonding electrons in sp^3 hybrid orbitals. These electrons should, therefore, be of the same energy. Experimentally we should not see them as occupying two different energy levels.

(c) Yes, lack of hybridization at the oxygen atom also results in two different energy levels for the two pairs of nonbonding electrons.

overlap of a $1s$ orbital
on hydrogen with a $2p$
orbital on oxygen

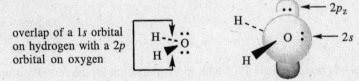

2.44 (cont)   An sp^2-hybridized oxygen atom would be expected to have a H—O—H bond angle of 120°. If oxygen were not hybridized, the H—O—H bond angle would be expected to be 90°. The experimentally determined value for the bond angle is 105°, between 120° and 90°.

    (d) The oxygen atom would have sp hybridization.

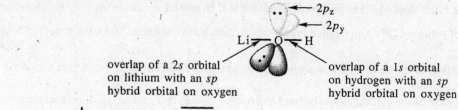

overlap of a 2s orbital
on lithium with an sp
hybrid orbital on oxygen

overlap of a 1s orbital
on hydrogen with an sp
hybrid orbital on oxygen

2.45 (a) E

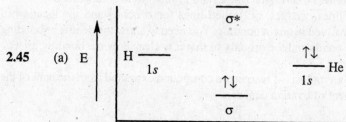

    (b) [HeH]⁻ would not be stable because it would have 4 electrons, which would fill both the bonding and antibonding molecular orbitals. It would be like the He₂ molecule; see Figure 2.7 (p. 45 in the text).

    (c)   He—H⁺

Supplemental Problems

S2.1   Circle and name all the functional groups in each of the following molecules.

(a) CH₃C≡CCCH₃
     with CH₃ and OH

(b) CH₃C=CHCCH₃
     with CH₃ and O

(c) benzene ring—CH₂NCH₃ with CH₃

(d) CH₃CH₂CHCH=CH₂
     with Br

(e) HOCCH₂CH₂COH (with two O)

(f) CH₃COCH₂CH₂OCH₃ (with O)

(g) benzene ring—CH with O

(h) CH₃CHCH₂CH with O
     with OH

(i) benzene ring—CH₂OH

(j) CH₃CH₂CH₂C≡N

S2.2   Diazene, N₂H₂, has the following connectivity: HNNH. It is a good reducing agent, giving up hydrogen atoms easily to form the stable molecule, N₂.

(a) Draw a Lewis structure for diazene in which all the atoms have closed shells.

(b) The bonding in diazene may also be described using the model of orbital hybridization and molecular orbitals. What is the hybridization of the nitrogen atoms in diazene?

(c) Draw a three-dimensional picture of one isolated nitrogen atom in diazene showing the hybrid orbitals and any other orbitals around it with the electrons occupying those orbitals clearly shown.

(d) Draw a three-dimensional structural formula for diazene using lines (—), dashed lines (- - -), and wedges (◄■■) to indicate σ bonds, and show the actual orbitals involved in π bonds.

S2.3   Dicyanocarbene has the following Lewis structure

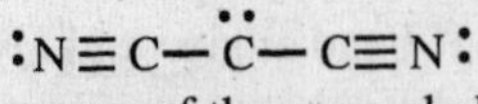

$$:N \equiv C - \overset{..}{C} - C \equiv N:$$

A carbene is a highly reactive neutral species because of the open shell on the central carbon atom.

(a) Predict the shape you would expect for dicyanocarbene from the structure given by drawing a three-dimensional representation, using lines, dashed lines, and wedges as needed. Nonbonding electrons may be shown in lobes (⌀), or at the end of wedges (➤), or dashed lines (⋰) to indicate direction. Show the expected bond angle on your drawing.

(b) What is the hybridization of the central carbon atom in dicyanocarbene according to the picture you drew in part (a)?

(c) What is the hybridization of the carbon atoms bonded to nitrogen (the cyano groups) in dicyanocarbene?

(d) Redraw the structure of dicyanocarbene using lines, wedges, or dashed lines for σ bonds and the location of nonbonding electrons, and show the p orbitals involved in any π bonding. You need to show the details of bonding for only *one* of the two cyano groups. Label the bonds and the orbitals so that it is clear how the bonding arises.

S2.4   The following two compounds are part of a synthesis of taxotere, a compound extracted from the bark of the Pacific yew tree and found to be useful in the treatment of ovarian cancer.

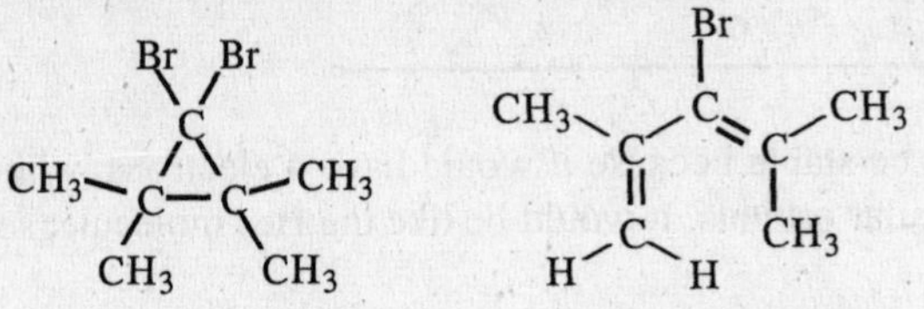

Compound A   Compound B

(a) What is the structural relationship, if any, of Compounds A and B? Some choices are (1) constitutional isomers, (2) stereoisomers, (3) resonance contributors, or (4) no structural relationship.

(b) Draw a structure for compounds that fit the descriptions given below.
  (1) A constitutional isomer of B that contains a triple bond.
  (2) A resonance contributor of B.
  (3) A constitutional isomer of B that contains a ring.
  (4) A constitutional isomer of B that has a different connectivity but no ring.

S2.5   An unusual structure, shown below, has been reported.

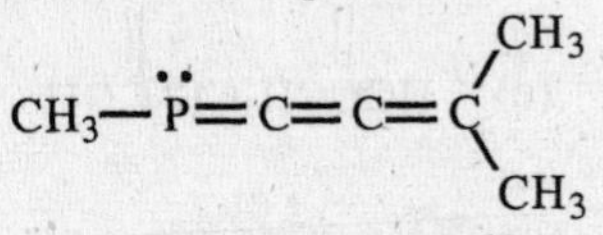

$$CH_3 - \overset{..}{P} = C = C = \overset{CH_3}{\underset{CH_3}{C}}$$

(a) What is the hybridization at each of the atoms in the compound?

(b) Draw a clear three-dimensional orbital picture of the π bonds in the compound. You may use CH_3 for the methyl groups, but do show how they, as well as the nonbonding electrons on phosphorus, point in space using wedges, dashes, or lines.

S2.6   Alkyl nitrile oxides (connectivity RCNO where R is a generic alkyl group) are a useful class of compounds for carrying out the preparation of more complex molecules. A relatively unimportant resonance contributor for methyl nitrile oxide is shown below.

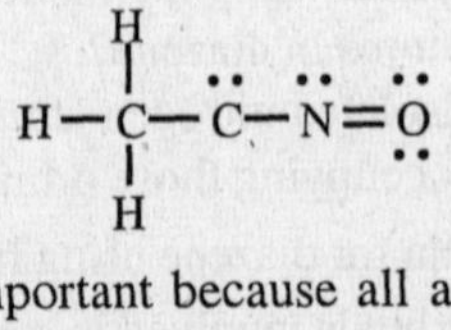

$$H - \overset{\overset{\displaystyle H}{|}}{\underset{\underset{\displaystyle H}{|}}{C}} - \overset{..}{C} - \overset{..}{N} = \overset{..}{O}$$

Two other resonance contributors are more important because all atoms have closed shells and some have formal charges of −1 and +1. What are they? Of the two forms that you have drawn, which one is the more important?

3 Reactions of Organic Compounds as Acids and Bases

Workbook Exercises

Much of Chapter 3 is devoted to learning how to rank acidities. It is important for you to be able to draw and recognize, both quickly and accurately, the form of a molecule resulting from the loss of a proton. Such a reaction is called a deprotonation reaction.

The connectivity change that occurs in deprotonation may be imagined in the following way:

$$(A = any\ atom) \quad A—H \longrightarrow A:^- \ + \ H^+$$

shared
electron pair

electron pair
with A

nucleus of a hydrogen
atom; a proton

EXERCISE I. Draw all the products that can result from a single deprotonation of the molecule shown below.

EXAMPLE

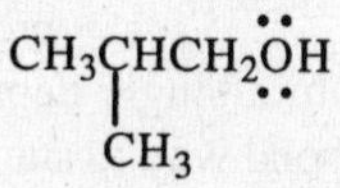

$$CH_3\overset{|}{C}HCH_2\overset{..}{O}H$$
$$\overset{|}{C}H_3$$

SOLUTION

(1) Redraw the compound as a complete Lewis structure and then decide, based on connectivity, how many different types of hydrogen atoms there are.

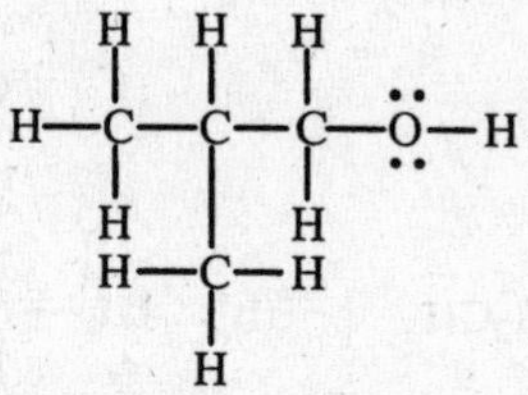

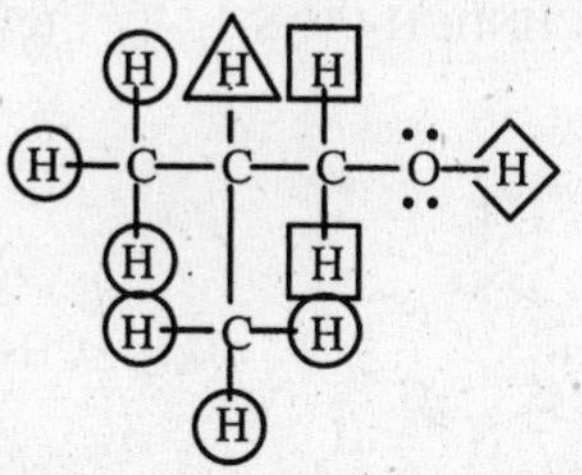

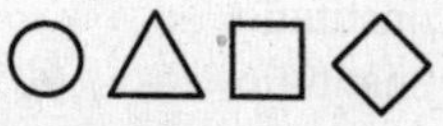

There are 4 different
types of hydrogen atoms.

(Note that regardless of how the molecule is drawn, the set of 6 hydrogens in circles are of the same type. That is, they are equally situated with respect to the other atoms in the molecule.)

(2) Four different deprotonations are possible.

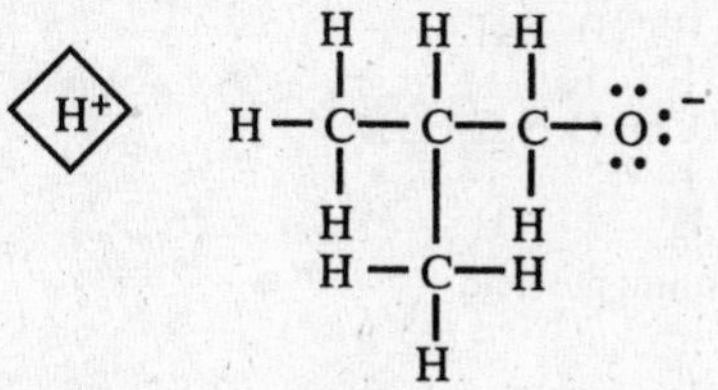

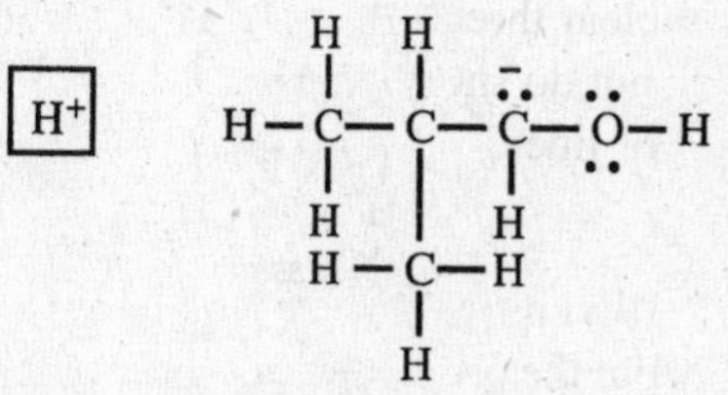

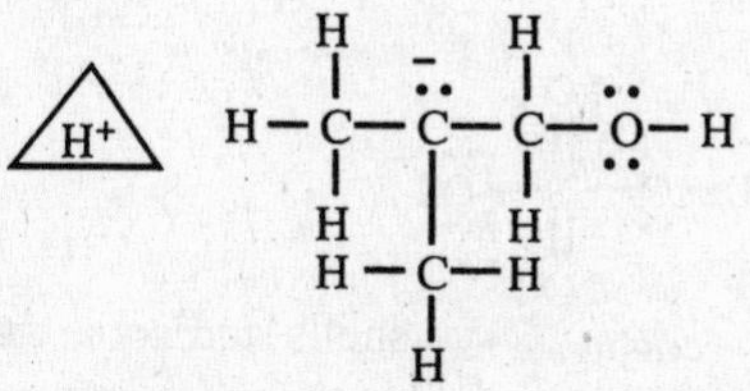

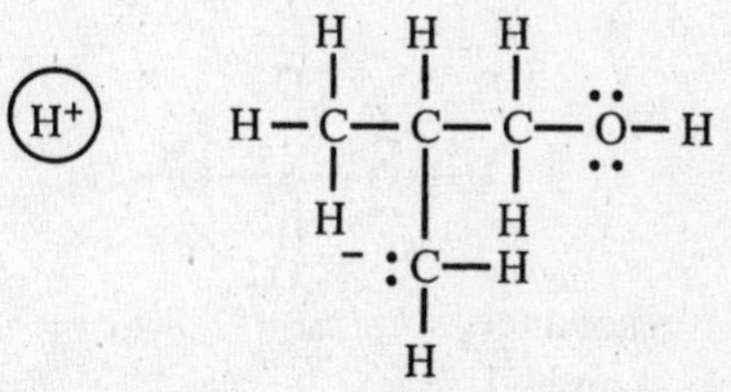

Workbook Exercises (cont)

PROBLEM I.   Draw all of the products that can result from a single deprotonation of the molecules shown below.

(a)

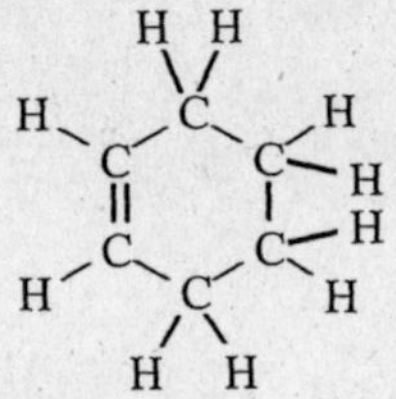

(b) $HOCCH_2CH_3$ (with O double bonded)

(c) H_2

(d) 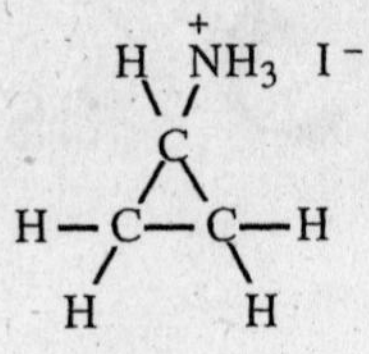 $H-C-C-H$ with $\overset{+}{NH_3}$ I^-

(e) $ClCH_2CN$

(f) $HC{\equiv}CCH_2CH$ (with O double bonded)

(g) $H_2\overset{+}{O}CH_2CH_3$

(h) $(CH_3)_3S^{+}\ ^{-}BF_4$

EXERCISE II.   The reverse of a deprotonation reaction is a protonation reaction. Protonation is the attaching of a proton, H⁺, to a molecule or ion. In order to form a bond with a proton, another atom must contribute both of the electrons to the bond (H⁺ has no electrons to contribute!). Nonbonding electron pairs from uncharged or negatively charged atoms are often the source of electrons for forming a new bond to a proton.

The connectivity change that occurs in protonation may be drawn in the following way:

(A = any atom)   uncharged $A\!:$ + a proton (H⁺) source $\longrightarrow$ $\overset{+}{A}\!-\!H$

anionic $A\!:^{-}$ + a proton (H⁺) source $\longrightarrow$ $A\!-\!H$

In both cases, the nonbonding electron pair
provides both electrons for the A—H bond.

Draw all of the products that can result from a single deprotonation of the molecules shown below. Locating the nonbonding electrons in each molecule is a good way to start.

(a) $CH_3CH_2CH_2O^- Na^+$

(b) $(CH_3)_2CHNHCH_2CH_2NH_2$

(c) CH_3OCNH_2

(d)

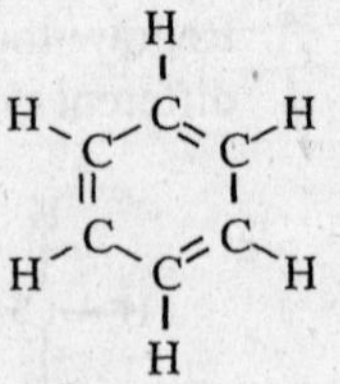

(e) $Li^+\ ^-C{\equiv}CCH_2CH_2O^- Li^+$

(f)

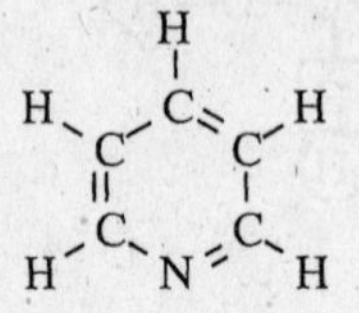

(g) 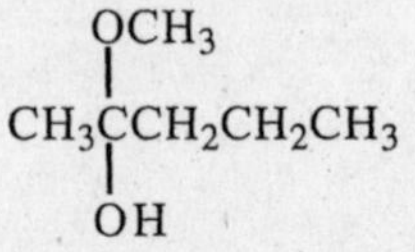 $CH_3CCH_2CH_2CH_3$ with OCH_3 above and OH below

(h) 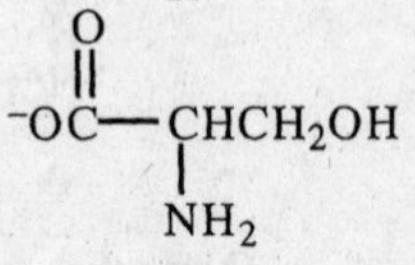 $^-OC-CHCH_2OH$ with O double bonded and NH_2 below

- -

3.1   (a) $(CH_3)_2O$ $(CH_3)_2OH^+$

base   conjugate acid

(b) $H_2SO_4 = (HO)_2SO_2$ $HSO_4^- = HOSO_3^-$

acid   conjugate base
                                              (one resonance contributor)

3.1   (c)

$^-NH_2$

H — N̈ — H

base

NH_3

H — N̈ — H

H (above N)

conjugate acid

(d)

$CH_3OH_2^+$

H   H

H — C — Ö⁺ — H

H

acid

CH_3OH

H

H — C — Ö — H

H

conjugate base

(e)

$H_2C=O$

:O:
‖
H — C — H

base

$H_2C=OH^+$

:Ö⁺ — H
‖
H — C — H

conjugate acid

(f)

CH_3OH

H

H — C — Ö — H

H

acid

CH_3O^-

H

H — C — Ö:⁻

H

conjugate base

- -

Concept Map 3.1   The Brønsted-Lowry theory of acids and bases.

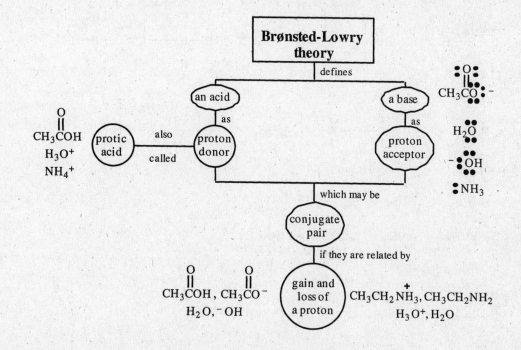

Concept Map 3.2   The Lewis theory of acids and bases.

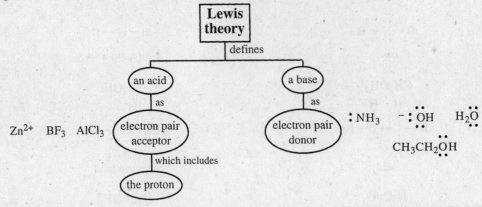

- -

3.2 (a)

$$Cu^{2+} + 4 \; H-\underset{\cdot\cdot}{N}-H \;\rightleftharpoons\; \left[\begin{array}{c} H \\ | \\ H-N-H \\ | \\ H-N-Cu-N-H \\ | \\ H-N-H \\ | \\ H \end{array}\right]^{2+}$$

Lewis acid   Lewis base

(b)

$$H-\underset{H}{\overset{H}{C}}-\underset{+}{\overset{\overset{H}{\overset{|}{H-C-H}}}{C}}-\underset{H}{\overset{H}{C}}-H \;+\; H-\underset{\cdot\cdot}{O}-H \;\rightleftharpoons\; H-\overset{H}{\underset{H}{C}}-\overset{\overset{H}{\overset{|}{H-C-H}}}{\underset{\underset{\cdot\cdot}{H-O-H}}{C}}-\overset{H}{\underset{H}{C}}-H$$

Lewis acid   Lewis base

(c)

$$\overset{\cdot\cdot}{\underset{\cdot\cdot}{:}F}-B-\overset{\cdot\cdot}{F:} \;+\; H-\overset{\overset{H}{\overset{|}{H-C-H}}}{\underset{H}{C}}-C=\overset{\cdot\cdot}{\underset{\cdot\cdot}{O}} \;\rightleftharpoons\; H-\overset{\overset{H}{\overset{|}{H-C-H}}}{\underset{H}{C}}-C=\overset{+}{O}-\overset{\overset{:F:}{|}}{B}-\overset{\cdot\cdot}{F:}^{-}$$

Lewis acid   Lewis base

(d)

$$Hg^{2+} \;+\; H-\overset{H}{\underset{H}{C}}-\overset{H}{\underset{H}{C}}-\overset{\cdot\cdot}{S}-H \;\rightleftharpoons\; \left[\begin{array}{c} H \quad H \quad Hg \\ | \quad | \quad | \\ H-C-C-S-H \\ | \quad | \\ H \quad H \end{array}\right]^{2+}$$

Lewis acid   Lewis base

3.3 (a) $CH_3\overset{\cdot\cdot}{N}HCH_3 + AlCl_3 \;\rightleftharpoons\; CH_3\overset{\overset{-AlCl_3}{\overset{|}{+}}}{\underset{\underset{H}{|}}{N}}CH_3$

$CH_3\overset{\cdot\cdot}{N}HCH_3 + H_2SO_4 \;\rightleftharpoons\; CH_3\overset{\overset{H}{\overset{|}{+}}}{\underset{\underset{H}{|}}{N}}CH_3 + HSO_4^-$

3.3 (b)

$$\overset{:O:}{\underset{}{\underset{HCH}{\|}}} + AlCl_3 \rightleftharpoons \overset{\overset{+}{:}O - \overset{-}{AlCl_3}}{\underset{HCH}{\|}}$$

$$\overset{:O:}{\underset{}{\underset{HCH}{\|}}} + H_2SO_4 \rightleftharpoons \overset{\overset{+}{:}O - H}{\underset{HCH}{\|}} + HSO_4^-$$

(c) $HC\equiv CH + AlCl_3 \rightleftharpoons \overset{}{\underset{}{HC}}\overset{\overset{-}{AlCl_3}}{\underset{}{=}}\overset{}{\underset{}{CH}}$ (with + on carbon)

$HC\equiv CH + H_2SO_4 \rightleftharpoons \overset{\overset{H}{|}}{\underset{}{HC}}=\overset{+}{CH} + HSO_4^-$

(d)

$$CH_3CH_2-\underset{\underset{\cdot\cdot}{\overset{\overset{CH_2CH_3}{|}}{P}}}{}-CH_2CH_3 + AlCl_3 \rightleftharpoons CH_3CH_2-\underset{\underset{AlCl_3}{\overset{-}{|}}}{\overset{\overset{CH_2CH_3}{|}}{\overset{+}{P}}}-CH_2CH_3$$

$$CH_3CH_2-\underset{\underset{\cdot\cdot}{\overset{\overset{CH_2CH_3}{|}}{P}}}{}-CH_2CH_3 + H_2SO_4 \rightleftharpoons CH_3CH_2-\underset{\underset{H}{\overset{}{|}}}{\overset{\overset{CH_2CH_3}{|}}{\overset{+}{P}}}-CH_2CH_3 + HSO_4^-$$

(e)

$$CH_3CH_2-\overset{\cdot\cdot}{\underset{\cdot\cdot}{O}}H + AlCl_3 \rightleftharpoons CH_3CH_2-\underset{\underset{\cdot\cdot}{}}{\overset{\overset{AlCl_3}{\overset{-}{|}}}{\overset{+}{O}}}H$$

$$CH_3CH_2-\overset{\cdot\cdot}{\underset{\cdot\cdot}{O}}H + H_2SO_4 \rightleftharpoons CH_3CH_2-\underset{\underset{\cdot\cdot}{}}{\overset{\overset{H}{\overset{}{|}}}{\overset{+}{O}}}H + HSO_4^-$$

(f)

$$CH_3CH_2-\overset{\cdot\cdot}{\underset{\cdot\cdot}{S}}-CH_2CH_3 + AlCl_3 \rightleftharpoons CH_3CH_2-\underset{\underset{\cdot\cdot}{}}{\overset{\overset{AlCl_3}{\overset{-}{|}}}{\overset{+}{S}}}-CH_2CH_3$$

$$CH_3CH_2-\overset{\cdot\cdot}{\underset{\cdot\cdot}{S}}-CH_2CH_3 + H_2SO_4 \rightleftharpoons CH_3CH_2-\underset{\underset{\cdot\cdot}{}}{\overset{\overset{H}{\overset{}{|}}}{\overset{+}{S}}}-CH_2CH_3 + HSO_4^-$$

3.4 (a) ethyl acetate

$$H-\overset{\overset{H}{|}}{\underset{\underset{H}{|}}{C}}-\overset{:O:}{\underset{}{\overset{\|}{C}}}-\overset{\cdot\cdot}{\underset{\cdot\cdot}{O}}-\overset{\overset{H}{|}}{\underset{\underset{H}{|}}{C}}-\overset{\overset{H}{|}}{\underset{\underset{H}{|}}{C}}-H$$

$$CH_3\overset{:O:}{\underset{}{\overset{\|}{C}}}-\overset{\cdot\cdot}{\underset{\cdot\cdot}{O}}-CH_2CH_3 + H_2SO_4 \rightleftharpoons CH_3\overset{\overset{+}{:}O-H}{\underset{}{\overset{\|}{C}}}-\overset{\cdot\cdot}{\underset{\cdot\cdot}{O}}-CH_2CH_3 + HSO_4^-$$

$$CH_3\overset{:O:}{\underset{}{\overset{\|}{C}}}-\overset{\cdot\cdot}{\underset{\cdot\cdot}{O}}-CH_2CH_3 + H_2SO_4 \rightleftharpoons CH_3\overset{:O:}{\underset{}{\overset{\|}{C}}}-\underset{\underset{H}{\overset{+}{|}}}{\overset{\cdot\cdot}{O}}-CH_2CH_3 + HSO_4^-$$

3.4   (b) 2-(*N*,*N*-dimethylamino)ethanol

$$H-\overset{\cdot\cdot}{\underset{\cdot\cdot}{O}}-CH_2CH_2-\overset{\overset{\displaystyle CH_3}{|}}{\underset{\cdot\cdot}{N}}-CH_3 \;+\; H_2SO_4 \;\rightleftharpoons\; H-\overset{\cdot\cdot}{\underset{\cdot\cdot}{O}}-CH_2CH_2-\overset{\overset{\displaystyle CH_3}{|}}{\underset{\underset{\displaystyle H}{|}}{\overset{+}{N}}}-CH_3 \;+\; HSO_4^-$$

$$H-\overset{\cdot\cdot}{\underset{\cdot\cdot}{O}}-CH_2CH_2-\overset{\overset{\displaystyle CH_3}{|}}{\underset{\cdot\cdot}{N}}-CH_3 \;+\; H_2SO_4 \;\rightleftharpoons\; H-\overset{+}{\underset{\underset{\displaystyle H}{|}}{O}}-CH_2CH_2-\overset{\overset{\displaystyle CH_3}{|}}{N}-CH_3 \;+\; HSO_4^-$$

(c) 2-propen-1-ol $H-C\!=\!C-C-\overset{\cdot\cdot}{\underset{\cdot\cdot}{O}}-H$

$$CH_2\!=\!CH_2CH_2-\overset{\cdot\cdot}{\underset{\cdot\cdot}{O}}-H \;+\; H_2SO_4 \;\rightleftharpoons\; CH_2\!=\!CH_2CH_2-\overset{+}{\underset{\underset{\displaystyle H}{|}}{O}}-H \;+\; HSO_4^-$$

$$CH_2\!=\!CHCH_2-\overset{\cdot\cdot}{\underset{\cdot\cdot}{O}}-H \;+\; H_2SO_4 \;\rightleftharpoons\; \underset{\underset{\displaystyle H}{|}}{CH_2}-\overset{+}{C}HCH_2-\overset{\cdot\cdot}{\underset{\cdot\cdot}{O}}-H \;+\; HSO_4^-$$

(d) methyl isocyanate

$$H-\overset{\overset{\displaystyle H}{|}}{\underset{\underset{\displaystyle H}{|}}{C}}-N\!=\!C\!=\!\overset{\cdot\cdot}{\underset{\cdot\cdot}{O}}$$

$$CH_3-\overset{\cdot\cdot}{N}\!=\!C\!=\!\overset{\cdot\cdot}{\underset{\cdot\cdot}{O}} \;+\; H_2SO_4 \;\rightleftharpoons\; CH_3-\overset{\overset{+}{}}{\underset{\underset{\displaystyle H}{|}}{N}}\!=\!C\!=\!\overset{\cdot\cdot}{\underset{\cdot\cdot}{O}} \;+\; HSO_4^-$$

$$CH_3-\overset{\cdot\cdot}{N}\!=\!C\!=\!\overset{\cdot\cdot}{\underset{\cdot\cdot}{O}} \;+\; H_2SO_4 \;\rightleftharpoons\; CH_3-N\!=\!C\!=\!\overset{+}{\underset{\cdot\cdot}{O}}-H \;+\; HSO_4^-$$

3.5   <u>base</u>   <u>conjugate acid</u>   <u>base</u>   <u>conjugate acid</u>

(a)

$$\underset{H}{\overset{CH_3CH_2}{\diagdown}}C\!=\!\underset{CH_2CH_3}{\overset{H}{\diagup}}C \qquad\qquad \underset{H}{\overset{CH_3CH_2}{\diagdown}}\overset{+}{C}-\underset{CH_2CH_3}{\overset{H}{|}}C-H$$

(b) $CH_3CH_2-\overset{\cdot\cdot}{\underset{\cdot\cdot}{O}}-CH_2CH_3$ $CH_3CH_2-\overset{+}{\underset{\underset{\displaystyle H}{|}}{O}}-CH_2CH_3$

(c) $CH_3CH_2\overset{\cdot\cdot}{N}CH_2CH_3$ $CH_3CH_2\overset{\overset{\displaystyle H}{\overset{|}{+}}}{N}CH_2CH_3$

   $\underset{\displaystyle H}{|}$ $\underset{\displaystyle H}{|}$

(d) $CH_3CH_2CH_2-\overset{\cdot\cdot}{\underset{\cdot\cdot}{S}}-CH_3$ $CH_3CH_2CH_2-\overset{+}{\underset{\underset{\displaystyle H}{|}}{S}}-CH_3$

(e) $CH_3CH_2\overset{\overset{\displaystyle :O:}{\|}}{C}H$ $CH_3CH_2\overset{\overset{\displaystyle :\overset{+}{O}-H}{\|}}{C}H$

(f) $CH_3CH_2CH_2-\overset{\cdot\cdot}{O}H$ $CH_3CH_2CH_2-\overset{\cdot\cdot}{\underset{+}{O}}H$

               $\underset{\displaystyle H}{|}$

(g) $CH_3CH_2\overset{\overset{\displaystyle :O:}{\|}}{C}CH_2CH_3$ $CH_3CH_2\overset{\overset{\displaystyle :\overset{+}{O}-H}{\|}}{C}CH_2CH_3$

3.6 Equations from Problem 3.3 rewritten:

(a)
$$CH_3\overset{\displaystyle CH_3}{\underset{\displaystyle H}{N:}} \quad \overset{\displaystyle :\ddot{C}l:}{\underset{\displaystyle :\ddot{C}l:}{Al-\ddot{C}l:}} \longrightarrow CH_3\overset{\displaystyle CH_3}{\underset{\displaystyle H}{\overset{+}{N}}}{-}\overset{-}{AlCl_3}$$

$$CH_3\overset{\displaystyle CH_3}{\underset{\displaystyle H}{N:}} \quad H{-}\ddot{O}{-}SO_3H \longrightarrow CH_3\overset{\displaystyle CH_3}{\underset{\displaystyle H}{\overset{+}{N}}}{-}H \quad \overset{-}{:}\ddot{O}{-}SO_3H$$

(b)
$$\underset{\displaystyle HCH}{\overset{\displaystyle :O:}{\parallel}} \quad \overset{\displaystyle :\ddot{C}l:}{\underset{\displaystyle :\ddot{C}l:}{Al-\ddot{C}l:}} \longrightarrow \underset{\displaystyle HCH}{\overset{\displaystyle \overset{+}{O}-\overset{-}{AlCl_3}}{\parallel}}$$

$$\underset{\displaystyle HCH}{\overset{\displaystyle :O:}{\parallel}} \quad H{-}\ddot{O}{-}SO_3H \longrightarrow \underset{\displaystyle HCH}{\overset{\displaystyle \overset{+}{O}-H}{\parallel}} \quad \overset{-}{:}\ddot{O}{-}SO_3H$$

(c)
$$HC\equiv CH \quad \overset{\displaystyle :\ddot{C}l:}{\underset{\displaystyle :\ddot{C}l:}{Al-\ddot{C}l:}} \longrightarrow \underset{\displaystyle HC=CH}{\overset{\displaystyle \overset{-}{AlCl_3}}{\overset{+}{}}}$$

$$HC\equiv CH \quad H{-}\ddot{O}{-}SO_3H \longrightarrow HC=CH_2 \quad \overset{-}{:}\ddot{O}{-}SO_3H$$

(d)
$$CH_3CH_2\overset{\displaystyle CH_3CH_2}{\underset{\displaystyle CH_3CH_2}{P:}} \quad \overset{\displaystyle :\ddot{C}l:}{\underset{\displaystyle :\ddot{C}l:}{Al-\ddot{C}l:}} \longrightarrow CH_3CH_2\overset{\displaystyle CH_3CH_2}{\underset{\displaystyle CH_3CH_2}{\overset{+}{P}}}{-}\overset{-}{AlCl_3}$$

$$CH_3CH_2\overset{\displaystyle CH_3CH_2}{\underset{\displaystyle CH_3CH_2}{P:}} \quad H{-}\ddot{O}{-}SO_3H \longrightarrow CH_3CH_2\overset{\displaystyle CH_3CH_2}{\underset{\displaystyle CH_3CH_2}{\overset{+}{P}}}{-}H \quad \overset{-}{:}\ddot{O}{-}SO_3H$$

(e)
$$CH_3CH_2{-}\ddot{O}H \quad \overset{\displaystyle :\ddot{C}l:}{\underset{\displaystyle :\ddot{C}l:}{Al-\ddot{C}l:}} \longrightarrow CH_3CH_2{-}\overset{+}{\underset{\displaystyle ..}{O}}\overset{\displaystyle \overset{-}{AlCl_3}}{H}$$

$$CH_3CH_2{-}\ddot{O}H \quad H{-}\ddot{O}{-}SO_3H \longrightarrow CH_3CH_2{-}\overset{\displaystyle H}{\underset{\displaystyle ..}{\overset{+}{O}}}H \quad \overset{-}{:}\ddot{O}{-}SO_3H$$

(f)
$$CH_3CH_2\ddot{S}{-}CH_2CH_3 \quad \overset{\displaystyle :\ddot{C}l:}{\underset{\displaystyle :\ddot{C}l:}{Al-\ddot{C}l:}} \longrightarrow CH_3CH_2\overset{\displaystyle \overset{-}{AlCl_3}}{\underset{\displaystyle ..}{\overset{+}{S}}}{-}CH_2CH_3$$

3.6 (f) (cont)

$$CH_3CH_2\overset{..}{\underset{..}{S}}-CH_2CH_3 \qquad H-\overset{..}{\underset{..}{O}}-SO_3H \longrightarrow CH_3CH_2\overset{H}{\overset{|+}{\underset{..}{S}}}-CH_2CH_3 \qquad {}^{-}:\overset{..}{\underset{..}{O}}-SO_3H$$

Equations from Problem 3.4 rewritten:

(a) $$\underset{\substack{|| \\ CH_3C-\overset{..}{\underset{..}{O}}-CH_2CH_3}}{:\overset{..}{O}:} \qquad H-\overset{..}{\underset{..}{O}}-SO_3H \longrightarrow \underset{\substack{|| \\ CH_3C-\overset{..}{\underset{..}{O}}-CH_2CH_3}}{:\overset{+}{\underset{}{O}}-H} \qquad {}^{-}:\overset{..}{\underset{..}{O}}-SO_3H$$

$$\underset{\substack{|| \\ CH_3C-\overset{..}{\underset{..}{O}}-CH_2CH_3}}{:\overset{..}{O}:}\;\;{\underset{\searrow H-\overset{..}{\underset{..}{O}}-SO_3H}{}} \longrightarrow \underset{\substack{|| \\ CH_3C-\overset{+}{\underset{|}{O}}-CH_2CH_3 \\ H}}{:\overset{..}{O}:} \qquad {}^{-}:\overset{..}{\underset{..}{O}}-SO_3H$$

(b) $$H-\overset{..}{\underset{..}{O}}-CH_2CH_2\overset{CH_3}{\underset{CH_3}{\overset{|}{N}}}: \qquad H-\overset{..}{\underset{..}{O}}-SO_3H \longrightarrow H-\overset{..}{\underset{..}{O}}-CH_2CH_2\overset{CH_3}{\underset{CH_3}{\overset{|+}{N}}}-H \qquad {}^{-}:\overset{..}{\underset{..}{O}}-SO_3H$$

$$H-\overset{..}{\underset{..}{O}}-CH_2CH_2\overset{CH_3}{\underset{\cdot\cdot}{\overset{|}{N}}}-CH_3 \;\; {\underset{\searrow H-\overset{..}{\underset{..}{O}}-SO_3H}{}} \longrightarrow H-\overset{..}{\underset{|+}{O}}-CH_2CH_2\overset{CH_3}{\underset{\cdot\cdot}{\overset{|}{N}}}-CH_3 \qquad {}^{-}:\overset{..}{\underset{..}{O}}-SO_3H$$

(c) $$CH_2{=}CHCH_2-\overset{..}{\underset{..}{O}}-H \;\; {\underset{\searrow H-\overset{..}{\underset{..}{O}}-SO_3H}{}} \longrightarrow CH_2{=}CHCH_2-\overset{..}{\underset{|+}{O}}-H \qquad {}^{-}:\overset{..}{\underset{..}{O}}-SO_3H$$

$$CH_2{=}CHCH_2-\overset{..}{\underset{..}{O}}-H \;\; {\underset{\searrow H-\overset{..}{\underset{..}{O}}-SO_3H}{}} \longrightarrow \overset{+}{CH_2}-\underset{\substack{| \\ H}}{CHCH_2}-\overset{..}{\underset{..}{O}}-H \qquad {}^{-}:\overset{..}{\underset{..}{O}}-SO_3H$$

(d) $$CH_3-\overset{..}{\underset{..}{N}}{=}C{=}\overset{..}{\underset{..}{O}} \;\; {\underset{\searrow H-\overset{..}{\underset{..}{O}}-SO_3H}{}} \longrightarrow CH_3-\overset{+}{\underset{\substack{| \\ H}}{N}}{=}C{=}\overset{..}{\underset{..}{O}} \qquad {}^{-}:\overset{..}{\underset{..}{O}}-SO_3H$$

$$CH_3-\overset{..}{\underset{..}{N}}{=}C{=}\overset{..}{\underset{..}{O}} \;\; {\underset{\searrow H-\overset{..}{\underset{..}{O}}-SO_3H}{}} \longrightarrow CH_3-\overset{..}{\underset{..}{N}}{=}C{=}\overset{..}{\underset{\substack{| \\ H}}{O}}{}^{+} \qquad {}^{-}:\overset{..}{\underset{..}{O}}-SO_3H$$

3.7 (a) $$\overset{CH_3CH_2}{\underset{H}{\overset{\diagdown}{:\overset{..}{O}:}}} \;\; {\underset{}{H-\overset{..}{\underset{..}{I}}:}} \longrightarrow CH_3CH_2-\overset{..}{\underset{\substack{|+ \\ H}}{O}}-H \qquad :\overset{..}{\underset{..}{I}}:^{-}$$

(b) $$CH_3-\overset{CH_3}{\underset{CH_3}{\overset{|}{N}}}: \;\; {\underset{}{H-\overset{..}{\underset{..}{Cl}}:}} \longrightarrow CH_3-\overset{CH_3}{\underset{CH_3}{\overset{|+}{N}}}-H \qquad :\overset{..}{\underset{..}{Cl}}:^{-}$$

3.7 (c)

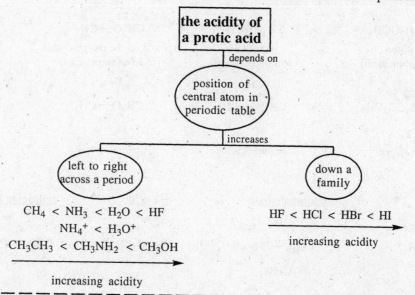

(d) $CH_3C\equiv CCH_3 \longrightarrow CH_3\overset{+}{C}=CCH_3$
$\quad\quad\quad\quad\quad\quad\quad\quad\quad\quad\quad\quad\quad\quad\quad\quad | $
$\quad\quad\quad\quad\quad\quad\quad\quad\quad\quad\quad\quad\quad\quad\quad\quad H$

(e) $CH_3-\overset{CH_3}{\underset{CH_2=CH_2}{\overset{|}{\underset{|}{C}}}}-CH_3 \longrightarrow CH_3-\overset{CH_3}{\underset{CH_2-CH_2}{\overset{|}{\underset{|}{C}}}}-CH_3$

3.8
(a)

most basic		least basic
CH_3NHCH_3	>	CH_3OCH_3
amine		ether
(nonbonding electrons on nitrogen)		(nonbonding electrons on oxygen)

(b)

CH_3^- > NH_2^- > OH^- > F^-

methyl anion   amide ion   hydroxide ion   fluoride ion

a carbanion

(nonbonding electrons on carbon; EN 2.5)

(nonbonding electrons on fluorine; EN 4.0)

EN = electronegativity

(c)

$CH_3\overset{-}{N}CH_3$ > CH_3O^-

amide anion   alkoxide anion

Concept Map 3.3   Relationship of acidity to position of the central element in the periodic table.

the acidity of a protic acid

depends on

position of central atom in periodic table

increases

left to right across a period   *down a family*

$CH_4 < NH_3 < H_2O < HF$ $HF < HCl < HBr < HI$
$\quad\quad NH_4^+ < H_3O^+$
$CH_3CH_3 < CH_3NH_2 < CH_3OH$ increasing acidity →

increasing acidity →

3.9 H_2S is the stronger acid.

$$H_2S + HO^- \rightleftharpoons HS^- + H_2O$$

Hydrogen sulfide ionizes more readily than does water because the hydrogen sulfide anion is more stable relative to its conjugate acid than is hydroxide ion. The hydrogen sulfide anion is a weaker base than hydroxide anion for the same reason that methanethiolate anion is a weaker base than the methoxide anion. The sulfur atom is larger than an oxygen atom. Therefore the negative charge on the hydrogen sulfide anion is more spread out. The anion is more stable relative to its conjugate acid than is hydroxide ion, and less likely to be protonated.

3.10 $CH_3CH_2SH + CH_3O^- \rightleftharpoons CH_3CH_2S^- + CH_3OH$

Ethanethiolate ion is a weaker base than methoxide ion because the negative charge is more spread out on the sulfur atom than on the oxygen atom. (See the answer to Problem 3.9.) In the reaction of ethanethiol and methoxide ion, the equilibrium lies far to the right.

3.11   (a)   One of the nonbonding electron pairs on the oxygen atom can react with the empty *p*-orbital of the carbocation.

an oxonium ion

(b)   The product formed is a protonated alcohol, an oxonium ion, which has a $pK_a \sim -2$. It will transfer a proton to the oxygen atom of water in an equilibrium reaction. With a large excess of water, equilibrium lies toward the right.

(excess)

3.12   (a)

least acidic					most acidic
$CH_3CH_2CH_3$	<	$CH_3CH_2NH_2$	<	CH_3CH_2OH	<	CH_3CH_2SH
alkene		amine		alcohol		thiol

(b)

				H
CH_3OCH_3	<	CH_3NHCH_3	<	$CH_3\overset{+}{O}-CH_3$
(proton on a carbon atom)		(proton on a nitrogen atom)		(proton on the oxygen atom of an oxonium ion)

(c)

		CH_3		H
$CH_3CH_2CH_3$	<	$CH_3\overset{+}{N}{-}CH_3$	<	$CH_3\overset{+}{O}{-}CH_3$
		H		
alkane		ammonium ion		oxonium ion

3.13

acid	conjugate base	acid	conjugate base

(a)

(b) $CH_3CH_2CH_2\overset{..}{\underset{..}{S}}H$ $CH_3CH_2CH_2\overset{..}{\underset{..}{S}}{:}^-$

(c)

(d)

3.13

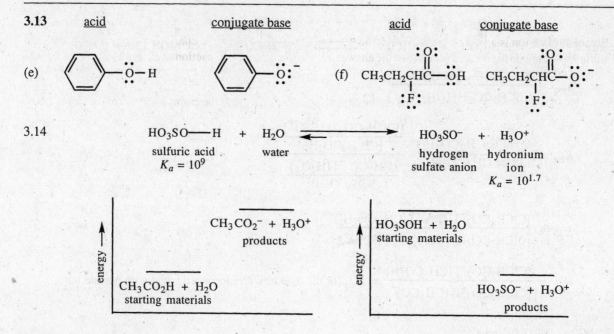

3.15   The acidity constant is

$$K_a = \frac{[H^+][\overset{\bullet}{\underset{\bullet}{}}B]}{[HB^+]}$$

where $\overset{\bullet}{\underset{\bullet}{}}B$ and HB^+ are the conjugate base and conjugate acid forms, respectively.

pK_a is defined as $-\log K_a$

For formic acid $\quad pK_a \quad = -\log(1.99 \times 10^{-4})$

$\qquad\qquad\qquad\quad = -(\log 1.99 + \log 10^{-4})$

$\qquad\qquad\qquad\quad = -(0.299 - 4)$

$\qquad\qquad pK_a \quad = -(-3.701)$

$\qquad\qquad pK_a \quad = +3.701$

For acetic acid $\quad pK_a = +4.76$

As the pK_a gets larger (more positive), K_a gets smaller, and the amount of dissociation of the acid decreases. Formic acid has the smaller pK_a, and thus the larger K_a, and is a stronger acid than acetic acid.

3.16

$$pK_a = -\log K_a$$
$$\log K_a = -7$$
$$K_a = 10^{-7}$$

base

conjugate acid
(one resonance contributor)

3.17

$$ClCH_2COH + H_2O \quad \rightleftharpoons \quad ClCH_2CO^- + H_3O^+$$

$$K_{eq} = \frac{[ClCH_2CO_2^-][H_3O^+]}{[ClCH_2CO_2H][H_2O]}$$

$$K_{eq}[H_2O] = K_a = \frac{[ClCH_2CO_2^-][H_3O^+]}{[ClCH_2CO_2H]}$$

3.17 (cont)

$$\underset{\text{ClCH}_2\overset{\displaystyle O}{\overset{\displaystyle \|}{\text{C}}}\text{OH}}{} + \underset{\text{CH}_3\overset{\displaystyle O}{\overset{\displaystyle \|}{\text{C}}}\text{O}^-}{} \rightleftharpoons \underset{\text{ClCH}_2\overset{\displaystyle O}{\overset{\displaystyle \|}{\text{C}}}\text{O}^-}{} + \underset{\text{CH}_3\overset{\displaystyle O}{\overset{\displaystyle \|}{\text{C}}}\text{OH}}{}$$

$$K_{eq} = \frac{[\text{ClCH}_2\text{CO}_2^-]\,[\text{CH}_3\text{CO}_2\text{H}]}{[\text{ClCH}_2\text{CO}_2\text{H}]\,[\text{CH}_3\text{CO}_2^-]}$$

$$K_{eq} = \frac{K_a \text{ of ClCH}_2\text{CO}_2\text{H}}{K_a \text{ of CH}_3\text{CO}_2\text{H}} = \frac{\dfrac{[\text{ClCH}_2\text{CO}_2^-]\,[\text{H}_3\text{O}^+]}{[\text{ClCH}_2\text{CO}_2\text{H}]}}{\dfrac{[\text{CH}_3\text{CO}_2^-]\,[\text{H}_3\text{O}^+]}{[\text{CH}_3\text{CO}_2\text{H}]}}$$

$$K_{eq} = \frac{[\text{ClCH}_2\text{CO}_2^-]\,\cancel{[\text{H}_3\text{O}^+]}\,[\text{CH}_3\text{CO}_2\text{H}]}{[\text{ClCH}_2\text{CO}_2\text{H}]\,[\text{CH}_3\text{CO}_2^-]\,\cancel{[\text{H}_3\text{O}^+]}}$$

$$K_{eq} = \frac{[\text{ClCH}_2\text{CO}_2^-]\,[\text{CH}_3\text{CO}_2\text{H}]}{[\text{ClCH}_2\text{CO}_2\text{H}]\,[\text{CH}_3\text{CO}_2^-]} = K_{eq} \text{ for the reaction of chloroacetic acid with acetate ion}$$

3.18

$$\Delta G° = -2.303\, RT \log K_{eq}$$

$$-2.59 \text{ kcal/mol} = (-2.303)(1.987 \times 10^{-3} \text{ kcal/mol·K})(298 \text{ K})(\log K_{eq})$$

$$\log K_{eq} = 1.899$$

$$K_{eq} = 10^{1.899} = 79.3$$

$$K_{eq} = \frac{K_a \text{ of ClCH}_2\text{CO}_2\text{H}}{K_a \text{ of CH}_3\text{CO}_2\text{H}}$$

$$79.3 = \frac{K_a \text{ of ClCH}_2\text{CO}_2\text{H}}{(1.75 \times 10^{-5})}$$

$$K_a \,(\text{ClCH}_2\text{CO}_2\text{H}) = (79.3)(1.75 \times 10^{-5}) = 1.39 \times 10^{-3}$$

$$pK_a \,(\text{ClCH}_2\text{CO}_2\text{H}) = -\log K_a = -\log (1.39 \times 10^{-3})$$

$$pK_a = -(-3 + 0.143) = 2.86$$

This is very close to the value, 2.8, given in the front cover of the textbook for chloroacetic acid.

- -

Concept Map 3.4   The relationship of pK_a to energy and entropy factors.

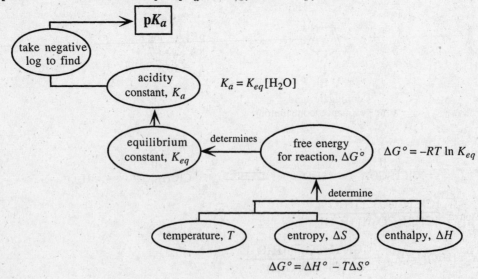

3.19

functional group	acid	pK_a

alkane

$$H-\underset{\underset{H}{|}}{\overset{\overset{H}{|}}{C}}-H \qquad 47$$

amine

$$H-\underset{\underset{H}{|}}{\overset{\cdot\cdot}{N}}-H \qquad 36$$

alcohol

$$CH_3CH_2-\overset{\cdot\cdot}{\underset{\cdot\cdot}{O}}-H \qquad 17$$

thiol

$$CH_3CH_2-\overset{\cdot\cdot}{\underset{\cdot\cdot}{S}}-H \qquad 10.5$$

increasing
acidity

ammonium ion

$$CH_3-\underset{\underset{CH_3}{|}}{\overset{\overset{CH_3}{|}}{\overset{+}{N}}}-H \qquad 9.8$$

carboxylic acid

$$\text{Ph}-\overset{\overset{\cdot\cdot}{\overset{O}{\parallel}}}{C}-\overset{\cdot\cdot}{\underset{\cdot\cdot}{O}}-H \qquad 4.2$$

oxonium ion

$$CH_3CH_2-\overset{\overset{H}{|}}{\underset{\cdot\cdot}{O}}{}^{+}-H \qquad -2.4$$

(Note that in this table acidity increases down the table, unlike the pK_a table inside the front cover of the textbook.)

3.20 (a) CH_3CH_2OH + K^+OH^- ⇌ $CH_3CH_2O^- K^+$ + H_2O

        pK_a 17   pK_a 15.7

(b) $CH_3CH_2\overset{+}{\underset{\underset{CH_3CH_2}{|}}{N}}H_2$ Cl^- + Na^+OH^- ⇌ $CH_3CH_2\underset{\underset{CH_3CH_2}{|}}{N}H$ + H_2O + Na^+Cl^-

        pK_a 10   pK_a 15.7

(c) $CH_2F\overset{\overset{O}{\parallel}}{C}OH$ + $Na^+HCO_3^-$ ⇌ $CH_2F\overset{\overset{O}{\parallel}}{C}O^- Na^+$ + H_2CO_3 $\left[⇌ H_2O + CO_2\uparrow\right]$

     pK_a 2   pK_a 6.5

(d) $CH_3-\langle\text{ring}\rangle-OH$ + $Na^+HCO_3^-$ ⇌ $CH_3-\langle\text{ring}\rangle-O^- Na^+$ + H_2CO_3

           pK_a 10   pK_a 6.5

(e) $\langle\text{ring}\rangle-NH_2$ + HCl ⇌ $\langle\text{ring}\rangle-\overset{+}{N}H_3$ Cl^-

                pK_a -7   pK_a 4.6

(f) $CH_3CH_2CH_2SH$ + $CH_3CH_2O^- Na^+$ ⇌ $CH_3CH_2CH_2S^- Na^+$ + CH_3CH_2OH

         pK_a 10.5   pK_a 17

(g) $CH_3-\langle\text{ring}\rangle-\overset{\overset{O}{\parallel}}{C}OH$ + Na^+CN^- ⇌ $CH_3-\langle\text{ring}\rangle-\overset{\overset{O}{\parallel}}{C}O^- Na^+$ + HCN

          pK_a 4   pK_a 9.1

3.20 (h)
$$\underset{\text{p}K_a\ 9.0}{CH_3\overset{O}{\overset{\|}{C}}CH_2\overset{O}{\overset{\|}{C}}CH_3} + CH_3O^-Na^+ \rightleftharpoons \underset{\underset{Na^+}{\overset{..}{}}}{CH_3\overset{O}{\overset{\|}{C}}\overset{}{C}HCH_3} + \underset{\text{p}K_a\ 15.5}{CH_3OH}$$

3.21 $\Delta G^\circ = -RT \ln K_{eq}$

  $= -2.303RT \log K_{eq}$

$\text{p}K_{eq} = -\log K_{eq}$

therefore $\Delta G^\circ = 2.303RT\,\text{p}K_{eq}$

At 298 K with $R = 1.987 \times 10^{-3}$ kcal/mol•K

$\Delta G^\circ = (2.303)(1.987 \times 10^{-3}\ \text{kcal/mol•K})(298\ \text{K})\,\text{p}K_{eq}$

$\Delta G^\circ = 1.36\ \text{kcal/mol} \times \text{p}K_{eq}$

3.22 $\text{p}K_{eq} = (\text{p}K_a\ \text{of starting acid}) - (\text{p}K_a\ \text{of product acid})$

starting materials	products
$\underset{\text{p}K_a\ \sim36}{CH_3CH_2^-} + CH_3NH_2$	$CH_3CH_3 + \underset{\text{p}K_a\ \sim49}{CH_3NH^-}$

$\text{p}K_{eq} = 36 - 49 = -13;\ \Delta G^\circ = 1.4\ \text{kcal/mol} \times (-13) = -18.2\ \text{kcal/mol}$
There is a large decrease in free energy so the reaction goes essentially to completion as written.

$\underset{\text{p}K_a\ 15.5}{CH_3NH^-} + CH_3OH$	$CH_3NH_2 + \underset{\text{p}K_a\ \sim36}{CH_3O^-}$

$\text{p}K_{eq} = 15.5 - 36 = -20.5;\ \Delta G^\circ = 1.4\ \text{kcal/mol} \times (-20.5) = -28.7\ \text{kcal/mol}$
There is a large decrease in free energy so the reaction goes essentially to completion as written.

$\underset{\text{p}K_a\ 15.5}{CH_3OH} + OH^-$	$CH_3O^- + \underset{\text{p}K_a\ 15.7}{H_2O}$

$\text{p}K_{eq} = 15.5 - 15.7 = -0.2;\ \Delta G^\circ = 1.4\ \text{kcal/mol} \times (-0.2) = -0.28\ \text{kcal/mol}$
There is only a small decrease in free energy. Approximately equal amounts of both starting material and products are present at equilibrium.

$\underset{\text{p}K_a\ \sim10.5}{CH_3SH} + OH^-$	$CH_3S^- + \underset{\text{p}K_a\ 15.7}{H_2O}$

$\text{p}K_{eq} = 10.5 - 15.7 = -5.2;\ \Delta G^\circ = 1.4\ \text{kcal/mol} \times (-5.2) = -7.3\ \text{kcal/mol}$
There is a large decrease in free energy so the reaction goes essentially to completion as written.

$\underset{\text{p}K_a\ 10.6}{CH_3NH_3^+} + OH^-$	$CH_3NH_2 + \underset{\text{p}K_a\ 15.7}{H_2O}$

$\text{p}K_{eq} = 10.6 - 15.7 = -5.1;\ \Delta G^\circ = 1.4\ \text{kcal/mol} \times (-5.1) = -7.1\ \text{kcal/mol}$
There is a large decrease in free energy so the reaction goes essentially to completion as written.

$\underset{\text{p}K_a\ \sim-1.7}{CH_3OH_2^+} + H_2O$	$CH_3OH + \underset{\text{p}K_a\ -1.7}{H_3O^+}$

$\text{p}K_{eq} = -1.7 - (-1.7) = 0;\ \Delta G^\circ = 1.4\ \text{kcal/mol} \times (0) = 0$
There is essentially no decrease in free energy. Approximately equal amounts of both starting material and products are present at equilibrium.

3.23 CF_3CH_2OH is more acidic than CH_3CH_2OH. CF_3CH_2OH ionizes more readily than CH_3CH_2OH because the $CF_3CH_2O^-$ ion is more stable relative to its conjugate acid than is the ethoxide ion. The electron-withdrawing effect (negative inductive effect) of the three fluorine atoms reduces the negative charge on the oxygen atom in 2,2,2-trifluoroethoxide ion relative to that in ethoxide ion. Withdrawal of negative charge from the oxygen atom stabilizes the anion.

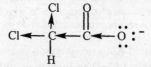

negative inductive effect of fluorine atoms reduces effective negative charge at the oxygen atom, stabilizing the anion

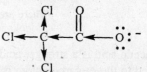

no electron-withdrawing groups to stabilize this anion

3.24

$ClCH_2\overset{O}{\overset{\|}{C}}OH$	$Cl_2CH\overset{O}{\overset{\|}{C}}OH$	$CCl_3\overset{O}{\overset{\|}{C}}OH$
pK_a 2.86	1.30	0.64

⟶ increasing acidity

Inductive effects are additive. The more electron-withdrawing chlorine atoms there are in the molecule, the more stable its conjugate base will be relative to the acid, and, therefore, the more acidic the conjugate acid will be.

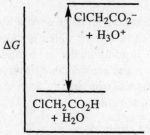

electron-withdrawing effect of one chlorine atom stabilizes the anion relative to acetate ion, making the conjugate acid more acidic than acetic acid (pK_a 4.8)

electron-withdrawing effect of two chlorine atoms leads to still further stabilization of the anion, and further strengthening of the conjugate acid

electron-withdrawing effect of three chlorine atoms makes this anion the most stable of the three compared here and, thus, its conjugate acid is the strongest

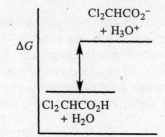

ΔG $ClCH_2CO_2^-$ + H_3O^+

$ClCH_2CO_2H$ + H_2O

large difference in energy between between acid and its conjugate base

ΔG $Cl_2CHCO_2^-$ + H_3O^+

Cl_2CHCO_2H + H_2O

ΔG $CCl_3CO_2^-$ + H_3O^+

CCl_3CO_2H + H_2O

much smaller difference in energy between acid and its conjugate base

3.25

:N≡C—C(H)—C(=O)—Ö—H + H_2O ⇌ :N≡C—C(H)—C(=O)—Ö:⁻ + H_3O^+

cyanoacetic acid   cyanoacetate anion

:N≡C—CH₂—C(=O)—Ö:⁻ ⟷ ⁻:N̈≡C—CH₂—C(=O)—Ö:⁻ ⟷ :N≡C—CH₂—C(—Ö)=Ö̈ ⟷ :N≡C—CH₂—C(—Ö⁻)=Ö:

3.25 (cont)

Cyanoacetic acid is stronger than acetic acid because its conjugate base is a weaker base than acetate ion. The atom attached to the α-carbon atom has a positive charge in one important resonance contributor. The cyano group is, therefore, electron withdrawing, reducing electron density at the carboxylate ion by its inductive effect. The withdrawal of charge stabilizes the cyanoacetate ion relative to its acid, shifting the equilibrium toward greater ionization for cyanoacetic acid than for acetic acid.

nitroacetic acid   nitroacetate anion

resonance involving the nitro group

resonance involving the carboxylate group
(There are three more resonance contributors in which the nitro group
looks as it does in the resonance contributor at the upper right.)

Nitroacetic acid is a stronger acid than acetic acid because its anion is a weaker base than acetate ion. The nitrogen attached to the α–carbon atom is more electronegative than hydrogen and always has a formal positive charge. The nitro group, with one nitrogen atom and two oxygen atoms, is a strongly electron-withdrawing group. Nitroacetic acid is a stronger acid than cyanoacetic acid because the atom attached to the α-carbon atom in nitroacetic acid carries a full positive charge (all resonance contributors have a positively-charged nitrogen atom). Only one resonance contributor in cyanoacetic acid has a positive carbon atom attached to the α-carbon atom.

3.26   In ethoxide anion, the basic oxygen atom is attached to an sp^3-hybridized carbon atom with no great polarity. In the acetate anion, the basic oxygen atom is attached to an sp^2-hybridized carbon atom which is more electronegative than an sp^3-hybridized carbon atom (see Section 2.7B in the text). In addition, this sp^2-hybridized carbon atom is attached to the still more electronegative oxygen atom, so it has considerable positive character. The positive nature of the carbon atom of the carbonyl group attracts electron density from the negatively charged oxygen atom of the acetate anion, resulting in stabilization of the anion.

3.27   (a)

pK_a ~10.6 pK_a 15.7
pK_{eq} = 10.6 – 15.7 = –5.1; $\Delta G°$ = 1.4 kcal/mol × (–5.1) = –7.1 kcal/mol
There is a large decrease in free energy so the reaction goes essentially to completion as written.

3.27 (b) H_2O + $CH_3CH_2\overset{+}{\underset{H}{O}}CH_2CH_3$ $\rightleftharpoons$ $CH_3CH_2OCH_2CH_3$ + H_3O^+

pK_a –3.6 pK_a –1.7

pK_{eq} = –3.6 – (–1.7) = –1.9; $\Delta G°$ = 1.4 kcal/mol × (–1.9) = –2.7 kcal/mol

There is a decrease in free energy. More product than starting material will be present at equilibrium.

(c) $CH_3\overset{O}{\overset{\|}{C}}OH$ + $CH_3CH_2S^-\,Na^+$ $\rightleftharpoons$ $CH_3\overset{O}{\overset{\|}{C}}O^-\,Na^+$ + CH_3CH_2SH

pK_a 4.8 pK_a 10.5

pK_{eq} = 4.8 – 10.5 = –5.7; $\Delta G°$ = 1.4 kcal/mol × (–5.7) = –8.0 kcal/mol

There is a large decrease in free energy so the reaction goes essentially to completion as written.

(d) $CCl_3\overset{O}{\overset{\|}{C}}OH$ + $Na^+\,HCO_3^-$ $\rightleftharpoons$ $CCl_3\overset{O}{\overset{\|}{C}}O^-\,Na^+$ + H_2CO_3 ($\rightleftharpoons$ H_2O + $CO_2\!\uparrow$)

pK_a 0.6 pK_a 6.5

pK_{eq} = 0.6 – 6.5 = –5.9; $\Delta G°$ = 1.4 kcal/mol × (–5.9) = –8.3 kcal/mol

There is a large decrease in free energy so the reaction goes essentially to completion as written.

Concept Map 3.5   The effects of structural changes on acidity and basicity.

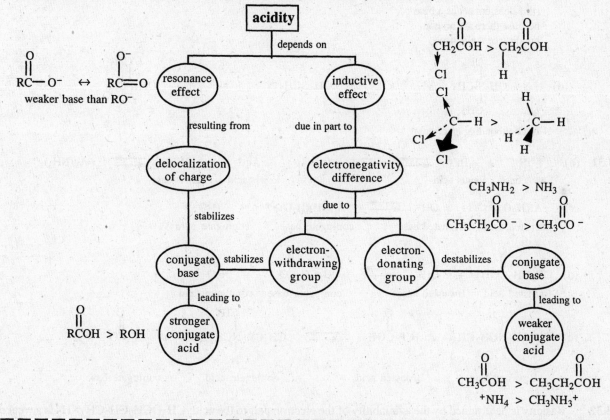

3.28   <u>least acidic</u>   <u>most acidic</u>

NaH < AlH_3 < H_2S < HCl

The order of relative acidity of the hydrogen atoms increases as the electronegativity of the atom to which the hydrogen atom is attached increases. This is due to the increasing stability of the conjugate base as the electronegativity of the central atom increases.

3.29   least acidic   most acidic

(a) $CH_3\overset{O}{\overset{\|}{C}}OH$ < $CCl_3\overset{O}{\overset{\|}{C}}OH$ < $CF_3\overset{O}{\overset{\|}{C}}OH$

(b) CH_3CH_2OH < $ClCH_2CH_2OH$ < FCH_2CH_2OH

(c) $CH_3CH_2NH_2$ < CH_3CH_2OH < $CH_3CH_2\overset{+}{S}\overset{O-H}{}$

(d) $CH_3\overset{O}{\overset{\|}{C}}OCH_3$ < $CH_3\overset{O}{\overset{\|}{C}}OH$ < $CH_3\overset{\overset{+}{O}-H}{\overset{\|}{C}}OH$

(e) $CH_3\overset{O}{\overset{\|}{C}}OCH_3$ < $CH_3\overset{O}{\overset{\|}{C}}NH_2$ < $CH_3\overset{O}{\overset{\|}{C}}OH$

3.30   least basic   most basic

(a) $CH_3\overset{..}{\underset{..}{O}}CH_3$ < $CH_3\overset{CH_3}{\underset{..}{N}}CH_3$ < $Cl_3C\overset{..}{:}{}^-$

(b) $:NF_3$ < $:NH_2\overset{..}{O}H$ < $:NH_3$

(c) $\overset{+}{N}H_4$ < $:NH_3$ < $^-\,\overset{..}{:}NH_2$

(this does not act as a base because there are no non-bonding or π electrons)

(d) $CH_3\overset{CH_3}{\underset{CH_3}{Si}}CH_3$ < $CH_3\overset{..}{\underset{..}{S}}CH_3$ < $CH_3\overset{CH_3}{\underset{..}{P}}CH_3$

(no nonbonding electrons)

3.31 (a) F^- + BF_3 $\rightleftharpoons$ BF_4^- (b) Ag^+ + $2NH_3$ $\rightleftharpoons$ $Ag(NH_3)_2^+$
Lewis base   Lewis acid   Lewis acid   Lewis base

(c) $Al(H_2O)_6^{3+}$ + OH^- $\rightleftharpoons$ $Al(OH)(H_2O)_5^{2+}$ + H_2O
Brønsted acid   Brønsted base   conjugate base   conjugate acid

(d) $CH_3CH_2\overset{O}{\overset{\|}{C}}OH$ + OH^- $\rightleftharpoons$ $CH_3CH_2\overset{O}{\overset{\|}{C}}O^-$ + H_2O
Brønsted acid   Brønsted base   conjugate base   conjugate acid

(e) $CH_3CH_2\overset{CH_3}{\underset{}{N}}CH_2CH_3$ + $CF_3\overset{O}{\overset{\|}{C}}OH$ $\rightleftharpoons$ $CH_3CH_2\overset{CH_3}{\underset{H}{\overset{+}{N}}}CH_2CH_3$ + $CF_3\overset{O}{\overset{\|}{C}}O^-$

Brønsted base   Brønsted acid   conjugate acid   conjugate base

3.32   Basicity is determined by the availability of the electron pair to the acid. $CH_3OCH_2CH_2CH_2NH_2$ is a weaker base than $CH_3CH_2CH_2CH_2NH_2$ because the electron-withdrawing oxygen atom of the methoxy group is pulling electron density away from the nitrogen atom. In $(CH_3O)_2CHCH_2NH_2$ there are two electron-withdrawing methoxy groups closer to the nitrogen atom than in the first case, which further reduce the availability of the electron pair. In $N\equiv CCH_2CH_2NH_2$ the cyano group is even more electron-withdrawing than the methoxy group because of the positive charge on the carbon atom of the cyano group in one resonance contributor (see Problem 3.25).

3.33

(a) CH_3CH_2-Br ⟶ $CH_3CH_2-C≡N:$

$^-:C≡N:$

$:Br:^-$

(b)

$CH_3-\overset{O}{\underset{|}{C}}-CH_3$ ⟶ $CH_3-\overset{O-H}{\underset{|}{C}}-CH_3$

$\overset{+}{O}H$

$\underset{H}{\overset{|}{O}}-CH_3$

$H-\overset{+}{O}-CH_3$
      H

$:\overset{H}{\underset{H}{O}}:$

(c) $CH_2=CH_2$ ⟶ $\overset{+}{C}H_2-CH_3$

$H-\overset{..}{O}-SO_3H$ $^-:\overset{..}{O}-SO_3H$

(d) $CH_3-\overset{CH_3}{\underset{CH_3}{C}}-Cl:$ ⟶ $CH_3-\overset{CH_3}{\underset{CH_3}{\overset{+}{C}}}$ $:Cl:^-$

(e)

$\overset{CH_3}{\underset{CH_3}{\overset{+}{C}}}-\overset{H}{\underset{H}{C}}-H$ ⟶ $\overset{CH_3}{\underset{CH_3}{C}}=\overset{H}{\underset{H}{C}}$

$:\overset{H}{\underset{H}{O}}:$ $H-\overset{+}{\underset{H}{O}}-H$

(f) $CH_3-\overset{CH_3}{\underset{:Cl:^-}{\overset{+}{C}}}-CH_3$ ⟶ $CH_3-\overset{CH_3}{\underset{:Cl:}{C}}-CH_3$

(g) $CH_3CH_2CH_2-\overset{..}{Br}:$ ⟶ $CH_3CH_2CH_2-\overset{+}{N}H_3$

$H-\overset{..}{N}-H$
      H

$:Br:^-$

(h) $H-\overset{H}{\underset{H}{C}}-\overset{:O:}{\underset{|}{C}}-CH_3$ ⟶ $H-C=\overset{:O:^-}{\underset{|}{C}}-CH_3$

$^-:\overset{..}{O}-H$ $H-\overset{..}{O}-H$

3.34

$Ph-\overset{:O:}{\underset{|}{C}}-\overset{\overset{..}{O}:}{\underset{|}{N}}-CH_2Ph$ ⟶ $Ph-\overset{:O:}{\underset{|}{C}}-N-CH_2Ph$ ⟶ $Ph-\overset{:O:}{\underset{|}{C}}-\overset{+}{N}=\overset{..}{O}-CH_2Ph$

$\overset{..}{O}:$
$\underset{C}{|}$
Ph
$:O:$

$H-\overset{+}{\underset{H}{O}}-H$ $:O:$

$:\overset{..}{O}:$
$\underset{H}{\overset{|}{O}}$

$\overset{..}{O}:$
$\underset{C}{|}$
Ph
$\overset{+}{O}:$
H

$\overset{..}{O}:$
$\underset{C}{|}$
Ph
$\overset{..}{O}:$
H

3.35   (a)

$\overset{O}{\underset{||}{CHCl_2COH}}$

$pK_a = 1.3$

$K_a = 10^{-pK_a} = 10^{-1.3}$

$\quad = 10^{+0.7} \times 10^{-2}$

$K_a = 5 \times 10^{-2}$

(b)

$CH_3\overset{+}{N}H_3$

$pK_a = 10.4$

$K_a = 10^{-pK_a} = 10^{-10.4}$

$\quad = 10^{+0.6} \times 10^{-11}$

$K_a = 4 \times 10^{-11}$

3.35   (c) CCl_3CH_2OH

$pK_a = 12.2$

$K_a = 10^{-pK_a} = 10^{-12.2}$

$\quad = 10^{+0.8} \times 10^{-13}$

$K_a = 6 \times 10^{-13}$

3.36   (a) CH_3CH_2OH

$K_a = 10^{-17} = 10^{-pK_a}$

$pK_a = -\log K_a = +17$

(b) CH_3CH_2SH

$K_a = 3.16 \times 10^{-11}$

$\quad = -\log(3.16 \times 10^{-11})$

$\quad = -(\log 3.16 + \log 10^{-11})$

$\quad = -(0.5 - 11) = -(-10.50)$

$pK_a = +10.50$

(c) $CH_3CH_2\overset{H}{\underset{+}{O}}CH_2CH_3$

$K_a = 3.98 \times 10^{+3}$

$pK_a = -\log(3.98 \times 10^{+3})$

$\quad = -(\log 3.98 + \log 10^{+3})$

$\quad = -(0.60 + 3) = -(3.60)$

$pK_a = -3.60$

3.37   (a) $H:^-$

(b) Na^+H^- + H_2O $\longrightarrow$ Na^+OH^- + $H_2\uparrow$

(c)   Hydride ion, H^-, is a base. Hydrogen, H_2, is its conjugate acid.

(d) CH_3CH_2OH + Na^+H^- $\longrightarrow$ $CH_3CH_2O^-Na^+$ + $H_2\uparrow$

3.38

(a)

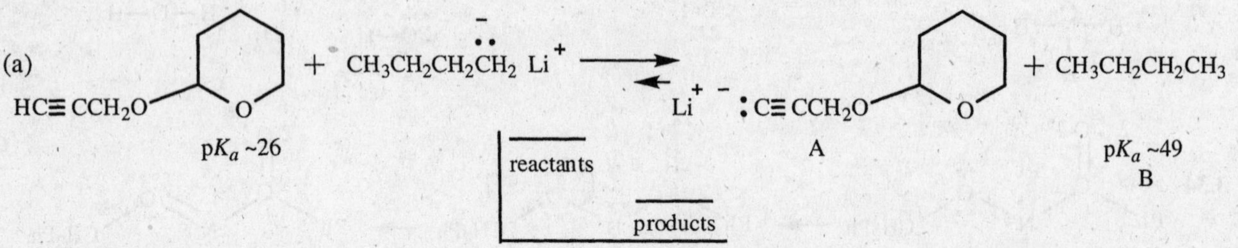

(b)

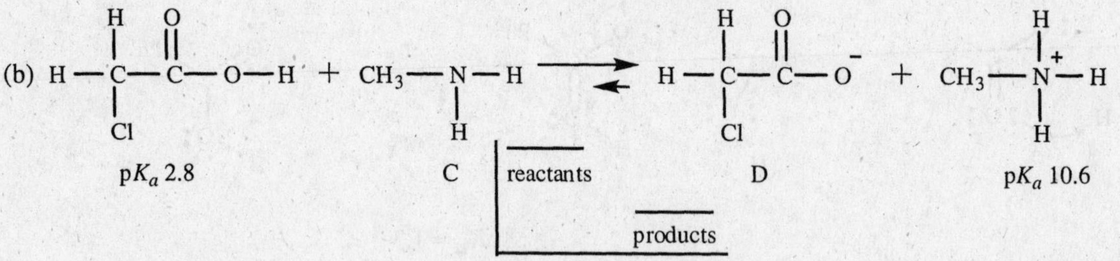

(c)

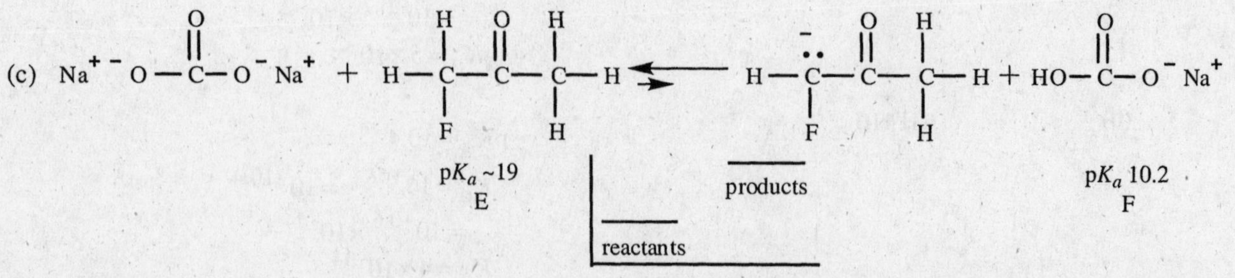

3.38 (cont)

(d) CH_3-O-CH_3 + HI $\rightleftharpoons$ $CH_3-\overset{H}{\underset{+}{O}}-CH_3$ + I^-

        G pK_a –9 $pK_a \sim$ –3.6   H

                     _____
                    reactants
                                 _____
                                 products

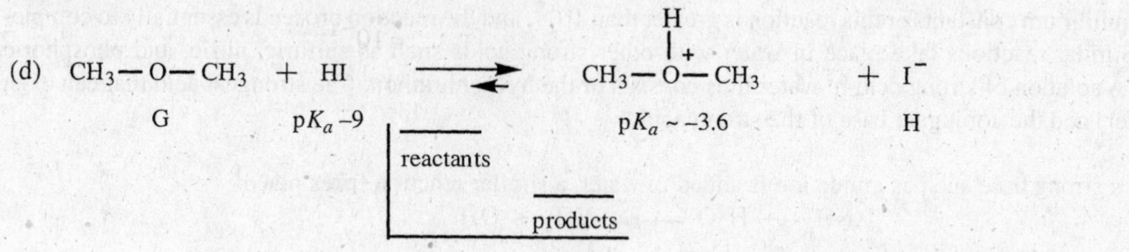

(e) [structure: salicylic acid] + Na^+ $^-O-\overset{O}{\overset{\|}{C}}-OH$ $\rightleftharpoons$ [structure: sodium salicylate] + $HO-\overset{O}{\overset{\|}{C}}-OH$

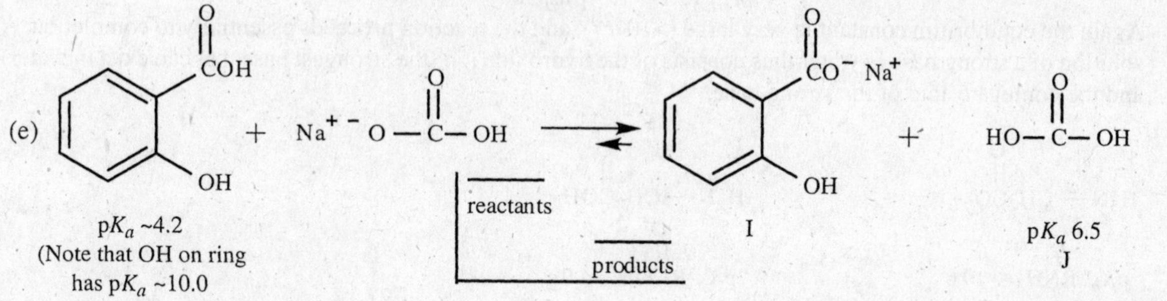

$pK_a \sim$4.2
(Note that OH on ring
has $pK_a \sim$10.0

                               _____
                             reactants   I pK_a 6.5
                                   _____   J
                                 products

3.39 (a) [Lewis structure of triflic acid: F_3C-SO_2-O-H]

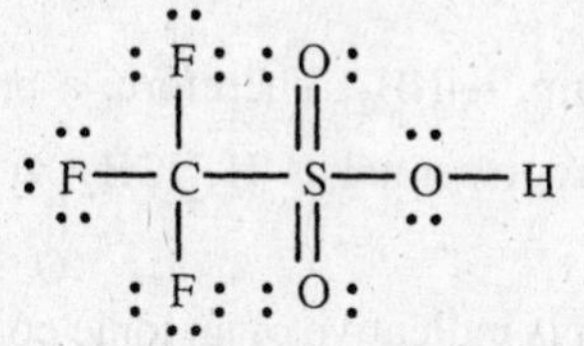

(b) A stronger acid because of the electron-withdrawing effects of the fluorine atoms, which would stabilize the conjugate base of triflic acid, relative to the stability of the conjugate base of benzenesulfonic acid

[structure of triflate anion with arrows showing inductive effect]   [structure of benzenesulfonate anion]

    strong inductive effect   resonance stabilization possible,
    in addition to resonance-   but not a significant inductive effect
    stabilization of the anion

(c) [structure of N-ethyl cyclic ketone] $\xrightarrow[\text{(1 molar equiv)}]{CF_3SO_3H}$ [structure of monocation A] $\xrightarrow[\text{(excess)}]{CF_3SO_3H}$ [structure of dication B]

                                        monocation A   dication B
                                           $pK_a \sim$10 $pK_a \sim$ –7
                                            +   +
                                      $CF_3SO_3^-$ another $CF_3SO_3^-$

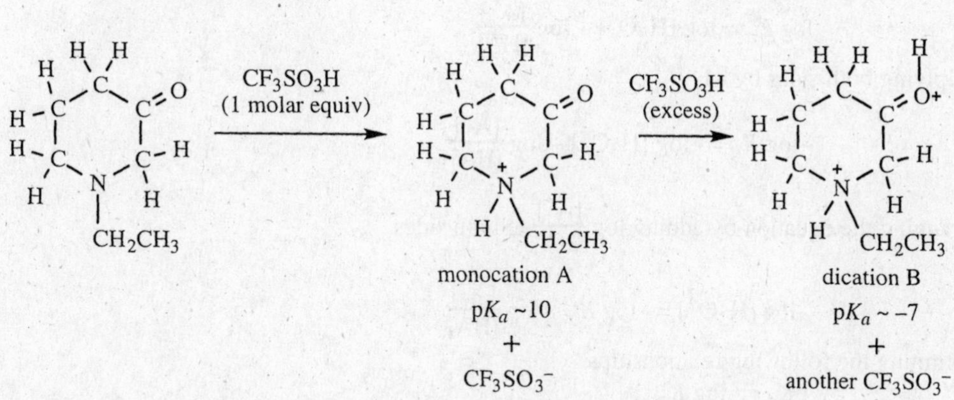

3.40   When a strong acid such as H—Cl is added to water, the following reaction takes place.

               H—Cl + H_2O $\longrightarrow$ Cl^- + H_3O^+

               pK_a –7 pK_a –1.7

3.40 (cont)

The equilibrium constant for this reaction is greater than 10^{+5}, and the reaction proceeds essentially to completion. Similar reactions take place in water with other strong acids such as sulfuric, nitric, and phosphoric acids. A solution of strong acid in water thus consists of the hydronium ion, (the strongest acid that can exist in water) and the conjugate base of the strong acid.

When a strong base such as amide ion is added to water, a similar reaction takes place.

$$NH_2^- + H_2O \longrightarrow NH_3 + OH^-$$
$$pK_a \ 15.7 \qquad pK_a \ 36$$

Again the equilibrium constant is very large ($>10^{+20}$), and the reaction proceeds essentially to completion. A solution of a strong base in water thus consists of the hydroxide ion (the strongest base that can exist in water) and the conjugate acid of the strong base.

3.41

$$\overset{+}{H_3N}\!\!-\!\!CH_2\overset{\displaystyle O}{\overset{\|}{C}}O^- \qquad\qquad\qquad H_2N\!\!-\!\!CH_2\overset{\displaystyle O}{\overset{\|}{C}}OH$$

$$pK_a \ \overset{+}{RNH_3} \sim 10.6 \qquad\qquad pK_a \ RCOH \sim 4.0$$

(a) The carboxylate anion, $-\overset{\displaystyle O}{\overset{\|}{C}}O^-$, is a weaker base than the amino group, $-NH_2$. Therefore, a proton will be transferred from the carboxyl group to the amino group. Glycine is best represented as $\overset{+}{H_3N}CH_2\overset{\|}{\underset{O}{C}}O^-$.

(b) The high melting point for a compound with such a low molecular weight is indicative of an ionic compound. The high water solubility is also indicative of strong electrostatic forces. Both observations are consistent with the charged structure, $\overset{+}{H_3N}CH_2\overset{\|}{\underset{O}{C}}O^-$, for glycine.

3.42 (a) $K_a = \dfrac{[A^-][H_3O^+]}{[HA]}$

taking the log of both sides

$$\log K_a = \log\left(\frac{[A^-][H_3O^+]}{[HA]}\right)$$

$$\log K_a = \log [H_3O^+] + \log \frac{[A^-]}{[HA]}$$

multiplying both sides by –1

$$-\log K_a = -\log [H_3O^+] - \log \frac{[A^-]}{[HA]}$$

rearranging the equation by adding $\log \dfrac{[A^-]}{[HA]}$ to both sides

$$-\log [H_3O^+] = -\log K_a + \log \frac{[A^-]}{[HA]}$$

substituting the following relationships
$$pK_a = -\log K_a$$
$$pH = -\log [H^+] = -\log [H_3O^+]$$
we get the Henderson-Hasselbalch equation

$$pH = pK_a + \log \frac{[A^-]}{[HA]}$$

3.42 (b) when pH = pK_a, $\log \dfrac{[A^-]}{[HA]} = 0$ and $\dfrac{[A^-]}{[HA]} = 1$

or the concentrations of the acid and its conjugate base are equal.

$[HA] = [A^-]$

(c) pH $\cong$ 6.5
$pK_a = 7$

$$pH = pK_a + \log \frac{[A^-]}{[HA]}$$

$$6.5 = 7 + \log \frac{[A^-]}{[HA]}$$

$$\log \frac{[A^-]}{[HA]} = -0.5$$

$$\frac{[A^-]}{[HA]} = 0.316$$

$[A^-] = 0.316\,[HA]$
and $[HA] = \sim 3\,[A^-]$

There is more protonated than unprotonated imidazole present at pH 6.5.

3.43

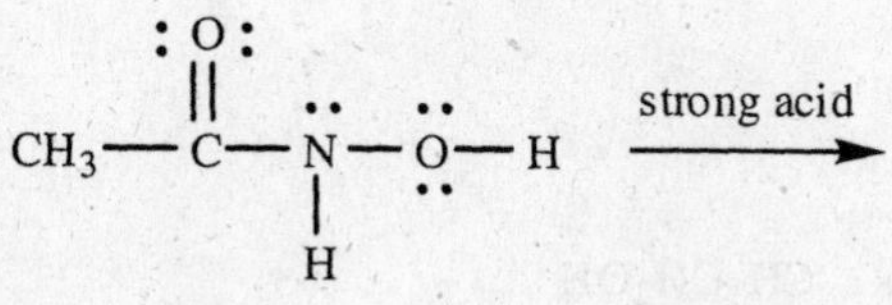

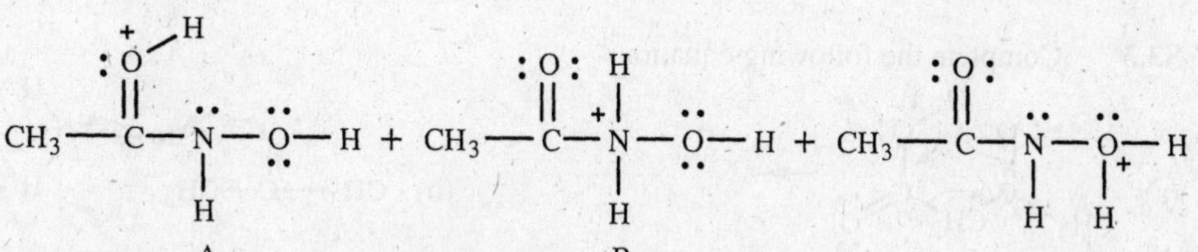

A is favored because the cation that is formed is stabilized by resonance.

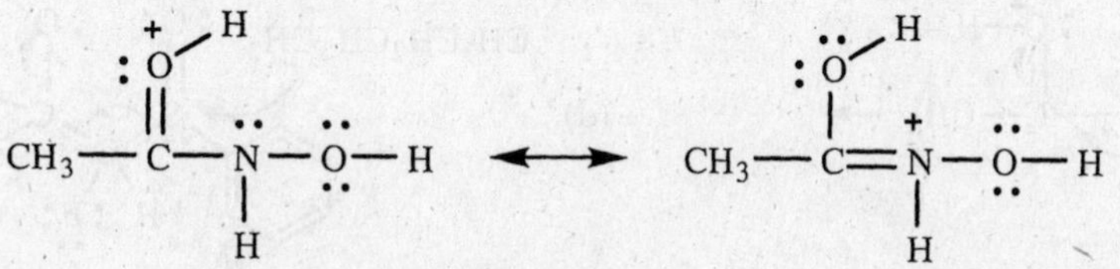

3.44 (a)

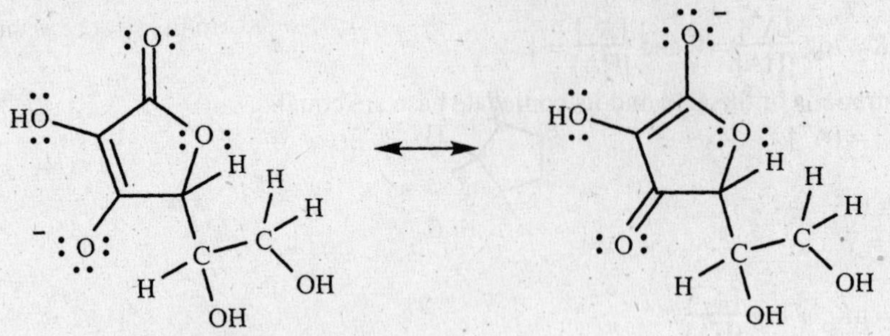

(b)

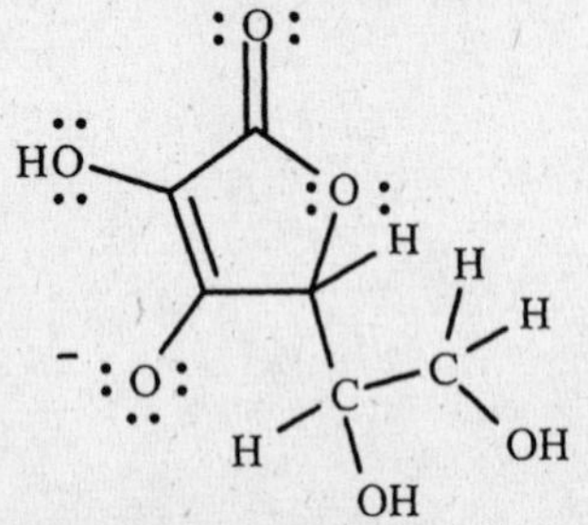

Supplemental Problems

S3.1   Complete the following equations. Use the table of pK_a values to decide where the equilibrium will lie by determining K_{eq} and whether there is a decrease or increase in free energy for each reaction.

(a) $CH_3CH_2\overset{\displaystyle O}{\overset{\|}{C}}OH$ + NH_3 $\longrightarrow$

(b) CH_3NO_2 + $CH_3CH_2O^-\ Na^+$ $\longrightarrow$

(c) CH_3CH_2SH + $NaNH_2$ $\longrightarrow$

(d) $HC\equiv CH$ + $Na^+\ HCO_3^-$ $\longrightarrow$

S3.2   For each acid-base reaction given below, label acids, bases, conjugate acids, and conjugate bases, and tell whether Brønsted acids or Lewis acids are present.

(a) $CH_3CH_2SCH_2CH_3$ + BF_3 $\longrightarrow$ $CH_3CH_2\underset{+}{\overset{\overset{\displaystyle \bar{B}F_3}{|}}{S}}CH_2CH_3$

(b) $H_2PO_4^-$ + OH^- $\longrightarrow$ HPO_4^{2-} + H_2O

(c) $ClCH_2\overset{\displaystyle O}{\overset{\|}{C}}OH$ + HCO_3^- $\longrightarrow$ $ClCH_2\overset{\displaystyle O}{\overset{\|}{C}}O^-$ + H_2CO_3

(d) $CH_3CH_2CH_2SH$ + $CH_3CH_2O^-$ $\longrightarrow$ $CH_3CH_2CH_2S^-$ + CH_3CH_2OH

S3.3   Complete the following equations.

(a)

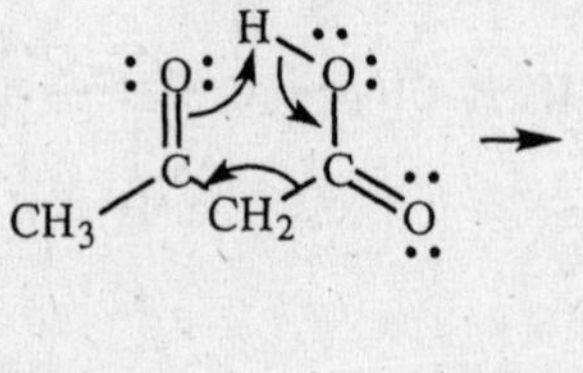

(b)

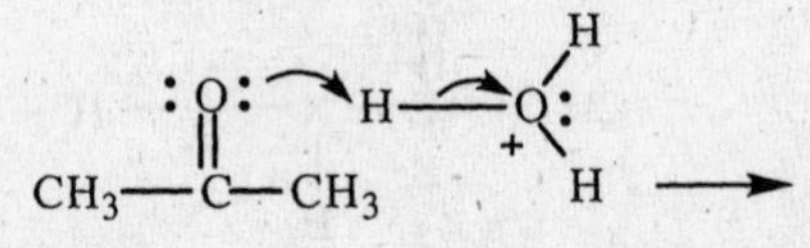

(c)

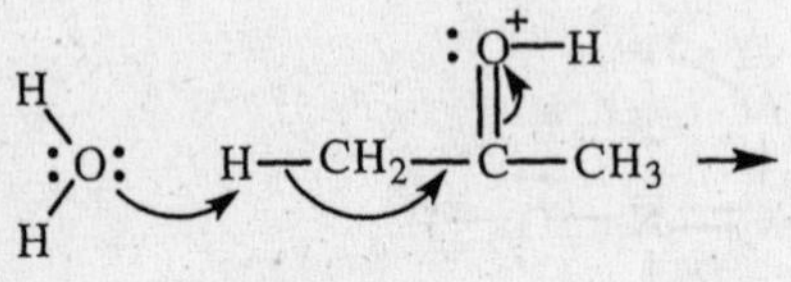

(d)

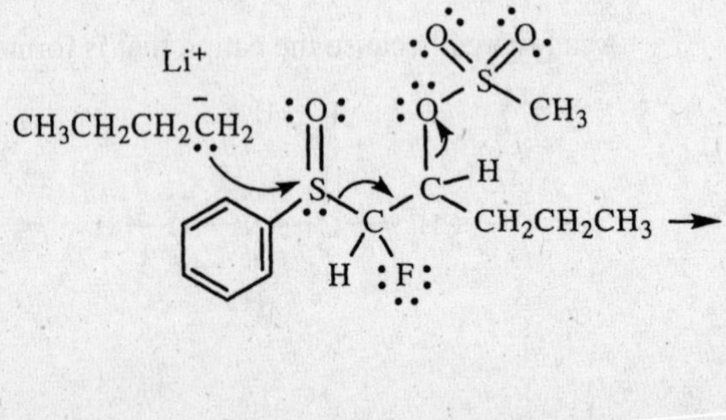

S3.4   1-Indanone has three different types of hydrogen atoms, labeled A, B, and C in the structure shown below.

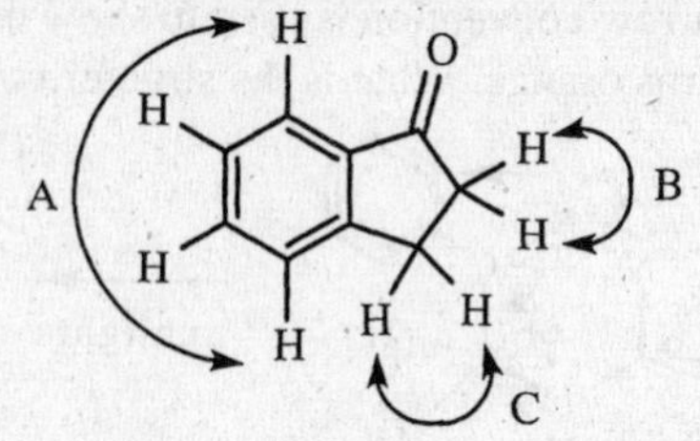

1-indanone

(a) Assign approximate pK_a values to the three different types of hydrogen atoms using exact values from analogous compounds in the pK_a table.

(b) Removing the most acidic proton from 1-indanone leads to the formation of an anion that is stabilized by resonance. Draw structural formulas for the conjugate base corresponding to this deprotonation and for its major resonance contributor.

(c) The conjugate base of 1-indanone can be protonated at two sites. One of these gives the 1-indanone structure; the other gives a new structure called the enol form of 1-indanone. What is the structure of the enol form of 1-indanone?

S3.5   The potency of an anesthetic has been found in recent research to be related to the extent to which it disrupts hydrogen bonding in nerve cell membranes. This capacity to disrupt hydrogen bonding is related to the *acidity of carbon-hydrogen bonds* and to the *basicity of oxygen atoms* in the compounds used as anesthetics. Three widely used anesthetics were examined. They are shown below.

$$CH_3CHClBr \qquad\qquad CH_3OCF_2CHCl_2 \qquad\qquad CHF_2OCF_2CHFCl$$

haloethane   methoxyflurane   enflurane

(a) Haloethane has a pK_a of 23.8. What is the approximate pK_a that you would expect for ethane? Explain in a few words and with structural representations why the acidity of haloethane differs from that of ethane.

(b) Methoxyflurane has two possible sites of deprotonation. The researchers found that only one site was deprotonated under the conditions of their reaction. Write a structural formula for the more stable of the two possible conjugate bases of methoxyflurane.

(c) Methoxyflurane is ten times more potent than enflurane as an anesthetic even though both compounds have approximately the same pK_a, ~26. According to the criteria listed at the beginning of this problem, why should methoxyflurane be more potent than enflurane? Explain in a few words and with structural representations.

S3.6   Lithium diisopropylamide is a strong Brønsted base used in many organic reactions. It is prepared by combining diisopropylamine with one of the following reagents. Which one would work and why?

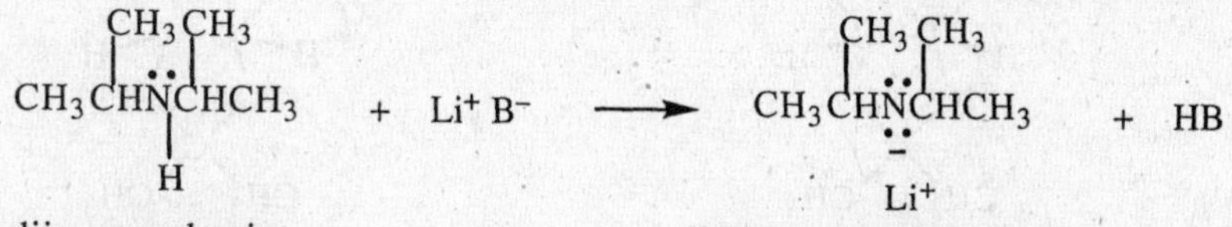

diisopropylamine   lithium diisopropylamide

Li^+ B^- may be butyllithium ($CH_3CH_2CH_2CH_2Li$), lithium methoxide ($LiOCH_3$), or lithium chloride ($LiCl$).

S3.7   The following data show that the cyano group (CN) increases the acidity of the hydroxyl group on phenol. Discuss how inductive effects and resonance may be used to rationalize these observations.

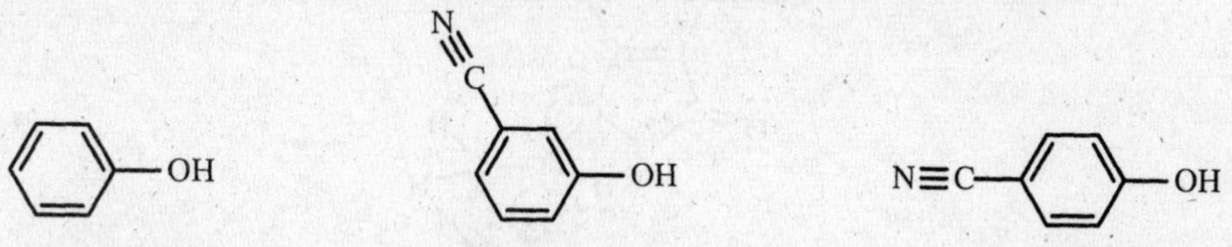

phenol   3-cyanophenol   4-cyanophenol
$pK_a = 10.0$ $pK_a = 8.8$ $pK_a = 7.95$

S3.8   Photochromic processes are those where a compound changes color when exposed to light. The following compound is photochromic. The curved arrow convention is used to show the changes in bonding that occur when the yellow compound on exposure to light turns orange. What is the structure of the orange species?

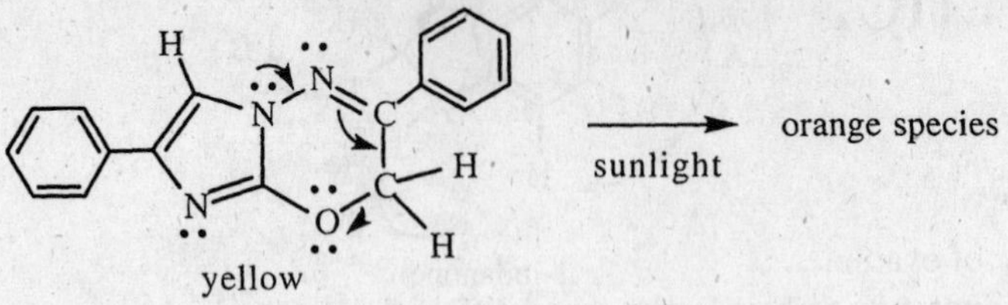

yellow

S3.9   The equation below shows a reaction known as a decarboxylation reaction, a reaction in which carbon dioxide, CO_2, is ejected from a molecule
  (a) Provide a mechanism for this reaction using the curved arrow convention.

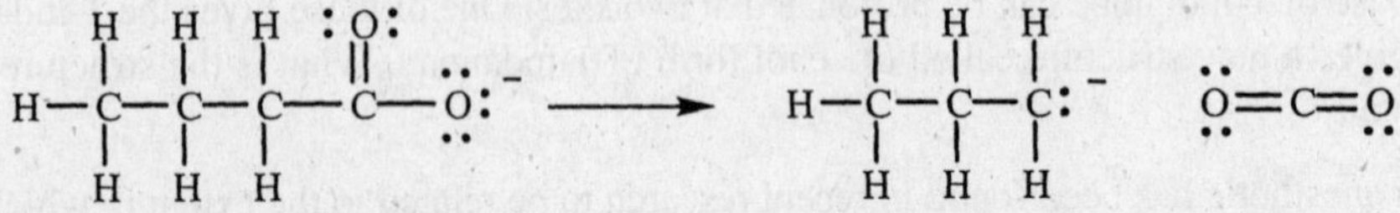

  (b) What is the product of the decarboxylation of the following compound? Is it more or less stable than the product of the reaction in part (a) and why?

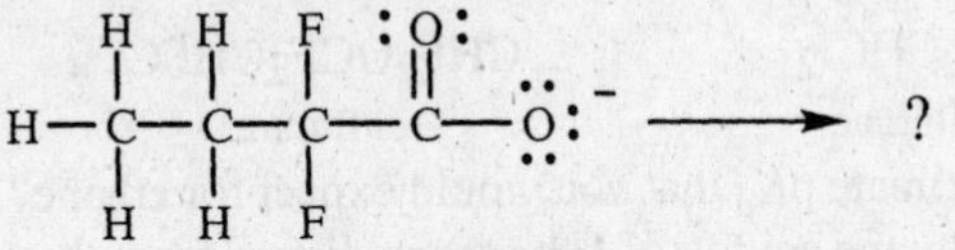

S3.10   Vitamin C is a naturally occurring antioxidant. The interaction of some substances related to vitamin C with oxygen have been studied.
  (a) Using the curved arrow convention, show the mechanism for the transformation shown below.

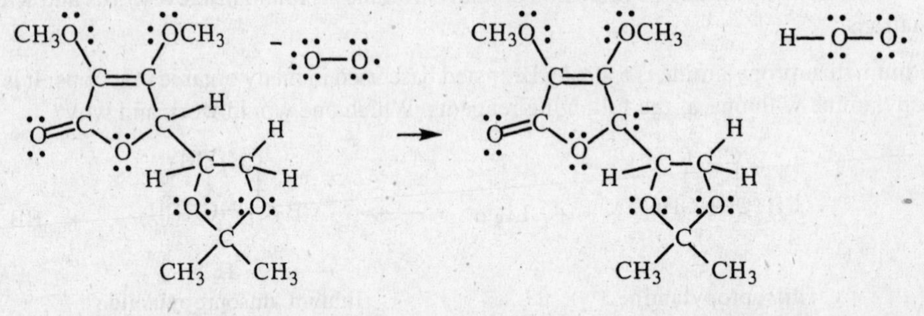

  (b) The next part of the transformation is represented by the mechanism shown below. What are the products implied by this mechanism?

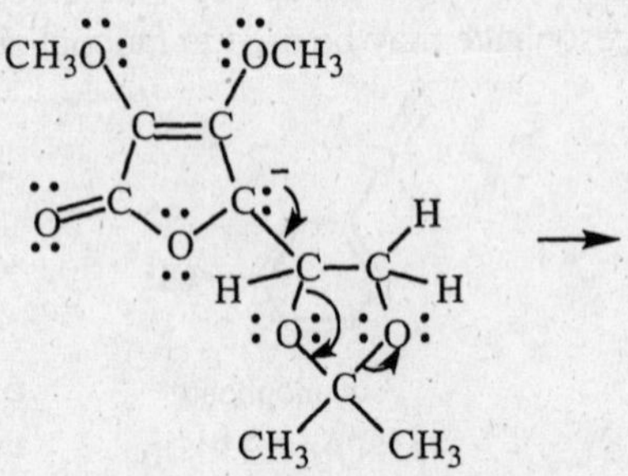

4 Reaction Pathways

Workbook Exercises

EXERCISE I. A number of examples of connectivity changes were given as Workbook Exercises I and II in Chapter 1. Use the curved-arrow notation that was introduced for mechanisms in Chapter 3 to restate the bonding changes you have already described with words.

EXERCISE II. Draw the structures of the products implied by each of the following mechanisms.

(a)

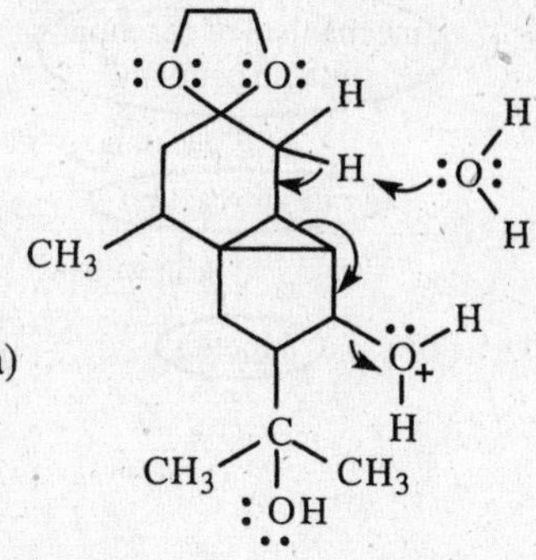

(b)

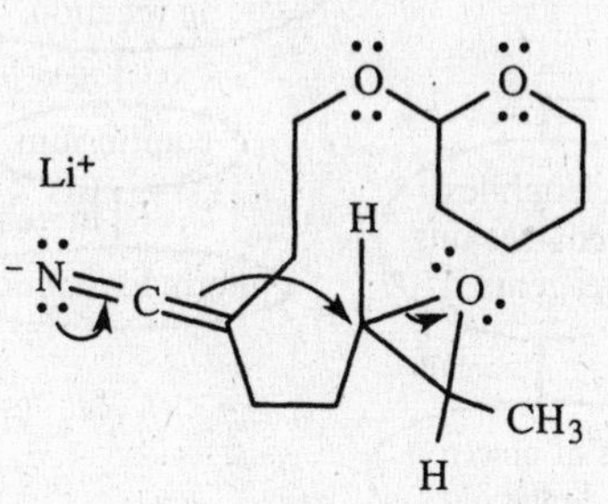

(c)

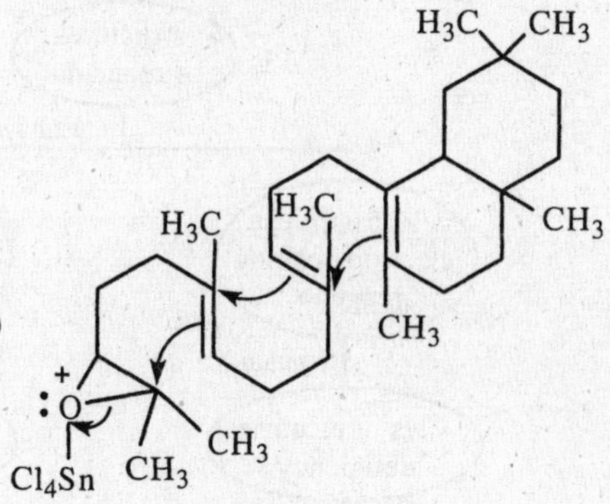

(d)

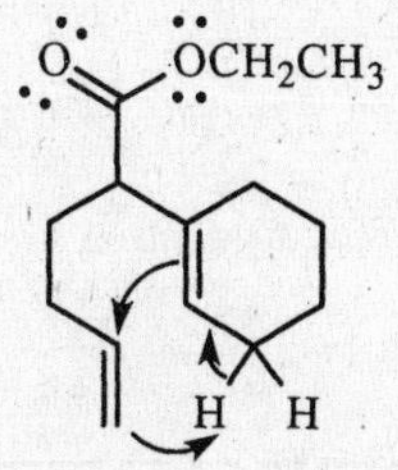

(e)

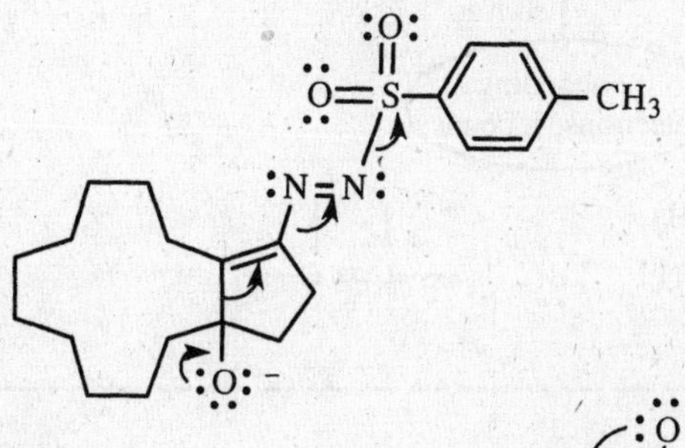

(f)

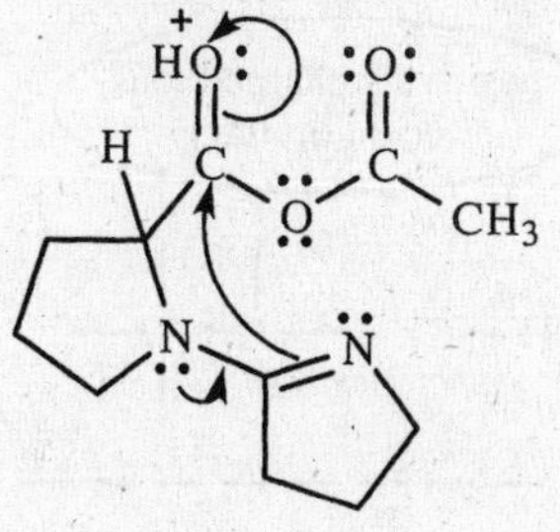

(g)

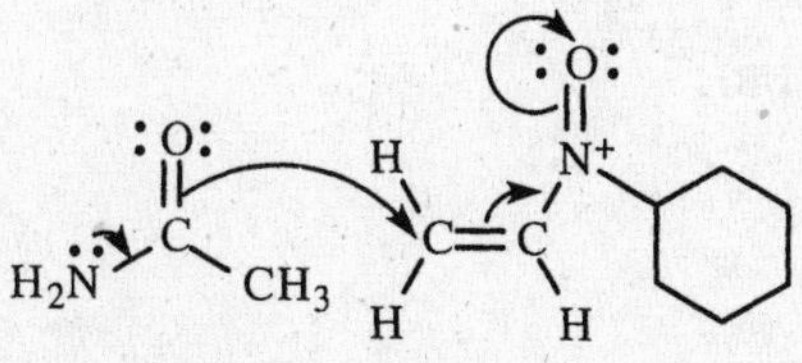

(h)

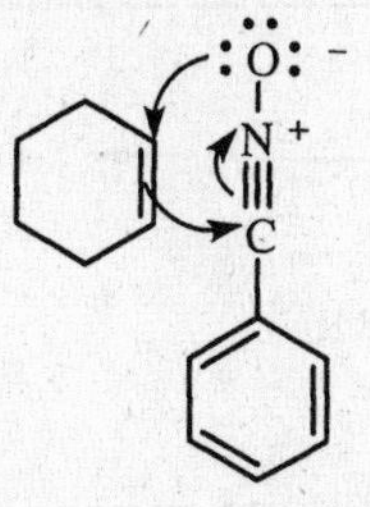

- -

Concept Map 4.1 (see p. 72)

- -

4.1 For those species in which more than one electrophilic or nucleophilic site is present, the stronger site is indicated. If a species contains both an electrophilic and nucleophilic site, the site that normally reacts is indicated.

(a) $CH_3-\ddot{O}{:}^-$
nucleophile

(b) ${:}PH_3$
nucleophile

(c) Cu^{2+}
electrophile

(d) $\overset{\delta+}{H}-\overset{\delta-}{\ddot{Br}{:}}$
electrophile

(e) $\overset{\delta+}{CH_3}-\overset{\delta-}{\ddot{Cl}{:}}$
electrophile

(f) $CH_3\ddot{N}H_2$
nucleophile

(g) $H{:}^-$
nucleophile

(h) $(CH_3)_3B$
electrophile

(i) $H\ddot{O}-NH_2$
nucleophile

(j) $HC{\equiv}CH$
nucleophile

4.1 (cont)

(k)   AlCl₃   (l)

$$\overset{H}{\underset{H}{\diagdown}}C \overset{\delta+}{=} \overset{\delta-}{\underset{\cdot\cdot}{O}}$$

(m)   :Ï:⁻   (n)   CH₃CH₂—S̈:⁻   (o)   Hg²⁺

     ↑   ↑   nucleophile   electrophile

electrophile   electrophile   nucleophile

Concept Map 4.1   The factors that determine whether a chemical reaction between a given set of reagents is likely.

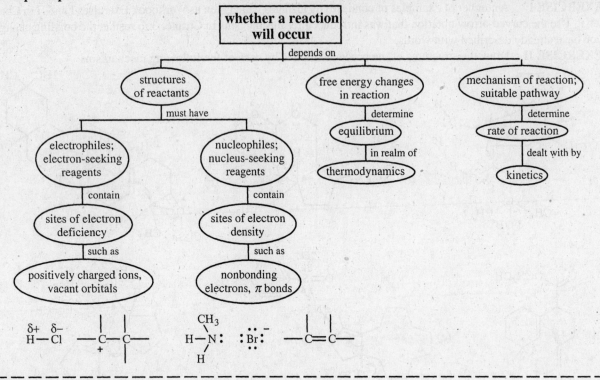

4.2   (a)   electrophile ---► CH₃⤻B̈r: ⟶   CH₃—Ö H   :B̈r:⁻

                nucleophile ---► ⁻:Ö–H

      (b)   electrophile ---► CH₃⤻Ï: ⟶   CH₃—Ö H   :Ï:⁻

                nucleophile ---► ⁻:Ö–H

      (c)   CH₃CH₂⤻C̈l: ⟶   CH₃CH₂—Ö H   :C̈l:⁻

               electrophile -´

           nucleophile ---► ⁻:Ö–H

      (d)   electrophile ---► CH₃⤻Ï: ⟶   CH₃—N⁺H₃   :Ï:⁻

                nucleophile ---► :NH₃

4.3   Sodium ion, Na$^+$, has a complete valence shell (see Problem 2.1 for the electronic configuration of elemental sodium), and the energies of the empty orbitals in the $n=3$ shell are too high to be available for accepting an electron pair. Both the mercury(II), Hg^{2+}, and copper(II), Cu^{2+}, ions are transition metals that have lower energy empty orbitals available to accept an electron pair.

4.4   (a)   Overall equation must be:

$$Cl_2 \; + \; NH_3 \longrightarrow \; H_3\overset{+}{N}-Cl \; + \; Cl^-$$

$$H_3\overset{+}{N}-Cl \; + \; Cl^- \; + \; Na^+\,OH^- \longrightarrow \; H_2N-Cl \; + \; Na^+\,Cl^- \; + \; H_2O$$

(b)   Mechanism

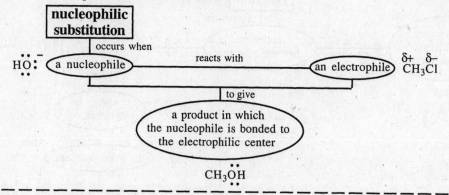

Concept Map 4.2   A nucleophilic substitution reaction.

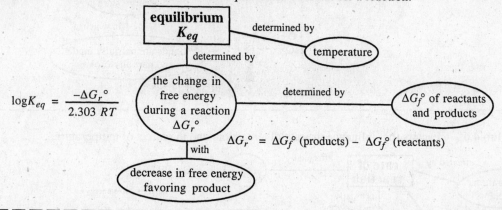

Concept Map 4.3   The factors that determine the equilibrium constant for a reaction.

equilibrium K_{eq}

determined by temperature

determined by

the change in free energy during a reaction ΔG_r°

$$\log K_{eq} = \frac{-\Delta G_r^\circ}{2.303\,RT}$$

determined by ΔG_f° of reactants and products

$$\Delta G_r^\circ \; = \; \Delta G_f^\circ \,(\text{products}) \; - \; \Delta G_f^\circ \,(\text{reactants})$$

with decrease in free energy favoring product

Concept Map 4.4   (see p. 74)

4.5   At 311 K, $k_r = 3.55 \times 10^{-5}$ L/mol·sec.
Initial rate $= k_r[CH_3Cl][OH^-]$ when $[OH^-]$ 0.05 M.
Initial rate $= (3.55 \times 10^{-5}$ L/mol·sec$)(0.003$ mol/L$)(0.05$ mol/L$)$.
            $= 5.33 \times 10^{-9}$ mol/L·sec, which is five times the original rate of 1.07×10^{-9} mol/L·sec.
When $[CH_3Cl] = 0.001$ M, the rate will be one-third the rate when $[CH_3Cl] = 0.003$ M, or 3.55×10^{-10} mol/L·sec.

Concept Map 4.4   The factors that influence the rate of a reaction as they appear in the rate equation.

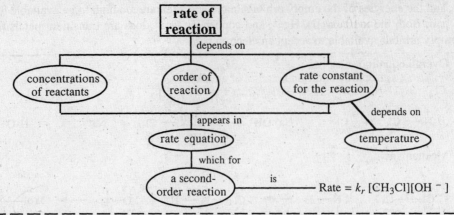

Concept Map 4.5   Some of the factors other than concentration that influence the rate of a reaction.

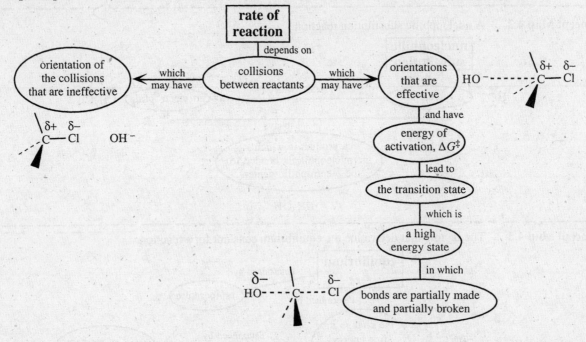

Concept Map 4.6   Factors that influence the rate of a reaction, and the effect of temperature.

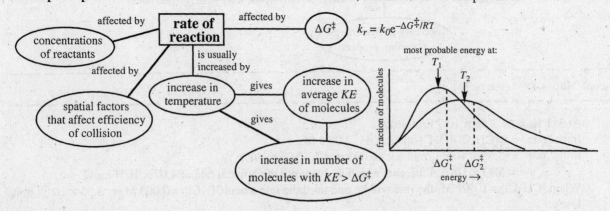

4.6   At 308 K, k_r $= 5.3 \times 10^{-4}$ L/mol·sec.
     Initial rate $= (5.3 \times 10^{-4}$ L/mol·sec)(0.001 mol/L)(0.01 mol/L)
                $= 5.3 \times 10^{-9}$ mol/L·sec

Concept Map 4.7   An electrophilic addition reaction of an alkene.

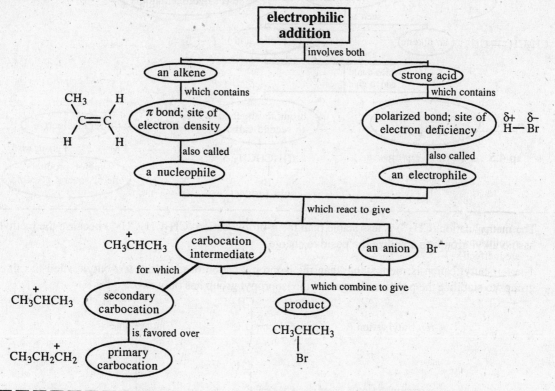

4.7   The double bond in propene is another possible nucleophile in the reaction mixture.

$$CHCH_3 = CH_2 \longrightarrow CH_3CH - CH_2 - \overset{\displaystyle CH_3}{\underset{+}{\overset{|}{C}HCH_3}}$$

$$\downarrow$$

$$\underset{+}{CH_3CHCH_3} \qquad \text{a new secondary carbocation}$$

Concept Map 4.8   (See p. 76)

4.8   (a) $CH_3\overset{\overset{\displaystyle CH_3}{|}}{C} = CH_2$ +   HBr $\longrightarrow$ $CH_3\overset{\overset{\displaystyle CH_3}{|}}{\underset{\underset{\displaystyle Br}{|}}{C}}CH_3$

 (b) $CH_3CH_2CH = CH_2$ +   HBr $\longrightarrow$ $CH_3CH_2\overset{}{\underset{\underset{\displaystyle Br}{|}}{C}HCH_3}$

 (c) $CH_3CH_2CH = CHCH_3$ + HBr $\longrightarrow$ $CH_3CH_2\underset{\underset{\displaystyle Br}{|}}{C}HCH_2CH_3$ + $CH_3CH_2CH_2\underset{\underset{\displaystyle Br}{|}}{C}HCH_3$

Concept Map 4.8   Markovnikov's Rule.

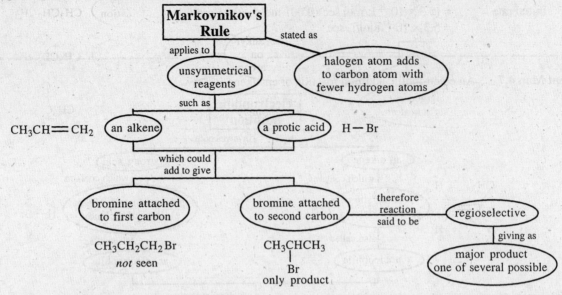

- -

4.9   The methyl cation (CH_3^+) is less stable than the *n*-propyl cation ($CH_3CH_2CH_2^+$) because the methyl cation has no alkyl groups to stabilize the positive charge.

4.10   The *tert*-butyl cation is more stable than the isopropyl cation because the *tert*-butyl cation has three alkyl groups to stabilize the positive charge and the isopropyl group has only two.

$$
\begin{array}{lll}
\textit{tert}\text{-butyl cation} & CH_3\!-\!\overset{\displaystyle CH_3}{\underset{\displaystyle CH_3}{C}}\!+ & \text{most stable} \\[3em]
\text{isopropyl cation} & CH_3\!-\!\overset{\displaystyle CH_3}{\underset{\displaystyle H}{C}}\!+ & \\[3em]
\textit{n}\text{-propyl cation} & CH_3CH_2\!-\!\overset{\displaystyle H}{\underset{\displaystyle H}{C}}\!+ & \\[3em]
\text{methyl cation} & H\!-\!\overset{\displaystyle H}{\underset{\displaystyle H}{C}}\!+ & \text{least stable}
\end{array}
$$

4.11 $CH_3 - \overset{+}{\underset{H}{C}} - \overset{..}{\underset{..}{F}} :$ $\longleftrightarrow$ $CH_3 - \underset{H}{C} = \overset{..}{\underset{..}{F}} \overset{+}{}$ Cation is stabilized by resonance.

all atoms closed shell

$CH_3 - \overset{+}{\underset{H}{C}} - C\!\equiv\!N :$ $\longleftrightarrow$ $CH_3 - \overset{+}{\underset{H}{C}} - C\!\equiv\!\overset{+}{N} : \overset{..}{}^-$ Resonance in the cyano group destabilizes the cation by putting two positive charges next to each other.

$CH_3 - \overset{+}{\underset{H}{C}} - H$ is a primary cation with no additional stabilizing or destabilizing factors.

Concept Map 4.9   Carbocations.

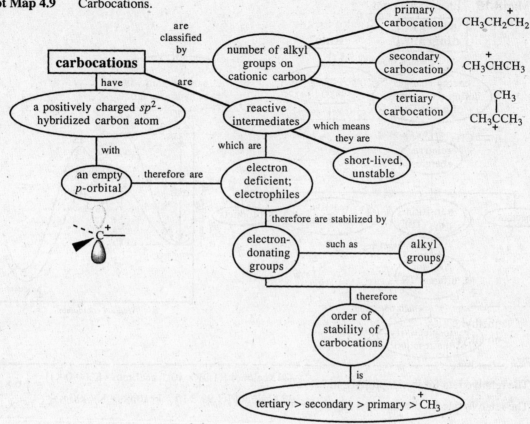

4.12 $\log K_{eq} = -(\Delta G°/2.303\ RT)$

At 25 °C (298 K),

   for 1-bromopropane

   $\log K_{eq} = -(-7.63\ \text{kcal/mol})/[(2.303)(1.99 \times 10^{-3}\ \text{kcal/mol·K})(298\ \text{K})]$
   $\log K_{eq} = 5.59$
   $K_{eq} = 10^{5.59} = 10^{0.59} \times 10^5 = 3.89 \times 10^5$

   for 2-bromopropane

   $\log K_{eq} = -(-8.77\ \text{kcal/mol})/[(2.303)(1.99 \times 10^{-3}\ \text{kcal/mol·K})(298\ \text{K})]$
   $\log K_{eq} = 6.42$
   $K_{eq} = 10^{6.42} = 10^{0.42} \times 10^6 = 2.63 \times 10^6$

Note that the answers that you get may not correspond exactly to the numbers obtained here, depending on whether you round off to the correct number of significant figures at each stage of the calculation or carry all figures through and round off at the end of the calculation.

4.13   (a) $CH_3CH = CH_2\ (g) + HI\ (g) \longrightarrow CH_3CHCH_3\ (g)$
                                                                        |
                                                                        I

   (b)   Rate $= k_r[CH_3CH=CH_2][HI]$

4.14   Concentration is defined as the number of moles of a substance in a given volume. For an ideal gas, the concentration at constant temperature and volume is proportional to the pressure.

   $$P = \frac{n}{V}\ RT$$

   $P_{\text{propene}} = P_{\text{HI}} = (45\ \text{mm Hg})/(760\ \text{mm Hg/atm})$
   Rate $= k_r(P_{\text{propene}})(P_{\text{HI}})$
   Rate $= (1.66 \times 10^{-6}/\text{atm·sec})(45/760)^2\text{atm}^2$
   Rate $= 5.8 \times 10^{-9}$ atm/sec

Concept Map 4.10   Energy diagrams.

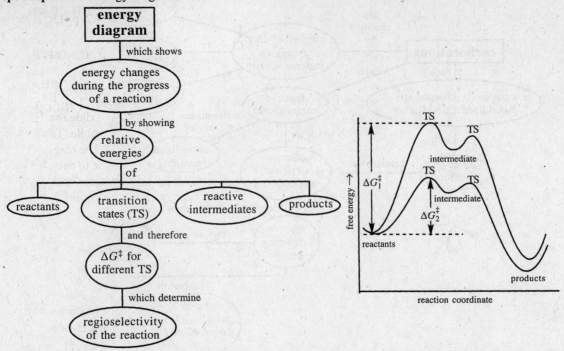

- -

4.15   The relative rate for the isopropyl cation $= e^{-22 \text{ kcal/mol}/[(1.99 \times 10^{-3} \text{ kcal/mol} \cdot \text{K})(490 \text{ K})]} = 1.6 \times 10^{-10}$

The relative rate for the *n*-propyl cation $= e^{-38 \text{ kcal/mol}/[(1.99 \times 10^{-3} \text{ kcal/mol} \cdot \text{K})(490 \text{ K})]} = 1.2 \times 10^{-17}$

$$\frac{\text{isopropyl}}{n\text{-propyl}} = \frac{1.6 \times 10^{-10}}{1.2 \times 10^{-17}} = 1.3 \times 10^{7}$$

4.16   (a)

$$CH_3CH_2 \rightarrow \ddot{\underset{\cdot\cdot}{Br}}\colon \longrightarrow CH_3CH_2 - \ddot{\underset{\cdot\cdot}{I}}\colon + \colon\ddot{\underset{\cdot\cdot}{Br}}\colon^{-}$$

electrophile

nucleophile $\longrightarrow \colon\ddot{I}\colon^{-}$

(b)

electrophile

$$\underset{nucleophile \longrightarrow \;^{-}\colon C \equiv N\colon}{CH_3CH \underset{\underset{\cdot\cdot}{\overset{CH_3}{|}}}{\rightleftharpoons} \ddot{\underset{\cdot\cdot}{Br}}\colon} \longrightarrow CH_3\overset{\overset{CH_3}{|}}{CH} - C \equiv N\colon + \colon\ddot{\underset{\cdot\cdot}{Br}}\colon^{-}$$

Note that even though both carbon and nitrogen have nonbonding electron pairs, the electrons on the carbon are more available for bonding in nucleophilic substitution reactions because carbon bears the negative charge and is less electronegative than nitrogen and, therefore, holds its electrons less tightly.

(c)

nucleophile   electrophile

$$CH_3CH_2CH = CH_2 \longrightarrow CH_3CH_2\overset{+}{CH} - \overset{\overset{H}{|}}{CH_2} \longrightarrow CH_3CH_2\underset{\underset{\colon\ddot{Cl}\colon}{|}}{CHCH_3}$$

electrophile $\longrightarrow H \rightleftharpoons \ddot{\underset{\cdot\cdot}{Cl}}\colon$ nucleophile $\longrightarrow \colon\ddot{\underset{\cdot\cdot}{Cl}}\colon^{-}$

(d)

electrophile

$$CH_3CH_2\overset{\overset{CH_3}{|}}{CH} \rightleftharpoons \ddot{\underset{\cdot\cdot}{Br}}\colon \longrightarrow CH_3CH_2\overset{\overset{CH_3}{|}}{CH} - \ddot{\underset{\cdot\cdot}{S}} - CH_3 \qquad + \colon\ddot{\underset{\cdot\cdot}{Br}}\colon^{-}$$

$$CH_3 - \ddot{\underset{\cdot\cdot}{S}}\colon \longleftarrow nucleophile$$

4.16 (e)
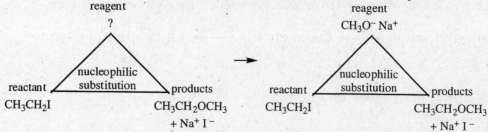

nucleophile   electrophile

CH_3CH=CH_2 + $\overset{\delta+}{Cl}$—$\overset{\delta-}{Cl}$ ⟶ CH_3CHCH_2—Cl (see Table 4.2, p. 136 in the text)
                                                                              |
                                                                              Cl

4.17   <u>reagent</u>   <u>type</u>   <u>reason for choice</u>
(a) $CH_3O^-\,Na^+$ nucleophile   The carbon atom bonded to the iodine atom in ethyl iodide is an electrophilic carbon and will react with a reagent that is a nucleophile. The reaction is a nucleophilic substitution. CH_3O^- is the nucleophile. The reagent is shown as also containing a sodium ion, because it is not possible to have a reagent with only anions in it. The sodium ions are called "spectator" ions. They do not participate directly in the reaction, but serve to neutralize the negative charges on the anions before and after the reaction takes place.

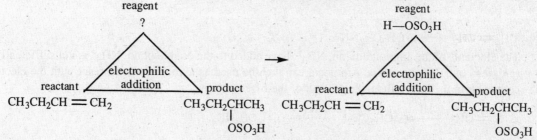

(b) H_2SO_4 electrophile   The double bond in 1-butene is a nucleophile and reacts with the electrophilic proton in sulfuric acid. The carbocation produced in the first step of the reaction is an electrophile and will react with the conjugate base of sulfuric acid, a nucleophile, which is also produced in the first step of the reaction. The overall reaction is an electrophilic addition.

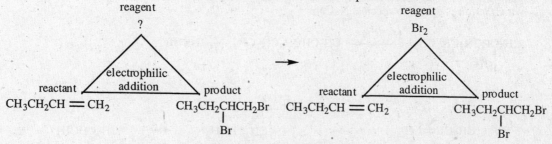

(c) Br_2 electrophile   The nucleophilic double bond in 1-butene will react with the electrophile bromine (see Chapter 8 for the mechanism of the addition of bromine to alkenes). The overall reaction is an electrophilic addition.

(d) NH_3 nucleophile   The carbon atom bonded to the bromine atom in *n*-propyl bromide is an electrophilic carbon and will react with a reagent that is a nucleophile. The reaction is a nucleophilic substitution.

4.17 (d) (cont)

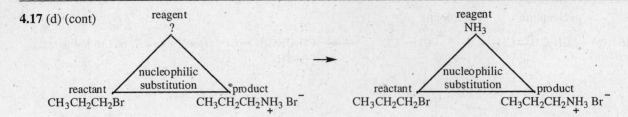

4.18 (a) $CH_3CH_2CH\!=\!CH_2 \xrightarrow{?} CH_3CH_2\underset{\underset{SH}{|}}{C}HCH_3$

The elements of hydrogen sulfide, H_2S, have added to the double bond. The double bond is a nucleophile and is protonated by strong Brønsted acids. H_2S, with a pK_a ~10.5, is not sufficiently acidic (hydrogen sulfide is similar to ethanethiol, CH_3CH_2SH, and can be expected to have a similar pK_a); a hydrogen halide such as HBr $(pK_a - 8)$ is. In the product that is formed, the bromine atom will be attached to an electrophilic carbon atom and will leave when the hydrogen sulfide anion is used as a nucleophile.

$$CH_3CH_2CH\!=\!CH_2 \xrightarrow{HBr} CH_3CH_2\underset{\underset{Br}{|}}{C}HCH_3 \xrightarrow{Na^+ SH^-} CH_3CH_2\underset{\underset{SH}{|}}{C}HCH_3 + Na^+ Br^-$$

(b) $CH_3CH\!=\!CHCH_3 \xrightarrow{?} CH_3\underset{\underset{Cl}{|}}{C}HCH_2CH_3$

The elements of hydrogen chloride, HCl, have added to the double bond. Unlike H_2S, hydrogen chloride is a strong enough acid $(pK_a -7)$ to protonate the double bond. The cation produced by protonation will then react with the conjugate base of hydrogen chloride to give the desired product.

$$CH_3CH\!=\!CHCH_3 \xrightarrow{HCl} CH_3\overset{+}{C}H\!-\!\underset{\underset{H}{|}}{C}HCH_3 + Cl^- \longrightarrow CH_3\underset{\underset{CH_3}{|}}{C}HCH_2CH_3$$

(c) $CH_2\!=\!CH_2 \xrightarrow{?} CH_3CH_2\overset{+}{N}H_3 \; I^-$

The elements of the ammonium ion, NH_4^+, have added to the double bond. NH_4^+ is not sufficiently acidic $(pK_a \; 9.4)$ but HI $(pK_a -9)$ is. Ammonia can then be used as a nucleophile to react with the electrophilic carbon atom in ethyl iodide, the product of the first reaction.

$$CH_2\!=\!CH_2 \xrightarrow{HI} CH_3CH_2I \xrightarrow{NH_3} CH_3CH_2\overset{+}{N}H_3 \; I^-$$

(d) $CH_3\underset{\underset{Br}{|}}{C}HCH_2CH_2CH_3 \xrightarrow{?} CH_3\underset{\underset{CN}{|}}{C}HCH_2CH_2CH_3$

Cyanide ion has substituted for bromide ion. This is a nucleophilic substitution reaction on 2-bromopentane, which requires only one species, NaCN.

$$CH_3\underset{\underset{Br}{|}}{C}HCH_2CH_2CH_3 \xrightarrow{Na^+ CN^-} CH_3\underset{\underset{CN}{|}}{C}HCH_2CH_2CH_3 + Na^+ Br^-$$

4.19 (a)

nucleophile ⋯   electrophile ⋯

$CH_3CH\!=\!CH_2 \longrightarrow CH_3\overset{+}{C}HCH_3 \longrightarrow CH_3\underset{\underset{:OSO_3H}{|}}{C}HCH_3$

electrophile ⋯➤ $H\!\rightarrow\!\overset{..}{\underset{..}{O}}\!-\!SO_3H$ nucleophile ⋯➤ $:\!\overset{..}{\underset{..}{O}}\!-\!SO_3H$

4.19 (b)

electrophile

$CH_3CH_2CH_2 \rightarrow \ddot{B}r: \rightarrow CH_3CH_2CH_2 \overset{+}{-}PH_3 \qquad :\ddot{B}r:^-$

nucleophile $\rightarrow :PH_3$

(c)

electrophile

$CH_3CH_2 \rightarrow \ddot{I}: \rightarrow CH_3CH_2-\ddot{S}-CH_2CH_3 \qquad :\ddot{I}:^-$

$CH_3CH_2-\ddot{S}:^- \leftarrow$ nucleophile

(d)

$\underset{\text{electrophile} \rightarrow H \rightarrow \ddot{C}l:}{CH_3\overset{\overset{CH_3}{|}}{C}=CH_2} \quad \xrightarrow{\text{nucleophile}\quad\text{electrophile}} \quad CH_3-\overset{\overset{CH_3}{|}}{\underset{}{C}}-CH_3 \quad \xrightarrow{\text{nucleophile} \rightarrow :\ddot{C}l:^-} \quad CH_3-\overset{\overset{CH_3}{|}}{\underset{\underset{:\ddot{C}l:}{|}}{C}}-CH_3$

(e) $CH_3CH_2CH=CHCH_3 + Br_2 \rightarrow CH_3CH_2\underset{Br}{\overset{|}{C}H}-\underset{Br}{\overset{|}{C}H}CH_3$

nucleophile   electrophile

4.20

reagent	type	reason for choice
(a) Na^+SH^-	nucleophile	The carbon atom bonded to the bromine atom in methyl bromide is an electrophilic carbon and will react with a reagent that is a nucleophile. The reaction is a nucleophilic substitution.
(b) Na^+CN^-	nucleophile	The carbon atom bonded to the bromine atom in ethyl bromide is an electrophilic carbon and will react with a reagent that is a nucleophile. The reaction is a nucleophilic substitution.

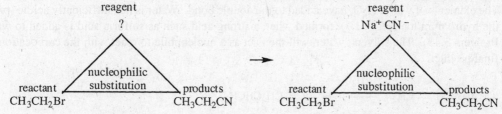

(c) Cl_2	electrophile	The nucleophilic double bond in ethylene will react with the electrophile chlorine (see Chapter 8 for the mechanism of the addition of chlorine to alkenes). The reaction is an electrophilic addition.
(d) $CH_3CH_2O^-Na^+$	nucleophile	The carbon atom bonded to the bromine atom in *n*-propyl bromide is an electrophilic carbon and will react with a reagent that is a nucleophile. The reaction is a nucleophilic substitution.
(e) HI	electrophile	The nucleophilic double bond in 1-butene will react with the electrophilic proton of hydrogen iodide and the electrophilic carbocation that is produced will then react with the iodide ion. The reaction is an electrophilic addition.

4.20 (e) (cont)

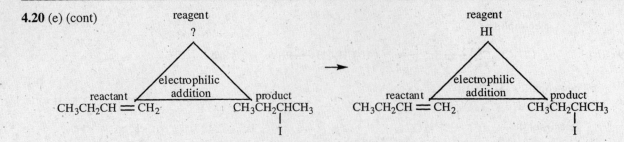

(f) $Na^+ I^-$ nucleophile   The carbon atom bonded to the bromine atom in isopropyl bromide
                                                          is an electrophilic carbon and will react with a reagent that is a nu-
                                                          cleophile. The reaction is a nucleophilic substitution.

4.21 (a) $CH_2 {=} CH_2 \xrightarrow{?} CH_3CH_2OCH_3$

The elements of methanol, CH_3OH, have added to the double bond. The double bond is a nucleophile and
is protonated by strong Brønsted acids. Methanol is not sufficiently acidic (pK_a 15.5), but a hydrogen
halide such as HBr (pK_a –8) is. In the product that is formed, the bromine atom will be attached to an
electrophilic carbon atom and will leave when sodium methoxide is used as a nucleophile.

$$CH_2 {=} CH_2 \xrightarrow{HBr} CH_3CH_2Br \xrightarrow{CH_3O^- Na^+} CH_3CH_2OCH_3$$

(b) $CH_3CH {=} CH_2 \xrightarrow{?} CH_3\underset{\underset{I}{|}}{C}HCH_3$

The elements of hydrogen iodide, HI, have added to the double bond. Hydrogen iodide is a strong enough
acid (pK_a –9) to protonate the double bond. The cation produced by protonation will then react with iodide
ion, the conjugate base of hydrogen iodide, to give the desired product.

$$CH_3CH {=} CH_2 \xrightarrow{HI} CH_3\underset{\underset{I}{|}}{C}HCH_3$$

(c) $CH_3CH_2CH {=} CH_2 \xrightarrow{?} CH_3CH_2\underset{\underset{OH}{|}}{C}HCH_3$

The elements of water, H_2O, have added to the double bond. Water is not sufficiently acidic (pK_a 15.7) but
the hydronium ion (pK_a –1.7), formed when a strong acid such as sulfuric acid is added to water, is (See
Problem 3.43). The solvent, water, will then act as a nucleophile to react with the carbocation to give the
final product.

$$CH_3CH_2CH {=} CH_2 \xrightarrow[H_2SO_4]{H_2O} CH_3CH_2\underset{\underset{OH}{|}}{C}HCH_3$$

(d) $CH_2 {=} CH_2 \xrightarrow{?} CH_3CH_2 {-} \overset{\overset{H}{|}}{\underset{\underset{H}{|}}{\overset{+}{N}}}CH_3 \ Br^-$

The elements of the methylammonium ion, $CH_3NH_3^+$, have added to the double bond. $CH_3NH_3^+$ is not
sufficiently acidic (pK_a 10.6) but HBr (pK_a –8) is. Methylamine can then be used as a nucleophile to react
with the electrophilic carbon atom in ethyl bromide, the product of the first reaction.

$$CH_2 {=} CH_2 \xrightarrow{HBr} CH_3CH_2Br \xrightarrow{CH_3NH_2} CH_3CH_2 {-} \overset{\overset{H}{|}}{\underset{\underset{H}{|}}{\overset{+}{N}}}CH_3 \ Br^-$$

4.22   (a)   electrophile ----▸ CH_3 ⇌ $\overset{..}{\underset{..}{Cl}}$: ⟶ :$\overset{..}{\underset{..}{I}}$—$CH_3$:$\overset{..}{\underset{..}{Cl}}$:⁻

nucleophile ----▸ :$\overset{..}{\underset{..}{I}}$:⁻

(b) CH_3—$\overset{\overset{\displaystyle CH_3}{|}}{\underset{\underset{\displaystyle CH_3}{|}}{C}}$—$\overset{..}{\underset{..}{O}}$—H   ⟶ CH_3—$\overset{\overset{\displaystyle CH_3}{|}}{\underset{\underset{\displaystyle CH_3}{|}}{C}}$⇌$\overset{..}{\underset{\underset{\displaystyle H}{|}}{O^+}}$—H   ⟶ CH_3—$\overset{\overset{\displaystyle CH_3}{|}}{\underset{\underset{\displaystyle CH_3}{|}}{C^+}}$:$\overset{..}{O}$—H

base ----▸ (on O)   acid ----▸ H ⇌ :$\overset{..}{\underset{..}{Br}}$:   :$\overset{..}{\underset{..}{Br}}$:⁻   (under O) H

(c)   :N≡C:⁻ $\overset{\overset{\displaystyle H}{|}}{\underset{\underset{\displaystyle H}{|}}{C}}$=$\overset{..}{\underset{..}{O}}$:   ⟶   :N≡C—$\overset{\overset{\displaystyle H}{|}}{\underset{\underset{\displaystyle H}{|}}{C}}$—$\overset{..}{\underset{..}{O}}$:⁻

nucleophile   electrophile

(d)   nucleophile, electrophile pyrrolidine/cyclohexene structure ⟶ CH_3 ⇌ :$\overset{..}{\underset{..}{I}}$:   ⟶ $^+$N= product with CH_3 :$\overset{..}{\underset{..}{I}}$:⁻

(e) CH_3—$\overset{\overset{\displaystyle :\overset{..}{O}:}{||}}{C}$—$\overset{..}{\underset{..}{Cl}}$:   ⟶ CH_3—$\overset{\overset{\displaystyle :\overset{..}{O}:⁻}{|}}{\underset{\underset{\displaystyle CH_3—\overset{+}{N}—H}{|}}{C^+}}$—$\overset{..}{\underset{..}{Cl}}$:   ⟶ CH_3—$\overset{\overset{\displaystyle :O:}{||}}{C}$—$\overset{\overset{\displaystyle H}{|}}{\underset{\underset{\displaystyle CH_3}{|}}{N^+}}$—$CH_3$ ⟶ CH_3—$\overset{\overset{\displaystyle :O:}{||}}{C}$—$\overset{\underset{\underset{\displaystyle CH_3}{|}}{N}}{|}$—$CH_3$

electrophile   electrophile   base ----▸ :$\overset{..}{\underset{..}{Cl}}$:⁻   H—$\overset{..}{\underset{..}{Cl}}$:   acid ----▸

$CH_3\overset{..}{N}$—H / $\overset{|}{CH_3}$ nucleophile

(If excess $(CH_3)_2NH$ is used, it will act as a base in the third step of the reaction.)

(f) CH_3CH CH_2=CH ⟶ CH_3CH—CH_2CH (with :$\overset{..}{O}$: and :$\overset{..}{O}$:⁻ groups)

:$\overset{..}{O}$:   :$\overset{..}{O}$:⁻   :$\overset{..}{O}$:⁻   :$\overset{..}{O}$:

electrophile   nucleophile

(g) CH_2=CH—$\overset{\overset{\displaystyle :\overset{..}{O}:}{||}}{C}$—$CH_3$ ⟶ CH_2—CH=$\overset{\underset{\underset{\displaystyle +NH_3}{}}{}}{C}$—$CH_3$ ⟷ $\overset{..}{CH_2}$—CH—$\overset{\overset{\displaystyle :\overset{..}{O}:}{||}}{C}$—$CH_3$

electrophile   :$\overset{..}{O}$:⁻

:NH_3 nucleophile $H_2\overset{+}{N}$—H   :NH_3 acid ----▸ ◂---- base

base

CH_2—CH—$\overset{\overset{\displaystyle :O:}{||}}{C}$—$CH_3$ ◂⟶ CH_2—CH—$\overset{\overset{\displaystyle :O:}{||}}{C}$—$CH_3$

$H_2\overset{..}{N}$:   H   :NH_3 H_2N: ⇌ H—$\overset{+}{N}H_3$ acid ----▸

4.23   Let us explore the two possible reactive intermediates for this reaction.

1° carbocation

2° carbocation destabilized by resonance

In this case the 1° carbocation is more stable than the 2° carbocation and is on the path with the lower $\Delta G^{\ddagger}$. The 1° carbocation forms faster than the 2° carbocation and most of the product comes from that pathway.

4.24   (a)   B, D, E   (b)   The steps by which C → G by way of transition state D.

   (c)   G   (d)   F

   (e) $\Delta G^{\ddagger}$ for C → F by way of E is +8.9 – (+3.0) = +5.9 kcal/mol.

   (f) ΔG_r° for the formation of G from A is –16.6 – (0) = –16.6 kcal/mol.

4.25   No, the rate of reaction will decrease as the reaction progresses. The rate is a function of the concentration of starting material and it decreases as starting material is converted into product.

4.26   At 308 K, $k = 5.3 \times 10^{-4}$ L/mol·sec. When the reaction is half over, the concentrations of hydroxide ion and of methyl bromide will each be half the initial concentrations.

$$
\begin{aligned}
\text{Rate} &= k\,[CH_3Br][OH^{-}] \\
&= k\,(0.5[CH_3Br]_{initial})(0.5[OH^{-}]_{initial}) \\
&= 0.25\,(\text{initial rate}) = 0.25(5.3 \times 10^{-9} \text{ mol/L·sec}) \\
&= 1.3 \times 10^{-9} \text{ mol/L·sec}
\end{aligned}
$$

4.27   When a reaction has gone to 99% completion, the final concentration of product will be 99% that of the initial starting material concentration.

$$[\text{product}]_{99\% \text{ completion}} = 0.99[\text{starting material}]_{initial}$$

and the concentration of starting material will be reduced to 1% of the initial concentration.

$$[\text{starting material}]_{99\% \text{ completion}} = 0.01[\text{starting material}]_{initial}$$

The reaction is over when it reaches equilibrium and

$$
\begin{aligned}
K_{eq} &= [\text{product}]_{eq}/[\text{starting material}]_{eq} = 0.99[\text{starting material}]_{initial}/0.01[\text{starting material}]_{initial} \\
&= 0.99/0.01 = 99
\end{aligned}
$$

$$\Delta G_r = -RT \ln K_{eq}$$

At 25 °C (298 K)

$$
\begin{aligned}
\Delta G_r &= -(1.99 \times 10^{-3} \text{ kcal/mol·K})(298 \text{ K}) \ln 99 \\
&= -2.7 \text{ kcal/mol}
\end{aligned}
$$

4.28   (a)

$$
\begin{aligned}
\Delta G_r^{\circ} &= \Delta H_r^{\circ} - T\Delta S_r^{\circ} \\
\Delta H_r^{\circ} &= -85.87 \text{ kcal/mol} - (-4.04 \text{ kcal/mol} - 68.32 \text{ kcal/mol}) = -13.51 \text{ kcal/mol} \\
\Delta S_r^{\circ} &= 46.30 \text{ cal/mol·K} - (16.72 \text{ cal/mol·K} + 70.17 \text{ cal/mol·K}) = -40.59 \text{ cal/mol·K} \\
&= -4.059 \times 10^{-2} \text{ kcal/mol·K} \\
\Delta G_r^{\circ} &= (-13.51 \text{ kcal/mol}) - (298 \text{ K})(-4.06 \times 10^{-2} \text{ kcal/mol·K}) \\
&= -1.41 \text{ kcal/mol} \\
\Delta G_r^{\circ} &= -RT \ln K_{eq}
\end{aligned}
$$

4.28 (a) (cont) $\ln K_{eq}$ $= -\Delta G_r°/RT$

K_{eq} $= \exp(-\Delta G_r°/RT) = \exp[-(-1.41 \text{ kcal/mol})/(1.99 \times 10^{-3} \text{ kcal/mol·K})(298 \text{ K})]$

$= 10.8$

(b)   When sulfuric acid is added to water, the proton is transferred completely to form the hydronium ion and the hydrogen sulfate anion.

$$H_2SO_4 + H_2O \longrightarrow H_3O^+ + HSO_4^-$$

$\text{p}K_a -9$ $\text{p}K_a -1.7$

In a dilute solution of H_2SO_4 in water, the electrophile is the hydronium ion, not sulfuric acid. The nucleophile is the double bond of the alkene, which requires a strong acid for protonation. H_2O (pK_a 15.7) is not strong enough, but H_3O^+ is. The hydronium ion is regenerated in the last step of the mechanism, so only a catalytic amount of sulfuric acid is necessary (see Section 8.4 for the mechanism).

4.29 (a)

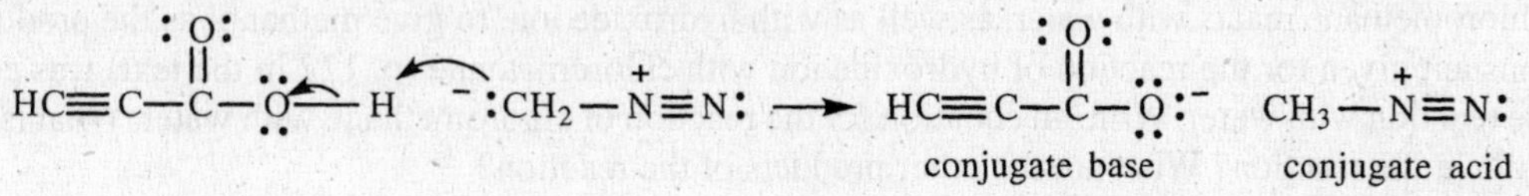

conjugate base   conjugate acid

(b)

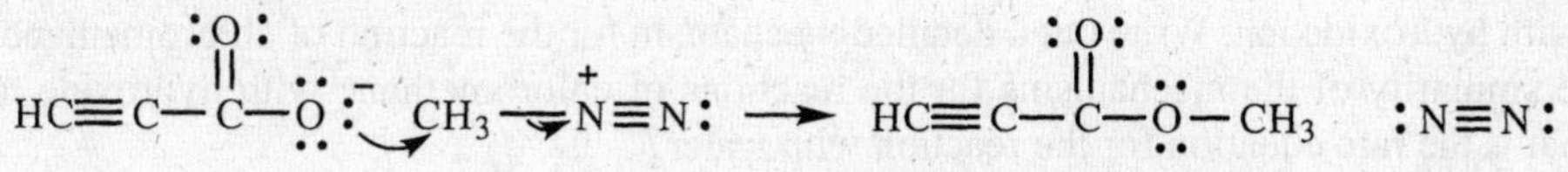

(c)

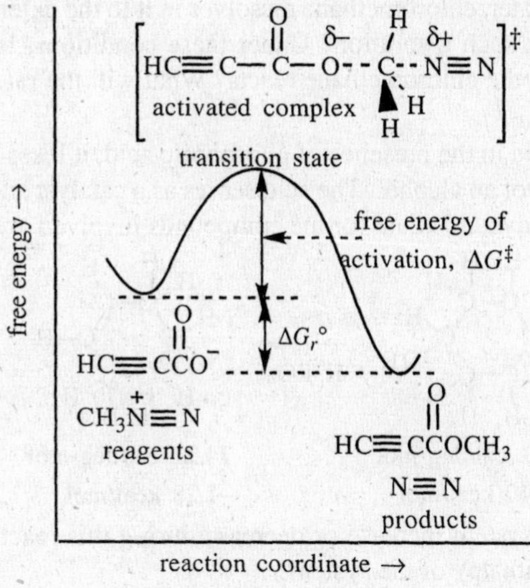

reaction coordinate →

4.30 (a)

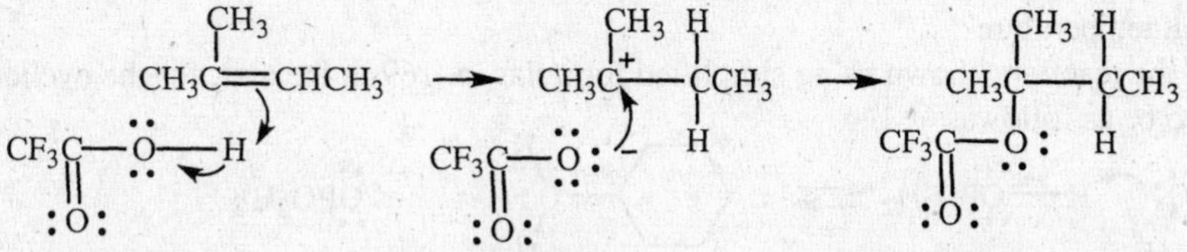

(b)

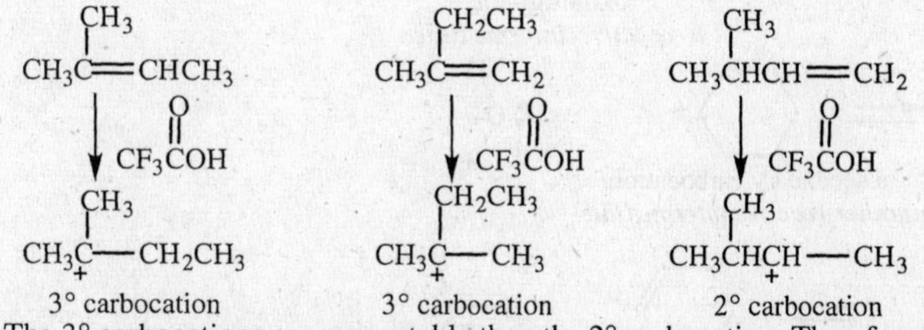

3° carbocation   3° carbocation   2° carbocation

The 3° carbocations are more stable than the 2° carbocation. They form with a lower $\Delta G^{\ddagger}$ and therefore at a faster rate than the 2° carbocation.

Supplemental Problems

S4.1   Supply a reagent that will give the chemical change shown in each of the equations below. Identify the reagents as electrophiles or nucleophiles and explain why you chose them. Constructing a concept triangle for each reaction may be helpful.

(a) $HOCH_2CH_2C \equiv CCH_2CH_2CH_2Cl \xrightarrow{?} HOCH_2CH_2C \equiv CCH_2CH_2CH_2CN$

(b)   (c) $CH_3CH_2OH \xrightarrow{?} CH_3CH_2O^- \ Na^+ \xrightarrow{?} CH_3CH_2OCH_3$

S4.2   (a)   Chloromethane reacts with water, as well as with hydroxide ion, to give methanol as the product. The rate constant given for the reaction of hydroxide ion with chloromethane (p. 122 in the text) was corrected for the reaction with water. Write an equation for the reaction of chloromethane with water. What is the nucleophile in the reaction? What are the other products of the reaction?

    (b)   The mechanism for the reaction of chloromethane with water is essentially the same as the one for the reaction with hydroxide ion. Write out a detailed mechanism for the reaction of chloromethane with water.

    (c)   Given the similarity of the mechanisms for the reactions of chloromethane with hydroxide ion and with water, what is the rate equation for the reaction with water?

    (d)   When bubbled through water, chloromethane dissolves in it to the extent of about 0.003 mol/L. What is the concentration of water in such a solution? Under these conditions, is it possible to see a change in the concentration of water as the chloromethane reacts? What will the rate equation look like for this experiment?

S4.3   When cyclohexanol is heated in the presence of phosphoric acid, it loses water to become cyclohexene. This reaction is known as the dehydration of an alcohol. The acid serves as a catalyst and is regenerated in the reaction. The equation for the reaction and thermodynamic data for the compounds involved are as follows:

S_f° (at 298 K)	78.32 cal/deg•mol	74.27 cal/deg•mol	45.11 cal/deg•mol
ΔH_f° (at 298 K)	−70.40 kcal/mol	−1.28 kcal/mol	−57.80 kcal/mol

(a)   Does the entropy of the system increase or decrease during this reaction? How can you account for this result? What about the enthalpy of the system?

(b)   Calculate the equilibrium constant for this reaction at 298 K. Will products be formed from the starting material at that temperature?

(c)   The stages of the reaction (shown using simplified formulas, p. 169 in the text, for the cyclic compounds) are believed to be as follows:

S4.3 (c) 4. H_3O^+ + $^-OPO_3H_2$ ⇌ H_2O + $HOPO_3H_2$

When cyclohexanol is mixed with phosphoric acid at room temperature, heat is evolved. What does this say about the relative energy levels of all the species involved in the first step of the reaction?

(d) No product is formed at room temperature. The reaction mixture must be heated for cyclohexene to be formed. Which step of this reaction do you think has the highest free energy of activation?

S4.4   Researchers from the University of North Carolina have reported that silica gel catalyzes the electrophilic addition reactions of hydrohalic acids (HCl, HBr, HI) to double and triple bonds.

(a) For the following example, use the general concept of the electrophilic addition reaction to predict the structure of the major product that results from the addition of one HBr to the triple bond.

$$CH_3CH_2CH_2CH_2CH_2C{\equiv}C{-}H \quad + \ HBr \quad \xrightarrow[\substack{\text{silica gel} \\ 24\ h}]{H_2O}$$
$$\text{1 equivalent}$$

(b) The addition of a hydrohalic acid to an alkene in the presence of silica gel is much faster than the addition to an alkyne. For example, the reaction with a simple alkene such as $CH_3CH_2CH_2CH_2CH_2CH{=}CH_2$ is complete in 1 hour in contrast to the 24 hours necessary for the alkyne shown above. The difference in rate between the reaction of an alkene and that of an alkyne is rationalized by considering the difference between reactive intermediates derived from the two starting materials.

1. Draw structural formulas for the reactive intermediates involved in the rate-determining step for the addition reaction of the alkyne and of the alkene.

2. How would you describe the difference between the two species shown in part 1, and why would this difference affect the relative rates of the two reactions?

S4.5   An electrophilic addition reaction is used in the preparation of ceralure, which is used to control the population of the Mediterranean fruit fly. The reaction shown below does not have high regioselectivity and gives two isomers. Draw their structural formulas and explain why the two compounds are formed in roughly the same amounts.

    + HI   →   A + B

S4.6   When the following compound is treated with aqueous acid, a reaction product is observed, the formation of which is believed to begin with the electrophilic addition of water to the double bond.

  $\xrightarrow[\substack{H_2O}]{H_3O^+\ Cl^-}$ →   + CH_3OH + $H_3O^+\ Cl^-$

$$A \qquad\qquad\qquad\qquad B \qquad\qquad C$$
$$+\ H_3O^+\ Cl^-$$

(a)   Using the curved arrow notation, draw a complete, stepwise mechanism for the transformation of A to B. (Note: strong bases, such as hydroxide ion, may not be used to describe the mechanism for reactions performed in acid, such as this one.)

(b)   Which of the following energy diagrams best describes the transformation of A to B?

(c)   On the energy diagram you selected in part (b), use an arrow to point to the transition state that corresponds to the rate-determining step.

5 Alkanes and Cycloalkanes

Workbook Exercises

EXERCISE I. Insight into molecular structure comes from a detailed examination of the numbers and kinds of atoms in a molecular formula. For each of the following molecular formulas, determine the units of unsaturation (see Section 1.5).

(a) $C_{10}H_{16}$, limonene from lemon oil

(b) $C_{17}H_{21}NO_4$, cocaine, a local anesthetic

(c) $C_{22}H_{37}NO_2$, anandiamide, a natural ligand for the receptor in the brain for tetrahydrocannabinol, the active ingredient of marijuana

(d) $C_{23}H_{46}$, muscalure, an insect attractant

(e) $C_5H_9NO_2$, proline, an amino acid

(f) $C_4H_{12}N_2$, putrescine, found in decayed tissues

(g) $C_6H_8O_6$, ascorbic acid

(h) C_3H_6O, propylene oxide, a sterilant

(i) $C_{43}H_{78}N_6O_{13}$, muroctasin, an immunostimulant

EXERCISE II. What are some structural features that could account for the units of unsaturation in each of the compounds in Exercise I? Try to give up to three examples of such structural features for each. For example, proline [part (e)], has two units of unsaturation. These units could be represented by two rings, a ring and a carbon-oxygen double bond, or a carbon-nitrogen triple bond. Other structural features are also possible based on the information given.

EXERCISE III. The structures for most of the compounds listed in Exercise I are found in chemistry textbooks. Find as many of these structures as you can, identify the units of unsaturation actually present, and compare this count to your answers to Exercise II.

5.1 (a)

(b)

(c)

(d)

5.1   (e)

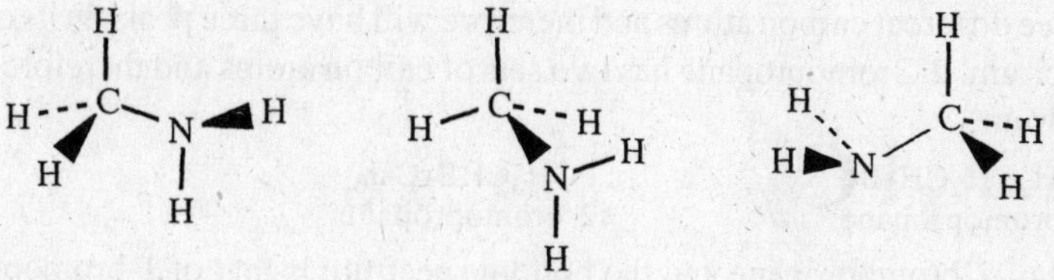

H—C⋯N◄H   (with H's around C and N)   H—C⋯H   (N below)   H⋯N—C⋯H etc.

5.2 $C_2H_3Cl_3$:

CH_3CCl_3 $ClCH_2CHCl_2$

Lewis structures

three-dimensional
representations

$C_2H_2Cl_4$:

$ClCH_2CCl_3$ $Cl_2CHCHCl_2$

Lewis structures

three-dimensional
representations

5.3   The carbon nuclear magnetic resonance spectrum tells us that there are two different carbon atoms in the molecule. The proton magnetic resonance spectrum tells us that there are two sets of hydrogens, in a ratio of 2:1 (16 mm to 7 mm from the integration curves). The two isomers with the molecular formula $C_2H_3Cl_3$ are shown below.

CH_3CCl_3 $ClCH_2CHCl_2$
1,1,1-trichloroethane   1,1,2-trichloroethane

The hydrogens in 1,1,1-trichloroethane are equivalent and would give only one peak in its proton magnetic resonance spectrum. The three hydrogens in 1,1,2-trichloroethane consist of two sets in a ratio of 2:1. The compound is 1,1,2-trichloroethane.

5.4   The carbon nuclear magnetic resonance spectrum of Compound A has two peaks and thus two different carbon atoms and that of Compound B has one peak and only one type of carbon atom. Compound A must be 1,1,1,2-tetrachloroethane and Compound B must be 1,1,2,2-tetrachloroethane

$ClCH_2CCl_3$ $Cl_2CHCHCl_2$
1,1,1,2-tetrachloroethane   1,1,2,2-tetrachloroethane

The proton magnetic resonance spectra confirm these assignments. The hydrogen atoms in Compound A are bonded to a carbon atom that has one chlorine attached. These hydrogen atoms absorb further to the right (closer to the TMS standard) than the hydrogen atoms in Compound B, which are bonded to carbon atoms bearing two chlorine atoms each.

5.5   (a) 1-Bromopropane has three different carbon atoms and therefore will have three peaks in its carbon nuclear magnetic resonance spectrum. 2-Bromopropane has two sets of carbon atoms and therefore will have two peaks in its carbon spectrum.

$$CH_3CH_2CH_2Br \qquad\qquad CH_3CHBrCH_3$$
1-bromopropane   2-bromopropane

The top spectrum is that of 2-bromopropane and the bottom spectrum is that of 1-bromopropane.

(b) 1-Bromopropane has three sets of equivalent hydrogen atoms in a ratio of 2:2:3. The hydrogen atoms (c) on the carbon bonded to the bromine will be furthest to the left. The hydrogen atoms (b) on carbon 2 will be closer to the TMS standard, and the methyl hydrogen atoms (a) will be closest to the TMS standard.

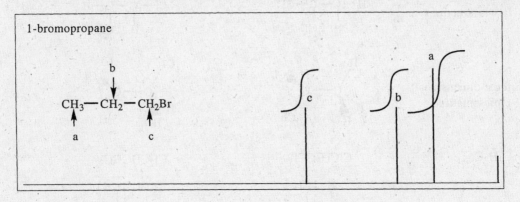

(c) 2-Bromopropane has two sets of equivalent hydrogen atoms in a ratio of 6:1. The hydrogen atom on carbon 2 (b) will be further to the left than the methyl hydrogen atoms (a).

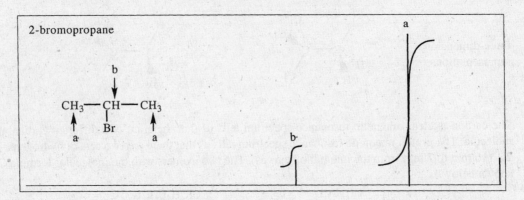

5.6 $C_3H_6Br_2$:
condensed formulas

(1)	$CH_3CH_2CHBr_2$		(2)	$CH_3CBr_2CH_3$
(3)	$CH_3CHBrCH_2Br$		(4)	$BrCH_2CH_2CH_2Br$

Lewis structures

5.6 (cont)

three-dimensional representations   (1)   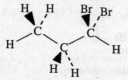   (2)

(3)   (4)

5.7   (a) *sec*-butyl   (b isobutyl   (c) isopropyl
        (d) propyl   (e) *tert*-butyl   (f) ethyl; butyl

5.8   (a)   Constitutional isomers   (b)   Not constitutional isomers; both formulas have the same connectivity.
          (c)   Constitutional isomers   (d)   Constitutional isomers
          (e)   Constitutional isomers   (f)   Constitutional isomers
          (g)   Not constitutional isomers; both formulas have the same connectivity.

Concept Map 5.1   Determining connectivity.

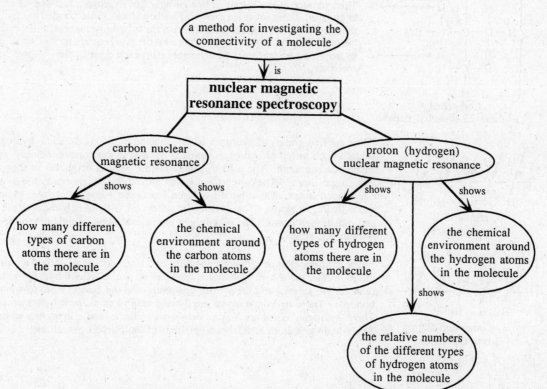

a method for investigating the
connectivity of a molecule

is

**nuclear magnetic
resonance spectroscopy**

carbon nuclear
magnetic resonance

proton (hydrogen)
nuclear magnetic resonance

shows   shows   shows   shows

how many different
types of carbon
atoms there are in
the molecule

the chemical
environment around
the carbon atoms
in the molecule

how many different
types of hydrogen
atoms there are in
the molecule

the chemical
environment around
the hydrogen atoms
in the molecule

shows

the relative numbers
of the different types
of hydrogen atoms
in the molecule

5.9   (a)

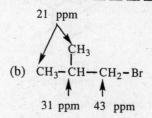

1.0 ppm

CH$_3$

CH$_3$—C—CH$_2$—Br

2.0 ppm →   H

3.3 ppm

(b)

21 ppm

CH$_3$

CH$_3$—CH—CH$_2$—Br

31 ppm   43 ppm

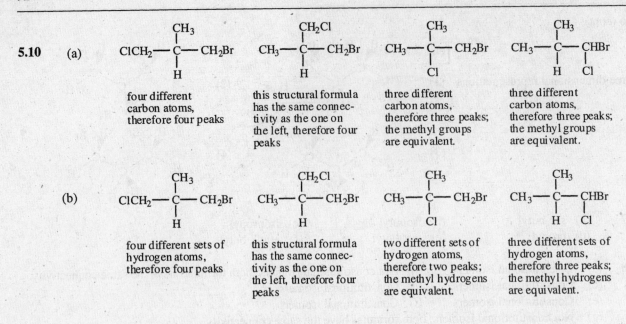

5.10 (a)

four different
carbon atoms,
therefore four peaks

this structural formula
has the same connec-
tivity as the one on
the left, therefore four
peaks

three different
carbon atoms,
therefore three peaks;
the methyl groups
are equivalent.

three different
carbon atoms,
therefore three peaks;
the methyl groups
are equivalent.

(b)

four different sets of
hydrogen atoms,
therefore four peaks

this structural formula
has the same connec-
tivity as the one on
the left, therefore four
peaks

two different sets of
hydrogen atoms,
therefore two peaks;
the methyl hydrogens
are equivalent.

three different sets of
hydrogen atoms,
therefore three peaks;
the methyl hydrogens
are equivalent.

5.11

Compound A
1-bromo-2,2-dimethylpropane

The hydrogen atoms belong in two groups, nine in group *a* and two
in group *b*. There are three groups of carbon atoms, those bonded to
the *a*-hydrogen atoms, the one bonded to the *b*-hydrogen atoms, and
the one labeled *c*, with no hydrogen atoms on it. There are no
hydrogen atoms that are 3-bond neighbors of each other in this
molecule.

Compound B
1-bromopentane

There are five groups of hydrogen atoms, labeled *a*-, *b*-, *c*-, *d*-, and *e*-hydrogen
atoms. Each group of hydrogen atoms is bonded to the correspondingly
labeled carbon atom. The *a*-hydrogen atoms are 3-bond neighbors to the
b-hydrogen atoms. The *b*-hydrogen atoms are 3-bond neighbors to the *a*- and
c-hydrogen atoms. The *c*-bond hydrogen atoms are 3-bond neighbors to the *b*-
and *d*-hydrogen atoms. The *d*-hydrogen atoms are 3-bond neighbors to the *c*-
and *e*-hydrogen atoms. The *e*-hydrogen atoms are 3-bond neighbors to the
d-hydrogen atoms.

Compound C
3-bromopentane

There are three groups of hydrogen atoms and three groups of carbon atoms in
the molecule, the *a*-hydrogen atoms (and carbon atoms), the *b*-hydrogen atoms
(and carbon atoms), and the *c*-hydrogen atom (and the carbon atom to which it is
bonded). The *a*-hydrogen atoms are 3-bond neighbors to the *b*-hydrogen atoms.
The *b*-hydrogen atoms are 3-bond neighbors to the *a*- and *c*-hydrogen atoms.
The *c*-hydrogen atom is a 3-bond neighbor to the *b*-hydrogen atoms.

5.12

perspective
formula

Newman
projection

staggered conformation

perspective
formula

Newman
projection

eclipsed conformation

Concept Map 5.2   **Conformation.**

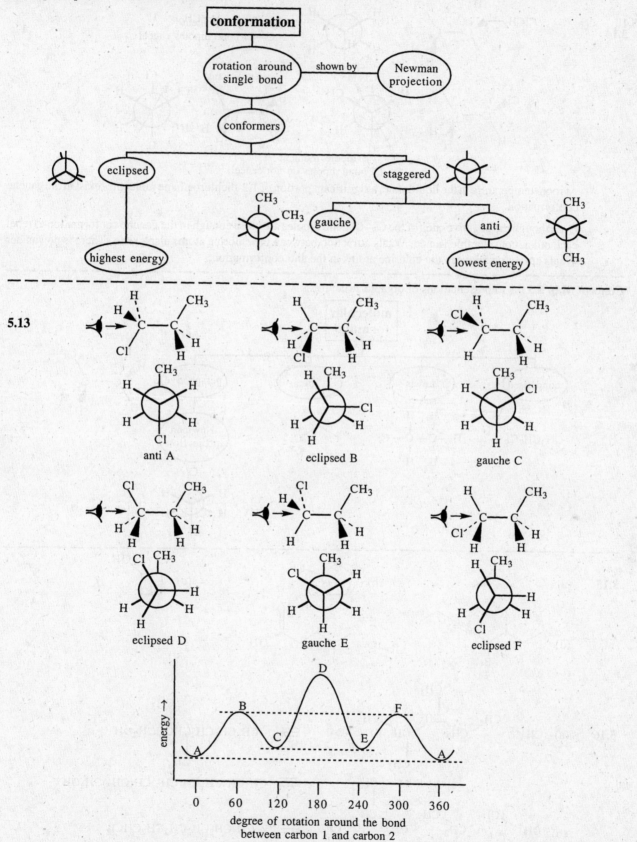

5.13

anti A

eclipsed B

gauche C

eclipsed D

gauche E

eclipsed F

degree of rotation around the bond
between carbon 1 and carbon 2

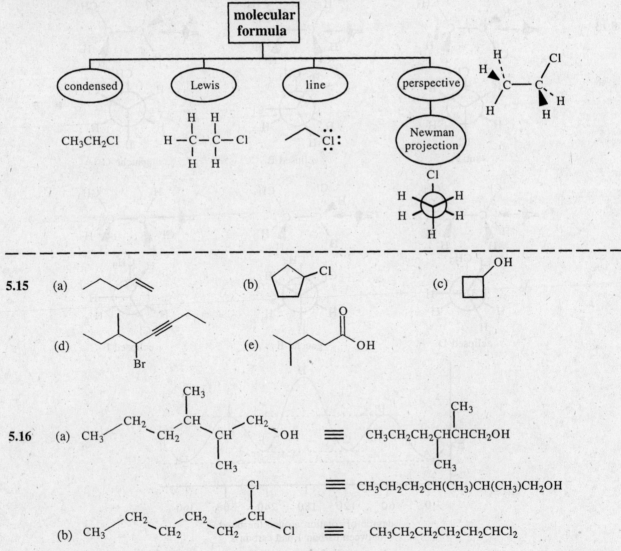

5.14

no dipole moment
C—Cl bond dipoles cancel

dipole moment
C—Cl bond dipoles do not cancel

At room temperature in the liquid state, a significant portion of 1,2-dichloroethane molecules exist in the gauche conformation.

Even though the negative ends of the C—Cl bond dipoles are close enough in the gauche conformation to repel each other, considerable van der Waals attraction between the chlorine atoms also exists. There is no van der Waals attraction between the chlorine atoms in the anti conformation.

Concept Map 5.3   Representations of organic structures.

molecular formula

condensed   Lewis   line   perspective

CH_3CH_2Cl

Newman projection

5.15 (a)   (b)   (c)

(d)   (e)

5.16 (a) $CH_3-CH_2-CH_2-CH(CH_3)-CH-CH_2-OH$ with CH_3 branch $\equiv$ $CH_3CH_2CH_2CHCHCH_2OH$ with CH_3 branch

$\equiv$ $CH_3CH_2CH_2CH(CH_3)CH(CH_3)CH_2OH$

(b) $CH_3-CH_2-CH_2-CH_2-CH_2-CH(Cl)-Cl$ $\equiv$ $CH_3CH_2CH_2CH_2CH_2CHCl_2$

5.16 (c)

$$CH_3CH_2CH_2CH_2CCHBrCH_3$$

$$\equiv CH_3CH_2CH_2CH_2COCHBrCH_3$$

(d)

$$\equiv CH_3CH_2CH_2CH_2CHCOCH_3$$

$$\equiv CH_3CH_2CH_2CH_2CH(CH_3)COOCH_3$$

(e)

5.17 (a) 2,2,6,7-tetramethyloctane (b) 5-ethyl-2,6-dimethyloctane (c) 2,3-dimethylpentane
(d) 4-*tert*-butylheptane (e) 2,2,5-trimethyl-6-propyldecane (f) 6-isopropyl-3-methylnonane

5.18 (a) $CH_3CCH_2CH_2CH_2CH_2CH_3$ (b) $CH_3CHCH_2CH_2CH_2CHCH_2CH_2CH_2CH_3$

(c) $CH_3CHCH_2CH_2CCH_2CH_2CHCH_3$ (d) $CH_3CCH_2CHCH_2CHCH_2CH_2CH_2CH_2CH_3$

5.19 Arrows connect equivalent carbon atoms. All hydrogen atoms on the same carbon atom are equivalent as are all hydrogen atoms on equivalent carbon atoms.

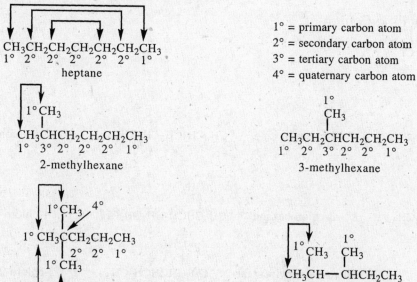

1° = primary carbon atom
2° = secondary carbon atom
3° = tertiary carbon atom
4° = quaternary carbon atom

5.19 (cont)

```
                    1°
                    CH₃
                 4° |
     CH₃CH₂——C——CH₂CH₃
     1°   2°  |        2°  1°
               CH₃
               1°
       3,3-dimethylpentane
```

```
        1°    1°
        CH₃   CH₃
        |     |
     CH₃CHCH₂CHCH₃
     1°  3° 2° 3° 1°

      2,4-dimethylpentane
   all 1° C and H are equivalent
   all 2° H are equivalent
   all 3° C and H are equivalent
```

```
        1°   1°
        CH₃  CH₃
        |    |
     CH₃C———CHCH₃
     1°  |4°  3° 1°
        CH₃
        1°
     2,2,3-trimethylbutane
```

```
           2°  1°
           CH₂CH₃
           |
     CH₃CH₂CHCH₂CH₃
     1°  2°  3° 2°  1°

        3-ethylpentane
   all 1° C and H are equivalent
   all 2° C and H are equivalent
```

5.20   (a) 2-bromo-2,3-dimethylpentane   (b) 3-pentanol   (c) 4-chloro-3-hexanol

(d) 4,4-dichloro-2,2-dimethylhexane   (e) 2,3-dibromo-2,3-dimethylbutane   (f) 4-*tert*-butyl-3-nonanol

5.21   (a)
```
     CH₃CHCH₂CH₂CH₂CH₂CH₂CH₃
        |
        I
```
(b)
```
     CH₃CH₂CHCH₂CH₂CH₃
           |
           OH
```
(c) CF₃CH₂OH

(d)
```
              CH₃
              |
     ClCH₂CHCH₂CCH₃
          |      |
          OH    CH₃
```
(e)
```
     HOCH₂CHCH₂OH
          |
          OH
```
(f) CH₃CH₂CH₂CH₂CH₂OH

(g)
```
        CH₂CH₃
        |
     CH₃CHCCH₂CH₂CH₂CH₃
        |  |
        Cl Cl
```

5.22   C₆H₁₃Cl

CH₃CH₂CH₂CH₂CH₂CH₂Cl   1° 1-chlorohexane
```
     CH₃CH₂CH₂CH₂CHCH₃
                    |
                    Cl
```
2° 2-chlorohexane

```
     CH₃CH₂CH₂CHCH₂CH₃
             |
             Cl
```
2° 3-chlorohexane
```
           CH₃
           |
     CH₃CHCH₂CH₂CH₂Cl
```
1° 1-chloro-4-methylpentane

```
     CH₃CH₂CH₂CHCH₂CH₃
             |
        CH₃  Cl
```

```
     CH₃CHCH₂CHCH₃
        |
        Cl
```
2° 2-chloro-4-methylpentane
```
           CH₃
           |
     CH₃CHCHCH₂CH₃
           |
           Cl
```
2° 3-chloro-2-methylpentane

5.22 (cont)

$$CH_3\overset{\overset{\displaystyle CH_3}{|}}{\underset{\underset{\displaystyle Cl}{|}}{C}}CH_2CH_2CH_3 \qquad 3° \quad \text{2-chloro-2-methylpentane}$$

$$ClCH_2\overset{\overset{\displaystyle CH_3}{|}}{CH}CH_2CH_2CH_3 \qquad 1° \quad \text{1-chloro-2-methylpentane}$$

$$CH_3CH_2\overset{\overset{\displaystyle CH_3}{|}}{CH}CH_2CH_2Cl \qquad 1° \quad \text{1-chloro-3-methylpentane}$$

$$CH_3CH_2\overset{\overset{\displaystyle CH_3}{|}}{\underset{\underset{\displaystyle Cl}{|}}{CH}}CHCH_3 \qquad 2° \quad \text{2-chloro-3-methylpentane}$$

$$CH_3CH_2\overset{\overset{\displaystyle CH_3}{|}}{\underset{\underset{\displaystyle Cl}{|}}{C}}CH_2CH_3 \qquad 3° \quad \text{3-chloro-3-methylpentane}$$

$$CH_3CH_2\overset{\overset{\displaystyle CH_2Cl}{|}}{CH}CH_2CH_3 \qquad 1° \quad \text{3-(chloromethyl)pentane or}$$
$$\text{1-chloro-2-ethylbutane}$$

$$CH_3\overset{\overset{\displaystyle CH_3}{|}}{\underset{\underset{\displaystyle CH_3}{|}}{C}}CH_2CH_2Cl \qquad 1° \quad \text{1-chloro-3,3-dimethylbutane}$$

$$CH_3\overset{\overset{\displaystyle CH_3}{|}}{\underset{\underset{\displaystyle CH_3}{|}}{C}}\!\!-\!\!\overset{}{\underset{\underset{\displaystyle Cl}{|}}{CH}}CH_3 \qquad 2° \quad \text{3-chloro-2,2-dimethylbutane}$$

$$CH_3\overset{\overset{\displaystyle CH_2Cl}{|}}{\underset{\underset{\displaystyle CH_3}{|}}{C}}CH_2CH_3 \qquad 1° \quad \text{1-chloro-2,2-dimethylbutane}$$

$$CH_3\overset{\overset{\displaystyle CH_3}{|}}{\underset{\underset{\displaystyle CH_3}{|}}{CH}}CHCH_2Cl \qquad 1° \quad \text{1-chloro-2,3-dimethylbutane}$$

$$CH_3\overset{\overset{\displaystyle CH_3}{|}}{CH}\!\!-\!\!\overset{\overset{\displaystyle CH_3}{|}}{\underset{\underset{\displaystyle Cl}{|}}{C}}CH_3 \qquad 3° \quad \text{2-chloro-2,3-dimethylbutane}$$

5.23 (1) 1,1-dichlorobutane (2) 2,2-dichlorobutane (3) 1,2-dichlorobutane
(4) 1,3-dichlorobutane (5) 2,3-dichlorobutane (6) 1,4-dichlorobutane
(7) 1,1-dichloro-2-methylpropane (8) 1,2-dichloro-2-methylpropane (9) 1,3-dichloro-2-methylpropane

5.24 (a) *sec*-butylbenzene (b) isobutylbenzene (c) isopropylbenzene
(d) propylbenzene (e) *tert*-butylbenzene (f) 1-butyl-4-ethylbenzene

Note that if the substituent groups are simple alkyl groups, the compounds are named as alkylbenzenes. When the alkyl groups become large or are complicated by having other substituents on them (Problem 5.25), the benzene ring is treated as one of the substituents with the substituent name phenyl.

5.25 (a) 1-chloro-3-phenylpropane (b) 3-iodo-3-phenylhexane (c) 1-isopropyl-2-methylbenzene
(d) 1,2-dichloro-1-phenylpropane (e) 2-methyl-2-phenylhexane (f) 3-phenyl-1-propanol

5.26 (a) $\overset{}{\underset{\underset{\displaystyle OH}{|}}{CH}}CH_2CH_2CH_2CH_3$

(b) $CH_2CH_2\overset{\overset{\displaystyle CH_3}{|}}{\underset{\underset{\displaystyle Br}{|}}{C}}CH_3$

5.26 (c) 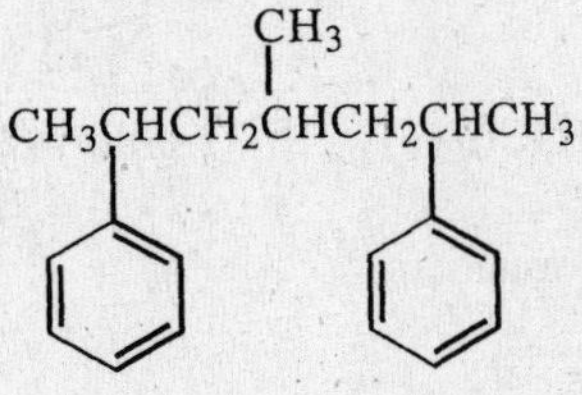

(d) 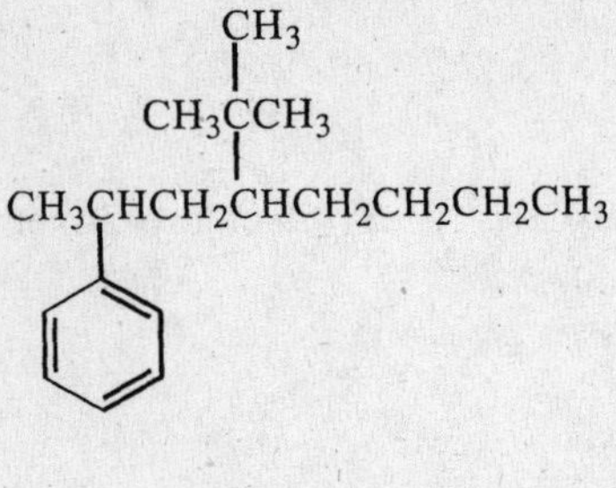

5.27
(a) 1-methylcyclobutanol (b) ethylcyclopropane (c) 1-ethyl-1-methylcyclopentane (d) iodocyclopentane

5.28 (a) 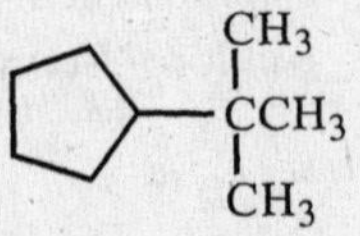

(b) 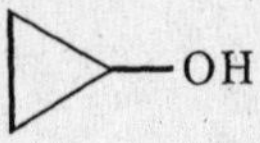

(c) 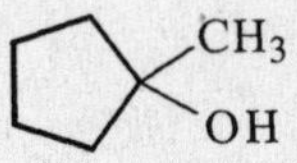

(d) 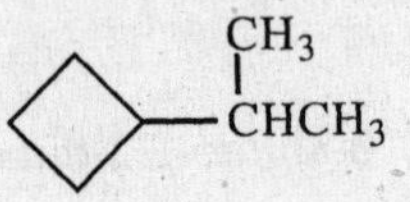

(e) 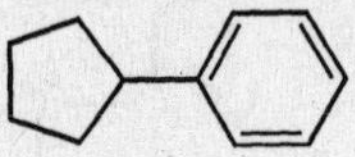

Concept Map 5.4 Conformation in cyclic compounds.

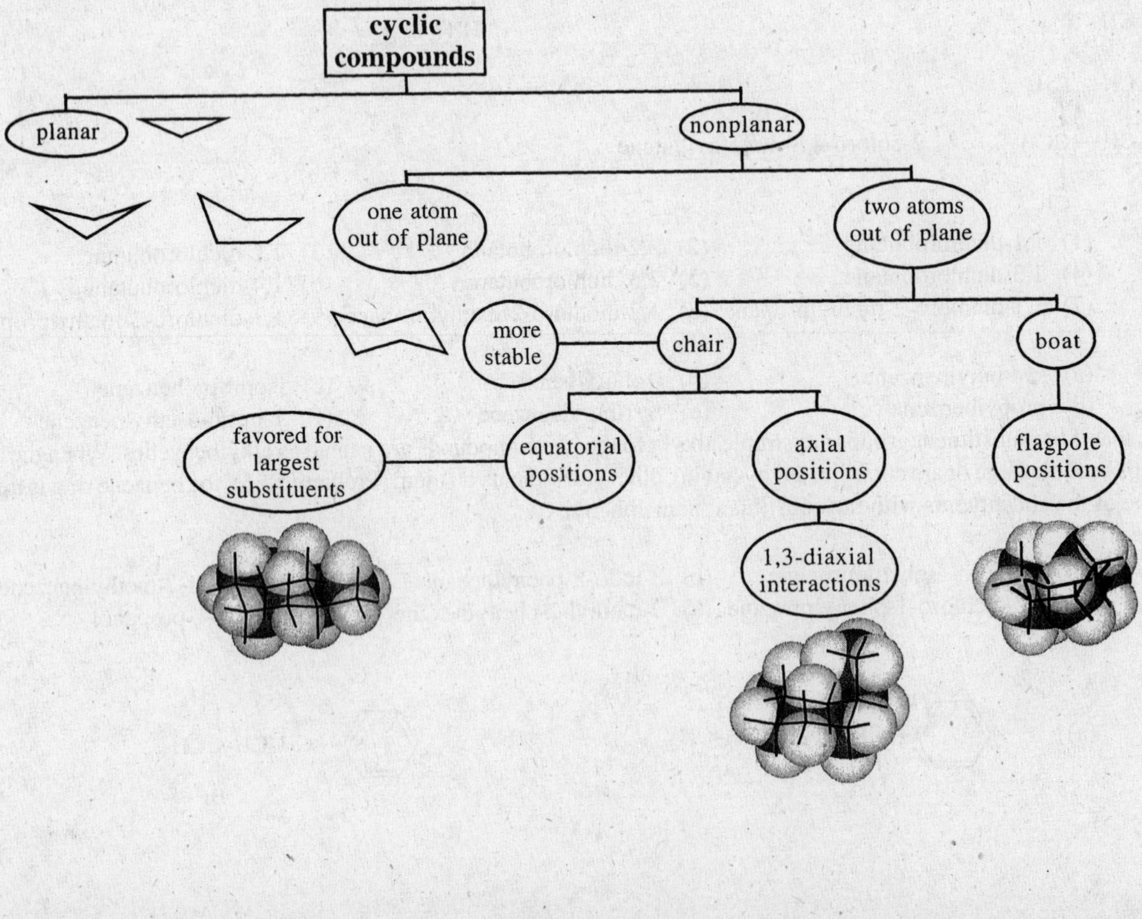

5.29

planar form   chair conformations

                axial chlorine   equatorial chlorine

planar form   chair conformations

            axial hydroxyl group   equatorial hydroxyl group

5.30

$$CH_3C\cdot \quad > \quad CH_3C\cdot \quad > \quad CH_3C\cdot \quad > \quad H-C\cdot$$

5.31

(a) 1-bromo-1-methylcyclopropane   (b) 2,2,5,5-tetramethylhexane   (c) 1,1-dimethylcyclobutane
(d) 2-chloro-2-methylpentane   (e) isopropylcyclohexane or 2-cyclohexylpropane (f) 1-ethyl-1-cyclopentanol
(g) 1-chloro-1-ethylcyclohexane   (h) 3-bromo-3-ethyl-1-heptanol

5.32

BrCHCF$_3$

Cl

2-bromo-2-chloro-1,1,1-trifluoroethane

5.33   Beginning with this problem, if the benzene ring is not involved in a reaction or in resonance or is bonded to only one substituent, it will be represented by Ph, the symbol for the phenyl group (Section 5.10D) [see (g)].

(a) CH$_3$CCH$_2$CH$_2$CH$_2$CH$_3$ with CH$_3$ and OH

(b) cyclobutane with CH$_3$ and Cl

(c) cyclopropane with I

(d) CH$_3$CH$_2$CH$_2$CHCH$_2$CH$_2$CH$_3$ with CH$_3$CCH$_3$ / CH$_3$ group

(e) CH$_3$CH$_2$CH$_2$CH$_2$CHCH$_2$CH$_2$CH$_3$ with cyclopropyl

(f) cyclopentane with CH$_2$CHCH$_3$ / CH$_3$

(g) CH$_3$CH$_2$CHCH$_2$CH$_2$CH$_2$CH$_3$ with Ph

(h) cyclohexane with CH$_3$ and Br

(i) CH$_3$CCH$_2$CH$_2$CH$_2$CH$_2$CCH$_3$ with CH$_3$ and Cl, and CH$_3$ and OH

5.34   Isomers of $C_3H_8O_2$

$$CH_3OCH_2OCH_3$$
two ether groups
no hydrogen bonding
bp 45 °C

$$CH_3OCH_2CH_2OH$$
ether, alcohol
hydrogen bonding
bp 124 °C

$$HOCH_2CH_2CH_2OH$$
2 hydroxyl groups
more hydrogen bonding
bp 215 °C

5.35   (a) C_5H_9Br

Each halogen atom counts as a hydrogen atom in determining the units of unsaturation.
Units of unsaturation = $[(2 \times 5) + 2 - 10]/2 = (12\text{-}10)/2 = 1$
Some structural formulas are shown below:

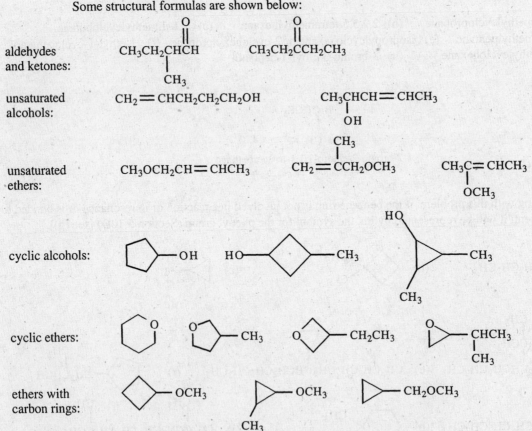

alkenes: $CH_3CH_2CH = CHCH_2Br$ $CH_3CH_2CH = \underset{\underset{Br}{|}}{C}CH_3$ $CH_3\underset{\underset{Br}{|}}{C} = \overset{\overset{CH_3}{|}}{C}CH_3$

cyclic
compounds:

(b) $C_5H_{10}O$

Oxygen atoms do not affect the calculation
Units of unsaturation = $[(2 \times 5) + 2 - 10]/2 = (12\text{-}10)/2 = 1$
Some structural formulas are shown below:

aldehydes
and ketones: $CH_3CH_2\underset{\underset{CH_3}{|}}{C}\overset{\overset{O}{||}}{H}$ $CH_3CH_2\overset{\overset{O}{||}}{C}CH_2CH_3$

unsaturated
alcohols: $CH_2 = CHCH_2CH_2CH_2OH$ $CH_3\underset{\underset{OH}{|}}{C}HCH = CHCH_3$

unsaturated
ethers: $CH_3OCH_2CH = CHCH_3$ $CH_2 = \underset{\underset{CH_2OCH_3}{}}{\overset{\overset{CH_3}{|}}{C}}$ $CH_3\underset{\underset{OCH_3}{|}}{C} = CHCH_3$

cyclic alcohols:

cyclic ethers:

ethers with
carbon rings:

5.35   (c) C_5H_8O

Units of unsaturation = $[(2 \times 5) + 2 - 8]/2 = (12-8)/2 = 2$

Some structural formulas are shown below:

unsaturated
aldehydes
and ketones:

$CH_3CH_2CH{=}CHCH$ $CH_2{=}CCH_2CH$ $CH_2{=}CCCH_3$ $CH_2{=}CHCCH_2CH_3$

unsaturated
alcohols
and ethers:

$CH_3C{\equiv}CCHCH_3$ $CH_2{=}CHCH{=}CHCH_2OH$ $CH_2{=}CCH{=}CH_2$

cyclic
compounds:

(d) Alkenes will react with bromine in carbon tetrachloride, therefore, the isomer cannot be an alkene. Cyclic compounds that have rings with four or more carbon atoms will not react easily with bromine in carbon tetrachloride. The ones with the molecular formula C_5H_9Br are:

(e) The hydrogenation data tells us that the original compound was an alkene with the carbon skeleton of 3-methyl-1-butanol.

$CH_3CHCH_2CH_2OH$ with CH_3 branch

The following structures are possible.

$CH_3C{=}CHCH_2OH$ with CH_3 branch $CH_2{=}CCH_2CH_2OH$ with CH_3 branch

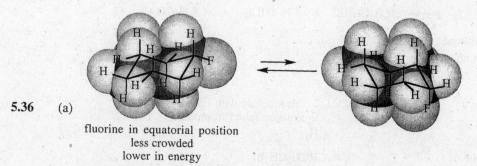

fluorine in equatorial position
less crowded
lower in energy

5.36   (a)

(b) The equilibrium constant, $K_{eq} = \dfrac{[\text{axial}]}{[\text{equatorial}]}$, is less than 1. The conformation with the fluorine in the equatorial position predominates.

5.36   (c)   The energy difference between the two conformations of bromocyclohexane would be expected to be larger than that for the fluoro compound. Bromine is considerably larger than fluorine and would experience even more crowding in the axial position than would fluorine.

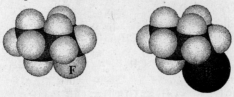

5.37   Ph = ⟨benzene ring⟩— in these answers

(a) $CH_3CH_2NH_2$ +   HCl $\longrightarrow$ $CH_3CH_2NH_3^+$ Cl^-
       base   acid

(b) $CH_3CH_2C\equiv CH$ + $NaNH_2$ $\longrightarrow$ $CH_3CH_2C\equiv C^- Na^+$ + NH_3
       acid   base

(c) $(CH_3CH_2CH_2)_2NH$ + CH_3CH_2Br $\longrightarrow$ $(CH_3CH_2CH_2)_2\overset{+}{N}HCH_2CH_3$ Br^-
     nucleophile   electrophile

(d) $CH_3CH_2CH=CHCH_2CH_3$ +   HBr $\longrightarrow$ $CH_3CH_2CH_2CHBrCH_2CH_3$
      nucleophile   electrophile

(e) $CH_3CH_2CH_2SH$ +   NaOH $\longrightarrow$ $CH_3CH_2CH_2S^- Na^+$ + H_2O
      acid   base

(f) $Ph\overset{+}{N}H_2CH_3$ Cl^- +   NaOH $\longrightarrow$ $PhNHCH_3$ +   NaCl   + H_2O
      acid   base

(g) $CH_3CH_2CH=CHCH_3$ + Br_2 $\longrightarrow$ $CH_3CH_2CHBrCHBrCH_3$
      nucleophile electrophile

(h)   PhSNa   + $CH_3CH_2CH_2Br$ $\longrightarrow$ $PhSCH_2CH_2CH_3$ +   NaBr
  nucleophile   electrophile

(i)   Br—⟨ring⟩—$\overset{\overset{\displaystyle O}{\parallel}}{C}OH$ + $NaHCO_3$ $\longrightarrow$ Br—⟨ring⟩—$\overset{\overset{\displaystyle O}{\parallel}}{C}O^- Na^+$ + H_2O + $CO_2\uparrow$
     acid   base

(j) CH_3CH_3 + Cl_2 $\xrightarrow{h\nu}$ CH_3CH_2Cl + CH_3CHCl_2 + $ClCH_2CH_2Cl$ +   HCl
  1 volume   1 volume

A pattern is not obvious for reaction (j).

                              electrophile with carbon
                              skeleton found in product

5.38   (a)   PhSH $\xrightarrow[\;\;H_2O\;\;]{NaOH}$ $PhS^- Na^+$ $\xrightarrow{CH_3\overset{\displaystyle CH_3}{\overset{\displaystyle |}{C}}HCH_2CH_2Br}$ $PhSCH_2CH_2\overset{\displaystyle CH_3}{\overset{\displaystyle |}{C}}HCH_3$
             lost in product   nucleophile
           therefore need OH^-

nucleophile

5.38   (b) $CH_3CH_2Br \xrightarrow{CH_3NH_2} CH_3CH_2\overset{+}{N}H_2CH_3$ Br^-

this portion of the product comes from the nucleophile;
the rest must be from the electrophile with which it reacts

(c) $CH_2{=\!\!=}CH_2 \xrightarrow{HBr} CH_3CH_2Br \xrightarrow{CH_3O^- Na^+} CH_3CH_2OCH_3$
nucleophile

looks as though CH_3OH has added; CH_3OH
is too weak an acid to add to double bond;
must use two consecutive reactions

(d) $CH_3CH_2CH_2CH_2NH_2 \xrightarrow{HCl} CH_3CH_2CH_2CH_2\overset{+}{N}H_3$ Cl^-
base

conjugate acid of base

(e) $CH_3CH_2C{\equiv}CH \xrightarrow[\text{electrophile}]{2\ Br_2} CH_3CH_2CBr_2CHBr_2$
nucleophile with carbon
skeleton of product;
triple bond will react two
times with Br_2

Br_2 added twice

electrophile with carbon
skeleton found in product

(f) $CH_3CH_2CH_2CH_2OH \xrightarrow{Na^+ NH_2^-} CH_3CH_2CH_2CH_2O^- \xrightarrow{CH_3Br} CH_3CH_2CH_2CH_2OCH_3$
nucleophile

proton lost in product,
therefore need base

(g) $CH_3CHCH_2CH{=\!\!=}CH_2 \xrightarrow[PtO_2]{H_2} CH_3CHCH_2CH_2CH_3$
       | |
       OH   OH

has two more hydrogens than starting
material, therefore reagent is H_2

(h) ⬠ $\xrightarrow[h\nu]{Br_2}$ ⬠—Br

hydrogen substituted by halogen,
therefore need halogen with light

5.39   (a)

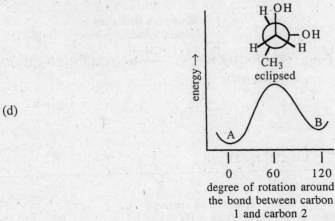

gauche   anti   gauche   anti

1,2-ethanediol   1,2-dimethoxyethane

Only one of the two gauche conformations is shown for each compound. The second gauche conformation can be obtained by rotating the front carbon atom in the anti conformation 120° in a clockwise direction.

(b) Hydrogen bonding can take place in the gauche conformation of 1,2-ethanediol but not in the anti conformation.

(c) The order of stability is A > C > B.

(d)

energy →

H OH
eclipsed

A   B

0   60   120
degree of rotation around
the bond between carbon
1 and carbon 2

5.40   (a)   1-chloro-1-methylcyclohexane

(b)

CH₃

CH₃

Cl

Cl

(c) The conformer in which the methyl group is axial must have larger 1,3-diaxial interactions between the axial group and the two axial hydrogens on the same side of the ring. The methyl group is effectively larger than the chlorine atom.

5.40 (c) (cont)

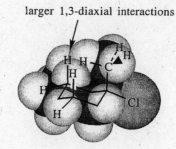

larger 1,3-diaxial interactions

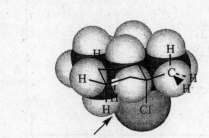

smaller 1,3-diaxial interactions

5.41

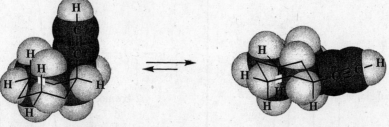

Linear shape of ethynyl group minimizes 1,3-diaxial interactions in axial conformation, therefore there is not as much difference in energy between the two forms.

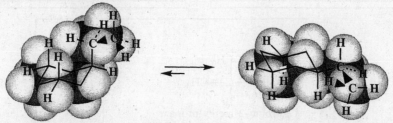

1,3-Diaxial interactions in the axial conformer makes it less stable relative to the equatorial conformer.

5.42   The three other isomers with molecular formula C_4H_9Br are given below:

$CH_3CH_2CH_2CH_2Br$ $CH_3CH_2CHBrCH_3$

$$CH_3 \atop CH_3CBr \atop CH_3$$

1-bromobutane
(*n*-butyl bromide)
four sets of hydrogen atoms
in a ratio of 2:2:2:3
and four different carbon atoms

2-bromobutane
(*sec*-butyl bromide)
four sets of hydrogen atoms
in a ratio of 1:2:3:3
and four different carbon atoms

2-bromo-2-methylpropane
(*tert*-butyl bromide)
one set of hydrogen atoms
and two sets of carbon atoms

The rough proton spectra are:

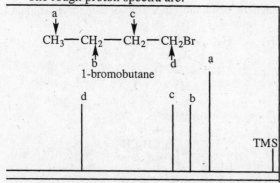

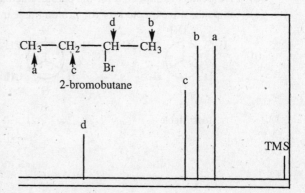

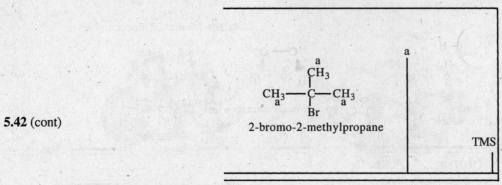

5.42 (cont)

The rough carbon spectra are:

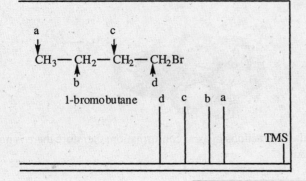

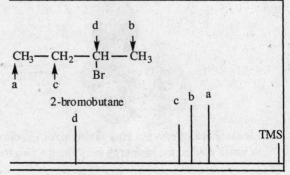

5.43   (a)   CH₃CH₂ — C (C) — OH

(b) Five peaks are expected for the carbon nuclear magnetic resonance spectrum of 2-methyl-1-butanol.

(c) Six peaks are expected for the proton nuclear magnetic resonance spectrum of 2-methyl-1-butanol.

(d)

A
least stable

B

C
most stable

perspective formula corresponding to C

5.43   (e)

D
highest energy
eclipsed conformation

perspective formula corresponding to D

Supplemental Problems

S5.1   Name the following compounds.

(a) $ClCCH_2CH_2CH_3$ (with Cl above and Cl below the first C)

(b) $CH_3CH_2CHCH_2CH_2CH_2CH_2CH_2CH_3$ (with I below the CH)

(c) $-CH_2CH_2CH_2-$

(d) $-CHCH_2CHCH_3$ (with CH_3 above and Cl below)

(e)   ▷$-Br$

(f)

(g) $CH_3CH_2CHCH_2CH_2CHCHCH_3$ (with CH_3CCH_3 / CH_3 group, and Cl Cl below)

(h) $-CH_2CH_2CHCH_3$ (with CH_3 above)

(i) $-OH$

S5.2   Draw structural formulas for the following compounds
(a) 4-*tert*-butyl-3-ethyloctane   (b) 1,2-ethanediol   (c) 4-bromo-3,3-dimethylheptane
(d) 2-bromo-3-ethylhexane   (e) 2-phenyl-1-ethanol   (f) chlorocyclopentane   (g) 2-cyclopentyloctane
(h) 1,3-diphenylpropane   (i) 1-chloro-3-methyl-1-phenylbutane

S5.3   For each molecular formula, draw structural formulas of all possible constitutional isomers. Name each one.
(a) $C_5H_{12}O$ (Ethers may be named for the two alkyl groups they contain: for example, $CH_3OCH_2CH_3$ is methyl ethyl ether.)
(b) $C_3H_5Cl_3$

S5.4   (a) A compound with the molecular formula C_7H_{12} has two units of unsaturation. Such a compound may contain one triple bond, two double bonds, a ring and a double bond, or two rings. For each of the four categories given as a possibility for the molecular formula C_7H_{12}, draw one structural formula.
(b) Adding hydrogen to the compound with the molecular formula C_7H_{12} gives a C_7H_{16} compound, which is identified as being 2-methylhexane. In light of this evidence, what structures are possible for the C_7H_{12} compound?

S5.5   The conformations of 1,1,1,4-tetrachlorobutane can be clearly depicted using Newman projections.
(a) Draw the structure of 1,1,1,4-tetrachlorobutane in its most stable conformation (conformation A) using a perspective formula and a Newman projection looking down the bond between carbon 2 and carbon 3.
(b) Show a gauche (conformation C) and one of the two more stable eclipsed forms (conformation B) of this molecule as Newman projections.
(c) Draw a potential energy diagram showing the relative energies of conformations A, B, and C, as one is transformed into the other. Label the "hills" and "valleys" with letters indicating the conformations that correspond to them.

S5.6   The ratios of different chair conformations at equilibrium for a large number of monosubstituted cyclohexanes are known. Examples of three of these are given below.

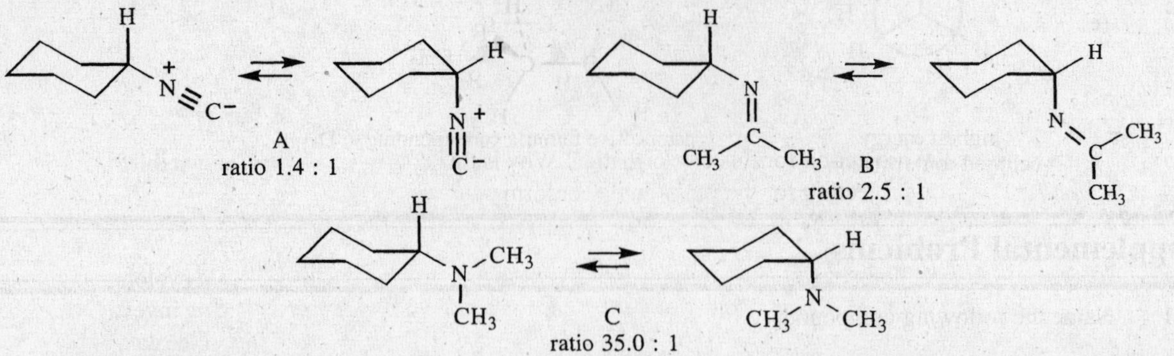

(a)   Which of the three equilibria (A, B, or C) will have the largest free energy difference ($\Delta G°$) between the two conformations?

(b)   What is responsible for the difference in observed ratios at equilibrium *(be very specific about structural features)?*

(c)   What is the value of the equilibrium constant, K_{eq}, for the equilibrium you selected in part (a)? Your answer may be expressed as a fraction.

(d)   When the compound in example C is treated with methyl iodide, it is converted to a new compound. Write an equation for this reaction. Show the two chair conformations for this new constant. Will the equilibrium constant, $K_{eq} = \dfrac{\text{[axial]}}{\text{[equatorial]}}$, for the conformational change for this new compound be smaller or larger than it was for example C? Explain your decision.

S5.7   2-Amino-3,3-difluoropentanedioic acid has been synthesized at the University of Michigan for research into the toxicity of compounds used in cancer chemotherapy. A perspective formula of the compound is shown below.

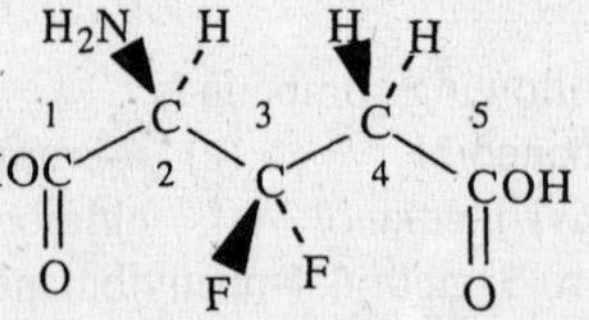

(a)   Draw a Newman projection of the conformation of this compound by looking along the bond between carbon 2 and carbon 3.

(b)   Rotate the front carbon atom by 120° of the Newman projection that you drew in part (a), and redraw the *perspective formula* corresponding to this new conformer of the compound.

S5.8   Experimental measurements of dipole moments have shown that for 1,2-dichloroethane in carbon tetrachloride solution at 25 °C, 70% of the molecules are in the anti and 30% in the gauche conformation. For 1,2-dibromoethane under the same conditions, 89% of the molecules have the anti and 11% the gauche conformation. Offer a rationalization for the differences in the percentages of anti and gauche conformers for the 1,2-dichloro- and 1,2-dibromoethanes. Pictures as well as words will be helpful.

S5.9   For the following molecule, the equilibrium constant, $\dfrac{\text{[equatorial]}}{\text{[axial]}}$, is less than 1. Draw an accurate picture of the more stable chair form and briefly state the reason for its stability.

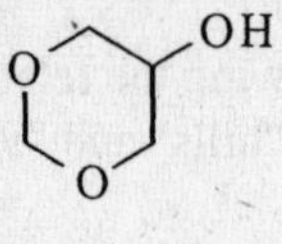

S5.10   If the oxygen atom in fluoromethanol is assumed to be sp^3-hybridized, the highest energy conformation for the compound when viewed along the O—C bond is shown below.

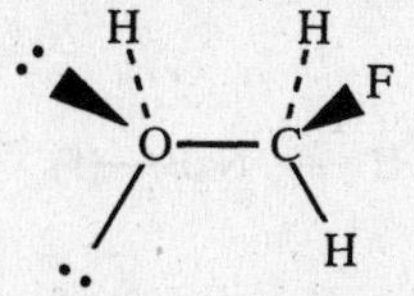

(a)   Draw a Newman projection for this conformation. Why is this conformation the least stable?

(b)   Draw a Newman projection for the most stable conformation.

S5.11   Researchers recently investigated the effect of different substituents at the 2-position of cyclohexanones on the equilibrium between the two chair conformers of the compounds. The equilibrium being investigated is shown below, with **X** representing the substituent at the 2-position on cyclohexanone. Some of the data obtained in this research are given in the table to the right of the equilibrium.

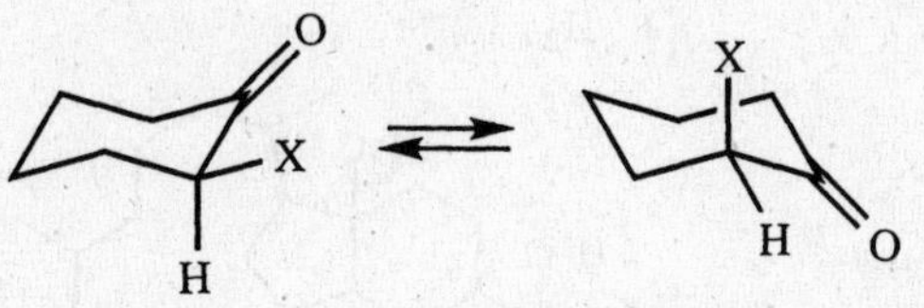

X	% of axial form at equilibrium
F	17
Cl	45
Br	71
I	88
CH$_3$O	28
CH$_3$S	85

(a)   Consider the equilibrium for 2-fluorocyclohexane. Will the equilibrium constant, $\frac{[\text{equatorial}]}{[\text{axial}]}$, for the conformational change shown above be greater than, equal to, or less than 1?

(b)   Draw an energy diagram showing the relative energies of the axial and the equatorial forms of 2-bromocyclohexanone.

(c)   Which 2-substituted cyclohexanone has the largest difference in free energy (ΔG) between its equatorial and axial forms?

(d)   According to the data given in the table, what factor or factors appear to be important in determining whether a 2-substituted cyclohexanone exists primarily in the equatorial or the axial form?

S5.12   The following represent chemical transformations showing pathways from starting materials to products. All of these equations use the types of reactions that you have learned involving electrophilic and nucleophilic species. Fill in the structural formulas for the starting materials, reagents, or products represented by the letters below, using changes in molecular connectivity as clues.

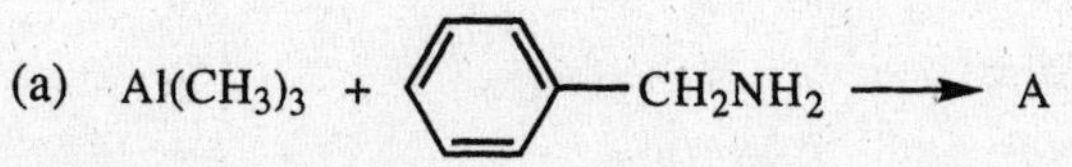

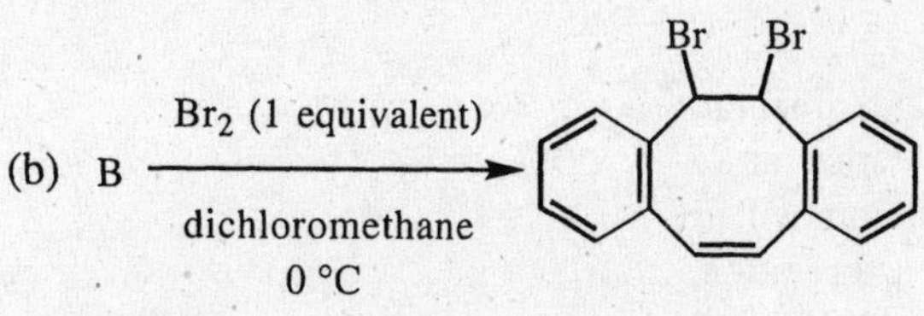

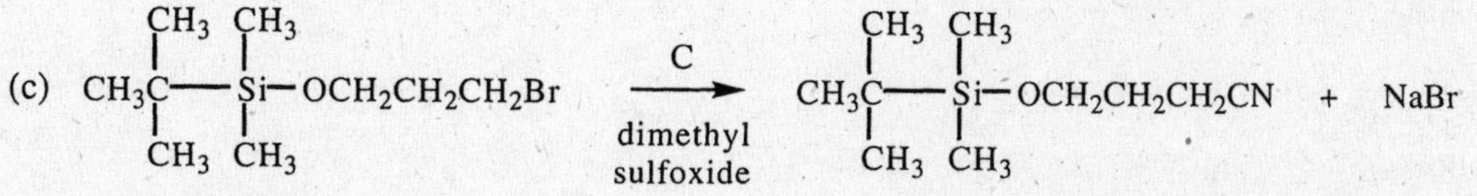

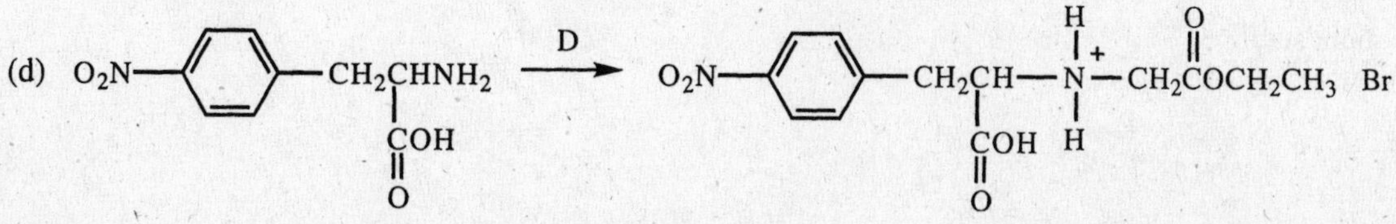

(a) $Al(CH_3)_3$ + [benzene]—CH$_2$NH$_2$ ⟶ A

(b) B $\xrightarrow[\substack{\text{dichloromethane} \\ 0\,°C}]{Br_2 \text{ (1 equivalent)}}$ [product]

(c) [silyl]—OCH$_2$CH$_2$CH$_2$Br $\xrightarrow[\substack{\text{dimethyl} \\ \text{sulfoxide}}]{C}$ [silyl]—OCH$_2$CH$_2$CH$_2$CN + NaBr

(d) O$_2$N—[benzene]—CH$_2$CHNH$_2$ $\xrightarrow{D}$ O$_2$N—[benzene]—CH$_2$CH—N$^+$—CH$_2$COCH$_2$CH$_3$ Br$^-$

S5.12 (cont)

(e) + Na$^+$ $^-$SCH$_2$CH$_3$ ⟶ E + NaBr

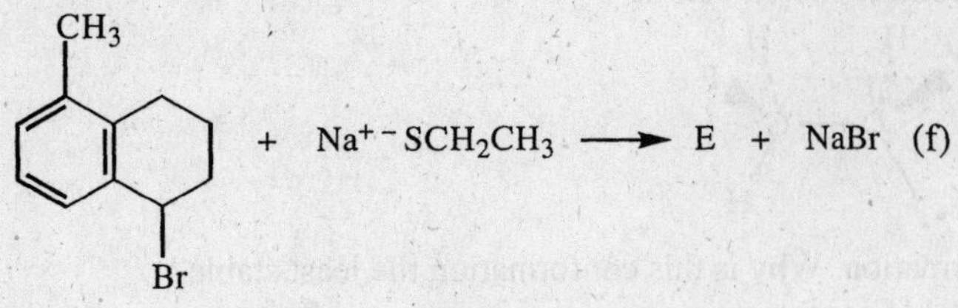

(f)

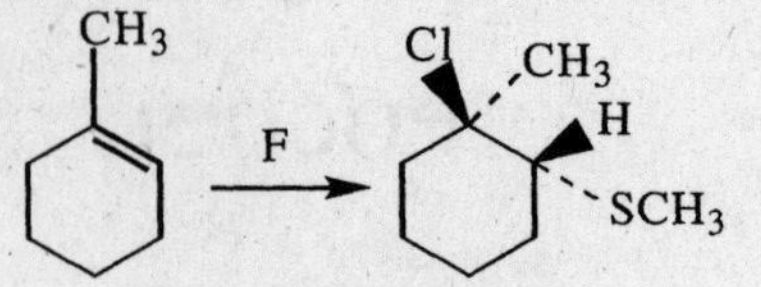

(Hint: a single reagent does this in one step. What must it be?)

(g)

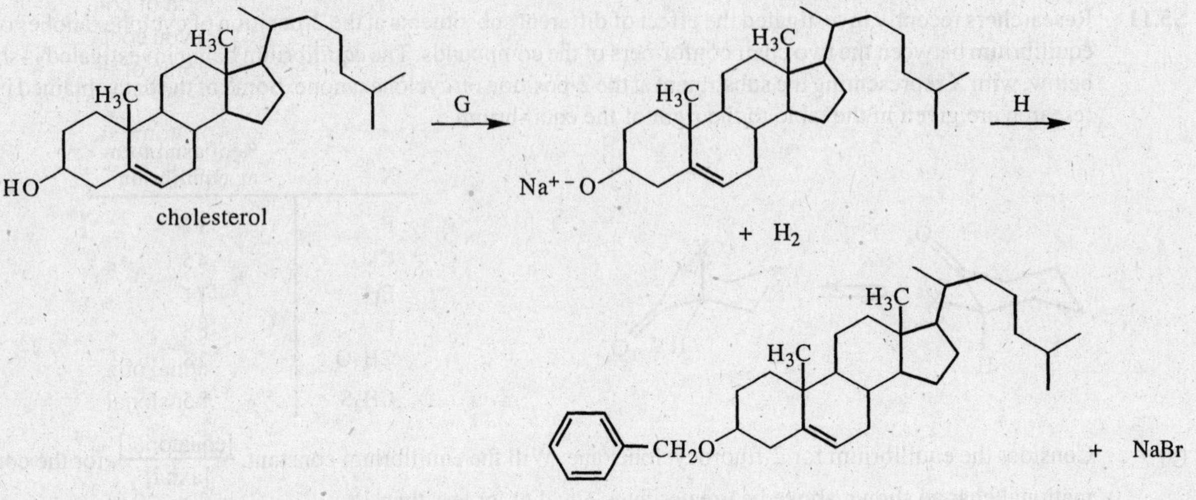

cholesterol

$\xrightarrow{\text{G}}$

+ H$_2$

$\xrightarrow{\text{H}}$

+ NaBr

6 Stereochemistry

Workbook Exercises

Before you begin to learn about some of the three-dimensional relationships that can exist in molecules, you should realize that these relationships are not unique to molecules, but also exist in the macroscopic world of common objects. To get the most benefit from the following exercises, you will need the two cardboard tetrahedra you constructed for the workbook exercises in Chapter 2. Use your model set to construct two tetrahedra with four atoms of different colors attached to each center.

In earlier chapters you learned the symbolism used to represent a tetrahedral arrangement of atoms around a central atom, for example, methane, CH_4.

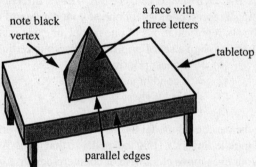

or

among many others

These are two-dimensional representations, called perspective drawings, of a three-dimensional object. Your study of stereochemistry in Chapter 6 will be easier if you become comfortable relating these two-dimensional pictures to the three-dimensional structures they represent.

EXERCISE I.

(a) Place one of the tetrahedra on the table so that the face with only letters on it is toward you, and one of the edges is parallel to an edge of the table.

note black vertex

a face with three letters

tabletop

parallel edges

(b) The overall arrangement of the three letters on the face of the cardboard tetrahedron pointed towards you should be

△ or ⅄ , if you use lines to point to each vertex from the center.

Add the letters A, B, and C
as they appear to you on the   △   ⅄
face of the tetrahedron:

(c) You can also use lines, dashes, and wedges to draw the arrangement of the four vertices of your cardboard model. If you imagine a point at the center of the cardboard model and lines originating from there to each of the four vertices, the picture would look as follows:

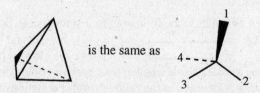

is the same as

111

Workbook Exercises (cont)

The face defined by the numbers 1-2-3,  , is still directed toward you.

Draw a representation of the cardboard model you have on the table using lines, dashes, and wedges, and add the A, B, C, and ● labels as they appear to you on the model.

(d) If you let your eyes travel from number 1 to number 2 to number 3 on the front face of the tetrahedron drawn in part (c), your eyes will be moving the way the hands of a clock do, or "clockwise."

counterclockwise (clock face) clockwise

Look at your own cardboard model and decide which direction your eyes must travel in going from A to B to C.

(e) Place your other cardboard model on the table with one of its edges parallel to the tabletop and with its A-B-C face toward you. How do your eyes travel in going from A to B to C? Draw the figure using the symbolism of lines, dashes, and wedges to indicate three dimensionality.

(f) Place a molecular model of a tetrahedron with four different attached atoms on the table so that two of the attached atoms line up with the edge of the tabletop in a way similar to the way the front edge of your cardboard model is situated on the table. The four atoms on the molecular model should correspond in space to the vertices of your cardboard model. Draw the molecular model using three-dimensional symbolism.

(g) Select one of your cardboard models. There are five other edges on the cardboard model besides the one parallel to the edge of the table. Reposition the model so that each edge, in turn, is parallel to the edge of the table. Draw a three-dimensional representation for each one. Note that all six pictures you will have drawn for the model represent the same object. For those with an A-B-C face toward you, is the A→B→C direction clockwise or counterclockwise?

(h) You now have six three-dimensional representations for one of your cardboard models and one representation for the second model. Are any of the six representations of one model an exact match for the second model?

EXERCISE II.   We will now arbitrarily assign a set of priorities to the four vertices on your cardboard models:

$$(1) = \text{highest priority} = B$$
$$(2) = \text{second highest} = A$$
$$(3) = \text{third highest} = ●$$
$$(4) = \text{lowest priority} = C$$

(a) With one of your hands, grasp the cardboard model by the lowest priority vertex (C). (Lightly pinch the vertex between your thumb, index and middle fingers.) Hold the model so that the B-A-● face that is opposite this vertex is facing you. On this face, do your eyes move clockwise or counterclockwise in looking from (1) to (2) to (3) (or B→A→ ●)? Using the same set of priorities, answer the same question for your other cardboard model. Now assign a set of priorities to the four colored atoms on one of your molecular models. Assign a clockwise or counterclockwise direction in going from (1) to (2) to (3) by grasping the group with priority (4) and holding the model so that the 1-2-3 face is pointing toward you.

(b) Using the same molecular model from part (a), disconnect and reconnect any two of the atoms, exchanging their positions. Keeping the same priorities that you assigned in part (a), decide whether your eyes move clockwise or counterclockwise when you look from (1) to (2) to (3) when atom (4) is held away from you. Exchange any two other atoms and examine the (1) to (2) to (3) direction again. Repeat this exercise once more. What generalization can you make about the number of different arrangements possible for four different atoms in a tetrahedral arrangement?

EXERCISE III.   When you hold the molecular model so that atom (4) is towards you, in which direction do your eyes move looking from (1) to (2) to (3)?

EXERCISE IV.   Arbitrarily assign priorities to the following four symbols: P, β, Σ, and q.

$$(1) = \text{highest priority} = \underline{\hspace{2cm}} \qquad (2) = \text{second highest} = \underline{\hspace{2cm}}$$
$$(3) = \text{third highest} = \underline{\hspace{2cm}} \qquad (4) = \text{lowest priority} = \underline{\hspace{2cm}}$$

Using your molecular models as a guide, decide whether your eyes move clockwise or counterclockwise going from (1) to (2) to (3) when the priority (4) symbol in the following representations is held away from you.

Workbook Exercises (cont)

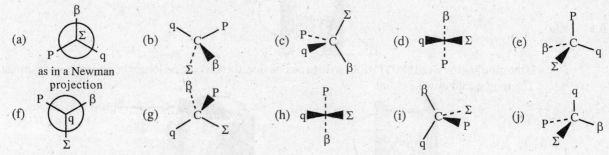

Which of the representations above correspond to the same arrangement in space?

- -

Concept Map 6.1   Stereochemical relationships.

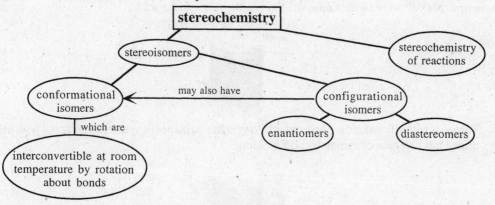

- -

Concept Map 6.2   Chirality.

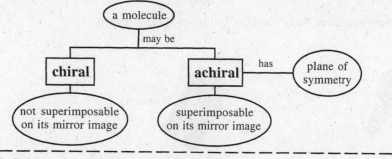

- -

Concept Map 6.3   Definition of enantiomers.

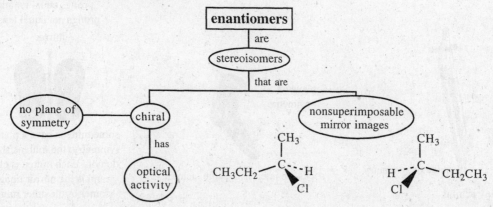

6.1   (a)

If the mug had the word "MOTHER" written on one side, there would no longer be any planes of symmetry. The mug would now be chiral.

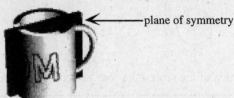plane of symmetry

If the mug had the word "MOM" instead of "MOTHER" written on one side, it would still be chiral. If the word "MOM" were written opposite the handle, it would be achiral.

(b)

Thread is helically wound around a spool. A helix has no planes of symmetry. A spool with helically wound thread has no plane of symmetry and is chiral.

(c)

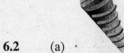 ← plane of symmetry

An empty spool has many planes of symmetry. One is shown.

6.2   (a)

helical thread; chiral

(b)   plane of symmetry

achiral

(c)

prongs equal length; achiral
prongs not equal length; chiral

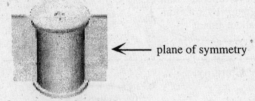

mirror

(d)

plane of symmetry

achiral

(e)   plane of symmetry

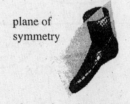

socks are interchangeable; achiral

(f)

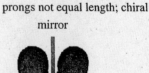

some mittens have a plane of symmetry; the mittens shown do not. Each mitten is chiral and is the mirror image isomer of the other mitten.

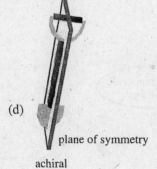

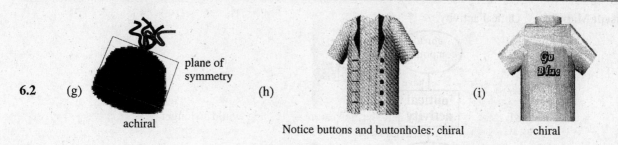

6.2 (g)   plane of
                     symmetry

         achiral

(h)   Notice buttons and buttonholes; chiral

(i)   chiral

6.3 (a)   No

The molecule has a plane of symmetry.
The atoms that are behind the plane
are shown as gray letters

(b)   Yes

(c)   Yes

(d)   No

The molecule has a plane of symmetry.
[See part (a)]

(e)   Yes

(f)   Yes

Concept Map 6.4   Optical activity.

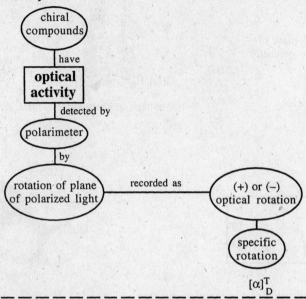

6.4 $$\frac{\text{optical rotation of unknown solution}}{\text{optical rotation of known concentration}} = \frac{+13.3°}{+53.2°} = \frac{1}{4}$$

therefore the concentration of the unknown solution = $\frac{1}{4}$ × the concentration of the original solution

$$= \frac{1}{4} \times (20 \text{ g}/100 \text{ mL}) = 5.0 \text{ g}/100 \text{ mL} = 0.05 \text{ g/mL}$$

6.5 $[\alpha]_{D}^{20°} = \dfrac{+2.46°}{(1.0 \text{ dm})(5 \text{ g}/100 \text{ mL})} = +49.2°$ (c 0.05, ethanol)

6.6   The equation for specific rotation can be rearranged.

$$\alpha = [\alpha]_{D}^{20°} \, lc$$

The actual rotation, α, is directly proportional to the concentration. If the solution is diluted, for example, to twice the original volume, the concentration, and therefore the optical rotation, will be cut in half. Thus if the original specific rotation is +90°, the actual rotation for the diluted solution will be 45° in a clockwise direction. If the specific rotation is –270°, the actual rotation will be –135°, or 225° in a clockwise direction. In general, decreasing the concentration will cause a positive rotation to decrease in the clockwise direction and a negative rotation to increase in the clockwise direction.

6.7   (a)   enantiomeric excess = $\dfrac{\text{actual rotation}}{\text{rotation of pure enantiomer}} \times 100$

enantiomeric excess = $\dfrac{-14.70°}{-39.5°} \times 100 = 37.2\%$

(b)

(S)-(–)-2-bromobutanoic acid   (R)-(+)-2-bromobutanoic acid
          68.6%   31.4%

The mixture containing a 37.2% enantiomeric excess of the levorotatory compound is 100 – 37.2 or 62.8% racemic. The racemic portion of the mixture contains equal amounts (62.8/2 = 31.4%) of each enantiomer. Therefore the solution contains 31.4% of the dextrorotatory and (31.4 + 37.2)% or 68.6% of the levorotatory isomer.

Concept Map 6.5   Formation of racemic mixtures in chemical reactions.

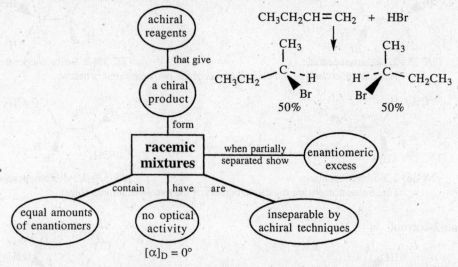

- -

6.8 (a) (*S*)-3-hexanol   (b) (*S*)-2-bromopentane   (c) (*R*)-3-iodo-3-methylhexane

(d) (*R*)-1-bromo-1-chloropropane   (e) (*R*)-1,2-dichlorobutane   (f) (*S*)-4-chloro-2-methyl-1-butanol

(g) (*S*)-1-chloro-3,4-dimethylpentane   (h) (*R*)-4-bromo-2-butanol

6.9 (a)   (b)   (c)

(d)   (e)   (f)

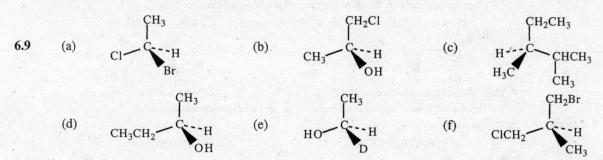

6.10 (a)

(*R*)-3-methylhexane   (*S*)-3-methylhexane

These are enantiomers

(b)

(2*R*, 4*R*)-2-bromo-4-methylhexane   (2*S*, 4*S*)-2-bromo-4-methylhexane
These enantiomers are the diastereomers of the two enantiomers below.

(2*R*, 4*S*)-2-bromo-4-methylhexane   (2*S*, 4*R*)-2-bromo-4-methylhexane
These enantiomers are the diastereomers of the two enantiomers above.

6.10   (c)

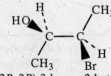

(2R, 3R)-2,3-dibromopentane   (2S, 3S)-2,3-dibromopentane

These enantiomers are the diastereomers of the two enantiomers below.

(2R, 3S)-2,3-dibromopentane   (2S, 3R)-2,3-dibromopentane

These enantiomers are the diastereomers of the two enantiomers above.

6.11   3-bromo-2-butanol

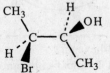

(2R, 3R)-3-bromo-2-butanol   (2S, 3S)-3-bromo-2-butanol

These enantiomers are the diastereomers of the two enantiomers below.

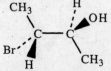

(2R, 3S)-3-bromo-2-butanol   (2S, 3R)-3-bromo-2-butanol

These enantiomers are the diastereomers of the two enantiomers above.

2-chloro-3-methylheptane

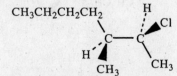

(2R, 3R)-2-chloro-3-methylheptane   (2S, 3S)-2-chloro-3-methylheptane

These enantiomers are the diastereomers of the two enantiomers below.

(2R, 3S)-2-chloro-3-methylheptane   (2S, 3R)-2-chloro-3-methylheptane

These enantiomers are the diastereomers of the two enantiomers above.

6.12   2,3-butanediol

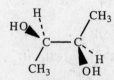

plane of symmetry

(2R, 3R)-   (2S, 3S)-   (2R, 3S)-
2,3-butanediol   2,3-butanediol   2,3-butanediol
                                                                          meso compound

These enantiomers are the diastereomers of the meso compound.

6.12 (cont)

2,3-pentanediol

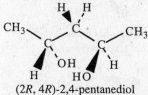

(2*R*, 3*R*)-2,3-pentanediol   (2*S*, 3*S*)-2,3-pentanediol

These enantiomers are the diastereomers of the two enantiomers below.

(2*R*, 3*S*)-2,3-pentanediol   (2*S*, 3*R*)-2,3-pentanediol

These enantiomers are the diastereomers of the two enantiomers above.

2,4-pentanediol

(2*R*, 4*R*)-2,4-pentanediol   (2*S*, 4*S*)-2,4-pentanediol   (2*R*, 4*S*)-2,4-pentanediol
These enantiomers are the diastereomers of the meso compound.   meso compound

Concept Map 6.6   Configurational isomers.

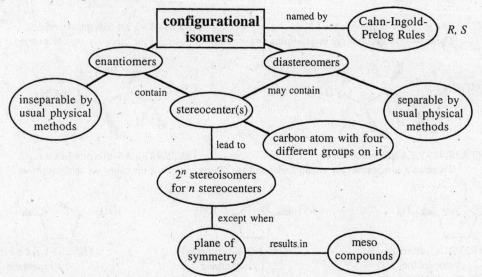

6.13

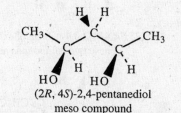

Note that the lowest priority group on carbon 3, H, is on a bond coming out of the plane toward you, thus the clockwise rotation you see is actually counterclockwise, and the designation is *S*, not *R*.

6.14   A compound with *n* stereocenters can have a maximum of 2^n stereoisomers. 2,3,4-Tribromohexane has 3 stereocenters and thus a maximum of 8 stereoisomers. The number of different isomers will be reduced if there are meso compounds. The easiest way to draw a set of stereoisomers is to start with one in which all stereocenters are the same and first reverse the designation of all the stereocenters to get the mirror image enantiomer. Then change one stereocenter at a time to generate the rest of the stereoisomers.

(2R,3R,4R)-2,3,4-tribromohexane   (2S,3S,4S)-2,3,4-tribromohexane

These two compounds are enantiomers and are diastereomers of the other six stereoisomers.

(2S,3R,4R)-2,3,4-tribromohexane   (2R,3S,4S)-2,3,4-tribromohexane

These two compounds are enantiomers and are diastereomers of the other six stereoisomers.

(2R,3S,4R)-2,3,4-tribromohexane   (2S,3R,4S)-2,3,4-tribromohexane

These two compounds are enantiomers and are diastereomers of the other six stereoisomers.

(2R,3R,4S)-2,3,4-tribromohexane   (2S,3S,4R)-2,3,4-tribromohexane

These two compounds are enantiomers and are diastereomers of the other six stereoisomers.

6.15

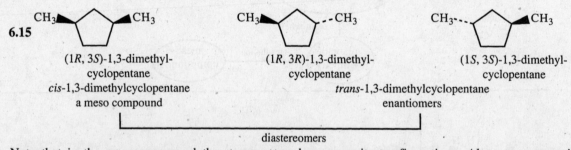

(1R, 3S)-1,3-dimethyl-   (1R, 3R)-1,3-dimethyl-   (1S, 3S)-1,3-dimethyl-
cyclopentane   cyclopentane   cyclopentane

cis-1,3-dimethylcyclopentane   *trans*-1,3-dimethylcyclopentane
a meso compound   enantiomers

diastereomers

Note that in the meso compound the stereocenters have opposite configurations. Also, once an assignment of configuration is carefully made for one of the stereoisomers, the configurations of stereocenters can be assigned by looking to see if they are the same, or opposite, in configuration. For example, once the stereocenters in *cis*-1,3-dimethylcyclopentane have been assigned (see explanation that follows), the other two stereoisomers are easy. The first trans isomer has the same configuration (methyl up) at carbon 1 and the opposite configuration (methyl down) at carbon 3 as the cis isomer. Therefore, it is the (1R, 3R) isomer.

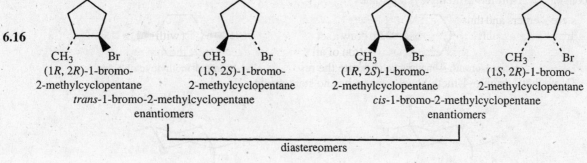

6.16

(1R, 2R)-1-bromo-
2-methylcyclopentane

(1S, 2S)-1-bromo-
2-methylcyclopentane

trans-1-bromo-2-methylcyclopentane
enantiomers

(1R, 2S)-1-bromo-
2-methylcyclopentane

(1S, 2R)-1-bromo-
2-methylcyclopentane

cis-1-bromo-2-methylcyclopentane
enantiomers

diastereomers

6.17   (a)

(b)

enantiomer of kainic acid

one diastereomer of kainic acid

6.18   1-bromo-2-methylcyclohexane

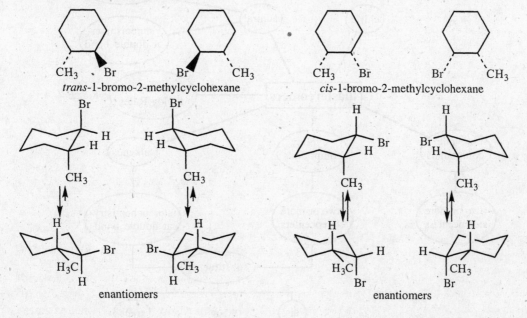

trans-1-bromo-2-methylcyclohexane

cis-1-bromo-2-methylcyclohexane

enantiomers

enantiomers

6.18 (cont)   1-bromo-4-methylcyclohexane

CH₃----⟨ ⟩◀Br   CH₃◀⟨ ⟩◀Br

trans-1-bromo-4-methylcyclohexane   *cis*-1-bromo-4-methylcyclohexane

Note that 1-bromo-4-methylcyclohexane has no stereocenters

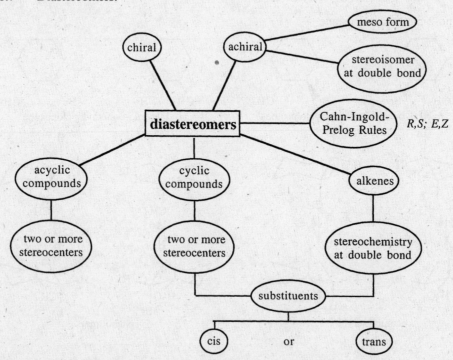

6.19   (a)   (b)

Concept Map 6.7   Diastereomers.

chiral   achiral ── meso form
                         ── stereoisomer
                            at double bond

diastereomers ── Cahn-Ingold-Prelog Rules   *R,S; E,Z*

acyclic compounds   cyclic compounds   alkenes

two or more stereocenters   two or more stereocenters   stereochemistry at double bond

substituents

cis   or   trans

6.20   (a)

CH₃CH₂   CH₃
      \   /
       C == C
      /   \
    H   CH₂CH₃

(*E*)-3-methyl-3-hexene

(b)

Br   CH₂Br
  \   /
   C == C
  /   \
Cl   CH₃

(*Z*)-1,3-dibromo-1-chloro-
2- methylpropene

(c)

                    O
                    ‖
ClCH₂CH₂   COH
        \   /
         C == C
        /   \
    CH₃CH₂   H

(Z)-5-chloro-3-ethyl-
2-pentenoic acid

(d)

                         H
                         |
                         C
                        ‖
                        C
CH₃CH₂   /
        \   /
         C == C
        /   \
       H   CH₂CH₃

(Z)-3-ethyl-3-hexen-1-yne

- -

Concept Map 6.8   The process of resolution.

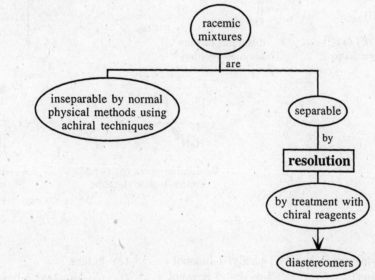

racemic mixtures

are

inseparable by normal physical methods using achiral techniques

separable

by

resolution

by treatment with chiral reagents

diastereomers

- -

6.21

CH₃
 |
Ph—C⋯H   •
 |
 ⁺NH₃

HO   OH
   \ |
    C   C
   /   / \\
⁻O—C   H   COH
    ‖
    O

water-soluble salt

→ NaOH / H₂O →

CH₃
 |
Ph—C⋯H   +
 |
 NH₂

water-insoluble amine

HO   OH
   \ |
    C   C
   /   / \\
Na⁺⁻O—C H   CO⁻Na⁺
      ‖   ‖
      O   O

water-soluble salt

6.22

      CH₃
       |
Ph —— C —— H
       |
      NH₂

(*R*)-(+)-1-amino-1-phenylethane

+

      CH₃
       |
  H —— C —— Ph
       |
     H₂N

(*S*)-(−)-1-amino-1-phenylethane

+

HO   OH
   \ |
    C   C
   / \\   / \\
HOC H   H   COH
 ‖
 O

(2*R*,3*R*)-(+)-tartaric acid

↓ methanol

6.22 (cont)

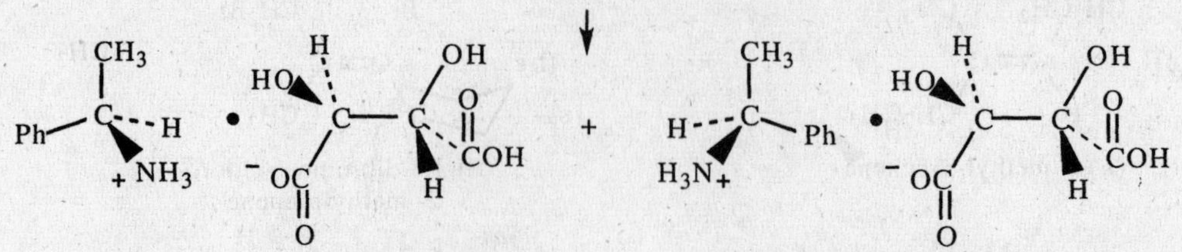

salt of (R)-(+)-1-amino-1-phenylethane
with (2R,3R)-(+)-tartaric acid; more
soluble in methanol; stays in solution

+

salt of (S)-(−)-1-amino-1-phenylethane
with (2R,3R)-(+)-tartaric acid; less
soluble in methanol; crystallizes out

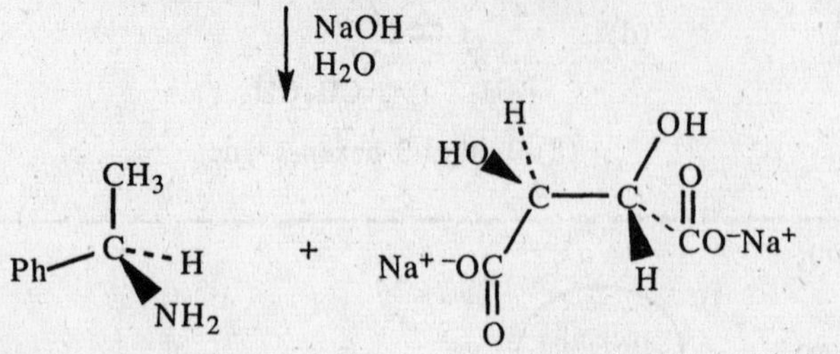

NaOH
H₂O

distillation gives (R)-(+)-1-
amino-1-phenylethane

disodium tartrate
stays in solution

NaOH
H₂O

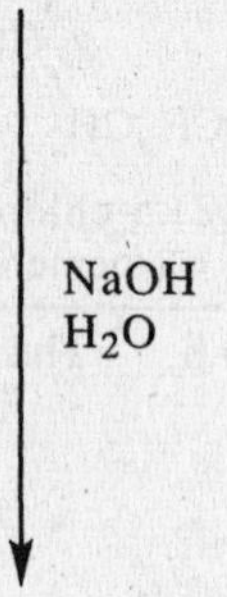

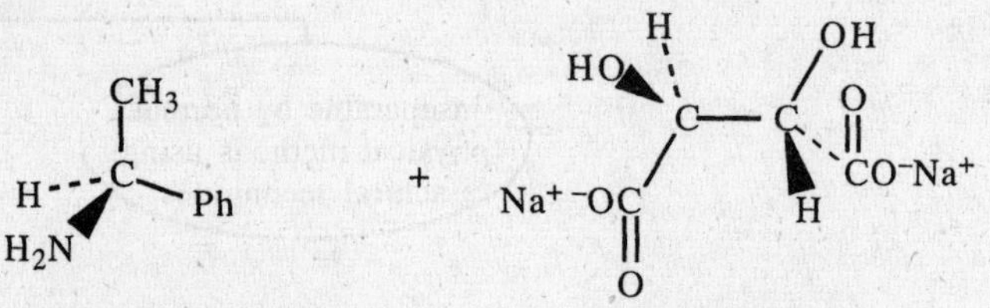

distillation gives (S)-(−)-1-
amino-1-phenylethane

disodium tartrate
stays in solution

6.23

(a) (S)-3-bromo-2-methylpentane   (b) (S)-1-phenyl-1-butanol   (c) butane
(d) (S)-1,2-dibromopentane   (e) (R)-3-methyl-3-hexanol   (f) (S)-3-methyl-3-hexanol
(g) (2R, 3S)-2,3-dibromobutane (*meso*-2,3-dibromobutane)   (h) (1S, 2S, 4S)-1,2-dibromo-4-methylcyclohexane

The assignment of configuration for (g) is easiest if the chair form is converted to the planar form using wedges and dashed
lines to indicate the stereochemistry of the substituents. Assignment of configuration at carbon-2, for example, is shown
below.

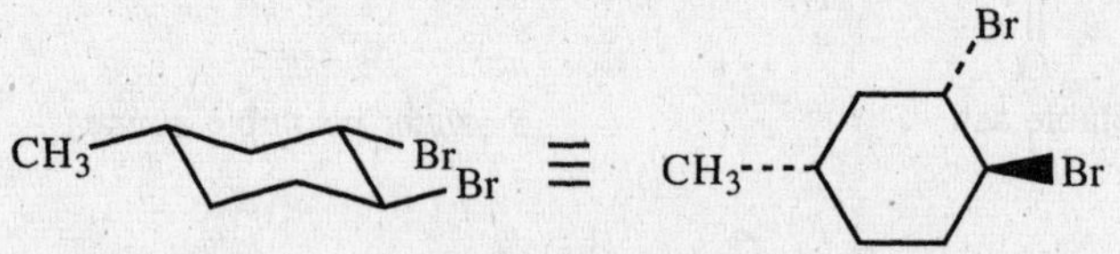

(1S, 2S, 4S)-1,2-dibromo-4-methylcyclohexane

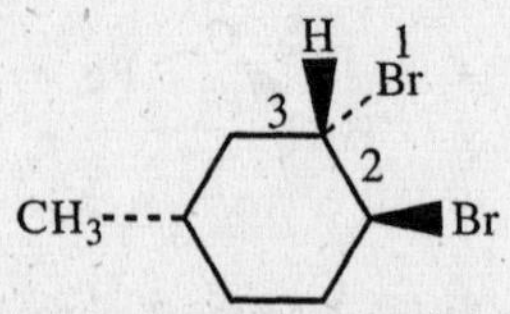

The bromine is pointing down, therefore the group of lowest priority, the hydrogen, is pointing up. Our eyes travel
clockwise from priority 1 to priority 2 to priority 3, therefore the stereocenter *looks R*, but is really *S* because we are
looking at it from the "wrong face" of the molecule.

6.24 b, c, e, h, and i have stereocenters.

(b)

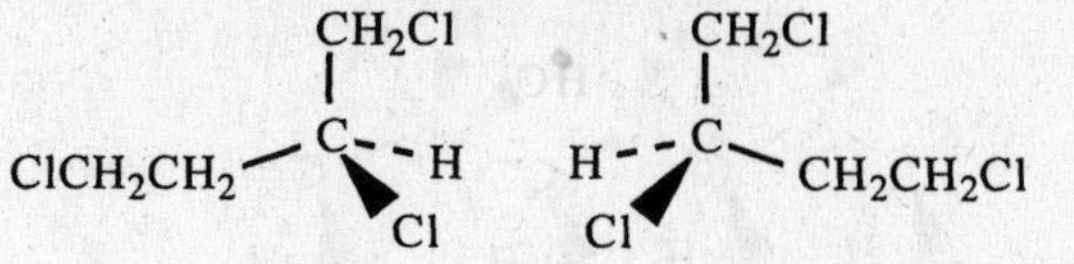

(c)

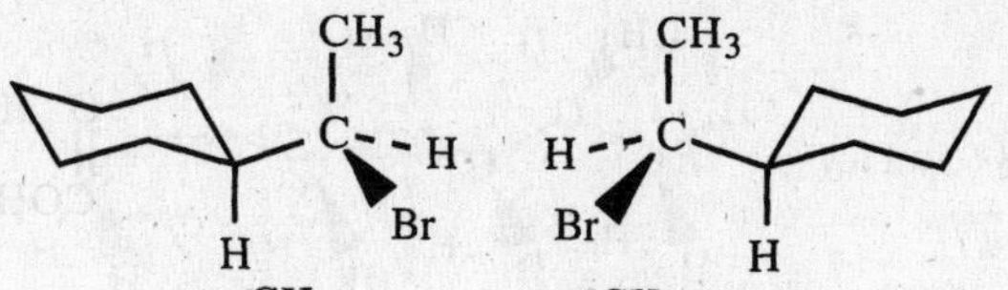

(e)

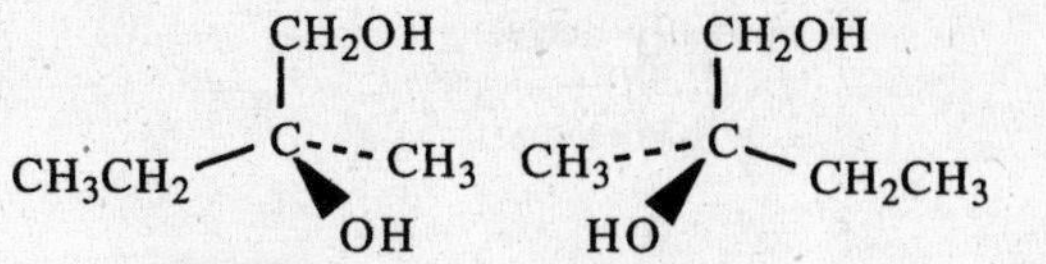

(h)

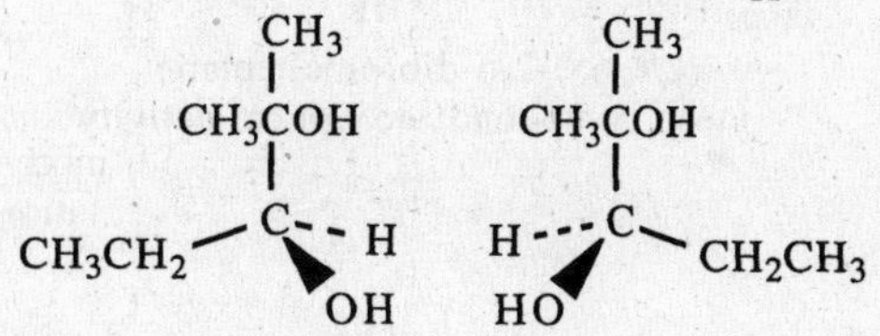

(i)

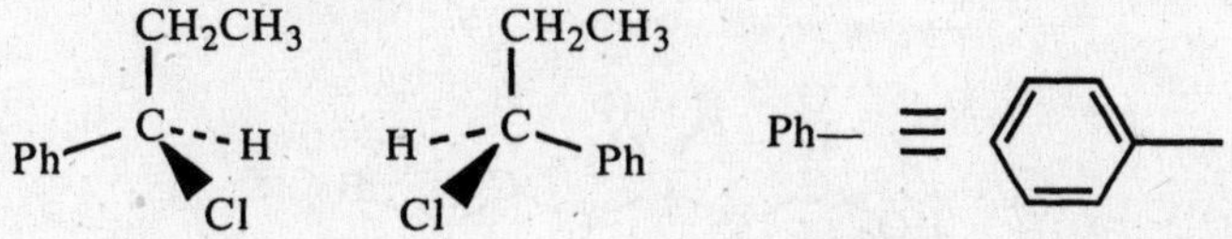

$Ph— \equiv$

6.25

(a) enantiomers — The molecule on the left has the S configuration, the one on the right, the R configuration.

(b) identical   (c) identical   (d) identical

(e) enantiomers — The molecule on the left has the R configuration; the other, the S configuration. Note also that the molecule on the right is not the mirror image of the molecule on the left but is a conformer of the mirror image.

(f) constitutional isomers — (R)-3-bromo-1-chlorobutane and $(2R,3S)$-2-bromo-3-chlorobutane

(g) identical — Both have the R configuration.

(h) conformers — The methyl group on carbon 2 is anti to the hydroxyl group on the left-hand molecule and gauche on the right-hand molecule.

(i) constitutional isomers — (R)-2-bromo-1-chlorobutane and (S)-3-bromo-1-chlorobutane

(j) conformers — The molecule on the left has an equatorial chlorine; the other has an axial chlorine.

6.26 (a)

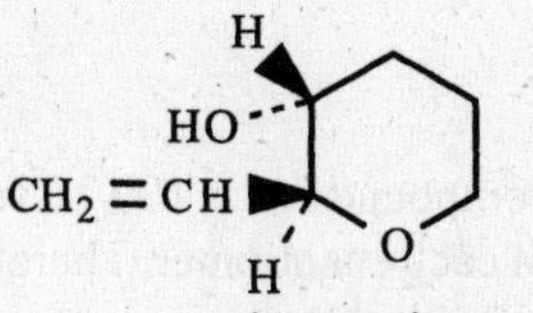

or the enantiomer

(b)

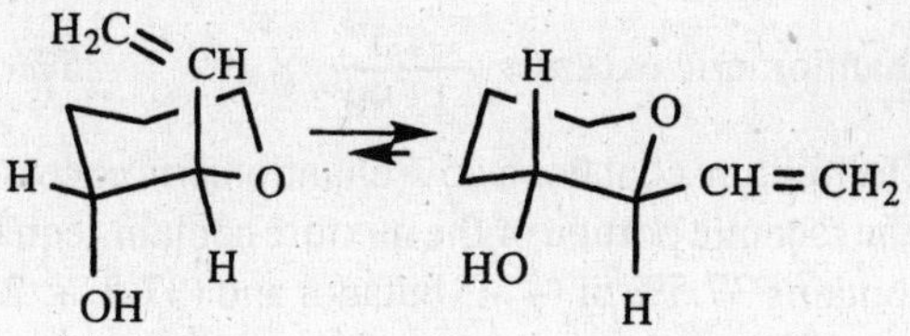

6.27

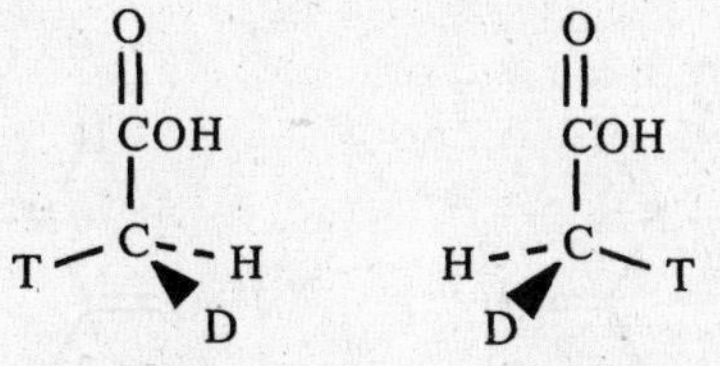

(S)-isomer (R)-isomer

sp^2-hybridized, planar; Br^- can attack from front face (a) or back face (b) of carbocation

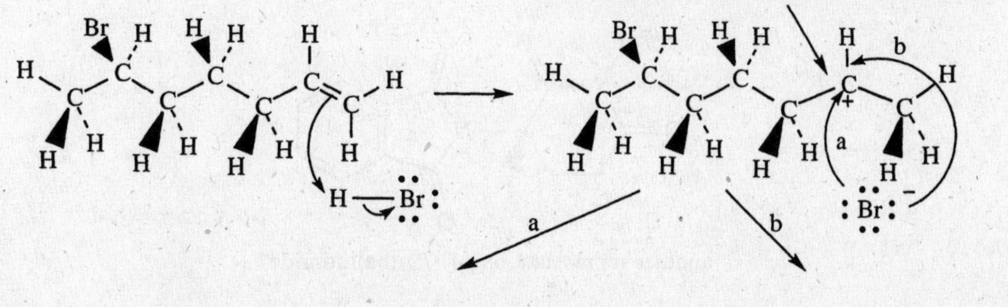

6.28

6.28 (cont)

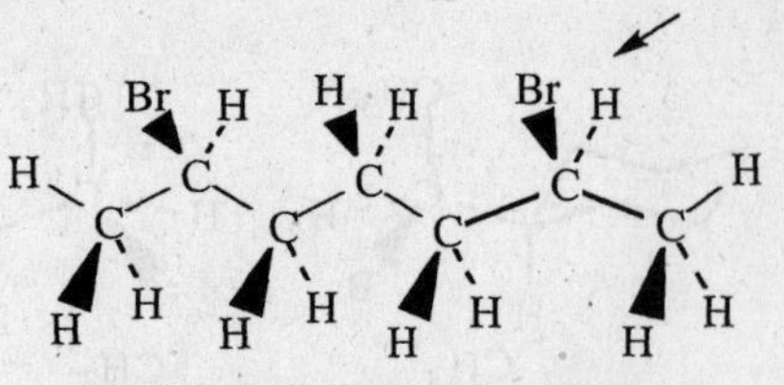

(2*R*, 6*S*)-2,6-dibromoheptane;
meso compound; no optical activity

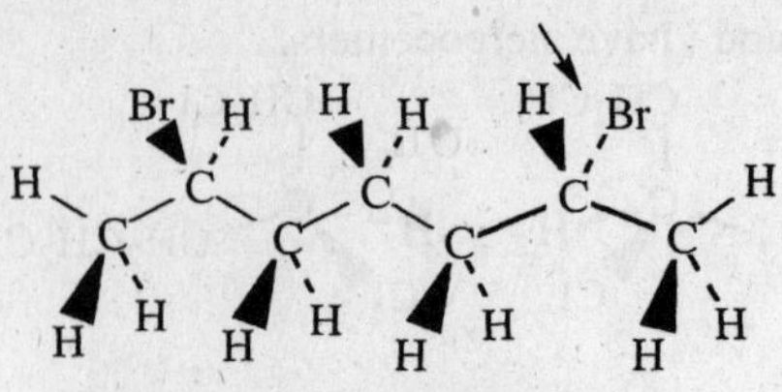

(2*R*, 6*R*)-2,6-dibromoheptane;
optically active

6.29   (a)

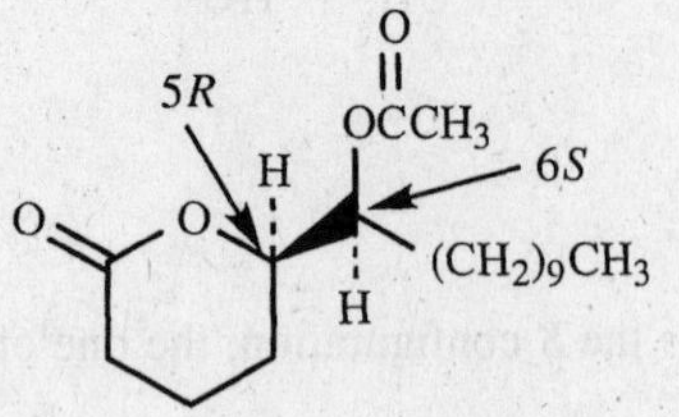

(b)

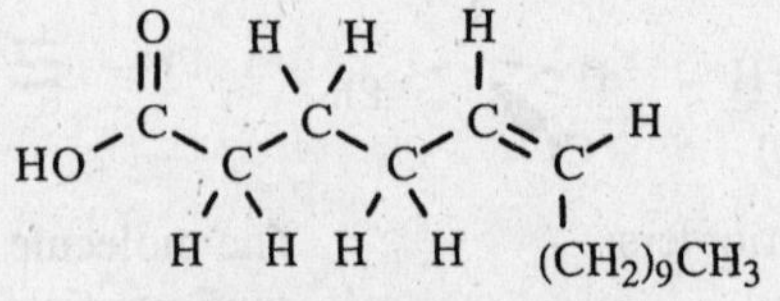

(*Z*)-5-hexadecenoic acid

6.30 $\alpha = [\alpha]_D^{20°} \, lc$

$\alpha = (-92°)(0.5 \text{ dm})(1 \text{ g}/100 \text{ mL})$

$\alpha = -0.46°$

6.31   enantiomeric excess $= \dfrac{\text{actual rotation}}{\text{rotation of pure enantiomer}} \times 100$

$[\alpha]_D^{20°}$ for pure (+) enantiomer $= +13.90°$, therefore $[\alpha]_D^{20°}$ for pure (–) enantiomer $= -13.90°$.

enantiomeric excess $= \dfrac{-3.5°}{-13.90°} \times 100 = 25\%$

The mixture contains a 25% enantiomeric excess of the levorotatory compound and is 100 – 25 or 75% racemic. The racemic portion of the mixture contains equal amounts (37.5%) of each enantiomer. Therefore the solution contains 37.5% of (+)-2-butanol and (37.5 + 25)% or 62.5% of (–)-2-butanol.

6.32

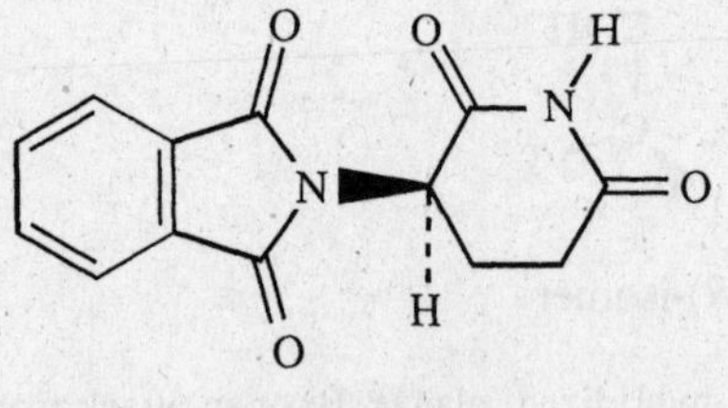

(*R*)-thalidomide 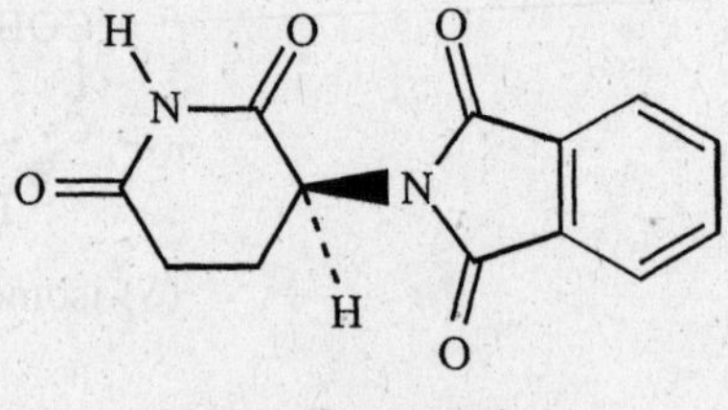   (*S*)-thalidomide

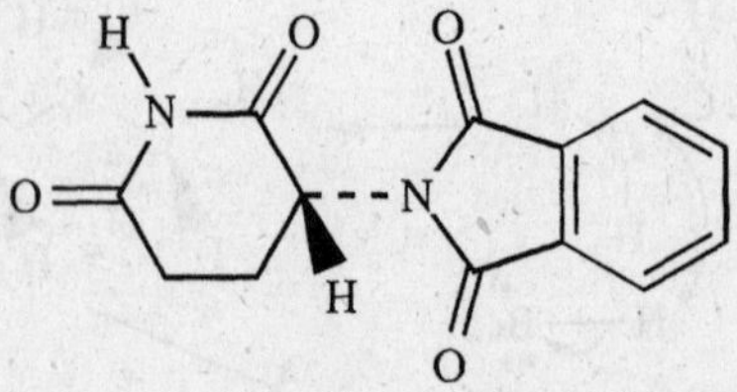

another representation of (*R*)-thalidomide

6.33

(2S, 3R)-4-dimethylamino-1,2-
diphenyl-3-methyl-2-butanol
medically useful form

(2R, 3S)-4-dimethylamino-1,2-
diphenyl-3-methyl-2-butanol

(2S, 3S)-4-dimethylamino-1,2-
diphenyl-3-methyl-2-butanol

(2R, 3R)-4-dimethylamino-1,2-
diphenyl-3-methyl-2-butanol

6.34

In *trans*-decalin, both of the methylene
groups on the ring junction are
equatorial in each case; more stable

In *cis*-decalin, one of the methylene
groups on the ring junction is
axial in each case; less stable

6.35 (a) Compound A has no enantiomers (meso). Compound B has an enantiomer.

(b) Compound A is not optically active. Compound B is optically active.

(c) *R* in Compound A and *R* in Compound B

(d) Compound A and Compound B each have two diastereomers.

(e)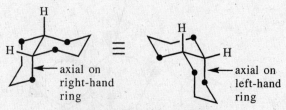

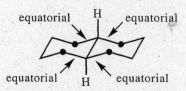

Conformer C   Conformer D
Conformer C is the more stable form.

6.36 (a) HO⬢⬡⬡⬡⬡⬡OCH₃   or   HO⬡⬡OCH₃

(b) HO⬢⬡⬡⬡⬡⬡OCH₃   or   HO⬡⬡H
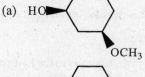

6.36   (c)

diequatorial conformer of *cis*-
1,3-dimethoxycyclohexane;
most stable conformer of the
four shown

diaxial conformer of *cis*-
1,3-dimethoxycyclohexane

both conformers of *trans*-1,3-dimethoxycyclohexane
have one axial and one equatorial methoxy group

6.37   (a)

(b) dextrorotatory

(c) The resulting compound would be a diastereomer of (+)-sandresolide A. We cannot tell what the optical
rotation would be.

6.38   (a) R   (b) P   (c) P, Q, R, T   (d) P, T
(e) Q   (f) T   (g) Q, R

Supplemental Problems

S6.1   Pasteur isolated the stereoisomers of tartaric acid, and the compound was used to establish absolute
configuration for the first time (Chapter 23). Tartaric acid has the formula

Its IUPAC name is 2,3-dihydroxybutanedioic acid. (+)-Tartaric acid is (2*R*, 3*R*)-2,3-dihydroxybutanedioic acid. Draw
(+)-tartaric acid and any other possible stereoisomers for tartaric acid.

S6.2   (a)   In a stereoisomer of 2-isopropyl-5-methylcyclohexanol, the methyl group is cis to the hydroxyl group,
and the isopropyl group is trans. Draw the two chair forms for the compound, and decide which
conformer would be more stable.

(b)   Draw the chair conformations of a stereoisomer of the compound described in part (a). There are several
possible stereoisomers.

S6.3   Name the following compounds.

(a)

(b)

S6.4   Which of the following compounds have stereocenters? Draw three-dimensional pictures of any that do, showing any enantiomers and diastereomers.

(a) $CH_3CHCH_2CH_2Br$
              |
             CH_3

(b) $CH_3CHCHCHCH_3$
              | |
             OH OH

       OH

S6.5   For each of the following pairs of structural formulas, tell whether the two represent identical molecular species, conformers of the same species, constitutional isomers, enantiomers, or diastereomers.

(a)

(b)

S6.6   Citronellol is a fragrant component of various plant oils, such as geranium oil and rose oil. A synthetic sample of (–)-citronellol with an enantiomeric excess of 88% has $[\alpha]_D^{20} = -4.1°$. What is the optical rotation of the pure enantiomer? (–)-Citronellol has the *S* configuration. The structure of citronellol is

$$CH_3C = CHCH_2CH_2CHCH_2CH_2OH$$
     | |
    CH_3 CH_3

Draw a stereochemically correct structural formula for (–)-citronellol.

S6.7   One method used to synthesize amino acids is a process known as reductive amination. This process, performed with 3-phenyl-2-oxopropanoic acid, results in a racemic mixture of the amino acid known as phenylalanine.

3-phenyl-2-oxopropanoic acid   racemic phenylalanine

Large quantities of stereochemically pure (*S*)-phenylalanine are required for the synthesis of the artificial sweetener, aspartame (NutraSweet). A Japanese firm is preparing (*S*)-phenylalanine by using an enzyme, phenylalanine dehydrogenase, from a bacterium, *Bacillus sphaericus,* in the presence of a biological reducing agent, NADH.

3-phenyl-2-oxopropanoic acid   (*S*)-phenylalanine

(a) Draw a stereochemically unambiguous structural formula for (*S*)-phenylalanine.
(b) Is it possible to say whether the sign of optical rotation of the product of the reductive amination (H_2/catalyst) will be +, –, or 0?
(c) Is the product from the enzymatic reduction (NADH/enzyme) optically active? Can you tell what the sign of rotation of the product of this reaction will be?

S6.7   (d) The structure of aspartame (not showing the stereochemistry of the phenylalanine unit) is drawn below. Find any other stereocenter(s) in the molecule, and assign configuration.

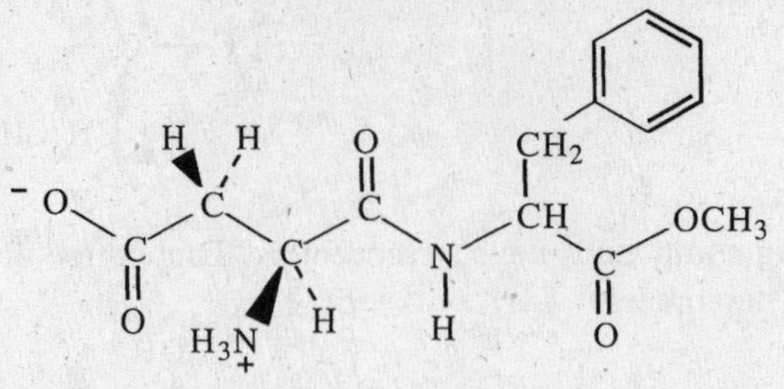

S6.8   Cyclic esters known as lactones are important components of natural products that have a wide range of antibiotic, antifungal, and antitumor activity. The biological activity of such compounds is dependent on their stereochemistry. Synthesis of such compounds must be designed in a way that gives the natural stereoisomer. One such synthesis starts with a racemic mixture of the methyllactone shown below and oxidizes its conjugate base with a chiral oxidizing agent. The major product of the oxidation is (2*S*, 3*R*)-(–)-2-hydroxy-3-methylpentanolide. The reactions are as shown.

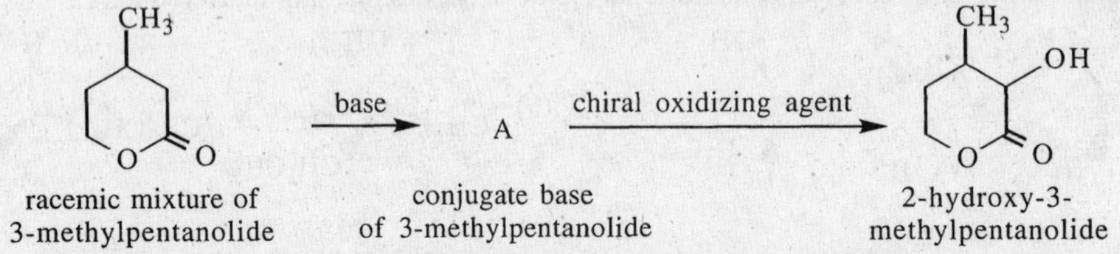

racemic mixture of
3-methylpentanolide

conjugate base
of 3-methylpentanolide

2-hydroxy-3-
methylpentanolide

(a) What is the structure of the most stable conjugate base of 3-methylpentanolide? Use the pK_a table, and explore resonance contributors for each possible conjugate base before making your decision.

(b) How many stereoisomers are possible for 2-hydroxy-3-methylpentanolide? Draw structural formulas for all of them. Which one is the major product of the reaction?

S6.9   A number of recently found small peptides with the potential for being useful as medications all contain the unusual amino acid (3*S*, 4*S*)-4-amino-3-hydroxy-6-methylheptanoic acid (common name statine) or one of its analogs. How biologically active the peptides are depends mainly on which statine-like amino acid is found in the peptide. Heptanoic acid has seven carbon atoms in a chain with the first carbon atom being a carboxylic acid. What is the stereochemically correct structural formula of the amino acid statine?

S6.10   Cyclotheonamide A, isolated from marine sponges, is a potent inhibitor of thrombin, which plays an important role in the clotting of blood. The structure of cyclotheonamide A is shown below.

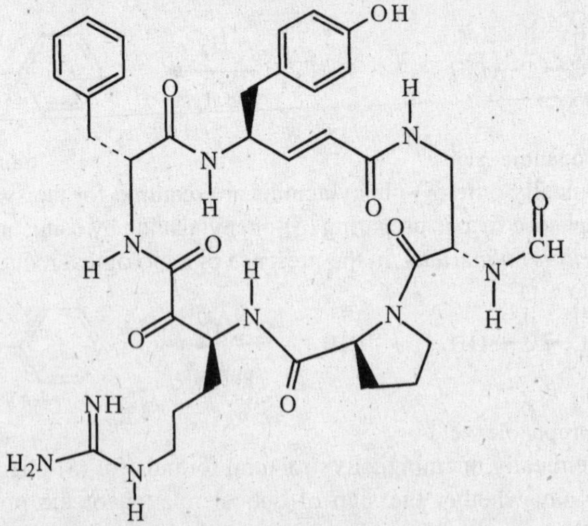

(a) How many stereocenters does cyclotheonamide A have?

(b) Does the compound contain any other source of stereoisomerism?

(c) Assign configuration to each source of stereoisomerism in the molecule.

S6.11 (a) The following compound, Compound **A**, was synthesized in a study of carbohydrate chemistry. Identify each stereocenter in the compound and assign its configuration (*R* or *S*).

$$AcO\!\!-\text{ is shorthand for } CH_3\overset{\displaystyle O}{\overset{\|}{C}}O\!\!-$$

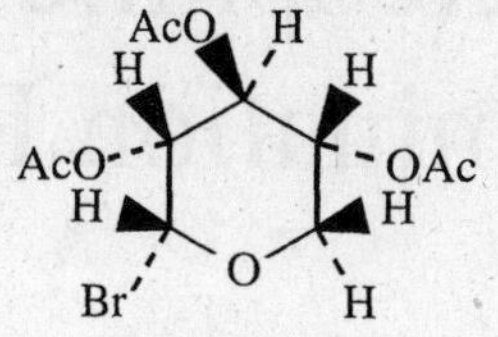

Compound **A**

(b) Use the skeletal chair form given below to draw a structural formula for the most stable chair form of Compound **A**.

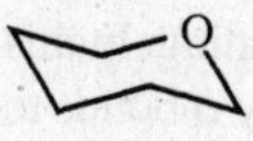

S6.12   A certain reaction produces the following two isomers, Compounds **B** and **C**, in almost equal amounts. The two isomers are observed to interconvert in an acidic solution to give an equilibrium mixture in which the major isomer is now 93% of the mixture.

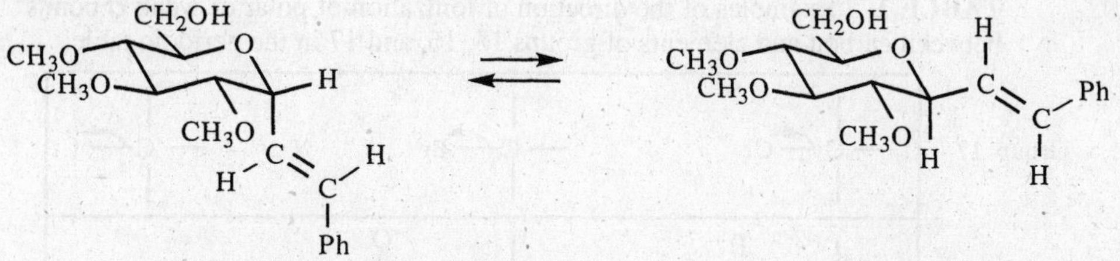

Compound **B**   Compound **C**

(a) Which is the major isomer at equilibrium and why?

(b) Draw an energy diagram showing the relative positions of isomers **B** and **C**.

S6.13

(a) Draw an accurate three-dimensional representation for the other chair conformation of the molecule shown below.

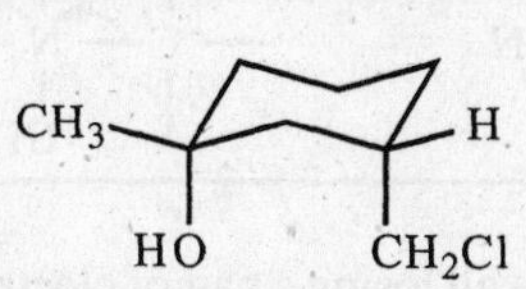

(b) Is the equilibrium constant (K_{eq}) for this conformational change greater than, less than, or equal to 1? Why?

(c) When this molecule is treated with methyllithium, the following products are observed.

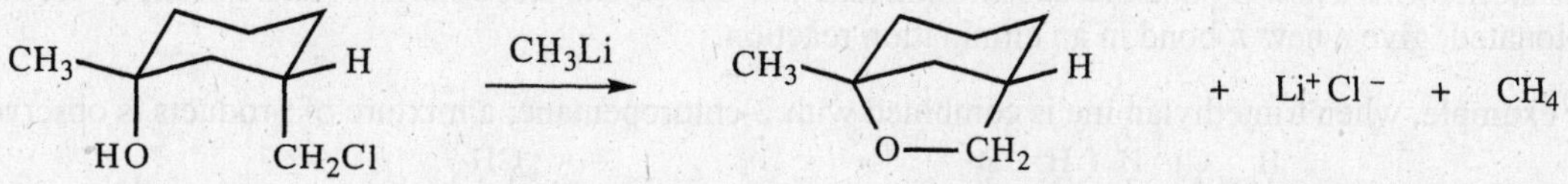

$$\xrightarrow{\text{CH}_3\text{Li}}\qquad\qquad +\ \ Li^+Cl^-\ \ +\ \ CH_4$$

Use the curved arrow convention to draw out the mechanism of the Brønsted acid/base reaction that is proposed to be the first step, the structure of the deprotonated intermediate (the conjugate base), and the mechanism of its conversion to the observed product.

7 Nucleophilic Substitution and Elimination Reactions

Workbook Exercises

The structures in this text contain single (σ), double ($\sigma + \pi$), and/or triple ($\sigma + 2\pi$) bonds. It is not surprising, then, that chemical reactions involve changes in σ and/or π bonding. In this chapter, you will be introduced to two important ways in which certain σ bonds can undergo changes. Many σ bonds to carbon are polar, or they are weak, which means they can be induced to break relatively easily. The ones shown below usually break heterolytically, in the same way that hydrogen chloride reacts with a base:

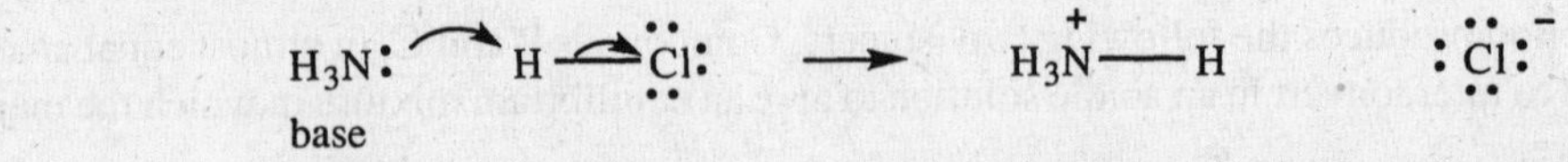

$$H_3N: \quad H \quad Cl: \quad \longrightarrow \quad H_3N^+ - H \qquad :Cl:^-$$

base

TABLE 1 Examples of the direction of ionization of polar or weak σ bonds between carbon and elements of groups 15, 16, and 17 in the periodic table.

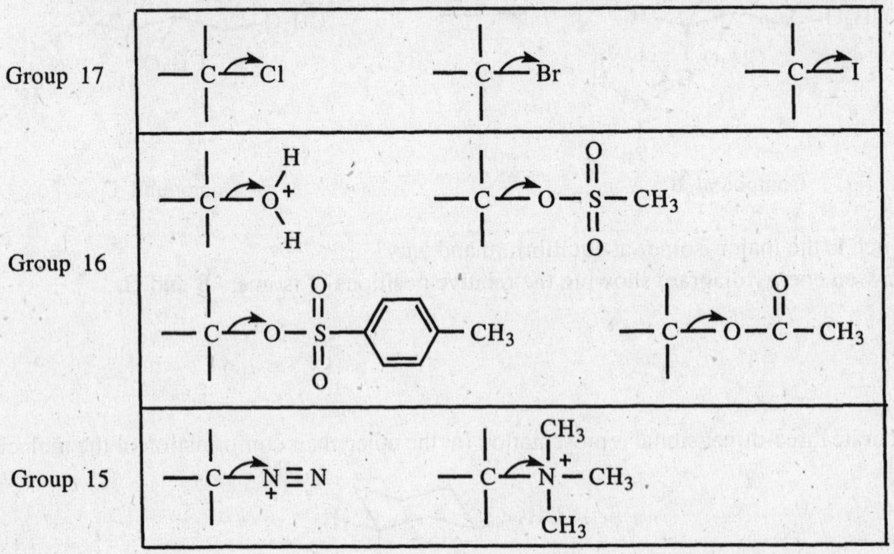

Group 17

Group 16

Group 15

The carbon atoms shown in the table above are all losing a pair of electrons to the group called the leaving group, which leaves when the bond breaks. To maintain or regain the closed shell configuration at these atoms, a new bond derived from an incoming electron pair is required. Two sources of such electron pairs are possible: (1) A nonbonding electron pair of an atom forms a new σ bond in a substitution reaction, and (2) the electrons from an adjacent C—H bond, which is deprotonated, give a new π bond in an elimination reaction.

For example, when trimethylamine is combined with 2-chloropentane, a mixture of products is observed:

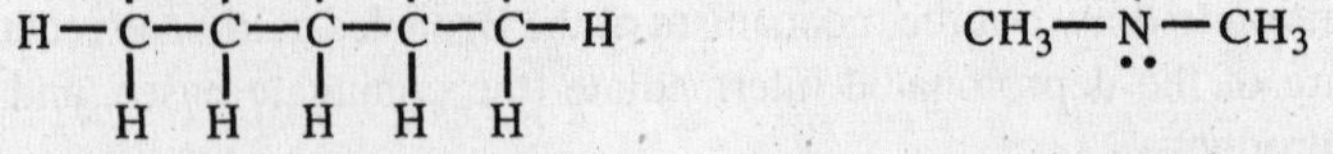

2-chloropentane
(the C—Cl bond can break
to form a chloride ion)

trimethylamine
(the N atom electron pair can be used to
(1) form a new C—N bond or (2)
deprotonate a C—H adjacent to the C—Cl)

Workbook Exercises (cont)

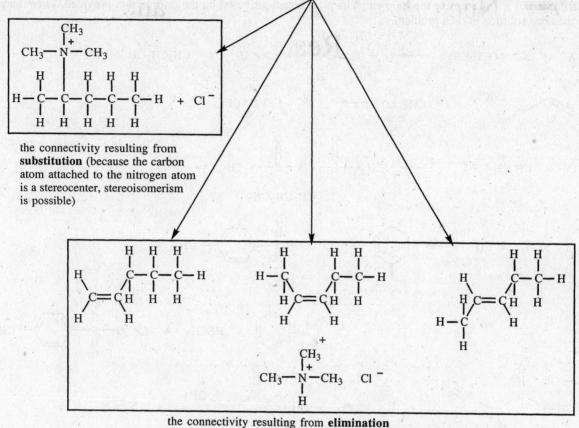

the connectivity resulting from
substitution (because the carbon
atom attached to the nitrogen atom
is a stereocenter, stereoisomerism
is possible)

the connectivity resulting from **elimination**
(loss of the chloride ion as well as the proton from an adjacent
C—H to give a π bond. Sometimes stereoisomers are possible

EXERCISE I.   Predict the connectivities of the possible substitution and elimination product(s) derived from the
following chemical reactions. Think about the possibility of stereoisomers forming, and show them when appropriate.
Use Table 1 to help you decide which atoms or groups can be lost easily.

(a)   (structure) + H₂O ⟶   (b)   (structure) + Na⁺ OH⁻ ⟶

(c)   (structure) + CH₃—C≡C⁻ Li⁺ ⟶   (d)   (structure) + Na⁺ ⁻SCH₂CH₃ ⟶

(e)   (structure) + (structure) ⟶   (f)   Na⁺ CN⁻ + (structure) ⟶

Workbook Exercises (cont)

EXERCISE II.   Complete the following chemical equations based on the information provided. There may be more than one solution to each problem.

(a)   A   + Na^+ $^-OCH_2CH_3$ $\longrightarrow$ $\underset{CH_3}{\overset{CH_3}{C}}{=}\underset{H}{\overset{H}{C}}$ + $Na^+\,Cl^-$ + CH_3CH_2OH

(b)   (allyl)—$SiCl_3$ +   3 $CH_3^-\,Li^+$ $\longrightarrow$ B   +   3 $Li^+\,Cl^-$

(c)   C   + $CH_3CH_2CH_2I$ $\longrightarrow$ $CH_3{-}\overset{O}{\overset{\|}{C}}{-}\underset{CH_2CH_2CH_3}{CH}{-}\overset{O}{\overset{\|}{C}}{-}OCH_3$ + $Na^+\,I^-$

(d) Li^+ $^-SCH_3$ +   D $\longrightarrow$ (cyclohexene with SCH₃)   + $Li^+\,{}^-O{-}\overset{O}{\underset{O}{\overset{\|}{\underset{\|}{S}}}}{-}$(C₆H₄)$CH_3$

$\Big($ +   (1,3-cyclohexadiene)   + $HSCH_3$ + $Li^+\,{}^-O{-}\overset{O}{\underset{O}{\overset{\|}{\underset{\|}{S}}}}{-}$(C₆H₄)$CH_3$ $\Big)$

(e) $CH_3O{-}\overset{O}{\overset{\|}{C}}{-}\overset{+}{\underset{CH_2}{\overset{}{C}}}\overset{N{\equiv}N}{\underset{H}{}}$ (with phenyl)   +   E $\longrightarrow$ $CH_3O{-}\overset{O}{\overset{\|}{C}}{-}\underset{CH_2}{\overset{\overset{+}{O}H_2}{C}}\underset{H}{}$ (with phenyl)   + $N{\equiv}N$

+   F (a stereoisomer of the substitution product)   +   G and H (two elimination products)

(f)   (bicyclic structure with CH_3, $Br{-}O{-}H$)   + $Na^+\,OH^-$ $\longrightarrow$ I   + H_2O + $Na^+\,Br^-$

(Hint: This is an elimination reaction in which bromide ion is the leaving group.)

- -

7.1   (a) $\underset{CH_3}{CH_3CHCH_2CH_2}{-}\ddot{\underset{\cdot\cdot}{I}}\!:$ $\longrightarrow$ $\underset{CH_3}{CH_3CHCH_2CH_2}{-}\ddot{\underset{\cdot\cdot}{O}}Ph$ Ph— $\equiv$ (phenyl)

Na^+ $^-\!:\!\ddot{\underset{\cdot\cdot}{O}}Ph$ Na^+ $:\!\ddot{\underset{\cdot\cdot}{I}}\!:^-$

(b)   (piperidine N–H) $\overset{}{\longrightarrow}$ (piperidinium $\overset{+}{N}$–H, CH_3) $:\!\ddot{\underset{\cdot\cdot}{I}}\!:^-$

$\overset{\cdot\cdot}{N}{-}CH_3{-}\ddot{\underset{\cdot\cdot}{I}}\!:$

first product;
acid with $pK_a \sim 10$

7.1 (b) (cont)

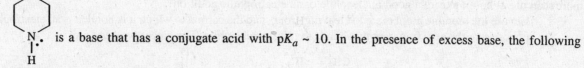

is a base that has a conjugate acid with $pK_a \sim 10$. In the presence of excess base, the following equilibrium will be established.

	base	acid	conjugate acid	conjugate base
	in large excess	$pK_a \sim 10$	$pK_a \sim 10$	

Because the two acids have roughly the same acidity, the relative amounts of the different species at equilibrium will be determined by the relative concentrations of the different reagents. Because there is a large excess of the amine used as a starting material in the reaction, the proton will spend most of its time with that amine and the product amine will be mostly in the unprotonated form.

(c) $CH_3CH_2CH_2 \underset{}{\overset{}{-}} \ddot{B}r: \longrightarrow CH_3CH_2CH_2 - \ddot{S} - CH_3$

$Na^+ \quad {}^- : \ddot{S} - CH_3 \qquad\qquad Na^+ \quad : \ddot{B}r : {}^-$

7.2   The answers to part (b) analyze the problem as shown in the *Art of Solving Problems* section of the chapter. The rest of the problems show only the equations, written so as to emphasize the necessity of reasoning backwards from the structure of the desired product to the structure of the starting material.

(a)
$$\underset{\underset{SCH_2CH_3}{|}}{CH_3CHCH_3} \xleftarrow{CH_3CH_2S^- Na^+} \underset{\underset{Br}{|}}{CH_3CHCH_3} \xleftarrow{HBr} CH_3CH = CH_2$$

(b) $CH_3CH_2CH_2CH_2CH_2Br \longrightarrow CH_3CH_2CH_2CH_2CH_2CH(\overset{\overset{O}{\|}}{C}OCH_2CH_3)_2$

1. What are the connectivities of the two compounds? How many carbon atoms does each contain? Are there any rings? What are the positions of branches and functional groups on the chains?

    The starting material has five carbon atoms in a row bonded to a bromine atom. In the product the bromine has been replaced by a carbon atom bearing two ester groups.

2. How do the functional groups change in going from starting material to product? Does the starting material have a good leaving group?

    The bromine atom in the starting material is a good leaving group. The alkyl halide function in the starting material has been converted to an extra carbon with two ester functions on it.

3. Is it possible to dissect the structures of the starting material and product to see which bonds must have been broken and which formed?

$CH_3CH_2CH_2CH_2CH_2 - \xi - Br$ $CH_3CH_2CH_2CH_2CH_2 - \xi - CH(\overset{\overset{O}{\|}}{C}OCH_2CH_3)_2$

        bond broken   bond formed

7.2 (b) (cont)

4.　New bonds are created when an electrophile reacts with a nucleophile. Do we recognize any part of the product molecule as coming from a good nucleophile or an electrophilic addition?

Because the bromine atom is a good leaving group, and the carbon to which it is bonded is an electrophile, we suppose that the new fragment that bonds to it is a nucleophile. The pieces we need are:

$$CH_3CH_2CH_2CH_2CH_2 \overset{\delta+}{\underset{}{—}} \overset{\delta-}{Br} \quad \text{and} \quad {}^-:CH(COCH_2CH_3)_2$$

electrophilic carbon　　nucleophilic carbon

We find the nucleophile in Table 7.1.

5.　What type of compound would be a good precursor to the product?

Displacement of the bromide ion by an incoming nucleophilic carbon atom will give us the product we want. We need a nucleophile with the structure shown above.

6.　After this step, do we see how to get from starting material to product? If not, do we need to analyze the structure obtained in step 5 by applying questions 4 and 5 to it?

We can now complete the synthesis.

$$CH_3CH_2CH_2CH_2CH_2CH(COCH_2CH_3)_2 \xleftarrow{\quad Na^{+-}:CH(COCH_2CH_3)_2 \quad} CH_3CH_2CH_2CH_2CH_2Br$$

(c)

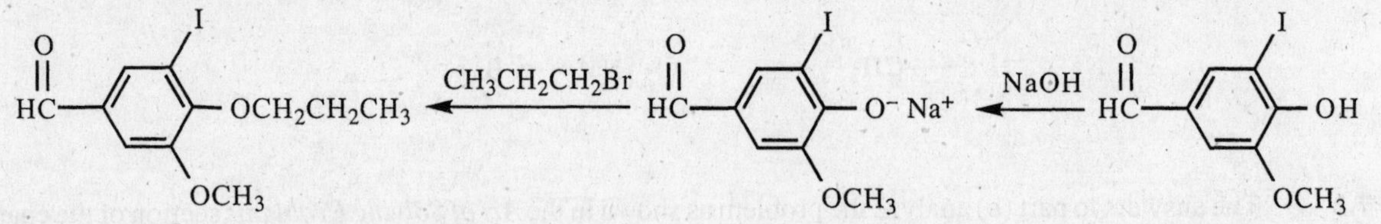

7.3　The initial rate of the reaction will be

rate　$=$　$k\,[CH_3CH_2Br]\,[OH^-]$

rate　$=$　$(1.7 \times 10^{-3}\ \text{L/mol·s})(0.05\ \text{mol/L})(0.07\ \text{mol/L})$

rate　$=$　$6 \times 10^{-6}\ \text{mol/L·s}$

If the concentration of bromoethane is doubled (increased from 0.05 mol/L to 0.1 mol/L), the rate will double. The new rate would be $\sim 1 \times 10^{-5}$ mol/L·s.

7.4　For the reaction

$$CH_3CH_2CH_2CH_2Br + K^+I^- \xrightarrow[\text{acetone}]{} CH_3CH_2CH_2CH_2I + K^+Br^-\downarrow$$

The rate at which *n*-butyl bromide is used up is:

rate　$=$　$k\,[CH_3CH_2CH_2CH_2Br]\,[I^-]$

rate　$=$　$(1.09 \times 10^{-1}\ \text{L/mol·min})(0.06\ \text{mol/L})(0.02\ \text{mol/L})$

rate　$=$　1.3×10^{-4} mol/L·min

7.5　For the reaction

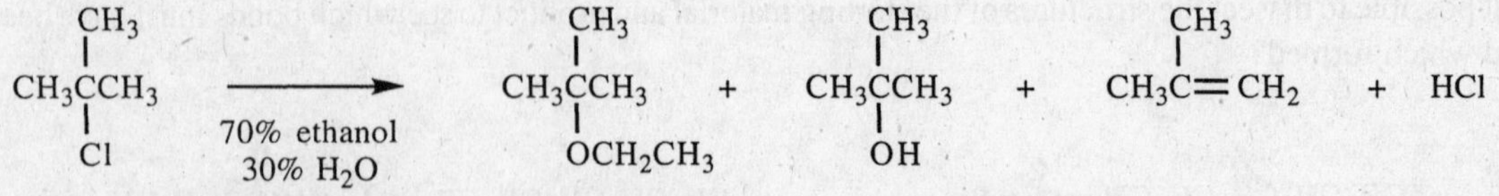

rate　$=$　$k\,[\textit{tert}\text{-butyl chloride}]$

rate　$=$　$(0.145/\text{h})(0.0824\ \text{mol/L})$

rate　$=$　1.19×10^{-2} mol/L·h

7.5 (cont)

The units in which this rate is expressed are mol/L·h. The rate constant will remain the same, but the rate of the reaction will decrease as the *tert*-butyl chloride is used up in the reaction. When half of the *tert*-butyl chloride is gone, the rate of the reaction will be half of what it was initially or 5.95×10^{-3} mol/L·h.

7.6   The following species can act as bases (electron donors).

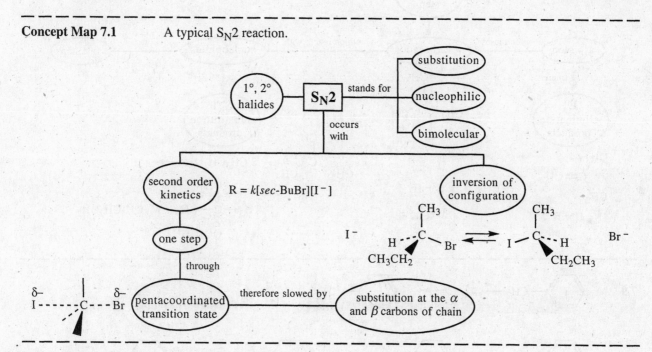

7.7   (*S*)-(+)-2-Bromobutane will racemize by undergoing reversible bimolecular nucleophilic substitution reactions with inversion of configuration. The optical rotation will gradually decrease to 0° as the system reaches equilibrium.

- -

Concept Map 7.1   A typical S_N2 reaction.

substitution

S_N2 stands for nucleophilic

1°, 2° halides

bimolecular

occurs with

inversion of configuration

second order kinetics $R = k[sec\text{-BuBr}][I^-]$

one step

through

pentacoordinated transition state ——— therefore slowed by ——— substitution at the α and β carbons of chain

- -

7.8   The fact that the rate of racemization is twice the rate of substitution of the radioactive iodine, I^*, shows that every substitution occurs with inversion of configuration.

(*S*)-(+)-2-iodooctane

(*R*)-(−)-2-iodooctane
inversion of configuration

Each S_N2 reaction incorporates *one* radioactive iodine atom into 2-iodooctane, but results in the loss of optical activity equal to that of two molecules of (*S*)-(+)-2-iodooctane: the one that had been converted to the (*R*)-isomer, and the other, the optical activity of which is cancelled by the presence of the (*R*)-isomer. Therefore the rate of racemization, the rate at which optical activity is lost, is *twice* the rate of the substitution reaction.

Concept Map 7.2   The S_N1 reaction.

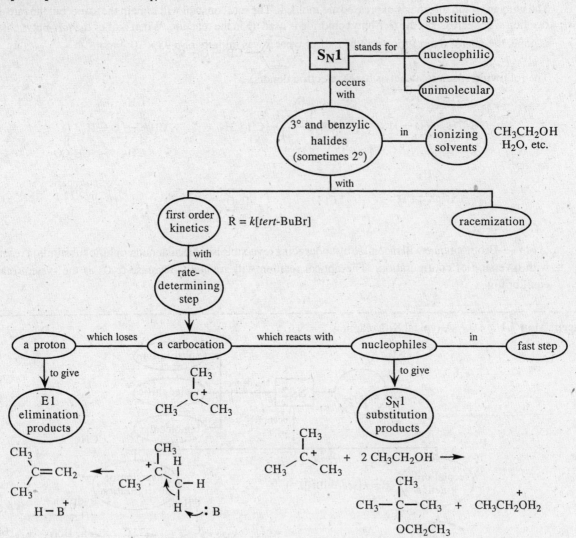

- -

7.9

7.10 S_N1 mechanism

7.10 (cont)

S$_N$2 mechanism

$$\text{C}_6\text{H}_5-\overset{\delta+}{\text{CH}_2}\overset{\longrightarrow}{\ \ }\overset{\delta-}{:\text{Cl}:} \longrightarrow \text{C}_6\text{H}_5-\text{CH}_2-\overset{..}{\text{O}}-\text{H} \qquad :\overset{..}{\underset{..}{\text{Cl}}}:^-$$

$$\overset{..}{:}\text{O}-\text{H}$$

7.11 (a) $\text{C}_6\text{H}_5-\text{S}^-$ > $\text{C}_6\text{H}_5-\text{O}^-$ (b) OH^- > CH_3CO^- ($\overset{O}{\parallel}$)

(Sulfur is more polarizable than oxygen.)

(The hydroxide ion is more basic than the acetate ion. See Problem 7.12.)

(c) OH^- > NO_3^-

(The hydroxide ion is more basic than the nitrate ion. See Problem 7.12.)

7.12

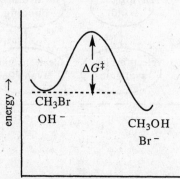

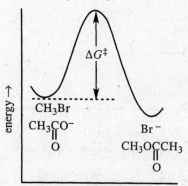

Hydroxide ion is more basic than acetate ion. The pK_a of water, the conjugate acid of hydroxide ion, is 15.7. The pK_a of acetic acid, the conjugate acid of acetate ion, is 4.8. The greater basicity of the oxygen atom in hydroxide ion means that its electrons are more loosely held than those in acetate ion. Bonding between hydroxide ion and the carbon atom in methyl bromide starts more easily, and it requires less energy to get to the transition state.

7.13 The increase in the relative rates of the substitution reaction in this series of solvents is related to the decreasing ability of the solvent to hydrogen bond to the nucleophile, chloride ion, and to the steric hindrance of the positive end of the dipole of the carbonyl group.

	$\overset{O}{\parallel}$	$\overset{O}{\parallel}$	$\overset{O}{\parallel}$
CH_3OH	HCNH_2	$\text{HCN(CH}_3)_2$	$\text{CH}_3\text{CN(CH}_3)_2$
methanol	formamide	*N,N*-dimethylformamide	*N,N*-dimethylacetamide

Methanol hydrogen bonds strongly and reduces the nucleophilicity of the chloride anion. Formamide has N—H bonds, so it also hydrogen bonds with chloride anion, but not as strongly as methanol does. Neither *N,N*-dimethylformamide nor *N,N*-dimethylacetamide can hydrogen bond at all, therefore the nucleophile behaves as a strong nucleophile in both solvents. The methyl substituent on the carbonyl group of *N,N*-dimethylacetamide prevents the chloride anion from getting as close to the positive end of the dipole as it can in *N,N*-dimethylformamide, making the chloride anion an even stronger nucleophile.

Concept Map 7.3   The factors that are important in determining nucleophilicity.

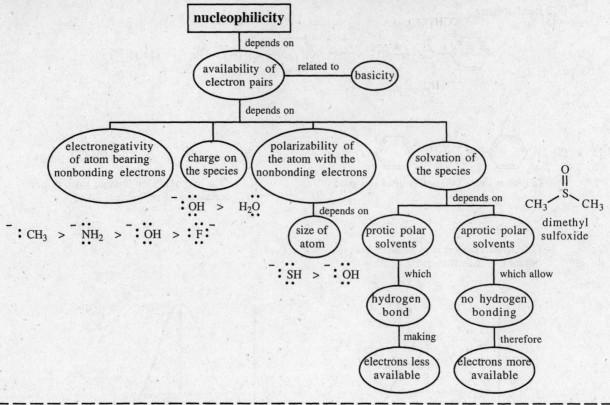

- -

7.14   (a)

$$H_2N-OH \xrightarrow{HCl} \overset{+}{H_3N}-OH \quad :Cl:^-$$

$$H_2N-OH \xrightarrow{CH_3I} CH_3\overset{\overset{H}{|}\,+}{\underset{H}{N}}-OH \quad :I:^-$$

(b)   The pK_as of the oxonium ions are all less than zero while the pK_as of the ammonium ions are all between 4 and 11. The pK_a of the conjugate acid of hydroxylamine falls in the range of values for the ammonium ion.

$$pK_a \qquad \overset{+}{NH_4} \qquad 9.4$$

$$pK_a \qquad \overset{+}{H_3NOH} \qquad 5.97$$

The presence of an electron-withdrawing oxygen atom on the nitrogen atom makes the nonbonding electrons on nitrogen less available. Hydroxylamine is therefore a weaker base than ammonia, and the conjugate acid of hydroxylamine is a stronger acid than ammonium ion.

7.15   The hydronium ion comes from protonation of the solvent, water, by the acid, HBr, that was added to the reaction mixture.

$$H_2O + HBr \longrightarrow H_3O^+ + Br^-$$

7.16   The other nucleophiles that are present are the starting alcohol and water.
Two additional substitution reactions are possible. The carbocation can react with water to regenerate the starting alcohol or with another alcohol molecule to produce an ether.

7.16 (cont)

Nucleophiles are also bases and can remove a proton from the carbocation to give an alkene, the elimination product.

7.17 CH_3NH_2 + $CH_3CH_2CH_2CH_2\overset{+}{O}\!-\!H$ (with H above) $\rightleftharpoons$ $CH_3\overset{+}{N}H_3$ + $CH_3CH_2CH_2CH_2O\!-\!H$

   base   acid   acid   conjugate base
                        $pK_a \sim -2.4$ pK_a 10.6

7.18   All of the nucleophiles in Table 7.1 are too strongly basic to be used with oxonium ions formed by protonating an alcohol.

nucleophile	conjugate acid of nucleophile	pK_a of conjugate acid
HO^-	H_2O	15.7
RO^-	ROH	~16
ArO^-	$ArOH$	~10
HS^-	H_2S	~10
RS^-	RSH	~10.5
ArS^-	$ArSH$	~7.8
NH_3	NH_4^+	9.4
RNH_2	RNH_3^+	~10.6
N_3^-	HN_3	4.7
CN^-	HCN	9.1
$RC\!\equiv\!C^-$	$RC\!\equiv\!CH$	~26
$(CH_3CH_2OC)_2CH^-$ (with =O above C)	$(CH_3CH_2OC)_2CH_2$ (with =O above C)	~11

7.19

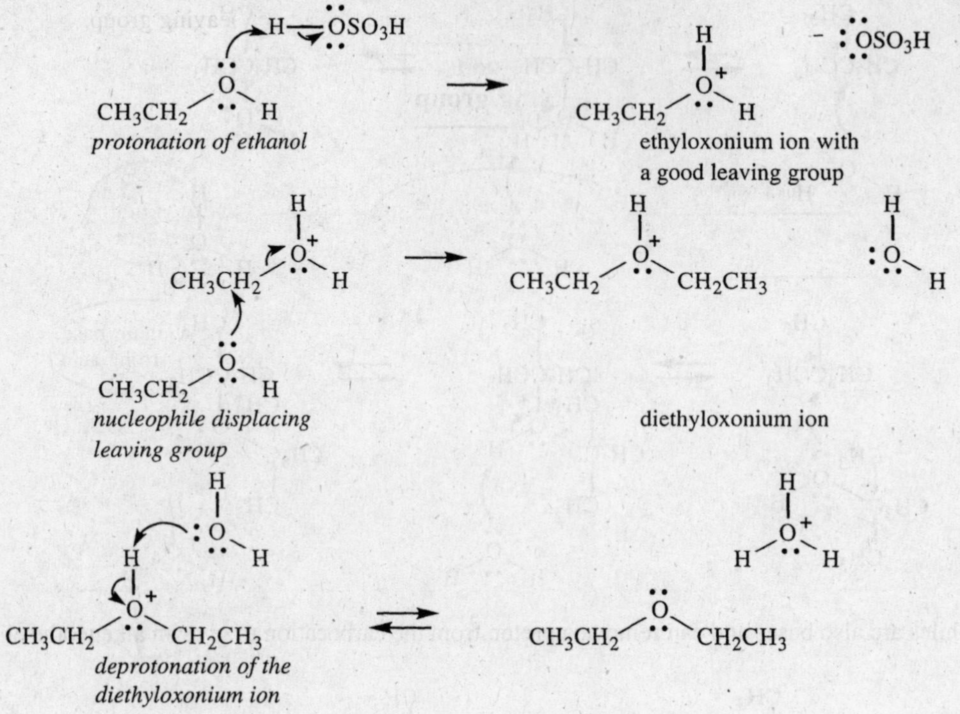

H——OSO₃H

CH₃CH₂—Ö—H

protonation of ethanol

⟶

CH₃CH₂—Ö⁺—H ⁻:OSO₃H
     |
     H

ethyloxonium ion with
a good leaving group

CH₃CH₂—Ö⁺—H
     |
     H

CH₃CH₂—Ö—H

*nucleophile displacing
leaving group*

⟶

CH₃CH₂—Ö⁺—CH₂CH₃
     |
     H

diethyloxonium ion

:Ö—H
  |
  H

:Ö—H
 |
 H

CH₃CH₂—Ö⁺—CH₂CH₃

*deprotonation of the
diethyloxonium ion*

⇌

H—Ö⁺—H
   |
   H

CH₃CH₂—Ö—CH₂CH₃

7.20

H——Ï:

CH₃CH₂—Ö—CH₂CH₃

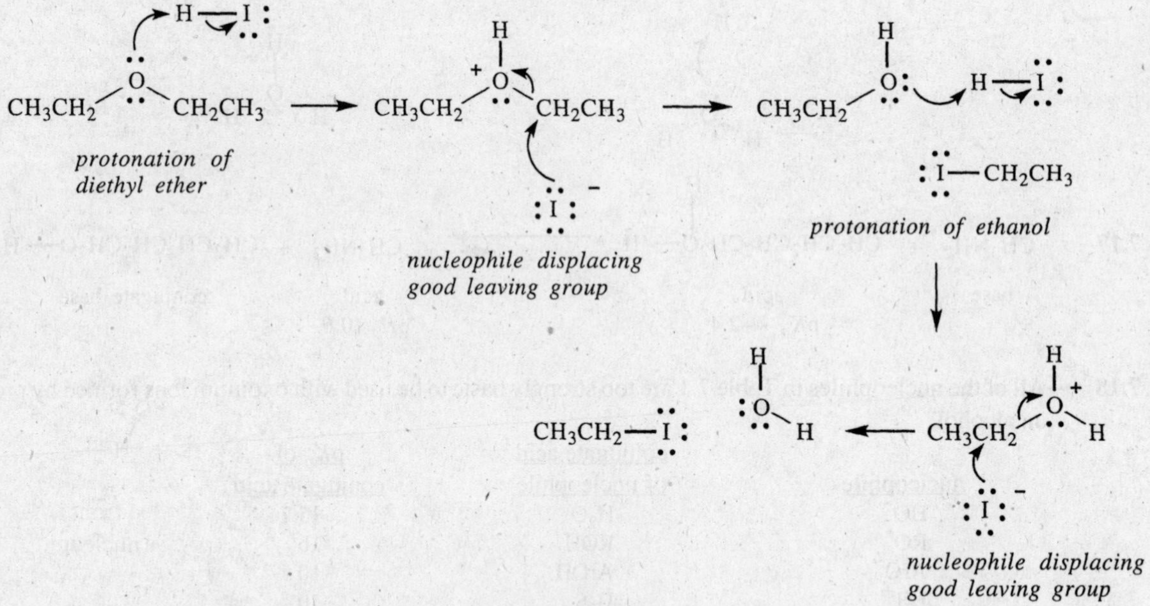

*protonation of
diethyl ether*

⟶

CH₃CH₂—Ö⁺—CH₂CH₃
     |
     H

:Ï:⁻

*nucleophile displacing
good leaving group*

⟶

CH₃CH₂—Ö: H——Ï:
        |
        H

:Ï—CH₂CH₃

protonation of ethanol

CH₃CH₂—Ï: :Ö—H   CH₃CH₂—Ö⁺—H
                | |
                H   H

:Ï:⁻

*nucleophile displacing
good leaving group*

7.21

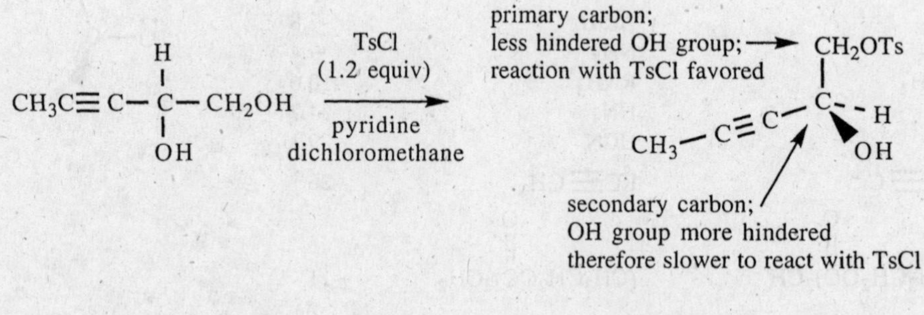

CH₃C≡C—C—CH₂OH
         |
         OH

TsCl
(1.2 equiv)
⟶
pyridine
dichloromethane

primary carbon;
less hindered OH group; ⟶ CH₂OTs
reaction with TsCl favored

CH₃—C≡C—C—H
           |
           OH

secondary carbon;
OH group more hindered
therefore slower to react with TsCl

Concept Map 7.4   The factors that determine whether a substituent is a good leaving group.

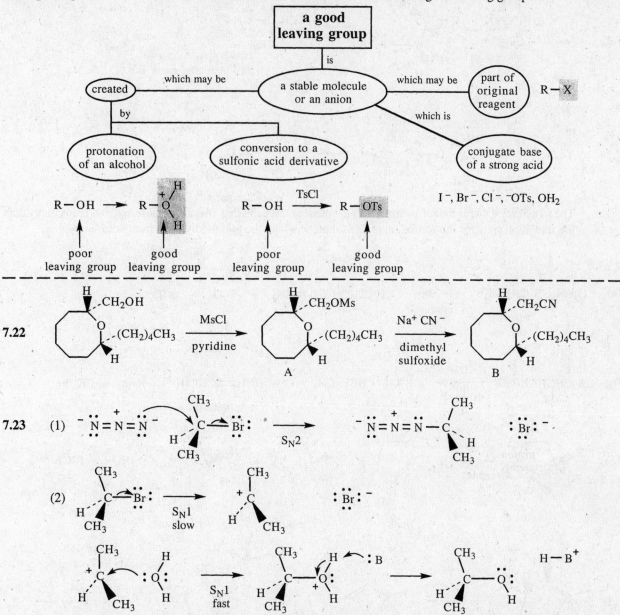

Note that attack of a nucleophile on the electrophilic carbocation can occur at either the front face (where the leaving group was attached) or at the back face. The next equation shows attack of a nucleophile at the back face.

7.23 (3)

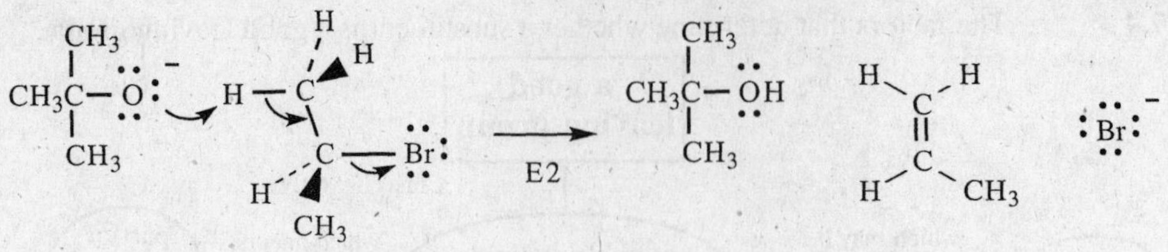

E2

7.24 Yes

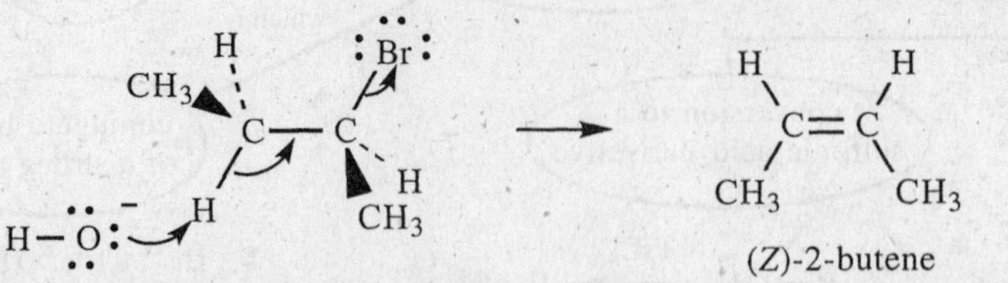

(Z)-2-butene

The transition state for the formation of (E)-2-butene is less crowded, more favorable, and lower in energy than the transition state for the formation of (Z)-2-butene where the two methyl groups are close together.

7.25

(a) $CH_3CH_2CH_2CH_2CH_2Br \xrightarrow[\Delta]{\text{KOH} \atop \text{ethanol}} CH_3CH_2CH_2CH=CH_2 \ + \ H_2O \ + \ K^+Br^-$

(b)

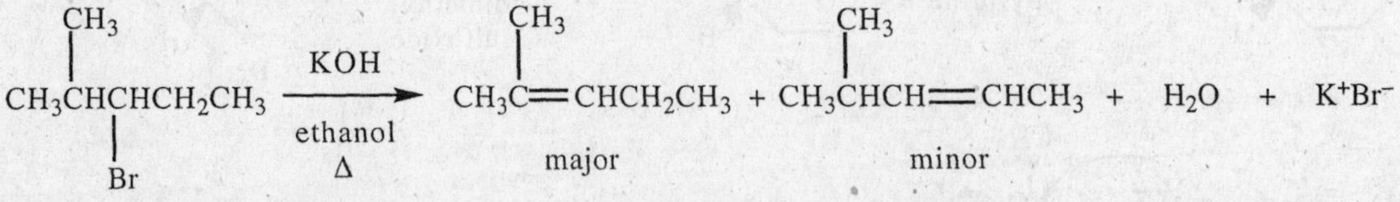

(c)

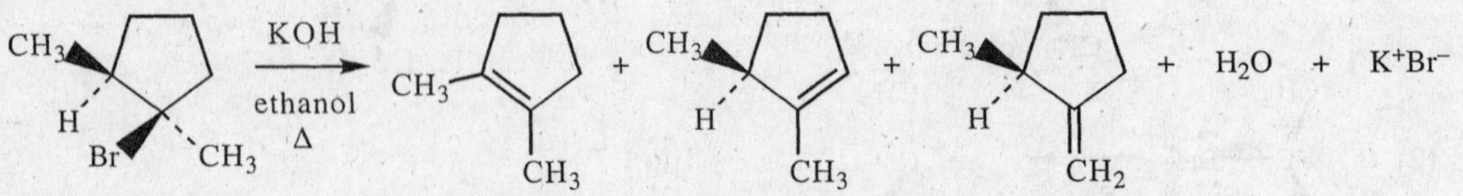

(d)

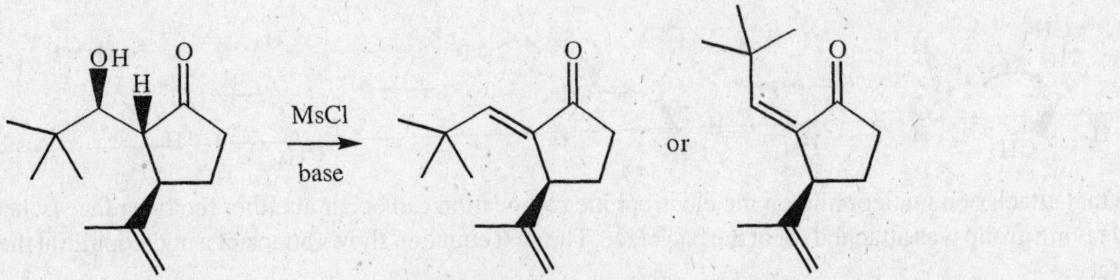

found experimentally

7.26 (a)

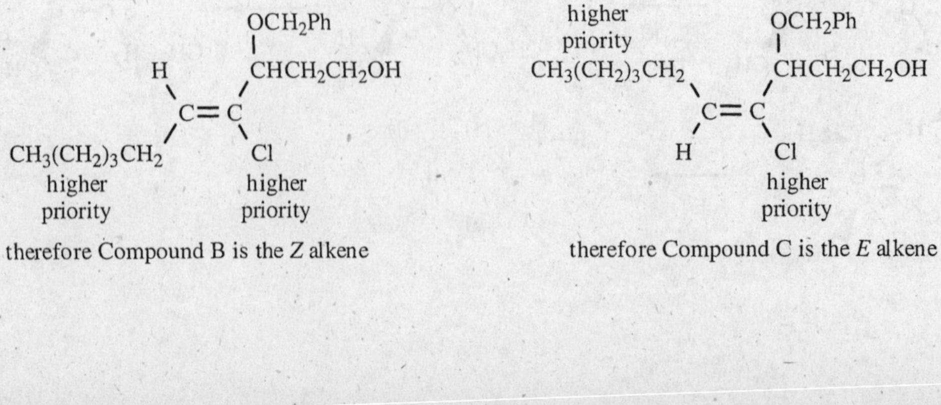

therefore Compound B is the Z alkene   therefore Compound C is the E alkene

7.26 (b)

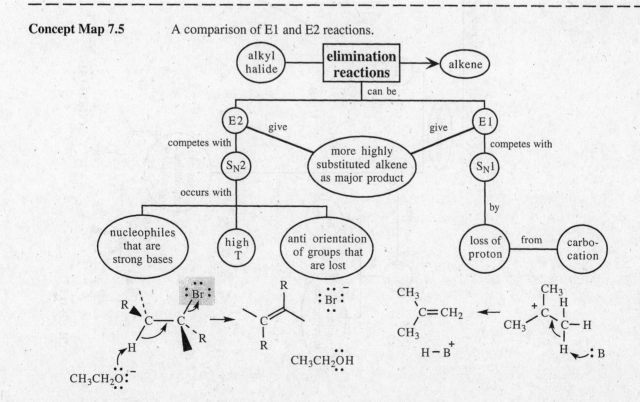

Newman projection for conformation
leading to the *Z*-alkene

Newman projection for conformation
leading to the *E*-alkene

(c) Compound B will be the major product. The conformation leading to the formation of Compound B has the large groups anti to each other, whereas they are gauche to each other in the conformation leading to Compound C. The $\Delta G^{\ddagger}$ for the formation of B will be lower than the $\Delta G^{\ddagger}$ for the formation of C; B will form faster, so most of the product will be B.

- -

Concept Map 7.5   A comparison of E1 and E2 reactions.

Concept Map 7.6   Overall view of nucleophilic substitution and elimination reactions

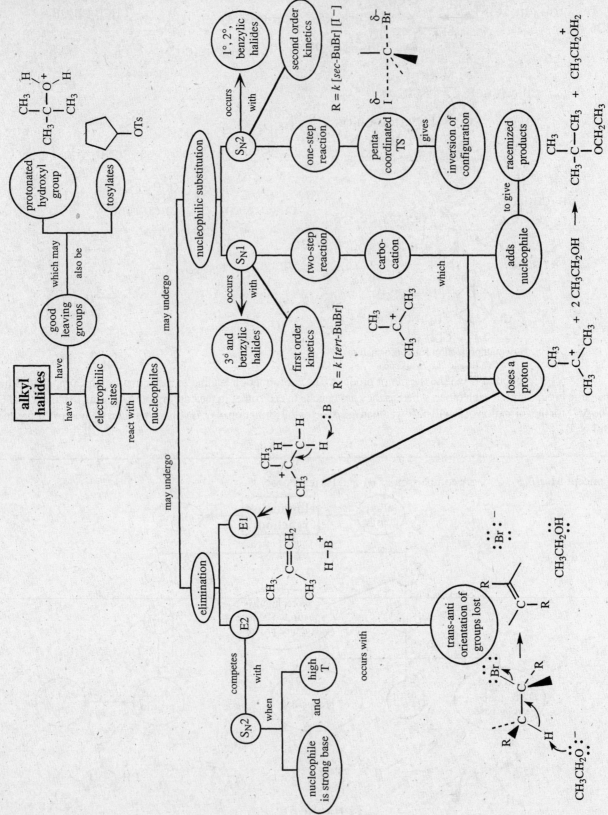

7.27

(a) $CH_3CH_2CH_2CH_2Br$ $\xrightarrow[\substack{\text{ethanol} \\ \Delta}]{\text{KOH}}$ $CH_3CH_2CH{=}CH_2$ + $CH_3CH_2CH_2CH_2OH$ + H_2O + K^+Br^-

with strong base at high temperature

major product

(b) $CH_3CH_2CH_2CH_2Br$ $\xrightarrow[\substack{\text{50\% aqueous} \\ \text{ethanol}}]{\text{0.1 M NaOH}}$ $CH_3CH_2CH_2CH_2OH$ + Na^+Br^-

(c) $CH_3CH_2CH_2CH_2Br$ $\xrightarrow{NH_3}$ $CH_3CH_2CH_2CH_2\overset{+}{N}H_3\ Br^-$ $\xrightarrow{NH_3\ (\text{excess})}$ $CH_3CH_2CH_2CH_2NH_2$ + $NH_4^+\ Br^-$

(d) $CH_3CH_2CH_2CH_2Br$ $\xrightarrow{Na^+N_3^-}$ $CH_3CH_2CH_2CH_2N_3$ + Na^+Br^-

(e) $CH_3CH_2CH_2CH_2Br$ $\xrightarrow{Na^+CN^-}$ $CH_3CH_2CH_2CH_2CN$ + Na^+Br^-

(f) $CH_3CH_2CH_2CH_2Br$ $\xrightarrow{CH_3CH_2S^-Na^+}$ $CH_3CH_2CH_2CH_2SCH_2CH_3$ + Na^+Br^-

(g) $CH_3CH_2CH_2CH_2Br$ $\xrightarrow{CH_3\overset{\displaystyle O}{\overset{\|}{C}}O^-Na^+}$ $CH_3CH_2CH_2CH_2O\overset{\displaystyle O}{\overset{\|}{C}}CH_3$ + Na^+Br^-

(h) $CH_3CH_2CH_2CH_2Br$ $\xrightarrow{CH_3CH_2C{\equiv}C^-Na^+}$ $CH_3CH_2CH_2CH_2C{\equiv}CCH_2CH_3$ + Na^+Br^-

(i) $CH_3CH_2CH_2CH_2Br$ $\xrightarrow{Na^+\ ^-CH(\overset{\displaystyle O}{\overset{\|}{C}}OCH_2CH_3)_2}$ $CH_3CH_2CH_2CH_2CH(\overset{\displaystyle O}{\overset{\|}{C}}OCH_2CH_3)_2$ + Na^+Br^-

7.28 (a)
leaving group

⬠—$CH_2CH_2CH_2OTs$ + $Na^+N_3^-$ ⟶ ⬠—$CH_2CH_2CH_2N_3$ + TsO^-Na^+

nucleophile

(b)
leaving group   nucleophile

$PhCH_2CH_2Cl$ + Na^+CN^- ⟶ $PhCH_2CH_2CN$ + $NaCl$

$^-{:}C{\equiv}N{:}$ ⟷ ${:}C{=}\overset{\cdot\cdot}{N}{:}^-$

more nucleophilic

(c)
leaving group   nucleophile

$CH_3CH_2CH_2CH_2Br$ + $CH_3CH_2S^-Na^+$ ⟶ $CH_3CH_2CH_2CH_2SCH_2CH_3$ + $NaBr$

(d)
leaving group   nucleophile

$CH_3CH_2CH_2Cl$ + $CH_3C{\equiv}C^-Na^+$ ⟶ $CH_3C{\equiv}CCH_2CH_2CH_3$ + $NaCl$

(e)
leaving group   nucleophile

$\overset{\displaystyle CH_3}{\underset{\displaystyle |}{}}$
$CH_3CHCH_2CH_2Br$ + $NH_3\ (\text{excess})$ ⟶ $CH_3CHCH_2CH_2NH_2$ + NH_4Br

7.28 (f) CH_3I + $Na^+{}^-CH(COCH_2CH_3)_2$ $\longrightarrow$ $CH_3CH(COCH_2CH_3)_2$ +   NaI

leaving group   nucleophile

7.29 (a) $SH^- > Cl^-$

(b) $(CH_3)_3P > (CH_3)_3B$
(The boron atom has an empty orbital and is an electrophile; the phosphorus atom has nonbonding electrons.)

(c) $CH_3NH^- > CH_3NH_2$

(d) $CH_3SCH_3 > CH_3OCH_3$

7.30 (a) $PhCHCH_2CH_3$ (Br) $\xrightarrow[\Delta]{\text{KOH} \atop \text{ethanol}}$ [alkene structures]   major   +   minor

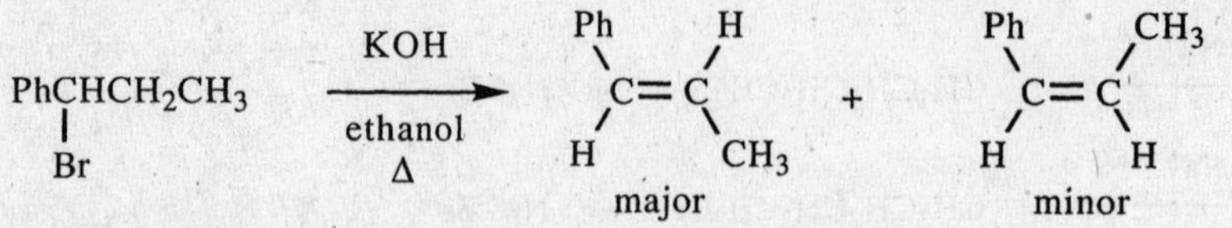

(b) $CH_3CH_2CCH_2CH_2CH_3$ (CH₃)(Br) $\xrightarrow[\Delta]{\text{KOH} \atop \text{ethanol}}$ [alkene structures] major

+ [alkene structures] minor

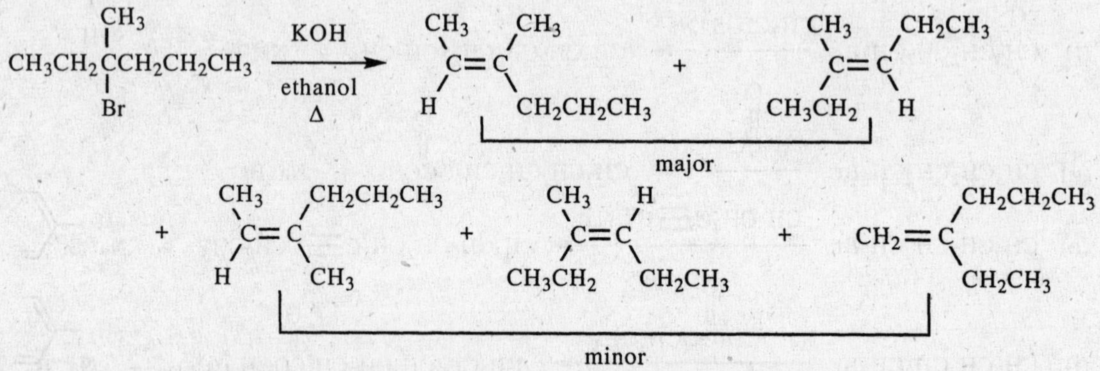

(c) [1-bromo-1-methylcyclohexane] $\xrightarrow[\Delta]{\text{KOH} \atop \text{ethanol}}$ [methylcyclohexene] $-CH_3$ major   +   [methylenecyclohexane] $=CH_2$ minor

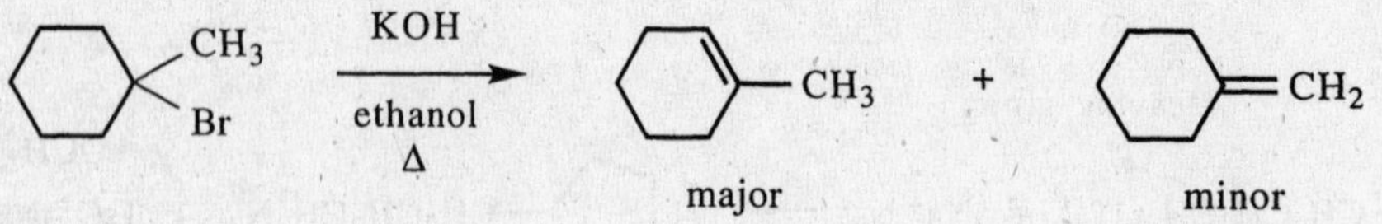

(d) $CH_3CH_2CH_2CHCOOH$ (Br) $\xrightarrow[\Delta]{\text{KOH} \atop \text{ethanol}}$ [alkene carboxylate] major   +   minor

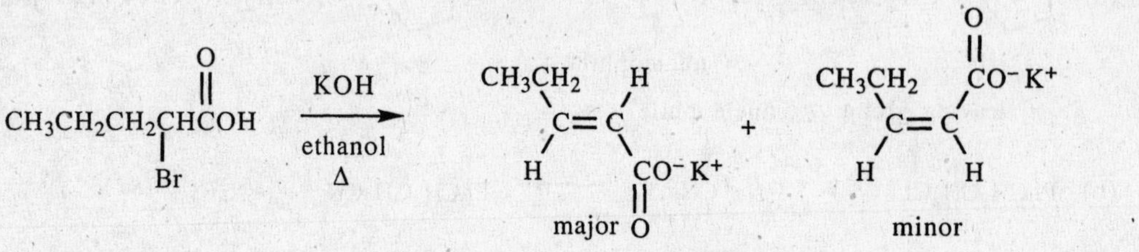

7.31 (a) [styrene derivative with OH] $\xrightarrow[\text{triethylamine}]{CH_3SCl(=O)_2}$ A $\xrightarrow{NaN_3}$ B

(b) [Ph–(CH₂)₈CH₂Br] $\xrightarrow{CH_3NH_2 \text{ (excess)}}$ [Ph–(CH₂)₈CH₂NHCH₃] + $CH_3\overset{+}{N}H_3\ Br^-$
                                                                 C

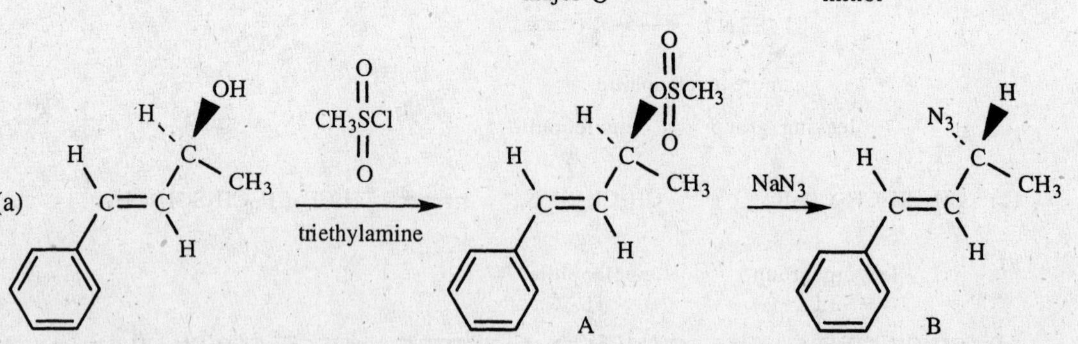

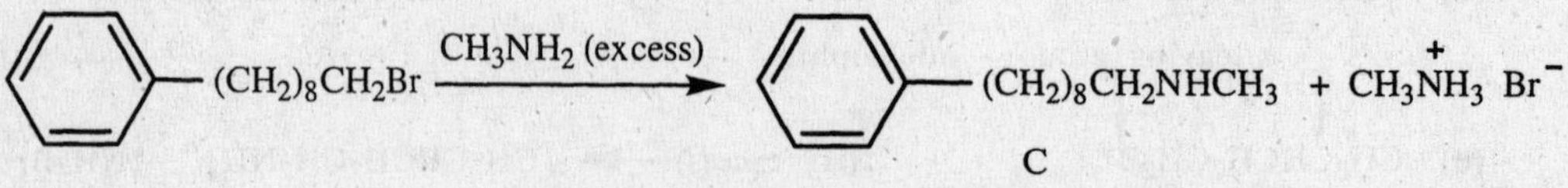

7.31 (c)

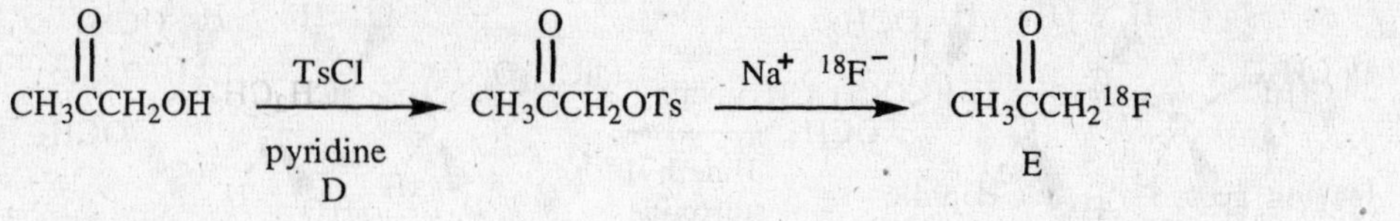

(d)

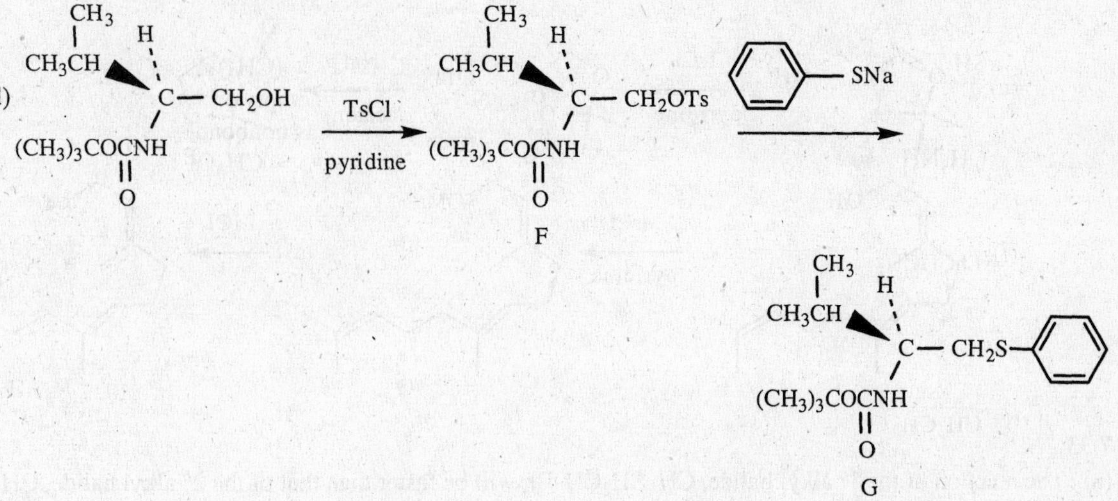

(e)

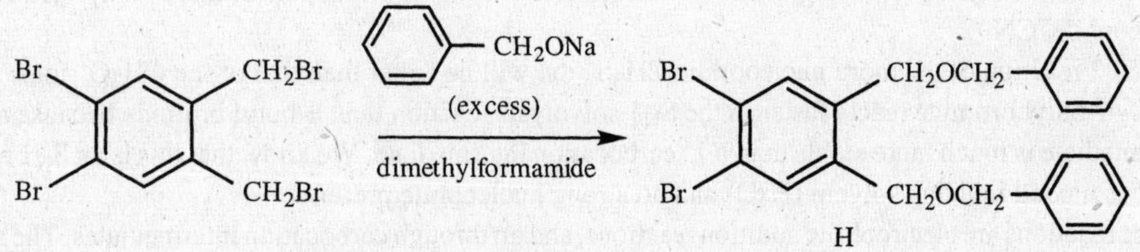

(f)

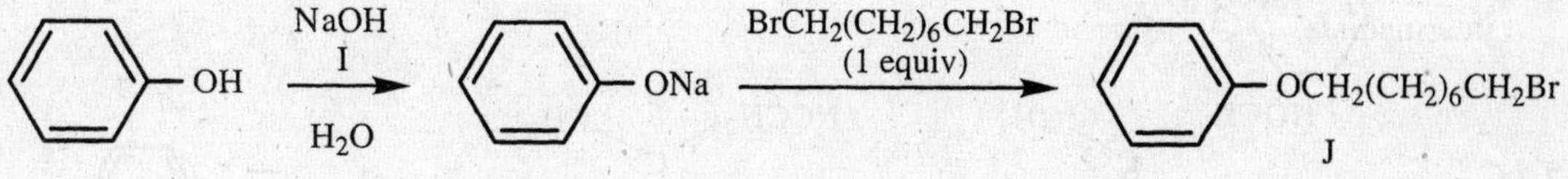

7.32 (a)

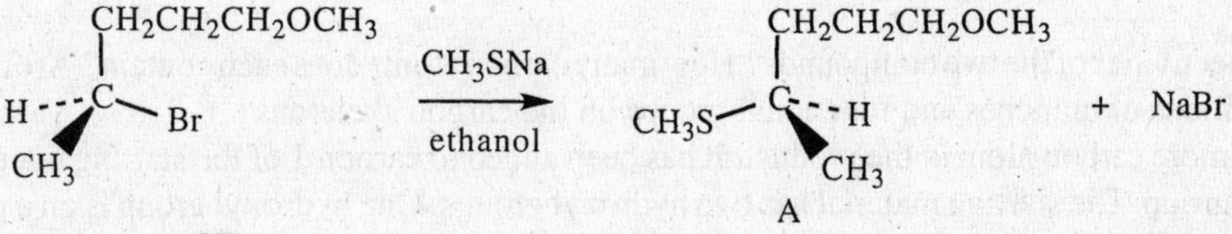

(b)

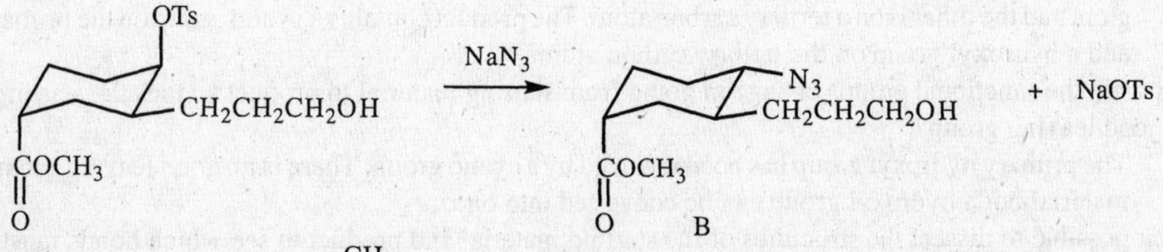

(c)

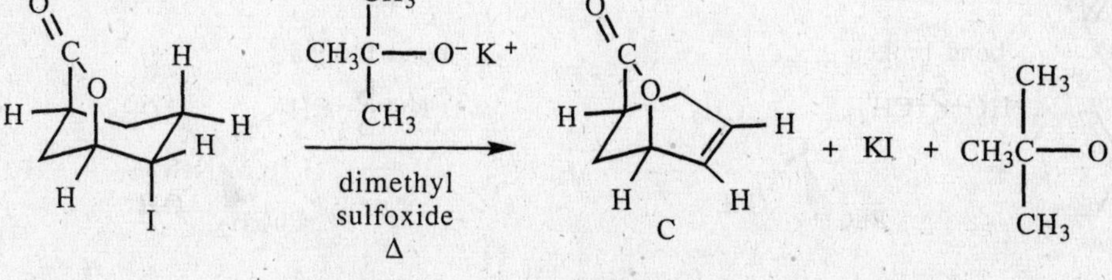

7.32   (d)

(e)

(f)

7.33

(a) The reaction of the 1° alkyl halide, $CH_3CH_2CH_2Br$, will be faster than that of the 2° alkyl halide, $CH_3CHBrCH_3$. There is greater steric crowding in the transition state for the S_N2 reaction of the 2° alkyl halide with the good nucleophile, CN^-.

(b) The S_N2 reaction of the more nucleophilic CH_3S^- ion will be faster than that of the CH_3O^- ion.

(c) The *tert*-butyl bromide reacts faster in the S_N1 solvolysis reaction than *n*-butyl bromide because a 3° carbocation intermediate is much more stable than a 1° carbocation intermediate. We know that this is an S_N1 reaction because there is a good ionizing solvent (H_2O) and no strong nucleophile present.

(d) These reactions are electrophilic addition reactions, and go through carbocation intermediates. The second reaction, which gives a 3° carbocation intermediate, goes faster than the first reaction, which gives a 2° carbocation intermediate.

7.34

1. What are the connectivities of the two compounds? How many carbon atoms does each contain? Are there any rings? What are the positions of branches and functional groups on the carbon skeletons?

   There is one more carbon atom in the product. It has been added to carbon 1 of the starting material and is part of the cyano group. The starting material has two hydroxyl groups. One hydroxyl group is on a primary carbon atom and the other is on a tertiary carbon atom. The product contains a cyano group on the primary carbon atom and a hydroxyl group on the tertiary carbon atom.

2. How do the functional groups change in going from starting material to product? Does the starting material have a good leaving group?

   The primary hydroxyl group has been replaced by a cyano group. There is no good leaving group in the starting material but a hydroxyl group can be converted into one.

3. Is it possible to dissect the structures of the starting material and product to see which bonds must be broken and which formed?

7.34 (cont)

4. New bonds are created when an electrophile reacts with a nucleophile. Do we recognize any part of the product molecule as coming from a good nucleophile or an electrophilic addition?

   The cyano group, —CN, comes from CN⁻, a good nucleophile.

5. What type of compound would be a good precursor to the product?

   Something with a good leaving group where the hydroxyl group is, such as water, a halide, or a tosylate group, would be a good precursor. We cannot use strong acid to convert the hydroxyl group into a better leaving group because the cyanide anion is a better base than the alcohol and would deprotonate the oxonium ion.

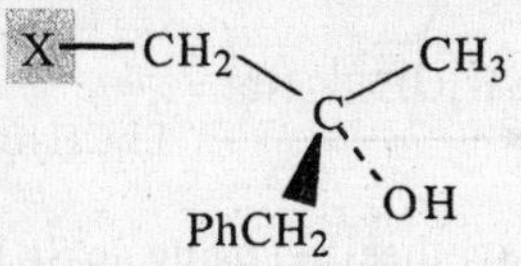

6. After this last step, do we see how to get from starting material to product? If not, we need to analyze the structure obtained in step 5 by applying questions 4 and 5 to it.

   The precursor containing the tosylate group can be made easily from the starting material and tosyl chloride. A primary alcohol will react faster with tosyl chloride than a tertiary alcohol (see Problem 7.21). We can now complete the problem.

Complete correct answer:

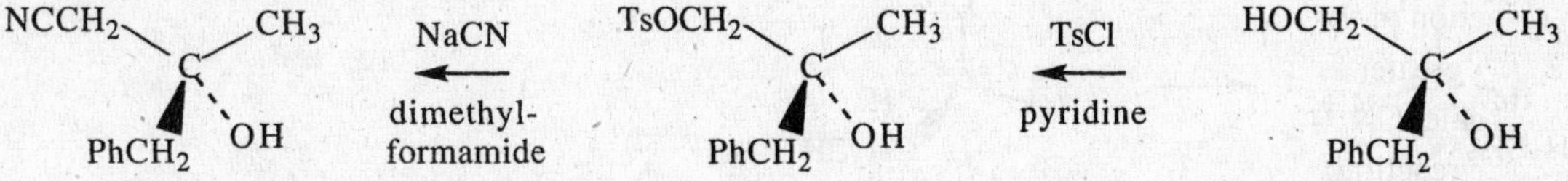

7.35 The answers to parts (a), (c), and (d) analyze the problems as shown in the *Art of Solving Problems* section of the chapter. The rest of the problems show only the equations, written to emphasize the necessity of reasoning backward from the structure of the desired product to the structure of the starting material. Ph is used to represent the benzene ring.

(a) $PhCH_2SCH_2CH_3$ ⟵$^?$— $PhCH_2OH$

1. What are the connectivities of the two compounds? How many carbon atoms does each contain? Are there any rings? What are the positions of branches and functional groups on the carbon skeletons?

   The product has two more carbon atoms than the starting material. These carbon atoms are on the sulfur atom. The thioether group in the product is on the same carbon atom to which the hydroxyl group was attached in the starting material.

2. How do the functional groups change in going from starting material to product? Does the starting material have a good leaving group?

   The hydroxyl group has been replaced by a thioether. The hydroxyl group is not a good leaving group but can be converted into one.

3. Is it possible to dissect the structures of the starting material and product to see which bonds must be broken and which formed?

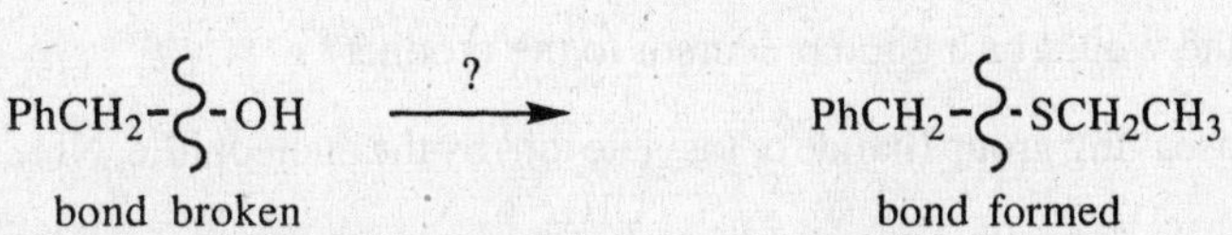

4. New bonds are created when an electrophile reacts with a nucleophile. Do we recognize any part of the product molecule as coming from a good nucleophile or an electrophilic addition?

   The ethanethiolate anion, $CH_3CH_2S^-$, is a good nucleophile.

7.35 (a) (cont)

5. What type of compound would be a good precursor to the product?

    The displacement of a leaving group (halide or tosylate ion) by the nucleophile will give us the product we want.

$$PhCH_2-X \qquad or \qquad PhCH_2-OTs$$

6. After this last step, do we see how to get from starting material to product? If not, we need to analyze the structure obtained in step 5 by applying questions 4 and 5 to it.

    We now can complete the synthesis.

$$PhCH_2SCH_2CH_3 \xleftarrow{\quad CH_3CH_2S^- \, Na^+ \quad} PhCH_2Br \xleftarrow[\quad H_2O \quad]{\quad HBr \quad} PhCH_2OH$$

    The compound could also have been synthesized using an S_N1 reaction with ethanethiol in an ionizing solvent such as water, but the product of the reaction with the solvent would also form. When possible, it is better to use the S_N2 reaction for synthesis, instead of an S_N1 reaction, because there will be fewer competing side reactions to lower the yield of the desired product.

(b) $CH_3CH_2CH_2CH_2CH_2I \xleftarrow[\quad acetone \quad]{\quad NaI \quad} CH_3CH_2CH_2CH_2CH_2OTs \xleftarrow[\quad pyridine \quad]{\quad TsCl \quad} CH_3CH_2CH_2CH_2CH_2OH$

(c)

1. What are the connectivities of the two compounds? How many carbon atoms does each contain? Are there any rings? What are the positions of branches and functional groups on the carbon skeletons?

    The carbon skeletons of the two compounds are the same. The compounds have opposite stereochemistry.

2. How do the functional groups change in going from starting material to product? Does the starting material have a good leaving group?

    A hydroxyl group has been replaced by an amino group. The hydroxyl group is not a good leaving group but can be converted into one.

3. Is it possible to dissect the structures of the starting material and product to see which bonds must be broken and which formed?

bond broken   bond formed

4. New bonds are created when an electrophile reacts with a nucleophile. Do we recognize any part of the product molecule as coming from a good nucleophile or an electrophilic addition?

    The amino group comes from ammonia, NH_3, a good nucleophile.

5. What type of compound would be a good precursor to the product?

    Displacement of a leaving group (halide or tosylate ion) by the nucleophile, NH_3, will give the product we want.

    This compound must have the same stereochemistry as the alcohol, so that an S_N2 reaction with ammonia gives the amine with the inverted configuration.

7.35 (c) (cont)

6. After this last step, do we see how to get from starting material to product? If not, we need to analyze the structure obtained in step 5 by applying questions 4 and 5 to it.

We know how to convert a hydroxyl group into a leaving group with retention of configuration. We can now finish the synthesis.

(d) $PhCH_2CH_2CHCH_3$ $\xleftarrow{?}$ $PhCH_2CH_2CH{=}CH_2$
                           |
                          N_3

1. What are the connectivities of the two compounds? How many carbon atoms does each contain? Are there any rings? What are the positions of branches and functional groups on the carbon skeleton?

The carbon skeletons are identical in starting material and products. The azide functional group in the product is on the second carbon atom of the chain, carbon two of the double bond in the starting material.

2. How do the functional groups change in going from the starting material to the product? Does the starting material have a good leaving group?

The double bond has disappeared. The product is an azide. There is no good leaving group in the starting material.

3. Is it possible to dissect the structure to see which bonds must be broken and which formed?

4. New bonds are created when an electrophile reacts with a nucleophile. Do we recognize any part of the product molecule as coming from a good nucleophile or an electrophilic addition?

The azide group, $-N_3$, comes from N_3^-, a good nucleophile.

5. What type of compound would be a good precursor to the product?

We need a compound that has a good leaving group where the azide group is in the product. Displacement of a leaving group (halide or tosylate ion) by the azide nucleophile will give us the product we want.

$$PhCH_2CH_2CHCH_3$$
$$|$$
$$X$$

6. After this last step, do we see how to get from starting material to product? If not, we need to analyze the structure obtained in step 5 by applying questions 4 and 5 to it.

Since the starting material does not have a leaving group, we need to repeat steps 4 and 5.

4. (again) New bonds are created when an electrophile reacts with a nucleophile. Do we recognize any part of the product molecule as coming from a good nucleophile or an electrophilic addition?

A good leaving group such as a halide can come from an electrophilic addition of a hydrogen halide to the double bond.

7.35 (d) (cont)

5.   (again) What type of compound would be a good precursor to the product?

The electrophilic addition of a hydrogen halide such as hydrogen bromide to the double bond will give us the intermediate we need. We can now complete the problem.

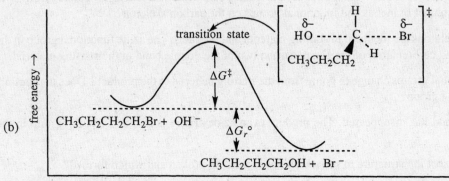

7.36   (a) Rate = k [CH$_3$CH$_2$CH$_2$CH$_2$Br] [NaOH]

(b)

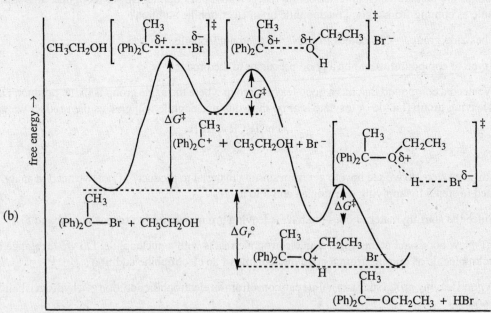

(c) The rate will be doubled.

(d) The rate will be halved.

(e) If some of the ethanol is replaced with water, there will be only a slight effect on the rate since there is only a slight build-up of positive charge in the transition state on the carbon atom undergoing the substitution.

7.37   (a) Rate = k [(Ph)$_2$CCH$_3$]   (with Br above C)

(b)

7.37   (c)   The rate will double.

(d)   Water is a better ionizing solvent than ethanol. The reaction involves the formation of a carbocation, resulting in the separation of charge in the rate-determining step, so an increase in the water content of the solvent will cause an increase in the rate of the reaction by stabilizing the activated complex at the transition state.

7.38

$$CH_3CH=CHCH_2-Cl: \longrightarrow \left[CH_3CH=CH-\overset{+}{C}H_2 \longleftrightarrow CH_3\overset{+}{C}H-CH=CH_2 \right] \quad :Cl:^-$$

resonance-stabilized carbocation, an allylic cation, has
partial positive character at carbon 1 and carbon 3

$$CH_3CH=CH-\overset{+}{C}H_2 \longrightarrow CH_3CH=CH-CH_2-\overset{+}{O} \begin{smallmatrix} H \\ \\ H \end{smallmatrix} \longrightarrow CH_3CH=CHCH_2OH$$

$$CH_3\overset{+}{C}H-CH=CH_2 \longrightarrow CH_3CH-CH=CH_2 \longrightarrow CH_3CH-CH=CH_2$$

7.39

most stable product;
interaction of *p* orbitals of
double bond and aromatic ring

For *cis*-2-phenylcyclohexyl tosylate, the most stable product arises from a conformation of the cyclohexane ring that has the phenyl group in the equatorial position, and the leaving group, the tosyl group, in the axial position required for the E2 reaction to occur (see Section 7.7C).

no reaction

In *trans*-2-phenylcyclohexyl tosylate the bulky phenyl ring prevents the molecule from achieving the diaxial conformation required for elimination, and no reaction takes place.

7.40   (a) $$CH_3CH_2O-CH_2-Cl: \longrightarrow \left[CH_3CH_2O-\overset{+}{C}H_2 \longleftrightarrow CH_3CH_2\overset{+}{O}=CH_2 \right]$$

$$:Cl:^-$$

Ionization of halide in a first-order reaction is the rate-determining step.

7.40   (b) The relative rates of this first-order reaction are determined by the rate at which the alkyl chlorides ionize, and this is determined by the relative stabilities of the carbocations that would form upon ionization. The carbocation from the ionization of chloroethoxymethane is resonance stabilized as shown in the answer to part (a).

$$CH_3CH_2CH_2CH_2 \overset{..}{\underset{..}{Cl}} : \longrightarrow CH_3CH_2CH_2CH_2^+ \quad : \overset{..}{\underset{..}{Cl}} :^-$$

A primary cation, much less stable than the cation from chloroethoxymethane, therefore much higher $\Delta G^{\ddagger}$, therefore much slower reaction.

$$CH_3\overset{..}{\underset{..}{O}} - CH_2 - CH_2 \overset{..}{\underset{..}{Cl}} : \longrightarrow CH_3\overset{..}{\underset{..}{O}} \leftarrow CH_2 \leftarrow CH_2^+ \quad : \overset{..}{\underset{..}{Cl}} :^-$$

The oxygen in this cation is in a position to destabilize the primary cation that forms still further by its inductive effect. This leads to a still higher $\Delta G^{\ddagger}$, and a still slower reaction.

7.41 $^- : \overset{..}{S} - C \equiv N : \longleftrightarrow \overset{..}{S} = C = \overset{..}{N} :^- \qquad R - \overset{..}{S} - C \equiv N :$

                                thiocyanate ion

$\overset{..}{O} = C = \overset{..}{N} :^- \longleftrightarrow {}^- : \overset{..}{O} - C \equiv N : \qquad R - \overset{..}{N} = C = \overset{..}{O}$

                                cyanate ion

In the thiocyanate ion, the electrons on the sulfur are more polarizable than those on the nitrogen, and that end of the ion is the better nucleophile. In the cyanate ion, the electrons on the nitrogen are more polarizable than those on the oxygen, and the nitrogen is a better nucleophile.

7.42   (a)

$$CH_3 \underset{H}{\overset{H}{\diagdown}} C \underset{Br}{\overset{H}{\diagup}} CH_3 \xrightarrow{NaBr} CH_3 \underset{Br}{\overset{H}{\diagdown}} C \underset{H}{\overset{H}{\diagup}} CH_3$$

(*S*)--2-bromopentane   (*R*)--2-bromopentane
                                          inversion of configuration

These compounds are enantiomers.

(b) The results of Experiment B, in which the reaction rate increases with an increase in the concentration of bromide ion, indicates that the reaction proceeds by an S_N2 reaction. The rate of an S_N2 reaction depends on the concentration of the nucleophile as well as on the concentration of the alkyl halide.

(c) The starting material is reversibly converted to product (Statement 3).

The starting material and the product are at the same level with respect to energy, therefore neither form is favored. The reaction proceeds randomly, converting each enantiomer to the other, until the mixture becomes optically inactive. The interconversion of the enantiomers continues, but can no longer be observed by a change in the optical activity of the solution.

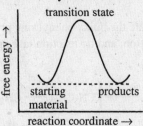

free energy →

transition state

starting material   products

reaction coordinate →

7.43

(a)

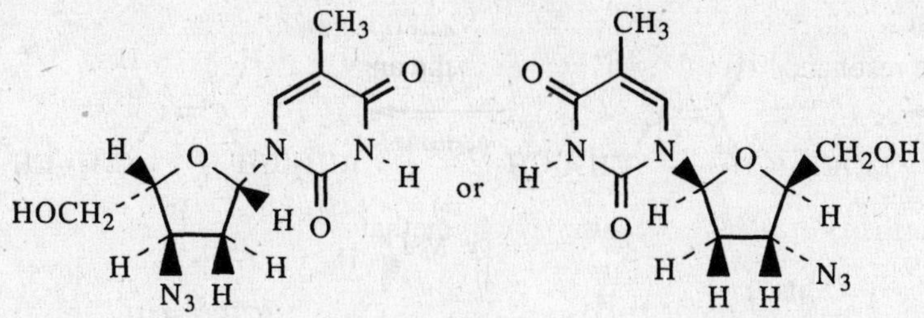

(–)-AZT

(b) HOCH$_2$

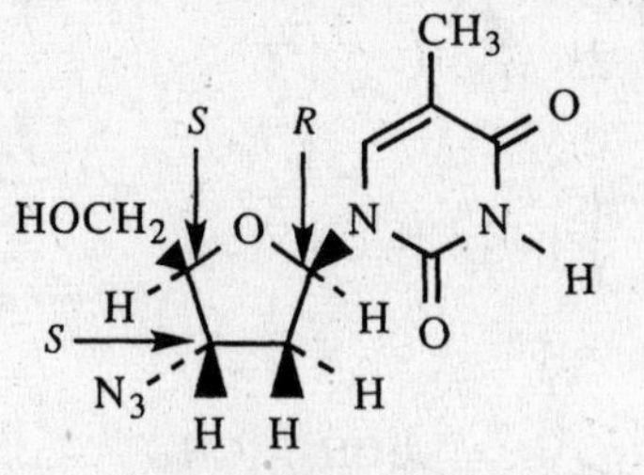

(+)-AZT

(c) –56°

(d) The number of stereoisomers possible for a compound with *n* stereocenters is 2^n. There are 3 stereocenters so the total number of stereoisomers possible is 2^3 or 8.

7.44   (a) Alkene **C** is the major product. It is the more highly substituted alkene. The transition state for its formation is of lower energy than the transition state for the formation of the other alkene.

(b)

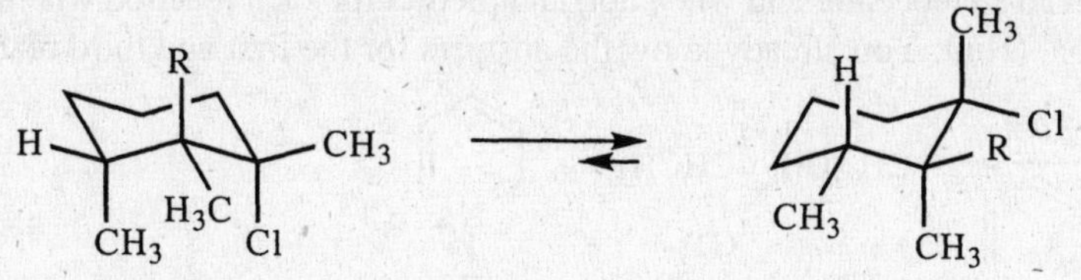

(c)

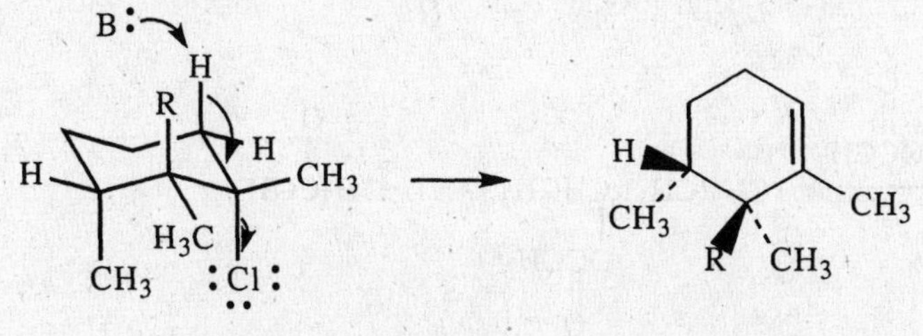

7.45   Compound **X** forms 56% of the mixture. **X** and **Y** are not formed in equal amounts at equilibrium because they are diastereomers of each other. Enantiomers, which are of equal energy, form in equal amounts at equilibrium (see Problem 7.42). Diastereomers have different energies and form in different amounts.

7.46

(a) The double bond has the *Z* configuration.

(b)

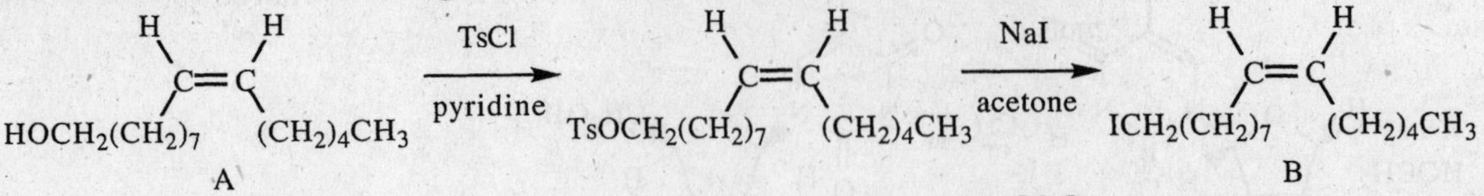

(c)

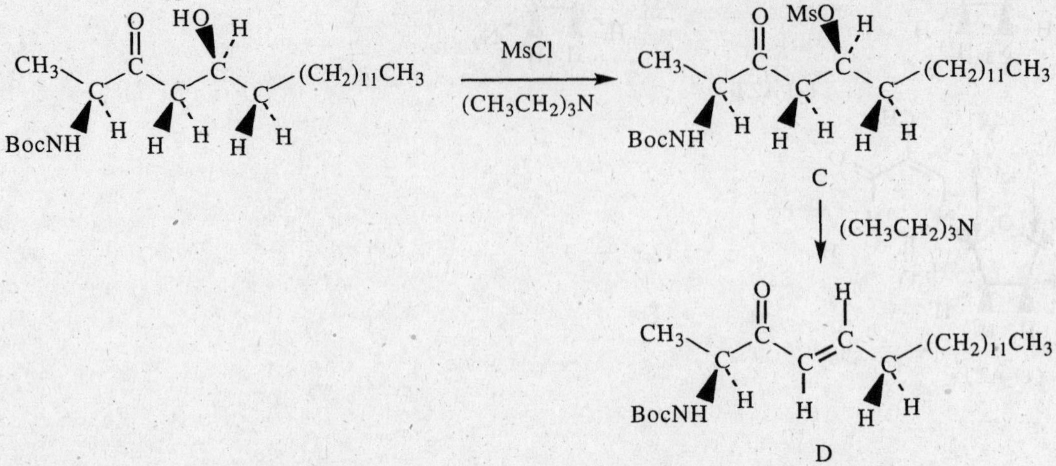

Supplemental Problems

S7.1   You can obtain experimental evidence for inversion of configuration by carefully working out the configurational relationships of different optically active compounds. One such cycle is outlined below. Look at all of the reactions, decide when bonds are being broken at a stereocenter and when not, and predict for each reaction whether it goes with inversion or retention of configuration. (Hint: You already know the answers for the first and third reactions.)

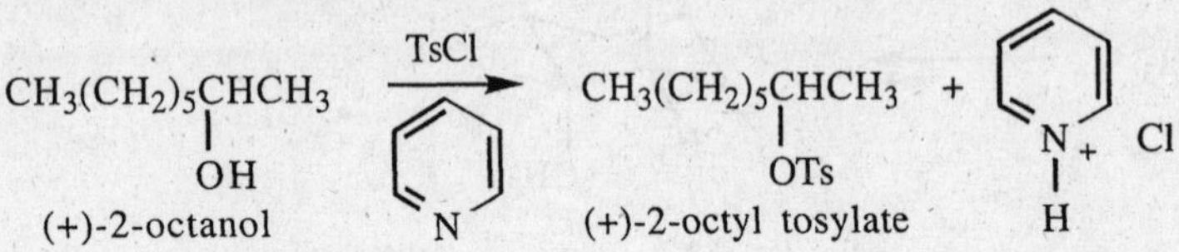

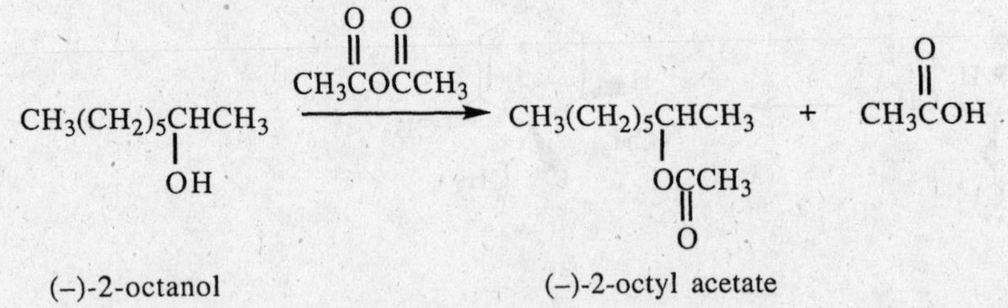

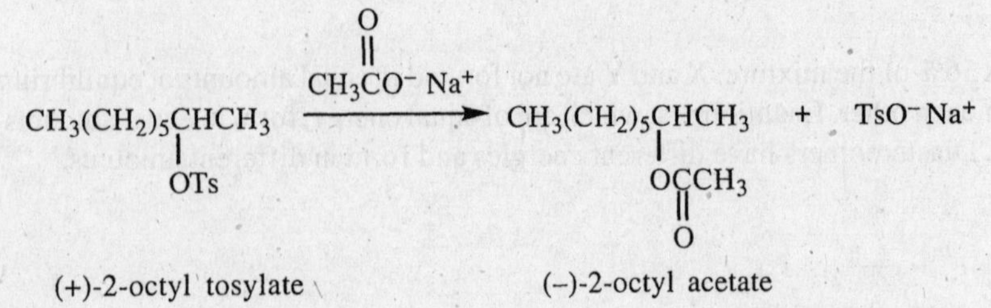

S7.2   The following equations represent some nucleophilic substitution reactions that are possible. Only primary alkyl groups are used, so the reactions will be S_N2, and competing elimination reactions will not be a problem. Complete the equations, showing the products that will form. In each case it will be helpful to label the leaving group and the incoming nucleophile.

(a) CH_3CH_2I + $P(CH_2CH_2CH_2CH_3)_3 \longrightarrow$ (b) $HOCH_2CH_2Cl$ + $CH_3S^- Na^+ \longrightarrow$

S7.3   The following incomplete reaction sequences give only a starting material and a product that can be prepared from it. Show the reagents that would be necessary to carry out each transformation and any major products that would be formed in the intermediate stages. There may be more than one correct way to complete a sequence. Ph is used to represent the benzene ring.

(a) $CH_3CH_2CH_2CH_2Br$ $\rightarrow$ $CH_3CH_2CH_2CH_2C{\equiv}CH$ (b) $PhSH$ $\rightarrow$ $PhSCH_2CH_2Ph$

S7.4   One of the classic experiments in which the stereochemistry of E2 reactions was explored involved menthyl chloride and neomenthyl chloride. These two compounds differ from each other only in the stereochemistry at the carbon atom to which the chlorine atom is attached. The following equations show the reactions they undergo without revealing their stereochemistry.

(a)   Using the information given, assign relative stereochemistry to menthyl chloride and neomenthyl chloride. Menthyl chloride is derived from natural sources, and the substituents on the cyclohexane ring are in the most stable configuration.

(b)   When the reaction with menthyl chloride is carried out in 80% aqueous ethanol with no added base, the following products are obtained. How do you explain these results?

S7.5   The following experimental observations were made. Even though the two reactions proceeded at very different rates, the relative amounts of the products obtained were practically identical.

S7.5 (cont)

The first reaction is 7.5 times faster than the second reaction.

(a) What do these data suggest about the mechanism of the reaction?

(b) Why are the rates of these two reactions different?

S7.6   Chromanones are compounds that are responsible for color in leaves and flowers. The following transformation was carried out as part of a recent synthesis of a chromanone.

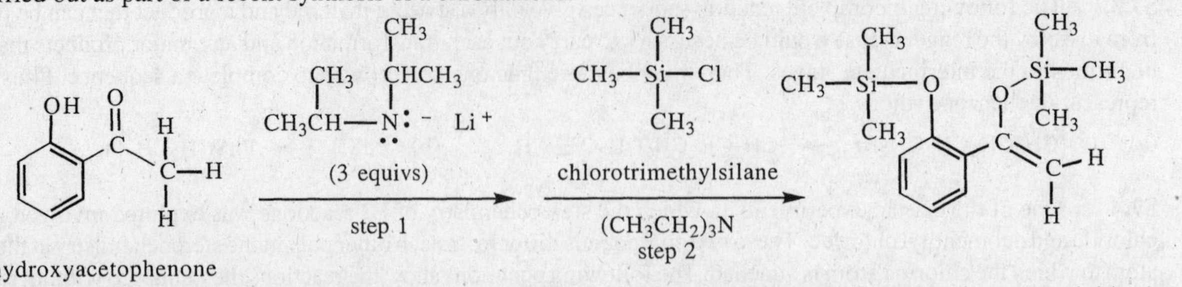

2-hydroxyacetophenone

(a) The first step is the deprotonation of the starting material, 2-hydroxyacetophenone. Examine the connectivities in the starting material and the product to identify where deprotonation has occurred. What is the base in the reaction? Use the pK_a table to assign an approximate pK_a value to the conjugate acid of this base.

(b) Which is the most acidic proton on the starting material and what is an approximate pK_a value for it? Draw a structural formula for the anion resulting from the first deprotonation of 2-hydroxyacetophenone.

(c) The second step of the synthesis involves reaction of the anion from the deprotonation of 2-hydroxyacetophenone with chlorotrimethylsilane. Silicon (Si) is just below carbon in the periodic table and resembles carbon in its ability to form covalent bonds. What general type of reaction has taken place in step 2 of the synthesis shown above? Write a mechanism using the curved-arrow symbolism to show how the anion formed in part (b) from 2-hydroxyacetophenone would react with chlorotrimethylsilane.

(d) Another deprotonation reaction takes place on the product of the reaction of the anion of 2-hydroxyacetophenone with chlorotrimethylsilane found in part (c). Use the pK_a table to determine the *second most acidic* proton in 2-hydroxyacetophenone. Resonance is important in stabilizing the anion that results from this second deprotonation. Draw a structural formula for this new anion and for its major resonance contributor.

S7.7   The metabolism of amino acids is often traced by synthesizing compounds that are labeled with isotopes of carbon or hydrogen at specific positions.

(a)   Provide the product of the following reaction used to synthesize arginine labeled with ^{13}C.

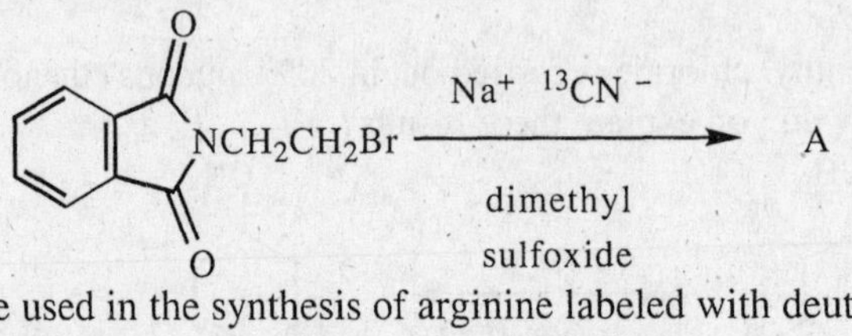

(b) The following reactions were used in the synthesis of arginine labeled with deuterium, ^2H or D. Write structural formulas for compounds B and C, the intermediates in the synthesis.

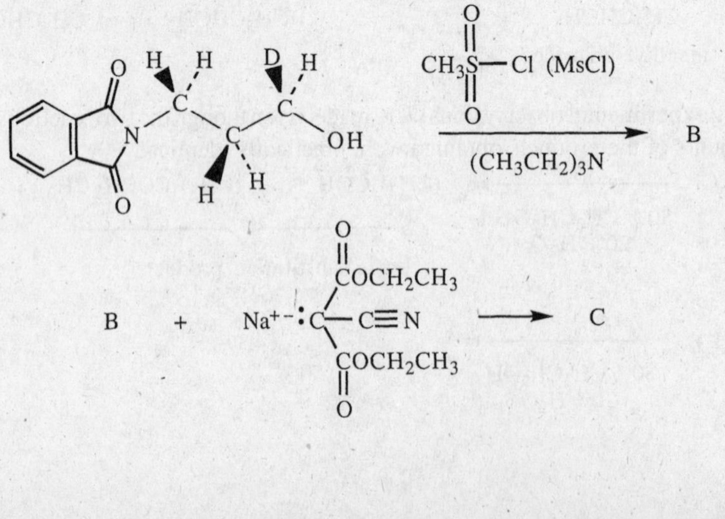

S7.8   Many compounds that are toxic to cells act by alkylating DNA. One such compound is chlorambucil, which is used in the chemotherapy of cancer, especially for leukemia. The compound reacts with the nitrogen at position 7 of guanine, one of the four bases that are part of the genetic code in DNA. The reaction is helped by participation of the nonbonding electrons on the nitrogen at position 9 in guanine. Draw a mechanism for the reaction of guanine with chlorambucil. Only one of the leaving groups is lost in the reaction. In DNA, chlorambucil can react twice, linking different parts of the DNA helix together, thus stopping reproduction of the cell.

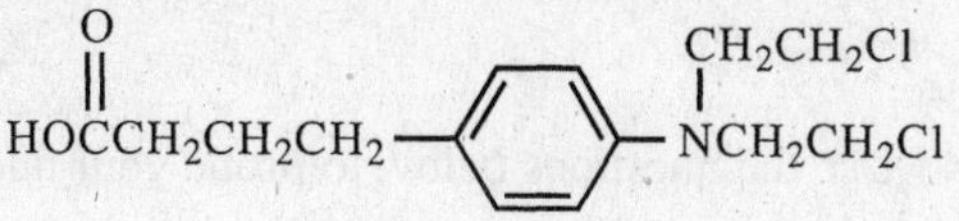

chlorambucil 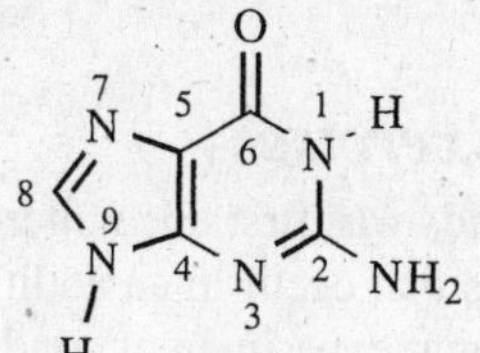   guanine

S7.9   The following equations represent chemical reactions carried out in research into the synthesis of useful and interesting compounds. Fill in structural formulas for the missing starting materials, products, or reagents.

(a)   → A (Is the configuration of the product *R* or *S*?)

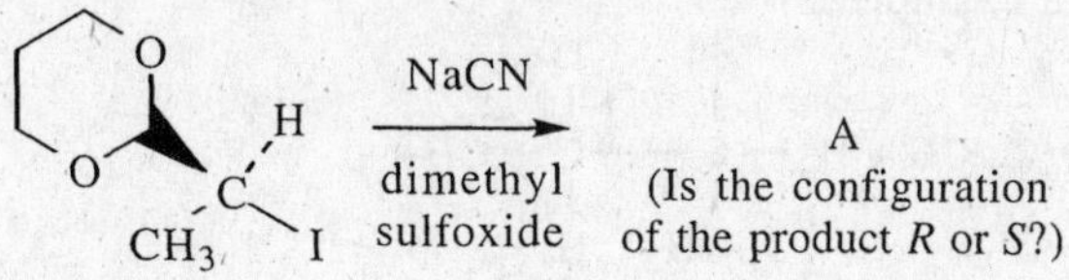

(b)   →   B

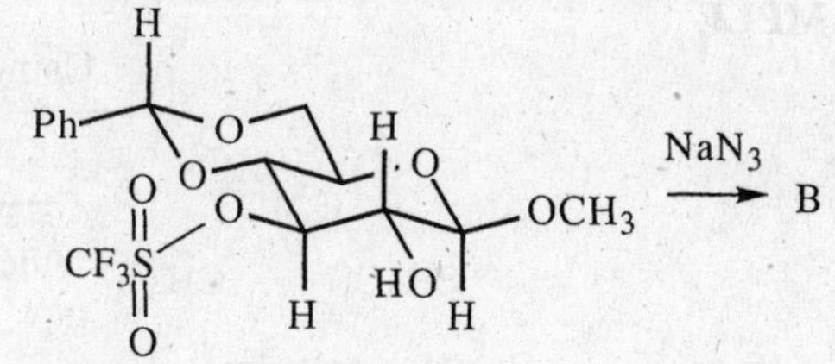

(c)   →   D

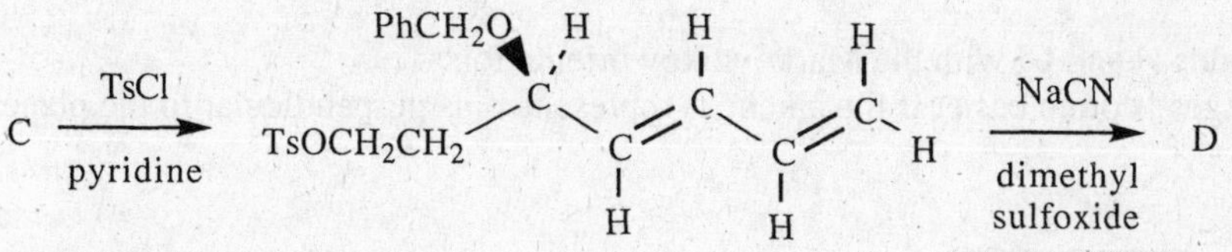

(d)

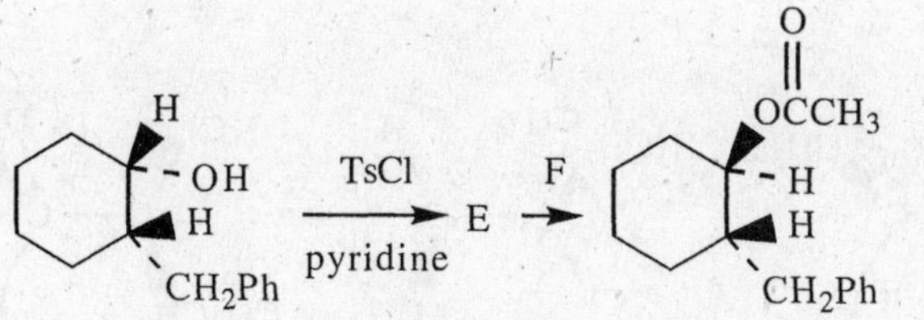

(e)

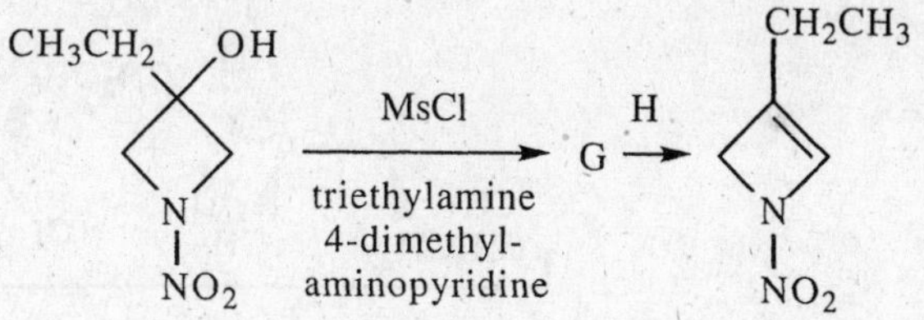

8 Alkenes

Workbook Exercises

Addition to π bonds was first encountered in Chapter 4. Use the questions below to probe your understanding of the structural changes that occur in an addition reaction.

Q1. Under what experimental conditions does the addition occur?

Q2. What molecule is adding; what are the groups that add?

Q3. What is the regioselectivity (for example, the Markovnikov orientation)?

Q4. What is the stereochemistry? The two groups can add on the same side of the π bond (called **syn** addition), or on opposite sides of the π bond (called **anti** addition).

EXAMPLE

Using a familiar example:

$$CH_3CH_2\text{--}C(CH_3)\text{=}C(D)(CH_3) \xrightarrow[\substack{\text{diethyl ether} \\ 0\,°C}]{HCl} CH_3CH_2\text{--}\overset{Cl}{\underset{CH_3}{C}}\text{--}\overset{D}{\underset{H}{C}}\text{--}CH_3$$

(D = deuterium, ^{2}H)

a mixture of stereoisomers

SOLUTION

Answers to Q1, Q2, and Q3: HCl adds H and Cl with the Markovnikov orientation.

Answer to Q4: Visualizing the changes is often easier if the alkene is represented as perpendicular to the plane of the paper.

$$CH_3CH_2, CH_3\text{C=C}D, CH_3 \xrightarrow[\substack{\text{diethyl ether} \\ 0\,°C}]{HCl}$$

a pair of enantiomers representing
syn (same side) addition

a pair of enantiomers representing
anti (opposite side) addition

An important skill for you to develop is to connect a written (and/or verbal) description of a structural change with a molecular representation of that transformation. The following example will help you develop this skill.

EXAMPLE

Draw the structure corresponding to the anti addition of bromine to cyclohexene (the six-membered hydrocarbon ring with one carbon-carbon double bond).

SOLUTION

Restatement of the verbal description in structural terms:

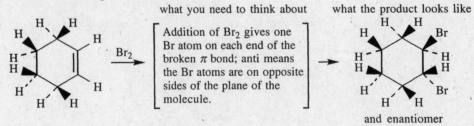

what you need to think about

[Addition of Br$_2$ gives one Br atom on each end of the broken π bond; anti means the Br atoms are on opposite sides of the plane of the molecule.]

what the product looks like

and enantiomer

Workbook Exercises (cont)

EXERCISE I.   In each case, provide structural formulas for the starting materials, reagents, or products needed to complete the equation. Examine all of the information carefully.

(a)

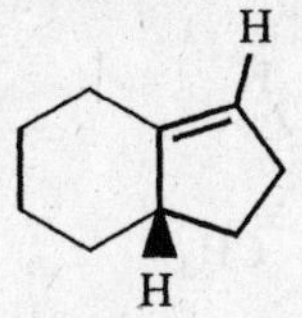

syn addition of H_2 → A + B

a pair of diastereomers
derived from the addition to
each face of the π bond

(b) C or D or E   Markovnikov addition of HBr → 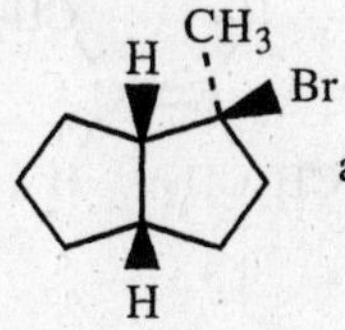   and other isomers

(c) F   syn addition of an oxygen atom → 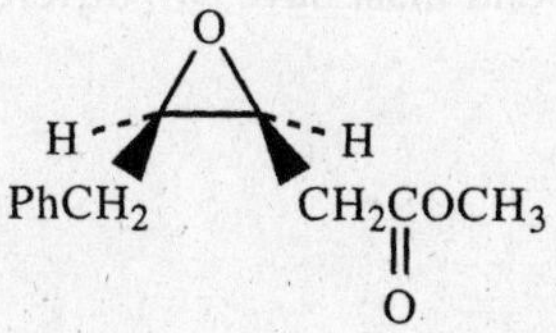   and enantiomer

EXERCISE II.   Give a complete description of the changes in connectivity that accompany each of the following reactions.

(a)

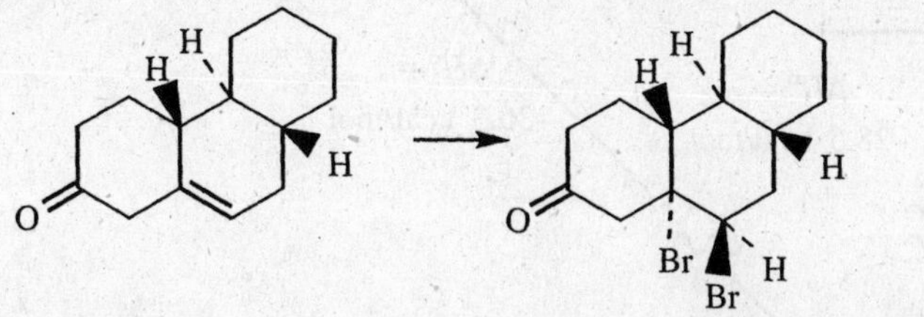

(b)

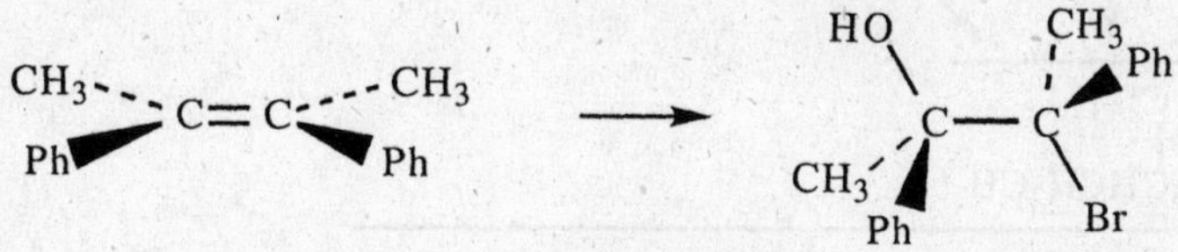

and enantiomer

(c)

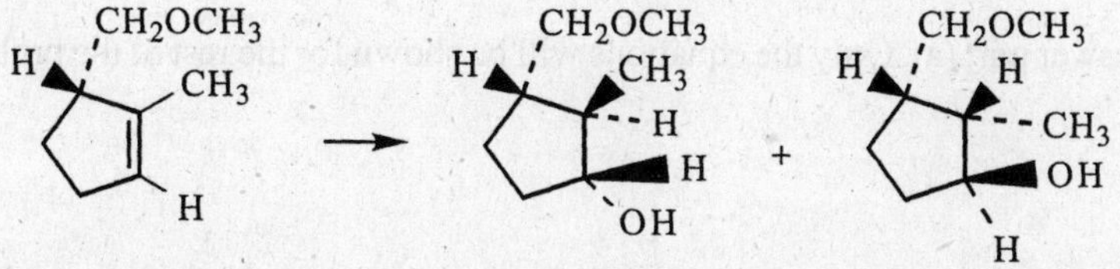

8.1 C_5H_{10}

(1) $CH_3CH_2CH_2CH=CH_2$

(2)

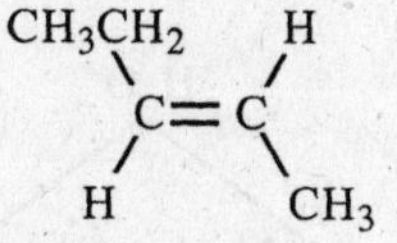

(3)

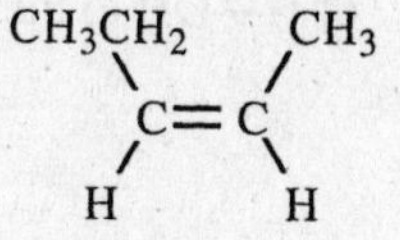

(4) 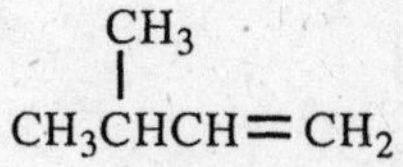 $CH_3CHCH=CH_2$

(5)

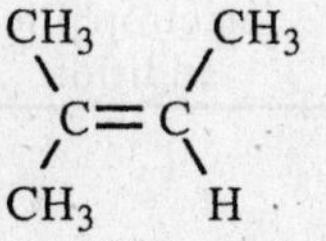

(6) 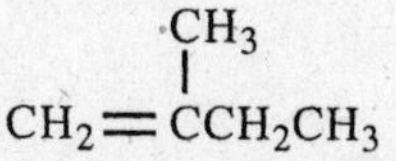 $CH_2=CCH_2CH_3$

8.2 (1) 1-pentene   (2) (*E*)-2-pentene   (3) (*Z*)-2-pentene
         (4) 3-methyl-1-butene   (5) 2-methyl-2-butene   (6) 2-methyl-1-butene

8.3 (a)   (b)   (c)

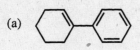

(d)   (e)   (f)

8.4 No. Each carbon atom of the double bond must have two different substituents for this type of isomerism to be possible.

8.5

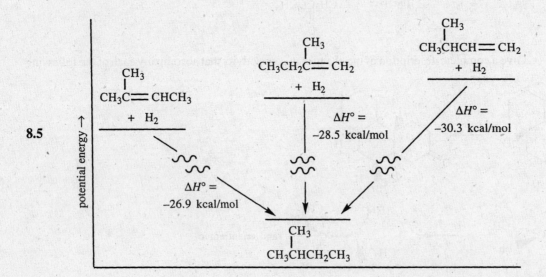

8.6 A concept triangle will be used to answer part (a). Only the equations will be shown for the rest of the problem.

(a) $\xrightarrow[\text{H}_2\text{SO}_4]{\text{H}_2\text{O}}$?

The starting material is an alkene. The reagent is a mixture of water with a strong acid; therefore, the most abundant species in solution will be the hydronium ion, H_3O^+, an electrophile. The type of reaction is an electrophilic addition to a double bond. Because the electrophile in H_2O is H_3O^+, the overall addition is that of H—OH to the double bond.

H_3O^+

Markovnikov electrophilic addition

8.6 (a) (cont)

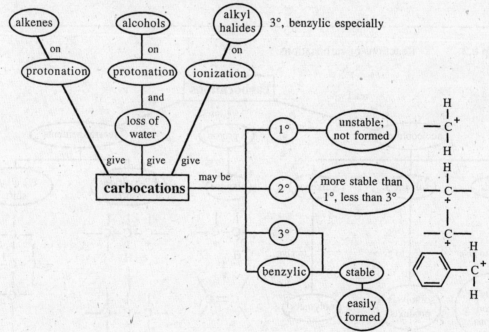

The alkene is symmetric, so we do not have to worry about regioselectivity in this case.

(b)

(c) $CH_3CH_2CH_2CH_2CH=CH_2$ $\xrightarrow[H_2SO_4]{H_2O}$ $CH_3CH_2CH_2CH_2CHCH_3$ with OH

(d) CH_3CCH_3 with CH$_3$ and OH $\xrightarrow[\Delta]{H_3PO_4}$ $CH_3C=CH_2$ with CH$_3$

(e) $CH_3C=CHCH_2CH_3$ with CH$_3$ $\xrightarrow[H_2SO_4]{H_2O}$ $CH_3CCH_2CH_2CH_3$ with CH$_3$ and OH

(f)

Concept Map 8.1   Reactions that form carbocations.

3°, benzylic especially

alkenes — on — protonation

alcohols — on — protonation — and — loss of water

alkyl halides — on — ionization

give   give   give

carbocations   may be

1° — unstable; not formed

2° — more stable than 1°, less than 3°

3°

benzylic — stable

easily formed

8.7

$CH_3C=CCH_3$ (with CH$_3$, CH$_3$)   → $CH_3C—CCH_3$ (with CH$_3$, CH$_3$, $+$, H)

$H—\overset{..}{\underset{..}{O}}—PO_3H_2$ $:\overset{..}{\underset{..}{O}}—PO_3H_2$

8.7 (cont)

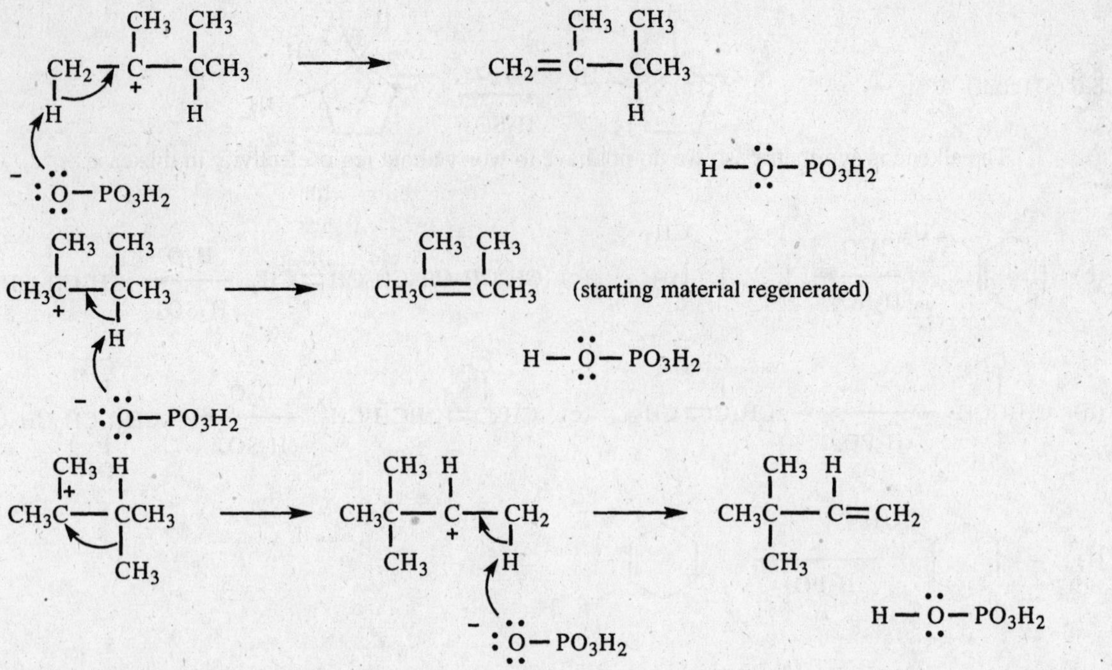

Note that very little of this product is seen. It arises from a secondary cation, which is less stable than a tertiary cation.

Concept Map 8.2 Reactions of carbocations.

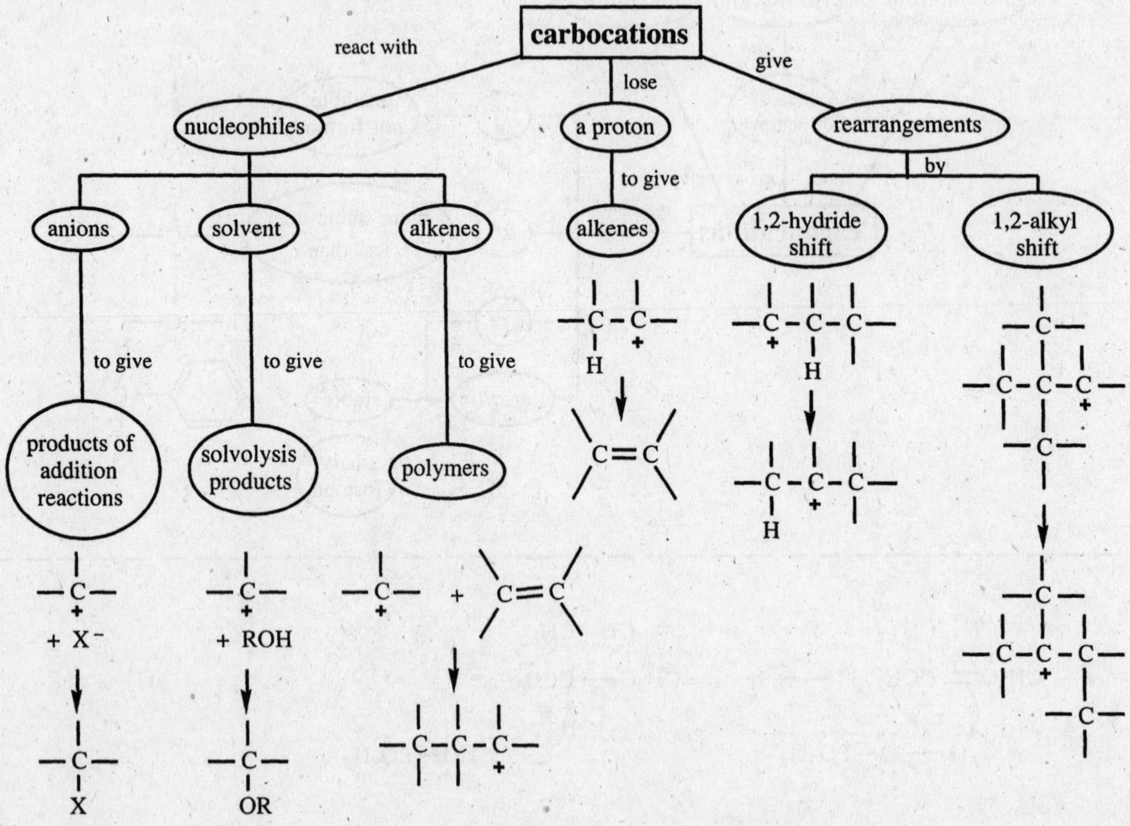

8.8

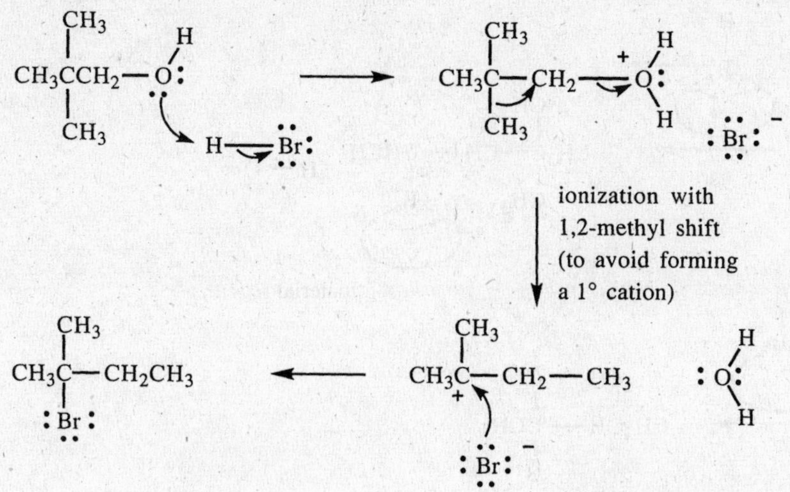

ionization with
1,2-methyl shift
(to avoid forming
a 1° cation)

8.9

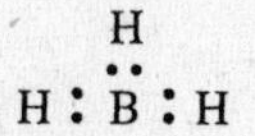

shape is trigonal planar;
hybridization of the
boron atom is *sp*^{2}

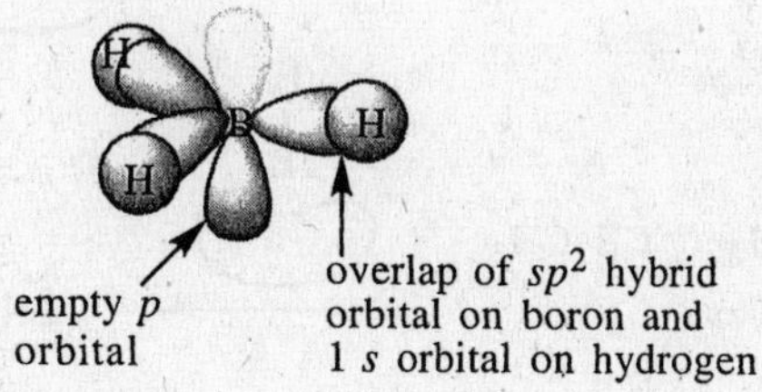

empty *p*
orbital

overlap of *sp*^{2} hybrid
orbital on boron and
1 *s* orbital on hydrogen

8.10

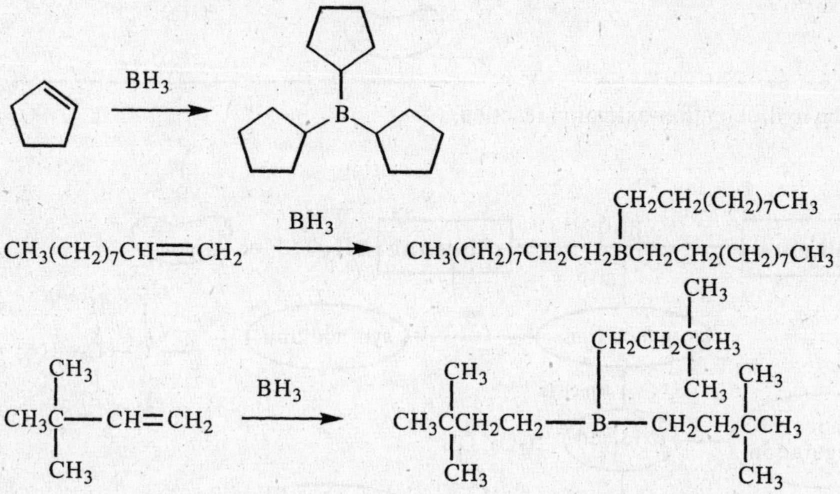

$CH_3(CH_2)_7CH = CH_2$ $\xrightarrow{BH_3}$ $CH_3(CH_2)_7CH_2CH_2BCH_2CH_2(CH_2)_7CH_3$

8.11 (a) $CH_3CH_2CH_2C = CH_2$ $\longrightarrow$ $CH_3CH_2CH_2CHCH_2 - B$

(b) $PhCH = CH_2$ $\longrightarrow$ $PhCH_2CH_2 - B$

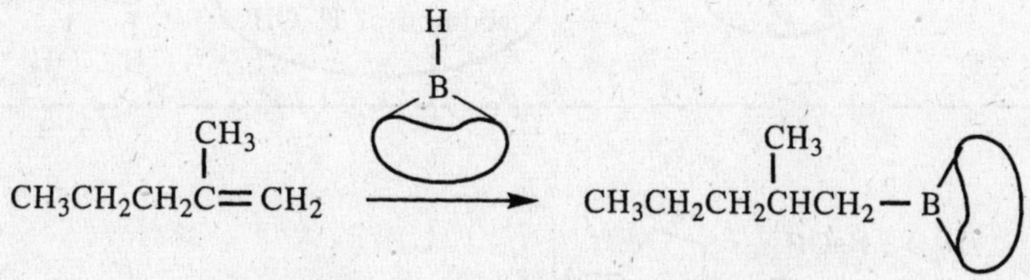

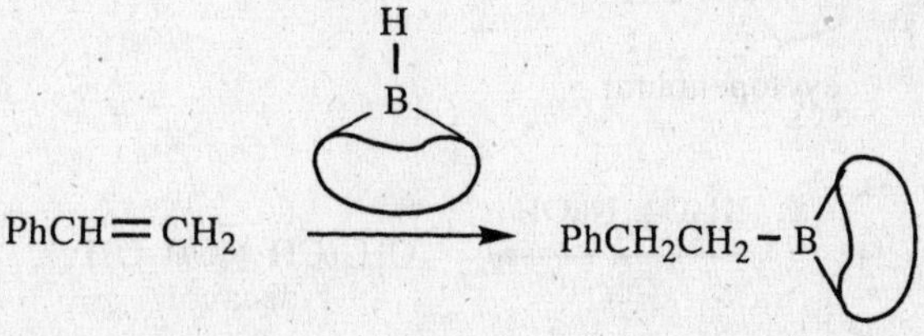

8.11   (c)

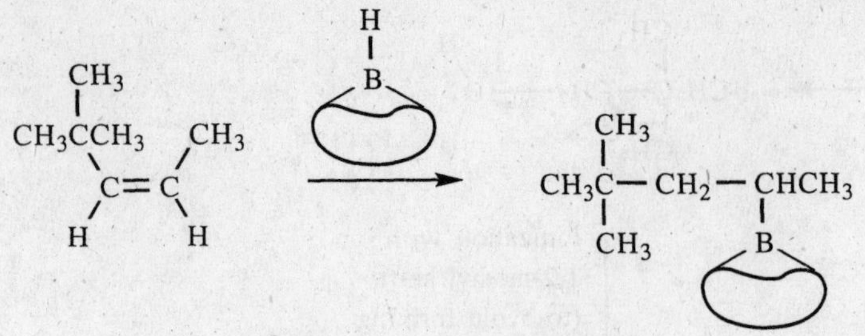

$$CH_3CCH_3 \quad CH_3$$

with diborane reagent gives

$$CH_3C-CH_2-CHCH_3$$
$$CH_3$$ (B)

(d)

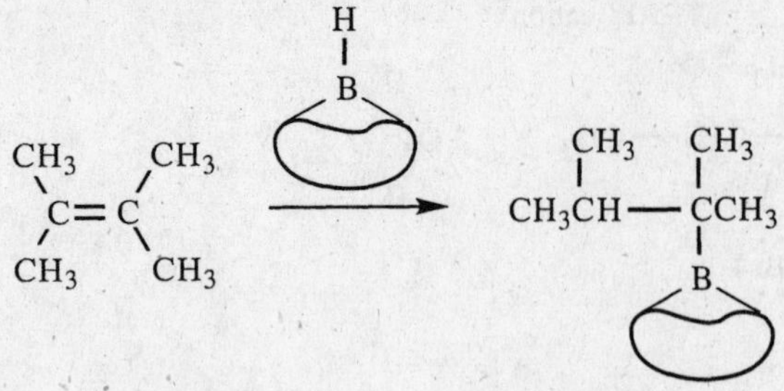

$$CH_3C=CCH_3 \quad (CH_3)_2$$

gives

$$CH_3CH-CCH_3$$

(e)

$$CH_3CH_2 \quad CH_2CH_3$$
$$C=C$$
$$H \qquad H$$

gives

$$CH_3CH_2CH_2CHCH_2CH_3$$
(B)

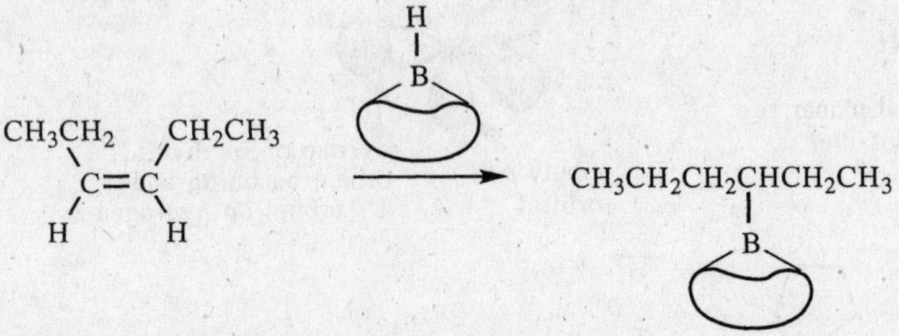

Concept Map 8.3   The hydroboration-oxidation reaction.

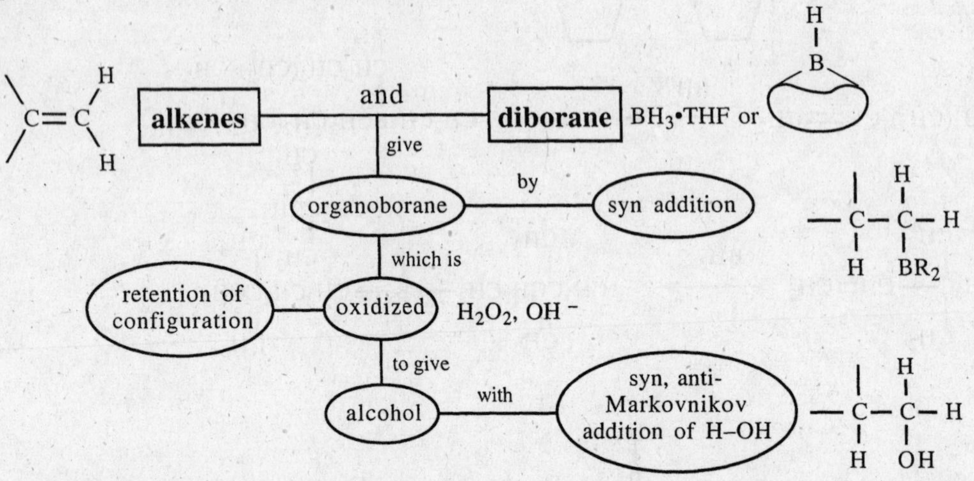

alkenes **and** **diborane** $BH_3 \cdot THF$ or

give

organoborane — by — syn addition

$$-\underset{H}{\overset{H}{C}}-\underset{BR_2}{\overset{H}{C}}-H$$

retention of configuration — oxidized H_2O_2, OH^-

to give

alcohol — with — syn, anti-Markovnikov addition of H–OH

$$-\underset{H}{\overset{H}{C}}-\underset{OH}{\overset{H}{C}}-H$$

8.12   (a)

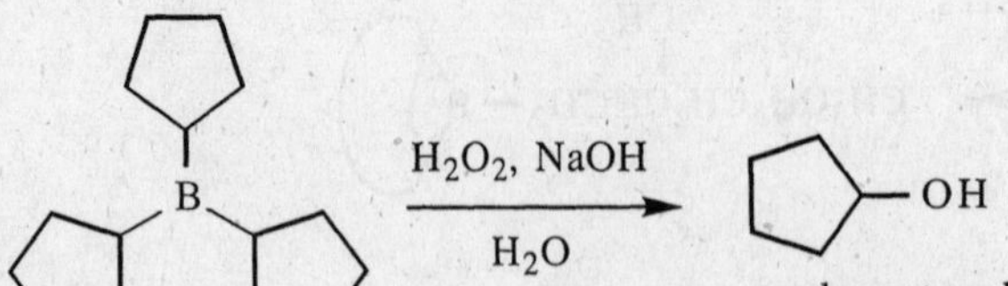

tricyclopentylborane $\xrightarrow[\text{H}_2\text{O}]{\text{H}_2\text{O}_2, \text{ NaOH}}$ cyclopentanol — OH

(b) $CH_2CH_2(CH_2)_7CH_3$

$CH_3(CH_2)_7CH_2CH_2BCH_2CH_2(CH_2)_7CH_3$ $\xrightarrow[\text{H}_2\text{O}]{\text{H}_2\text{O}_2, \text{ NaOH}}$ $CH_3(CH_2)_8CH_2OH$
1-decanol

8.12 (c)

CH$_3$CHCH$_2$CH$_2$—B—CH$_2$CH$_2$CCH$_3$ $\xrightarrow[\text{H}_2\text{O}]{\text{H}_2\text{O}_2, \text{ NaOH}}$ CH$_3$CCH$_2$CH$_2$OH

where the boron bears CH$_2$CH$_2$CCH$_3$ (with CH$_3$), CH$_3$ substituents, CH$_3$ and CH$_3$ groups

3,3-dimethyl-1-butanol

(a') CH$_3$CH$_2$CH$_2$CHCH$_2$—B $\xrightarrow[\text{H}_2\text{O}]{\text{H}_2\text{O}_2, \text{ NaOH}}$ CH$_3$CH$_2$CH$_2$CHCH$_2$OH
       with CH$_3$ on the CH   with CH$_3$ on the CH

2-methyl-1-pentanol

(b') PhCH$_2$CH$_2$—B $\xrightarrow[\text{H}_2\text{O}]{\text{H}_2\text{O}_2, \text{ NaOH}}$ PhCH$_2$CH$_2$OH

2-phenylethanol

(c') CH$_3$C—CH$_2$—CHCH$_3$ $\xrightarrow[\text{H}_2\text{O}]{\text{H}_2\text{O}_2, \text{ NaOH}}$ CH$_3$C—CH$_2$—CHCH$_3$
      with CH$_3$, CH$_3$ on C and B on CH   with CH$_3$, CH$_3$ and OH

4,4-dimethyl-2-pentanol

(d') CH$_3$CH—CCH$_3$ $\xrightarrow[\text{H}_2\text{O}]{\text{H}_2\text{O}_2, \text{ NaOH}}$ CH$_3$CH—CCH$_3$
      with CH$_3$, CH$_3$ and B   with CH$_3$, CH$_3$ and OH

2,3-dimethyl-2-butanol

(e') CH$_3$CH$_2$CH$_2$CHCH$_2$CH$_3$ $\xrightarrow[\text{H}_2\text{O}]{\text{H}_2\text{O}_2, \text{ NaOH}}$ CH$_3$CH$_2$CH$_2$CHCH$_2$CH$_3$
       with B   with OH

3-hexanol

Concept Map 8.4   Catalytic hydrogenation reactions.

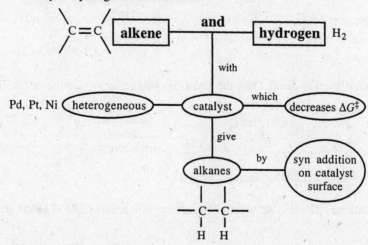

8.13

(a)

$$CH_3CCH=CH_2 \xrightarrow[\text{Ni}]{H_2} CH_3CCH_2CH_3$$

with CH₃ groups above and below, product labeled **A**

(b)

$$CH_3CH_2CH=CHCH_2CH_3 \xrightarrow[\text{Pt}]{H_2} CH_3CH_2CH_2CH_2CH_2CH_3$$

labeled **B**

(c)

$$CH_3CH_2CCH_2CH_3 \text{ (with CH}_3\text{ and OH)} \xrightarrow[\Delta]{H_2SO_4} CH_3CH=CCH_2CH_3 \text{ (C)} + CH_3CH_2CCH_2CH_3 \text{ (D)} \xrightarrow[\text{Pt}]{H_2} CH_3CH_2CHCH_2CH_3 \text{ (E)}$$

(d) cyclobutene—CH₃ $\xrightarrow[\text{Pd/C}]{H_2}$ cyclobutane—CH₃ (**F**)

(e) benzene ring $\xrightarrow[\text{Pd/C}]{H_2 \text{ (excess)}}$ cyclohexane (**G**)

8.14   1.   Attack at carbon 3 of the enantiomeric bromonium ions derived from (Z)-2-pentene.

(2S, 3S)-2,3-dibromopentane   (2R, 3R)-2,3-dibromopentane

This is the same mixture as that obtained by attack at carbon 2 shown in the text.

2.   Attack at carbon 2 of the enantiomeric bromonium ions derived from (E)-2-pentene.

(2S, 3R)-2,3-dibromopentane   (2R, 3S)-2,3-dibromopentane

Attack at carbon 3 of the enantiomeric bromonium ions derived from (E)-2-pentene.

(2R, 3S)-2,3-dibromopentane   (2S, 3R)-2,3-dibromopentane

This is the same mixture as that obtained by attack at carbon 2.

8.15   Part (a) will be analyzed in detail. Only the equations will be shown for the rest of the problem.

(a)

$$CH_3CH_2, H \atop C=C \atop H, CH_2CH_3 \xrightarrow[\text{carbon tetrachloride}]{Br_2} \quad ?$$

1.   To what functional group class do the reactants belong?

One reactant is an alkene. The double bond of the alkene is a nucleophile. The other reagent is a halogen, an electrophile.

8.15 (a) (cont)

2.   Does either reactant have a good leaving group?

Yes, Br⁻ from the bromine molecule is a good leaving group.

3.   Are any of the reactants good acids, bases, nucleophiles, or electrophiles? Is there an ionizing solvent present?

Bromine, above the arrow, is an electrophile. The solvent, carbon tetrachloride, is not an ionizing solvent.

4.   What is the most likely first step for the reaction? Most common reactions can be classified either as protonation-deprotonation, or as reactions of a nucleophile with an electrophile.

Attack by the nucleophilic double bond on the electrophilic bromine leading to the formation of a bromonium ion is the first step.

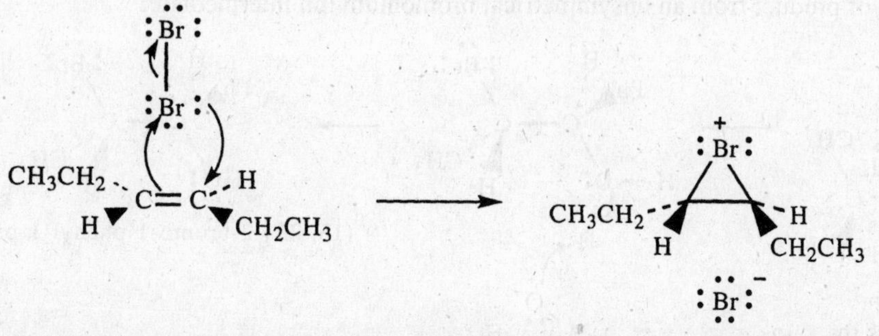

5.   What are the properties of the species present in the reaction mixture after the first step? Is any further reaction likely?

The species present in the reaction mixture are the bromonium ion, an electrophile, and bromide ion, a nucleophile. The bromonium ion will react with bromide ion to give the dibromide product.

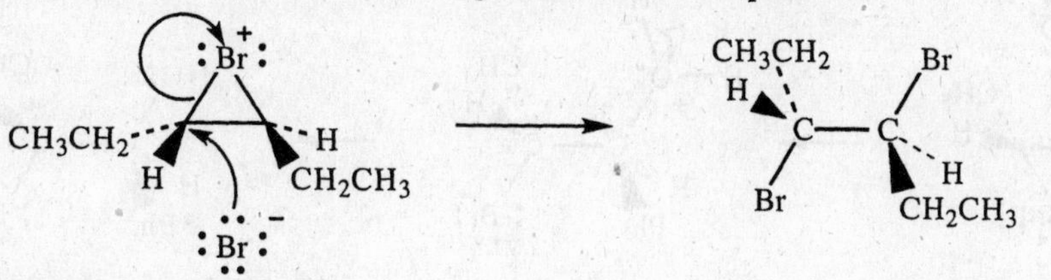

6.   What is the stereochemistry of the reaction?

The opening of the bromonium ion by bromide leads to an overall anti addition and the creation of two stereocenters. The product, however, has a plane of symmetry and is therefore a meso compound.

Complete correct answer:

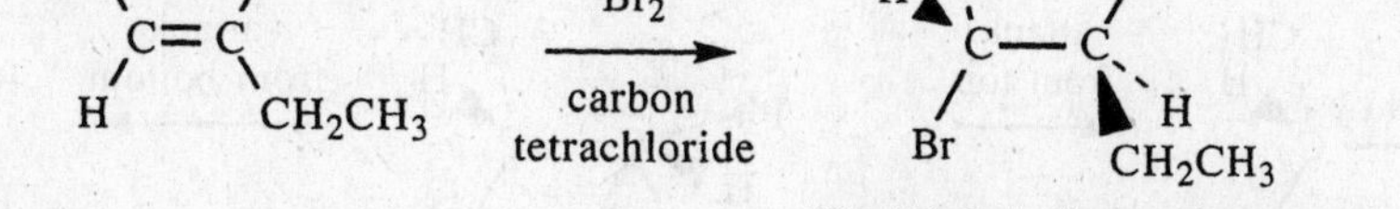

(3R,4S)-3,4-dibromohexane
meso-3,4-dibromohexane

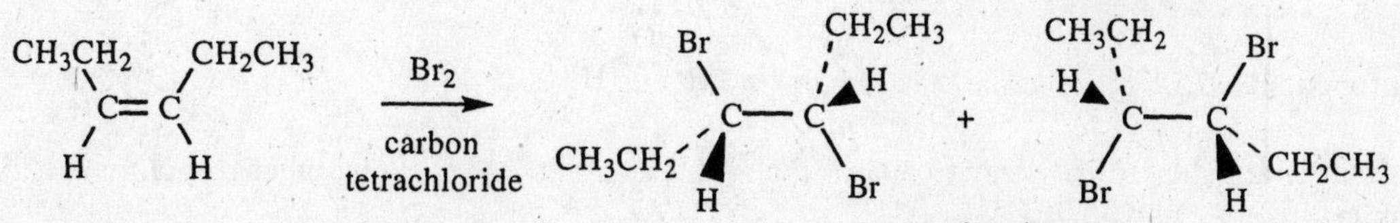

racemic mixture

8.15

(c)

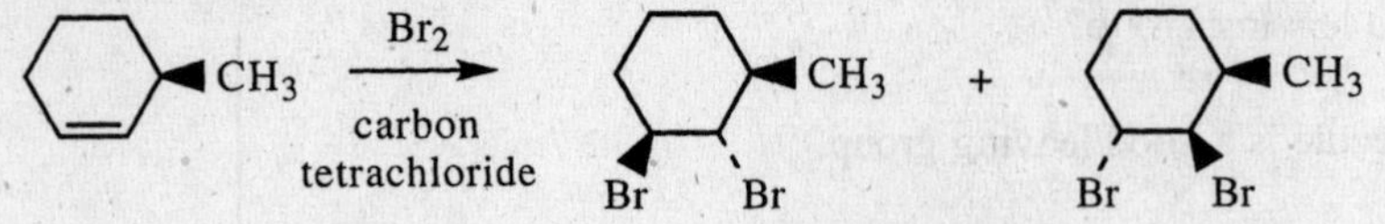

diastereomers

(d)

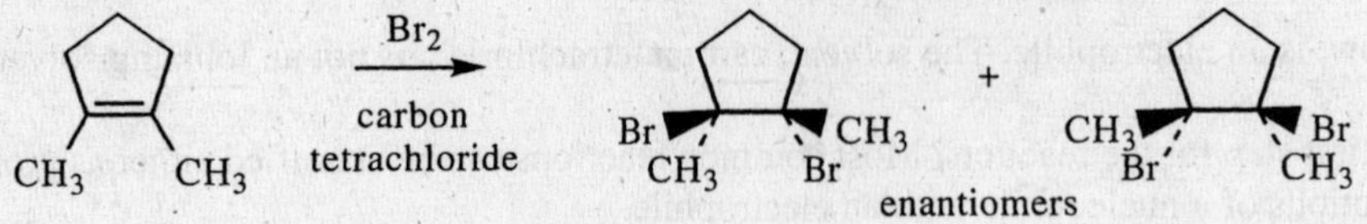

enantiomers

8.16   Formation of product from an unsymmetrical bromonium ion intermediate.

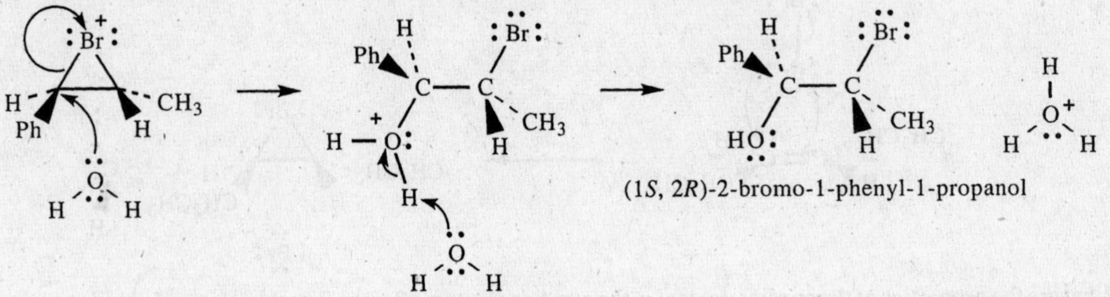

(1*S*, 2*R*)-2-bromo-1-phenyl-1-propanol

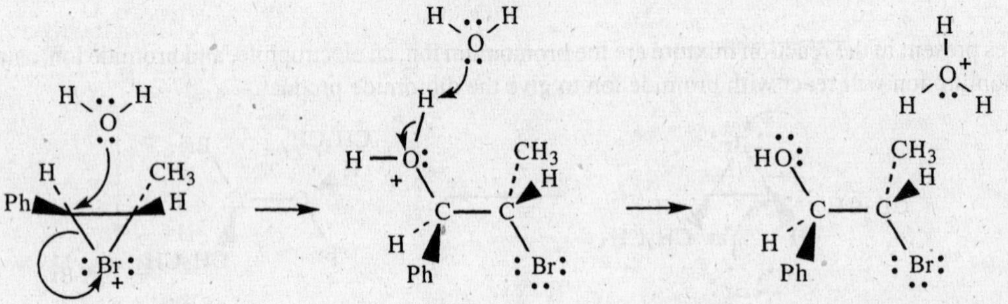

(1*R*, 2*S*)-2-bromo-1-phenyl-1-propanol

Formation of products from a carbocation.

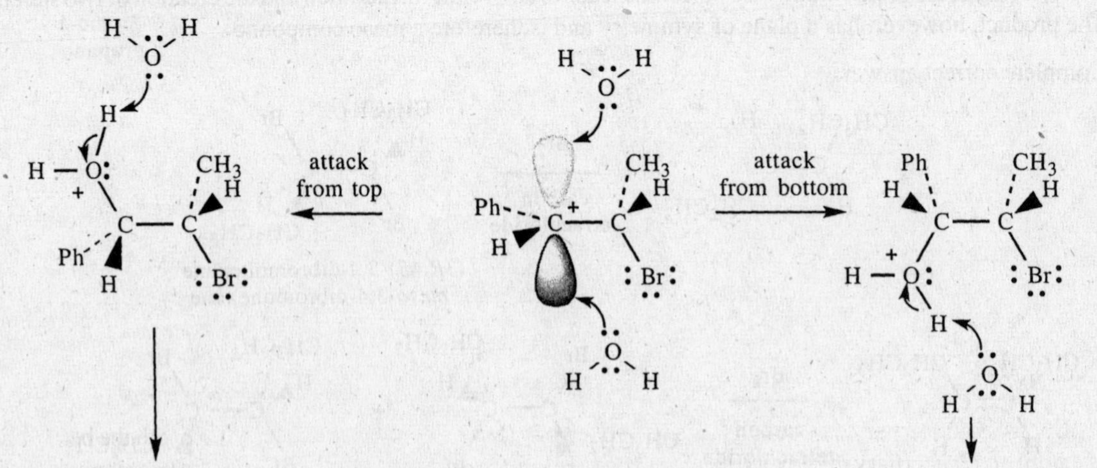

attack
from top

attack
from bottom

8.16 (cont)

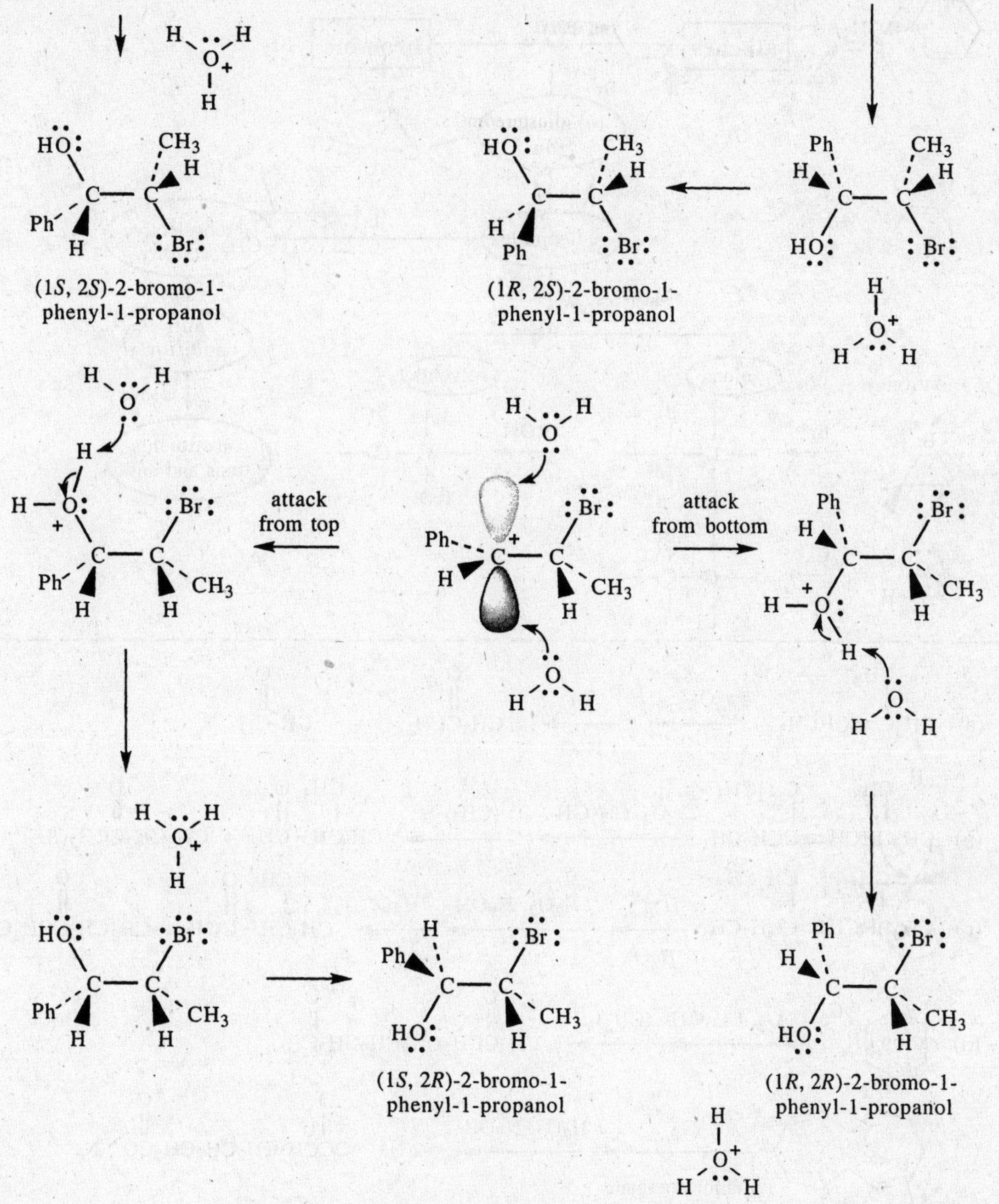

(1*S*, 2*S*)-2-bromo-1-
phenyl-1-propanol

(1*R*, 2*S*)-2-bromo-1-
phenyl-1-propanol

attack
from top

attack
from bottom

(1*S*, 2*R*)-2-bromo-1-
phenyl-1-propanol

(1*R*, 2*R*)-2-bromo-1-
phenyl-1-propanol

8.17

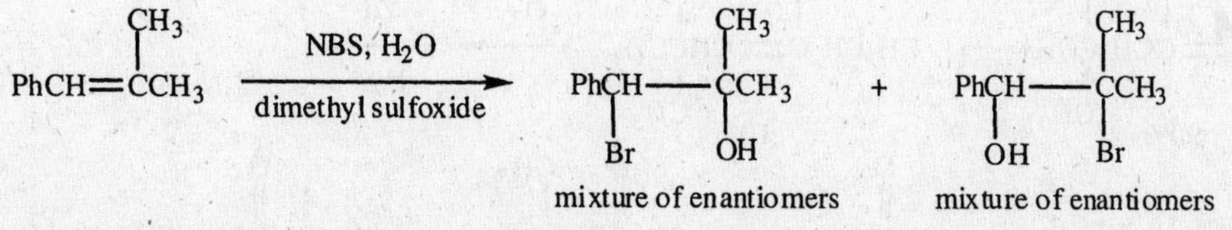

$$PhCH=CCH_3 \xrightarrow[\text{dimethyl sulfoxide}]{\text{NBS, H}_2\text{O}} PhCH-CCH_3 \ + \ PhCH-CCH_3$$

mixture of enantiomers   mixture of enantiomers

Benzylic and tertiary carbocations are both relatively stable intermediates. Therefore, both the benzylic carbon atom and the tertiary carbon atom have about the same amount of partial charge in the bromonium ion. Attack by the nucleophile can occur at either carbon.

Concept Map 8.5   The addition of bromine to alkenes.

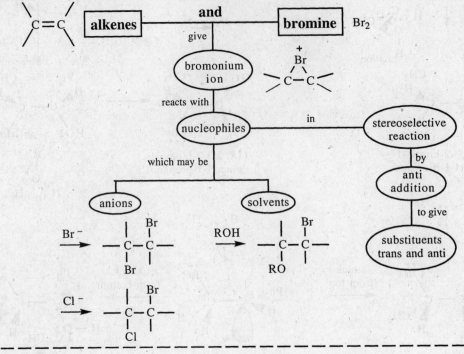

- -

8.18   (a) $\underset{\displaystyle CH_3C=CHCH_3}{\overset{\displaystyle CH_3}{|}} \xrightarrow{O_3} \xrightarrow[H_2O]{Zn} CH_3\overset{O}{\overset{||}{C}}CH_3 + CH_3\overset{O}{\overset{||}{C}}H$

(b) $\underset{\displaystyle CH_3CHCH=CCH_2CH_3}{\overset{CH_3 \quad\quad CH_2CH_3}{|\quad\quad\quad\;|}} \xrightarrow{O_3,\ CH_3OH} \xrightarrow{(CH_3)_2S} \underset{CH_3\ O}{CH_3CH-CH} + CH_3CH_2CCH_2CH_3$

(c) $\underset{\displaystyle CH_3CHCH=CCH_2CH_3}{\overset{CH_3 \quad\quad CH_2CH_3}{|\quad\quad\quad\;|}} \xrightarrow[H_2O]{O_3} \xrightarrow{H_2O_2,\ NaOH} \xrightarrow{H_3O^+} CH_3CH-COH + CH_3CH_2CCH_2CH_3$

(d) $\xrightarrow{O_3,\ CH_3OH} \xrightarrow{(CH_3)_2S} CH_3CCH_2CH_2CH_2CH_2CH$

(e) $\xrightarrow[\substack{\text{dichloromethane}\\ 0\,°C}]{O_3} \xrightarrow{H_2O_2,\ NaOH} 2\,Na^+{}^-OCCH_2CH_2CH_2CH_2CO^-Na^+$

8.19 $\underset{90\%}{CH_3CH=CCH_2CCH_3} + \underset{10\%}{CH_3CH_2C=CHCCH_3} \xrightarrow{O_3} \xrightarrow[H_2O]{Zn}$

$\underset{\sim45\%}{CH_3CH} + \underset{\sim45\%}{CH_3CH_2CCH_2CCH_3} + \underset{\sim5\%}{CH_3CH_2CCH_2CH_3} + \underset{\sim5\%}{CH_3C-CH}$

8.20

HO   OH
H—C—C—CH₂CH₂CH₃
CH₃CH₂CH₂   H
eclipsed conformation

⟶ ⟸

HO   OH
H—C—C—CH₂CH₂CH₃
CH₃CH₂CH₂   CH₂CH₂CH₃
one gauche conformation

⇅

HO   CH₂CH₂CH₃
H—C—C—OH
CH₃CH₂CH₂   H
the other gauche conformation

⟸ ⟶

HO   H
H—C—C—CH₂CH₂CH₃
CH₃CH₂CH₂   OH
anti conformation

8.21

CH₃CH₂CH₂   CH₂CH₂CH₃
       C=C
H   H
(Z)-4-octene

$\xrightarrow[\substack{\textit{tert}\text{-butyl alcohol}\\ \text{tetraethylammonium hydroxide}}]{\text{OsO}_4\text{ (cat), (CH}_3)_3\text{COOH}}$ $\xrightarrow[\text{H}_2\text{O}]{\text{Na}_2\text{SO}_3}$

CH₃CH₂CH₂   CH₂CH₂CH₃
H—C—C—H
HO   OH
(4S, 5R)-octane-4,5-diol
the R,S isomer, a meso compound
and diastereomer of the compounds
shown on p. 315 in the text.

⟶ ⟸

CH₃CH₂CH₂   CH₂CH₂CH₃
H—C—C—OH
HO   H
one gauche conformation

8.22

(a)   CH₃(CH₂)₇CH=CH₂ $\xrightarrow[\substack{\textit{tert}\text{-butyl alcohol}\\ \text{tetraethylammonium hydroxide}}]{\text{OsO}_4\text{ (cat), (CH}_3)_3\text{COOH}}$ $\xrightarrow[\text{H}_2\text{O}]{\text{Na}_2\text{SO}_3}$

CH₃(CH₂)₇   H
H—C—C—H
HO   OH
and enantiomer
A

⟶ ⟸

CH₃(CH₂)₇   H
H—C—C—OH
HO   H

(b)   [cyclooctadiene] $\xrightarrow[\substack{\text{pyridine}\\ \text{cyclopentane}}]{\text{OsO}_4\text{ (1 molar equivalent)}}$ $\xrightarrow[\text{H}_2\text{O}]{\text{Na}_2\text{SO}_3}$ [cyclooctene with two OH]
B

(c)

H   COOH
   C=C
Ph   H

$\xrightarrow[\textit{tert}\text{-butyl alcohol}]{\text{OsO}_4\text{ (cat), H}_2\text{O}_2}$ $\xrightarrow{\text{H}_2\text{O}}$

H   H
Ph—C—C—COOH
HO   OH
and enantiomer
C

⟶ ⟸

H   COOH
Ph—C—C—OH
HO   H

Hydrogen peroxide, H₂O₂, is the oxidizing agent used to reoxidize the catalytic osmium tetroxide.

Concept Map 8.6   The oxidation reactions of alkenes.

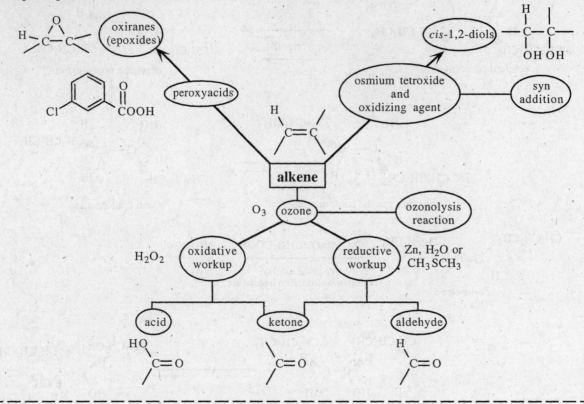

8.23   (a)   CH₃C=CCH₂CH=CH₂ with CH₃ substituents, reacting with 3-chloroperoxybenzoic acid (one molar equivalent), chloroform, 25 °C → epoxide product and enantiomer

(b)   cyclohexene with CH=CH₂ substituent + 3-chloroperoxybenzoic acid (one molar equivalent), chloroform, 10 °C → epoxide product with CH=CH₂

Note that in parts (a) and (b), the more highly substituted double bond is more nucleophilic and therefore reacts faster with the peroxyacid than the less substituted double bond.

(c)   CH₃CH₂ and CH₃ on C=C with H and H + 3-chloroperoxybenzoic acid (one molar equivalent), dichloromethane, 25 °C → epoxide CH₃CH₂----CH₃ with H and H and enantiomer

8.23   (d)

$$CH_3CH_2 \quad H$$
$$\diagdown C = C \diagup$$
$$H \diagup \quad \diagdown CH_3$$

Cl — [benzene ring] — COOH (with C=O)
(one molar equivalent)

dichloromethane
25 °C

→

$$CH_3CH_2 - [epoxide: O] - H$$
$$H \quad CH_3$$
and enantiomer

8.24

$$CH_3 \quad H$$
$$\diagdown C = C \diagup$$
$$H \diagup \quad \diagdown CH_3$$

? →

$$CH_3 - C - C - OH$$
$$HO \quad H$$
$$H \quad CH_3$$

1.   What are the connectivities of the two compounds? How many carbon atoms does each contain? Are there any rings? What are the positions of branches and functional groups on the carbon skeletons?

$$CH_3 \quad H$$
$$\diagdown C = C \diagup$$
$$H \diagup \quad \diagdown CH_3$$
(*E*)-2-butene

$$HO \quad OH$$
$$C - C$$
$$CH_3 \quad H$$
$$H \quad CH_3$$
(2*S*, 3*S*)-2,3-butanediol
drawn in the eclipsed
conformation. No plane
of symmetry is present.

$$H \quad OH$$
$$CH_3 \quad C - C$$
$$HO \quad H$$
$$CH_3$$
anti conformation of
(2*S*, 3*S*)-2,3-butanediol

$$CH_3 \quad OH$$
$$HO \quad C - C$$
$$H \quad H$$
$$CH_3$$
gauche conformation of
(2*S*, 3*S*)-2,3-butanediol;
more stable than anti because
of hydrogen bonding

conformations of one of the enantiomeric products

The carbon skeletons of the two compounds are identical. Each contains four carbon atoms. The two carbon atoms that were doubly bonded in the starting material have hydroxyl groups on them in the product. The methyl groups in the starting alkene and in the eclipsed conformation of the product diol are on the opposite side of the two reacting carbon atoms.

2.   How do the functional groups change in going from starting material to product? Does the starting material have a good leaving group?

The starting material is an alkene. It contains a double bond. The product is a diol with the two hydroxyl groups on adjacent carbon atoms. There is no leaving group present in the starting material.

3.   Is it possible to dissect the structures of the starting material and product to see which bonds must be broken and which formed?

bond broken

$$CH_3 \quad H$$
$$C \diagup \diagdown C$$
$$H \quad CH_3$$

bonds formed

$$HO \quad OH$$
$$C - C$$
$$CH_3 \quad H$$
$$H \quad CH_3$$

4.   New bonds are created when an electrophile reacts with a nucleophile. Do we recognize any part of the product molecule as coming from a nucleophile or an electrophilic addition reaction?

A double bond is converted to a diol by reaction with an oxidizing agent (an electrophile) such as osmium tetroxide. The reaction of osmium tetroxide is a syn addition so would give the stereochemistry we need to get from an alkene with methyl groups on opposite sides of the double bond to a diol, which also has methyl groups on the opposite sides of the molecule in its eclipsed conformation.

5.   Do we see how to get from starting material to the product?

The starting material can be converted directly to product.

Complete correct answer:

$$CH_3 \quad H$$
$$\diagdown C = C \diagup$$
$$H \diagup \quad \diagdown CH_3$$

OsO$_4$, (CH$_3$)$_3$COOH →

$$[O = Os = O]$$
$$O \quad O$$
$$C - C$$
$$CH_3 \quad H$$
$$H \quad CH_3$$
and enantiomer

Na$_2$SO$_3$
water →

$$HO \quad OH$$
$$C - C$$
$$CH_3 \quad H$$
$$H \quad CH_3$$
and enantiomer

→

$$CH_3 \quad H$$
$$C - C$$
$$HO \quad OH$$
$$H \quad CH_3$$
and enantiomer

8.25 (a) (*E*)-3-methyl-3-heptene   (b) 3-ethyl-2,5-dimethyl-2-hexene   (c) (*E*)-3,4-dichloro-3-heptene
     (d) (*Z*)-3,4-dimethyl-3-heptene   (e) (*R*)-1,3-dichlorobutane   (f) (*Z*)-4-bromo-3-ethyl-2-pentene
     (g) 3-chloro-2-methyl-2-pentene   (h) *cis*-1,2-dibromocyclobutane
     (i) (2*R*, 3*S*)-2-bromo-3-chlorobutane

8.26 (a)
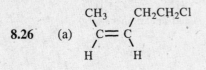

(b) and enantiomer (c)

(d) $CH_3CH_2CHCH=CH_2$ with Cl   (e)   (f)

8.27 (a) ...

(b) ...

(c) ...

(d) ...

(e) ...

(f) ...

(g) ...

8.28 (*Z*)-6-dodecen-1-ol

8.29

(a) $\quad \xrightarrow[Pt]{H_2} \quad CH_3CH_2CHCH_2CH_3$

(b) $\quad \xrightarrow[H_2SO_4]{H_2O} \quad CH_3CH_2CCH_2CH_3$ with OH

8.29 (cont)

(c)

$$\underset{\underset{H}{|}}{\overset{\overset{CH_3}{|}}{C}}=\underset{\underset{CH_2CH_3}{|}}{\overset{\overset{CH_3}{|}}{C}} \xrightarrow{HBr} CH_3CH_2\underset{\underset{Br}{|}}{\overset{\overset{CH_3}{|}}{C}}CH_2CH_3$$

(d)

$$\underset{\underset{H}{|}}{\overset{\overset{CH_3}{|}}{C}}=\underset{\underset{CH_2CH_3}{|}}{\overset{\overset{CH_3}{|}}{C}} \xrightarrow[\text{chloroform}]{} CH_3\cdots\!\overset{O}{\triangle}\!\cdots CH_3$$

(m-chlorobenzoic acid, Cl—⬡—COOH)

H   CH_2CH_3

and enantiomer

(e)

$$\underset{\underset{H}{|}}{\overset{\overset{CH_3}{|}}{C}}=\underset{\underset{CH_2CH_3}{|}}{\overset{\overset{CH_3}{|}}{C}} \xrightarrow{O_3, CH_3OH} \xrightarrow{(CH_3)_2S} \underset{}{\overset{O}{\overset{\|}{C}}}CH_3CH \;+\; CH_3\overset{O}{\overset{\|}{C}}CH_2CH_3$$

(f)

$$\underset{\underset{H}{|}}{\overset{\overset{CH_3}{|}}{C}}=\underset{\underset{CH_2CH_3}{|}}{\overset{\overset{CH_3}{|}}{C}} \xrightarrow[\text{diethyl ether}]{OsO_4, (CH_3)_3COOH} \xrightarrow[\text{H}_2\text{O}]{Na_2SO_3}$$

HO   CH_2CH_3

CH_3 ─ C ─ C ─ CH_3

H   OH

and enantiomer

(g)

$$\underset{\underset{H}{|}}{\overset{\overset{CH_3}{|}}{C}}=\underset{\underset{CH_2CH_3}{|}}{\overset{\overset{CH_3}{|}}{C}} \xrightarrow[\text{water}]{O_3} \xrightarrow{H_2O_2, NaOH} \xrightarrow{H_3O^+} CH_3\overset{O}{\overset{\|}{C}}OH \;+\; CH_3\overset{O}{\overset{\|}{C}}CH_2CH_3$$

(h)

$$\underset{\underset{H}{|}}{\overset{\overset{CH_3}{|}}{C}}=\underset{\underset{CH_2CH_3}{|}}{\overset{\overset{CH_3}{|}}{C}} \xrightarrow[\text{carbon}\atop\text{tetrachloride}]{Br_2}$$

Br   CH_3

CH_3 ─ C ─ C ─ CH_2CH_3

H   Br

and enantiomer

(i)

$$\underset{\underset{H}{|}}{\overset{\overset{CH_3}{|}}{C}}=\underset{\underset{CH_2CH_3}{|}}{\overset{\overset{CH_3}{|}}{C}} \xrightarrow{Br_2, H_2O}$$

Br   CH_3   Br   CH_3

CH_3 ─ C ─ C ─ CH_2CH_3   +   CH_3 ─ C ─ C ─ CH_2CH_3

H   Br   H   OH

and enantiomer   and enantiomer

8.30 (a)

□ (cyclobutene) $\xrightarrow[\text{carbon}\atop\text{tetrachloride}]{Br_2}$ (cyclobutane with Br, Br)

and enantiomer

(b)

$$\underset{\underset{H}{|}}{\overset{\overset{CH_3}{|}}{C}}=\underset{\underset{H}{|}}{\overset{\overset{CH_3}{|}}{C}} \xrightarrow[\text{carbon}\atop\text{tetrachloride}]{Br_2}$$

Br   H

H ─ C ─ C ─ CH_3

CH_3 Br

and enantiomer

(c)

$$CH_2\!=\!CCH\!=\!CH_2 \;(\text{with } CH_3)$$

$$\xrightarrow[\text{H}_2\text{O}]{\overset{N-Br}{\text{(1 molar equiv.)}}}$$

HO   CH_3

BrCH_2 ─ C ─ CH=CH_2

and enantiomer

8.30   (d)

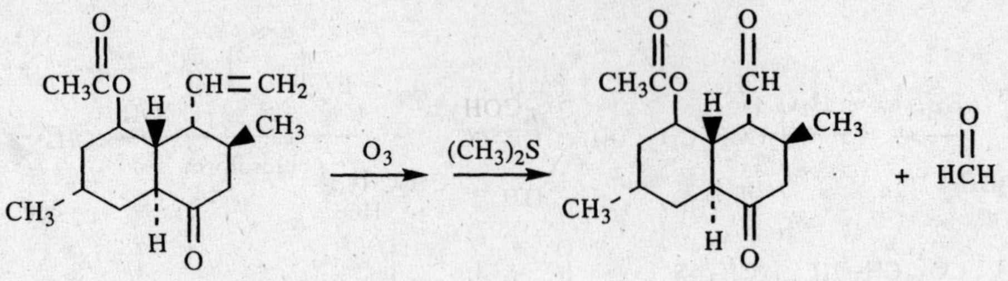

(e)

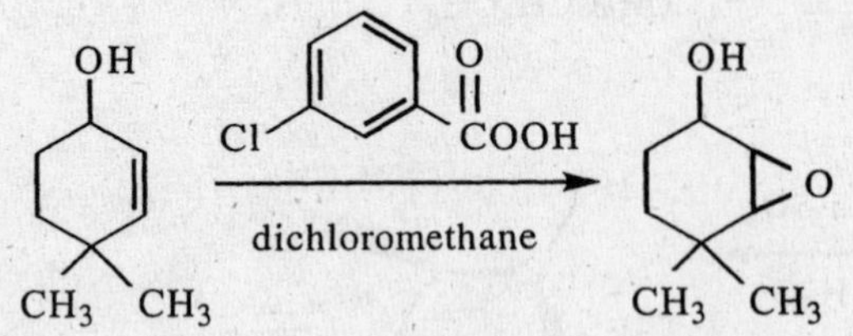

(f)

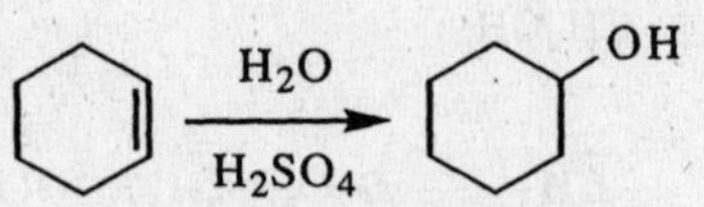

8.31

(a)

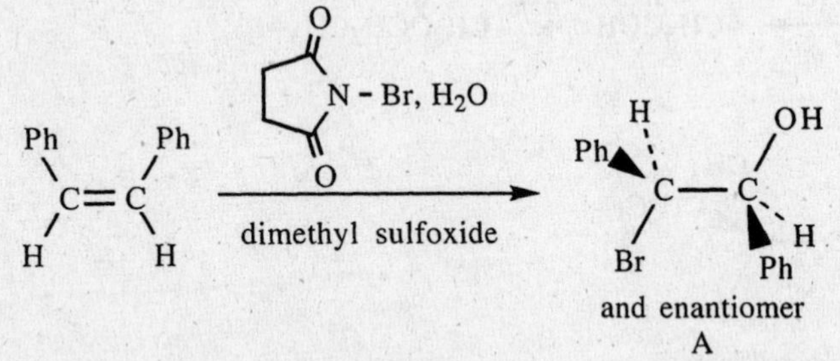

and enantiomer
A

(b)

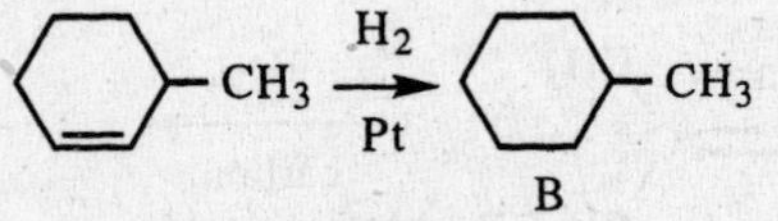

(c)

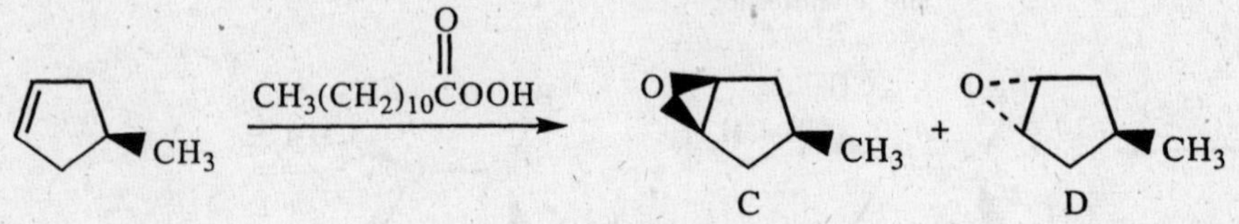

(d)

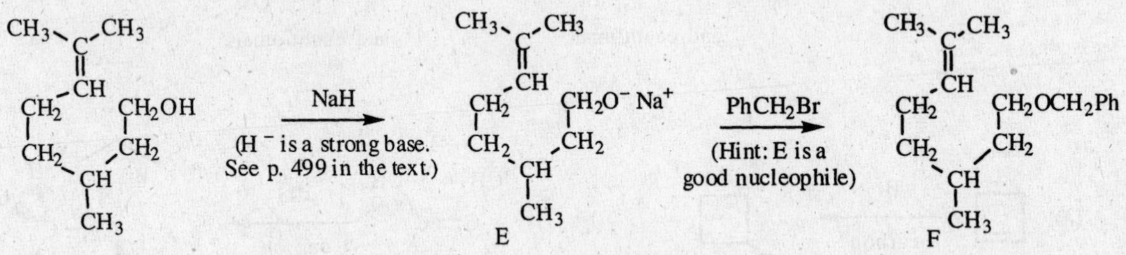

(e)

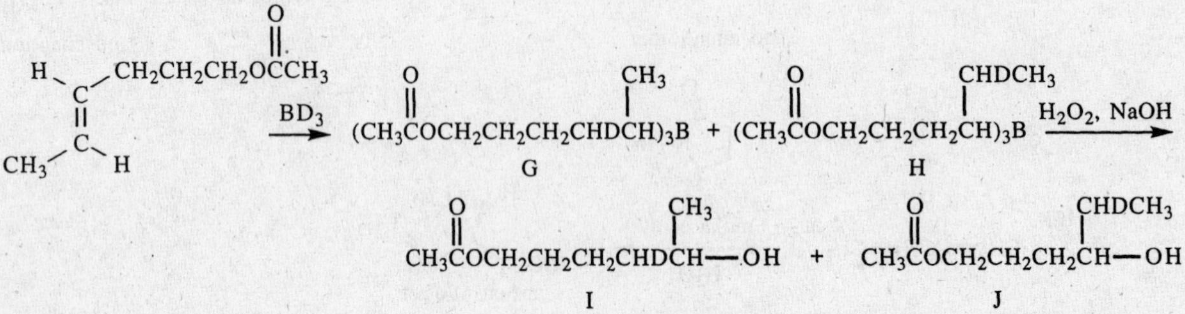

8.31 (cont)

(f)

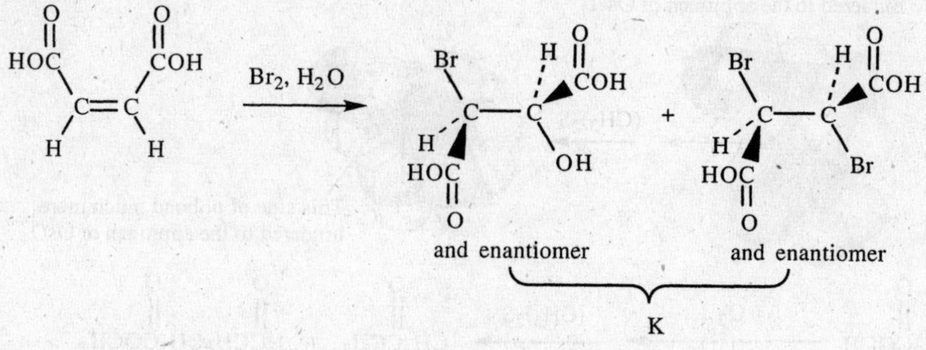

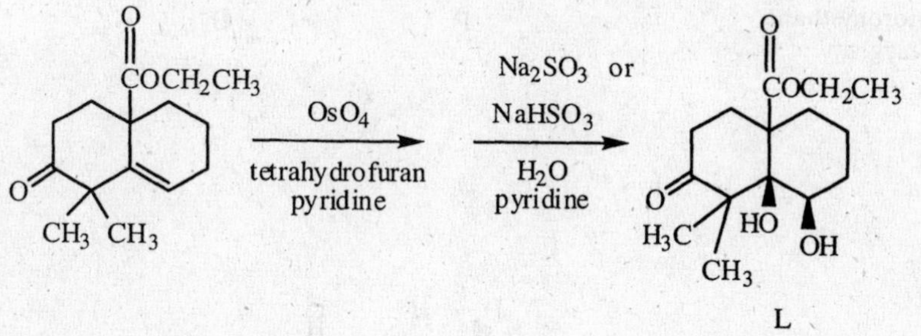

8.32

(a)

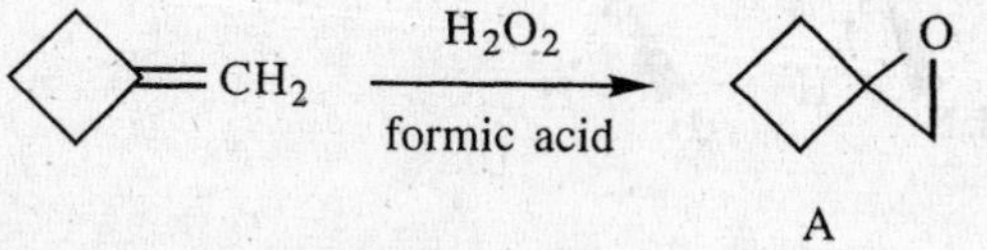

(b)

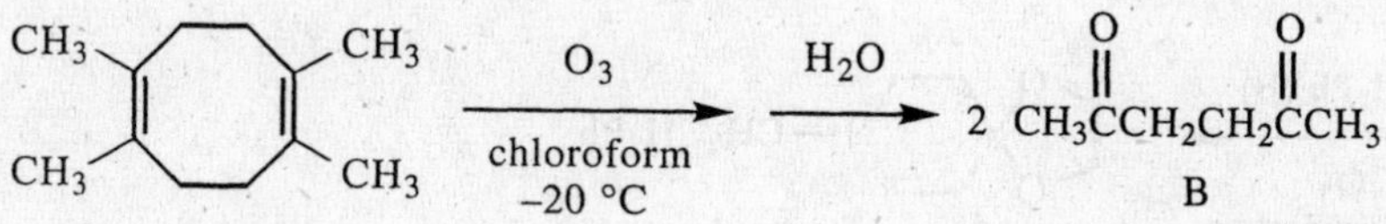

(c)

C
optically active

D 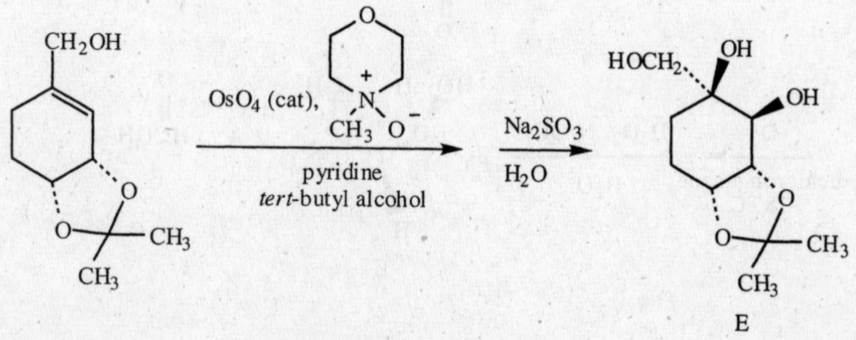
optically inactive

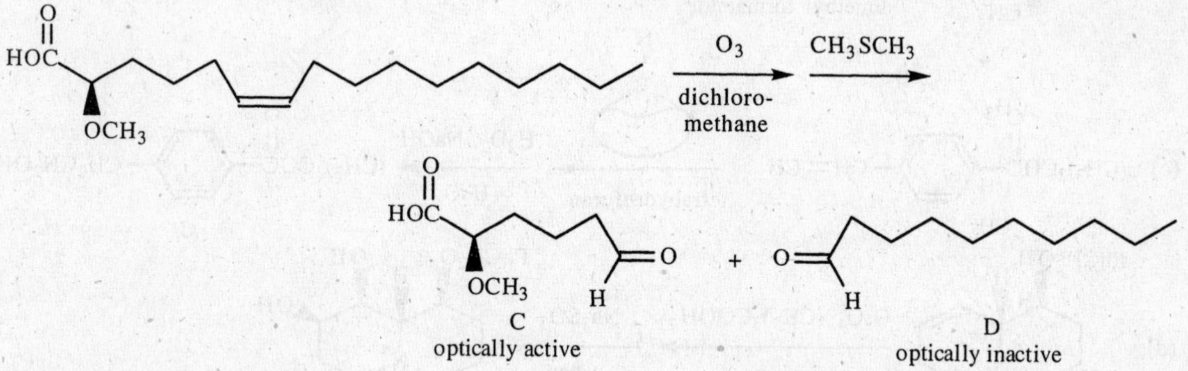

(d)

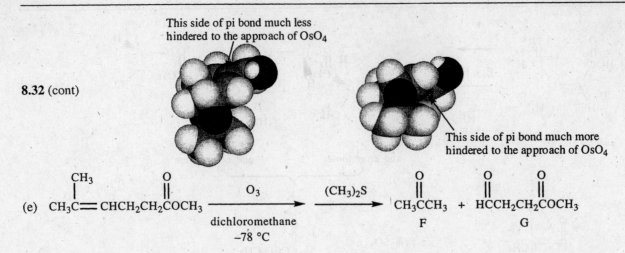

This side of pi bond much less
hindered to the approach of OsO₄

This side of pi bond much more
hindered to the approach of OsO₄

8.32 (cont)

(e) $CH_3C \!\!=\!\! CHCH_2CH_2\overset{\displaystyle O}{\overset{\|}{C}}OCH_3$ $\xrightarrow[\substack{\text{dichloromethane} \\ -78\ ^\circ C}]{O_3}$ $\xrightarrow{(CH_3)_2S}$ $CH_3\overset{\displaystyle O}{\overset{\|}{C}}CH_3$ + $H\overset{\displaystyle O}{\overset{\|}{C}}CH_2CH_2\overset{\displaystyle O}{\overset{\|}{C}}OCH_3$

with CH_3 above the first carbon

F   G

8.33

(a) $PhCH_2\cdots\overset{H}{\underset{H_2N}{C}}\!\!-\!\!\overset{H}{\underset{\|}{C}}\!\!=\!\!\overset{}{\underset{H}{C}}\!\!-\!\!\overset{\displaystyle O}{\overset{\|}{C}}OCH_3$ $\xrightarrow[\substack{Pd/C \\ methanol}]{H_2}$ PhCH₂...C—C—C—COCH₃ with H₂N, H, H substituents

A

(b) (dioxolane–spiro–piperidine) N—H $\xrightarrow[\substack{K_2CO_3 \\ dimethyl\ formamide}]{BrCH_2CH_2Ph\ (B)}$ N—CH₂CH₂Ph

(c) $(CH_3)_3C\overset{\displaystyle O}{\overset{\|}{C}}O$ —⟨benzene ring⟩— $CH\!\!=\!\!CH_2$ $\xrightarrow[tetrahydrofuran]{\underset{B}{\overset{H}{|}}}$ $\xrightarrow[\substack{H_2O}]{H_2O_2,\ NaOH}$ $(CH_3)_3C\overset{\displaystyle O}{\overset{\|}{C}}O$ —⟨benzene ring⟩— CH₂CH₂OH

C

(d) (bicyclic PhCH₂O-substituted oxazinone, H) $\xrightarrow{OsO_4,\ (CH_3)_3COOH}$ $\xrightarrow[water]{Na_2SO_3}$ (diol product with PhCH₂O, H, OH, OH)

D

(e) $Ph\!\!-\!\!\overset{HO\ \ H}{\underset{}{C}}\!\!-\!\!\overset{H}{\underset{H}{C}}\!\!-\!\!\overset{H}{\underset{H}{C}}\!\!=\!\!\overset{H}{\underset{H}{C}}$ $\xrightarrow[\substack{\text{dichloromethane} \\ -78\ ^\circ C}]{O_3}$ $\xrightarrow[\substack{H_2O}]{H_2O_2,\ NaOH}$ $Ph\!\!-\!\!\overset{HO\ \ H}{\underset{}{C}}\!\!-\!\!\overset{H}{\underset{H}{C}}\!\!-\!\!\overset{\displaystyle OH}{\overset{}{C}}\!\!\overset{\displaystyle O}{}$ + $H\overset{\displaystyle O}{\overset{\|}{C}}OH$

E   F

(f)

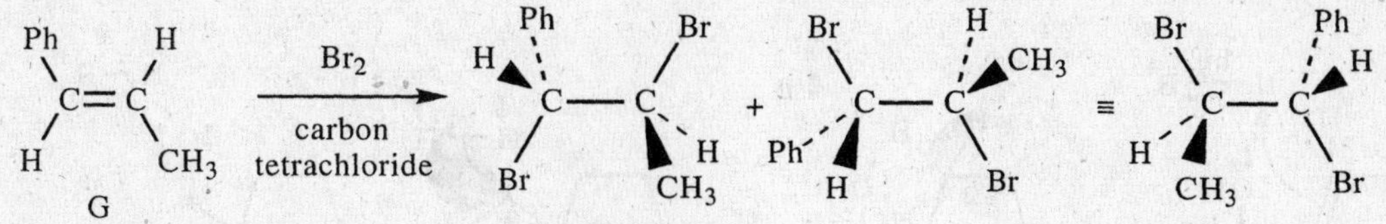

(g) $CH_3(CH_2)_6CH=CH_2$ $\xrightarrow{\text{HBr} \atop (H)}$ $CH_3(CH_2)_6CHCH_3$
$\underset{\displaystyle Br}{|}$

(h)

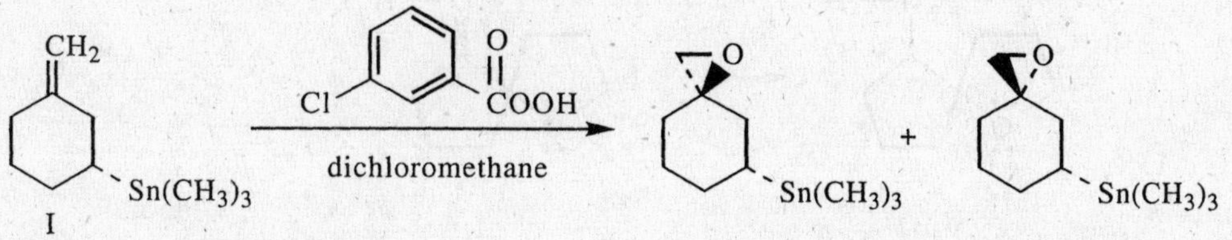

(i)

(j)

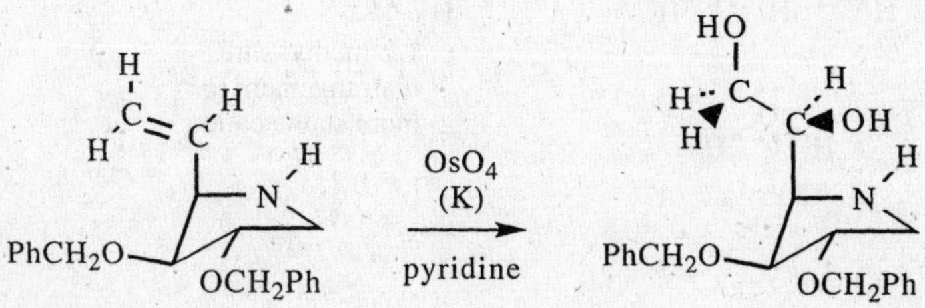

8.34

1,2-shift of sigma
bonding electrons
of a ring carbon;
another type of
alkyl shift

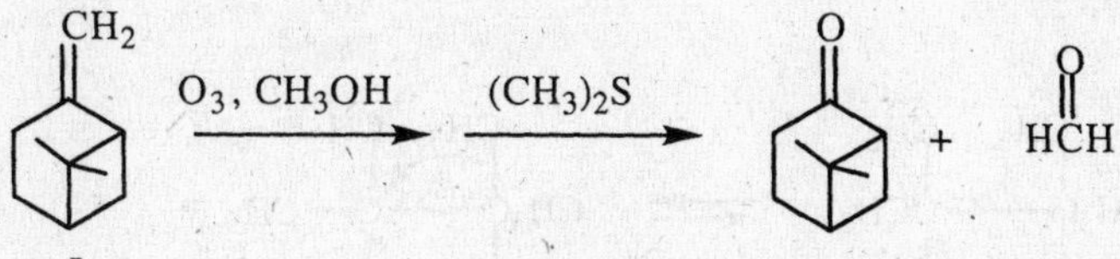

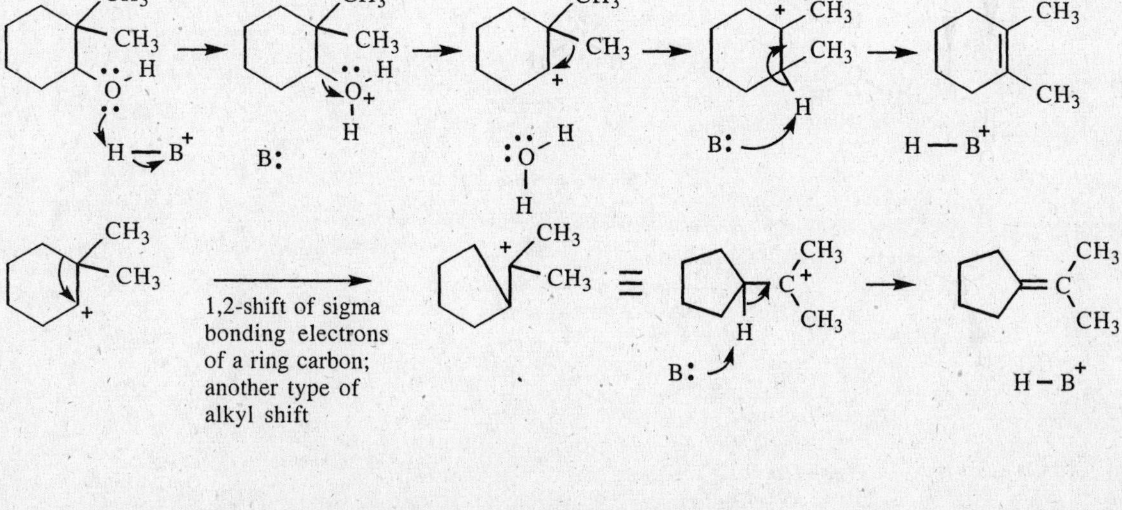

8.35 (a)

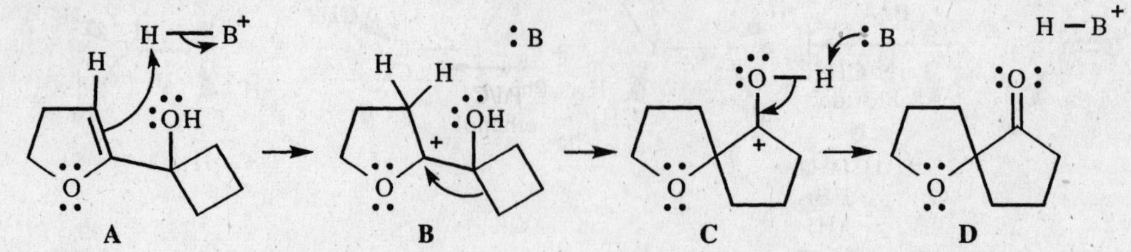

A → B → C → D

(b) Both B and C have the positive charge on a carbon adjacent to oxygen, which allows for resonance stabilization of the cation.

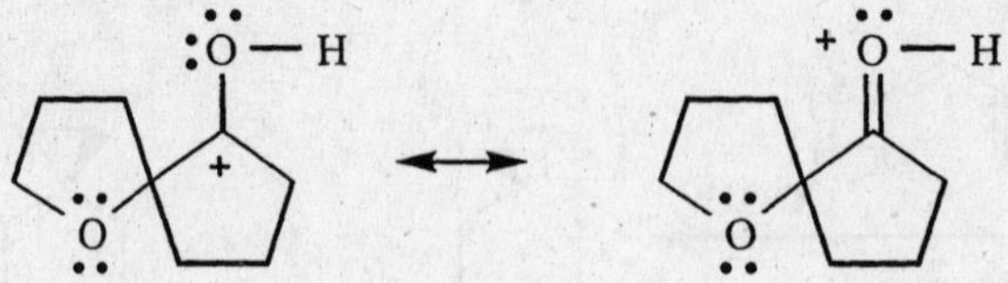

8.36   Formation of pinacolone from pinacol

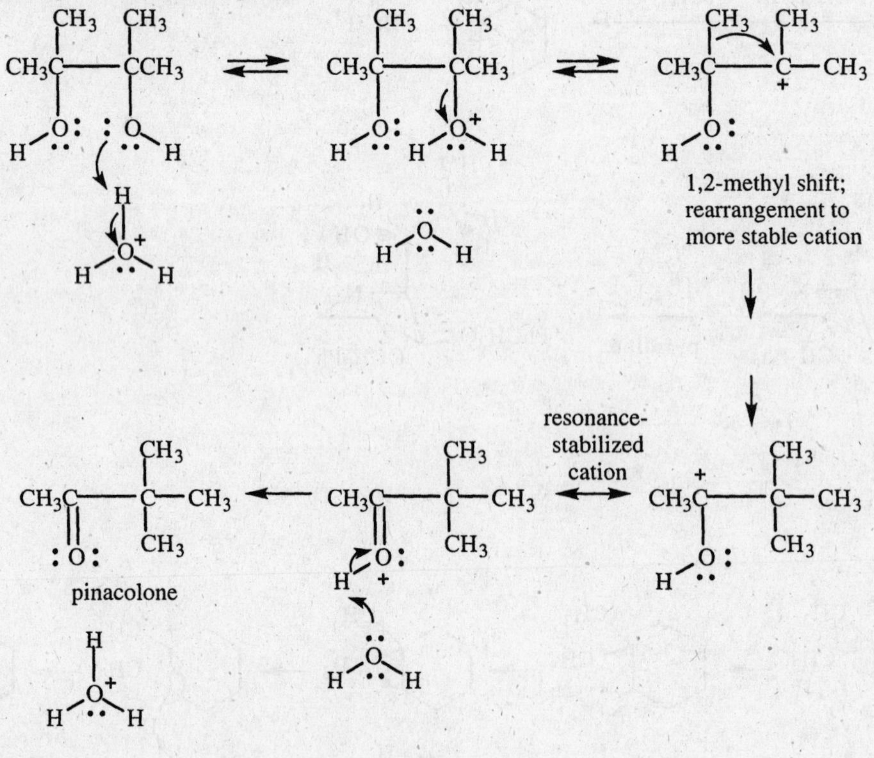

1,2-methyl shift;
rearrangement to
more stable cation

resonance-
stabilized
cation

pinacolone

8.37

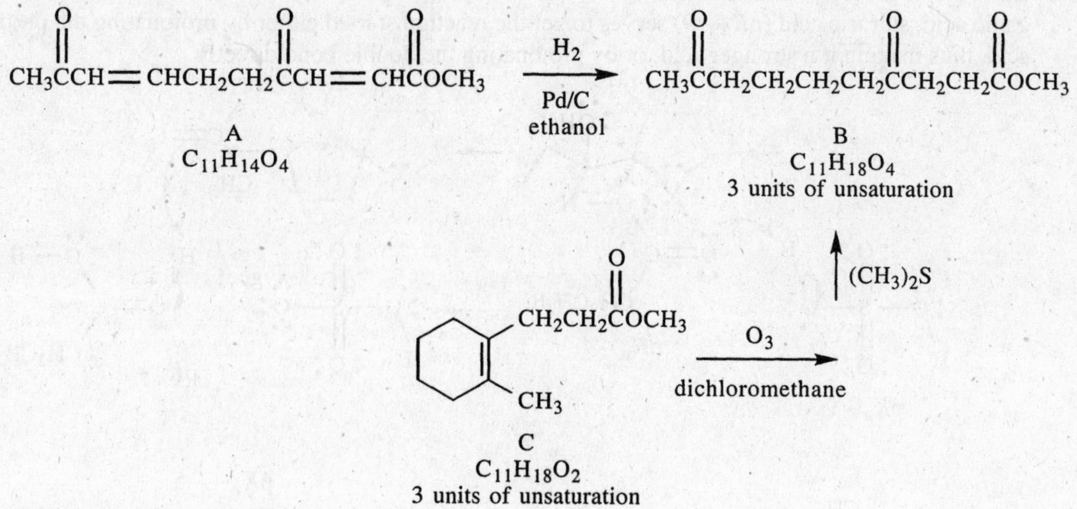

8.38   (a)

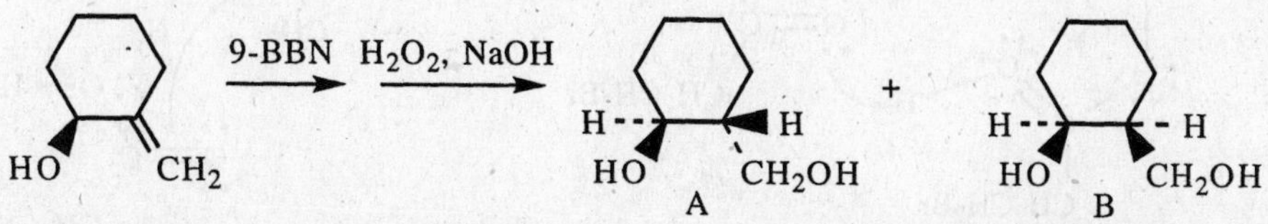

Compound A, in which the large groups are trans to each other on the cyclohexane ring, is expected to be thermodynamically more stable. Compound B, however, is formed by way of a sterically less hindered transition state.

(b)

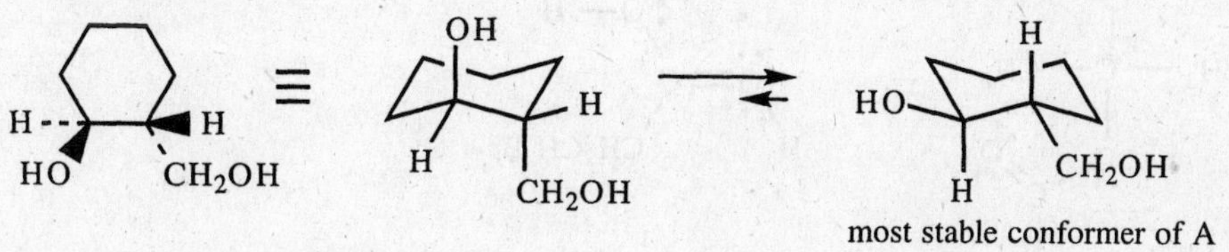

most stable conformer of A

8.39   The reaction causes a change in the connectivities of the carbon atoms. In the starting material the phenyl ring is attached to an unlabeled carbon atom. In the product, it is bonded to the C-14-labeled carbon atom. A rearrangement has taken place. Such rearrangements require carbocation intermediates.

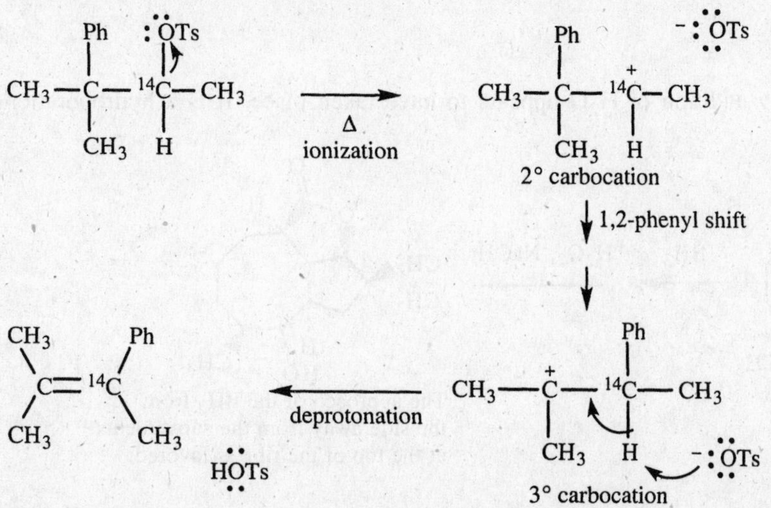

8.40   (a)   3-Bromopropanoic acid is not a strong enough acid ($pK_a \sim 4$) to protonate the double bond. The inorganic acid, sulfuric acid ($pK_a - 9$) serves to get the reaction started either by protonating the propanoic acid, thus making it a stronger acid, or by protonating the double bond directly.

(b)   There will be five distinct peaks in the carbon nuclear magnetic resonance spectrum of the product.

(c)   There will be three distinct peaks in the hydrogen nuclear magnetic resonance spectrum of the product.

8.41   (a)   An anti-Markovnikov addition of H_2O appears to have taken place. It is a hydroboration-oxidation reaction.

(b)

The approach of the BH_3 from the side away from the substituents at the top of the ring is favored.

8.42   (a)

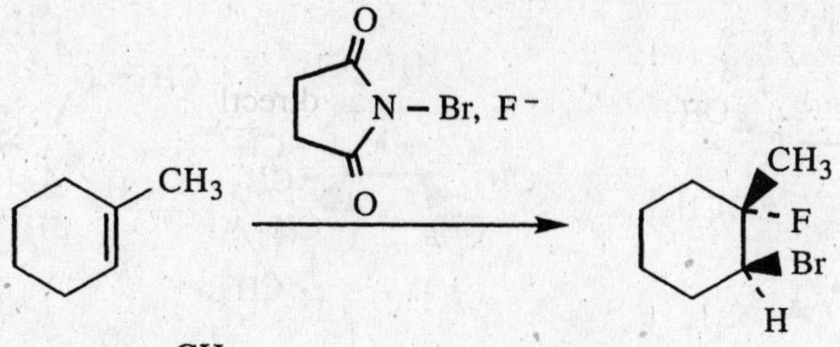

and enantiomer

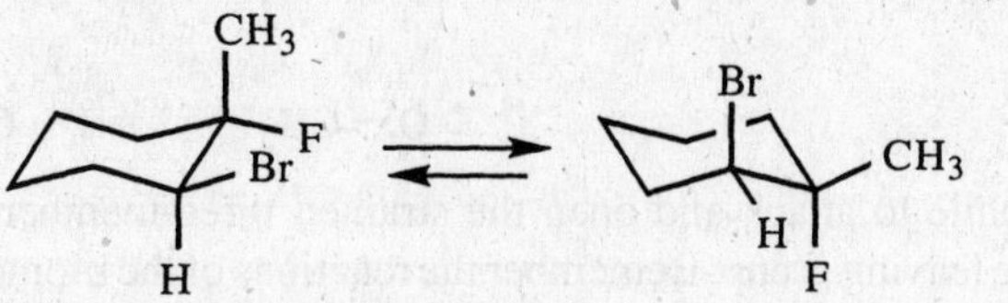

(b) *S* at each stereocenter
(1*S*, 2*S*)-2-bromo-1-fluoro-1-methylcyclohexane

(c)

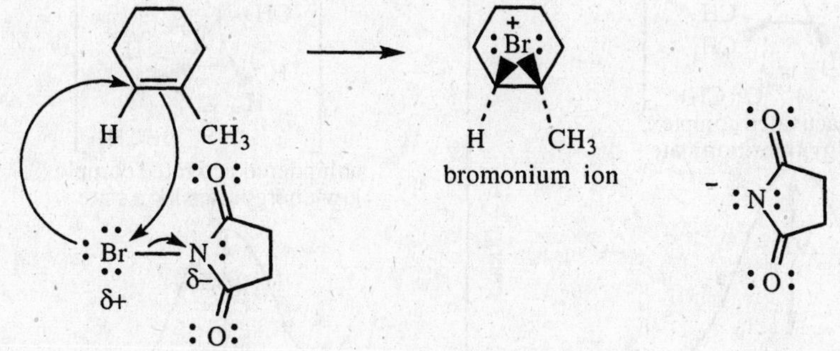

bromonium ion

If a bromonium ion with little charge on the carbon atoms were the intermediate in the reaction, we would expect attack of fluoride ion at the less hindered carbon atom, which is not the regioselectivity observed. Therefore, the intermediate must be the partially open bromonium ion with cationic character at the 3° carbon atom.

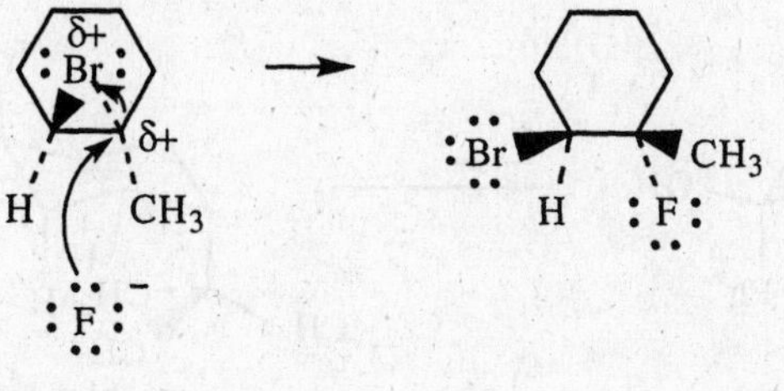

8.43

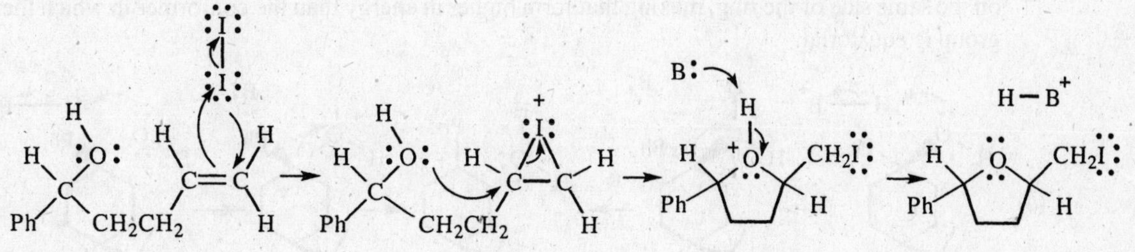

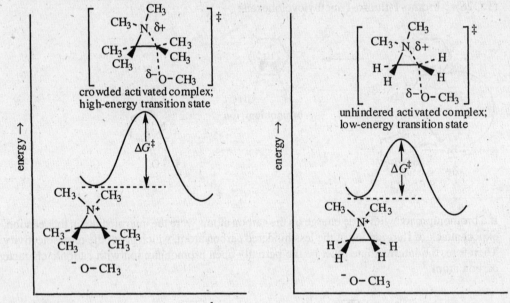

8.44 (a)

(b) Methanol is a strong enough nucleophile to attack and open the strained three-membered ring with a positively-charged nitrogen atom as the leaving group. Remember the reactions of the bromonium ion with water or alcohols. Methanol is not a strong enough base to deprotonate a carbon atom to form the alkene.

(c) The rate of the reaction increases as crowding at the carbon atom undergoing the S_N2 reaction decreases.

crowded activated complex;
high-energy transition state

unhindered activated complex;
low-energy transition state

$\Delta G^{\ddagger}$

$\Delta G^{\ddagger}$

energy →

energy →

8.45 (a)

lower energy

higher energy

(b) The *tert*-butyl group in an axial position has severe 1,3-diaxial interactions with the axial hydrogen atoms on the same side of the ring, making that form higher in energy than the conformer in which the *tert*-butyl group is equatorial.

(c)

same cation whether we start with *cis*
or *trans* isomer of starting material

8.45 (c) (cont)

Supplemental Problems

S8.1   A mixture of two primary alcohols obtained from natural sources, 3-methyl-1-butanol and 2-methyl-1-butanol, gives chiefly 2-methyl-2-butene when dehydrated with acid. 2-Methyl-2-butene is converted to 2-methyl-2-butanol with 90% yield by treatment with 50% aqueous sulfuric acid at 0 °C. Write equations showing the details of the conversion of the mixture of primary alcohols into a single tertiary alcohol.

S8.2   Complete the following equations, predicting the starting material or products indicated by letter and showing stereochemistry wherever possible. If a racemic mixture is formed, show one enantiomer, and say "+ enantiomer."

S8.2 (cont)

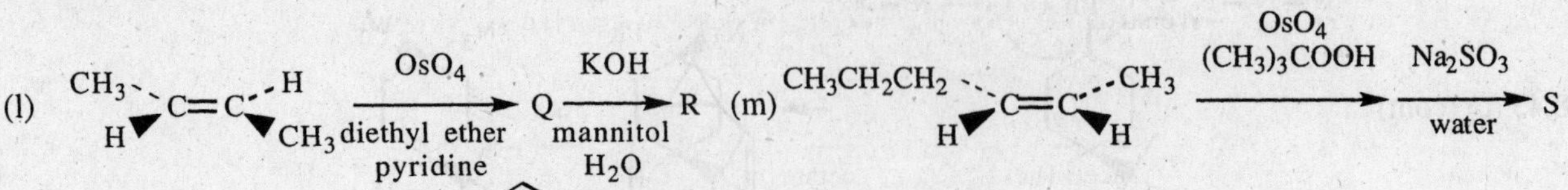

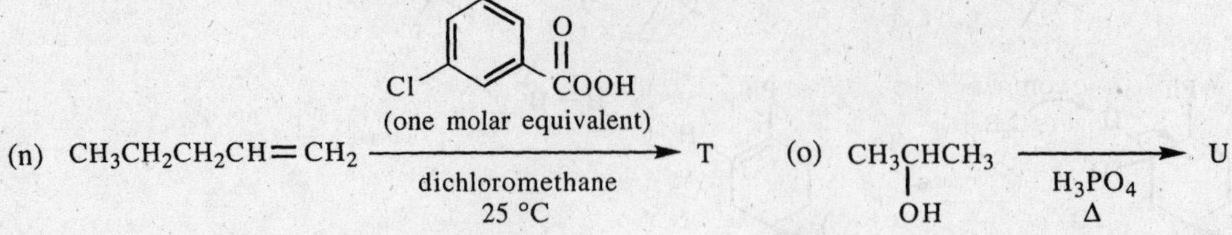

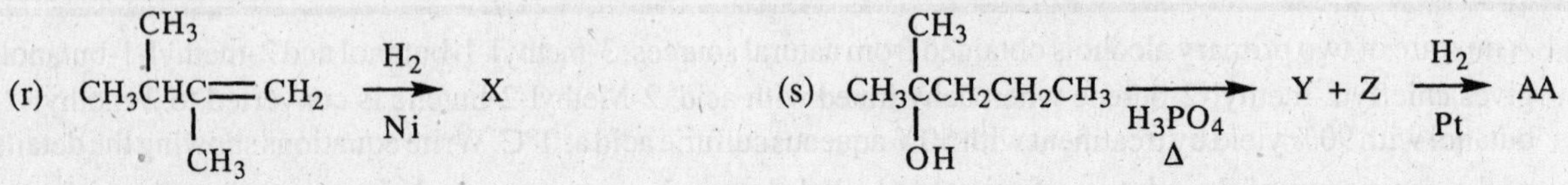

(l) CH₃ C=C H $\xrightarrow{\text{OsO}_4}$ diethyl ether pyridine $\xrightarrow{\text{KOH}}$ Q mannitol H₂O R

(m) CH₃CH₂CH₂ C=C CH₃ $\xrightarrow[\text{water}]{\text{OsO}_4 \ (\text{CH}_3)_3\text{COOH} \ \ \text{Na}_2\text{SO}_3}$ S

(n) CH₃CH₂CH₂CH=CH₂ $\xrightarrow[\substack{\text{dichloromethane} \\ 25\,°C}]{\begin{array}{c}\text{Cl} \quad \overset{O}{\overset{\|}{C}}\text{OOH} \\ \text{(one molar equivalent)}\end{array}}$ T

(o) CH₃CHCH₃ with OH $\xrightarrow[\Delta]{\text{H}_3\text{PO}_4}$ U

(p) CH₂=CH₂ $\xrightarrow[\text{H}_2\text{SO}_4]{\text{H}_2\text{O}}$ V

(q) CH₃C=CCH₃ with CH₃ (top) and CH₃ (bottom) $\xrightarrow[\text{Ni}]{\text{H}_2}$ W

(r) CH₃CHC=CH₂ with CH₃ (top) and CH₃ (bottom) $\xrightarrow[\text{Ni}]{\text{H}_2}$ X

(s) CH₃CCH₂CH₂CH₃ with CH₃ (top) and OH (bottom) $\xrightarrow[\Delta]{\text{H}_3\text{PO}_4}$ Y + Z $\xrightarrow[\text{Pt}]{\text{H}_2}$ AA

S8.3 A tertiary alcohol was dehydrated to give an alkene. Treatment of the alkene with ozone, then with water and zinc, gave equal amounts of 3-pentanone and acetaldehyde. What is the structure of the alcohol?

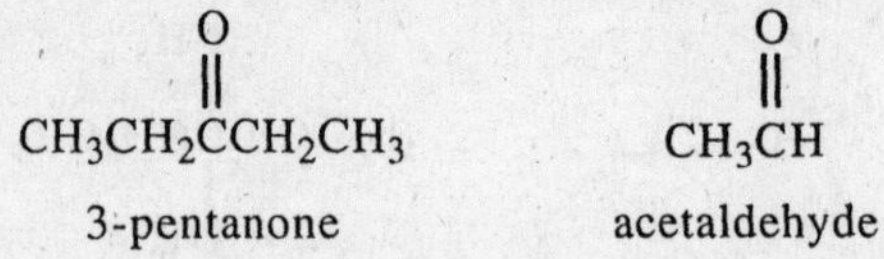

$$\underset{\text{3-pentanone}}{CH_3CH_2\overset{\overset{\displaystyle O}{\|}}{C}CH_2CH_3} \qquad \underset{\text{acetaldehyde}}{CH_3\overset{\overset{\displaystyle O}{\|}}{C}H}$$

S8.4 Name each of the following compounds, assigning the correct configuration to any for which stereochemistry is shown.

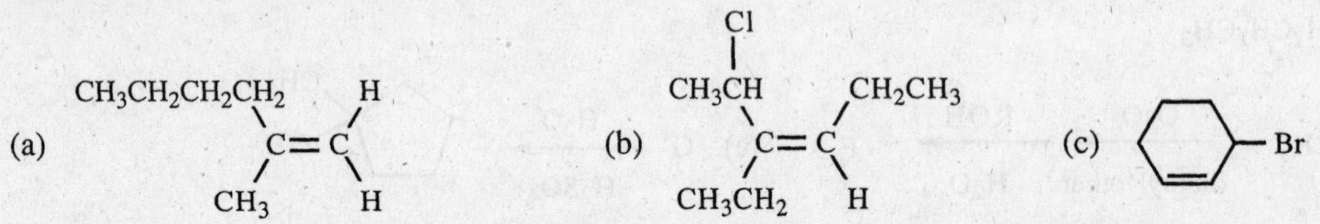

(a) CH₃CH₂CH₂CH₂ \ / H C=C CH₃ / \ H

(b) Cl | CH₃CH \ CH₂CH₃ C=C CH₃CH₂ / \ H

(c) ⬡—Br

S8.5 Propose a mechanism for the following reaction that accounts for the regioselectivity that is observed.

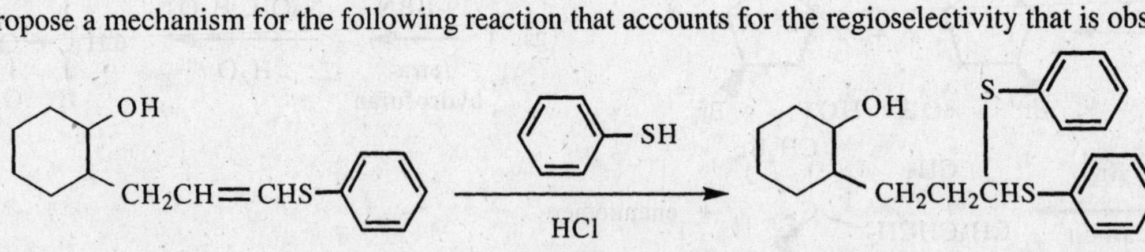

⬡ with OH and CH₂CH=CHS—⬡ $+$ ⬡—SH $\xrightarrow{\text{HCl}}$ ⬡ with OH and CH₂CH₂CHS—⬡ with S—⬡

S8.6 Neopentyl bromide reacts in 50% aqueous ethanol at a high temperature (125 °C) in a first-order reaction. Under these conditions, the major products of the reaction are 2-ethoxy-2-methylbutane and 2-methyl-2-butene.

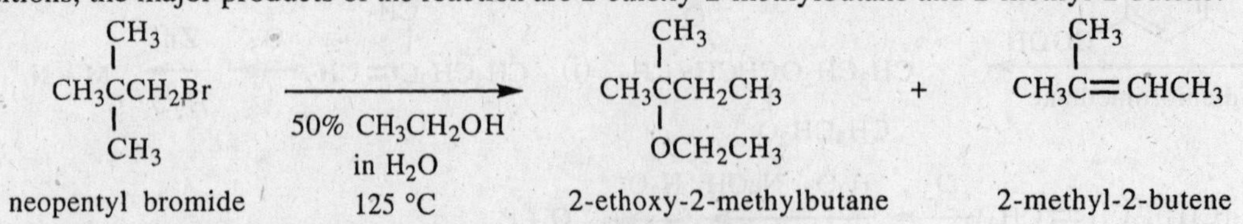

CH₃ | CH₃CCH₂Br | CH₃ $\xrightarrow[\substack{\text{in H}_2\text{O} \\ 125\,°C}]{50\%\ \text{CH}_3\text{CH}_2\text{OH}}$ CH₃ | CH₃CCH₂CH₃ | OCH₂CH₃ + CH₃ | CH₃C=CHCH₃

neopentyl bromide   2-ethoxy-2-methylbutane   2-methyl-2-butene

Write a mechanism for the reaction that explains the observed products.

S8.7   (a) The structure of the following bicyclic compound was investigated by subjecting it to hydrogenation and ozonolysis. Give structural formulas for the products of the reactions shown.

$$B + C \xleftarrow[\substack{\text{acetic acid} \\ \Delta}]{H_2O_2} \xleftarrow[\substack{\text{ethyl acetate} \\ -70\ °C}]{O_3} \qquad \xrightarrow[\substack{\text{Pd/C} \\ \text{tetrahydrofuran} \\ 24\ °C,\ 1\ \text{atm}}]{H_2\ (2\ \text{molar equiv})} A$$

(b) Another bicyclic compound, compound D, C_7H_{10}, also gives compound B after ozonolysis under the same conditions. What is the structure of compound D?

S8.8   Reactions of *cis*-4,5-dimethylcyclohexene and 3-*tert*-butylcyclohexene with *m*-chloroperbenzoic acid give predominantly one product in each case. Show the conformation of each starting alkene, and predict, by showing the reaction with the peroxyacid, the structure of the expected product. (Hint: Using molecular models will help.)

S8.9   Recently pheromones of both the Japanese beetle and the mosquito were synthesized in British Columbia.

(a) The structure of the pheromone of the Japanese beetle is shown below *without* any stereochemical information. To attract the beetle, the compound has to be the (*R*)-(–)-(*Z*)-isomer. As little as 5% of the (*S*)-isomer lowers the attractiveness of the compound to beetles. Draw a structural formula that clearly shows the correct stereochemistry for the (*R*)-(–)-(*Z*)-stereoisomer of this compound.

$$\text{(structure)} \qquad CH = CH(CH_2)_7CH_3$$

pheromone of the Japanese beetle

(b) The synthesis of this pheromone involved preparation of a chiral alcohol from a ketone by the use of baker's yeast. The stereochemical structure of this chiral alcohol is shown below. Name the compound. (Hint: the name of a compound containing two hydroxyl groups ends in "diol.") Locate any sources of stereoisomerism in the compound, and assign configuration, *R* or *S*, *E* or *Z*, to those centers. What is the correct name for the compound, including its stereochemistry?

(c) In a later step of the synthesis, the double bond in the molecule had to be removed to create an aldehyde function. A partial equation for the reaction that was used is shown below. Fill in structural formulas for the reagents (A and B) that were used and for compound C, which is the other product of the reaction.

S8.10   The following reaction would be expected to produce the bromohydrin shown below on the left. However, the reaction actually gives the allylic bromide shown on the right.

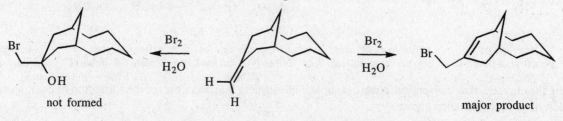

not formed   major product

S8.10   (a)   What are the structures for the two different bromonium ion intermediates that could result from the reaction of the alkene with bromine?

(b)   Only one of these bromonium ions is proposed to form in this reaction. Which one is it, and what is wrong with the other one?

(c)   Write a mechanism that rationalizes the formation of the allylic bromide from the bromonium ion that you selected as the one that forms in the reaction.

S8.11   Research into the reactions of alkenes substituted at the double bond with a good leaving group and with groups that stabilize carbocations show that they undergo S_N1 reactions. For example, the following reaction was observed.

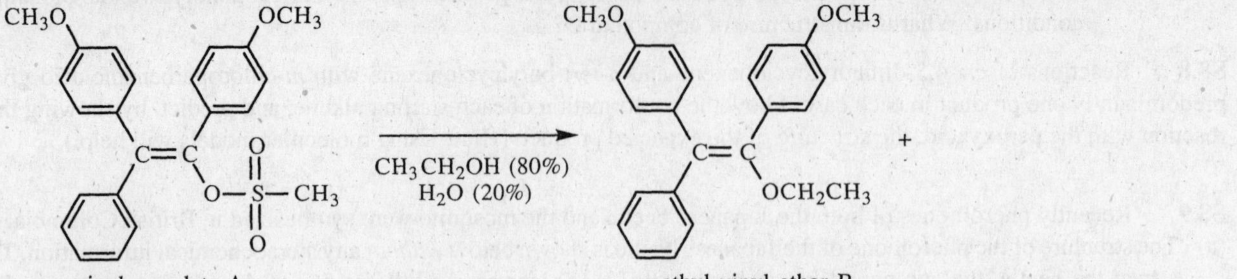

vinyl mesylate A   ethyl vinyl ether B

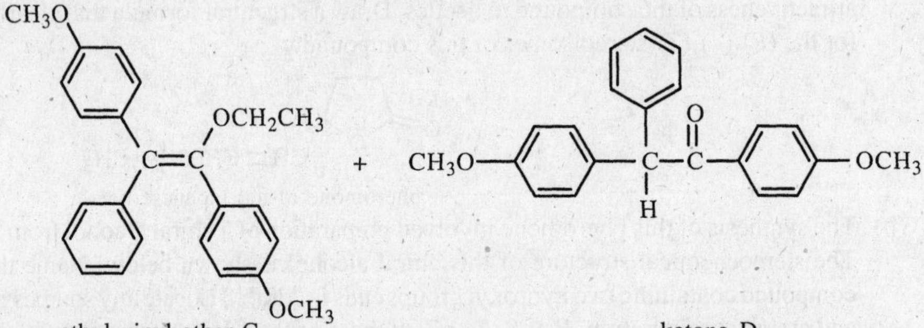

ethyl vinyl ether C   ketone D

In aqueous ethanol, vinyl mesylate gives a mixture of stereoisomeric ethyl vinyl ethers B and C, and ketone D.

(a)   What is the configuration, *E* or *Z*, of ethyl vinyl ether B?

(b)   What is the structure of the carbocation that is the reactive intermediate in the reaction shown above? What is the hybridization of the carbon with the positive charge in this cation? What is the bond angle between the two nearest neighbors of the cationic carbon?

(c)   Using the curved arrow convention, draw the complete mechanism for the conversion of vinyl mesylate A to ketone D. Think carefully about the participation of the solvent in this reaction. A shorthand version of the structures for A and B are given below. You may use these to save time in writing your mechanism. You may also use HB^+ and B: as general acids and bases if you wish.

vinyl mesylate A   ketone D

S8.12   When the tosylate of 3-methyl-2-butanol is dissolved in trifluoroacetic acid, the major product is the trifluoroacetic acid ester of 2-methyl-2-butanol.

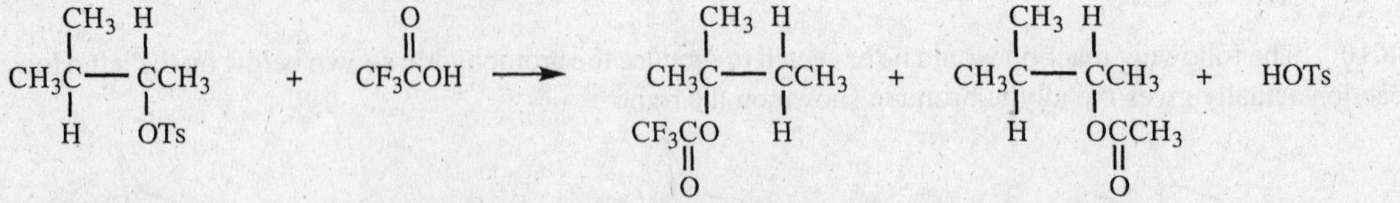

| tosylate of 3-methyl-2-butanol | trifluoroacetic acid, also serves as solvent | trifluoroacetate of 2-methy-2-butanol 98.5% | trifluoroacetate of 3-methy-2-butanol 1.5% | *p*-toluene-sulfonic acid |

Using the curved arrow convention, write a complete mechanism that accounts for the formation of the major product as described in the equation above.

S8.13 The following transformation has been observed.

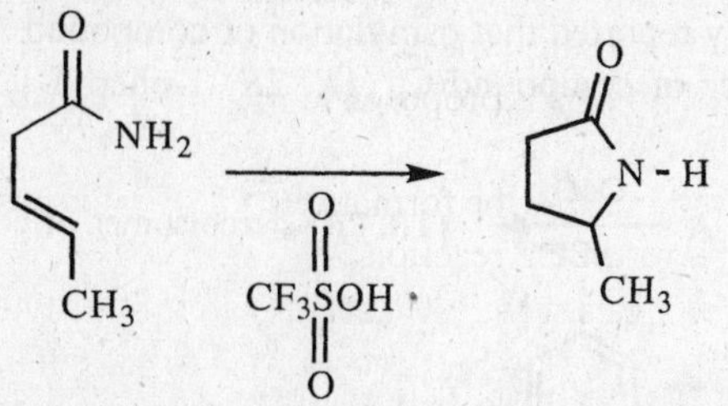

The reaction is catalyzed by trifluoromethylsulfonic acid, which is regenerated during the reaction.
(a) Propose an approximate pK_a for trifluoromethylsulfonic acid.
(b) Using the curved arrow convention, draw a mechanism for the conversion shown above.

S8.14 The following reactions are all used by chemists in the synthesis of natural products that display interesting biological effects. Supply structural formulas for the missing reagents or products indicated by letters. Show stereochemistry wherever it is known.

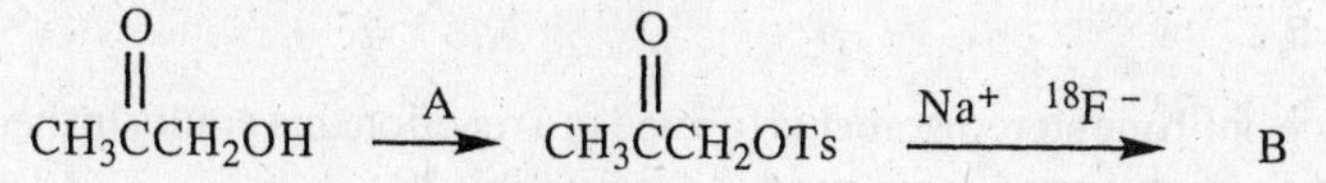

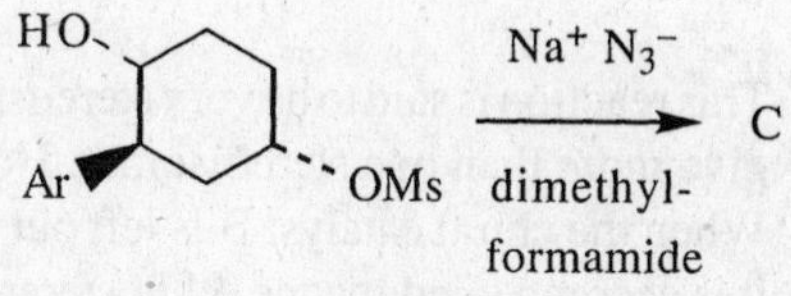

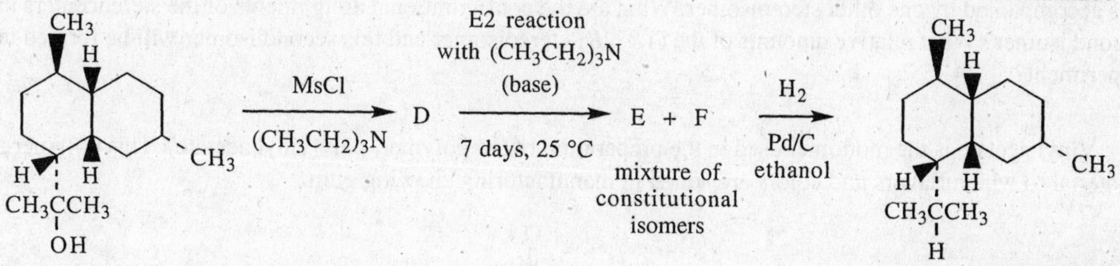

S8.15 The following solvolysis reactions were studied.
(a) Both are first-order reactions. Reaction B is ~10^4 times faster than reaction A. Draw a mechanism using the curved arrow convention for the rate-determining step of reaction A.

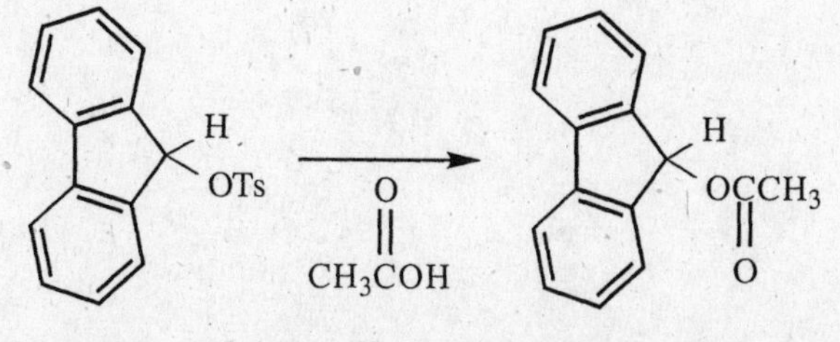

(b) Why is reaction B faster than reaction A? Use both words and pictures in your answer.

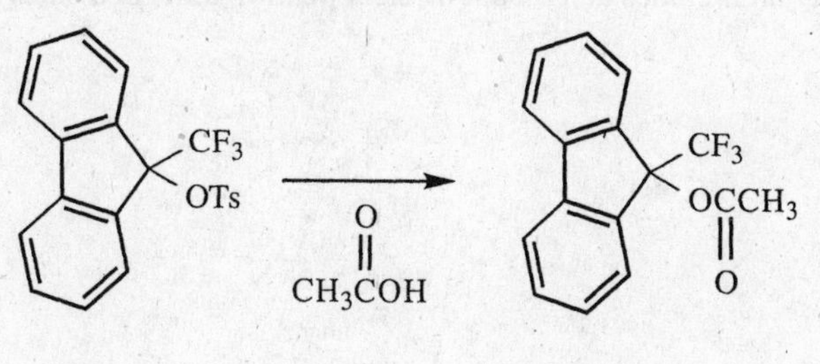

S8.16

(a) Professor Barry Sharpless recently reported that osmylation of compound A in the presence of the chiral catalyst B produced a single stereoisomer of compound C, (1R, 2R)-1-phenyl-1,2-propanediol. Provide structures for compounds A and C.

compound A $\xrightarrow[\text{Cl}]{\text{OsO}_4}$ (1R, 2R)-stereoisomer of compound C

catalyst B

(b) This reaction is said to be very stereospecific, since only one stereoisomer is formed in a reaction that could, in theory, give more than one stereoisomer. How many stereoisomers of compound C are possible?

(c) When the chiral catalyst B is left out of the reaction, (1R, 2R)-1-phenyl-1,2-propanediol is once again formed, and it is accompanied by one other stereoisomer. What are the configurational assignments of the stereocenters in this second isomer? What relative amounts of the (1R, 2R)-stereoisomer and this second isomer will be formed in this experiment?

S8.17   Vinyl acetate is the monomer used in the preparation of the polymer, poly(vinyl acetate). This polymer is the base material to which flavors and colors are added in manufacturing chewing gum.

vinyl acetate

Poly(vinyl acetate) forms an atactic (stereochemically nonregular) polymeric structure. Draw a representation of a poly(vinyl acetate) segment that includes at least six of the repeating units and shows the atactic nature of the structure.

9 Alkynes

Workbook Exercises

Your study of the structure and reactivity of organic compounds has included a number of different ways to represent a molecule based upon the type of information that is known or needed. One way of organizing this information is under the general heading of "isomerism," which can be viewed according to the hierarchy presented below.

(1) **Isomerism** (or ways to represent molecules)

(2) A **molecular formula** provides the possibility of constitutional isomers that differ in their connectivity (see Chapter 1).

(3) A particular **connectivity** provides the possibility that there are stereoisomers that differ in the three-dimensional arrangement of bonded atoms (see Chapter 6).

(4) A particular **stereoisomer** (or a particular connectivity, if no stereoisomers are possible) may have conformational isomers that differ in nonbonded three-dimensional arrangements (see Chapter 5).

EXERCISE I. For each of the three molecular formulas (a, b, and c) shown below,

(a) $C_6H_{12}O_2$ (b) C_7H_{10} (c) $C_5H_{11}NO$

1. Create 10 different constitutional isomers (for a total of 30 molecules). The molecules you create should consist of uncharged, closed-shell atoms.

2. Decide how many stereoisomers are possible for each constitutional isomer that you create.

3. Choose five examples of single bonds between any two tetrahedral atoms in the isomers you create. Draw Newman projections for each example showing different possible conformations. For any cyclohexane or closely related six-member ring, create the two appropriate chair forms.

(Note: This is an ideal problem to work with other students. You should also save your responses to part 1 of this exercise for the workbook exercise in Chapter 11.)

9.1   (a) 6,6-dichloro-2-methyl-3-heptyne   (b) 3,3-dimethyl-1-pentyne

         (c) 4-bromo-2-hexyne   (d) 2,6-dimethyl-3-octyne

9.2 (a) $\underset{\underset{\displaystyle Cl}{|}}{\overset{\overset{\displaystyle CH_3}{|}}{CH_3CH_2CH_2C}}\equiv CCCH_3$ (b) $CH_3C\equiv CCH_2\underset{\underset{\displaystyle Ph}{|}}{CH}CH_2CH_2CH_3$ (c) $CH_3\underset{\underset{\displaystyle CH_3}{|}}{\overset{\overset{\displaystyle CH_3}{|}}{C}}\!-\!C\equiv CCH_2CH_3$

Concept Map 9.1   Outline of the synthesis of a disubstituted alkyne from a terminal alkyne.

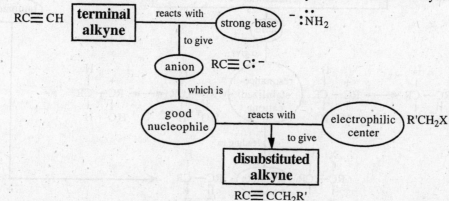

9.3   (a) $CH_3CH_2CH_2C \equiv CH$ $\xrightarrow[\text{NH}_3 \text{ (liq)}]{\text{NaNH}_2}$ $CH_3CH_2CH_2C \equiv C{:}^- Na^+$ $\xrightarrow{CH_3CH_2CH_2Br}$
                                                                                                      A

$CH_3CH_2CH_2C \equiv CCH_2CH_2CH_3$
                                          B

(b) $HC \equiv C{:}^- Na^+$ $\xrightarrow[\text{NH}_3 \text{ (liq)}]{\overset{\displaystyle CH_3}{\overset{\displaystyle |}{CH_3CHCH_2CH_2Br}}}$ $\overset{\displaystyle CH_3}{\overset{\displaystyle |}{CH_3CHCH_2CH_2C \equiv CH}}$
       C   D

(c) $CH_3CH_2CH_2C \equiv CH$ $\xrightarrow[\text{NH}_3 \text{ (liq)}]{\text{NaNH}_2}$ $CH_3CH_2CH_2C \equiv C{:}^- Na^+$ $\xrightarrow{(CH_3)_2SO_4}$ $CH_3CH_2CH_2C \equiv CCH_3$
                                                  E   F   G

The leaving group in dimethyl sulfate, $(CH_3)_2SO_4$, is the methyl sulfate anion, $^-OSO_3CH_3$.

9.4   (a) [cyclopentene with Cl] $\xrightarrow{HBr}$ [cyclopentane with Cl and Br]

(b) $CH_3C \equiv CCH_3$ $\xrightarrow{HBr \text{ (excess)}}$ $CH_3CH_2\overset{\displaystyle Br}{\underset{\displaystyle Br}{\overset{\displaystyle |}{\underset{\displaystyle |}{C}}}}CH_3$

(c) $CH_3CH_2CH_2CH = CH_2$ $\xrightarrow{HCl}$ $CH_3CH_2CH_2\overset{\displaystyle |}{\underset{\displaystyle Cl}{C}}HCH_3$

(d) [cyclohexene with CH$_3$] $\xrightarrow{HI}$ [cyclohexane with CH$_3$ and I]

9.5 $CH_3C \equiv C(CH_2)_7COH$ $\xrightarrow[\text{H}_2\text{SO}_4]{80\%}$ $CH_3\overset{O}{\overset{\|}{C}}CH_2(CH_2)_7\overset{O}{\overset{\|}{C}}OH$ + $CH_3CH_2\overset{O}{\overset{\|}{C}}(CH_2)_7\overset{O}{\overset{\|}{C}}OH$

- -

Concept Map 9.2   Electrophilic addition of acids to alkynes.

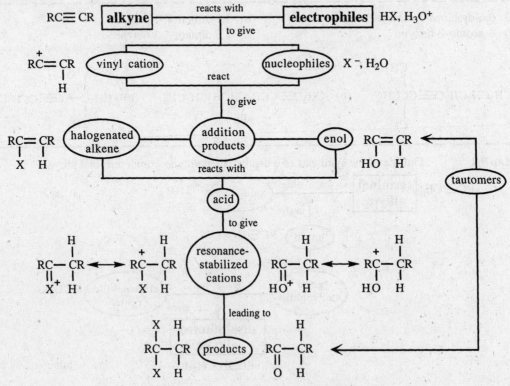

9.6 $CH_3CH_2C\equiv CCH_3$ $\longrightarrow$ $CH_3CH_2\overset{+}{C}=CHCH_3$ $\longrightarrow$ $CH_3CH_2C=CHCH_3$ $\longrightarrow$ $CH_3CH_2C=CHCH_3$

$CH_3CH_2CCH_2CH_3$ $\longleftarrow$ $\left[\; CH_3CH_2C-CH_2CH_3 \longleftrightarrow CH_3CH_2\overset{+}{C}-CH_2CH_3 \;\right]$

9.7 $HC\equiv CH$ $\longrightarrow$ $CH_2=\overset{+}{C}H$ $\longrightarrow$ $CH_2=CH$ $\longrightarrow$ $CH_2=CH$

CH_3CH $\longleftarrow$ $\left[\; CH_3-CH \longleftrightarrow CH_3-\overset{+}{C}H \;\right]$

9.8

1. What are the connectivities of the starting material and the products? How many carbon atoms does each contain? Are there any rings? What are the positions of branches and functional groups on the carbon skeletons?

$$CH_3(CH_2)_{11}C\equiv CH \quad \overset{?}{\longrightarrow} \quad CH_3(CH_2)_{11}C\equiv CCH_2(CH_2)_7CH_3$$

1-tetradecyne   10-tricosyne

The starting material has a linear chain of 14 carbon atoms; the product has a linear chain of 23 carbon atoms. The starting material is a 1-alkyne, and the product is a 10-alkyne.

2. How do the functional groups change in going from starting material to product? Does the starting material have a good leaving group?
A nonyl group in the product has replaced the terminal hydrogen in the starting material.

3. Is it possible to dissect the structures of the starting material and product to see which bonds must be broken and which formed?

9.8 (cont)

bond broken   bond formed

$CH_3(CH_2)_{11}C \equiv C - \{ - H$ $CH_3(CH_2)_{11}C \equiv C - \{ - CH_2(CH_2)_7CH_3$

Pieces that we need: $CH_3(CH_2)_{11}C \equiv C$——   and   ——$CH_2(CH_2)_7CH_3$

4.   New bonds are created when an electrophile reacts with a nucleophile. Do we recognize any part of the product molecule as coming from an electrophile or a nucleophile?
The new carbon-carbon bond must have come from a nucleophilic substitution reaction. The terminal alkyne can be converted into a nucleophile. The alkyl group, attached to a leaving group, must supply the electrophile.

$CH_3(CH_2)_{11}C \equiv C{:}^-$ $\xrightarrow{CH_3(CH_2)_7CH_2Br}$ $CH_3(CH_2)_{11}C \equiv CCH_2(CH_2)_7CH_3$

5.   After this last step, do we see how to get from starting material to product? If not, we need to analyze the structure obtained in step 4 by applying questions 3 and 4 to it.

Complete correct answer:

$CH_3(CH_2)_{11}C \equiv CH$ $\xrightarrow[NH_3 \text{ (liq)}]{NaNH_2}$ $CH_3(CH_2)_{11}C \equiv C{:}^- Na^+$ $\xrightarrow{CH_3(CH_2)_7CH_2Br}$ $CH_3(CH_2)_{11}C \equiv CCH_2(CH_2)_7CH_3$

9.9 $CH_3(CH_2)_{11}C \equiv CCH_2(CH_2)_7CH_3$ $\xrightarrow[\substack{PtO_2 \\ \text{hexane}}]{H_2 \text{ (1 equivalent)}}$

$CH_3(CH_2)_{11}$ $CH_2(CH_2)_7CH_3$

$\underset{H}{\overset{}{C}} = \underset{H}{\overset{}{C}}$ $+$ $CH_3(CH_2)_{21}(CH_3)$ $+$ $CH_3(CH_2)_{11}C \equiv CCH_2(CH_2)_7CH_3$

alkene   alkane   alkyne

In the presence of one equivalent of hydrogen, the first product to form will be the alkene. As the concentration of alkene builds up in the reaction mixture, it will start to react with hydrogen to give the alkane. To the extent that this second reaction occurs, some of the alkyne will remain unchanged, because the system will run out of hydrogen before it runs out of alkyne.

- -

Concept Map 9.3   Reduction reactions of alkynes.

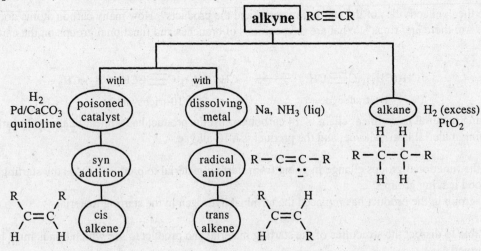

9.10   (a) $CH_3CH_2CH_2CH_2C \equiv CH$ $\xrightarrow[\text{quinoline}]{\text{H}_2 \atop \text{Pd/BaSO}_4}$ $CH_3CH_2CH_2CH_2CH = CH_2$

(b) $CH_3CH_2CH_2CH_2C \equiv CCH_3$ $\xrightarrow[\text{quinoline}]{\text{H}_2 \atop \text{Pd/BaSO}_4}$

(c) $CH_3CH_2CH_2CH_2C \equiv CCH_3$ $\xrightarrow[\text{NH}_3 \text{ (liq)}]{\text{Na}}$

(d) $CH_3CH_2CH_2CH_2C \equiv CCH_3$ $\xrightarrow[\text{acetic acid}]{\text{H}_2 \text{ (excess)} \atop \text{Pt}}$ $CH_3CH_2CH_2CH_2CH_2CH_2CH_3$

9.11 $CH_3C \equiv CCH_3$ $\xrightarrow[\text{Pd/Al}_2\text{O}_3]{\text{D}_2}$

9.12

(a) $CH_3CH_2CH_2\underset{\underset{Br}{|}}{C}HCH_3$ $\xleftarrow{\text{HBr}}$ $CH_3CH_2CH_2CH = CH_2$ $\xleftarrow[\text{quinoline}]{\text{H}_2 \atop \text{Pd/BaSO}_4}$ $CH_3CH_2CH_2C \equiv CH$ $\xleftarrow{CH_3CH_2CH_2Br}$

$Na^{+-}:C \equiv CH$ $\xleftarrow[\text{NH}_3 \text{ (liq)}]{\text{NaNH}_2}$ $HC \equiv CH$

(b) $CH_3CH_2CH_2\overset{\overset{O}{||}}{C}CH_2CH_3$ $\xleftarrow[\text{H}_2\text{SO}_4]{\text{H}_2\text{O}}$ $CH_3CH_2C \equiv CCH_2CH_3$ $\xleftarrow{CH_3CH_2Br}$

$CH_3CH_2C \equiv C:^- \ Na^+$ $\xleftarrow[\text{NH}_3 \text{ (liq)}]{\text{NaNH}_2}$ $CH_3CH_2C \equiv CH$

(c)

and enantiomer $\xleftarrow[\text{H}_2\text{O}]{\text{OsO}_4}$

$\xleftarrow[\text{quinoline}]{\text{H}_2 \atop \text{Pd/BaSO}_4}$

$CH_3CH_2CH_2C \equiv CCH_2CH_3$ $\xleftarrow{CH_3CH_2Br}$ $CH_3CH_2CH_2C \equiv C:^- \ Na^+$ $\xleftarrow[\text{NH}_3 \text{ (liq)}]{\text{NaNH}_2}$ $CH_3CH_2CH_2C \equiv CH$

9.13   (a)   (*R*)-2-chloro-4-octyne   (b)   (*Z*)-3-methyl-3-heptene
          (c)   2-methyl-4-octyn-2-ol   (d)   *trans*-1-methyl-2-vinylcyclohexane

9.14 (a)

$$
\underset{\substack{\\ \text{Cl} \quad\quad \text{H} \\[-2pt] }}{\overset{\substack{CH_3 \\ CH_3CCH_2 \\[-2pt]}}{C}} = \underset{\substack{\\ H \quad\quad CH_2CH_2CH_3}}{C}
$$

(b)

$$
\underset{\substack{Br}}{\overset{\substack{CH_2CH_2CH_3}}{H--C}}--CH_2C\equiv CH
$$

(c)

$$
HO\text{——}\langle\;\rangle\text{——}CH_3
$$
and enantiomer

(d) $CH_3CH_2C\equiv CCH_2CH_2OH$

9.15 (a) $CH_3CH_2C\equiv CCH_3 \xrightarrow[\substack{Pd/CaCO_3 \\ quinoline}]{H_2}$ $\underset{\substack{H \quad\quad H}}{\overset{\substack{CH_3CH_2 \quad CH_3}}{C}}=C$

(b) $CH_3CH_2C\equiv CCH_3 \xrightarrow[\substack{Pt}]{H_2\ (excess)} CH_3CH_2CH_2CH_2CH_3$

(c) $CH_3CH_2C\equiv CCH_3 \xrightarrow[\substack{NH_3\ (liq)}]{Na} \underset{\substack{H \quad\quad CH_3}}{\overset{\substack{CH_3CH_2 \quad H}}{C}}=C$

(d) $\underset{\substack{H \quad\quad H}}{\overset{\substack{CH_3CH_2 \quad CH_3}}{C}}=C \xrightarrow[\substack{dichloromethane}]{\substack{Cl\text{—}C_6H_4\text{—}COOH}} CH_3CH_2\overset{O}{\triangle}CH_3$ and enantiomer

(e) $\underset{\substack{H \quad\quad CH_3}}{\overset{\substack{CH_3CH_2 \quad H}}{C}}=C \xrightarrow[\substack{dichloromethane}]{\substack{Cl\text{—}C_6H_4\text{—}COOH}} CH_3CH_2\overset{O}{\triangle}H\ CH_3$ and enantiomer

(f) $CH_3CH_2C\equiv CCH_3 \xrightarrow{HBr\ (1\ molar\ equiv)}$ $\underset{\substack{H \quad\quad CH_3}}{\overset{\substack{CH_3CH_2 \quad Br}}{C}}=C$ + $\underset{\substack{Br \quad\quad CH_3}}{\overset{\substack{CH_3CH_2 \quad H}}{C}}=C$

major products

(g) $CH_3CH_2C\equiv CCH_3 \xrightarrow{HBr\ (2\ molar\ equiv)} CH_3CH_2CBr_2CH_2CH_3 + CH_3CH_2CH_2CBr_2CH_3$

(h) $CH_3CH_2C\equiv CCH_3 \xrightarrow{Br_2\ (1\ molar\ equiv)} \underset{\substack{Br \quad\quad CH_3}}{\overset{\substack{CH_3CH_2 \quad Br}}{C}}=C$

(i) $CH_3CH_2C\equiv CCH_3 \xrightarrow{Br_2\ (2\ molar\ equiv)} CH_3CH_2CBr_2CBr_2CH_3$

(j) $CH_3CH_2C\equiv CCH_3 \xrightarrow[\substack{H_2SO_4 \\ HgSO_4}]{H_2O} CH_3CH_2\overset{O}{\overset{\|}{C}}CH_2CH_3 + CH_3CH_2CH_2\overset{O}{\overset{\|}{C}}CH_3$

9.16   (a) 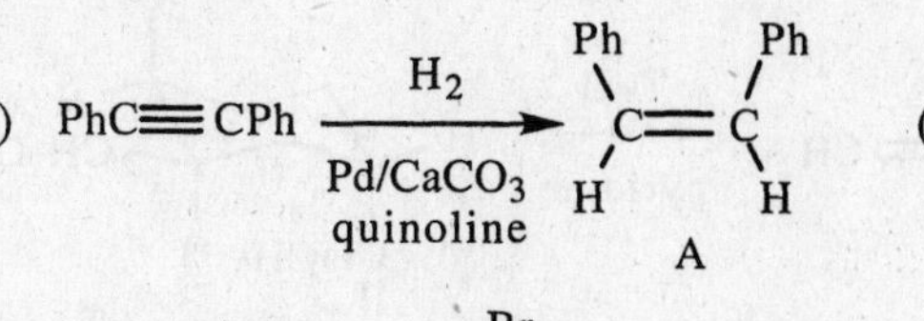 PhC≡CPh $\xrightarrow[\substack{Pd/CaCO_3 \\ quinoline}]{H_2}$ Ph—C=C—Ph (with H's) **A**

(b) HC≡CCH₂COH (with =O) $\xrightarrow[]{\substack{HI\ (1\ molar \\ equiv)}}$ CH₂=CCH₂COH (with =O) **I**
B

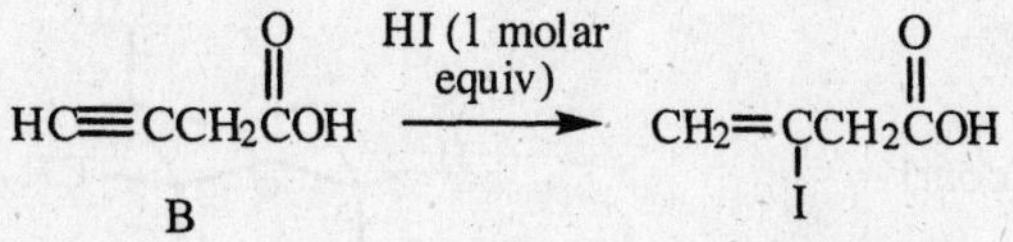

(c) 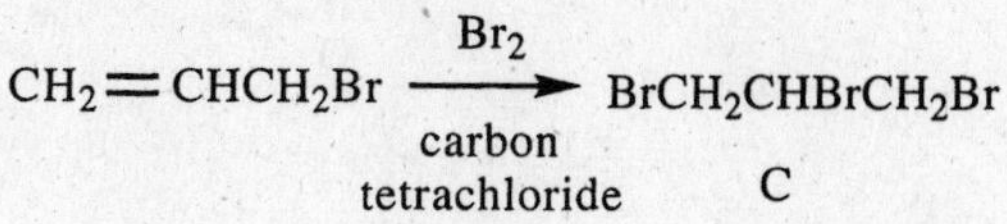 CH₂=CHCH₂Br $\xrightarrow[\substack{carbon \\ tetrachloride}]{Br_2}$ BrCH₂CHBrCH₂Br **C**

(d) PhC≡CCH₂CH₃ $\xrightarrow[\substack{NH_3\ (liq)}]{Na}$ Ph—C=C—H / CH₂CH₃ **D**

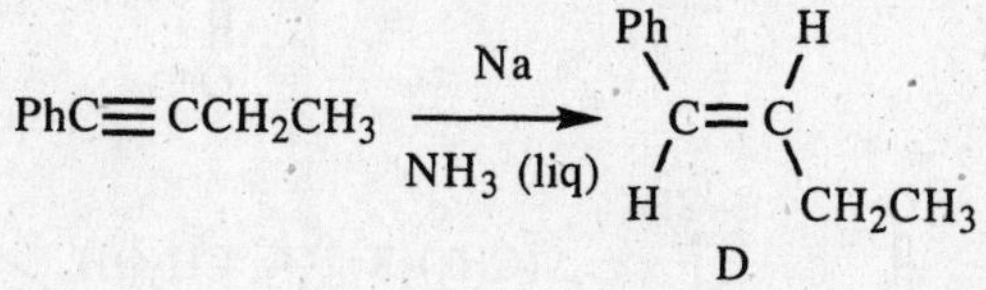

(e)

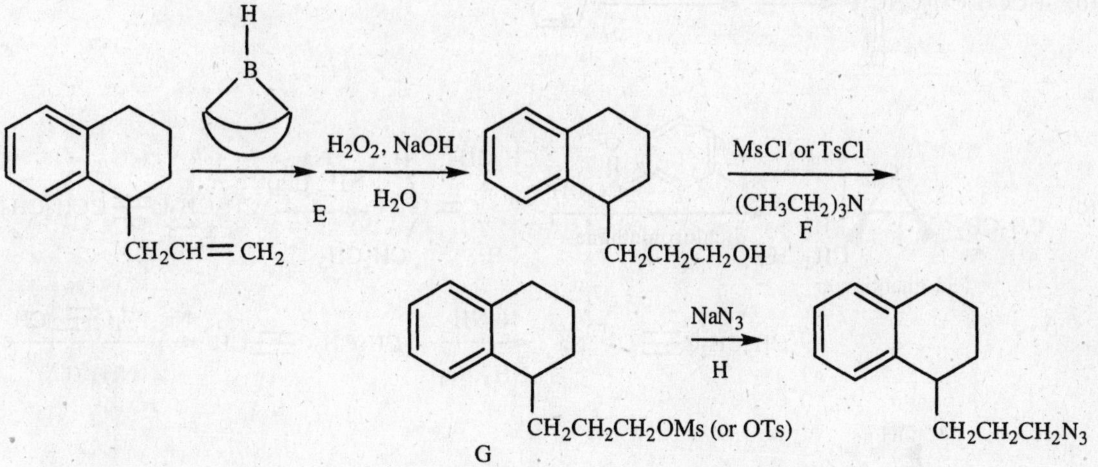

(f)

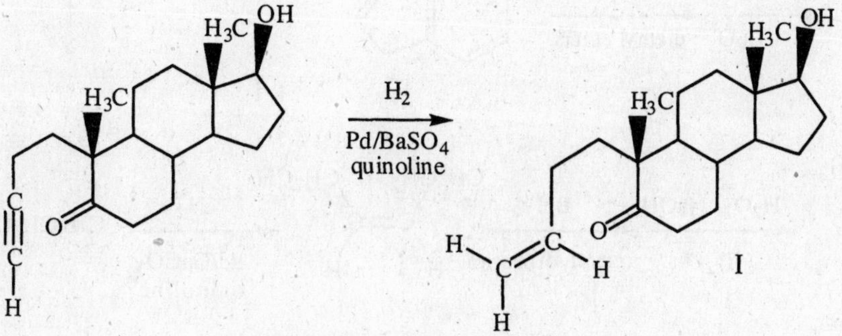

(g) 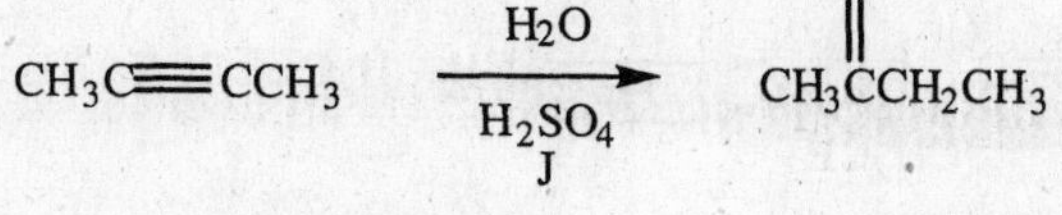 CH₃C≡CCH₃ $\xrightarrow[H_2SO_4]{H_2O}$ CH₃CCH₂CH₃ (with =O)
J

(h) HC≡CH $\xrightarrow[K]{Na^+NH_2^-}$ HC≡C:⁻ Na⁺ $\xrightarrow[M]{PhCH_2OCH_2CH_2Br}$ HC≡CCH₂CH₂OCH₂Ph
L

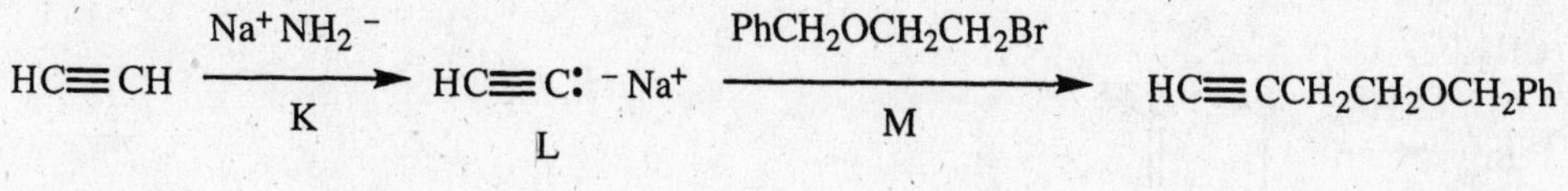

9.17

(a)

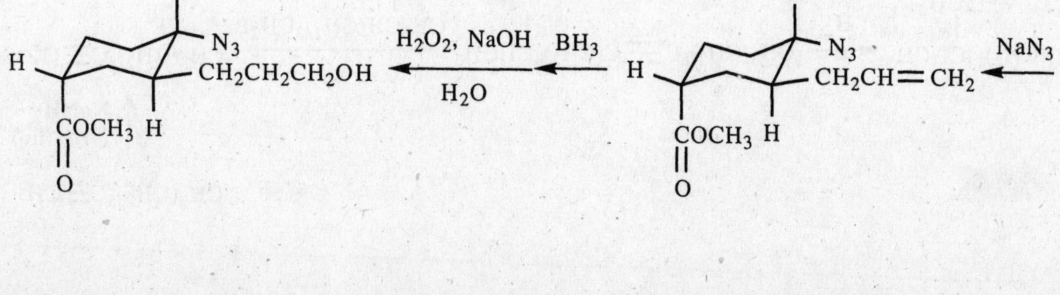

9.17 (a) (cont)

$$\text{(structure with OTs, H, CH}_2\text{CH=CH}_2\text{, COCH}_3\text{, O)} \xleftarrow[\text{pyridine}]{\text{TsCl}} \text{(structure with OH, H, CH}_2\text{CH=CH}_2\text{, COCH}_3\text{, O)}$$

(b)

$$\underset{\underset{O}{\parallel}}{\text{HCCH}_2\text{CH}_2}\underset{\underset{O}{\underset{\parallel}{}}}{\overset{\overset{\displaystyle CH_3}{\mid}}{\text{CHCH}}} \xleftarrow{(CH_3)_2S \quad O_3,\ CH_3OH} \text{(methylcyclopentene)}$$

(c)

$$\underset{\substack{H \quad\quad CH_2CH_3 \\ \text{and enantiomer}}}{\overset{O}{CH_3CH_2\cdots\bigtriangleup\cdots H}} \xleftarrow[\text{dichloromethane}]{\text{Cl} \ \ \text{COOH (m-chlorobenzoic acid)}} \underset{\substack{H \qquad CH_2CH_3}}{\overset{CH_3CH_2 \qquad H}{C=C}} \xleftarrow[\text{NH}_3\ (liq)]{Na} CH_3CH_2C\equiv CCH_2CH_3 \xleftarrow{CH_3CH_2Br}$$

$$CH_3CH_2C\equiv C\!:^-\ Na^+ \xleftarrow[\text{NH}_3\ (liq)]{NaNH_2} CH_3CH_2C\equiv CH \xleftarrow[\text{NH}_3\ (liq)]{Na^+\ {}^-\!:C\equiv CH} CH_3CH_2Br$$

(d)

$$\underset{\text{and enantiomer}}{\text{(indane-1,2-diol, OH, OH)}} \xleftarrow[\text{H}_2O]{\text{KOH}} \xleftarrow[\text{diethyl ether}]{OsO_4} \text{(indene)}$$

(e)

$$\underset{\substack{OH \\ \text{and enantiomer}}}{\overset{\overset{\displaystyle CH_2CH_3}{\mid}}{CH_3CH_2CH_2\overset{}{\underset{}{C}}\text{-}H}} \xleftarrow[\text{H}_2O]{H_2O_2,\ NaOH} \xleftarrow{BH_3} \underset{\substack{H \qquad\quad H}}{\overset{CH_3CH_2 \qquad CH_2CH_3}{C=C}} \xleftarrow[\substack{Pd/CaCO_3 \\ \text{quinoline}}]{H_2} CH_3CH_2C\equiv CCH_2CH_3$$

$$\xleftarrow[\text{tetrahydrofuran}]{}$$

An alternate synthesis is given below.

$$\underset{\substack{OH \\ \text{and enantiomer}}}{\overset{\overset{\displaystyle CH_2CH_3}{\mid}}{CH_3CH_2CH_2\overset{}{\underset{}{C}}\cdots H}} \xleftarrow[\text{H}_2SO_4]{H_2O} \underset{\substack{H \qquad\quad CH_2CH_3}}{\overset{CH_3CH_2 \qquad H}{C=C}} \xleftarrow[\text{NH}_3\ (liq)]{Na} CH_3CH_2C\equiv CCH_2CH_3$$

(f)

$$\underset{}{\overset{\overset{\displaystyle CH_3}{\mid}}{\text{(1-bromo-1-methylcyclohexane), Br}}} \xleftarrow{\text{HBr}} \text{(1-methylcyclohexene)}$$

9.18

$$\underset{\text{Pt}}{\overset{\overset{\displaystyle CH_3}{\mid}}{CH_3(CH_2)_{14}CHCH_3}} \xleftarrow{H_2} CH_3(CH_2)_8C\equiv C(CH_2)_4\overset{\overset{\displaystyle CH_3}{\mid}}{CHCH_3} \xleftarrow{CH_3\overset{\overset{\displaystyle CH_3}{\mid}}{CH}(CH_2)_3CH_2Br} CH_3(CH_2)_8C\equiv C\!:^-\ Na^+$$

$$\Big\uparrow \substack{NaNH_2 \\ NH_3\ (liq)}$$

$$CH_3(CH_2)_8C\equiv CH$$

9.19

(a)

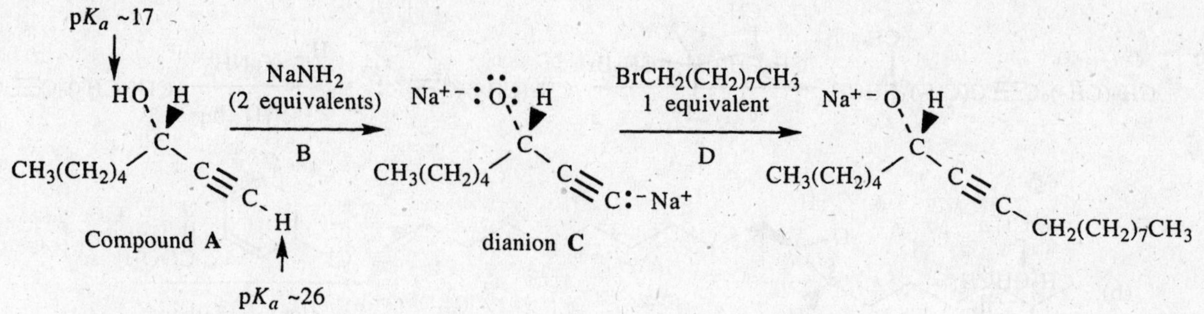

Compound **A**

dianion **C**

(b)   The nucleophilic sites are the oxygen anion, $-\ddot{\underset{\cdot\cdot}{O}}:^-$, and the carbon anion, $-C\equiv C:^-$. The reason for the regioselectivity of this reaction is that the carbon anion is much more nucleophilic than the oxygen anion because carbon is less electronegative than oxygen.

(c)   (*S*)-1-octyn-3-ol

9.20

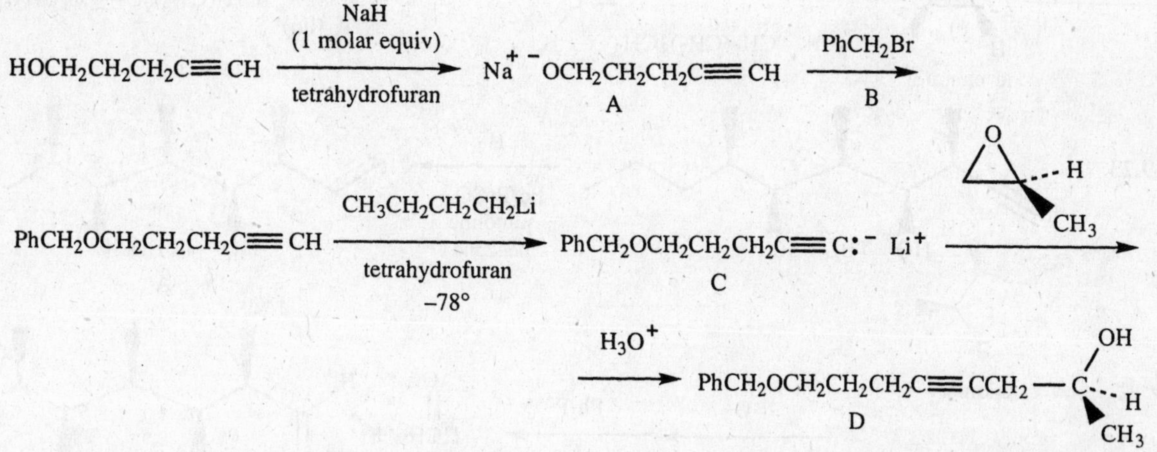

9.21   (a)

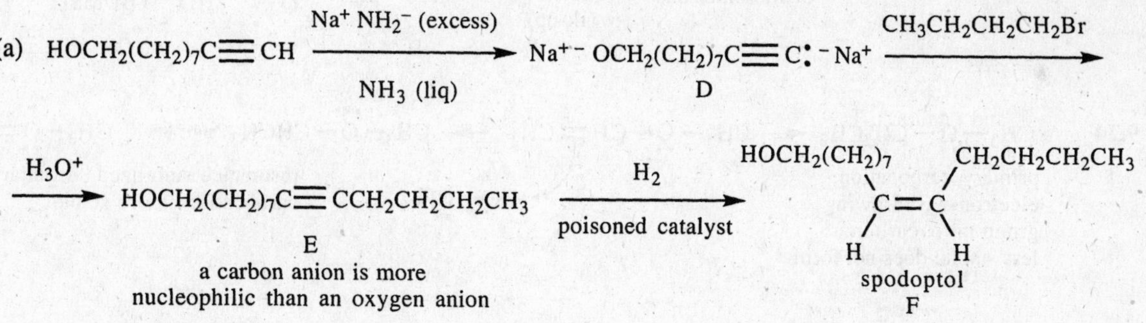

a carbon anion is more
nucleophilic than an oxygen anion

spodoptol
F

(b)

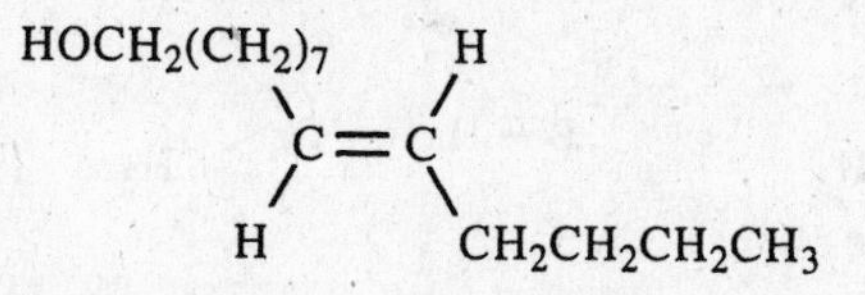

(c) $CH_3CH_2CH_2CH_2\overset{\overset{O}{\|}}{C}H$ $HOCH_2(CH_2)_7\overset{\overset{O}{\|}}{C}H$

Aldehyde A   Aldehyde B

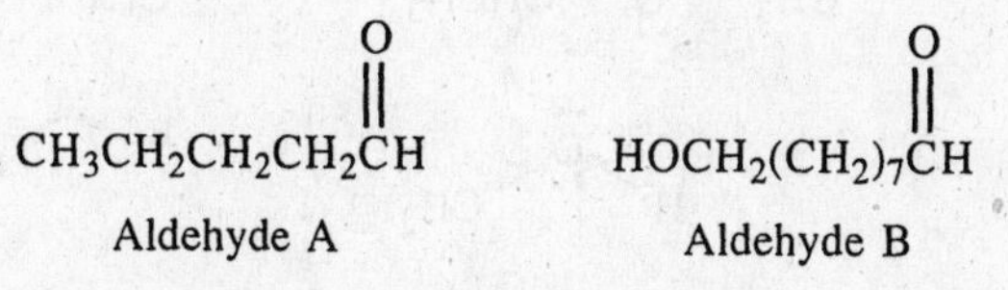

9.22   (a)

disparlure

dichloromethane

H_2
Pd/CaCO$_3$
quinoline

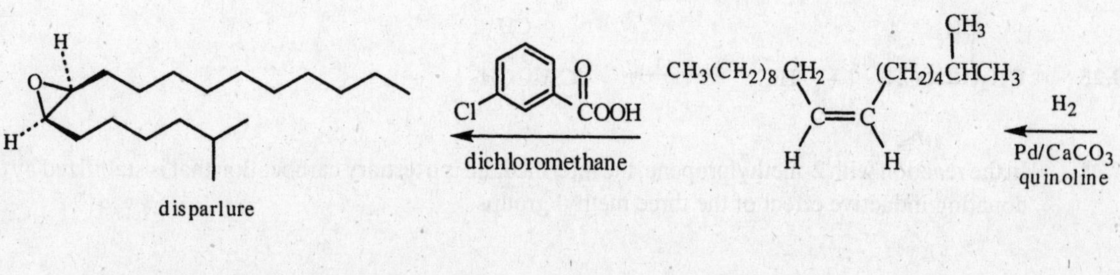

9.22 (a) (cont)

$$CH_3(CH_2)_9C\equiv C(CH_2)_4\overset{\overset{\displaystyle CH_3}{|}}{CH}CH_3 \xleftarrow{\quad CH_3\overset{\overset{\displaystyle CH_3}{|}}{CH}(CH_2)_3CH_2Br \quad} CH_3(CH_2)_9C\equiv C:^-Na^+ \xleftarrow[NH_3 \ (liq)]{NaNH_2} CH_3(CH_2)_9C\equiv CH$$

(b)

$$\xleftarrow[\text{dichloromethane}]{Cl-\!\!\!\!\bigcirc\!\!\!\!-COOH}$$

$$\xleftarrow[NH_3 \ (liq)]{Na} CH_3(CH_2)_9C\equiv C(CH_2)_4\overset{\overset{\displaystyle CH_3}{|}}{CH}CH_3$$

9.23

$$\xrightarrow[\substack{Pd/BaSO_4 \\ quinoline \\ benzene}]{H_2}$$

A

$$\xrightarrow[\text{dichloromethane}]{O_3} \xrightarrow[\substack{(\text{reductive} \\ \text{work-up})}]{Ph_3P} \quad \underset{\text{B}}{\overset{\displaystyle O}{\underset{\displaystyle \|}{HCH}}} \ + \quad \text{C}$$

9.24

$$CH_3-\overset{..}{\underset{..}{O}}-CH_2\overset{+}{C}H_2 \longleftarrow CH_3-\overset{..}{\underset{..}{O}}-CH=CH_2 \longrightarrow CH_3-\overset{..}{\underset{..}{O}}-\overset{+}{C}HCH_3 \longleftrightarrow CH_3-\overset{+}{\underset{..}{O}}=CHCH_3$$

primary carbocation;
electron-withdrawing
group on α-carbon;
less stable; does not form

$H\overset{\frown}{}\overset{..}{\underset{..}{Cl}}:$

resonance stabilized secondary
carbocation; more stable

$$CH_3-\overset{..}{\underset{\underset{\displaystyle H}{\overset{..}{\underset{..}{O}}}\,CH_3}{\overset{+}{O}}}-CHCH_3 \longrightarrow CH_3-\overset{..}{\underset{..}{O}}-CHCH_3 \longrightarrow CH_3-\overset{..}{\underset{..}{O}}-CHCH_3$$

$$B:\overset{\frown}{}H\overset{\frown}{}\overset{+}{\underset{..}{O}}\,CH_3 \qquad\qquad ^+B-H \qquad :\overset{..}{\underset{..}{O}}\,CH_3$$

9.25

$$\overset{\overset{\displaystyle CH_3}{|}}{CH_3C}=CH_2 \ + \ HCl \longrightarrow \overset{\overset{\displaystyle CH_3}{|}}{\underset{\underset{\displaystyle Cl}{|}}{CH_3CCH_3}}$$

In the reaction with 2-methylpropene, the intermediate is a tertiary carbocation that is stabilized by the electron-donating inductive effect of the three methyl groups.

9.25 (cont)

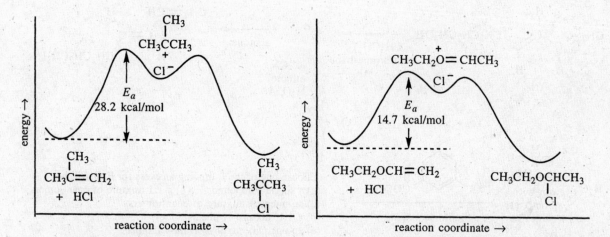

tertiary carbocation; less stable than the resonance-
stabilized carbocation from ethyl vinyl ether

$$CH_3CH_2OCH=CH_2 \;+\; HCl \;\longrightarrow\; CH_3CH_2OCHCH_3$$
$$\underset{Cl}{|}$$

In the reaction with ethyl vinyl ether, the intermediate is a carbocation in which the positive charge can be delocalized over the carbon atom and the adjacent oxygen atom.

$$CH_3CH_2\ddot{O}-CH=CH_2 \;\longrightarrow\; \left[\; CH_3CH_2\ddot{O}-\overset{+}{C}HCH_3 \;\longleftrightarrow\; CH_3CH_2\overset{+}{O}=CHCH_3 \;\right] \quad :\ddot{Cl}:^-$$

major contributor

delocalized carbocation; more stable

$$CH_3CH_2\ddot{O}-\overset{+}{C}HCH_3 \;\longrightarrow\; CH_3CH_2\ddot{O}-CHCH_3$$
$$:\ddot{Cl}:^- \qquad\qquad\qquad\qquad :\ddot{Cl}:$$

The reactive intermediate formed in this reaction is more highly stabilized, and therefore, closer in energy to the energy level of the starting materials than is the case for 2-methylpropene. The E_a for this reaction is therefore lower, and the reaction is faster.

9.26 (a)

Compound **X**   Compound **Y**
                           tetronic acid

9.26   (b)

$$H-CH-O-C=O \longleftrightarrow H-CH-O-C-O^-$$

Supplemental Problems

S9.1   The following equations represent chemical transformations showing pathways from starting materials to products. All of these use the types of reactions that you have learned involving electrophilic and nucleophilic species. Fill in the structural formulas for the starting materials, reagents, or products represented by the letters, using changes in molecular connectivity as clues. Be sure to show stereochemistry when it is known.

(a) $CH_3C \equiv CCH_2CNH_2 \xrightarrow[\text{quinoline}]{\text{H}_2 \atop \text{Pd/BaSO}_4}$ A

(b) $\xrightarrow[\substack{\text{tetrahydrofuran} \\ -78\,°C}]{C_6H_{13}C \equiv C:^- Li^+ \\ (2\ \text{equivalents})}$ B

(c) $CH_2=C-C=CH_2CH_2CH_2OH \xrightarrow[\substack{(CH_3CH_2)_3N \\ \text{dichloromethane}}]{\text{MsCl}}$ C $\xrightarrow[\text{acetone}]{E}$ $CH_2=CH$ C=C H, CH_2CH_2CH_2I

$C \xrightarrow[\substack{\text{ethanol} \\ H_2O,\ \Delta}]{K^+\ CN^-}$ D

(d) $H_2C=C-C-H \xrightarrow[\text{dichloromethane}]{\text{Cl}\ C_6H_4\ COOH}$ F

Choose one of the following answers for F. It will be:
(a) a single compound   (b) a 1:1 mixture of enantiomers
(c) an unequal mixture of enantiomers
(d) a 1:1 mixture of diastereomers
(e) an unequal mixture of diastereomers

(e) $(CH_3)_3SiCH_2C \equiv CCH_2CH_2CHPh \xrightarrow[\substack{\text{Lindlar catalyst} \\ \text{(a poisoned Pd catalyst)}}]{\text{H}_2}$ G
          |
          OH

(f) $H-C \equiv C-C-H \xrightarrow{H}$ CH_3CH_2-C-H
          NHBoc   NHBoc

(g) $\xrightarrow[NH_3\ (liq)]{\text{Na}}$ I

Boc is a group used to protect the amino group.
It is unaffected by this reaction.

S9.1

(h) [structure: N-methyl dihydropyridinone with CH₂Br and CH₃CCH₃/H substituents] $\xrightarrow{\text{PhC}\equiv\text{C}^- \text{Li}^+}$ J

(i) [structure: H—C(Ph)(CH₃)—CH₂OH] $\xrightarrow[\text{pyridine}]{\text{TsCl}}$ K $\xrightarrow[\text{acetone}]{\text{L}}$ [structure: H—C(Ph)(CH₃)—CH₂Br]

(j) $CH_3(CH_2)_4C\equiv CCH_2COCH_3$ $\xrightarrow{\text{M}}$ [structure: cis alkene CH₃(CH₂)₄ and CH₂COCH₃ on C=C with H, H]

(k) stereochemistry unknown [structure: cyclohexenol with OH (wavy) and R,H] $\xrightarrow{\text{N}}$ [structure O: epoxide] + [structure P: epoxide]

R is a large peptide chain and is not changed in the reaction.

O   P

Which one of the oxirane stereoisomers, O or P, would you expect to be formed in the larger amount and why?

(l) $HC\equiv CH$ $\xrightarrow{\text{HBr (excess)}}$ Q

(m) $PhCH_2CH\!=\!CH_2$ $\xrightarrow{\text{HBr}}$ R

(n) $PhC\equiv C - COCH_2CH_3$ $\xrightarrow[\substack{\text{Pd/BaSO}_4 \\ \text{quinoline}}]{\text{H}_2}$ S

S9.2   How would you carry out each of the following transformations? More than one step may be necessary.

(a) [bicyclic structure with H, H] $\longrightarrow$ [structure: HOCCH₂ and COH diacid with H, H]

(b) $CH_3CHCH_2CH_2Br$ (with CH₃ substituent) $\longrightarrow$ CH_3CHCH (with CH₃ and =O)

(c) [cyclohexane with CH₃, H, HO, CH₃] $\longrightarrow$ [cyclohexane with CH₃, CH₃, O epoxide]

(d) CH_3CH_2Br $\longrightarrow$ [structure: (CH₃CH₂)(H)C=C(H)(CH₂CH₃)]

(e) $CH_3CH_2C\equiv CCH_2CH_3$ $\longrightarrow$ [epoxide: H, CH₃CH₂, CH₂CH₃, H]

S9.3   Substitution of hydrogen with fluorine often causes significant changes in the chemical and physical properties of compounds while allowing them to be taken up by cells and to interact with key enzyme systems. For this reason, there is much interest in the synthesis of such compounds. Some steps from an investigation of trifluoromethyl compounds are shown on the next page.

S9.3   (a)   A racemic mixture of the following compound, compound A, was resolved by treatment with a mixture of vinyl acetate and an enzyme that converted the *S*-enantiomer to an acetate ester, and left the *R*-enantiomer untouched. The equation for this reaction is given below with no stereochemistry shown.

$$CF_3C \equiv CCHCH_2OCH_2Ph \quad + \quad \begin{matrix} H \\ \diagdown \\ C=C \\ \diagup \\ H \end{matrix} \begin{matrix} OCCH_3 \\ \diagup \\ \diagdown \\ H \end{matrix} \xrightarrow[\text{hexane}]{\text{Novozym® 435}}$$

racemic compound A   vinyl acetate

$$CF_3C \equiv CCHCH_2OCH_2Ph \quad + \quad CF_3C \equiv CCHCH_2OCH_2Ph \quad + \quad \begin{bmatrix} H \diagup \diagdown OH \\ C=C \\ H \diagup \diagdown H \end{bmatrix}$$

(*R*)-enantiomer of compound A   acetate ester of   compound B
                                               the (*S*)-enantiomer
                                               of compound A

Draw a correct structural formula for the (*R*)-enantiomer of compound A, and show how it reacts with hydrogen in the presence of a poisoned palladium catalyst.

(b)   Fill in the structures of the ozonolysis products of compound C below, and draw a Newman projection for it looking down the bond indicated.

compound C

$\xrightarrow{\text{ozonolysis}}$ optically active product + optically inactive product

(c)   Compound B was shown in brackets in the equation in part (a) because it is not recovered from the mixture in the form written, but as acetaldehyde. What is the relationship between acetaldehyde and compound B?

(d)   Use the curved arrow convention to show how compound B is converted to acetaldehyde in the reaction mixture. Use HB$^+$ and B: as acids and bases as needed.

S9.4   Polyunsaturated fatty acids labeled with deuterium (D, an isotope of hydrogen, sometimes shown as ^{2}H,) are needed to study the metabolism of such acids. A compound used in the synthesis of such fatty acids is $HOCH_2C \equiv CCD_2CD_3$. 2-Propyn-1-ol (propargyl alcohol), $HOCH_2C \equiv CH$, is readily available as a starting material. Show how you would convert 2-propyn-1-ol to the desired compound. The synthesis involves several steps. Pay special attention to the relative amounts of the reagents you need to use. Solvents may be ignored. Your pK_a table will be useful in choosing reagents and predicting the structures of reactive intermediates. You may assume that you can buy any necessary deuterated reagents.

S9.5   The following transformation is proposed to occur in a single, bimolecular step. Provide the curved arrows for the mechanism.

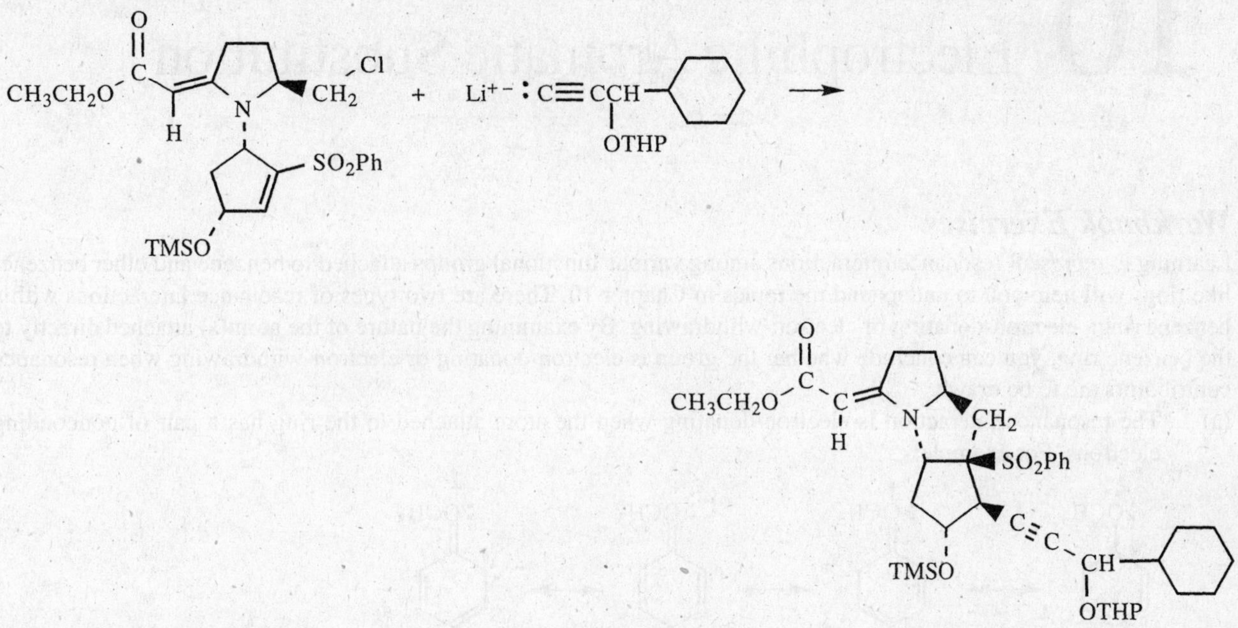

TMS- and THP- are groups that protect alcohols from reactions (pp. 845 and 847 in the text). They do not themselves react with the reagents shown.

S9.6   Chemists are interested in how natural products are constructed in biological systems. In research into how some plant products are synthesized, the following reaction pathway was proposed. Use the curved arrow convention to show the electronic and bonding changes that occur as the starting material is converted to the final cation shown.

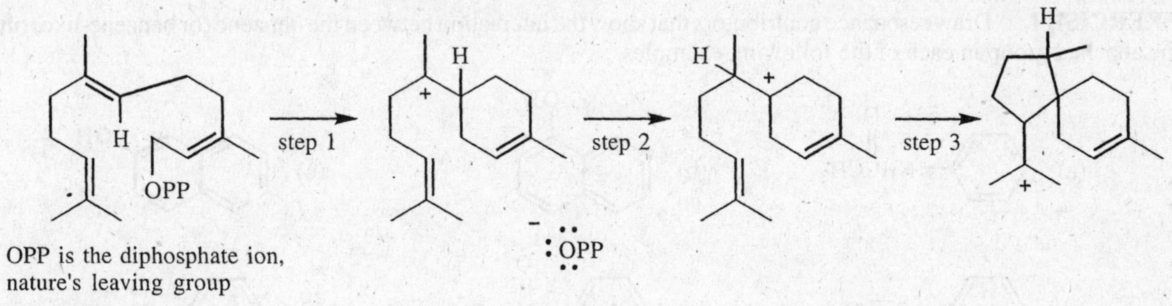

OPP is the diphosphate ion, nature's leaving group

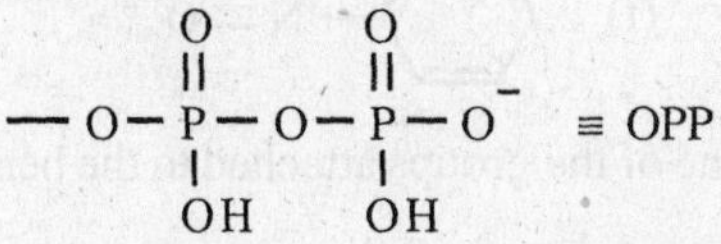

$$-O-\overset{\overset{O}{\|}}{P}-O-\overset{\overset{O}{\|}}{P}-O^- \equiv OPP$$
$$\quad\quad\;\; | \quad\quad\quad | $$
$$\quad\quad OH \quad\quad OH$$

What is the transformation shown in step 2 called?

10 The Chemistry of Aromatic Compounds. Electrophilic Aromatic Substitution

Workbook Exercises

Learning to represent resonance interactions among various functional groups attached to benzene and other benzene-like rings will help you to understand the topics in Chapter 10. There are two types of resonance interactions with a benzene ring: electron-donating or electron-withdrawing. By examining the nature of the atom(s) attached directly to the benzene ring, you can conclude whether the group is electron-donating or electron-withdrawing when resonance contributors are to be drawn.

(a) The resonance interaction is electron-donating when the atom attached to the ring has a pair of nonbonding electrons. For example:

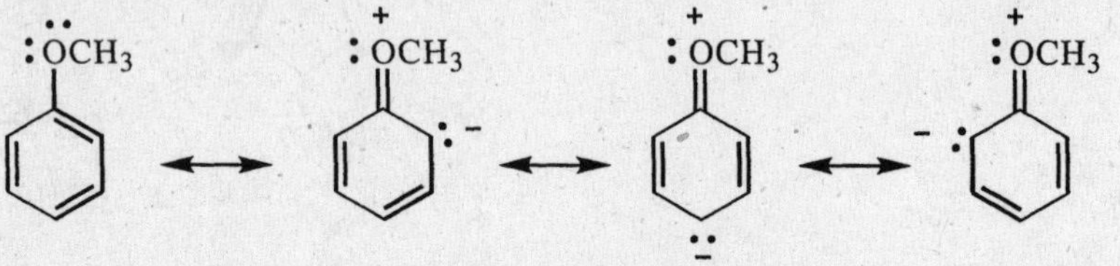

(b) The resonance interaction is electron-withdrawing when the atom attached to the ring is involved with a multiple bond to an electronegative atom such as oxygen or nitrogen. For example:

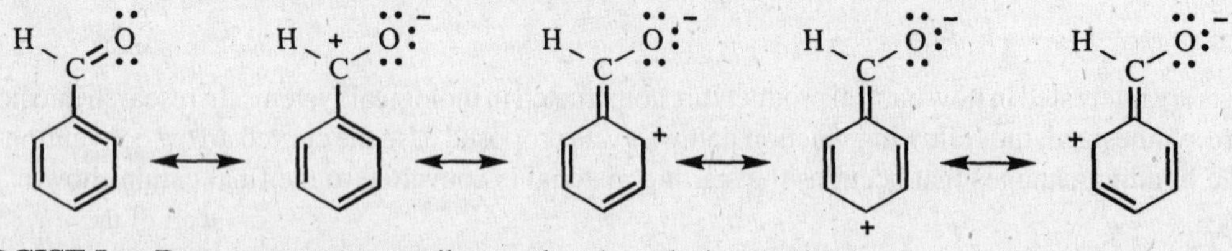

EXERCISE I. Draw resonance contributors that show the interaction between the benzene (or benzene-like) ring and the attached group in each of the following examples.

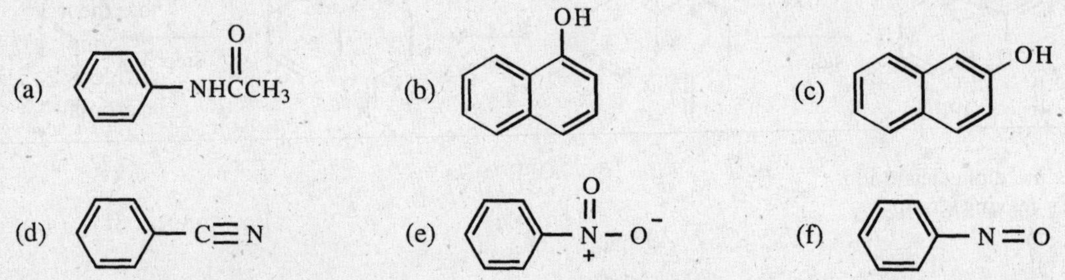

EXERCISE II. Which of the following ions will be resonance stabilized by one of the groups attached to the benzene ring?

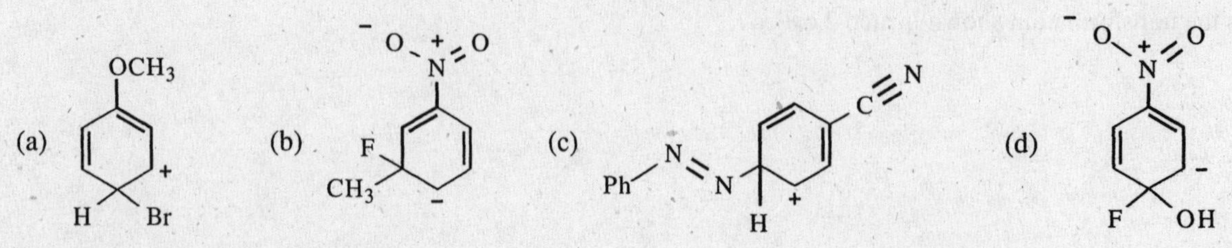

10.1   The lone pair electrons that are counted in determining aromaticity are those that are in the plane of the conjugated cyclic π system and bring the total number of π electrons to $4n + 2$.

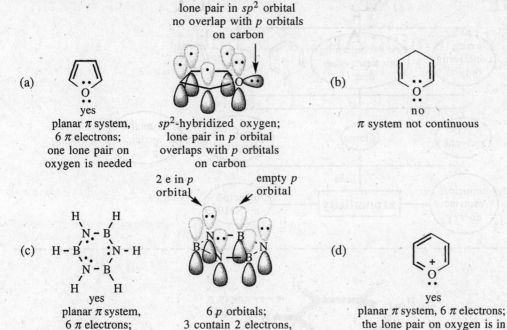

(a)
yes
planar π system,
6 π electrons;
one lone pair on
oxygen is needed

lone pair in sp^2 orbital
no overlap with p orbitals
on carbon

sp^2-hybridized oxygen;
lone pair in p orbital
overlaps with p orbitals
on carbon

(b)
no
π system not continuous

(c)
yes
planar π system,
6 π electrons;
all three lone pairs on
nitrogen are needed

2 e in p
orbital

empty p
orbital

6 p orbitals;
3 contain 2 electrons,
3 have room to
accept electrons

(d)
yes
planar π system, 6 π electrons;
the lone pair on oxygen is in
sp^2 hybrid orbital

(e)
yes; planar π system, 6 π electrons;
the lone pairs on the nitrogen in sp^2 hybrid orbitals,
do not overlap with π bonds

(f)
yes; planar π system, 6 π electrons;
the lone pair on the nitrogen bonded to the
hydrogen is in a p orbital and is part of the π system,
the lone pair on the other nitrogen is not.

10.2   The hydrogen atoms on the sp^2-hybridized carbon atoms in thiophene absorb close to where the hydrogen atoms in benzene do (Figure 10.2 in the text). Those bound to the sp^2-hybridized atoms in methyl vinyl sulfide absorb more in the region seen for 1,3,5,7-cyclooctatetraene (Figure 10.2) indicating that the sulfide is an alkene. The chemical shifts of the thiophene hydrogen atoms fall in the region where hydrogen atoms on aromatic rings absorb, pointing to the presence of a ring current, which is one of the criteria for aromaticity.

thiophene
$(4n + 2)$ π electrons with $n = 1$, therefore
an aromatic system with delocalization
of electrons over the ring

$CH_3 - \overset{\cdot\cdot}{\underset{\cdot\cdot}{S}} - CH = CH_2$

methyl vinyl sulfide

10.3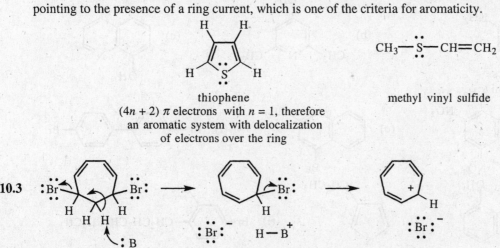

Concept Map 10.1   Aromaticity.

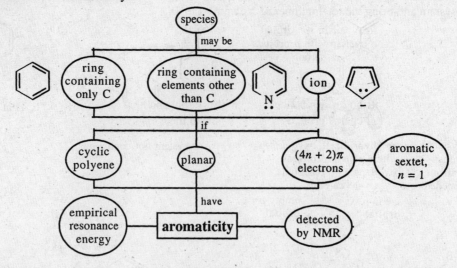

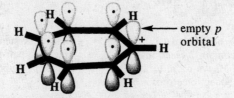

10.4

empty *p* orbital

σ bonds between the carbon atoms and
between the carbon and hydrogen atoms

10.5   (a) 1,3-dinitrobenzene
               m-dinitrobenzene
           (b) 2-methylnaphthalene
               β-methylnaphthalene
           (c) 3-ethylnitrobenzene
               m-ethylnitrobenzene
           (d) 4-chlorophenol
               p-chlorophenol
           (e) 2,4-dichlorotoluene
           (f) 4-*tert*-butylnitrobenzene
               p-*tert*-butylnitrobenzene

10.6   (a)   (b)   (c)

           (d)   (e)   (f) Br——⬡—⬡——Br

           (g)   (h)   (i)

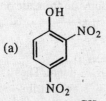

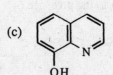

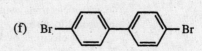

10.7

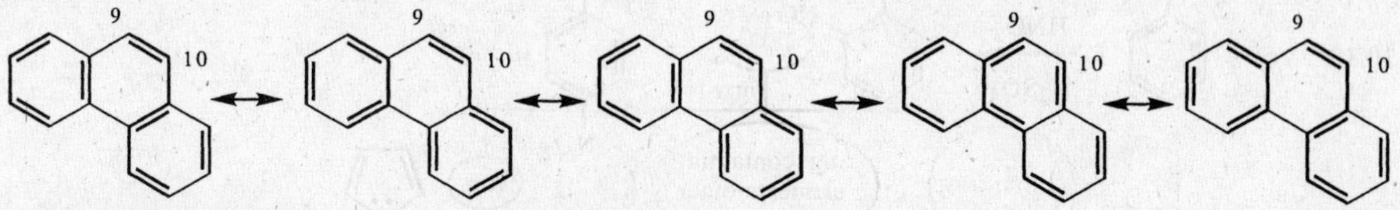

Four of the five resonance contributors of phenanthrene have a double bond at the 9,10 position. That bond, therefore, has more localized double bond character than the other bonds in the molecule.

10.8

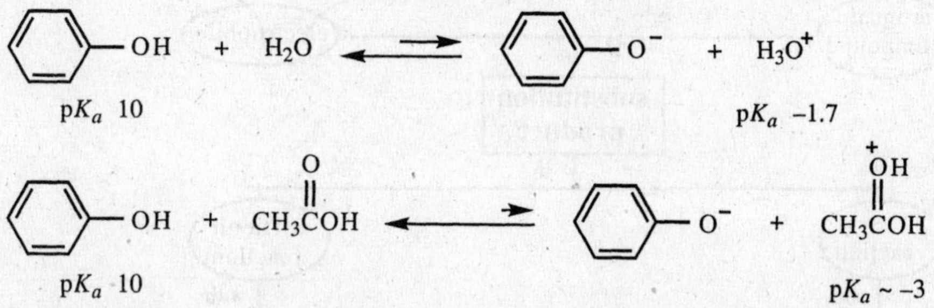

Phenol is much less likely to transfer a proton to acetic acid than it is to water. The water solution will contain a higher concentration of phenolate ion than will the acetic acid solution.

10.9

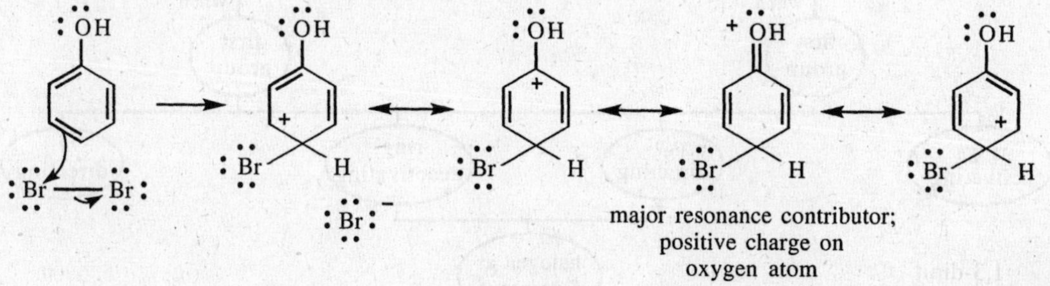

major resonance contributor;
positive charge on
oxygen atom

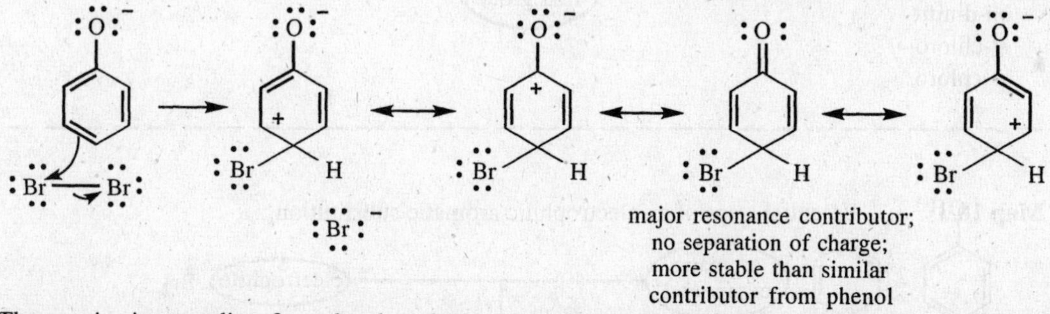

major resonance contributor;
no separation of charge;
more stable than similar
contributor from phenol

The reactive intermediate from the phenolate ion is better stabilized than the one from phenol. The phenolate ion gives rise to a reactive intermediate that has no separation of charge in its major resonance contributor, and hence is more stable than the reactive intermediate from phenol. A more stable reactive intermediate corresponds to a lower transition state and, therefore, to a faster reaction.

10.10   (a)

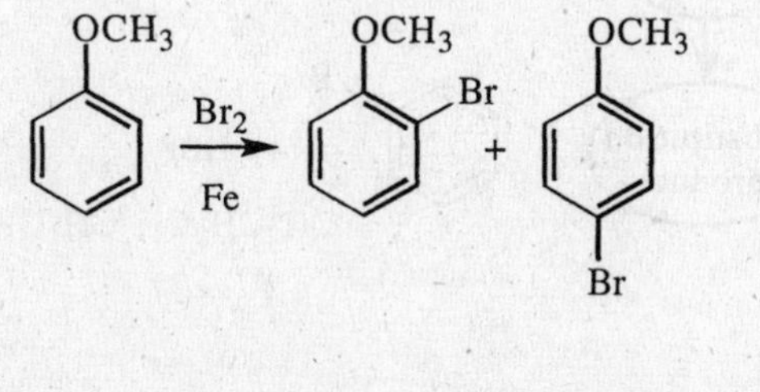

(b)

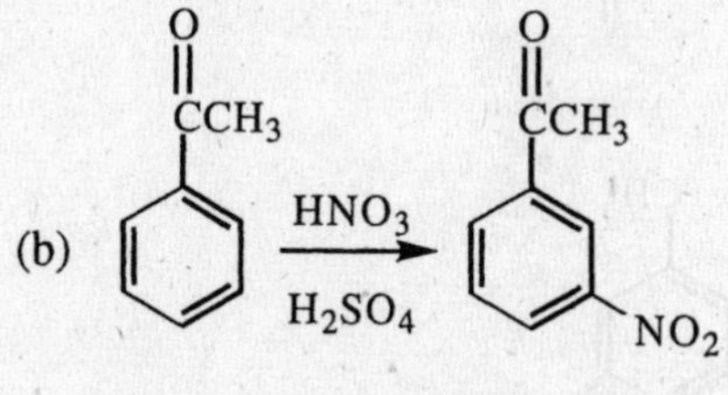

10.10   (c)

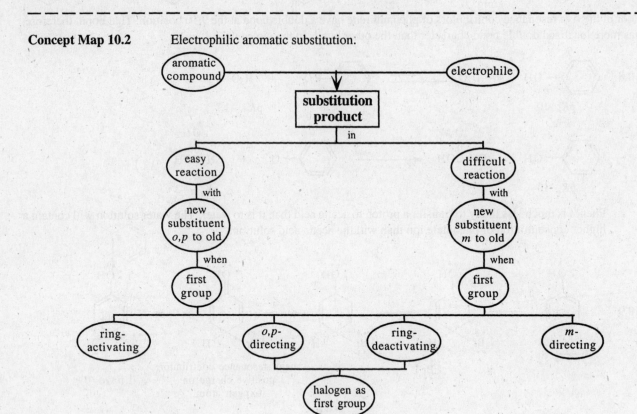

Concept Map 10.2   Electrophilic aromatic substitution.

aromatic compound —————— electrophile

substitution product

in

easy reaction

with

new substituent *o,p* to old

when

first group

difficult reaction

with

new substituent *m* to old

when

first group

ring-activating   *o,p*-directing   ring-deactivating   *m*-directing

halogen as first group

Concept Map 10.3   Essential steps of an electrophilic aromatic substitution.

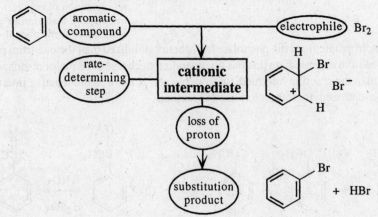

aromatic compound —————— electrophile   Br₂

rate-determining step —— **cationic intermediate**

loss of proton

substitution product

Concept Map 10.4   Reactivity and orientation in electrophilic aromatic substitution.

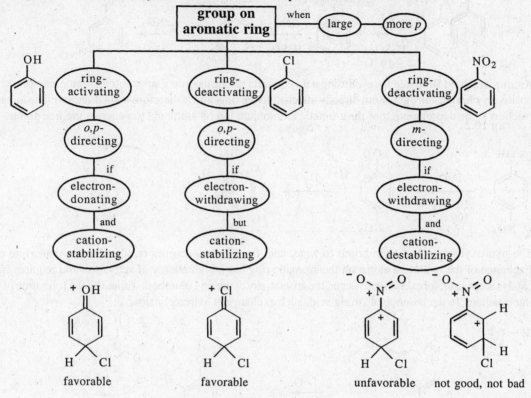

- -

10.11 If electrophilic substitution occurs at the para position of the ring, the intermediate will have the following resonance contributors:

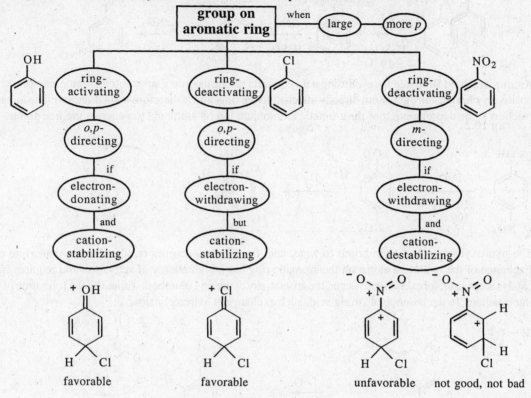

In this resonance contributor
the positive charge on the ring
is next to the positively charged
nitrogen atom; unfavorable

A similar set of resonance contributors can be drawn for ortho substitution. If electrophilic substitution occurs at the meta position, the positive charge on the ring is separated from the positive charge on the nitrogen atom by at least one carbon atom in all resonance contributors.

The intermediate formed on electrophilic attack at the meta position is, therefore, of lower energy than the intermediate formed when reaction occurs at the ortho or para positions.

10.12

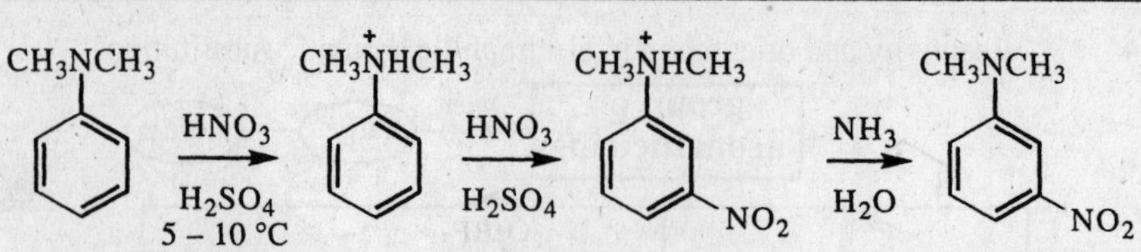

Amines are good Brønsted bases. Strong acid converts the amine into a substituted ammonium ion, which has a positively charged nitrogen atom directly attached to the ring, and is therefore meta directing. The last step of the reaction is the deprotonation of the aromatic ammonium ion by ammonia to generate the free amine.

10.13

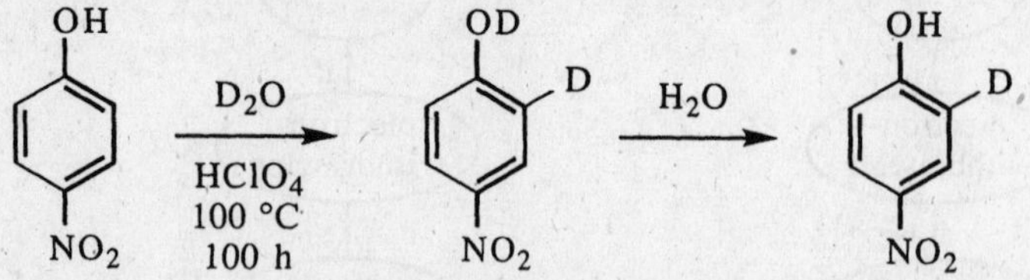

The hydroxyl group hydrogen bonds to water and, therefore, undergoes rapid hydrogen-deuterium exchange. Exchange of the hydrogen atoms on the aromatic ring has a high energy of activation and requires strong acid catalysis because, for exchange to occur, the aromaticity of the ring must be disrupted by the formation of a cationic intermediate. In the absence of strong acid, such exchange is extremely slow.

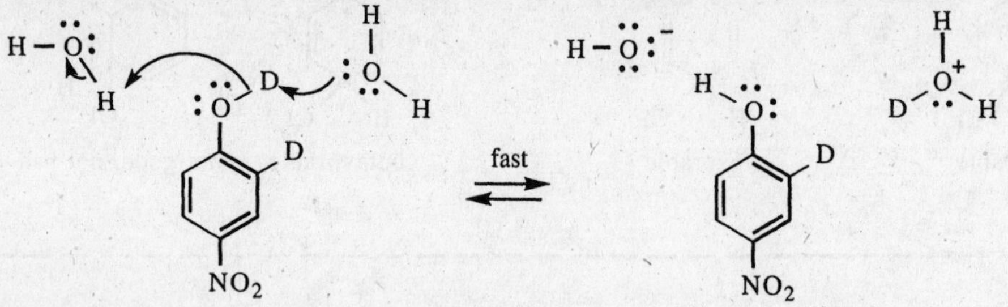

- -

Concept Map 10.5 (See p. 218)

- -

10.14   (a)

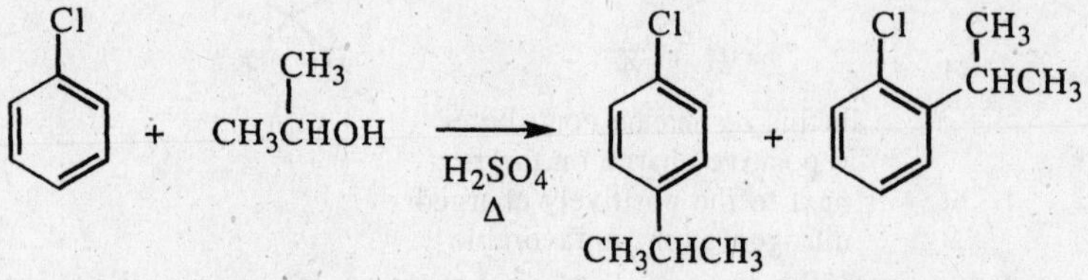

(b)

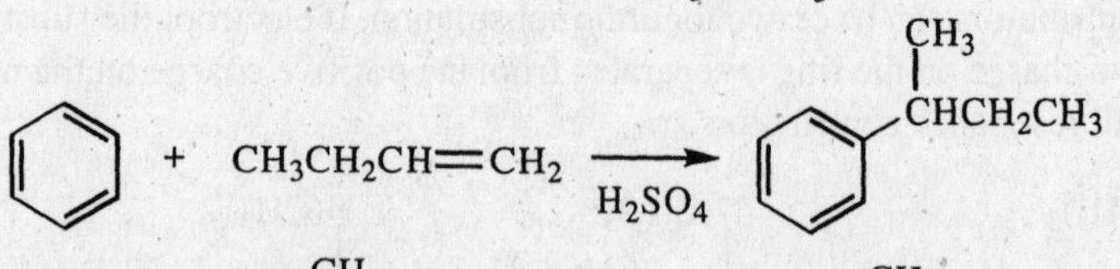

(c)

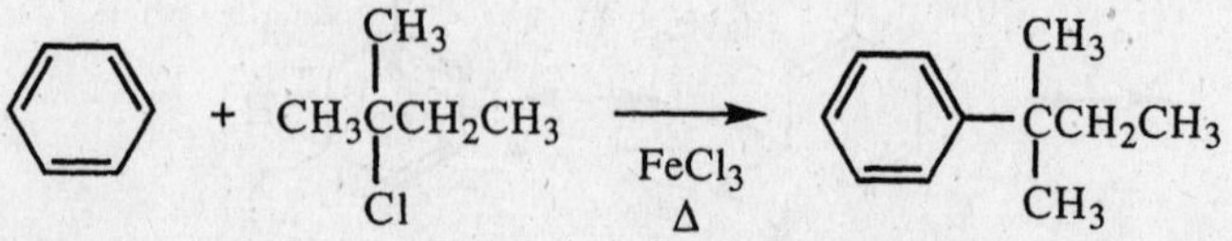

(d)

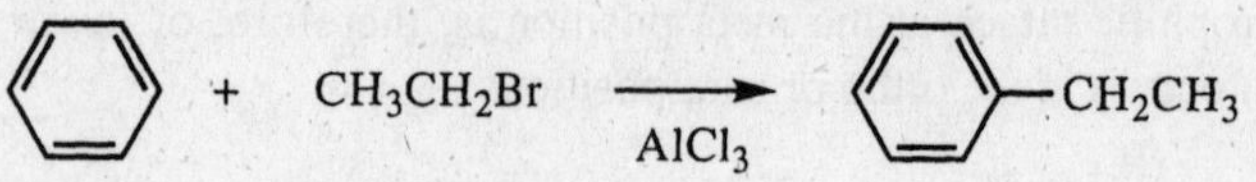

10.15

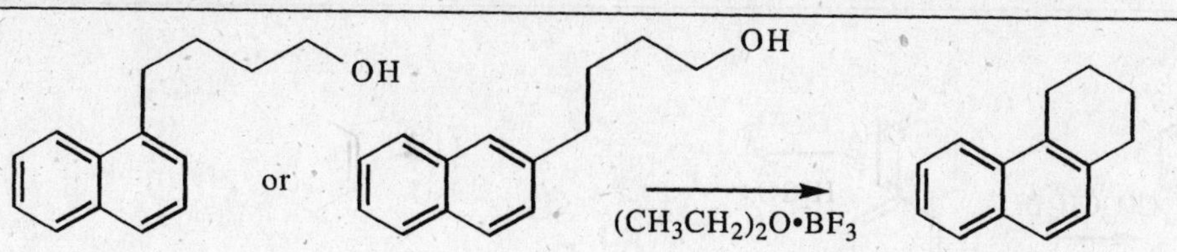

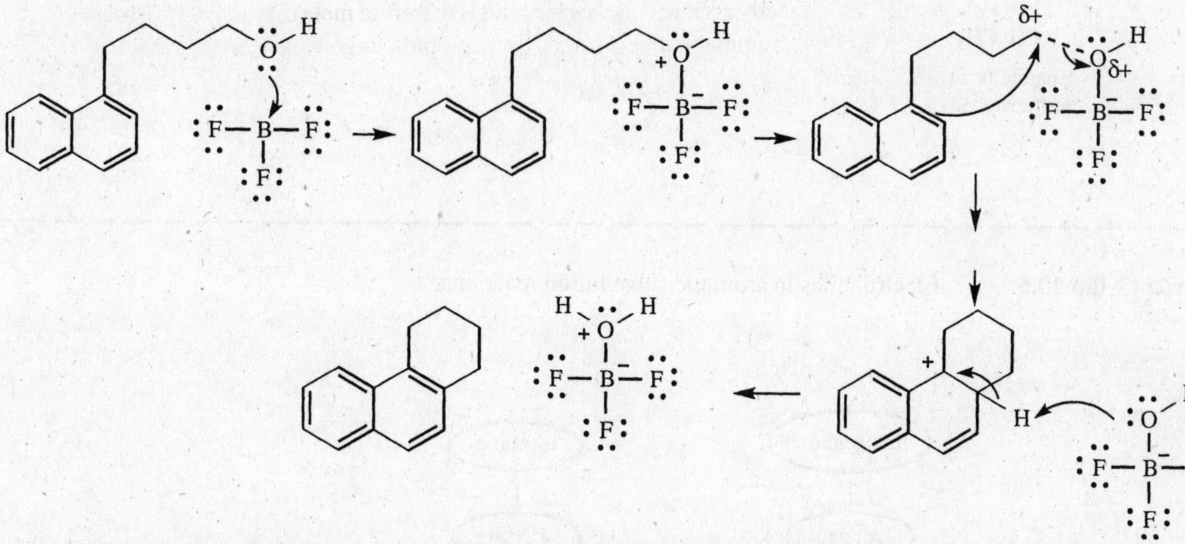

The same process would take place with the other alcohol except that the point of attack of the chain would be different. Attack would be at carbon 1 of the ring instead of carbon 2.

10.16   (a)

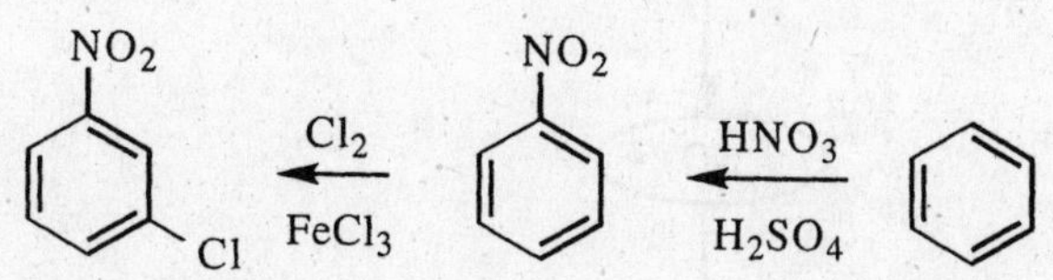

(Note: reversing the order in which these reagents are used would give ortho- and para-substituted products.)

(b)

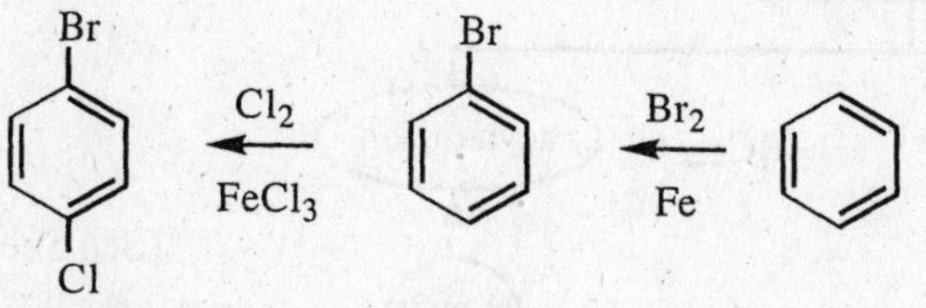

separate from
ortho isomer

(c)

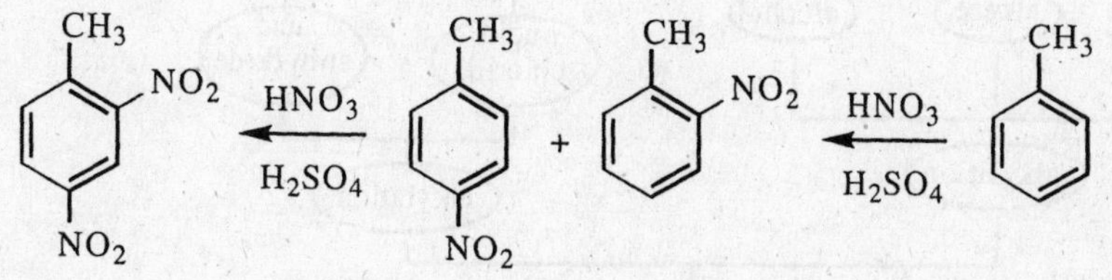

10.16

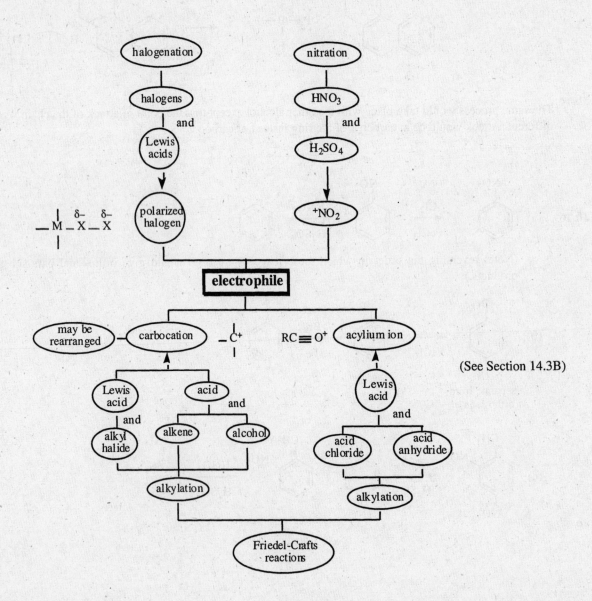

(d) $\xleftarrow[\text{AlCl}_3]{\text{CH}_3\text{CH}_2\text{Br}}$

separate from
ortho isomer

- -

Concept Map 10.5   Electrophiles in aromatic substitution reactions.

(See Section 14.3B)

10.17

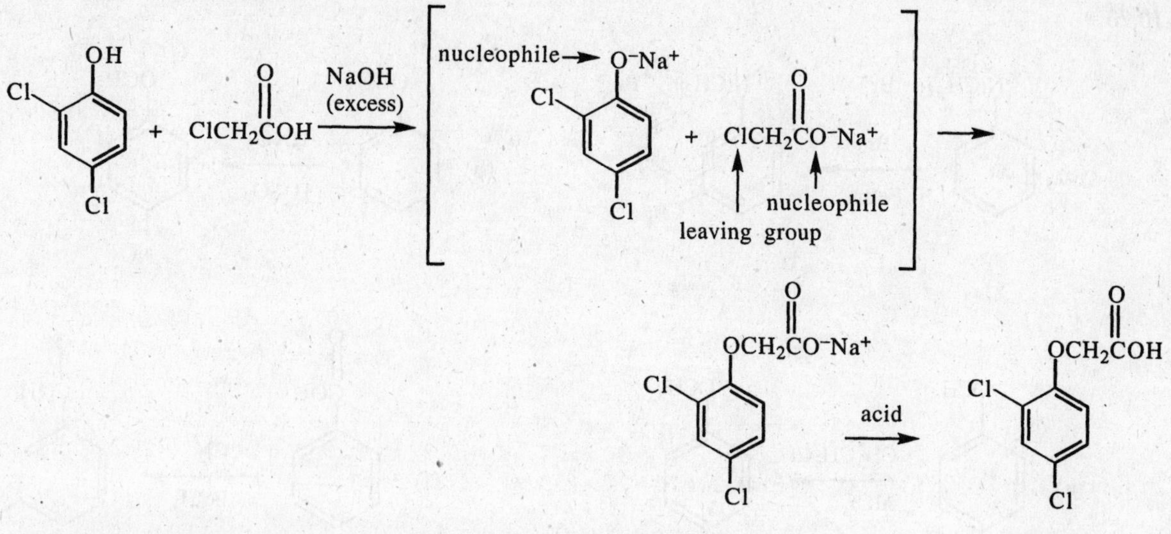

2,4-D

The phenolate anion is a stronger base and a better nucleophile than the chloroacetate anion. The oxygen atom of the phenolate anion will displace the chloride leaving group of the chloroacetate anion. The chlorine atoms on the aromatic ring of the dichlorophenolate ion are not displaced in nucleophilic substitution reactions.

10.18 (a) 3,4-dimethylnitrobenzene   (b) 4-methyl-1-phenylpentane (c) 2-chloro-6-ethylnaphthalene
     (d) 2,4-dinitrotoluene   (e) *m*-bromobenzoic acid   (f) methyl 3,5-dimethylbenzoate
     (g) 1-(4-methylphenyl)-1-butanone (h) *p*-bromobenzaldehyde (4-bromobenzaldehyde)

10.19 (a)

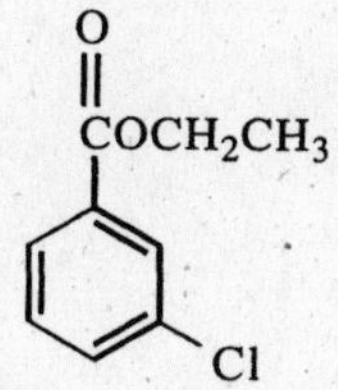

(b)

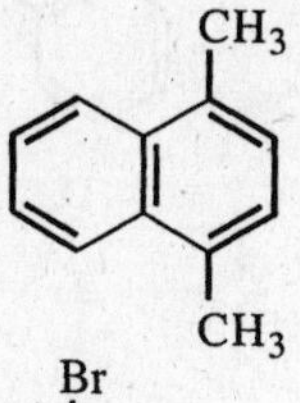

(c)

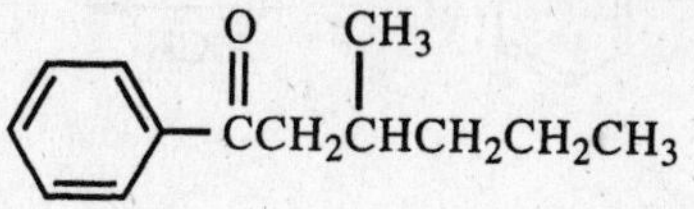

(d)

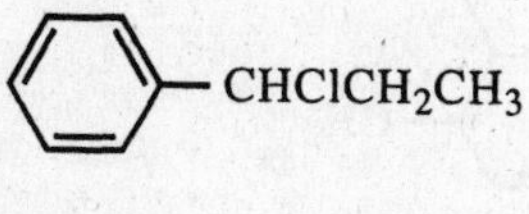

(e)

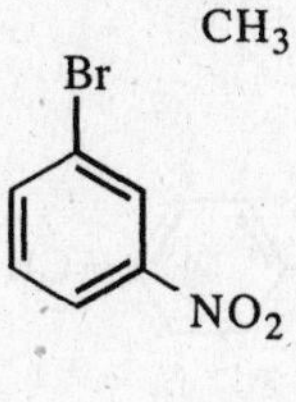

(f)

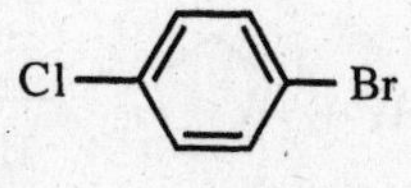

10.20 (a)

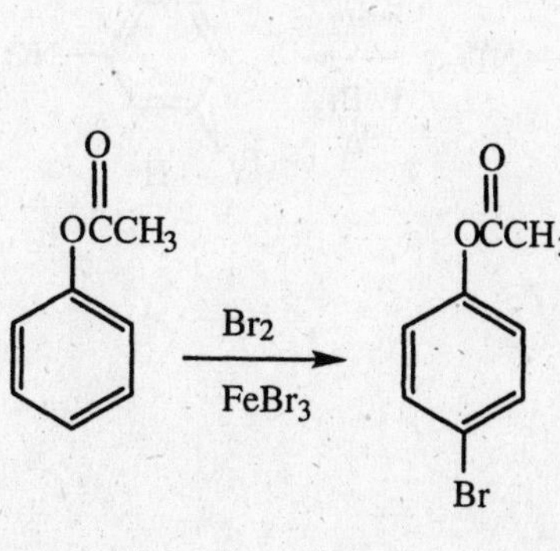

(b)

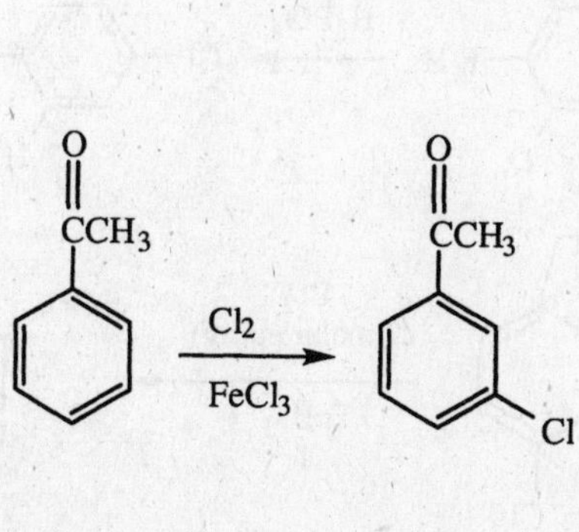

10.20

(c)

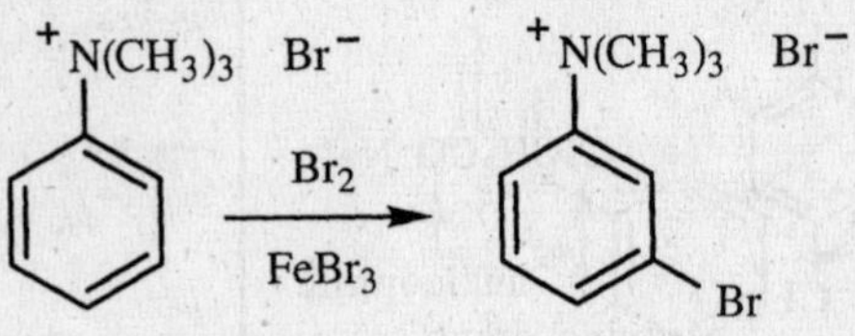

(d)

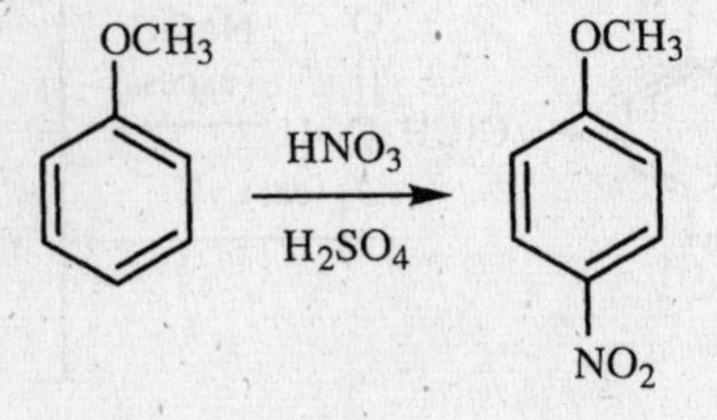

(e)

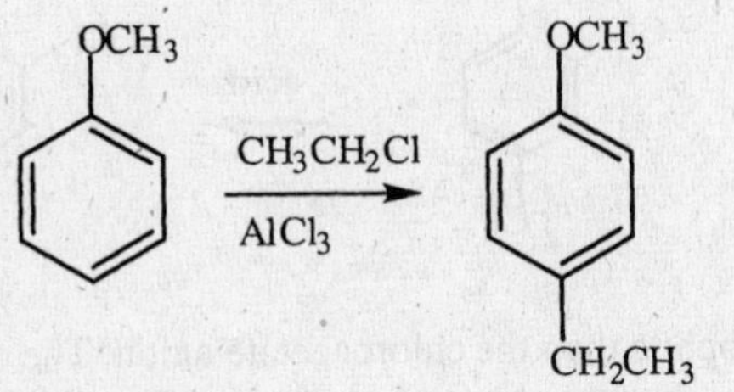

(f)

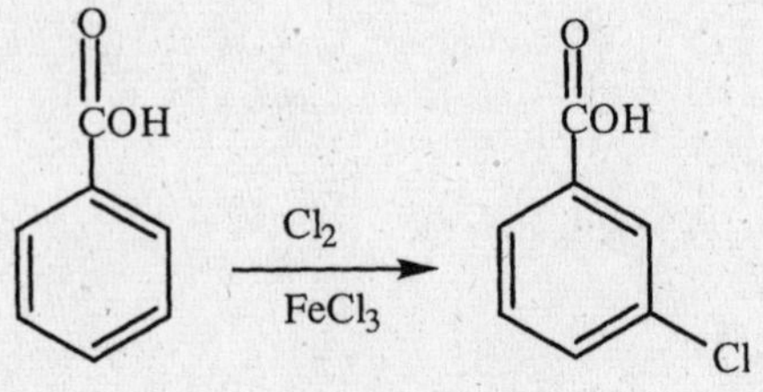

10.21

(a)

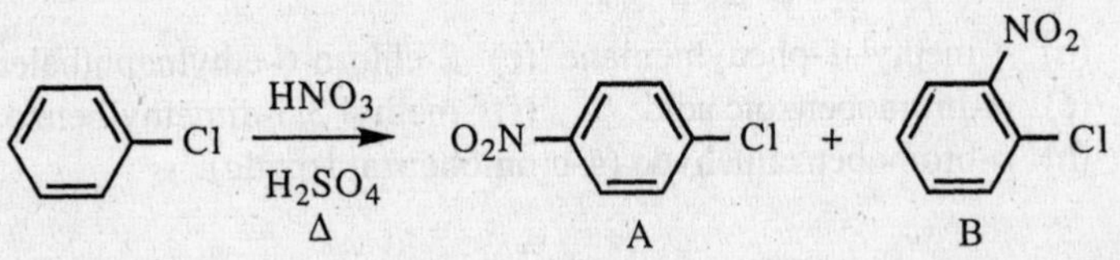

(b)

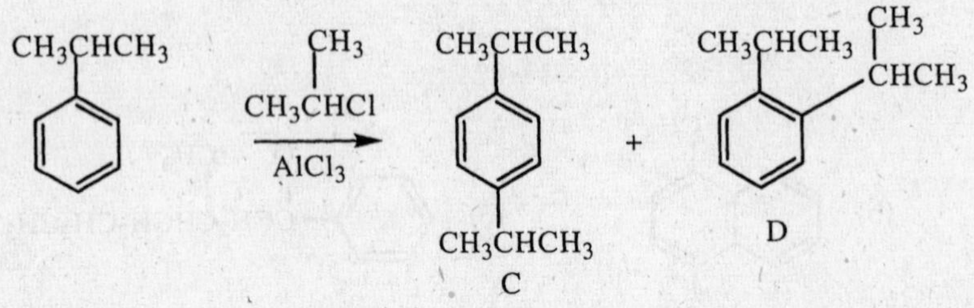

(c)

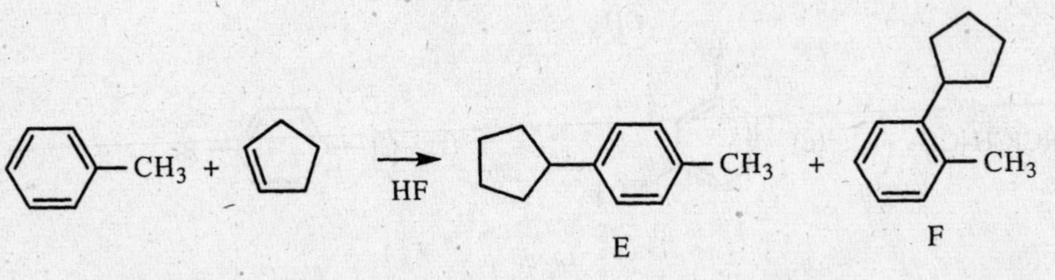

(d)

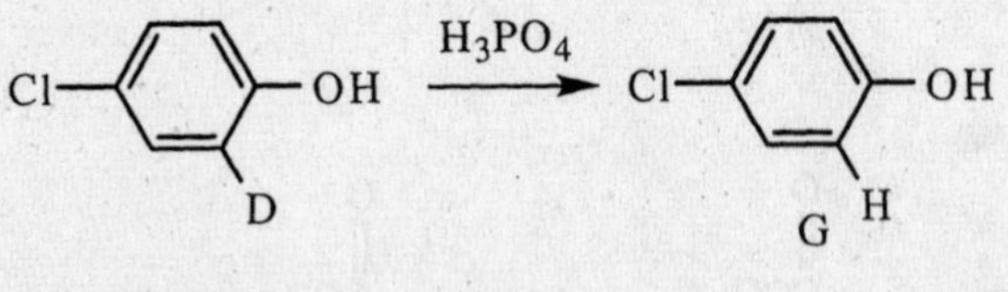

(e)

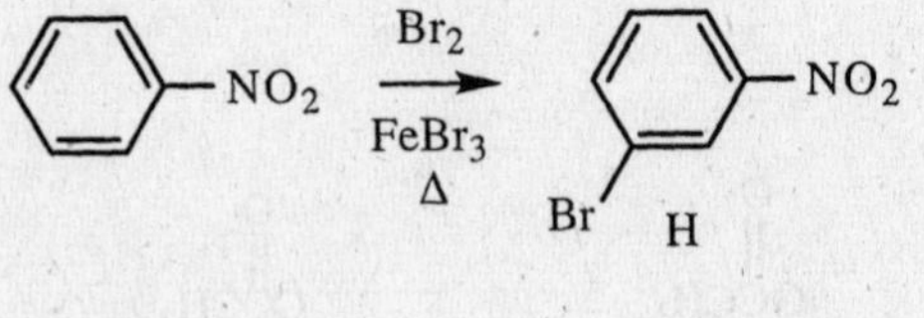

(f)

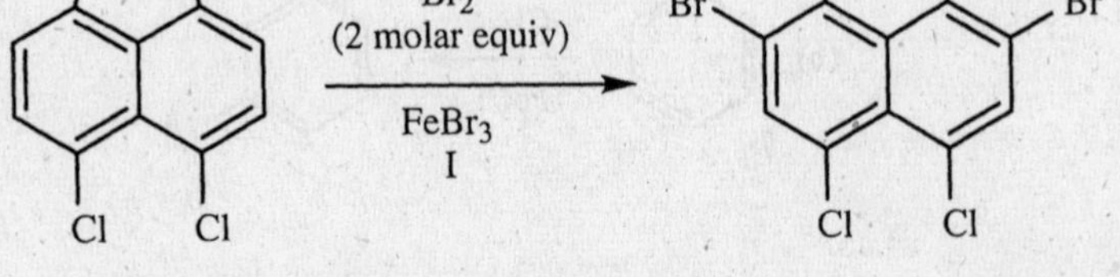

10.21 (cont)

(g)

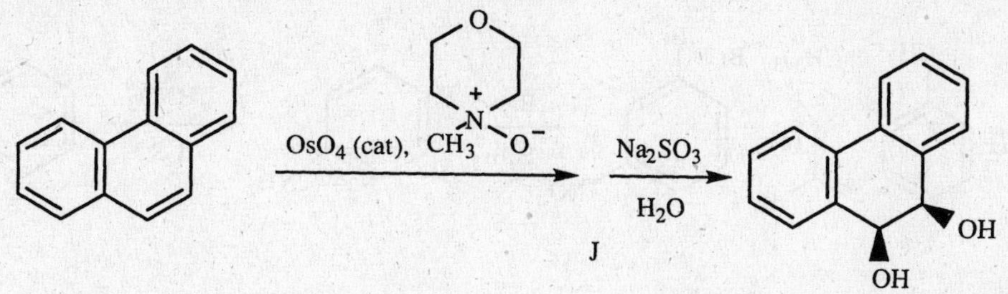

J

(h)

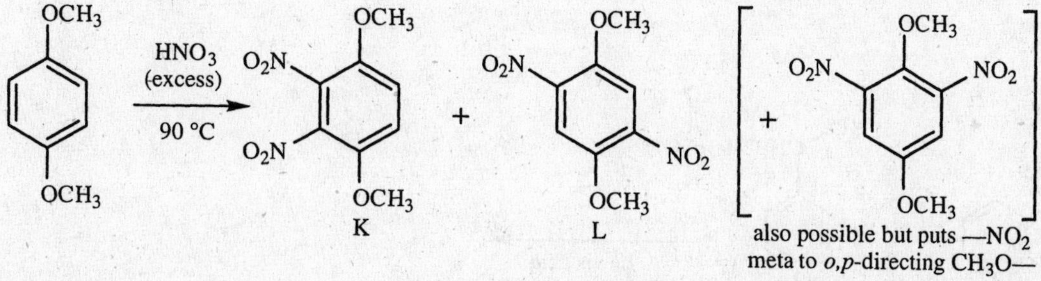

K   L

also possible but puts —NO$_2$
meta to *o,p*-directing CH$_3$O—

(i)

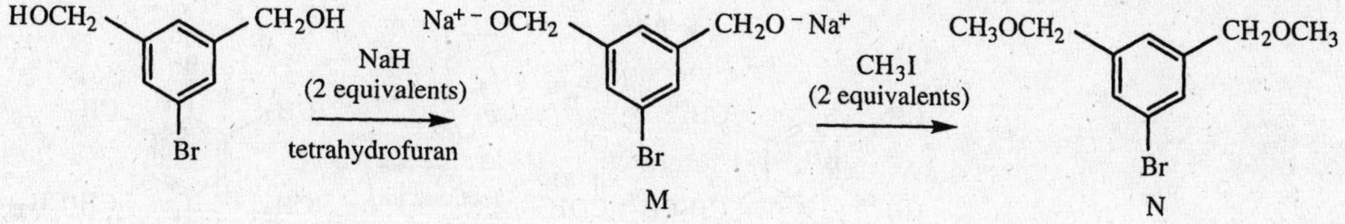

M   N

10.22

(a)

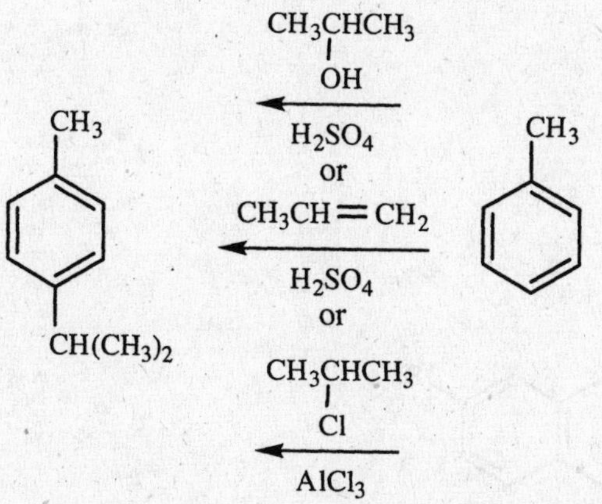

(b)

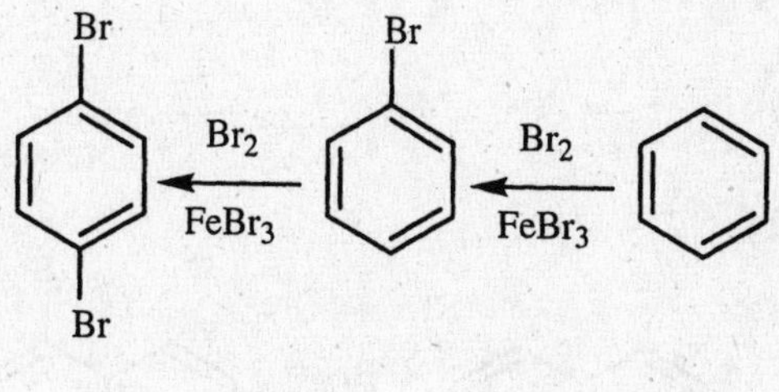

(c)

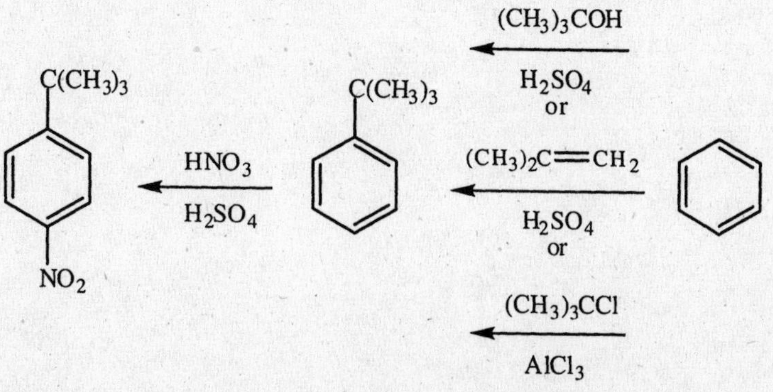

10.22

(d)

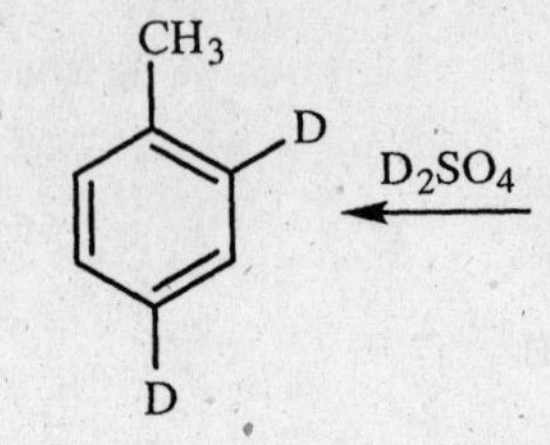

$$\text{(structure with CH}_3\text{, D, D)} \xleftarrow{\text{D}_2\text{SO}_4} \text{(toluene)}$$

(e)

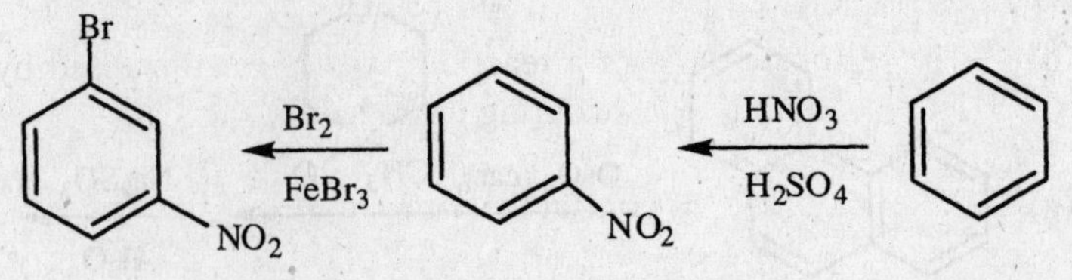

$$\text{(bromo-nitrobenzene)} \xleftarrow[\text{FeBr}_3]{\text{Br}_2} \text{(nitrobenzene)} \xleftarrow[\text{H}_2\text{SO}_4]{\text{HNO}_3} \text{(benzene)}$$

10.23

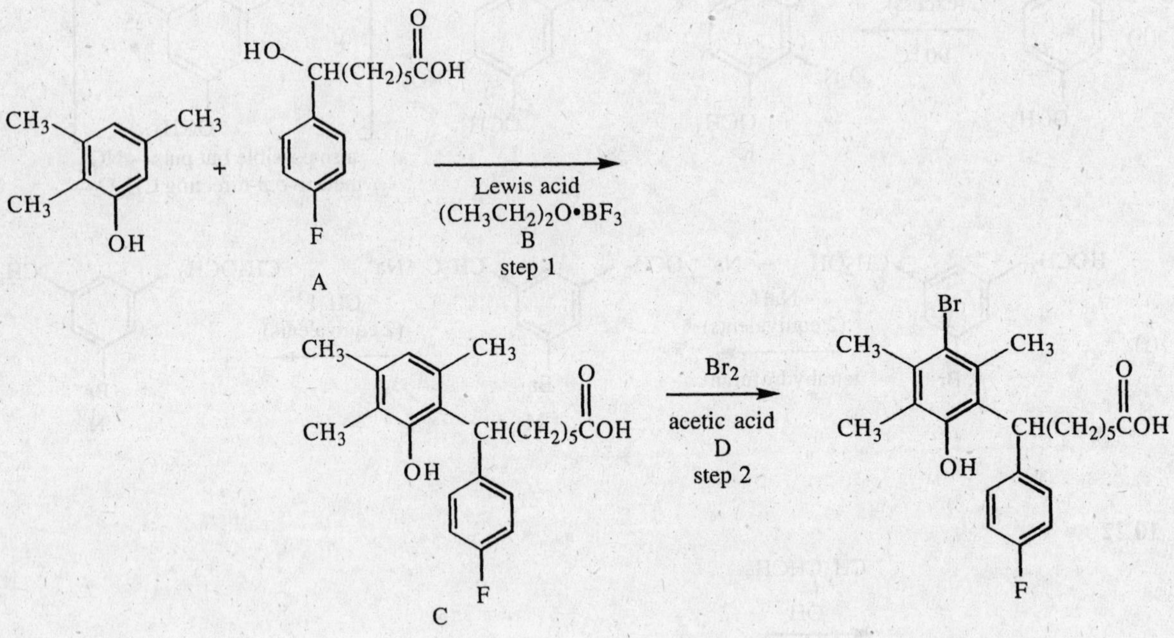

$$\text{A} + \text{(phenol with CH}_3\text{ groups)} \xrightarrow[\substack{\text{(CH}_3\text{CH}_2)_2\text{O}\bullet\text{BF}_3 \\ \text{B} \\ \text{step 1}}]{\text{Lewis acid}}$$

$$\text{C} \xrightarrow[\substack{\text{acetic acid} \\ \text{D} \\ \text{step 2}}]{\text{Br}_2}$$

10.24

(a)

naphthalene $\xrightarrow[\text{AlBr}_3]{\text{Br}_2}$ 1-bromonaphthalene 99% + 2-bromonaphthalene 1%

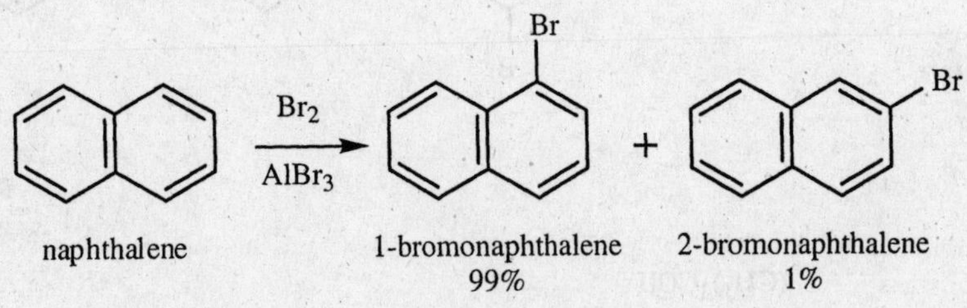

10.24

(b)   The regioselectivity of a reaction is usually rationalized by looking at the relative stabilities of the reactive intermediates that form during the reaction.

Attack of the bromine at the 1-position:

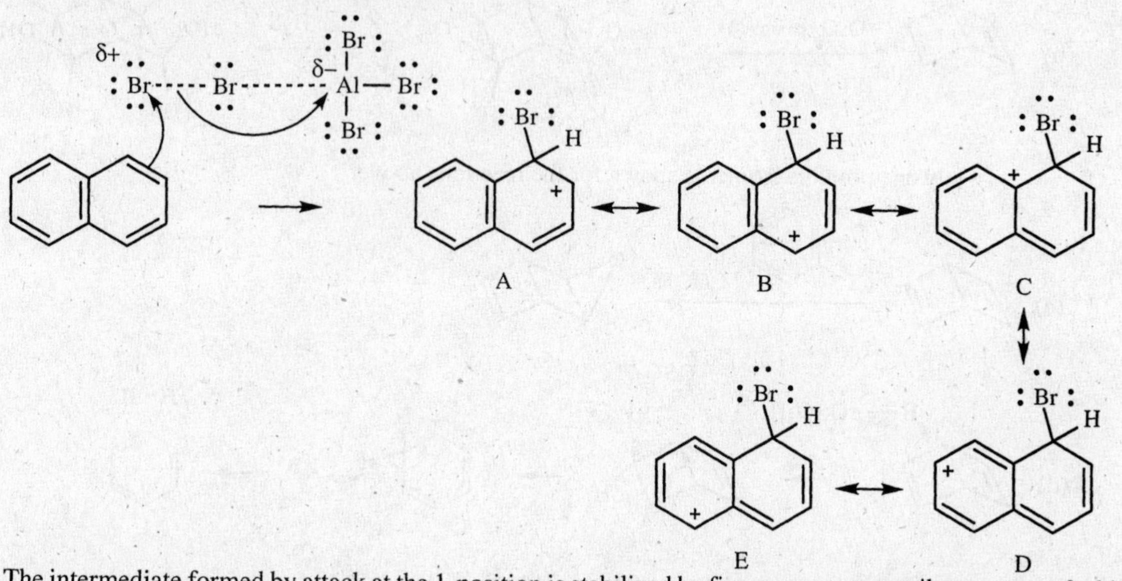

The intermediate formed by attack at the 1-position is stabilized by five resonance contributors, two of which, A and B, have fully aromatic rings.

Attack of bromine at the 2-position:

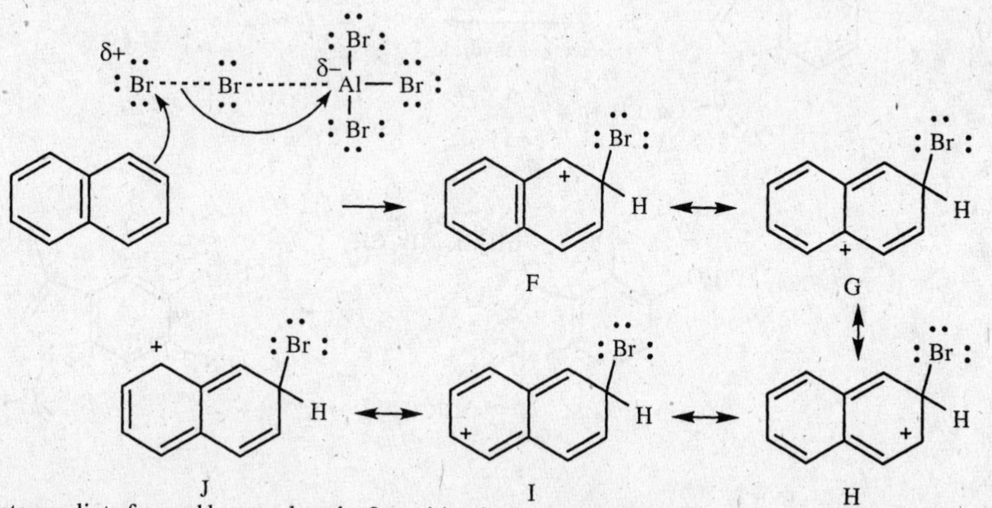

The intermediate formed by attack at the 2-position is also stabilized by five resonance contributors, but only one, F, has a fully aromatic ring.

10.25 (a)

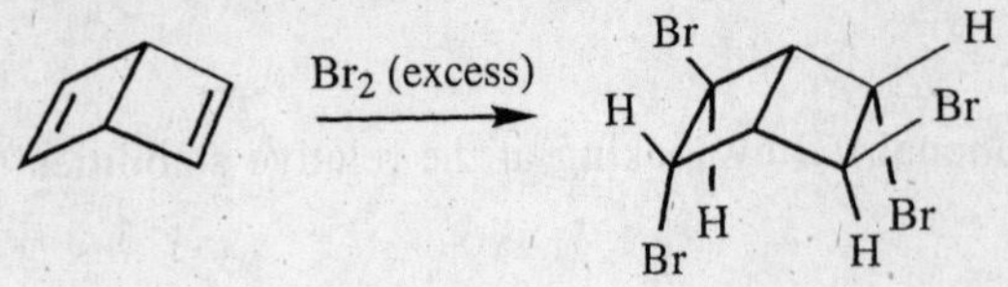

(b)

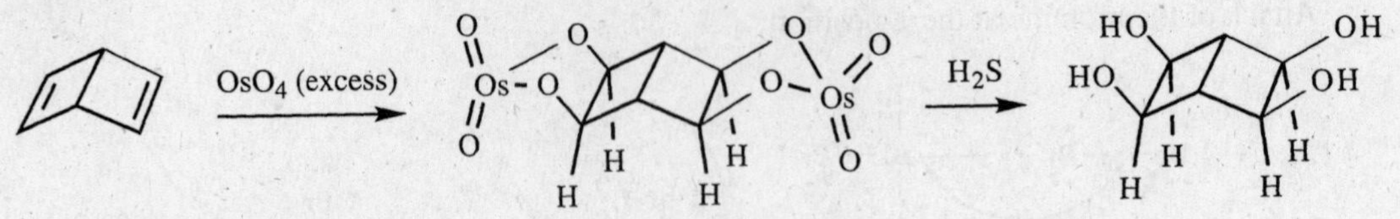

Only one possible isomer is shown for the reactions above.

(c)

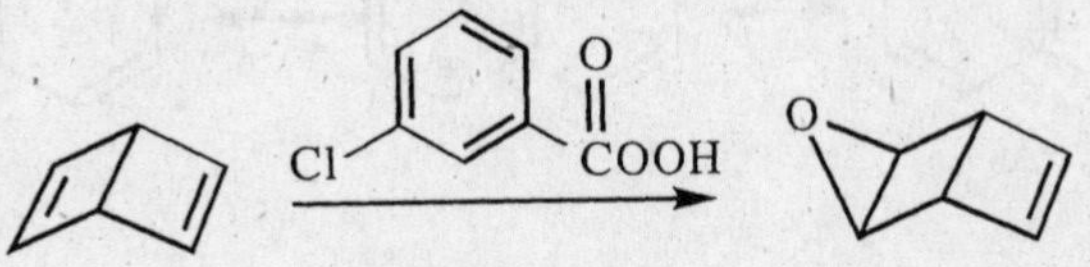

(d)

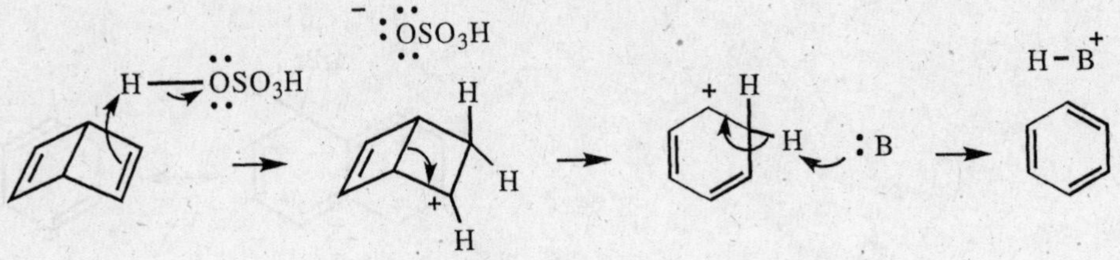

10.26 (a)

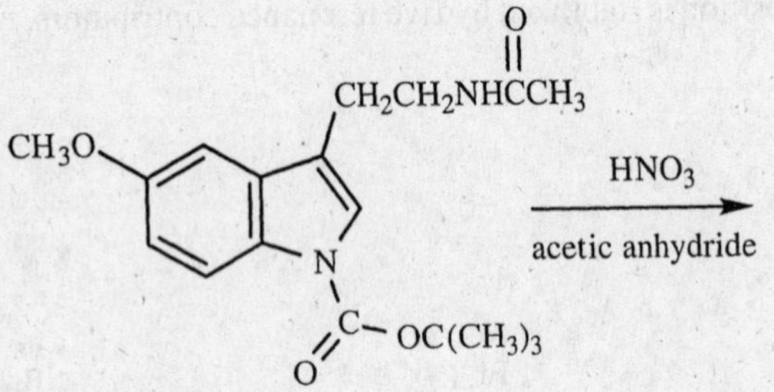

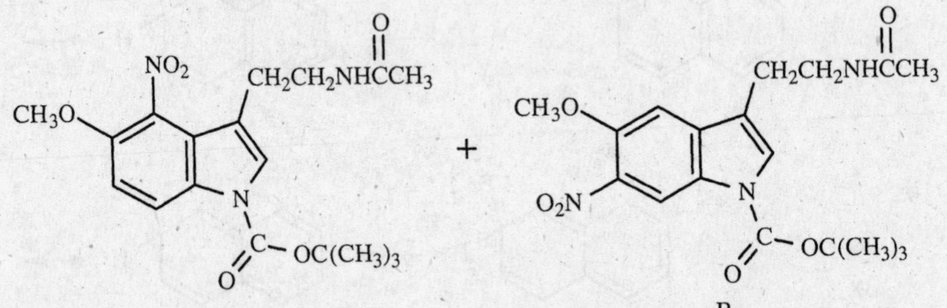

A   +   B

(b)

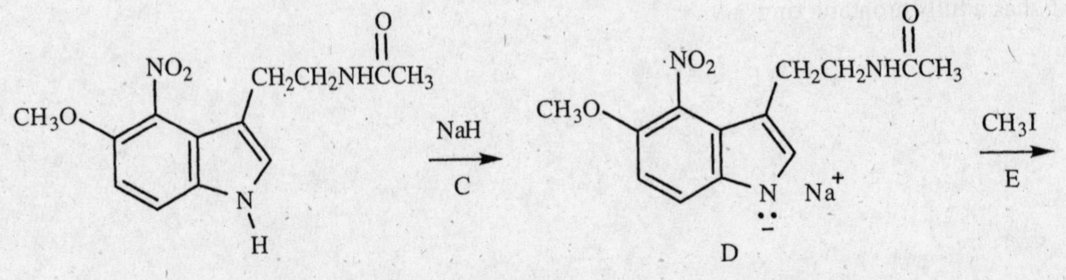

10.26 (b) (cont)

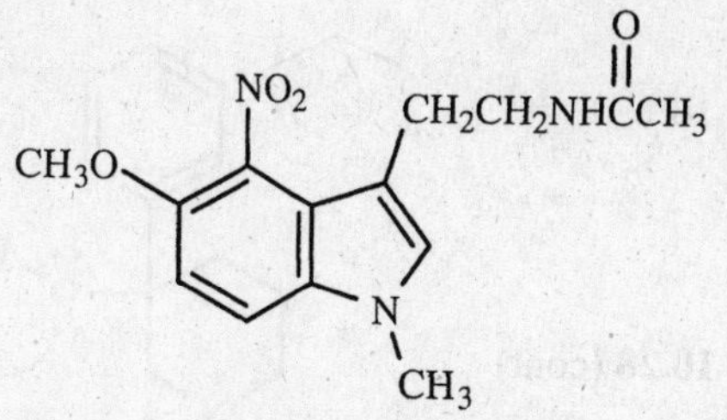

10.27

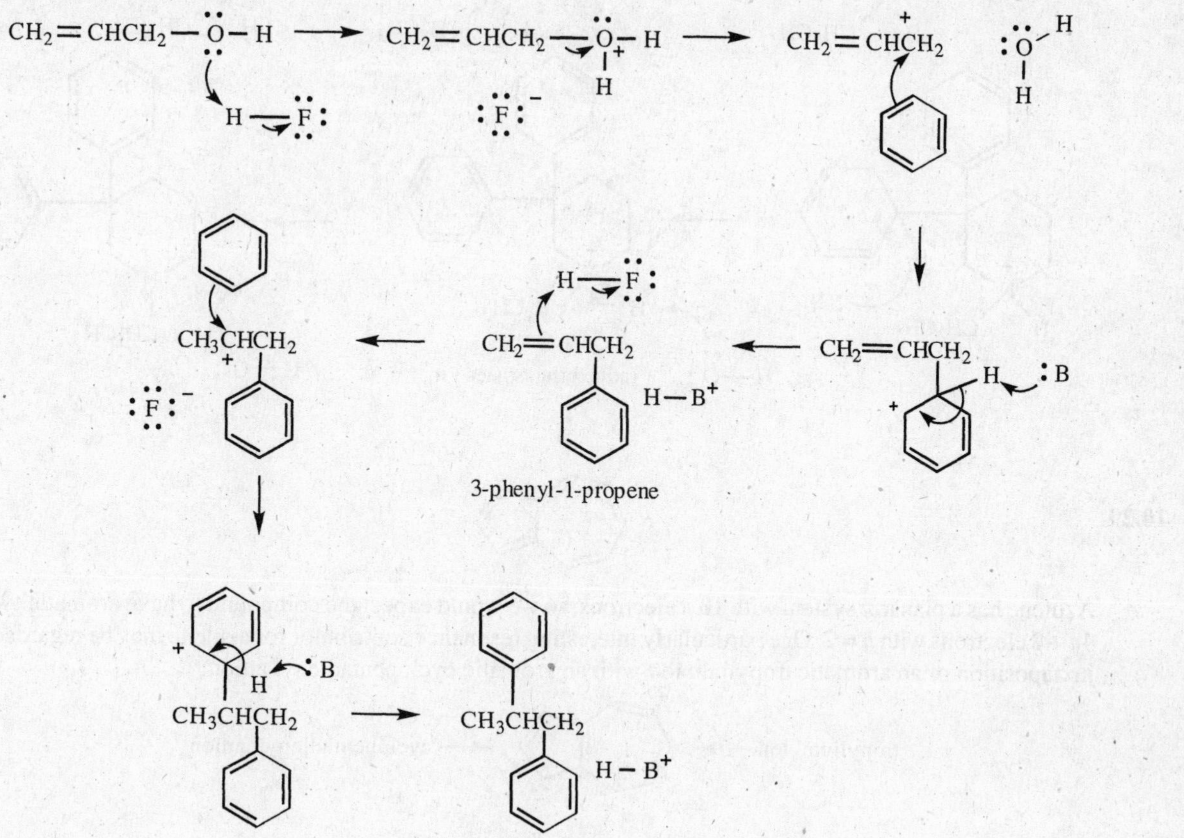

3-phenyl-1-propene

1,2-diphenylpropane

10.28

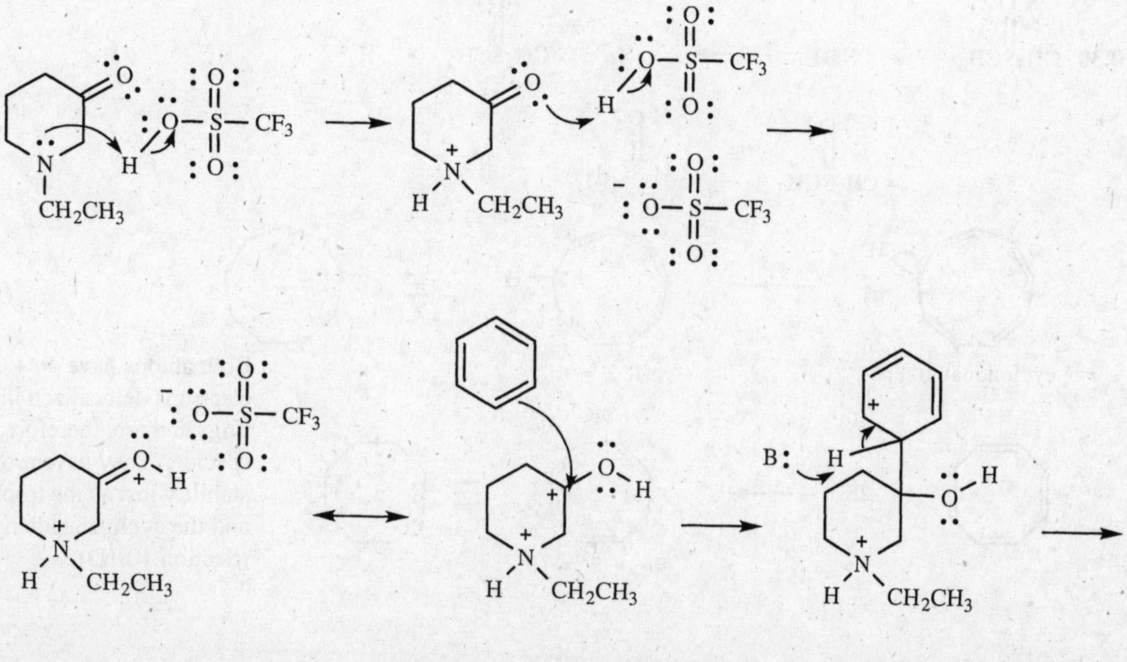

10.28 (cont)

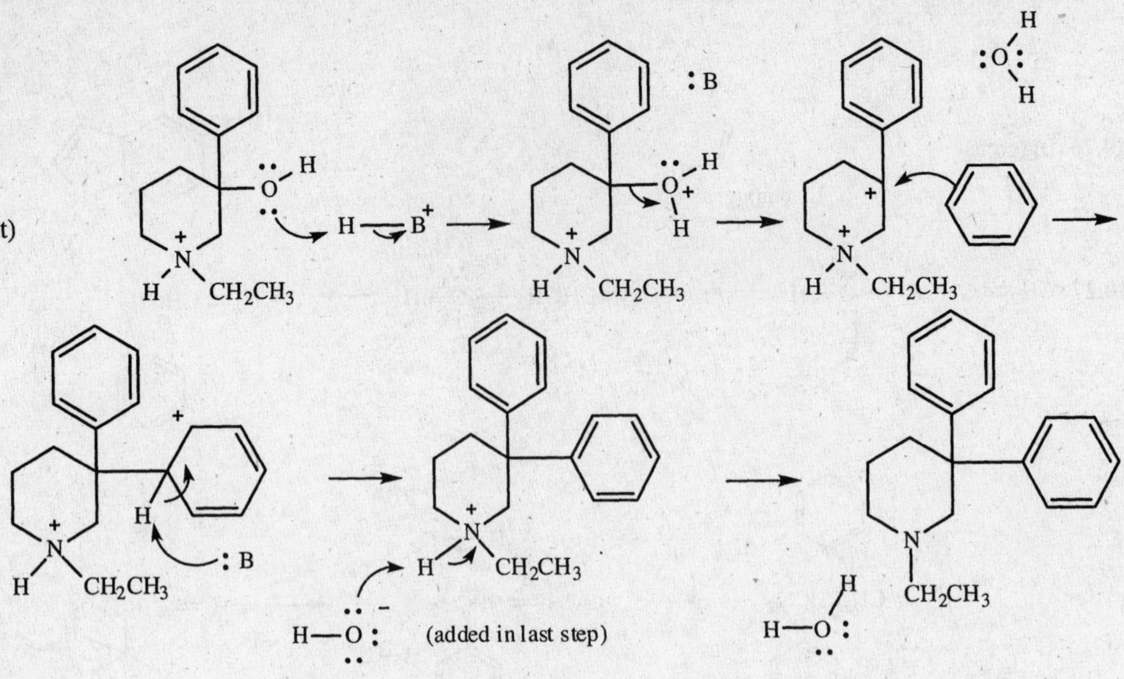

10.29

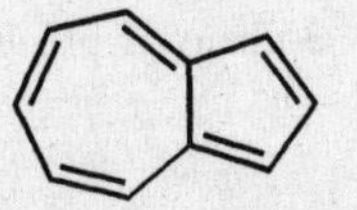

Azulene has a planar π system with 10 π electrons, so we would expect the compound to have aromaticity. It has $4n + 2$ electrons with $n = 2$. One particularly interesting resonance contributor for azulene may be regarded as the juxtaposition of an aromatic tropylium ion with an aromatic cyclopentadienyl anion.

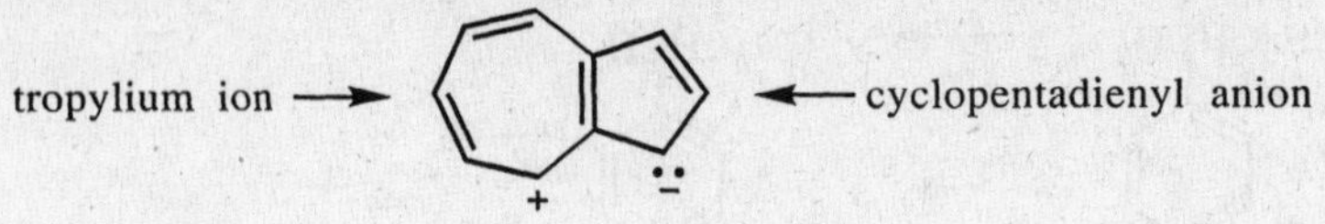

tropylium ion ⟶ ⟵ cyclopentadienyl anion

10.30 $CH_3\overset{\overset{\ddot{O}\!:}{\|}}{\underset{\ddot{}}{S}}CH_3$ +   NaH   ⟶ $Na^+ \; {}^- : CH_2\overset{\overset{\ddot{O}\!:}{\|}}{S}CH_3$ + $H_2\uparrow$

cyclononatetraene $4n + 2 = 10$

                       $2K^+$

$4n + 2 = 10$

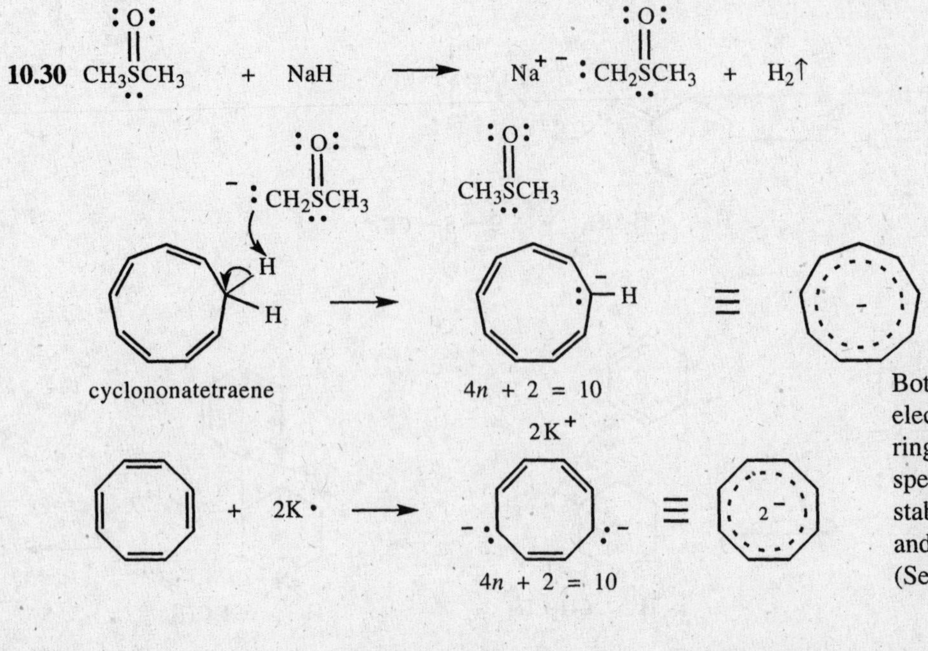

Both anions have $4n + 2 = 10$ π electrons delocalized in planar rings and are, therefore, aromatic species. They have aromatic stability just as the tropylium ion and the cyclopentadienyl anion do (Section 10.1D).

10.31

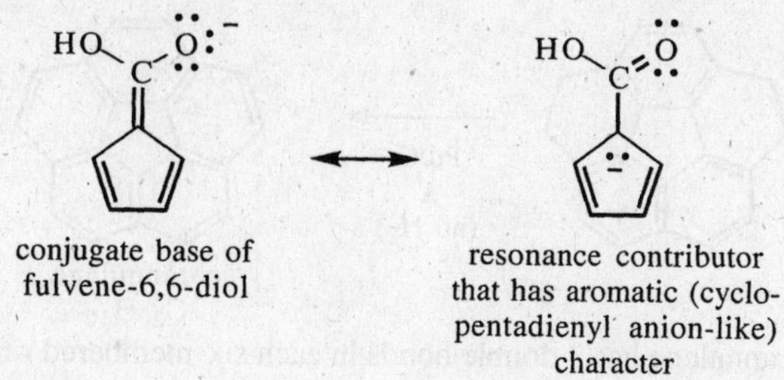

conjugate base of
fulvene-6,6-diol

resonance contributor
that has aromatic (cyclo-
pentadienyl anion-like)
character

10.32

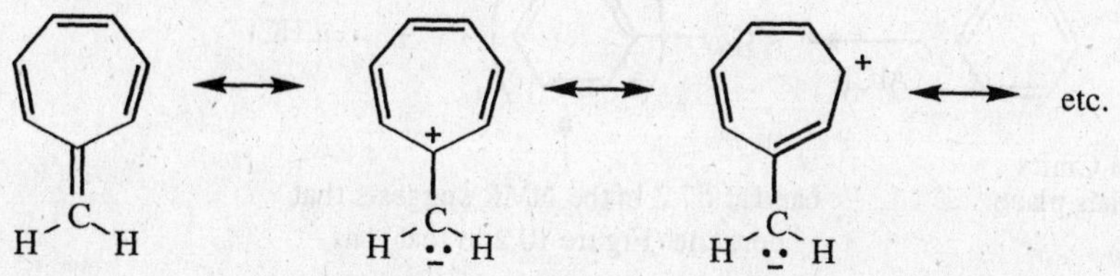

etc.

Resonance contributors for heptafulvene have the positive charge delocalized over the seven-membered ring in what is a tropylium ion, but the negative charge is localized on a carbon atom. In the dicyano compound, the negative charge can be delocalized onto the two nitrogen atoms. The two electron-withdrawing cyano groups thus stabilize the negative character at the carbon atom outside the ring.

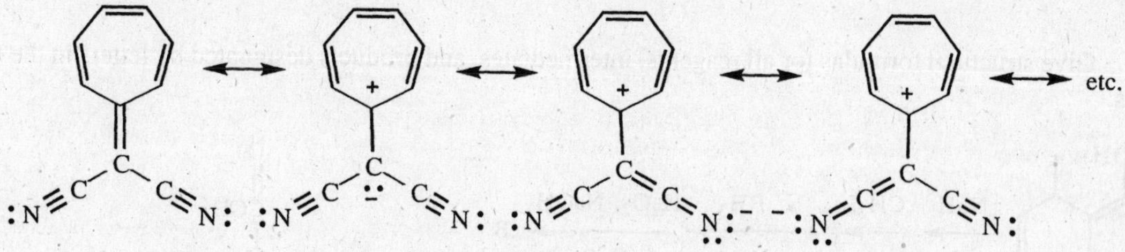

etc.

10.33

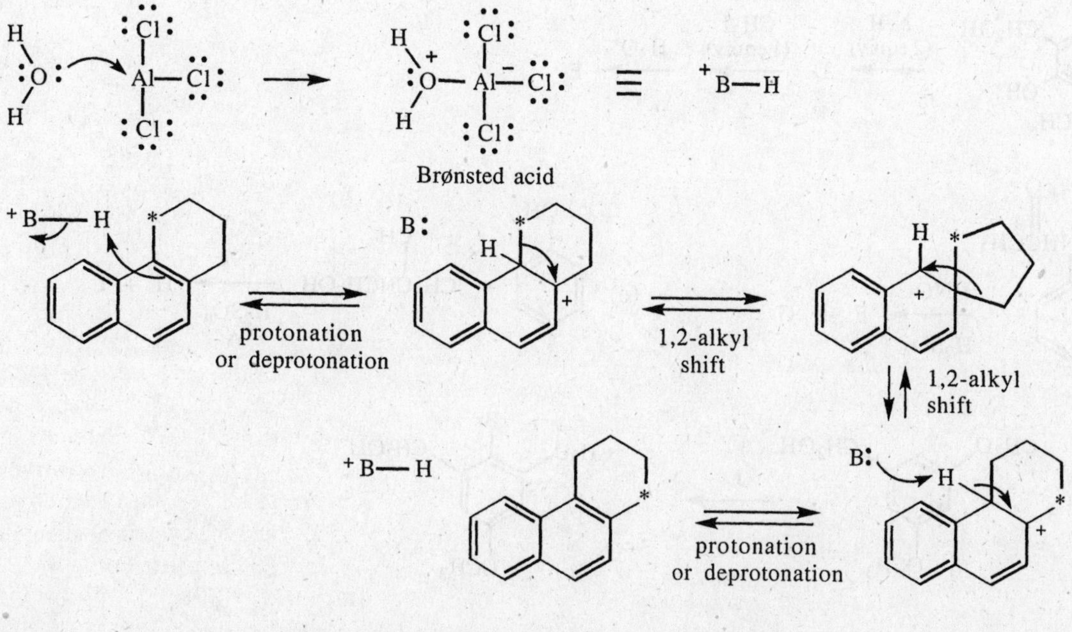

Brønsted acid

protonation
or deprotonation

1,2-alkyl
shift

1,2-alkyl
shift

protonation
or deprotonation

10.34

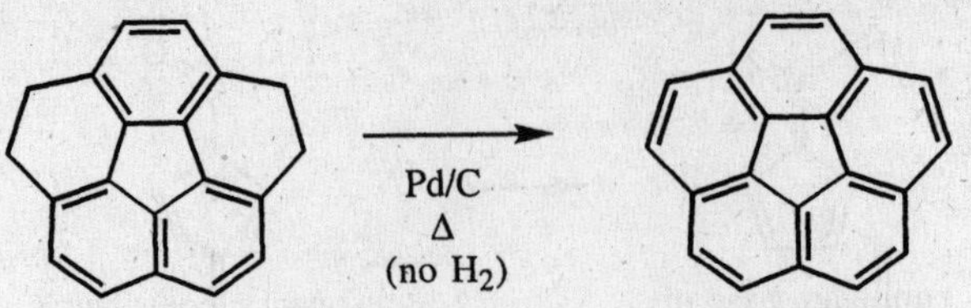

corannulene
$C_{20}H_{10}$

Corannulene has 3 double bonds in each six-membered ring.

10.35

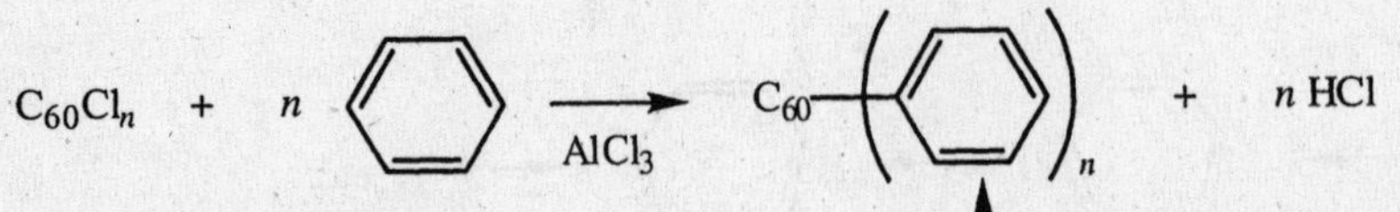

$C_{60}Cl_n$ + n ⬡ $\xrightarrow{AlCl_3}$ C_{60}(⬡)$_n$ + n HCl

A Friedel-Crafts
reaction takes place

band at $\delta 7.2$ in the NMR suggests that
benzene (Figure 10.2 in text) has
become part of the molecule

Supplemental Problems

S10.1   Give structural formulas for all reagents, intermediates, and products designated by letters in the following equations.

(a)

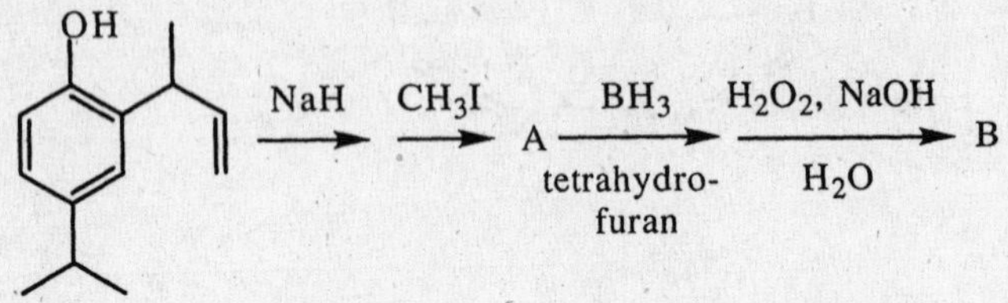

(b)

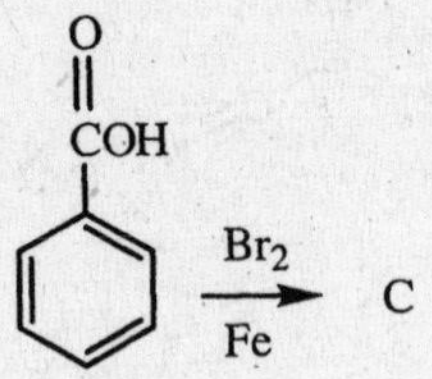

(c)

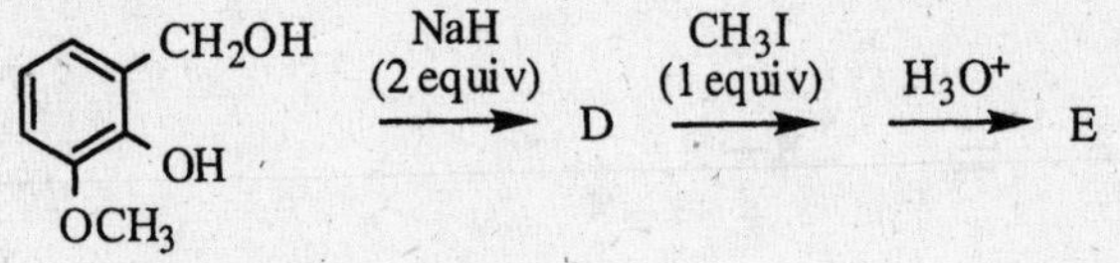

(d)

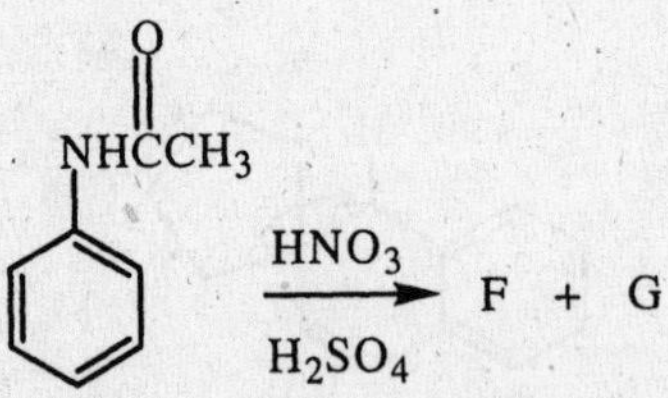

(e)

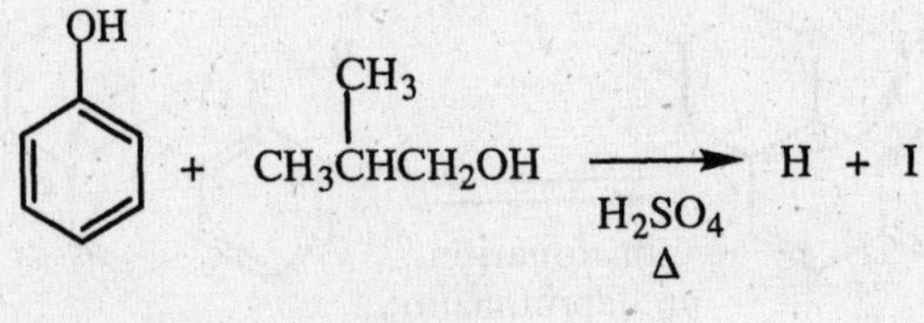

(f)

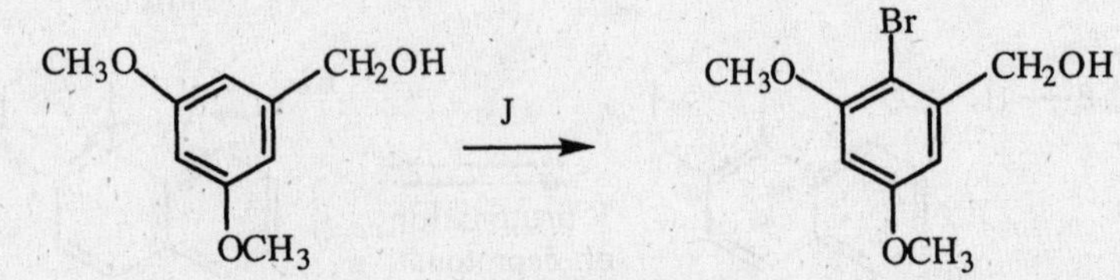

S10.2   Propose a complete, stepwise mechanism for the following transformation using the curved arrow convention.

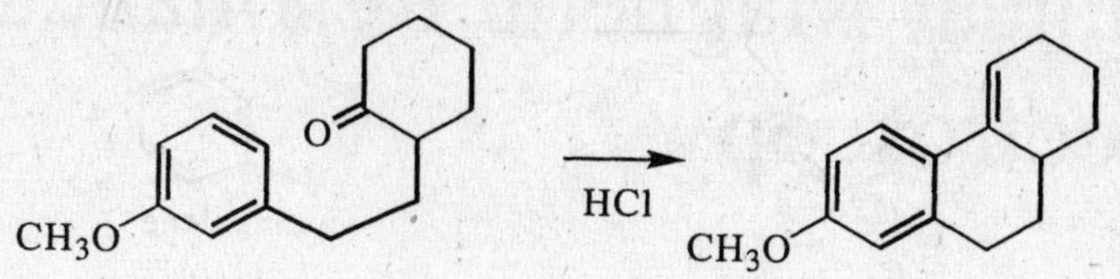

S10.3   When 2-hydroxybenzoic acid is heated with isobutyl alcohol in the presence of sulfuric acid, compound A is formed. The same product is obtained when *tert*-butyl alcohol and sulfuric acid are used. What is the structure of compound A? Write a mechanism that accounts for these experimental facts.

S10.4   When 3,7-di-*tert*-butylnaphthalene is treated with two equivalents of bromine in the presence of ferric bromide, the major product has three sets of equivalent hydrogen atoms in its proton magnetic resonance spectrum in a ratio of 9:1:1. What is the most likely structure for this product? (Hint: Be sure to take steric factors into consideration in your answer.)

S10.5   The heterocyclic compounds pyrrole and indole, shown below, are not very basic. Pyrrole has $pK_a \sim 15$ and thus has considerable acidity. Explain why pyrrole is so much less basic (so much more acidic) than ammonia ($pK_a \sim 36$).

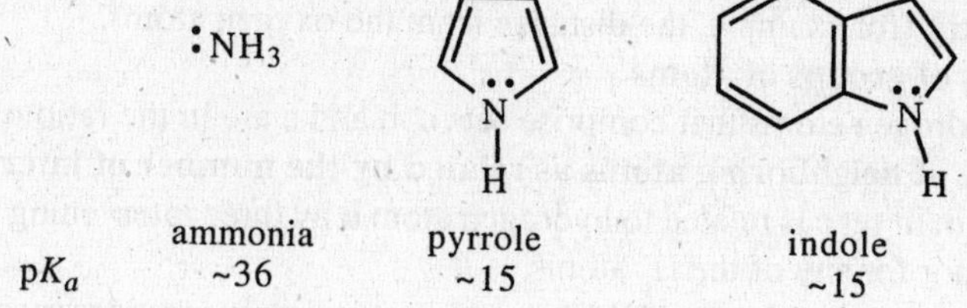

	ammonia	pyrrole	indole
pK_a	~36	~15	~15

11 Nuclear Magnetic Resonance Spectroscopy

Workbook Exercises

In Sections 5.4 and 5.7 of your text, you encountered nuclear magnetic resonance (NMR) spectroscopy as a tool that allows us to observe the equivalence of atoms and groups in molecular structures, in other words, to detect symmetry in structures. You solved problems in which you used NMR spectra to make decisions about the connectivity of molecules. In this chapter, we will learn more about NMR and its uses in determining the structure of molecules.

A brief review of the ideas and terminology introduced in Sections 5.4 and 5.7
(You should re-examine those sections of the text for a more complete look at this topic.)

There are three structural descriptions for this molecule that relate to NMR spectroscopy:

I.   **The number of sets of atoms and their relationships within a molecule**
There are three distinct, nonequivalent sets, or groups, of hydrogen atoms (a, b and c) based on overall molecular symmetry (for example, the distance from the oxygen atom).

II.   **The relative sizes of groups of atoms**
The total of 10 hydrogen atoms that comprise sets a, b and c are in the ratio of 3:1:6, respectively.

III.   **The environment of neighboring atoms as related by the number of intervening bonds**
Any hydrogen atom in set c is related to hydrogen atom b by three intervening bonds (H_b—C—C—H_c). H_b is a 3-bond neighbor for any of the H_c atoms.

As you begin to study Chapter 11, remember that NMR spectroscopy simply provides graphical information that can be interpreted in terms of:

1.   the number of groups of equivalent atoms and their connectivity relationships within a molecule.
2.   the relative numbers of atoms within each of the groups.
3.   the environment of groups of atoms based on the number of bonds intervening between them and other atoms.

The following two workbook exercises are extensions of problems from Chapter 5. They will focus your attention on structural ideas important to your study of NMR spectroscopy.

EXERCISE I.   You created a number of molecules with different connectivities for part (1) of the workbook exercise in Chapter 9. Identify the number and relative size of the sets of equivalent hydrogen atoms for each of your molecules. Do the same for sets of carbon atoms.

EXERCISE II.   In this problem, groups of equivalent atoms are ranked according to their proximity to an electronegative group in the molecule (for example, of groups a, b and c, group a is closest to the most electronegative group).

Molecules A, B, and C have the same molecular formula, $C_6H_{12}O_2$. Molecule A has a carbonyl group but no hydroxyl group. Molecule A has five sets of equivalent hydrogen atoms (a, b, c, d, and e) in the ratio of 3:2:2:2:3, respectively. Set a has no nonequivalent 3-bond hydrogen neighbors (hereafter referred to as "neighbors" unless a different description is required). Set b has two neighbors, set c has four, set d has five, and set e has two. Molecule A has six sets of carbon atoms.

Workbook Exercises (cont)

Molecule B has a carboxylic acid functional group. Molecule B has four groups of equivalent hydrogen atoms (a, b, c, and d) in the ratio of 1:2:6:3, respectively. Set a has no neighbors. Set b has three neighbors. Set c has no neighbors, and set d has two neighbors. There are five groups of carbon atoms in molecule B.

Molecule C has no alcohol, carbon-carbon double bond, or carbonyl functional groups. There are three groups of hydrogen atoms in molecule C (a,b, and c) in the ratio of 2:1:3 respectively. Set a has two neighbors. Set b has four neighbors. Set c has no neighbors. There are four groups of carbon atoms in molecule C.

Draw structural formulas for molecules A, B, and C that are compatible with the spectral data (and the molecular formula) that you have for them.

- -

11.1   Replacement of any one of the *a* hydrogen atoms with a methyl group gives the same connectivity and, therefore, the same compound. Replacement of a *b* hydrogen atom gives a different connectivity.

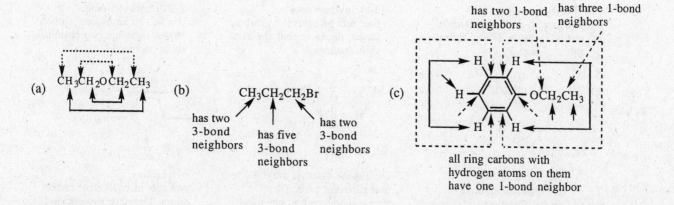

11.2   Solid arrows point to chemical-shift equivalent hydrogen atoms; dashed arrows point to chemically equivalent carbon atoms. Hydrogen atoms on the same methyl group are always chemical-shift equivalent. Hydrogen atoms on the same methylene group are usually, but not always, chemical-shift equivalent. The carbon atoms to which the indicated hydrogen atoms are attached are also equivalent. In addition, there are two carbons with no hydrogen atoms on them, one in (c) and one in (e) that are different from the others.

11.2

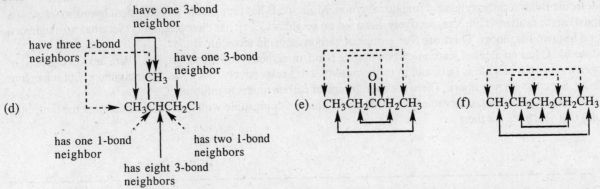

(d)

have three 1-bond neighbors

have one 3-bond neighbor

have one 3-bond neighbor

CH₃

→ CH₃CHCH₂Cl

has one 1-bond neighbor

has eight 3-bond neighbors

has two 1-bond neighbors

(e) CH₃CH₂CCH₂CH₃

(f) CH₃CH₂CH₂CH₂CH₃

11.3 The first compound has two sets of equivalent carbon atoms and, therefore, will have two chemical shifts in the ¹³C nuclear magnetic resonance spectrum. The second compound has three sets of equivalent carbon atoms and, therefore, three chemical shifts in the ¹³C nuclear magnetic resonance spectrum. The bottom spectrum is that of tetrahydrofuran, and the top spectrum is that of propanoic anhydride. The chemical shift assignments are given below.

26 ppm

H₂C — CH₂

H₂C   CH₂
       O

68 ppm
tetrahydrofuran

29 ppm

O   O
‖   ‖
CH₃CH₂COCCH₂CH₃

170 ppm   9 ppm
propanoic anhydride

11.4 (a) $^{28}_{14}$Si   even atomic and mass number; no nuclear magnetic moment

       (b) $^{15}_{7}$N   odd mass and atomic number; has nuclear magnetic moment

       (c) $^{19}_{9}$F   odd mass and atomic number; has nuclear magnetic moment

       (d) $^{31}_{15}$P   odd mass and atomic number; has nuclear magnetic moment

       (e) $^{11}_{5}$B   odd mass and atomic number; has nuclear magnetic moment

       (f) $^{32}_{16}$S   even atomic and mass number; no nuclear magnetic moment

11.5

CH₃
|
CH₃– CH — CH₂ — CH₂ — CH₃

2-methylpentane;
five sets of equivalent carbon atoms, therefore five chemical shifts; isomer 3

        CH₃
        |
CH₃– CH₂ — CH — CH₂ — CH₃

3-methylpentane;
four sets of equivalent carbon atoms, therefore four chemical shifts; isomer 2

CH₃   CH₃
| |
CH₃– CH — CH — CH₃

2,3-dimethylbutane;
two sets of equivalent carbon atoms, therefore two chemical shifts; isomer 1

11.6

δ 200

δ 130, 152

δ 24, 26, 38

2-cyclohexenone;
six different carbon atoms, therefore six chemical shifts; bottom spectrum

δ 30   δ 38
        O
        ‖
    CH₃CCH₂CH₃

δ 210   δ 8

2-butanone (methyl ethyl ketone);
four different carbon atoms, therefore four chemical shifts; top spectrum

δ 26

δ 48 →   ← δ 48
        N
        |
        H

pyrrolidine;
two sets of equivalent carbon atoms, therefore two chemical shifts; middle spectrum

11.7   Compound A, C_5H_8O, has two units of unsaturation. The chemical shift at δ 219.6 is that of the carbon of a ketone, which uses up one unit of unsaturation. There are two more chemical shifts and four carbon atoms left. The molecule must be symmetric with two additional sets of two carbon atoms. There are no peaks in the alkene region, therefore the other unit of unsaturation must be a ring. Only cyclopentanone fits the given ^{13}C data.

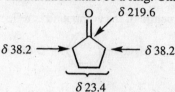

11.8   The peak that moves progressively upfield as the solution of the alcohol is diluted is the peak for the hydrogen atom of the hydroxyl group. As the alcohol is diluted, the amount of hydrogen bonding decreases.

Hydrogen bonding further deshields a hydrogen atom because the hydrogen atom involved in hydrogen bonding is partially bonded to another electronegative atom (O, N, or F) and has even less electron density around it than a hydrogen atom that is not involved in hydrogen bonding.

R — O — H   R — O — H - - - - - O — R
                                                                              /
                                                                            H

dilute solution; no hydrogen bonding;   concentrated solution; hydrogen bonding;
the hydrogen atom is less deshielded   the hydrogen atom is more deshielded
and absorbs farther upfield   and absorbs farther downfield

11.9

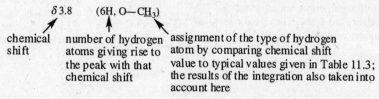

The absorption peaks between δ 1.0 and 3.0 in the spectrum of cyclohexene fall in the region for hydrogen atoms that are bonded to sp^3-hybridized carbon atoms. The four hydrogen atoms absorbing farther downfield (δ ~2) are the allylic hydrogen atoms on carbon atoms 3 and 6. These hydrogens are more deshielded than the hydrogens on carbon atoms 4 and 5 because they are closer to the double bond. The four hydrogen atoms farther upfield (δ ~1.6) are those on carbon atoms 4 and 5.

11.10   In this problem, in which we assign the structure of a compound from its molecular formula and spectral data, the proton magnetic resonance data are analyzed in two ways. The chemical shift values for groups of different hydrogen atoms are read off the spectrum, these values are compared with the chemical shift values given in Table 11.3 for different types of hydrogen atoms, and a tentative assignment of partial structure is made. The integration of the spectrum is examined by measuring the relative heights of the steps in the integration curves. The two types of information are summarized in the answers to these problems in the following way:

δ 3.8   (6H, O—CH_3)

chemical   number of hydrogen   assignment of the type of hydrogen
shift   atoms giving rise to   atom by comparing chemical shift
                the peak with that   value to typical values given in Table 11.3;
                chemical shift   the results of the integration also taken into
                                       account here

Compound B, $C_8H_{10}O_2$, has four units of unsaturation. Integration of the spectrum gives a ratio of 11:7.5 or 3:2 for the relative areas under the peaks. The molecular formula tells us that there are ten hydrogen atoms that absorb so the actual ratio is 6:4. Therefore, we have six hydrogen atoms that absorb at δ 3.8 and four that absorb at δ 6.8. The results of this reasoning and of our inspection of Table 11.3 are summarized on the next page.

11.10 (cont) δ 3.8   (6H, —OC$\underline{H}_3$)

                   δ 6.8   (4H, Ar$\underline{H}$)

The compound is 1,4-dimethyoxybenzene.

Note that the symmetry of the structure of the compound is reflected in the simplicity of the spectrum.

$$CH_3O-\langle\!\!\!\bigcirc\!\!\!\rangle-OCH_3$$

Compound C, C_7H_7Cl, has four units of unsaturation. (Each halogen atom counts as one hydrogen atom in determining units of unsaturation.) Integration of the spectrum gives a ratio of 10:22 or 2:5 (for a total of seven hydrogen atoms in C) for the relative areas under the peaks.

           δ 4.6   (2H, Y—C$\underline{H}_2$—X, where both X and Y deshield the methylene group)

           δ 7.4   (5H, Ar$\underline{H}$)

The compound is benzyl chloride.

$$\langle\!\!\!\bigcirc\!\!\!\rangle-CH_2Cl$$

Note that in using Table 11.3, we must realize that the values given will not match exactly the chemical shift values that we get from the spectra. We must put all the information together to make assignments of structure. In this problem, once we know that an aryl ring is present, we understand why the hydrogen atoms of the methylene group absorb at lower field than a typical RCH$_2$X group (δ 2.6 - 4.3 in Table 11.3).

Compound D, C_8H_8O, has five units of unsaturation. Integration of the spectrum gives a ratio of 9:15 or 3:5 (for a total of eight hydrogen atoms in D) for the relative areas under the peaks.

                               O
                               ‖

        δ 2.6   (3H, — CC$\underline{H}_3$)

       δ 7.4 – 8.0   (5H, Ar$\underline{H}$)

The compound is acetophenone (1-phenylethanone).

$$\langle\!\!\!\bigcirc\!\!\!\rangle-\overset{\overset{\displaystyle O}{\|}}{C}CH_3$$

The presence of the carbonyl group in the molecule was deduced from the chemical shift of the methyl group, the total number of carbon atoms in the molecule, and the one unit of unsaturation that had to be accounted for once the aromatic ring, with four units of unsaturation, was identified.

11.11   The spectrum of 1,1-dichloroethane has two groups of peaks with an integration giving a ratio of 30:10 or 3:1 for the relative areas under the groups of small peaks.

These hydrogen atoms are adjacent to one hydrogen atom. They give rise to a doublet centered at δ 2.05.

$$\begin{array}{ccc} & H & Cl \\ & | & | \\ H - & C - & C - Cl \\ & | & | \\ & H & H \end{array}$$

This hydrogen atom is adjacent to three chemical-shift equivalent hydrogen atoms. It gives rise to a quartet centered at δ 5.92.

                       ↓↑↑   ↓↓↑

                       ↑↓↑   ↓↑↓

              ↑↑↑   ↑↑↓   ↑↓↓   ↓↓↓

There are eight possible spin orientations of the three hydrogen atoms of the methyl group. Three orientations in the two middle sets are indistinguishable. The magnetic field sensed by the adjacent hydrogen atom varies slightly with a probability of 1:3:3:1, giving rise to similar relative intensities of the peaks in the quartet.

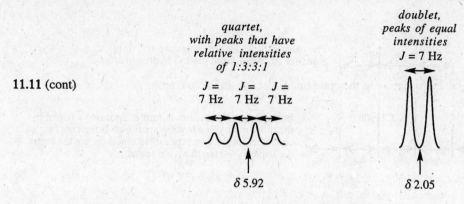

11.11 (cont)

The data from the spectrum may be summarized as follows:

δ 2.05   (3H, d, J = 7 Hz, C$\underline{H}_3$CHCl$_2$)

δ 5.92   (1H, q, J = 7 Hz, CH$_3$C$\underline{H}$Cl$_2$—)

(s = singlet, d = doublet, t = triplet, q = quartet)

11.12   C$_4$H$_{10}$O has no units of unsaturation. The proton-decoupled spectrum tells us that four different carbon atoms are present. Three of the carbon atoms are alkyl carbons bonded to other carbon atoms (δ 10, 23, and 32). The fourth carbon atom (δ 69) is bonded to the oxygen atom.

The proton-coupled spectrum tells us how many hydrogen atoms are bonded to the carbon atoms. Two of the carbons are quartets (at δ 10 and 23), which tells us that these two carbon atoms are methyl carbons. The carbon atom at δ 32 is a triplet and thus has two hydrogens bonded to it. The carbon atom bonded to the oxygen (δ 69) is a doublet and has only one hydrogen bonded to it. The compound is 2-butanol.

$$\text{CH}_3\!-\!\underset{\underset{\text{OH}}{|}}{\text{CH}}\!-\!\text{CH}_2\!-\!\text{CH}_3$$

11.13

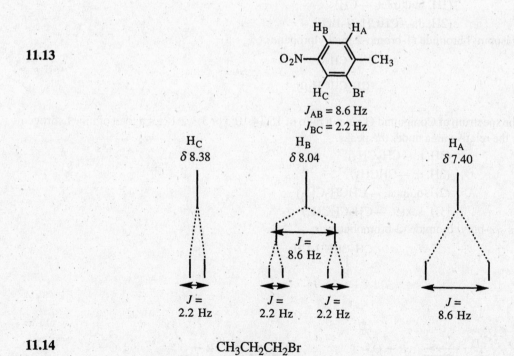

11.14   CH$_3$CH$_2$CH$_2$Br
                             a   b   c
                    J_{ab} = J_{bc} = 7 Hz

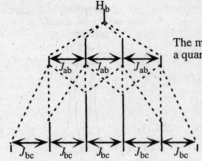

11.14 (cont)

The methylene hydrogen atoms are split into a quartet by the adjacent methyl hydrogen atoms.

Each peak of the quartet is further split into a triplet by the other methylene hydrogen atoms. If the coupling constants for the two types of interactions are the same, the triplets overlap to give a sextet.

11.15   CH₃—⟨benzene ring⟩—C≡N   CH₃-substituted benzene—C≡N   CH₃-substituted benzene—C≡N

six sets of carbon atoms   eight sets of carbon atoms   eight sets of carbon atoms

The spectrum is that of 4-methylbenzonitrile. The symmetrical splitting pattern for the hydrogen atoms on the aromatic ring, similar in appearance to a doublet of doublets, is typical of an aromatic ring that is substituted in the para positions with two substituents of differing electronegativity.

11.16   Compounds F and G, C_4H_9Br, have no units of unsaturation. Integration of the spectrum of Compound F gives a ratio of 37:7:12 or 6:1:2 for the relative area under the peaks. The molecular formula tells us that there are nine hydrogen atoms that absorb so we have accounted for all the hydrogen atoms. The chemical shifts and coupling information are summarized below.

δ 1.0   (6H, d, —CH(C$\underline{H}_3$)₂)

δ 1.95   (1H, multiplet, —C$\underline{H}$)

δ 3.3   (2H, d, —CHC$\underline{H}_2$—Br)

Compound F is isobutyl bromide (1-bromo-2-methylpropane).

$$CH_3\overset{\overset{\displaystyle CH_3}{|}}{CH}CH_2Br$$

Integration of the spectrum of Compound G gives a ratio of 15:14:10:5 or 3:3:2:1 (for a total of nine hydrogen atoms in G) for the relative area under the peaks.

δ 1.0   (3H, t, —CH₂C$\underline{H}_3$)

δ 1.7   (3H, d, —CHC$\underline{H}_3$)

δ 1.8   (2H, quintet, —CHC$\underline{H}_2$CH₃)

δ 4.1   (1H, sextet, —CH₂C$\underline{H}$CH₃)

Compound G is *sec*-butyl bromide (2-bromobutane).

$$CH_3\underset{\underset{\displaystyle Br}{|}}{CH}CH_2CH_3$$

11.17   R—Ö: ⤴ D—Ö—D ⇌ R—Ö⁺—D ⇌ R—Ö: :Ö—D
           | | | |
           H   H :Ö—D   H   H

11.18   Compound H, $C_4H_8O_2$, has one unit of unsaturation. Its proton magnetic resonance spectrum has peaks at

δ 1.25	(3H, t, C$\underline{H}_3$CH$_2$)
δ 2.05	(3H, s, C$\underline{H}_3$C=O)
δ 4.15	(2H, q, CH$_3$C$\underline{H}_2$O)

The compound is ethyl acetate (ethyl ethanoate)

$$\overset{\overset{\displaystyle O}{\|}}{CH_3C}OCH_2CH_3$$

Note the characteristic triplet at ~δ 1 and quartet at ~δ 4 that are characteristic of an ethyl group bonded to an electronegative atom, in this case oxygen.

Compound I, $C_5H_8O_2$, has two units of unsaturation. The proton magnetic resonance spectrum has peaks at

δ 1.9	(3H, s, C$\underline{H}_3$C=C)
δ 2.2	(3H, s, C$\underline{H}_3$C=O)
δ 5.7	(1H, s, C$\underline{H}$=C)
δ 12.1	(1H, broad s, $\underline{H}$OC=O)

The compound is 3-methyl-2-butenoic acid.

$$\begin{array}{ccc}
CH_3 & & \overset{\overset{\displaystyle O}{\|}}{C}OH \\
\diagdown & & \diagup \\
& C=C & \\
\diagup & & \diagdown \\
CH_3 & & H
\end{array}$$

At first glance the two peaks around δ 2.0 appear to be a doublet, but since there are no other peaks in the spectrum showing the same separation, the two peaks cannot have arisen because of coupling to another hydrogen atom. We assign the peaks to two nonequivalent methyl groups on a double bond.

Compound J, $C_7H_{12}O_3$, has two units of unsaturation. The proton magnetic resonance spectrum has peaks at

δ 1.25	(3H, t, C$\underline{H}_3$CH$_2$)
δ 2.20	(3H, s, C$\underline{H}_3$C=O)
δ 2.55	(2H, t, CH$_2$C$\underline{H}_2$C=O)
δ 2.75	(2H, t, O=CC$\underline{H}_2$CH$_2$)
δ 4.15	(2H, q, CH$_3$C$\underline{H}_2$O)

Compound J has an isolated ethyl group bonded to oxygen and an isolated methyl group next to a carbonyl group. Each methylene group is also next to a carbonyl group. The compound is ethyl 4-oxopentanoate.

$$CH_3CH_2O\overset{\overset{\displaystyle O}{\|}}{C}CH_2CH_2\overset{\overset{\displaystyle O}{\|}}{C}CH_3$$

Compound K, C_4H_8O, has one unit of unsaturation. The ^{13}C magnetic resonance spectrum has peaks at

δ 26	(t, C$\underline{H}_2$)
δ 68	(t, C$\underline{H}_2$O)

Only two carbons are seen in the spectrum, but Compound K has four carbon atoms. Each peak must correspond to two equivalent carbon atoms. There are no peaks above 100 ppm, therefore the unit of unsaturation must be assigned to a ring and not a carbonyl group or an alkene. The splitting pattern tells us that each carbon atom is bonded to two hydrogen atoms. The compound is tetrahydrofuran.

11.19

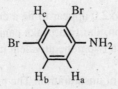

$\delta\, 0.8$ (3H, t)

These are the methyl hydrogen atoms labeled a. They are split into a triplet by the two hydrogen atoms on the adjacent methylene group.

$\delta\, 1.2$ (3H, d)

These are the methyl hydrogen atoms labeled b. They are split into a doublet by the hydrogen atom on the adjacent benzylic carbon atom.

$\delta\, 1.6$ (2H, quintet, $J_{ac} = J_{cd}$)

These are the methylene hydrogen atoms labeled c. J_{ac} and J_{cd} are equal. The methylene hydrogens have four 3-bond neighbors and, therefore, are split into a quintet by the hydrogen atom on the adjacent benzylic carbon atom and the hydrogen atoms on the adjacent methyl group.

$\delta\, 2.6$ (1H, sextet, $J_{bd} = J_{cd}$)

This is the benzylic hydrogen atom labeled d. J_{bd} and J_{cd} are equal, and the benzylic hydrogen, with five 3-bond neighbors, is split into a sextet by the hydrogen atoms on the adjacent methylene group and the hydrogen atoms on the adjacent methyl group.

$\delta\, 7.2$ (5H, multiplet)

These are the aromatic hydrogen atoms labeled e. They are not chemical-shift equivalent and couple to each other in a complex pattern.

11.20

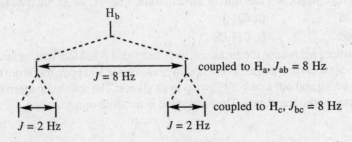

The peaks are analyzed below:

$\delta\, 4.10$	(2H, broad multiplet, —N$\underline{H}_2$)
$\delta\, 6.65$	(1H, d, $J_{ab} = 8$ Hz, Ar$\underline{H}_a$)
$\delta\, 7.20$	(1H, doublet of doublets, $J_{ab} = 8$ Hz, $J_{bc} = 2$ Hz, Ar$\underline{H}_b$)
$\delta\, 7.55$	(1H, d, $J_{bc} = 2$ Hz, Ar$\underline{H}_c$)

H_a is the farthest upfield of the aryl hydrogen atoms because it is next to the amino group, which is strongly electron donating (see Section 16.1C in the text) and distant from the electron-withdrawing bromine atoms. It is split into a doublet by H_b with $J = 8$ Hz.

H_b has the next highest chemical shift. It is next to only one bromine atom. A diagram showing the splitting pattern for H_b is shown below.

H_c is the most deshielded of the aryl hydrogen atoms because it is between two bromine atoms. It is also split into a doublet by H_b with $J = 2$ Hz.

11.21   The signal at δ 178 is that of a carbonyl carbon atom. It is split into a quartet, with a very small coupling constant, by three adjacent methyl hydrogen atoms that are three-bond neighbors of the carbonyl carbon atom. The signal of the methyl carbon at δ 20 is also split into a quartet, with a very large coupling constant, by the methyl hydrogen atoms that are one-bond neighbors to it.

11.22   Compound L, $C_4H_8Cl_2$, has no units of unsaturation. We know that there are only two different types of carbon atoms in X because the ^{13}C magnetic resonance spectrum has only two peaks. X must be 1,4-dichlorobutane.

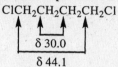

2,3-Dichlorobutane also has the symmetry required for this spectrum, but we would expect the methyl groups to absorb at higher field than the chemical shift values observed (Table 11.2, p. 403 in the text).

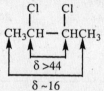

Also note that the carbon atoms bearing the chlorine atoms in this compound are even more highly substituted than in 1,4-dichlorobutane. We would expect them to absorb at lower field.

11.23   There are four possible isomers with the molecular formula C_4H_9Br.

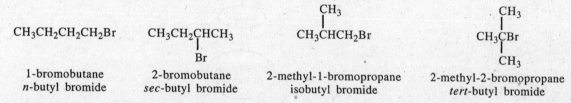

We know that there are four different types of carbon atoms in the two spectra shown because the ^{13}C magnetic resonance spectra each have four peaks. That eliminates isobutyl bromide (3 sets of carbon atoms) and *tert*-butyl bromide (2 sets of carbon atoms) as possible structures.

Tentative assignments can be made by looking at the carbon atom bonded to the bromine. The chemical shifts given in Table 11.2 (p. 403 in the text) show that the more highly substituted a carbon atom is, the further downfield it absorbs. Primary alkyl bromide carbon atoms, such as the one in *n*-butyl bromide absorb between 28 and 35 ppm and secondary alkyl bromine carbon atoms, such as in *sec*-butyl bromide will absorb farther downfield. We can assign the top spectrum to *n*-butyl bromide because it has no carbon atoms absorbing below 40 ppm. The bottom spectrum would be that of *sec*-butyl bromide, since it has one carbon atom at 54 ppm, suggesting that this carbon atom is more highly substituted.

Further confirmation could be made on the basis of the proton-coupled ^{13}C spectra. *n*-Butyl bromide would show three triplets and one quartet and *sec*-butyl bromide would show one doublet, one triplet, and two quartets. The proton magnetic resonance spectra can also be used for confirmation.

12.1

more electronegative oxygen,
higher energy, shorter
wavelength $n \rightarrow \pi^*$ transition

less electronegative sulfur,
lower energy, longer
wavelength $n \rightarrow \pi^*$ transition

Concept Map 12.1   Visible and ultraviolet spectroscopy.

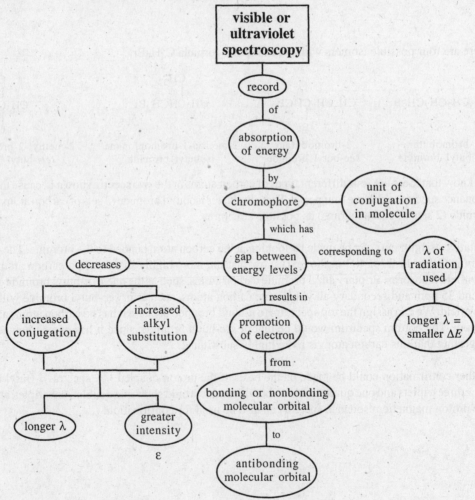

12.2   The compound with the most extended conjugation would be expected to absorb at the longest wavelength.

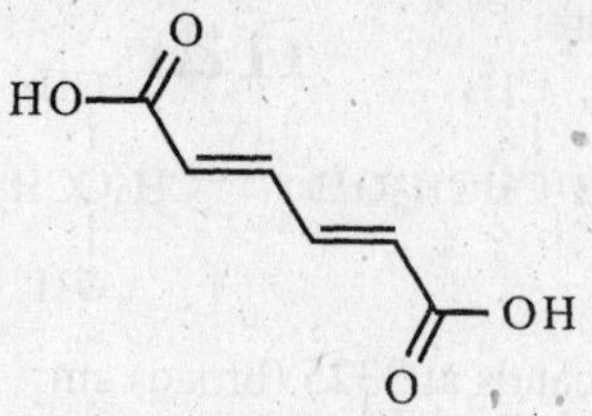

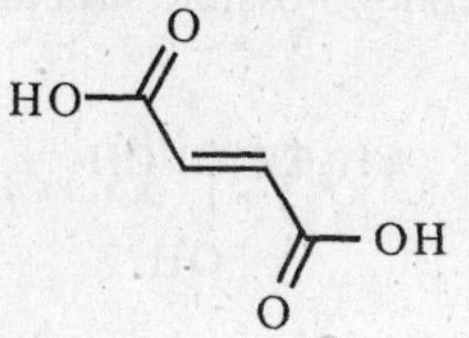

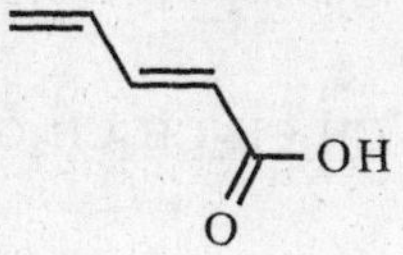

The compound above has the most extensive conjugation and would absorb at the longest wavelength.

12.3

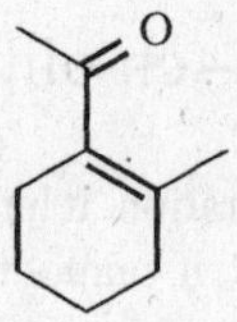

1-Acetyl-2-methyl-1-cyclohexene has two absorption bands, $\lambda_{max}^{ethanol}$ 247 nm (ε 6000) and 305 nm (ε 100). The band at 247 nm corresponds to the $\pi \to \pi^*$ transition of the conjugated π system and the band at 305 nm, to the $n \to \pi^*$ transition of the carbonyl group.

12.4

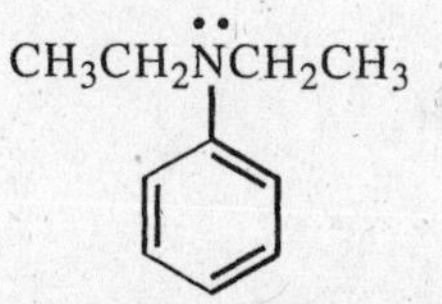

$CH_3CH_2\overset{..}{N}CH_2CH_3$

This compound has the nonbonding electrons on the nitrogen atom in conjugation with the aromatic ring. It is an arylamine and should have an ultraviolet spectrum in which bands are shifted to longer wavelengths than those in the spectrum of toluene.

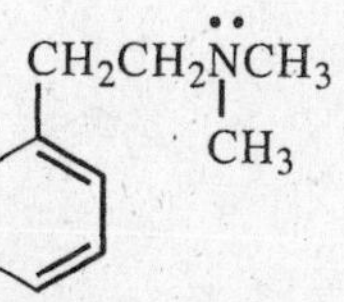

$CH_2CH_2\overset{..}{N}CH_3$
$\qquad\qquad\quad CH_3$

In this compound the nitrogen atom is separated from the aromatic ring by sp^3-hybridized carbon atoms. The chromophore in this compound is an alkylbenzene, like toluene.

A comparison of the spectrum in Figure 12.9 with that of toluene (Figure 12.8) identifies Compound A as *N,N*-diethylaniline. The chromophore is an arylamine in which the nonbonding electrons on the nitrogen atom can interact with the aromatic ring.

12.5   A shift of λ_{max} to a longer wavelength when potassium hydroxide is added to the solution points to a phenol. There are three possible phenols with the molecular formula C_7H_8O. Compound B may be:

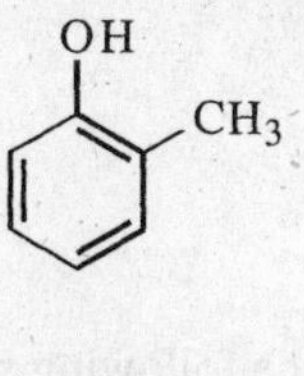

o-cresol 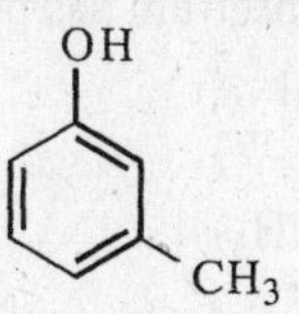   *m*-cresol 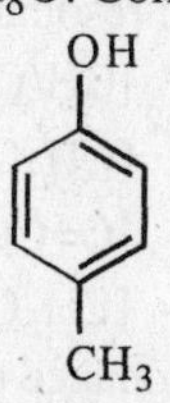   *p*-cresol

12.6   Compound C has bands at 3090 cm^{-1} (=C—H) and 1650 cm^{-1} (C=C); therefore compound C is an alkene.
Compound D has bands at 3300 cm^{-1} (≡C—H) and 2100 cm^{-1} (C≡C); therefore compound D is an alkyne.
Compound E has a broad band at 3330 cm^{-1} (—O—H) and a band at 1060 cm^{-1} (C—O); therefore compound E is an alcohol.
Compound F has bands at 3080 cm^{-1} (=C—H) and 1640 cm^{-1} (C=C); therefore compound F is an alkene.

12.7   Compound G, $C_4H_{10}O$, has no units of unsaturation. It has bands at 3325 (broad) cm^{-1} (—O—H) and 1040 cm^{-1} (C—O); therefore it is an alcohol. Possible structures are:

$$CH_3CH_2CH_2CH_2OH \qquad \underset{\underset{OH}{|}}{CH_3CH_2CHCH_3} \qquad \underset{\overset{|}{CH_3}}{CH_3CHCH_2OH} \qquad \underset{\underset{OH}{|}}{\overset{\overset{CH_3}{|}}{CH_3CCH_3}}$$

Compound H, $C_6H_{12}O$, has one unit of unsaturation. It has bands at 3325 (broad) cm^{-1} (—O—H) and 1060 cm^{-1} (C—O); therefore it must be an alcohol. Since there is one unit of unsaturation and no alkene stretching frequency, it must be a cyclic alcohol. Possible structures are:

Compound I, $C_6H_{12}O$, has one unit of unsaturation. It has a band at 1700 cm^{-1} (C=O). The unit of unsaturation corresponds to the carbonyl group; therefore, it cannot have a ring. There is no band at 2700 cm^{-1}; therefore it must be a ketone. Possible structures are:

Compound J, C_4H_8O, has one unit of unsaturation. It has a band at 2700 cm^{-1} (O=C—H); the band at 2900 cm^{-1} is hidden in the alkane band. In addition there is a band at 1740 cm^{-1} (C=O). Possible structures are:

12.8   Compound K, $C_5H_{10}O$, has one unit of unsaturation. It has bands at 1715 cm^{-1} (C=O). The absorption at 1715 cm^{-1} indicates that it has a carbonyl group. With only one unit of unsaturation, it cannot contain a ring. No absorption at ~2700 cm^{-1} indicates that it is not an aldehyde, but a ketone.
Possible structures are:

The proton nuclear magnetic resonance spectrum has peaks at

       $\delta\,1.11$ [6H, d, $(C\underline{H}_3)CH$]

       $\delta\,2.15$ [3H, s, $C\underline{H}_3C=O$]

       $\delta\,2.60$ [1H, septet, $O=CC\underline{H}(CH_3)_2$] (The two outer peaks are very difficult to see.)

The ^{13}C nuclear magnetic resonance spectrum has peaks at

       $\delta\,18.15$ [$CH(\underline{C}H_3)_2$]

       $\delta\,27.43$ [$\underline{C}H_3C=O$]

       $\delta\,41.65$ [$O=C\underline{C}H(CH_3)_2$]

       $\delta\,212.47$ [$CH_3\underline{C}=O$]

The nuclear magnetic resonance data allows us to assign the following structure to K:

Concept Map 12.2   Infrared spectroscopy.

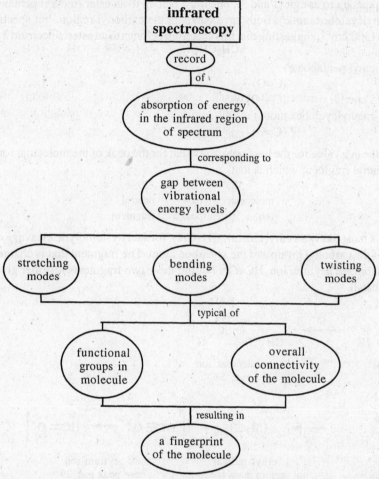

- -

12.9   The best approach to identifying which spectrum belongs to which compound is to first determine the functional groups that are present in each compound and then to look for the specific bands in the spectra that correspond to the functional groups.

There is only one alcohol, which should have a broad band at ~3350 cm^{-1} and no band in the carbonyl region. The only spectrum that fits is spectrum 4. Spectrum 4, therefore, must belong to 1-octanol.

3325 (broad) cm^{-1}	(—O—H)
1060 cm^{-1}	(C—O)

One of the compounds is an acid and an alkene. Acids should have a broad band at ~3300–2400 cm^{-1} and a band in the carbonyl region. Alkenes should have a band at ~1600 cm^{-1}. Spectrum 5 has a band at 1645 cm^{-1}, as well as those corresponding to a carboxylic acid, therefore, spectrum 5 belongs to 3-butenoic acid.

3100–2500 (broad) cm^{-1}	(—O—H)
1715 cm^{-1}	(C=O)
1645 cm^{-1}	(C=C)

One of the compounds is an aldehyde, which should have a band in the carbonyl region and a band at 2700 cm^{-1}. Spectrum 2 has these two bands and must, therefore, belong to 10-undecenal, an unsaturated aldehyde. Spectrum 2 also has a band at ~1600 cm^{-1}, which confirms the presence of an alkene.

2715 cm^{-1}	(O=C—H)
1725 cm^{-1}	(C=O)
1640 cm^{-1}	(C=C)

12.9 (cont)

Of the last two spectra to assign (1 and 3), one must belong to an ester (methyl pentanoate) and the other to a ketone (3-methylcyclohexanone). Both have bands in the carbonyl region, but spectrum 1 has broad bands between 1000-1300 cm^{-1}, suggesting that this spectrum belongs to an ester. Spectrum 3 must, therefore, belong to the ketone.

Spectrum 1 (methyl pentanoate)

    1740 cm^{-1} (C=O)

    1260, 1175 cm^{-1} (C—O)

Spectrum 3 (3-methylcyclohexanone)

    1710 cm^{-1} (C=O)

12.10 If we subtract the m/z value for the base peak from that for the peak of the molecular ion, we can determine the mass of the neutral fragment which is lost.

$$58 \; - \; 29 \; = \; 29$$

    molecular   base   neutral

       ion   peak   fragment

The fragment is most likely an ethyl radical, CH_3CH_2, formed by homolytic cleavage of the bond between the carbon atom of the carbonyl group and the α-carbon atom. The fragment that is charged, with an m/z of 29, is a resonance-stabilized acylium ion, $HC\equiv O^+$. Putting these two fragments together gives us the identity of the unknown, propanal.

propanal   molecular ion m/z 58

M‡ ethyl radical   an acylium ion

m/z 58   not seen in mass spectrum   base peak m/z 29

12.11 A hydrogen atom is lost in going from the molecular ion at m/z 46 to the peak at m/z 45. The loss of a hydroxyl radical (m/z 17) from the molecular ion leads to the base peak at m/z 29. Both of these fragments come from homolytic cleavage of the two bonds to the carbonyl group. Putting these fragments together gives us the structure of the unknown acid, which is formic acid.

formic acid   molecular ion m/z 46

M‡ hydrogen   an acylium ion

m/z 46   atom m/z 45

M‡ hydroxyl   another acylium ion

m/z 46   radical   base peak m/z 29

12.12   Two possible cleavages of the molecular ion of 3-pentanone at the bonds to the α-carbon atom are shown below:

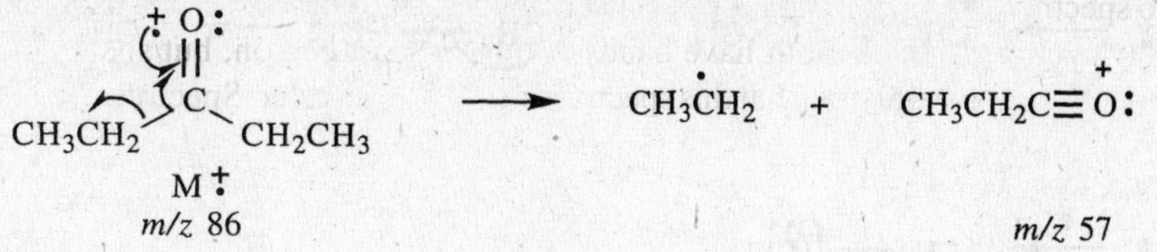

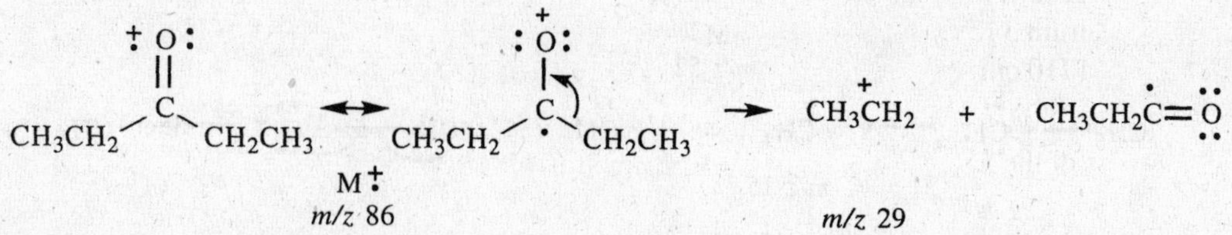

$$CH_3CH_2 \quad \overset{\overset{+O:}{\|}}{C} \quad CH_2CH_3 \quad \longrightarrow \quad CH_3\overset{\cdot}{C}H_2 \quad + \quad CH_3CH_2C\equiv\overset{+}{O}:$$

M⁺·
m/z 86   *m/z* 57

$$CH_3CH_2 \quad \overset{\overset{+O:}{\|}}{C} \quad CH_2CH_3 \quad \longleftrightarrow \quad CH_3CH_2 \quad \overset{\overset{+}{:O:}}{\underset{\cdot}{C}} \quad CH_2CH_3 \quad \longrightarrow \quad CH_3\overset{+}{C}H_2 \quad + \quad CH_3CH_2\overset{\cdot}{C}=\overset{\cdot\cdot}{\underset{\cdot\cdot}{O}}$$

M⁺·
m/z 86   *m/z* 29

The expected cleavages of the molecular ion of 2-pentanone at the bonds to the two different α-carbon atoms are shown below:

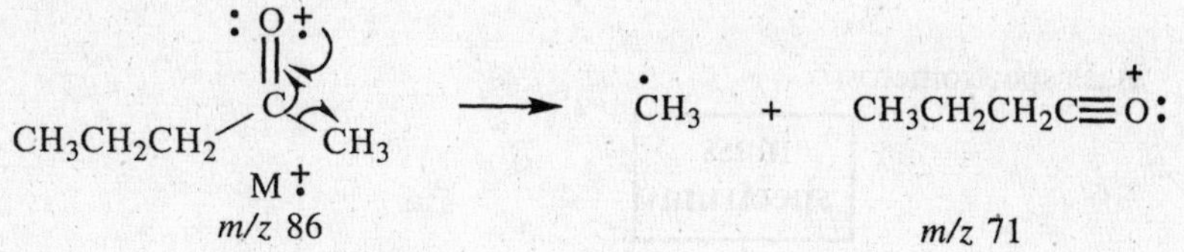

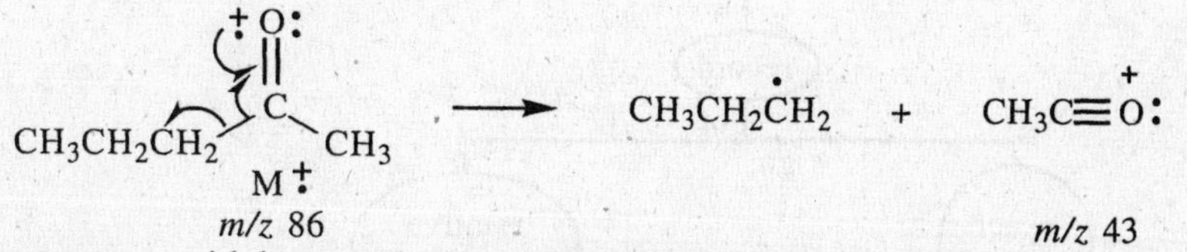

$$CH_3CH_2CH_2 \quad \overset{\overset{:O\overset{+}{\cdot}}{\|}}{C} \quad CH_3 \quad \longrightarrow \quad \overset{\cdot}{C}H_3 \quad + \quad CH_3CH_2CH_2C\equiv\overset{+}{O}:$$

M⁺·
m/z 86   *m/z* 71

$$CH_3CH_2CH_2 \quad \overset{\overset{+O:}{\|}}{C} \quad CH_3 \quad \longrightarrow \quad CH_3CH_2\overset{\cdot}{C}H_2 \quad + \quad CH_3C\equiv\overset{+}{O}:$$

M⁺·
m/z 86   *m/z* 43

The mass spectrum with base peaks at *m/z* 29 and *m/z* 57 is the spectrum of 3-pentanone. The mass spectrum with the base peak at *m/z* 43 and a small peak at *m/z* 71 is the spectrum of 2-pentanone.

12.13   The formation of the base peak in the spectrum of 2,2-dimethylpropanal from the molecular ion is shown below:

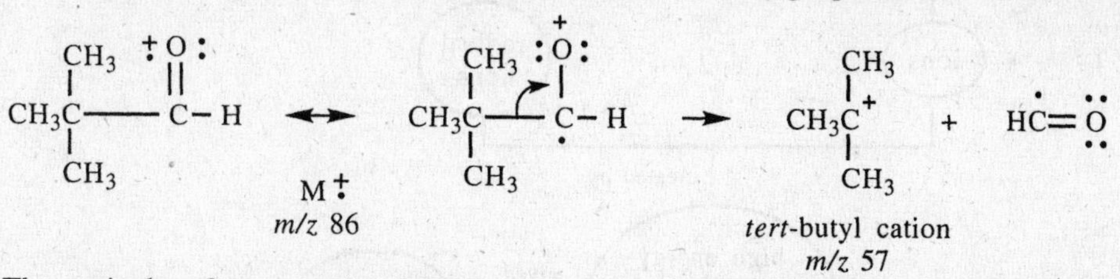

$$\overset{CH_3}{\underset{CH_3}{CH_3C}}\overset{\overset{+O:}{\|}}{---C}-H \quad \longleftrightarrow \quad \overset{CH_3}{\underset{CH_3}{CH_3C}}\overset{\overset{:\overset{+}{O}:}{\underset{\cdot}{C}}}{---C}-H \quad \longrightarrow \quad \overset{CH_3}{\underset{CH_3}{CH_3\overset{+}{C}}} \quad + \quad H\overset{\cdot}{C}=\overset{\cdot\cdot}{\underset{\cdot\cdot}{O}}$$

M⁺·
m/z 86   *tert*-butyl cation
                                                        m/z 57

The equivalent fragmentation from the molecular ion of acetaldehyde is shown below:

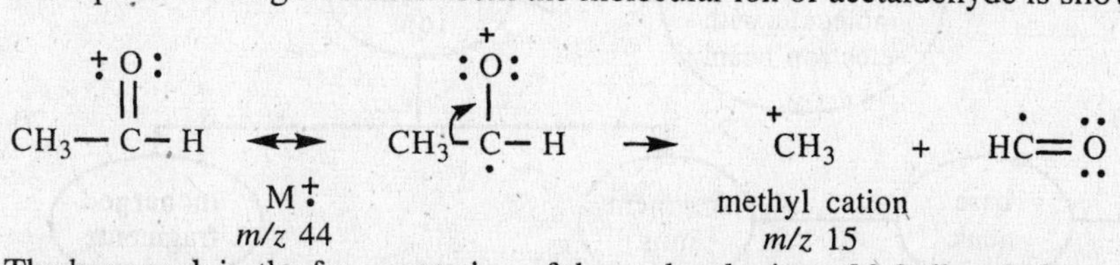

$$CH_3-\overset{\overset{+O:}{\|}}{C}-H \quad \longleftrightarrow \quad CH_3\overset{\overset{:\overset{+}{O}:}{\underset{\cdot}{C}}}{}-H \quad \longrightarrow \quad \overset{+}{C}H_3 \quad + \quad H\overset{\cdot}{C}=\overset{\cdot\cdot}{\underset{\cdot\cdot}{O}}$$

M⁺·
m/z 44   methyl cation
                                                   m/z 15

The base peak in the fragmentation of the molecular ion of 2,2-dimethylpropanal is the stable tertiary cation, the *tert*-butyl cation. The comparable species in the fragmentation of the molecular ion of acetaldehyde would be the unstable methyl cation, *m/z* 15, which does not appear in the spectrum of acetaldehyde.

12.14 $$CH_3-\overset{\cdot\cdot}{\underset{\cdot\cdot}{^{35}Cl}}: \quad \xrightarrow{-e^-} \quad CH_3-^{35}\overset{\cdot\cdot}{\underset{\cdot\cdot}{Cl}}⁺·$$

M⁺·
m/z 50

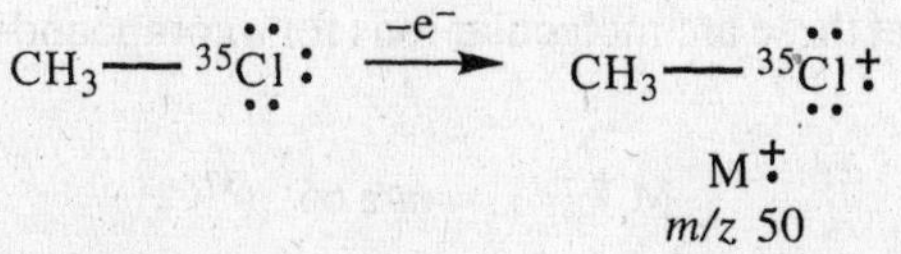

12.14 (cont)

$CH_3 \longrightarrow {}^{35}\ddot{C}l^+ \longrightarrow {}^+CH_3 + {}^{35}\cdot\ddot{C}l^{..}$

           m/z 15

$CH_3 \longrightarrow {}^{35}\ddot{C}l^+ \longrightarrow \cdot CH_3 + {}^{35}\ddot{\overset{..}{C}l}^+$

                         m/z 35
                        not evident in
                        the spectrum

$CH_3 \longrightarrow {}^{37}\ddot{C}l^{..} \xrightarrow{-e^-} CH_3 \longrightarrow {}^{37}\ddot{C}l^+$

                            M$^+$
                        m/z 52

$CH_3 \longrightarrow {}^{37}\ddot{C}l^+ \longrightarrow {}^+CH_3 + {}^{37}\cdot\ddot{C}l^{..}$

           m/z 15

$CH_3 \longrightarrow {}^{37}\ddot{C}l^+ \longrightarrow \cdot CH_3 + {}^{37}\ddot{\overset{..}{C}l}^+$

                         m/z 37
                        not evident in
                        the spectrum

Concept Map 12.3   Mass spectrometry.

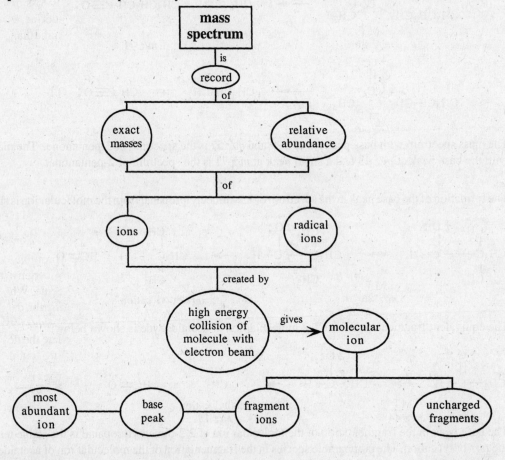

12.15   The two ions with the highest *m/z* values appear in this spectrum at *m/z* 64 and *m/z* 66. The one at the higher mass is one-third the intensity of the other, suggesting that these are molecular ions for a compound containing chlorine.

        M$^+$ *m/z* 64 (^{35}Cl)   M$^+$ *m/z* 66 (^{37}Cl)

12.15 (cont)

An important fragmentation of the molecular ion gives rise to the peak at m/z 29. This is the ethyl cation and results from loss of chlorine from the molecular ion. Putting these fragments together, we identify the haloalkane as ethyl chloride (chloroethane).

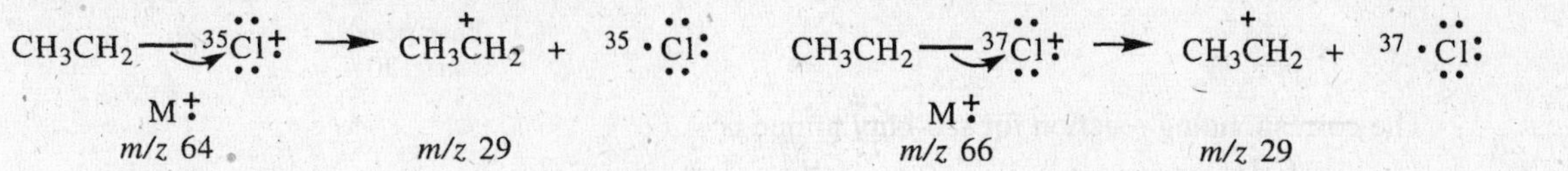

$$CH_3CH_2 \overset{\curvearrowright}{-\!\!-} {}^{35}\ddot{\underset{\cdot\cdot}{Cl}}{\overset{+}{\cdot}} \longrightarrow CH_3\overset{+}{C}H_2 + {}^{35} \cdot \ddot{\underset{\cdot\cdot}{Cl}} \qquad CH_3CH_2 \overset{\curvearrowright}{-\!\!-} {}^{37}\ddot{\underset{\cdot\cdot}{Cl}}{\overset{+}{\cdot}} \longrightarrow CH_3\overset{+}{C}H_2 + {}^{37} \cdot \ddot{\underset{\cdot\cdot}{Cl}}$$

$$\underset{m/z\ 64}{M^{+}} \qquad\qquad\qquad \underset{m/z\ 29}{} \qquad\qquad\qquad \underset{m/z\ 66}{M^{+}} \qquad\qquad\qquad \underset{m/z\ 29}{}$$

12.16   The high mass region of this mass spectrum has a pair of ions with equal intensity differing by 2 mass units (m/z 148, 150). This suggests that one of the two halogens in the unknown is bromine. The presence of bromine is confirmed by the two peaks of equal intensity at m/z 79 and 81.

The first fragmentation, m/z 148 to m/z 129, (or m/z 150 to m/z 131) is the loss of a species with a mass of 19, which corresponds to the atomic weight of fluorine. The base peak, which is at m/z 69, results from the loss of a bromine atom. The compound can have only one bromine atom, because the molecular weight is too low for more than one. Therefore, there must be three fluorine atoms. The compound is bromotrifluoromethane, CF_3Br.

The fragmentations are shown below. The molecular ion is shown as being formed by loss of a nonbonding electron from bromine. Because the electronegativity of bromine is lower than that of fluorine, and the bromine atom is much larger than a fluorine atom, it would be easier to lose the electron from bromine than from fluorine.

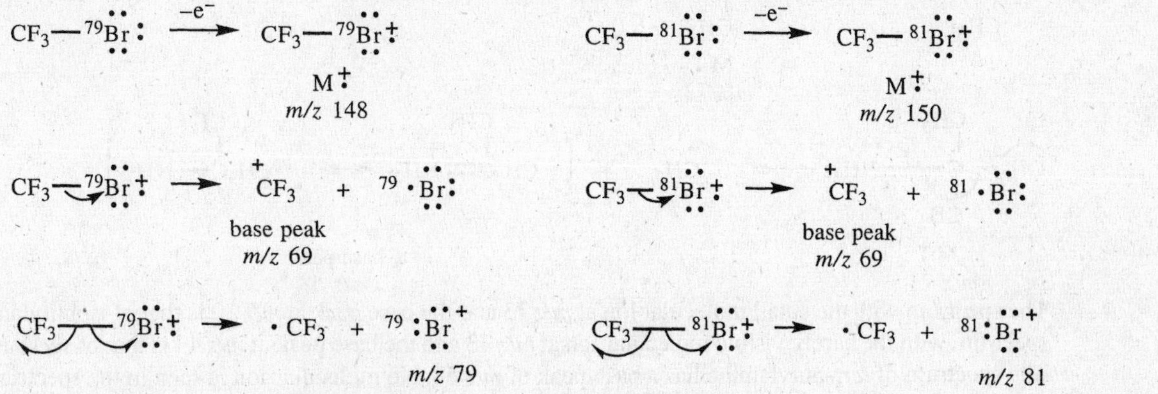

$$CF_3 \!-\! {}^{79}\ddot{\underset{\cdot\cdot}{Br}}\!: \overset{-e^-}{\longrightarrow} CF_3 \!-\! {}^{79}\ddot{\underset{\cdot\cdot}{Br}}{\overset{+}{\cdot}} \qquad\qquad CF_3 \!-\! {}^{81}\ddot{\underset{\cdot\cdot}{Br}}\!: \overset{-e^-}{\longrightarrow} CF_3 \!-\! {}^{81}\ddot{\underset{\cdot\cdot}{Br}}{\overset{+}{\cdot}}$$

$$\underset{m/z\ 148}{M^{+}} \qquad\qquad\qquad\qquad\qquad\qquad \underset{m/z\ 150}{M^{+}}$$

$$CF_3 \overset{\curvearrowright}{=\!\!\!=} {}^{79}\ddot{\underset{\cdot\cdot}{Br}}{\overset{+}{\cdot}} \longrightarrow \overset{+}{C}F_3 + {}^{79} \cdot \ddot{\underset{\cdot\cdot}{Br}}\!: \qquad CF_3 \overset{\curvearrowright}{=\!\!\!=} {}^{81}\ddot{\underset{\cdot\cdot}{Br}}{\overset{+}{\cdot}} \longrightarrow \overset{+}{C}F_3 + {}^{81} \cdot \ddot{\underset{\cdot\cdot}{Br}}\!:$$

$$\underset{\substack{\text{base peak}\\ m/z\ 69}}{} \qquad\qquad\qquad\qquad\qquad \underset{\substack{\text{base peak}\\ m/z\ 69}}{}$$

$$CF_3 \overset{\curvearrowleft}{=\!\!\!=} {}^{79}\ddot{\underset{\cdot\cdot}{Br}}{\overset{+}{\cdot}} \longrightarrow \cdot CF_3 + {}^{79} \overset{\cdot\cdot\,+}{\underset{\cdot\cdot}{Br}} \qquad CF_3 \overset{\curvearrowleft}{=\!\!\!=} {}^{81}\ddot{\underset{\cdot\cdot}{Br}}{\overset{+}{\cdot}} \longrightarrow \cdot CF_3 + {}^{81} \overset{\cdot\cdot\,+}{\underset{\cdot\cdot}{Br}}$$

$$\underset{m/z\ 79}{} \qquad\qquad\qquad\qquad\qquad\qquad \underset{m/z\ 81}{}$$

Even though the majority of the molecular ions result from loss of an electron from the bromine atom, a certain number of the molecular ions will result from loss of an electron from a fluorine atom. A better way of thinking about this, of course, is to remember that all of the electrons belong to the whole molecule. When an electron is lost, the fragments that result from this molecular ion may be rationalized by picturing the deficiency of the electron as being localized at one site or another. The relative intensities of the bands at m/z 69, 129, and 131 in the spectrum are a measure of the ease with which bromine and fluorine accommodate the deficiency of an electron. Localization of the deficiency of an electron on fluorine leads to the following fragmentations:

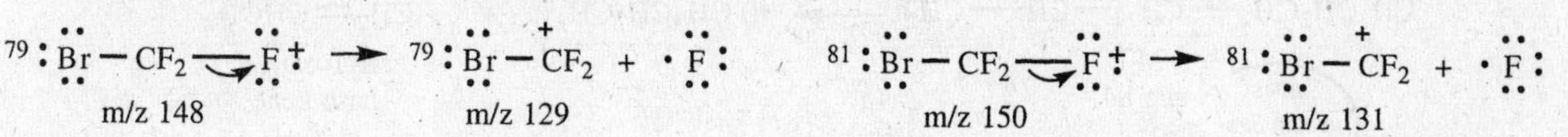

$$ {}^{79}\ddot{\underset{\cdot\cdot}{Br}} \!-\! CF_2 \overset{\curvearrowright}{=\!\!\!=} \ddot{\underset{\cdot\cdot}{F}}{\overset{+}{\cdot}} \longrightarrow {}^{79}\ddot{\underset{\cdot\cdot}{Br}} \!-\! \overset{+}{C}F_2 + \cdot \ddot{\underset{\cdot\cdot}{F}}\!: \qquad {}^{81}\ddot{\underset{\cdot\cdot}{Br}} \!-\! CF_2 \overset{\curvearrowright}{=\!\!\!=} \ddot{\underset{\cdot\cdot}{F}}{\overset{+}{\cdot}} \longrightarrow {}^{81}\ddot{\underset{\cdot\cdot}{Br}} \!-\! \overset{+}{C}F_2 + \cdot \ddot{\underset{\cdot\cdot}{F}}\!:$$

$$\underset{m/z\ 148}{} \qquad\qquad\quad \underset{m/z\ 129}{} \qquad\qquad\qquad \underset{m/z\ 150}{} \qquad\qquad\quad \underset{m/z\ 131}{}$$

12.17   The formation of the molecular ion for isobutylamine and the pathway for its fragmentation to the base peak are shown below:

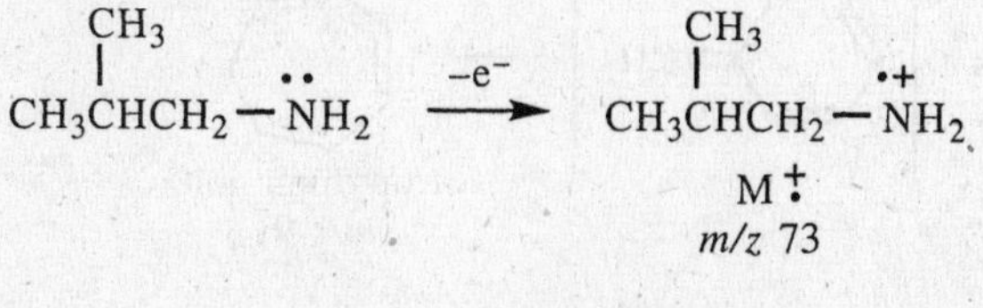

$$\begin{array}{c} CH_3 \\ | \\ CH_3CHCH_2 \!-\! \ddot{N}H_2 \end{array} \overset{-e^-}{\longrightarrow} \begin{array}{c} CH_3 \\ | \\ CH_3CHCH_2 \!-\! \overset{\cdot\,+}{N}H_2 \end{array}$$

$$\underset{m/z\ 73}{M^{+}}$$

12.17 (cont)

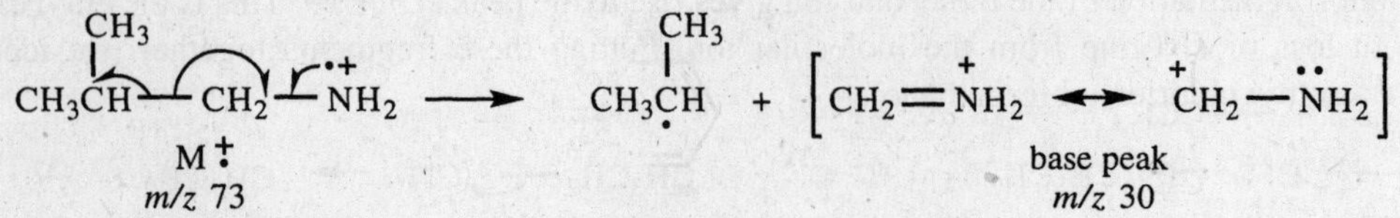

$$CH_3\overset{CH_3}{\underset{|}{CH}}-CH_2 \overset{\bullet+}{NH_2} \longrightarrow CH_3\overset{CH_3}{\underset{\bullet}{CH}} + \left[CH_2=\overset{+}{NH_2} \longleftrightarrow \overset{+}{CH_2}-\overset{\bullet\bullet}{NH_2}\right]$$

$$\underset{m/z\ 73}{M\overset{+}{\bullet}} \qquad\qquad \underset{\substack{\text{base peak}\\ m/z\ 30}}{}$$

The corresponding reaction for *sec*-butylamine is:

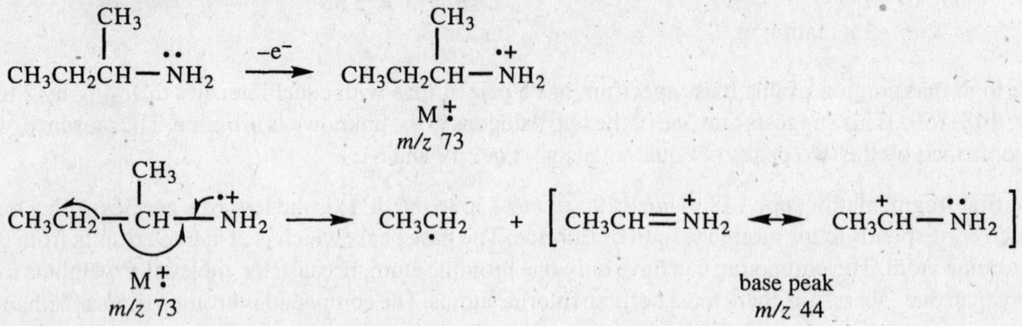

$$CH_3CH_2\overset{CH_3}{\underset{|}{CH}}-\overset{\bullet\bullet}{NH_2} \xrightarrow{-e^-} CH_3CH_2\overset{CH_3}{\underset{|}{CH}}-\overset{\bullet+}{NH_2}$$

$$\underset{m/z\ 73}{M\overset{+}{\bullet}}$$

$$CH_3CH_2-\overset{CH_3}{\underset{|}{CH}}\overset{\bullet+}{NH_2} \longrightarrow CH_3\overset{\bullet}{CH_2} + \left[CH_3CH=\overset{+}{NH_2} \longleftrightarrow CH_3\overset{+}{CH}-\overset{\bullet\bullet}{NH_2}\right]$$

$$\underset{m/z\ 73}{M\overset{+}{\bullet}} \qquad\qquad \underset{\substack{\text{base peak}\\ m/z\ 44}}{}$$

Similarly, for tert-butylamine:

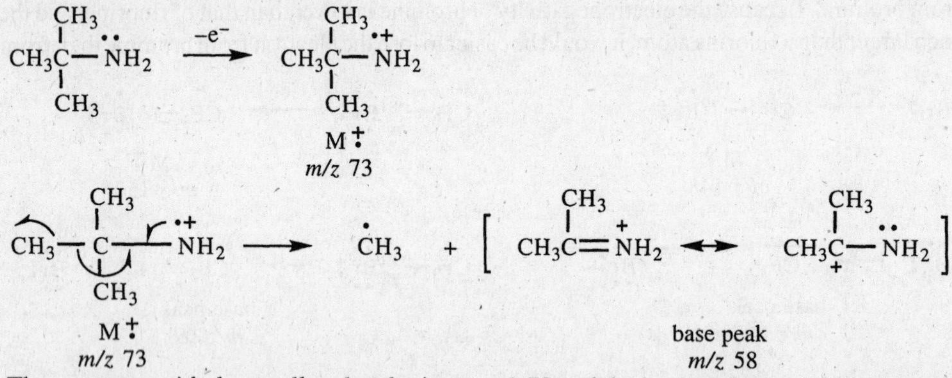

$$CH_3\overset{CH_3}{\underset{\underset{CH_3}{|}}{\overset{|}{C}}}-\overset{\bullet\bullet}{NH_2} \xrightarrow{-e^-} CH_3\overset{CH_3}{\underset{\underset{CH_3}{|}}{\overset{|}{C}}}-\overset{\bullet+}{NH_2}$$

$$\underset{m/z\ 73}{M\overset{+}{\bullet}}$$

$$CH_3-\overset{CH_3}{\underset{\underset{CH_3}{|}}{\overset{|}{C}}}\overset{\bullet+}{NH_2} \longrightarrow \overset{\bullet}{CH_3} + \left[CH_3\overset{CH_3}{\underset{|}{C}}=\overset{+}{NH_2} \longleftrightarrow CH_3\overset{CH_3}{\underset{|}{\overset{+}{C}}}-\overset{\bullet\bullet}{NH_2}\right]$$

$$\underset{m/z\ 73}{M\overset{+}{\bullet}} \qquad\qquad \underset{\substack{\text{base peak}\\ m/z\ 58}}{}$$

The spectrum with the small molecular ion at *m/z* 73 and the base peak at *m/z* 30 is that of isobutylamine. The spectrum with the barely visible molecular ion at *m/z* 73 and the base peak at *m/z* 44 is that of *sec*-butylamine. The spectrum of *tert*-butylamine has a base peak of *m/z* 58. No molecular ion is seen in the spectrum of *tert*-butylamine.

12.18 $$CH_3CH_2CH_2CH_2CH=CH_2 \xrightarrow{-e^-} CH_3CH_2CH_2CH_2CH\overset{\bullet+}{-}CH_2$$

$$\underset{\text{1-hexene}}{} \qquad\qquad \underset{m/z\ 84}{M\overset{+}{\bullet}}$$

$$CH_3CH_2CH_2-CH_2-CH\overset{\bullet+}{-}CH_2 \longrightarrow CH_3CH_2\overset{\bullet}{CH_2} + CH_2=CH\overset{+}{CH_2}$$

$$\underset{m/z\ 84}{M\overset{+}{\bullet}} \qquad\qquad \underset{\substack{\text{allyl cation}\\ \text{base peak}\\ m/z\ 41}}{}$$

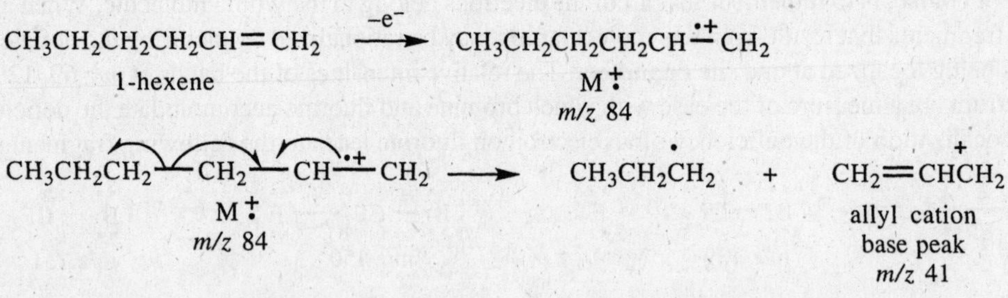

12.19 The expected fragmentations of the molecular ion from 2,2-dimethyl-1-phenylpropane are shown below:

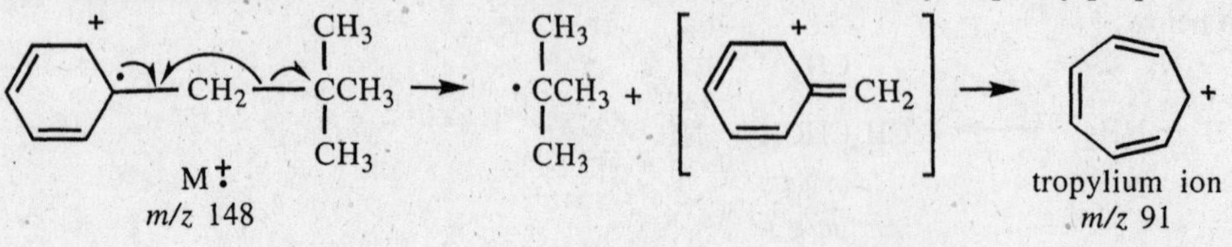

$$\underset{\underset{m/z\ 148}{M\overset{+}{\bullet}}}{\text{Ph}^+}-CH_2-\overset{CH_3}{\underset{\underset{CH_3}{|}}{\overset{|}{C}}}CH_3 \longrightarrow \overset{CH_3}{\underset{\underset{CH_3}{|}}{\overset{|}{\bullet C}}}CH_3 + \left[\text{Ph}=CH_2\right]^+ \longrightarrow \underset{\substack{\text{tropylium ion}\\ m/z\ 91}}{\text{tropylium}^+}$$

12.19 (cont)

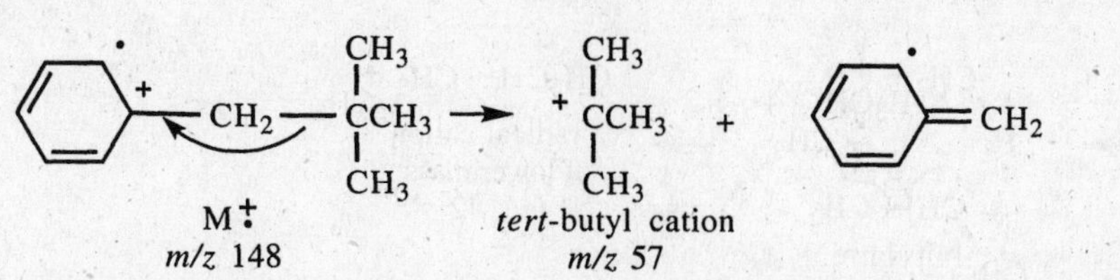

The expected fragmentation of the molecular ion from 2-methyl-3-phenylbutane is shown below:

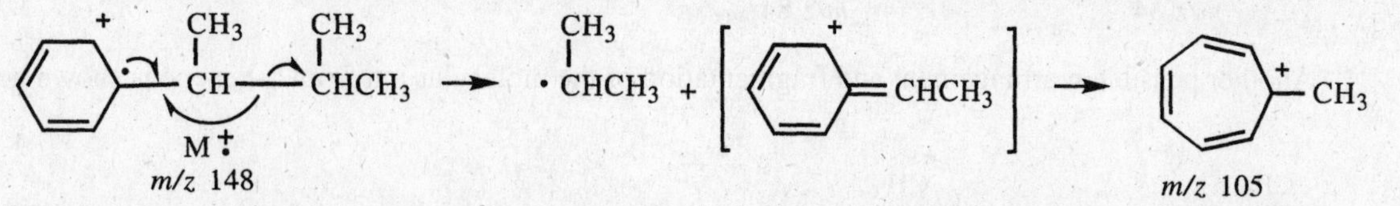

The spectrum with a peak at *m/z* 91 and the base peak at *m/z* 57 is that of 2,2-dimethyl-1-phenylpropane. The spectrum with the base peak at *m/z* 105 is that of 2-methyl-3-phenylbutane.

12.20

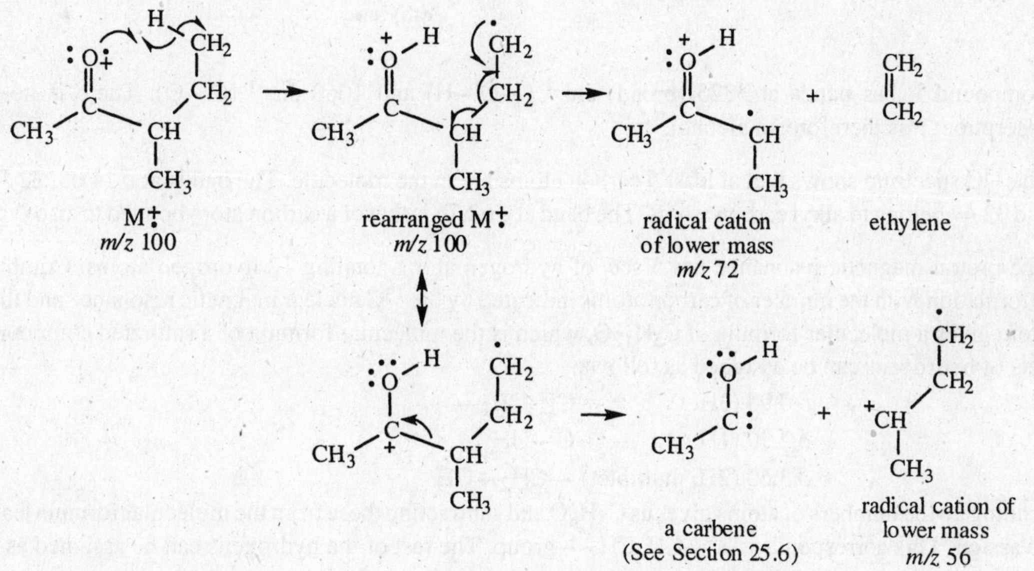

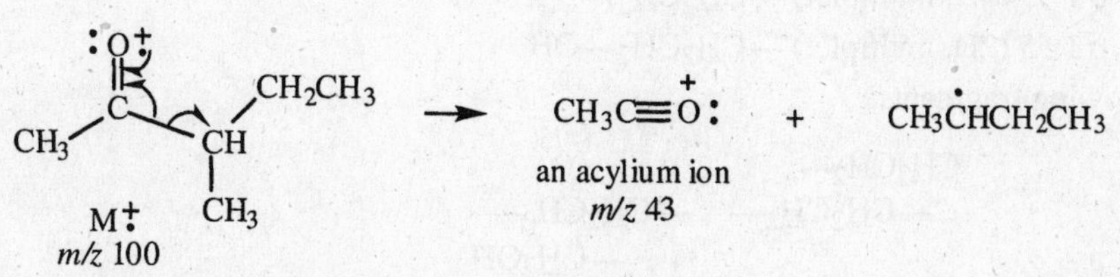

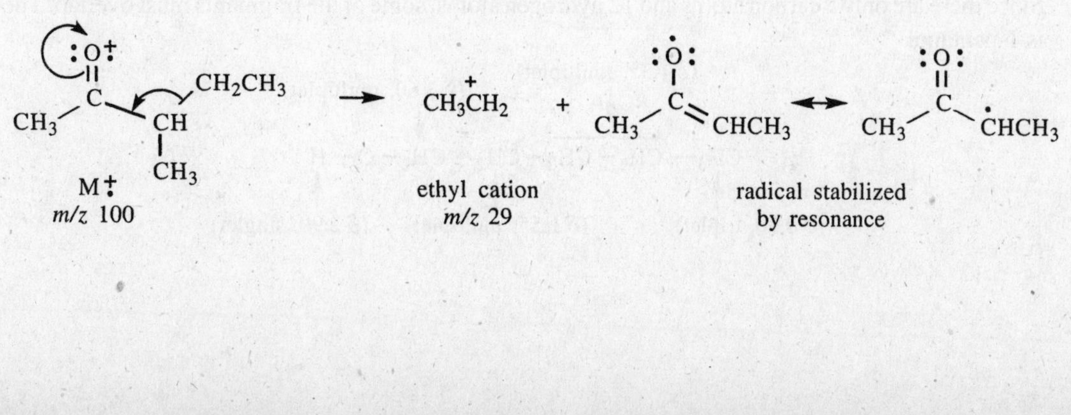

12.21 One possible rearrangement and fragmentation of the molecular ion from 1-hexene is shown below:

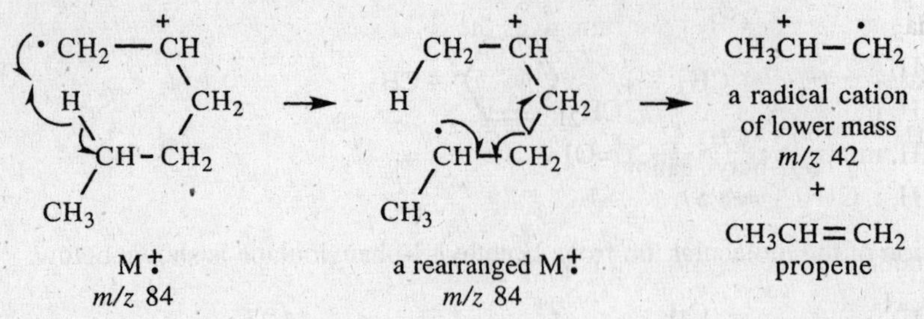

Another possible rearrangement and fragmentation of the molecular ion from 1-hexene is shown below:

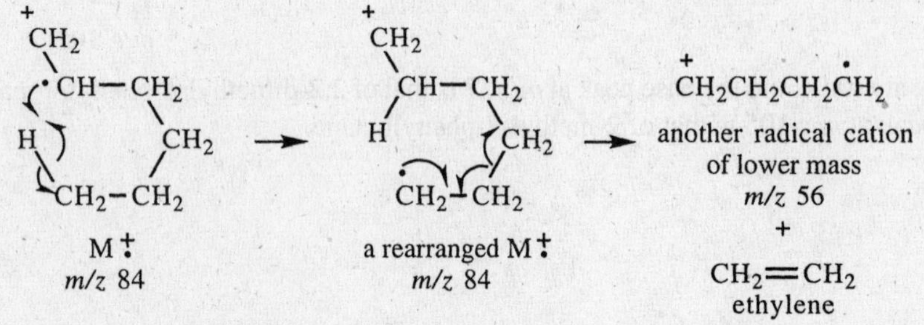

12.22 Compound L has bands at 3325 (broad) cm^{-1} (—O—H) and 1060 cm^{-1} (C—O). There is no carbonyl absorption; it is therefore an alcohol.

The ^{13}C spectrum shows that at least 5 carbon atoms are in the molecule. The bands at δ 14.05, 22.56, 28.02, and 32.47 belong to alkyl carbon atoms. The band at δ 62.76 is that of a carbon atom bonded to an oxygen atom.

The proton magnetic resonance has 5 sets of hydrogen atoms totaling 12 hydrogen atoms. Combining this information with the number of carbon atoms indicated by the ^{13}C nuclear magnetic resonance and the oxygen atom gives a molecular formula of $C_5H_{12}O$, which is the molecular formula of a saturated compound. Three sets of hydrogens can be assigned as follows:

δ 0.90 (3H, t)   C$\underline{H}_3$CH$_2$—

δ 2.90 (1H, s)   C—O$\underline{H}$

δ 3.60 (2H, multiplet) —C$\underline{H}_2$—OH

Adding up the numbers of atoms gives us C_3H_8O and subtracting these from the molecular formula leaves C_2H_4 to assign. This corresponds to a —CH$_2$CH$_2$— group. The rest of the hydrogens can be assigned as follows:

δ 1.35 (4H, multiplet) —C$\underline{H}_2$C$\underline{H}_2$—

δ 1.55 (2H, multiplet) —C$\underline{H}_2$CH$_2$—OH

We now have the following fragments:

CH$_3$CH$_2$—

—C$\underline{H}_2$C$\underline{H}_2$—   —C$\underline{H}_2$CH$_2$—

—C$\underline{H}_2$OH

Since there are only 5 carbon atoms and 12 hydrogen atoms, some of the fragments must overlap. The molecule is 1-pentanol.

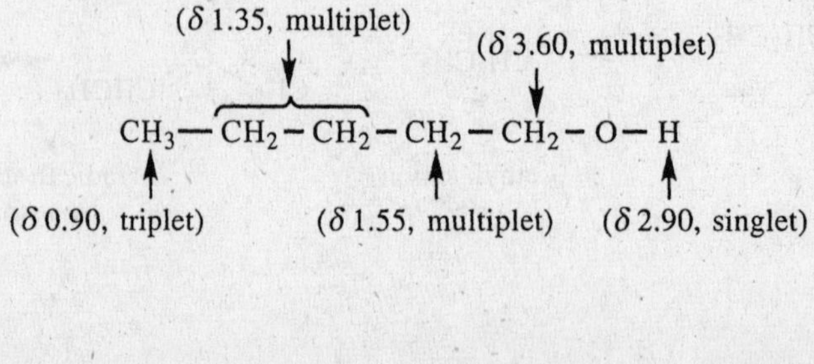

12.23   Compound M has bands at 2720 cm^{-1} (O=C—H) and 1730 cm^{-1} (C=O). It is therefore an aldehyde.

Analysis of the proton magnetic resonance spectrum gives the following:

δ 0.95   [6H, d, (C$\underline{H}_3$)$_2$CH]
δ 2.20   [1H, multiplet, (CH$_3$)$_2$C$\underline{H}$CH$_2$]
δ 2.30   [2H, multiplet, CHC$\underline{H}_2$CH=O]
δ 9.75   [1H, t, CH$_2$C$\underline{H}$=O]

Analysis of the ^{13}C magnetic resonance spectrum is given below:

δ 22.57   [($\underline{C}$H$_3$)$_2$CH]
δ 23.50   [(CH$_3$)$_2$$\underline{C}HCH_2$]
δ 52.59   [$\underline{C}$H$_2$CH=O]
δ 202.65   [$\underline{C}$H=O]

Compound M is 3-methylbutanal, (CH$_3$)$_2$CHCH$_2\overset{\displaystyle O}{\overset{\|}{C}}$H .

12.24

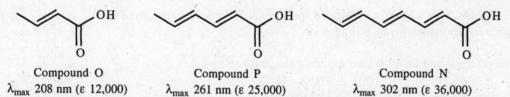

Compound O   Compound P   Compound N
λ_{max} 208 nm (ϵ 12,000) λ_{max} 261 nm (ϵ 25,000) λ_{max} 302 nm (ϵ 36,000)

The wavelength of maximum absorption and the intensity of absorption increase with increasing conjugation.

12.25   (a) A compound with a molecular formula of C$_4$H$_6$O has two units of unsaturation.

(b) The observed λ_{max} 219 nm (ϵ 16,600) is due to the $\pi \rightarrow \pi^*$ transition of a conjugated system, while λ_{max} 318 nm (ϵ 30) is due to the $n \rightarrow \pi^*$ transition of a carbonyl group. Only three structures fit both the molecular formula and the ultraviolet spectrum.

(c) Some other structures that are not compatible with the ultraviolet spectral data are:

12.26   (+)-17-Methyltestosterone

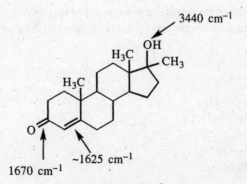

12.27   The absence of bands above 3000 cm^{-1} in the infrared spectrum eliminates both a phenol and an alcohol. The absence of bands in the region between 2000 and 1600 cm^{-1} eliminates a carbonyl compound such as an aldehyde or ketone. The ultraviolet spectrum is very similar to that of methoxybenzene (Figure 12.8), suggesting that Compound Q is a substituted aromatic ether. Possible structures are:

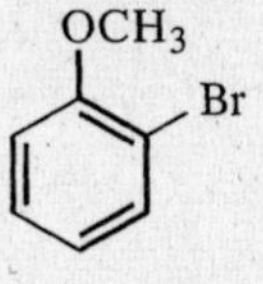

| *o*-bromomethoxy-benzene | *m*-bromomethoxy-benzene | *p*-bromomethoxy-benzene |

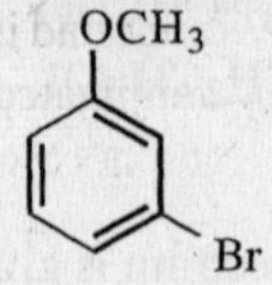

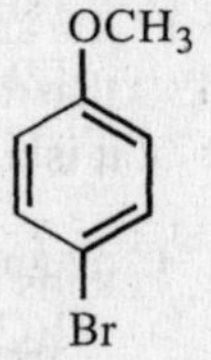

12.28   The mass spectrum of Compound R has two molecular ions at *m/z* 60 and 62 having a ratio of intensities of 3:1. This ratio tells us that the halogen in Compound R is chlorine. If we subtract 35 from 60, we get a mass of 25, which corresponds to the molecular formula C_2H, and the structure $HC\equiv C$—. Compound R is chloroethyne, $HC\equiv CCl$.

12.29   Compound S has a band in its infrared spectrum at 1685 cm^{-1} (conjugated C=O). According to the infrared spectral information given, S is a ketone.

The proton magnetic resonance spectrum has peaks at

δ 1.20   (6H, d, (C$\underline{H}_3$)$_2$CH—)

δ 3.57   (1H, septet, (CH$_3$)$_2$C$\underline{H}$—)

δ 7.50   (3H, multiplet, Ar$\underline{H}$)

δ 7.95   (2H, multiplet, Ar$\underline{H}$)

The chemical shift of the methine hydrogen atom and the fact that the aryl hydrogen atoms are split indicate that the carbonyl group is between the aryl and the isopropyl groups. This is confirmed by the presence of a band for a conjugated carbonyl group in the infrared. Compound S is 2-methyl-1-phenyl-1-propanone

(isobutyrophenone),

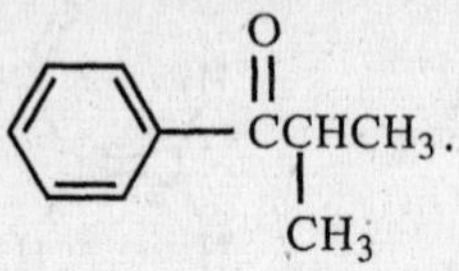

12.30   The two peaks of equal size in the mass spectrum of Compound T at *m/z* 79 and 81 tell us that bromine is present. The fact that the molecular ion fragments before it can be recorded suggests that the compound gives rise to a very stable cation, *m/z* 43. An acylium ion, $CH_3C\equiv O^+$, fits this description. Compound T is acetyl bromide.

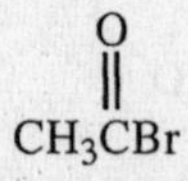

(Other possibilities that may come to mind are isopropyl or *n*-propyl bromide. Neither alkyl cation is as stable as the resonance-stabilized acylium ion, and we would therefore expect to see the molecular ion for either of these compounds.)

12.31   Compound U has bands at

3335 (broad) cm^{-1} (—O—H)

3080 cm^{-1} (C=C—H)

1650 cm^{-1} (C=C)

1048 cm^{-1} (C—O)

It is an alcohol and an alkene.

12.31 (cont)

Compound V has bands at

2200 cm^{-1} (C≡C)

1675 cm^{-1} (C=O)

No band is seen at 3300 cm^{-1}; therefore, the compound is not a terminal alkyne. The low value for the carbonyl absorption suggests that the carbonyl group is conjugated with the multiple bond. The absence of a band at 2700 cm^{-1} indicates that it is a conjugated ketone and not a conjugated aldehyde.

Compound W has bands at

3460 (broad) cm^{-1} (—O—H)

1710 cm^{-1} (C=O)

There is no band at 2700 cm^{-1}; therefore, the compound is a ketone and not an aldehyde. The other functional group in the compound is an alcohol.

Compound X has bands at

3100 cm^{-1} (C=C—H)

1740 cm^{-1} (C=O, not conjugated with alkene)

1640 cm^{-1} (C=C)

1240, 1040 cm^{-1} (C—O)

The bands at 1240 and 1040 cm^{-1}, along with the presence of a carbonyl, indicate an ester. The other functional group is an alkene.

12.32   The absorption bands reported for the first compound are comparable to those for an alkylbenzene (compare with the spectrum of toluene, Figure 12.8). This suggests that the double bond in this compound is not conjugated with the aromatic ring. The first compound is 3-phenylpropene. In the spectrum of the second compound, the absorption bands are shifted to longer wavelengths and are also more intense, suggesting that this compound has a more extensive system of conjugation than the first one. The second compound is, therefore, 1-phenylpropene.

$$\text{C}_6\text{H}_5\text{—CH}_2\text{CH}=\text{CH}_2 \qquad \text{C}_6\text{H}_5\text{—CH}=\text{CHCH}_3$$

3-phenylpropene   1-phenylpropene
no conjugation of the   double bond in con-
double bond with the ring   jugation with the ring

12.33

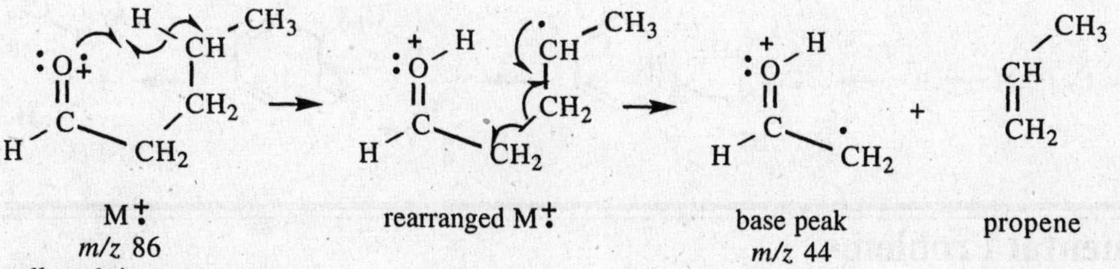

M⁺·   rearranged M⁺·   base peak   propene
m/z 86   *m/z* 44
small peak in spectrum

12.34   The infrared spectral information tells us that Compound Y is an aldehyde (1703 cm^{-1} is the C=O stretching frequency, and 2730 cm^{-1} distinguishes an aldehyde group from a ketone group).

$\delta\, 2.45$ (3H, s, ArC$\underline{\text{H}}_3$)

$\delta\, 7.32, 7.78$ (4H, para substitution pattern, Ar$\underline{\text{H}}$)

$\delta\, 9.95$ (1H, s, —C$\underline{\text{H}}$=O)

12.34 (cont)

These fragments add up to C_8H_8O corresponding to a molecular weight of 120, which is the *m/z* of the molecular ion of Compound Y. Putting the subunits from the proton magnetic resonance spectrum together gives us *p*-tolualdehyde. The first base peak, *m/z* 119, results from the loss of a hydrogen atom from the molecular ion, giving an acylium ion. The second base peak, *m/z* 91, is the tropylium ion, which most likely comes from a rearrangement of the methylphenyl cation formed by loss of carbon monoxide from the acylium ion.

an acylium ion
m/z 119

M$\overset{+}{\cdot}$
m/z 120

m/z 119

tropylium ion
m/z 91

The ^{13}C nuclear magnetic resonance is analyzed below:

134.4 ppm
singlet

145.3 ppm
singlet

21.6 ppm
split into
quartet by three
hydrogen atoms

129.6 ppm
each split into
doublet by one
hydrogen atom

191.4 ppm
split into
doublet by one
hydrogen atom

12.35 (a) $C_{60} = 60 \times 12$ amu $= 720$ amu

1656 amu $- 720$ amu $= 936$ amu

MW of benzene $= 78$ amu

936 amu/78 amu $= 12$ benzene units,

therefore MW 1656 corresponds to $C_{60}(C_6H_6)_{12}$.

(b) The proton nuclear magnetic resonance spectrum shows that the hydrogen atoms on benzene have been converted into two kinds of hydrogens. Five of them remain as aromatic hydrogens; one has been converted into another type of hydrogen, perhaps more typical of alkene hydrogens.

(c)

Supplemental Problems

S12.1   Of the following three isomeric compounds, which one will absorb at the longest wavelength?

S12.2   Compound Z is an ester. It has a singlet in its proton magnetic resonance spectrum at δ 3.69. Identify the molecular ion from its mass spectrum, shown in Figure S12.1. Assign a structure to that ion and to the ions having an intensity greater than 50%.

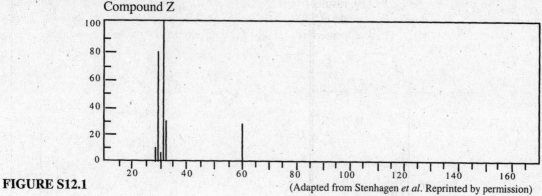

Compound Z

FIGURE S12.1

(Adapted from Stenhagen *et al*. Reprinted by permission)

S12.3   The mass spectra of two isomeric alcohols, compounds AA and BB, are given in Figure S12.2. The molecular ion shows up in the spectrum of one but not the other. Assign structures to the two alcohols, and write equations for the formation of the ion responsible for the base peak in each case.

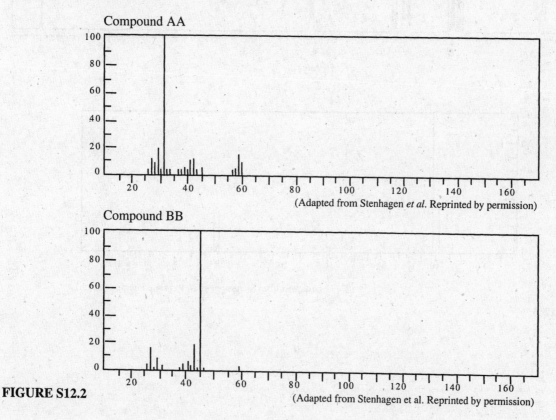

Compound AA

(Adapted from Stenhagen *et al*. Reprinted by permission)

Compound BB

FIGURE S12.2

(Adapted from Stenhagen et al. Reprinted by permission)

S12.4   Compound CC has the proton magnetic resonance and mass spectra shown in Figure S12.3. There is a peak for the molecular ion in the mass spectrum. The ^{13}C nuclear magnetic resonance spectrum of the compound has peaks at δ 33, 120, and 135. Tell which spectral peaks belong to each carbon and each hydrogen atom in the compound. Write an equation for the formation of the ion responsible for the base peak in the mass spectrum.

S12.4 (cont)
Compound CC

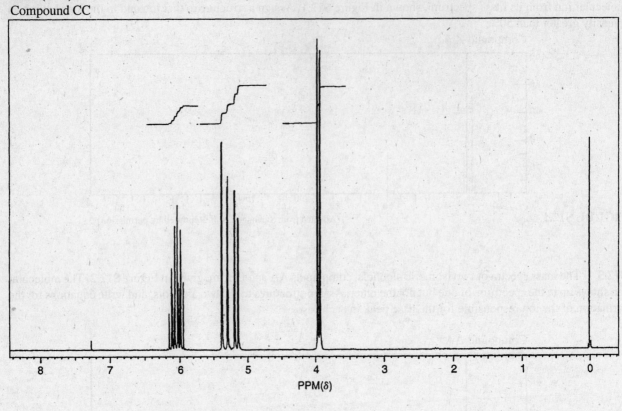

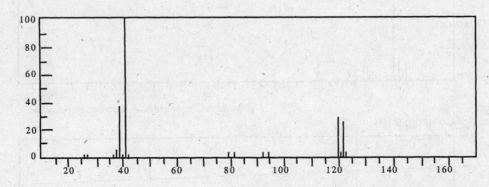

(Adapted from Stenhagen *et al.* Reprinted by permission)

FIGURE S12.3

13 Alcohols, Diols, and Ethers

13.1   1,2-Ethanediol (ethylene glycol) is a very polar compound and hydrogen bonds strongly to water. It is infinitely soluble in water. A solution of ethylene glycol in water depresses the freezing point of water so that the mixture does not freeze at temperatures usually seen in the winter time. This keeps pipes and radiators from bursting because of the expansion of the water as it forms ice. The high boiling point of ethylene glycol ensures that it will not boil out of the radiator of a car when the engine is hot.

13.2

Even though it has a hydroxyl group, cholesterol is not very soluble in water. The rest of the molecule is nonpolar and too large to be brought into solution by the small polar hydroxyl group (Section 1.10C).

13.3   Cholesterol is an alkene as well as an alcohol.

13.4 (a) The location of the hydroxyl group on the chain affects the solubility in water of the isomeric alcohols. *n*-Butyl alcohol has the polar hydroxyl group at one end of the molecule and a long hydrophobic chain stretching away from it. In *sec*-butyl alcohol all parts of the hydrocarbon portion of the molecule are closer to, and more under the influence of, the polar hydroxyl group. In *tert*-butyl alcohol there is no large hydrophobic chain that is far away from the polar, water-soluble portion of the molecule. The hydroxyl group is able to carry the rest of the molecule into solution with it.

(b) Hydrogen bonding of the alcohol is the most important factor in determining water solubility. The bromo compound is less soluble than the slightly more polar chloro compound. The dipole is weaker for the bromo compound, so even those interactions are not helpful for the bromo compound.

(c) Hydrogen bonding among molecules of the diol give it a very high boiling point. The alcohol also has hydrogen bonding, but with only one hydroxyl group to participate in the hydrogen bonding, it has a lower boiling point than the diol. The ether cannot have hydrogen bonding among its molecules, therefore it has the lowest boiling point of the three.

13.5 (a) 3-methyl-1-butanol (b) (*E*)-4-hexen-3-ol (c) 2-pentyn-1-ol
(d) 4,5-dichloro-1-hexanol (e) 1-chloro-3-ethoxypropane (f) diphenyl ether (phenoxybenzene)
(g) cyclohexanol (h) (1*R*, 2*R*)-2-methylcyclopentanol (i) *cis*-4-*tert*-butylcyclohexanol
(j) (*R*)-5-hexen-3-ol (k) 1,3-dimethoxypropane (l) 1,2-dimethoxybenzene

13.6 (a)

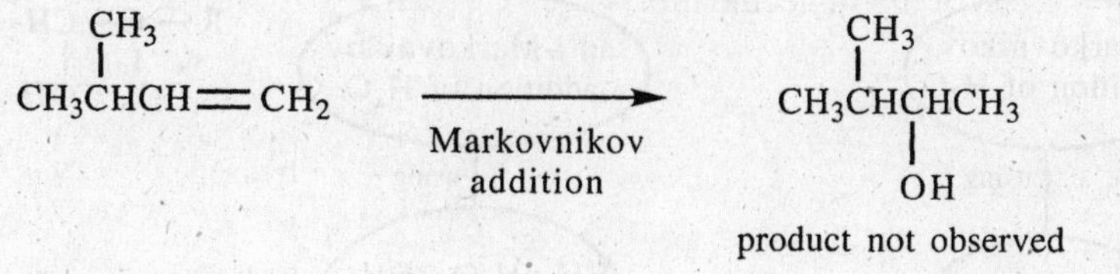

$$CH_3CHCH=CH_2 \xrightarrow[\text{addition}]{\text{Markovnikov}} CH_3CHCHCH_3$$

with CH_3 substituent on left carbon; product CH_3CHCHCH_3 with OH

product not observed

(b)

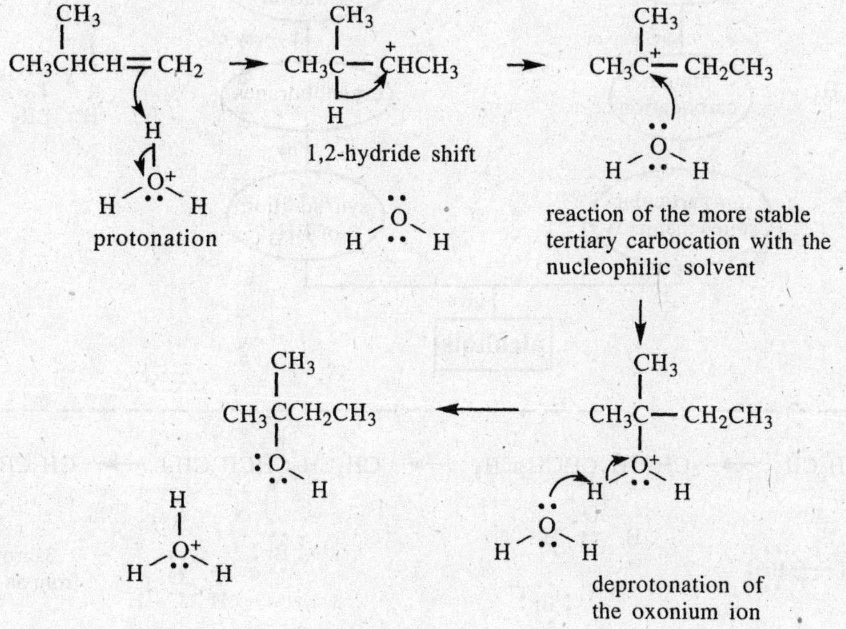

1,2-hydride shift

protonation

reaction of the more stable tertiary carbocation with the nucleophilic solvent

deprotonation of the oxonium ion

13.7

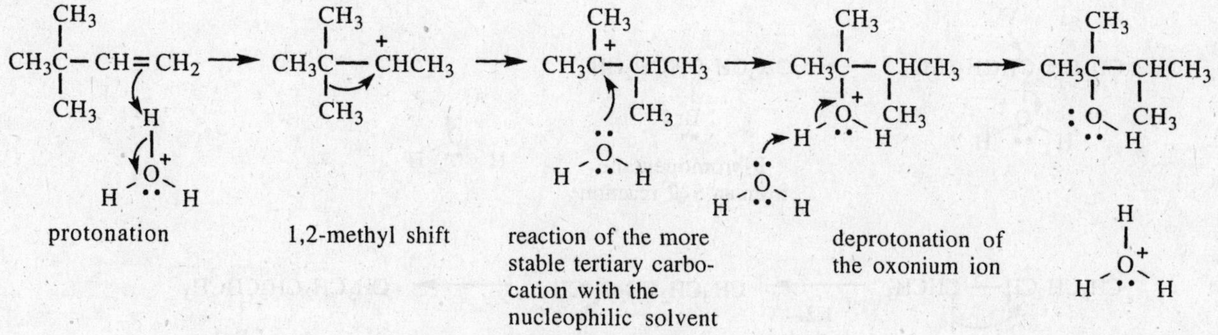

protonation   1,2-methyl shift   reaction of the more stable tertiary carbocation with the nucleophilic solvent   deprotonation of the oxonium ion

13.8

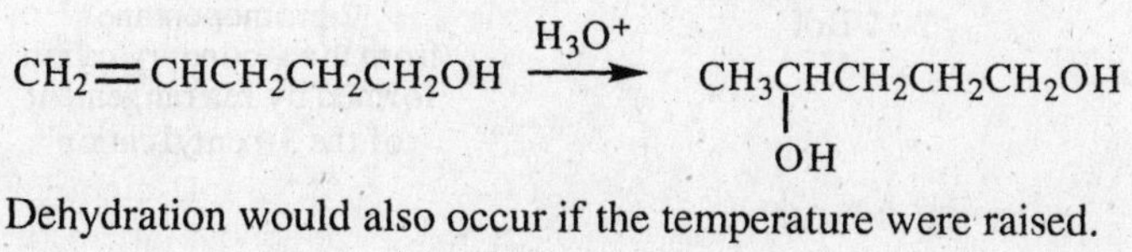

$$CH_2=CHCH_2CH_2CH_2OH \xrightarrow{H_3O^+} CH_3CHCH_2CH_2CH_2OH$$

with OH on second carbon

Dehydration would also occur if the temperature were raised.

Concept Map 13.1   Conversion of alkenes to alcohols.

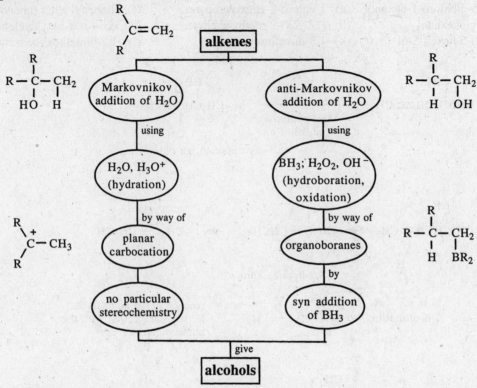

13.9

3-bromopentane
from an S_N1 reaction

3-bromopentane
from an S_N2 reaction

2-bromopentane
from the secondary cation
formed by rearrangement
of the 3-pentyl cation

1,2-
hydride
shift

13.10

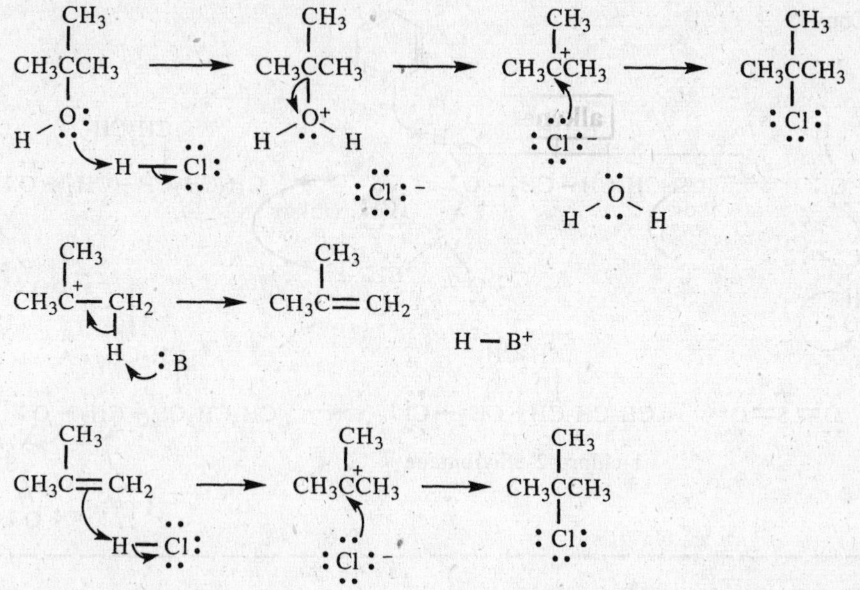

13.11 (1) With zinc chloride in concentrated hydrochloric acid, we have reaction conditions that allow for ionization of the alkyl halides that are formed as products and for equilibrium among different ionic species. As we saw in the *One Small Step*, p. 495 in the text, zinc chloride assists in the ionization of alkyl halides.

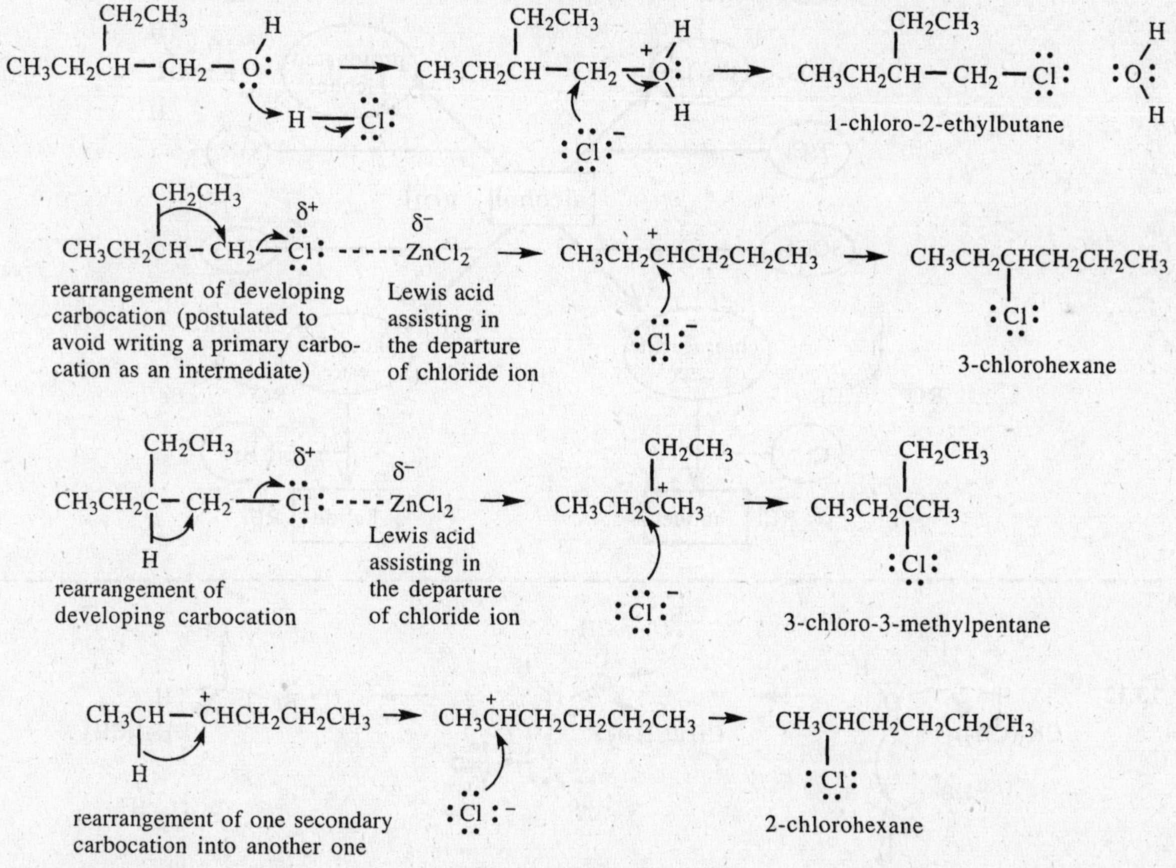

1-chloro-2-ethylbutane

rearrangement of developing carbocation (postulated to avoid writing a primary carbocation as an intermediate)

Lewis acid assisting in the departure of chloride ion

3-chlorohexane

rearrangement of developing carbocation

Lewis acid assisting in the departure of chloride ion

3-chloro-3-methylpentane

rearrangement of one secondary carbocation into another one

2-chlorohexane

(2) With thionyl chloride in pyridine, an S_N2 reaction takes place. The alcohol is converted into the corresponding halide without rearrangement.

13.11 (2) (cont)

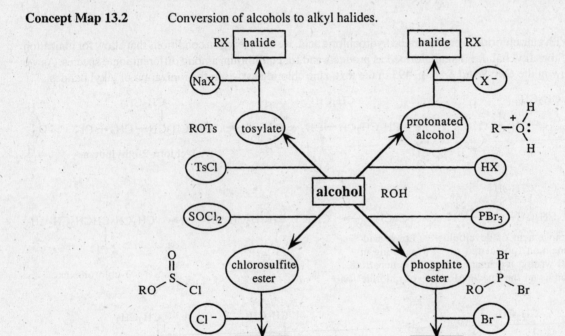

1-chloro-2-ethylbutane

───

Concept Map 13.2   Conversion of alcohols to alkyl halides.

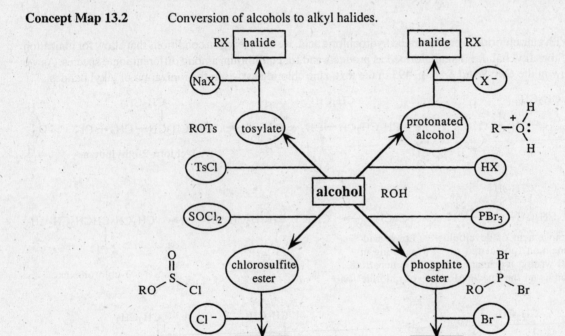

───

13.12

nucleophilic substitution
on phosphorus

intermediate with a good leaving
group undergoing a nucleophilic
reaction at the chiral carbon atom
with inversion of configuration

13.13

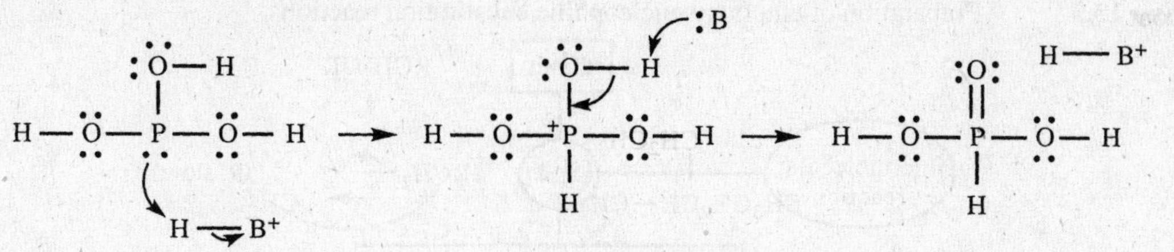

13.14

(a)

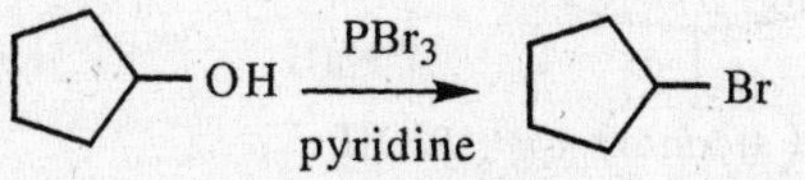

(b)

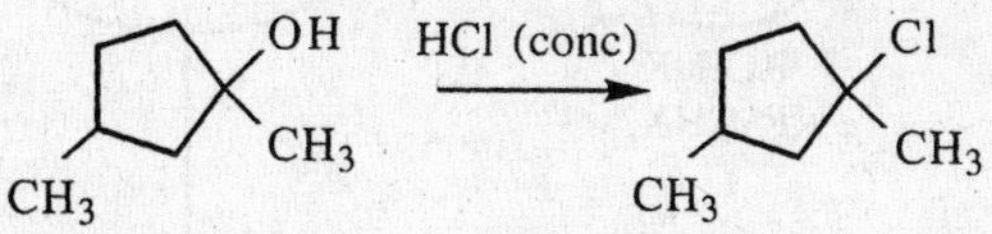

(c) $CH_3(CH_2)_9CH_2OH \xrightarrow[\Delta]{SOCl_2} CH_3(CH_2)_9CH_2Cl$

(d) $CH_3(CH_2)_8CH_2OH \xrightarrow[\Delta]{HBr} CH_3(CH_2)_8CH_2Br$

(e) $CH_3SCH_2CH_2OH \xrightarrow[chloroform]{SOCl_2} CH_3SCH_2CH_2Cl$

(f) $CH_3CH_2OCH_2CH_2OH \xrightarrow[pyridine]{PBr_3} CH_3CH_2OCH_2CH_2Br$

(g)

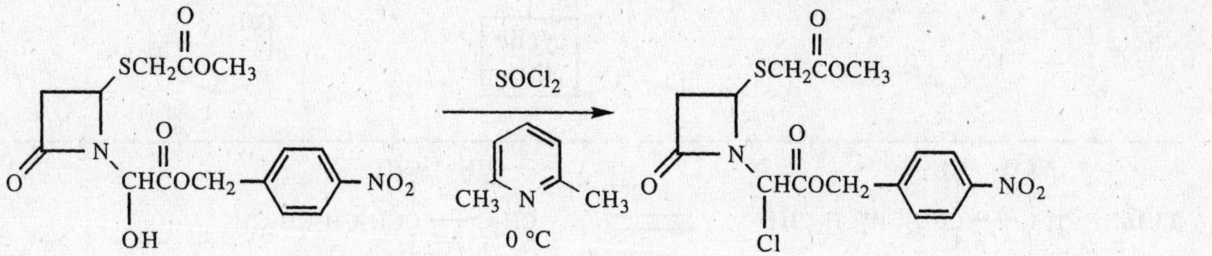

13.15

(a)

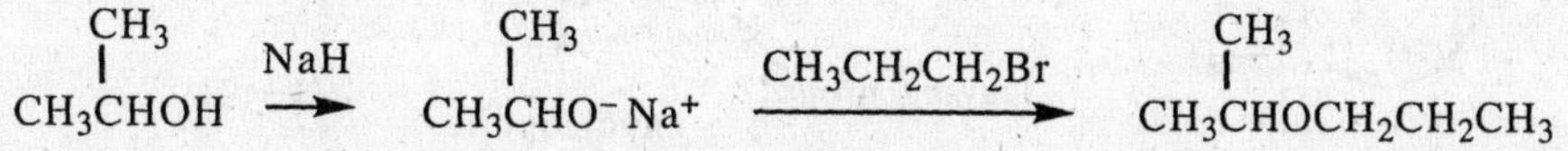

(b)

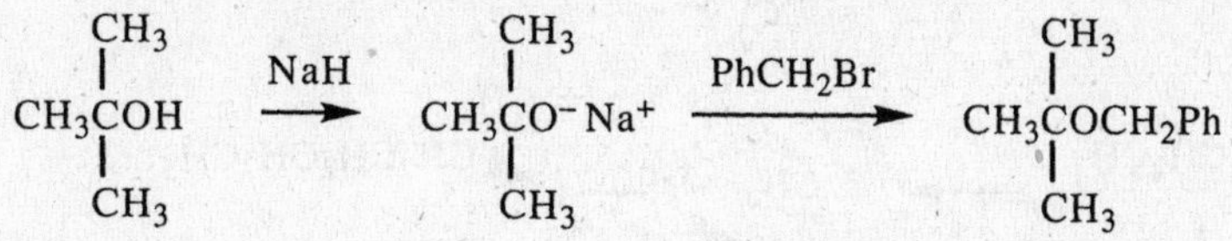

(c) $CH_3CH_2OH \xrightarrow{NaH} CH_3CH_2O^-Na^+ \xrightarrow{CH_2=CHCH_2Br} CH_3CH_2OCH_2CH=CH_2$

or $CH_2=CHCH_2OH \xrightarrow{NaH} CH_2=CHCH_2O^-Na^+ \xrightarrow{CH_3CH_2Br} CH_3CH_2OCH_2CH=CH_2$

(d) $CH_3CH_2CH_2CH_2OH \xrightarrow{NaH} CH_3CH_2CH_2CH_2O^-Na^+ \xrightarrow{CH_3I} CH_3CH_2CH_2CH_2OCH_3$

or $CH_3OH \xrightarrow{NaH} CH_3O^-Na^+ \xrightarrow{CH_3CH_2CH_2CH_2Br} CH_3CH_2CH_2CH_2OCH_3$

(e) $CH_3CH_2CH_2CH_2CH_2CH_2OH \xrightarrow{NaH} CH_3CH_2CH_2CH_2CH_2CH_2O^-Na^+ \xrightarrow{CH_3CH_2Br}$

$CH_3CH_2CH_2CH_2CH_2CH_2OCH_2CH_3$

or $CH_3CH_2OH \xrightarrow{NaH} CH_3CH_2O^-Na^+ \xrightarrow{CH_3CH_2CH_2CH_2CH_2CH_2Br}$

$CH_3CH_2CH_2CH_2CH_2CH_2OCH_2CH_3$

(f)

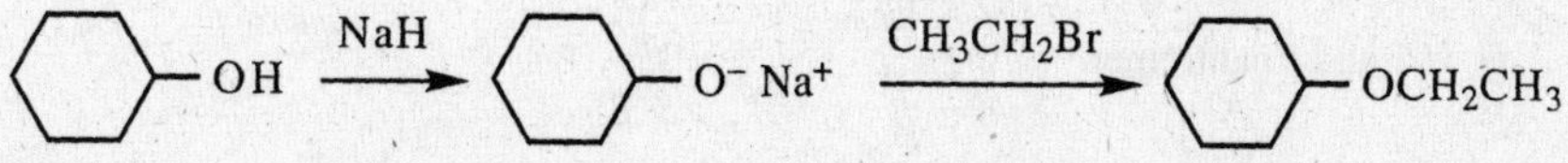

Concept Map 13.3   Preparation of ethers by nucleophilic substitution reactions.

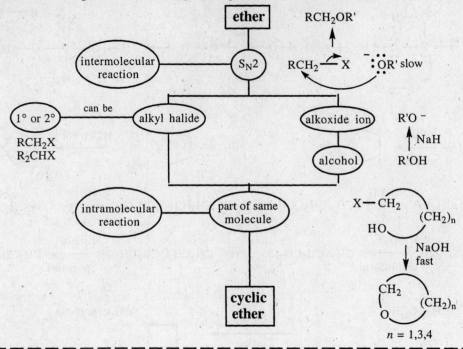

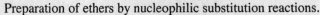

13.16

trans, anti orientation of the
nucleophile and the leaving group
that is beginning to ionize from
a tertiary carbon atom

transition state for an intramolecular substitution reaction

one of two possible enantiomers

13.17 (a)

13.17 (b)

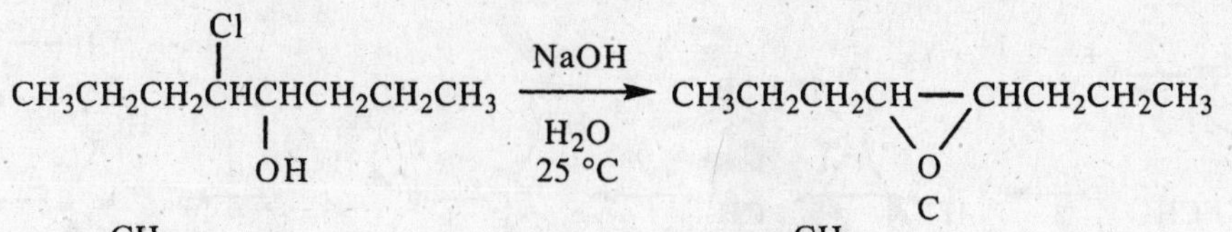
NaOH
H_2O
25 °C
C

(c)

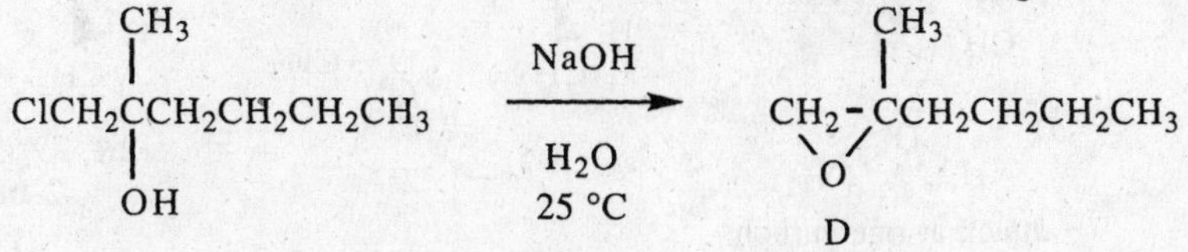
NaOH
H_2O
25 °C
D

(d)

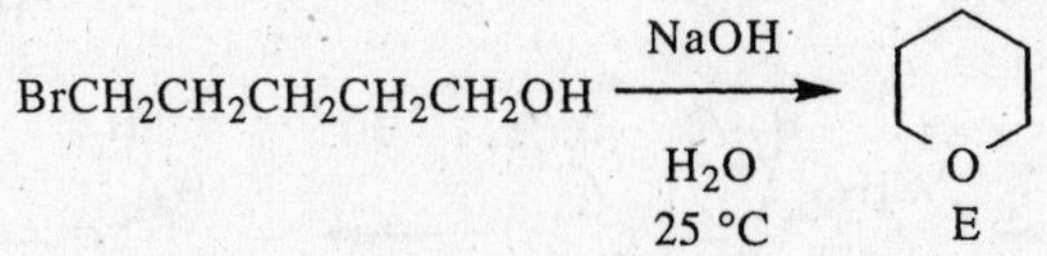
NaOH
H_2O
25 °C
E

(e)

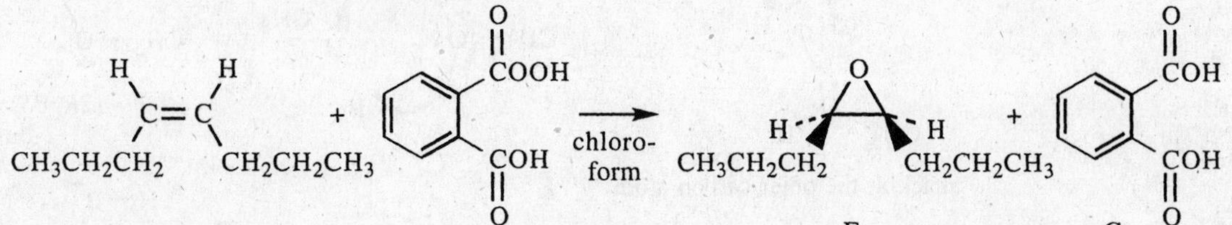
chloro-
form
F
G

13.18

(1)

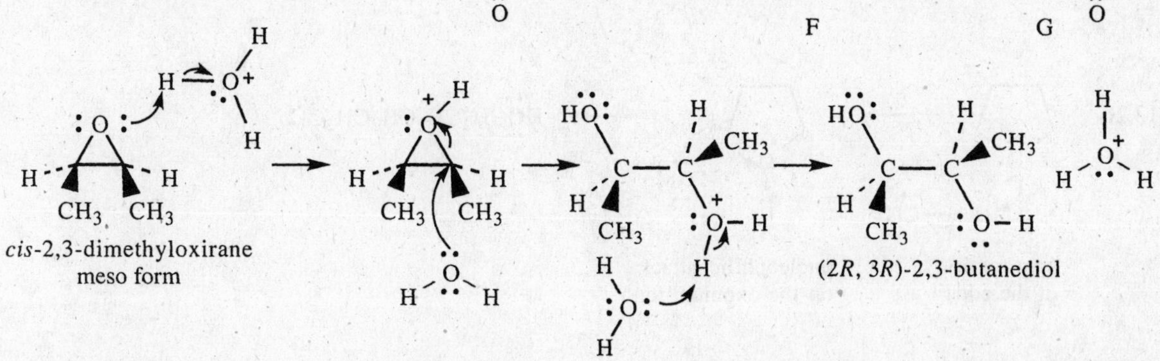
cis-2,3-dimethyloxirane
meso form
(2*R*, 3*R*)-2,3-butanediol

(2)

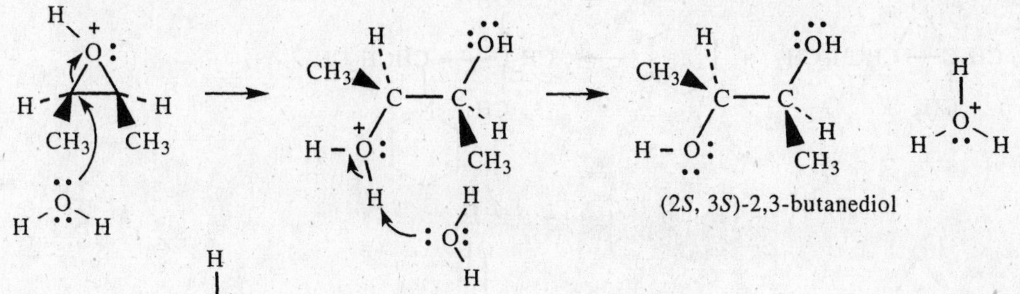
(2*S*, 3*S*)-2,3-butanediol

(3)

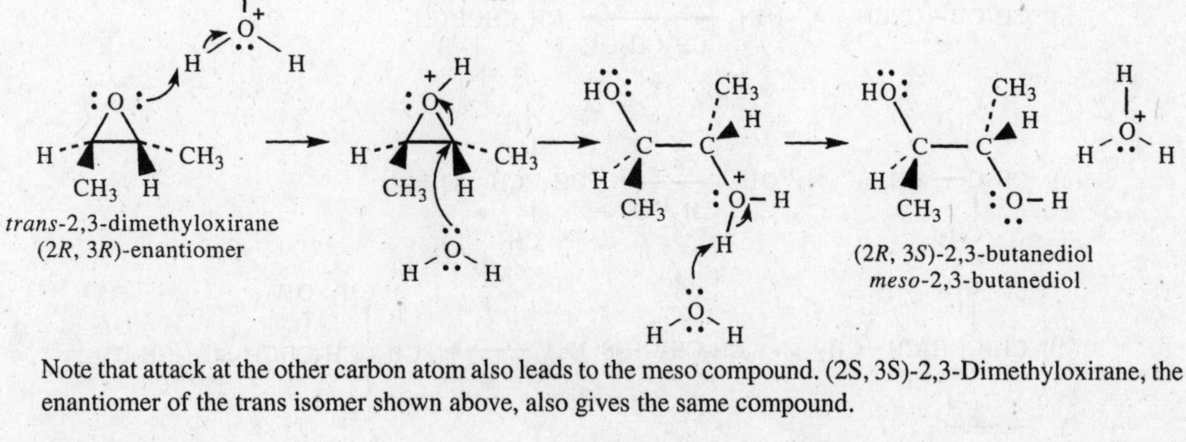
trans-2,3-dimethyloxirane
(2*R*, 3*R*)-enantiomer
(2*R*, 3*S*)-2,3-butanediol
meso-2,3-butanediol

Note that attack at the other carbon atom also leads to the meso compound. (2S, 3S)-2,3-Dimethyloxirane, the enantiomer of the trans isomer shown above, also gives the same compound.

13.19

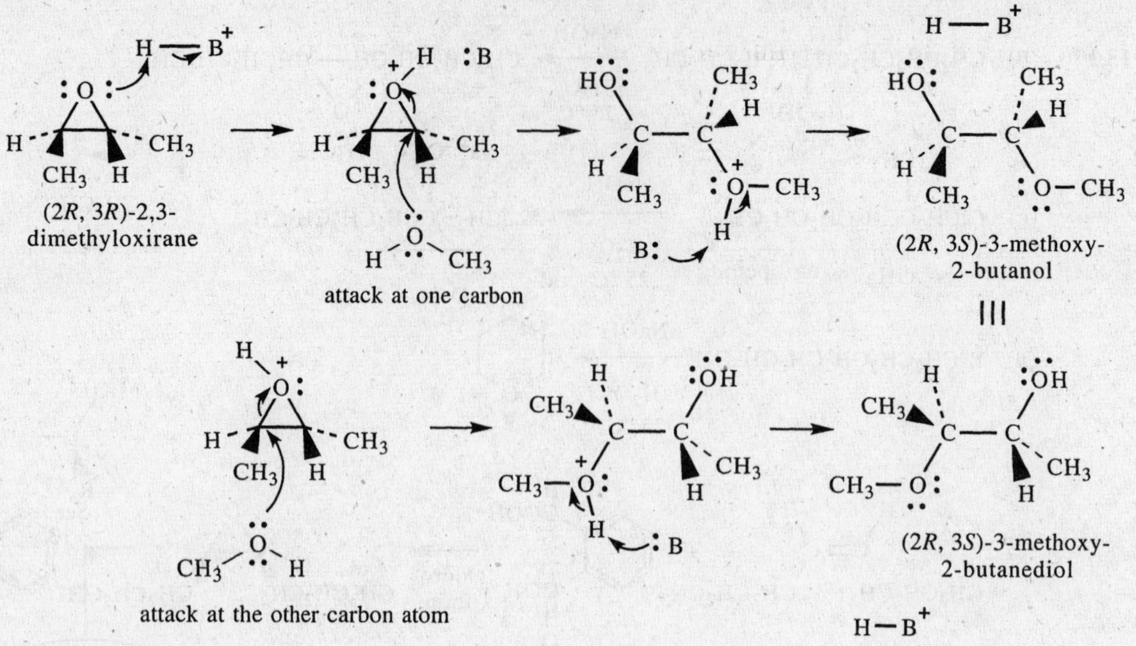

(2R, 3R)-2,3-
dimethyloxirane

attack at one carbon

(2R, 3S)-3-methoxy-
2-butanol

III

attack at the other carbon atom

(2R, 3S)-3-methoxy-
2-butanediol

H—B$^+$

13.20

HOCH$_2$CH$_2$CH$_2$CH$_2$Cl:

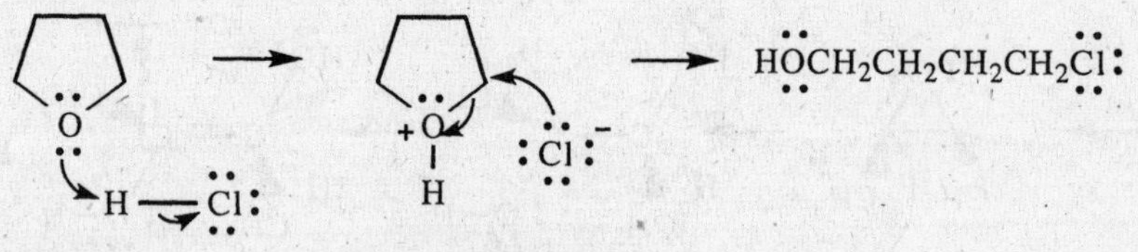

protonation
of the ether

nucleophilic attack
on the oxonium ion

13.21 (a)

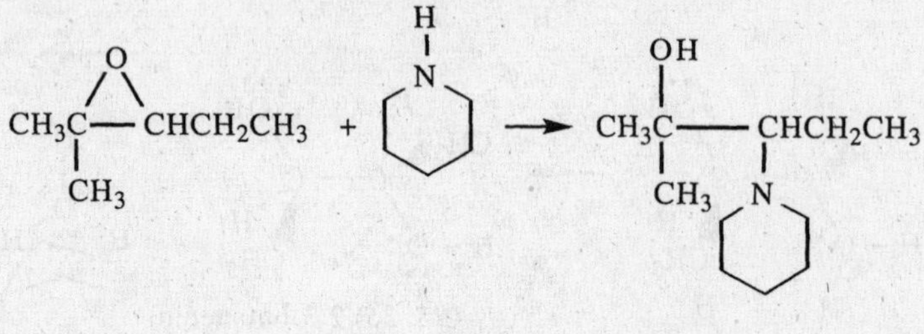

(b)

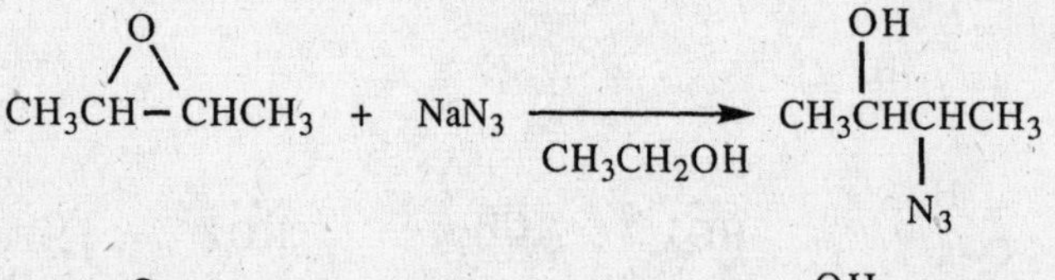

(c)

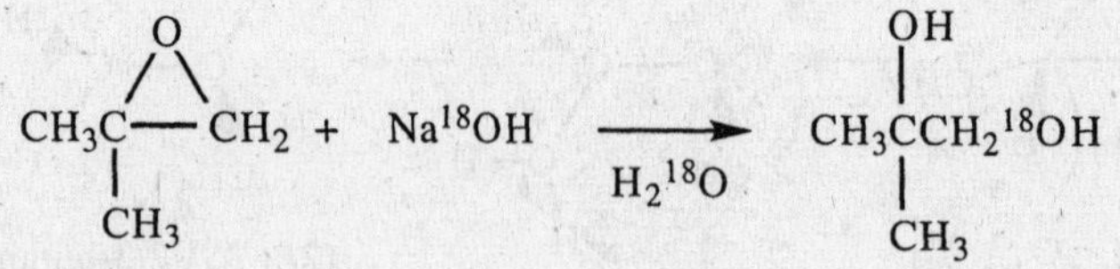

(d)

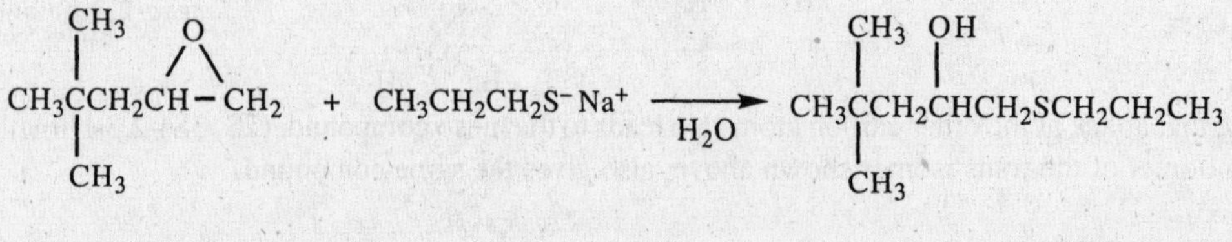

13.21 (e) CH_3CH-CH_2 + HBr $\longrightarrow$ CH_3CHCH_2Br + CH_3CHCH_2OH

(with epoxide O over CH–CH$_2$; OH over the middle carbon of first product; Br over the CH of second product)

(f) CH_2-CH_2 + $Na^+ {}^-:C \equiv CCH_2CH_3$ $\longrightarrow$ $Na^+ {}^-OCH_2CH_2C \equiv CCH_2CH_3$ $\xrightarrow{H_3O^+}$

$HOCH_2CH_2C \equiv CCH_2CH_3$

- -

Concept Map 13.4   Ring-opening reactions of oxiranes.

- -

13.22 CH_3SH + $CH_3CH_2O^- Na^+$ $\longrightarrow$ $CH_3S^- Na^+$ + CH_3CH_2OH

13.23   (a)

13.23 (b)

and enantiomer

1. What are the connectivities of the two compounds? How many carbon atoms does each contain? Are there any rings? What are the positions of branches and functional groups on the carbon skeletons?

   The carbon skeletons are the same. The alkene has two methyl groups cis to each other on the double bond. The methyl groups are also on the same side of the plane of the page that contains the hydroxyl groups, carbon-2, and carbon-3.

   The two hydroxyl groups are anti to each other.

2. How do the functional groups change in going from starting material to product? Does the starting material have a good leaving group?

   The double bond of the alkene has disappeared. Two hydroxyl groups have been attached to the carbon atoms that were part of the double bond.

3. Is it possible to dissect the structures of the starting material and product to see which bonds must be broken and which formed?

   bond broken   bonds formed

   A π bond was broken; two carbon-oxygen bonds were formed anti to each other. The anti orientation of the two hydroxyl groups suggests the opening of an oxirane ring by hydroxide ion or water.

4. New bonds are created when an electrophile reacts with a nucleophile. Do we recognize any part of the product molecule as coming from a good nucleophile or an electrophilic addition?

   An alkene can be converted into a compound containing one oxygen by oxidation with the electrophilic reagent, *m*-chloroperoxybenzoic acid. A hydroxyl group can come from water, a nucleophile.

5. What type of compound would be a good precursor to the product?

   An oxirane can be converted to the 1,2-diol.

6. After this last step, do we see how to get from starting material to product? If not, we need to analyze the structure obtained in step 5 by applying questions 4 and 5 to it.

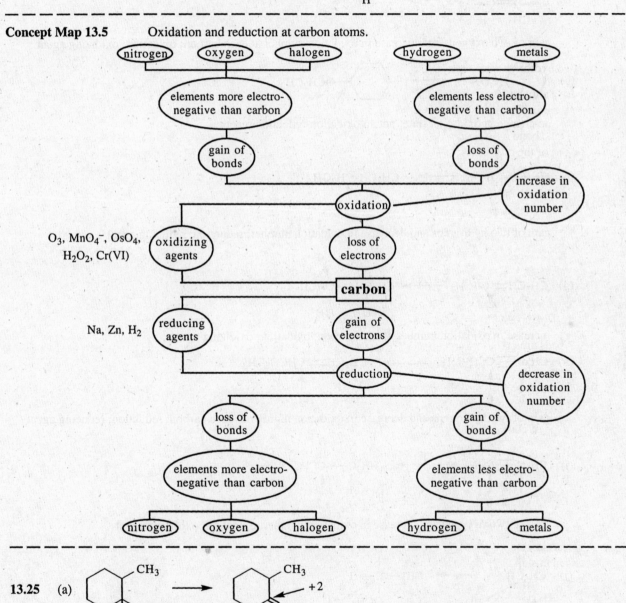

13.24 (1) ... (2)

Concept Map 13.5   Oxidation and reduction at carbon atoms.

13.25 (a)

loss of two hydrogens and increase in oxidation number; oxidation; oxidizing agent

13.25 (b)

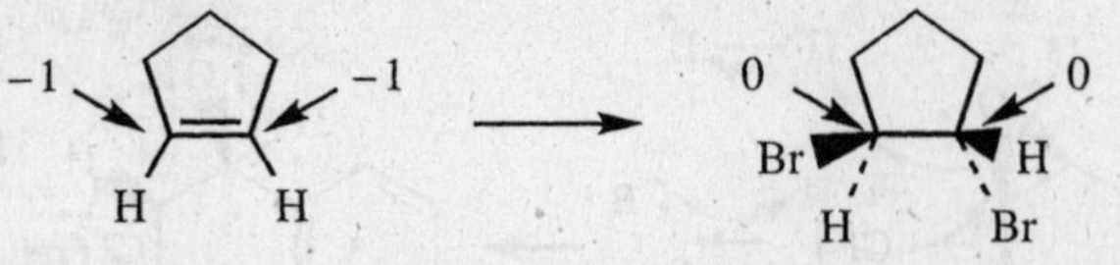

an increase in oxidation number at each carbon; oxidation; oxidizing agent

(c) $CH_3CH_2Br \longrightarrow CH_3CH_2NH_2$

-1 -1

no change in oxidation state; not an oxidation-reduction reaction

(d)

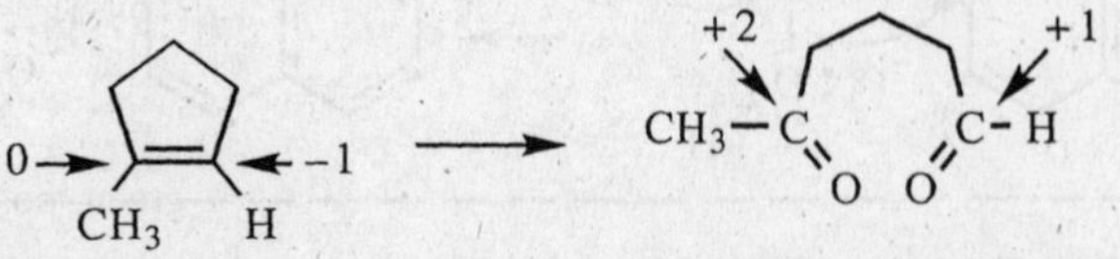

gain of two oxygens and increase in oxidation number at each carbon; oxidation; oxidizing agent

(e)

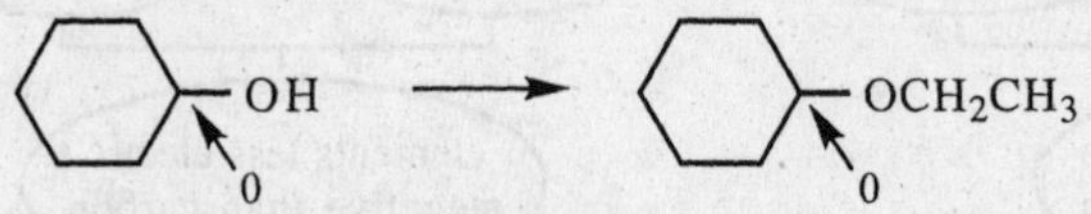

no change in oxidation state; not an oxidation-reduction reaction

(f)

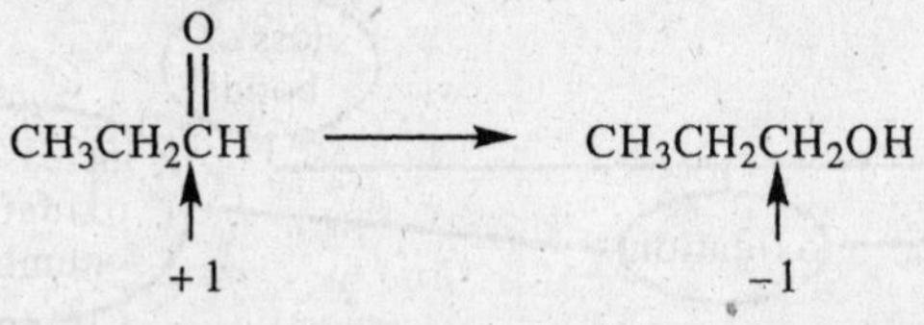

gain of two hydrogens and decrease in oxidation number; reduction; reducing agent

(g) $CH_3CH = CHCH_3 \longrightarrow$

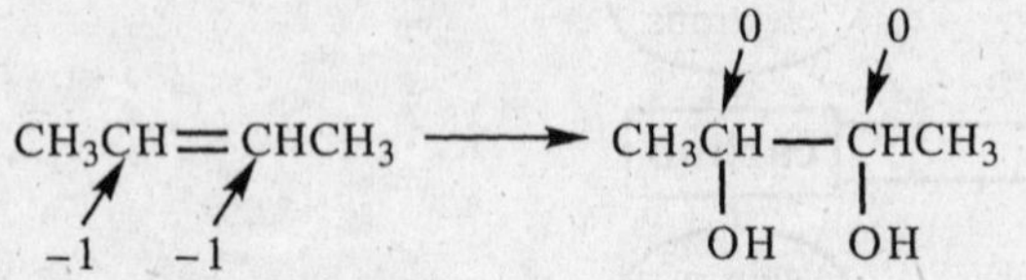

increase in oxidation number at each carbon; oxidation; oxidizing agent

(h) $CH_3C \equiv CCH_2CH_3 \longrightarrow CH_3CH = CHCH_2CH_3$

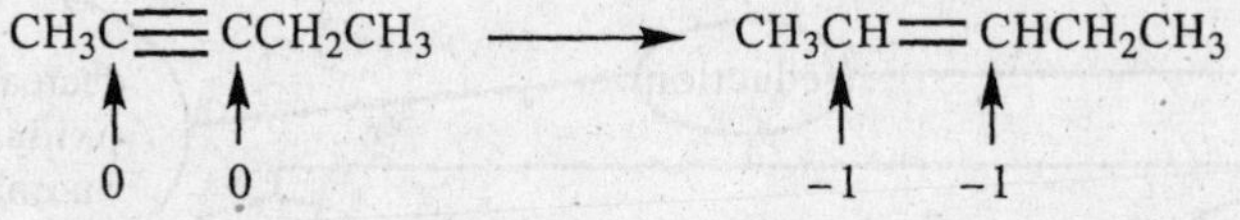

gain of two hydrogens and decrease in oxidation number at each carbon; reduction; reducing agent

(i) $CH_3C \equiv CCH_3 \longrightarrow$

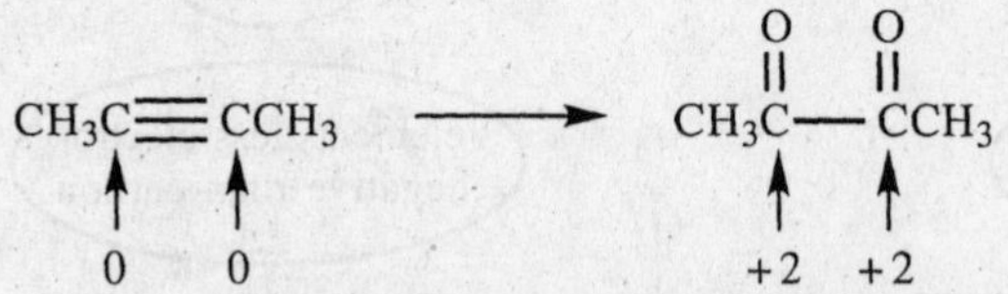

gain of two oxygens and increase in oxidation number; oxidation; oxidizing agent

(j)

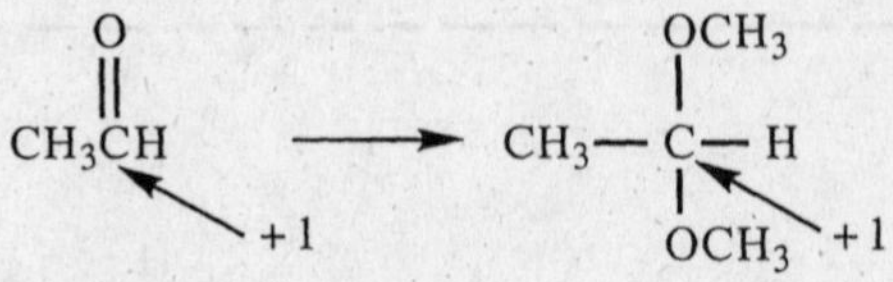

no change in oxidation state; not an oxidation-reduction reaction

13.26

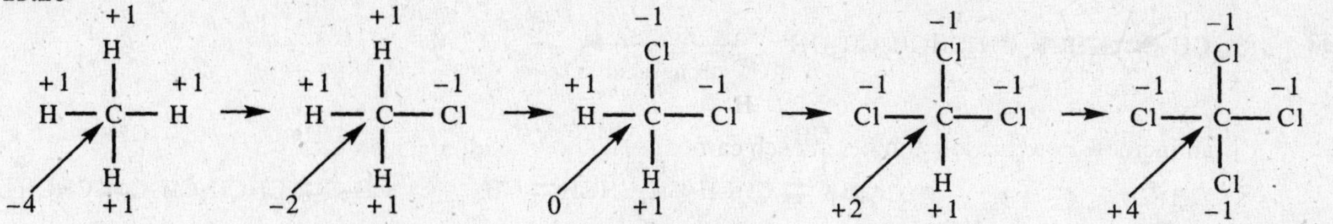

methane   chloromethane   dichloromethane   trichloromethane   tetrachloromethane
                 methyl chloride   methylene chloride   chloroform   carbon tetrachloride

13.27 (a)

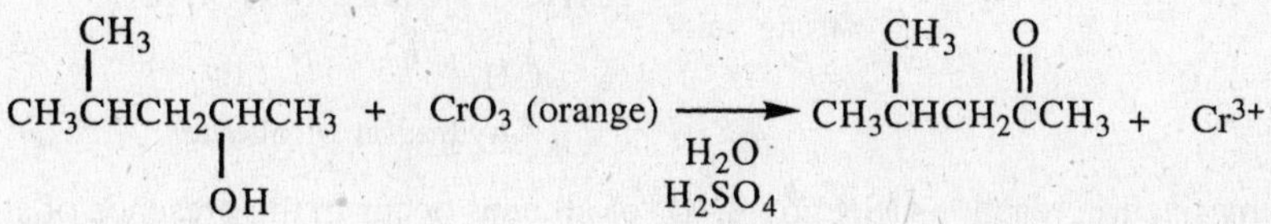

$$CH_3CHCH_2CHCH_3 + CrO_3 \text{ (orange)} \xrightarrow[H_2SO_4]{H_2O} CH_3CHCH_2CCH_3 + Cr^{3+} \text{ (blue-green, positive test)}$$

(b)

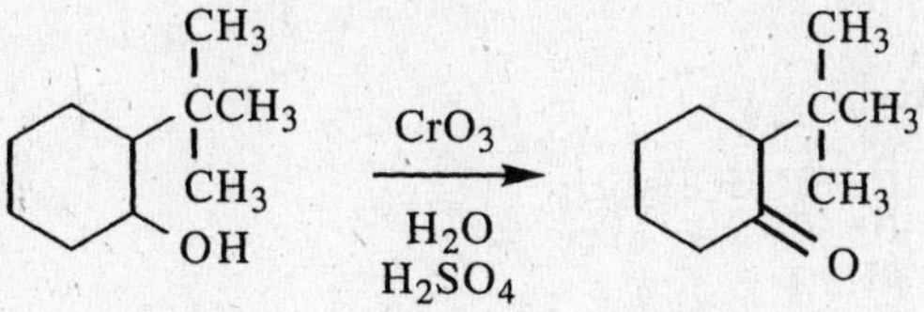

$+ Cr^{3+}$ (positive test)

(c)

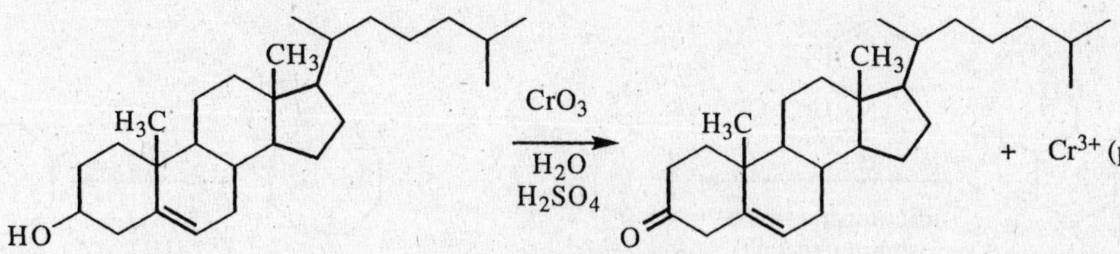

$+ Cr^{3+}$ (positive test)

(d)

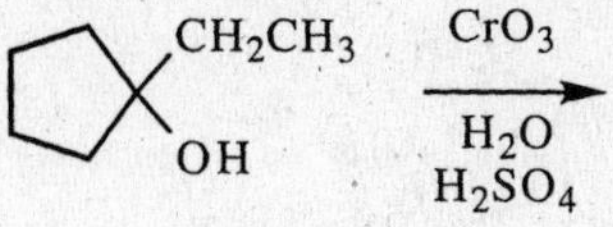

$\xrightarrow[H_2SO_4]{CrO_3 \atop H_2O}$ 3° alcohol, solution remains orange (negative test)

(e) $CH_3CH_2CH_2CH_2CH_2OH \xrightarrow[H_2SO_4]{CrO_3 \atop H_2O} CH_3CH_2CH_2CH_2COH + Cr^{3+}$ (positive test)

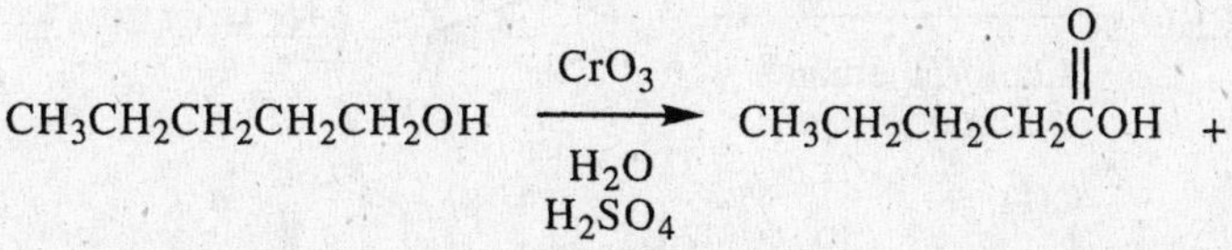

(f)

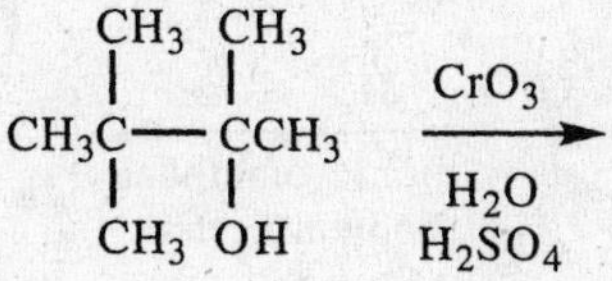

$\xrightarrow[H_2SO_4]{CrO_3 \atop H_2O}$ 3° alcohol, solution remains orange (negative test)

13.28 (1)

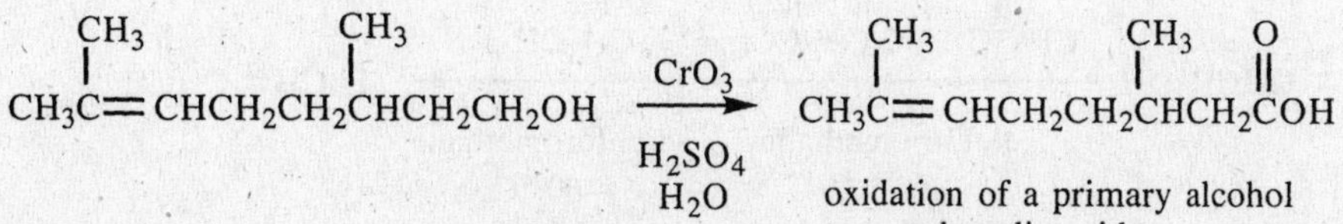

oxidation of a primary alcohol
to a carboxylic acid

13.28 (2) $\underset{\substack{| \\ CH_3}}{CH_3C} = CHCH_2CH_2\underset{\substack{| \\ CH_3}}{CHCH_2CH_2OH}$ $\xrightarrow{\quad\text{strong acid}\quad}$

$\underset{\substack{| \\ CH_3}}{CH_3C} = CHCH_2CH_2\underset{\substack{| \\ CH_3}}{CHCH} = CH_2$ + $\underset{\substack{| \\ CH_3}}{CH_3C} = CHCH_2CH_2\underset{\substack{| \\ CH_3}}{C} = CHCH_3$

dehydration of alcohol

(3) $\underset{\substack{| \\ CH_3}}{CH_3C} = CHCH_2CH_2\underset{\substack{| \\ CH_3}}{CHCH_2CH_2OH}$ $\xrightarrow{\quad\text{strong acid}\quad}$ $CH_3\underset{\substack{| \\ OH}}{C}CH_2CH_2CH_2\underset{\substack{| \\ CH_3}}{CHCH_2CH_2OH}$

hydration of alkene

In addition, polymerization reactions (Section 8.4B) of alkenes occur in the presence of strong acids.

13.29 (a) $CH_3\underset{\substack{| \\ OH}}{CH}\underset{\substack{| \\ CH_3}}{CH}\underset{\substack{| \\ CH_3}}{CH}CHCH_2CH_3$ $\xrightarrow[\substack{H_2SO_4 \\ H_2O \\ \text{acetone}}]{CrO_3}$ $CH_3\underset{\substack{| \\ CH_3}}{CH}\underset{\substack{\| \\ O}}{C}\underset{\substack{| \\ CH_3}}{CH}CHCH_2CH_3$

(b)

(c)

(d)

(e)

(f)

13.30 (a) $PhCH = CHCH_2OH$ $\xrightarrow[\substack{\text{dichloromethane} \\ -60\ °C}]{\substack{\| \\ CH_3SCH_3,\ ClC-CCl}}$ $\xrightarrow[\substack{\text{dichloromethane} \\ -60\ °C}]{(CH_3CH_2)_3N}$ $PhCH = CHC\overset{\displaystyle O}{\underset{\displaystyle \|}{}}H$

13.30 (b)

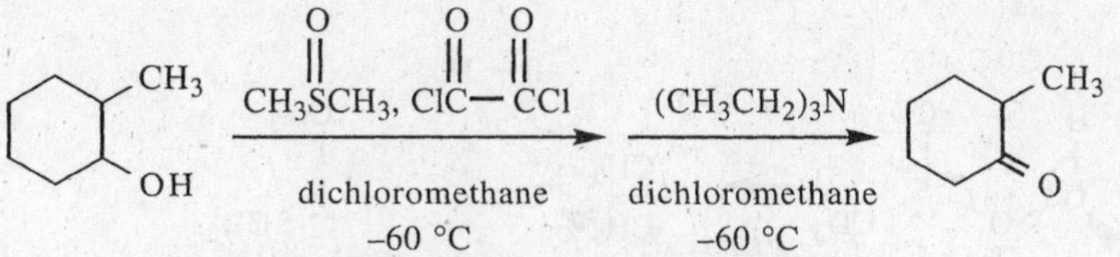

CH₃SCH₃, ClC—CCl
dichloromethane
−60 °C

(CH₃CH₂)₃N
dichloromethane
−60 °C

(c)

CH₃SCH₃, ClC—CCl
dichloromethane
−10 °C

(CH₃CH₂)₃N
dichloromethane
−10 °C

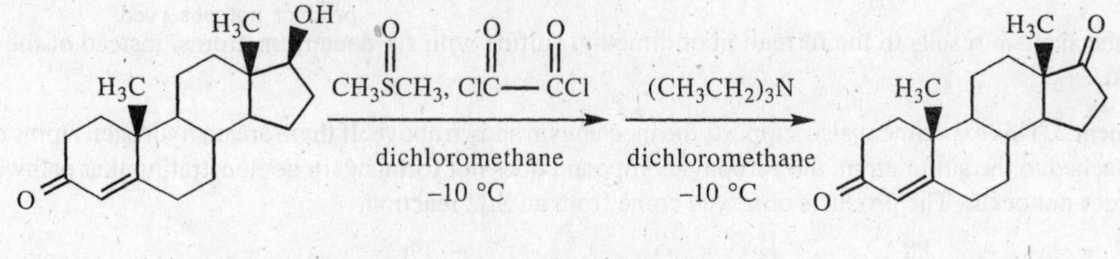

Concept Map 13.6   Reactions of alcohols with oxidizing agents.

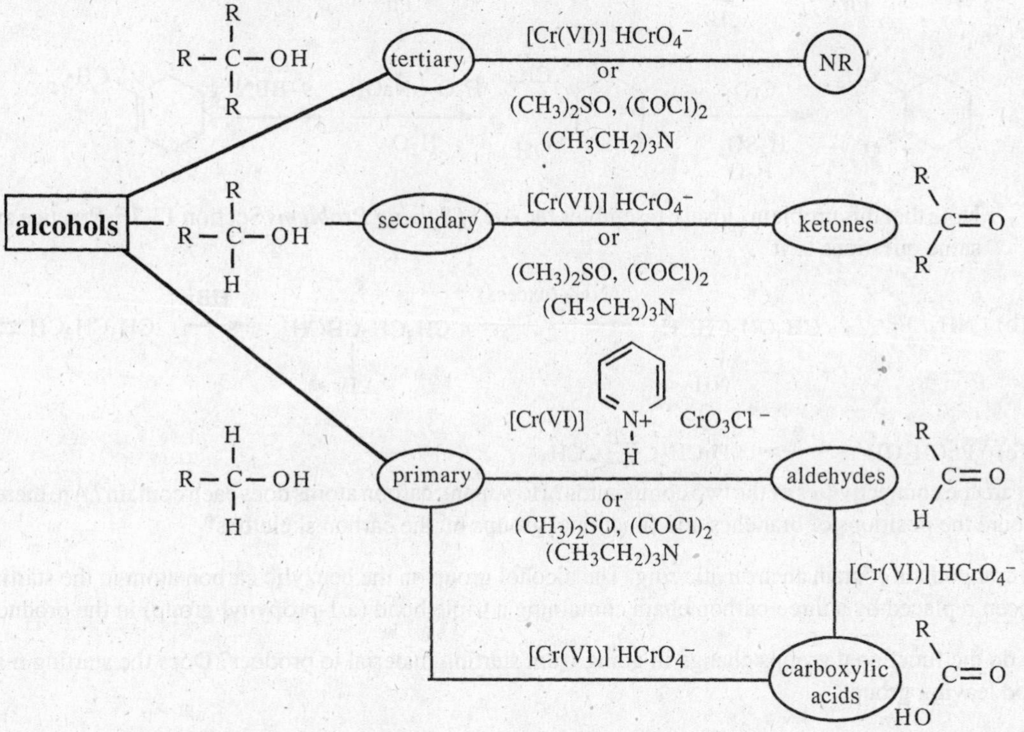

alcohols

$$R-\underset{\underset{R}{|}}{\overset{\overset{R}{|}}{C}}-OH \quad \text{tertiary}$$

[Cr(VI)] HCrO₄⁻
or
(CH₃)₂SO, (COCl)₂
(CH₃CH₂)₃N

NR

$$R-\underset{\underset{H}{|}}{\overset{\overset{R}{|}}{C}}-OH \quad \text{secondary}$$

[Cr(VI)] HCrO₄⁻
or
(CH₃)₂SO, (COCl)₂
(CH₃CH₂)₃N

ketones

$$\underset{R}{\overset{R}{>}}C=O$$

$$R-\underset{\underset{H}{|}}{\overset{\overset{H}{|}}{C}}-OH \quad \text{primary}$$

[Cr(VI)] ⟨N⁺—H⟩ CrO₃Cl⁻
or
(CH₃)₂SO, (COCl)₂
(CH₃CH₂)₃N

aldehydes

$$\underset{H}{\overset{R}{>}}C=O$$

[Cr(VI)] HCrO₄⁻

[Cr(VI)] HCrO₄⁻

carboxylic acids

$$\underset{HO}{\overset{R}{>}}C=O$$

13.31

Experiment 1. The mechanism for the second stage of the Swern reaction has the base deprotonating the carbon atom next to the sulfur atom and not the one next to the oxygen atom. Dimethyl sulfoxide with deuterium substituting for hydrogen makes it possible to see the location of the deprotonation.

(a)

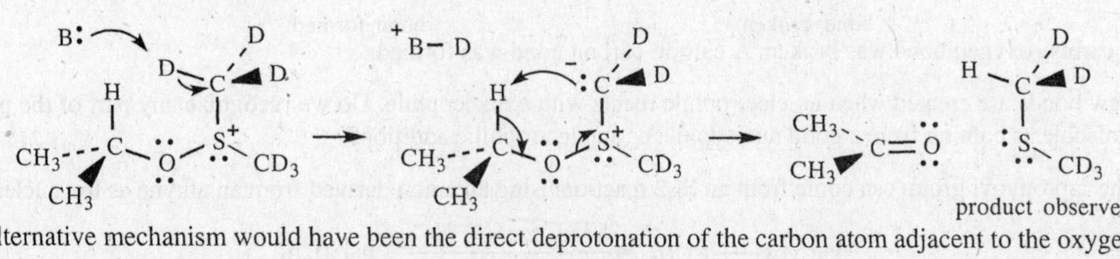

product observed

An alternative mechanism would have been the direct deprotonation of the carbon atom adjacent to the oxygen atom.

13.31 (cont)

(b)

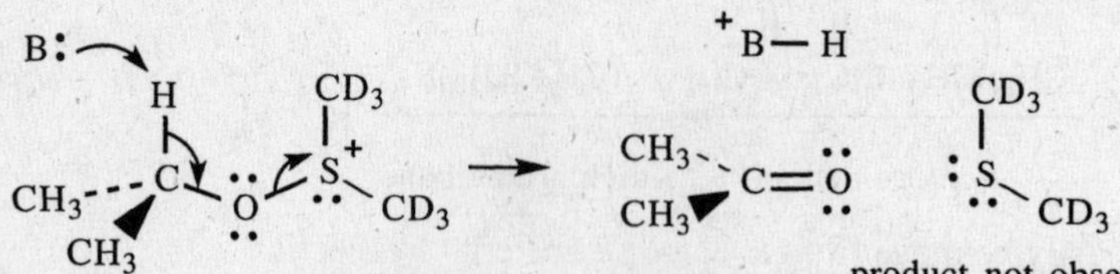

Such a mechanism results in the formation of dimethyl sulfide with six deuterium atoms, instead of the five actually observed.

Experiment 2. This experiment also supports the mechanism shown above. If there are no hydrogen atoms on the carbon atom attached to the sulfur atom, the carbonyl compound does not form, again demonstrating that pathway (b) shown above does not occur. The products observed come from an S_N2 reaction.

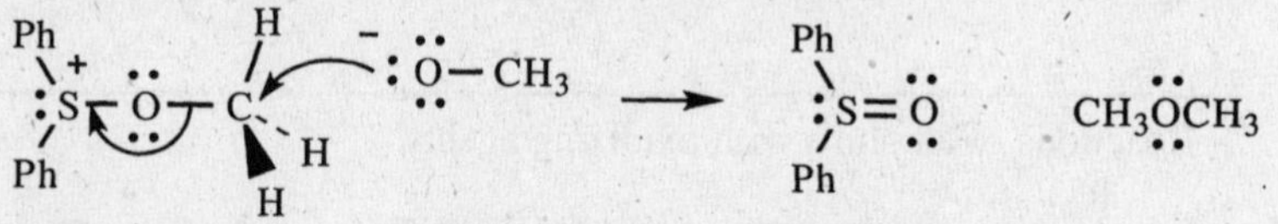

13.32 (a)

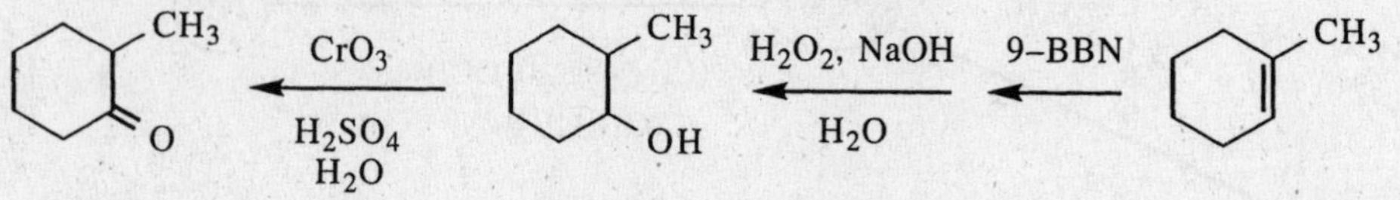

Note that this problem closely resembles the *Art of Solving Problems* Section 13.7F. Practice applying the same questions to it.

(b) 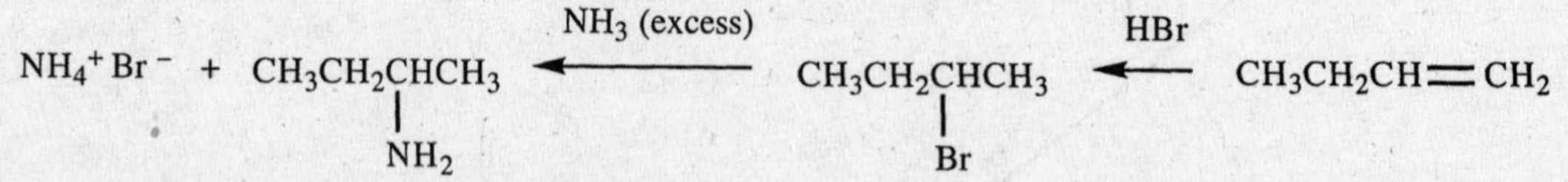 $NH_4^+ Br^-$ + $CH_3CH_2CHCH_3$ $\xleftarrow{NH_3 \text{ (excess)}}$ $CH_3CH_2CHCH_3$ $\xleftarrow{HBr}$ $CH_3CH_2CH=CH_2$

with NH_2 below first, Br below second.

(c) $PhCH_2OH$ $\xrightarrow{?}$ $PhCH_2C\equiv CCH_3$

1. What are the connectivities of the two compounds? How many carbon atoms does each contain? Are there any rings? What are the positions of branches and functional groups on the carbon skeletons?

Both compounds contain an aromatic ring. The alcohol group on the benzylic carbon atom in the starting material has been replaced by a three-carbon chain containing a triple bond (a 1-propynyl group) in the product.

2. How do the functional groups change in going from starting material to product? Does the starting material have a good leaving group?

The benzylic carbon atom is bonded to the first carbon atom of the triple bond. There is no good leaving group present in the starting material.

3. Is it possible to dissect the structures of the starting material and product to see which bonds must be broken and which formed?

$PhCH_2-\xi-OH$ (bond broken) $PhCH_2-\xi-C\equiv CCH_3$ (bond formed)

A carbon-oxygen bond was broken. A carbon-carbon bond was formed.

4. New bonds are created when an electrophile reacts with a nucleophile. Do we recognize any part of the product molecule as coming from a good nucleophile or an electrophilic addition?

The 1-propynyl group can come from an S_N2 reaction using an anion derived from an alkyne as the nucleophile.

$PhCH_2C\equiv CCH_3$ $\xleftarrow{CH_3C\equiv C:^- Na^+}$ $PhCH_2Br$

13.32 (c) (cont)

5. What type of compound would be a good precursor to the product?

A compound with a good leaving group, such as an alkyl halide or a tosylate, would be a good precursor to the product.

$$PhCH_2 \text{—} \boxed{X}$$

An alkyl bromide can easily be made from the starting material by reaction with concentrated hydrobromic acid.

$$PhCH_2Br \xleftarrow{\text{HBr (conc)}} PhCH_2OH$$

6. After this last step, do we see how to get from starting material to product? If not, we need to analyze the structure obtained in step 5 by applying questions 4 and 5 to it.

$$PhCH_2C \equiv CCH_3 \xleftarrow{CH_3C \equiv C\colon^- Na^+} PhCH_2Br \xleftarrow[\text{(conc)}]{HBr} PhCH_2OH$$

Concept Map 13.7   Summary of the preparation and reactions of alcohols and ethers.

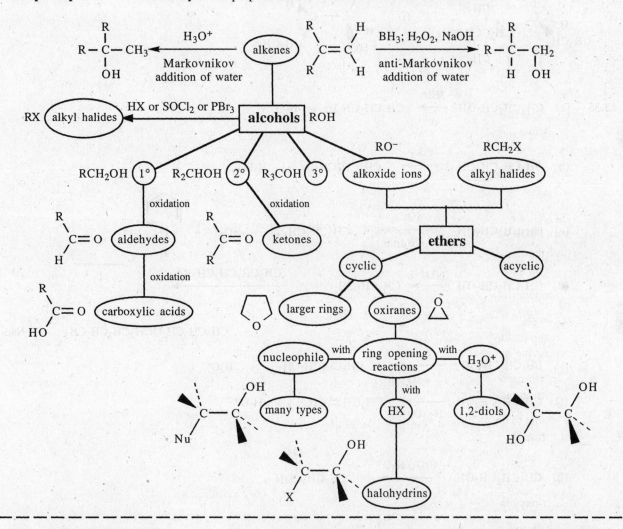

13.33   (a) 1,3-propanediol   (b) 1,2-dimethoxyethane   (c) 3-ethoxy-2,2-dimethyl-1-propanol

(d) 1-bromo-2-methoxyethane   (e) ethoxybenzene (ethyl phenyl ether)   (f) 2-ethoxyethanol

(g) cyclobutanol   (h) (2*S*)(*E*)-5-methyl-4-hepten-2-ol   (i) (1*R*, 2*R*)-2-bromocyclopentanol

(j) benzyloxyphenylmethane (dibenzyl ether)   (k) (1*S*, 2*S*)-2-methoxycyclohexanol

(l) 1,1,1-triethoxyethane

13.34

(a) $CH_3CH_2OC=CH_2$
   $\quad\quad\quad\quad\quad\; |$
   $\quad\quad\quad\quad\quad Cl$

(b)
$$CH_3CH_2CH_2O\quad\; H$$
$$\quad\quad\quad\quad\; C=C$$
$$\quad\quad\; H\quad\quad OCH_3$$

(c) $ClCH_2CHCH_2Cl$
   $\quad\quad\quad\; |$
   $\quad\quad\quad OH$

(d) ⬦—CH_2OH

(e)
$$\quad\quad\; OH$$
$$\quad\quad\; |$$
$$CH_3\cdots C$$
$$\quad\quad / \backslash$$
$$CH_3CH_2\quad CH_2CH=CH_2$$

(f)
$$\quad\quad CH_2OH$$
$$\quad\quad\; |$$
$$H\cdots C$$
$$\quad / \backslash$$
$$CH_3\quad Cl$$

(g)
$$\quad\quad\; H$$
$$\quad\quad\; |||$$
$$\quad\quad\; C$$
$$\quad\quad\; |||$$
$$\quad\quad\; C$$
$$\quad\quad\; |$$
$$CH_3\cdots C$$
$$\quad\quad / \backslash$$
$$CH_3CH_2\quad OH$$

(h) $O_2NCH_2CH_2OH$

(i)
$$\quad\quad\; OH$$
$$\quad\quad\; |$$
$$\quad\quad\; C\quad\quad\quad CH_3$$
$$H\cdots / \backslash\quad\quad |$$
$$CH_3CH_2\quad CH_2CCH_2CH_3$$
$$\quad\quad\quad\quad\quad\; |$$
$$\quad\quad\quad\quad\quad CH_3$$

(j)
$$CH_3CH_2\quad\quad CH_3$$
$$\quad\quad\; \backslash\quad\quad / \quad H$$
$$\quad\quad\; C\quad\quad C$$
$$\quad / \quad\quad \backslash$$
$$H\cdots\quad\quad CH_2CH_3$$
$$HO$$

13.35 (a) $CH_3CH_2CH_2OH \xrightarrow[\Delta]{HBr} CH_3CH_2CH_2Br + H_2O$

(b) $CH_3CH_2CH_2OH \xrightarrow{PBr_3} CH_3CH_2CH_2Br + H_3PO_3$

(c) $CH_3CH_2CH_2OH \xrightarrow[\text{pyridine}]{SOCl_2} CH_3CH_2CH_2Cl + SO_2\uparrow + $ ⬠N^+—H Cl^-

(d) $CH_3CH_2CH_2OH \xrightarrow{NaH} CH_3CH_2CH_2O^-Na^+ \xrightarrow{CH_3CH_2CH_2CH_2Br}$
$\quad\quad\quad\quad\quad\quad\quad + H_2\uparrow$

$\quad\quad\quad\quad\quad\quad\quad\quad\quad\quad\quad\quad CH_3CH_2CH_2OCH_2CH_2CH_2CH_3 + NaBr$

(e) $CH_3CH_2CH_2OH \xrightarrow[\text{cold}]{H_3O^+} CH_3CH_2CH_2\overset{+}{O}H_2 + H_2O$

(f) $CH_3CH_2CH_2OH \xrightarrow[\substack{H_2SO_4 \\ \Delta}]{} CH_3CH=CH_2 + H_2O$

(g) $CH_3CH_2CH_2OH \xrightarrow[H_2O]{CrO_3, H_3O^+} CH_3CH_2\overset{O}{\overset{||}{C}}OH$

(h) $CH_3CH_2CH_2OH \xrightarrow[\text{dichloromethane}]{\text{⬠}N^+\text{—H } CrO_3Cl^-} CH_3CH_2\overset{O}{\overset{||}{C}}H$

13.35

(i) $CH_3CH_2CH_2OH$ $\xrightarrow{\underset{CH_3SCH_3,\ ClC—CCl}{\overset{O\quad O\quad O}{}}}$ $\xrightarrow{(CH_3CH_2)_3N}$ $CH_3CH_2\overset{O}{\overset{\|}{C}}H$

13.36

(a) $CH_3CH_2CH_2CH_2\underset{OH}{CH}CH_3$ $\xrightarrow[\Delta]{HBr}$ $CH_3CH_2CH_2CH_2\underset{Br}{CH}CH_3$ + $CH_3CH_2CH_2\underset{Br}{CH}CH_2CH_3$

major
+ H_2O + some elimination products

(b) $CH_3CH_2CH_2CH_2\underset{OH}{CH}CH_3$ $\xrightarrow[\Delta]{PBr_3}$ $CH_3CH_2CH_2CH_2\underset{Br}{CH}CH_3$ + H_3PO_3

(c) $CH_3CH_2CH_2CH_2\underset{OH}{CH}CH_3$ $\xrightarrow[pyridine]{SOCl_2}$ $CH_3CH_2CH_2CH_2\underset{Cl}{CH}CH_3$ + $SO_2\uparrow$ + pyridinium Cl^-

(d) $CH_3CH_2CH_2CH_2\underset{OH}{CH}CH_3$ $\xrightarrow{NaH}$ $CH_3CH_2CH_2CH_2\underset{O^-Na^+}{CH}CH_3$ + $H_2\uparrow$

(e) $CH_3CH_2CH_2CH_2\underset{OH}{CH}CH_3$ $\xrightarrow[ZnCl_2]{HCl}$ $CH_3CH_2CH_2\underset{Cl}{CH}CH_3$ + $CH_3CH_2CH_2\underset{Cl}{CH}CH_2CH_3$

(f) $CH_3CH_2CH_2CH_2\underset{OH}{CH}CH_3$ $\xrightarrow[\Delta]{H_2SO_4}$ $CH_3CH_2CH_2CH=CHCH_3$ +

$CH_3CH_2CH_2CH_2CH=CH_2$ + $CH_3CH_2CH=CHCH_2CH_3$ + H_2O

(g) $CH_3CH_2CH_2CH_2\underset{OH}{CH}CH_3$ $\xrightarrow[H_2O]{CrO_3,\ H_3O^+}$ $CH_3CH_2CH_2CH_2\overset{O}{\overset{\|}{C}}CH_3$

(h) $CH_3CH_2CH_2CH_2\underset{OH}{CH}CH_3$ $\xrightarrow[dichloromethane]{\text{pyridinium } CrO_3Cl^-}$ $CH_3CH_2CH_2CH_2\overset{O}{\overset{\|}{C}}CH_3$

(i) $CH_3CH_2CH_2CH_2\underset{OH}{CH}CH_3$ $\xrightarrow{\underset{CH_3SCH_3,\ ClC—CCl}{\overset{O\quad O\quad O}{}}}$ $\xrightarrow{(CH_3CH_2)_3N}$ $CH_3CH_2CH_2CH_2\overset{O}{\overset{\|}{C}}CH_3$

13.37   (a)   The amino alcohol with two hydrogen bonding groups is most soluble in water. The alcohol, which can act as a donor as well as an acceptor of hydrogen bonds, is more soluble in water than the ester, which is only an acceptor of hydrogen bonds.

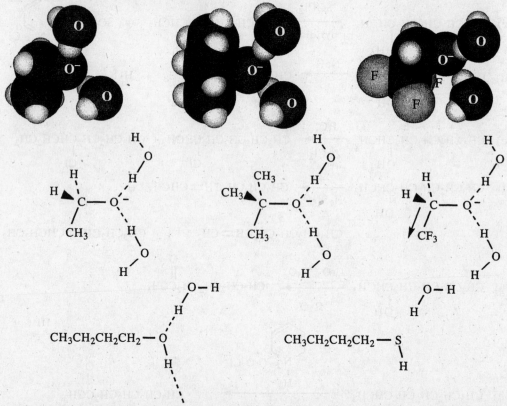

(b)   The more stable the conjugate base (an anion in this example), the stronger is the acid (the alcohol). The *tert*-butoxide anion is destabilized by the bulk of the methyl groups, which interferes with solvation (and therefore stabilization) of the anion. The trifluoroethoxide anion, by contrast, is stabilized by the electron-withdrawing effect of the fluorine atoms.

(c)   The compound that has the highest boiling point and the highest solubility in water is the one with two hydroxyl groups, which lead to increased hydrogen bonding, both between molecules of the same kind (giving rise to a higher boiling point) and between the compound and water (giving rise to higher solubility in water). Of the compounds with only one hydroxyl group, the one with the higher molecular weight and the longer carbon chain has the higher boiling point (from increased van der Waals attraction) and the lower solubility in water (because the hydrophobic part of the molecule is too large to be brought into solution by the polar hydroxyl group). See Sections 1.10B and 1.10C for review.

13.37 (c) (cont)

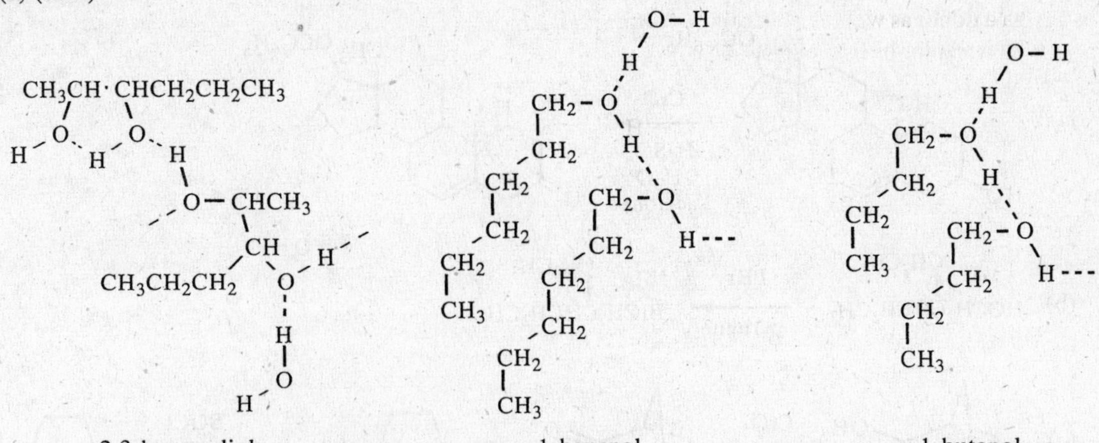

2,3-hexanediol   1-hexanol   1-butanol

(d) The hydroxyl group hydrogen bonds more strongly than the thiol group, and the alcohol is therefore more soluble than the corresponding sulfur compound.

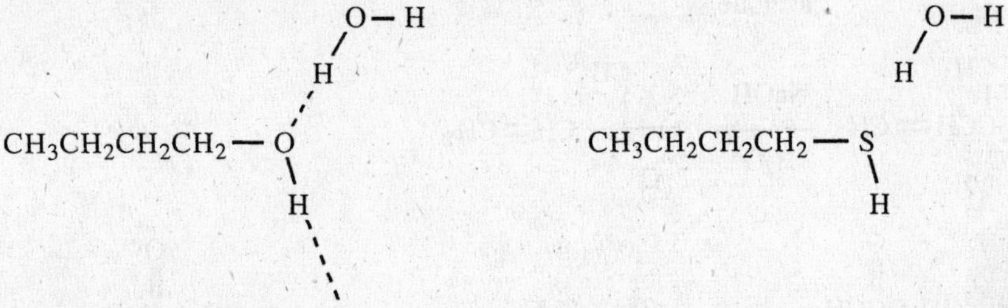

13.38 (a)

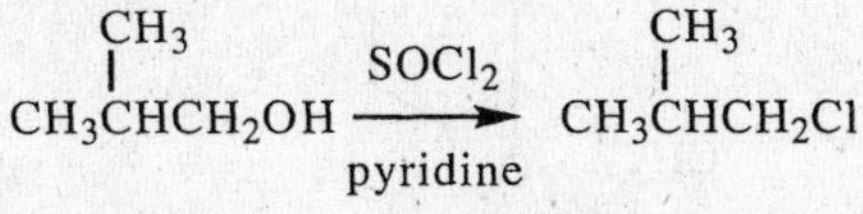

(b)

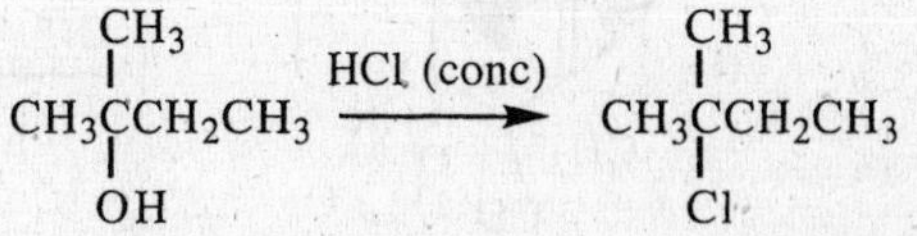

(c)

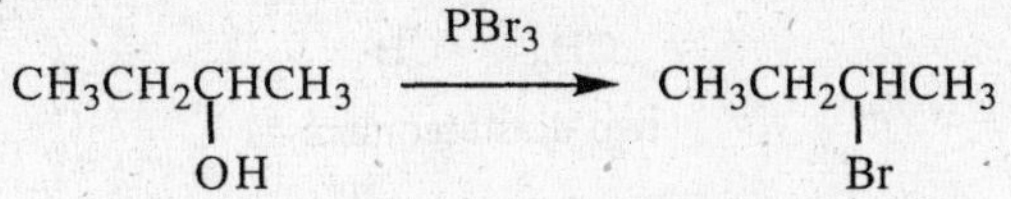

(d)

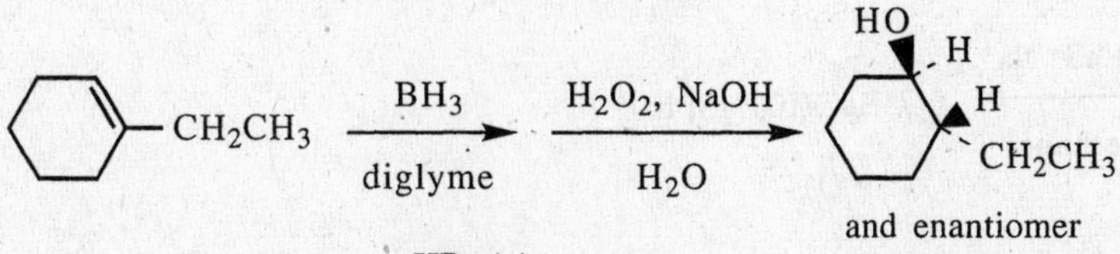

and enantiomer

(e) HOCH$_2$(CH$_2$)$_4$CH$_2$OH $\xrightarrow{\text{HBr (g)} \atop \text{(excess)}}$ BrCH$_2$(CH$_2$)$_4$CH$_2$Br

(f)

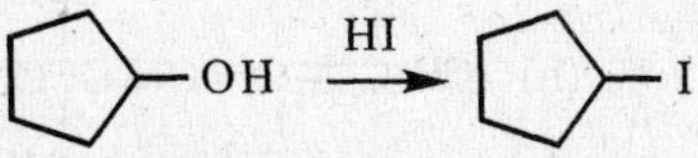

(g)

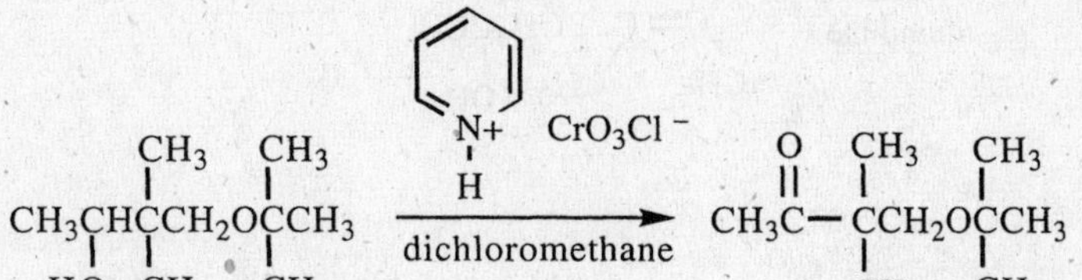

(h)

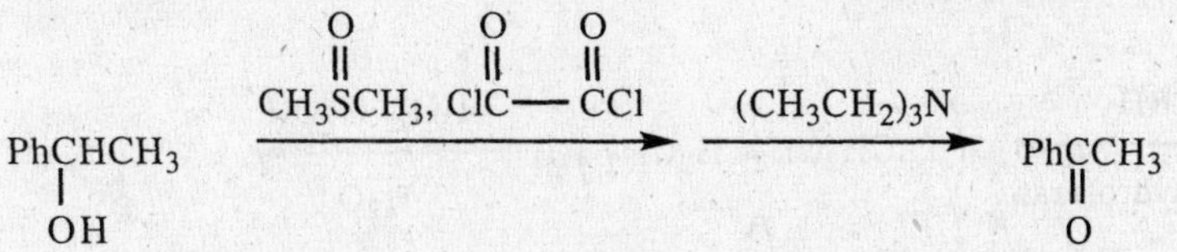

13.39 (a)

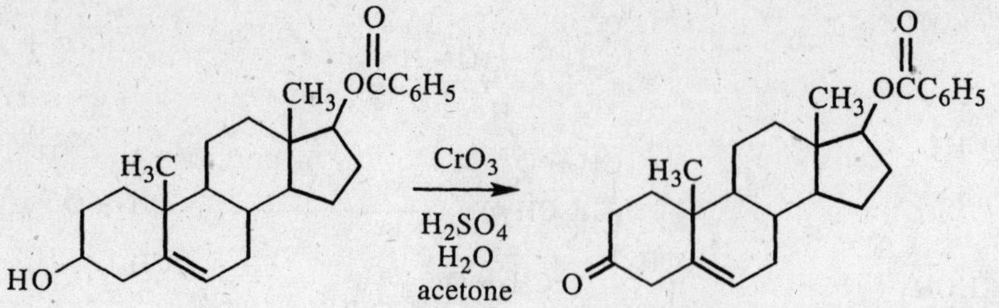
CH3 OCC6H5
H3C
HO
CrO3
H2SO4
H2O
acetone
O

(b)

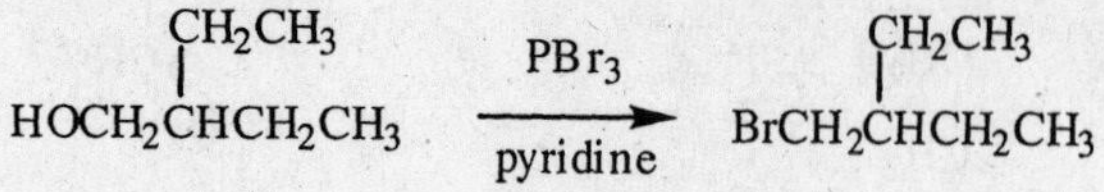
CH2CH3
HOCH2CHCH2CH3
PBr3
pyridine
BrCH2CHCH2CH3

(c)

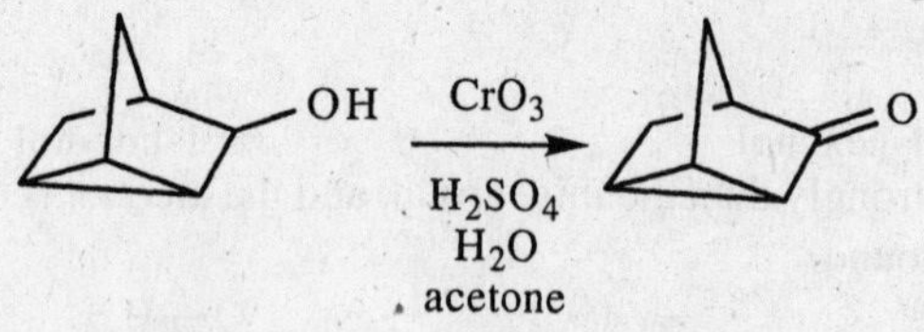
OH
CrO3
H2SO4
H2O
acetone
O

(d)

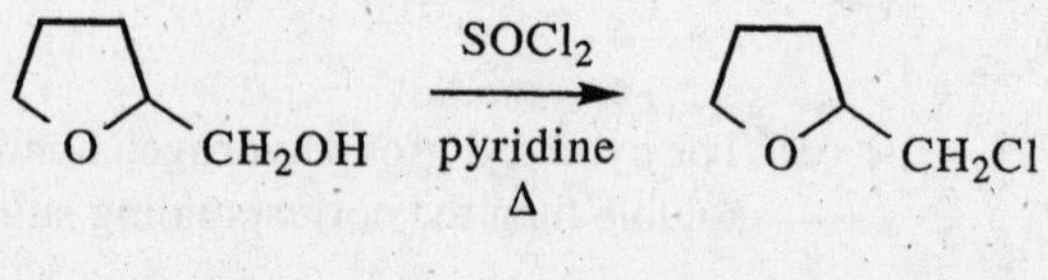
O
CH2OH
SOCl2
pyridine
Δ
CH2Cl

(e)

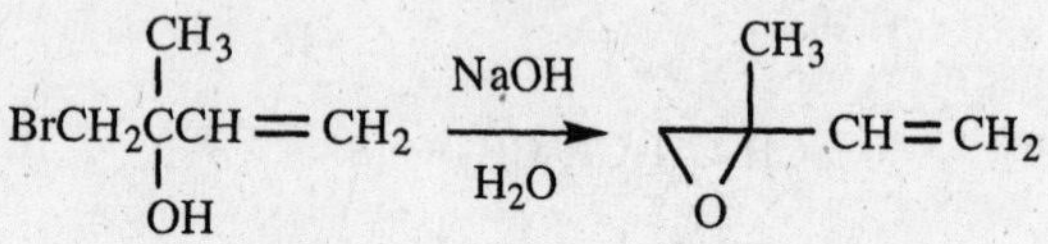
CH3
BrCH2CCH=CH2
OH
NaOH
H2O
CH=CH2

(f)

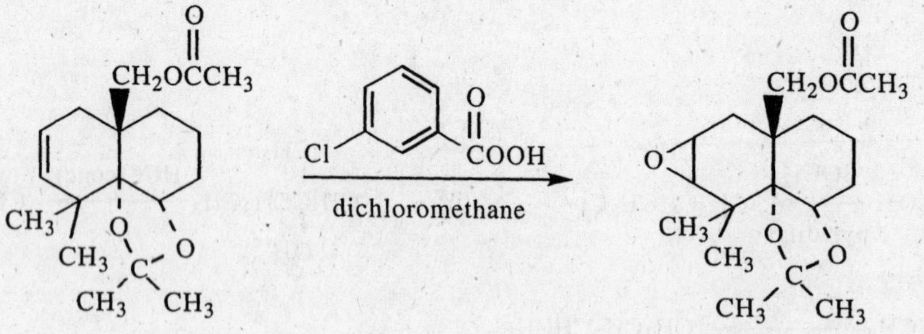
CH2OCCH3
Cl
COOH
dichloromethane
CH3
O
C

two diastereomers

(g)

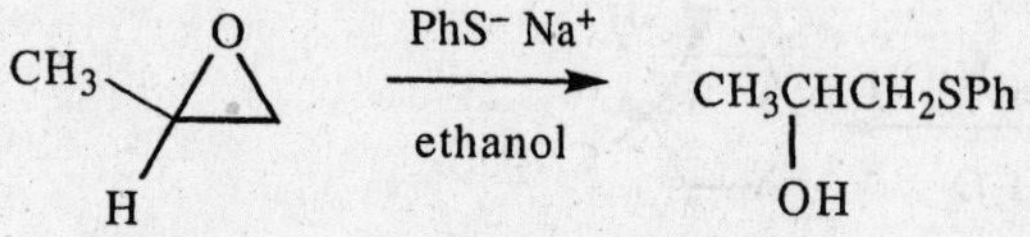
CH3
H
O
PhS− Na+
ethanol
CH3CHCH2SPh
OH

(h)

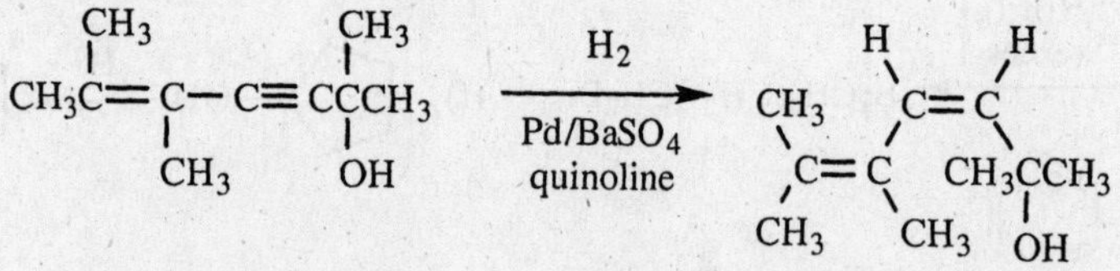
CH3C=C−C≡CCCH3
H2
Pd/BaSO4
quinoline
CH3CCH3
OH

13.40

(a)

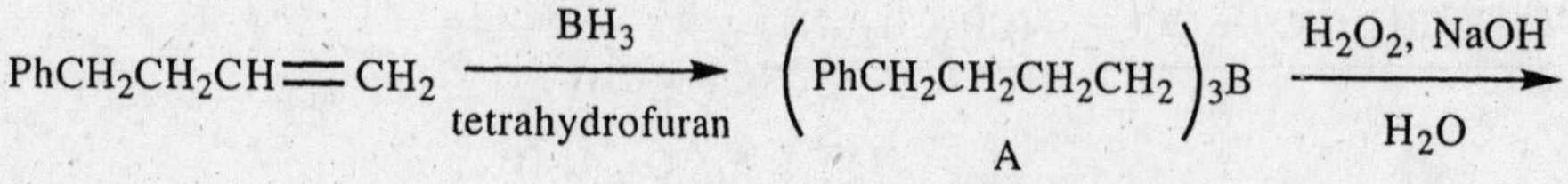
PhCH2CH2CH=CH2
BH3
tetrahydrofuran
(PhCH2CH2CH2CH2)3B
A
H2O2, NaOH
H2O

13.40 (a) (cont)

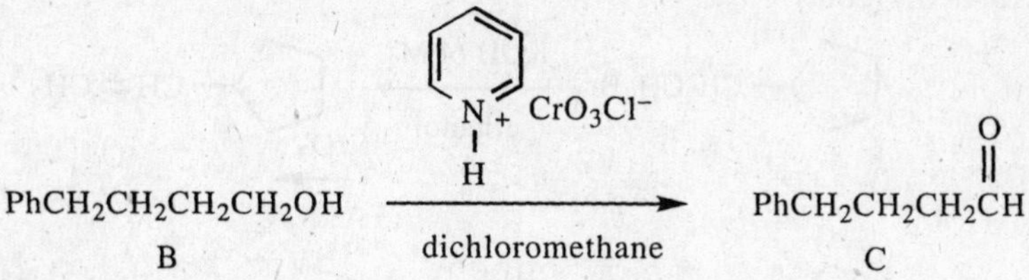

$$PhCH_2CH_2CH_2CH_2OH \xrightarrow[\text{dichloromethane}]{\text{(pyridinium chlorochromate)}} PhCH_2CH_2CH_2CH$$

B → C

(b)

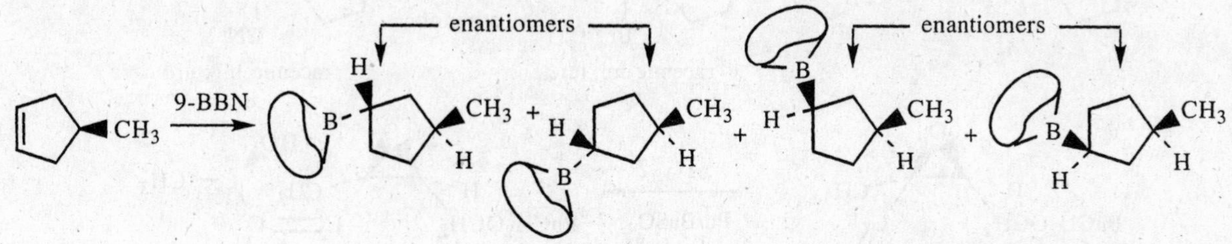

D
major isomers from less
hindered transition state

E
minor isomers

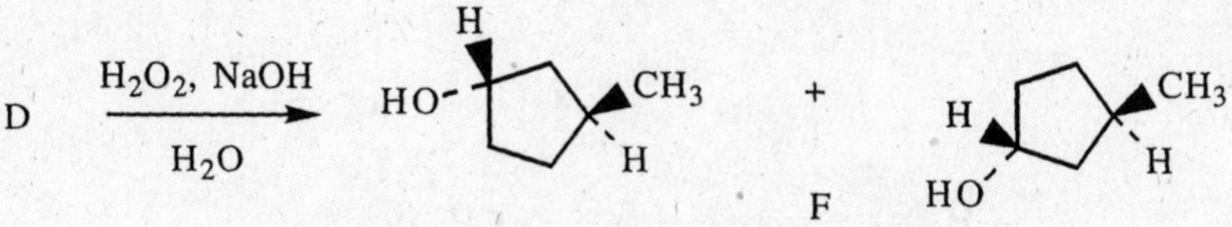

$$D \xrightarrow[\text{H}_2\text{O}]{\text{H}_2\text{O}_2, \text{ NaOH}} $$

F
enantiomers

$$E \xrightarrow[\text{H}_2\text{O}]{\text{H}_2\text{O}_2, \text{ NaOH}} $$

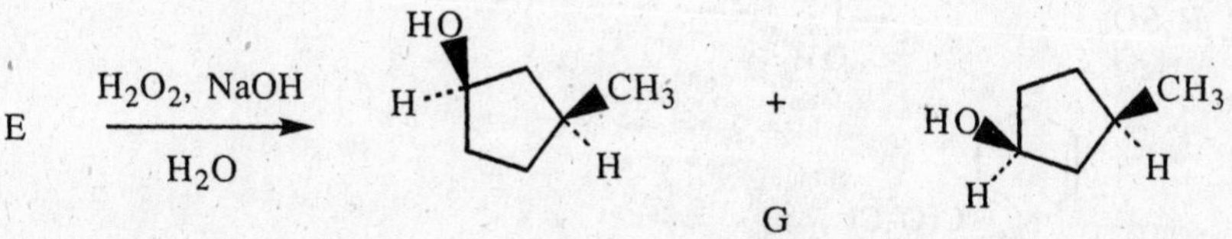

G
enantiomers

(c)

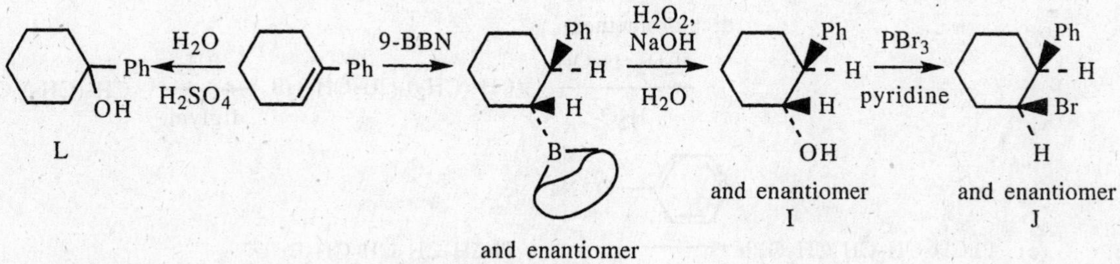

L

and enantiomer
H

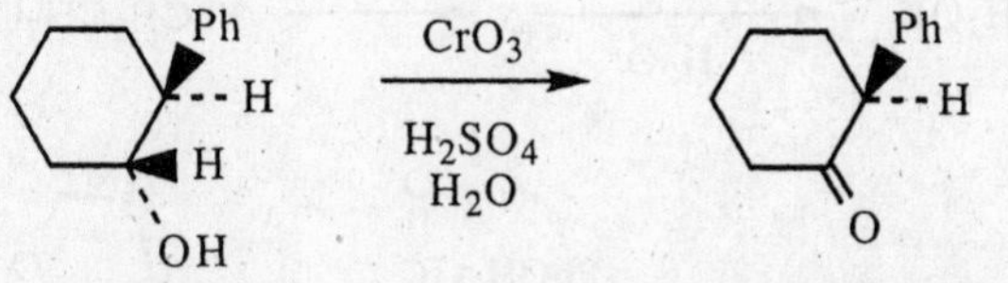

and enantiomer
I

and enantiomer
J

and enantiomer
I

$$\xrightarrow[\text{H}_2\text{O}]{\text{CrO}_3 \atop \text{H}_2\text{SO}_4}$$

and enantiomer
K

(d)

$$\text{—CH}_2\text{CH}_2\text{Br} \xrightarrow[\text{NH}_3 \text{ (liq)}]{\text{HC}\equiv\text{C}:^-\text{Na}^+} \text{—CH}_2\text{CH}_2\text{C}\equiv\text{CH} \xrightarrow[\text{H}_2\text{SO}_4 \atop \text{HgSO}_4]{\text{H}_2\text{O}} \text{—CH}_2\text{CH}_2\text{CCH}_3$$

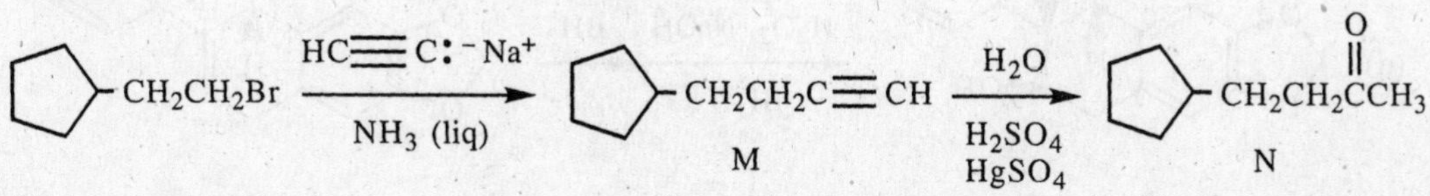

M → N

13.40 (d) (cont)

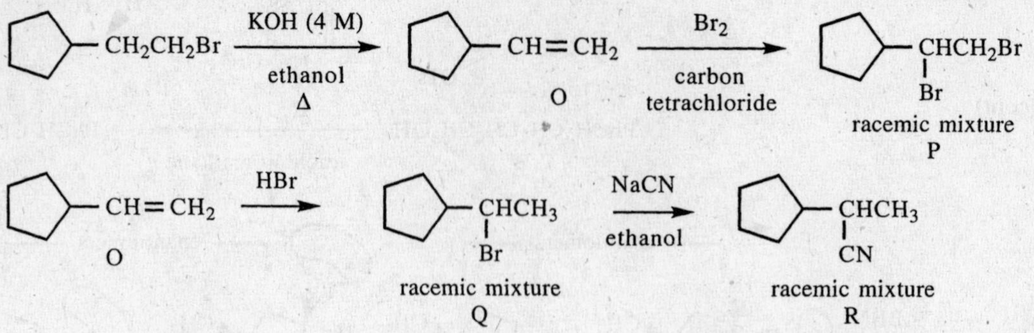

(e)

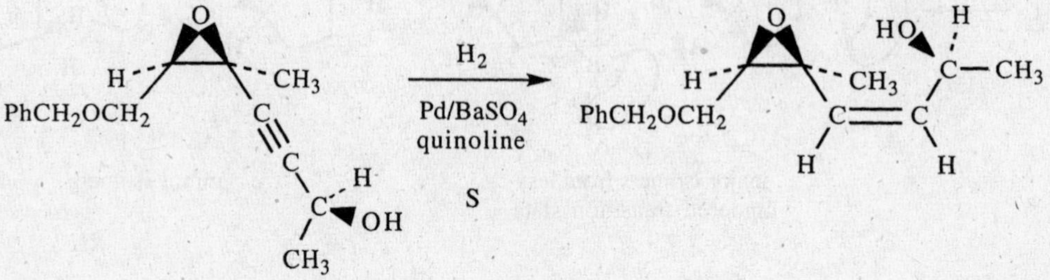

13.41   (a)

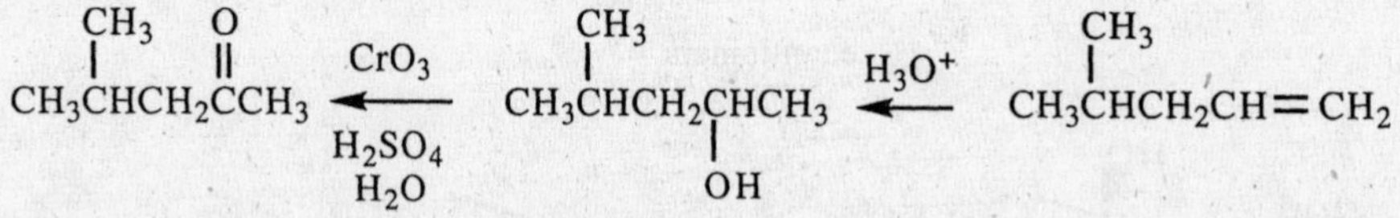

(b)

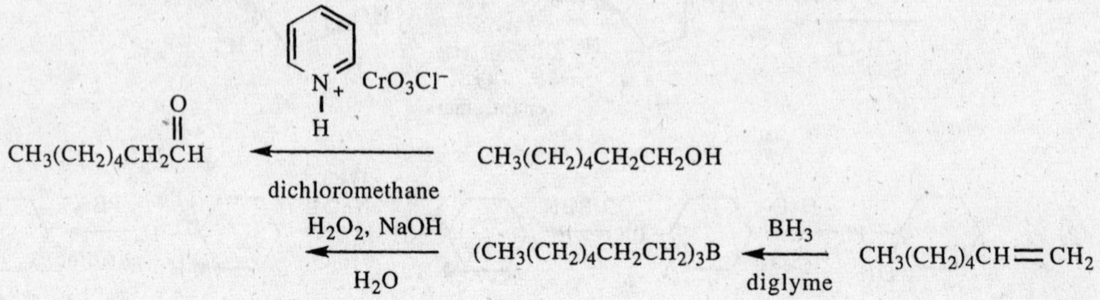

(c)

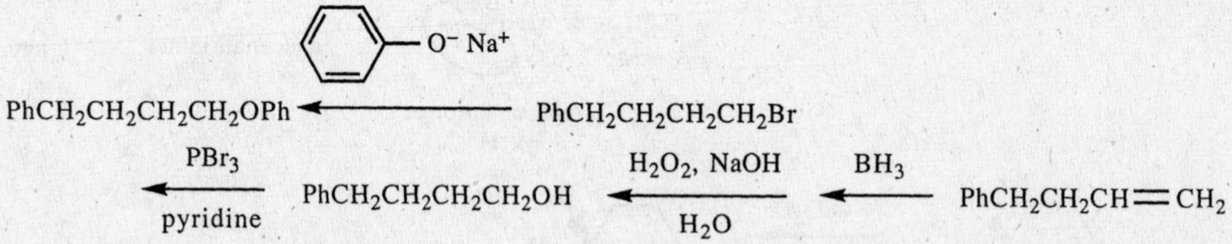

(d)

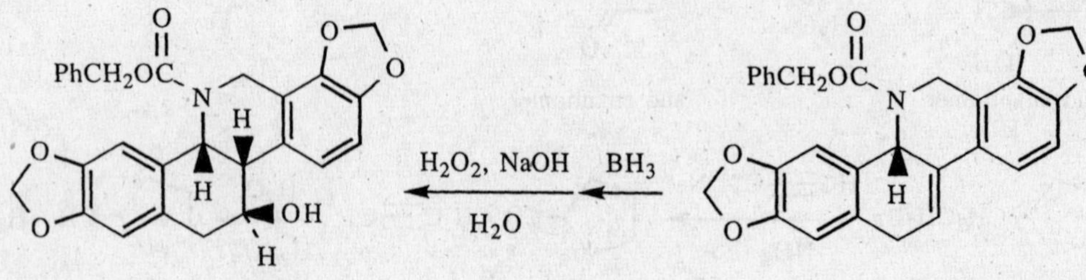

13.41 (e)

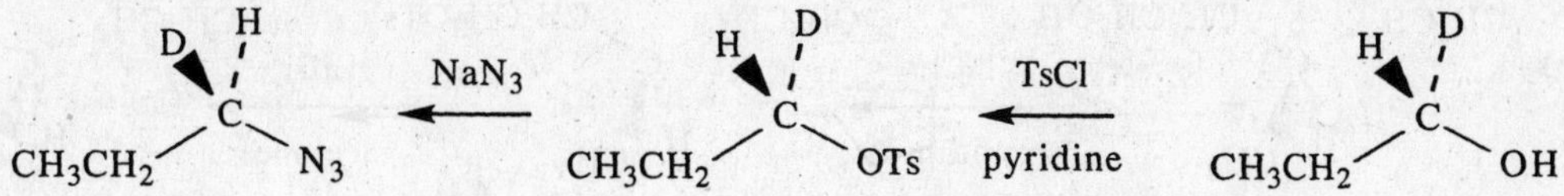
NaN3
TsCl
pyridine
CH3CH2
N3
OTs
OH

(f)

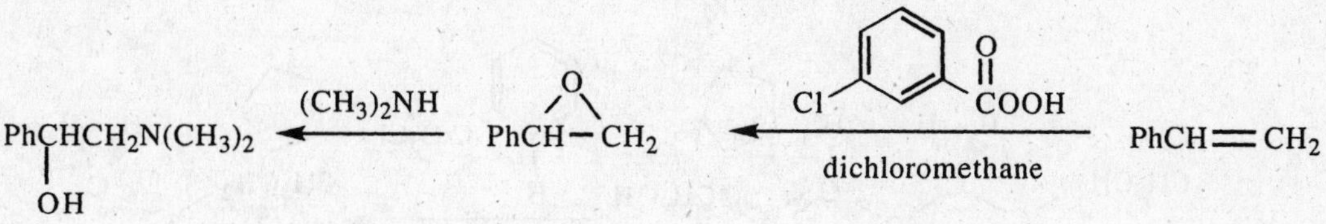
(CH3)2NH
PhCHCH2N(CH3)2
OH
PhCH—CH2
Cl
COOH
dichloromethane
PhCH=CH2

13.42

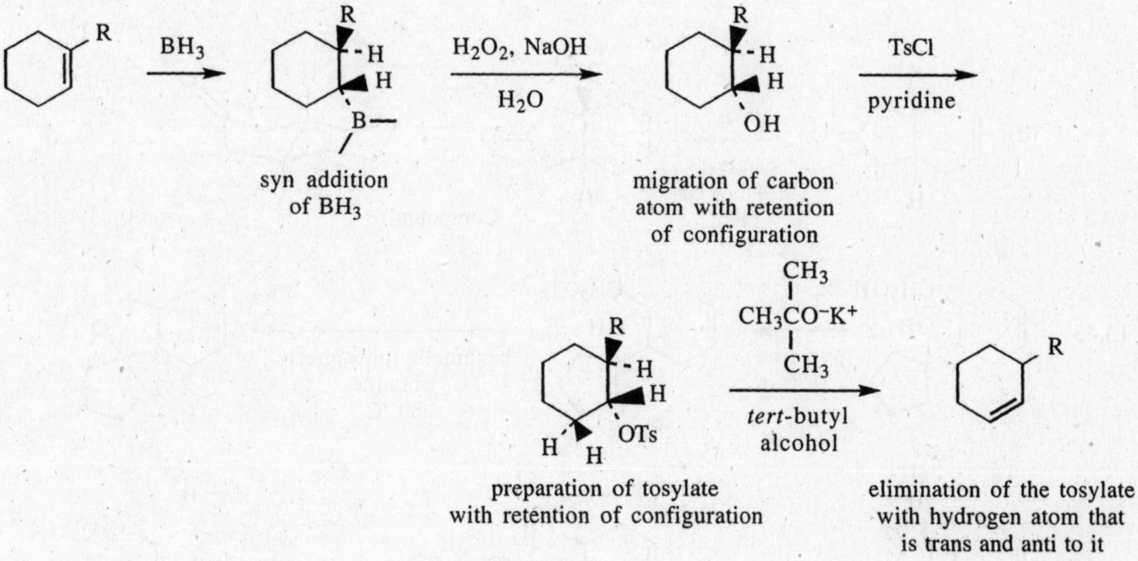
BH3
H2O2, NaOH
H2O
TsCl
pyridine
syn addition
of BH3
migration of carbon
atom with retention
of configuration
CH3CO−K+
tert-butyl
alcohol
preparation of tosylate
with retention of configuration
elimination of the tosylate
with hydrogen atom that
is trans and anti to it

13.43 (a)

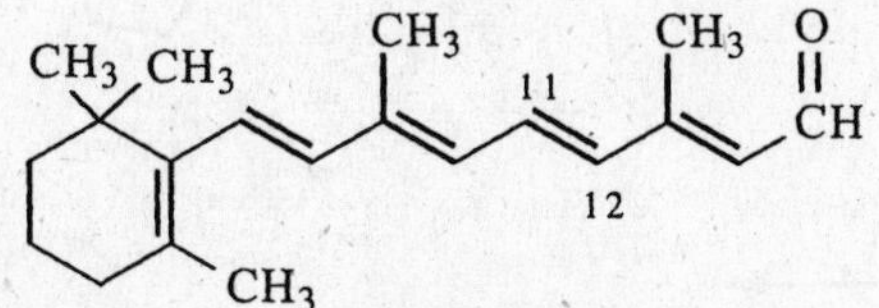

(11*E*)-retinal

(b)

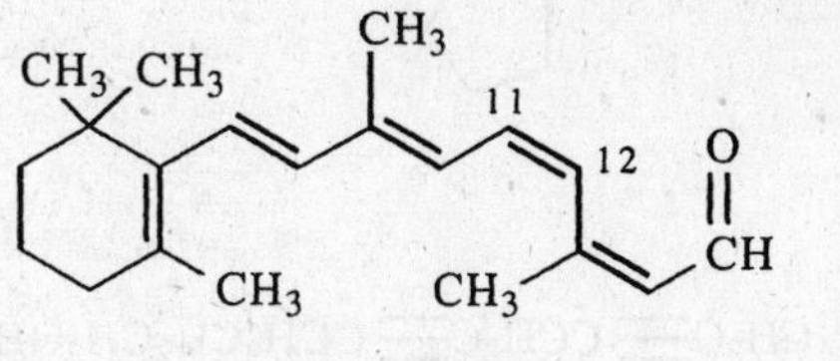

(11*Z*)-retinal

(c)

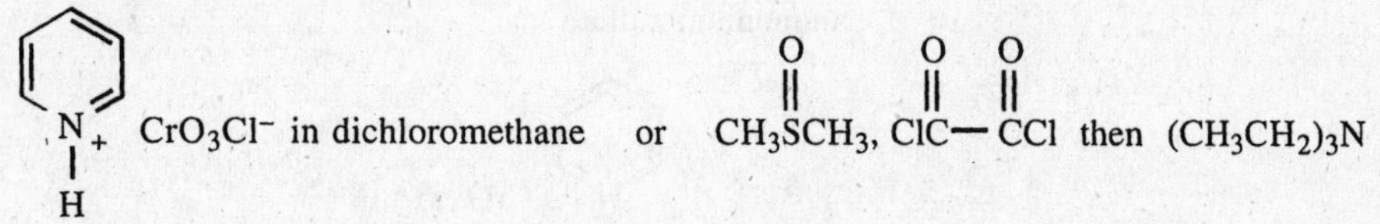
CrO3Cl− in dichloromethane or CH3SCH3, ClC—CCl then (CH3CH2)3N

13.44 (a)

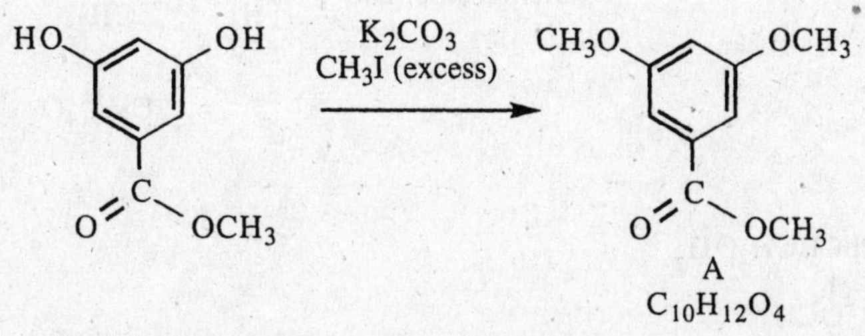
K2CO3
CH3I (excess)
A
C10H12O4

13.44 (b)

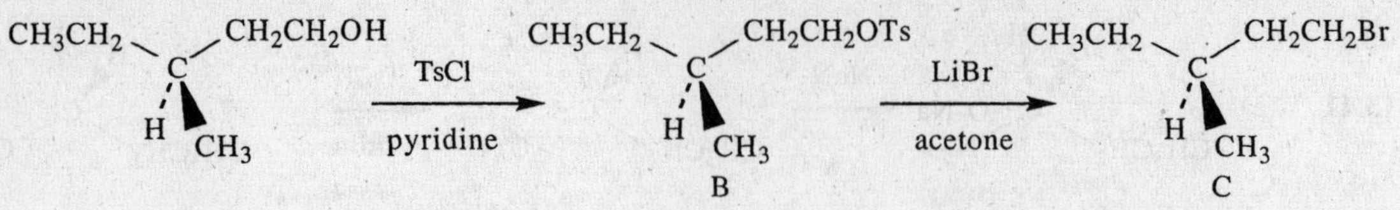

$$CH_3CH_2\overset{\displaystyle CH_2CH_2OH}{\underset{\underset{CH_3}{H}}{C}} \xrightarrow[\text{pyridine}]{TsCl} CH_3CH_2\overset{\displaystyle CH_2CH_2OTs}{\underset{\underset{CH_3}{H}}{C}} \xrightarrow[\text{acetone}]{LiBr} CH_3CH_2\overset{\displaystyle CH_2CH_2Br}{\underset{\underset{CH_3}{H}}{C}}$$

B   C

(c)

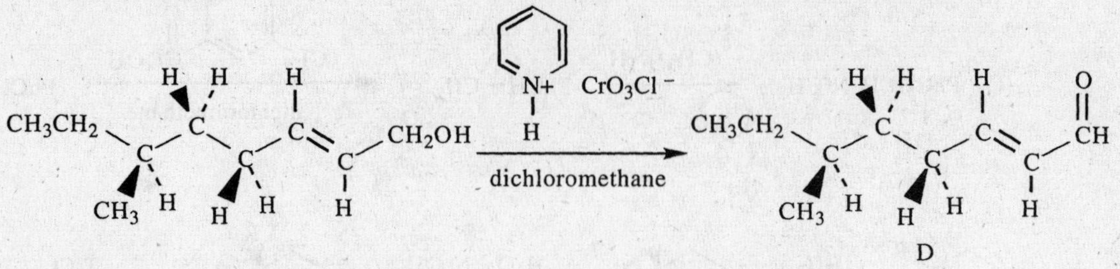

(d)

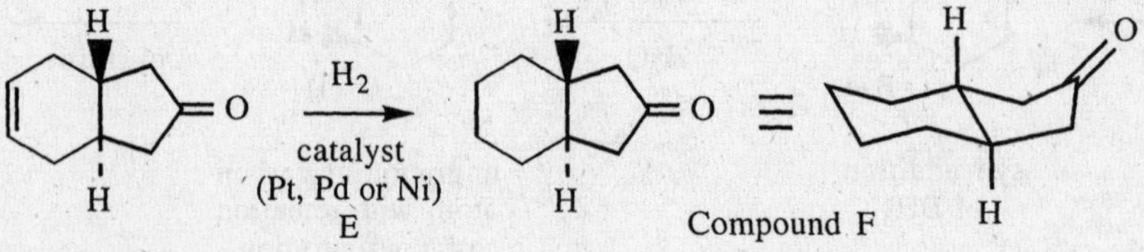

Compound F

13.45

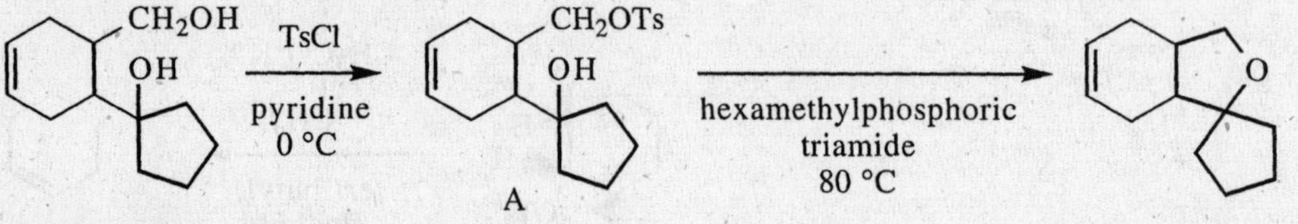

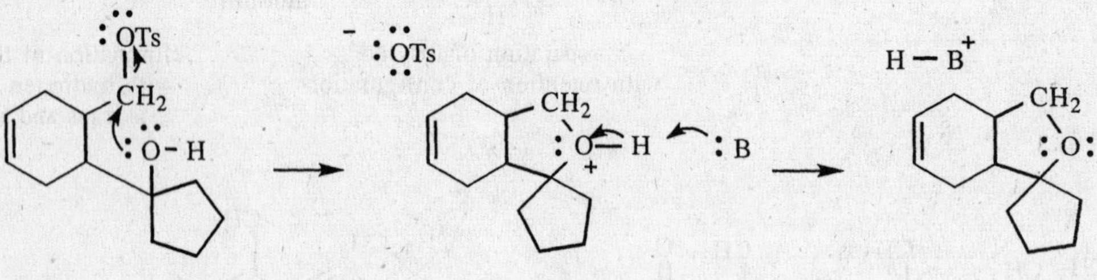

A

13.46 $CH_3C \equiv CCH_2C \equiv CCH_2CH_2CH_2OH \xrightarrow[\substack{NH_3 \text{ (liq)} \\ \textit{tert}\text{-butyl alcohol} \\ \text{ammonium sulfate}}]{Li}$

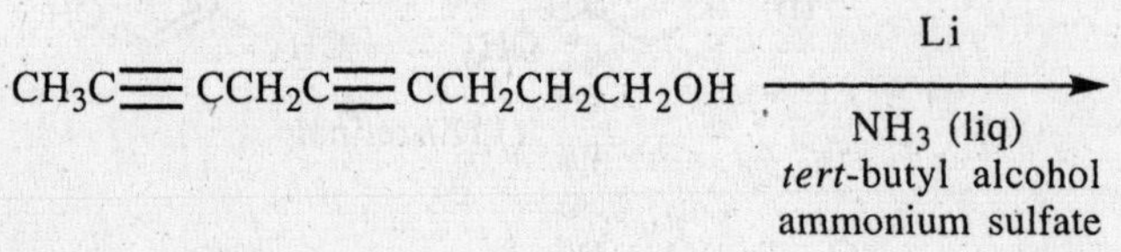

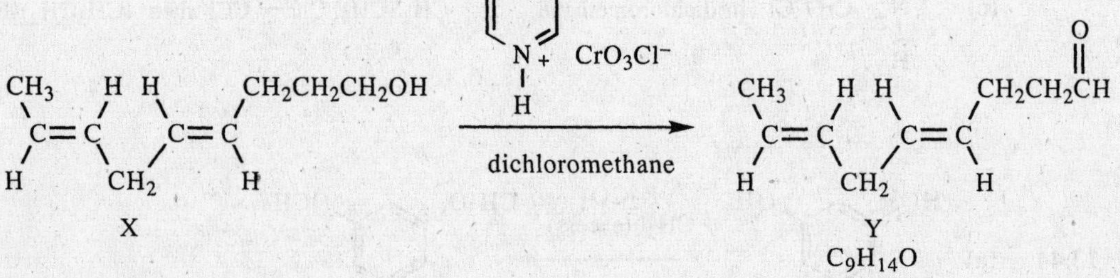

X   Y
$C_9H_{14}O$

13.47 $\underset{O}{PhCH-CH_2} \xrightarrow[H_2SO_4]{CH_3OH} \underset{\underset{OCH_3}{|}}{PhCHCH_2OH}$

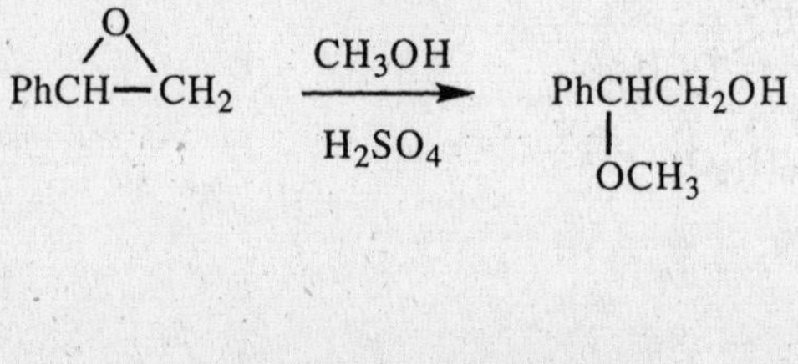

13.47 (cont)

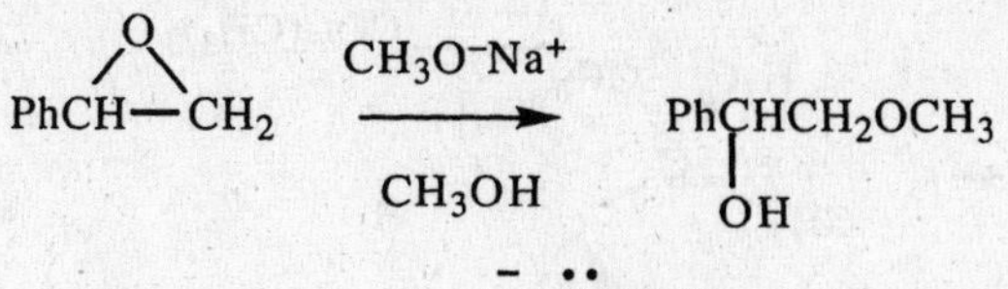
PhCH—CH2
CH3O−Na+
CH3OH
PhCHCH2OCH3
OH

(1)

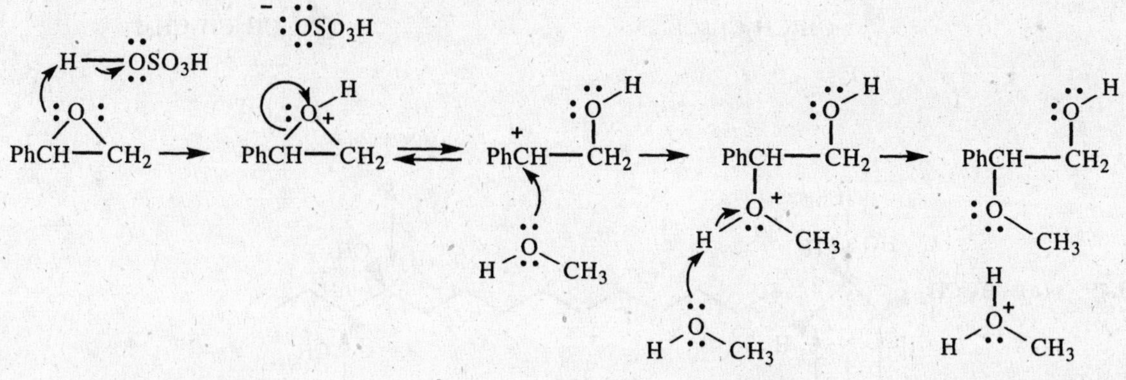
H—OSO3H
OSO3H
PhCH—CH2
H
CH3

(2)

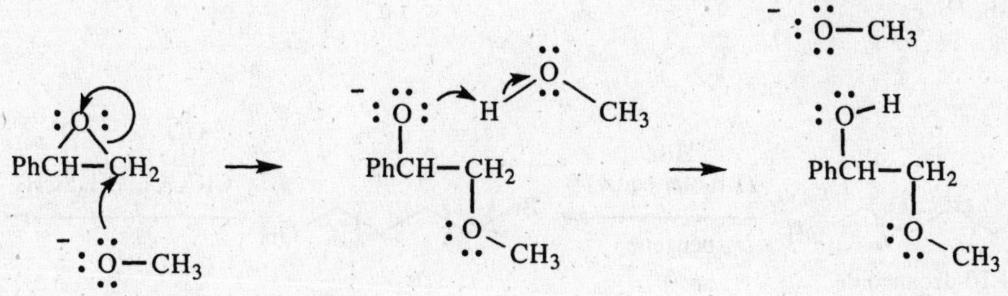
PhCH—CH2
O—CH3
H
CH3

13.48 (a)

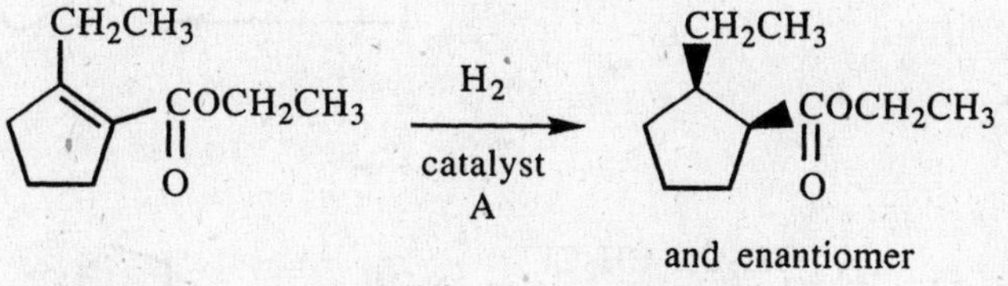
CH2CH3
COCH2CH3
O
H2
catalyst
A
and enantiomer

(b)

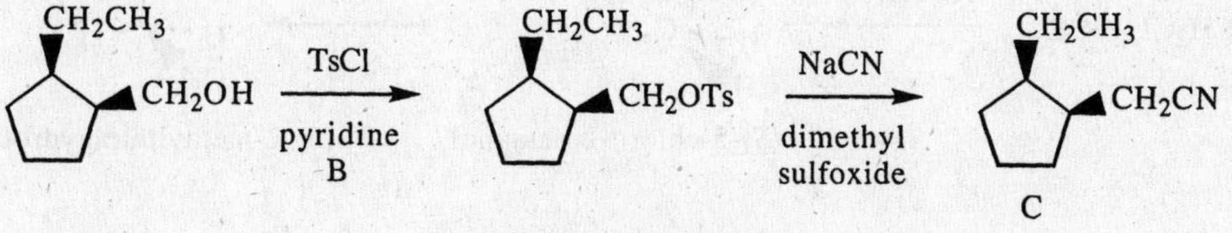
CH2CH3
CH2OH
TsCl
pyridine
B
CH2OTs
NaCN
dimethyl
sulfoxide
CH2CN
C

(c)

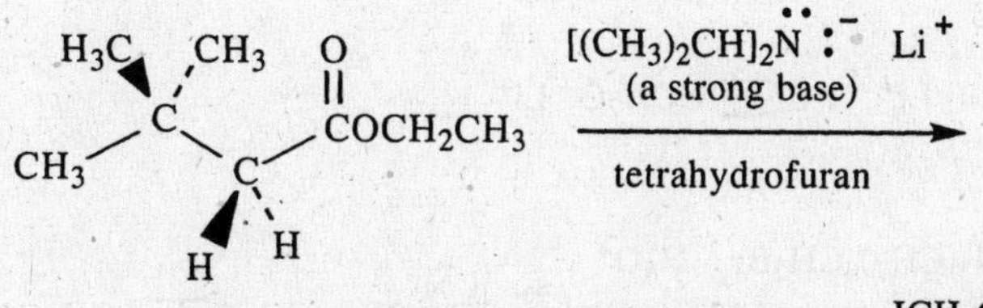
H3C CH3 O
COCH2CH3
CH3
H
[(CH3)2CH]2N:− Li+
(a strong base)
tetrahydrofuran

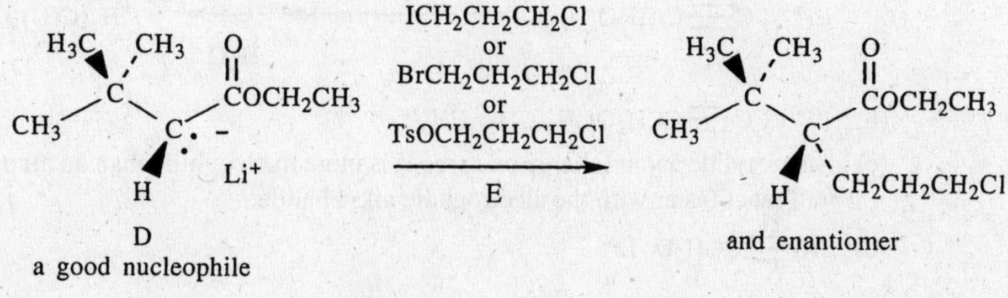
H3C CH3 O
COCH2CH3
CH3
Li+
H
D
a good nucleophile
ICH2CH2CH2Cl
or
BrCH2CH2CH2Cl
or
TsOCH2CH2CH2Cl
E
CH2CH2CH2Cl
and enantiomer

13.48 (d)

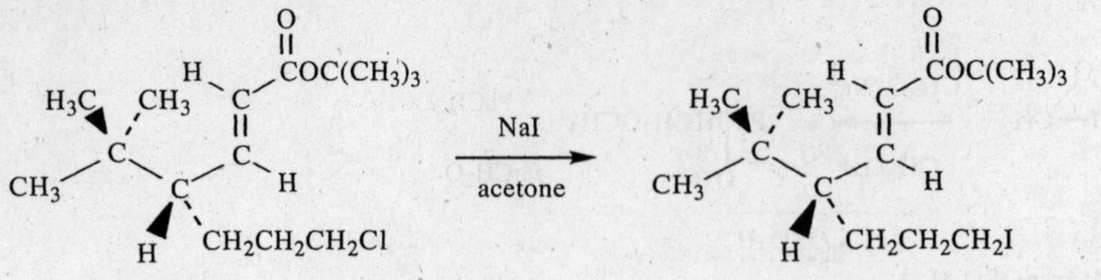

F

13.49 (a)

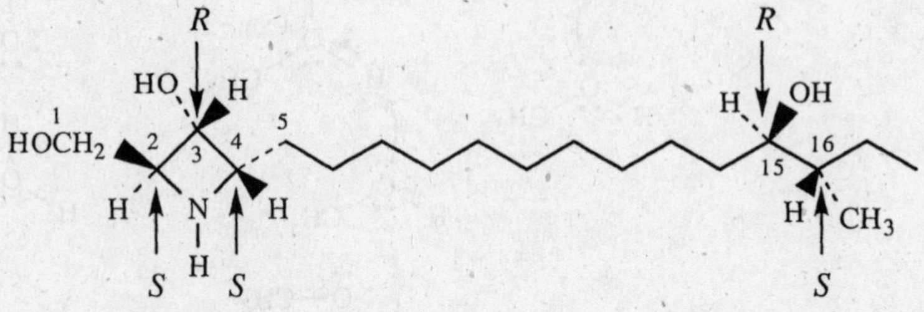

(b)

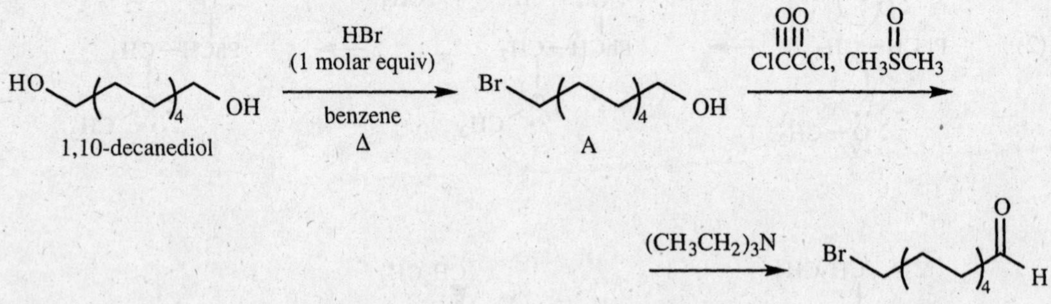

1,10-decanediol A B

13.50

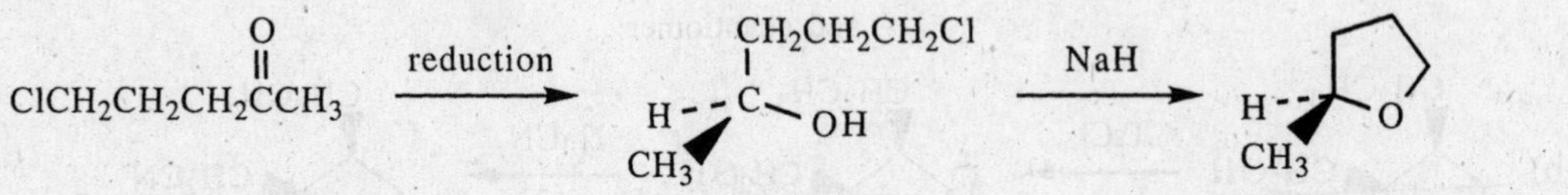

(*S*)-5-chloro-2-pentanol (*S*)-2-methyltetrahydrofuran

13.51 (a) $HC{\equiv}CCH_2OH \xrightarrow[NH_3\ (liq)]{2\ LiNH_2} Li^{+}\ {}^{-}{:}C{\equiv}CCH_2O^{-}\ Li^{+}$ (A)

$Li^{+}\ {}^{-}{:}C{\equiv}CCH_2O^{-}\ Li^{+}$ (A) $\xrightarrow{CH_3(CH_2)_3CH_2Br} \xrightarrow[H_2O]{H_3O^{+}} CH_3(CH_2)_3CH_2C{\equiv}CCH_2OH$ (2-octyn-1-ol)

(b) $Li^{+}\ {}^{-}{:}C{\equiv}CCH_2OCH_2(CH_2)_3CH_3$

(c) An acetylide anion (charge on carbon) is more nucleophilic than an alkoxide ion (charge on oxygen) and will react faster with the electrophilic alkyl halide.

(d) $HC{\equiv}CCH_2O^{-}\ Li^{+}$

13.52

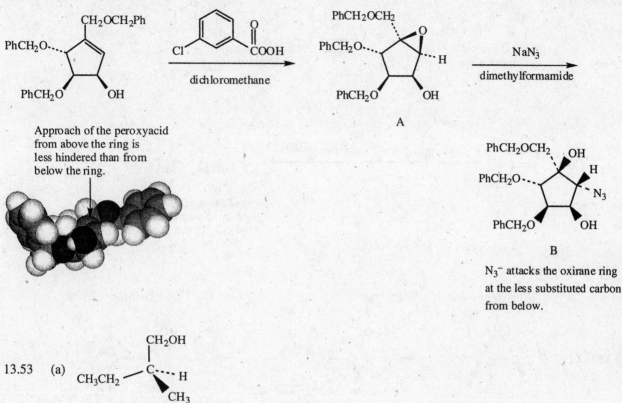

Approach of the peroxyacid from above the ring is less hindered than from below the ring.

A

N_3^- attacks the oxirane ring at the less substituted carbon from below.

B

13.53   (a)

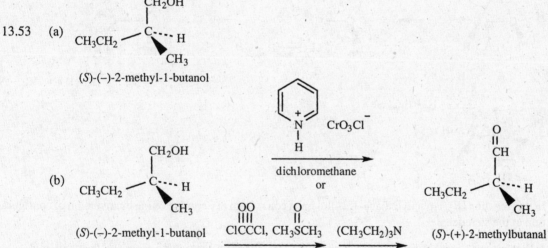

(S)-(−)-2-methyl-1-butanol

(b)

(S)-(−)-2-methyl-1-butanol

(S)-(+)-2-methylbutanal

13.54   (a)

looks *R* because we are looking at the stereocenter from the side on which the group of lowest priority comes up at us, therefore it is really *S*.

(2S, 5S)-2,5-dimethyl-1-boracyclopentane

The numbers in the circles indicate priorities of groups on the left-hand stereocenter for the purpose of assigning configuration; the numbers in the squares indicate priority of groups on the right-hand stereocenter. Boron has a lower atomic number than carbon and so comes third in the order of priorities.

13.54   (b)

(S)-(+)-2-methyl-
cyclopentanone

or

(R)-(−)-2-methyl-
cyclopentanone

(c)

H_2O_2, NaOH

syn addition of H—OH to give
the R configuration at the carbon
atom bearing the methyl group

13.55   (a)

(R)-styrene oxide   (R)-1-phenylethane-1,2-diol

(b)

In this reaction (R)-1-phenylethane-1,2-diol must come from (S)-styrene oxide by inversion of configuration at the stereocenter.

13.56   The following equation can be used to calculate the free energy of activation from the heat and entropy of activation.

$$\Delta G^{\ddagger} = \Delta H^{\ddagger} - T\Delta S^{\ddagger}$$

For the formation of oxirane

$$\Delta G^{\ddagger} = (23.2 \text{ kcal/mol} - (303 \text{ K})(9.9 \times 10^{-3} \text{ kcal/mol·K})$$

$$\Delta G^{\ddagger} = 20.2 \text{ kcal/mol}$$

For the formation of tetrahydrofuran

$$\Delta G^{\ddagger} = (19.8 \text{ kcal/mol} - (303 \text{ K})(-5 \times 10^{-3} \text{ kcal/mol·K})$$

$$\Delta G^{\ddagger} = 21.3 \text{ kcal/mol}$$

The formation of oxirane has the smaller free energy of activation and therefore the faster rate. Even though the heat of activation is smaller for the formation of tetrahydrofuran than for oxirane, the reaction giving rise to tetrahydrofuran has a negative (unfavorable) entropy of activation. The entropy of activation for the formation of oxirane is positive, contributing to a smaller overall free energy of activation for the reaction.

13.57   Compound A, $C_7H_{16}O$, has no units of unsaturation. The infrared spectrum shows the presence of an alcohol (3300 cm^{-1} for the O—H stretch and 1110 cm^{-1} for the C—O stretch).

13.57 (cont)

Compound B, $C_7H_{14}O$, has one unit of unsaturation. The infrared spectrum shows the presence of an aldehyde (2700 cm^{-1} for the C—H stretch and 1725 cm^{-1} for the C=O stretch).

Analysis of the proton magnetic resonance spectrum of Compound A is shown below:

δ 0.9	(3H, t, C$\underline{H}_3$CH$_2$—)
δ 1.3	(10H, m, —(C$\underline{H}_2$)$_5$—) (A linear chain of methylene groups that have similar chemical shifts and are at least one carbon away from electron-withdrawing atoms or π bonds will usually appear as a complex multiplet in this region.)
δ 2.2	(1H, s, —O$\underline{H}$)
δ 3.6	(2H, t, —CH$_2$C$\underline{H}_2$—O—)

The ^{13}C magnetic resonance spectrum shows six alkyl carbon atoms (δ 14.2, 23.1, 26.4, 29.7, 32.4, and 33.2) suggesting a linear chain. The seventh carbon atom (δ 62.2) falls in the range of a carbon atom bonded to an oxygen atom. Compound A is 1-heptanol, $CH_3(CH_2)_5CH_2OH$. It must be oxidized by pyridinium chlorochromate to heptanal.

Analysis of the proton magnetic resonance spectrum of Compound B confirms that it is heptanal.

δ 0.9	(3H, t, C$\underline{H}_3$CH$_2$—)
δ 1.3	(8H, m, —(C$\underline{H}_2$)$_4$—)
δ 2.4	(2H, m, —C$\underline{H}_2$C=O)
δ 9.8	(1H, t, —C$\underline{H}_2$C$\overset{\displaystyle O}{\overset{\displaystyle \|}{H}}$)

The ^{13}C magnetic resonance spectrum confirms the presence of an aldehyde functional group (δ 202.2).

$$CH_3(CH_2)_5CH_2OH \qquad\qquad CH_3(CH_2)_5\overset{\displaystyle O}{\overset{\displaystyle \|}{C}}H$$
$$\text{A} \qquad\qquad\qquad\qquad\qquad \text{B}$$

13.58 Compound C, $C_5H_{12}O$, has no units of unsaturation. The infrared spectrum shows the presence of an alcohol (3400 cm^{-1} for the O—H stretch). Compound C does not react with chromium trioxide so it must be a tertiary alcohol. Only one structural formula is possible for a five-carbon tertiary alcohol. Compound C must be 2-methyl-2-butanol (*tert*-amyl alcohol).

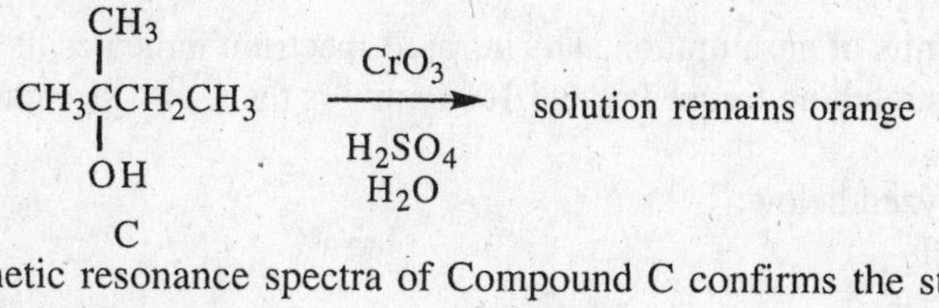

$$\underset{\displaystyle \text{C}}{\overset{\displaystyle CH_3}{\underset{\displaystyle OH}{CH_3CCH_2CH_3}}} \quad \xrightarrow[\substack{H_2SO_4 \\ H_2O}]{CrO_3} \quad \text{solution remains orange}$$

Analysis of the magnetic resonance spectra of Compound C confirms the structure. The proton magnetic resonances are given below:

δ 0.9	(m, (C$\underline{H}_3$CH$_2$—)
δ 1.2	(s, (C$\underline{H}_3$)$_2$C—)
δ 1.5	(m, —C$\underline{H}_2$CH$_3$)

There is no peak for a hydrogen atom on a carbon bonded to an oxygen atom (δ ~4), which also confirms that Compound C is a tertiary alcohol. The proton on the oxygen atom is hidden in one of the multiplets.

The ^{13}C magnetic resonance spectrum shows only four types of carbon atoms. Three of the types are alkyl carbon atoms (δ 8.8, 28.9, and 36.8) and one is a tertiary carbon atom bonded to an oxygen atom (δ 70.6).

Compound D, $C_5H_{12}O$, also has no units of unsaturation. The infrared spectrum again shows the presence of an alcohol (3400 cm^{-1} for the O—H stretch). A positive reaction with chromium trioxide points to a secondary or primary alcohol. The possible structures are shown on the next page:

13.58 (cont)

Primary alcohols:

$$CH_3CH_2CH_2CH_2CH_2OH$$

$$\underset{\overset{|}{CH_3}}{CH_3CHCH_2CH_2OH}$$

$$\underset{\overset{|}{CH_3}}{CH_3CH_2CHCH_2OH}$$

$$\underset{\overset{|}{CH_3}}{\overset{\overset{CH_3}{|}}{CH_3CCH_2OH}}$$

Secondary alcohols:

$$\underset{\overset{|}{OH}}{CH_3CH_2CH_2CHCH_3}$$

$$\underset{\overset{|}{OH}}{CH_3CH_2CHCH_2CH_3}$$

$$\underset{\overset{|}{OH}}{\overset{\overset{CH_3}{|}}{CH_3CHCHCH_3}}$$

The ^{13}C magnetic resonance spectrum shows five peaks. The three alcohols that have five different types of carbon atoms are 1-pentanol, 2-pentanol, and 2-methyl-1-butanol. Of these, only 2-pentanol will have the observed doublet for the hydrogen atoms on carbon-1 and a multiplet for the hydrogen atom on the carbon atom bearing the hydroxyl group.

Analysis of the rest of the magnetic resonance spectra of Compound D confirms the structure. The proton magnetic resonances are given below:

δ 0.9-1.4   (complex multiplet, C$\underline{H}_3$C$\underline{H}_2$C$\underline{H}_2$—)
δ 1.2   (d, C$\underline{H}_3$CH—)
δ 3.7   (m, — CH$_2$C$\underline{H}$CH$_3$)
$$\underset{\overset{\overset{|}{O}}{\,}}{\,}$$
$$O—$$

$$CH_3CH_2CH_2CHCH_3 \xrightarrow[\substack{H_2SO_4 \\ H_2O}]{CrO_3} CH_3CH_2CH_2\overset{\overset{O}{\parallel}}{C}CH_3$$
$$\underset{\overset{|}{OH}}{\,}$$
$$D$$

Note that the hydrogen atom on the hydroxyl group is not seen as a separate peak in the spectrum of either Compound C or Compound D.

13.59   Compound E has 4 units of unsaturation. The infrared spectrum indicates the presence of an alcohol (3338 cm^{-1} is the O—H stretching frequency and 1031 cm^{-1} is the C—O stretching frequency).

The ^{13}C spectrum is analyzed below:
δ 32.02   (t, —C$\underline{H}_2$—)
δ 34.13   (t, —C$\underline{H}_2$—)
δ 61.95   (t, —C$\underline{H}_2$OH)
δ 125.74   (d, sp^2-C$\underline{H}$)
δ 128.29   (d, sp^2-C$\underline{H}$)
δ 128.33   (d, sp^2-C$\underline{H}$)
δ 141.78   (s, sp^2-$\underline{C}$)

The proton magnetic resonance spectrum is analyzed below:
δ 1.8   (2H, quintet, —CH$_2$C$\underline{H}_2$CH$_2$—)
δ 2.6   (1H, s, —O$\underline{H}$)
δ 2.8   (2H, t, ArC$\underline{H}_2$CH$_2$—)
δ 3.6   (2H, multiplet, —CH$_2$C$\underline{H}_2$OH) (see next page)
δ 7.2   (5H, multiplet, Ar$\underline{H}$)

The compound is 3-phenyl-1-propanol. The assignments are shown on the next page.

13.59 (cont).

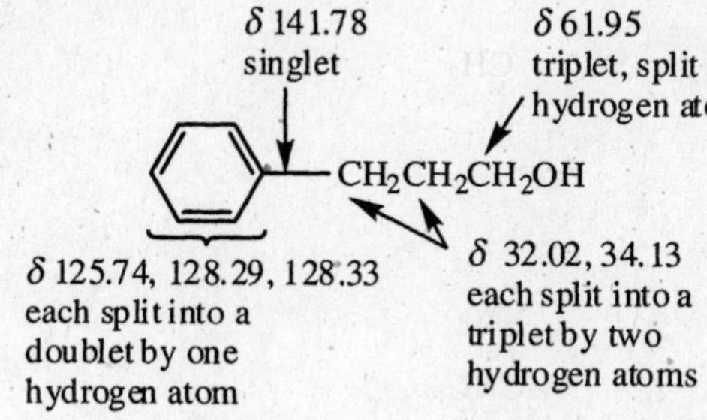

δ 141.78
singlet

δ 61.95
triplet, split by two
hydrogen atoms

$\text{CH}_2\text{CH}_2\text{CH}_2\text{OH}$

δ 125.74, 128.29, 128.33
each split into a
doublet by one
hydrogen atom

δ 32.02, 34.13
each split into a
triplet by two
hydrogen atoms

^{13}C nuclear magnetic resonance

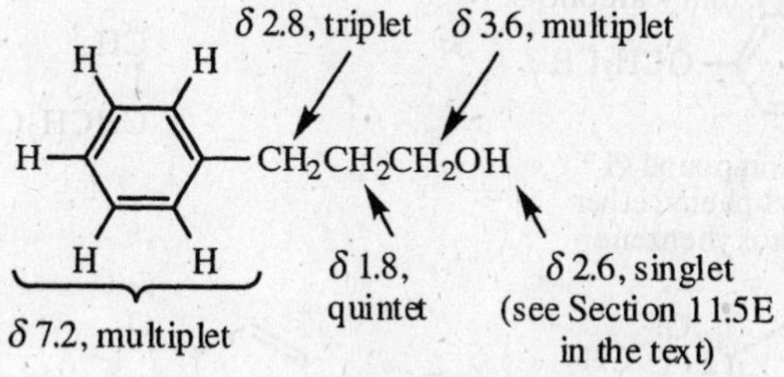

δ 2.8, triplet   δ 3.6, multiplet

$\text{CH}_2\text{CH}_2\text{CH}_2\text{OH}$

δ 7.2, multiplet

δ 1.8,
quintet

δ 2.6, singlet
(see Section 11.5E
in the text)

proton magnetic resonance

A molecular formula of $C_9H_{12}O$ corresponds to a molecular weight of 136. The molecular ion in the mass spectrum is at m/z 136. The base peak (m/z 91) is the tropylium ion, which results from loss of a C_2H_5O unit and is the result of the following fragmentation:

$\cdot\text{CH}_2\text{—}\text{CH}_2\text{CH}_2\ddot{\text{O}}\text{H} \longrightarrow \cdot\text{CH}_2\text{CH}_2\ddot{\text{O}}\text{H} \quad + \quad [\quad =\text{CH}_2] \longrightarrow \quad +$

M⁺
m/z 136

tropylium ion
m/z 91

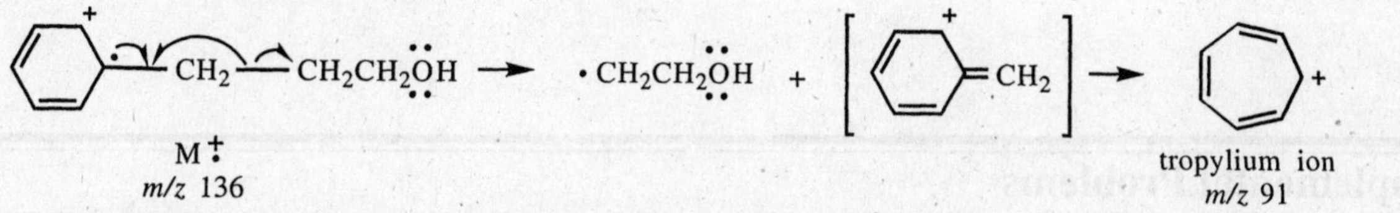

13.60 The infrared spectrum tells us that Compound F is an alcohol (3329 cm^{-1} is the O—H stretching frequency and 1046 cm^{-1} is the C—O stretching frequency), and Compound G is an ether (1245 and 1049 cm^{-1} are the C–O stretching frequencies). The proton magnetic resonance spectra are analyzed below:

Compound F:

δ 2.5   (2H, t, J = 7 Hz, —C$\underline{\text{H}}_2$CH$_2$—)

δ 3.4   (2H, t, J = 7 Hz, —OC$\underline{\text{H}}_2$CH$_2$—)

δ 3.7   (1H, s, —O$\underline{\text{H}}$)

δ 6.7   (5H, s, Ar$\underline{\text{H}}$)

Compound G:

δ 1.3   (3H, t, J = 7 Hz, —CH$_2$C$\underline{\text{H}}_3$)

δ 3.7   (2H, q, J = 7 Hz, —OC$\underline{\text{H}}_2$CH$_3$)

δ 6.3 – 7.0 (5H, m, Ar$\underline{\text{H}}$)

The fragments for both Compound F and Compound G add up to $C_8H_{10}O$, which corresponds to a molecular weight of 122, which is where the molecular ion of each compound appears in the mass spectrum. We thus have the molecular formulas for Compounds F and G. Compound F is 2-phenylethanol. The base peak for Compound F, m/z 91, is the tropylium ion, which results from the following fragmentation:

$\text{—CH}_2\text{CH}_2\text{OH}$

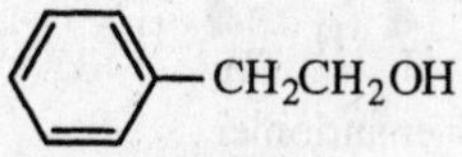

Compound F
2-phenylethanol

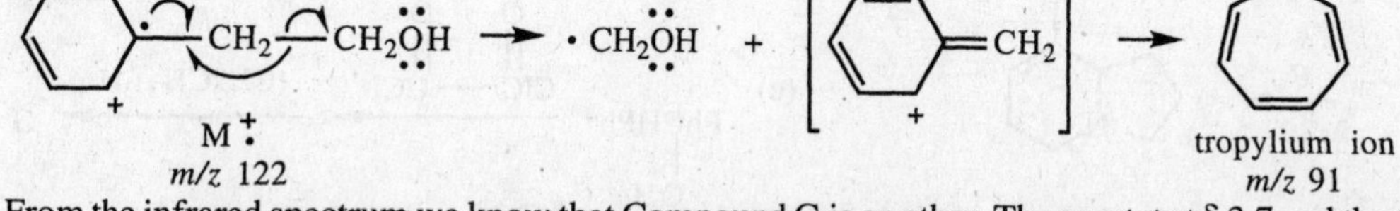

$\cdot\text{CH}_2\text{—}\text{CH}_2\ddot{\text{O}}\text{H} \longrightarrow \cdot\text{CH}_2\ddot{\text{O}}\text{H} \quad + \quad [\quad =\text{CH}_2] \longrightarrow \quad +$

M⁺
m/z 122

tropylium ion
m/z 91

From the infrared spectrum we know that Compound G is an ether. The quartet at δ 3.7 and the triplet at δ 1.3 in the proton magnetic resonance spectrum tell us that there is an isolated ethyl group in which the methylene group is attached to an oxygen atom. Compound G is ethyl phenyl ether (ethoxybenzene). The base peak in Compound G, m/z 94, is 28 mass units less than the molecular ion and corresponds to a rearrangement in which ethylene, C_2H_4, is lost.

13.60 (cont)

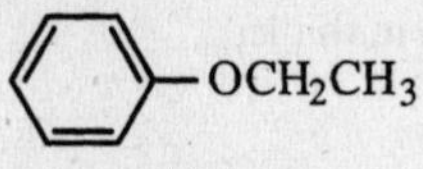

Compound G
ethyl phenyl ether
ethoxybenzene

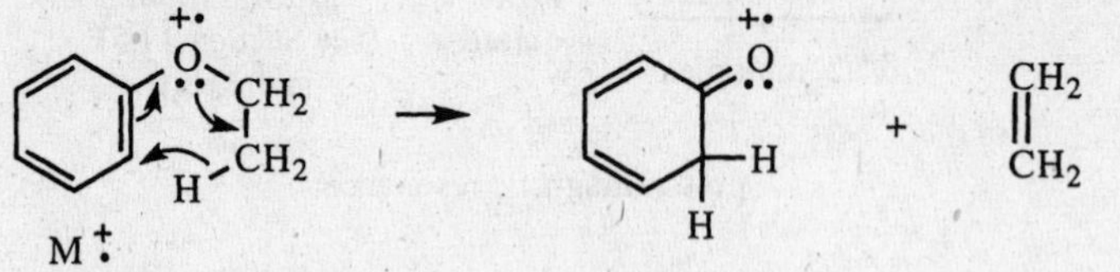

M⁺
m/z 122

m/z 94

Supplemental Problems

S13.1   Find the products or reagents corresponding to each of the capital letters. Show stereochemistry by appropriate conventions. When a racemic mixture forms, show the structure for one enantiomer and write "and enantiomer" below it.

(a)

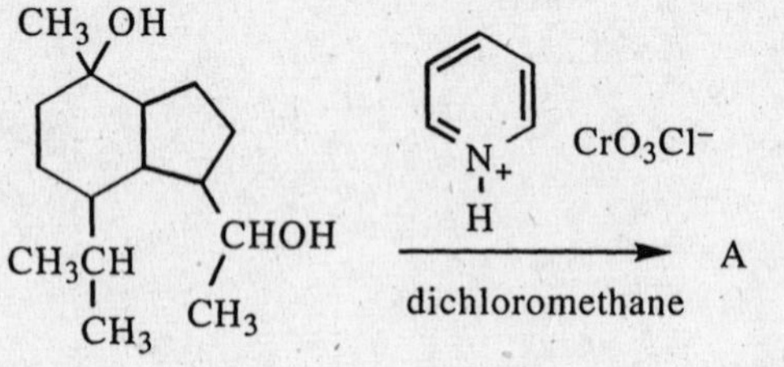

(b)

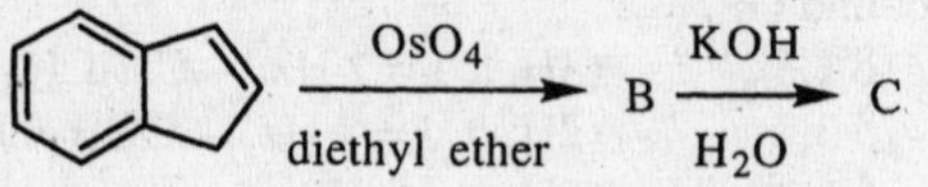

(c)

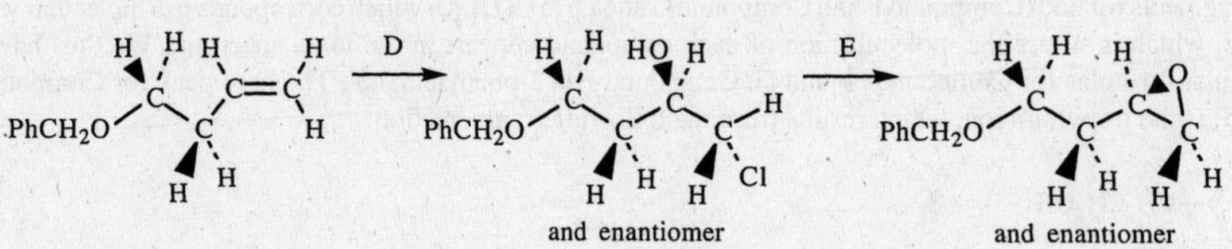

and enantiomer   and enantiomer

(d)

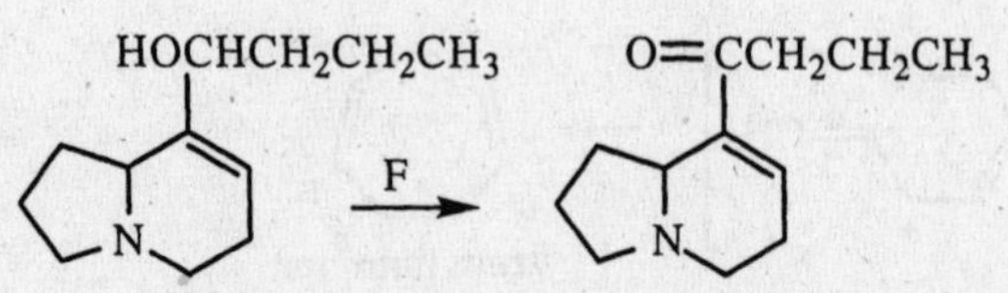

(e)

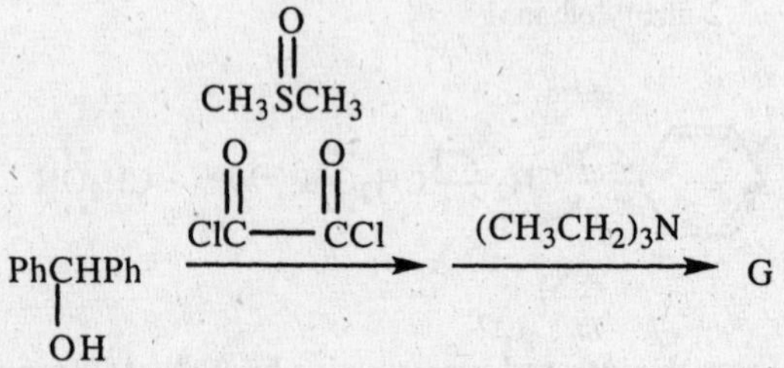

S13.2   Supply the reagents that are necessary and the intermediate compounds that will form in the following transformations. There may be more than one good way to carry out each synthesis.

(a) $HOCH_2(CH_2)_6CH_2OH$ ⟶ $PhCH_2OCH_2(CH_2)_6CH$ (with =O)

(b)
$$HOCH_2\underset{\underset{NH_3\ Br^-}{+}}{\overset{\overset{CH_2CH_3}{|}}{C}}CH_2OH \longrightarrow BrCH_2\underset{\underset{NH_3\ Br^-}{+}}{\overset{\overset{CH_2CH_3}{|}}{C}}CH_2Br$$

(c) 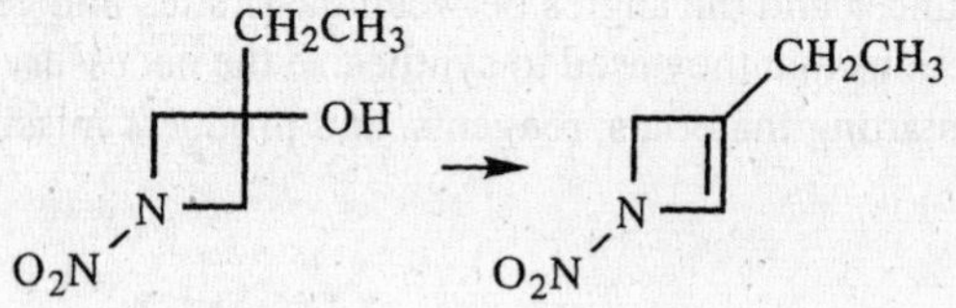 [structure: azetidine ring with CH_2CH_3 and OH on carbon, N bearing O_2N] ⟶ [azetine ring with CH_2CH_3, $N-O_2N$]

(d) 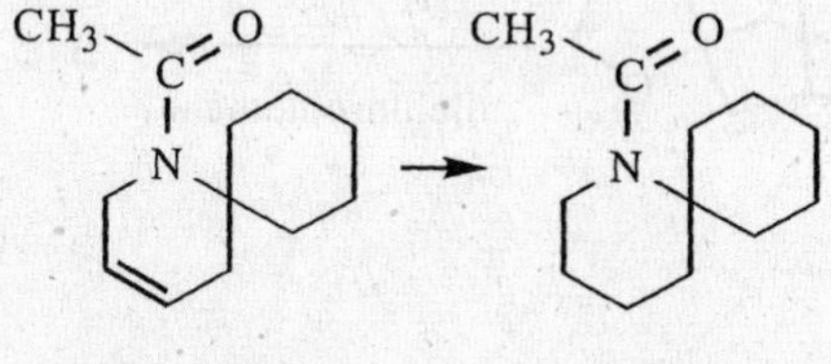 [spirocyclic acetyl enamine] ⟶ [spirocyclic acetyl amine, reduced ring]

(e) 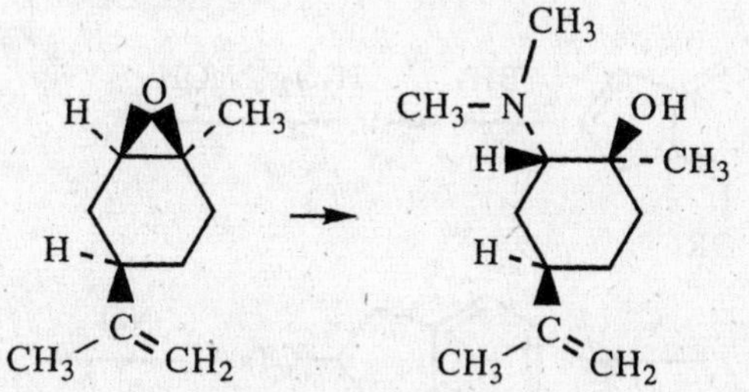 [epoxide cyclohexane with H, CH_3, isopropenyl $CH_3-C=CH_2$] ⟶ [CH_3-N, OH, CH_3 substituted cyclohexane with isopropenyl group]

(f) 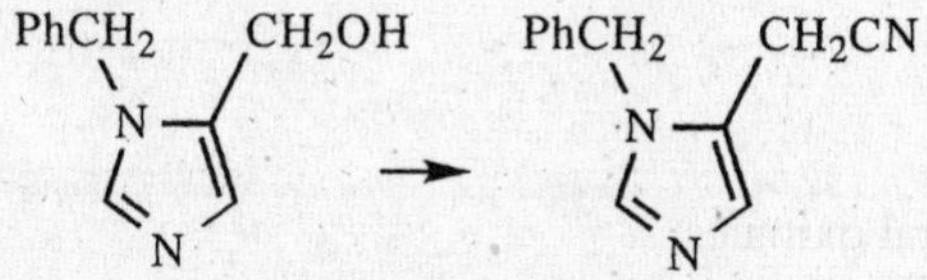 [imidazole with $PhCH_2$ on N and CH_2OH] ⟶ [imidazole with $PhCH_2$ on N and CH_2CN]

(g) $CH_3(CH_2)_3CH{=}CH_2$ ⟶ $CH_3(CH_2)_4CH_2OCH_3$

S13.3   (a)   The following reactions were carried out in research leading to the synthesis of a compound isolated from the lichens of the Western Himalayas. Supply structures for the major products of the reactions shown below.

(1) 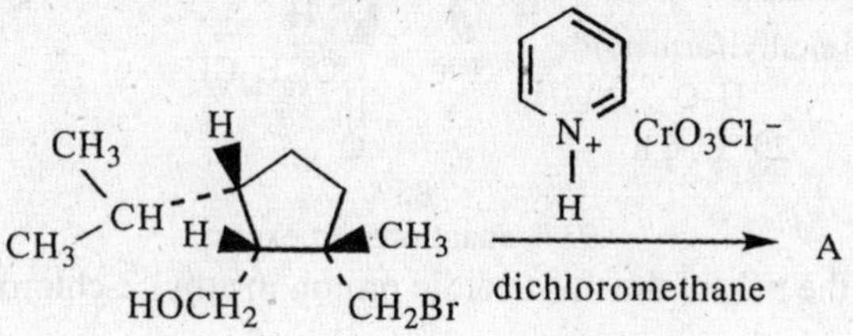 [cyclopentane with $CH_3CH(CH_3)-CH$, H, $HOCH_2$, CH_2Br, CH_3 substituents] $\xrightarrow[\text{dichloromethane}]{\text{pyridinium } CrO_3Cl^-}$ A

(2) [cyclopentene with isopropyl, CH_3, and $CH_2O-Si(CH_3)_2CC(CH_3)_3$ (t-butyldimethylsilyl) groups] $\xrightarrow[\text{tetrahydrofuran}]{BH_3}$ $\xrightarrow{H_2O_2,\ NaOH}$ B + C

(3) [cyclopentane with $CH_3-CH(CH_3)$, H, CH_3, $=CH_2$, CH_2OH substituents] $\xrightarrow{O_3,\ CH_3OH}$ $\xrightarrow{CH_3SCH_3}$ D + E

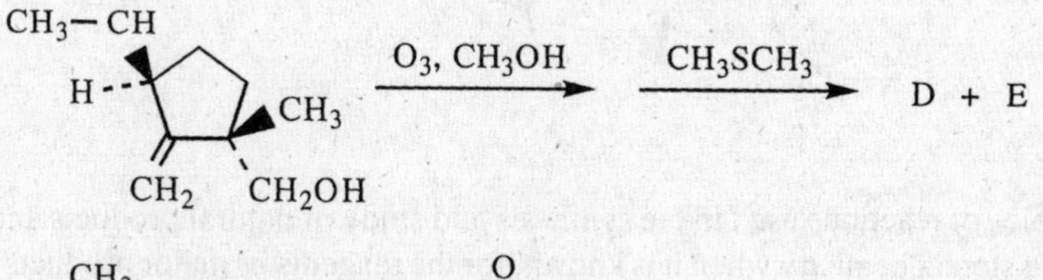

(4) 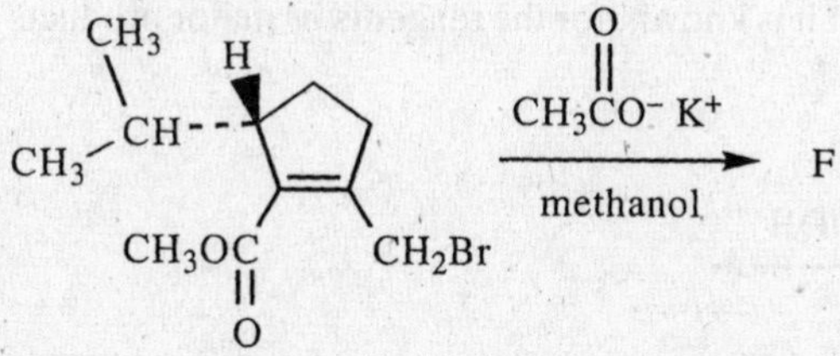 [cyclopentene with $CH_3CH(CH_3)-$, H, $CH_3OC(=O)$, CH_2Br substituents] $\xrightarrow[\text{methanol}]{CH_3CO^-\ K^+\ (=O)}$ F

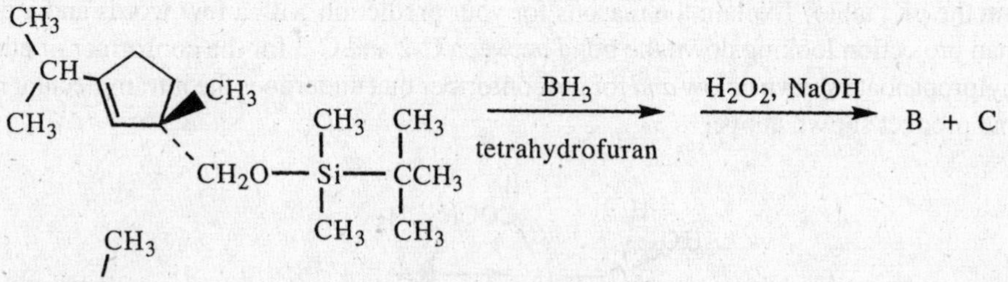

S13.3   (b) The starting material for the first reaction was found to be quite unstable and to lose HBr spontaneously to give a cyclic ether. Propose a structure for the ether and show, using the curved arrow representation, how it would form.

S13.4   In research into the stereochemistry of proton transfer reactions, researchers synthesized compounds in which the distances and the angles between basic sites and certain protons in the same molecules would be known. Some of the reactions that they used to synthesize the necessary compounds are shown below. Draw correct structural formulas for the starting materials, reagents, and products missing from the equations.

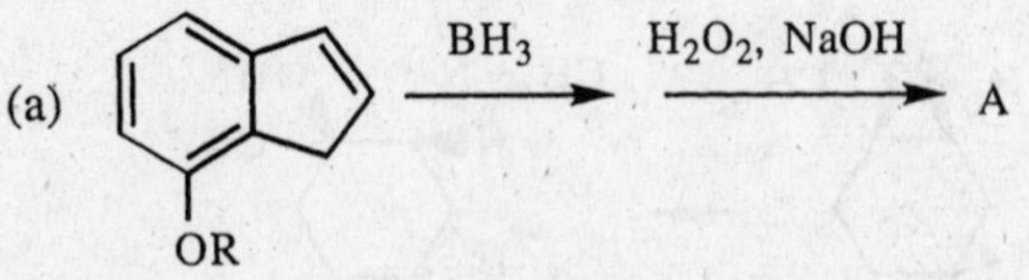

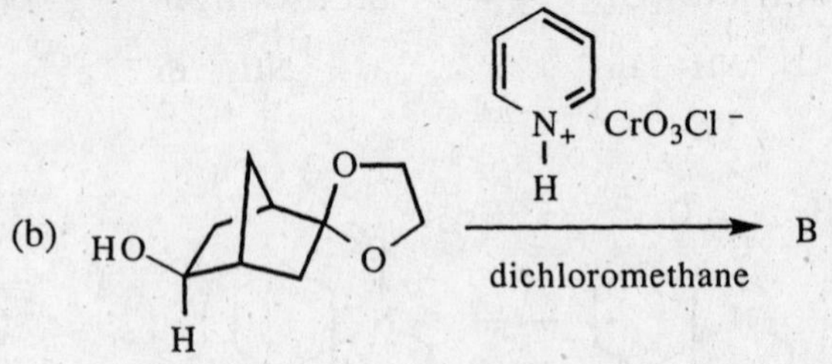

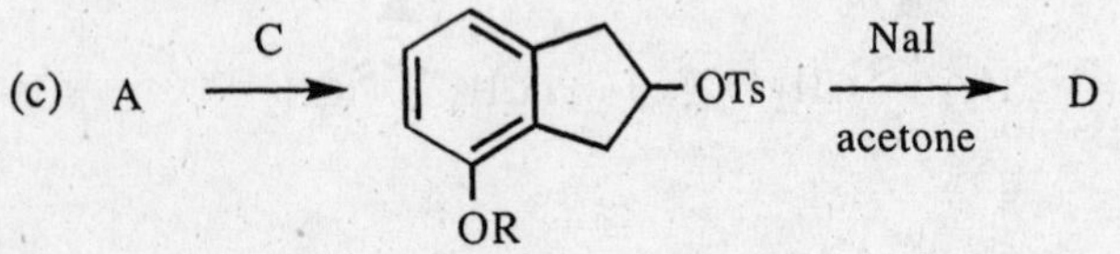

S13.5   The following reaction was observed in the synthesis of chiral oxiranes.

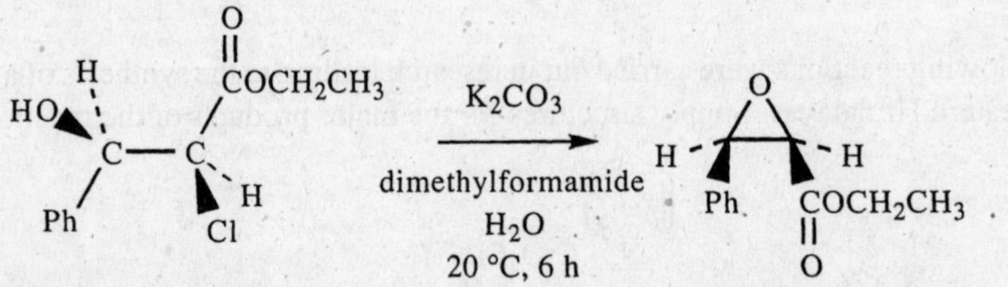

ethyl 2-chloro-3-hydroxy-
3-phenylpropanoate

85%
91% enantiomeric excess

(a) What is the best analogy from the pK_a table for the pK_a of the most acidic proton in ethyl 2-chloro-3-phenylpropanoate?
(b) Do you expect ethyl 2-chloro-3-hydroxy-3-phenylpropanoate to be a weaker or stronger acid than the compound you selected from the pK_a table? Explain the reasons for your prediction with a few words and a picture.
(c) Draw the Newman projection looking down the bond between C-2 and C-3 for the conformer of ethyl 2-chloro-3-hydroxy-3-phenylpropanoate shown below *and* for the conformer that undergoes the intramolecular reaction giving rise to the oxirane product shown above.

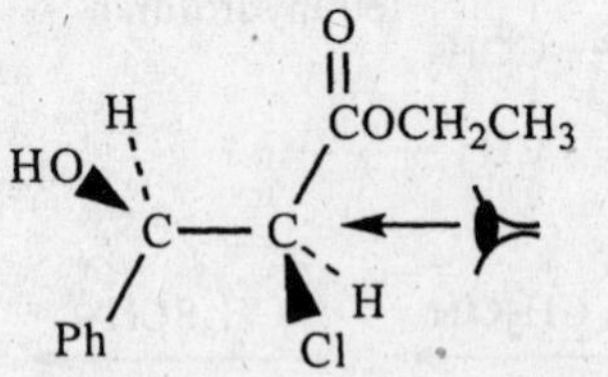

S13.6   The following are examples of reactions used in the synthesis and study of natural products and their analogs. Supply precise formulas, including stereochemistry when it is known, for the reagents or major products as indicated by the capital letters.

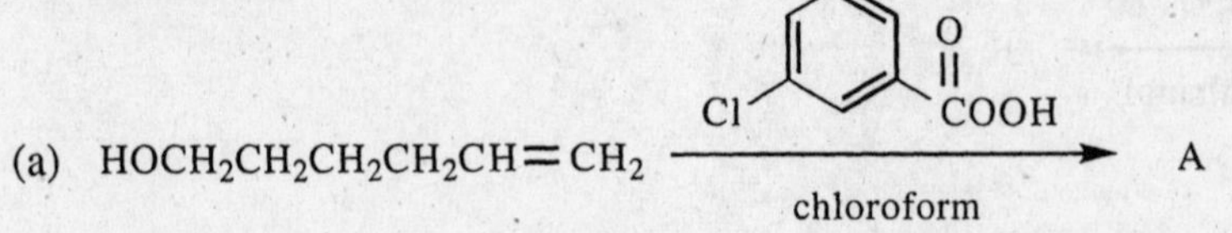

(a) $HOCH_2CH_2CH_2CH_2CH=CH_2$ $\xrightarrow{\text{chloroform}}$ A

S13.6

(b) $\xrightarrow{\text{NH}_3 \text{ (excess)}}$ B

(c) $\xrightarrow[\substack{\text{chloroform} \\ -78\ ^\circ\text{C}}]{\substack{\text{Br}_2 \\ \text{(1 equivalent)}}}$ C $\xrightarrow[\text{tetrahydrofuran}]{\text{OsO}_4}$ $\xrightarrow{\text{H}_2\text{O}}$ D

(d) $\xrightarrow[\substack{\text{dichloromethane} \\ -78\ ^\circ\text{C}}]{\text{O}_3}$ $\xrightarrow[\substack{\text{(behaves like} \\ \text{CH}_3\text{SCH}_3)}]{\text{Ph}_3\text{P}}$ E + F

(e) CH_3O——$\text{CH}_2\text{CH}_2\text{CH}=\text{CH}_2$ $\xrightarrow{\text{G}}$ $\xrightarrow{\text{H}}$ CH_3O——$\text{CH}_2\text{CH}_2\text{CH}_2\text{CH}_2\text{OH}$

(f) $\xrightarrow{\text{I}}$ $\xrightarrow{\text{J}}$ $\xrightarrow[\text{(CH}_3\text{CH}_2)_3\text{N}]{\text{CH}_3\text{SCH}_3,\ \text{ClC}-\text{CCl}}$ K

$\text{Tr} = \text{C(Ph)}_3$

a protecting
group for alcohols

S13.7   Provide structural formulas for the starting materials, reagents, or products indicated by capital letters.

(a) $\xrightarrow[\text{(CH}_3\text{CH}_2)_3\text{N}]{\substack{\text{O} \quad\quad \text{O}\ \text{O} \\ \text{CH}_3\text{SCH}_3,\ \text{ClC}-\text{CCl}}}$ A

(b) $\xrightarrow[\text{hexane}]{\text{CH}_3\text{CH}_2\text{CH}_2\text{CH}_2\text{Li}}$ B
product of an
E2 elimination

(c) $\xrightarrow[\text{Pd-C}]{\text{H}_2}$ C $\xrightarrow[\text{pyridine}]{\text{TsCl}}$ D $\xrightarrow{\text{NaBr}}$ E $\xrightarrow[\substack{\text{(a good source} \\ \text{of nucleophilic} \\ \text{hydride, H}:^-)}]{\text{NaBH}_3\text{CN}}$ F

S13.7

(d) $CH_2=CHCH_2CH_2C-\overset{CH_3}{\underset{CH_3}{C}}-COCH_3$ (with two C=O groups) $\xrightarrow{O_3}$ $\xrightarrow{(CH_3)_2S}$ G + H

(e) $CH_3CH_2C\equiv CCH_2CH_3$ $\xrightarrow{I}$ J $\xrightarrow{K}$

(f) $\xrightarrow[\text{dimethyl-formamide}]{Na^+ H:^-}$ L $\xrightarrow[\text{dimethyl-formamide}]{PhCH_2Br}$ M

(g) $BrCH_2CH_2-\overset{CH_2CH_2Br}{\underset{CH_2CH_2Br}{C}}-CH_2CH_2Br$ $\xrightarrow[\text{acetonitrile}\ \Delta]{K^+ CN^- \text{ (excess)}}$ N

S13.8   Provide structural formulas for the starting materials, reagents, or products indicated by capital letters.

(a) $\xrightarrow{A}$ $\xrightarrow{B}$

(b) $\xrightarrow[\text{pyridine}]{\text{TsCl}\ (1\text{ equivalent})}$ $\xrightarrow{\text{NaCN}}$ C

(c) D $\xrightarrow{CH_3COOH}$ $\xrightarrow{NaN_3}$ $\xrightarrow{H_3O^+}$ E

(d) $\xrightarrow[\substack{H_2O,\ H_2SO_4\\ \text{acetone}}]{CrO_3}$ F

(e) $C\equiv CCH_2CH_3$ $\xrightarrow[\substack{Pd/BaSO_4\\ \text{quinoline}}]{H_2}$ G $\xrightarrow[H_2O]{OsO_4}$ H

(f) $\underset{C_6H_8}{I}$ $\xrightarrow{O_3}$ $\xrightarrow{(CH_3)_2S}$ $2\ HCCH_2CH$ (with two C=O groups)

(g) CH_3OC $NHPG$ $\xrightarrow[\text{ethyl acetate}]{H_2\ \ Pt/C}$ J + K (stereoisomers of each other)

PG is a protecting group for the amine. It does not participate in this reaction.

S13.9   When *trans*-1-bromo-2-methoxycyclopentane is treated with a dilute solution of aqueous sodium hydroxide, only *cis*-2-methoxy-1-cyclopentanol is obtained. When *trans*-2-bromocyclopentanol is subjected to the same conditions, only *trans*-1,2-cyclopentanediol is formed. Draw mechanisms for these two reactions that account for the experimental observations described above.

S13.10   2,5-Dimethyl-1,5-hexadiene is converted to 2,2,5,5-tetramethyltetrahydrofuran when treated with aqueous sulfuric acid, as shown below.

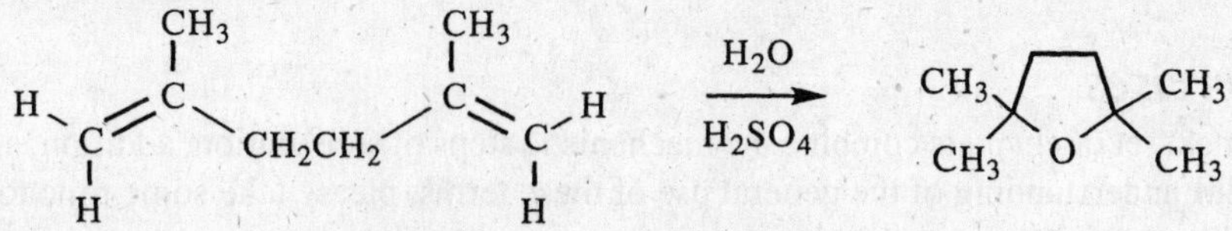

Write a clear, rational mechanism for this transformation using the curved-arrow convention.

S13.11   In a key step in the synthesis of ciguatoxin, a neurotoxin responsible for occurrences of seafood poisonings, the following reactions were observed.

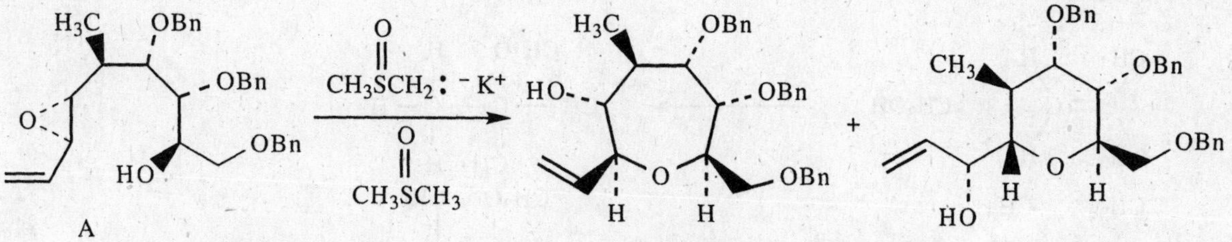

(a) The base used in this reaction is shown over the arrow in the equation given above. It is the conjugate base of dimethyl sulfoxide, shown below the arrow. Use your pK_a table to assign an approximate pK_a value to dimethyl sulfoxide. To which compound in the pK_a table is it most analogous?

(b) Write an equation for the reaction of compound A (abbreviated as ROH) with the conjugate base of dimethyl sulfoxide, and indicate by the relative lengths of the equilibrium arrows whether equilibrium will lie to the left or the right of the equation.

(c) Choose *one* of the two cyclic ethers, B or C, shown at the beginning of this problem, and write a mechanism using the curved arrow convention that rationalizes its formation from alcohol A in the presence of base. You may use B: and HB⁺ to indicate bases and acids in the mechanism.

14 Aldehydes and Ketones. Addition Reactions at Electrophilic Carbon Atoms

Workbook Exercises

The reactions in the next set of chapters combine the mechanistic steps of substitution, addition, and elimination. If you feel unsure about your understanding of the general use of these terms, please take some time to review.

Like carbon-carbon π bonds, carbon-oxygen and carbon-nitrogen π bonds undergo addition reactions. Notice the similar changes in connectivity associated with the addition of methanol, under acidic conditions, to a $\diagdown C = C \diagup$, a $\diagdown C = N \diagup$,

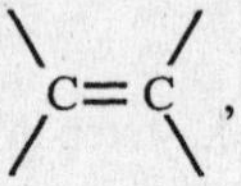

and a $\diagdown C = O$ double bond.

$$CH_3CH=CHCH_3\ (C=C\ arrangement) + CH_3OH \longrightarrow CH_3-C(CH_3)(OCH_3)-CH(H)-H$$

$$CH_3C(CH_3)=N-CH_3\ arrangement + CH_3OH \longrightarrow CH_3-C(OCH_3)(CH_3)-N(CH_3)-H$$

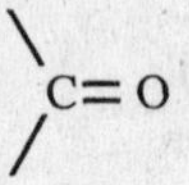

$$CH_3C(CH_3)=O + CH_3OH \longrightarrow CH_3-C(OCH_3)(H)-O-CH_3$$

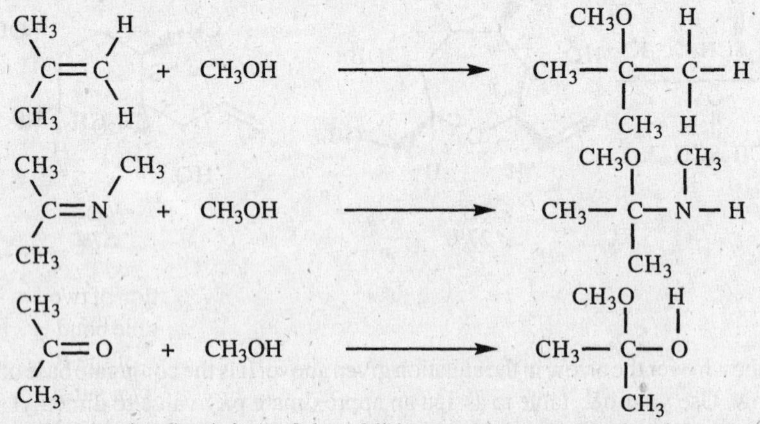

The orientation (regioselectivity) of the addition reaction to π bonds where one of the atoms is nitrogen or oxygen is highly predictable in the direction shown above. The reaction is also reversible.

$$CH_3-C(CH_3)=eN \underset{\text{elimination of NuH}}{\overset{\text{addition of NuH}}{\rightleftarrows}} CH_3-C(Nu)(CH_3)-eN(H)$$

$$eN = NR, O \qquad Nu = NR_2, OR, SR$$

The simple addition reaction products shown above are generally not observed when the nucleophilic group is a nitrogen (RNH_2), an alcohol (ROH), or a thiol (RSH). Instead, the observed product, in many cases, can be understood in terms of an overall loss of water (H_2O) between the reactants:

Connectivity changes that show the overall stoichiometry:
For nucleophiles involving RNH_2:

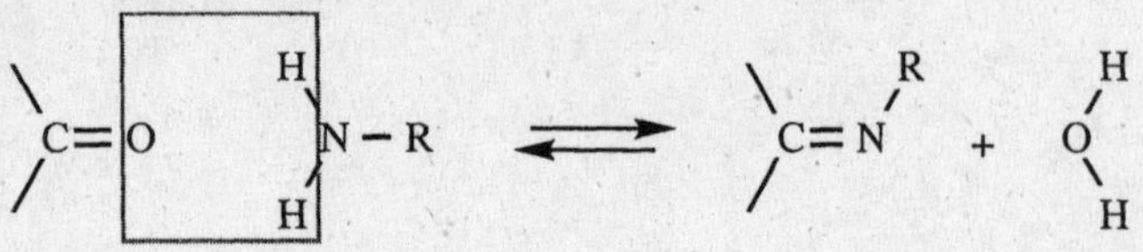

$$\diagdown C = O \quad + \quad H-N(H)-R \rightleftarrows \diagdown C = N-R \quad + \quad H-O-H$$

Workbook Exercises (cont)

For nucleophiles involving ROH or RSH:

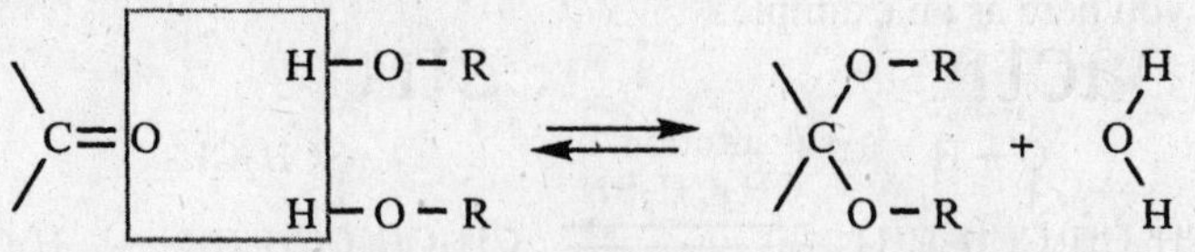

The ways of visualizing the reactions shown above are useful for quickly predicting structural relationships.

EXERCISE I. Predict the product of the reaction, where water is also formed, when each of the compounds shown below is combined in separate reactions with (1) ethanol (CH_3CH_2OH), (2) methylamine (CH_3NH_2), (3) hydrazine (NH_2NH_2), and (4) ethanethiol (CH_3CH_2SH).

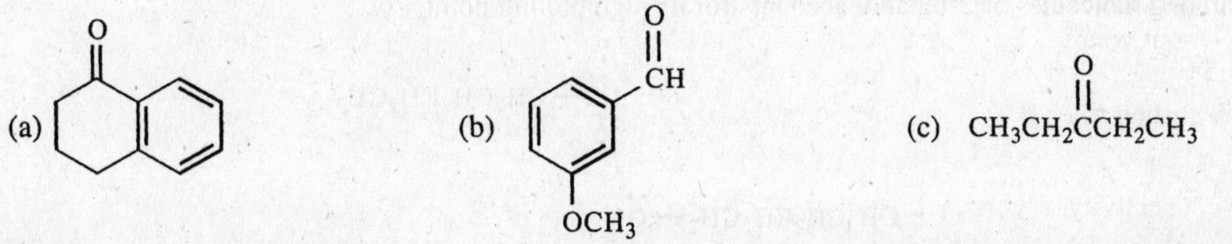

(a)

(b)

(c) $CH_3CH_2CCH_2CH_3$

EXERCISE II. The reactions shown above are reversible when an excess of water is present. What are the structures of the carbonyl compounds (the compounds with a carbon-oxygen π bond) and the nucleophile(s) that result from re-introducing water into the compounds shown below?

(a)

(b)

(c)

(d)

(e)

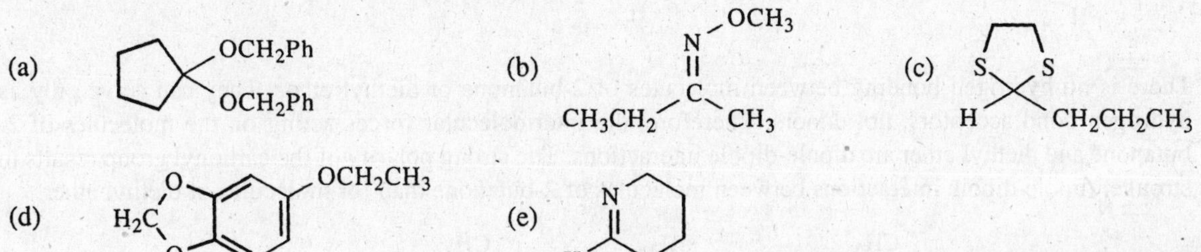

The transformations shown above and on the preceding page may be seen in detail as the combination of two fundamental mechanistic steps. In the case of the RNH_2 nucleophiles, the formation of the carbon-nitrogen double bond is understood to be an addition to the carbon-oxygen π bond followed by the elimination of water. In the case of the ROH and RSH nucleophiles, the first step is also an addition to the carbon-oxygen π bond. This is followed by a substitution of the hydroxyl group, which, under acidic conditions, leaves as a molecule of water. The reverse reactions also have addition-elimination or substitution-elimination steps.

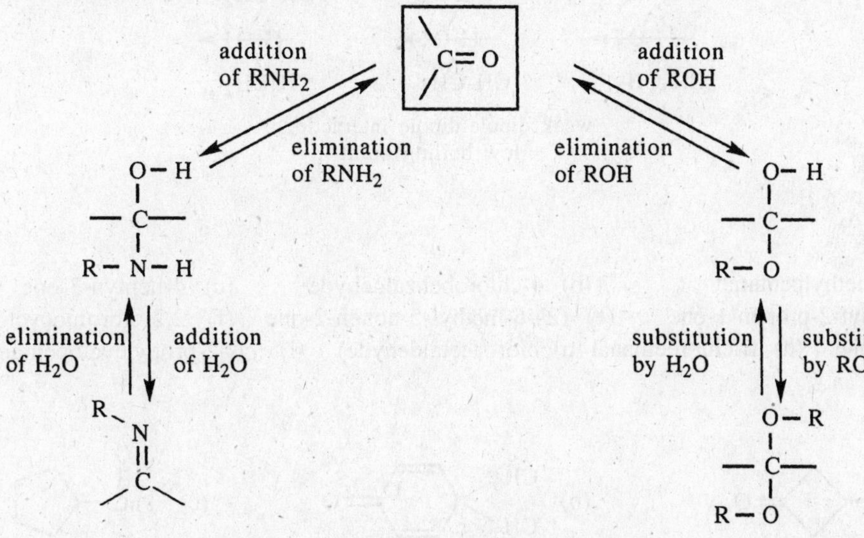

Workbook Exercises (cont)

EXERCISE III.   Redo Workbook Exercises I and II using the mechanistic scheme outlined on the previous page. Problem I (c), using ethanethiol, is solved for you here as an example.

SOLUTION

$$
\underset{CH_3CH_2CCH_2CH_3}{\overset{\overset{\displaystyle O}{\|}}{}}
\underset{\text{elimination of}\atop CH_3CH_2SH}{\overset{\text{addition of}\atop CH_3CH_2SH}{\rightleftharpoons}}
\underset{\underset{CH_3CH_2S}{CH_3CH_2CCH_2CH_3}}{\overset{\overset{\displaystyle O-H}{|}}{}}
\underset{\text{substitution by}\atop H_2O}{\overset{\text{substitution by}\atop CH_3CH_2SH}{\rightleftharpoons}}
\underset{\underset{CH_3CH_2S}{CH_3CH_2CCH_2CH_3}}{\overset{\overset{\displaystyle SCH_2CH_3}{|}}{}} + \; H_2O
$$

— —

14.1   2-Butanol is an alcohol and can serve as both a hydrogen bond donor and an acceptor. Strong hydrogen bonding among molecules of 2-butanol accounts for its high boiling point.

$$
\ddot{O} - CH_2CH_2CH_2CH_3
$$
$$
\overset{|}{H}
$$
$$
CH_3CH_2CH_2CH_2 - \ddot{O}\!:
$$
$$
\overset{|}{H}
$$
$$
\ddot{O} - CH_2CH_2CH_2CH_3
$$
$$
\overset{|}{H}
$$

There is no hydrogen bonding between molecules of 2-butanone or diethyl ether. They can serve only as hydrogen bond acceptors, not donors. Therefore, the intermolecular forces acting on the molecules of 2-butanone and diethyl ether are dipole-dipole interactions. The strong polarity of the carbonyl group results in stronger dipole-dipole interactions between molecules of 2-butanone than for molecules of diethyl ether.

$$
\underset{CH_3CH_2}{\overset{CH_3}{>}}\!C\!=\!\ddot{O} \rightarrow \quad
\underset{CH_3CH_2}{\overset{CH_3}{>}}\!C\!=\!\ddot{O} \rightarrow \quad
\underset{CH_3CH_2}{\overset{CH_3}{>}}\!C\!=\!\ddot{O} \rightarrow
$$

strong dipole-dipole interactions,
high boiling point

$$
\underset{CH_3CH_2}{\overset{CH_3CH_2}{>}}\!\ddot{O}\!: \rightarrow \quad
\underset{CH_3CH_2}{\overset{CH_3CH_2}{>}}\!\ddot{O}\!: \rightarrow \quad
\underset{CH_3CH_2}{\overset{CH_3CH_2}{>}}\!\ddot{O}\!: \rightarrow
$$

weak dipole-dipole interactions,
low boiling point

14.2

(a) 3-hydroxy-2-methylpentanal   (b) 4-chlorobenzaldehyde   (c) 4-heptyn-3-one

(d) (*E*)-1,3-diphenyl-2-propen-1-one   (e) (*Z*)-6-methyl-5-nonen-2-one   (f) 2,2-dibromocyclohexanone

(g) 2,4-pentanedione   (h) trichloroethanal (trichloroacetaldehyde)   (i) *cis*-2-propylcyclopentanecarbaldehyde

14.3   (a) CH₃—⬦=O   (b) (CH₃)₂ ring =O   (c) PhC(=O)-cyclopentane

14.3 (d)

$$CH_3CH_2 \quad H$$
$$C=C$$
$$H \quad CH_2CH_2CH$$
(with C=O on the CH group)

(e) $O_2N-\langle\text{benzene ring}\rangle-CH$ (with C=O)

(f) $CH_3CH_2CCH_2CH_2OH$ (with C=O)

(g) $HCCH_2CH_2CH_2CH_2CH$ (with C=O at both ends)

(h) $PhCHCPh$ (with C=O, and OH below)
$$OH$$

(i) $O=\langle\text{cyclohexane ring}\rangle=O$

Concept Map 14.1   Some ways to prepare aldehydes and ketones.

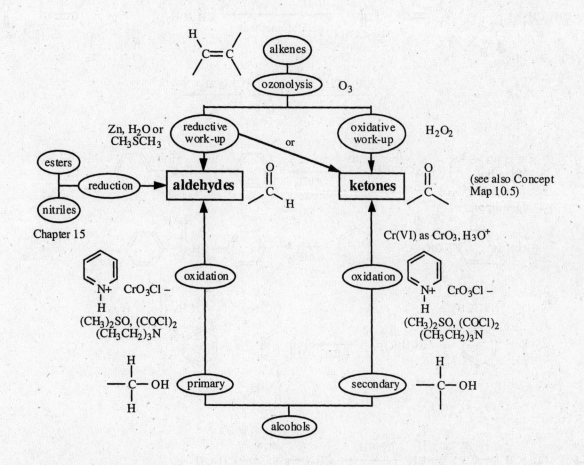

14.4

(a) $HC\equiv CCH_2CH_2CH=CHCH_2OH \xrightarrow[\quad]{\underset{\displaystyle ClC-CCl}{\overset{\displaystyle \overset{O}{\underset{\|}{CH_3SCH_3}}}{\overset{O\ \ \ \ O}{\|\quad\ \ \|}}}} \xrightarrow{(CH_3CH_2)_3N} HC\equiv CCH_2CH_2CH=CHCH$

$$A$$

14.4

(b)

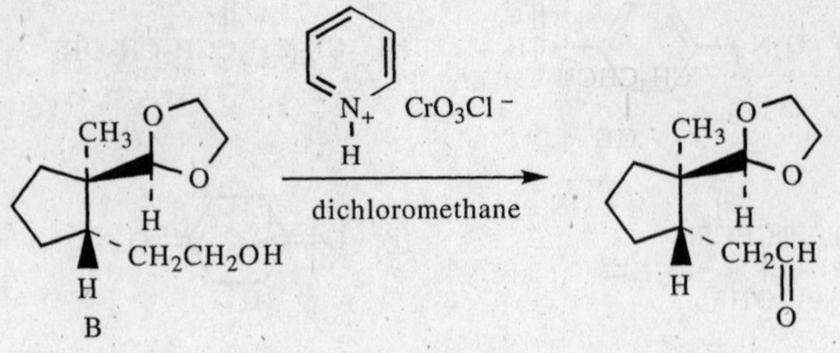

(c)

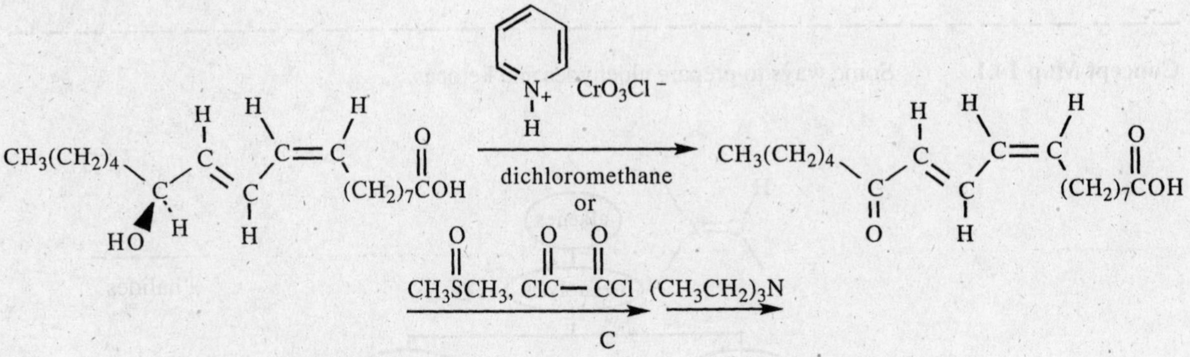

14.5 (a)

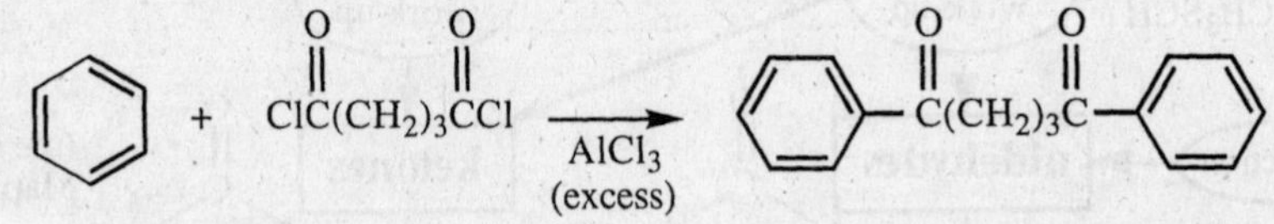

(b)

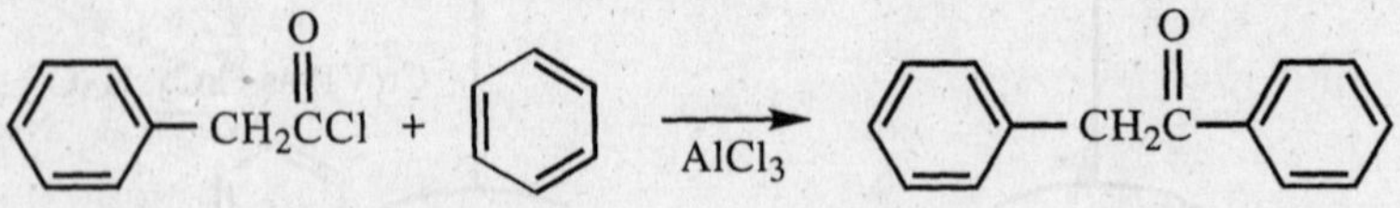

(c)

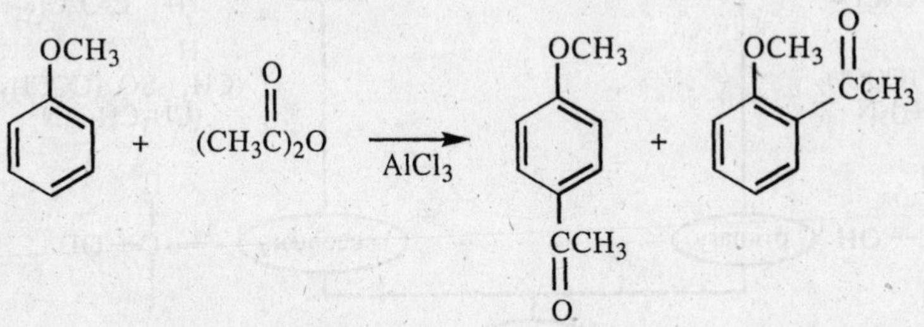

14.6 (a)

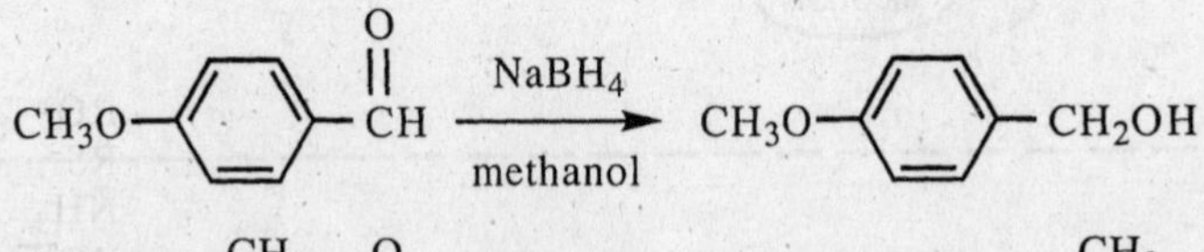

$$CH_3O-\langle\rangle-\underset{O}{\overset{\parallel}{C}}H \xrightarrow[\text{methanol}]{NaBH_4} CH_3O-\langle\rangle-CH_2OH$$

(b)

$$CH_3(CH_2)_5\underset{CH_3}{\overset{|}{C}}HCH_2\underset{O}{\overset{\parallel}{C}}H \xrightarrow[\text{diethyl ether}]{LiAlH_4} \xrightarrow{H_3O^+} CH_3(CH_2)_5\underset{CH_3}{\overset{|}{C}}HCH_2CH_2OH$$

(c) $PhCH_2\overset{O}{\overset{\parallel}{C}}CH_3 \xrightarrow[\substack{\text{diethyl}\\\text{ether}}]{LiAlH_4} \xrightarrow{H_3O^+} PhCH_2\underset{OH}{\overset{|}{C}}HCH_3$ (d) 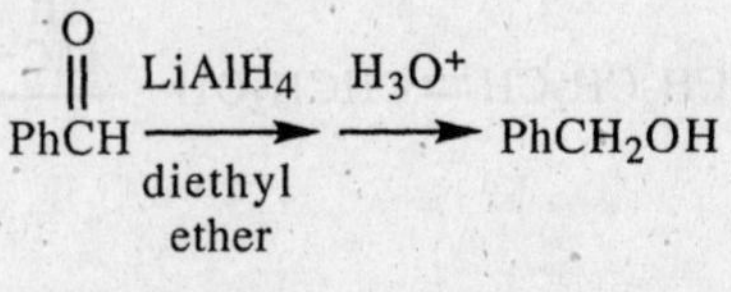 $Ph\overset{O}{\overset{\parallel}{C}}H \xrightarrow[\substack{\text{diethyl}\\\text{ether}}]{LiAlH_4} \xrightarrow{H_3O^+} PhCH_2OH$

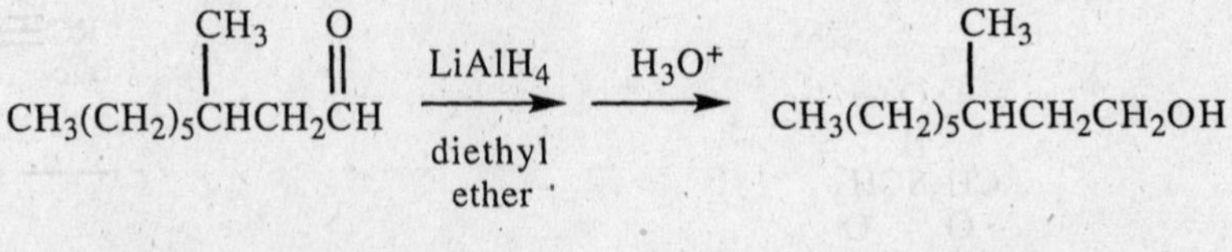

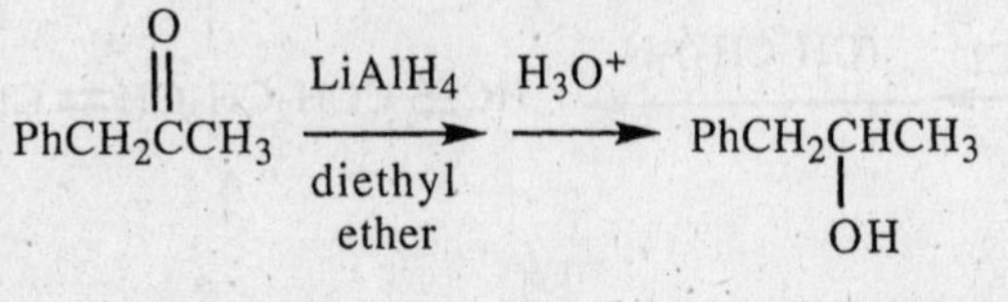

14.6   (e)

$$\underset{\text{diethyl}}{\overset{\text{LiAlH}_4}{\longrightarrow}}\ \overset{\text{H}_3\text{O}^+}{\longrightarrow}$$

O   CH₃
‖ |
CH₃CCH₂CHCH₃ $\xrightarrow[\text{ether}]{}$ CH₃CHCH₂CHCH₃

CH₃
|
CH₃CHCH₂CHCH₃
|
OH

(f)

O
‖
CH₃CH=CHCCH₃ $\xrightarrow[\substack{\text{CeCl}_3 \cdot 7\,\text{H}_2\text{O} \\ \text{methanol}}]{\text{NaBH}_4}$ CH₃CH=CHCHCH₃
|
OH

Concept Map 14.2   The relationship between carbonyl compounds, alcohols, and alkyl halides.

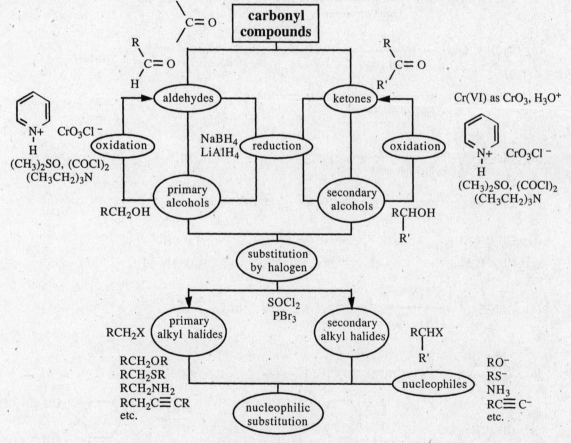

14.7

$$\underset{\underset{CH_3}{}}{\overset{H}{}}C=O \quad + \quad {}^-C\equiv N \quad \longrightarrow \quad H-\overset{O^-}{\underset{CH_3}{C}}-C\equiv N \quad + \quad CH_3-\overset{H}{\underset{O^-}{C}}-C\equiv N$$

attack by cyanide ion can occur at the top or the bottom of the carbonyl group

R   *S*

racemic mixture

↓ HB⁺

$$H-\overset{O-H}{\underset{CH_3}{C}}-\underset{O}{\overset{}{C}}OH \quad + \quad CH_3-\overset{H}{\underset{O-H}{C}}-\underset{}{\overset{O}{C}}OH \quad \xleftarrow[\Delta]{H_3O^+} \quad H-\overset{OH}{\underset{CH_3}{C}}-C\equiv N \quad + \quad CH_3-\overset{H}{\underset{OH}{C}}-C\equiv N$$

(*R*)-lactic acid   (*S*)-lactic acid   *R*   *S*

also racemic mixture   racemic mixture

Note that the hydrolysis of each nitrile proceeds with retention of configuration, but hydrolysis of the racemic mixture of nitriles gives rise to a racemic mixture of hydroxyacids

14.8   (a) $CH_3CH_2CH_2CH_2Br \xrightarrow[\text{tetrahydrofuran}]{Li} CH_3CH_2CH_2CH_2Li + Li^+Br^-$

(b) $PhCH_2CH_2Br \xrightarrow[\text{diethyl ether}]{Mg} PhCH_2CH_2MgBr$

(c) $CH_2=CHCl \xrightarrow[\underset{\Delta}{\text{tetrahydrofuran}}]{Mg} CH_2=CHMgCl$

(d) $PhC\equiv CH \xrightarrow[\text{diethyl ether}]{CH_3CH_2MgBr} PhC\equiv CMgBr + CH_3CH_3\uparrow$

(e) $PhCH_2Br \xrightarrow[\text{tetrahydrofuran}]{Li} PhCH_2Li + Li^+Br^-$

(f) $CH_3CH_2CH_2Li + NH_3 \longrightarrow CH_3CH_2CH_3\uparrow + Li^+NH_2^-$

(g) $CH_3CH_2MgI + CH_3OH \longrightarrow CH_3CH_3\uparrow + CH_3O^-Mg^{2+}I^-$

(h) $PhC\equiv CH \xrightarrow[\text{NH}_3 \text{ (liq)}]{Na^+NH_2^-} PhC\equiv C^-Na^+ + NH_3$

14.9   (a)   (cyclohexanone)$=O + CH_3CH_2CH_2CH_2Li \longrightarrow$

$$\text{(cyclohexane ring)}\overset{CH_2CH_2CH_2CH_3}{\underset{O^-Li^+}{}} \xrightarrow{H_3O^+} \text{(cyclohexane ring)}\overset{CH_2CH_2CH_2CH_3}{\underset{OH}{}}$$

A   B

(b) $PhBr \xrightarrow[\underset{\text{ether}}{\text{diethyl}}]{Mg} PhMgBr \xrightarrow{\overset{O}{\overset{/\backslash}{CH_2-CH_2}}} PhCH_2CH_2O^-Mg^{2+}Br^- \xrightarrow{H_3O^+} PhCH_2CH_2OH$

      C   D   E

14.9 (c) $\text{PhBr} \xrightarrow[\text{tetrahydrofuran}]{\text{Li}} \text{PhLi} \xrightarrow{\overset{\displaystyle O}{\overset{\|}{CH_3CH_2CH_2CH}}} \underset{\underset{\displaystyle G}{O^-Li^+}}{\overset{|}{PhCHCH_2CH_2CH_3}} \xrightarrow{H_3O^+} \underset{\underset{\displaystyle H}{OH}}{\overset{|}{PhCHCH_2CH_2CH_3}}$

$\underset{\displaystyle F}{}$

(d) $\text{CH}_3\text{CH}_2\text{C} \equiv \text{CH} \xrightarrow[\text{NH}_3 \text{ (liq)}]{\text{NaNH}_2} \underset{\displaystyle I}{\text{CH}_3\text{CH}_2\text{C} \equiv \text{C}^- \text{ Na}^+} \xrightarrow{\text{diethyl ether}}$

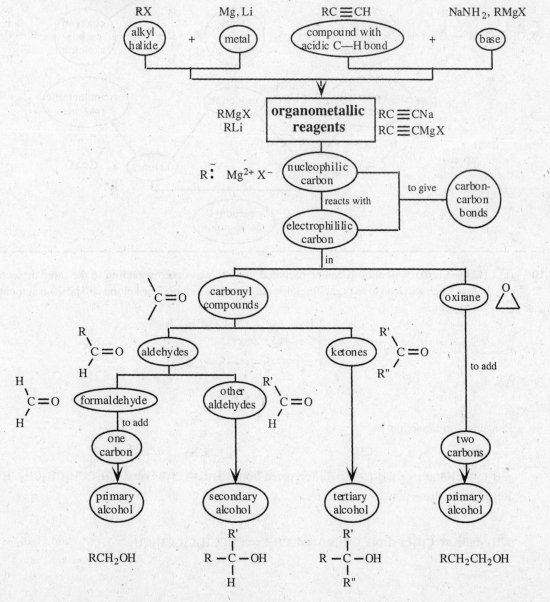

- -

Concept Map 14.3   Organometallic reagents and their reactions with compounds containing electrophilic carbon atoms.

Concept Map 14.4   Some ways to prepare alcohols.

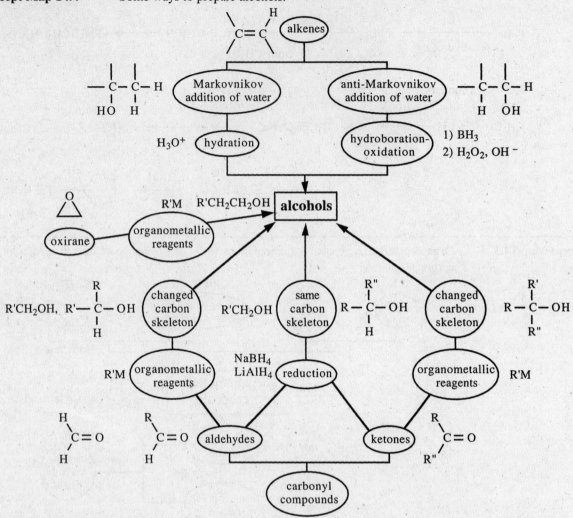

- -

14.10 (a) This is a tertiary alcohol. It can be prepared in three ways corresponding to the three dissections shown below. Each will lead to an organometallic reagent and a ketone. The ketone will be the fragment containing the oxygen.

$$CH_3C\!=\!CH\!-\!\overset{\displaystyle CH_3}{\underset{\displaystyle OH}{\overset{\textcircled{\scriptsize 3}}{\underset{\textcircled{\scriptsize 1}}{C}}}}\!\overset{\displaystyle CH_2CH_2CH_2CH_3}{\underset{\textcircled{\scriptsize 2}}{}}\!-\!CH_3$$

Reagents for dissection 1:

$$\overset{\displaystyle CH_3}{CH_3C\!=\!CHMgBr} \text{ or } \overset{\displaystyle CH_3}{CH_3C\!=\!CHLi} \text{ (prepared from } \overset{\displaystyle CH_3}{CH_3C\!=\!CHBr}) \text{ and } \overset{\displaystyle O}{CH_3CCH_2CH_2CH_2CH_3}$$

Reagents for dissection 2:

$$CH_3MgBr \text{ or } CH_3Li \text{ (from } CH_3Br) \text{ and } \overset{\displaystyle CH_3 \quad O}{CH_3C\!=\!CHCCH_2CH_2CH_2CH_3}$$

14.10 (a) (cont)

Reagents for dissection 3:

$$CH_3CH_2CH_2CH_2MgBr \text{ or } CH_3CH_2CH_2CH_2Li \text{ (from } CH_3CH_2CH_2CH_2Br \text{) and } CH_3\overset{\overset{\textstyle CH_3}{|}}{C}=CHC\overset{\overset{\textstyle O}{||}}{C}CH_3$$

(b) This is a secondary alcohol. Two dissections are possible leading to an aldehyde and an organometallic reagent. A third dissection indicates that a reduction reaction, hydride addition, will also give this alcohol.

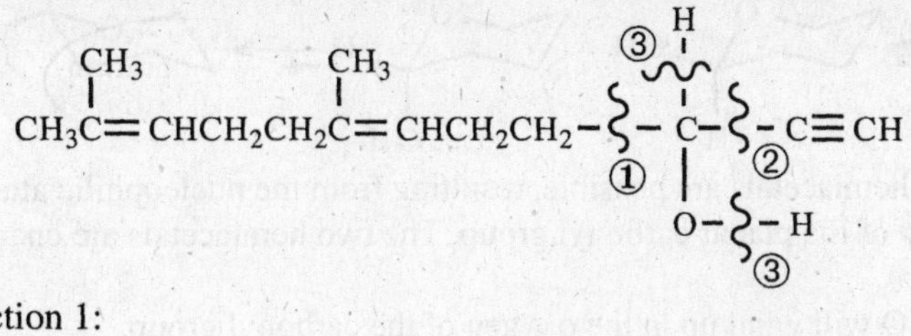

$$CH_3\overset{\overset{\textstyle CH_3}{|}}{C}=CHCH_2CH_2\overset{\overset{\textstyle CH_3}{|}}{C}=CHCH_2CH_2-\underset{①}{\xi}-\overset{\overset{\textstyle ③\ H}{|}}{\underset{\underset{\textstyle O-\underset{③}{\xi}-H}{|}}{C}}-\underset{②}{\xi}-C\equiv CH$$

Reagents for dissection 1:

$$CH_3\overset{\overset{\textstyle CH_3}{|}}{C}=CHCH_2CH_2\overset{\overset{\textstyle CH_3}{|}}{C}=CHCH_2CH_2MgBr \text{ or } CH_3\overset{\overset{\textstyle CH_3}{|}}{C}=CHCH_2CH_2\overset{\overset{\textstyle CH_3}{|}}{C}=CHCH_2CH_2Li \text{ (prepared from}$$

$$CH_3\overset{\overset{\textstyle CH_3}{|}}{C}=CHCH_2CH_2\overset{\overset{\textstyle CH_3}{|}}{C}=CHCH_2CH_2Br \text{) and } HC\overset{\overset{\textstyle O}{||}}{C}\equiv CH. \text{ This aldehyde may give problems depending on}$$

whether nucleophilic addition to the carbonyl group will compete with deprotonation of the alkyne.

Reagents for dissection 2:

$$HC\equiv CMgBr \text{ or } HC\equiv CNa \text{ (from } HC\equiv CH \text{) and } CH_3\overset{\overset{\textstyle CH_3}{|}}{C}=CHCH_2CH_2\overset{\overset{\textstyle CH_3}{|}}{C}=CHCH_2CH_2\overset{\overset{\textstyle O}{||}}{C}H$$

This is the better choice for an aldehyde and an organometallic reagent.

Reagents for dissection 3:

$$NaBH_4 \text{ and } CH_3\overset{\overset{\textstyle CH_3}{|}}{C}=CHCH_2CH_2\overset{\overset{\textstyle CH_3}{|}}{C}=CHCH_2CH_2\overset{\overset{\textstyle O}{||}}{C}-C\equiv CH$$

(c) This is a primary alcohol. A primary alcohol is prepared 1) by reduction of an aldehyde, 2) by addition of an organometallic reagent to formaldehyde or, sometimes, by addition of an organometallic reagent to oxirane. The dissections corresponding to these three routes are shown below.

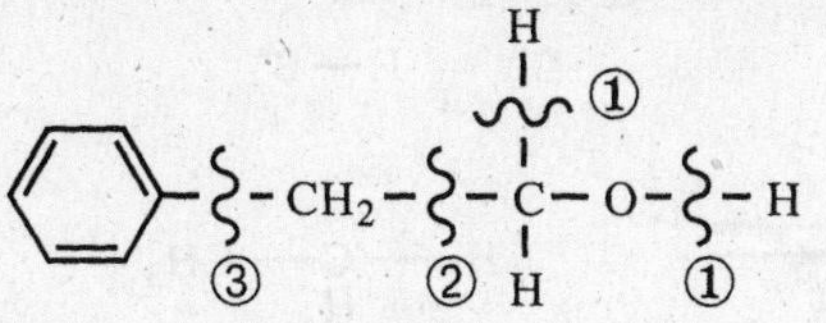

$$\text{Ph}-\underset{③}{\xi}-CH_2-\underset{②}{\xi}-\overset{\overset{\textstyle ①\ H}{|}}{\underset{\underset{\textstyle H}{|}}{C}}-O-\underset{①}{\xi}-H$$

Reagents for dissection 1:

$$NaBH_4 \text{ and } \text{Ph}-CH_2\overset{\overset{\textstyle O}{||}}{C}H$$

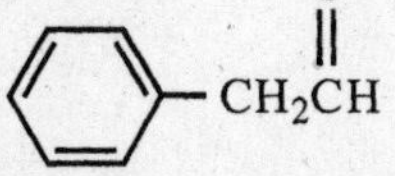

Reagents for dissection 2:

$$\text{Ph}-CH_2MgBr \text{ or } \text{Ph}-CH_2Li \text{ (from } \text{Ph}-CH_2Br \text{) and } H\overset{\overset{\textstyle O}{||}}{C}H$$

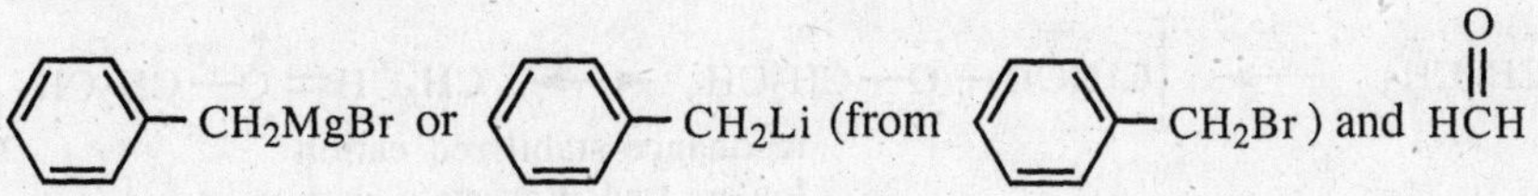

14.10 (c) (cont)

Reagents for dissection 3:

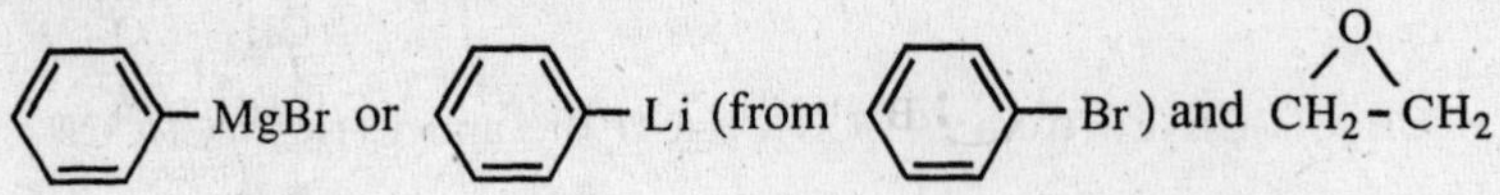
MgBr or
Li (from
Br) and CH2–CH2
O

14.11

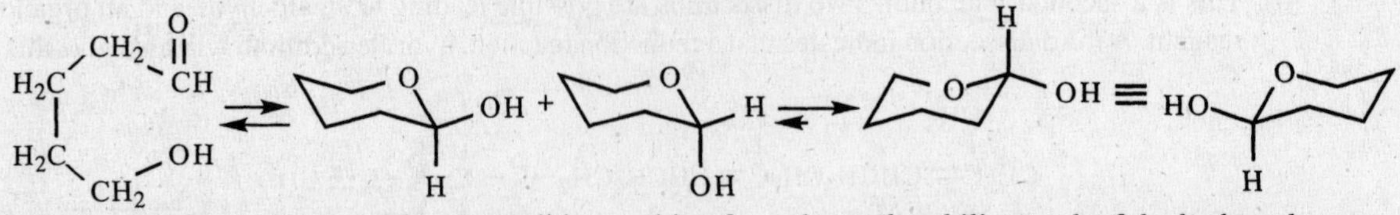

Two stereoisomeric hemiacetals are possible, resulting from the nucleophilic attack of the hydroxyl group at the bottom or the top of the planar carbonyl group. The two hemiacetals are enantiomers of each other.

14.12 Essentially all the ^{18}O will wind up in the oxygen of the carbonyl group.

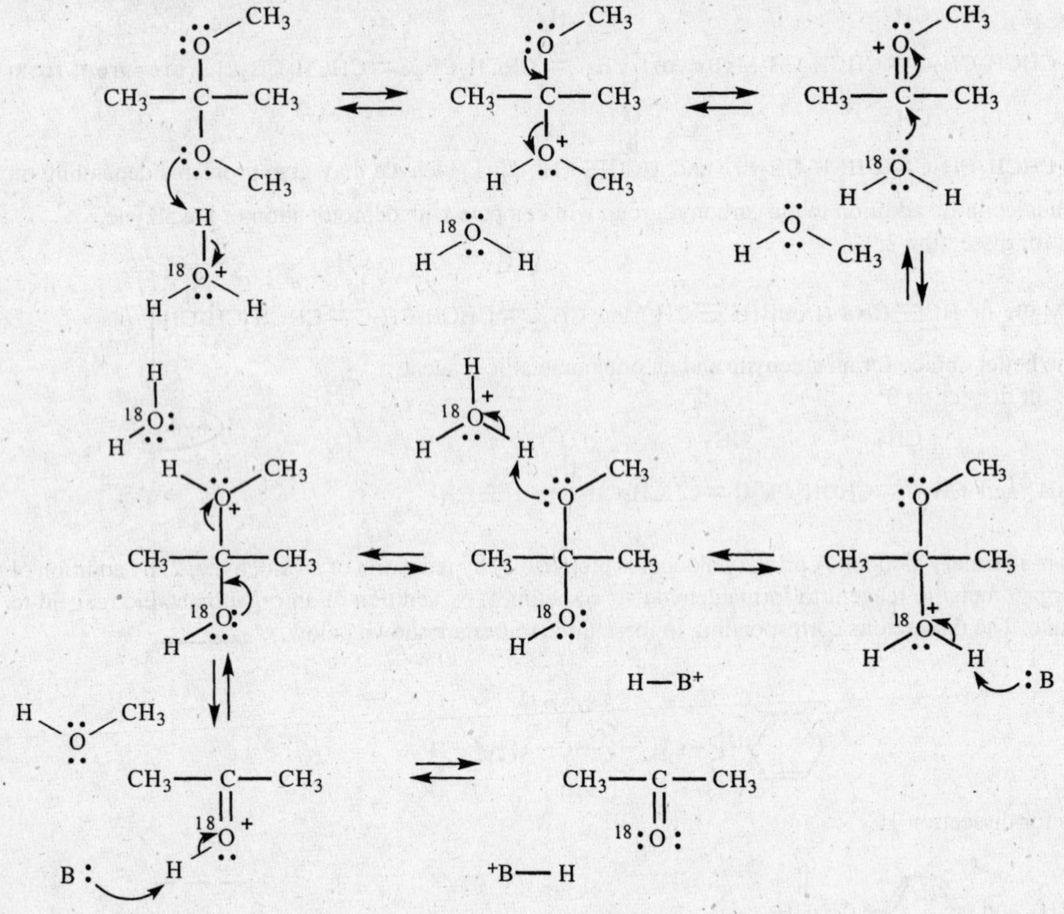

14.13

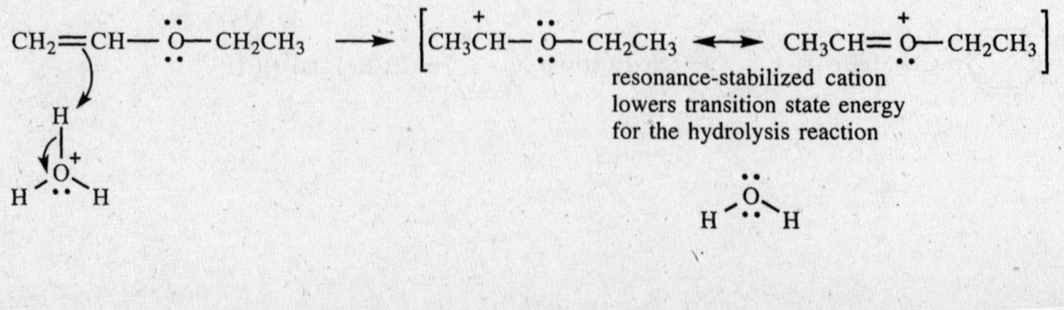
resonance-stabilized cation
lowers transition state energy
for the hydrolysis reaction

14.13 (cont)

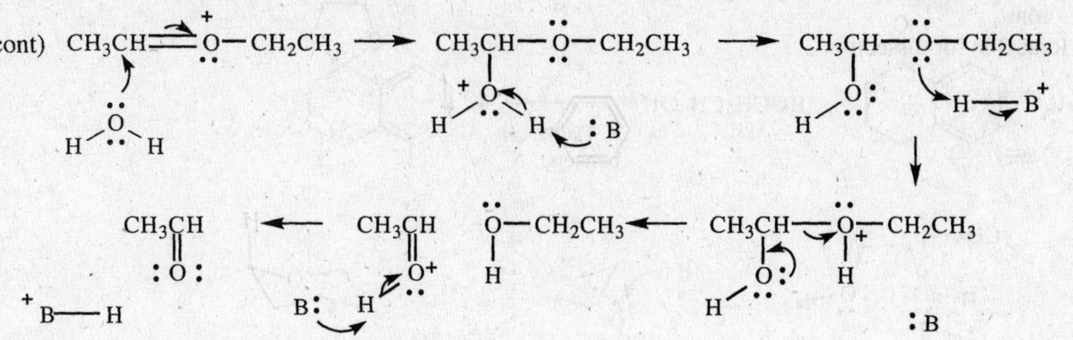

$$CH_3CH = \overset{+}{\underset{\cdot\cdot}{O}} - CH_2CH_3 \longrightarrow CH_3CH - \overset{\cdot\cdot}{\underset{+}{O}} - CH_2CH_3 \longrightarrow CH_3CH - \overset{\cdot\cdot}{O} - CH_2CH_3$$

Concept Map 14.5   Hydrates, acetals, and ketals.

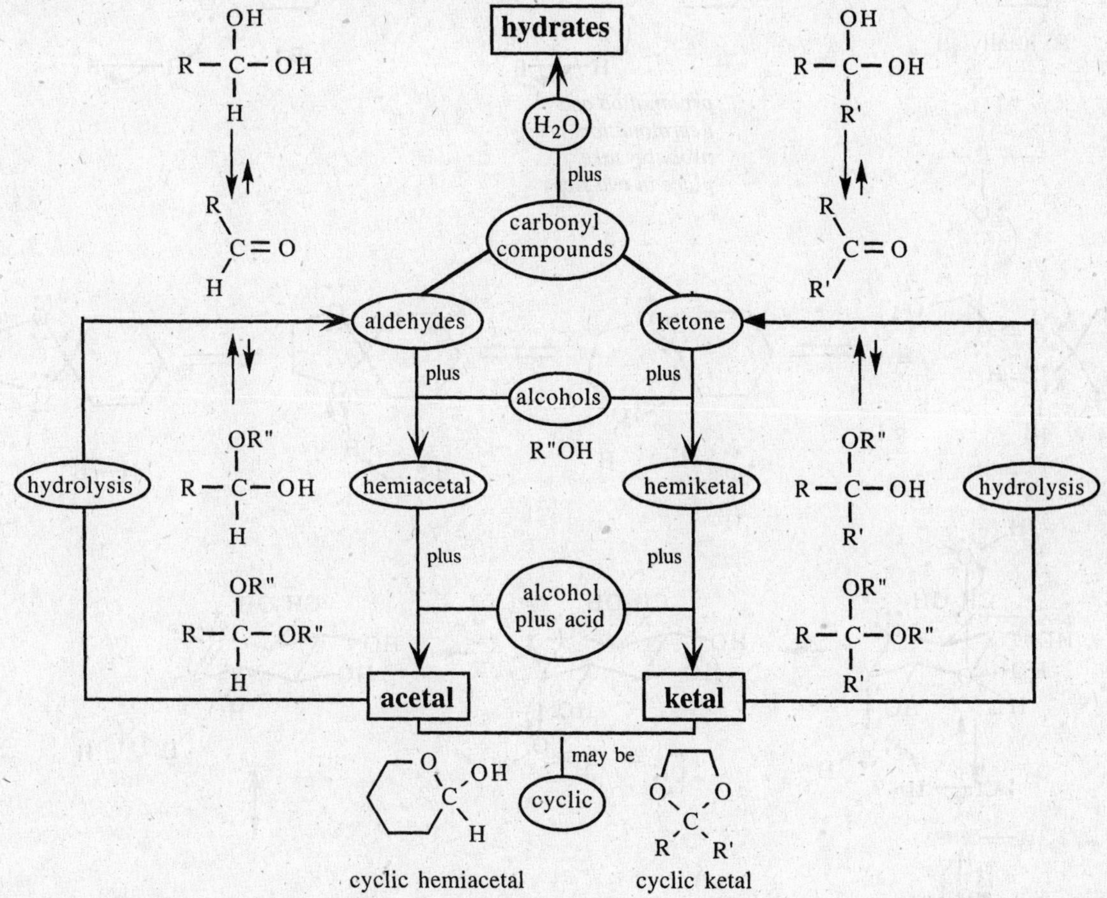

14.14 (a)

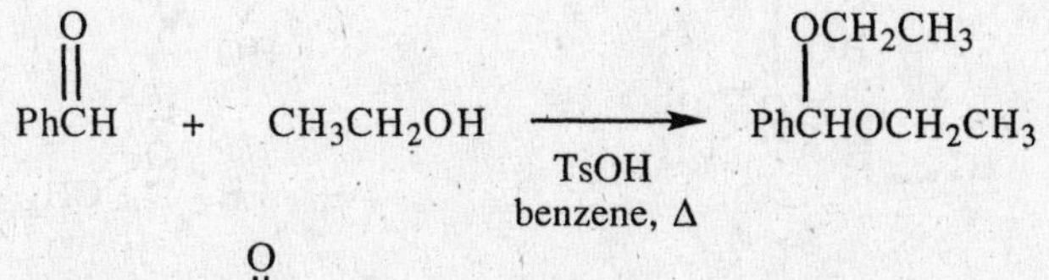

$$\underset{\text{O}}{\text{PhCH}} + CH_3CH_2OH \xrightarrow[\text{benzene, }\Delta]{\text{TsOH}} \underset{\text{OCH}_2\text{CH}_3}{\text{PhCHOCH}_2CH_3}$$

(b)

$$\underset{\text{O}}{CH_3CH_2CH_2CH} + CH_3OH \xrightarrow[\text{benzene, }\Delta]{\text{TsOH}} \underset{\text{OCH}_3}{CH_3CH_2CH_2CHOCH_3}$$

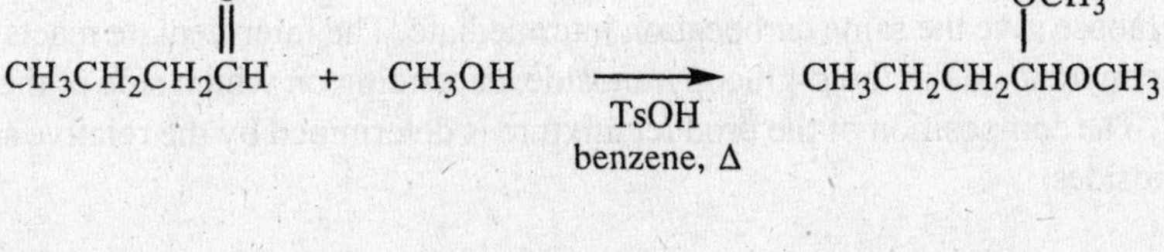

14.14 (c)

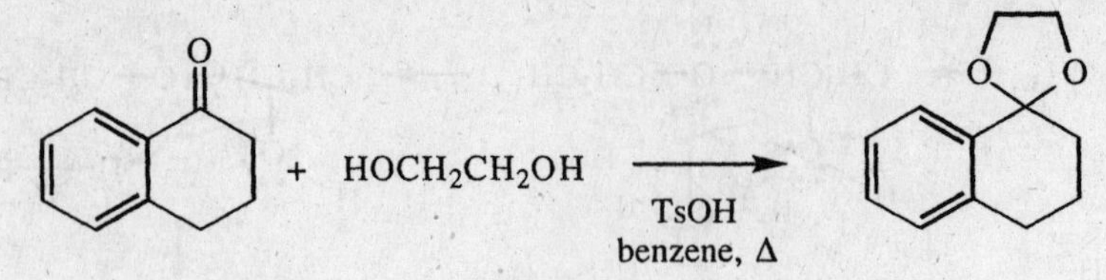

+ HOCH2CH2OH
TsOH
benzene, Δ

14.15

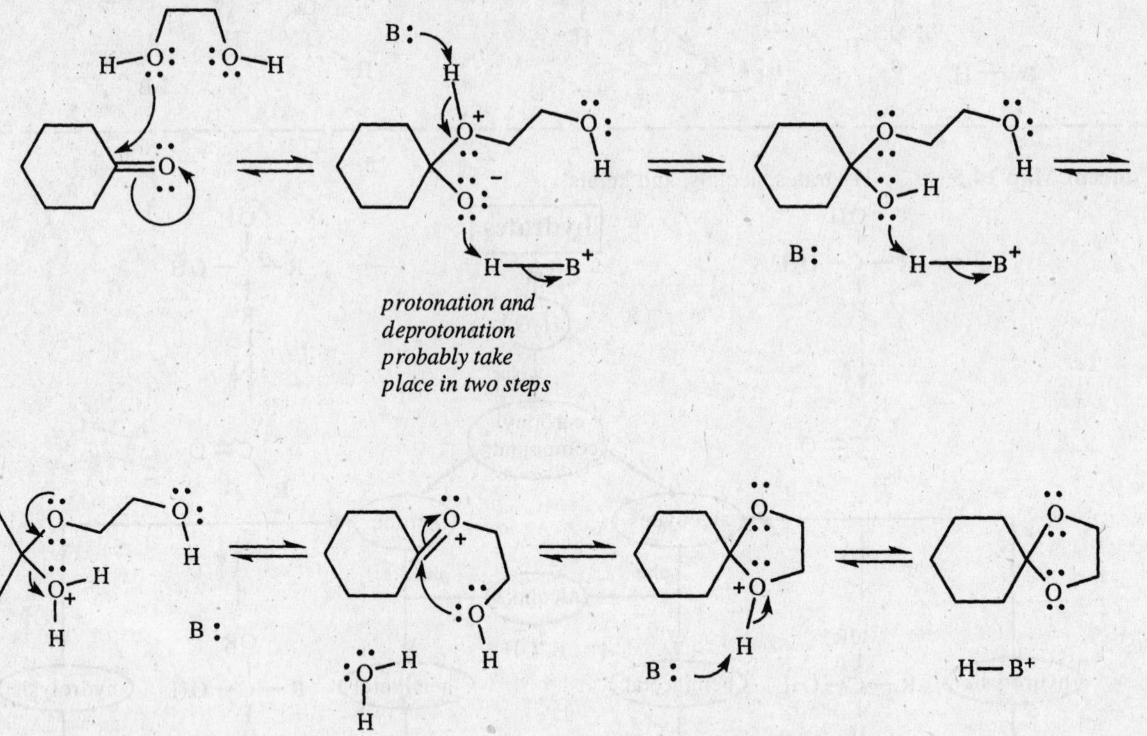

protonation and deprotonation probably take place in two steps

14.16

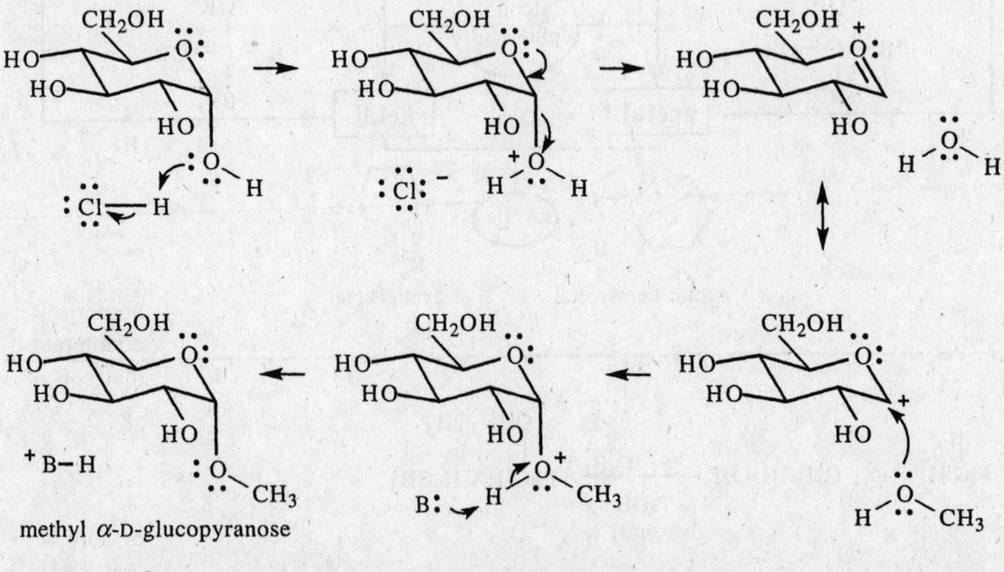

methyl α-D-glucopyranose

Both α- and β-glucopyranose give the same carbocation intermediate. The intermediate reacts with methanol to give either the α-glucopyranoside or the β-glucopyranoside, depending on which side of the carbocation is attacked by the alcohol. The composition of the product mixture is determined by the relative stabilities of the two methyl glucopyranosides.

14.17

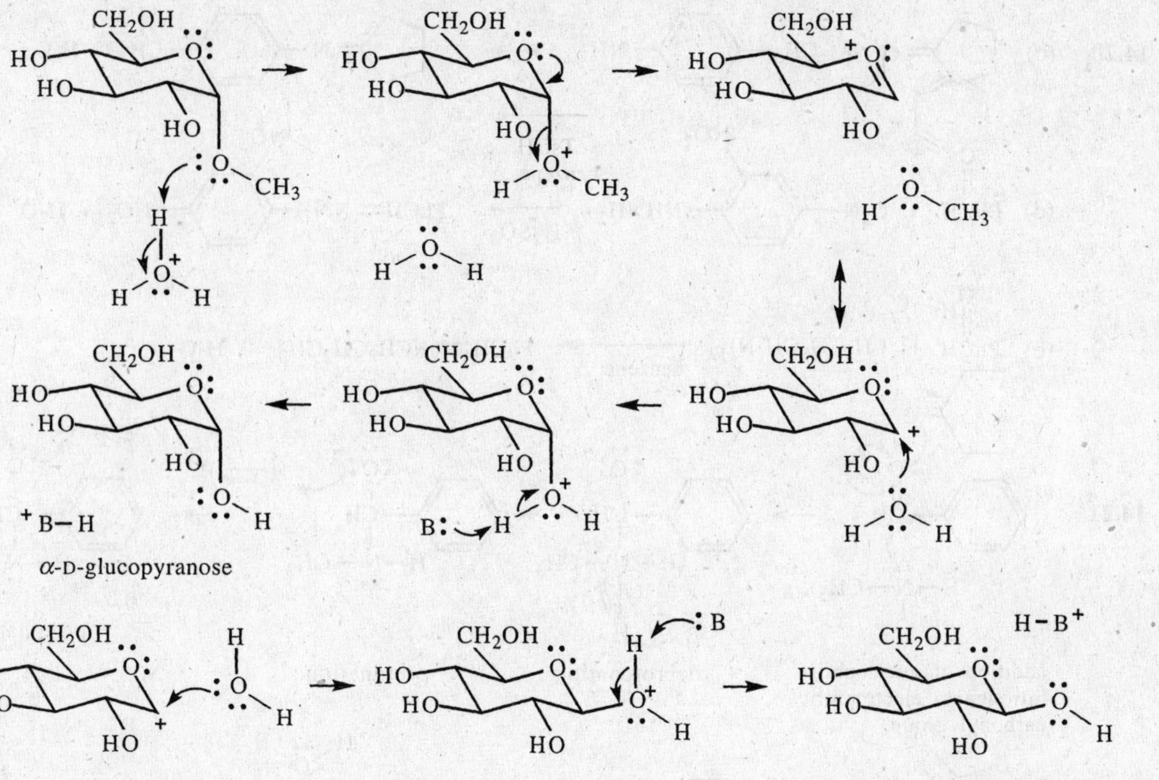
α-D-glucopyranose
β-D-glucopyranose

14.18

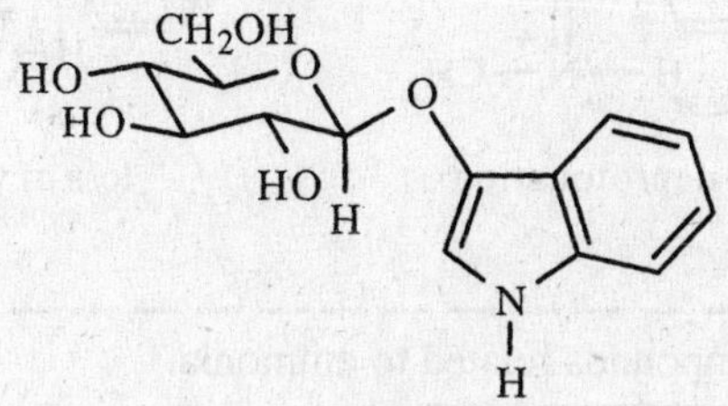

indican

14.19

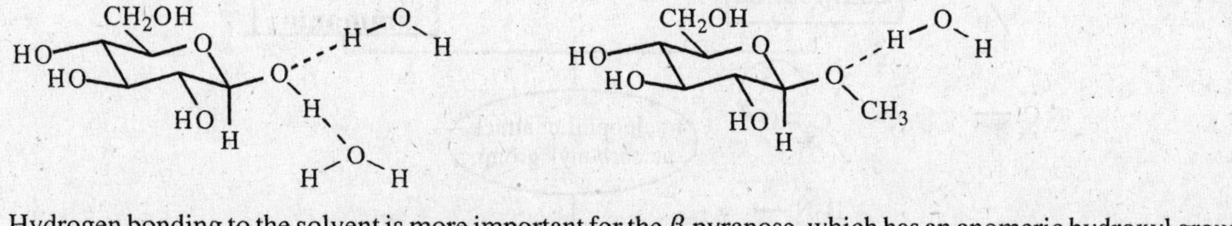

Hydrogen bonding to the solvent is more important for the β-pyranose, which has an anomeric hydroxyl group, than for the β-pyranoside, which has an anomeric methoxyl group. Hydrogen bonding counteracts the anomeric effect, stabilizing the equatorial hydroxyl group relative to the axial one, more than it affects the equilibrium between the forms with the equatorial and axial methoxyl groups.

14.20 (a)

$$CH_3\overset{CH_3}{\overset{|}{C}}HCH_2CH_2\overset{O}{\overset{\|}{C}}H + HO\overset{+}{N}H_3\ Cl^- \xrightarrow{CH_3\overset{O}{\overset{\|}{C}}O^-Na^+} CH_3\overset{CH_3}{\overset{|}{C}}HCH_2CH_2CH{=}NOH + H_2O$$

(b)

$$CH_3\overset{O}{\overset{\|}{C}}CH_2CH_2CH_3 + PhNHNH_2 \xrightarrow[\text{acetic acid}]{} CH_3CH_2CH_2\overset{CH_3}{\overset{|}{C}}{=}NNHPh + H_2O$$

14.20 (c)

cyclopentanone =O + CH₃—⟨aryl⟩—NH₂ →(Δ) cyclopentanone =N—⟨aryl⟩—CH₃ + H₂O

(d) PhCH(=O) + O₂N—⟨aryl with NO₂⟩—NHNH₂ —(H₂SO₄)→ PhCH=NNH—⟨aryl with NO₂⟩—NO₂ + H₂O

(e) PhCH(=O) + CH₃CH₂CH₂NH₂ —(benzene, Δ)→ PhCH=NCH₂CH₂CH₃ + H₂O

14.21

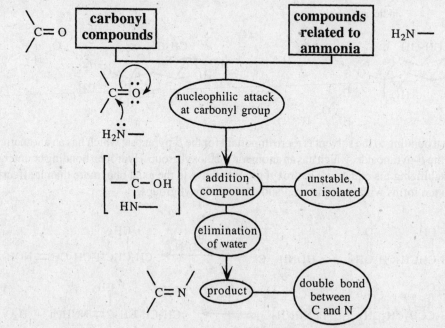

addition of nucleophilic nitrogen to electrophilic carbonyl group   deprotonation   protonation   protonation

loss of water

deprotonation

Concept Map 14.6   Reactions of carbonyl compounds with compounds related to ammonia.

carbonyl compounds   compounds related to ammonia   H₂N—

nucleophilic attack at carbonyl group

addition compound — unstable, not isolated

[—C—OH / HN—]

elimination of water

C=N product — double bond between C and N

14.22

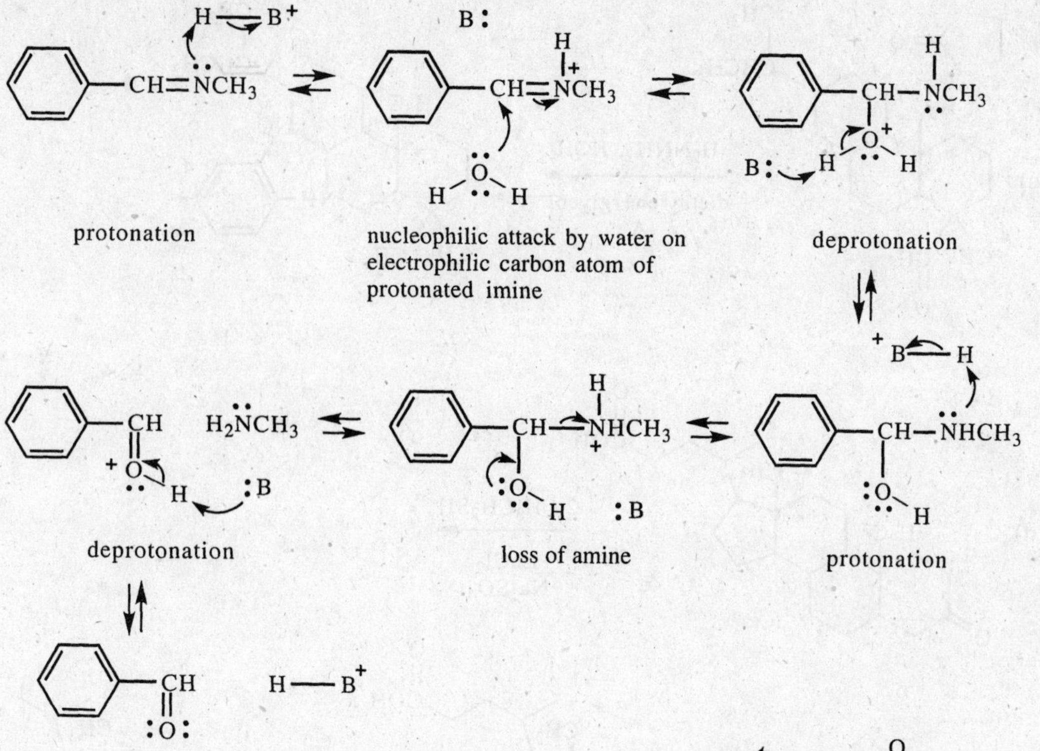

protonation

nucleophilic attack by water on electrophilic carbon atom of protonated imine

deprotonation

deprotonation

loss of amine

protonation

14.23

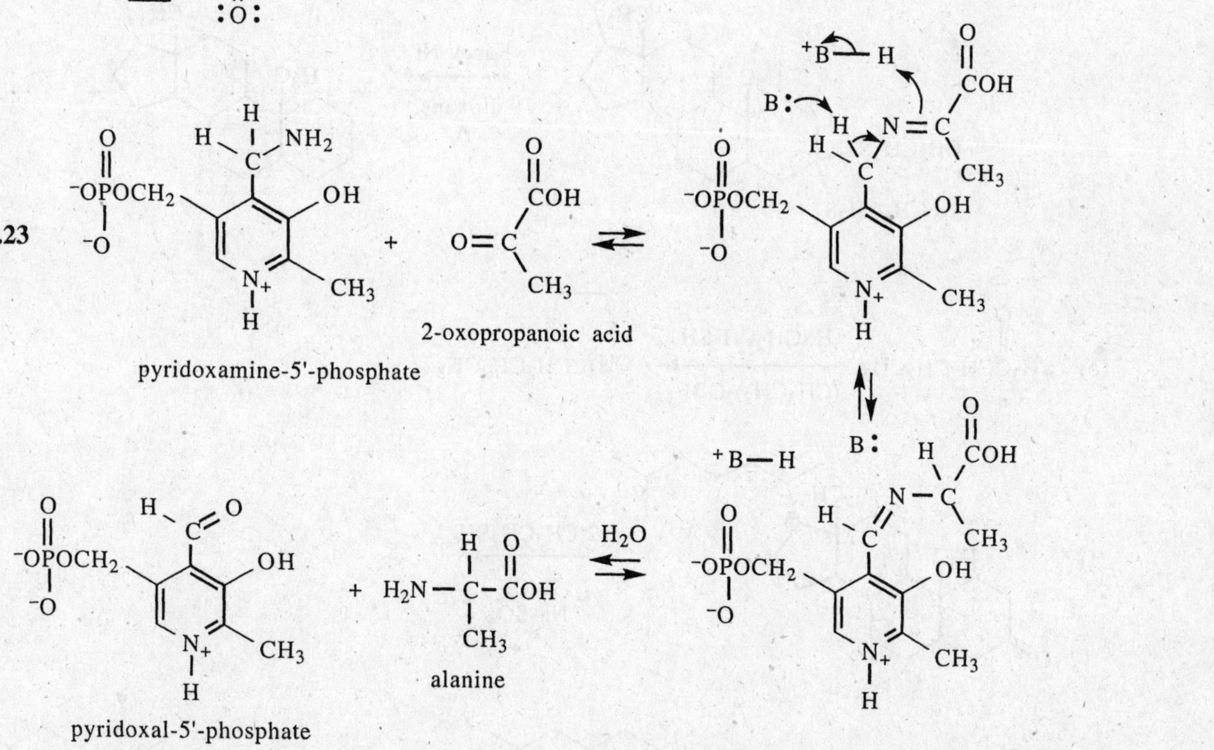

pyridoxamine-5'-phosphate

2-oxopropanoic acid

pyridoxal-5'-phosphate

alanine

In the presence of an enzyme, which is a chiral reagent, these reactions occur stereoselectively to give (S)-alanine.

14.24 (a) cyclohexanone $\xrightarrow[\text{diethylene glycol} \atop \Delta]{\text{H}_2\text{NNH}_2, \text{NaOH}}$ cyclohexane   (b) PhCH=O $\xrightarrow[\text{diethylene glycol} \atop \Delta]{\text{H}_2\text{NNH}_2, \text{NaOH}}$ PhCH$_3$

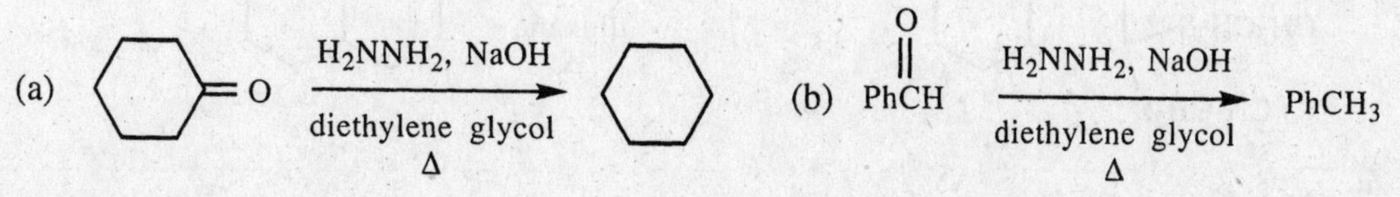

14.24 (c)

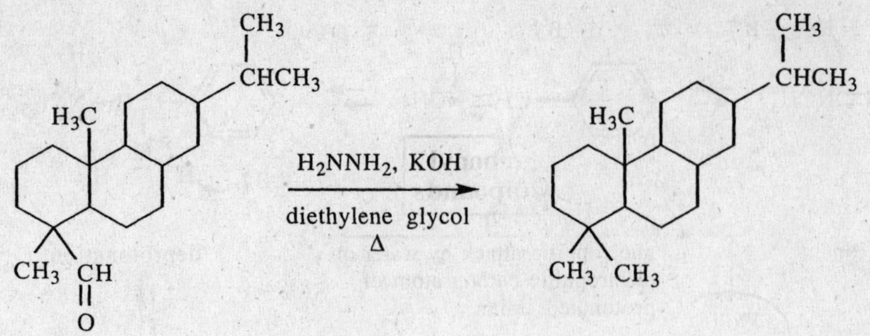
CH3
CHCH3
H3C
H2NNH2, KOH
diethylene glycol
Δ
CH3 CH
O
CH3
CHCH3
H3C
CH3 CH3

(d)

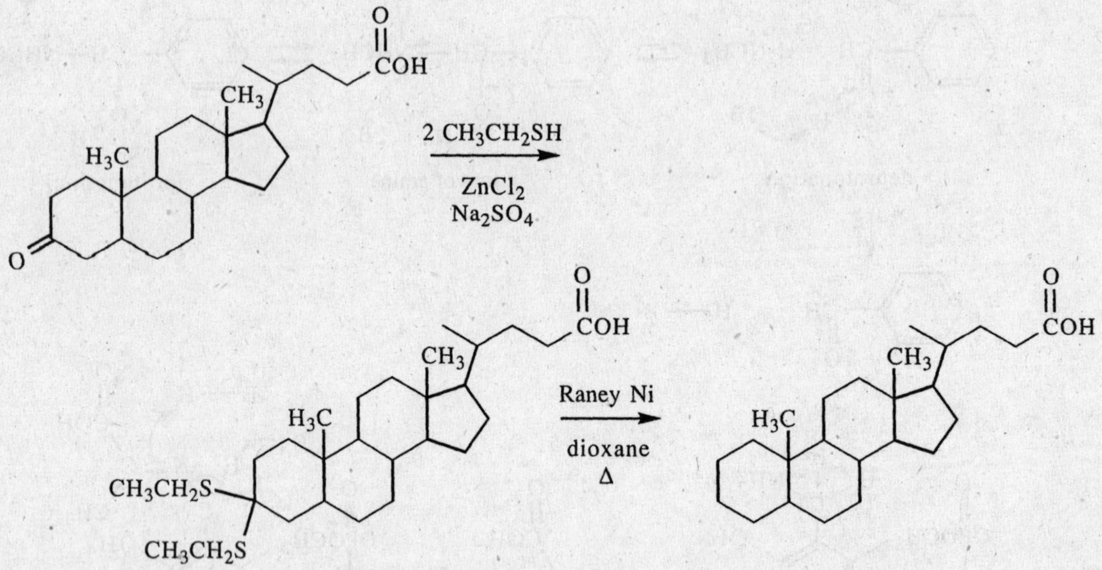
O
COH
CH3
H3C
2 CH3CH2SH
ZnCl2
Na2SO4
O
O
COH
CH3
H3C
CH3CH2S
CH3CH2S
Raney Ni
dioxane
Δ
O
COH
CH3
H3C

(e)

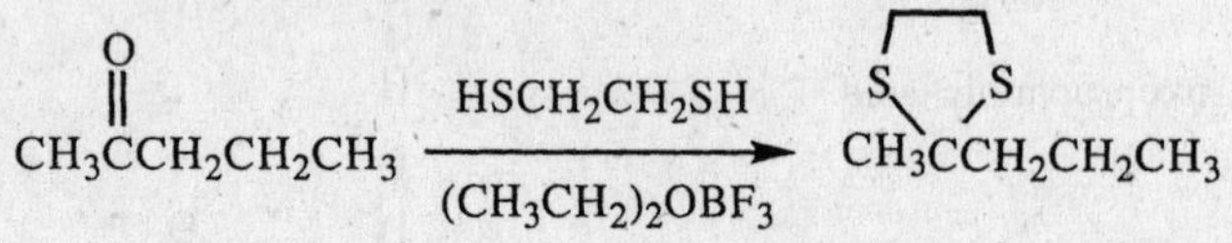
O
CH3CCH2CH2CH3
HSCH2CH2SH
(CH3CH2)2OBF3
S S
CH3CCH2CH2CH3

(f)

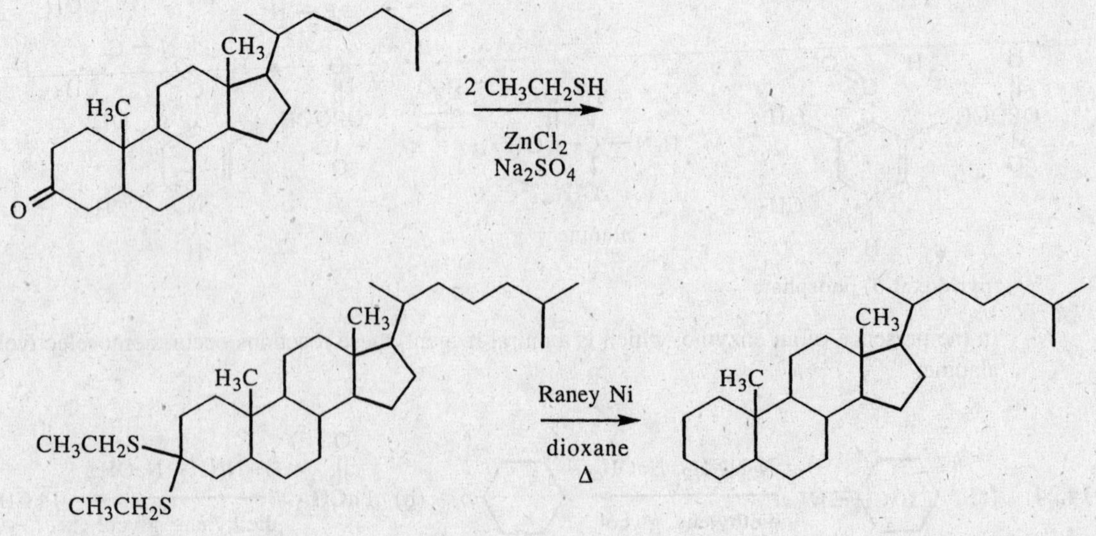
CH3
H3C
2 CH3CH2SH
ZnCl2
Na2SO4
O
CH3
H3C
CH3CH2S
CH3CH2S
Raney Ni
dioxane
Δ
CH3
H3C

Concept Map 14.7   Reduction of carbonyl groups to methylene groups.

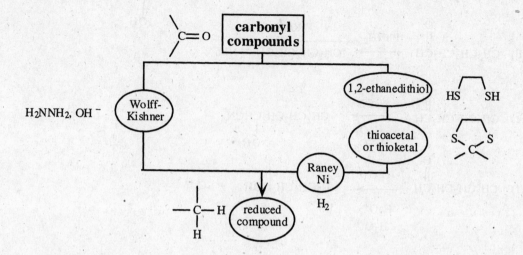

14.25

(a) cyclopentanecarbaldehyde   (b) cyclobutanone   (c) (*E*)-2-hexenal   (d) (*S*)-2-hydroxypropanal

(e) 2-methylcyclohexanone   (f) 5-methyl-2-hexanone   (g) 4-methoxybenzaldehyde (anisaldehyde)

(h) 3-methyl-2-cyclohexenone   (i) (4*S*, 5*S*)-4-chloro-5-methyl-2-heptanone

14.26

(a) 3,3-dimethylcyclopentanone   (b) $CH_3CH_2CH_2CH_2CCH_2Br$ with =O   (c) 3-chlorobenzaldehyde   (d) 4-methylacetophenone

(e) $CH_2=CHCH=CHCCH_3$ (f) $CH_3CH_2CH_2CHCH$ with CH_3 and =O   (g) cyclobutanone with Cl   (h) structure with CH_2CH=O, H, Br, CH_3

14.27

(a) $CH_3CH_2CH_2CH$ with =O $\xrightarrow[\text{H}_2\text{O}]{\text{NaBH}_4}$ $CH_3CH_2CH_2CH_2OH$

(b) $CH_3CH_2CH_2CH$ with =O $\xrightarrow[\text{H}_2\text{SO}_4]{}$ $CH_3CH_2CH_2CH=NNH$ (2,4-dinitrophenyl)

14.27

(c) $CH_3CH_2CH_2CH$ (with =O) $\xrightarrow{CH_3CH_2CH_2CH_2Li}$ $\xrightarrow{H_3O^+}$ $CH_3CH_2CH_2CHCH_2CH_2CH_2CH_3$ with OH below

(d) $CH_3CH_2CH_2CH$ (with =O) $\xrightarrow[\Delta]{PhNH_2}$ $CH_3CH_2CH_2CH=NPh$

(e) $CH_3CH_2CH_2CH$ (with =O) $\xrightarrow[\substack{H_2O \\ H_2SO_4}]{NaCN}$ $CH_3CH_2CH_2CHCN$ with OH below

(f) $CH_3CH_2CH_2CH$ (with =O) $\xrightarrow[\substack{H_3O^+ \\ H_2O}]{CrO_3}$ $CH_3CH_2CH_2COH$ (with =O) $+$ Cr^{+3}

(g) $CH_3CH_2CH_2CH$ (with =O) $\xrightarrow[\substack{CH_3CO^-Na^+ \\ (C=O)}]{HONH_3^+Cl^-}$ $CH_3CH_2CH_2CH=NOH$

(h) $CH_3CH_2CH_2CH$ (with =O) $\xrightarrow[\text{diethyl ether}]{CH_3MgI}$ $CH_3CH_2CH_2CHCH_3$ with $O^-Mg^{2+}I^-$ below $\xrightarrow{H_3O^+}$ $CH_3CH_2CH_2CHCH_3$ with OH below

(i) $CH_3CH_2CH_2CH$ (with =O) $\xrightarrow[\text{diethyl ether}]{LiAlH_4}$ $\xrightarrow{H_3O^+}$ $CH_3CH_2CH_2CH_2OH$

(j) $CH_3CH_2CH_2CH$ (with =O) $\xrightarrow[\substack{\text{diethylene glycol} \\ \Delta}]{H_2NNH_2, KOH}$ $CH_3CH_2CH_2CH_3$

14.28

(a) $CH_3CH_2CH_2CCH_3$ (with =O) $\xrightarrow[H_2O]{NaBH_4}$ $CH_3CH_2CH_2CHCH_3$ with OH below

(b) 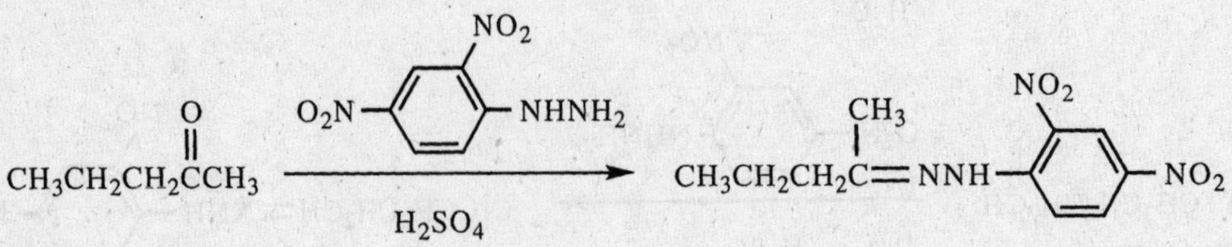 $CH_3CH_2CH_2CCH_3$ (with =O) $\xrightarrow[H_2SO_4]{O_2N-C_6H_3(NO_2)-NHNH_2}$ $CH_3CH_2CH_2C(CH_3)=NNH-C_6H_3(NO_2)_2$

(c)
$$CH_3CH_2CH_2\overset{\overset{\displaystyle O}{\|}}{C}CH_3 \xrightarrow[\Delta]{PhNH_2} CH_3CH_2CH_2\overset{\overset{\displaystyle CH_3}{|}}{C}=NPh$$

(d)
$$CH_3CH_2CH_2\overset{\overset{\displaystyle O}{\|}}{C}CH_3 \xrightarrow[\text{tetrahydrofuran}]{PhLi} CH_3CH_2CH_2\overset{\overset{\displaystyle CH_3}{|}}{\underset{\underset{\displaystyle O^-Li^+}{|}}{C}}Ph \xrightarrow[H_2O]{NH_4Cl} CH_3CH_2CH_2\overset{\overset{\displaystyle CH_3}{|}}{\underset{\underset{\displaystyle OH}{|}}{C}}Ph$$

(e)
$$CH_3CH_2CH_2\overset{\overset{\displaystyle O}{\|}}{C}CH_3 \xrightarrow[\substack{H_2O \\ H_2SO_4}]{NaCN} CH_3CH_2CH_2\overset{\overset{\displaystyle CH_3}{|}}{\underset{\underset{\displaystyle OH}{|}}{C}}CN$$

(f)
$$CH_3CH_2CH_2\overset{\overset{\displaystyle O}{\|}}{C}CH_3 \xrightarrow[\substack{CH_3CO^-Na^+ \\ \|\\ O}]{HONH_3^+Cl^-} CH_3CH_2CH_2\overset{\overset{\displaystyle CH_3}{|}}{C}=NOH$$

(g)
$$CH_3CH_2CH_2\overset{\overset{\displaystyle O}{\|}}{C}CH_3 \xrightarrow[\text{diethyl ether}]{CH_3CH_2MgI} CH_3CH_2CH_2\overset{\overset{\displaystyle CH_3}{|}}{\underset{\underset{\displaystyle O^-Mg^{2+}Br^-}{|}}{C}}CH_2CH_3 \xrightarrow{H_3O^+} CH_3CH_2CH_2\overset{\overset{\displaystyle CH_3}{|}}{\underset{\underset{\displaystyle OH}{|}}{C}}CH_2CH_3$$

(h)
$$CH_3CH_2CH_2\overset{\overset{\displaystyle O}{\|}}{C}CH_3 \xrightarrow{CH_3C\equiv C^-Na^+} CH_3CH_2CH_2\overset{\overset{\displaystyle CH_3}{|}}{\underset{\underset{\displaystyle O^-Na^+}{|}}{C}}C\equiv CCH_3 \xrightarrow[H_2O]{NH_4Cl} CH_3CH_2CH_2\overset{\overset{\displaystyle CH_3}{|}}{\underset{\underset{\displaystyle OH}{|}}{C}}C\equiv CCH_3$$

(i)
$$CH_3CH_2CH_2\overset{\overset{\displaystyle O}{\|}}{C}CH_3 \xrightarrow[\substack{\text{diethylene glycol} \\ \Delta}]{H_2NNH_2,\ KOH} CH_3CH_2CH_2CH_2CH_3$$

14.29 (a)
cyclohexanone $+$ PhNHNH$_2$ $\xrightarrow{\text{acetic acid}}$ cyclohexanone phenylhydrazone (=NNHPh)

(b)
$\xrightarrow[\substack{TsOH \\ \text{benzene} \\ \Delta}]{\substack{HOCH_2CH_2OH \\ \text{(1 equivalent)}}}$

(c)
$$CH_2=CHC\overset{\overset{\displaystyle O}{\|}}{}H \xrightarrow[\text{diethyl ether}]{PhLi} \xrightarrow{H_3O^+} CH_2=CHCHPh$$
$$\underset{\underset{\displaystyle OH}{|}}{}$$

14.29 (cont)

(d)

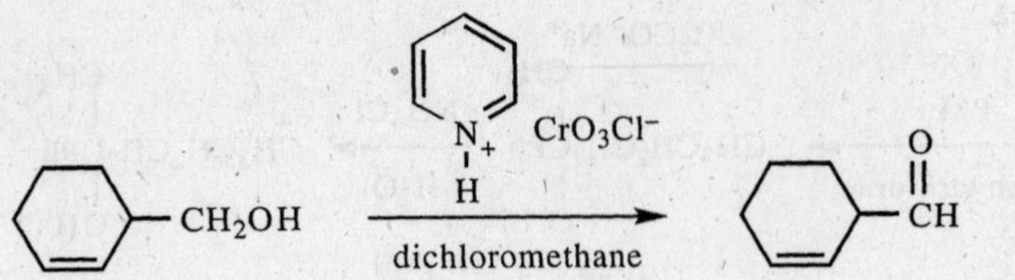

(e)

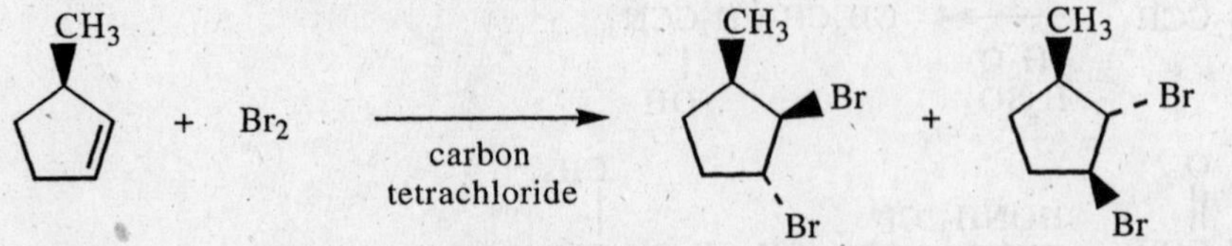

(f)

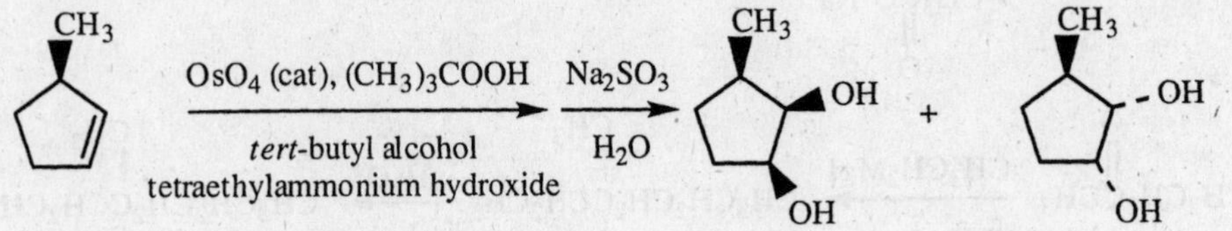

(g)

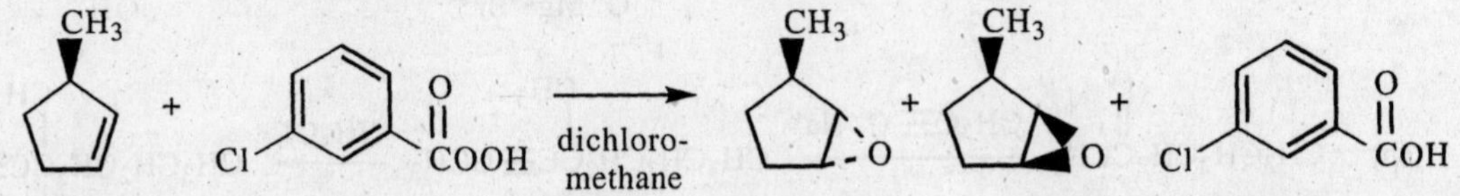

(h)

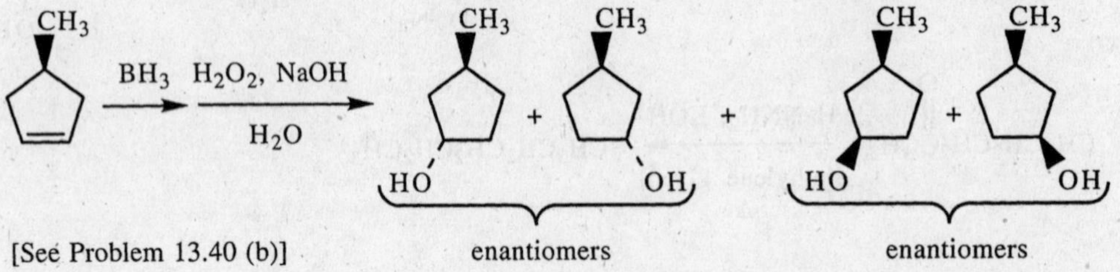

[See Problem 13.40 (b)]   enantiomers   enantiomers

(i)

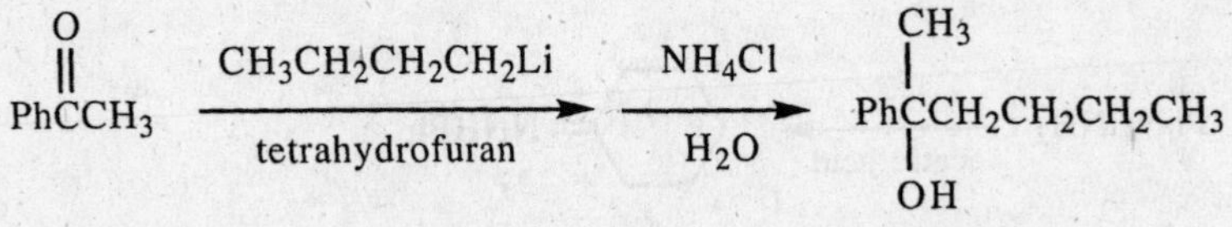

(j)

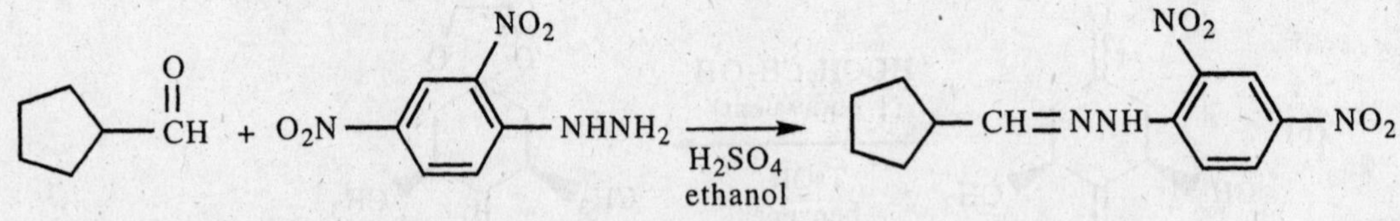

(k)

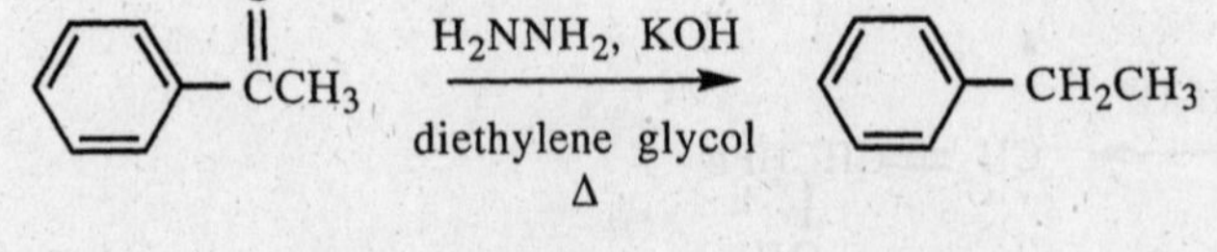

14.30 (a) [structure] + $HONH_3^+ Cl^-$ $\xrightarrow{CH_3CO^- Na^+}$ [structure]

(b) [furan-CHO structure] + [amino alcohol structure] $\xrightarrow{\text{benzene} \; \Delta}$ [imine structure]

(c) $CH_3CH_2CH_2C\equiv CH + NaNH_2 \xrightarrow{NH_3 \text{ (liq)}} CH_3CH_2CH_2C\equiv C^- Na^+ + NH_3$

(d) $CH_3CH_2CH_2C\equiv CH + CH_3CH_2MgBr \xrightarrow{\text{diethyl ether}} CH_3CH_2CH_2C\equiv CMgBr + CH_3CH_3\uparrow$

(e) [cyclohexanone] $\xrightarrow[\text{diethyl ether}]{CH_3CH_2MgBr} \xrightarrow[H_2O]{NH_4Cl}$ [1-ethylcyclohexanol]

(f) CH_3—[benzene]—$\overset{O}{\overset{\|}{C}}CH_2CH_3 + HOCH_2CH_2OH \xrightarrow[\substack{\text{TsOH} \\ \text{benzene} \\ \Delta}]{}$ CH_3—[benzene]—[dioxolane]CCH_2CH_3

(g) [cyclohexanone] $O + HSCH_2CH_2OH \xrightarrow{(CH_3CH_2)_2OBF_3}$ [oxathiolane spiro structure]

(h) $CH_3CH_2C\equiv C\overset{OTHP}{\overset{|}{C}}HCH_2CH_2CH=CHCCH_3 \xrightarrow[\text{base}]{CH_3ONH_3^+ Cl^-} CH_3CH_2C\equiv C\overset{OTHP}{\overset{|}{C}}HCH_2CH_2CH=CHC\overset{NOCH_3}{\overset{\|}{C}}CH_3$

(i) $HOCH_2CH_2\overset{CH_3}{\underset{CH_3}{\overset{|}{\underset{|}{C}}}}CH_2CH=CH\overset{O}{\overset{\|}{C}}CH_3 \xrightarrow[\text{dichloromethane}]{[\text{pyridinium chlorochromate}]} H\overset{O}{\overset{\|}{C}}CH_2\overset{CH_3}{\underset{CH_3}{\overset{|}{\underset{|}{C}}}}CH_2CH=CH\overset{O}{\overset{\|}{C}}CH_3$

(j) [phenyl ketone structure] $\overset{O}{\overset{\|}{C}}CH_2CH_2CH_3 \xrightarrow[\substack{BF_3 \\ \Delta}]{HSCH_2CH_2SH}$ [dithiolane structure] $CH_2CH_2CH_3 \xrightarrow[\substack{\text{ethanol} \\ \Delta}]{\text{Raney Ni}}$ [phenyl structure] $CH_2CH_2CH_2CH_3$

14.31

(a)

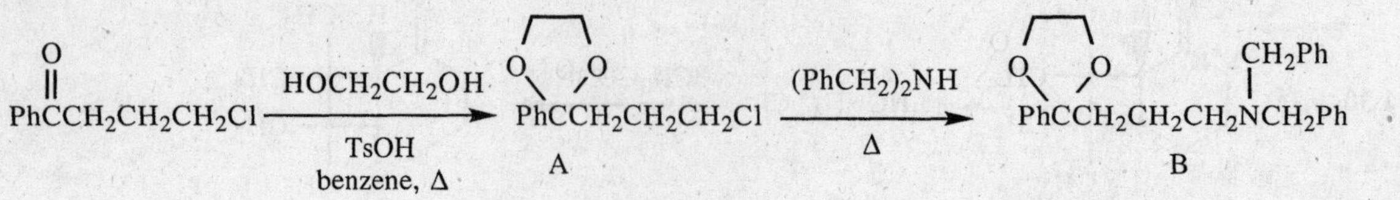

$$PhCCH_2CH_2CH_2Cl \xrightarrow[\text{benzene, } \Delta]{\text{HOCH}_2\text{CH}_2\text{OH} \atop \text{TsOH}} PhCCH_2CH_2CH_2Cl \xrightarrow[\Delta]{(PhCH_2)_2NH} PhCCH_2CH_2CH_2NCH_2Ph$$

A   B

(b)

$$CH_3CH_2CH_2CCH_3 \xrightarrow[\text{H}_2\text{O}]{\text{NaBH}_4} CH_3CH_2CH_2CHCH_3 \xrightarrow[\text{pyridine}]{\text{TsCl}} CH_3CH_2CH_2CHCH_3 \xrightarrow[\text{ethanol}]{\text{KOH (4 M)}}$$

C (OH)   D (OTs)

$$CH_3CH_2CH=CHCH_3 \xrightarrow[\text{dichloromethane}]{Cl\text{-}C_6H_4\text{-}COOH} CH_3CH_2CH-CHCH_3$$

major product E   F

(c)

cyclopentyl-CH(=O) $\xrightarrow[\text{tetrahydrofuran}]{PhLi}$ cyclopentyl-CHPh(O$^-$Li$^+$) G $\xrightarrow{H_3O^+}$ cyclopentyl-CHPh(OH) H $\xrightarrow[\text{acetone}]{CrO_3, H_3O^+}$

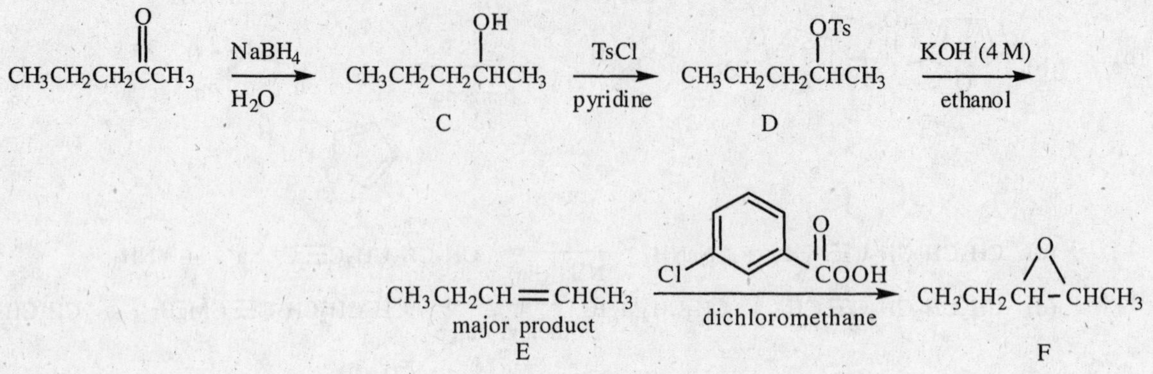

cyclopentyl-CPh(=O) I $\xrightarrow[\text{acetic acid}]{PhNHNH_2}$ cyclopentyl-C(Ph)=NNHPh J

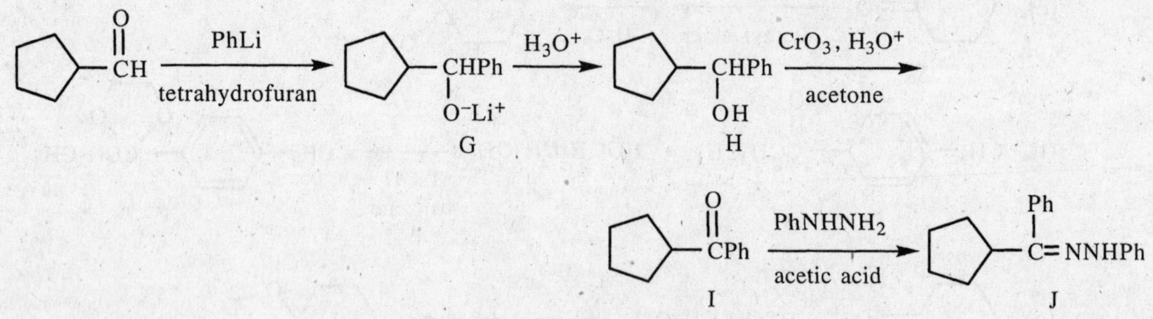

(d)

bicyclo-ketone (=O) $\xrightarrow{H_2NNH_2}$ bicyclo (=NNH$_2$) K $\xrightarrow[\Delta]{\text{(CH}_3\text{)}_3\text{CO}^- \text{K}^+ \atop \text{dimethyl sulfoxide}}$ bicyclo L

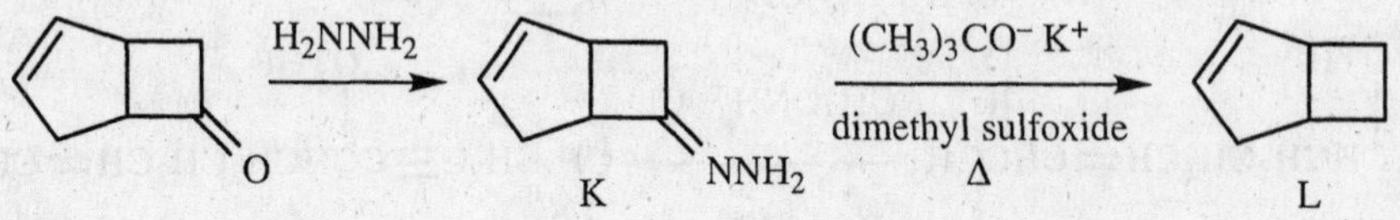

(e)

dimethoxybenzene $\xrightarrow[\text{AlCl}_3]{CH_3COCCH_3}$ M

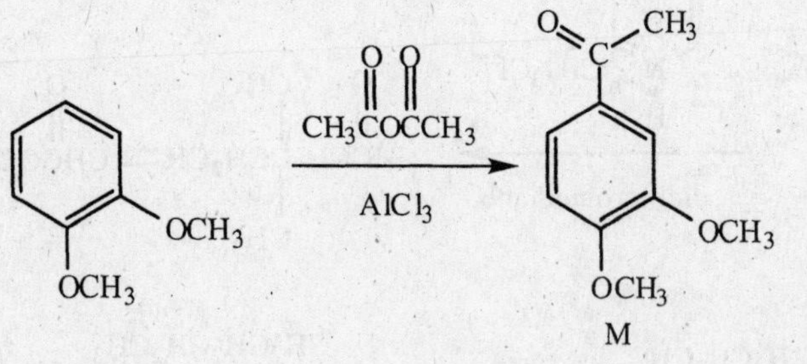

(f)

alkenol-OH $\xrightarrow{CH_3SCH_3, \ ClC\text{---}CCl \ (CH_3CH_2)_3N}$ N (aldehyde)

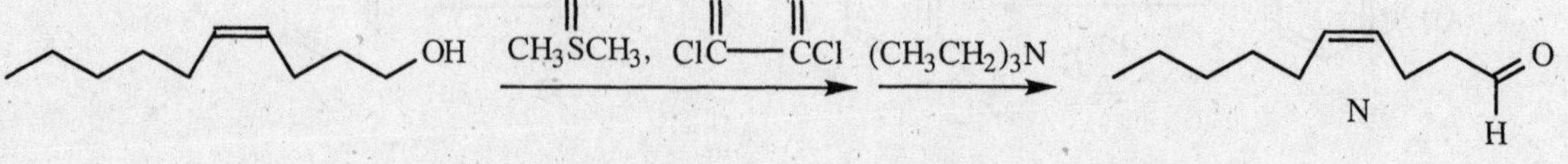

14.32

(a) 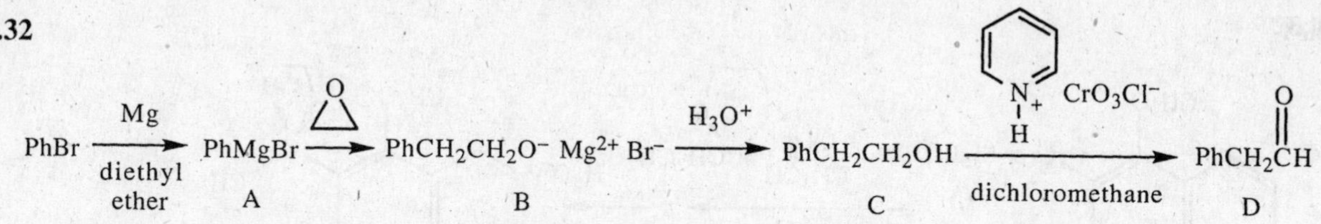 $PhBr \xrightarrow[\substack{\text{diethyl} \\ \text{ether}}]{Mg} PhMgBr \xrightarrow{\triangle\!O} PhCH_2CH_2O^- \ Mg^{2+} Br^- \xrightarrow{H_3O^+} PhCH_2CH_2OH \xrightarrow[\text{dichloromethane}]{} PhCH_2\overset{O}{\overset{\|}{C}H}$

A · B C D

(b)

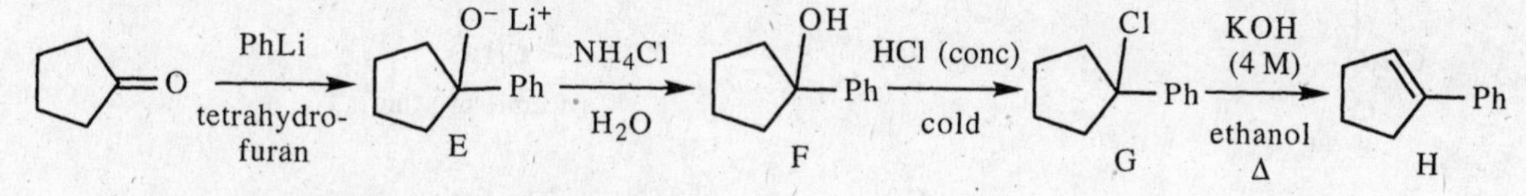

(c)

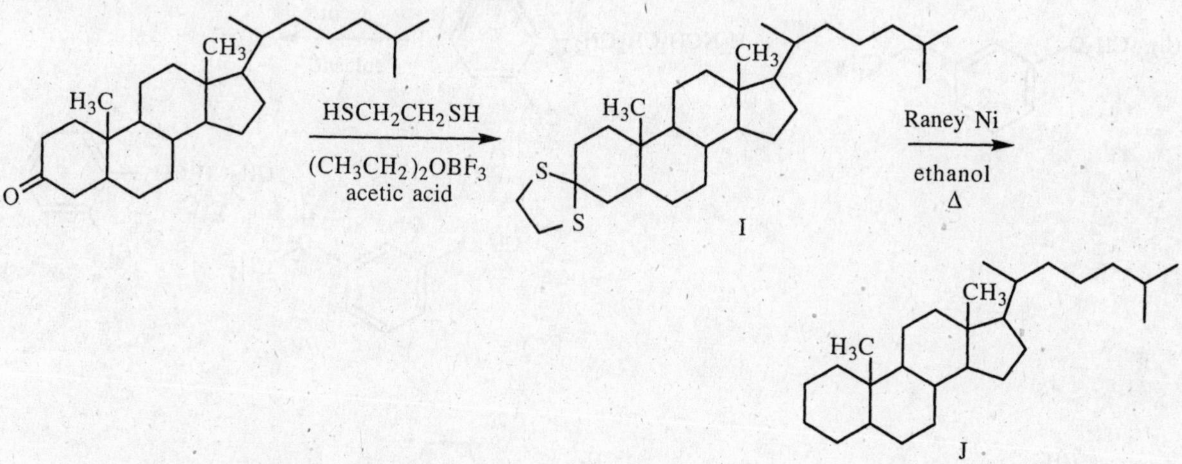

(d)

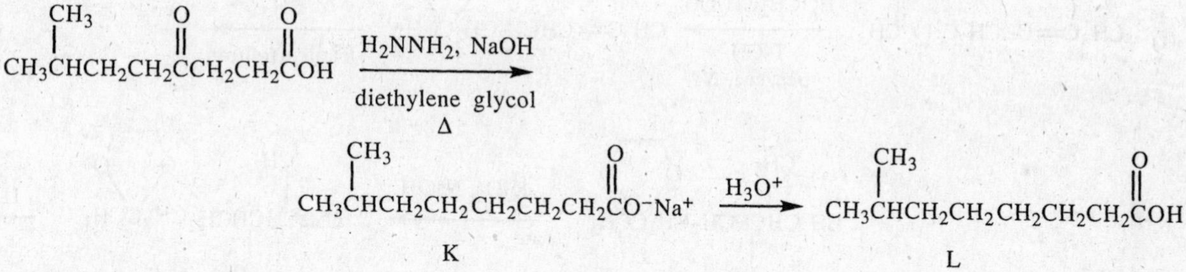

(e)

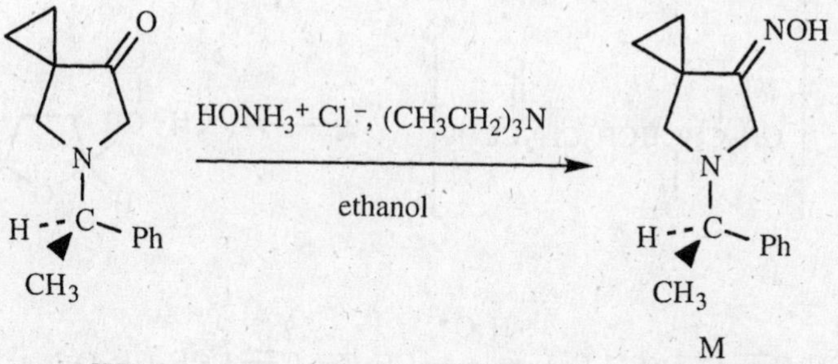

(f)

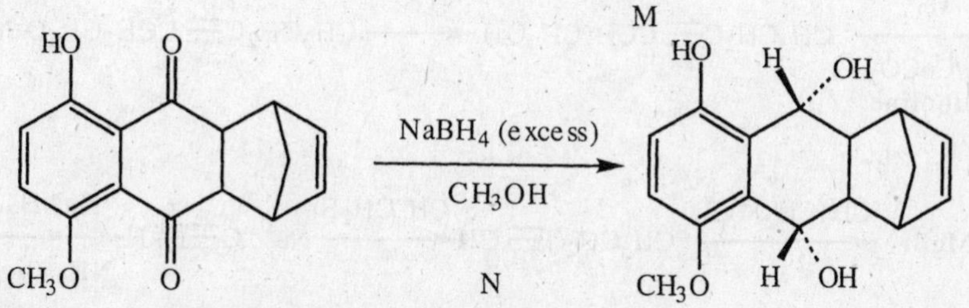

(Note: The top of the molecule is more open to attack by borohydride than the bottom.)

14.32

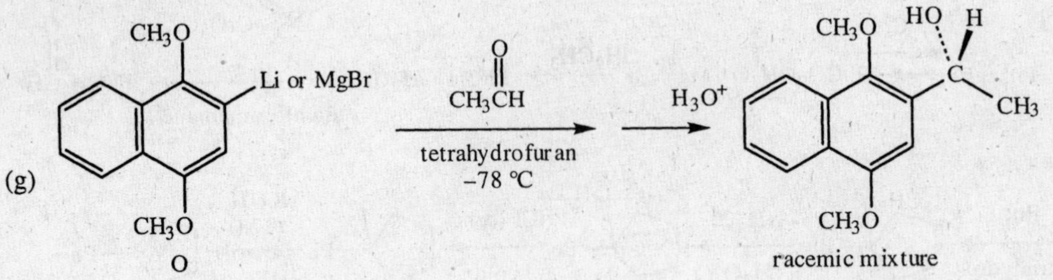
(g)
CH_3O
Li or MgBr
CH_3O
O
CH_3CH
tetrahydrofuran
−78 °C
H_3O^+
HO H
CH_3
racemic mixture

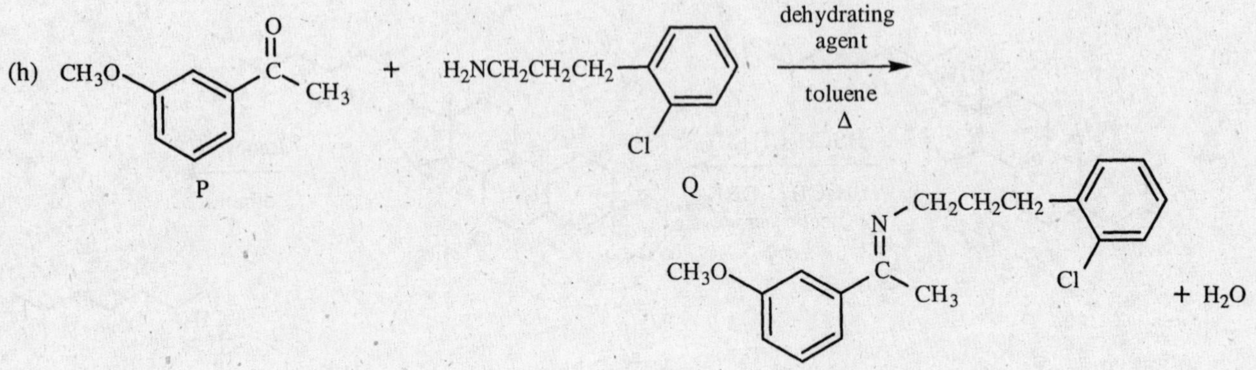
(h) CH_3O
CH_3
+ $H_2NCH_2CH_2CH_2$
Cl
P
Q
dehydrating agent
toluene
Δ
$CH_2CH_2CH_2$
N
C
+ H_2O

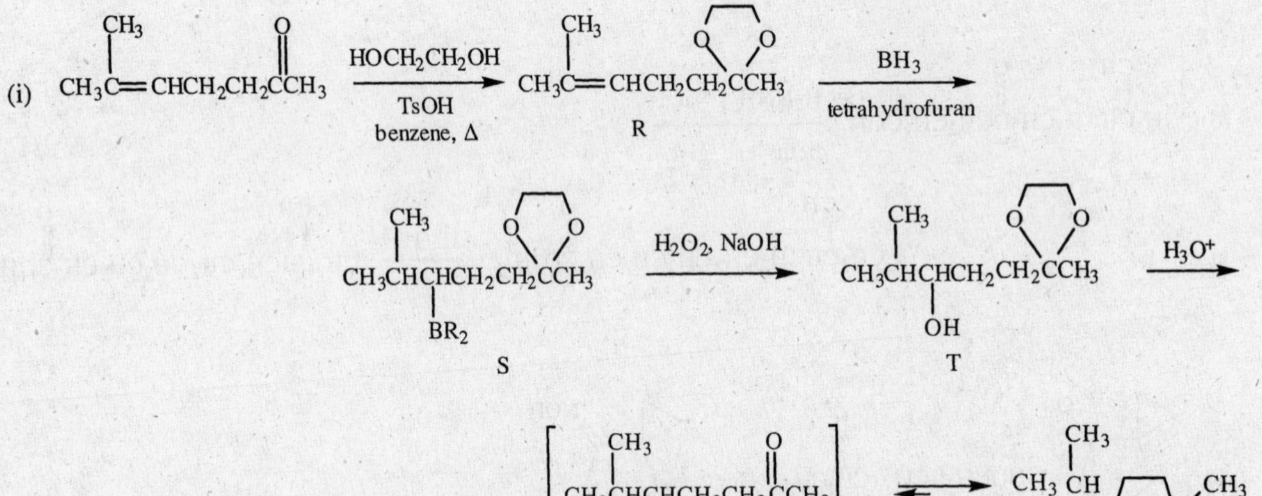
(i) $CH_3C=CHCH_2CH_2CCH_3$
$HOCH_2CH_2OH$
TsOH
benzene, Δ
R
BH_3
tetrahydrofuran
$CH_3CHCHCH_2CH_2CCH_3$
BR_2
S
H_2O_2, NaOH
OH
T
H_3O^+
U

14.33

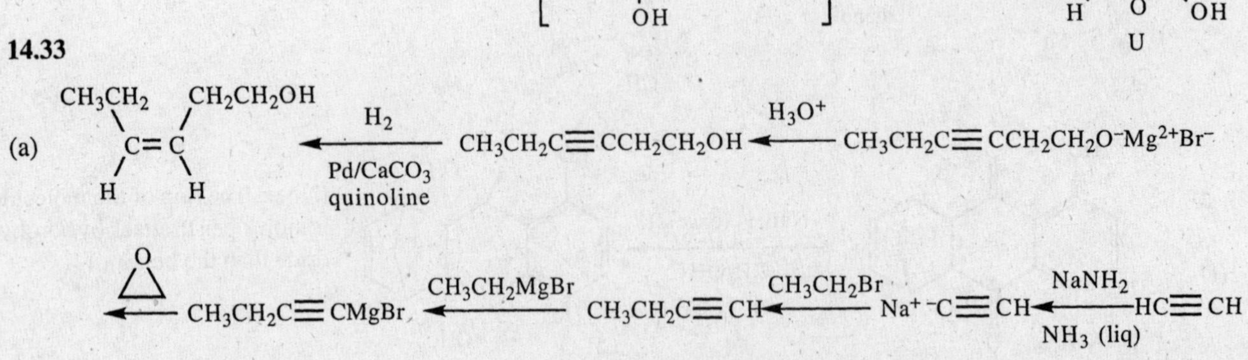
(a)
CH_3CH_2 CH_2CH_2OH
H_2
Pd/$CaCO_3$
quinoline
$CH_3CH_2C\equiv CCH_2CH_2OH$
H_3O^+
$CH_3CH_2C\equiv CCH_2CH_2O^-Mg^{2+}Br^-$
$CH_3CH_2C\equiv CMgBr$
CH_3CH_2MgBr
$CH_3CH_2C\equiv CH$
CH_3CH_2Br
$Na^+ {}^-C\equiv CH$
$NaNH_2$
NH_3 (liq)
$HC\equiv CH$

14.33

(b)

$$\underset{\text{CH}_3\text{CH}_2\text{CH}_2\overset{\displaystyle O}{\overset{\|}{\text{C}}}\text{CH}_2\text{CH}_3}{} \xleftarrow[\text{H}_3\text{O}^+]{\text{CrO}_3} \underset{\text{OH}}{\overset{}{\text{CH}_3\text{CH}_2\text{CH}_2\overset{}{\text{CH}}\text{CH}_2\text{CH}_3}} \xleftarrow{\text{H}_3\text{O}^+}$$

$$\underset{\text{O}^-\text{Mg}^{2+}\text{Br}^-}{\text{CH}_3\text{CH}_2\text{CH}_2\overset{}{\text{CH}}\text{CH}_2\text{CH}_3} \xleftarrow{\overset{\displaystyle O}{\overset{\|}{\text{CH}_3\text{CH}_2\text{CH}}}} \text{CH}_3\text{CH}_2\text{CH}_2\text{MgBr} \xleftarrow[\substack{\text{diethyl}\\\text{ether}}]{\text{Mg}} \text{CH}_3\text{CH}_2\text{CH}_2\text{Br}$$

(c)

$$\underset{\substack{\text{major product}}}{\underset{\text{CH}_3}{\overset{\text{CH}_3}{\text{CH}_3\text{C}=\text{CHCH}_2\text{CH}_2\text{CH}_3}}} \xleftarrow[\substack{\text{H}_3\text{PO}_4\\\Delta}]{} \underset{\substack{\text{CH}_3\\\text{OH}}}{\text{CH}_3\overset{}{\text{C}}\text{CH}_2\text{CH}_2\text{CH}_2\text{CH}_3} \xleftarrow{\text{H}_3\text{O}^+} \underset{\substack{\text{CH}_3\\\text{O}^-\text{Mg}^{2+}\text{Br}^-}}{\text{CH}_3\overset{}{\text{C}}\text{CH}_2\text{CH}_2\text{CH}_2\text{CH}_3} \xleftarrow{\overset{\displaystyle O}{\overset{\|}{\text{CH}_3\text{C}\text{CH}_3}}}$$

$$\text{CH}_3\text{CH}_2\text{CH}_2\text{CH}_2\text{MgBr} \xleftarrow[\substack{\text{diethyl}\\\text{ether}}]{\text{Mg}} \text{CH}_3\text{CH}_2\text{CH}_2\text{CH}_2\text{Br} \xleftarrow[\text{pyridine}]{\text{PBr}_3} \text{CH}_3\text{CH}_2\text{CH}_2\text{CH}_2\text{OH} \xleftarrow{\text{H}_3\text{O}^+}$$

$$\text{CH}_3\text{CH}_2\text{CH}_2\text{CH}_2\text{O}^-\text{Mg}^{2+}\text{Br}^- \xleftarrow{\triangle} \text{CH}_3\text{CH}_2\text{MgBr} \xleftarrow[\substack{\text{diethyl}\\\text{ether}}]{\text{Mg}} \text{CH}_3\text{CH}_2\text{Br}$$

(d)

$$\text{PhCH}_2\text{CH}_2\text{Br} \xleftarrow[\text{pyridine}]{\text{PBr}_3} \text{PhCH}_2\text{CH}_2\text{OH} \xleftarrow{\text{H}_3\text{O}^+} \text{PhCH}_2\text{CH}_2\text{O}^-\ \text{Mg}^{2+}\ \text{Br}^- \xleftarrow{\triangle} \text{PhMgBr} \xleftarrow[\substack{\text{diethyl}\\\text{ether}}]{\text{Mg}} \text{PhBr}$$

(e)

$$\xleftarrow[\substack{\text{carbon}\\\text{tetrachloride}}]{\text{Br}_2} \underset{}{\overset{\text{CH}_3\text{CH}_2\text{CH}_2}{\text{C}=\text{C}}} \xleftarrow[\text{NH}_3\ (\text{liq})]{\text{Na}} \text{CH}_3\text{CH}_2\text{CH}_2\text{C}\equiv\text{CCH}_3 \xleftarrow{\text{CH}_3\text{I}}$$

$$\text{CH}_3\text{CH}_2\text{CH}_2\text{C}\equiv\text{C}^-\ \text{Na}^+ \xleftarrow[\text{NH}_3\ (\text{liq})]{\text{NaNH}_2} \text{CH}_3\text{CH}_2\text{CH}_2\text{C}\equiv\text{CH} \xleftarrow{\text{CH}_3\text{CH}_2\text{CH}_2\text{Br}} \text{Na}^+\ ^-\text{C}\equiv\text{CH} \xleftarrow[\text{NH}_3\ (\text{liq})]{\text{NaNH}_2} \text{HC}\equiv\text{CH}$$

(f)

14.33 (f) (cont)

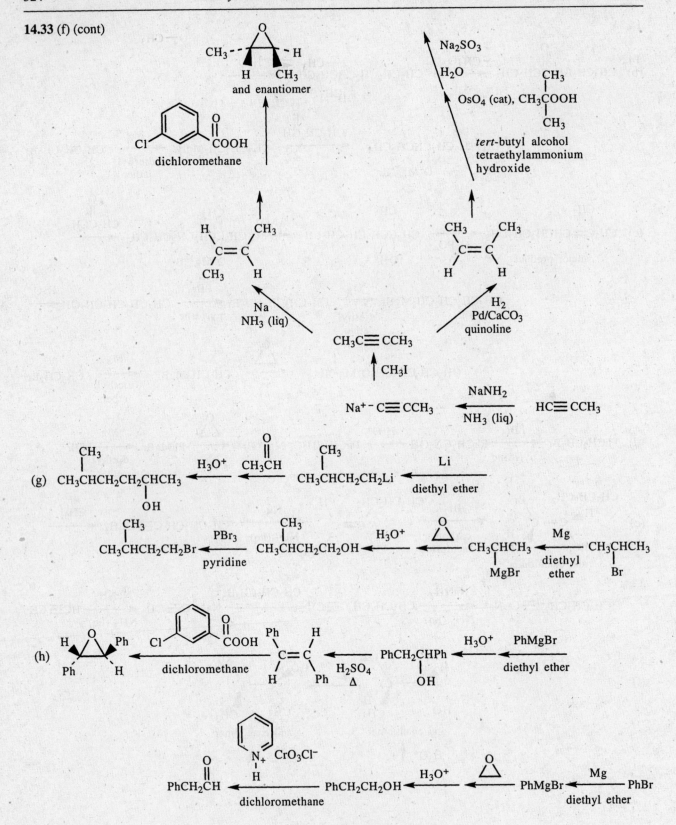

14.34

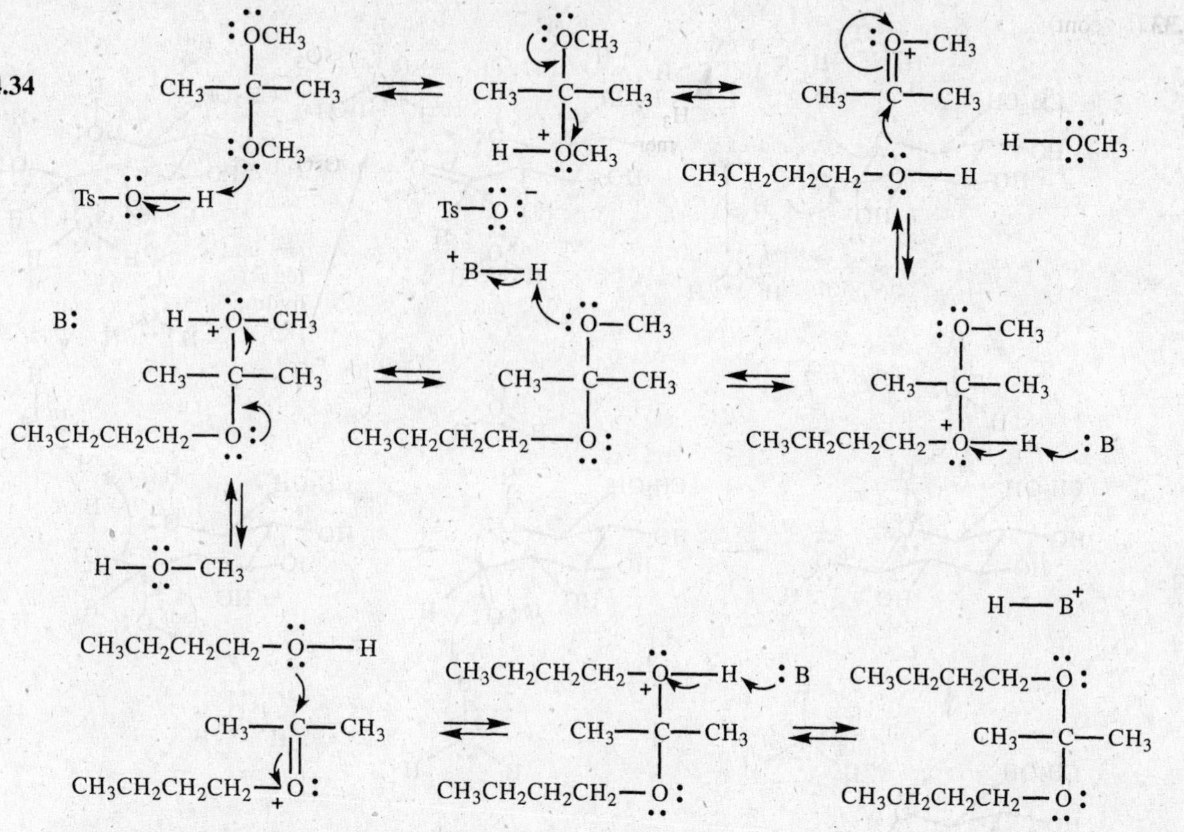

The reaction proceeds by a series of steps, each of which is an equilibrium. Removal by distillation of methanol, the lowest boiling component of the mixture, as it forms, will push the equilibrium towards the dibutoxypropane.

14.35

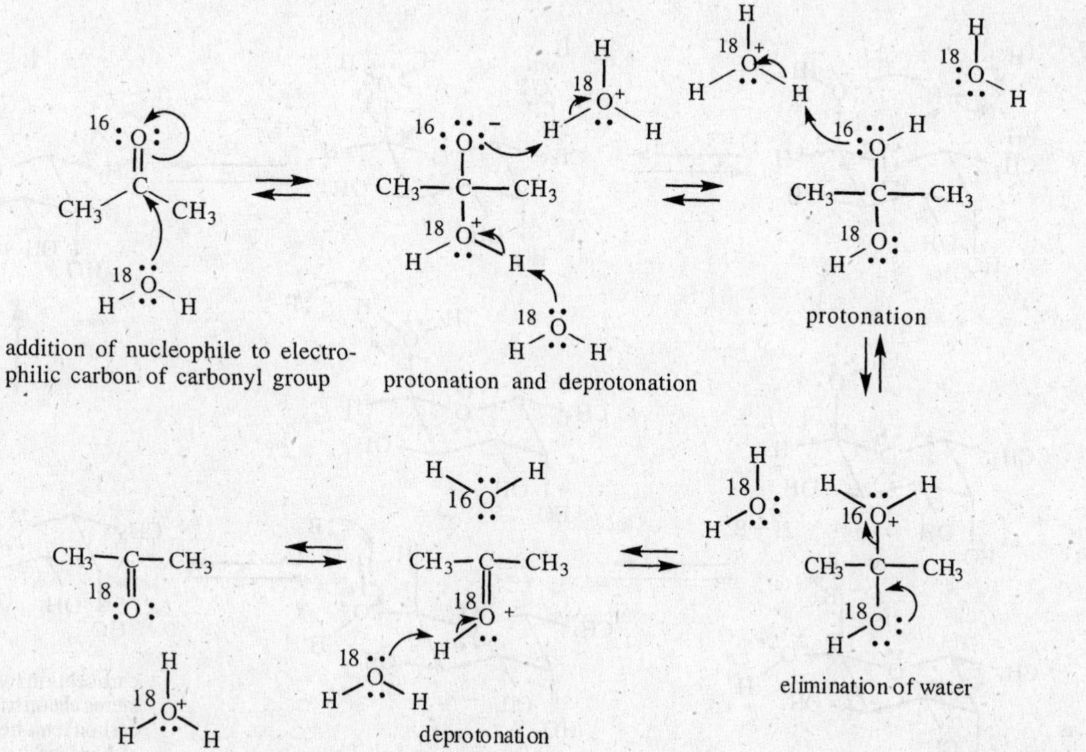

addition of nucleophile to electro-
philic carbon of carbonyl group   protonation and deprotonation   protonation

elimination of water

deprotonation

14.36

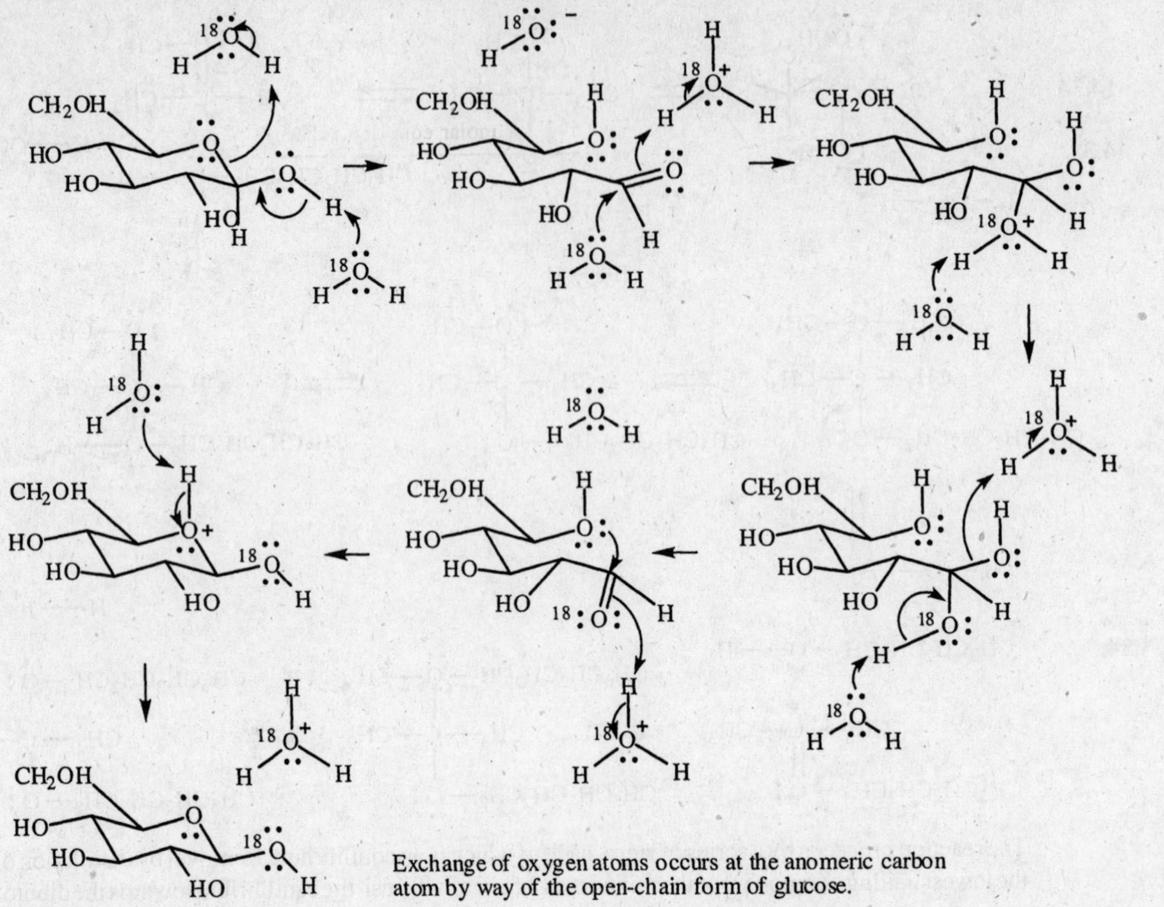

Exchange of oxygen atoms occurs at the anomeric carbon atom by way of the open-chain form of glucose.

14.37

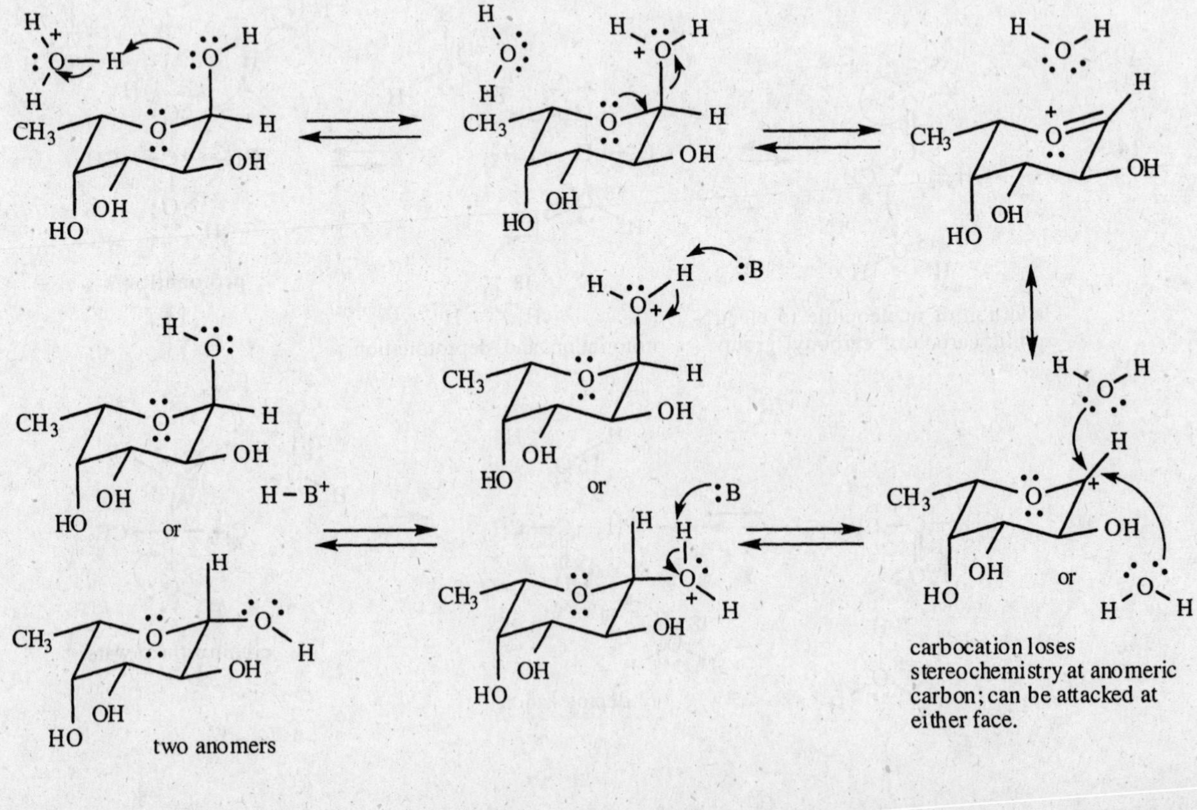

two anomers

carbocation loses stereochemistry at anomeric carbon; can be attacked at either face.

14.38

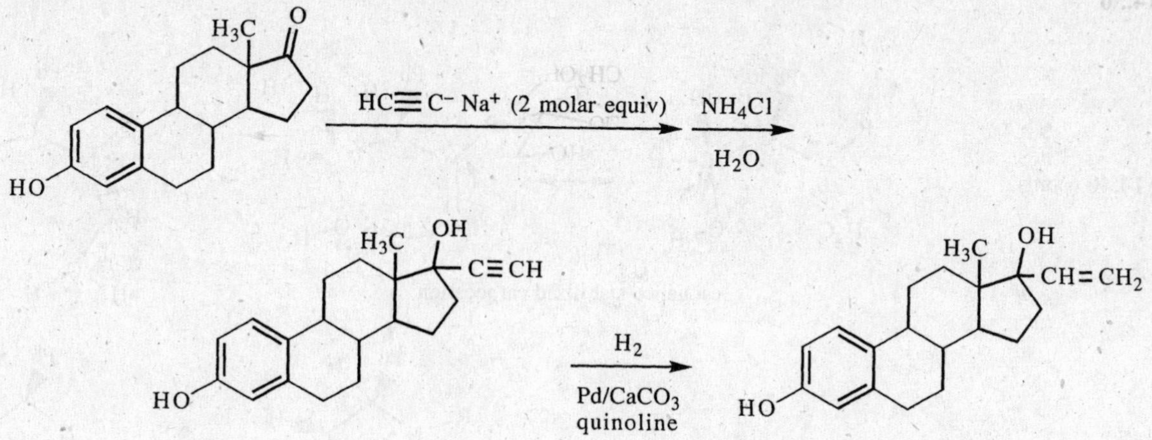

14.39

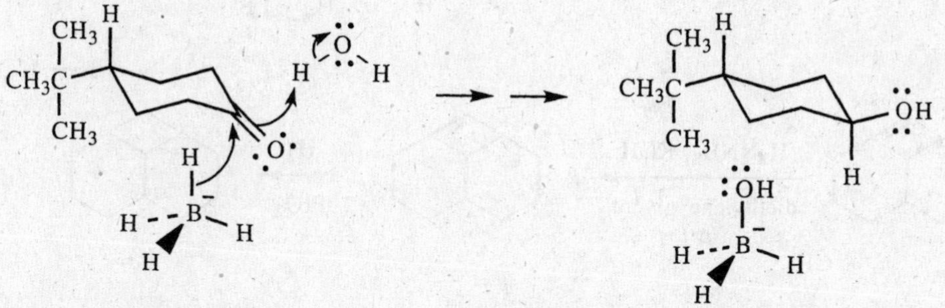

The large *tert*-butyl group occupies an equatorial position on the ring (Section 5.11C). Reduction takes place so that the major product has the hydroxyl group also in the equatorial position. This means that the hydride is delivered to the carbonyl group from the axial direction.

48%   52%

product from axial approach   product from equatorial approach
of the borohydride anion   of the borohydride anion

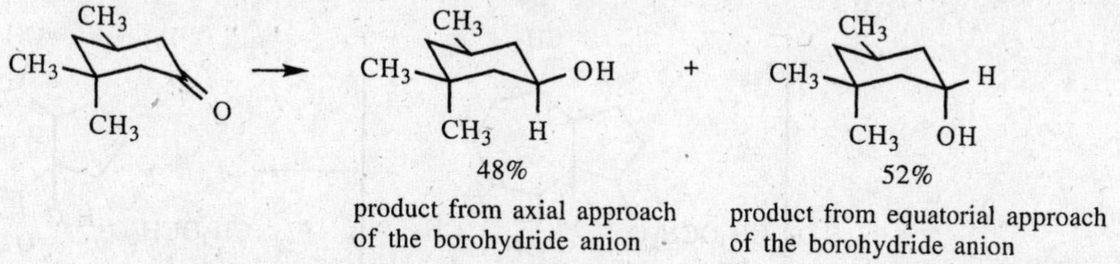

In 3,3,5-trimethylcyclohexanone, one of the methyl groups on carbon 3 is axial and hinders the approach of the borohydride anion from the axial direction. The major product is now the result of reduction of the carbonyl group from the less hindered side of the molecule.

14.40

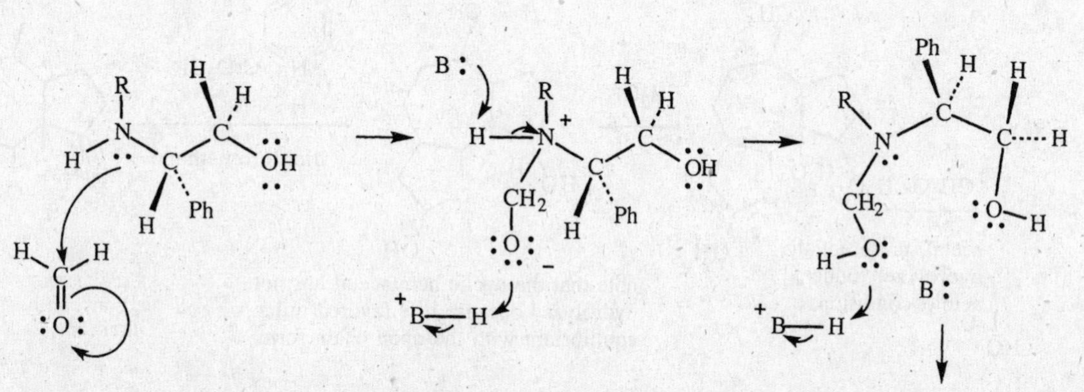

14.40 (cont)

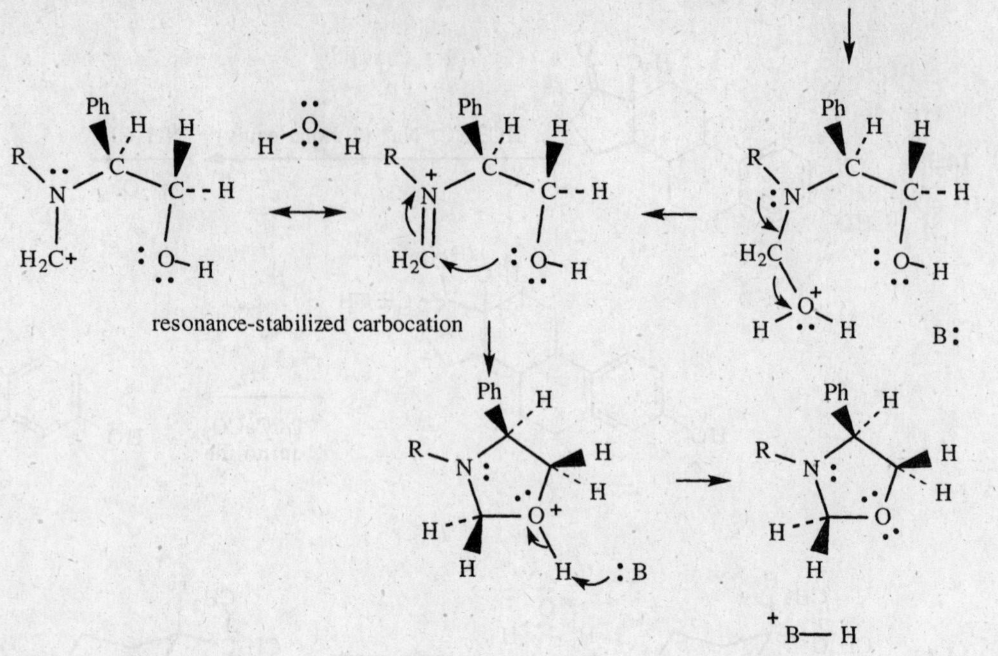

resonance-stabilized carbocation

14.41 (a)

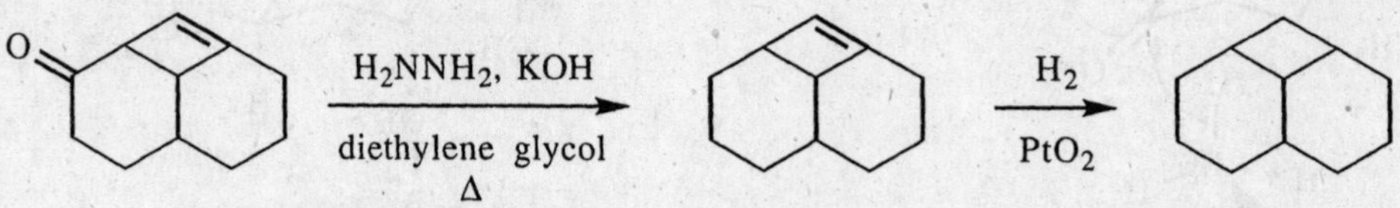

(b)

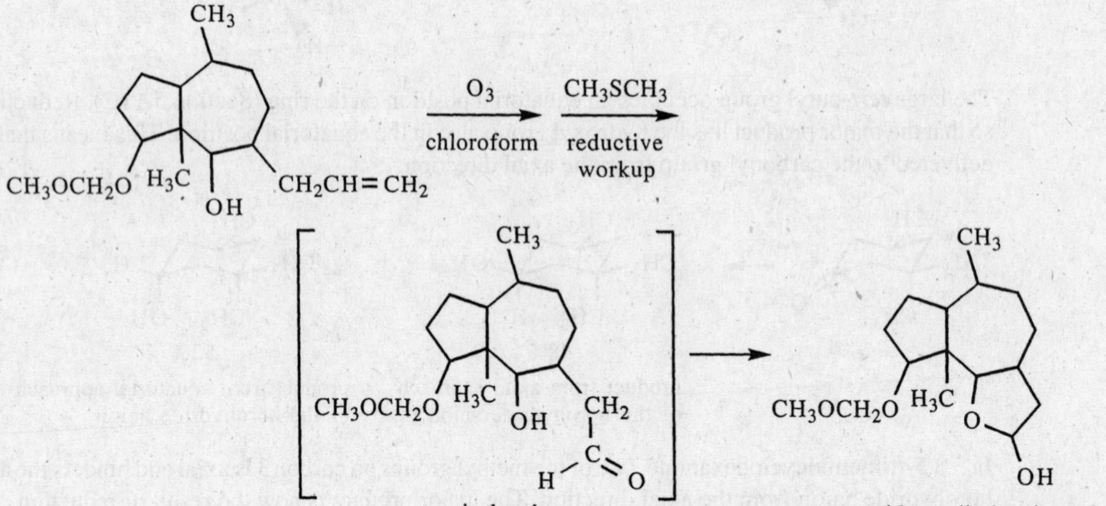

not isolated

stable cyclic hemiacetal

(c)

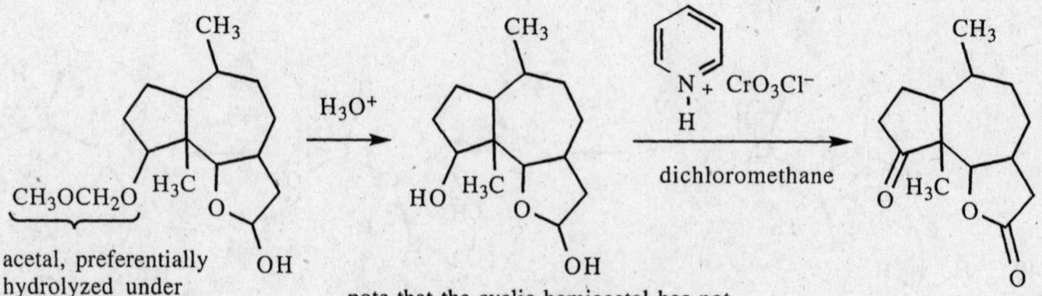

acetal, preferentially
hydrolyzed under
acidic conditions

note that the cyclic hemiacetal has not
hydrolyzed because it is favored in its
equilibrium with the open chain form

14.41   (d)

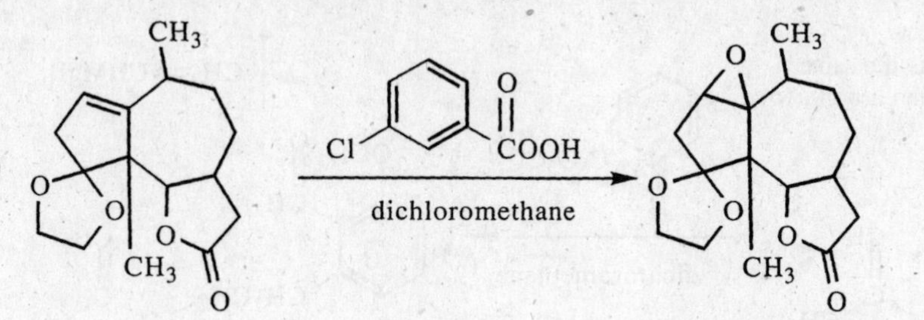

14.42   (a)

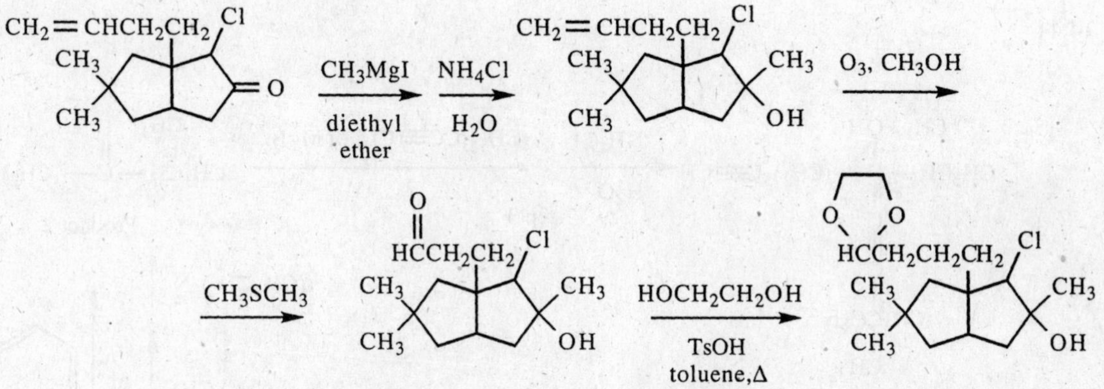

(b)

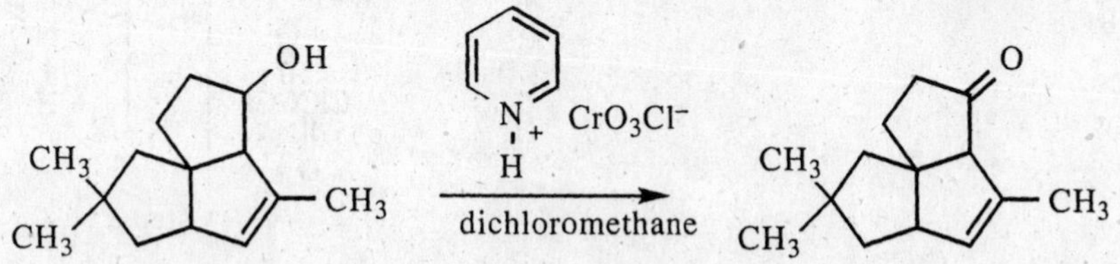

14.43   The benzylic carbon atom in the starting material that has a hydroxyl group on it is the carbon atom of a carbonyl group in the product. It is also bonded to another carbon atom. The formation of a carbon-carbon σ bond suggests the reaction of an organometallic reagent containing a nucleophilic center with the electrophilic carbon atom of a carbonyl compound. A ketone is formed by oxidation of a secondary alcohol. A secondary alcohol is formed by reaction of an organometallic reagent with an aldehyde. Therefore, we need an aldehyde function where the benzylic alcohol group is in the starting material.

this carbon is the same as the benzylic carbon with the hydroxyl group in the starting material shown below

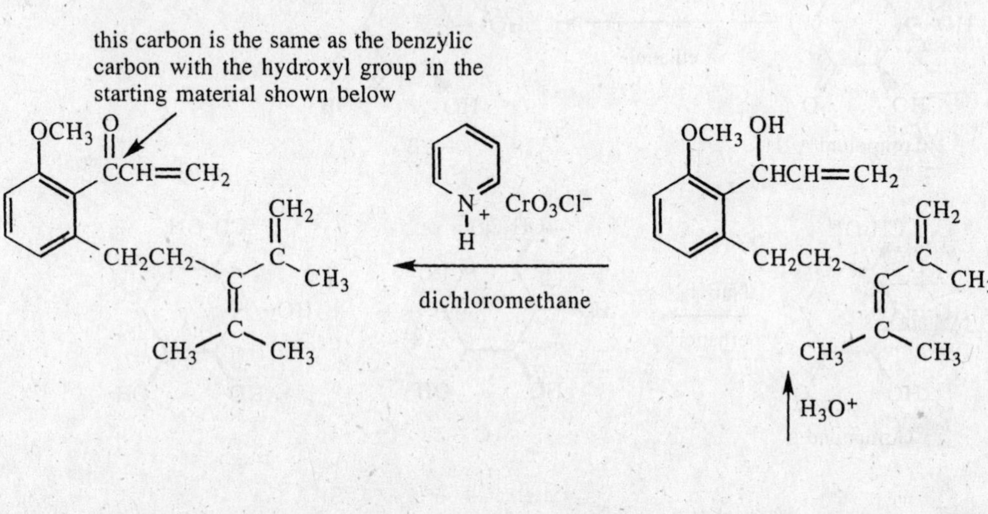

14.43 (cont)

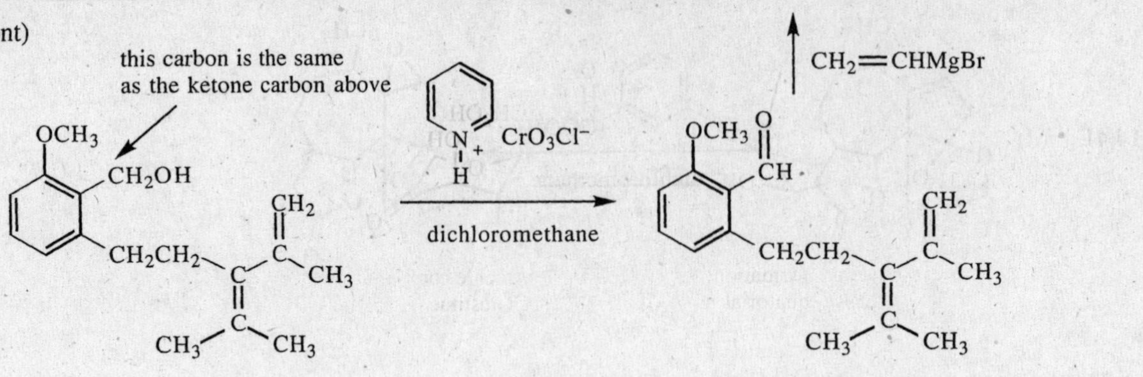
this carbon is the same
as the ketone carbon above
OCH3
CH2OH
CH2
CH2CH2
C
CH3
CH3
CH3
CrO3Cl−
dichloromethane
CH2=CHMgBr
CH

14.44

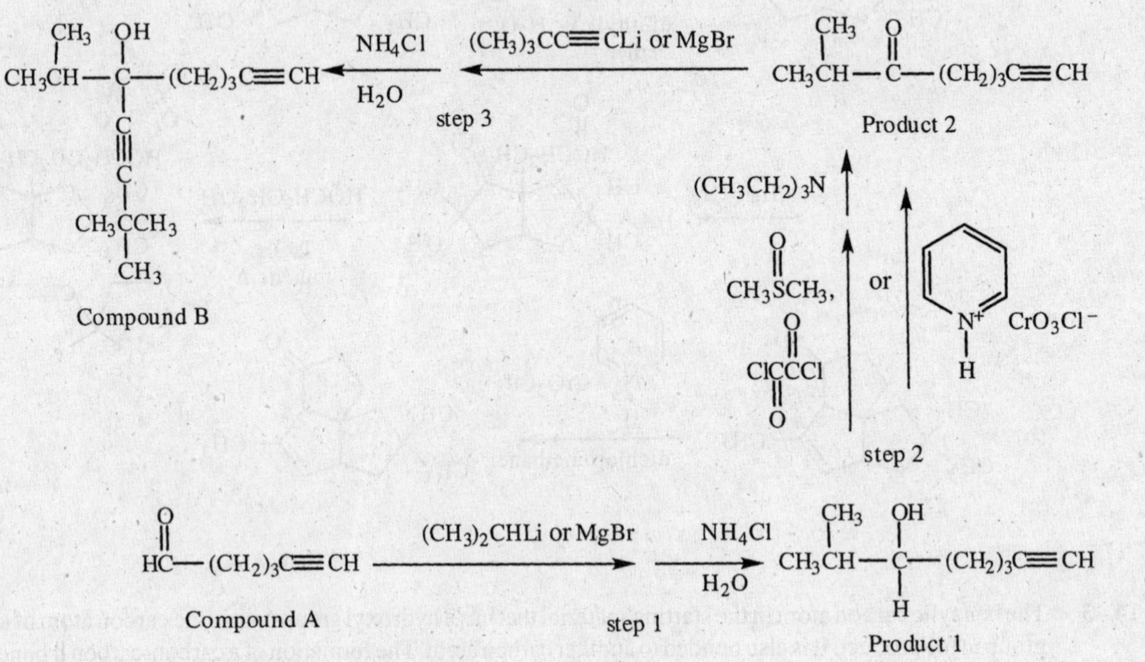
NH4Cl
(CH3)3CC≡CLi or MgBr
H2O
step 3
Compound B
Product 2
(CH3CH2)3N
CH3SCH3,
ClCCCl
or
CrO3Cl−
step 2
Compound A
(CH3)2CHLi or MgBr
step 1
Product 1

14.45 (a)

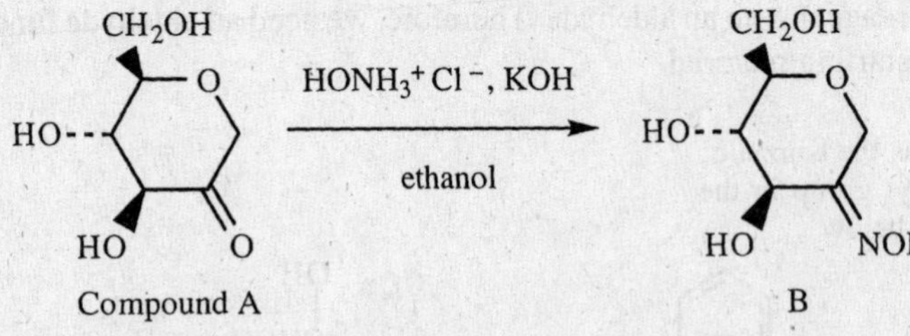
CH2OH
HONH3+ Cl−, KOH
ethanol
Compound A
B
NOH

(b)

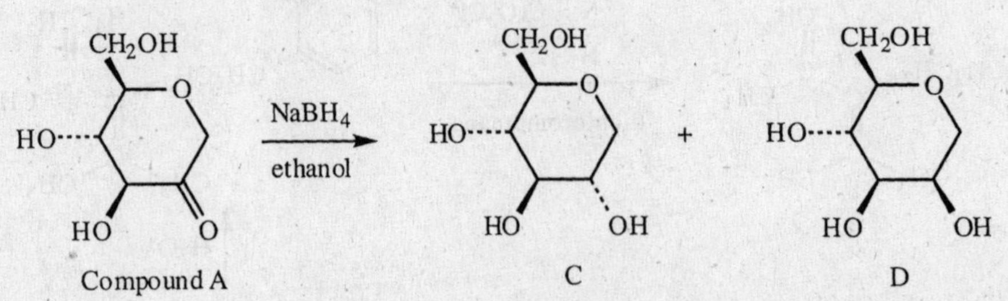
NaBH4
ethanol
Compound A
C
D

attack from this face of
carbonyl gives major product

14.45 (c)

more stable conformation;
substituents equatorial

less stable conformation;
substituents axial

less sterically hindered
side of carbonyl

14.46

14.47

14.47 (cont)

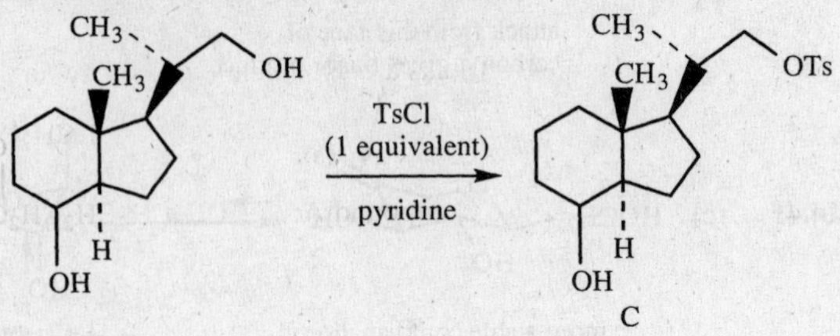

14.48

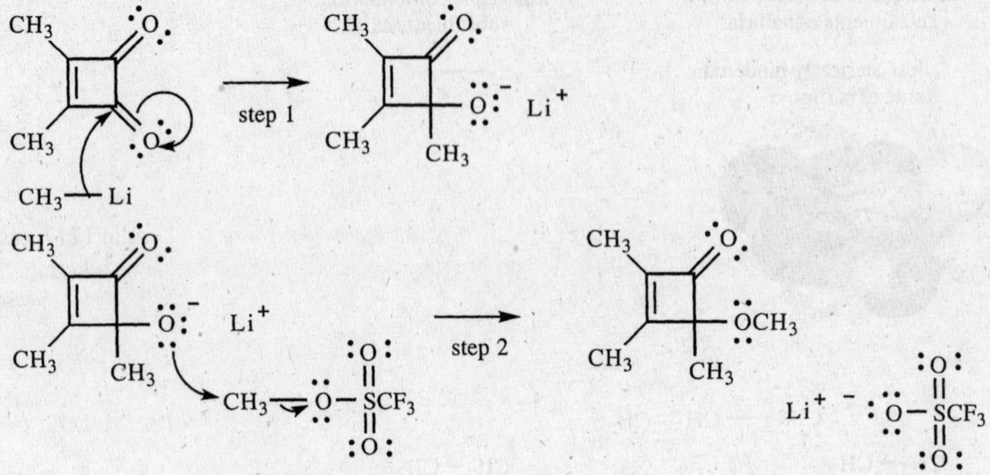

14.49

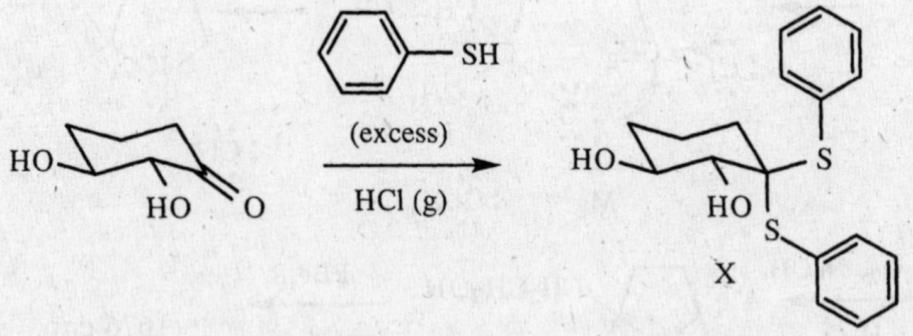

14.50   Compounds A and B, molecular formula $C_6H_{12}O$, each have one unit of unsaturation. The infrared spectra of the two compounds show the presence of a ketone (C=O at 1717 cm^{-1} and no bands at 2900 and 2700 cm^{-1}, which tells us that the compounds are not aldehydes). The presence of a ketone is confirmed by the ^{13}C magnetic resonance spectra in which the ketone carbon atom can be seen for A and B (A at δ 208.88 and B at δ 214.21). The carbonyl group of a ketone accounts for one unit of unsaturation, therefore neither A nor B can have a ring or a double bond. There are four ketones that have the molecular formula $C_6H_{12}O$.

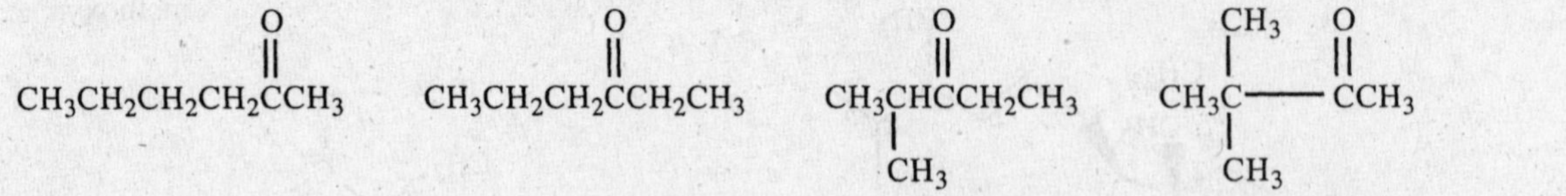

2-hexanone   3-hexanone   2-methyl-3-pentanone   3,3-dimethyl-2-butanone

The ^{13}C spectra tell us how many types of carbon atoms each compound has. Compound A has six different carbon atoms. Compound B has four sets of carbon atoms, and therefore must have three which are the same. 3,3-Dimethyl-2-butanone has four sets of carbon atoms and, therefore, must be Compound B. This is confirmed by the proton magnetic resonance spectrum. Both 2-hexanone and 3-hexanone have six different carbon atoms and could be Compound A. The proton magnetic resonance spectrum allows us to choose the correct structure for Compound A.

14.50 (cont)

The structural formulas are shown below, along with the proton magnetic resonance data for each one.

$$\delta\,0.92 \qquad \delta\,1.60$$

Compound A: $CH_3CH_2CH_2CH_2CCH_3$

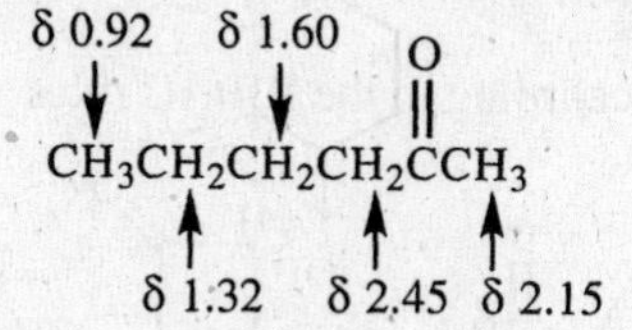

$$\delta\,1.32 \qquad \delta\,2.45 \quad \delta\,2.15$$

Compound B: $\delta\,1.05$
$$\begin{cases} CH_3 & O \\ | & \| \\ CH_3C\!\!-\!\!-\!\!-\!\!-CCH_3 \\ | & \\ CH_3 & \delta\,2.15 \end{cases}$$

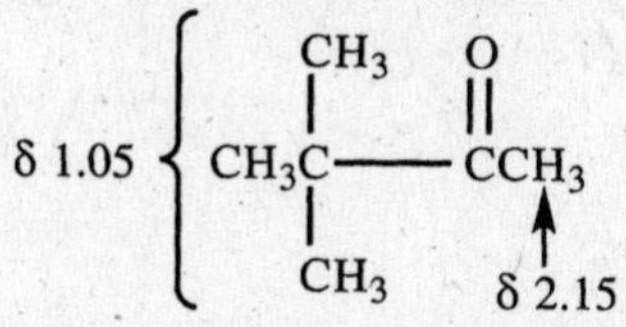

14.51 Compound C, $C_{10}H_{12}O$, has five units of unsaturation. The infrared spectrum has bands at 1717 cm^{-1} (C=O). There are no bands in the infrared at about 2700 cm^{-1}; therefore, the compound must be a ketone and not an aldehyde.

The proton magnetic resonance spectrum has bands at

$\delta\,2.13$	(3H, s, C$\underline{H}_3$C=O)
$\delta\,2.76$	(2H, multiplet, —C$\underline{H}_2$CH$_2$—)
$\delta\,2.90$	(2H, multiplet, —CH$_2$C$\underline{H}_2$—)
$\delta\,7.2$	(5H, multiplet, Ar$\underline{H}$)

The compound is 4-phenyl-2-butanone, $\bigcirc\!\!-CH_2CH_2\overset{\text{O}}{\overset{\|}{C}}CH_3$.

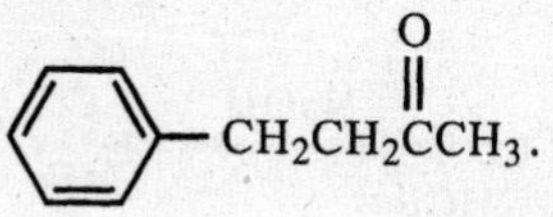

14.52 The infrared spectral information tells us that Compound D is a conjugated ketone (1676 cm^{-1} is the C=O stretching frequency for a conjugated carbonyl group). The proton magnetic resonance spectrum is analyzed below:

$\delta\,2.57$	(3H, s, C$\underline{H}_3$C=O)
$\delta\,3.88$	(3H, s, —OC$\underline{H}_3$)
$\delta\,6.95, 7.95$	(4H, para substitution pattern, Ar$\underline{H}$)

The fragments add up to $C_9H_{10}O_2$, which corresponds to a molecular weight of 150. The molecular ion of Compound D is at m/z 150. Therefore, all atoms are accounted for. Compound D is *p*-methoxyacetophenone.

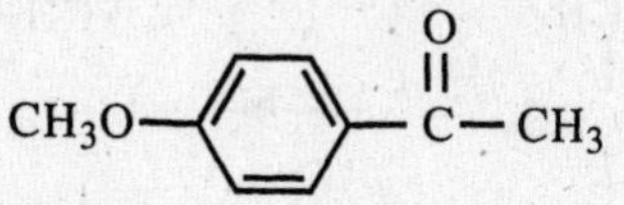

$$CH_3O\!\!-\!\!\bigcirc\!\!-\overset{\text{O}}{\overset{\|}{C}}\!-CH_3$$

The base peak, m/z 135, results from loss of a methyl radical from the molecular ion, leaving a resonance-stabilized acylium ion with m/z 135.

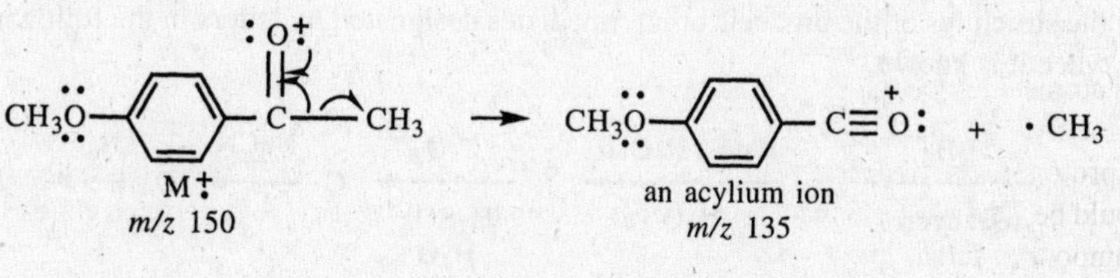

$$CH_3\overset{..}{\underset{..}{O}}\!\!-\!\!\bigcirc\!\!-\overset{:\overset{+}{O}:}{\overset{\|}{C}}\!\!-\!CH_3 \;\longrightarrow\; CH_3\overset{..}{\underset{..}{O}}\!\!-\!\!\bigcirc\!\!-C\!\equiv\!\overset{+}{O}: \;+\; \cdot CH_3$$

M^{+}
m/z 150

an acylium ion
m/z 135

Supplemental Problems

S14.1   Name the following compounds according to the IUPAC rules.

(a) $CH_3CCH_2CH_2CH_3$ (with =O on the C)

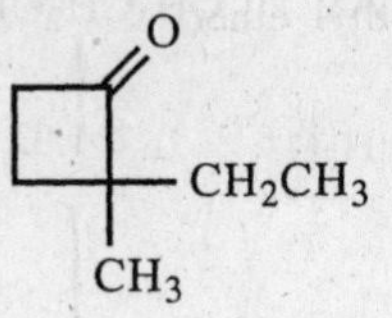

(b) [cyclohexane ring with two CH$_3$ groups and a CH=O substituent]

(c) [cyclobutanone ring with CH$_2$CH$_3$ and CH$_3$ substituents]

(d) 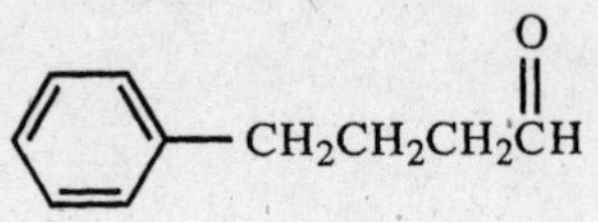 [benzene ring]$-CH_2CH_2CH_2CH$ (with =O)

S14.2   Complete the following equations.

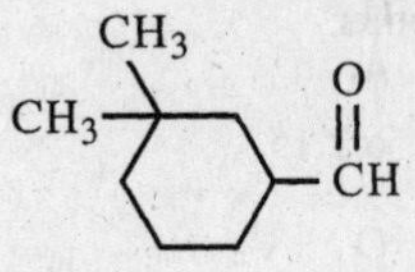

(a) [bicyclic diketone] $\xrightarrow[\substack{CeCl_3 \cdot 7\,H_2O \\ ethanol}]{NaBH_4\ (excess)}$ A

(b) [cyclobutanone, =O] $\xrightarrow[\substack{diethyl \\ ether}]{LiAlH_4 \quad H_3O^+}$ B

(c) [dimethyl cyclohexenone] $\xrightarrow[\substack{diethyl \\ ether}]{CH_3Li \quad H_3O^+}$ C $\longrightarrow$ D

(d) $PhCCH_2CH_2CH_3 + H_2NNH_2 \xrightarrow{\Delta}$ E (the PhC has =O)

(e) $CH_3CCH_2CH_3 + CH_3CHCH_2NH_2 \xrightarrow[HCl]{} \xrightarrow{NaOH}$ F (the first C has =O, the second has CH$_3$)

(f) $CH_3CHCH_2CH_2CH + CH_3OH \xrightarrow[\substack{TsOH \\ benzene \\ \Delta}]{}$ G (CH$_3$ on first C, =O on last CH)

(g) $CH_3CHCH_2CCH_2CHCH_3 \xrightarrow[\substack{diethylene\ glycol \\ \Delta}]{H_2NNH_2,\ NaOH}$ H (CH$_3$ groups on carbons 2 and 6, =O on central C)

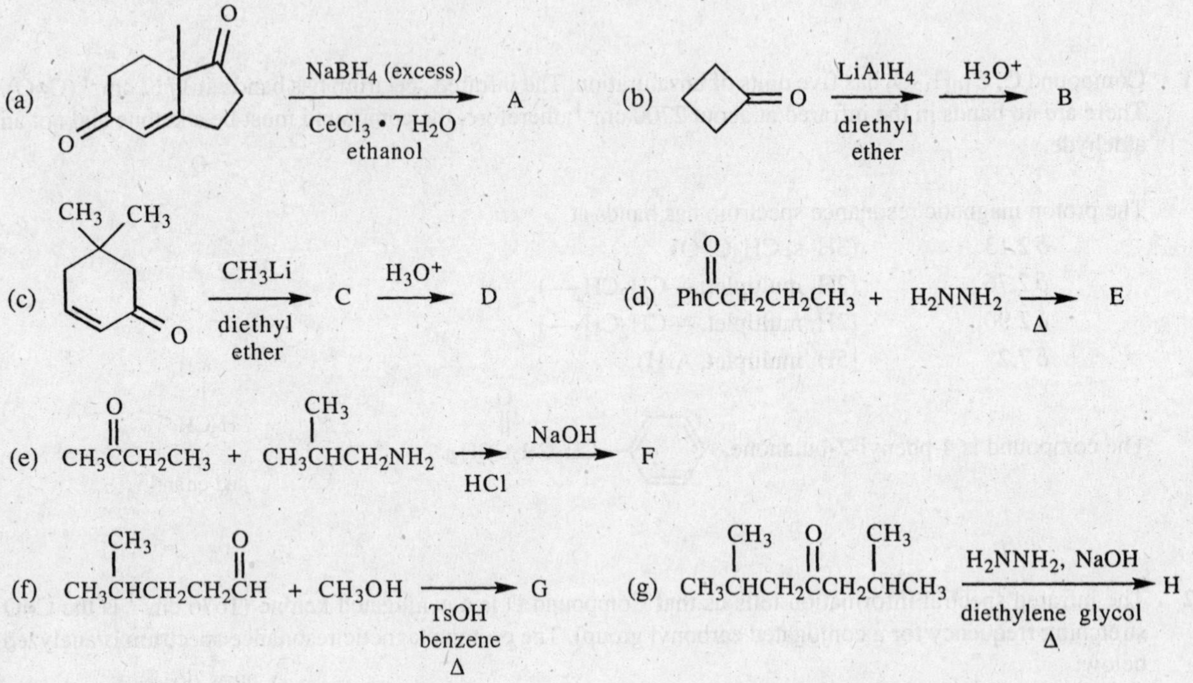

S14.3   A synthesis of talaromycin, a toxic metabolite of a fungus that grows on chicken litter, involved the transformation shown below. How would you carry it out?

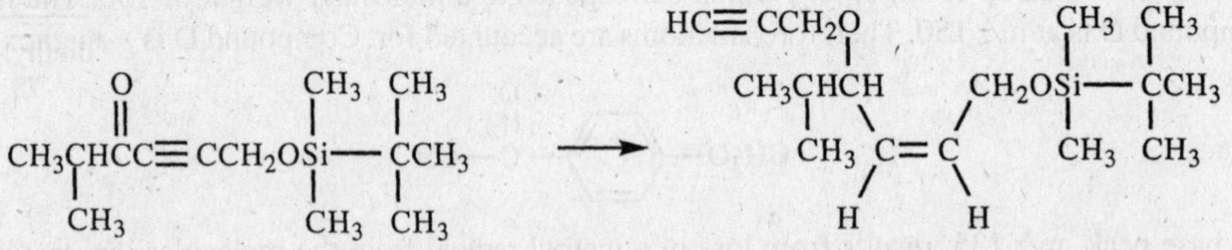

$CH_3CHCC\equiv CCH_2OSi-CCH_3$ [with CH$_3$ groups and Si(CH$_3$)$_3$] $\longrightarrow$ [product with $HC\equiv CCH_2O$, CH_3CHCH, $CH_2OSi-CCH_3$, and C=C alkene]

S14.4   Predict the structures of the products or intermediates designated by letters in the following equations. Show stereochemistry when it is known.

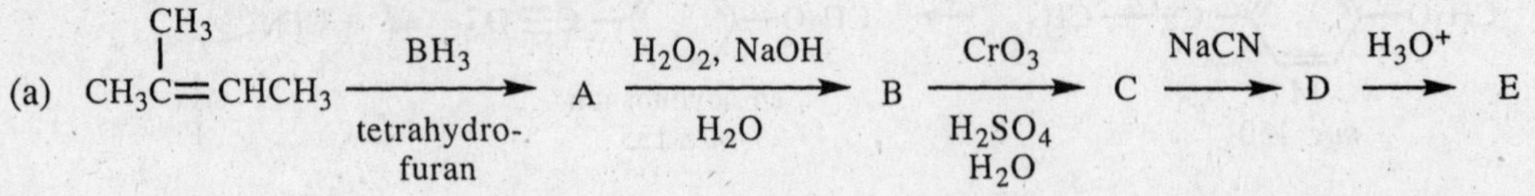

(a) $CH_3C\equiv CHCH_3$ (CH$_3$ on second C) $\xrightarrow[\substack{tetrahydro- \\ furan}]{BH_3}$ A $\xrightarrow[H_2O]{H_2O_2,\ NaOH}$ B $\xrightarrow[\substack{H_2SO_4 \\ H_2O}]{CrO_3}$ C $\xrightarrow{NaCN}$ D $\xrightarrow{H_3O^+}$ E

S14.4

(b) 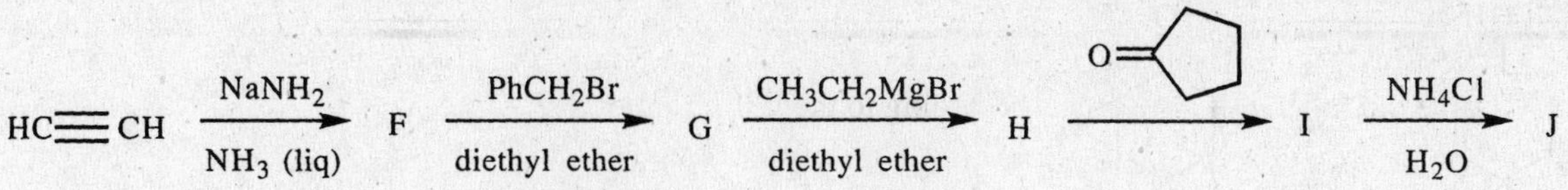 $HC \equiv CH \xrightarrow[\text{NH}_3 \text{ (liq)}]{\text{NaNH}_2} F \xrightarrow[\text{diethyl ether}]{\text{PhCH}_2\text{Br}} G \xrightarrow[\text{diethyl ether}]{\text{CH}_3\text{CH}_2\text{MgBr}} H \longrightarrow I \xrightarrow[\text{H}_2\text{O}]{\text{NH}_4\text{Cl}} J$

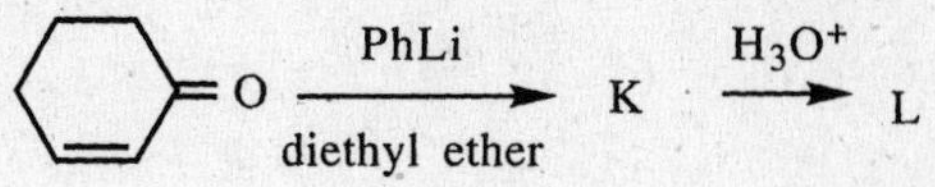

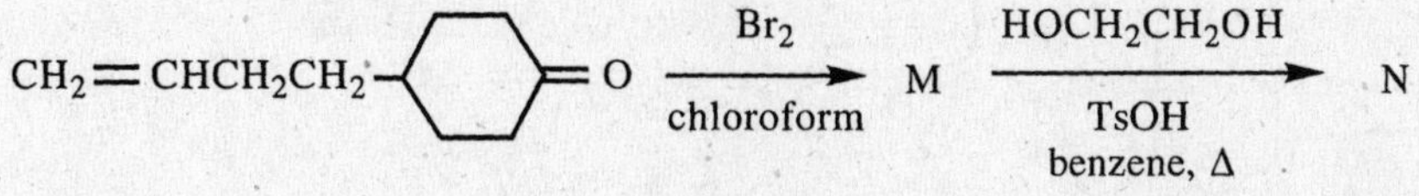

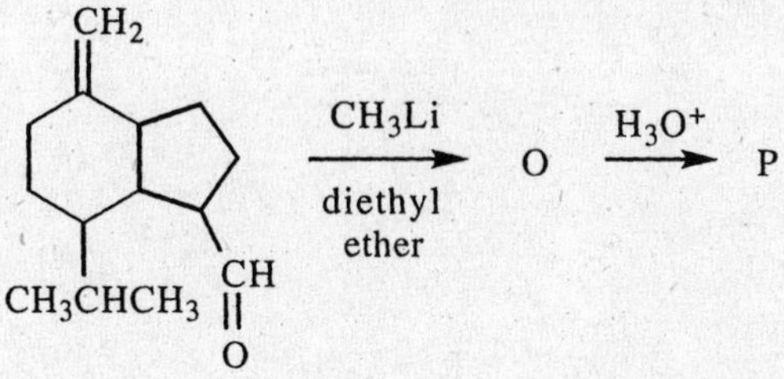

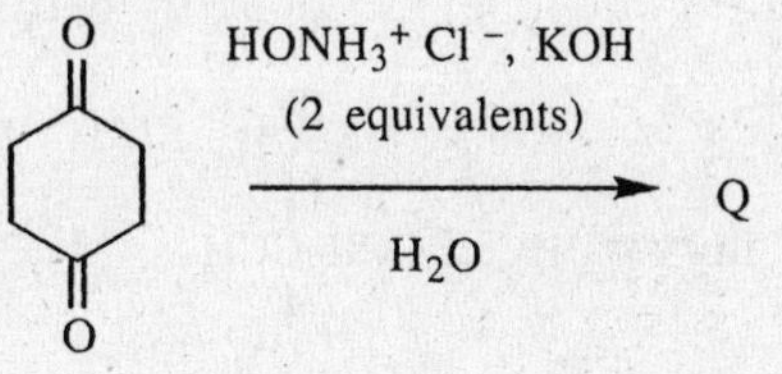

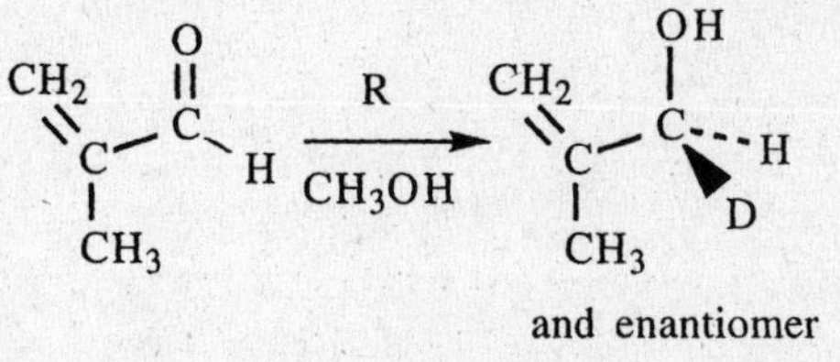

(c) $\xrightarrow[\text{diethyl ether}]{\text{PhLi}} K \xrightarrow{\text{H}_3\text{O}^+} L$

(d) $CH_2 = CHCH_2CH_2 - \bigcirc = O \xrightarrow[\text{chloroform}]{\text{Br}_2} M \xrightarrow[\substack{\text{TsOH} \\ \text{benzene, } \Delta}]{\text{HOCH}_2\text{CH}_2\text{OH}} N$

(e) $\xrightarrow[\substack{\text{diethyl} \\ \text{ether}}]{\text{CH}_3\text{Li}} O \xrightarrow{\text{H}_3\text{O}^+} P$

(f) $\xrightarrow[\text{H}_2\text{O}]{\substack{\text{HONH}_3^+ \text{Cl}^-, \text{ KOH} \\ (2 \text{ equivalents})}} Q$

(g) $\xrightarrow[\text{CH}_3\text{OH}]{R}$ and enantiomer

(h) $S + T \longrightarrow U \xrightarrow[\text{H}_2\text{O}]{\text{NH}_4\text{Cl}}$ and enantiomer

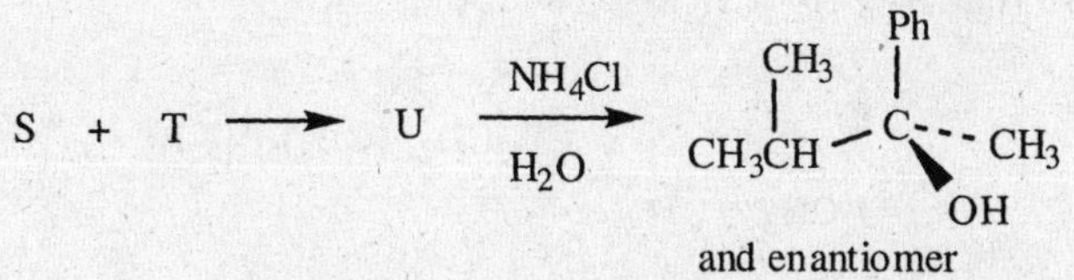

S14.5 Imines behave toward Grignard reagents in much the same way that carbonyl groups do. The following transformations were carried out. Supply the reagents that are necessary to carry out the reactions observed.

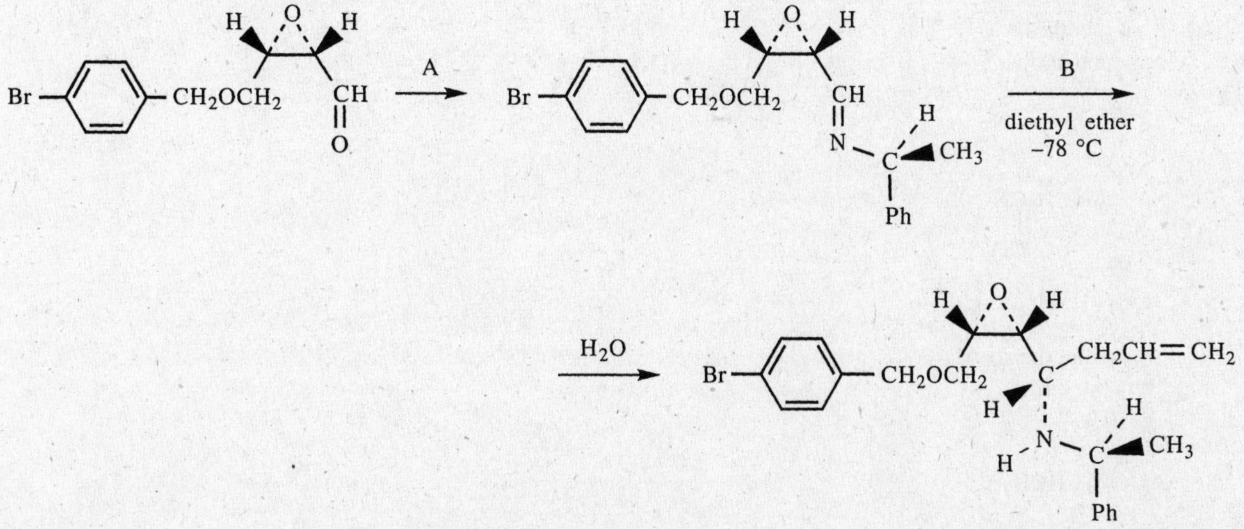

$\xrightarrow{A}$

$\xrightarrow[\substack{\text{diethyl ether} \\ -78\,°C}]{B}$

$\xrightarrow{\text{H}_2\text{O}}$

S14.6   Supply reagents for each of the following transformations. More than one step may be necessary for some.

(a)

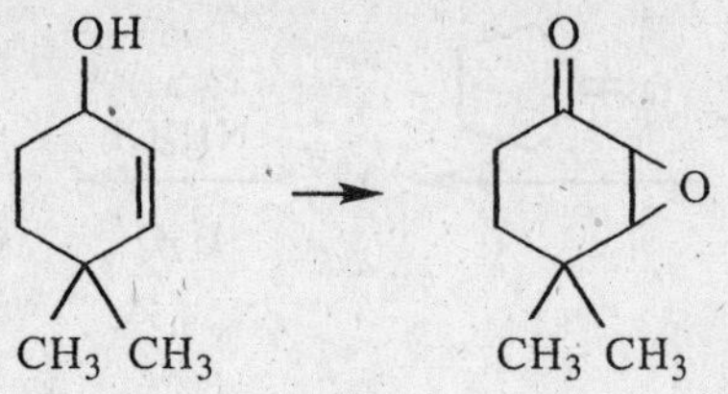

(b)

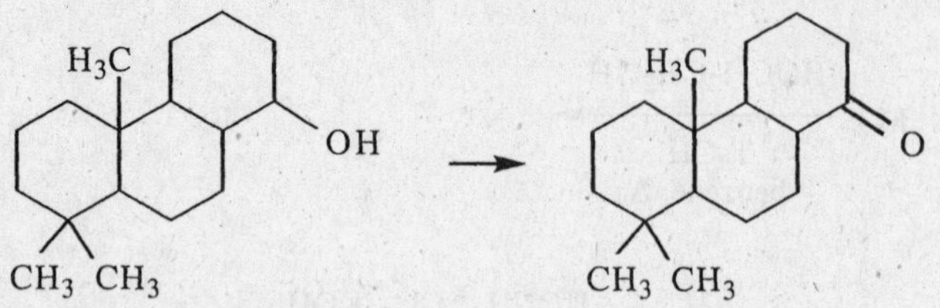

(c)   HC≡CCH₂Br ⟶ HC≡CCH₂CCH₃ with CH₃ and OH substituents

$HC{\equiv}CCH_2Br \longrightarrow HC{\equiv}CCH_2\underset{\underset{OH}{|}}{\overset{\overset{CH_3}{|}}{C}}CH_3$

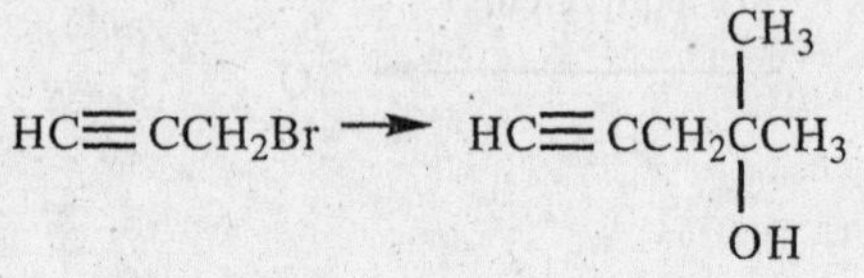

(d) $HOCH_2(CH_2)_5CH_2Br \longrightarrow$

$CH_3CH_2 \quad (CH_2)_5CH_2Br$

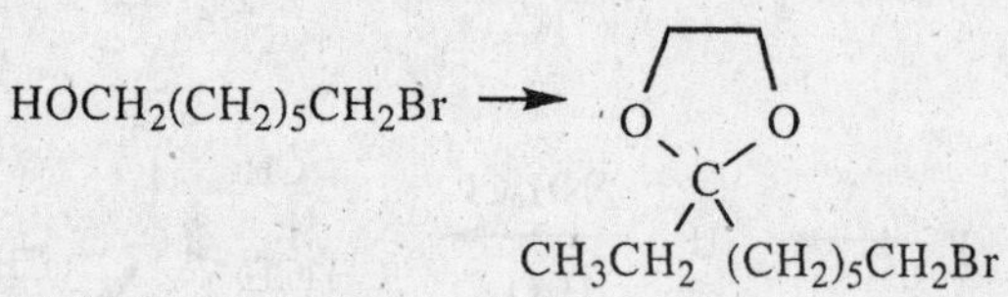

15 Carboxylic Acids and Their Derivatives. Acyl-Transfer Reactions

Workbook Exercises

There are two common classifications of functional groups that contain the carbonyl group ($-\overset{\overset{\displaystyle O}{\|}}{C}-$). These two classifications are based on an observed difference in reactivity. In the case of aldehydes and ketones (see Chapter 14), the fundamental reaction is addition. As demonstrated by the formation of imines and ketals, among others, the addition products of aldehydes and ketones can undergo further reactions. In Chapter 15, the second type of carbonyl chemistry is presented. In this case, groups other than simple alkyl or aryl groups (as in ketones) or hydrogen (as in aldehydes) are attached to the carbon atom of a carbonyl group. Instead, halogen, oxygen, nitrogen, or sulfur atoms (symbolized below as the generic electronegative group eN) are bonded to the carbonyl group. The overall reaction of these carbonyl compounds is not addition, as in the case of aldehydes and ketones, but rather substitution of the eN group.

$$R-\overset{\overset{\displaystyle O}{\|}}{C}-eN \quad \xrightarrow{\text{nucleophiles} \atop (Nu)} \quad R-\overset{\overset{\displaystyle O}{\|}}{C}-Nu \;+\; eN$$

overall reaction = substitution at the sp^2-hybridized carbon atom

(eN = many O, N, S, and halogen groups, among others)

The functional unit symbolized by $R-\overset{\overset{\displaystyle O}{\|}}{C}-$ is called an **acyl group.** The S_N1 and S_N2 substitution reactions of alkyl halides by nucleophiles (R—LG → R—Nu) are called alkylations of the nucleophile, or alkyl transfer reactions. The substitution of the eN atom group of $R-\overset{\overset{\displaystyle O}{\|}}{C}-eN$ by a nucleophile to give $R-\overset{\overset{\displaystyle O}{\|}}{C}-Nu$ is called an **acylation** of the nucleophile, or an **acyl transfer reaction.**

A new term, such as acylation, is created in order to express a new idea. The process of acylation can be described as addition followed by elimination. The reference to ideas previously encountered is designed to make an unfamiliar process, such as acylation, more familiar to you. If the reference to addition and elimination reactions is not meaningful to you, your best strategy is to refamiliarize yourself with these previous types of reactions before attempting to master new ones. These types of analogies can be used in many ways to communicate meaning. When referring to the overall change in connectivity in an acylation, the term substitution is used to describe the transformation at the sp^2-hybridized carbon atom.

EXAMPLE
Compare the four examples of transformations using water as a nucleophile shown on the following page. Each one is accompanied by an equation describing the overall change in connectivity and the mechanistic rationalization of the change observed.

Workbook Exercises (cont)

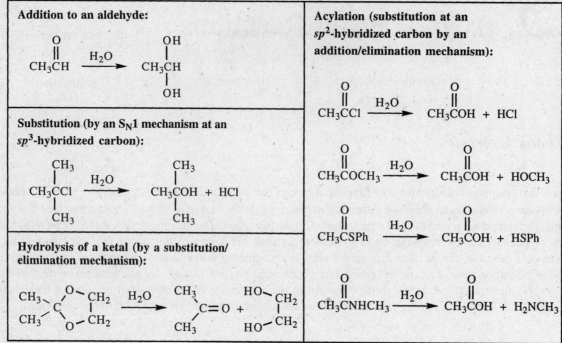

EXERCISE I.   Complete the following acylation reactions.

(a) $PhCOCH_2CH_3$ + NH_3 $\longrightarrow$ A + B   (b) (lactone ring) O + CH_3OH $\longrightarrow$ C

(c) $HCNCH_3$ (with CH_3) + D $\longrightarrow$ $HCNCH_2CH_2CH_2CH_3$ (with H) + E   (d) F + G $\longrightarrow$ CH_3COCCH_3 + NaCl

(e) (cyclopentane ring with CH_3 and $CCCl_3$, =O) + NaOH $\longrightarrow$ H* + I*

(*The trichloromethyl group, —CCl_3, behaves as a leaving group in this reaction. Under the aqueous base conditions of this reaction, the leaving group is protonated to give an uncharged product, while the acylated nucleophile is deprotonated to give its conjugate base.)

A MORE DETAILED LOOK AT THE ACYLATION MECHANISM

The overall transformation in an acylation reaction is viewed as the combination of two fundamental mechanistic steps: addition to the carbonyl group followed by an elimination. This is the same way that the reactions of the carbonyl group in Chapter 14 were characterized.

$$R-C-eN \;\; \underset{\text{elimination of Nu-H}}{\overset{\text{addition of Nu-H}}{\rightleftharpoons}} \;\; R-C-Nu \;\; \underset{\text{addition of H-eN}}{\overset{\text{elimination of H-eN}}{\rightleftharpoons}} \;\; R-C-Nu$$

alcohol intermediate

Workbook Exercises (cont)

If the nucleophile is an anion, Nu$^-$, the intermediate will be an alkoxide ion, $R-\overset{\overset{\displaystyle O^-}{|}}{\underset{\underset{\displaystyle eN}{|}}{C}}-Nu$, rather than an alcohol.

EXAMPLE

This scheme, applied to one of the molecules shown on the previous page, gives the following more detailed description.

previously shown: $CH_3CSPh \xrightarrow{H_2O} CH_3COH + HSPh$

more detailed view: $CH_3-\overset{\overset{\displaystyle O}{||}}{C}-SPh$ $\underset{\substack{\text{elimination}\\\text{of }H_2O}}{\overset{\substack{\text{addition}\\\text{of }H_2O}}{\rightleftharpoons}}$ $CH_3-\overset{\overset{\displaystyle OH}{|}}{\underset{\underset{\displaystyle SPh}{|}}{C}}-OH$ $\underset{\substack{\text{addition}\\\text{of PhSH}}}{\overset{\substack{\text{elimination}\\\text{of PhSH}}}{\rightleftharpoons}}$ $CH_3-\overset{\overset{\displaystyle O}{||}}{C}-OH$

EXERCISE II.   Redo Exercise I using the mechanistic scheme outlined above.

15.1

progress of reaction for reaction
of aldehyde or ketone with amine

progress of reaction for reaction
of acid derivative with amine

The carbonyl group of an aldehyde or a ketone is not as highly stabilized as the carbonyl group of an ester is. The free energy of the aldehyde or ketone is thus higher than that of the ester. If we assume that the tetrahedral intermediates, in which we no longer have resonance, are approximately the same energy, then $\Delta G^\ddagger$ for an aldehyde or ketone is lower than for an ester. Therefore, an aldehyde or ketone reacts faster with the nucleophile than the ester does.

15.2   (a) $CH_3CH_2-\overset{\overset{\displaystyle O}{||}}{C}-OH$ $CH_3CH_2-\overset{\overset{\displaystyle O}{||}}{C}-Cl$ $CH_3CH_2-\overset{\overset{\displaystyle O}{||}}{C}-O-\overset{\overset{\displaystyle O}{||}}{C}-CH_2CH_3$

$CH_3CH_2-\overset{\overset{\displaystyle O}{||}}{C}-OCH_2CH_3$ $CH_3CH_2-\overset{\overset{\displaystyle O}{||}}{C}-NH_2$

15.2   (b) $CH_3CH_2-\overset{O}{\overset{\|}{C}}-\overset{+}{OH_2} \; HSO_4^- \;\underset{H_2SO_4}{\overset{}{\rightleftharpoons}}\; CH_3CH_2-\overset{O}{\overset{\|}{C}}-OH \;\underset{}{\overset{H_2SO_4}{\rightleftharpoons}}\; CH_3CH_2-\overset{\overset{+}{O}-H}{\overset{\|}{C}}-OH \; HSO_4^-$

major product

$CH_3CH_2-\overset{O}{\overset{\|}{C}}-\overset{\overset{+}{OCH_2CH_3}}{\underset{H}{}} \; HSO_4^- \;\underset{H_2SO_4}{\overset{}{\rightleftharpoons}}\; CH_3CH_2-\overset{O}{\overset{\|}{C}}-OCH_2CH_3 \;\underset{}{\overset{H_2SO_4}{\rightleftharpoons}}\; CH_3CH_2-\overset{\overset{+}{O}-H}{\overset{\|}{C}}-OCH_2CH_3 \; HSO_4^-$

major product

$CH_3CH_2-\overset{O}{\overset{\|}{C}}-\overset{+}{NH_3} \; HSO_4^- \;\underset{H_2SO_4}{\overset{}{\rightleftharpoons}}\; CH_3CH_2-\overset{O}{\overset{\|}{C}}-NH_2 \;\underset{}{\overset{H_2SO_4}{\rightleftharpoons}}\; CH_3CH_2-\overset{\overset{+}{O}-H}{\overset{\|}{C}}-NH_2 \; HSO_4^-$

major product

(c)   The good leaving groups are shaded in the formulas shown in parts (a) and (b).

Concept Map 15.1   Relative reactivities in nucleophilic substitutions.

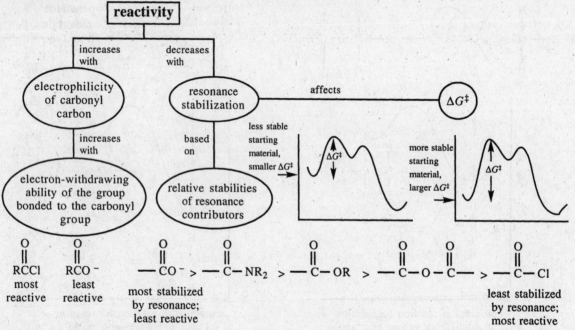

15.3 $CH_3CH_2\overset{:O:}{\overset{\|}{C}}-\overset{..}{\underset{..}{O}}-H \longleftrightarrow CH_3CH_2\overset{:\overset{..}{O}:^-}{\overset{|+}{C}}-\overset{..}{\underset{..}{O}}-H \longleftrightarrow CH_3CH_2\overset{:\overset{..}{O}:^-}{\overset{|}{C}}=\overset{..}{\underset{..}{O}}-H$

$CH_3CH_2\overset{:O:}{\overset{\|}{C}}-\overset{..}{\underset{..}{Cl}}: \longleftrightarrow CH_3CH_2\overset{:\overset{..}{O}:^-}{\overset{|+}{C}}-\overset{..}{\underset{..}{Cl}}: \longleftrightarrow CH_3CH_2\overset{:\overset{..}{O}:^-}{\overset{|}{C}}=\overset{+}{\underset{..}{Cl}}:$

$CH_3CH_2\overset{:O:}{\overset{\|}{C}}-\overset{..}{\underset{..}{O}}-\overset{:O:}{\overset{\|}{C}}CH_2CH_3 \longleftrightarrow CH_3CH_2\overset{:\overset{..}{O}:^-}{\overset{|+}{C}}-\overset{..}{\underset{..}{O}}-\overset{:O:}{\overset{\|}{C}}CH_2CH_3 \longleftrightarrow CH_3CH_2\overset{:\overset{..}{O}:^-}{\overset{|}{C}}=\overset{+}{\underset{..}{O}}-\overset{:O:}{\overset{\|}{C}}CH_2CH_3 \longleftrightarrow$

15.3 (cont)

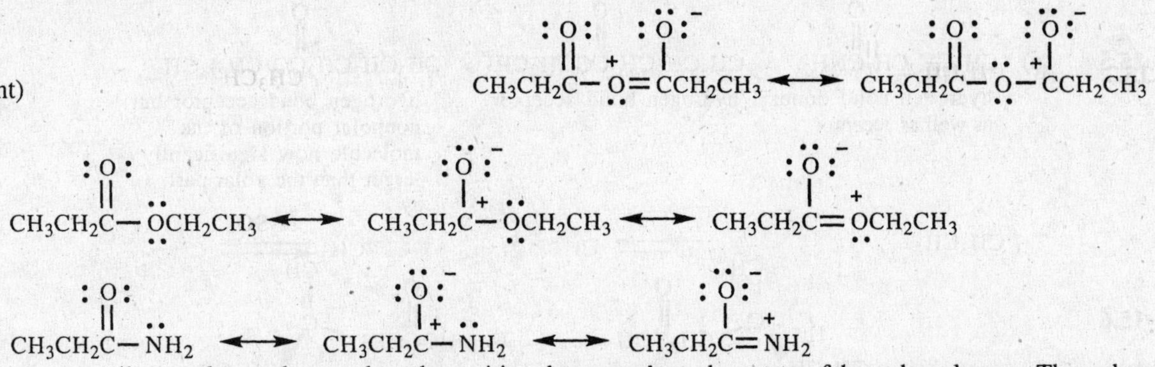

$$CH_3CH_2\overset{\overset{\displaystyle :\ddot{O}:}{||}}{C}-\overset{+}{\underset{\cdot\cdot}{O}}=CCH_2CH_3 \longleftrightarrow CH_3CH_2\overset{\overset{\displaystyle :\ddot{O}:^-}{|}}{C}-\overset{\cdot\cdot}{\underset{\cdot\cdot}{O}}-\overset{+}{C}CH_2CH_3$$

$$CH_3CH_2\overset{\overset{\displaystyle :O:}{||}}{C}-\overset{\cdot\cdot}{O}CH_2CH_3 \longleftrightarrow CH_3CH_2\overset{\overset{\displaystyle :\ddot{O}:^-}{|}}{C}-\overset{+}{\underset{\cdot\cdot}{O}}CH_2CH_3 \longleftrightarrow CH_3CH_2\overset{\overset{\displaystyle :\ddot{O}:^-}{|}}{C}=\overset{+}{O}CH_2CH_3$$

$$CH_3CH_2\overset{\overset{\displaystyle :O:}{||}}{C}-\overset{\cdot\cdot}{N}H_2 \longleftrightarrow CH_3CH_2\overset{\overset{\displaystyle :\ddot{O}:^-}{|}}{C}-\overset{+}{\underset{\cdot\cdot}{N}}H_2 \longleftrightarrow CH_3CH_2\overset{\overset{\displaystyle :\ddot{O}:^-}{|}}{C}=\overset{+}{N}H_2$$

The resonance contributors shown above reduce the positive charge on the carbon atom of the carbonyl group. The carbon atom of the carbonyl group in acid derivatives is thus less electrophilic than the carbon atom of the carbonyl group in aldehydes and ketones. The compounds in which the resonance contributors involve oxygen or nitrogen are more stable than the one in which chlorine is involved. The orbitals of oxygen and nitrogen (second row elements of smaller size) overlap with those of carbon better than those of chlorine, therefore their resonance contributors are more important in the stabilization of the compounds. The amide is the most stable of all, because nitrogen is less electronegative than oxygen, and therefore bears a positive charge more easily than oxygen does.

15.4   Propanamide is much less basic than propylamine because the nitrogen atom in propanamide has a partial positive charge (see the resonance contributors for propanamide in Problem 15.3). This means that there is less electron density on the nitrogen atom in propanamide than in propylamine, and, therefore, lower basicity for the amide.

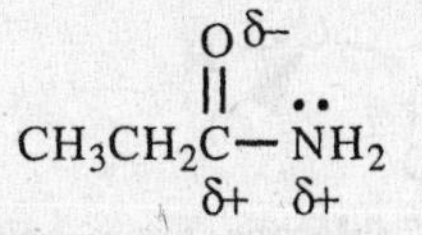

$$CH_3CH_2\overset{\overset{\displaystyle O^{\delta-}}{||}}{\underset{\delta+\ \ \delta+}{C}}-\overset{\cdot\cdot}{N}H_2$$

The electron pair on the nitrogen atom is less available; the amide nitrogen is less basic

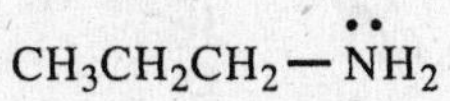

$$CH_3CH_2CH_2-\overset{\cdot\cdot}{N}H_2$$

The electron pair on the nitrogen atom is more available; the amine nitrogen is more basic

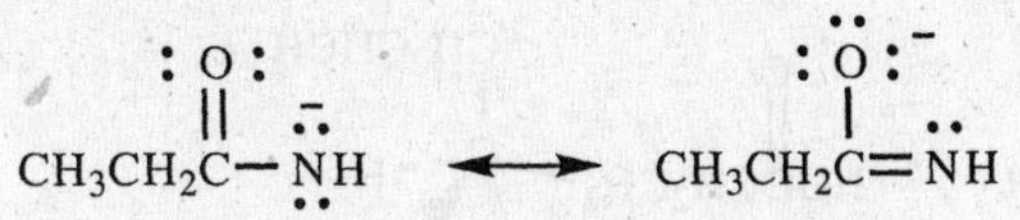

$$CH_3CH_2\overset{\overset{\displaystyle :O:}{||}}{C}-\overset{\cdot\cdot}{\underset{\cdot\cdot}{N}}H \longleftrightarrow CH_3CH_2\overset{\overset{\displaystyle :\ddot{O}:^-}{|}}{C}=\overset{\cdot\cdot}{N}H$$

delocalization of charge in conjugate base of propanamide; hence stabilization of base relative to its conjugate acid

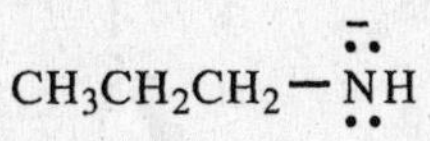

$$CH_3CH_2CH_2-\overset{\cdot\cdot}{\underset{\cdot\cdot}{N}}H$$

no stabilization of negative charge; therefore strong base

15.5   decreasing solubility →

(a) $CH_3CH_2CH_2CH_2\overset{\overset{\displaystyle O}{||}}{C}O^-Na^+$ > $CH_3CH_2CH_2CH_2\overset{\overset{\displaystyle O}{||}}{C}OH$ > $CH_3CH_2\overset{\overset{\displaystyle O}{||}}{C}OCH_2CH_3$

   ionic

covalent, but hydrogen bond donor as well as acceptor

only hydrogen bond acceptor

(b) $CH_3CH_2CH_2\overset{\overset{\displaystyle O}{||}}{C}OH$ > $CH_3CH_2CH_2CH_2OH$ > $CH_3CH_2\overset{\overset{\displaystyle O}{||}}{C}OCH_2CH_3$

can hydrogen bond at the carbonyl group as well as at the hydroxyl group

only one site for hydrogen bonding

only hydrogen bond acceptor

15.5   (c) $CH_3CH_2CH_2\overset{\overset{\displaystyle O}{\|}}{C}NH_2$ > $CH_3CH_2CH_2\overset{\overset{\displaystyle O}{\|}}{C}OCH_2CH_3$ > $CH_3CH_2CH_2\overset{\overset{\displaystyle O}{\|}}{C}O(CH_2)_4CH_3$

hydrogen bond donor   hydrogen bond acceptor   hydrogen bond acceptor but
as well as acceptor   nonpolar portion of the
                                                                  molecule now significantly
                                                                   larger than the polar part

15.6

(R)-3-methylpentanoic acid   (S)-3-methylpentanoic acid

(S)-2-hydroxypropanoic acid   (R)-2-hydroxypropanoic acid

(S)-2-chlorohexanoic acid   (R)-2-chlorohexanoic acid

(R)-4-hydroxy-6-methylheptanoic acid   (S)-4-hydroxy-6-methylheptanoic acid

(S)-alanine   (R)-alanine

(2R, 3S)-tartaric acid   (2S, 3S)-tartaric acid   (2R, 3R)-tartaric acid
meso tartaric acid

15.6 (cont)

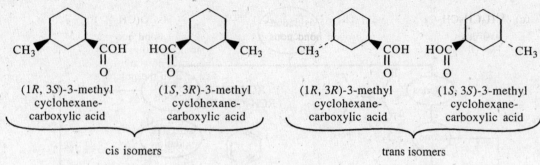

(1R, 3S)-3-methyl
cyclohexane-
carboxylic acid

(1S, 3R)-3-methyl
cyclohexane-
carboxylic acid

(1R, 3R)-3-methyl
cyclohexane-
carboxylic acid

(1S, 3S)-3-methyl
cyclohexane-
carboxylic acid

cis isomers   trans isomers

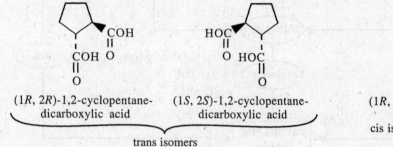

(1R, 2R)-1,2-cyclopentane-
dicarboxylic acid

(1S, 2S)-1,2-cyclopentane-
dicarboxylic acid

(1R, 2S)-1,2-cyclopentane-
dicarboxylic acid
cis isomer; meso compound

trans isomers

15.7 (a) undecanoic acid   (b) (E)-2-pentenoic acid   (c) (S)-2-hydroxybutanoic acid
(d) m-bromobenzoic acid   (e) 2-chloro-4-methylpentanoic acid (f) 3-hydroxyhexanedioic acid

15.8 (a) hexanoyl chloride (b) butanoic anhydride   (c) methyl (Z)-4-hexenoate (d) isobutyl p-chlorobenzoate

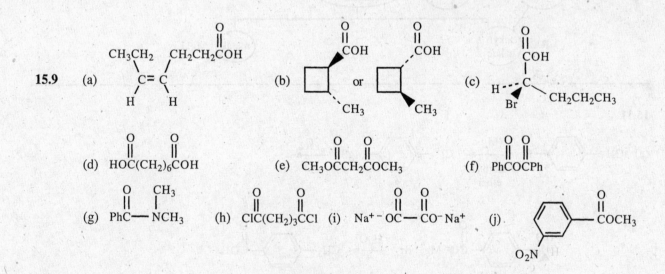

15.10 (a) ethyl 6-oxooctanoate   (b) decanoic acid   (c) m-bromobenzoyl chloride
(d) sodium octadecanoate (sodium stearate)   (e) N-methylacetamide   (f) p-methylbenzamide
(g) (1R, 2R)-2-bromocyclopentanecarboxylic acid   (h) (E)-2-butenoic acid   (i) (R)-3-hydroxybutanoic acid
(j) heptanenitrile   (k)   3-methylbutanoic anhydride   (l)   N-phenylbutanamide (butananilide)

Concept Map 15.2   Preparation of carboxylic acids.

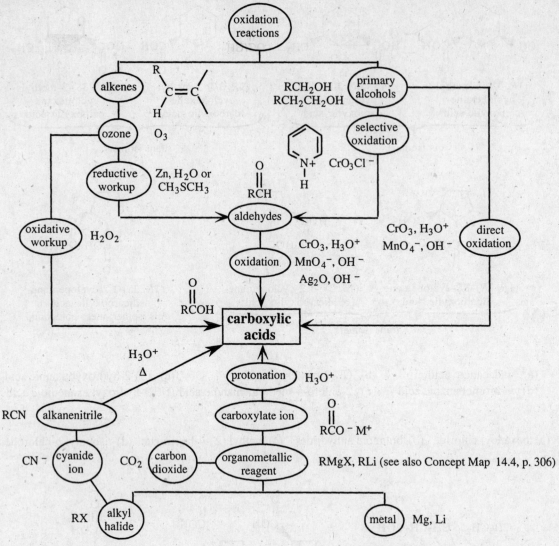

- -

15.11

(a) CH_3—⟨benzene⟩—Br $\xrightarrow[\substack{\text{diethyl} \\ \text{ether}}]{Mg}$ CH_3—⟨benzene⟩—MgBr $\xrightarrow{CO_2}$

   A

CH_3—⟨benzene⟩—$\overset{\overset{O}{\|}}{C}O^-$ Mg^{2+} Br^- $\xrightarrow{H_3O^+}$ CH_3—⟨benzene⟩—$\overset{\overset{O}{\|}}{C}OH$

   B   C

(b) $\underset{\underset{CH_2CH_2CH_2}{}}{CH_3CH_2CH_2}\overset{\overset{CH_3}{|}}{C}HCH_2CH_2OH \xrightarrow[\substack{H_2O \\ \Delta}]{KMnO_4,\ NaOH} CH_3CH_2CH_2\overset{\overset{CH_3}{|}}{C}HCH_2\overset{\overset{O}{\|}}{C}O^-\ Na^+ \xrightarrow{H_3O^+} CH_3CH_2CH_2\overset{\overset{CH_3}{|}}{C}HCH_2\overset{\overset{O}{\|}}{C}OH$

                                                                                    D   E

15.11

(c) [cyclohexene] $\xrightarrow[\text{chloroform}]{O_3}$ [ozonide F] $\xrightarrow[\substack{H_2O \\ \Delta}]{Ag_2O,\ NaOH}$ Na$^+$ $^-$OC(CH$_2$)$_4$CO$^-$ Na$^+$ $\xrightarrow{H_3O^+}$ HOC(CH$_2$)$_4$COH

F   G   H

(d) PhCH$_2$CH$_2$CH ($=$O) $\xrightarrow[\substack{H_2O \\ \Delta}]{Ag_2O,\ NaOH}$ PhCH$_2$CH$_2$CO$^-$ Na$^+$ $\xrightarrow{H_3O^+}$ PhCH$_2$CH$_2$COH

                                                    I   J

(e) CH$_3$(CH$_2$)$_9$CH$=$CH$_2$ $\xrightarrow[\text{chloroform}]{O_3}$ $\xrightarrow{H_2O_2,\ NaOH}$ $\xrightarrow{H_3O^+}$ CH$_3$(CH$_2$)$_9$COH + HCOH

                                                                    K   L

15.12 (a) PhCH$_2$CH$_2$COH $\xrightarrow[\Delta]{SOCl_2}$ PhCH$_2$CH$_2$CCl   (b) CH$_3$CH$_2$CH$_2$COH $\xrightarrow{PCl_3}$ CH$_3$CH$_2$CH$_2$CCl

(c) CH$_3$CH$_2$CH$_2$CCl + Na$^+$ $^-$OCCH$_2$CH$_2$CH$_3$ $\longrightarrow$ CH$_3$CH$_2$CH$_2$COCCH$_2$CH$_2$CH$_3$

(d) HOCCH$_2$CH$_2$COH $\xrightarrow{CH_3COCCH_3}$ [succinic anhydride]   (e) CH$_3$CH$_2$COH $\xrightarrow[\substack{\Delta \\ 650\ °C}]{clay}$ CH$_3$CH$_2$COCCH$_2$CH$_3$

(f) HOC(CH$_2$)$_5$COH $\xrightarrow[\Delta]{\text{excess }SOCl_2}$ ClC(CH$_2$)$_5$CCl

15.13

(a) [pyrrolidine-N-CCH$_2$CH$_3$] $\longleftarrow$ [pyrrolidine NH] + ClCCH$_2$CH$_3$ (b) [piperidine N-CCH$_2$CH$_3$] $\longleftarrow$ [piperidine NH] + ClCCH$_2$CH$_3$

15.14 [mechanism: CH$_3$-C(=O)-Cl with H$_2$O $\longrightarrow$ tetrahedral intermediate $\longrightarrow$ $\longrightarrow$ CH$_3$-C(=O)-OH + Cl$^-$]

15.15

$$\underset{\text{p}K_a\ 2.8}{\text{ClCH}_2\overset{\overset{\displaystyle O}{\|}}{\text{C}}\text{OH}} \quad + \quad {}^-\text{C}\!\equiv\!\text{N} \quad \longrightarrow \quad \text{ClCH}_2\overset{\overset{\displaystyle O}{\|}}{\text{C}}\text{O}^- \quad + \quad \underset{\text{p}K_a\ 9.3}{\text{HC}\!\equiv\!\text{N}}$$

Chloroacetic acid is a much stronger acid than hydrogen cyanide. The cyanide anion will, therefore, be converted to hydrogen cyanide and will not be available to act as a nucleophile if chloroacetic acid is not first neutralized with base. In addition, even small amounts of hydrogen cyanide are lethal to humans.

Concept Map 15.3   Hydrolysis reactions of acid derivatives.

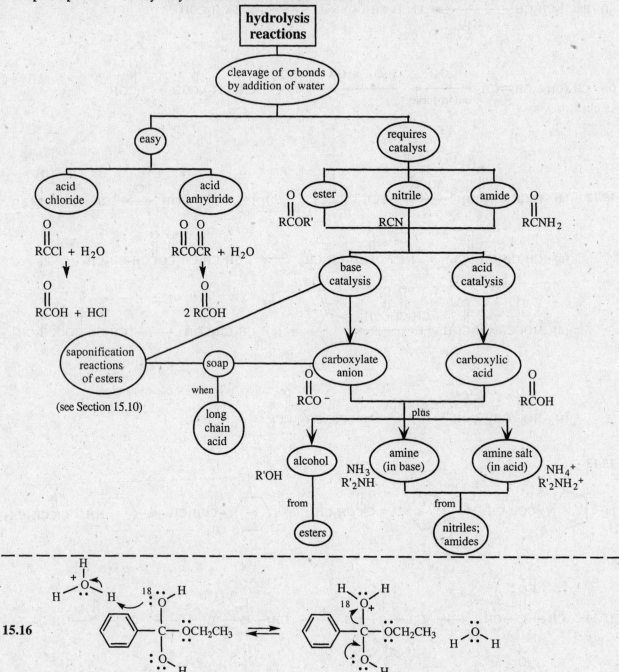

15.16

15.16 (cont)

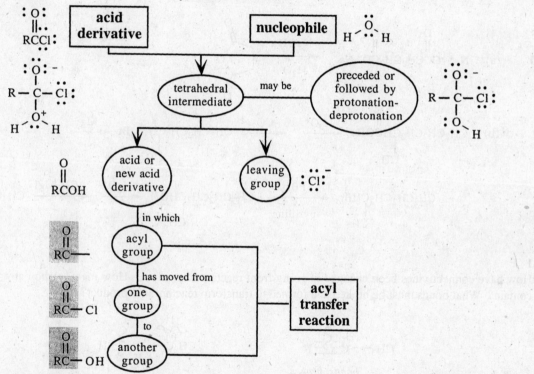

15.17   The two oxygen atoms are equivalent and cannot be distinguished from each other (Section 1.6A).

Concept Map 15.4   Mechanism of acyl transfer reactions.

15.18

15.19

For the carbon-nitrogen bond to break in the basic hydrolysis of acetamide, the leaving group must be the conjugate base of ammonia, amide anion. In basic solution, it is unlikely that protonation of the nitrogen in the tetrahedral intermediate will take place before the carbon-nitrogen bond begins to break. However, amide anion is such a strong base that it is improbable that it has any real existence in water. The mechanism accounts for this by showing the amide ion taking a proton from water as the tetrahedral intermediate breaks up.

15.20   (a)

      (b)

15.21

1. How have connectivities been changed in going from reactant to product? How many carbon atoms does each contain? What bonds must be broken and formed to transform reactant into product?

bonds broken

bonds formed

The reactant and the product have the same number of carbon atoms. Two of the bonds between carbon and nitrogen have been broken. A carbon-oxygen double bond and two nitrogen-hydrogen single bonds have been formed.

2. What reagents are present? Are they good acids, bases, nucleophiles, or electrophiles?

H_3O^+ is an acid, H_2O is a weak nucleophile. The carbon atom of the nitrile is electrophilic, and the nitrogen atom has nonbonding electrons on it.

electrophile   nonbonding electrons

3. What is the most likely first step for the reaction: protonation or deprotonation, ionization, attack by a nucleophile, or attack by an electrophile?

15.21 (3) (cont)

An acid is present; therefore, protonation is the most likely first step. Water is a weak nucleophile and is unlikely to attack the unprotonated nitrile.

4. What are the properties of the species present in the reaction mixture after the first step? What is likely to happen next?

The electrophilicity of the carbon bonded to nitrogen is greatly increased by the protonation of the nitrile; this carbon will now react with water.

4 (repeated). What are the properties of the species present in the reaction mixture after this step? What is likely to happen next?

The oxonium ion resulting from the attack by water will be deprotonated by the solvent.

Comparing this new species to the desired product shows that it has a carbon-oxygen bond and a nitrogen-hydrogen bond in the right locations. To get to the product requires protonation at nitrogen and a shifting of the double bond.

| *bonds broken* | *bonds formed* |

This is a tautomerization.

The complete mechanism is shown on the next page:

15.21 (cont)

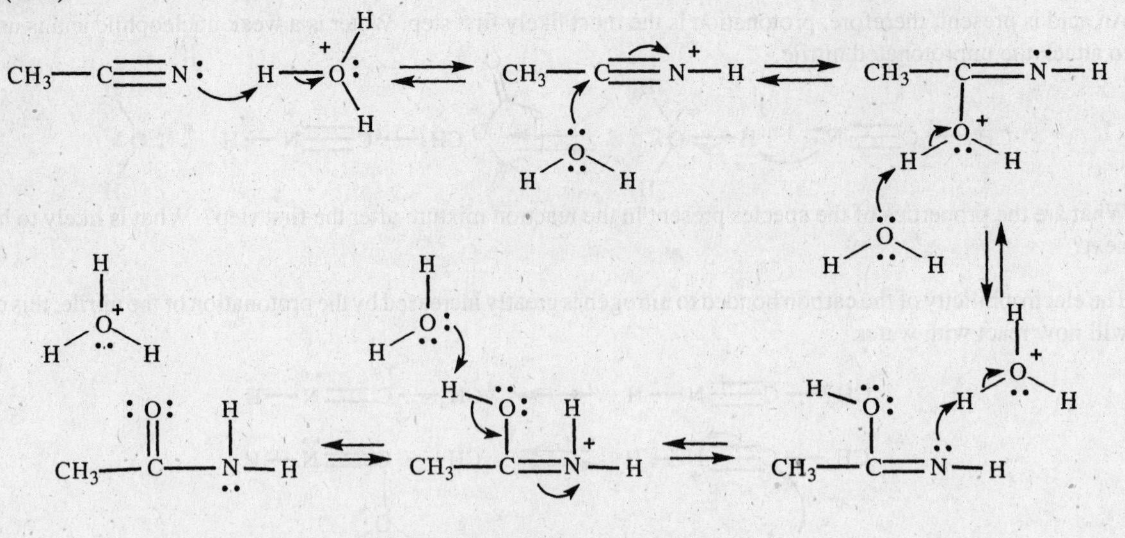

15.22

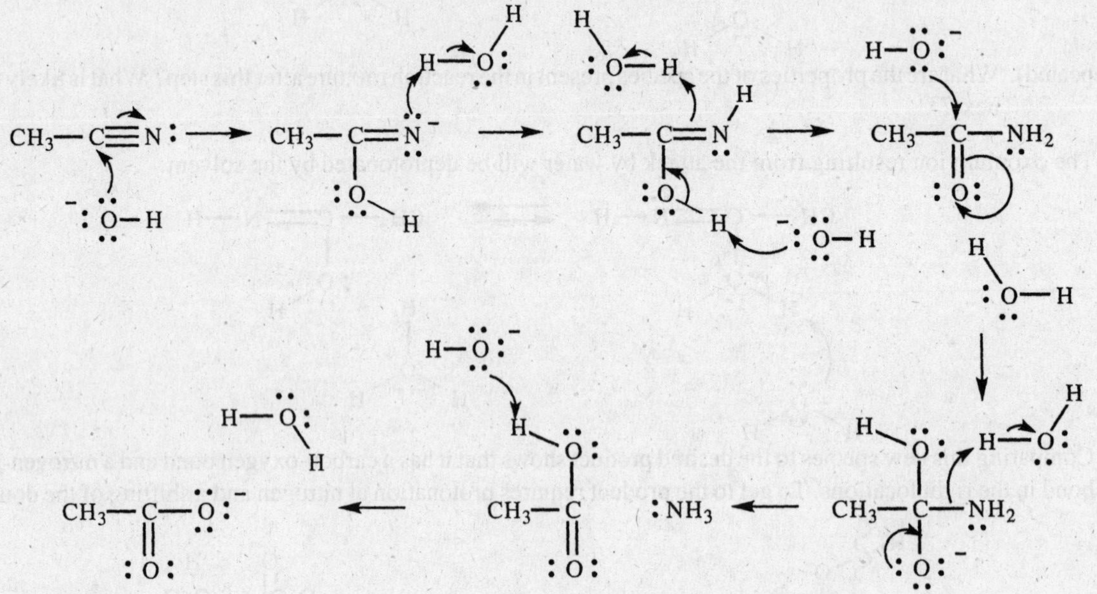

15.23

$$\text{CH}_3\text{CH}_2\text{CH}_2\text{CH}_2\text{OH} + \text{Ph}\overset{\overset{\displaystyle\text{O}}{\|}}{\text{C}}\text{Cl} \xrightarrow[\text{H}_2\text{O}]{\text{10\% NaOH}} \text{Ph}\overset{\overset{\displaystyle\text{O}}{\|}}{\text{C}}\text{OCH}_2\text{CH}_2\text{CH}_2\text{CH}_3$$

The other organic product formed would be sodium benzoate.

$$\text{Ph}\overset{\overset{\displaystyle\text{O}}{\|}}{\text{C}}\text{Cl} \xrightarrow[\text{H}_2\text{O}]{\text{10\% NaOH}} \text{Ph}\overset{\overset{\displaystyle\text{O}}{\|}}{\text{C}}\text{O}^- \text{Na}^+$$

15.24

(a) $$\text{CH}_3\underset{\underset{\displaystyle\text{CH}_3}{|}}{\overset{\overset{\displaystyle\text{CH}_3}{|}}{\text{C}}}\text{OH} + \text{CH}_3\overset{\overset{\displaystyle\text{O}}{\|}}{\text{C}}\text{O}\overset{\overset{\displaystyle\text{O}}{\|}}{\text{C}}\text{CH}_3 \longrightarrow \text{CH}_3\overset{\overset{\displaystyle\text{O}}{\|}}{\text{C}}\text{O}\underset{\underset{\displaystyle\text{CH}_3}{|}}{\overset{\overset{\displaystyle\text{CH}_3}{|}}{\text{C}}}\text{CH}_3 + \text{CH}_3\overset{\overset{\displaystyle\text{O}}{\|}}{\text{C}}\text{OH}$$

15.24

(b) $PhCCl$ + ⬡—OH → $PhCO$—⬡ (c) [maleic anhydride] + CH_3CH_2OH (1 equivalent) → [product with HOC and $COCH_2CH_3$ groups on $C=C$]

15.25

(a) $HOC(CH_2)_9COH$ $\xrightarrow[\substack{H_2SO_4 \\ \Delta}]{CH_3OH \text{ (excess)}}$ $CH_3OC(CH_2)_9COCH_3$ **A**

(b) $PhCHCOH$ (with OH) $\xrightarrow[\substack{HCl \text{ (g)} \\ \Delta}]{CH_3CH_2OH}$ $PhCHCOCH_2CH_3$ (with OH) **B** $\xrightarrow{SOCl_2}$ $PhCHCOCH_2CH_3$ (with Cl) **C**

(c) [cyclohexene anhydride] $\xrightarrow[\substack{TsOH \\ \Delta}]{CH_3CH_2OH \text{ (excess)}}$ [cyclohexene with two $COCH_2CH_3$ groups] **D**

(d) $CH_3CH=CHCOH$ $\xrightarrow[\substack{H_2SO_4 \\ \text{benzene} \\ \Delta}]{\substack{CH_3CH_2CHCH_3 \\ | \\ OH}}$ $CH_3CH=CHCOCHCH_2CH_3$ (with CH_3) **E**

15.26 (a)

$PhCOH$ + CH_3OH $\underset{H_2SO_4}{\rightleftharpoons}$ $PhCOCH_3$ + H_2O

mp 121° C
insoluble in
water; soluble in
diethyl ether

bp 66° C
miscible with
water; soluble in
diethyl ether

bp 198° C
insoluble in
water; soluble
in diethyl ether

(b) The reaction mixture consists of excess methanol, unreacted benzoic acid, sulfuric acid, and the products, water and methyl benzoate.

Step 1: Add more water to the reaction mixture; most of the methanol and the sulfuric acid will go into the water layer. Methyl benzoate is insoluble in water and will appear as an oily layer. Unreacted benzoic acid will remain dissolved in the ester. Add diethyl ether, bp 37 °C, d 0.7, to dissolve the organic compounds. Separate the organic layer from the water layer.

Step 2: Shake the organic layer (diethyl ether, methyl benzoate, benzoic acid) with saturated aqueous sodium bicarbonate.

H_2SO_4 + $NaHCO_3$ → $CO_2\uparrow$ + H_2O + SO_4^{2-} + $2\,Na^+$

(traces that
remain in the
methyl benzoate)

(excess)

soluble in water;
insoluble in
diethyl ether

15.26 (step 2 cont)

$$PhC\overset{O}{\overset{\|}{}}OH \quad + \quad NaHCO_3 \longrightarrow \quad PhC\overset{O}{\overset{\|}{}}O^-Na^+ \quad + \quad CO_2\uparrow \quad + \quad H_2O$$

insoluble in water;   soluble in water;
soluble in diethyl ether   insoluble in diethyl ether

The mixture will separate into two layers. The upper layer will be the ether solution containing methyl benzoate, some of the methanol, and small amounts of water. The lower layer will contain sodium sulfate (the salt formed by the reaction of the remaining sulfuric acid with the sodium bicarbonate), sodium benzoate (the salt from the reaction of benzoic acid with sodium bicarbonate), and some methanol.

Step 3: Separate the ether layer from the water layer. The ether and remaining methanol can be removed by distillation, leaving the higher boiling methyl benzoate behind. This is usually done after the ether layer is dried by adding a solid, anhydrous inorganic salt to it.

15.27   (a) – (d)

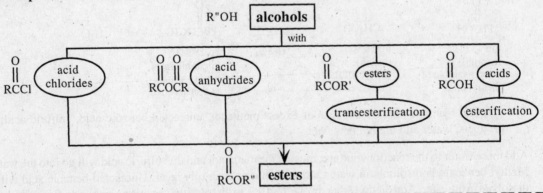

Concept Map 15.5   Reactions of acids and acid derivatives with alcohols.

15.29

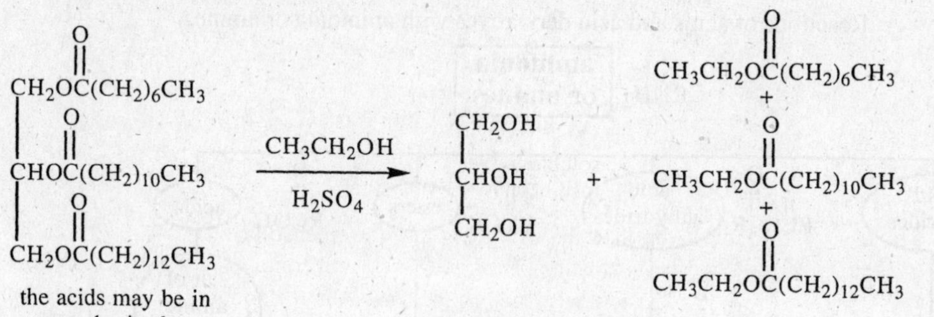

CH₂OC(CH₂)₆CH₃ ... CHOC(CH₂)₁₀CH₃ ... CH₂OC(CH₂)₁₂CH₃

the acids may be in any order in the triglyceride

$\xrightarrow[\text{H}_2\text{SO}_4]{\text{CH}_3\text{CH}_2\text{OH}}$

CH₂OH · CHOH · CH₂OH

+

CH₃CH₂OC(CH₂)₆CH₃ + CH₃CH₂OC(CH₂)₁₀CH₃ + CH₃CH₂OC(CH₂)₁₂CH₃

This is a transesterification reaction.

15.30

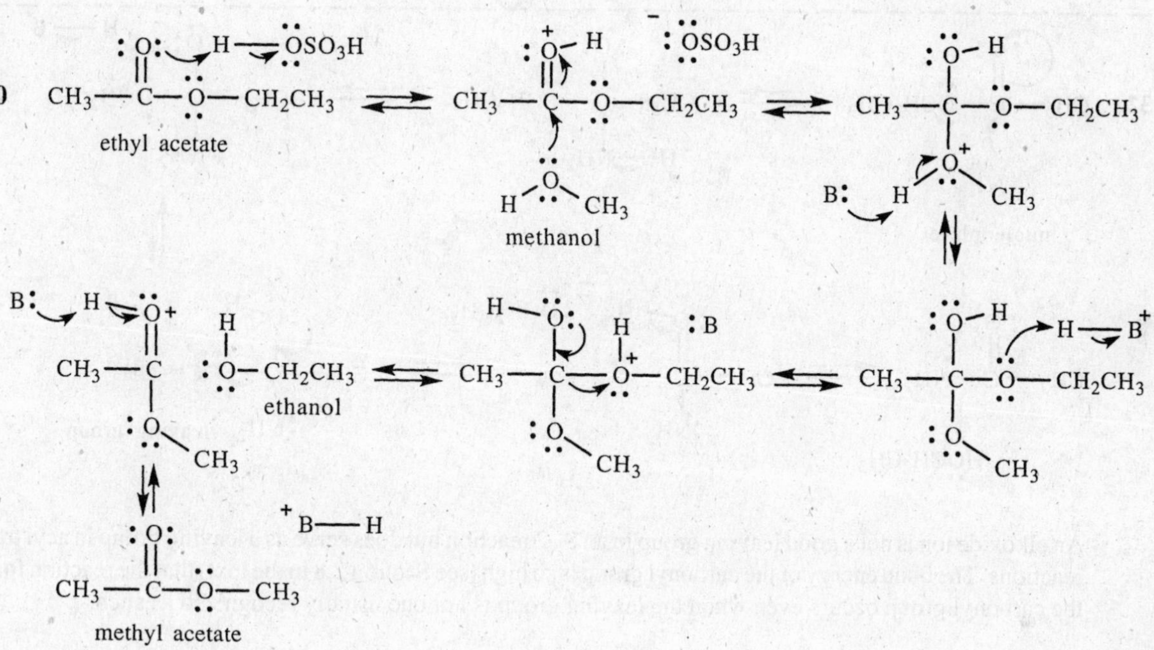

ethyl acetate

methanol

ethanol

methyl acetate

15.31

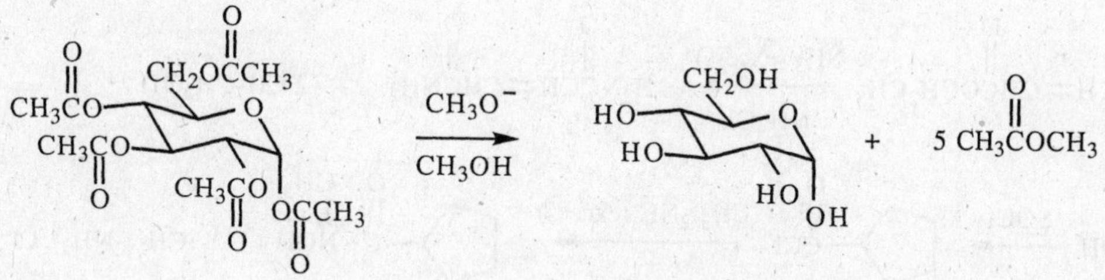

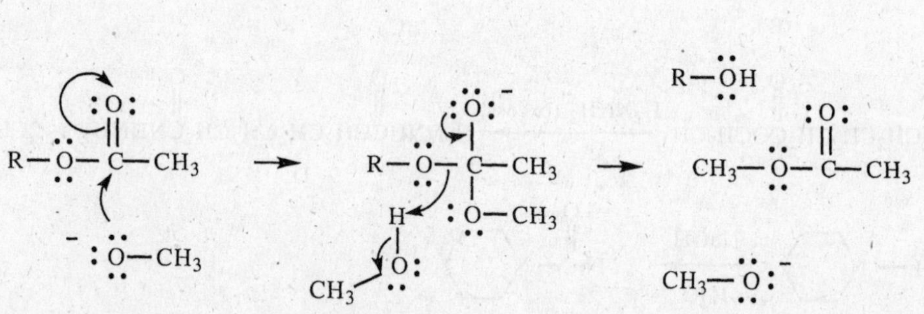

$\xrightarrow[\text{CH}_3\text{OH}]{\text{CH}_3\text{O}^-}$

+ 5 CH₃COCH₃

Concept Map 15.6   Reactions of acids and acid derivatives with ammonia or amines.

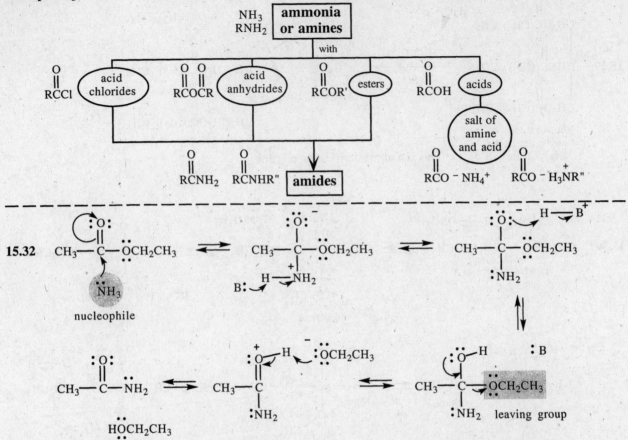

An alkoxide ion is not a good leaving group in an S_N2 reaction but does serve as a leaving group in acyl transfer reactions. The bond energy of the carbonyl group is so high (see Section 2.8 in the text) that the reaction forming the carbonyl group occurs even when the leaving group is not one usually recognized as such.

15.33

(a) $CH_3CH_2OCCH=CHCOCH_2CH_3 \xrightarrow[\substack{NH_4Cl \\ H_2O}]{NH_3 \text{ (excess)}} H_2NCCH=CHCNH_2 + 2 CH_3CH_2OH$
                                                                                        A

(b)   cyclopentane–COH $\xrightarrow[\Delta]{SOCl_2}$ cyclopentane–CCl $\xrightarrow{(CH_3)_2NH \text{ (excess)}}$ cyclopentane–C–NCH_3 + $(CH_3)_2NH_2^+$ Cl$^-$
                                        B   C

(c) $CH_3CH_2OCCH_2CH_2CH_2CH_2COCH_2CH_3 \xrightarrow{H_2NNH_2 \text{ (excess)}} H_2NNHCCH_2CH_2CH_2CH_2CNHNH_2 + 2CH_3CH_2OH$
                                                                                                    D

(d)   PhCCl + H—N(piperidine) $\xrightarrow[H_2O]{NaOH}$ PhC—N(piperidine)
                                                          E

15.34

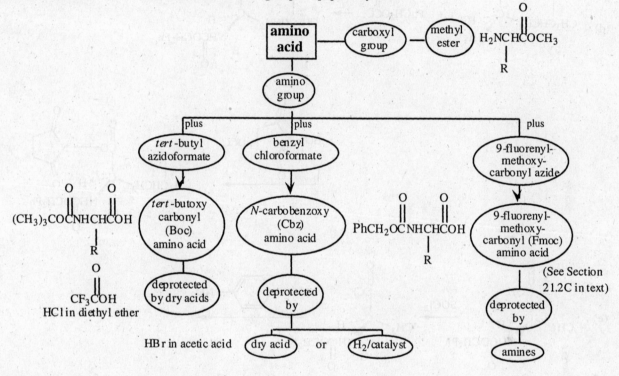

peptide linkage — alanylalanyl

peptide linkage — glycylglycine

Concept Map 15.7   Protection of functional groups in peptide synthesis.

Concept Map 15.8   (See page 357)

15.35

15.35

(c)

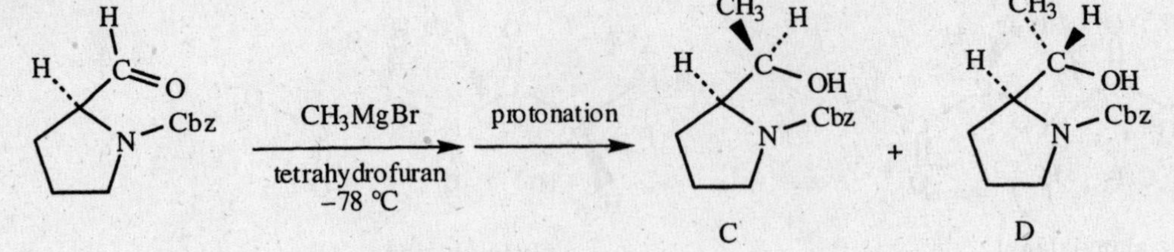
CH3MgBr
tetrahydrofuran
−78 °C
protonation
C
D

(d)

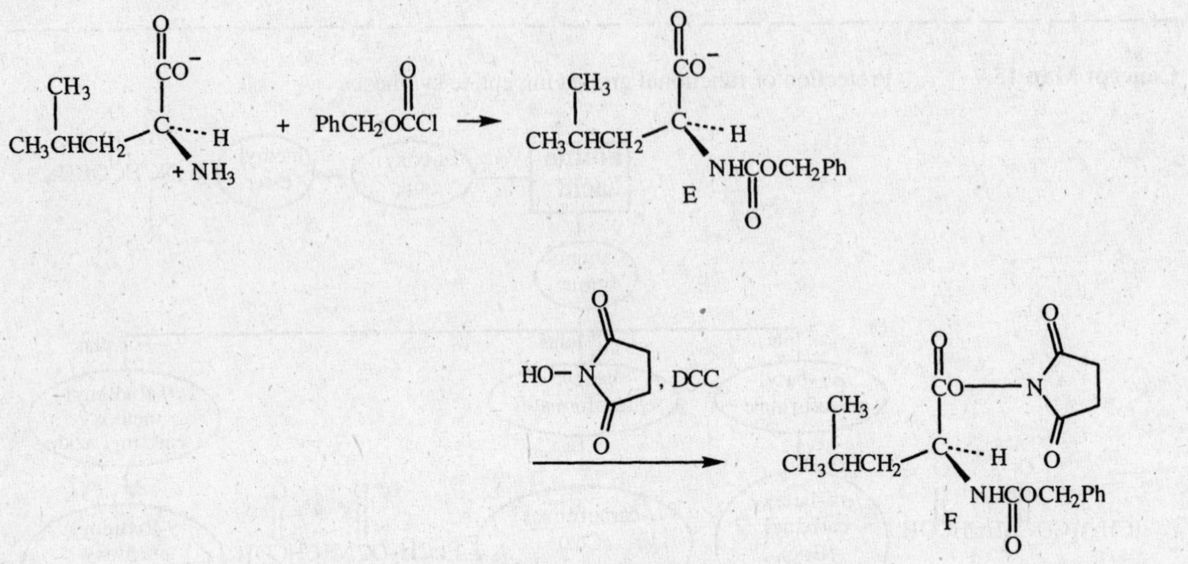
PhCH2OCCl
E
DCC
F

(e)

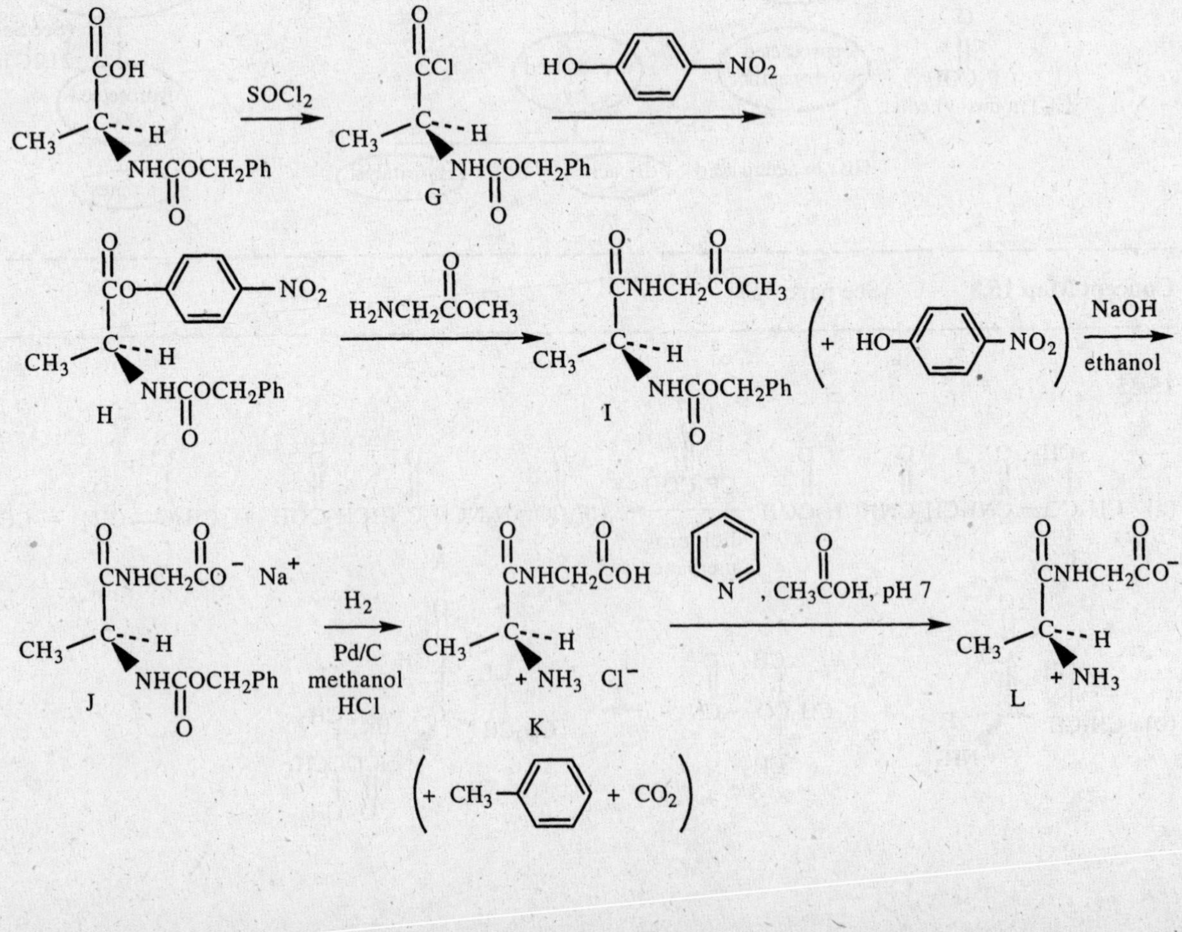
SOCl2
G
H
H2NCH2COCH3
I
NaOH
ethanol
J
H2
Pd/C
methanol
HCl
K
CH3COH, pH 7
L

Concept Map 15.8   Activation of the carboxyl group in peptide synthesis.

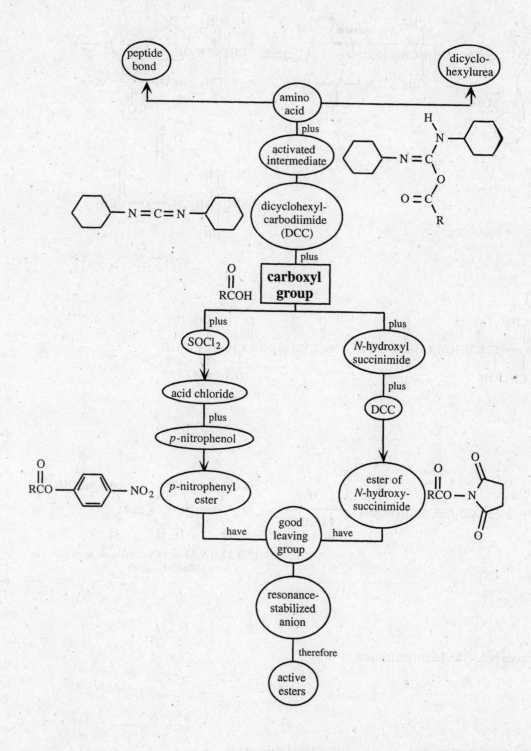

15.36

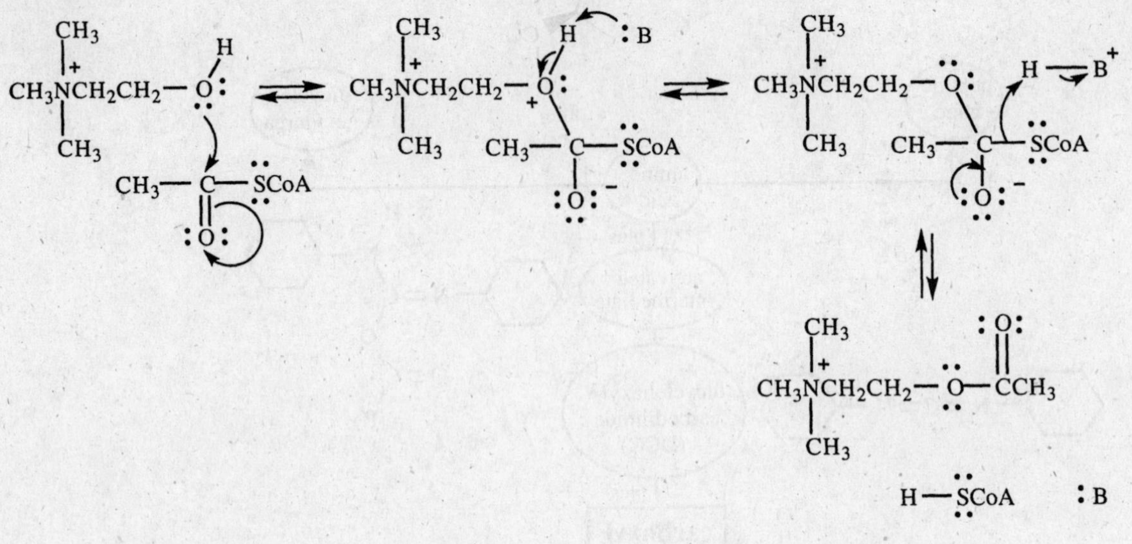

15.37

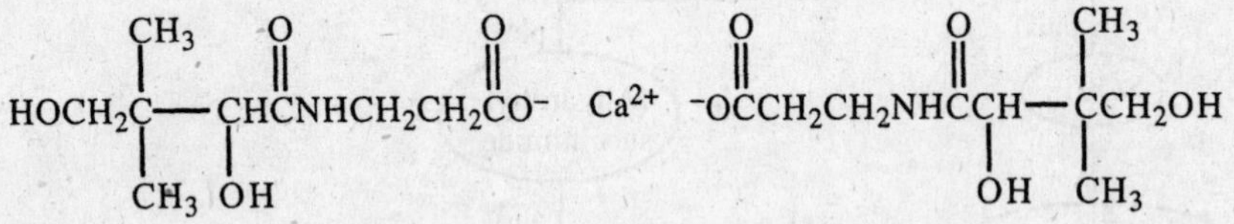

$$HOCH_2C\overset{\underset{\displaystyle CH_3}{|}}{\underset{\underset{\displaystyle CH_3}{|}}{C}}\overset{\overset{\displaystyle O}{||}}{\underset{\underset{\displaystyle OH}{|}}{CHCNHCH_2CH_2CO^-}} \quad Ca^{2+} \quad {}^-OCCH_2CH_2NHCCH-CCH_2OH$$

(with structure: CH_3, $\overset{O}{||}$, OH, CH_3 groups as drawn)

15.38

$$CH_3(CH_2)_4C\equiv CCH_2C\equiv C(CH_2)_7COH \xrightarrow[\substack{Pd/BaSO_4 \\ quinoline}]{H_2}$$

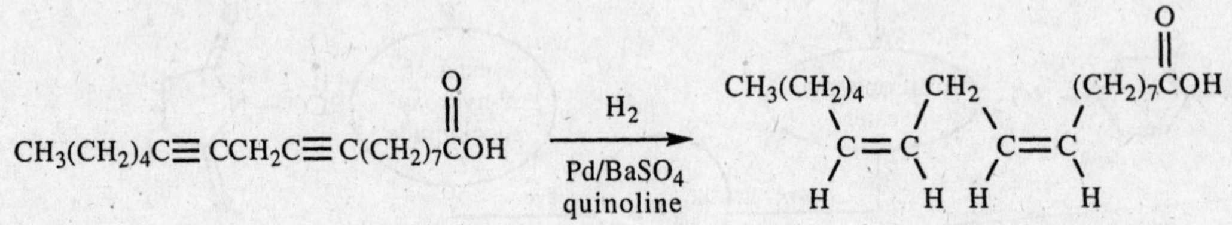

(9Z, 12Z)-9,12-octadecadienoic acid
(linoleic acid)

15.39   Compounds b, c, e, and f are surfactants.

15.40   (a)   2,4-dichlorobenzoic acid   (b)   4-bromohexanoic acid

        (c)   2-methylbutanedioic anhydride (2-methylsuccinic anhydride)   (d)   2-methylpentanenitrile

        (e)   ethyl (1R, 2R)-2-ethylcyclohexanecarboxylate   (f)   N,N-dimethylpentanamide

        (g)   (R)-2-hydroxypentanoic acid   (h)   2-chloropentanoyl chloride   (i)   4-ethylbenzanilide

        (j)   ethyl 2-aminobenzoate   (k)   (E)-4-octenoic acid   (l)   (1R,2R)-cyclobutanedicarboxylic acid

15.41

(a)
$$\begin{array}{c} CH_2COCH_3 \\ \text{(with C=O)} \end{array}$$
structure: central C bonded to CH_2COCH_3, H (dashed), CH_3CH_2, and Br

(b) cyclopentane with HO and COOH (C=O) substituents, *or enantiomer*

(c) 2,4-dibromobenzamide: Br-substituted benzene ring with Br and CNH_2 (C=O)

(d)
$$\begin{array}{c} CH_3CH_2 \\ \end{array} C=C \begin{array}{c} CH_2COCH_2CH_3 \\ H \end{array}$$
with H on lower carbons

(e) $CH_3CH_2CH_2CH_2CH_2CH_2C\equiv N$

(f) $CH_3CH_2CH_2CH_2CH_2CNCH_2CH_3$ (C=O), with CH_2CH_3 on N

(g) $CH_3CH_2OCCH_2CH_2C(CH_3)_2COCH_2CH_3$ (two C=O groups, central C with two CH_3)

(h) O_2N—benzene ring—$COCH_3$ (C=O)

(i) Br—benzene ring—COC (two C=O)—benzene ring—Br

(j) CH_3O—benzene ring—CCl (C=O)

15.42

(a) CH_3O—benzene ring—NH_2 $\xrightarrow[\begin{array}{c}\text{acetic acid}\\ H_2O \\ 0\text{-}5\,°C\end{array}]{CH_3COCCH_3 \text{ (acetic anhydride)}}$ CH_3O—benzene ring—$NHCCH_3$ (C=O) **A** $+\ CH_3COH$

(b) $PhCOH$ (C=O) $\xrightarrow[\Delta]{PBr_3}$ $PhCBr$ (C=O) **B**

(c)
$$\begin{array}{c} CH_2COH \\ CH_2 \\ CH_2COH \end{array}$$ (two C=O, i.e. glutaric acid) $\xrightarrow[\Delta]{CH_3COCCH_3}$ cyclic anhydride **C** $+\ CH_3COH$

(d)
3,4-dimethoxyphenyl–CH(H,H)–CH(NH$_2$)(Ar)(H) $\xrightarrow[\begin{array}{c}\text{base}\\ \text{dichloromethane}\end{array}]{CH_3CCl \text{ (C=O)}}$ 3,4-dimethoxyphenyl–CH(H,H)–CH(NHCCH$_3$)(Ar)(H) **D**

(e) $PhCHCNH_2$ (C=O) with CH_2CH_3 $\xrightarrow[\begin{array}{c}H_2SO_4 \\ \Delta\end{array}]{H_2O}$ $PhCHCOH$ (C=O) with CH_2CH_3 **E**

(f) 2-methylphenyl–CN $\xrightarrow[\begin{array}{c}H_2SO_4 \\ \Delta \\ 5\ h\end{array}]{H_2O}$ 2-methylphenyl–COH (C=O) **F**

(g) CH_3CH_2CCl (C=O) $\xrightarrow{NH_3\ (\text{excess})}$ $CH_3CH_2CNH_2$ (C=O) **G**

(h) $PhCOCH_2CH_3$ (C=O) $\xrightarrow{H_2NOH}$ $PhCNHOH$ (C=O) **H** $+\ CH_3CH_2OH$

15.42

(i)

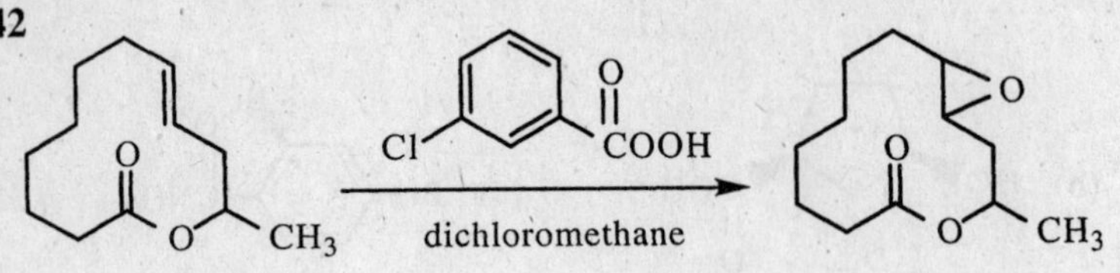

I

(j)

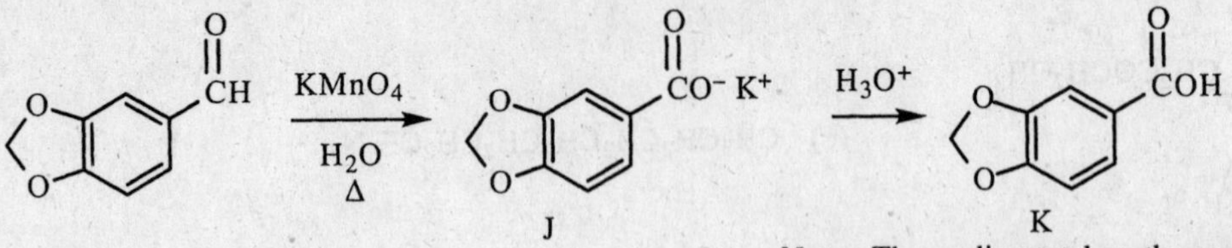

J   **K**

Note: The cyclic acetal on the aromatic ring is stable and does not open during the protonation of the carboxylate anion.

(k)

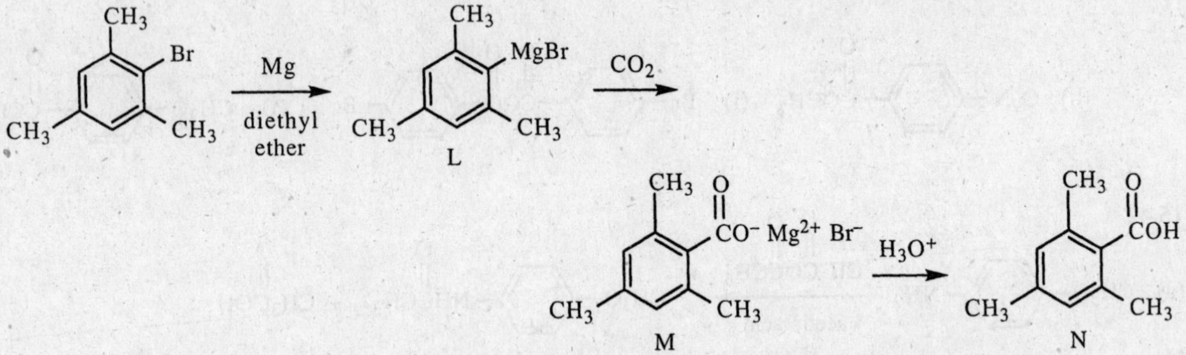

L

M   **N**

15.43   (a)

$$CH_3COCH_2C=CHCH_2NHCHCH_2OH \xrightarrow[\text{base}]{\underset{\text{(excess)}}{\text{PhCCl}}}$$

with CH₃ and CH=CH₂ substituents

$$CH_3COCH_2C=CHCH_2NCHCH_2OCPh$$

A   PhC=O

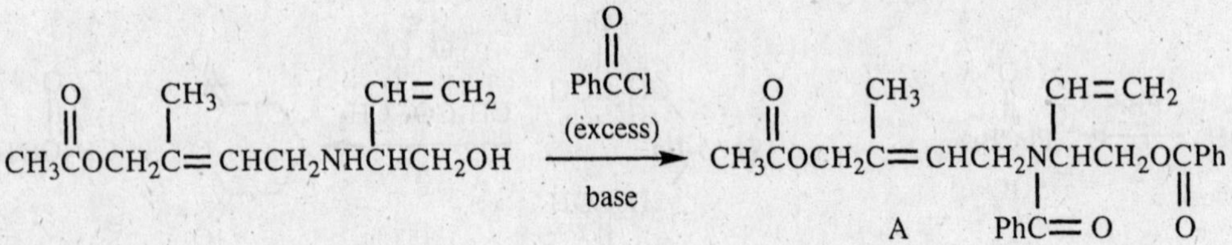

(b)

$$CH_3(CH_2)_{10}COCH_2(CH_2)_{10}CH_3 + CH_3OH \xrightarrow[\Delta]{H_2SO_4} CH_3(CH_2)_{10}COCH_3 + CH_3(CH_2)_{10}CH_2OH$$

    B   **C**

(c)

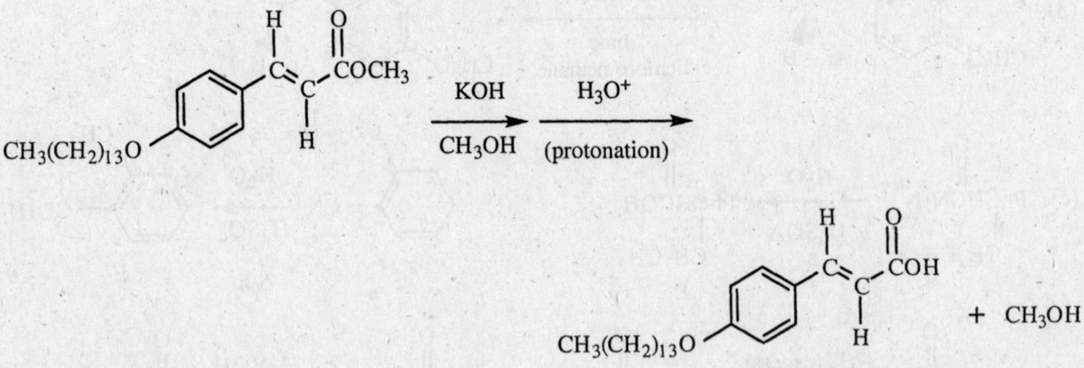

$$CH_3(CH_2)_{13}O-\text{...}-C=C-COCH_3 \xrightarrow[CH_3OH]{KOH} \xrightarrow[\text{(protonation)}]{H_3O^+}$$

    D   + CH₃OH   **E**

15.43 (d)

$$\text{(cyclohexenyl)}-\overset{\overset{\displaystyle O}{\|}}{C}H \xrightarrow[\text{H}_2\text{O}]{\text{Ag}_2\text{O}} \text{(cyclohexenyl)}-\overset{\overset{\displaystyle O}{\|}}{C}O^-Ag^+ \xrightarrow{\text{H}_3\text{O}^+} \text{(cyclohexenyl)}-\overset{\overset{\displaystyle O}{\|}}{C}OH$$

F   G

(e)

$$\text{HOC(CH}_2)_4\text{COH} \xrightarrow[]{\text{SOCl}_2 \text{ (excess)}} \text{ClC(CH}_2)_4\text{CCl} \xrightarrow[\underset{\text{PhNCH}_3}{\overset{\text{CH}_3}{|}}]{\overset{\text{CH}_3}{\underset{|}{\text{CH}_3\text{COH (excess)}}}} \text{CH}_3\overset{\text{CH}_3}{\underset{\text{CH}_3}{\overset{|}{C}}}O-\overset{O}{\overset{\|}{C}}(\text{CH}_2)_4\overset{O}{\overset{\|}{C}}-O\overset{\text{CH}_3}{\underset{\text{CH}_3}{\overset{|}{C}}}\text{CH}_3$$

H   I

(f)

$$\text{O}_2\text{N}-\text{(phenyl)}-\overset{\overset{\displaystyle O}{\|}}{C}\text{CH}_3 \xrightarrow[\underset{\Delta}{\text{H}_2\text{O}}]{\text{NaOH}} \text{O}_2\text{N}-\text{(phenyl)}-\overset{O}{\underset{\|}{C}}O^- \text{Na}^+ + \text{CH}_3\text{OH}$$

J   K

(g)

$$\xrightarrow[\substack{\text{NaHCO}_3 \\ \text{dichloromethane}}]{\substack{\text{PhCCl} \\ \text{(1 equivalent)}}}$$

L
major product

M
minor product

Nitrogen is more nucleophilic than oxygen, so the major product comes from acylation at the nitrogen atom.

(h) $\text{CCl}_3\text{CH}_2\text{CH}_2\text{CH}_2\text{CH}_2\text{OH} \xrightarrow[\underset{\Delta}{\text{H}_2\text{O}}]{\text{KMnO}_4} \text{CCl}_3\text{CH}_2\text{CH}_2\text{CH}_2\overset{O}{\overset{\|}{C}}\text{O}^-\text{K}^+ \xrightarrow{\text{H}_3\text{O}^+} \text{CCl}_3\text{CH}_2\text{CH}_2\text{CH}_2\overset{O}{\overset{\|}{C}}\text{OH}$

N   O

(i)

$\overset{+}{\underset{\text{H}}{\text{N}}}$ (pyridinium) is an acid with pK_a 5.2, close enough to the pK_a of the carboxylic acid (pK_a ~ 4.8) to maintain an equilibrium concentration of the acid. The carboxylic acid then forms the lactone under conditions in which water is removed from the reaction mixture.

15.44   (a)

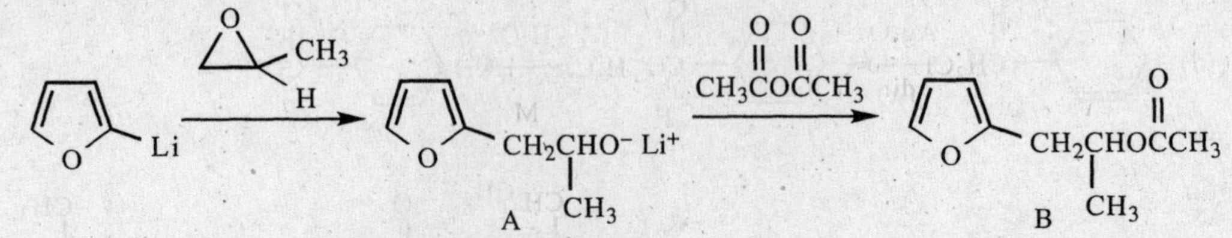

A
nucleophilic attack at the
less hindered carbon atom

(b)

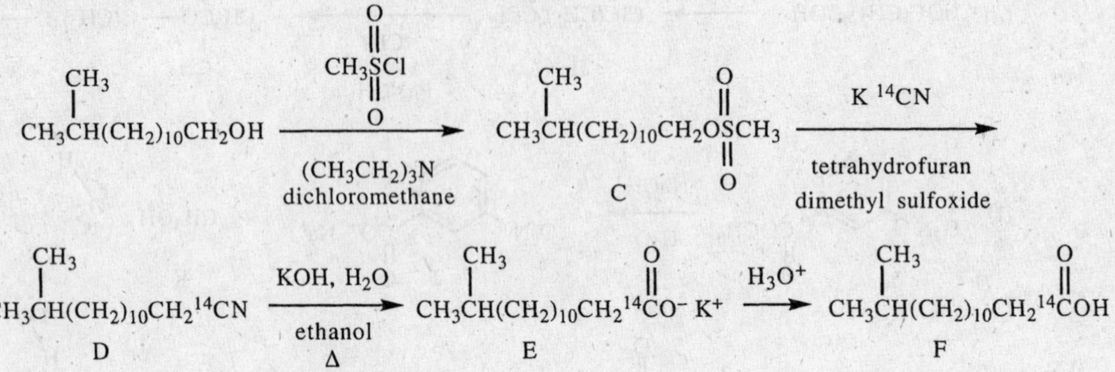

(c)

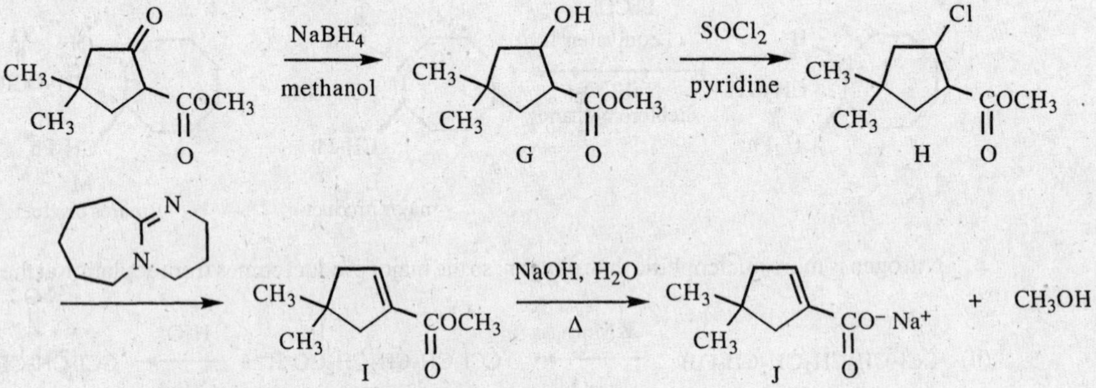

(A double bond conjugated with a carbonyl
group is favored. See Section 17.5)

(d)

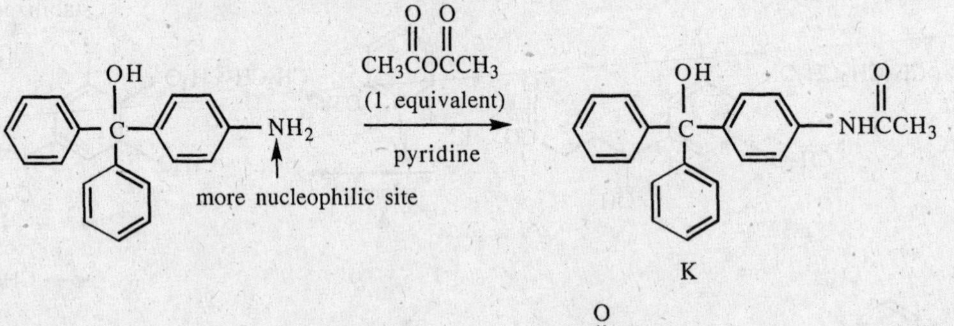

(e)

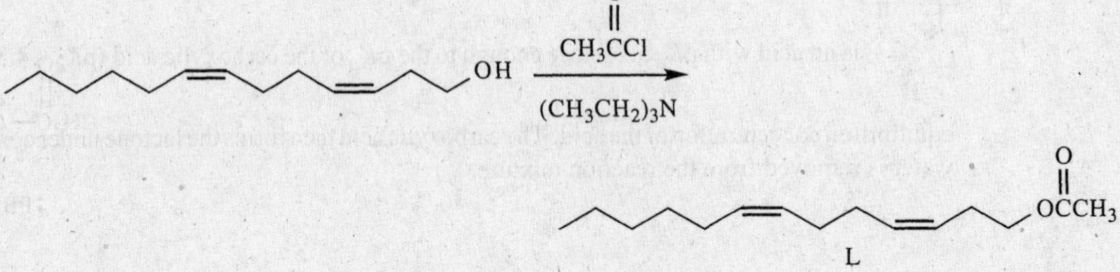

15.44 (f)

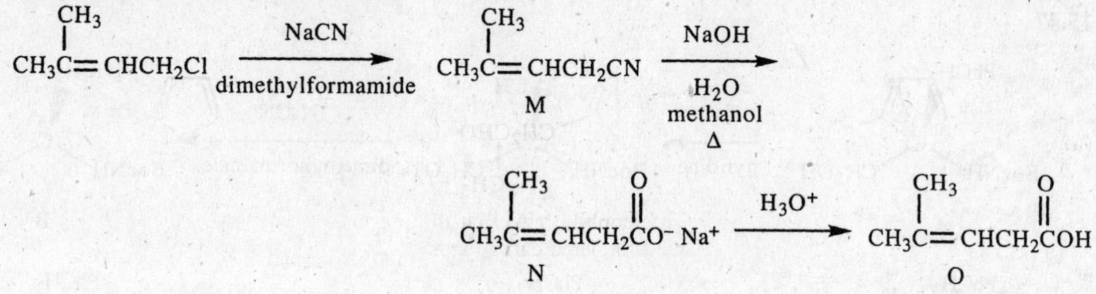

$$CH_3C{=}CHCH_2Cl \xrightarrow[\text{dimethylformamide}]{NaCN} CH_3C{=}CHCH_2CN \text{ (M)} \xrightarrow[\substack{H_2O \\ \text{methanol} \\ \Delta}]{NaOH}$$

with CH_3 on the carbon.

$$CH_3C{=}CHCH_2CO^- Na^+ \text{ (N)} \xrightarrow{H_3O^+} CH_3C{=}CHCH_2COH \text{ (O)}$$

15.45

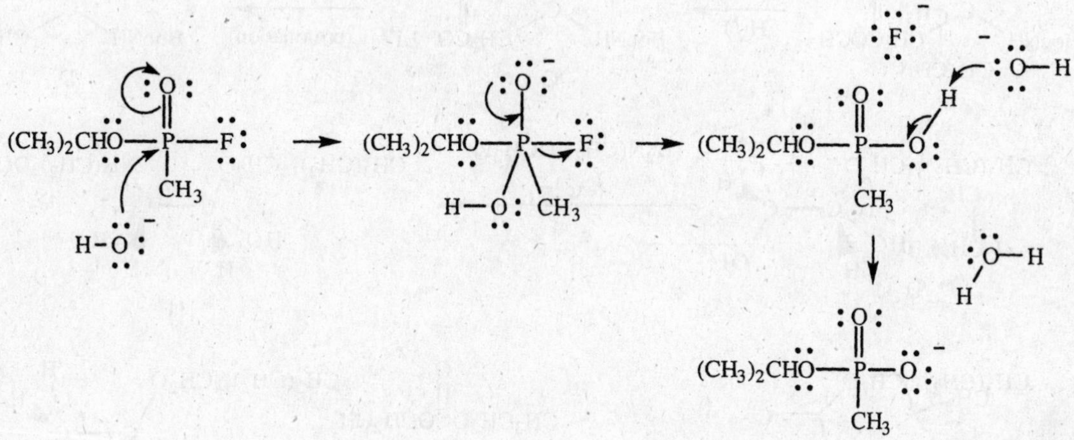

(CH₃)₂CHO—P(=O)(CH₃)—F → → → (CH₃)₂CHO—P(=O)(CH₃)—O⁻

15.46 (a)

$$CH_3C(=O){-}O{-}CCH_3 \longrightarrow CH_3C{-}O{-}CCH_3 \longrightarrow CH_3C{-}PBu_3^+ \quad {^-}O{-}CCH_3$$
$$Bu_3P:$$

new acyl transfer agent

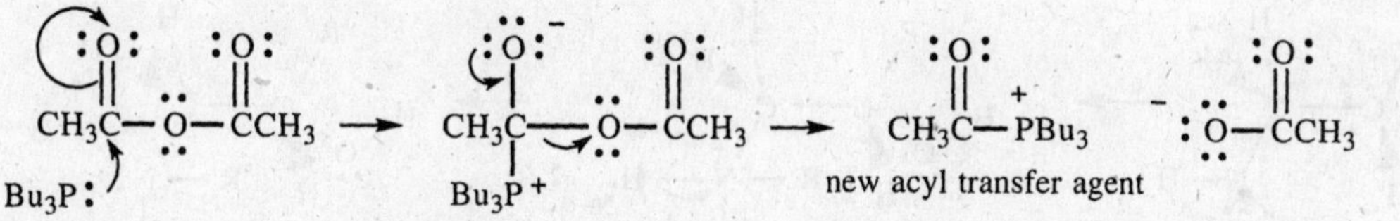

The new acyl transfer agent is more reactive than acetic anhydride because it is less stabilized by resonance.

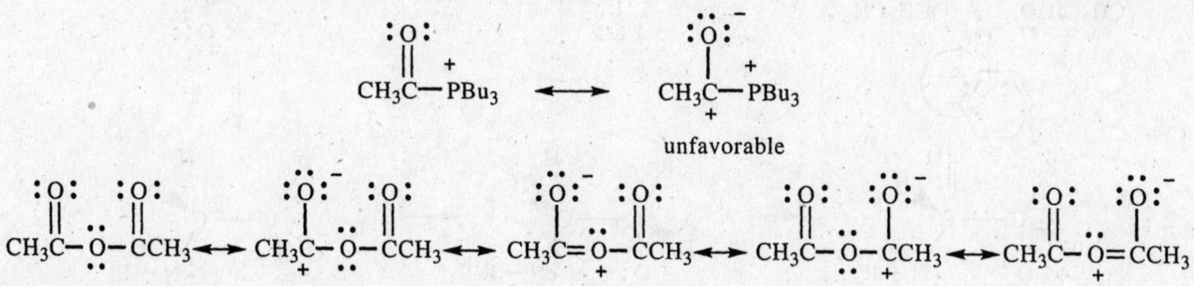

$$CH_3C(=O){-}PBu_3^+ \longleftrightarrow CH_3C(-O^-){-}\overset{+}{P}Bu_3$$

unfavorable

$$CH_3C{-}O{-}CCH_3 \longleftrightarrow CH_3C{-}O{-}CCH_3 \longleftrightarrow CH_3C{=}O{-}CCH_3 \longleftrightarrow CH_3C{-}O{-}CCH_3 \longleftrightarrow CH_3C{-}O{=}CCH_3$$

(b)

$$CH_3C(=O){-}PBu_3^+ \longrightarrow CH_3C(-O^-){-}PBu_3^+ \longrightarrow CH_3C(-O^-){-}PBu_3^+ \longrightarrow CH_3C{-}OCH_3$$
$$H{-}O{-}CH_3 \qquad H{-}\overset{+}{O}{-}CH_3 \qquad :O{-}CH_3 \qquad :PBu_3$$
$$N(CH_2CH_3)_3 \qquad HN(CH_2CH_3)_3$$

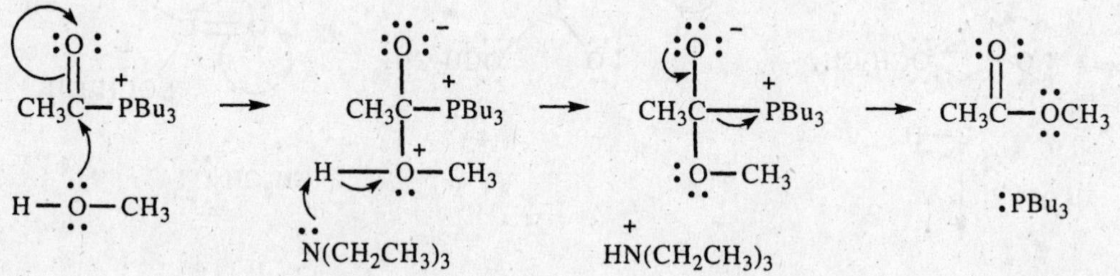

15.47

(a)

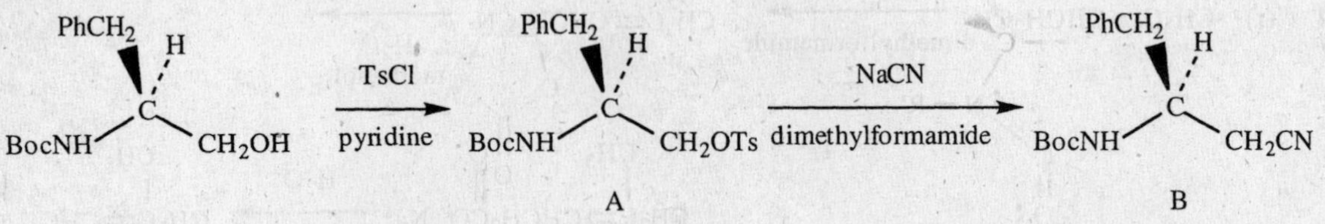

$$\text{PhCH}_2,\ H \quad \xrightarrow[\text{pyridine}]{\text{TsCl}} \quad \text{PhCH}_2,\ H \quad \xrightarrow[\text{dimethylformamide}]{\text{NaCN}} \quad \text{PhCH}_2,\ H$$

BocNH–C–CH$_2$OH   BocNH–C–CH$_2$OTs   BocNH–C–CH$_2$CN

A   B

(b)

$$\xrightarrow[\text{H}_2\text{O}]{\text{LiOH}} \qquad \xrightarrow[\text{(protonation)}]{\text{H}_3\text{O}^+}$$

BocNH–C–CH$_2$COCH$_3$ BocNH–C–CH$_2$CO$^-$ Li$^+$ BocNH–C–CH$_2$COH

C   D

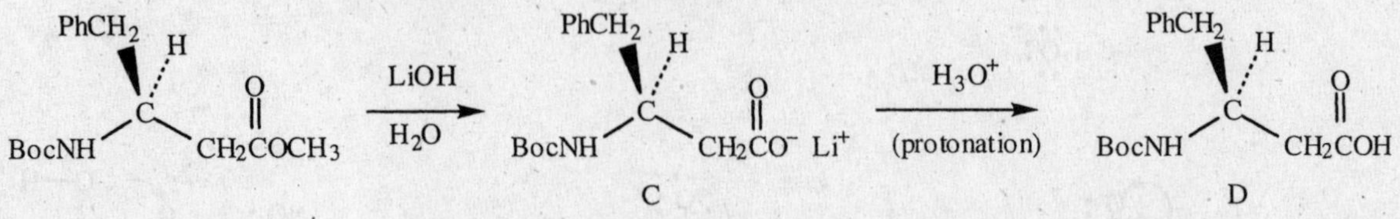

15.48

$$\text{CH}_3(\text{CH}_2)_{14}\text{CH}_2\text{O} \quad \xrightarrow[\text{(excess)}]{\text{H}_2\text{N}(\text{CH}_2)_4\text{OCH}_2\text{Ph}} \quad \text{CH}_3(\text{CH}_2)_{14}\text{CH}_2\text{O} \qquad \text{NH}(\text{CH}_2)_4\text{OCH}_2\text{Ph}$$

III

$$\text{CH}_3(\text{CH}_2)_{14}\text{CH}_2\text{O} \qquad \xleftarrow[\text{CH}_3\text{CH}_2\text{OCOCH}_2\text{CH}_3}]{} \qquad \text{CH}_3(\text{CH}_2)_{14}\text{CH}_2\text{O}$$

N(CH$_2$)$_4$OCH$_2$Ph   NH(CH$_2$)$_4$OCH$_2$Ph

B   A

CH$_3$CH$_2$O–C–OCH$_2$CH$_3$

CH$_3$CH$_2$O–C–OCH$_2$CH$_3$

H–B$^+$

CH$_3$CH$_2$O–C–OCH$_2$CH$_3$

CH$_3$CH$_2$O–C–OCH$_2$CH$_3$

$^+$B–H

CH$_3$CH$_2$O$^-$

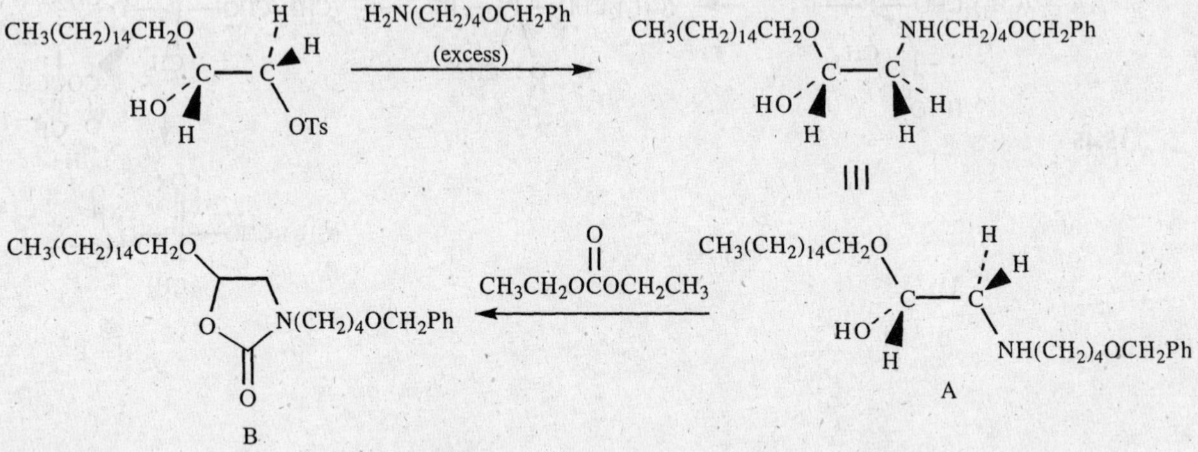

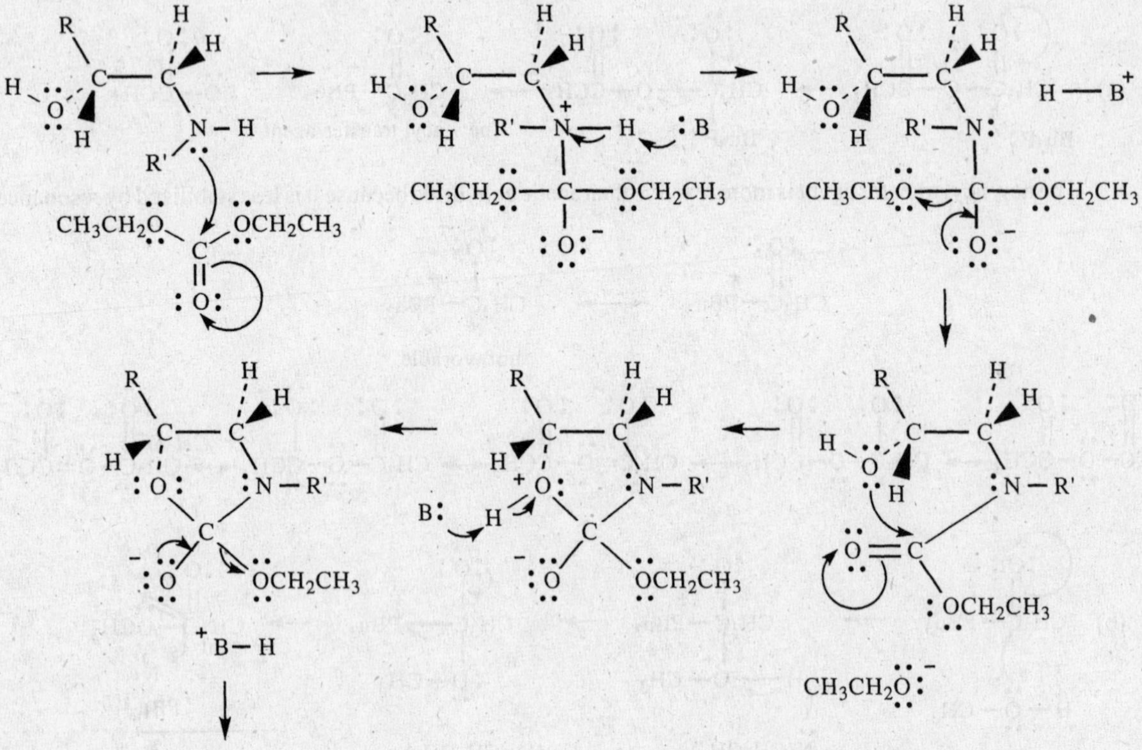

15.48 (b) (cont)

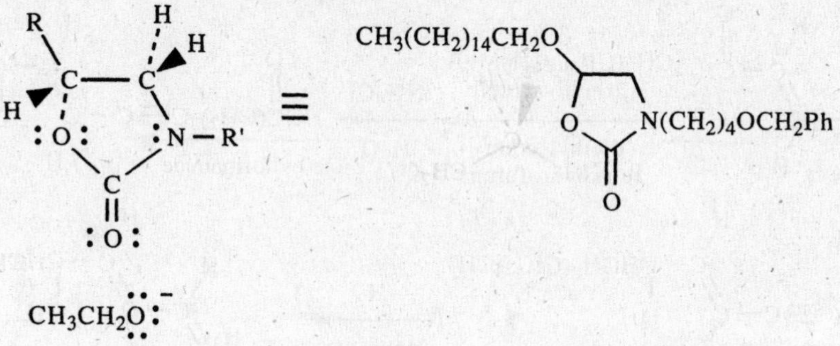

15.49

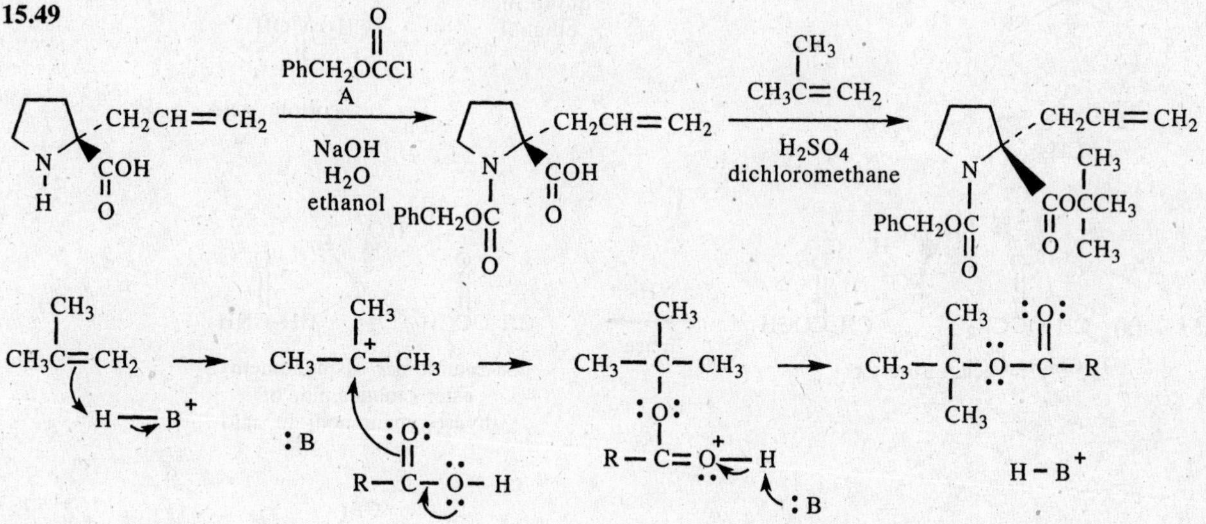

15.50 (a) $HO(CH_2)_8OH \xrightarrow[\text{H}_2\text{O}]{\text{HBr (48\%) (1 eq)}} Br(CH_2)_8OH \xrightarrow[\substack{\text{H}_2\text{SO}_4 \\ \text{acetone}}]{\text{CrO}_3,\ \text{H}_2\text{O}} Br(CH_2)_7\overset{\displaystyle O}{\overset{\|}{C}}OH$

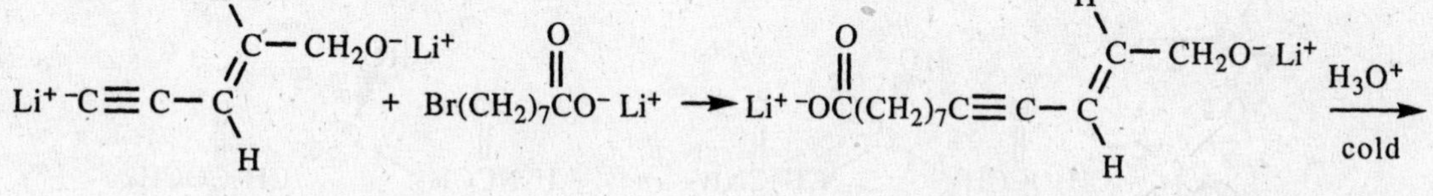

A   B

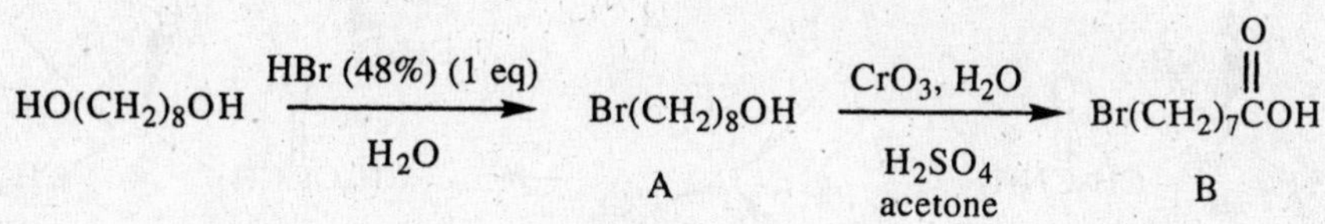

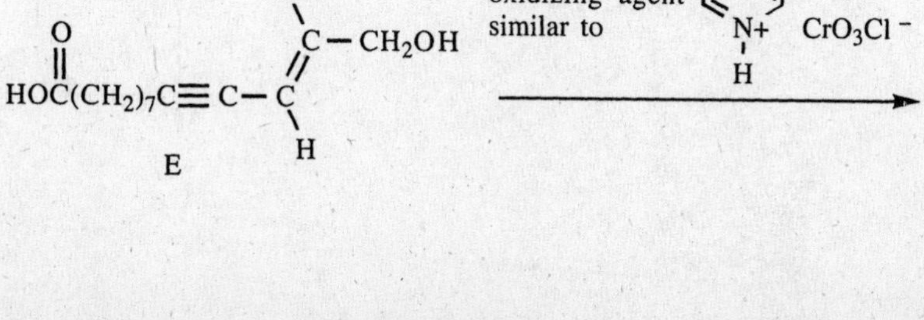

(b)

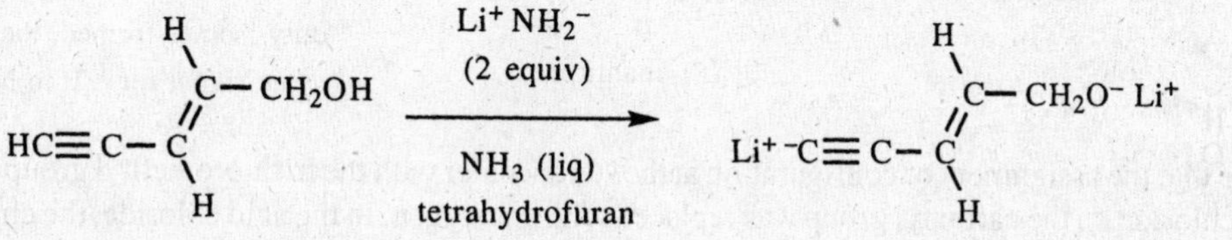

(c)

15.50 (c) (cont)

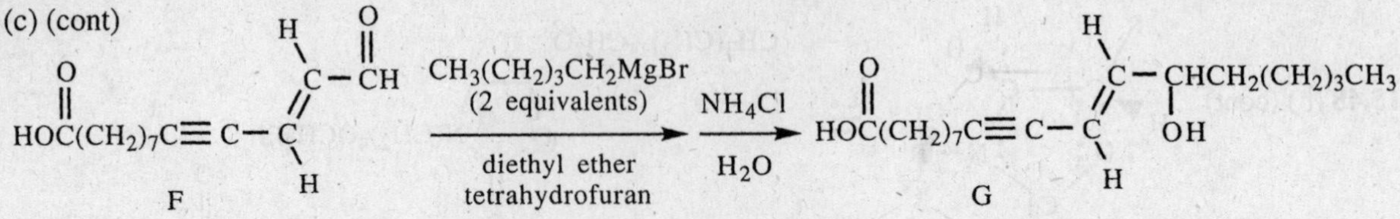

(d)

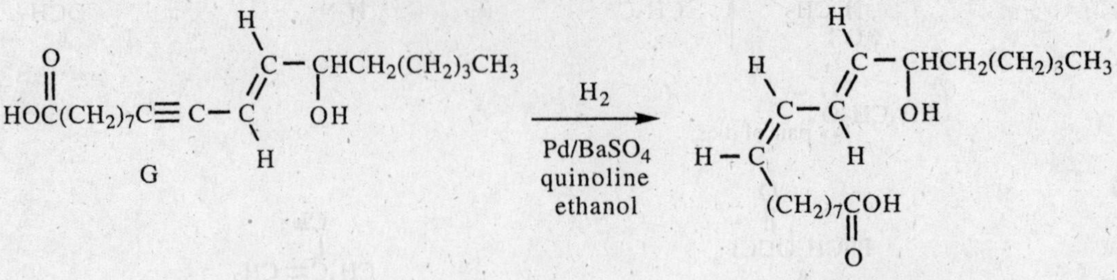

coriolic acid

15.51 (a)

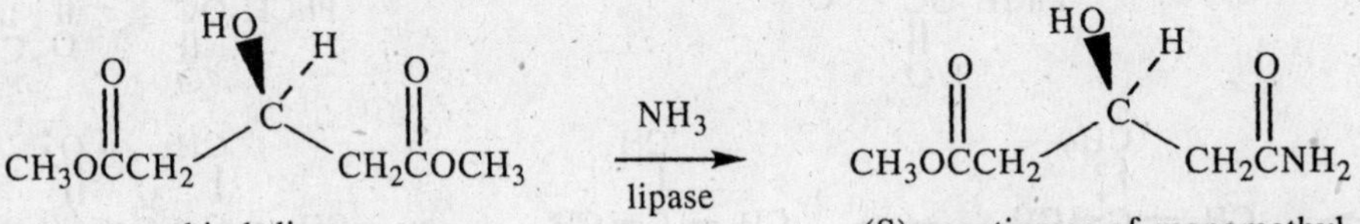

achiral diester

(S)-enantiomer of monomethyl ester - monoamide of 3-hydroxypentanedioic acid

(b)

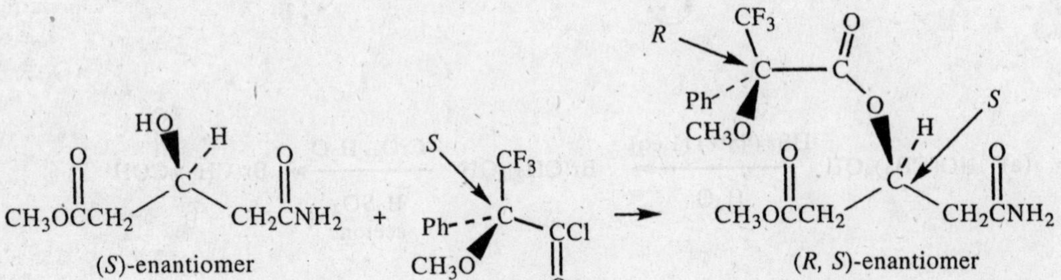

(*S*)-enantiomer

(*S*)-enantiomer

(*R, S*)-enantiomer only isomer formed, therefore one signal for ^{19}F in NMR

Note that the assignment of configuration at the stereocenter with the trifluoromethyl group changed when the chlorine on the carbonyl group was replaced with an oxygen. In the acid chloride, the chloroacyl group has a higher priority than the trifluoromethyl group; the priorities change when the ester is formed.

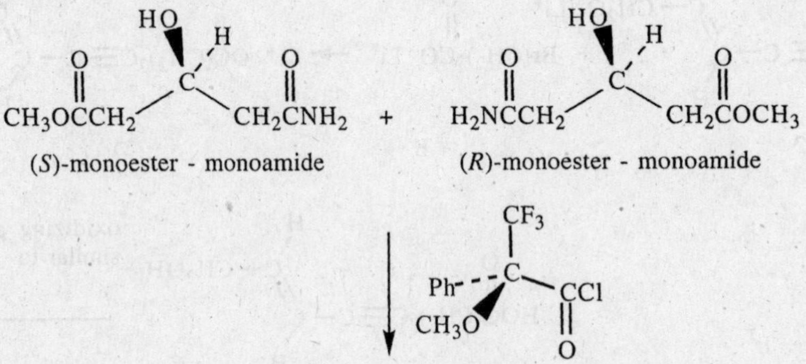

(*S*)-monoester - monoamide

(*R*)-monoester - monoamide

15.51 (b) (cont)

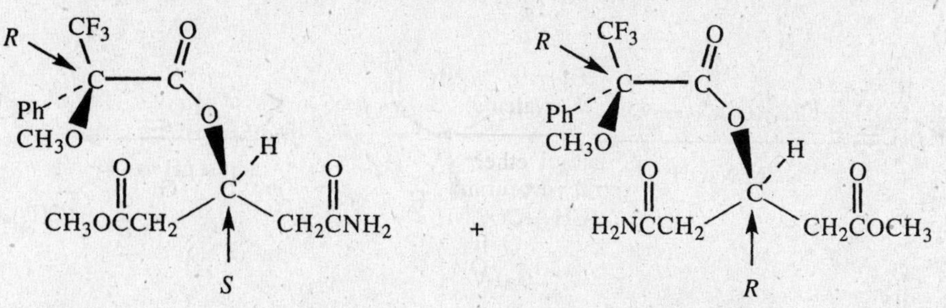

pair of diastereomers, therefore two signals for ^{19}F in NMR

15.52

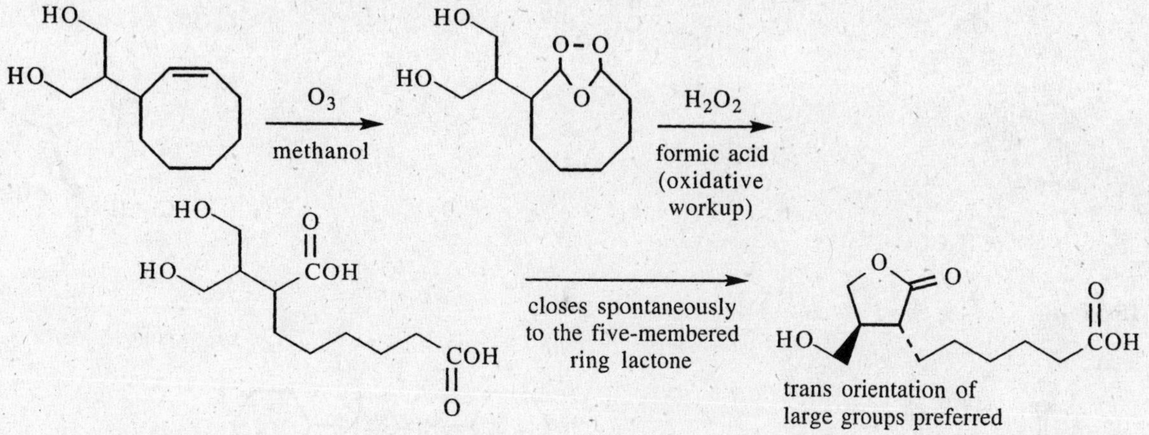
O_3
methanol
H_2O_2
formic acid
(oxidative
workup)
closes spontaneously
to the five-membered
ring lactone
trans orientation of
large groups preferred

15.53 Let $R\ddot{N}(H)-H$ stand for the amine.

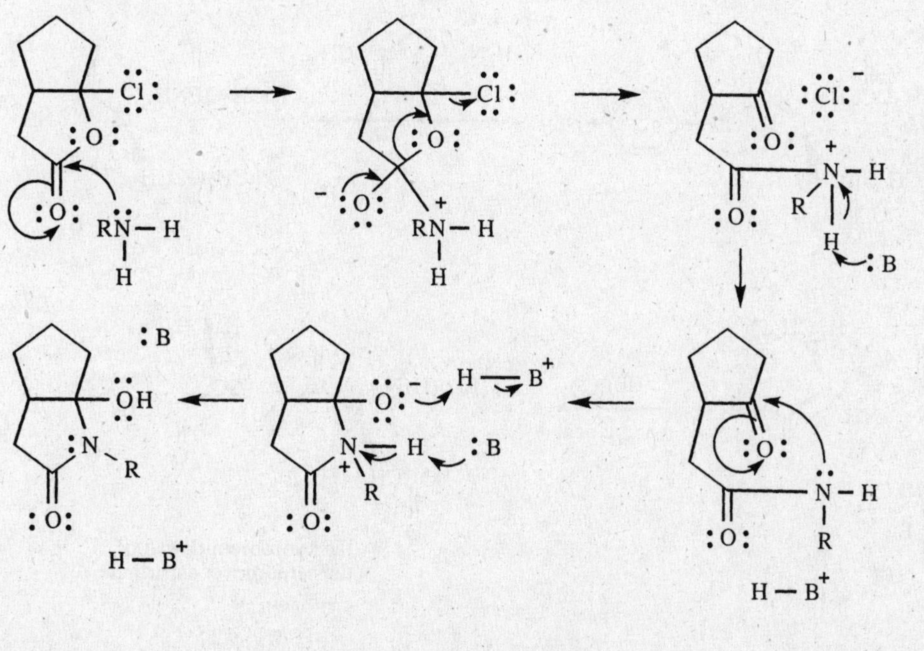

15.54

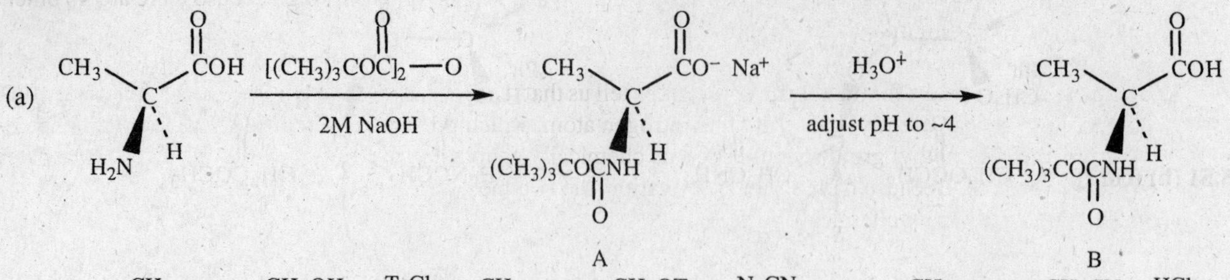

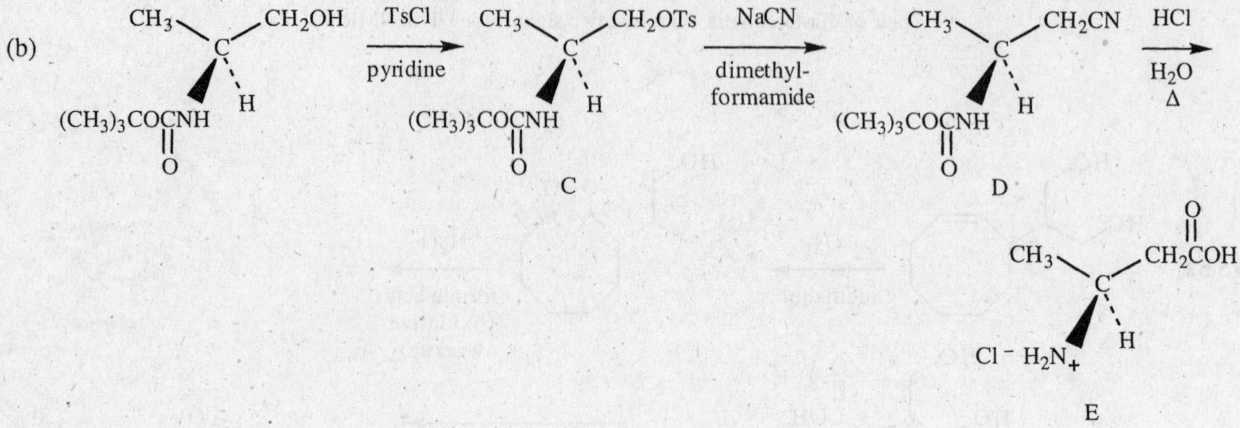

(a)

(b)

15.55

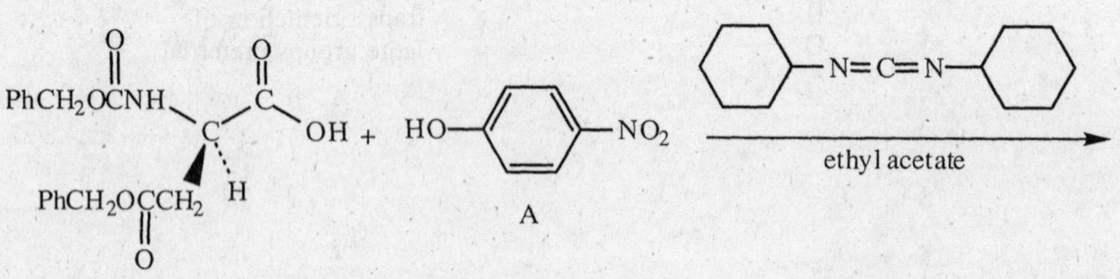

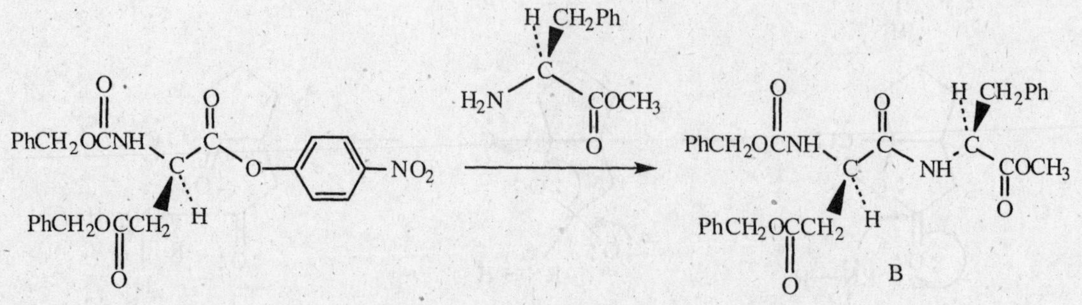

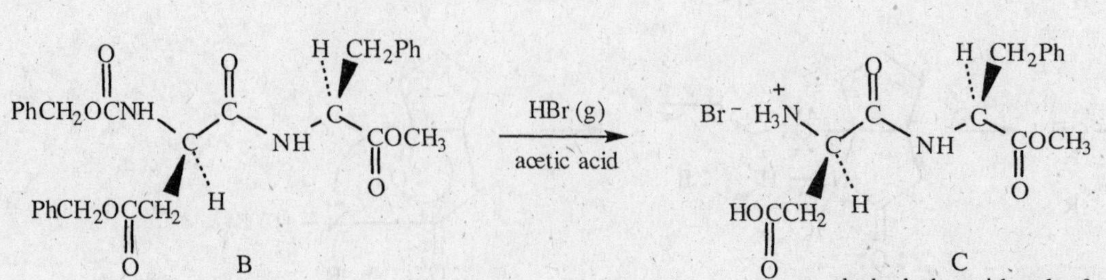

the hydrobromide salt of
a diastereomer of aspartame

15.56   The two peaks observed for the methyl groups cannot be due to spin-spin coupling because there are no other peaks in the spectrum that also show splitting.

The two separate peaks for the two methyl groups tell us that there is partial double bond character between the carbon atom of the carbonyl group and the nitrogen atom, which prevents free rotation. If free rotation were possible, the two methyl groups would become chemically equivalent, and only one chemical shift would be observed for both. Because there is not free rotation, the two methyl groups are in different environments and thus have two different chemical shifts.

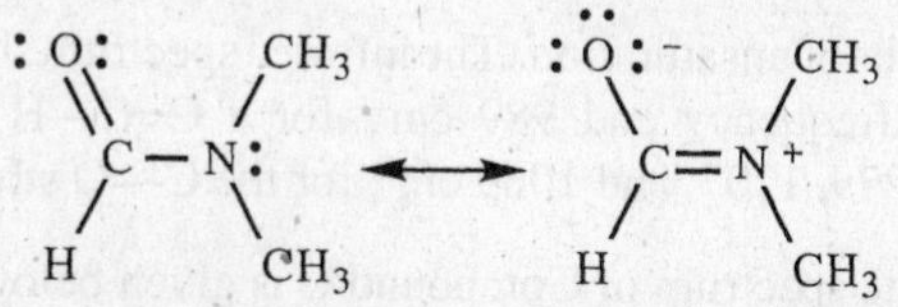

Resonance contributors for the amide group showing the existence of partial double bond character for the bond between the carbon atom of the carbonyl group and the nitrogen atom.

15.57   Compound A, $C_9H_{10}O_2$, has five units of unsaturation. When a compound has four or more units of unsaturation, we should look for an aromatic ring, which can be seen in both the proton and ^{13}C magnetic resonance spectra. The infrared spectrum shows the presence of an ester (1743 cm^{-1} for the C=O stretch and 1229 and 1118 cm^{-1} for the C—O stretch).

Analysis of the proton spectrum of Compound A is shown below:

δ 2.2   (3H, s, C$\underline{H}_3$C=O

δ 5.2   (2H, s, ArC$\underline{H}_2$—O—)

δ 7.4   (5H, s, Ar$\underline{H}$)

The absence of splitting for the aromatic hydrogen atoms tells us that the group bonded to the aromatic ring does not differ significantly from the aryl carbon atoms in electronegativity and in magnetic anisotropy. A tetrahedral carbon atom meets this requirement.

The ^{13}C magnetic resonance spectrum shows two alkyl carbon atoms (δ 20.7 and 66.1). There are three types of aromatic carbon atoms (δ 128.1, 128.4, and 136.2) and one carbonyl carbon atom (δ 170.5). Compound A is benzyl acetate.

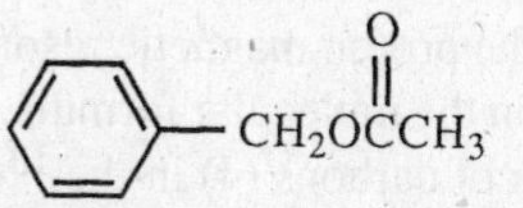

Note that the compound cannot be methyl phenylacetate because the band for the methyl singlet in the proton magnetic resonance spectrum of a methyl ester would be at ~δ 3.7.

15.58   Compound B, $C_9H_{10}O_3$, has five units of unsaturation. The infrared spectrum shows the presence of a carboxylic acid (3300-2600 cm^{-1} for the O—H and 1719 cm^{-1} for the C=O stretching frequencies).

Analysis of the proton magnetic spectrum of Compound B is given below:

δ 3.56   (2H, s, —C$\underline{H}_2$—)

δ 3.78   (3H, s, C$\underline{H}_3$—O—)

δ 6.85   (2H, multiplet, Ar$\underline{H}$)

δ 7.18   (2H, multiplet, Ar$\underline{H}$)

δ 11.80   (1H, s, O=CO$\underline{H}$)

The splitting pattern in the aromatic region is typical of para substituted rings.

Compound B is (4-methoxyphenyl)acetic acid. Analysis of the ^{13}C nuclear magnetic spectrum of Compound B is shown on the next page:

15.58 (cont)

δ 114.00, 125.24
130.35, and 158.75

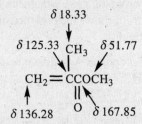

$$CH_3O-\langle\bigcirc\rangle-CH_2COH$$

δ 55.19   δ 40.14   δ 178.50

15.59   Compound C, $C_5H_8O_2$, has 2 units of unsaturation. The infrared spectrum shows the presence of an alkene (1635 cm⁻¹ for the C=C stretching frequency and 989 cm⁻¹ for a C=C—H bending frequency), and an ester (1731 cm⁻¹ for the C=O and 1279, 1207, and 1069 cm⁻¹ for the C—O stretching frequencies).

Analysis of the proton magnetic spectrum of Compound C is given below:
  δ 1.95   (3H, multiplet, C$\underline{H}_3$C=CH)
  δ 3.75   (3H, s, C$\underline{H}_3$O—)
  δ 5.55   (1H, multiplet, C=C$\underline{H}$)
  δ 6.20   (1H, multiplet, C=C$\underline{H}$)
Compound C is methyl 2-methylpropenoate (methyl methacrylate).
Analysis of the ¹³C nuclear magnetic spectrum of Compound C is shown below:

δ 18.33

δ 125.33 CH₃   δ 51.77

$$CH_2 = CCOCH_3$$

δ 136.28   O   δ 167.85

The allylic and vinyl hydrogen atoms are coupled to each other through the double bond.

15.60   Compound D, $C_8H_{14}O_4$, has 2 units of unsaturation. The infrared spectrum shows the presence of an ester (1736 cm⁻¹ for the C=O, and 1160 and 1032 cm⁻¹ for the C—O stretching frequencies).

The number of hydrogen atoms in the proton magnetic resonance spectrum of Compound D adds up to 7, or half the number of hydrogen atoms in the molecular formula. This points to symmetry in the molecule. This is confirmed by looking at the number of carbons (4) in the ¹³C nuclear magnetic spectrum.

Analysis of the proton magnetic spectrum of Compound D is given below:
  δ 1.26   (6H, t, 2 C$\underline{H}_3$CH₂—)
  δ 2.62   (4H, s, O=CC$\underline{H}_2$C$\underline{H}_2$C=O)
            these hydrogen atoms are chemical shift equivalent and do not couple
  δ 4.15   (4H, q, 2 CH₃C$\underline{H}_2$O—)
Compound D is diethyl butanedioate (diethyl succinate).

Analysis of the ¹³C nuclear magnetic spectrum of Compound D is shown below:

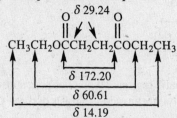

O δ 29.24 O

$$CH_3CH_2OCCH_2CH_2COCH_2CH_3$$

δ 172.20

δ 60.61

δ 14.19

15.61 Compounds E and F show evidence for an ester functional group in the infrared spectra (C=O stretch at 1743 cm^{-1} for E and 1741 cm^{-1} for F and C—O stretch at 1243 and 1031 cm^{-1} for E and 1198 and 1097 cm^{-1} for F). Analysis of the proton magnetic spectrum of Compound E is shown below:

δ 0.95   (3H, t, C$\underline{H}_3$CH$_2$—)
δ 1.40   (2H, sextet, CH$_3$C$\underline{H}_2$CH$_2$—)
δ 1.65   (2H, quintet, —CH$_2$C$\underline{H}_2$CH$_2$—)
δ 2.05   (3H, s, C$\underline{H}_3$C=O)
δ 4.10   (2H, t, CH$_3$C$\underline{H}_2$—O—)

Compound E is butyl acetate. Analysis of the ^{13}C nuclear magnetic spectrum of Compound E is shown below:

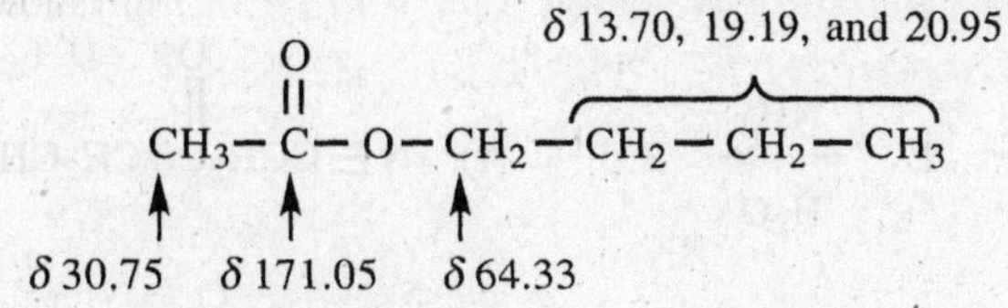

Analysis of the proton magnetic spectrum of Compound F is shown below:

δ 1.00   [3H, t, C$\underline{H}_3$CH$_2$—]
δ 1.65   [2H, sextet, CH$_3$C$\underline{H}_2$CH$_2$—]
δ 2.30   [2H, t, (C$\underline{H}_2$CH$_2$C=O]
δ 3.70   (3H, s, C$\underline{H}_3$—O—)

Compound F is methyl butanoate (methyl butyrate). Analysis of the ^{13}C nuclear magnetic spectrum of Compound F is shown below:

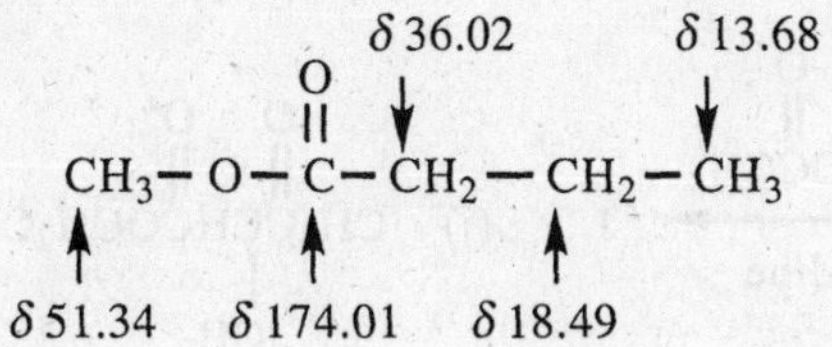

Supplemental Problems

S15.1   Name the following compounds, including an indication of the stereochemistry where appropriate.

(a) CH$_3$CCH$_2$CH$_2$CH$_2$COH

(b)

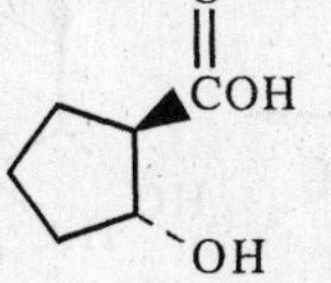

(c)

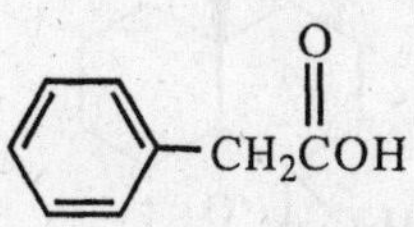

(d)

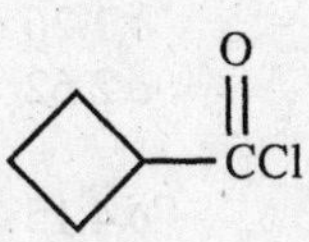

(e)

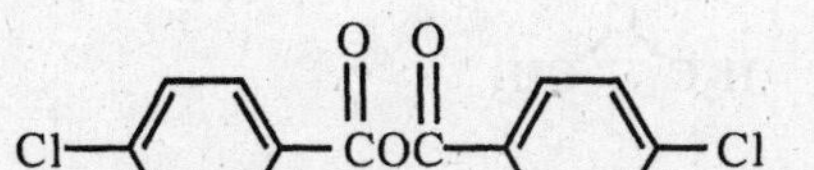

(f)

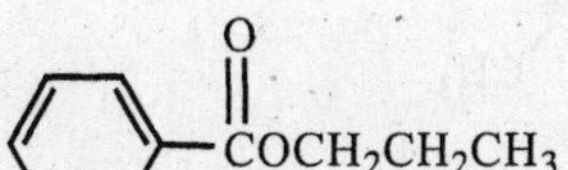

(g) CH$_3$CH$_2$CCH$_2$COCH$_2$CH$_3$

(h) CH$_3$CHCH$_2$CHCOCH$_3$ (CH$_3$ and Br substituents)

(i) HOCCHCH$_2$CH$_2$CH$_2$COH (CH$_3$ substituent)

(j) (CH$_3$CH$_2$CH$_2$CH$_2$CH$_2$C)$_2$O

S15.2   Give structural formulas for all of the reagents, organic intermediates, and products indicated by letters in the following equations.

(a) $CH_3(CH_2)_{10}COH \xrightarrow[\Delta]{SOCl_2} A$

(b) $CH_3CH_2CH_2CH_2OH \; + \; CH_3CH_2CH_2CCl \longrightarrow B$

(c) $HOCC\equiv CCOH \xrightarrow[\substack{H_2SO_4 \\ \Delta}]{CH_3OH \,(excess)} C$

(d) $BrCH_2COH \xrightarrow[\substack{H_2SO_4 \\ \Delta}]{CH_3CH_2OH} D$

(e) $CH_3(CH_2)_4\underset{\underset{CH_2CH_3}{|}}{CH}COH \xrightarrow[\Delta]{SOCl_2} E \xrightarrow[H_2O]{NH_3} F$

(f) $N\equiv CCH_2COCH_2CH_3 \xrightarrow[\substack{H_2O \\ 0\,°C}]{NH_3} G \; + \; CH_3CH_2OH$

(g) $CH_3(CH_2)_5CCl \xrightarrow{CH_3(CH_2)_5CO^-Na^+} H$

(h)   (structure with CH_2COH and COH on benzene ring) $\xrightarrow[\Delta]{(CH_3C)_2O} I \; + \; 2\,CH_3COH$

(i) $\xrightarrow[pyridine]{CH_3COCCH_3} J$

(j) $CH_3CCHCOCH_2CH_3$ (with CH_3 substituent) $\xrightarrow[\substack{base, \\ toluene \\ \Delta}]{K}$ product $+ \; CH_3CH_2OH$

S15.3   The following transformation was observed. Propose a mechanism for it.

S15.4   Dimethyl (2*R*,3*S*)-2-acetoxy-3-bromobutanedioate is converted into a chiral alcohol on treatment with methanol in the presence of a trace of hydrogen chloride. The chiral alcohol, in turn, reacts with potassium carbonate in acetone to give a chiral oxirane. Provide stereochemically correct structures for the ester, alcohol, and oxirane, and show how the transformations described occur.

S15.5   Compounds labeled with radioactive isotopes are needed for the medical imaging technique known as positron emission tomography (PET scanning). Such compounds have been synthesized to visualize the areas of the brain where cocaine (see section 22.8A in the text) is localized. A cocaine analog labeled with radioactive iodine on the aromatic ring was synthesized in the following way. Give structural formulas for compounds A and B.

S15.6   The following conversion was carried out. Work out the conditions for this transformation that requires several steps. There is more than one way to solve this problem.

S15.7   The reactions used in the determination of the structure of musk ambrette (p. 625 in the text) serve as a review of the chemistry of alkenes, alcohols, aldehydes, and acid derivatives. Some of the reactions are given in the form of the incomplete equations below. Complete the equations, giving structures for all compounds indicated by Roman numerals.

(a)   musk ambrette $\xrightarrow[\substack{H_2O \\ \Delta}]{NaOH}$ $\xrightarrow{H_3O^+}$ I

(b)   I $\xrightarrow[\substack{Pt \\ ethanol}]{H_2}$ II $\xrightarrow{(CH_3C)_2O}$ III ($C_{18}H_{34}O_4$)

(c)   II $\xrightarrow[\substack{H_2SO_4 \\ acetic\ acid}]{CrO_3}$ IV ($C_{16}H_{30}O_4$)

(d)   I $\xrightarrow{O_3}$ $\xrightarrow{\substack{reductive \\ work\ up}}$ V + VI

(e)   VI $\xrightarrow[H_2O]{KMnO_4}$ $\xrightarrow{H_3O^+}$ VII ($C_9H_{18}O_3$)

(f)   VII $\xrightarrow[H_2SO_4]{CrO_3}$ VIII ($C_9H_{16}O_4$)

(g)   V $\xrightarrow[Na_2CO_3]{HO\overset{+}{N}H_3\ Cl^-}$ IX ($C_7H_{13}NO_3$)

(h)   IX $\xrightarrow[\Delta]{(CH_3C)_2O}$ X ($C_7H_{11}NO_2$)

(i)   X $\xrightarrow[\substack{H_2O \\ \Delta}]{NaOH}$ $\xrightarrow{H_3O^+}$ heptanedioic acid

16 Structural Effects in Acidity and Basicity Revisited. Enolization

16.1

anion of *ortho*-nitrobenzoic acid

positive charge on the carbon atom that is adjacent to the carboxylate group; stabilization of the anion

anion of *meta*-nitrobenzoic acid

There is no resonance contributor of *meta*-nitrobenzoic acid in which there is a positive charge adjacent to the carboxylate group.

16.2 (a) Resonance contributors
  1. *m*-nitrophenolate anion

The negative charge is stabilized by the inductive effect of the electron-withdrawing nitro group, so *m*-nitrophenol is a stronger acid than phenol.

16.2 (a) 2.   *p*-nitrophenolate anion

The negative charge of the *p*-nitrophenolate anion is stabilized not only by the inductive effect but also by a resonance effect. Additional delocalization of charge, beyond what is available in the phenolate anion and in the *m*-nitrophenolate anion, is possible for the *p*-nitrophenolate anion. This extra stabilization of the conjugate base is reflected in the greater acidity of *p*-nitrophenol.

(b)

Picric acid is a sufficiently strong acid to protonate the weak base bicarbonate anion. Although it is a phenol, the effect of the three nitro groups stabilizes the conjugate base of picric acid and thus increases the acidity of the compound so that it is more acidic than a simple carboxylic acid.

16.3   Compound A is the stronger acid. Its conjugate base is better stabilized by resonance. In A the negative charge is delocalized to both nitro groups.

16.3 (cont)   In B the negative charge is delocalized to only one nitro group.

16.4   (a) $K_{diss} = \dfrac{[R_3N:][BR'_3]}{[R_3N^+:^-BR'_3]}$

(b) The dissociation constants for the amine-borane complexes become smaller as the number of substituents on the nitrogen atom increases from none for ammonia to two for dimethylamine. When there are three substituents on the nitrogen atom, the dissociation constant is larger than it is for the other two complexes. The dissociation constant measures the strength of the bond between the atoms of the Lewis base, the amine, and the Lewis acid, the borane. The bond strength is influenced by two factors: (1) electronic factors, mainly the availability of the nonbonding electron pair on the nitrogen atom and (2) steric factors. The values observed for the dissociation constants may be interpreted to mean that alkyl substitution on the nitrogen atom increases the availability of the nonbonding electrons on the nitrogen atom. The presence of three substituents on nitrogen leads to steric hindrance that interferes with bonding in the complex between trimethylamine and trimethylborane. The steric factor is more important in this case than in the reaction of the amine with a Brønsted-Lowry acid because trimethylborane is larger than a proton.

16.5   The acidity of a conjugate acid is related to the availability of the nonbonding electron pair on the base. The observed trend for the conjugate acids of the given amines is due to the decreasing availability of the electron pair on the nitrogen. This decreasing availability is caused by the electron-withdrawing ether and cyano substituents which pull electron density away by induction. The inductive effect is greatest when the electron-withdrawing group is closest to the amino group. The cyano group is more electron-withdrawing than the oxygen because resonance within the cyano group puts a partial positive charge on its carbon atom.

16.6

16.6 (cont)

16.7   All of the cyanoanilines are weaker bases than aniline itself because the cyano group is an electron-withdrawing group. The effect of the cyano group is greatest when it is ortho, and smallest when it is meta, to the amino group. Resonance effects are important for the ortho- and para-substituted compounds, but only the inductive effect is important in *m*-cyanoaniline.

nonbonding electrons of the amino group are delocalized to the ring and to the cyano group

no resonance delocalization of the nonbonding electrons of the amino group to the cyano group; only the inductive effect

delocalization of the nonbonding electrons of the amino group to the ring and to the cyano group; inductive effect of the cyano group in the para position is weaker than when it is in the ortho position.

16.8

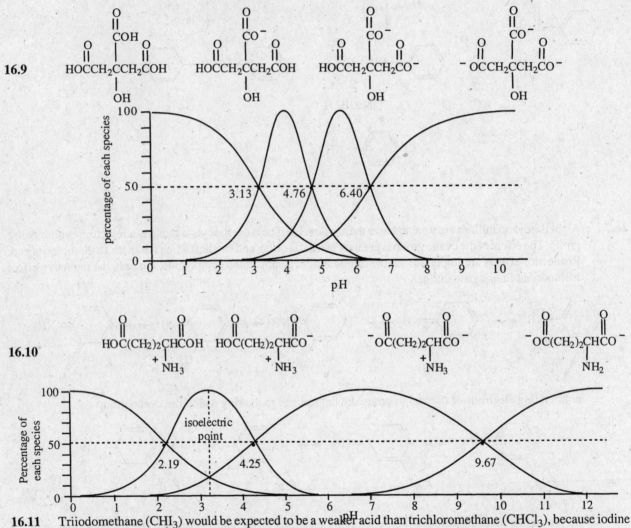

The carboxylic acid group on carbon 3 is likely to have the lowest pK_a value since it is closest to the electron withdrawing effect of the hydroxyl group.

16.9

16.10

16.11 Triiodomethane (CHI_3) would be expected to be a weaker acid than trichloromethane ($CHCl_3$), because iodine is less electronegative than chlorine and cannot stabilize the conjugate base of triiodomethane as effectively as chlorine stabilizes the trichloromethyl anion.

The strongly electronegative (3.0) chlorine atoms withdraw electron density from the carbon atom, stabilizing the anion.

The iodine atoms have about the same electronegativity as carbon (2.4 vs. 2.5), and, therefore, do not really have a negative inductive effect. Any stabilization comes from the presence of atoms larger than hydrogen, which help to spread the charge away from the carbon and over the whole ion in a way that the small hydrogen cannot.

16.12 (a)

$$HO-CH_2-C(=O)-O-H \quad :O-H^- \longrightarrow \left[HO-CH_2-C(=O)-O:^- \longleftrightarrow HO-CH_2-C=O \right]$$

$$+ H_2O$$

The anion produced by removal of the proton of the carboxyl group is stabilized by resonance. The anion that would be produced by removal of the proton on the hydroxyl group would not be stabilized by resonance.

(b)

$$CH_3CH_2CH-\overset{+}{N}(=O)(-O:^-) \quad H \quad :O-H^- \longrightarrow \left[CH_3CH_2CH-\overset{+}{N}(=O)(-O:^-) \longleftrightarrow CH_3CH_2CH=\overset{+}{N}(-O:^-)(-O:^-) \right]$$

$$+ H_2O$$

The anion produced by the removal of the α-hydrogen atom is stabilized by resonance.

(c)

$$HOCH_2CH_2CH_2-S-H \quad :O-H^- \longrightarrow HOCH_2CH_2CH_2-S:^- \quad + \quad H_2O$$

The removal of the proton from the thiol group is easier than the removal of a proton from the hydroxyl group. The larger size of the sulfur atom means that the negative charge is more spread out and, therefore, better stabilized than if it were on an oxygen atom.

(d)

$$:Br-C(Br)(=O)-CCH_3 \quad H \quad :O-H^- \longrightarrow \left[:Br-C(Br)(=O)-CCH_3 \longleftrightarrow :Br-C(Br)(-O:^-)=CCH_3 \right]$$

$$+ H_2O$$

The hydrogen atom on the α-carbon atom with the bromine atoms is more acidic than the hydrogen on the other α-carbon atom because the electron-withdrawing effect of the bromine atoms, as well as resonance, stabilizes the anion that results from its loss.

16.13 (a)

tautomers

(b)

resonance contributors

(c)

$$CH_2=\overset{+}{N}(-O:^-)(-O:^-) \longleftrightarrow :CH_2-\overset{+}{N}(=O)(-O:^-)$$

resonance contributors

(d)

$$H-C(H)(H)-N^{2+}(-O:^-)(-O:^-) \rightleftharpoons H-C(H)=\overset{+}{N}(-O-H)(-O:^-)$$

tautomers

Concept Map 16.1   Enolization.

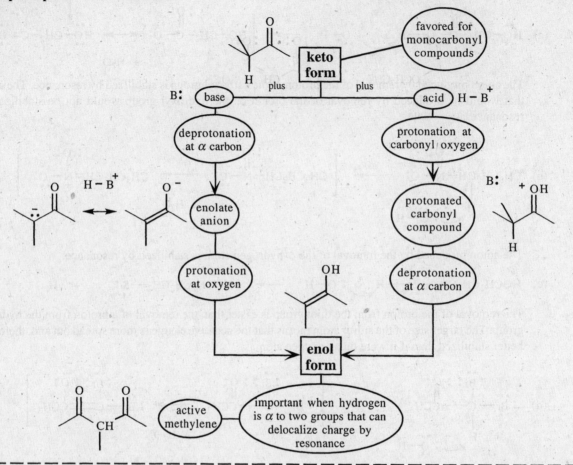

- -

16.14   (a) ethyl acetate

ethyl acetate
pK_a 23

(b) ethyl acetoacetate

an active
methylene group

ethyl acetoacetate
pK_a 11.0

16.14 (b) (cont)

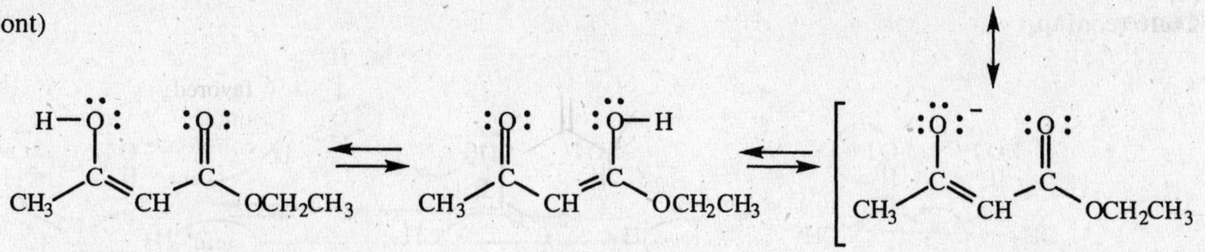
CH3
OCH2CH3
CH

(c) ethyl nitroacetate

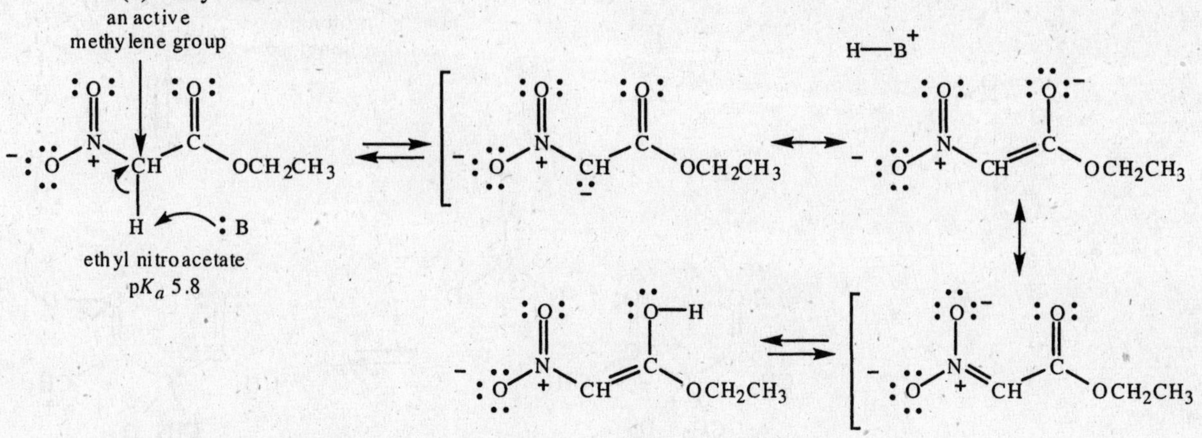
an active
methylene group
ethyl nitroacetate
pKa 5.8
H—B+
OCH2CH3

(d) ethyl cyanoacetate

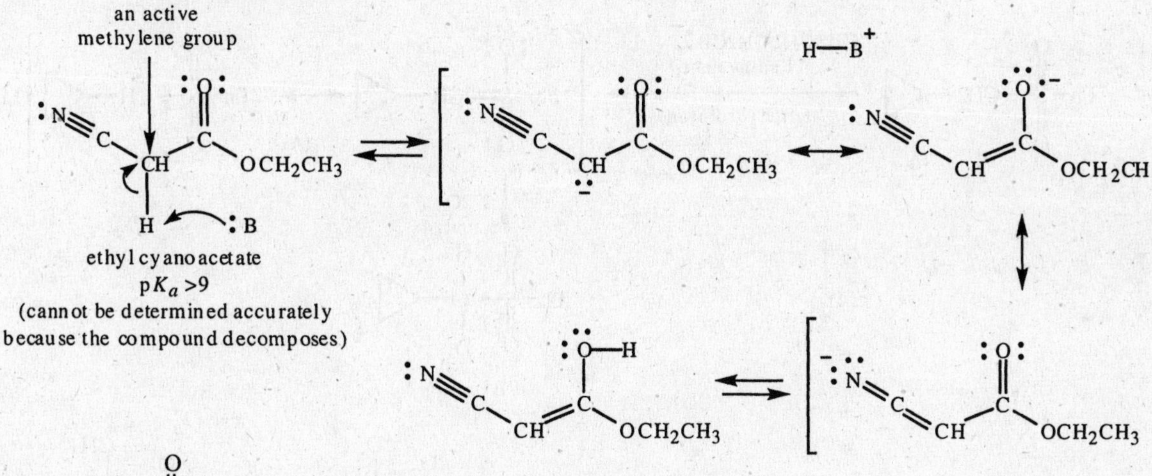
an active
methylene group
ethyl cyanoacetate
pKa >9
(cannot be determined accurately
because the compound decomposes)
H—B+
OCH2CH3

16.15 (a)

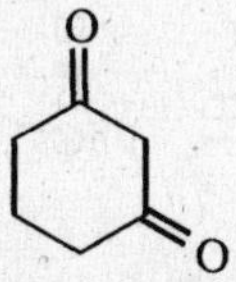

(active methylene)

(b) $PhC(=O)CH_2C(=O)CH_3$ active methylene

(c) $CH_3C(=O)OCH_2CH_3$ (more electrophilic carbonyl carbon atom)

(d) $CH_3OC(=O)CH_2C(=O)OCH_3$ (active methylene)

16.16

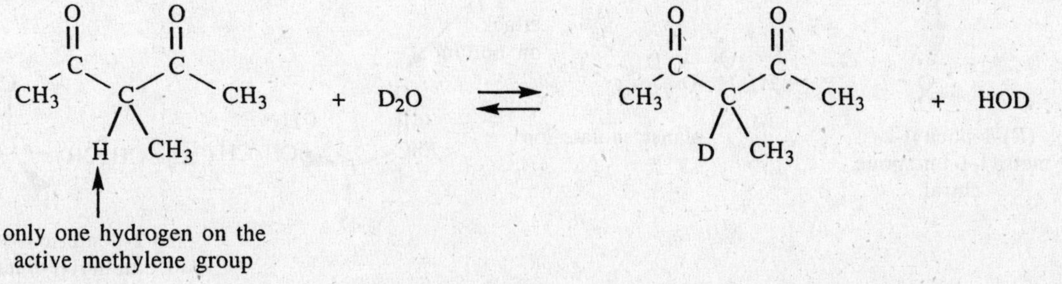
CH3
H
CH3
+ D2O
D
+ HOD
only one hydrogen on the
active methylene group

16.16 (cont)

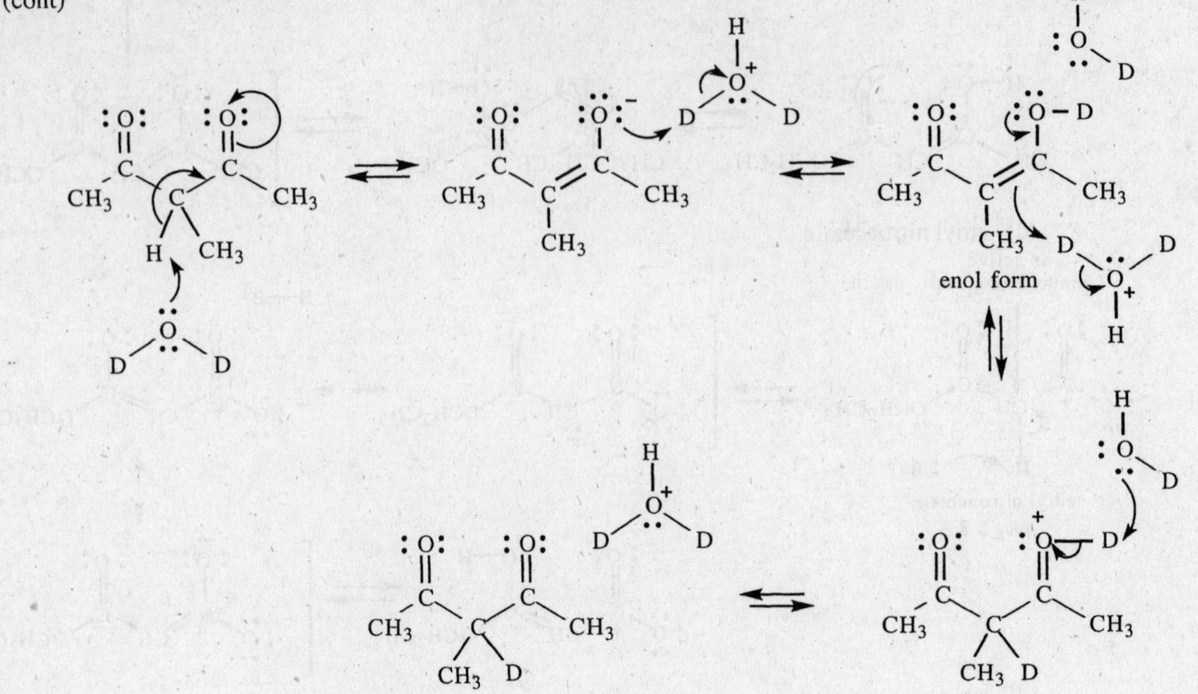
enol form

16.17

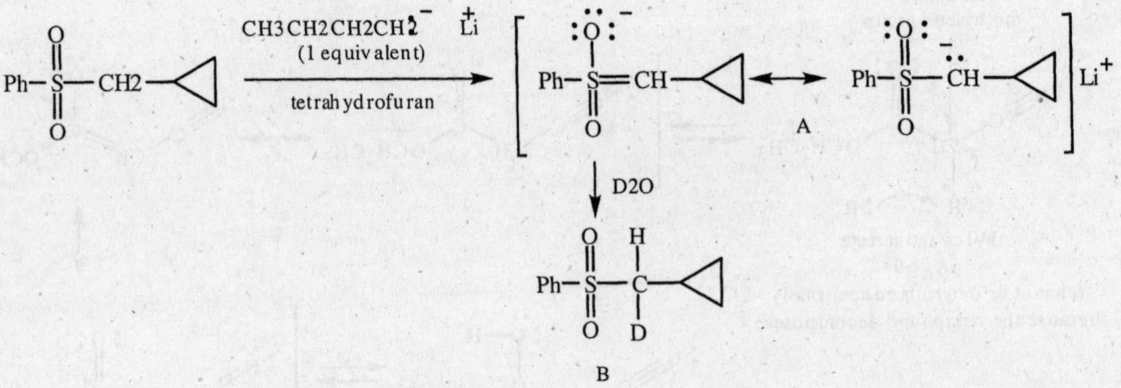
CH3CH2CH2CH2⁻ Li⁺
(1 equivalent)
tetrahydrofuran
A
D2O
B

16.18

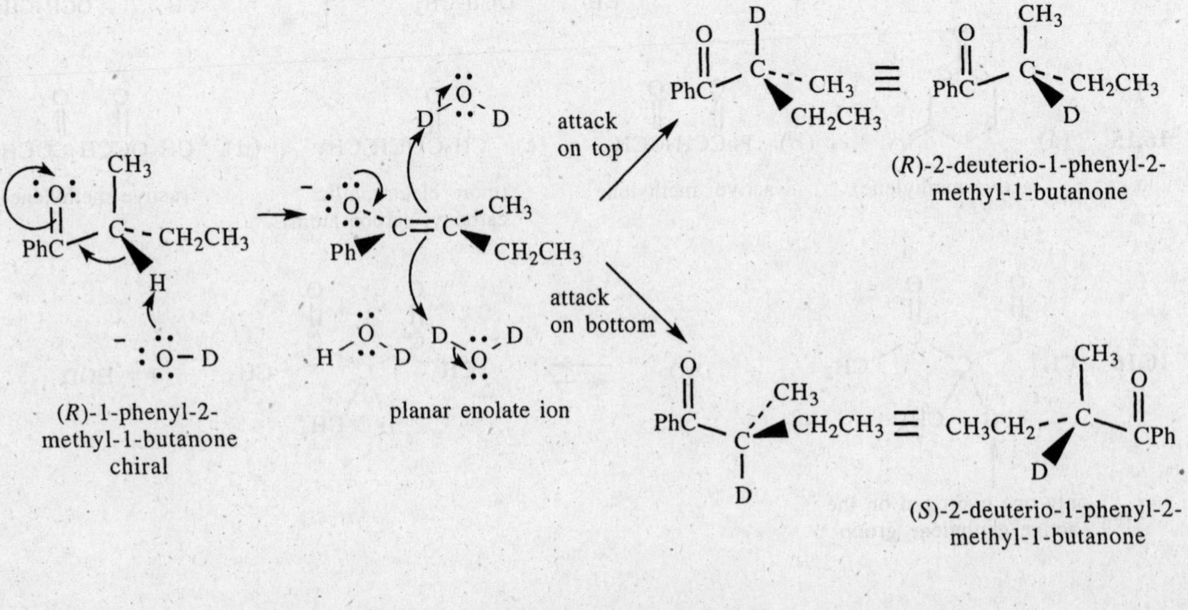
attack on top
attack on bottom
(R)-1-phenyl-2-methyl-1-butanone chiral
planar enolate ion
(R)-2-deuterio-1-phenyl-2-methyl-1-butanone
(S)-2-deuterio-1-phenyl-2-methyl-1-butanone

16.19

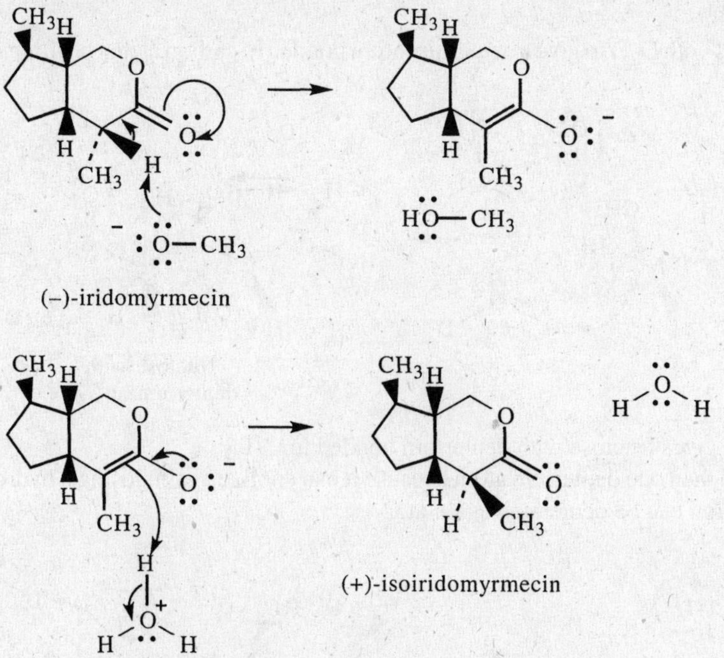

(–)-iridomyrmecin

(+)-isoiridomyrmecin

16.20   (a)

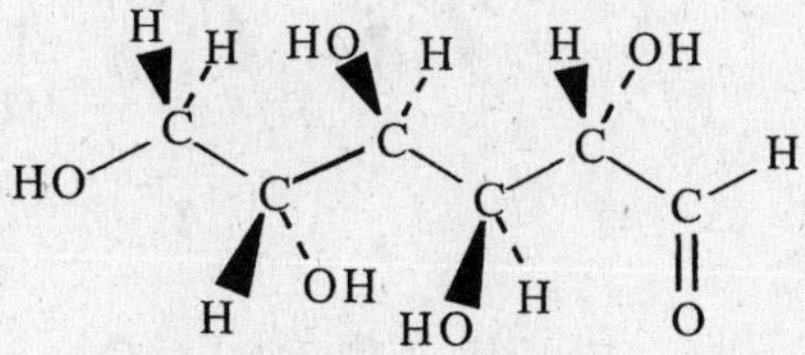

mannose

(b)

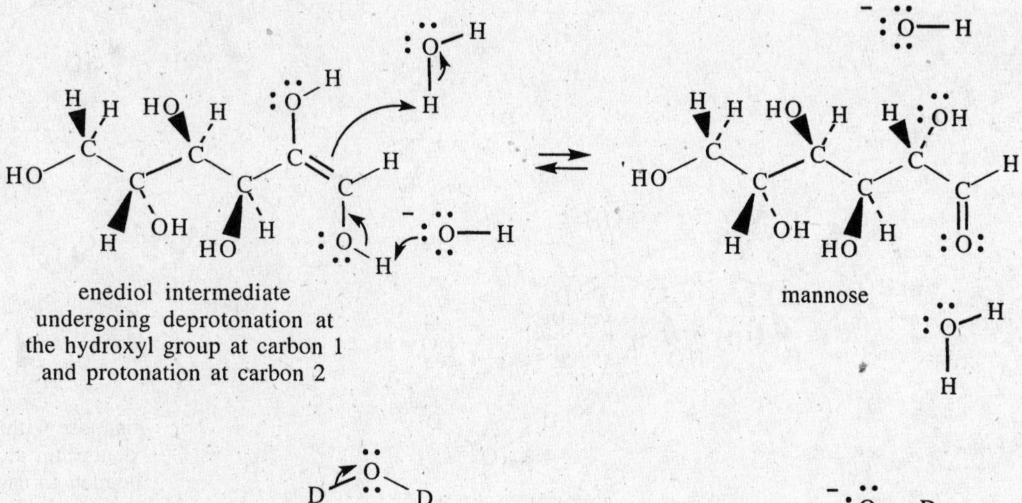

enediol intermediate
undergoing deprotonation at
the hydroxyl group at carbon 1
and protonation at carbon 2

mannose

16.21   (a)

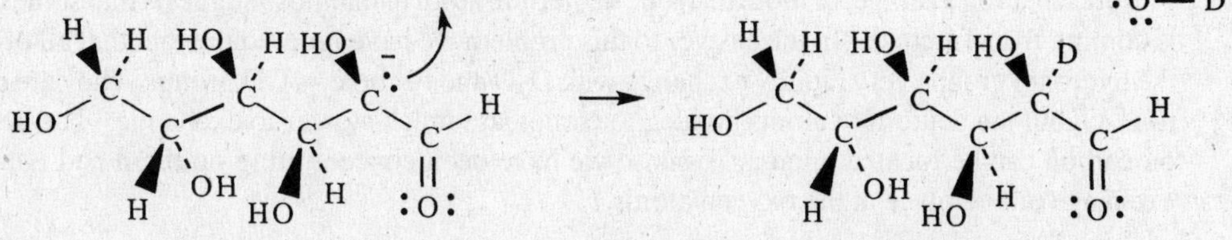

glucose with
deuterium at C-2

16.21 (a) (cont)

The carbanion at C-2 reacts with D_2O to give a deuterium atom bonded to carbon at that position in glucose or mannose.

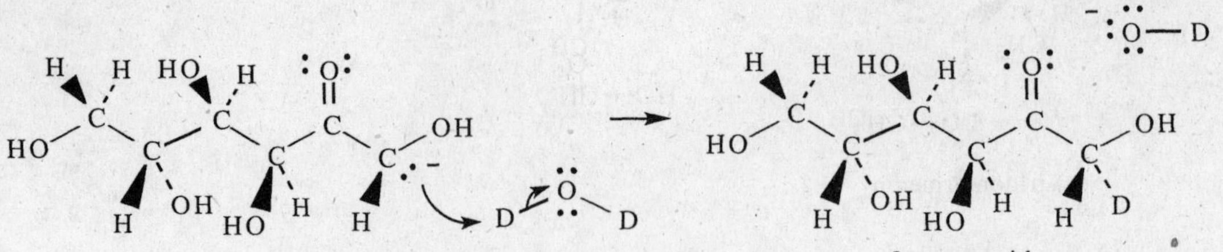

fructose with
deuterium at C-1

The carbanion at C-1 gives fructose with deuterium bonded to C-1.

(b) Fructose can pick up more than one deuterium atom because it can enolize again, losing a hydrogen atom from C-1. This enolate anion can be deuterated again at C-1.

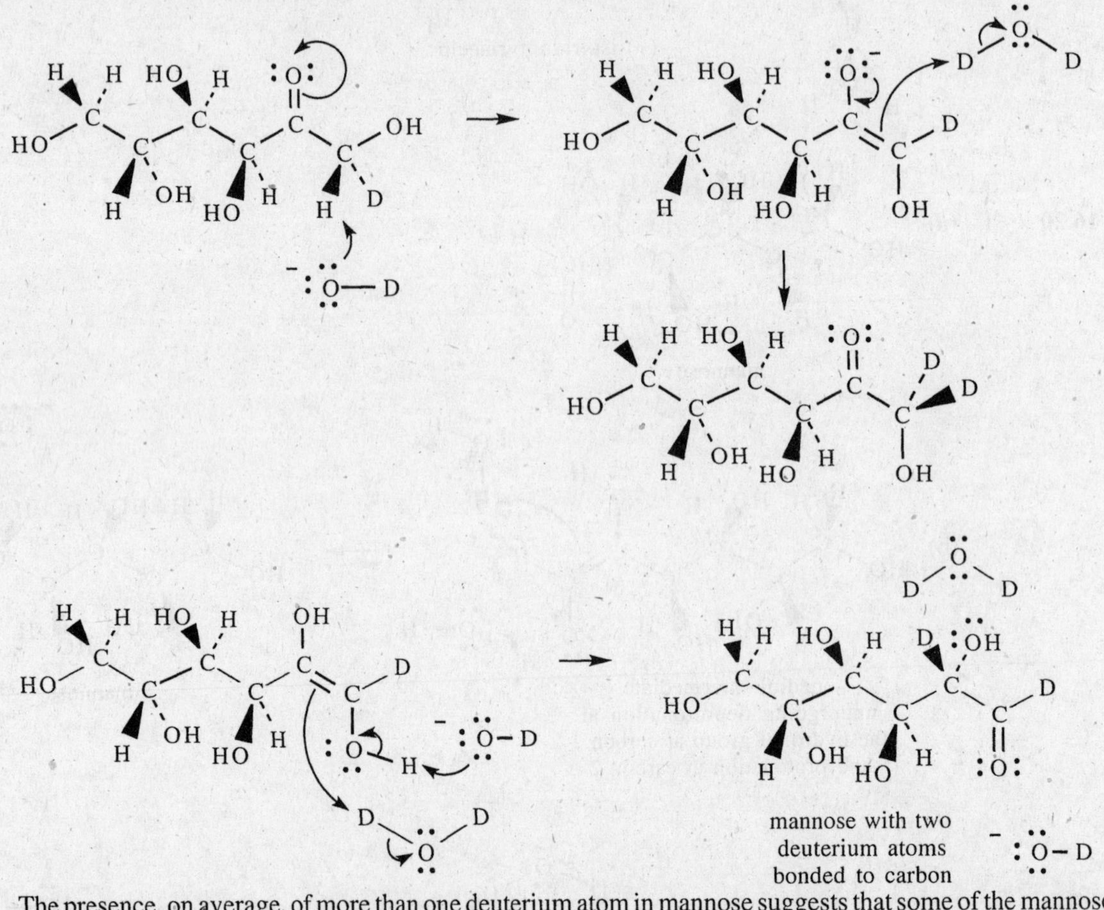

mannose with two
deuterium atoms
bonded to carbon

The presence, on average, of more than one deuterium atom in mannose suggests that some of the mannose is coming from fructose. (In the answer to this problem we have ignored the fact that all of the protons on the hydroxyl groups very rapidly exchange with D_2O and become —OD groups. They are converted back to H_2O, but the deuterium atoms bonded to carbon are much slower to exchange. The deuterium atoms on carbon can be located upon analysis so we have been concentrating on them and ignoring the rapid exchange taking place at the oxygen atoms.)

$$ROH \ + \ D_2O \ \rightleftharpoons \ ROD \ + \ HOD$$

$$ROD \ + \ H_2O \ \rightleftharpoons \ ROH \ + \ HOD$$

16.22 (a) $CH_3CCH_2CCF_3$ > $CH_3CCH_2CCH_3$ (b) >

(c) $CH_3CCH_2COCH_2CH_3$ > $CH_3CCHCOCH_2CH_3$
                                          |
                                        CH_2CH_3

(d) $O_2NCH_2NO_2$ > $O_2NCH_2CCH_3$ > $O_2NCH_2COCH_2CH_3$

(e) $HCCH_2CH$ > $CH_3CH_2OCCH_2COCH_2CH_3$ > $CH_3CH_2OCCHCOCH_2CH_3$
                                                                      |
                                                                   CH_2CH_3

16.23 The conjugate base of nitromethane is stabilized by resonance and by the positive charge next to the carbanion site. It is a much stronger acid than methane (pK_a 49), which has no additional stabilization of its conjugate base.

pK_a 10.2

Dinitromethane, $CH_2(NO_2)_2$, has two nitro groups to stabilize the carbanion and is, therefore, a stronger acid (more stable conjugate base) than nitromethane. Trinitromethane, $CH(NO_2)_3$, has three nitro groups to stabilize the carbanion and is an even stronger acid.

16.24 The conjugate bases of both diones are stabilized by resonance. The conjugate base of 1,1,1-trifluoro-2,4-pentanedione is also stabilized by the electron-withdrawing effect of the three fluorine atoms. Stabilization of the conjugate base increases the acidity of the corresponding acid.

$pK_a = 9.0$

$pK_a = 4.7$

16.25

(a)

2-(phenylsulfonyl)ethanol

16.25

(b)

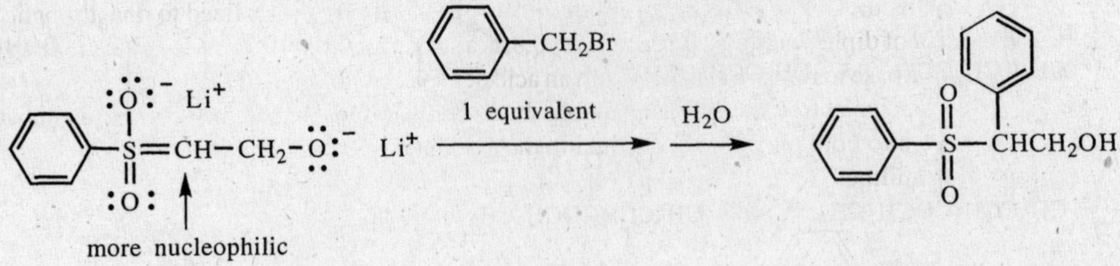

16.26

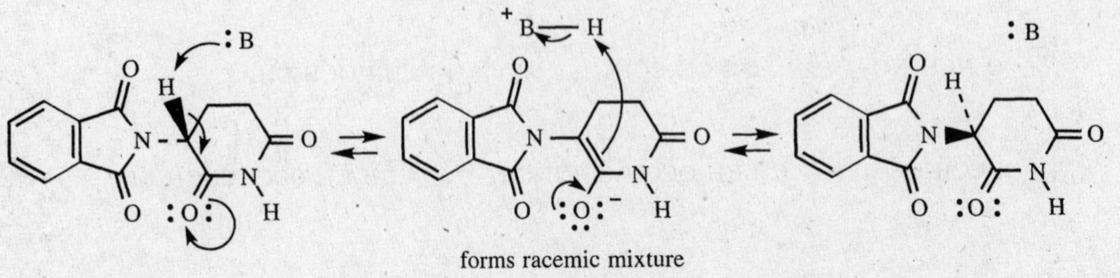

forms racemic mixture

16.27 (a) *p*-Toluidine is a slightly stronger base than aniline. This is attributed to the electron-donating effect of the methyl group. 4-Aminobenzophenone is a weaker base than aniline because the carbonyl group is an electron-withdrawing group.

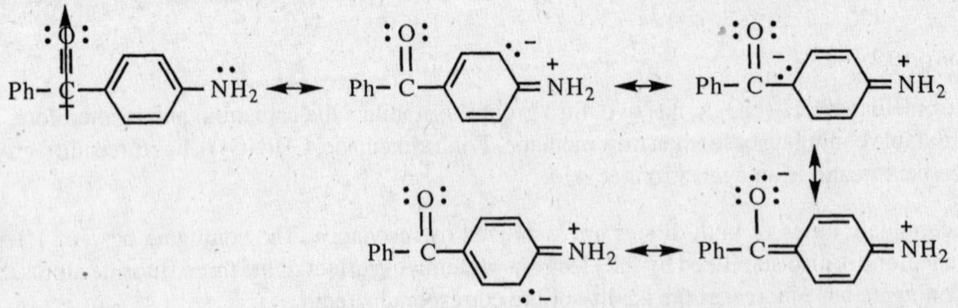

The carbonyl group decreases the availability of the nonbonding electrons on the amino group by the inductive effect and by resonance. In both series of compounds the trends can be rationalized by looking at the availability of the nonbonding electrons on the amino group in the conjugate base. The more available the electrons, the stronger the base, the weaker is the conjugate acid, and the larger the pK_a.

(b) The aromatic amine is less basic than the alkyl amines. The nonbonding electrons on the nitrogen atom are delocalized to the aromatic ring in *p*-toluidine; they are localized on the nitrogen atom in diethylamine and ethylamine.

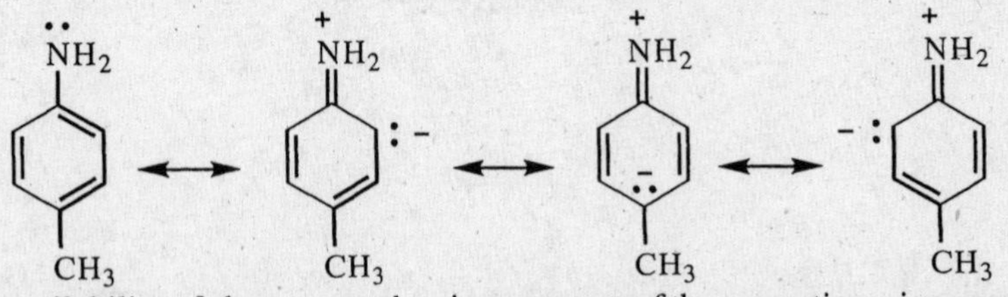

The lower availability of electrons on the nitrogen atom of the aromatic amine compared to the alkyl amine is used to rationalize its lower basicity and the increased acidity of its conjugate acid. Diethylamine, with two alkyl groups on the nitrogen, is slightly more basic than ethylamine, which has only one alkyl group on the nitrogen.

16.27 (c) The less positive (or more negative) a pK_a value is, the stronger is the acid, and the weaker is its conjugate base. In aniline, the nonbonding electrons of the nitrogen atom are delocalized to one aromatic ring. The conjugate acid of diphenylamine, with a pK_a of 0.8, is a strong acid, and diphenylamine is thus a weak base. The ability of a base to remove a proton from an acid depends on how available the electron pair of the base is, i.e., how much electron density is available. In diphenylamine, the electron pair on the nitrogen atom is delocalized into both phenyl rings, reducing the amount of electron density on the nitrogen atom even more than in aniline.

16.28 (a) pK_a 10.12 $\qquad$ CH$_3$—N—CH$_2$COH ⟵ pK_a 2.21

(b) The fully protonated structure of sarcosine will predominate below pH 2.21. The zwitterion structure will predominate at physiological pH.

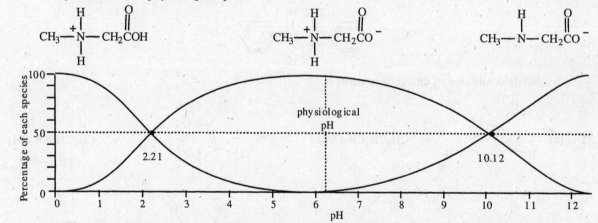

(c) The most abundant species at pH 13 is the anionic form, shown to the right on the plot above.

16.29 (a) H$_2$N—C(=NH)—NH—CH$_2$CH$_2$CH$_2$CHNH$_3^+$, CO$_2^-$ $\xrightarrow{\text{protonation}}$ ...

16.29 (b) Arginine is more basic than lysine because the conjugate acid of arginine is stabilized by resonance. The positive charge is delocalized to three nitrogen atoms in arginine. In lysine it is localized on a single nitrogen atom.

16.30 (a)

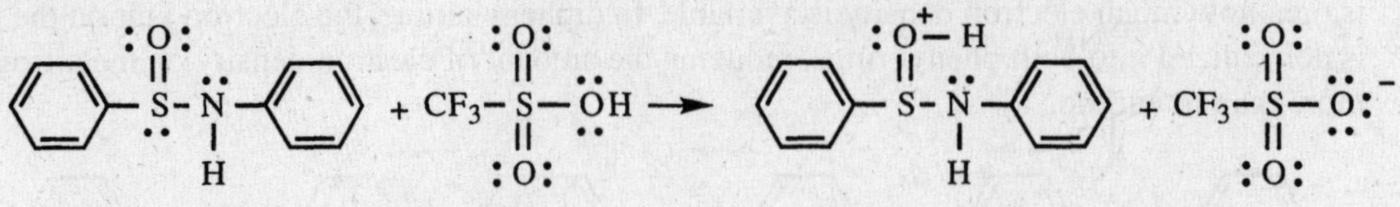

$pK_a \sim -1$

(b)

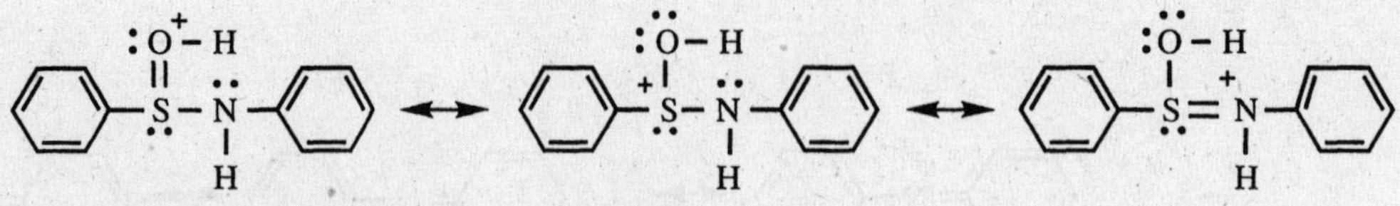

Resonance delocalization of charge is possible.

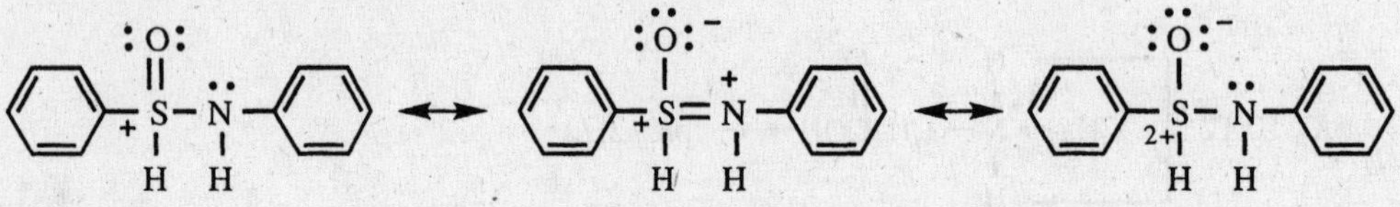

If protonation occurs at the sulfur atom, resonance delocalization of charge is possible only with separation of charge.

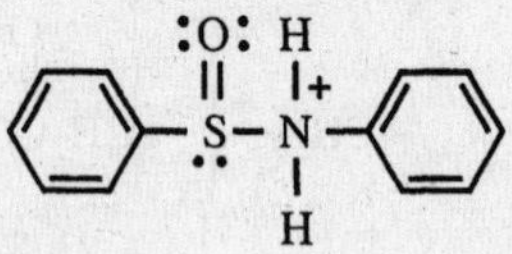

No delocalization of charge is possible.

16.31 (a)

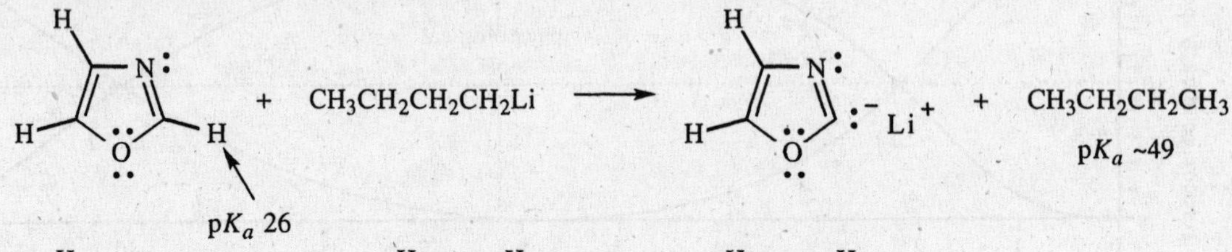

pK_a 26

$pK_a \sim 49$

(b)

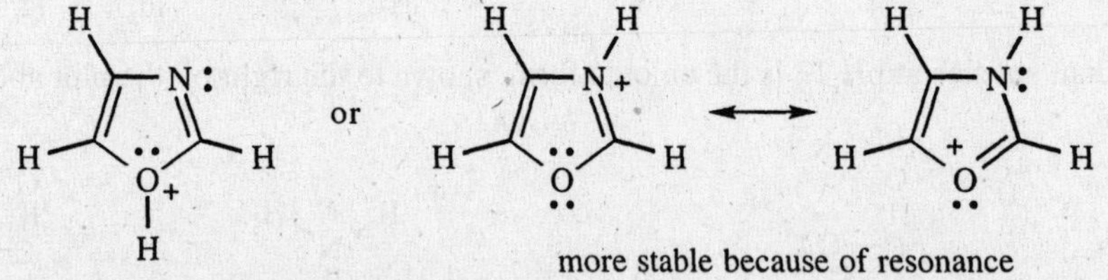

more stable because of resonance

16.32 At pH 4.15, the alkyl amine is fully protonated (pK_a of conjugate acid ~10), and is therefore not a nucleophile, while the aromatic amine (pK_a of conjugate acid ~4.5) is not, so it serves as the nucleophile in the reaction.

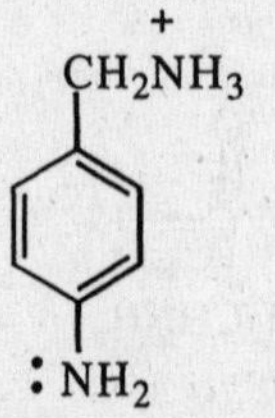

nucleophile at pH 4.15

16.32 (cont)

At pH 11.25, the alkyl amine, as well as the aromatic amine, is free to serve as the nucleophile. The alkyl amine is more nucleophilic than the aromatic amine because of the resonance delocalization of the pair of electrons on the nitrogen to the aromatic ring.

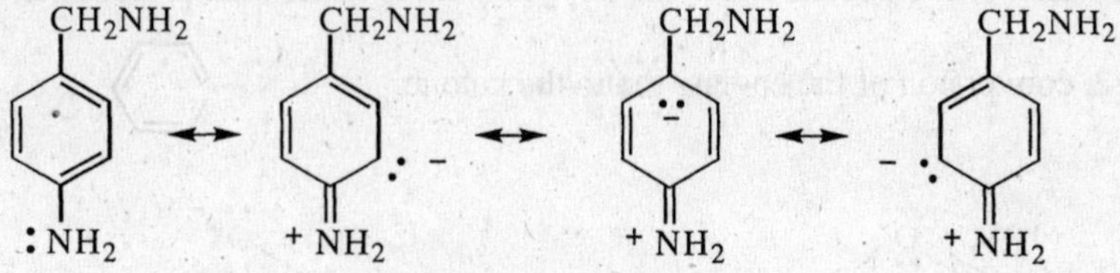

nucleophile at pH 11.25

16.33   The left hand compound is more acidic than the right hand compound. Its conjugate base is better stabilized by resonance.

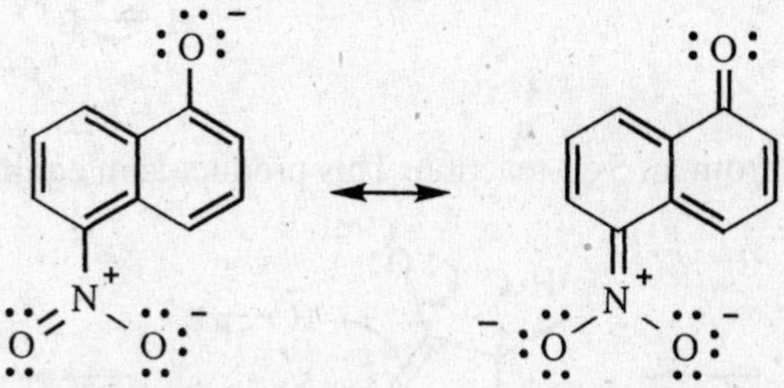

Delocalization of the negative charge onto the nitro group is not possible with the compound on the right.

16.34   (a) Diisopropylamine has pK_a ~36, therefore only butyllithium, (2), will deprotonate it almost completely.

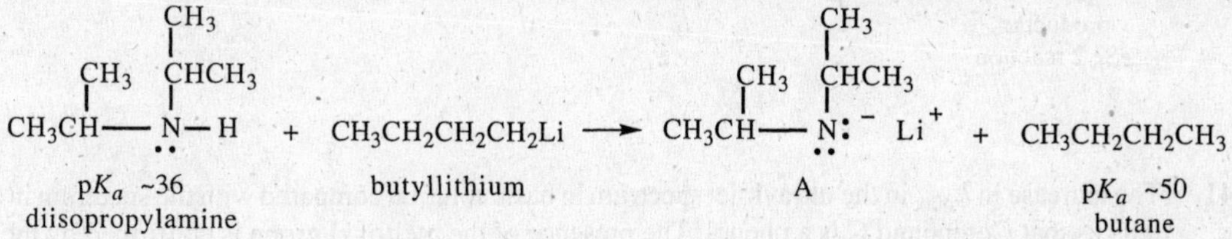

(b)

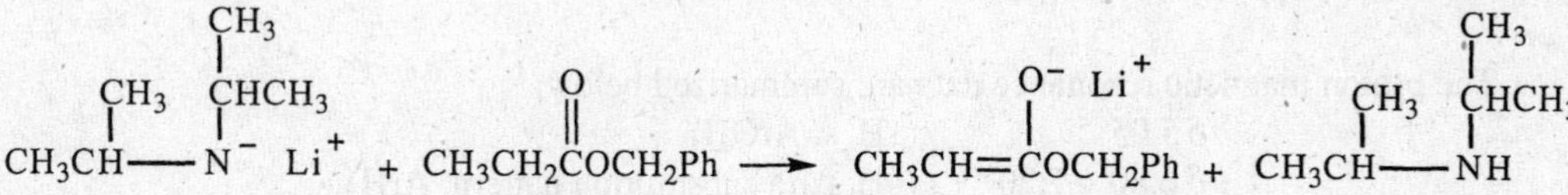

16.35

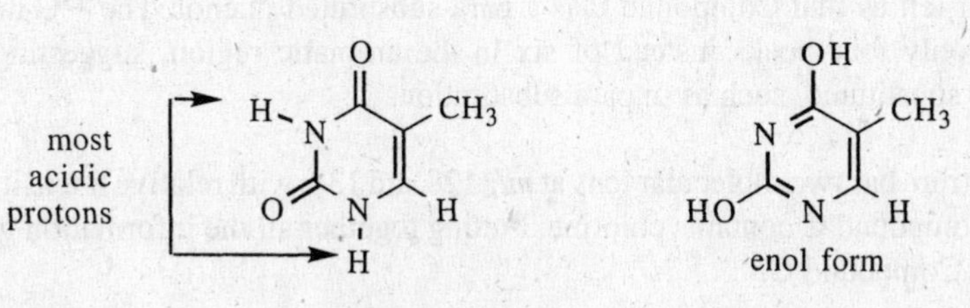

16.36   Cytochrome c will have a net positive charge at pH 6.

16.37   pyridine ring
pK_a of conjugate acid ~5

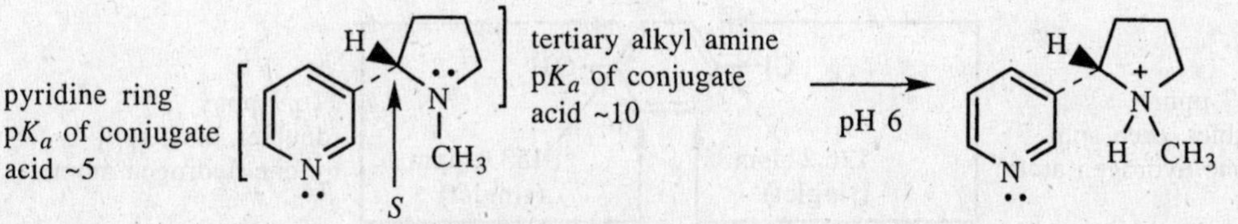

At physiological pH, nicotine will be fully protonated at the alkyl amine site but not at the pyridine site.

16.38   (a) $\Delta G_r^\circ = E - A$

   (b) $\Delta G_r^\circ < 0$

   (c) $K_{eq} > 1$

   (d) $\Delta G^\ddagger = B - A$

   (e) Protonation of the enolate ion at the oxygen atom is faster than protonation at carbon.

   (f) D

   (g) Step 2, conversion of the enolate ion to the ketone.

16.39

16.40   The initial product comes from an S_N2 reaction. This product is in equilibrium with its epimer.

product of
S_N2 reaction

16.41   The increase in λ_{max} in the ultraviolet spectrum in basic solution compared with the spectrum in acidic solution tells us that Compound C is a phenol. The presence of the hydroxyl group is confirmed by the broad band in the infrared spectrum at 3400 cm^{-1}.

The proton magnetic resonance data are summarized below:

     δ 5.95   (1H, s, ArO<u>H</u>)

    δ 6.80 – 7.20   (4H, para substitution pattern, Ar<u>H</u>)

The data so far tell us that Compound C is a para-substituted phenol. The ^{13}C nuclear magnetic resonance spectrum has only four peaks instead of six in the aromatic region, suggesting that the benzene ring is symmetrically substituted, such as in para substitution.

The mass spectrum has two molecular ions at *m/z* 128 and 130, with relative intensities of 3:1. These intensities suggest that Compound C contains chlorine. Putting together all the information gives us *p*-chlorophenol for the identity of Compound C.

The ^{13}C nuclear magnetic resonance spectrum is analyzed below:

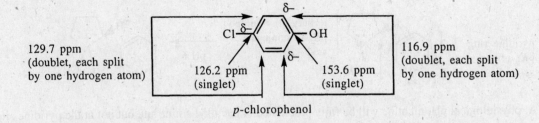

129.7 ppm
(doublet, each split
by one hydrogen atom)

126.2 ppm
(singlet)

153.6 ppm
(singlet)

116.9 ppm
(doublet, each split
by one hydrogen atom)

p-chlorophenol

16.42 In aqueous sulfuric acid, the anilinium ion, the conjugate acid of aniline will be present in solution. At pH 8, aniline will be present and the nonbonding electrons on the nitrogen are in conjugation with the aromatic ring. Extended conjugation shifts the ultraviolet absorption to longer wavelength. The spectrum on the left is that of aniline and the one on the right is that of its conjugate acid, the anilinium ion.

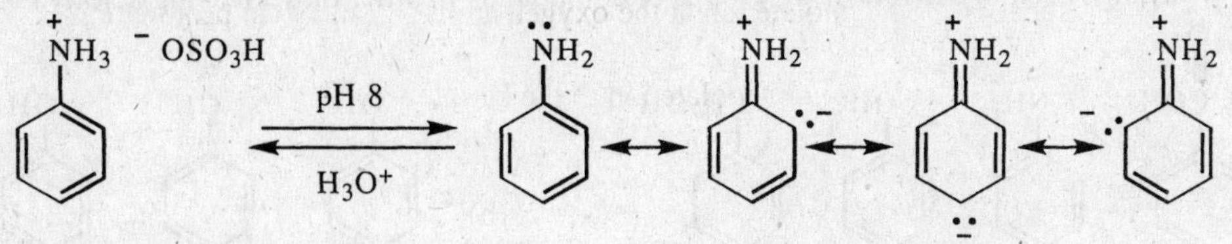

no interaction of nonbonding
electrons with aromatic ring

interaction of nonbonding
electrons with aromatic ring;
absorption at longer λ_{max}

16.43 Integration of the peaks in the proton magnetic resonance spectrum gives a ratio of 4:2:2:3. Disappearance of the broad absorption at $\delta 3.60$ when Compound D is shaken with D_2O tells us that those hydrogens must be acidic protons attached to an oxygen or a nitrogen atom. The decrease in λ_{max} in the ultraviolet spectra in acid solution compared with basic solution tells us that Compound D is an aryl amine. The two bands in the infrared spectrum at 3480 and 3376 cm^{-1} point to the presence of a primary amine, and the band at 1621 cm^{-1} confirms the presence of the aromatic ring.

The proton magnetic resonance data are summarized below:

$\delta 1.25$ (3H, t, $J = 7$ Hz, —CH$_2$CH$_3$)
$\delta 2.50$ (2H, q, $J = 7$ Hz, ArCH$_2$CH$_3$)
$\delta 3.60$ (2H, broad s, —NH$_2$)
$\delta 6.60 - 7.20$ (4H, multiplet, ArH)

The data so far tell us that Compound D is an ethylaniline. The aromatic region in the proton magnetic resonance spectrum does not show the typical splitting pattern for para substitution. Therefore, Compound D must be either ortho- or meta-substituted. The carbon-13 nuclear magnetic resonance spectrum has eight peaks, which also tells us that Compound D cannot be *p*-ethylaniline, which should have only six peaks in the ^{13}C spectrum because of the symmetry of the molecule. The data, however, do not allow us to distinguish between *o*-ethylaniline and *m*-ethylaniline.

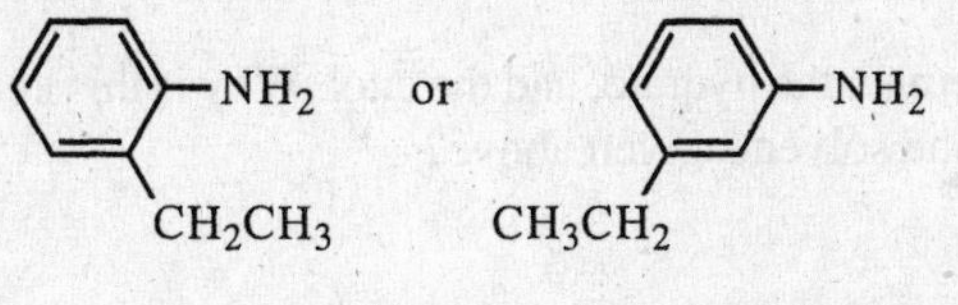

Supplemental Problems

S16.1   Arrange each of the following groups of compounds in order of decreasing acidity.

(a)

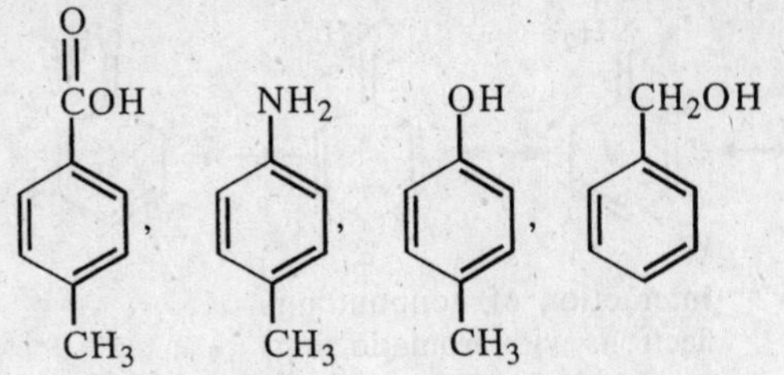

(b)

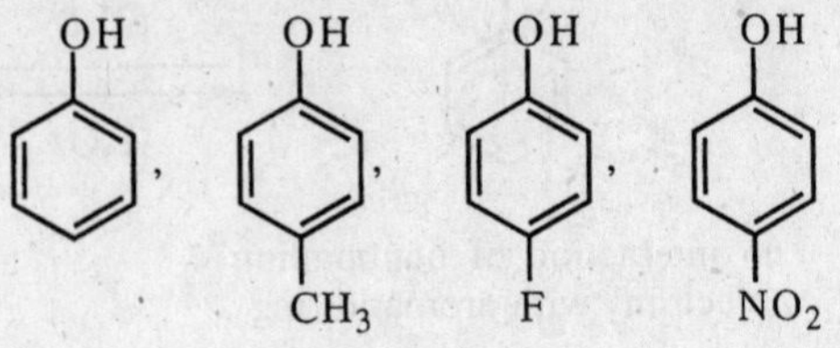

(c)

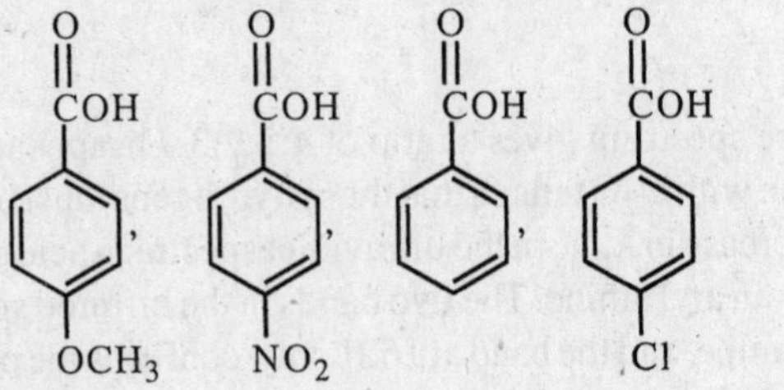

S16.2   The most acidic protons in the following compound have pK_a 13. Identify those protons, and show how you rationalize their acidity.

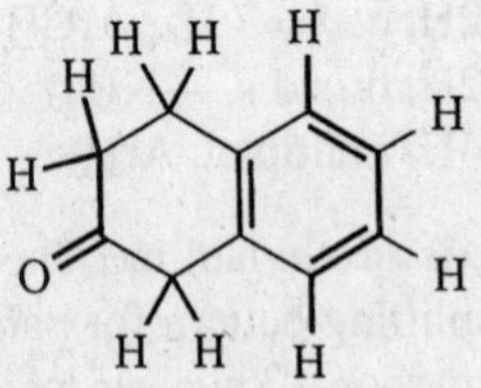

S16.3   2,3,3,4,4-Pentafluorocyclobutanone has been found to enolize completely in solvents such as acetonitrile, tetrahydrofuran, or diethyl ether. The keto form also is extremely hygroscopic, meaning that it spontaneously picks up moisture from the air to form a hydrate, which can be isolated as white crystals.

Draw structural formulas for the ketone, the hydrate, and the enol. Why is this ketone so reactive? Why is the enol form more stable than the keto form in the solvents given above?

S16.4

(a)   When Compound A is treated with a catalytic amount of sodium methoxide in methanol, Compound B is formed. This is an equilibrium process, where B is favored for steric reasons. Draw a complete mechanism, using the curved arrow convention, for the conversion of A to B.

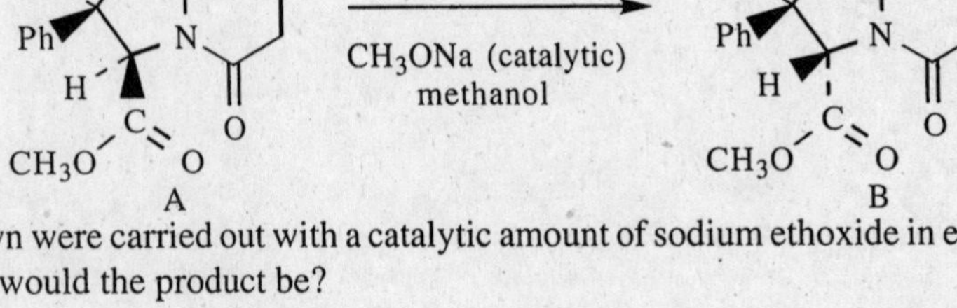

CH₃ONa (catalytic)
methanol

(b)   If the reaction shown were carried out with a catalytic amount of sodium ethoxide in ethanol instead of the reagent given above, what would the product be?

S16.5   The following reactions were carried out in the synthesis of vitamin D_3 derivatives that are deuterated. Provide structural formulas for the products obtained. In these products, all readily exchangeable hydrogen atoms have been substituted by deuterium.

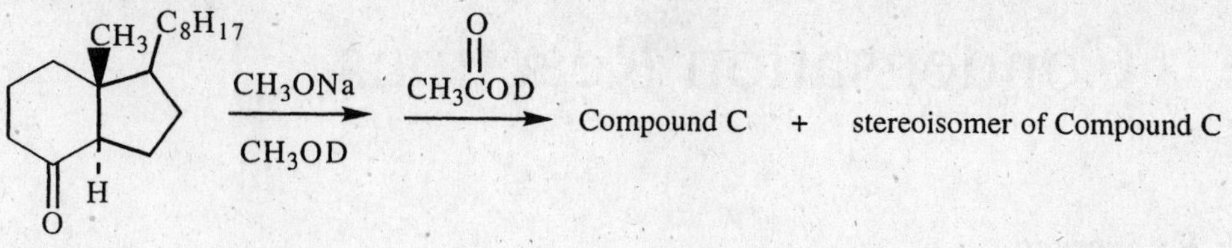

$$\xrightarrow[\text{CH}_3\text{OD}]{\text{CH}_3\text{ONa} \quad \text{CH}_3\overset{\text{O}}{\overset{\|}{\text{C}}}\text{OD}} \text{Compound C} \quad + \quad \text{stereoisomer of Compound C}$$

S16.6   Research into compounds that prevent the clotting of blood involve the following reactions. Provide structural formulas for the missing compounds.

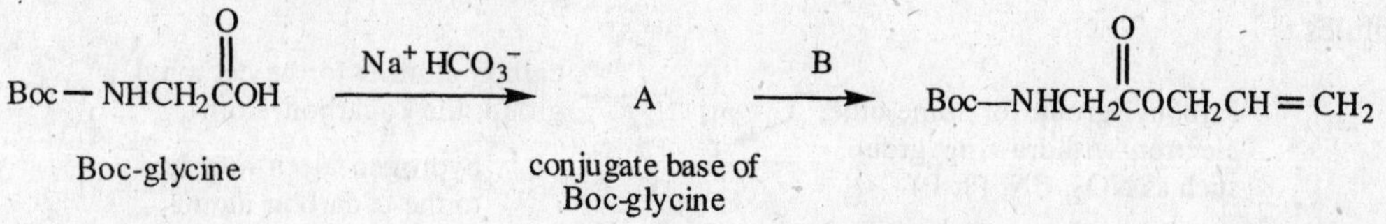

$$\text{Boc}-\text{NHCH}_2\overset{\text{O}}{\overset{\|}{\text{C}}}\text{OH} \xrightarrow{\text{Na}^+ \text{HCO}_3^-} \text{A} \xrightarrow{\text{B}} \text{Boc}-\text{NHCH}_2\overset{\text{O}}{\overset{\|}{\text{C}}}\text{OCH}_2\text{CH}=\text{CH}_2$$

Boc-glycine   conjugate base of
                                             Boc-glycine

S16.7   Phosphorus-oxygen double bonds, $-\overset{|}{\underset{|}{\text{P}}}=\text{O}$, and sulfur-oxygen double bonds, $\diagdown\text{S}=\text{O}$, behave like

carbonyl groups (carbon-oxygen double bonds, $\diagdown\text{C}=\text{O}$) in stabilizing anions on carbons adjacent to them. Chemists take advantage of this fact in the synthesis of interesting compounds.

(a)   Given this information, complete the following equation and predict whether the equilibrium will lie to the left or to the right by how you draw the equilibrium arrows. The pK_a table will be useful.

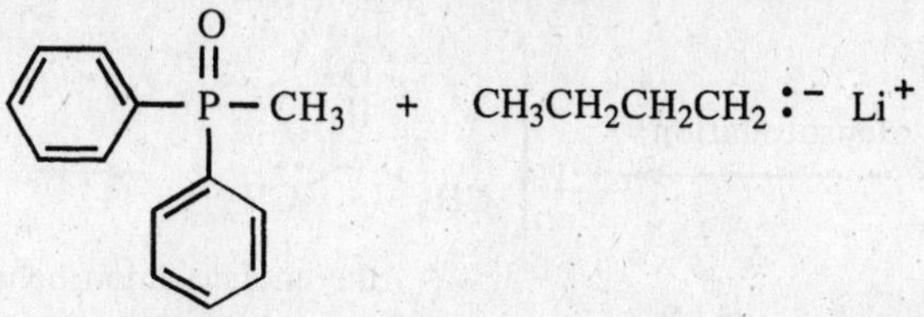

$$\text{P}-\text{CH}_3 \quad + \quad \text{CH}_3\text{CH}_2\text{CH}_2\text{CH}_2\!:^- \text{Li}^+$$

(b)   How do chemists rationalize the stability of the conjugate base of the phosphorus compound?

17 Enols and Enolate Anions as Nucleophiles. Alkylation and Condensation Reactions

Workbook Exercises

The most efficient way for you to understand the subject matter in Chapter 17 is to seek out the appropriate analogies in your earlier work. There are two fundamental ideas in this chapter that build on concepts from Chapter 16:

1. Carbon atoms connected to carbonyl groups, as well as to other electron-withdrawing groups, can behave as nucleophiles.

carbonyl group (or some other electron-withdrawing group, such as NO_2, CN, Ph_3P)

carbon atom α to the carbonyl group; the α-carbon atom

hydrogen atom attached to the α-carbon atom; the α-hydrogen atom

The carbon atom α to a carbonyl group can behave as a nucleophile in two ways:

as the uncharged tautomer; the enol form

as the anionic conjugate base; the enolate form

2. These carbon nucleophiles undergo the characteristic reactions that you already associate with nucleophiles: substitution, addition to carbonyl groups, and acylation. As before, reviewing these concepts from earlier chapters will benefit your learning.

EXAMPLE

deprotonation

the enolate anion behaves primarily as a carbon nucleophile

$ICH_2CH_2CH_2CH_3$

(S_N2, see Ch. 7)

$CH_3 \quad CH_2CH_2CH_2CH_2CH_3$ + I^-

Ph H

(addition, see Ch. 14)

$CH_3 \quad CH_2CHPh$ O$^-$

protonation

$CH_3 \quad CH_2CHPh$ OH

Ph OCH$_3$

(acylation, see Ch. 15)

$CH_3 \quad CH_2 \quad Ph$ + $^-OCH_3$

394

Workbook Exercises (cont)

EXERCISE I. For each of the following compounds, draw the structure of the enol and two resonance forms for the enolate anion . If there is more than one enol/enolate anion possible, show all of the possibilities.

(a) 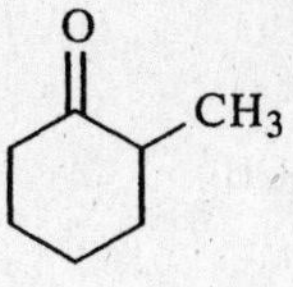

(b) PhC(=O)CH$_2$CH$_3$

(c) CH$_3$CH$_2$C(=O)OCH$_3$

(d) 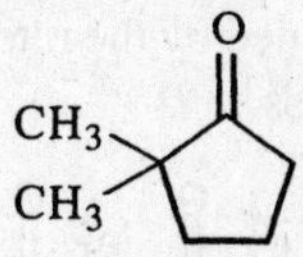

(e) PhCH$_2$C(=O)H

(f) CH$_3$C(=O)CH$_2$C(=O)CH$_3$

EXERCISE II. Use each of the enolate nucleophiles you created in Exercise I in substitution, addition, and acylation reactions with the three electrophiles shown in the example on the previous page ($CH_3CH_2CH_2CH_2I$, PhCHO, and $PhCO_2CH_3$). You have seen the mechanisms for each of these types of reactions in earlier chapters, so you should be able to draw mechanisms for the reactions you write.

Concept Map 17.1 Reactions of enols and enolates with electrophiles.

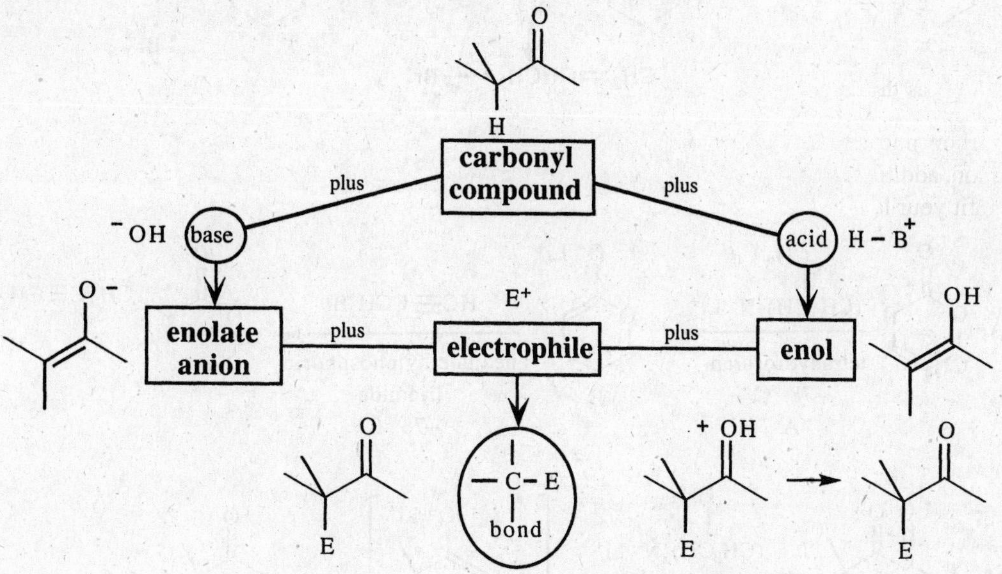

E^+ can be D_2O, X_2, RX, RC(=O)X, RC(=O)OR', RC(=O)OC(=O)R

17.1 (a) 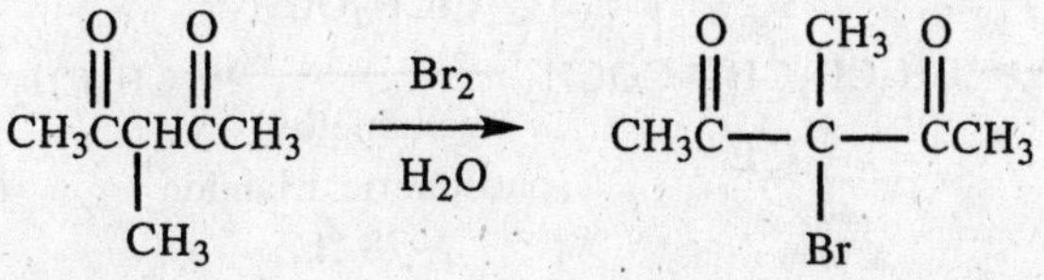

17.1   (b) 3-Methyl-2,4-pentanedione will enolize in water. This reaction is an equilibrium reaction, and after a day, the equilibrium concentration of enol is present. The reaction with bromine is fast because the electrophile reacts at the double bond of the enol until the equilibrium concentration of the enol is used up.

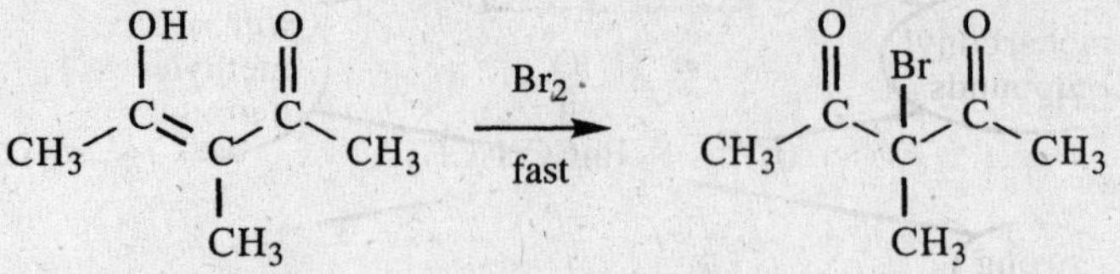

Once the enol has been used up, the rate of the overall reaction is determined by the rate at which the enol is formed. That is the slow, rate-determining step of this two-step process.

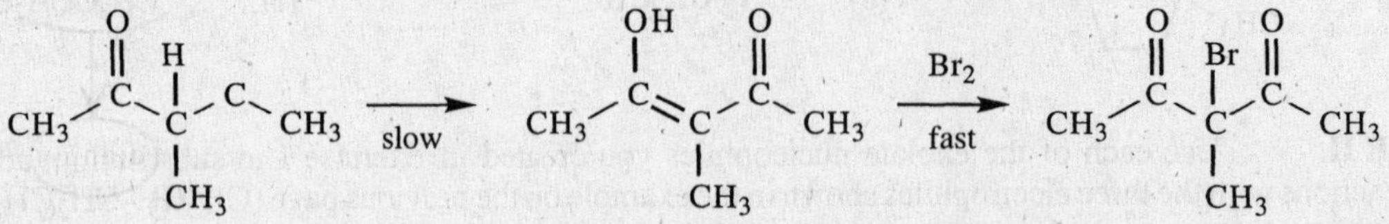

17.2

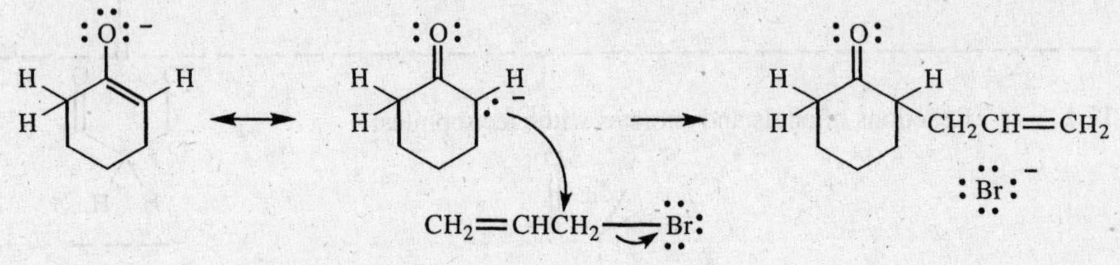

17.3   (a)

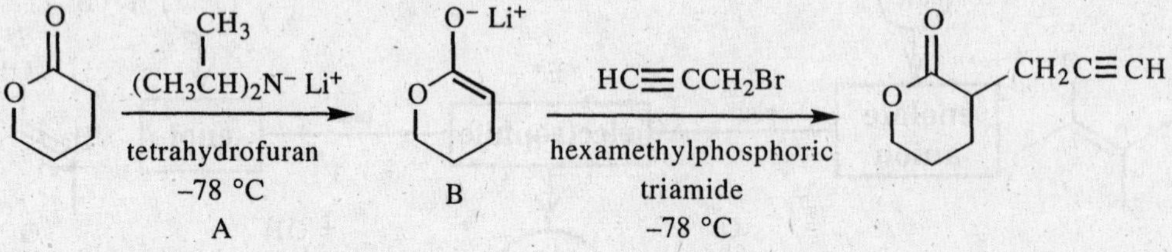

(b)

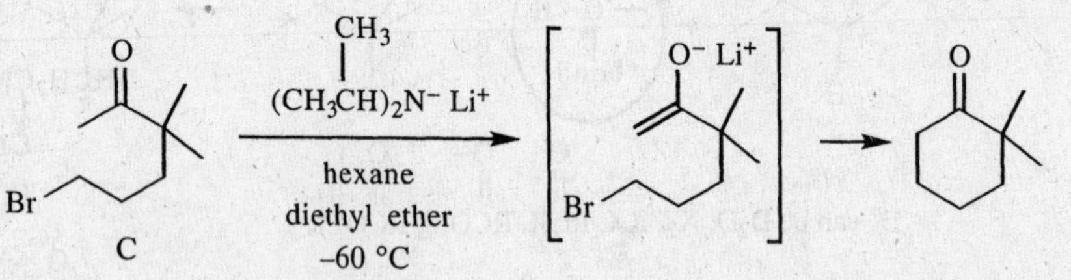

(c) $CH_3CH_2CH_2COCH_3$

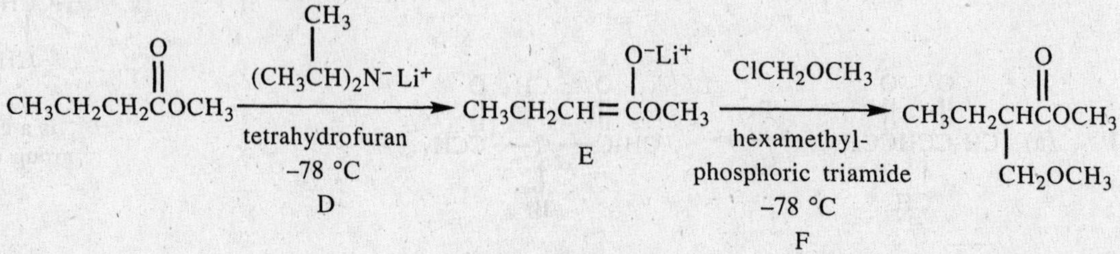

Concept Map 17.2   Alkylation reactions.

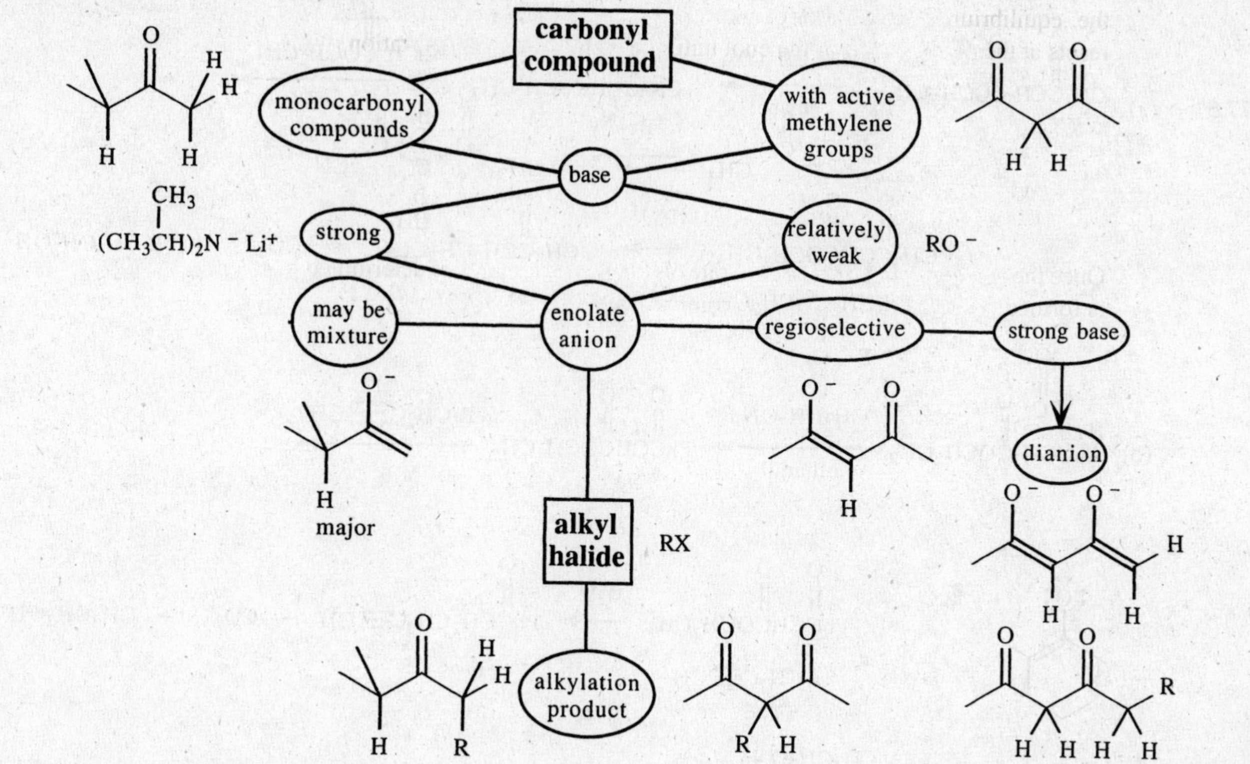

carbonyl compound

monocarbonyl compounds

with active methylene groups

base

strong

relatively weak   RO⁻

$(CH_3CH)_2N^- Li^+$

may be mixture

enolate anion

regioselective

strong base

major

alkyl halide   RX

dianion

alkylation product

(Br⁻ is a better leaving group than Cl⁻)

17.4 (a) $CH_2(COCH_2CH_3)_2$ $\xrightarrow[\text{ethanol}]{CH_3CH_2ONa}$ $Na^+ \ ^-CH(COCH_2CH_3)_2$ $\xrightarrow{CH_3CHCH_3 \ (Br)}$ $CH_3CHCH(COCH_2CH_3)_2 \ (CH_3)$

A

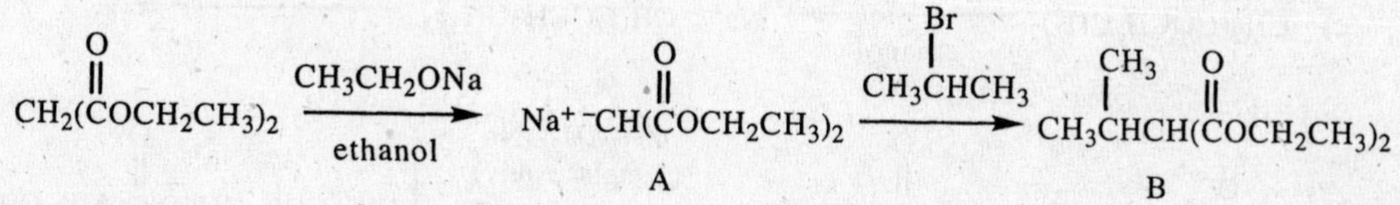

B

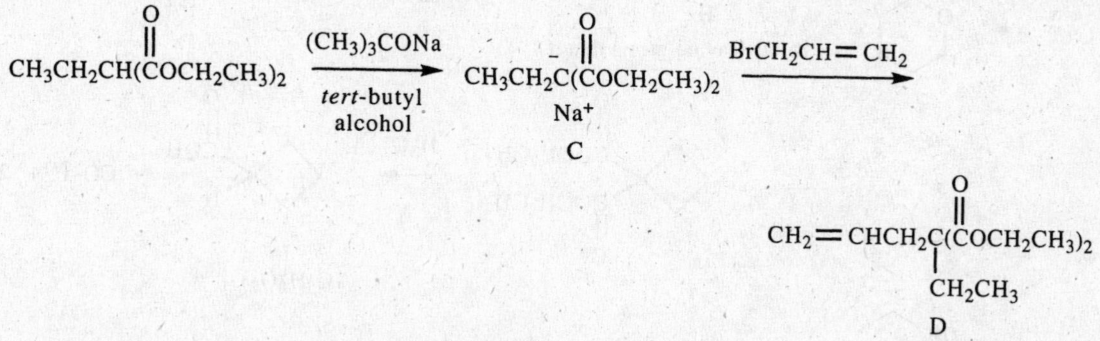

(b) $CH_3CH_2CH(COCH_2CH_3)_2$ $\xrightarrow[\substack{\textit{tert}\text{-butyl} \\ \text{alcohol}}]{(CH_3)_3CONa}$ $CH_3CH_2C(COCH_2CH_3)_2 \ Na^+$ $\xrightarrow{BrCH_2CH=CH_2}$

C

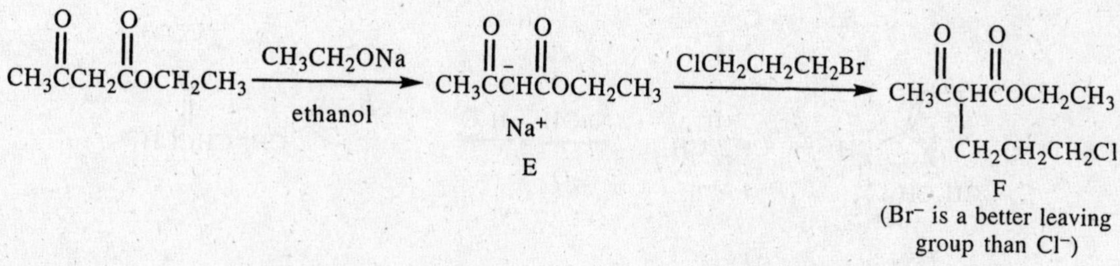

$CH_2=CHCH_2C(COCH_2CH_3)_2$
$\qquad\qquad\qquad\quad CH_2CH_3$

D

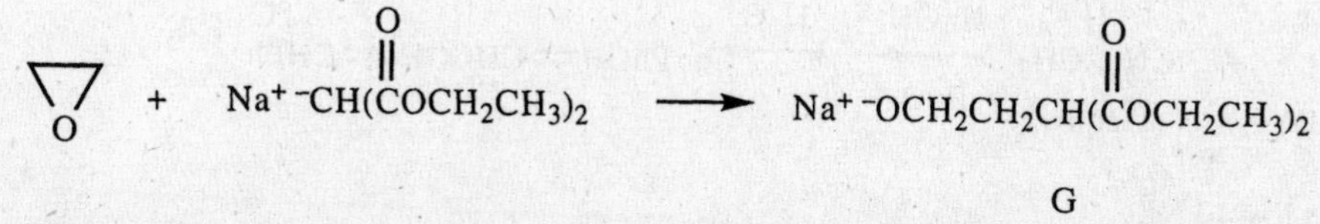

(c) $CH_3CCH_2COCH_2CH_3$ $\xrightarrow[\text{ethanol}]{CH_3CH_2ONa}$ $CH_3CCHCOCH_2CH_3 \ Na^+$ $\xrightarrow{ClCH_2CH_2CH_2Br}$ $CH_3CCHCOCH_2CH_3$
$\qquad\qquad\qquad\qquad\qquad\qquad\qquad\qquad\qquad\qquad\qquad\qquad\qquad\qquad CH_2CH_2CH_2Cl$

E

F
(Br⁻ is a better leaving group than Cl⁻)

(d) ▷O + $Na^+ \ ^-CH(COCH_2CH_3)_2$ $\longrightarrow$ $Na^+ \ ^-OCH_2CH_2CH(COCH_2CH_3)_2$

G

17.5 (a)

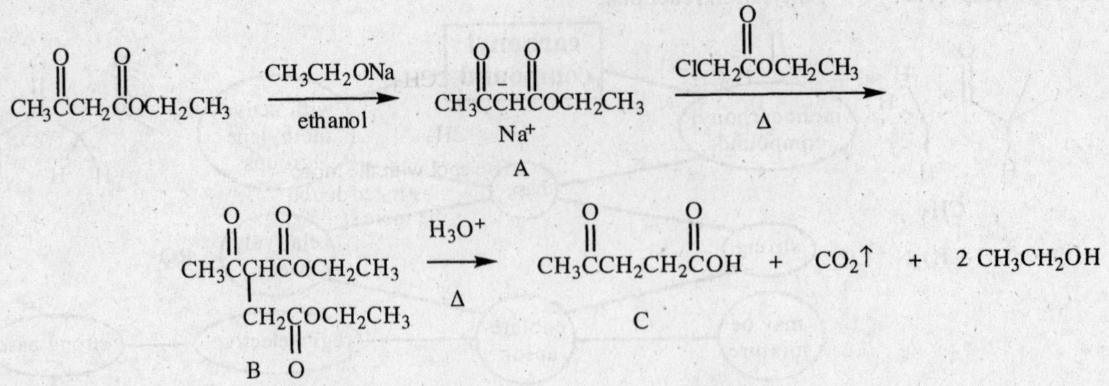

(b)

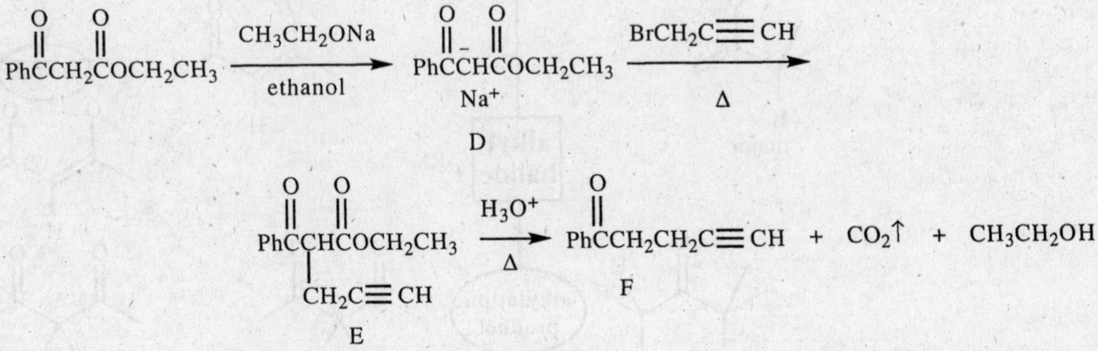

(c)

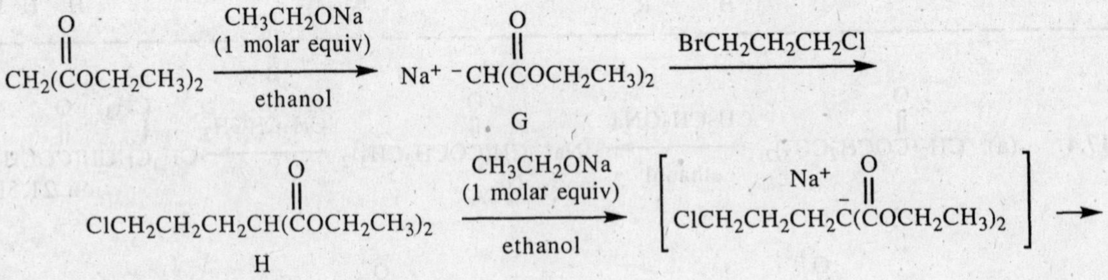

(Br $^-$ is a better leaving group than Cl $^-$)

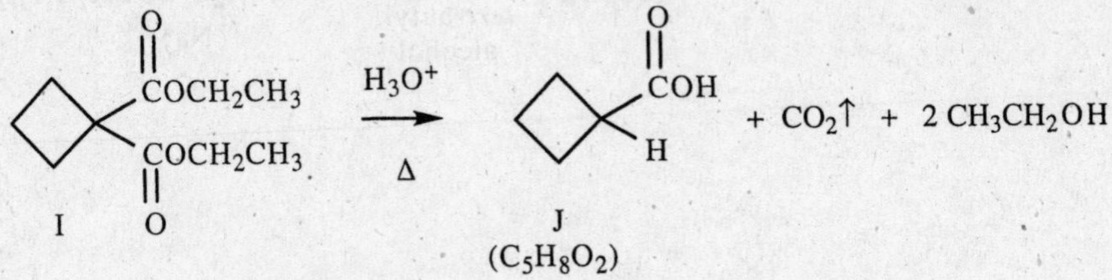

17.6 (a)

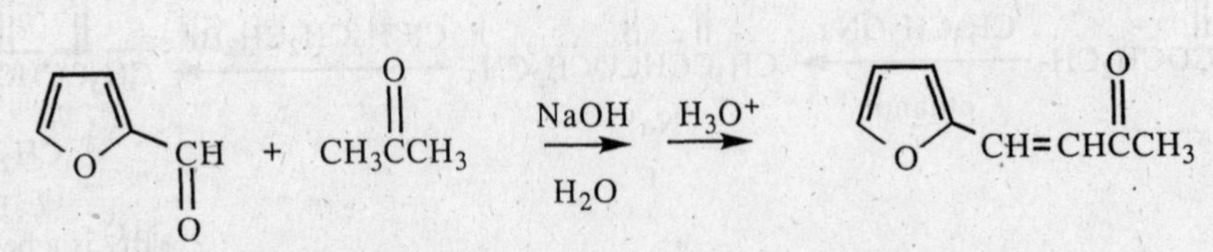

(b)

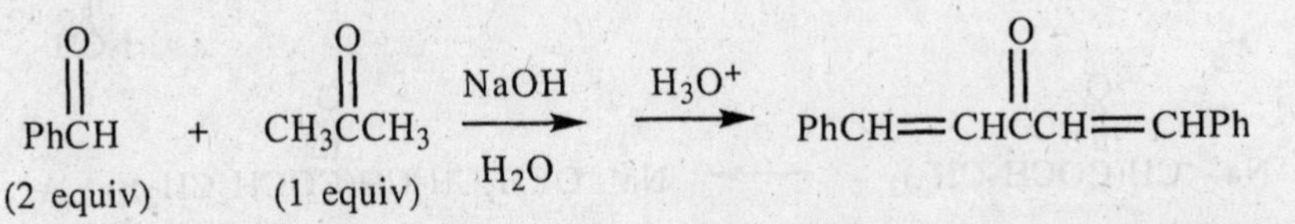

17.6   (c)

$$\underset{PhCH}{\overset{O}{\|}} + \underset{CH_3CCH_2CH_3}{\overset{O}{\|}} \xrightarrow{HCl} PhCH = \underset{\underset{CH_3}{|}}{\overset{O}{\|}}CCCH_3$$

(The enol with the more
highly substituted double
bond is the more stable one.)

(d) $\underset{(2\ equiv)}{\overset{O}{\underset{}{\|}}{PhCH_2CH}}$ $\xrightarrow[\text{H}_2\text{O}]{\text{NaOH}}$ $\underset{}{\overset{\overset{O}{\|}}{\overset{CH}{\|}}}$ PhC$=$CHCH$_2$Ph

----

Concept Map 17.3   The aldol condensation.

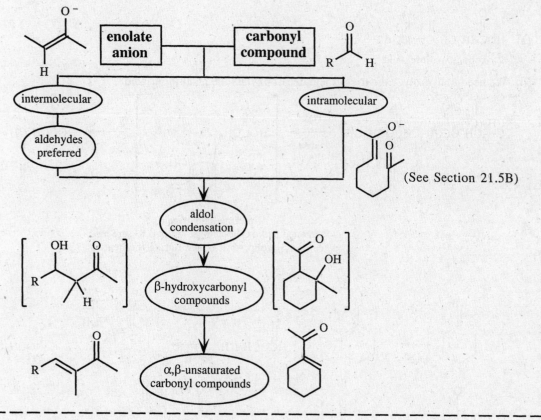

17.7   (a)

17.7 (b)

$$CH_3CH_2CH_2CH_2\overset{\displaystyle HO}{\underset{|}{CH}}\!-\!\!\!\!\{\!-\!\!\underset{\underset{\textstyle CH_2CH_2CH_3}{|}}{CH}\overset{\displaystyle O}{\overset{\|}{C}}H \longleftarrow 2\ CH_3CH_2CH_2CH_2\overset{\displaystyle O}{\overset{\|}{C}}H$$

(c)

$$\text{(cyclohexene ring with } \overset{O}{\overset{\|}{C}}\!-\!C(CH_3)_3 \text{ and Ph substituents)} \longleftarrow PhCCH_2CH_2CH_2CH_2CH_2C\overset{\displaystyle O}{\overset{\|}{}}\!-\!\underset{\underset{\textstyle CH_3}{|}}{\overset{\overset{\textstyle CH_3}{|}}{C}}CH_3$$

(d)

$$\underset{\underset{\textstyle H}{|}}{\overset{\overset{\textstyle Ph}{|}}{C}}=C=\underset{\underset{\textstyle CPh}{|}}{\overset{\overset{\textstyle H}{|}}{C}} \longleftarrow PhCH \overset{O}{\overset{\|}{}} + CH_3CPh \overset{O}{\overset{\|}{}}$$

17.8 (a)

$$HO\overset{O}{\overset{\|}{C}}CH_2CH_2\overset{O}{\overset{\|}{C}}\!-\!\overset{O}{\overset{\|}{C}}OH$$

2-oxopentanedioic acid

(b) We need to hydrolyze the triester and decarboxylate the resulting triacid.

$$CH_3CH_2O\overset{O}{\overset{\|}{C}}CH_2\overset{}{\underset{\underset{\textstyle \overset{O}{\overset{\|}{C}}\!-\!\overset{O}{\overset{\|}{C}}OCH_2CH_3}{|}}{CH}}\overset{O}{\overset{\|}{C}}OCH_2CH_3 \xrightarrow[\substack{H_2SO_4 \\ \Delta}]{H_2O} \left[HO\overset{O}{\overset{\|}{C}}CH_2\underset{\underset{\textstyle \overset{O}{\overset{\|}{C}}\!-\!COH}{|}}{CH}\!-\!COH \right] \longrightarrow HO\overset{O}{\overset{\|}{C}}CH_2CH_2\overset{}{\underset{O}{\underset{\|}{C}}}\!-\!\underset{O}{\underset{\|}{C}}OH$$

intermediate triacid;
screened carboxylic acid group is β to screened
carbonyl group, so it is lost during the heating

17.9 (a)

(cyclohexanone) $\xrightarrow{\text{base}}$ (cyclohexenolate) $+$ $HCOCH_2CH_3 \overset{O}{\overset{\|}{}} \longrightarrow$ (2-formylcyclohexanone) $+\ {}^-OCH_2CH_3$

(b) (cyclohexenolate) $+\ CH_3CH_2O\overset{O}{\overset{\|}{C}}\!-\!\overset{O}{\overset{\|}{C}}OCH_2CH_3 \longrightarrow$ (cyclohexanone with $\overset{O}{\overset{\|}{C}}\!-\!\overset{O}{\overset{\|}{C}}OCH_2CH_3$) $+\ {}^-OCH_2CH_3$

(c) (cyclohexenolate) $+\ CH_3CH_2O\overset{O}{\overset{\|}{C}}OCH_2CH_3 \longrightarrow$ (cyclohexanone with $\overset{O}{\overset{\|}{C}}OCH_2CH_3$) $+\ {}^-OCH_2CH_3$

17.10   (a)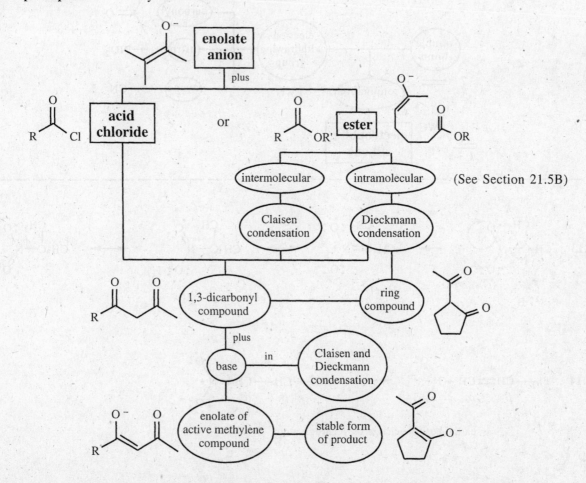

(b)

(c)

17.11

Concept Map 17.4   Acylation reactions of enolates.

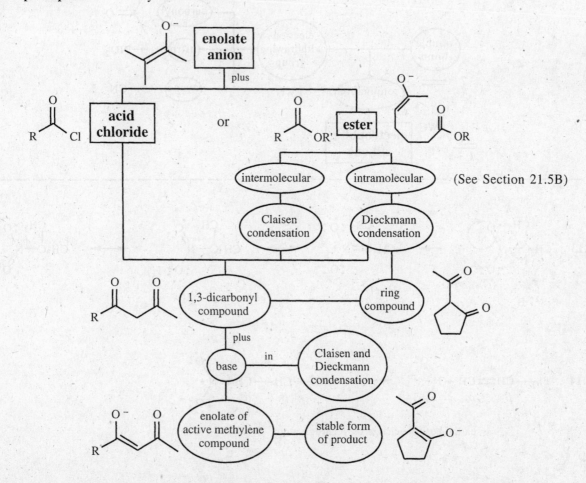

17.12 (a)

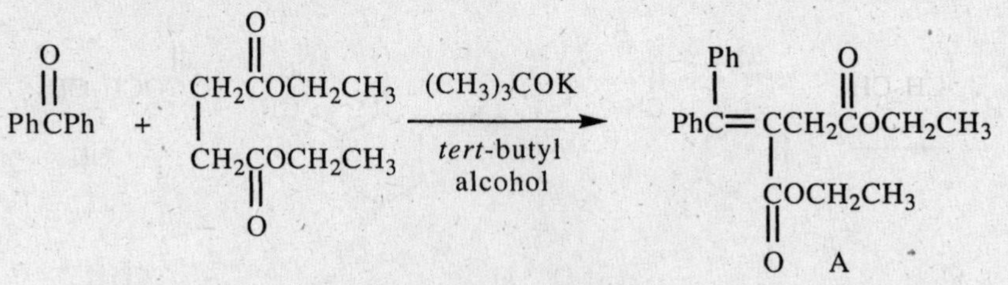

(b)

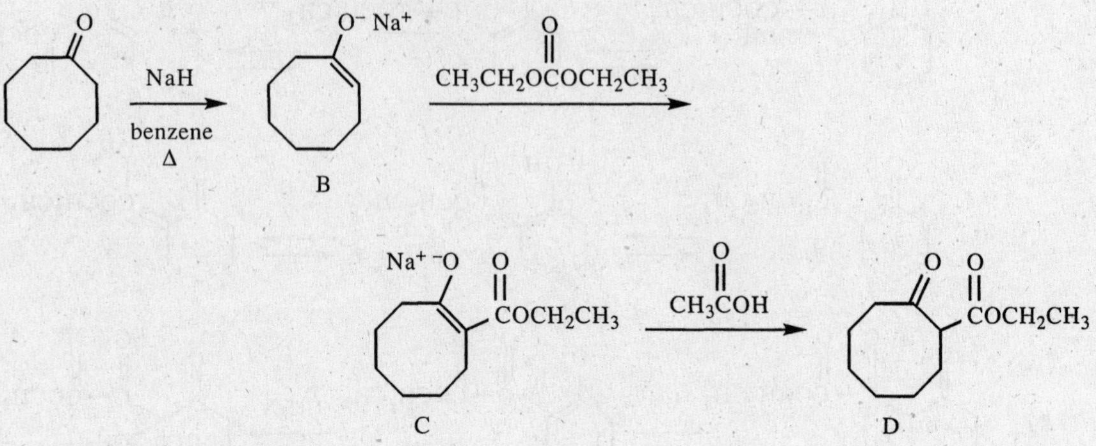

Concept Map 17.5   Electrophilic alkenes.

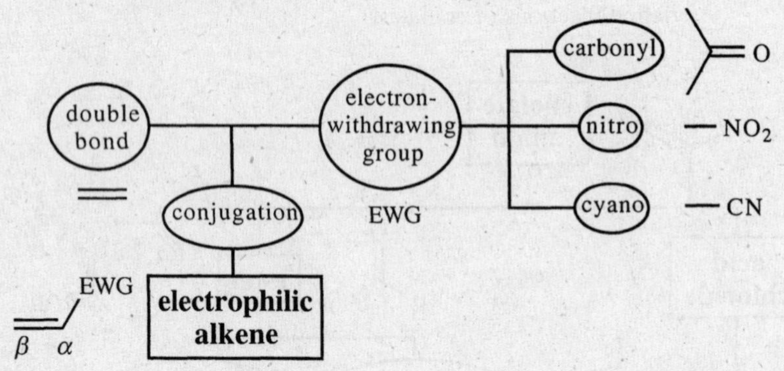

17.13

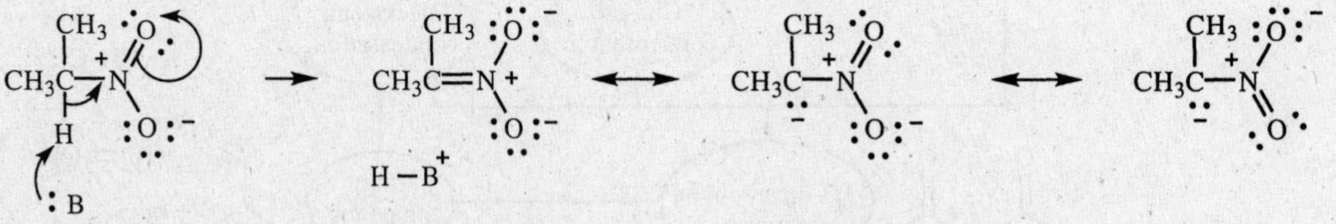

17.14

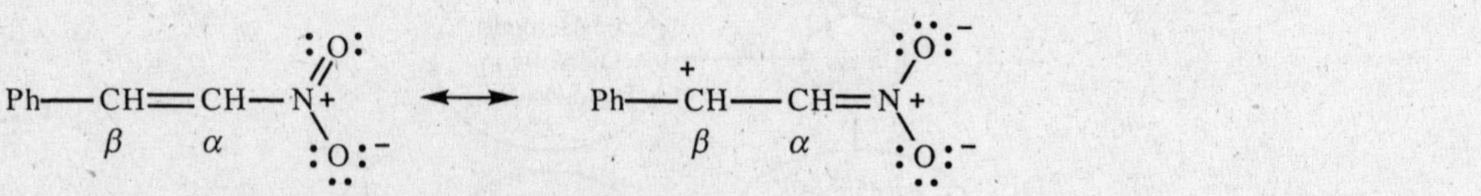

partial positive charge on β-carbon atom

17.15 (a) CH₃CHCCH₂COCH₂CH₃ + CH₂=CCCH₃ $\xrightarrow[\text{ethanol}]{\text{KOH}}$ CH₃CHCCHCOCH₂CH₃

(b) N≡CCH₂COCH₃ + CH₂=CHCPh $\xrightarrow[\text{methanol}]{\text{CH}_3\text{ONa}}$ PhCCH₂CH₂CHCOCH₃

(c) $\xrightarrow{\text{NaNH}_2}$ $\xrightarrow{\text{CH}_2=\text{CHC}\equiv\text{N}}$

(The more highly substituted enolate anion and the one in which the double bond is conjugated with the aromatic ring is the more stable one.)

(d) CH₃NO₂ + CH₃C=CHCCH₃ $\xrightarrow{\text{CH}_3\text{CH}_2\text{NHCH}_2\text{CH}_3}$ O₂NCH₂CCH₂CCH₃

17.16

17.17 (a)

$$CH_3C{=}CHCCH_3 + CH_3NHCH_3 \longrightarrow CH_3N{-}CCH_2CCH_3$$

A   B

(b)

$$CH_2{=}CHC{\equiv}N + NH_3 \xrightarrow{H_2O} N{\equiv}CCH_2CH_2NHCH_2CH_2C{\equiv}N$$

(2 equivalents)   C

(c)

$$PhSH + CH_2{=}CHCOCH_3 \xrightarrow{CH_3ONa} PhSCH_2CH_2COCH_3$$

D   E

(d)

$$PhC{\equiv}CCPh + CH_3CH_2NHCH_2CH_3 \xrightarrow[\text{ether}]{\text{diethyl}} PhC{=}CHCPh$$

(1 equiv)

F

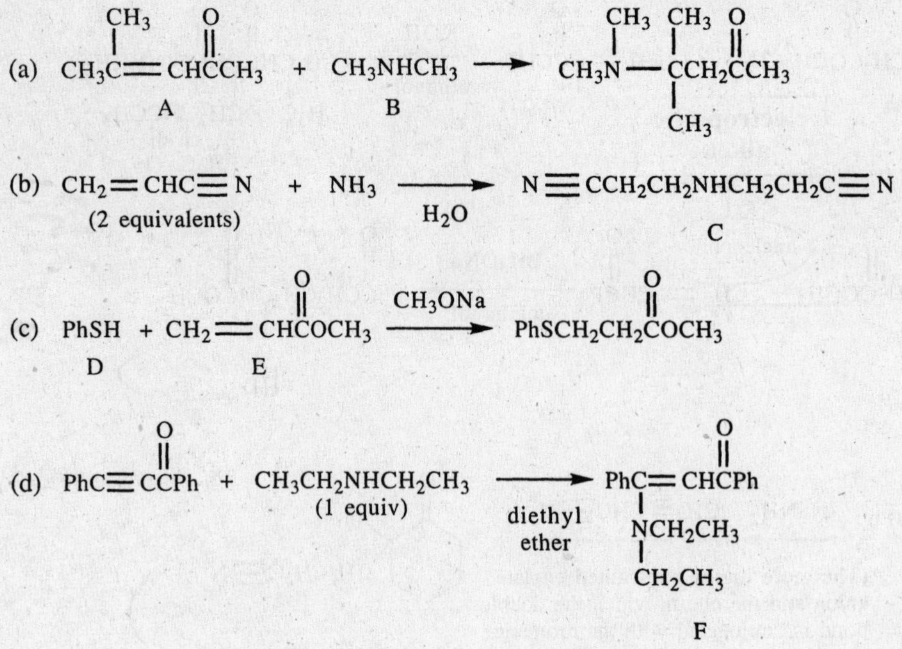

17.18

(E)--2-butenedioate
fumarate

attack at C-2 from top   attack at C-2 from bottom   attack at C-3 from top   attack at C-3 from bottom

(S)-(–)-malate   (R)-(+)-malate   (S)-(–)-malate   (R)-(+)-malate

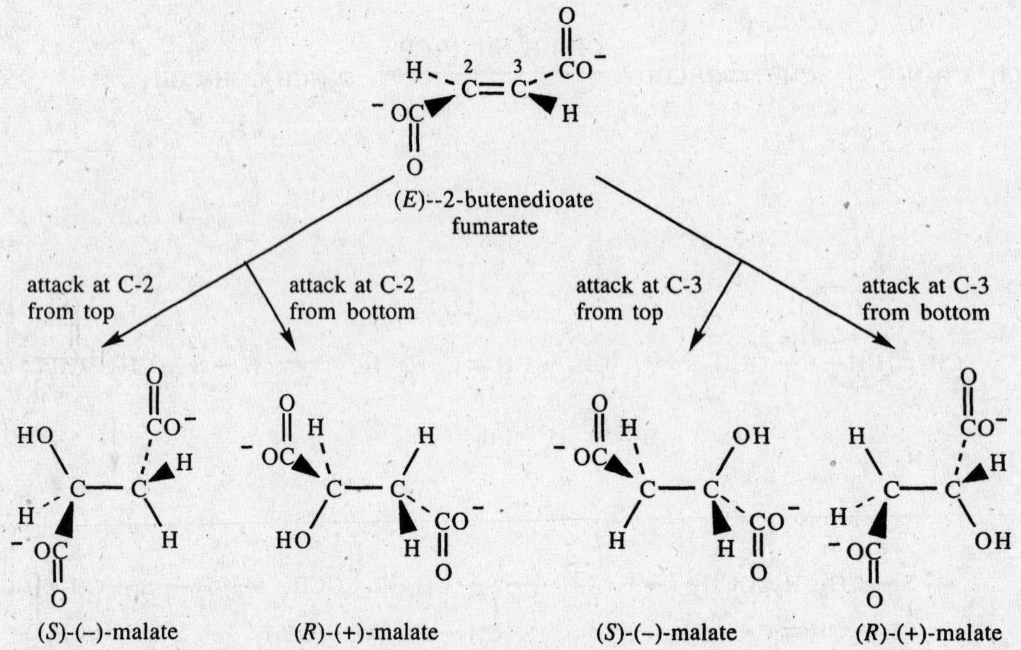

If the reaction were taking place in an achiral environment, we would expect a racemic mixture as a result of all four of these processes occurring with equal probability.

17.19 $$RSH + CH_3(CH_2)_8CCH{=}CH_2 \longrightarrow CH_3(CH_2)_8CCH_2CH_2SR$$

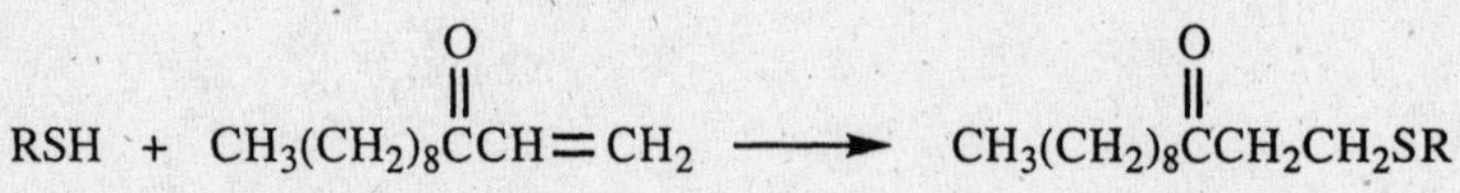

Concept Map 17.6   Reactions of electrophilic alkenes.

17.20   1.

The carbon atom of the carbonyl group in the carboxylate anion is much less electrophilic than the carbon atom of the ketone carbonyl.

17.20

The same result can be obtained by hydrolyzing the enol ether (Problem 14.13) and then dehydrating the alcohol.

17.21 The aldol condensation product must be reverting to an aldehyde and an enolate ion, which then reacts with the other aldehyde in the reaction mixture.

17.22

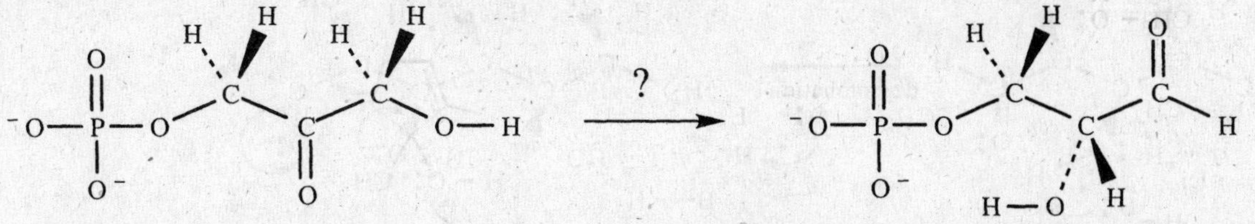

1.   How have connectivities changed in going from reactant to product? How many carbon atoms does each contain? What bonds must be broken and formed to transform starting material into product?

   Both reactant and product have three carbon atoms. There is an alcohol function on carbon 1 and a ketone function on carbon 2 in the reactant. There is an aldehyde function on carbon 1 and an alcohol function on carbon 2 in the product. Both have a phosphate ester on carbon 3. The bonds to be broken and formed are shown below.

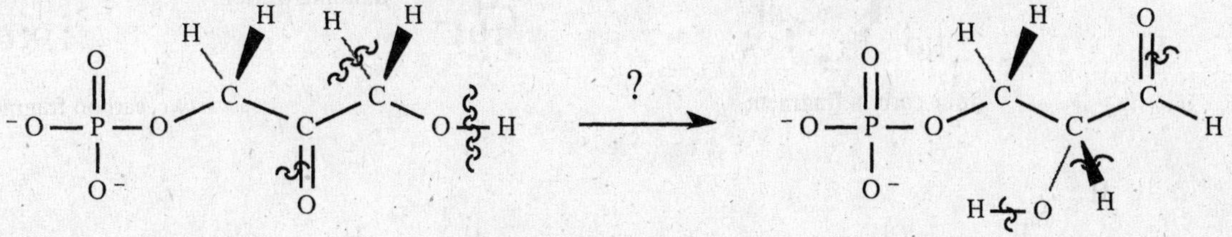

2.   What reagents are present? Are they good acids, bases, nucleophiles, or electrophiles? Is there an ionizing solvent present?

   The enzyme, triosephosphate isomerase, can act as an acid and a base.

3.   What is the most likely first step for the reaction: protonation or deprotonation, ionization, attack by a nucleophile, or attack by an electrophile?

   Protonation of the carbonyl group of the ester with later deprotonation of the α hydrogen on carbon 1 is the most likely first step.

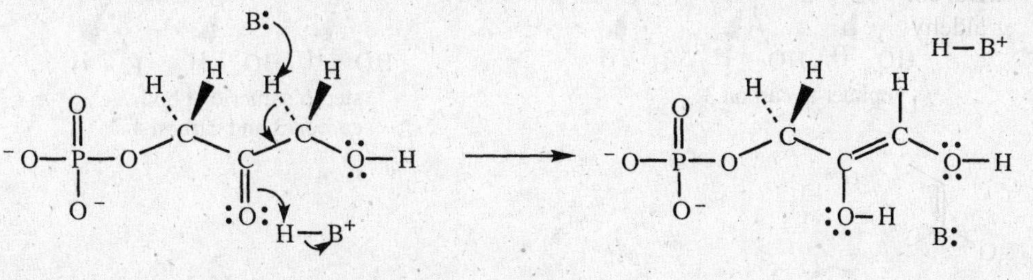

4.   What are the properties of the species present in the reaction mixture after the first step? What is likely to happen next?
   An enediol, which will tautomerize to the product, is formed.

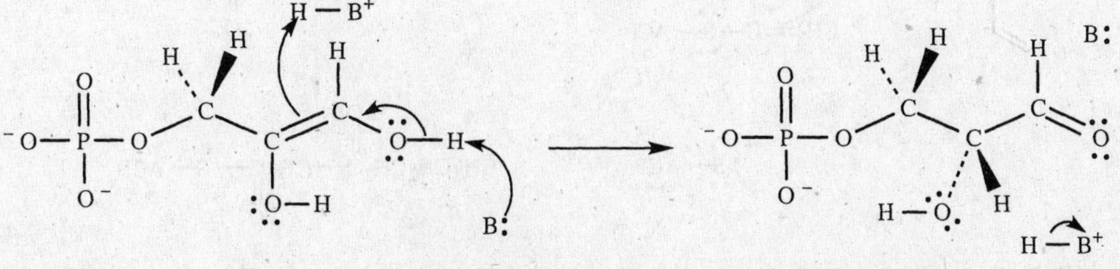

17.23

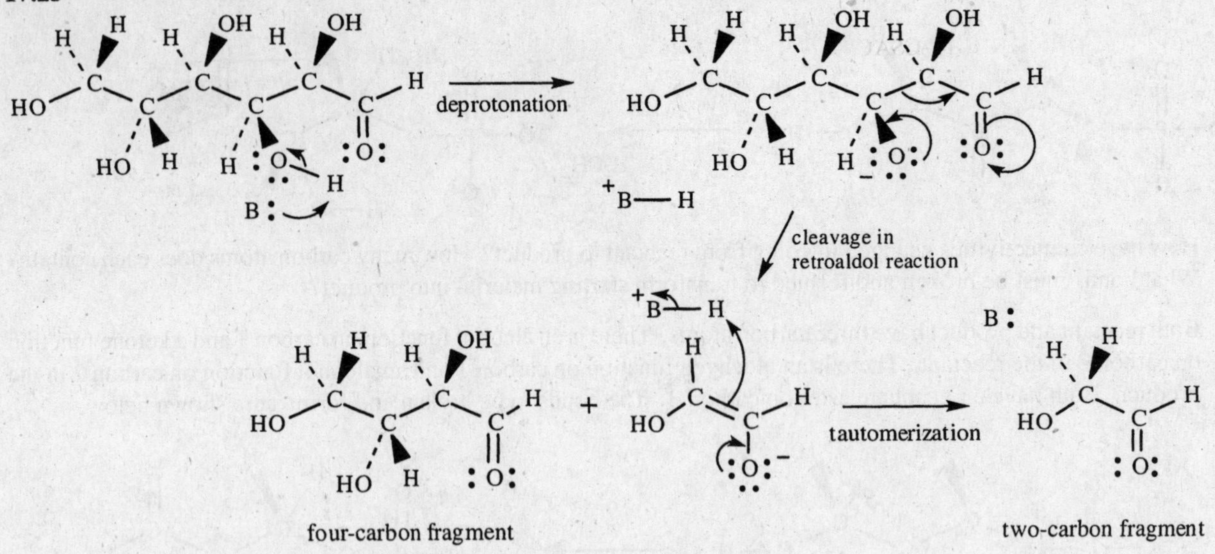
deprotonation
cleavage in retroaldol reaction
tautomerization
four-carbon fragment
two-carbon fragment

17.24

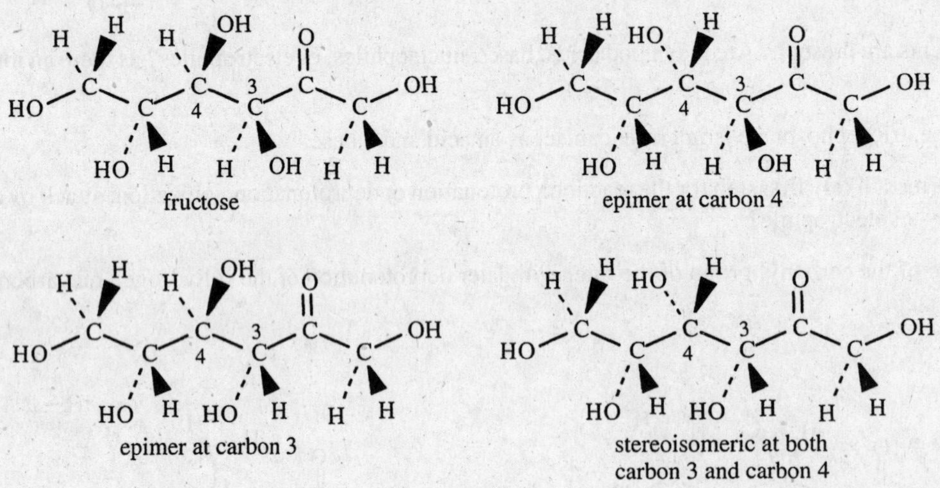
fructose
epimer at carbon 4
epimer at carbon 3
stereoisomeric at both carbon 3 and carbon 4

17.25

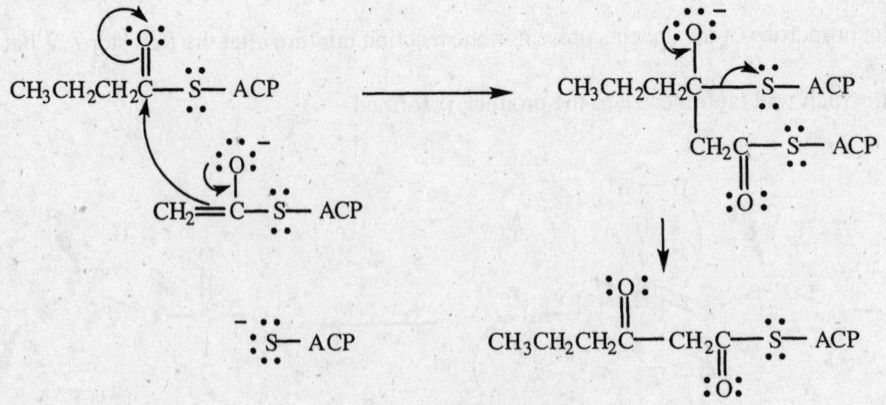

17.25 (cont)

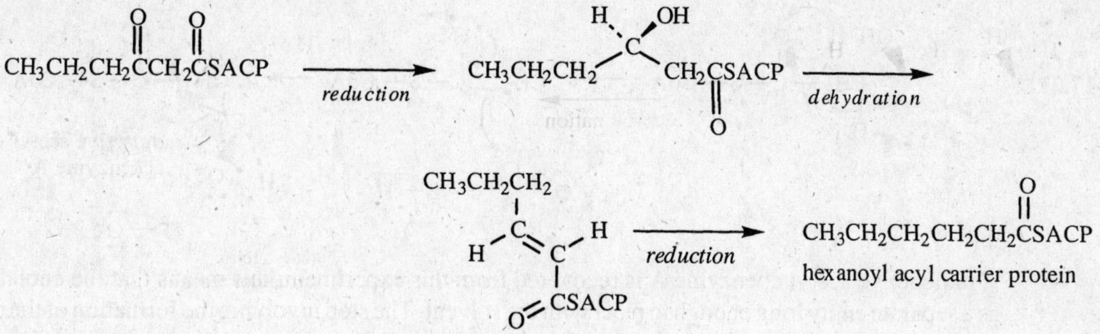

17.26

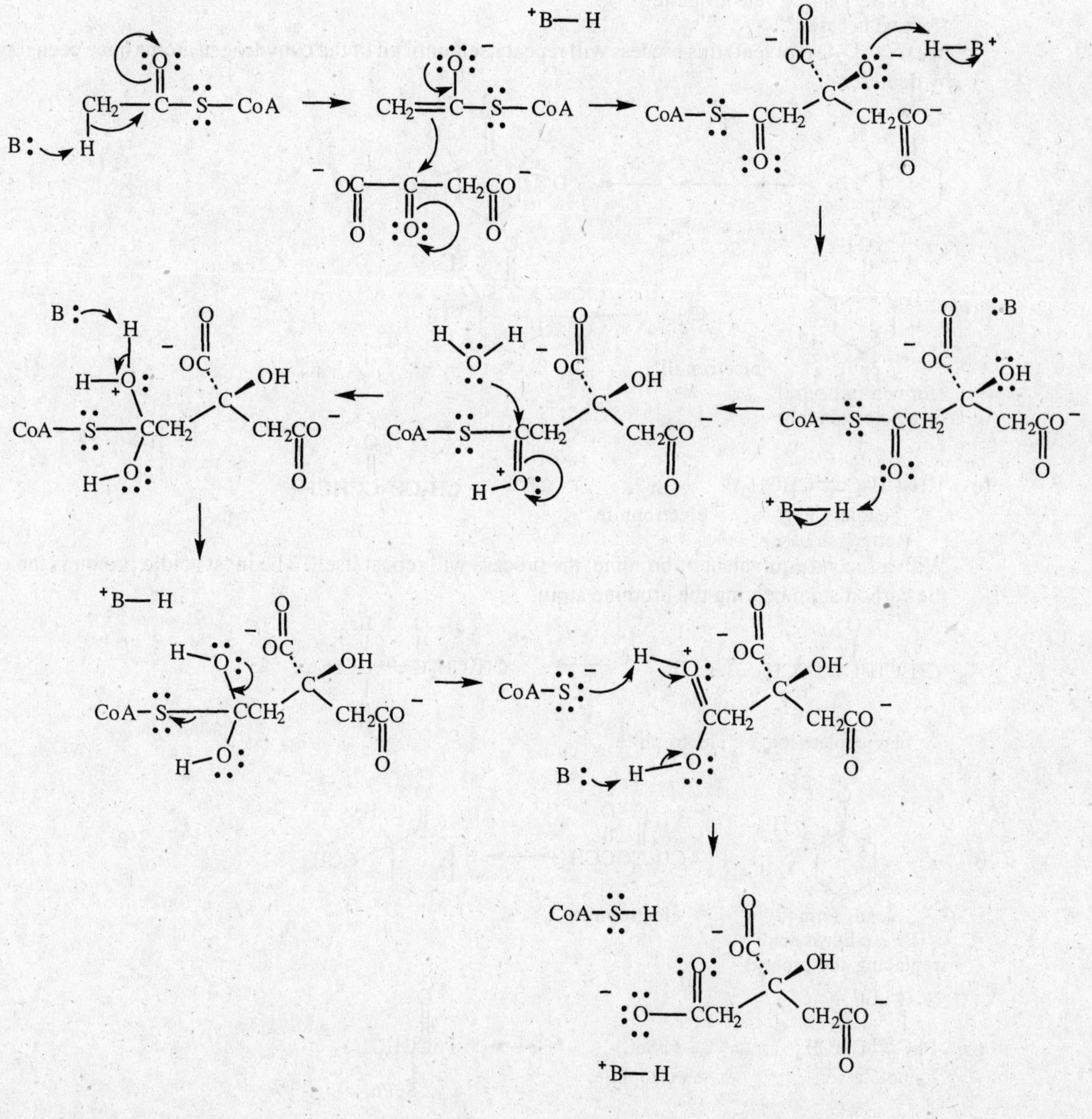

17.27

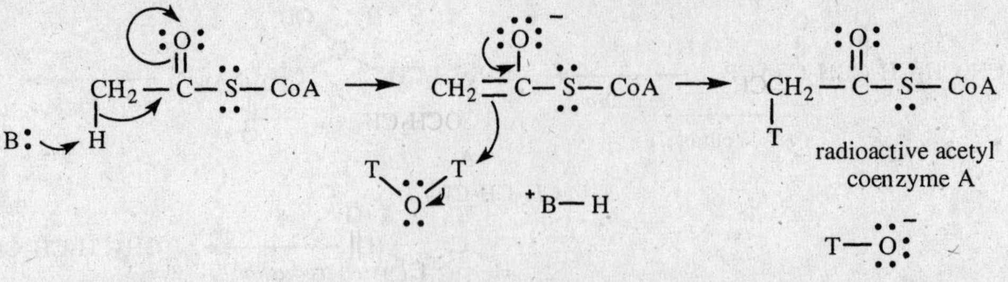

If radioactive acetyl coenzyme A is recovered from this experiment, this means that the enolate ion survives as a separate entity long enough to react with the solvent. The step involving the formation of the carbon-carbon bond is a separate second step.

17.28   (a)

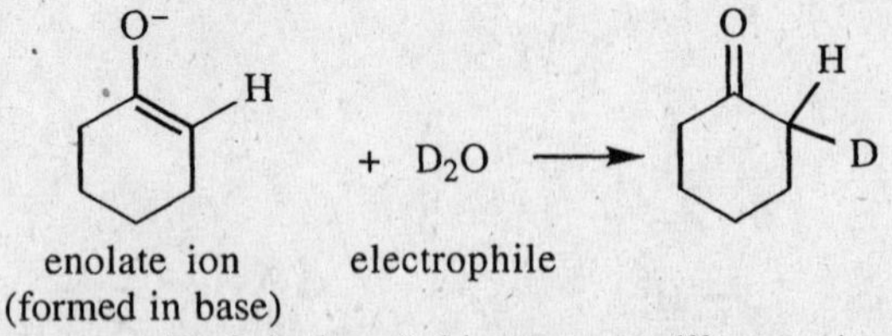

In excess D_2O (solvent) this process will repeat itself until all of the α-hydrogen atoms have been replaced by deuterium.

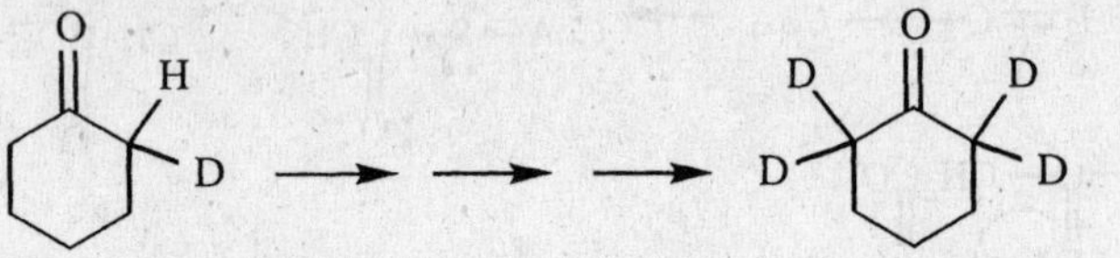

(b)

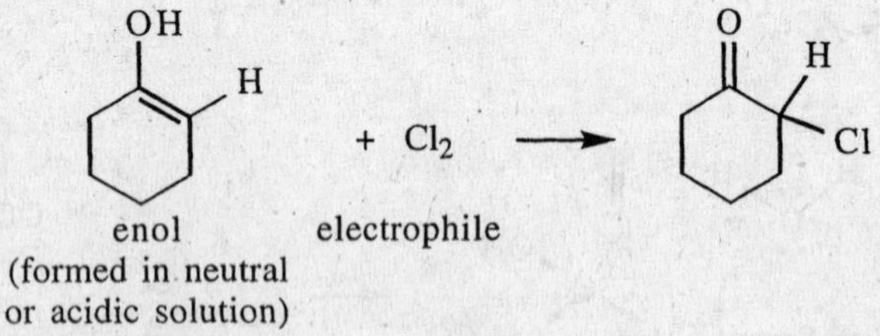

(c) $CH_3CH_2\overset{\overset{\displaystyle O^-}{|}}{C}=CHCH_3$ + Br_2 ⟶ $CH_3CH_2\overset{\overset{\displaystyle O}{||}}{C}\underset{\underset{\displaystyle Br}{|}}{C}HCH_3$

enolate ion   electrophile
(formed in base)

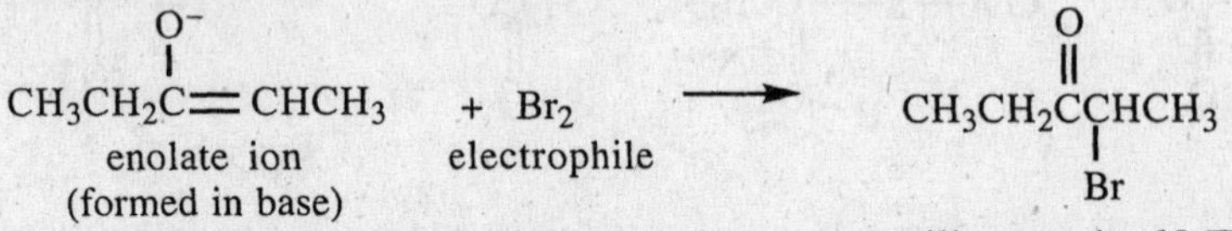

With a second equivalent of bromine, the process will repeat itself. The most acidic proton is the one on the carbon atom bearing the bromine atom.

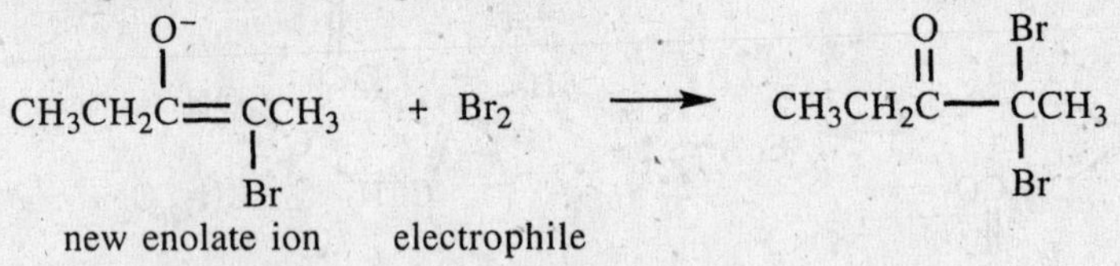

$CH_3CH_2\overset{\overset{\displaystyle O^-}{|}}{C}=\underset{\underset{\displaystyle Br}{|}}{C}CH_3$ + Br_2 ⟶ $CH_3CH_2\overset{\overset{\displaystyle O}{||}}{C}-\overset{\overset{\displaystyle Br}{|}}{\underset{\underset{\displaystyle Br}{|}}{C}}CH_3$

new enolate ion   electrophile

(d)

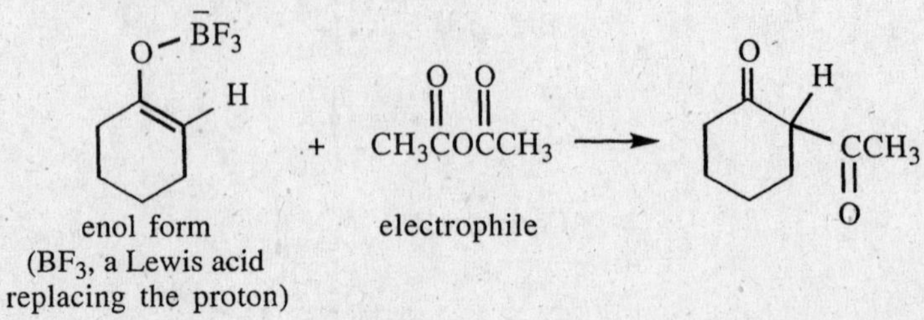

(e) $Ph\overset{\overset{\displaystyle O^- Li^+}{|}}{C}=CHCH_3$ + PhSeCl ⟶ $Ph\overset{\overset{\displaystyle O}{||}}{C}\underset{\underset{\displaystyle SePh}{|}}{C}HCH_3$

enolate ion   electrophile

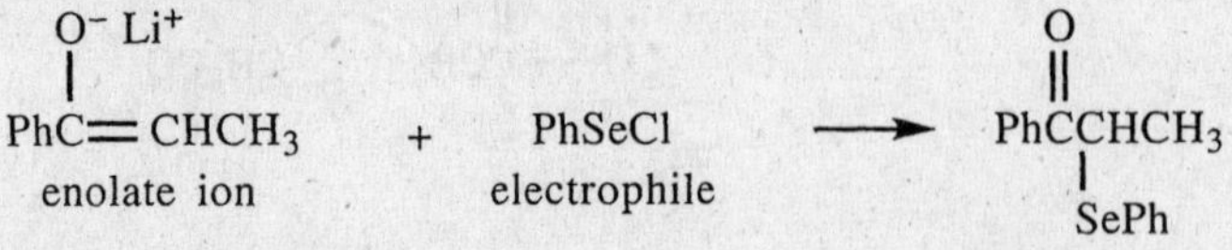

17.29 (a)

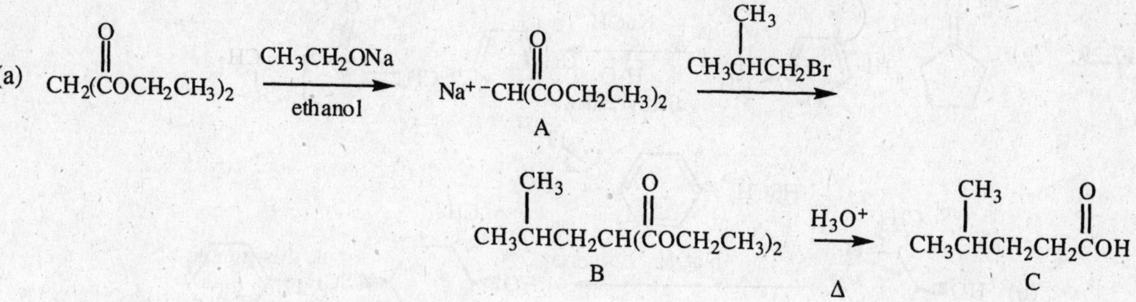

$$CH_2(COCH_2CH_3)_2 \xrightarrow[\text{ethanol}]{CH_3CH_2ONa} Na^+ \, {}^-CH(COCH_2CH_3)_2 \xrightarrow{CH_3CHCH_2Br}$$

A

$$CH_3CHCH_2CH(COCH_2CH_3)_2 \xrightarrow[\Delta]{H_3O^+} CH_3CHCH_2CH_2COH$$

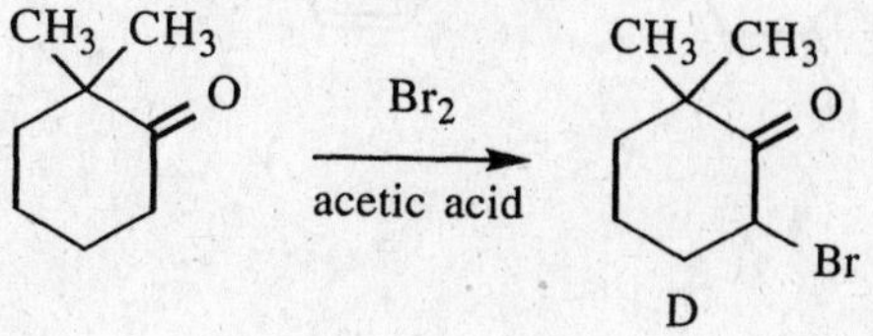

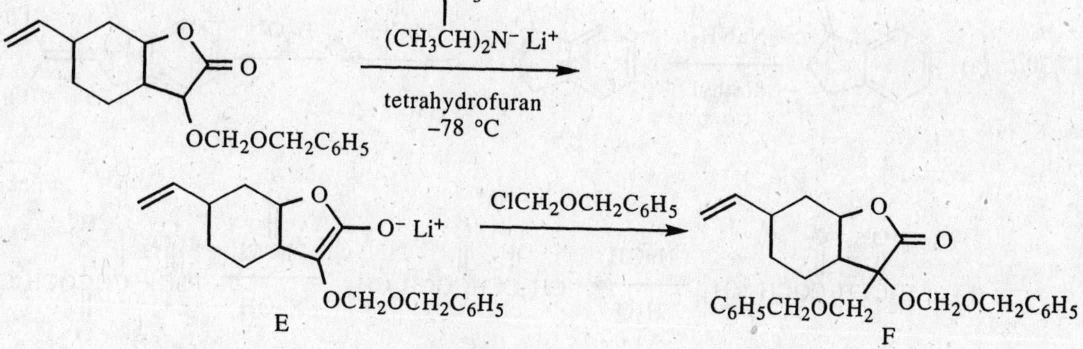

B C

(b)

$$\xrightarrow[\text{acetic acid}]{Br_2}$$

D

(c)

$$\xrightarrow[\substack{\text{tetrahydrofuran} \\ -78\,°C}]{(CH_3CH)_2N^- \, Li^+}$$

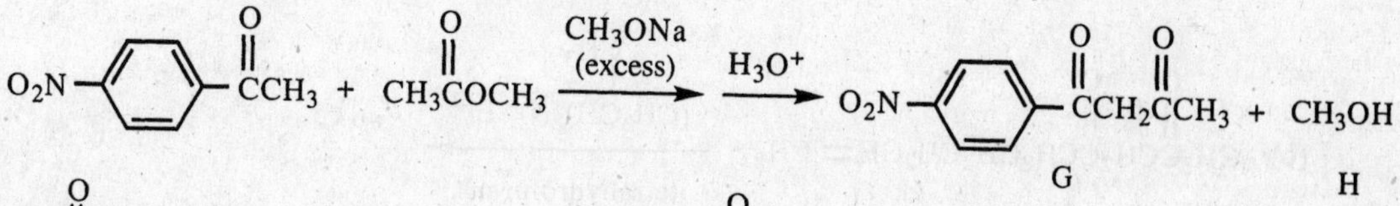

E $\xrightarrow{ClCH_2OCH_2C_6H_5}$ F

(d)

$$O_2N-\!\!\!\!\!\!\bigcirc\!\!\!\!\!\!-CCH_3 + CH_3COCH_3 \xrightarrow[\text{(excess)}]{CH_3ONa} \xrightarrow{H_3O^+} O_2N-\!\!\!\!\!\!\bigcirc\!\!\!\!\!\!-CCH_2CCH_3 + CH_3OH$$

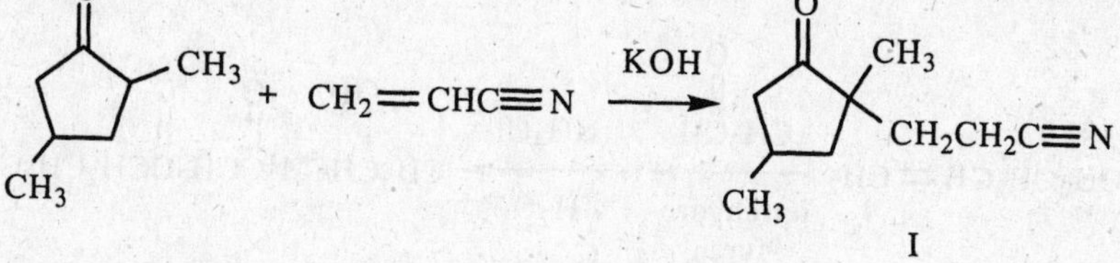

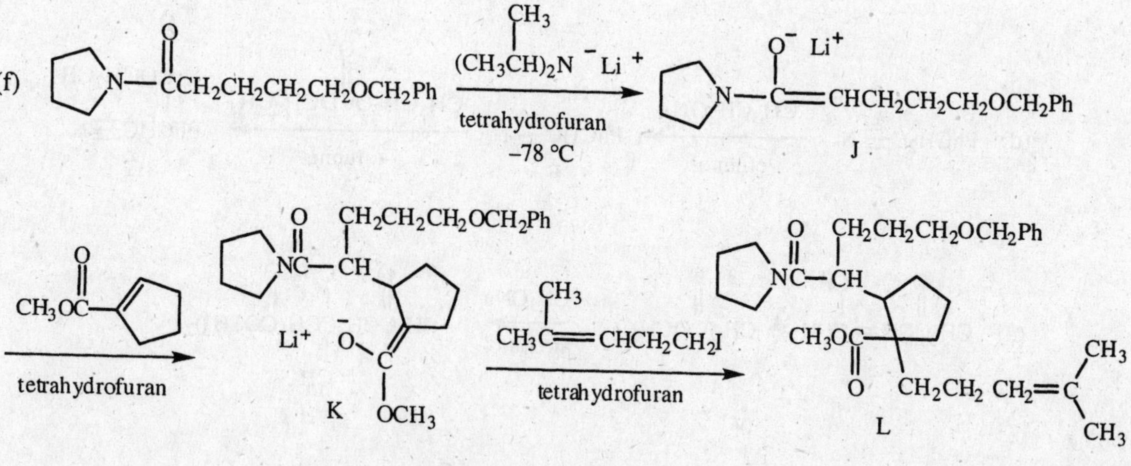

G H

(e) $+ CH_2=CHC\equiv N \xrightarrow{KOH}$ I

(f) $\xrightarrow[\substack{\text{tetrahydrofuran} \\ -78\,°C}]{(CH_3CH)_2N^- \, Li^+}$ J

17.29 (g)

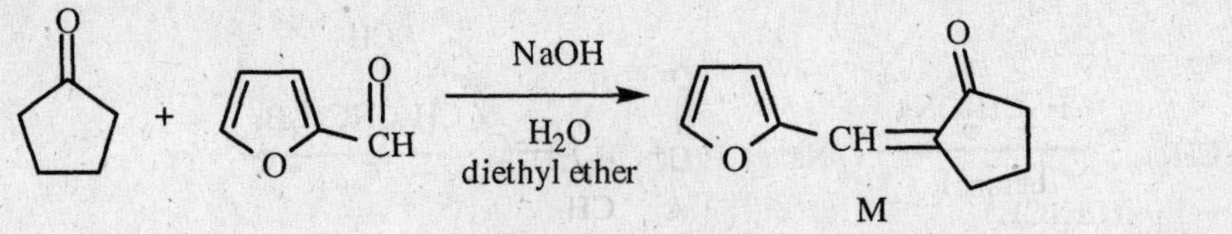
NaOH
H2O
diethyl ether
CH
CH=
M

(h)

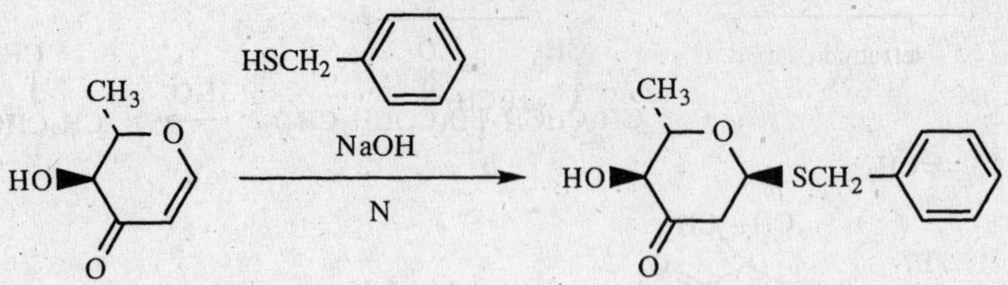
HSCH2
CH3
HO
NaOH
N
SCH2

17.30 (a)

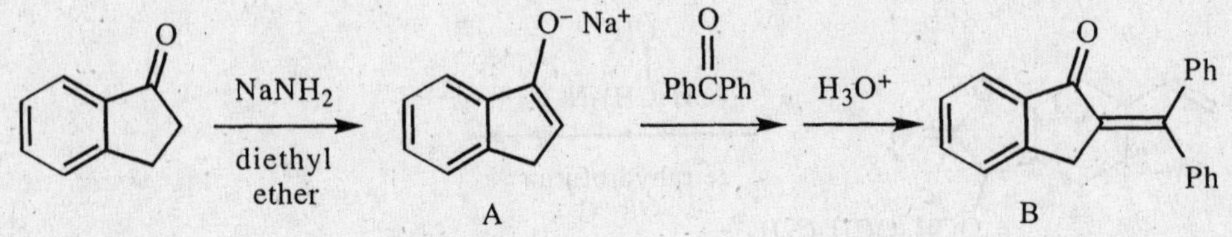
NaNH2
diethyl
ether
O⁻ Na⁺
A
PhCPh
H3O⁺
Ph
Ph
B

(b)

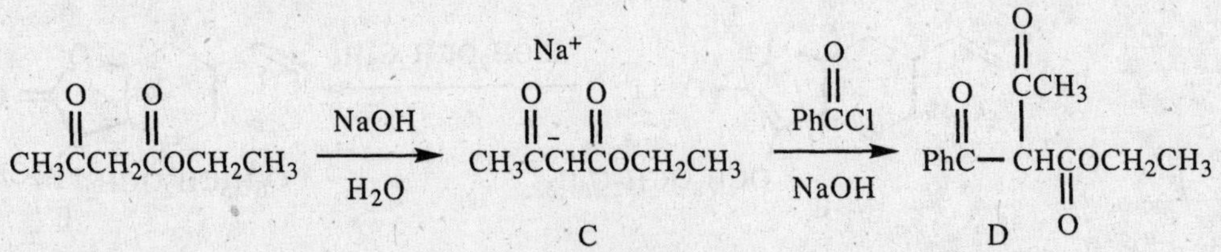
CH3CCH2COCH2CH3
NaOH
H2O
Na⁺
CH3CCHCOCH2CH3
C
PhCCl
NaOH
CCH3
PhC— CHCOCH2CH3
D

(c)

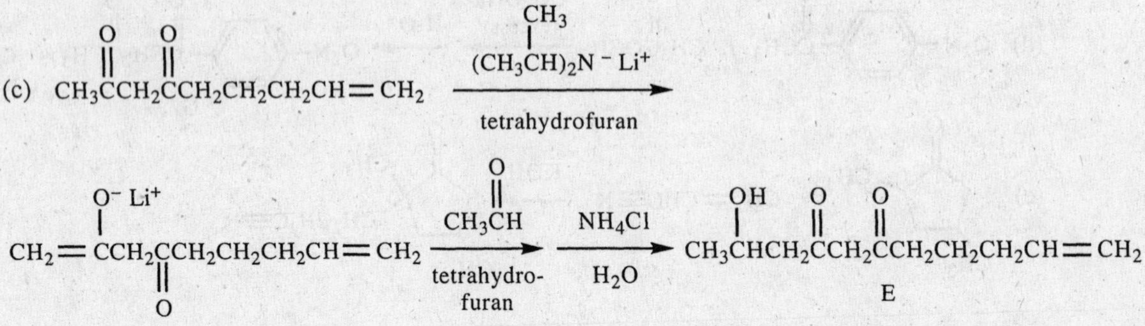
CH3CCH2CCH2CH2CH2CH=CH2
CH3
(CH3CH)2N⁻ Li⁺
tetrahydrofuran
O⁻ Li⁺
CH2=CCH2CCH2CH2CH2CH=CH2
CH3CH
tetrahydro-
furan
NH4Cl
H2O
OH
CH3CHCH2CCH2CCH2CH2CH2CH=CH2
E

(d)

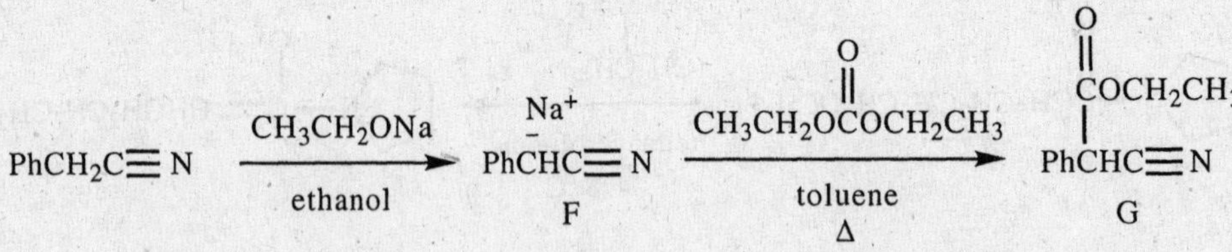
PhCH2C≡N
CH3CH2ONa
ethanol
Na⁺
PhCHC≡N
F
CH3CH2OCOCH2CH3
toluene
Δ
COCH2CH3
PhCHC≡N
G

(e)

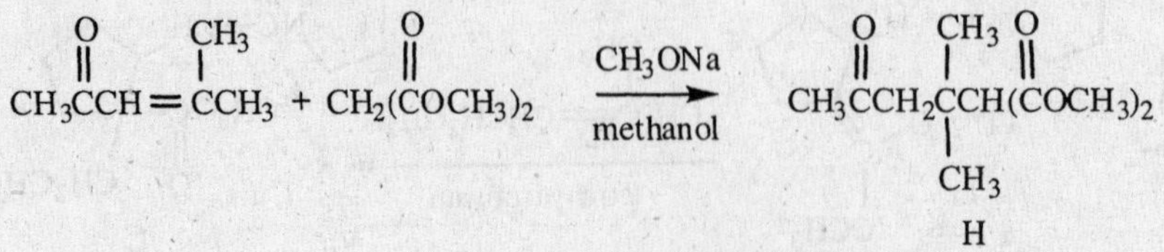
CH3CCH=CCH3 + CH2(COCH3)2
CH3
CH3ONa
methanol
CH3CCH2CCH(COCH3)2
CH3
H

17.30 (f)

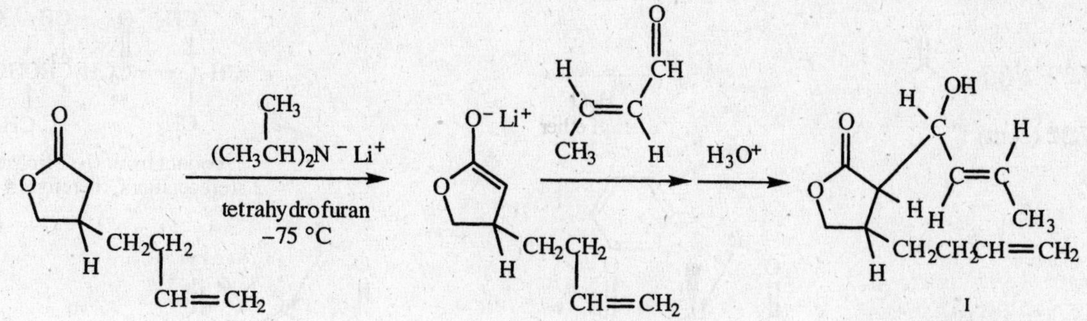
tetrahydrofuran
−75 °C
H3O+

I
mixture of diastereomers

(g)

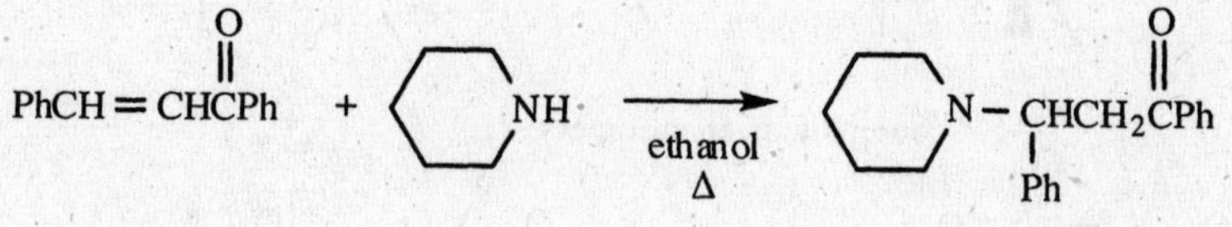
ethanol
Δ

J

(h)

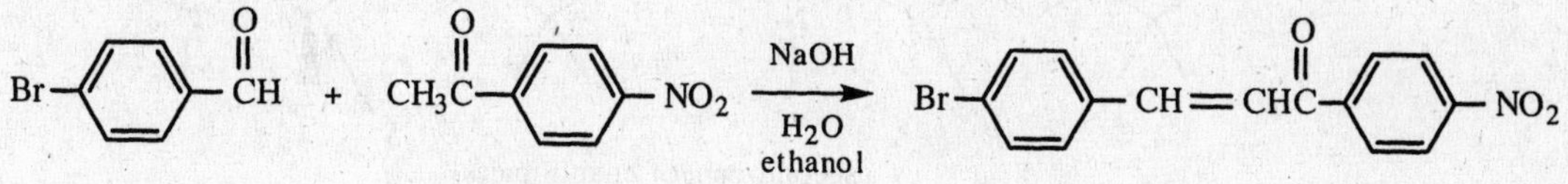
NaOH
H2O
ethanol

K

17.31 (a)

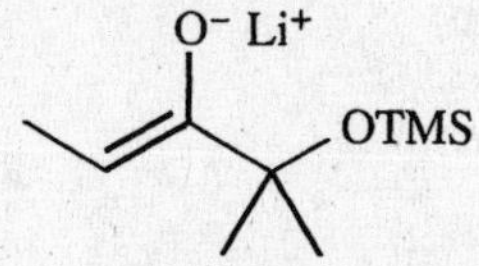

(*Z*)-enolate

(b)

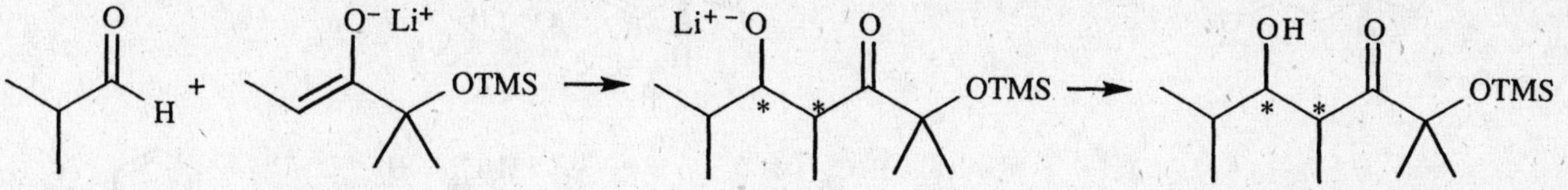

aldol product with two stereocenters

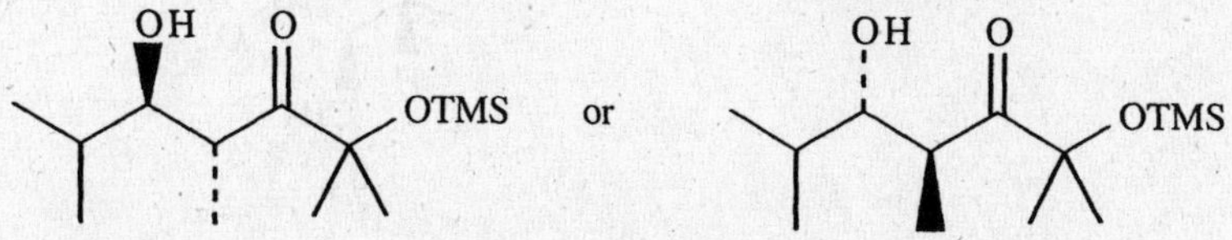

17.32

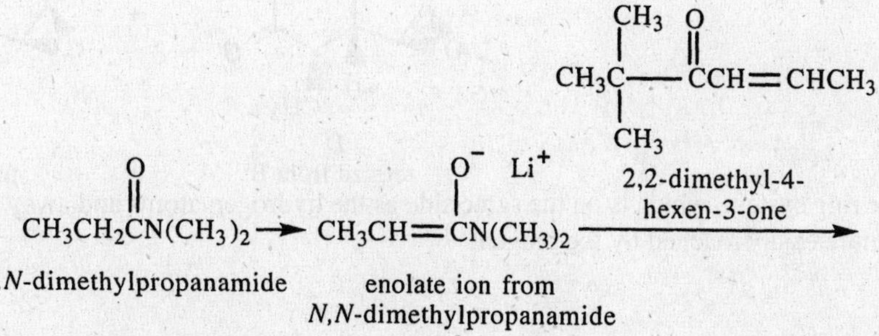
2,2-dimethyl-4-
hexen-3-one
N,N-dimethylpropanamide
enolate ion from
N,N-dimethylpropanamide

17.32 (cont)

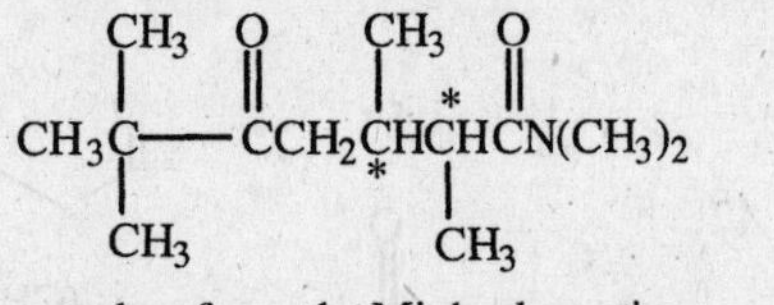

product from the Michael reaction;
2 stereocenters, therefore 4 stereoisomers

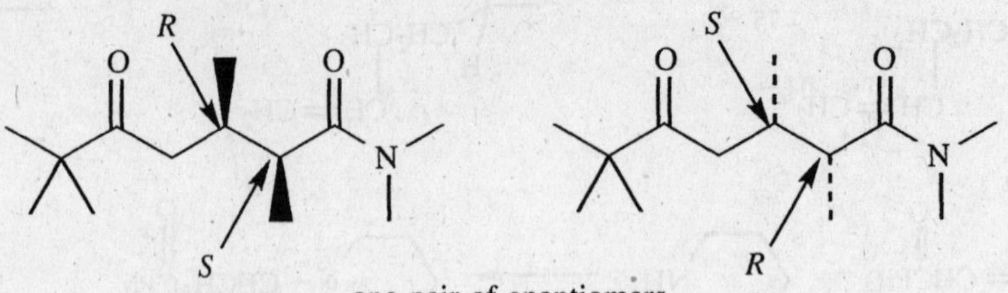

one pair of enantiomers

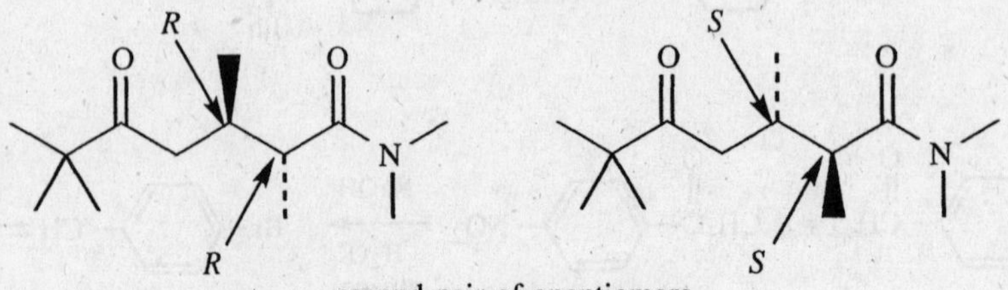

second pair of enantiomers

17.33

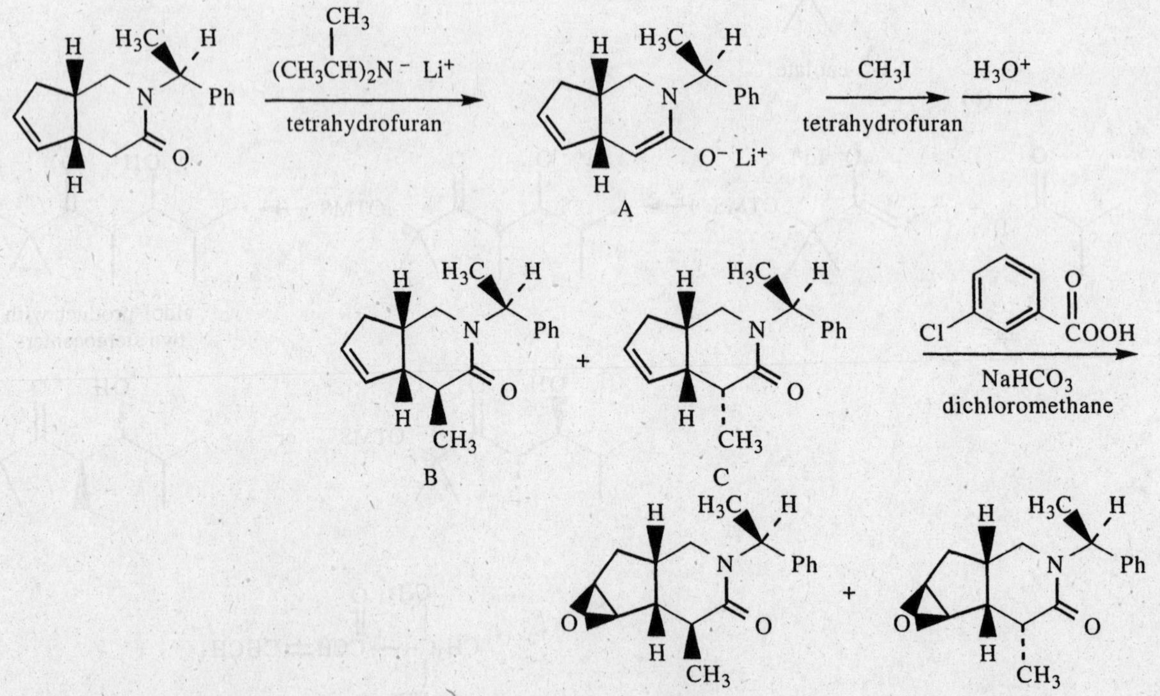

formed from B   formed from C

The top face of the ring system, which is on the same side as the hydrogen atoms and away from the bonds to the other ring, is more easily reached by the reagent.

17.34

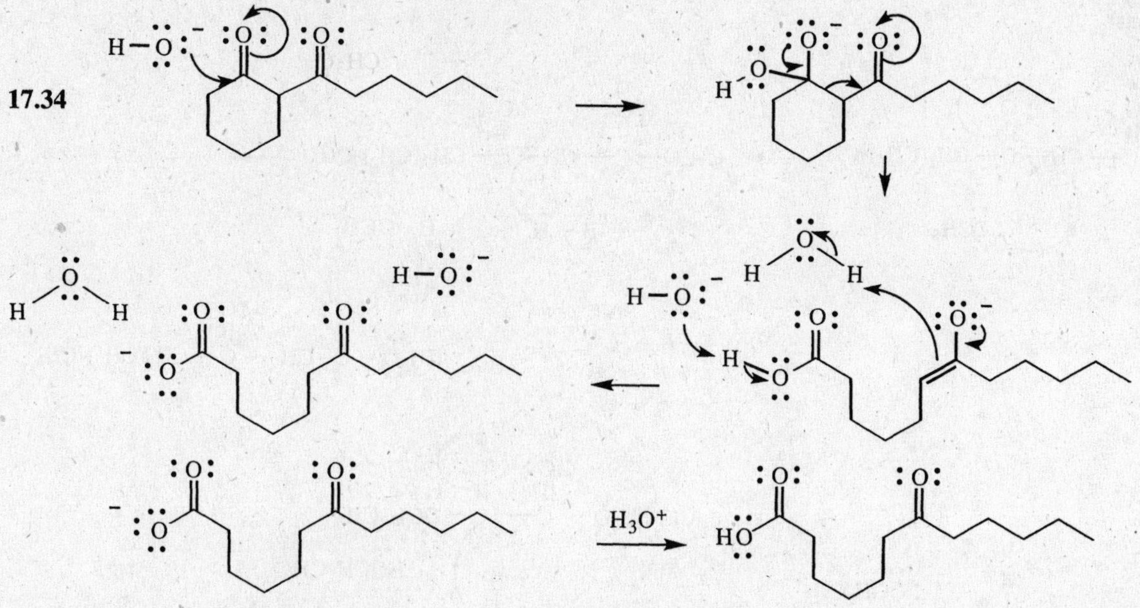
H₃O⁺

17.35

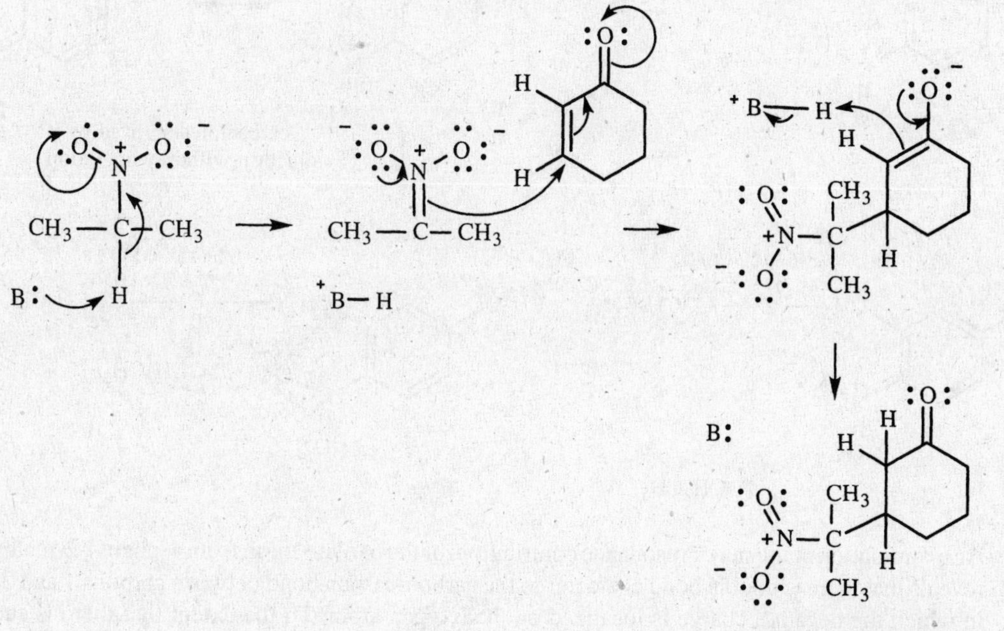
only one enantiomer formed
in presence of chiral catalyst

17.36

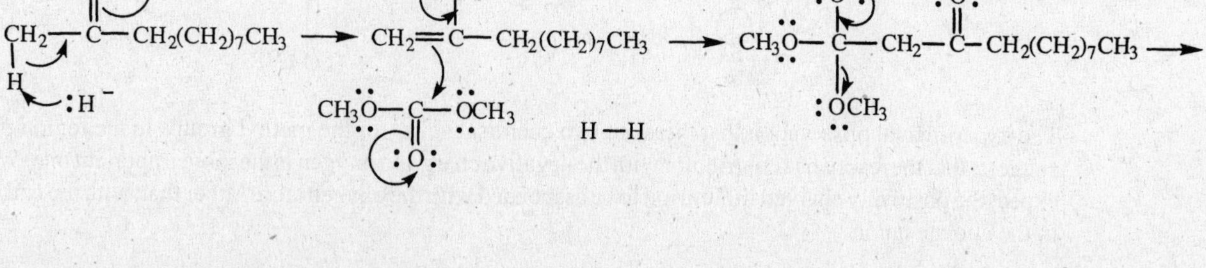

17.36 (cont)

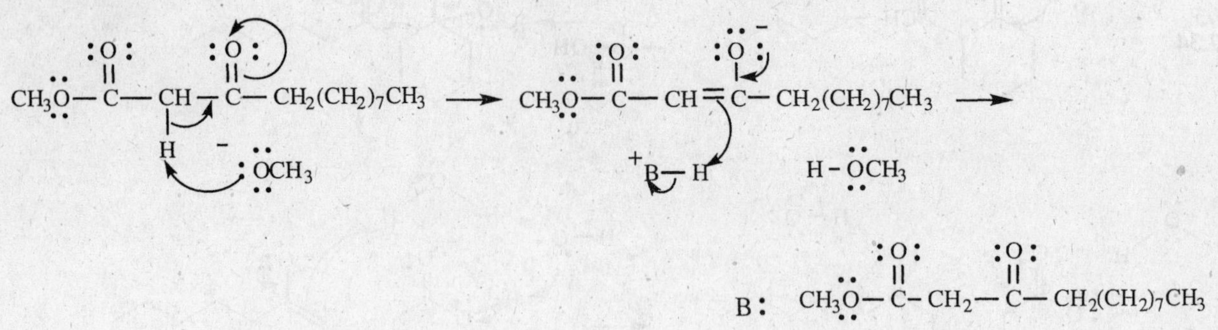

17.37

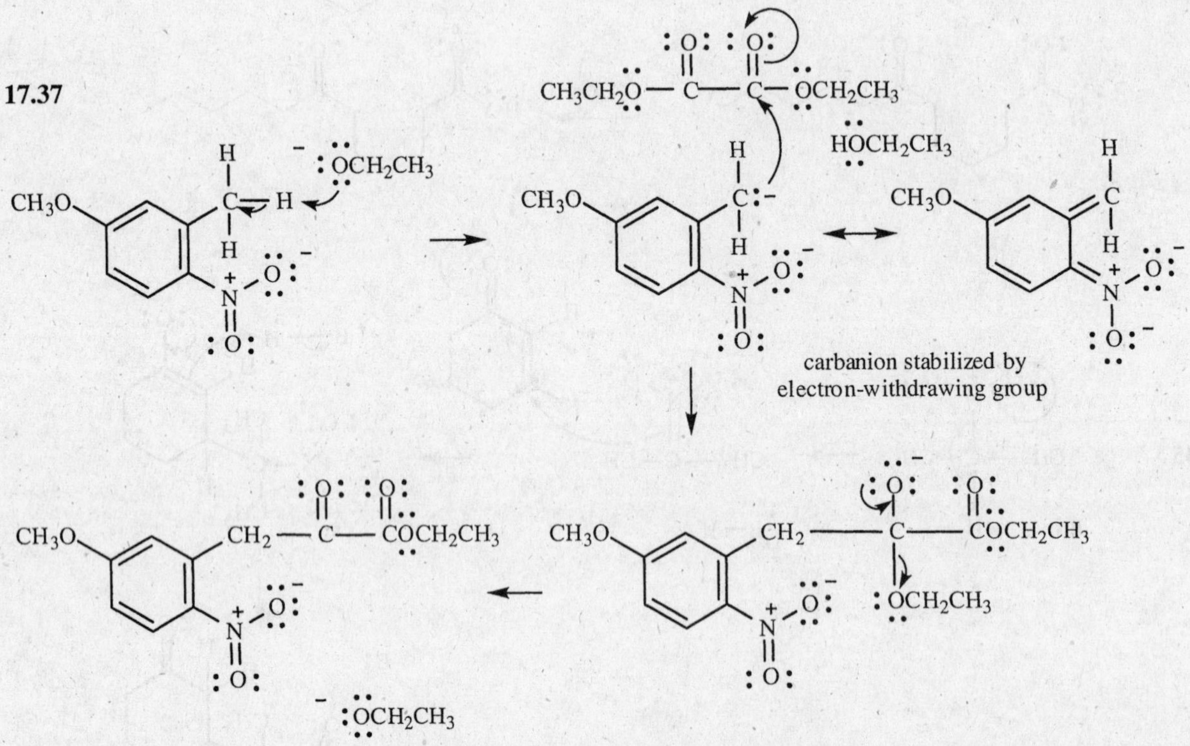

carbanion stabilized by
electron-withdrawing group

17.38   An examination of the major resonance contributors of the enolate anion from 1-phenyl-2-methyl-1-propanone reveals that there is double bond character to the carbon-carbon bond between carbons 1 and 2 in the structure in which the negative charge is localized on the oxygen atom. To the extent that there is such double bond character, free rotation around that bond is inhibited, and the two methyl groups are no longer chemical-shift equivalent (for example, one is trans and the other cis to the oxygen atom) and would be expected to have different chemical shifts.

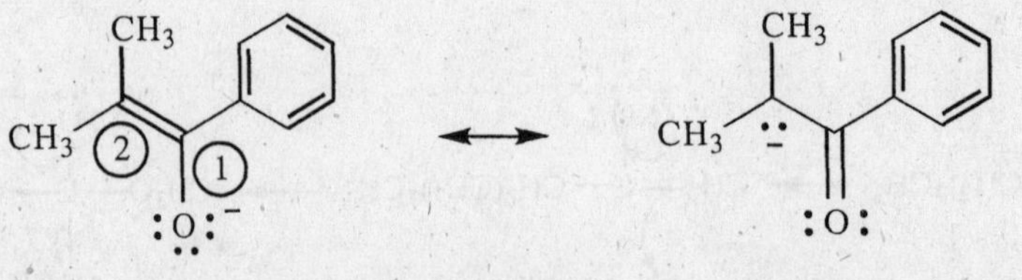

The experimental observation that there are two chemical shifts for the methyl groups in the region δ 1.0–2.0 suggests that the resonance contributor with the negative charge on oxygen is the more important one. We would expect the positively charged lithium ion to be associated with the oxygen atom rather than with the carbon atom in the enolate anion.

17.39

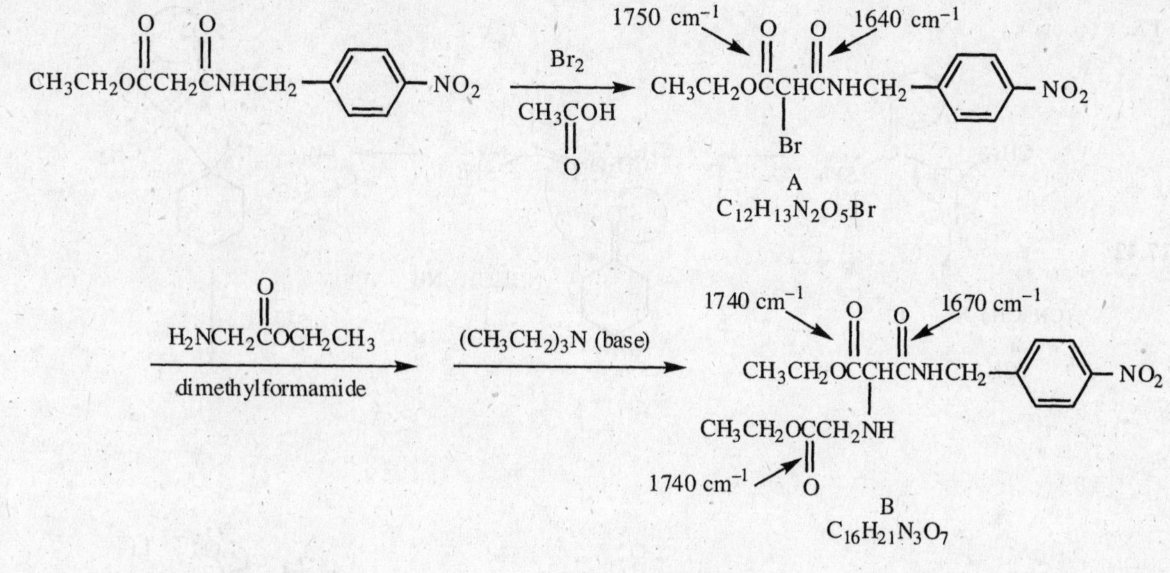

$$CH_3CH_2OCCH_2CNHCH_2 \quad \text{—} \quad NO_2 \xrightarrow[CH_3COH]{Br_2} CH_3CH_2OCCHCNHCH_2 \quad \text{—} \quad NO_2$$

1750 cm^{-1} 1640 cm^{-1}

Br

A
C$_{12}$H$_{13}$N$_2$O$_5$Br

$$\xrightarrow[\text{dimethylformamide}]{H_2NCH_2COCH_2CH_3} \xrightarrow{(CH_3CH_2)_3N \text{ (base)}}$$

1740 cm^{-1} 1670 cm^{-1}

$$CH_3CH_2OCCHCNHCH_2 \quad \text{—} \quad NO_2$$

$$CH_3CH_2OCCH_2NH$$

1740 cm^{-1}

O

B
C$_{16}$H$_{21}$N$_3$O$_7$

17.40

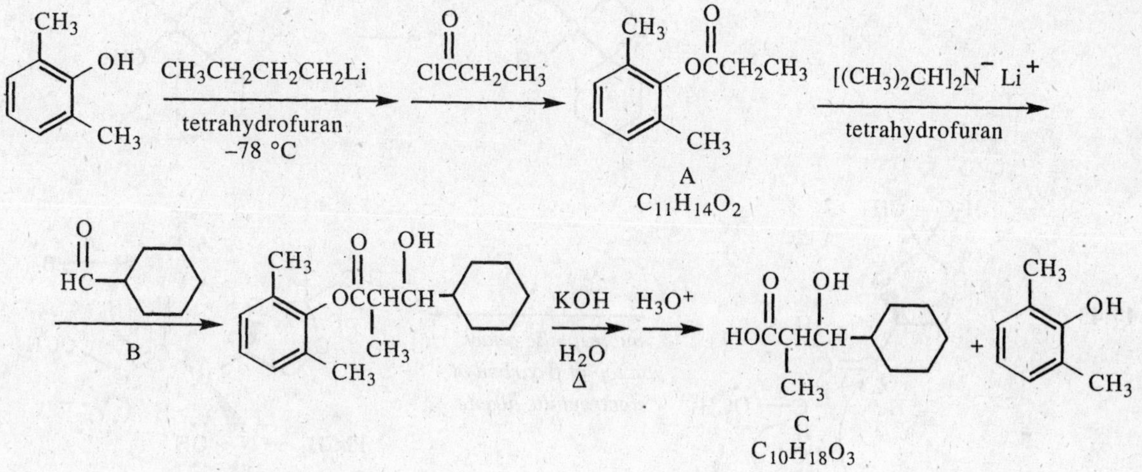

$$\xrightarrow[\substack{\text{tetrahydrofuran} \\ -78\,°C}]{CH_3CH_2CH_2CH_2Li} \xrightarrow{ClCCH_2CH_3}$$

A
C$_{11}$H$_{14}$O$_2$

$$\xrightarrow{[(CH_3)_2CH]_2N^- Li^+}$$

tetrahydrofuran

$$\xrightarrow{B}$$

$$\xrightarrow[\substack{H_2O \\ \Delta}]{KOH} \xrightarrow{H_3O^+}$$

C
C$_{10}$H$_{18}$O$_3$

17.41

$$CH_3OC \quad \text{...} \quad CH_2CH_3 \xrightarrow[A]{NaH} \xrightarrow{B} \xrightarrow{Cs_2CO_3 \text{ (base)}}$$

$$CH_2{=}CHC{\equiv}N \xrightarrow{} \quad C \quad \xrightarrow[\substack{H_2O \\ \Delta}]{KOH} \xrightarrow[\text{(protonation)}]{H_3O^+} \xrightarrow{\Delta}$$

D
C$_{17}$H$_{20}$N$_4$O

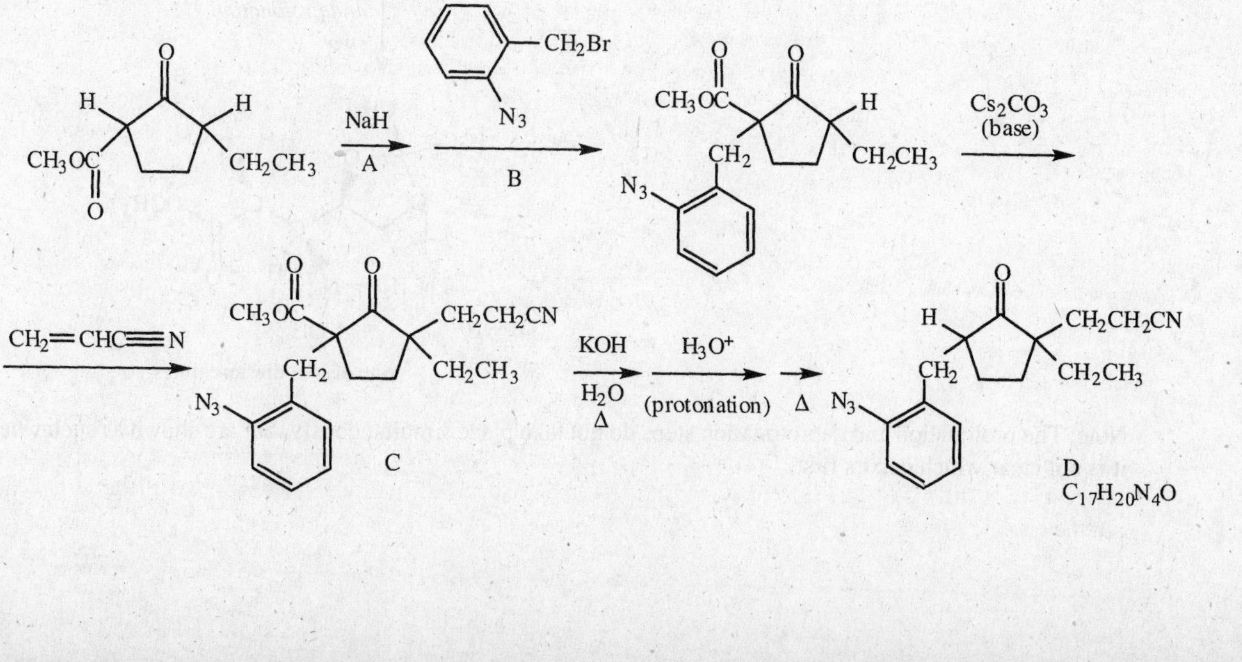

17.42

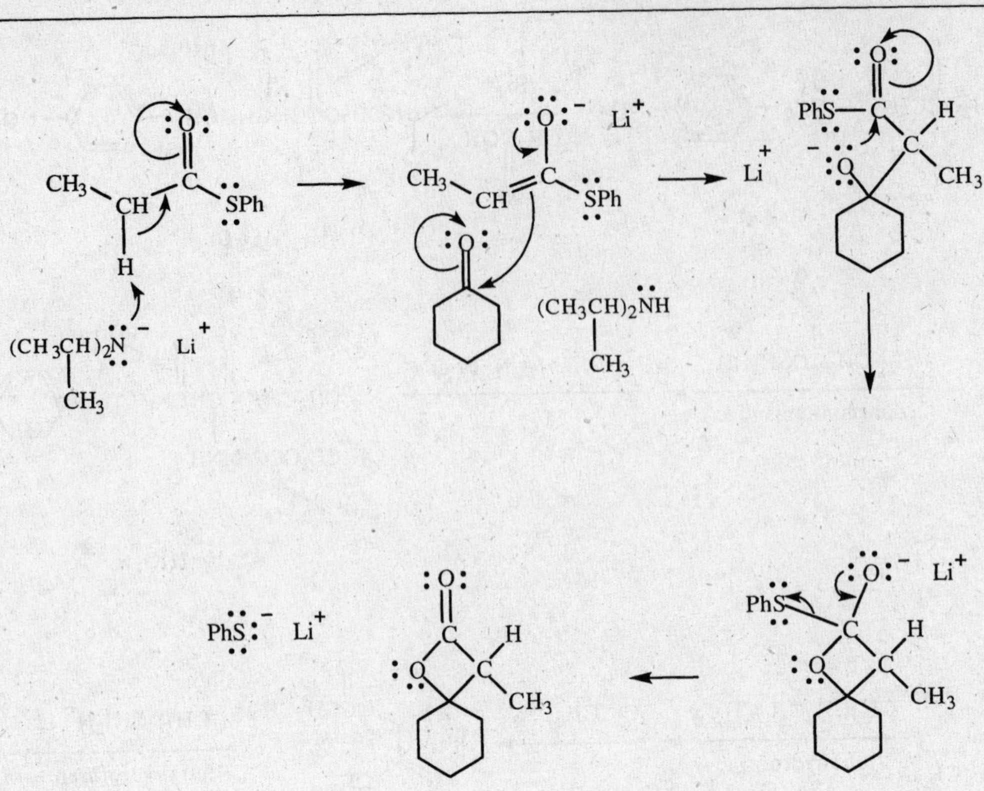

17.43

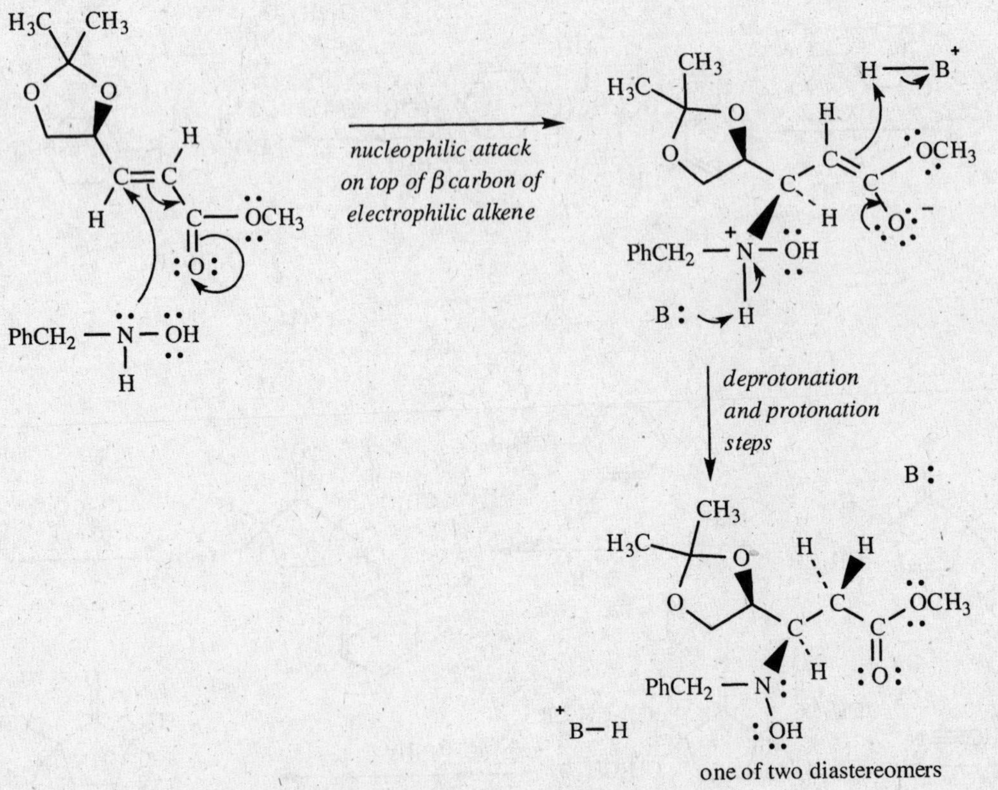

nucleophilic attack on top of β carbon of electrophilic alkene

deprotonation and protonation steps

one of two diastereomers

Note: The protonation and deprotonation steps do not take place simultaneously, but are shown as such when it is not clear which occurs first.

17.44

17.45   The kinetic product (the one that forms faster) of the reaction is the 1,2-addition product. It forms almost exclusively when the reaction is run for a short time at low temperatures. The 1,4-addition product is the thermodynamically more stable product. It is the major product when the reaction is run at higher temperatures and for a long period of time allowing for equilibrium to be established.

The reaction products must be formed reversibly, otherwise they could not equilibrate.

reversal of the
1,2-addition reaction

reversal of the
1,4-addition reaction

17.46   The reaction is an aldol condensation reaction between two glyceraldehyde molecules. The enediol intermediate that is necessary for the condensation reaction is formed faster from dihydroxyacetone (in which deprotonation occurs at an O—H bond) than from glyceraldehyde (in which deprotonation must occur at a C—H bond). The stereochemistry of the products is determined by the side of the carbonyl group that is attacked by the enediol. The reaction is shown as taking place with the polar and bulky hydroxyl group of the enediol pointing away from the oxygen atom of the carbonyl group undergoing nucleophilic attack.

17.46 (cont)

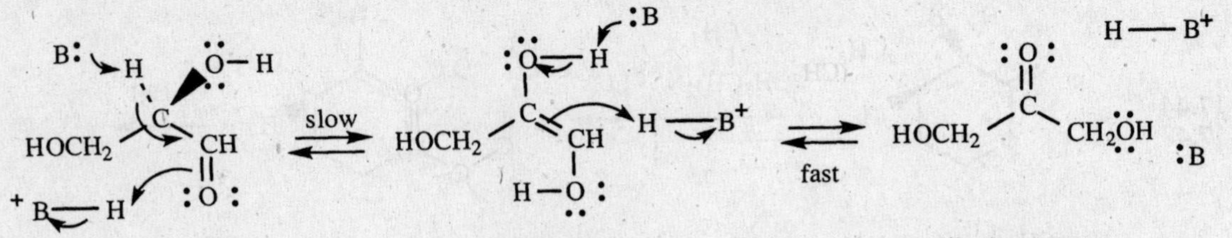

(R)-glyceraldehyde   endiol intermediate   dihydroxyacetone

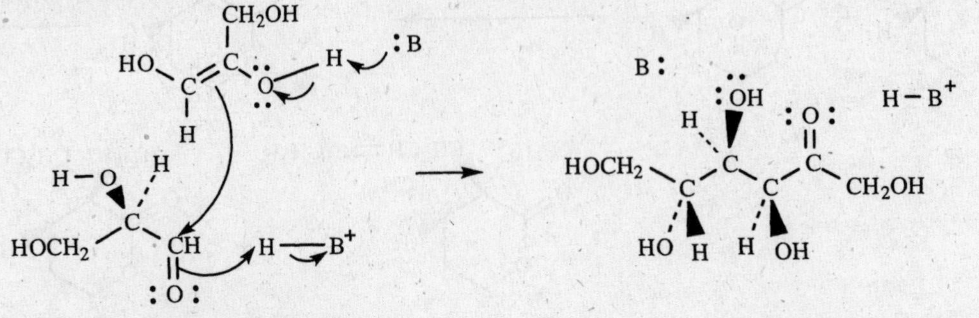

attack on one face of the carbonyl
group of (R)-glyceraldehyde by the
enediol intermediate; an aldol condensation   fructose

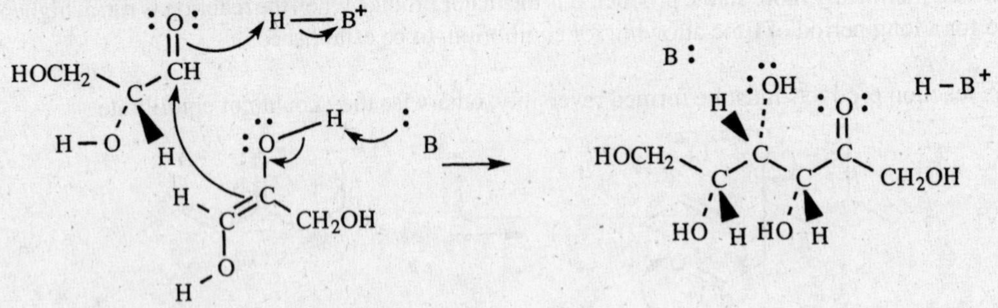

attack on the other face of the carbonyl
group of (R)-glyceraldehyde by the enediol
intermediate; an aldol condensation   sorbose

17.47 (a)

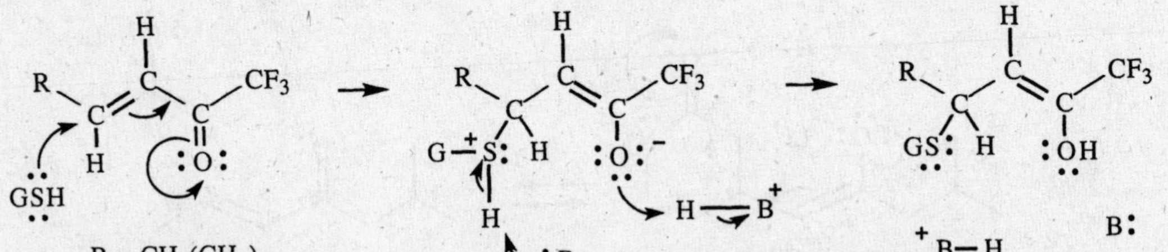

R = CH_3(CH_2)_7—

(b) The electron-withdrawing effect of the fluorine atoms stabilizes the enolate ion intermediate, lowering the
energy of activation for the reaction. It also increases the electrophilicity of the β-carbon atom, making it
a more attractive target for a nucleophilic attack, more so than for —CH_3 ketone.

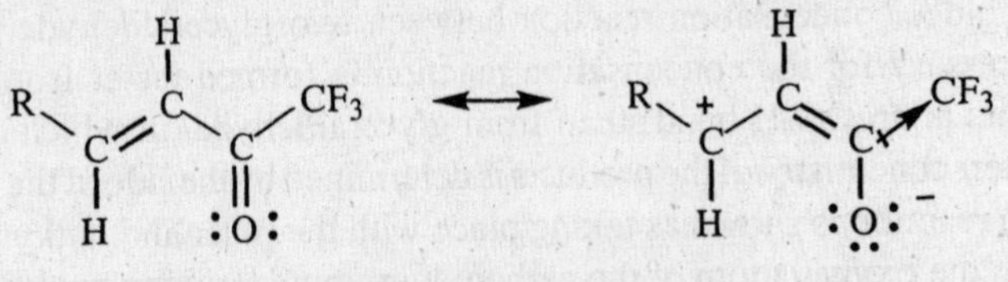

17.48

$CH_3OC(O)$—$C(H)(H)$—$COCH_3$ $\xrightarrow[A]{\substack{CH_3 \\ (CH_3CH)_2N^- Li^+ \\ \text{or NaH}}}$ $CH_3OC(O)$—$C(H)(:^-)$—$COCH_3$ $\xrightarrow[B]{BrCH_2CH=CH_2}$

$CH_3OC(O)$—$C(H)(CH_2CH=CH_2)$—$COCH_3$ $\xrightarrow[C]{\substack{CH_3 \\ (CH_3CH)_2N^- Li^+ \\ \text{or NaH}}}$ $CH_3OC(O)$—$C(:^-)(CH_2CH=CH_2)$—$COCH_3$ $\xrightarrow[D]{BrCH_2C\equiv CH}$

$CH_3OC(O)$—$C(HC\equiv CCH_2)(CH_2CH=CH_2)$—$COCH_3$ $\xrightarrow[\substack{\text{methanol} \\ \text{tetrahydrofuran}}]{KOH}$ $\xrightarrow[\Delta]{H_3O^+ Cl^-}$ $HC\equiv CCH_2$—$C(H)(CH_2CH=CH_2)$—$COH(O)$
E

infrared assignments

$3300\ cm^{-1}$ ← H—$C\equiv CCH_2$ ... $1708\ cm^{-1}$ → $C=O$ (CO—H) ← $3000\ cm^{-1}$

$2150\ cm^{-1}$ ... $CH_2CH=CH_2$ ← $1642\ cm^{-1}$

proton nuclear magnetic assignments

2.70 ppm → H ; 10.8 ppm → CO—H ; 5.10 ppm → CH₂CH=CH₂

H—$C\equiv CCH_2$ 2.01 ppm ; 2.37–2.60 ppm → CH₂CH=CH₂ ; 5.75 ppm

17.49

$\delta\,4.25$ (2H, q, $J = 7$ Hz)

$\delta\,1.32$ (3H, t, $J = 7$ Hz)

$\delta\,7.68$ (1H, d, $J = 14$ Hz) → H

$COCH_2CH_3$

$C=C$

$\delta\,7.35 - 7.50$ (5H, m)

H ← $\delta\,6.43$ (1H, d, $J = 14$ Hz)

Each vinylic hydrogen atom is coupled only to the other vinylic hydrogen atom, and each appears as a doublet with $J \sim 14$ Hz.

position of unsplit vinylic hydrogen

|← 14 Hz →|

The normal range for the chemical shift of vinylic hydrogen atoms is about $\delta\,5$–6. Both vinylic hydrogen atoms in ethyl (E)-3-phenyl propenoate are shifted farther downfield, indicating that they are being deshielded by something other than the carbon-carbon double bond. The higher field ($\delta\,6.43$) vinylic hydrogen atom is deshielded by the carbonyl group of the ester function. The lower field vinylic hydrogen atom ($\delta\,7.68$) is even more deshielded because it is β to the carbonyl group and adjacent to the phenyl ring.

17.49 (cont)

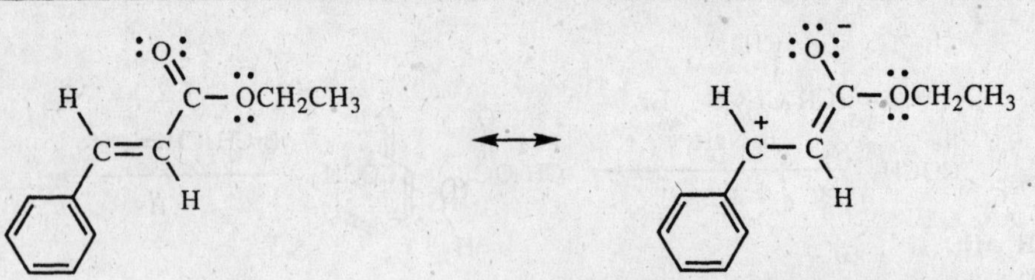

The chemical shift of the hydrogen atom on the β-carbon atom reflects the lower electron density at this carbon, symbolized by the positive charge shown in the resonance contributor above.

17.50

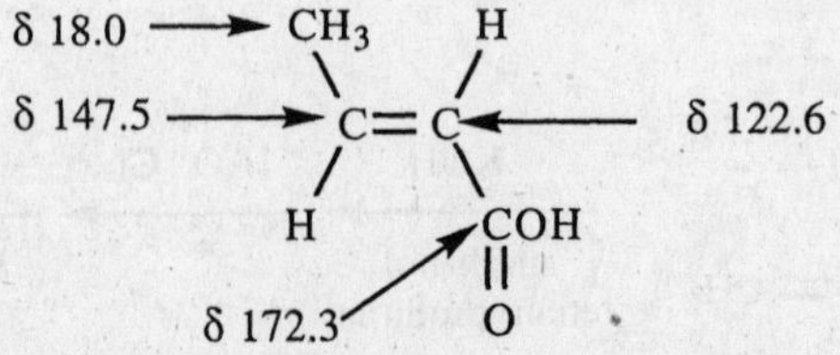

$$\delta\ 18.0 \longrightarrow CH_3 \qquad H$$
$$\delta\ 147.5 \longrightarrow C=C \longleftarrow \delta\ 122.6$$
$$H \qquad COH$$
$$\delta\ 172.3 \qquad O$$

Chemical shifts in ^{13}C magnetic resonance spectra have essentially the same kind of dependence on electron density as proton chemical shifts. A carbon atom attached to an electron-withdrawing group or atom will be shifted downfield (deshielded) relative to a carbon atom attached to an electron-donating group or atom. The carbon atom of the carboxylic acid group, attached to two oxygens, absorbs farthest downfield and the methyl group farthest upfield. Of the two carbon atoms of the double bond, carbon 3 absorbs farther downfield than carbon 2 because it is β to the carbonyl group.

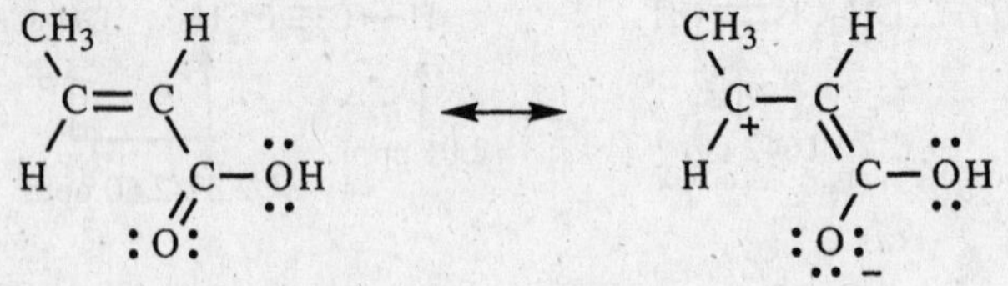

Supplemental Problems

S17.1   Give structural formulas for all species symbolized by letters in the following equations.

(a) 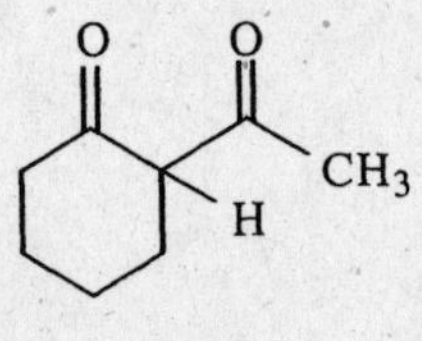 [cyclohexanone with CH₃ and H substituents] + $D_2O \longrightarrow$ A

(b) $CH_2(COCH_2CH_3)_2 \xrightarrow[\text{ethanol}]{CH_3CH_2ONa}$ B $\xrightarrow{}$ C   [benzyl chloride shown above arrow]

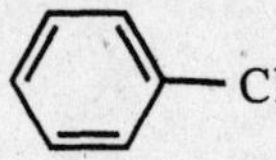

(c) $CH_3CCH_2COCH_2CH_3 \xrightarrow[\text{ethanol}]{CH_3CH_2ONa}$ D $\xrightarrow{BrCH_2CH_2COCH_2CH_3}$ E

(d) $CH_3CCH_2COCH_2CH_3 \xrightarrow[\text{ethanol}]{CH_3CH_2ONa}$ F $\xrightarrow{CH_2=CHCH_2CH_2Br}$ G $\xrightarrow[\text{toluene}]{NaH}$ H $\xrightarrow{CH_3CHCH_3}$ I

S17.1

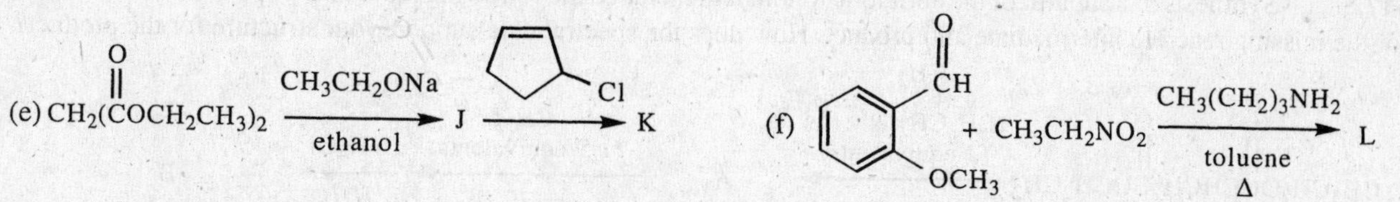

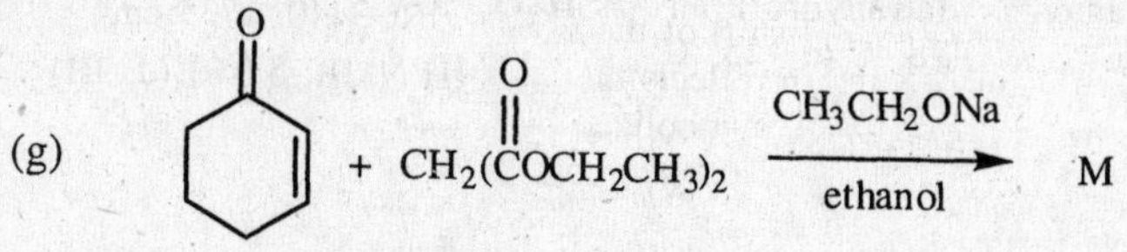

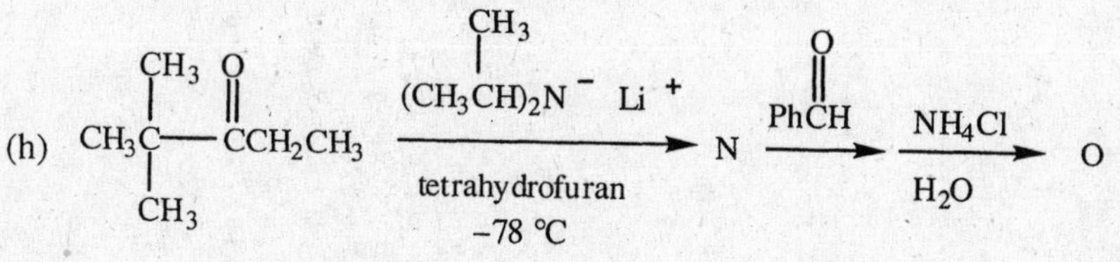

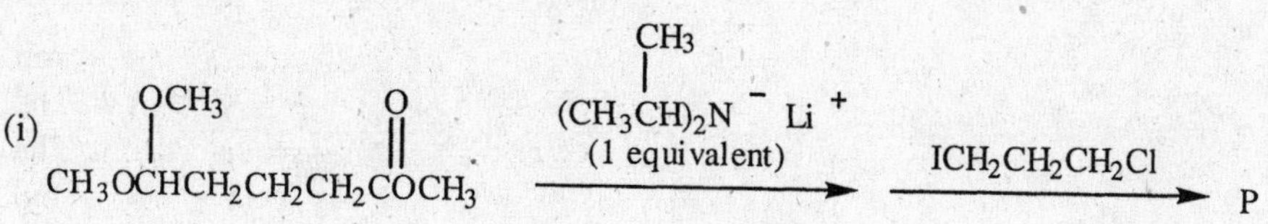

(e) $CH_2(COCH_2CH_3)_2$ $\xrightarrow[\text{ethanol}]{CH_3CH_2ONa}$ J $\xrightarrow{}$ K

(f) $\quad + CH_3CH_2NO_2$ $\xrightarrow[\substack{\text{toluene}\\\Delta}]{CH_3(CH_2)_3NH_2}$ L

(g) $\quad + CH_2(COCH_2CH_3)_2$ $\xrightarrow[\text{ethanol}]{CH_3CH_2ONa}$ M

(h) $CH_3\underset{\underset{CH_3}{|}}{\overset{\overset{CH_3}{|}}{C}}-CCH_2CH_3$ $\xrightarrow[\substack{\text{tetrahydrofuran}\\-78\ °C}]{(CH_3CH)_2N^-Li^+}$ N $\xrightarrow{PhCH}$ $\xrightarrow[H_2O]{NH_4Cl}$ O

(i) $CH_3OCHCH_2CH_2CH_2COCH_3$ $\xrightarrow[\substack{(CH_3CH)_2N^-Li^+\\(1\ \text{equivalent})}]{}$ $\xrightarrow{ICH_2CH_2CH_2Cl}$ P

S17.2   Write a mechanism for the decarboxylation of malonic acid, $HOCCH_2COH$.

S17.3   When the ketolactone shown below is heated with concentrated hydrochloric acid, the product is 5-chloro-2-pentanone. Treatment of 5-chloro-2-pentanone with aqueous sodium hydroxide gives cyclopropyl methyl ketone in about 80% yield. How would you rationalize these experimental observations?

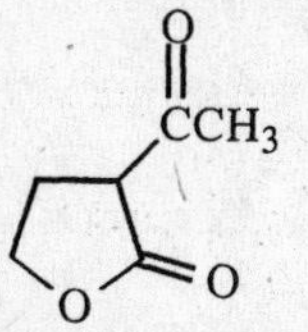

ketolactone used
as starting material

S17.4   Compounds other than aldehydes and ketones may be used to provide components of aldol-type reactions. For example, the following reaction was observed.

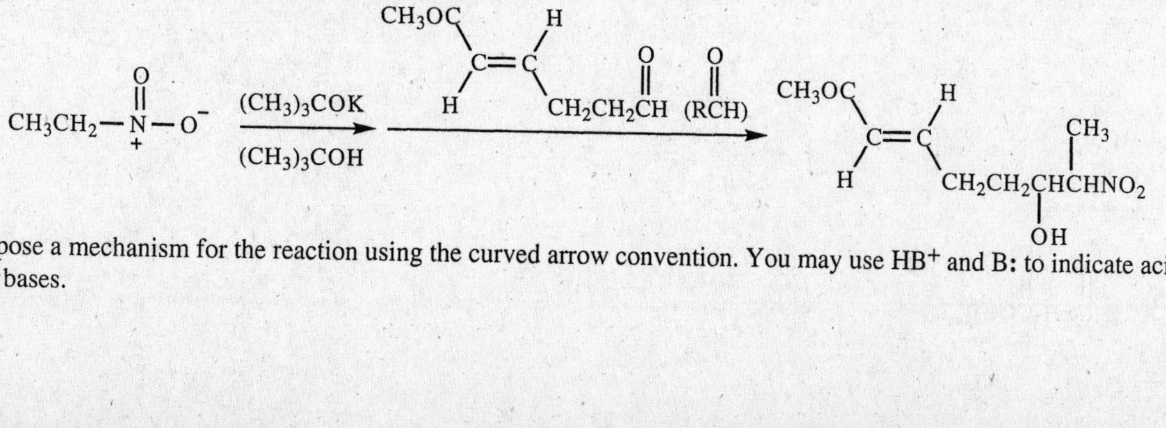

$CH_3CH_2-\overset{+}{N}\overset{O}{\underset{O^-}{\parallel}}$ $\xrightarrow[(CH_3)_3COH]{(CH_3)_3COK}$ $\xrightarrow{}$

Propose a mechanism for the reaction using the curved arrow convention. You may use HB^+ and $B:$ to indicate acids and bases.

S17.5   Synthesis of a subunit of the antibiotic ionomycin included the following reactions. Supply structural formulas for the missing reactive intermediate and product. How does the spectral data support your structure for the product?

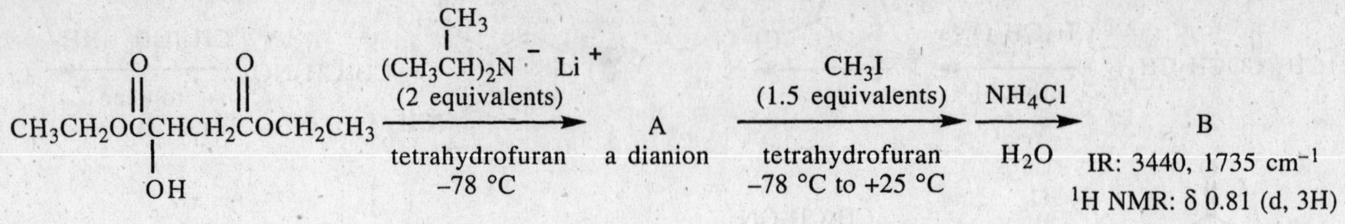

$$
\underset{\underset{OH}{|}}{CH_3CH_2OCCHCH_2COCH_2CH_3} \xrightarrow[\substack{\text{tetrahydrofuran} \\ -78\ ^\circ C}]{\substack{CH_3 \\ | \\ (CH_3CH)_2N^-\ Li^+ \\ (2\ \text{equivalents})}} \underset{\text{a dianion}}{A} \xrightarrow[\substack{\text{tetrahydrofuran} \\ -78\ ^\circ C\ \text{to}\ +25\ ^\circ C}]{\substack{CH_3I \\ (1.5\ \text{equivalents})}} \xrightarrow[H_2O]{NH_4Cl} \underset{\substack{\text{IR: 3440, 1735 cm}^{-1} \\ {}^1\text{H NMR: } \delta\ 0.81\ (d, 3H)}}{B}
$$

18 Polyenes

18.1   (a) 1,4-hexadiyne (b) 4-hexen-2-yne (see note) (c) 1,4-cyclohexadiene
(d) 1,5-cyclooctadiene (e) (*E*)-1-phenyl-1,4-pentadiene (f) 1-chloro-2,4-heptadiyne

Note: the lower number goes to the group that is the end of the name. When an alkene and an alkyne functional group are present in a compound, the alkyne is the at the end of the name and thus gets the lower number.

18.2   (a) $CH_2 = CHCH_2CH_2\overset{\displaystyle O}{\overset{\displaystyle \|}{C}}CH_3$ (b) $HC \equiv CCH_2CH_2C \equiv CH$ (c) $CH_2 = CHCH_2CH = CH_2$

(d) $CH_3C \equiv CCH_2C \equiv CCH_3$

(e) $\underset{H}{\overset{CH_3}{C}} = \underset{\underset{\underset{CH_3}{|}}{\overset{|}{CH_2CCH_2CH = CH_2}}}{\overset{H}{\overset{|}{\underset{|}{C}}}} \; CH_3$

(f) $HC \equiv CCH = CHCH = CH_2$

Concept Map 18.1   Different relationships between multiple bonds.

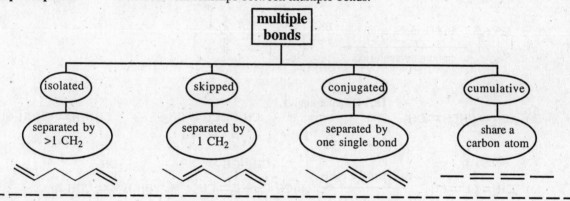

Concept Map 18.2   A conjugated diene.

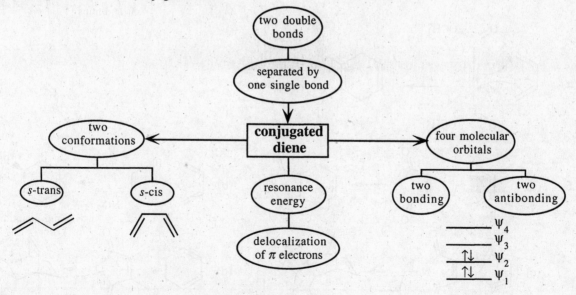

Concept Map 18.3   Addition to dienes.

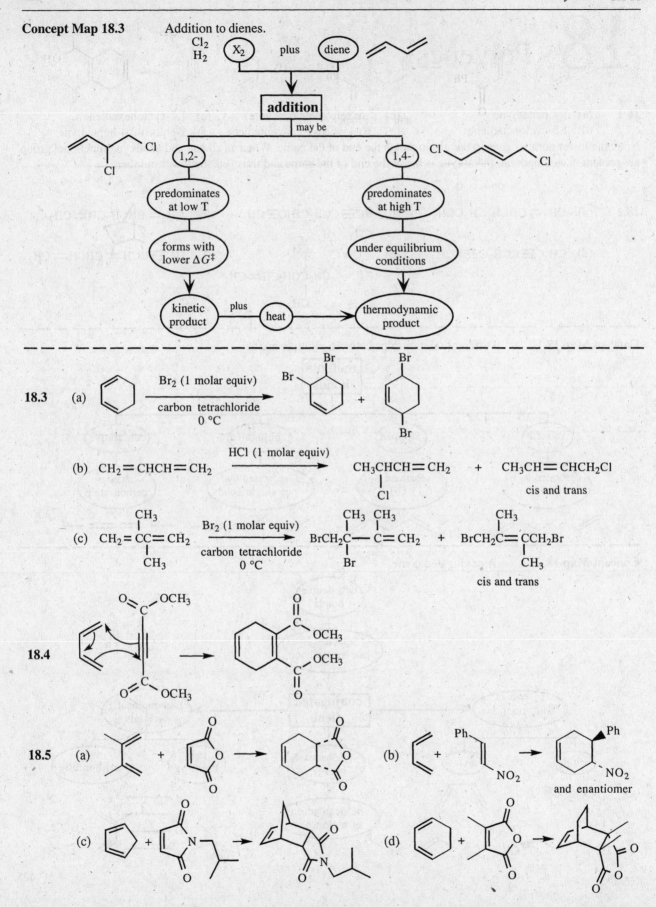

18.5   (e)   (f)

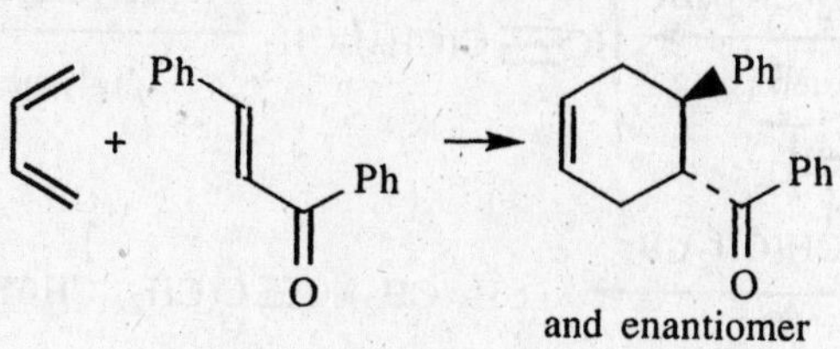

and enantiomer

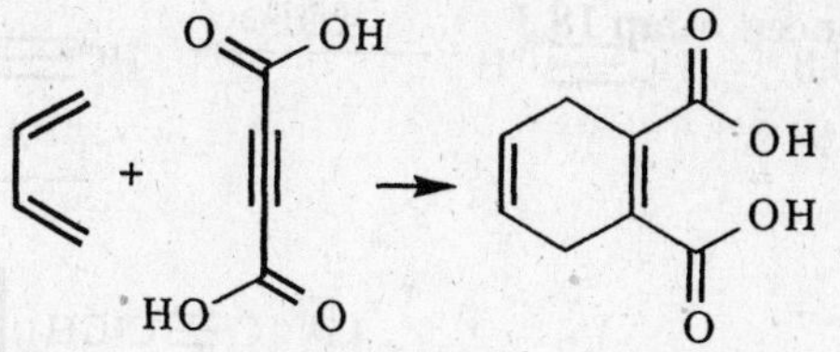

Concept Map 18.4   The Diels-Alder reaction.

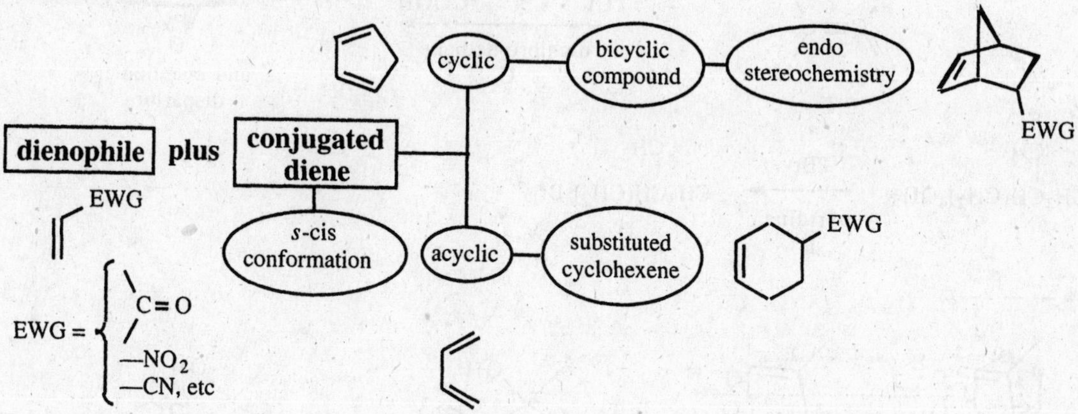

18.6

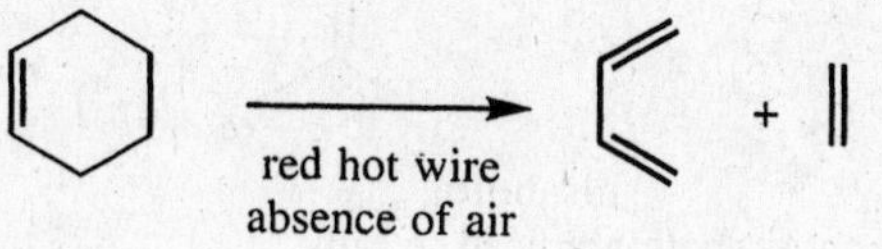

red hot wire
absence of air

18.7

(a)

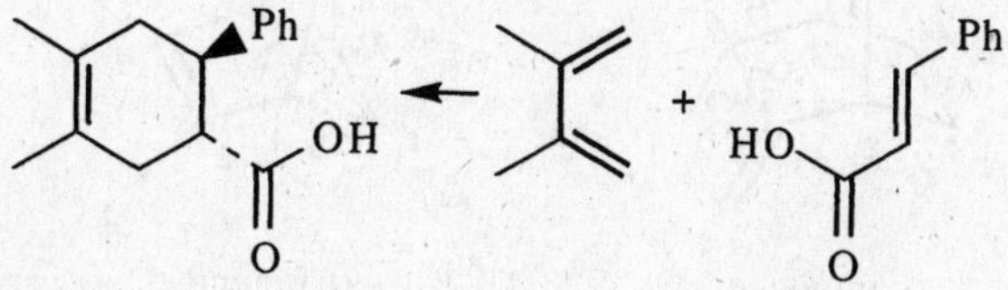

(b)

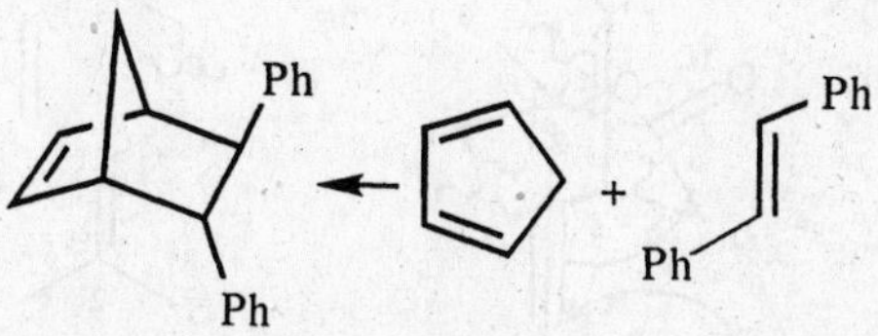

(c)

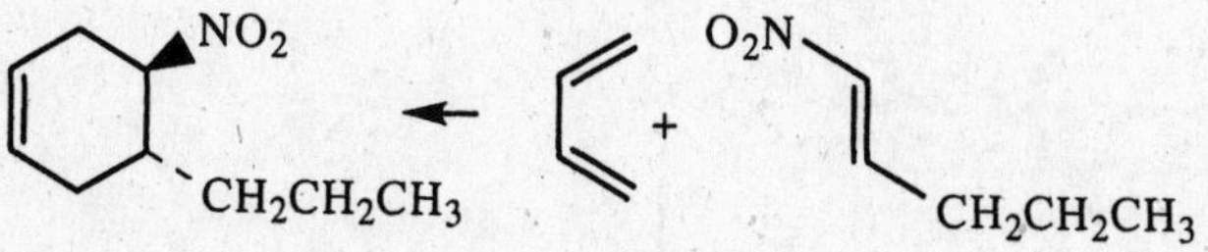

(d)

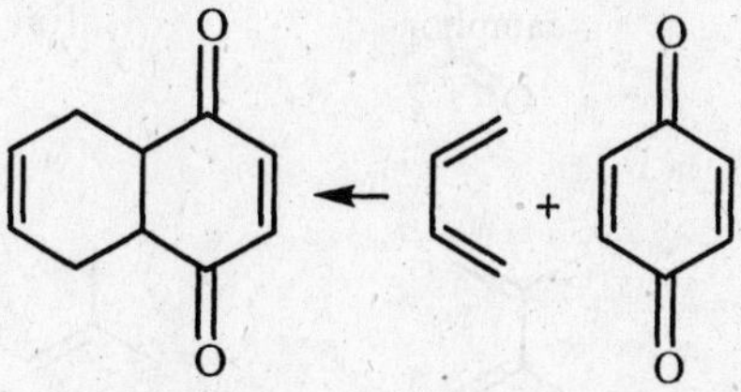

(e)

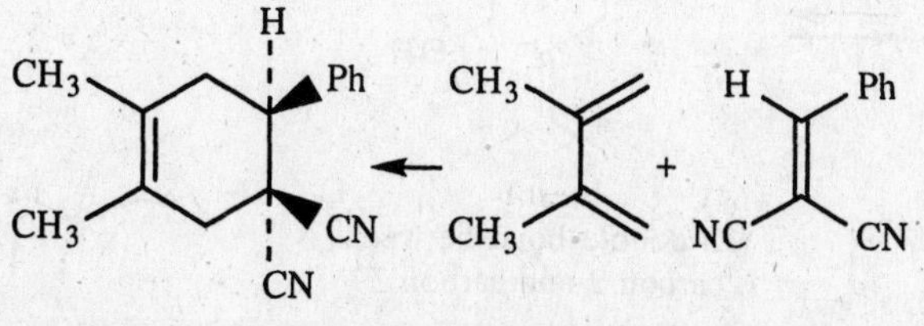

(f)

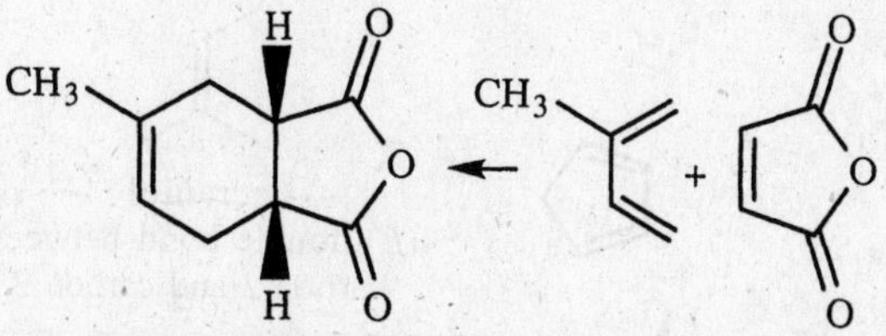

18.8 $HC \equiv CH$ $\xrightarrow{NaNH_2}$ $HC \equiv C^- Na^+$ $\xrightarrow{CH_3(CH_2)_9Br}$ $HC \equiv C(CH_2)_9CH_3$ $\xrightarrow[\text{diglyme}]{CH_3CH_2CH_2CH_2Li}$
                                                          A   B

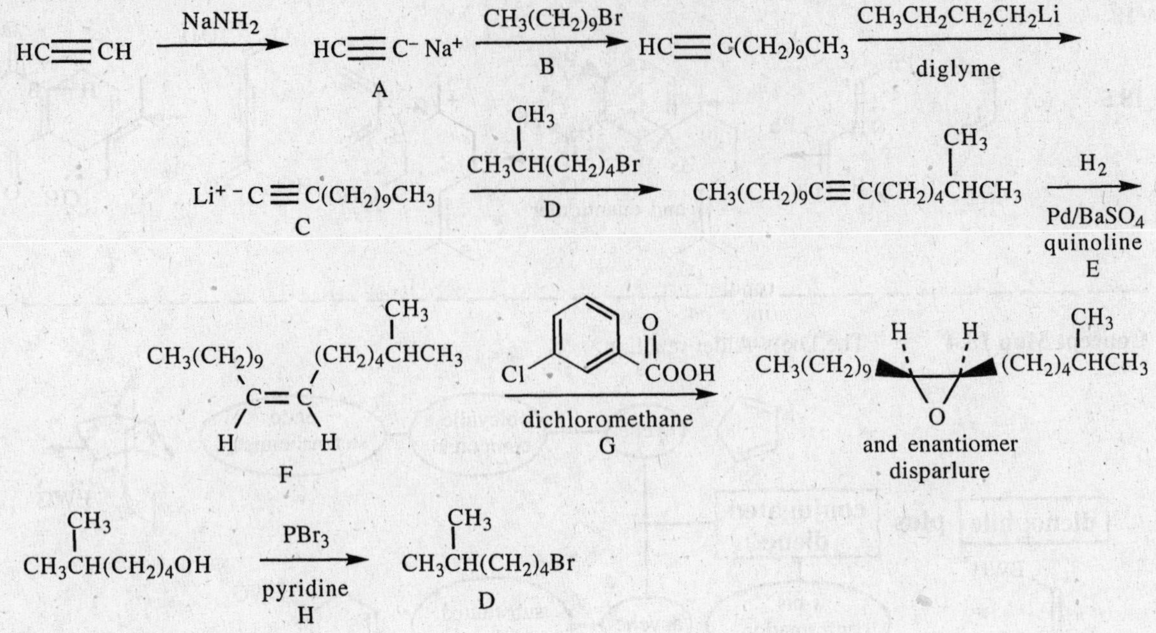

$Li^+ {}^- C \equiv C(CH_2)_9CH_3$ $\xrightarrow[D]{CH_3CH(CH_2)_4Br}$ $CH_3(CH_2)_9C \equiv C(CH_2)_4CHCH_3$ $\xrightarrow[\substack{Pd/BaSO_4 \\ quinoline \\ E}]{H_2}$
          C

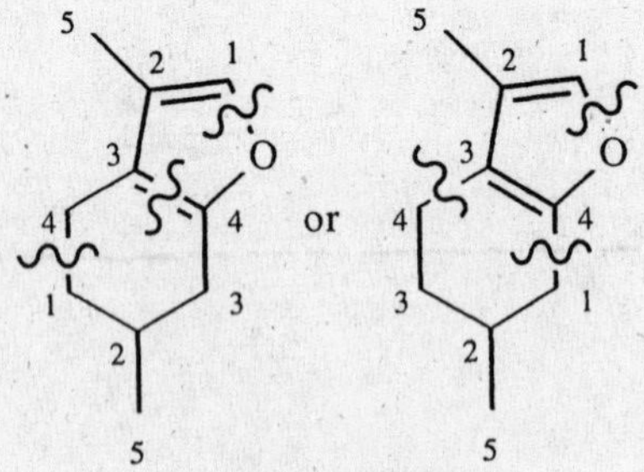

menthofuran
both are head to tail

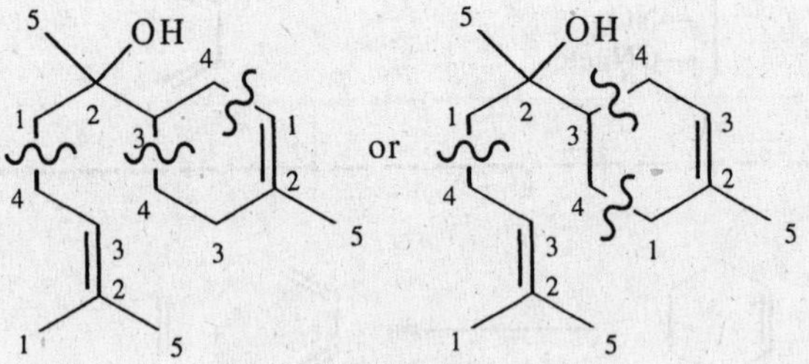

bisabolol
both are head to tail

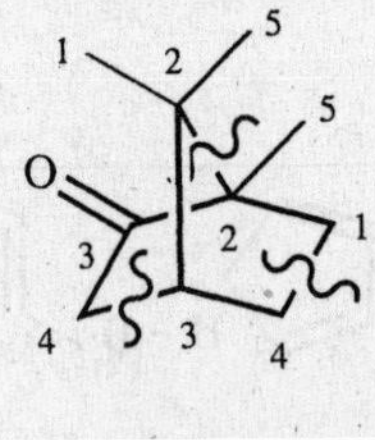

camphor 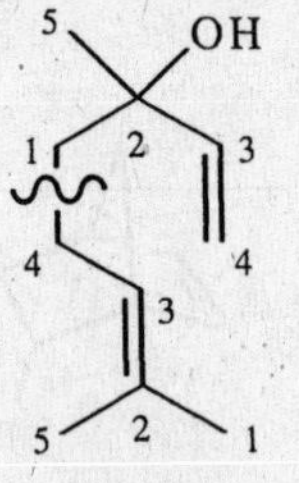   linaloöl 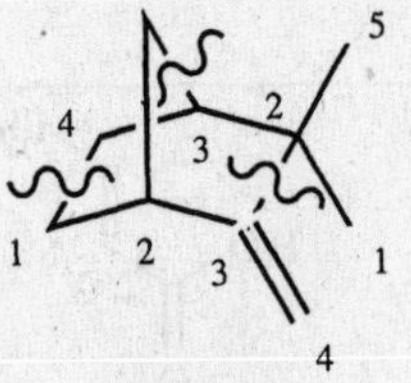   camphene 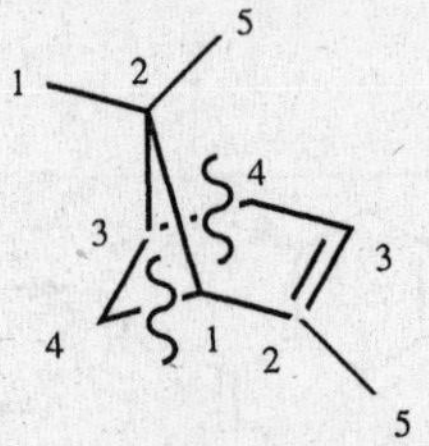   α-pinene

18.10

(a)

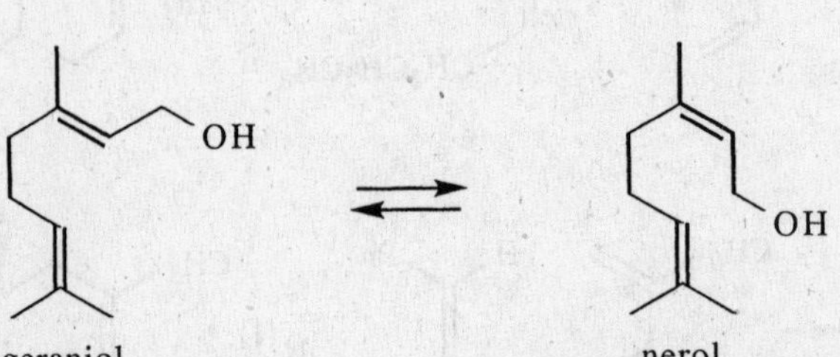

geraniol
(*E*)-double bond between
carbon 2 and carbon 3

nerol
(*Z*)-double bond between
carbon 2 and carbon 3

18.10 (a) (cont)

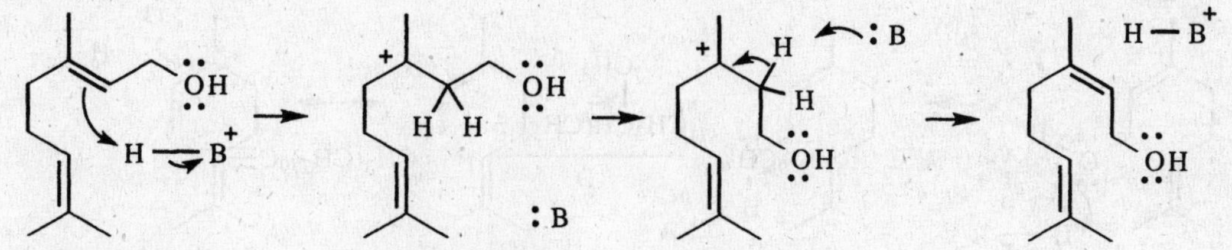

protonation of   rotation around the   deprotonation
the double bond   single bond between
                        carbon 2 and carbon 3

(b)

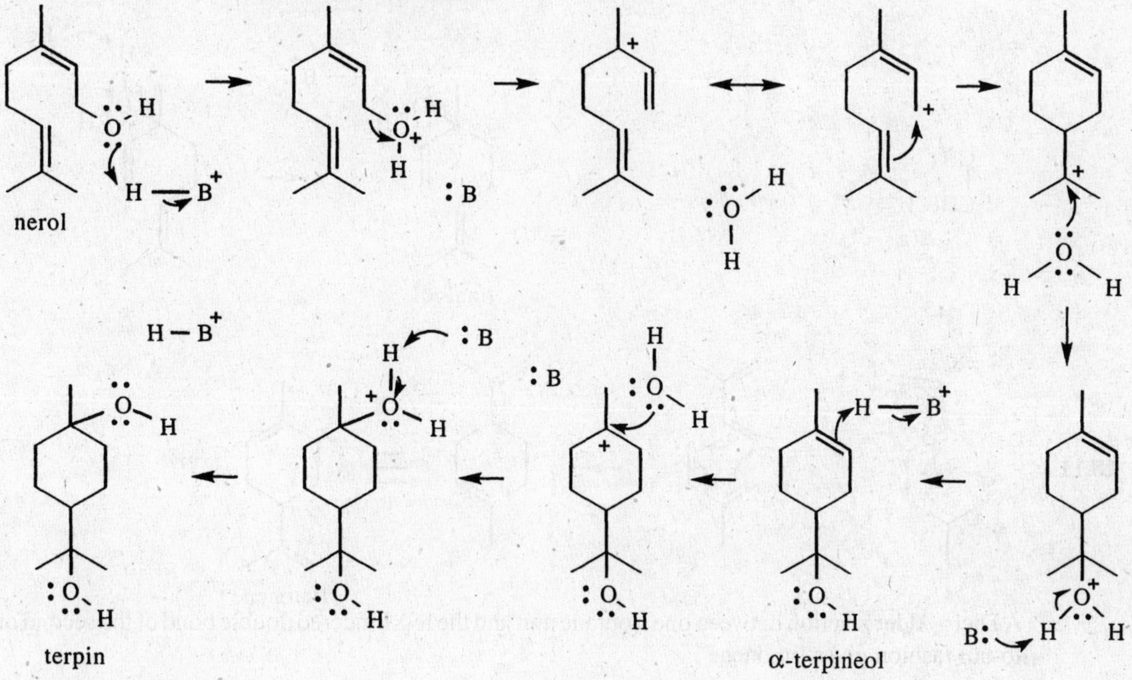

nerol

terpin   α-terpineol

(c)

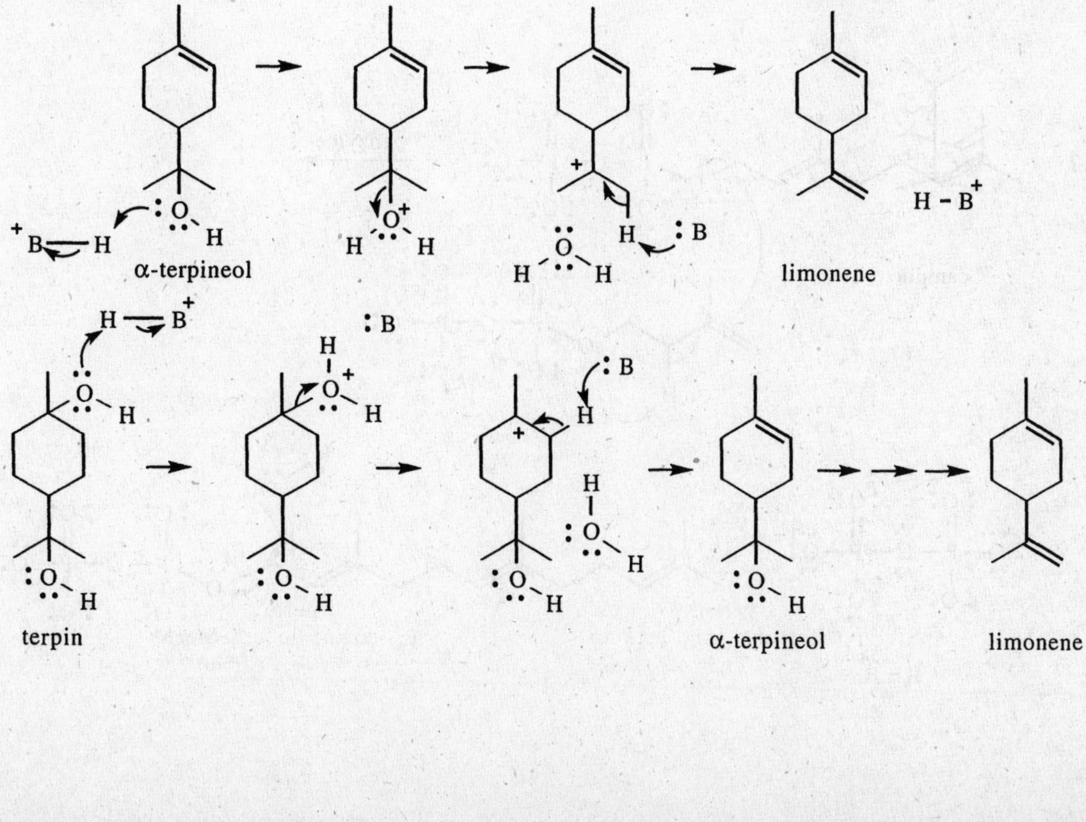

α-terpineol   limonene

terpin   α-terpineol   limonene

18.10

(d)

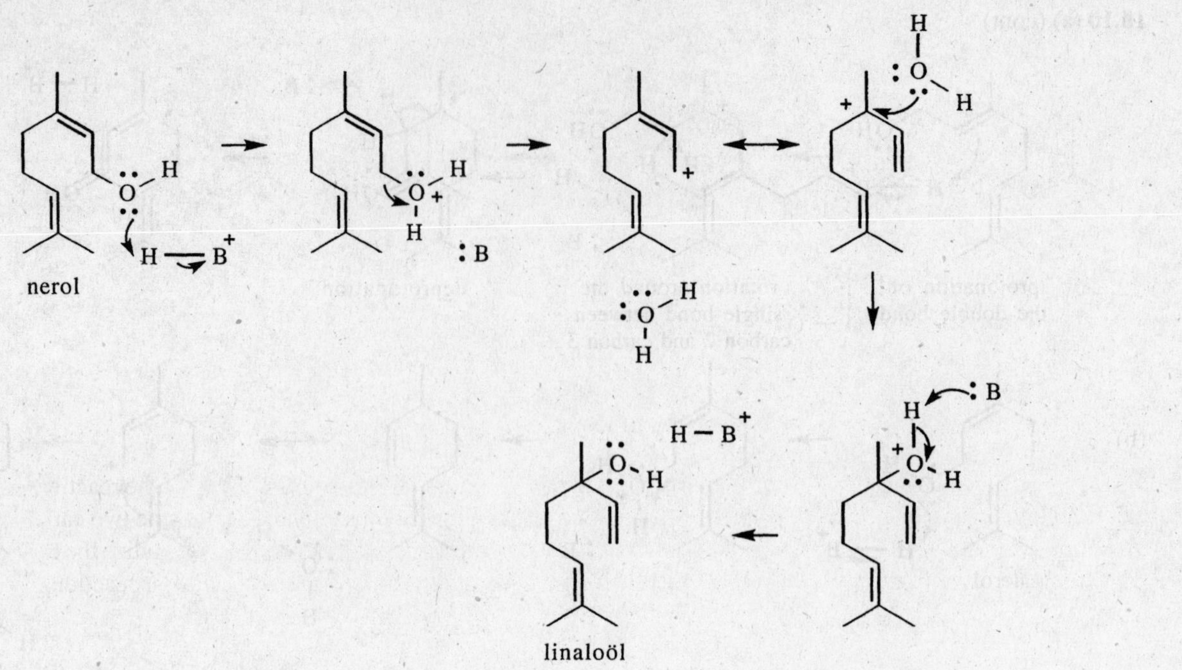

nerol

linaloöl

18.11

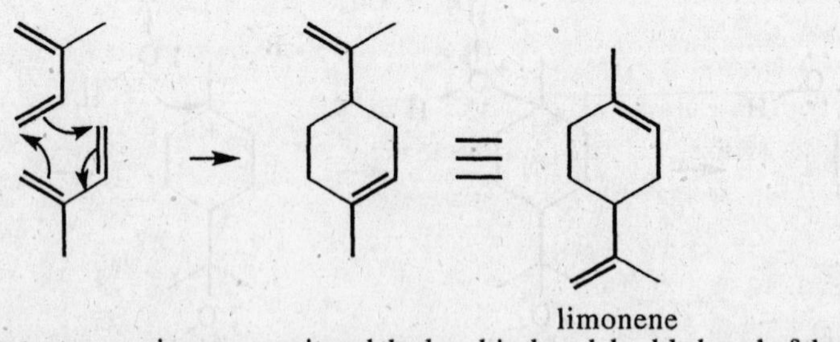

limonene

A Diels-Alder reaction between one isoprene unit and the less hindered double bond of the second one in a head-to-tail fashion gives limonene.

18.12

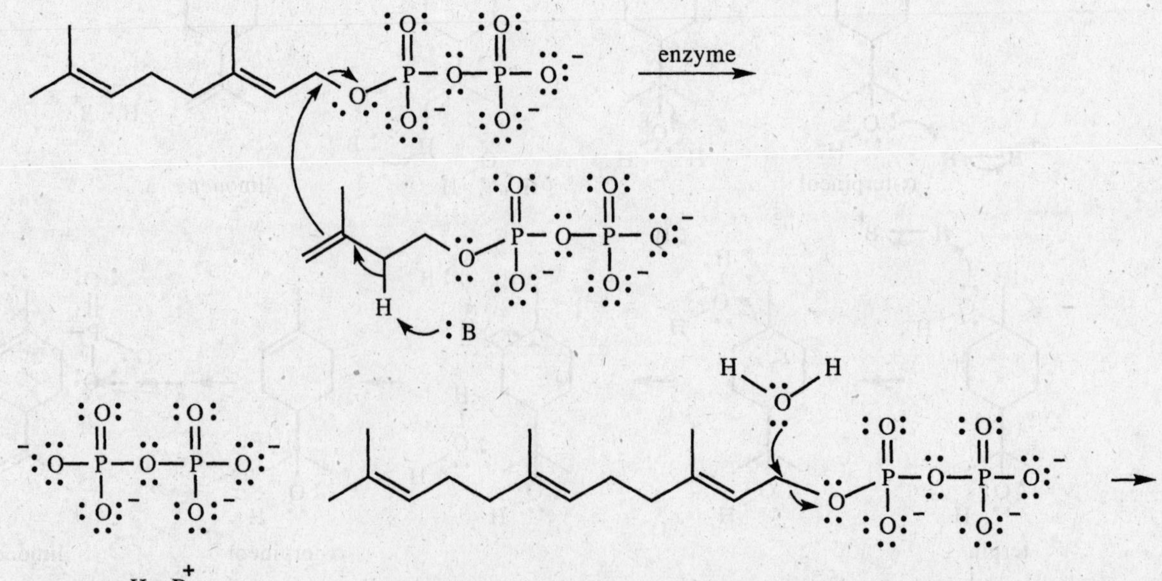

enzyme

H—B$^+$

18.12 (cont)

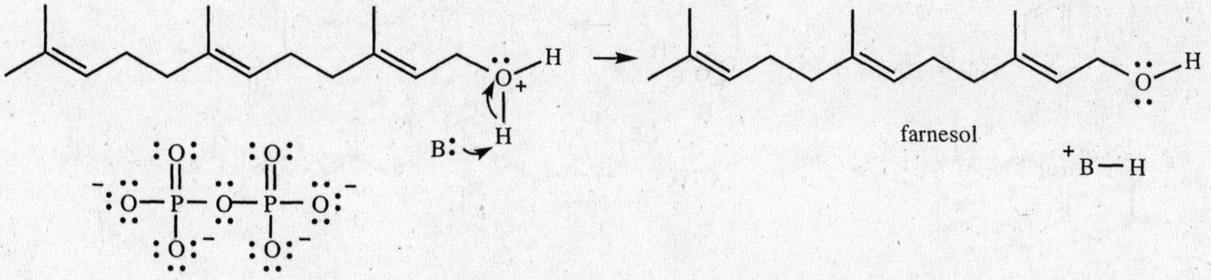

farnesol

18.13   If a carbocation is the reactive intermediate in the biosynthesis of terpenes, the rate of the reaction will depend on the stability of the cation, and hence $\Delta G^{\ddagger}$ for the formation of the cation. A comparison of the two cations, one with a fluorine substituent and the other without shows that we can expect the fluorine atom to destabilize the cation because of its negative inductive effect. This will result in a higher energy transition state, and a slower reaction.

electron-withdrawing effect of fluorine
destabilizes this cation relative to the
one substituted by H; higher $\Delta G^{\ddagger}$,
slower reaction expected.

18.14

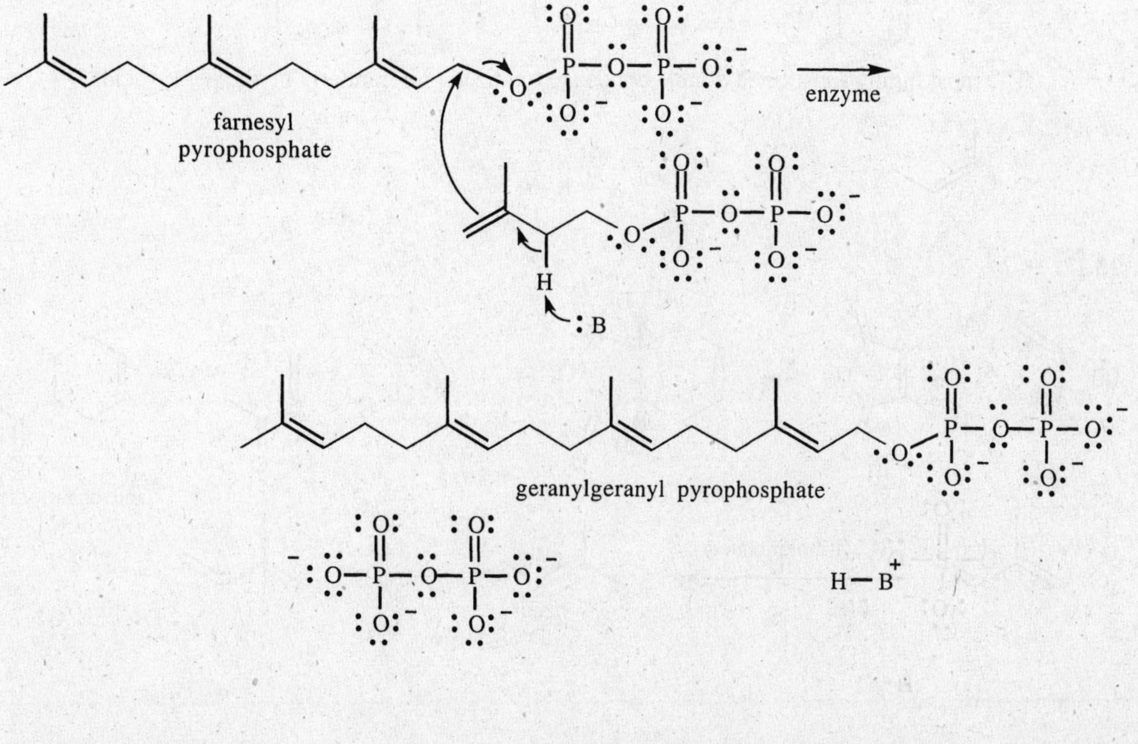

farnesyl
pyrophosphate

enzyme

geranylgeranyl pyrophosphate

18.15

(a)

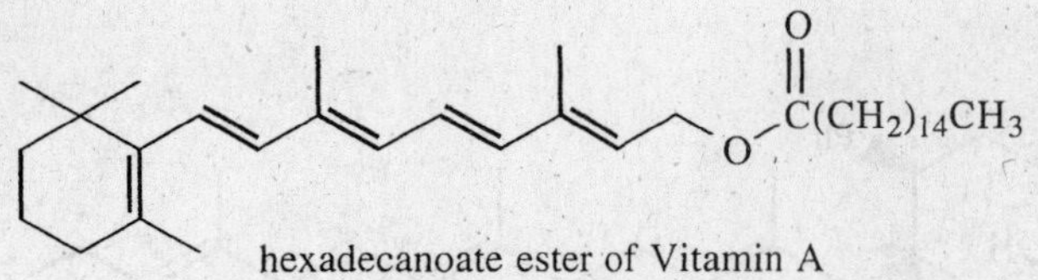

hexadecanoate ester of Vitamin A

(b)

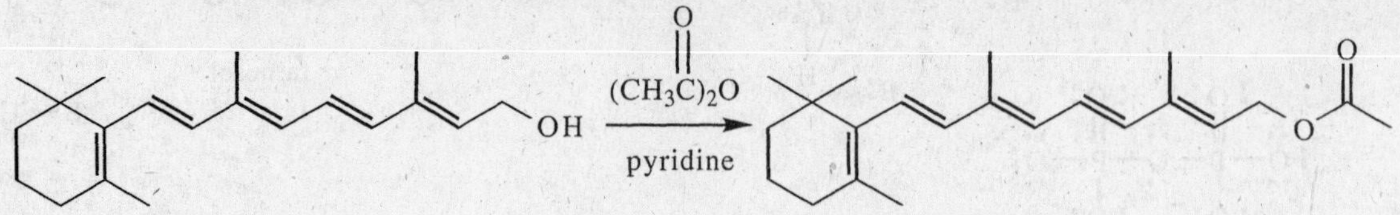

18.16

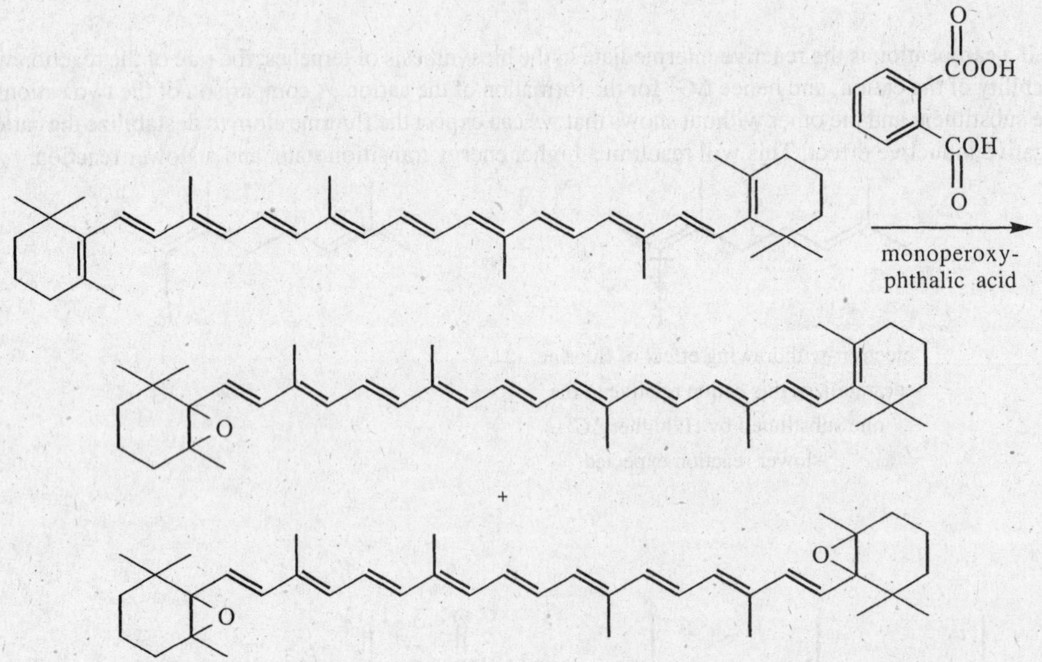

The most highly substituted double bonds are oxidized preferentially by the peroxyacid.

18.17

(a)

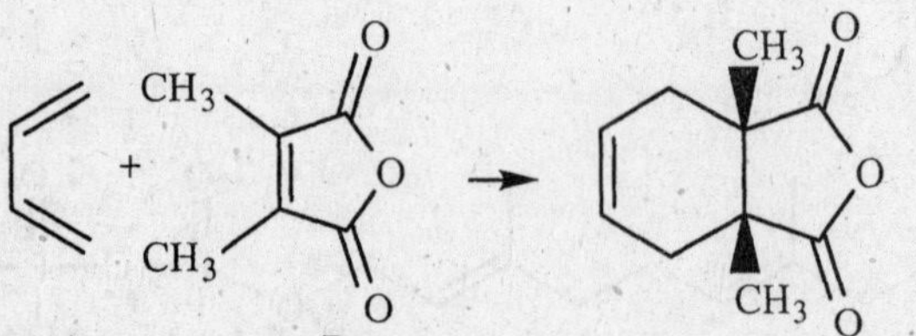

(b)

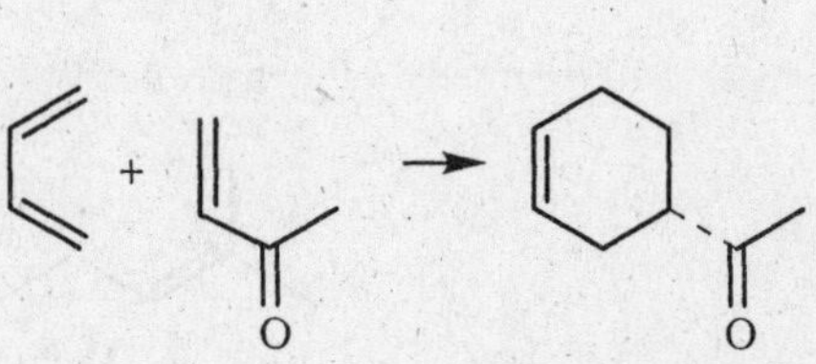

and enantiomer
C

(c)

HCl (1 molar equiv)

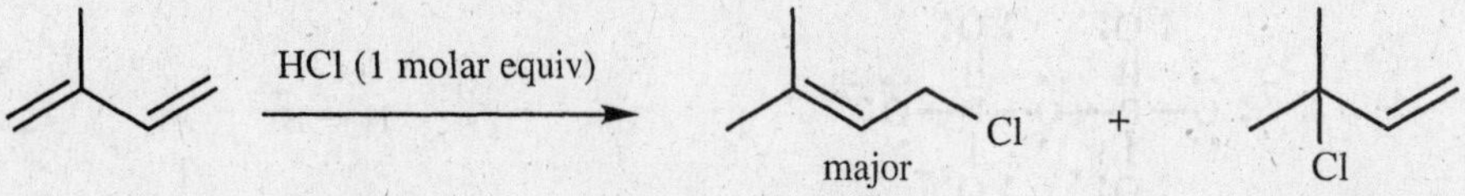

major
D

+

E

18.17

(d)

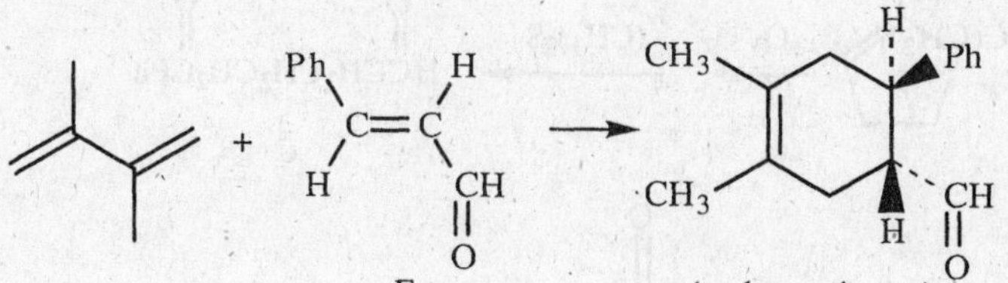

F   (and enantiomer)

(e) $CH_3C \equiv CCH_2C \equiv CCH_2CH_3$ $\xrightarrow[\text{quinoline}]{\substack{H_2 \\ Pd/BaSO_4}}$

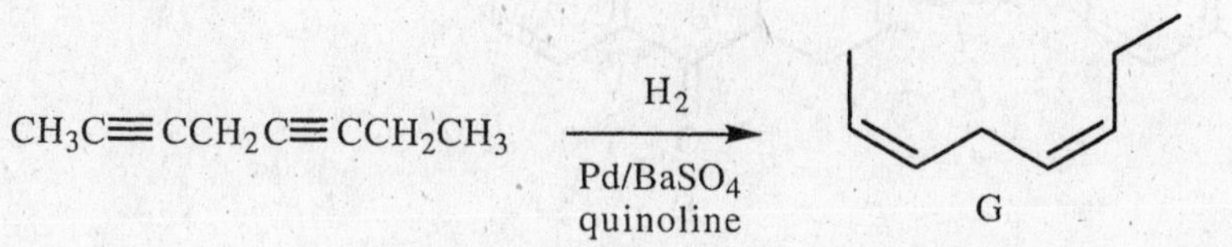

G

(f)

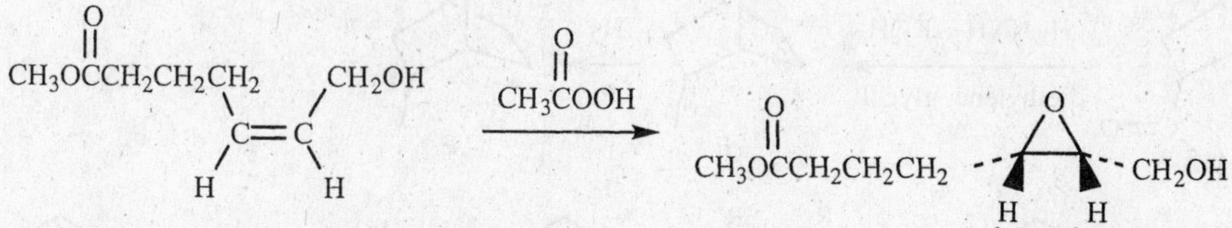

CH_3COOH

and enantiomer

(g)

$\xrightarrow{\Delta}$

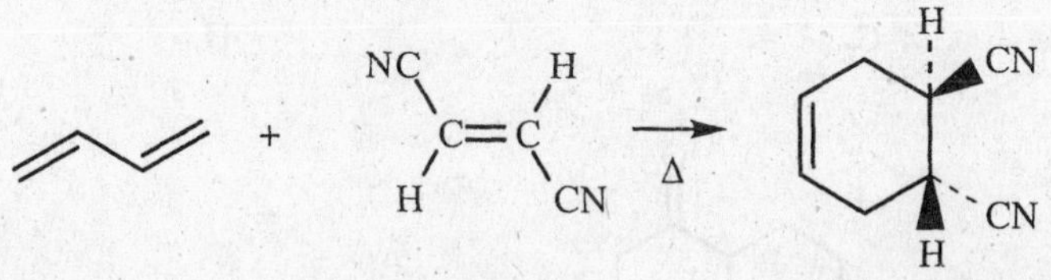

(and enantiomer)
I

(h)

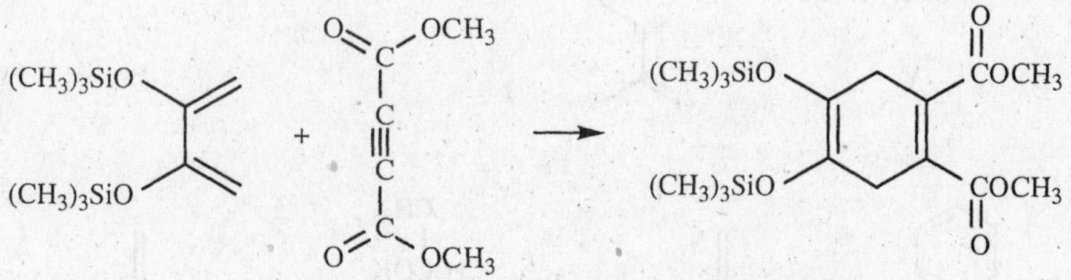

J

(i)

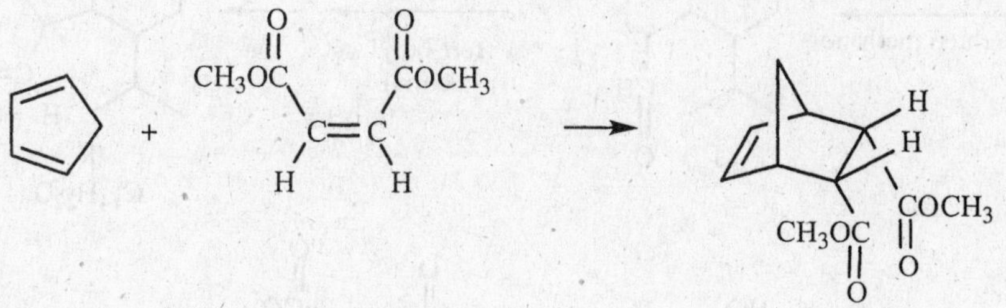

K

18.18 (a)

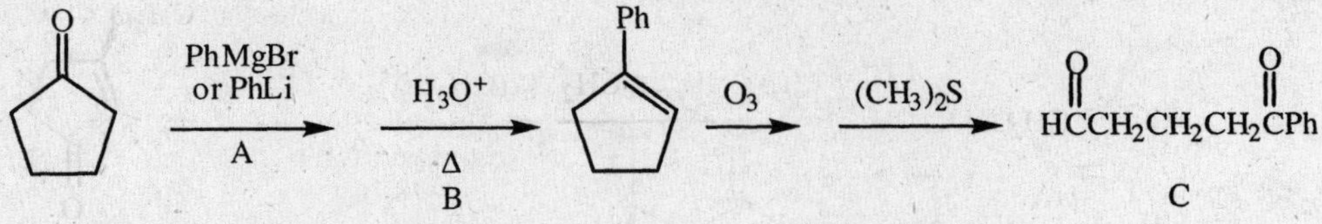

(b)

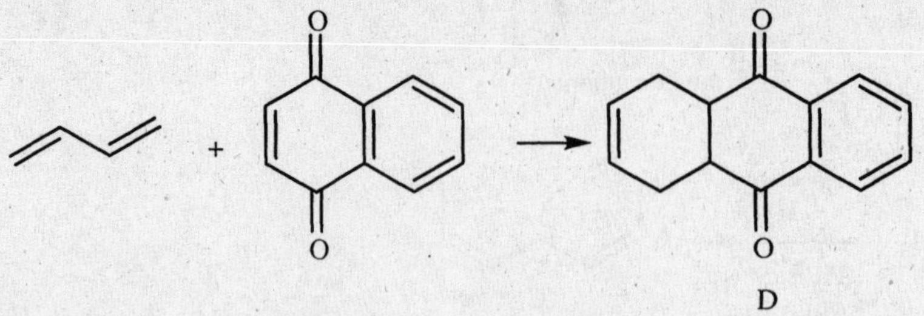

(c)

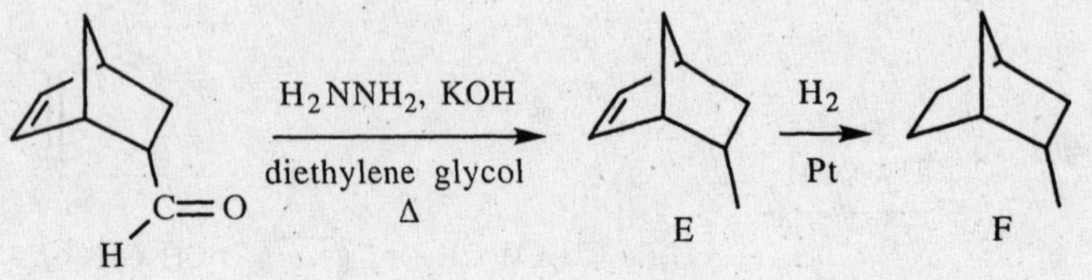

(d)

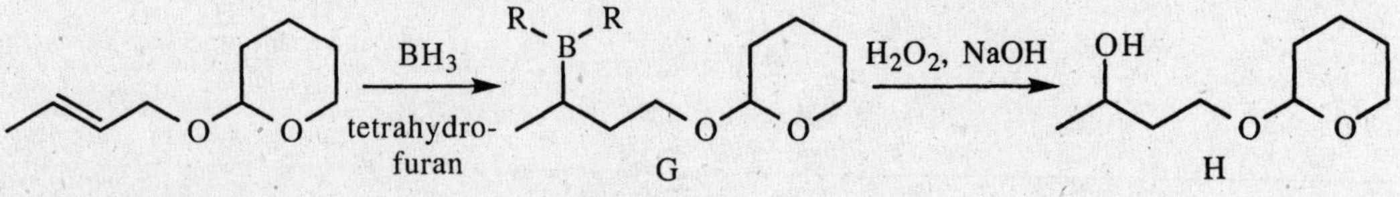

(e)

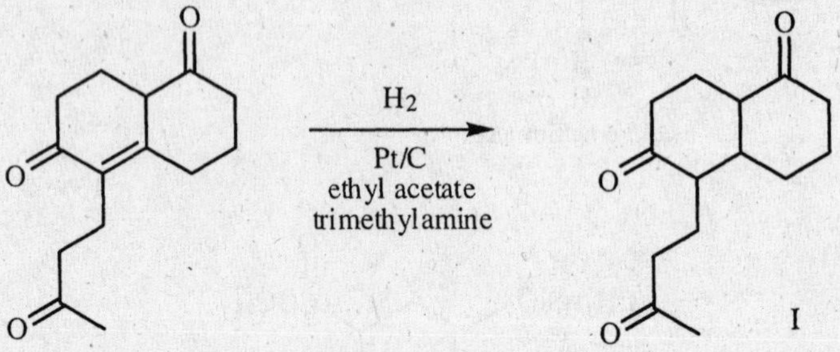

(f)

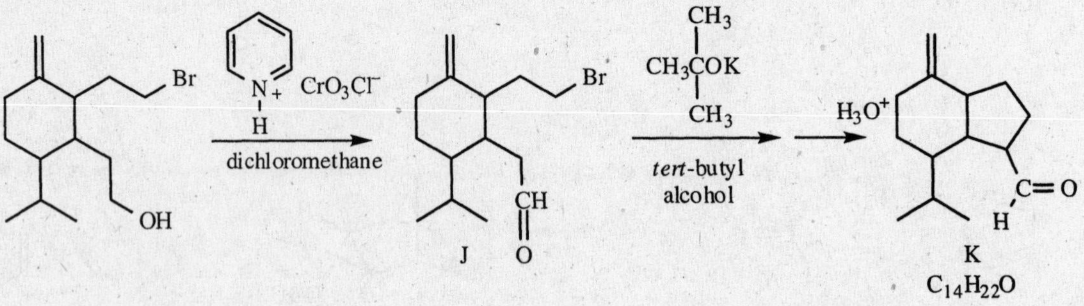

(g)

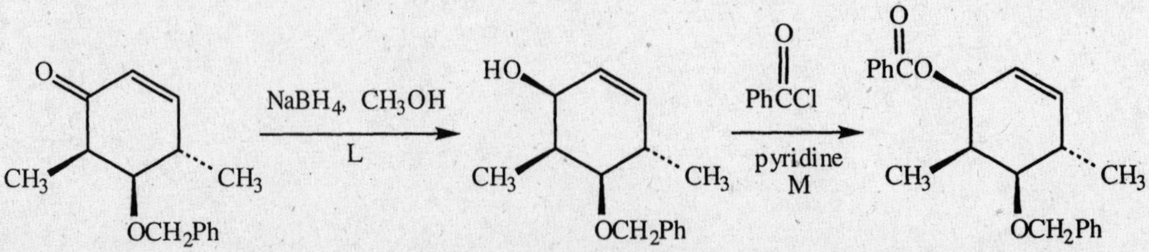

18.19   (a)

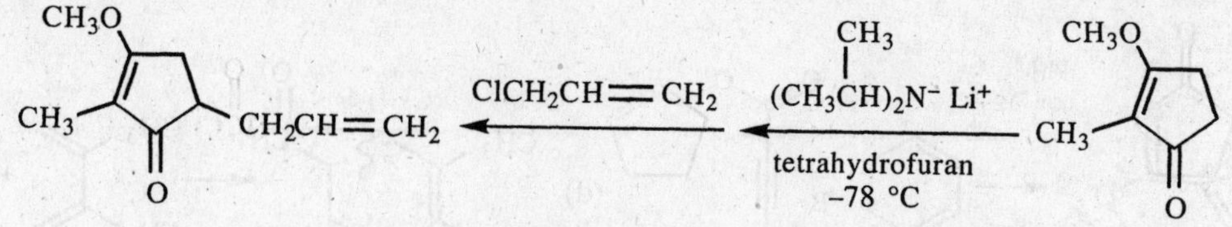

(b)

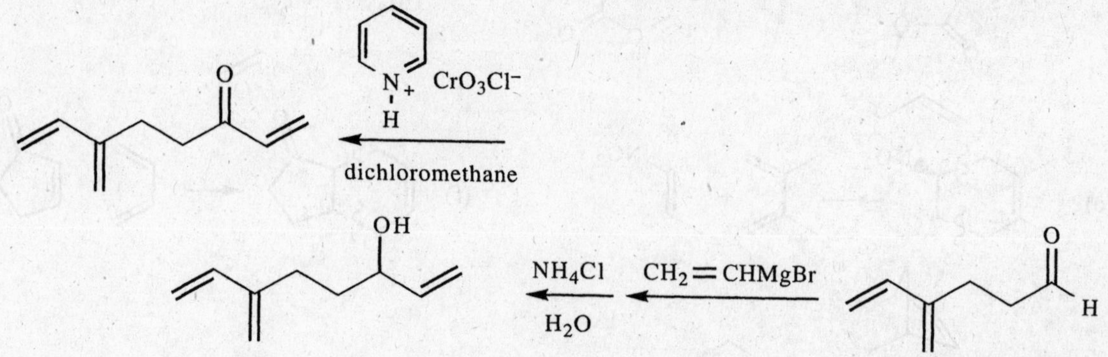

(c)

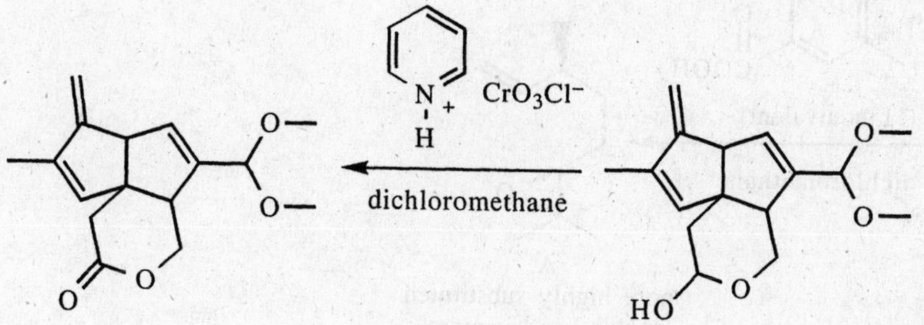

(d)

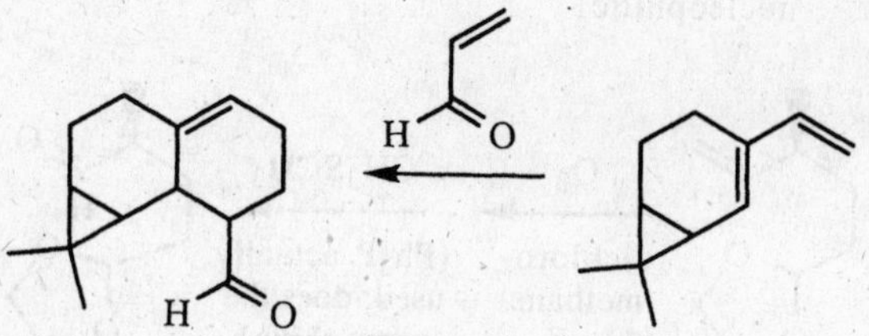

18.20   σ bonds formed   σ bonds formed

(a)

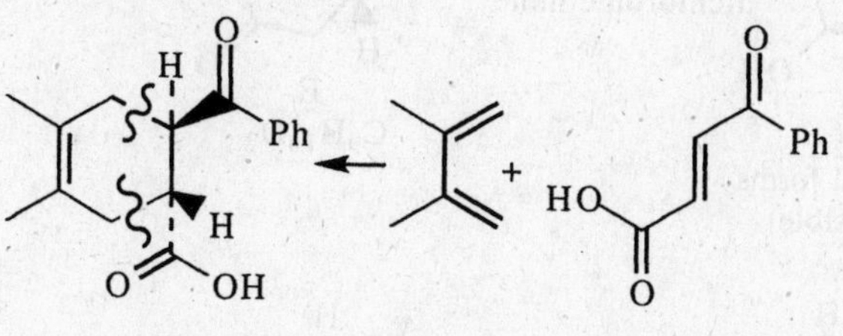

(b)

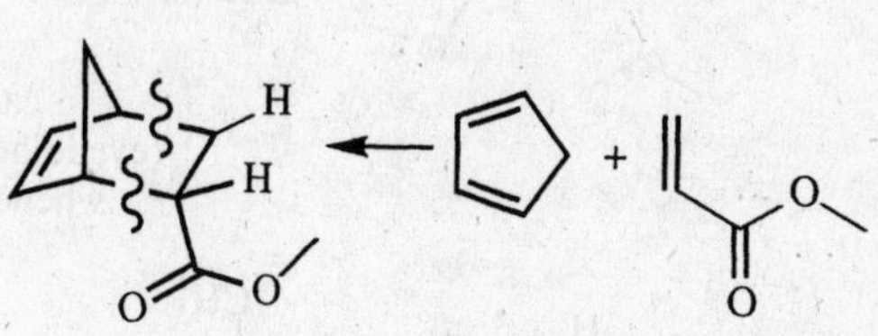

18.20

(c)

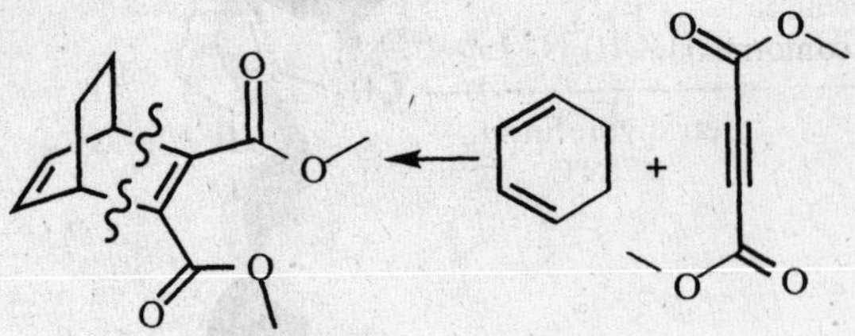

(d)

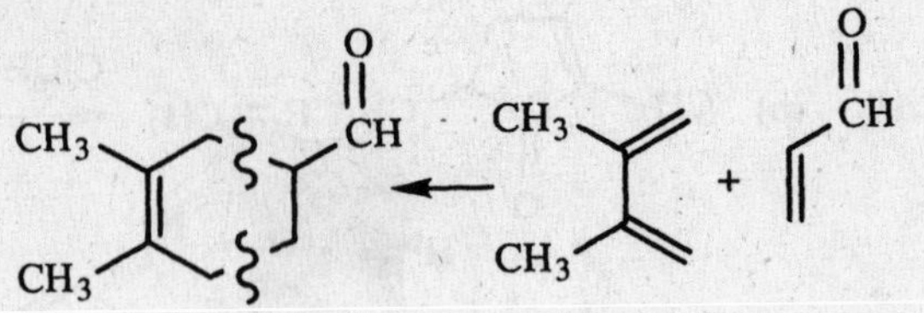

(e)

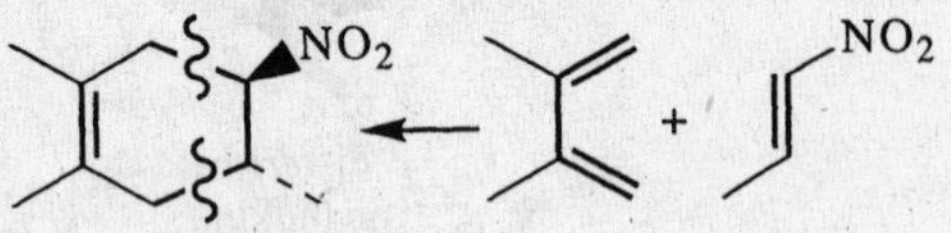

(f)

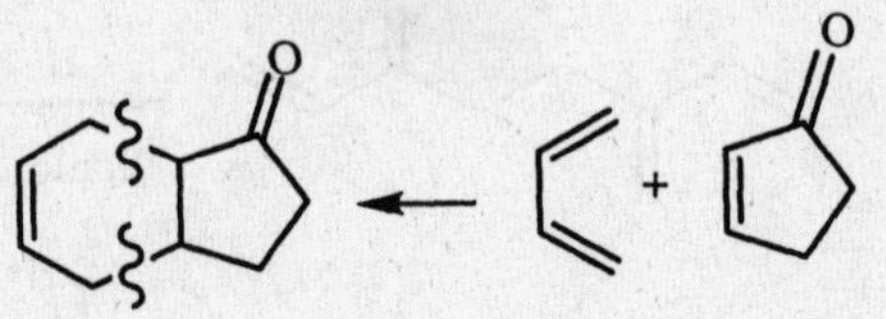

18.21 (a)

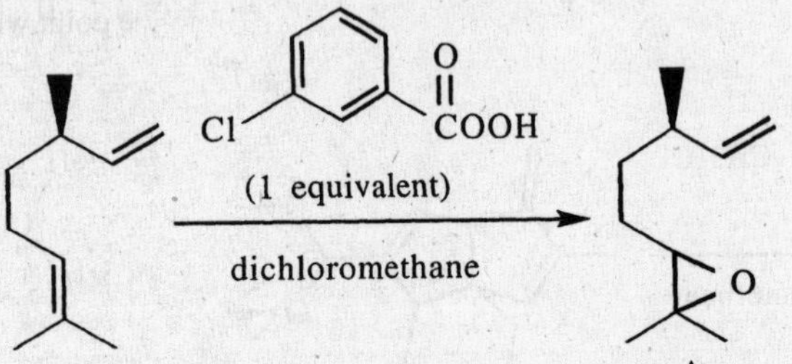

A
(more highly substituted
double bond is more
nucleophilic)

(b)

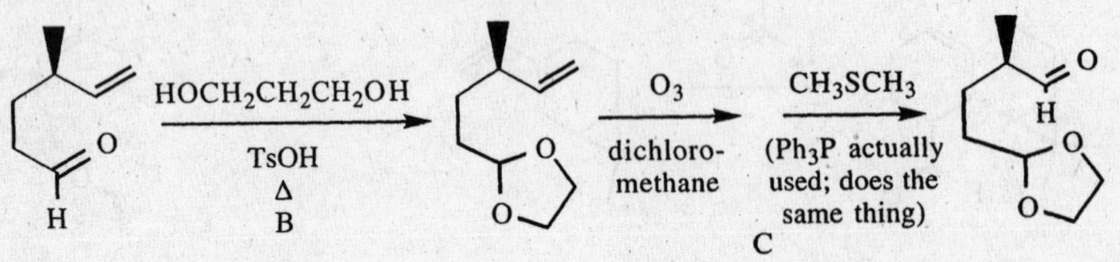

(c)

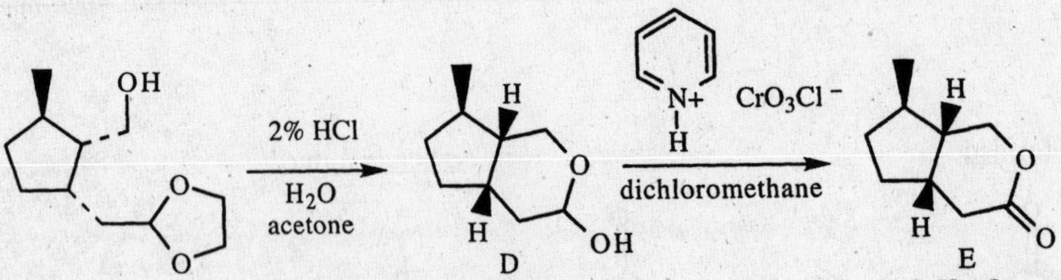

D
$C_9H_{16}O_2$
(cyclic hemiacetal forms
whenever possible)

E
$C_9H_{14}O_2$

(d)

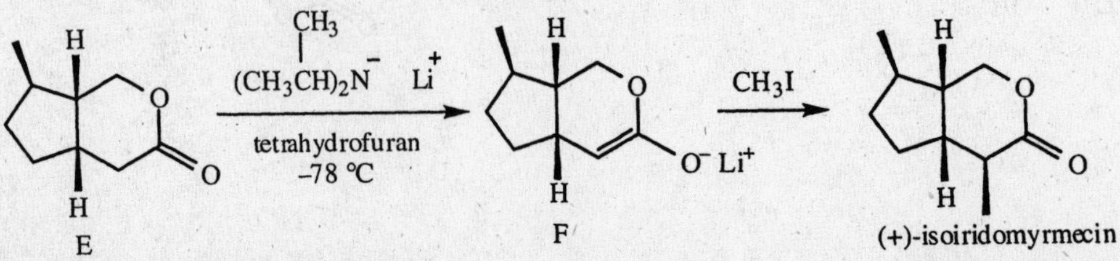

E

F

(+)-isoiridomyrmecin

18.22

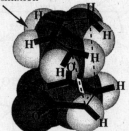

no steric hindrance;
diene can achieve
s-cis conformation

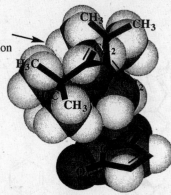

steric hindrance;
difficult for diene to
achieve *s*-cis conformation

transition state for the addition
of 1,3-butadiene to maleic anhydride

transition state for the addition of 2,3-di-*tert*-butyl-
1,3-butadiene to maleic anhydride. The bulkiness of
the *tert*-butyl groups prevents the diene from achieving
the *s*-cis conformation needed for the Diels-Alder reaction.
Good overlap of the π system is also difficult, raising
the energy of the transition state to the point where the
reaction does not proceed.

18.23

18.24 (a)

(R)-(+)-acetate intermediate

(b)

(c)

stereoisomer of the
(R)-(+)-acetate

18.25 (a)

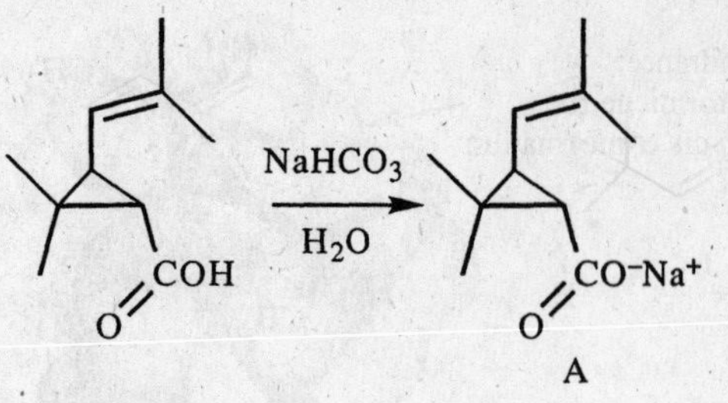
NaHCO$_3$
H$_2$O
COH
CO$^-$Na$^+$
A

(b)

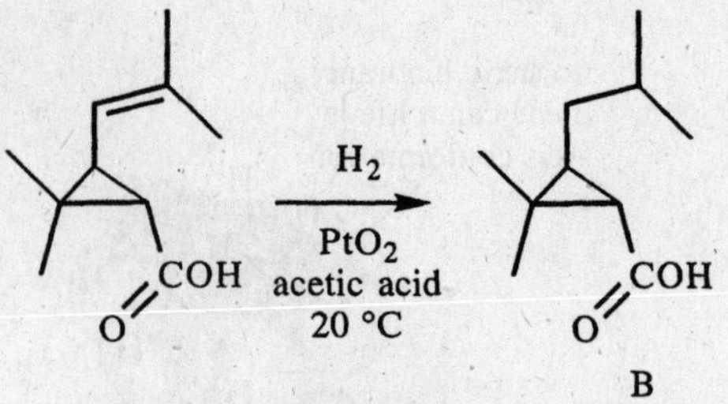
H$_2$
PtO$_2$
acetic acid
20 °C
COH
B

(c)

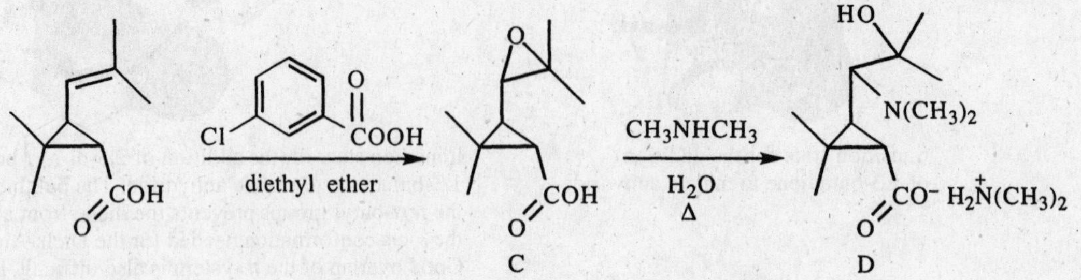
Cl
COOH
diethyl ether
CH$_3$NHCH$_3$
H$_2$O
Δ
HO
N(CH$_3$)$_2$
COH
CO$^-$ H$_2\overset{+}{N}$(CH$_3$)$_2$
C
D

(d)

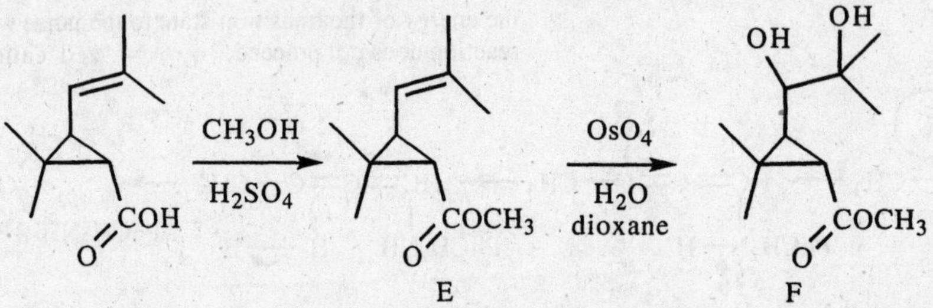
CH$_3$OH
H$_2$SO$_4$
OsO$_4$
H$_2$O
dioxane
OH
OH
COH
COCH$_3$
E
F

(e)

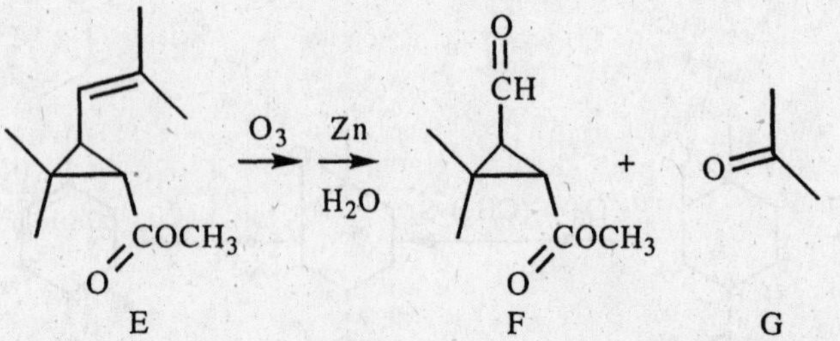
O$_3$
Zn
H$_2$O
CH
COCH$_3$
+
E
F
G

18.26 (a) Bisabolene has the molecular formula $C_{15}H_{24}$. The corresponding saturated alkane would have the formula $C_{15}H_{32}$. Bisabolene has four units of unsaturation. Hydrogenation of bisabolene gives Compound X, $C_{15}H_{30}$. Therefore, bisabolene contains one ring and three π bonds:

(b)

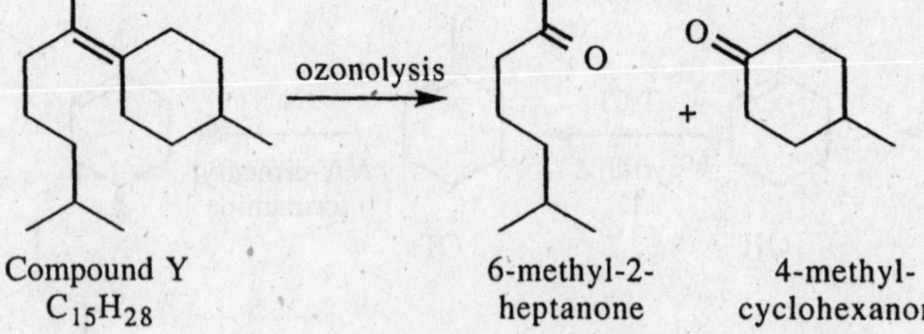
ozonolysis
+
Compound Y
C$_{15}$H$_{28}$
6-methyl-2-
heptanone
4-methyl-
cyclohexanone

(c)

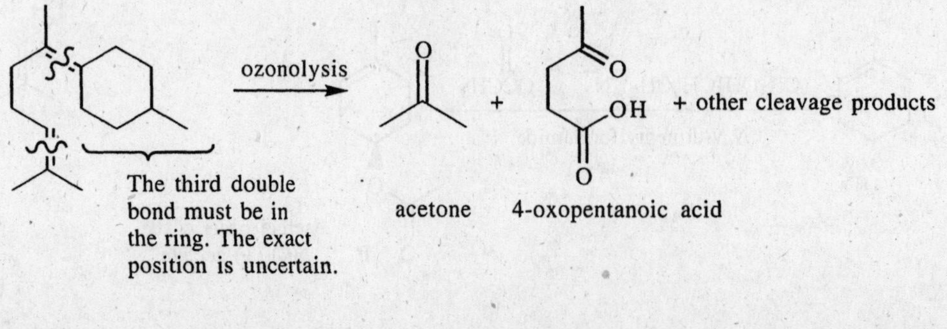
ozonolysis
+
OH
+ other cleavage products
The third double
bond must be in
the ring. The exact
position is uncertain.
acetone
4-oxopentanoic acid

18.26   (d)

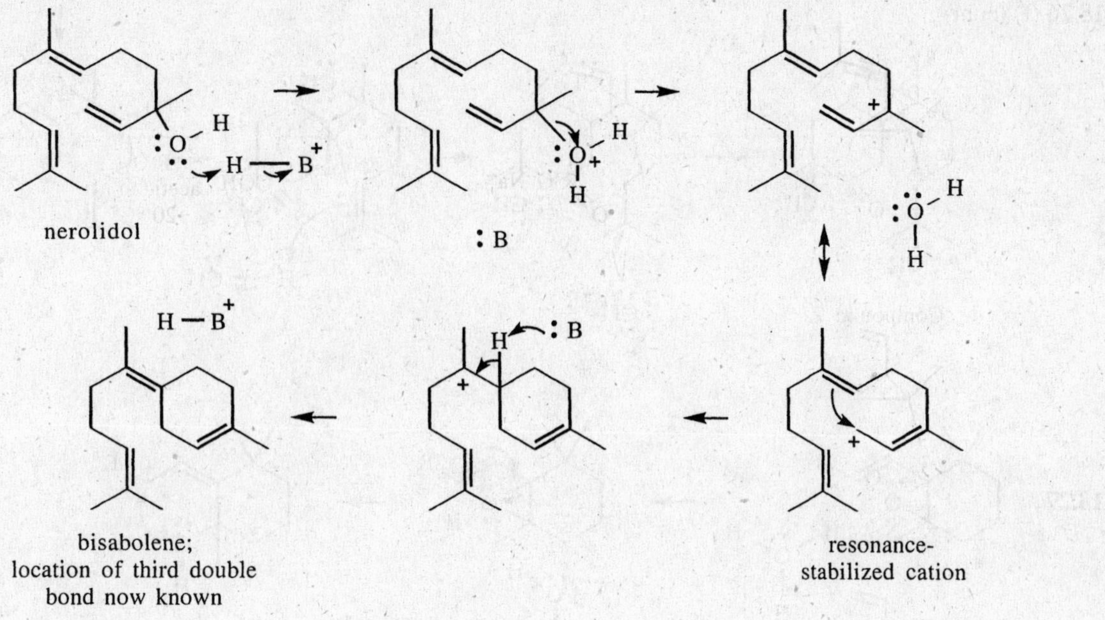

nerolidol

: B

resonance-
stabilized cation

bisabolene;
location of third double
bond now known

(e)

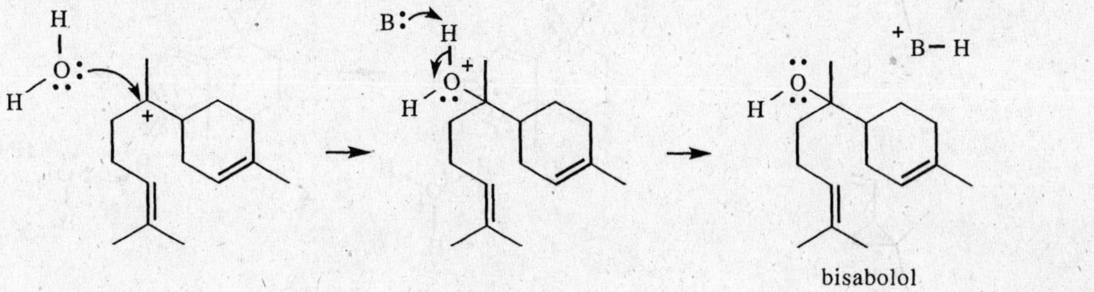

bisabolol

(f)

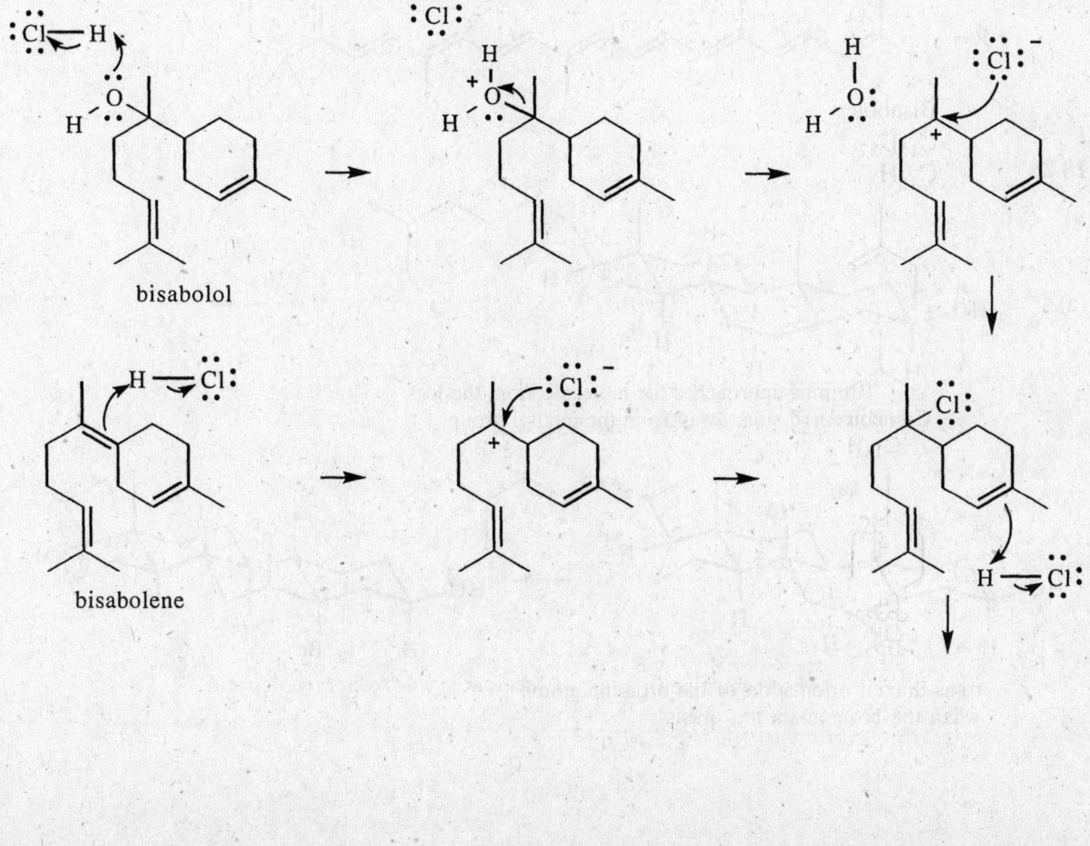

bisabolol

bisabolene

18.26 (f) (cont)

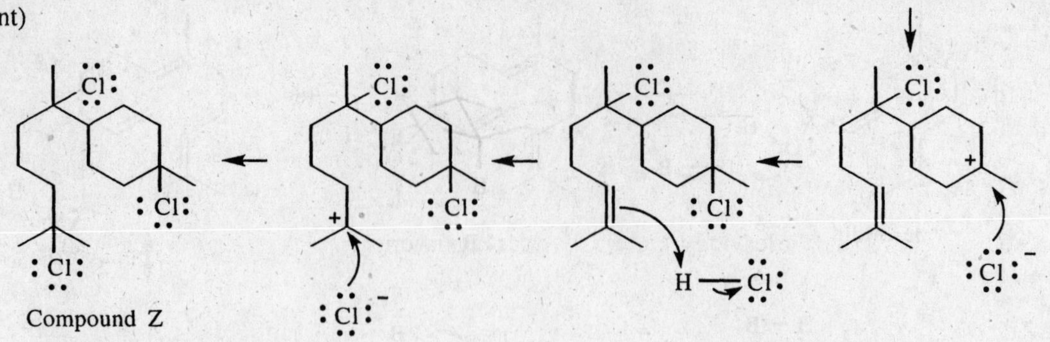

Compound Z

18.27

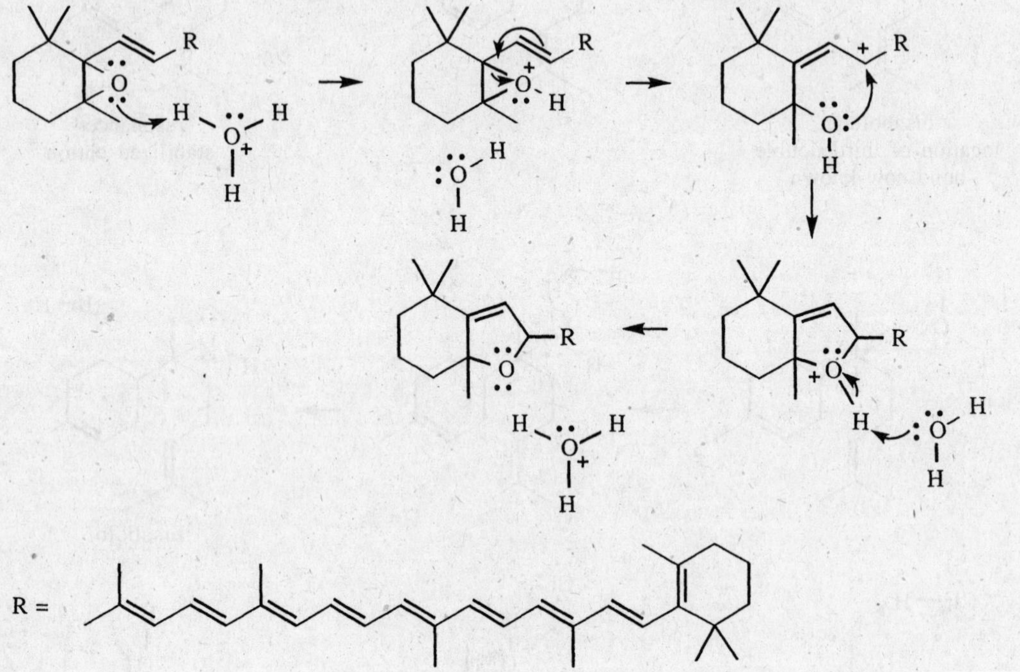

R =

18.28

(a)

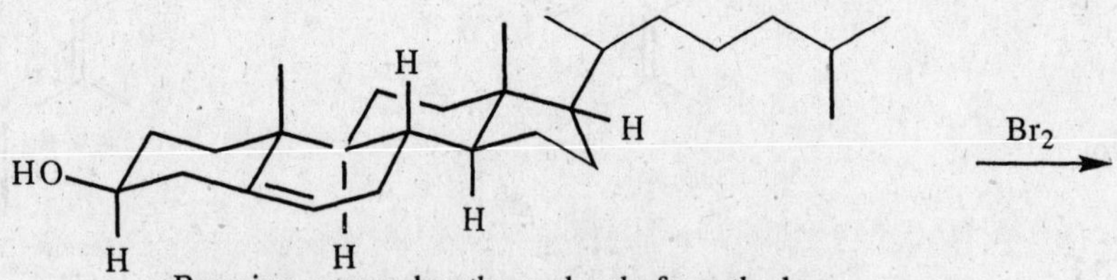

Br₂ →

Bromine approaches the molecule from the less
hindered side, away from the methyl group.

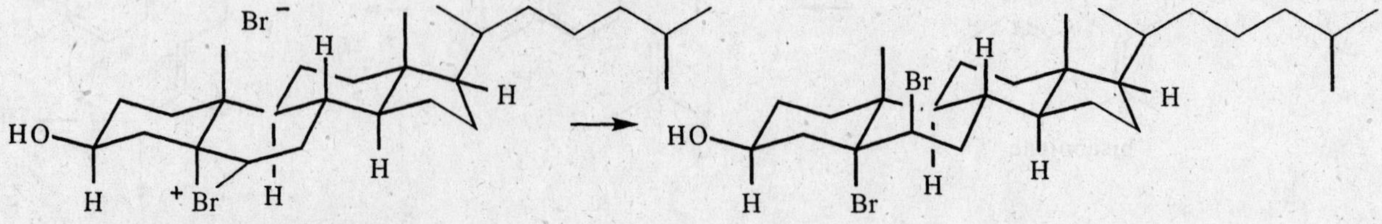

trans-diaxial orientation of the bromine atoms
when the bromonium ion opens

18.28

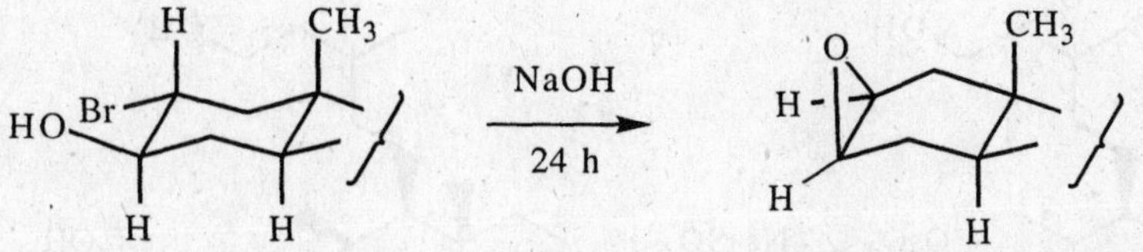

(b)

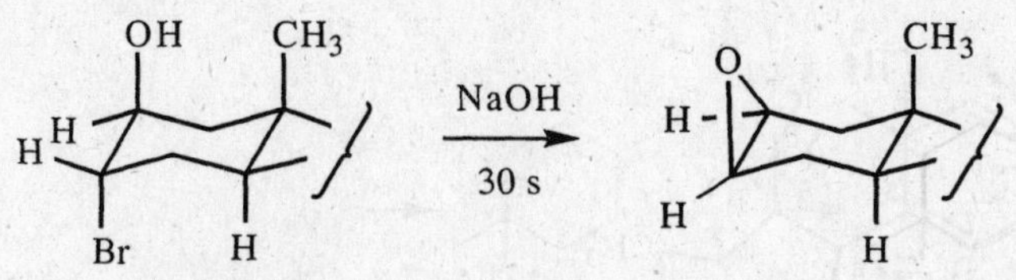

The nucleophile and the leaving group are in a trans-diaxial orientation that is favored for an intramolecular S_N2 reaction.

The nucleophile and the leaving group are in a trans-diequatorial orientation, which is less favorable for an intramolecular S_N2 reaction. Because of the rigidity of the fused ring system, the two groups can have the trans-diaxial orientation to each other only when the cyclohexane ring is in a boat or twist conformation (Section 5.11C). Such a conformation is higher in energy than the chair conformation allowed for the reaction of Compound A, so the rate of reaction is much slower for B than for A.

18.29

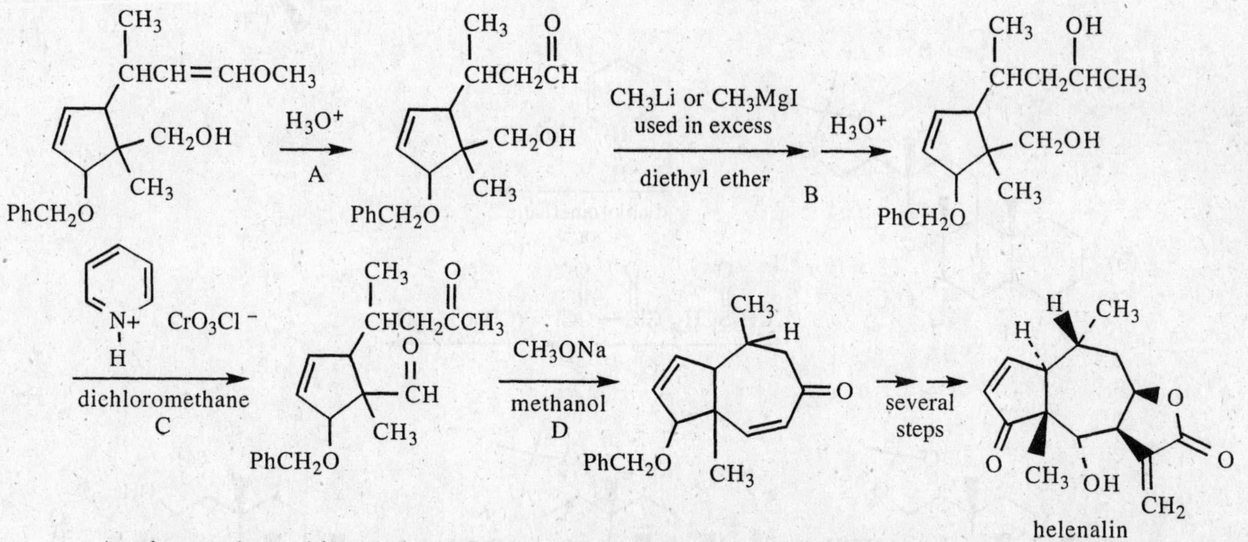

Another product with two fused five-membered rings is theoretically possible. The structure shown for helenalin in the text indicates that the bonds joining the five-membered ring to the seven-membered ring are trans to each other. Such trans stereochemistry at the junction of two fused five-membered rings would create a lot of strain, but is all right at the junction of a five- and seven-membered ring. You may wish to prove this to yourself by building molecular models of both systems.

18.30   Santonin follows the isoprene rule. It has three isoprene units attached head-to-tail.

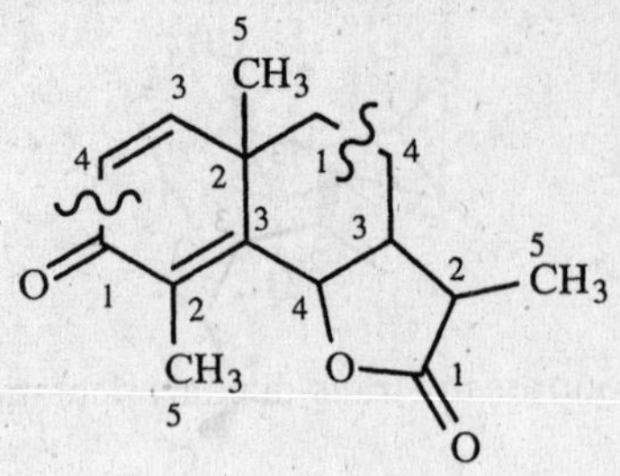

18.31   (a)

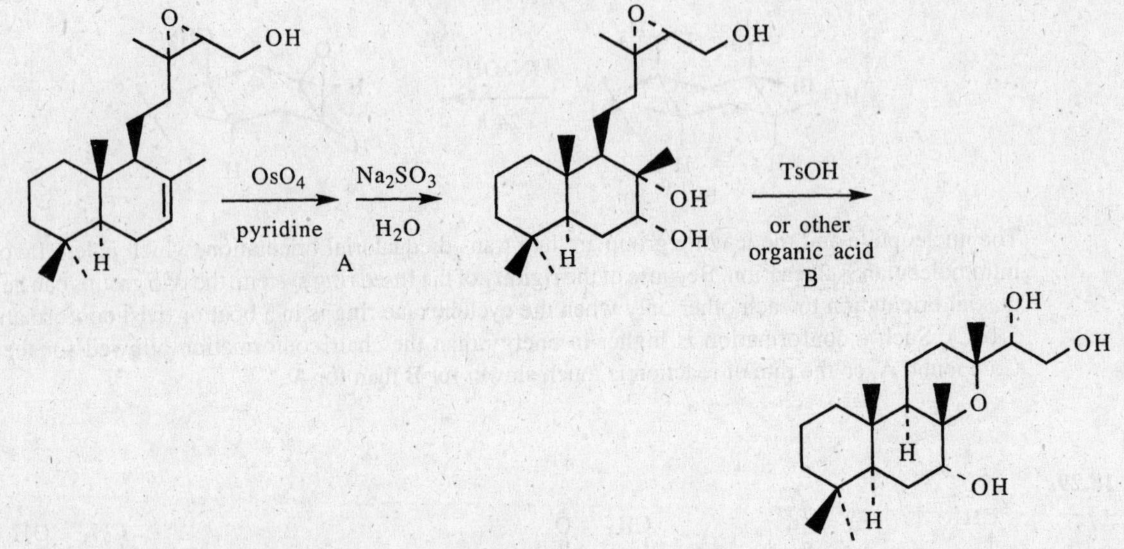

(b)

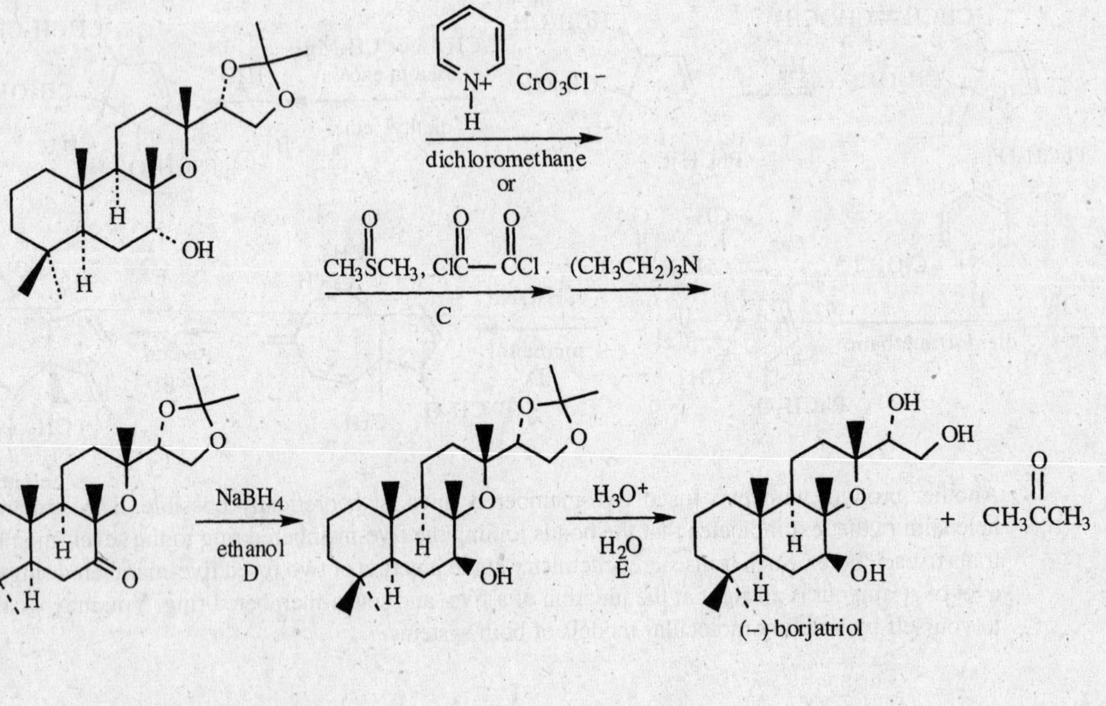

conformation of (−) borjatriol:

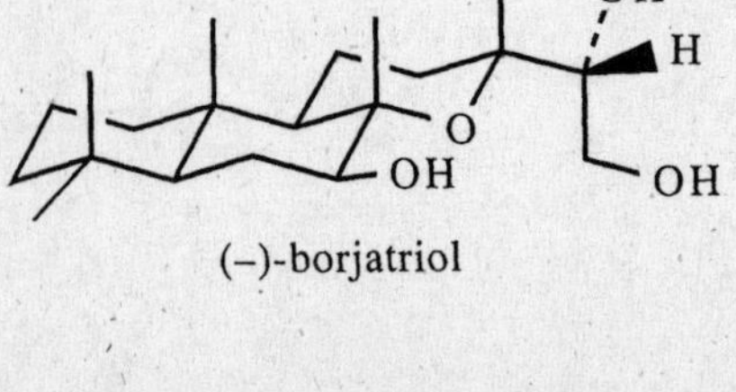

(−)-borjatriol

18.32   (a) Guaiol is a terpene, with head-to-tail connections of isoprene units as shown below.

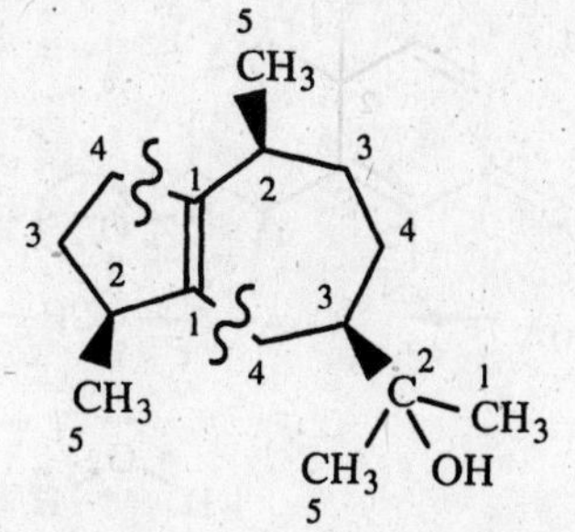

(b) The two rings in guaiazulene are planar with a 10 π electron system around the periphery. As such, the system fits Hückel's Rule, having $(4n + 2)$ π electrons with $n = 2$.

18.33   (a) The reaction is a Claisen condensation.

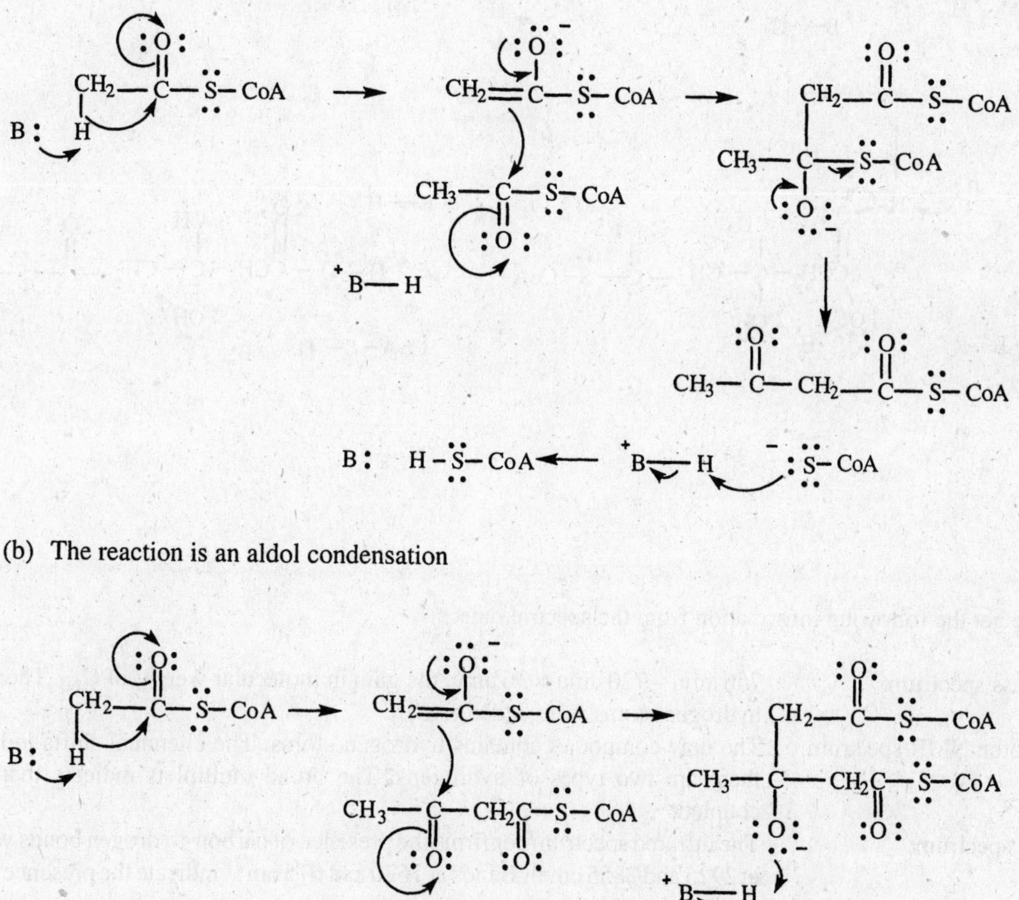

(b) The reaction is an aldol condensation

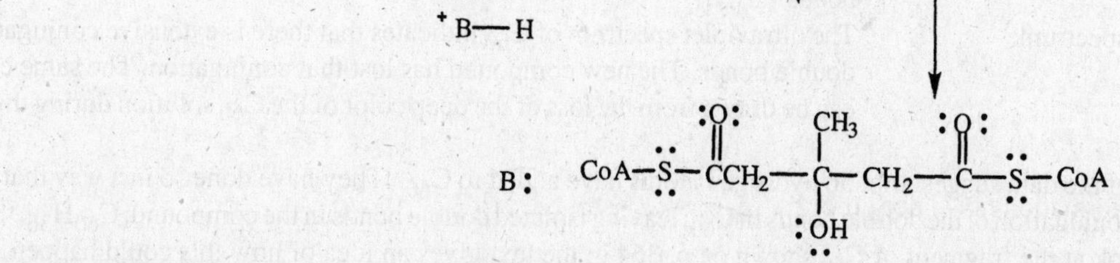

18.33   (c) The reaction is the hydrolysis of a thioester.

18.34   We get the following information from the spectral data:

Mass spectrum:	756 amu − 720 amu = 36 amu, the gain in molecular weight of C_{60}. Therefore, 36 hydrogen atoms were added to C_{60}.
Proton NMR spectrum:	The new compound contains hydrogen atoms. The chemical shifts indicate that there are two types of hydrogens. The broad multiplets indicate that they are coupled.
IR spectrum:	The infrared spectrum confirms the presence of carbon-hydrogen bonds with bands at 2925 and 2855 cm^{-1}. Bands at 1620 and 675 cm^{-1} indicate the presence of double bonds.
UV spectrum:	The ultraviolet spectrum of C_{60} indicates that there is extensive conjugation of the double bonds. The new compound has lost that conjugation. The same conclusion can be drawn from the loss of the deep color of the C_{60} solution during the reaction.

All of the data suggest that 36 hydrogen atoms have added to C_{60}. They have done so in a way that breaks up the conjugation of the double bonds in C_{60}, leaving isolated double bonds in the compound, $C_{60}H_{36}$, that results. A look at the fragment of C_{60} shown on p. 364 in the text gives an idea of how this could happen.

18.34 (cont)

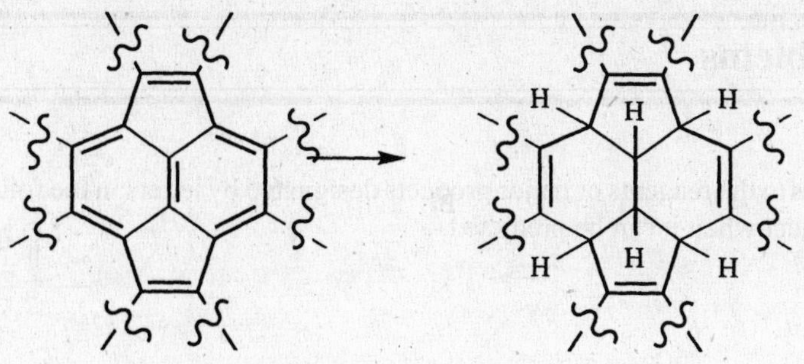

18.35

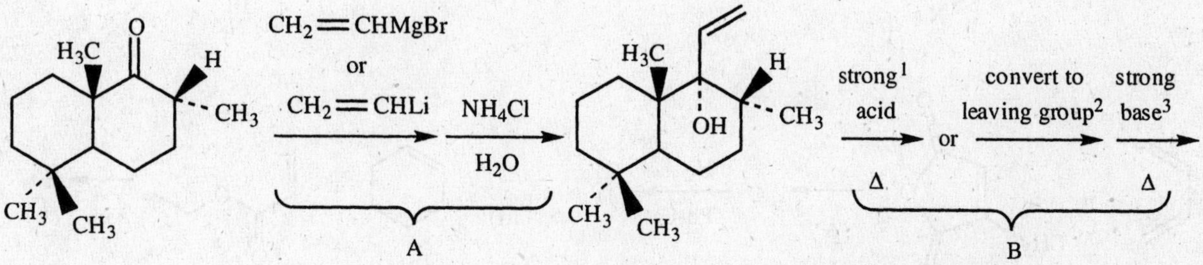

$^1H_2SO_4$, H_3PO_4, etc.
^2TsCl, $SOCl_2$, PBr_3, etc.
^3KOH, CH_3CH_2ONa, $(CH_3)_3COK$, etc.

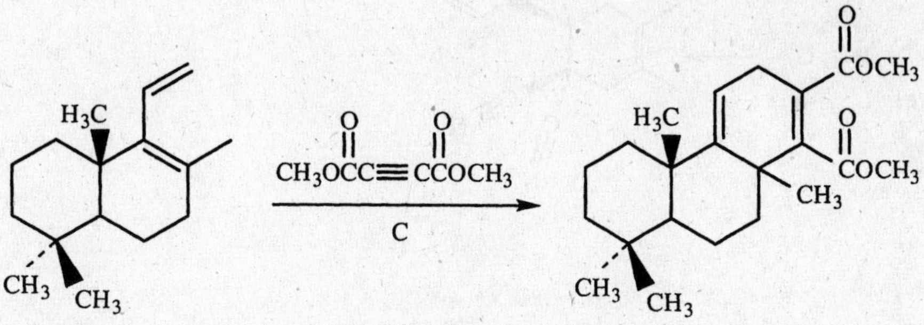

infrared assignments

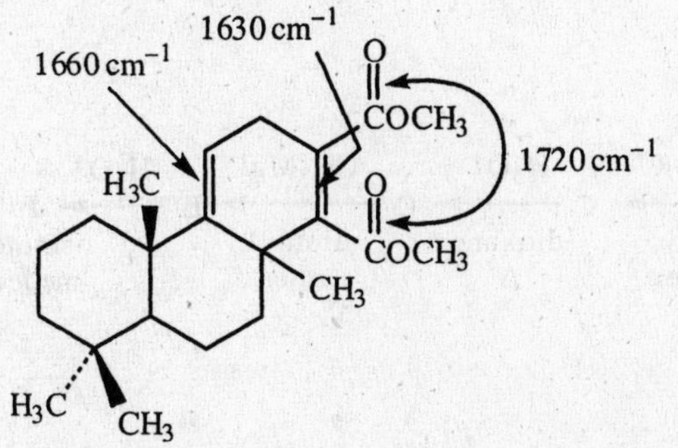

1660 cm^{-1}
1630 cm^{-1}
1720 cm^{-1}

^1H NMR assignments

δ 3.14, 2.77
δ 5.64
δ 3.81, 3.73
δ 1.31 – 0.85, 4 singlets

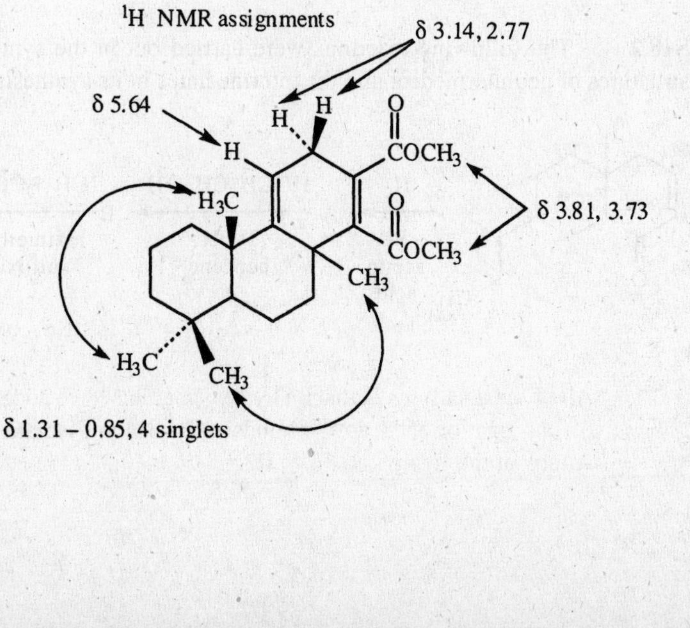

Supplemental Problems

S18.1   Assign structures to the reagents or major products designated by letters in the following reactions. Show the stereochemistry of the product when it can be predicted.

(a) 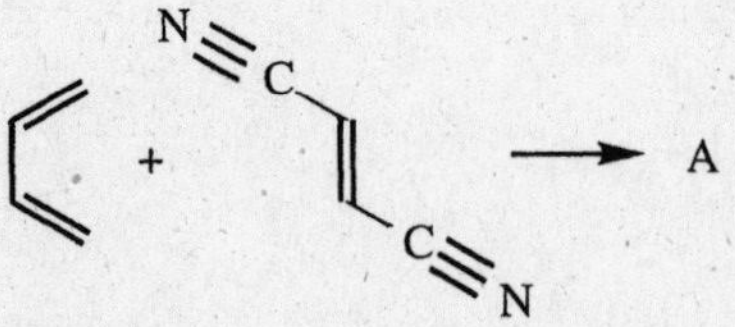 → A

(b) → B

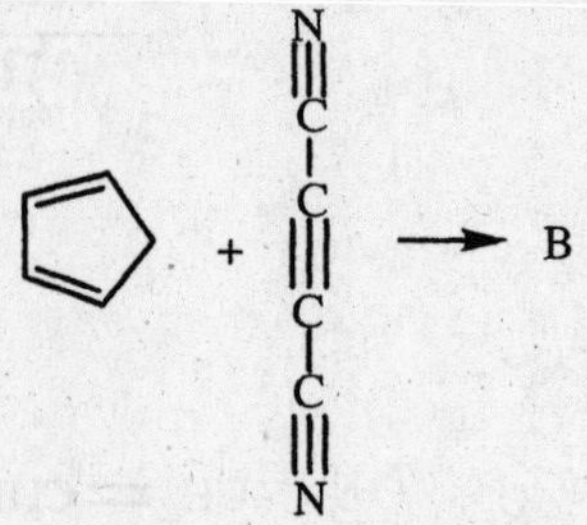

(c) 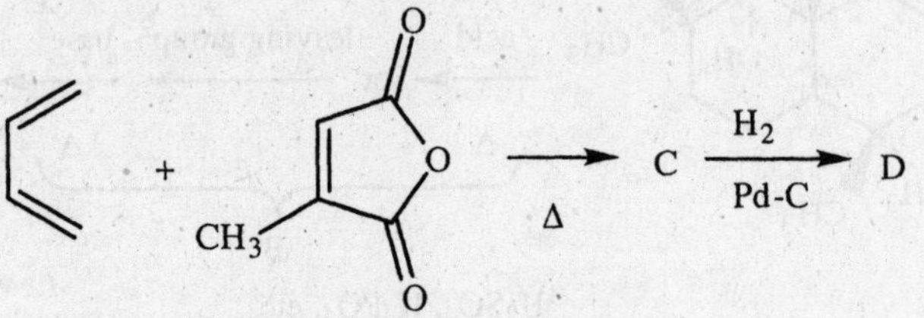 → C $\xrightarrow[\text{Pd-C}]{\text{H}_2}$ D

(d) 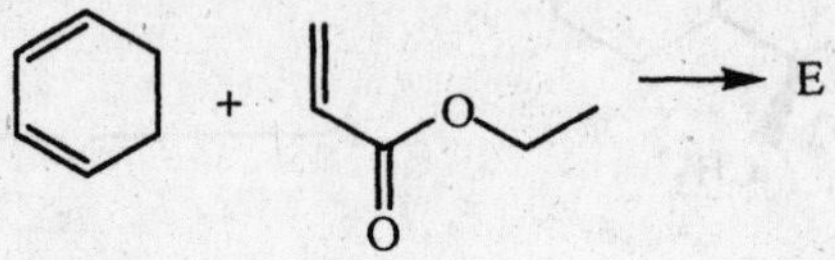 → E

(e) F + G ⟶

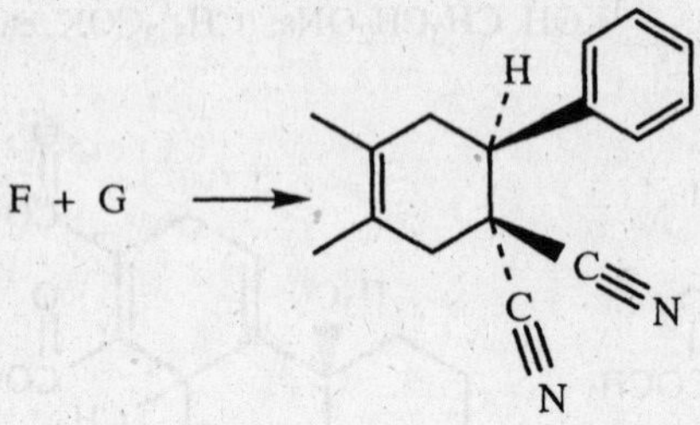

S18.2   The following reactions were carried out in the synthesis of the terpene neointermedeol. What are the structures of neointermedeol and the intermediates in its synthesis?

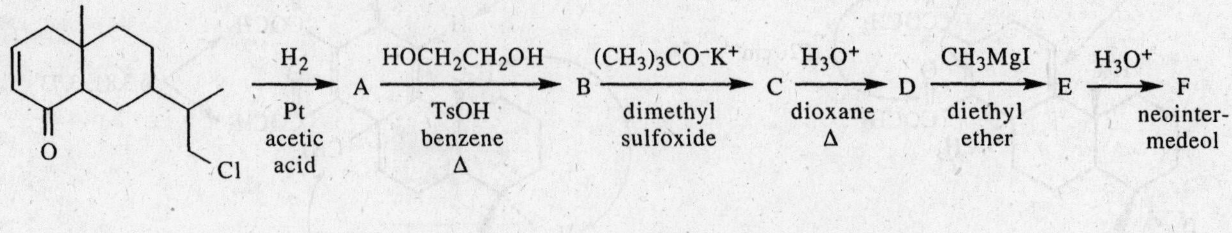

S18.3   An intermediate in the synthesis of quadrone, a sesquiterpene with possible antitumor activity, was prepared by the following reactions. Write structural formulas for the compounds represented by letters.

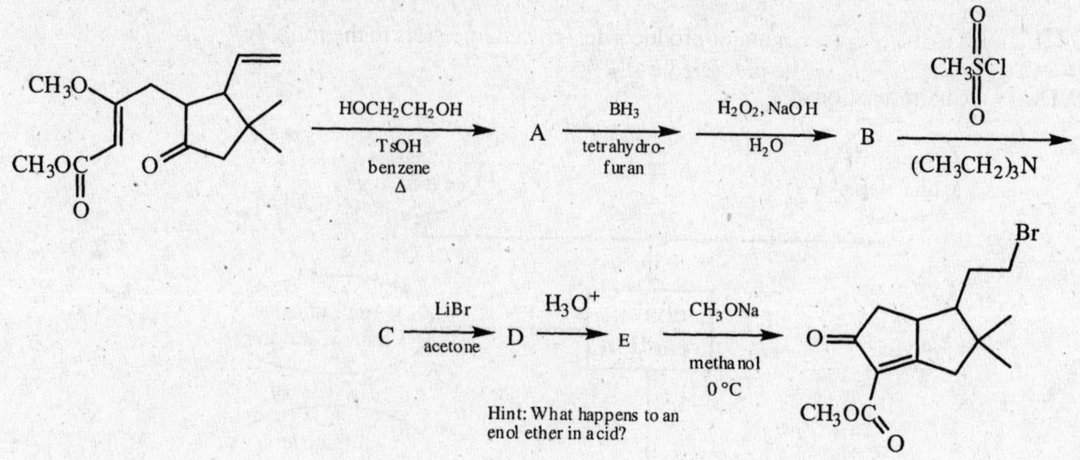

Hint: What happens to an enol ether in acid?

S18.4   Write structural formulas for the products and intermediates designated by letters in the following sequence of reactions.

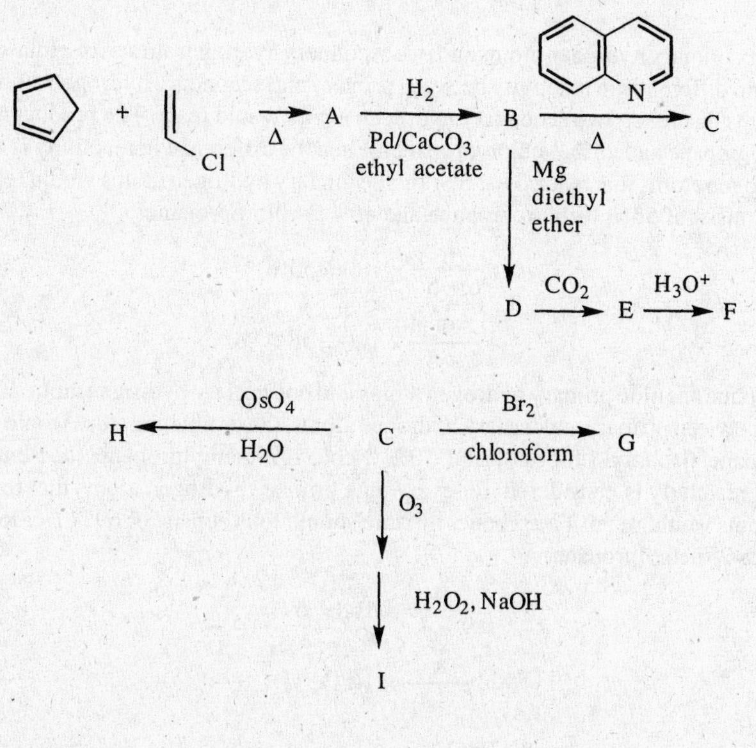

19 Free Radicals

Concept Map 19.1   Chain reactions.

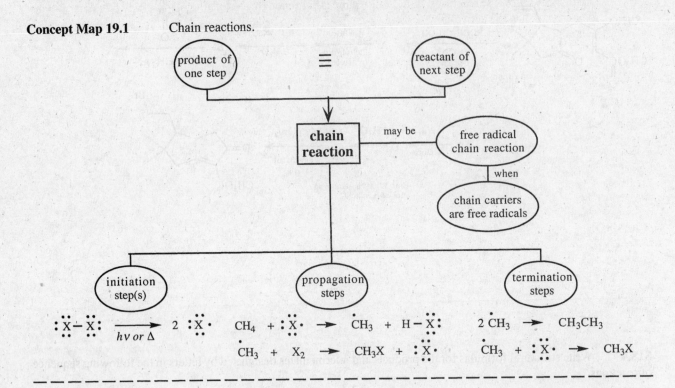

- -

19.1   Propane has six primary hydrogen atoms and two secondary hydrogen atoms for a total of eight hydrogen atoms. If there were no difference in reactivity between primary and secondary hydrogen atoms, for every six primary hydrogen atoms that react, two secondary hydrogen atoms would react. The product mixture would consist of 75% 1-chloropropane and 25% 2-chloropropane. When the difference in reactivity is considered, for every six primary hydrogen atoms that react, 2×2.5 or five secondary hydrogen atoms would react. The product mixture should thus consist of 55% 1-chloropropane and 45% 2-chloropropane.

$$\frac{6}{6+5} \times 100 = 55\%$$

$$\frac{5}{6+5} \times 100 = 45\%$$

2-Methylpropane has nine primary hydrogen atoms and one tertiary hydrogen atom. If there were no difference in reactivity between primary and tertiary hydrogen atoms, the product mixture would consist of 90% 1-chloro-2-methylpropane (isobutyl chloride) and 10% 2-chloro-2-methylpropane (*tert*-butyl chloride). When the difference in reactivity is considered, for every nine primary hydrogen atoms that react, 1×4 or four tertiary hydrogen atoms would react. The product mixture should thus consist of 69% 1-chloro-2-methylpropane and 31% 2-chloro-2-methylpropane.

$$\frac{9}{9+4} \times 100 = 69\%$$

$$\frac{4}{9+4} \times 100 = 31\%$$

448

19.2   The selectivity observed at 300 °C depends on small differences between energies of activation for the abstraction of primary, secondary, and tertiary hydrogen atoms. As the temperature increases, the average kinetic energy of chlorine atoms and alkane molecules increases. The number of collisions that have energies in excess of the energy of activation for the abstraction of the different kinds of hydrogen atoms also increases. Differences in energies of activation for different reactions become less and less relevant. We see a leveling off of the selectivity of the reactions. If the temperature is high enough, almost all molecules will have energy equal to or greater than the activation energies for any hydrogen to be abstracted, and the rates of reaction would be the same.

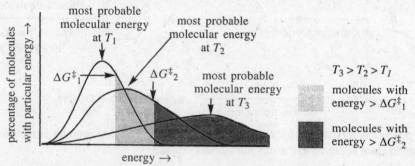

Concept Map 19.2   Halogenation of alkanes.

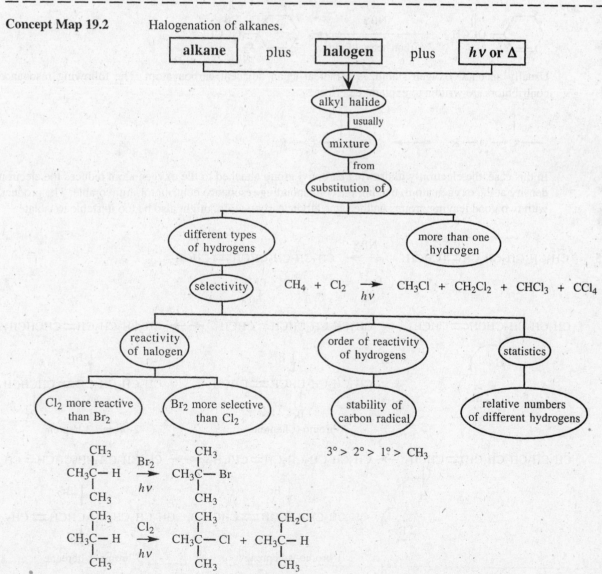

19.3   Neopentane (2,2-dimethylpropane) has only primary hydrogen atoms. The enthalpy for the abstraction of a primary hydrogen atom by a chlorine atom is –5 kcal/mol. The activation energy for this exothermic step is small (~1 kcal/mol), and the rate is correspondingly fast. The enthalpy for the abstraction of a primary hydrogen atom by a bromine atom, on the other hand, is +12 kcal/mol. The activation energy must be at least as large as the enthalpy, and the rate is correspondingly slow.

19.4   (a)

(b)

(c)

This product is reported to form in 58% yield. Another possible product is the one in which the allylic hydrogen atom adjacent to the oxygen atom is replaced.

Usually an ether oxygen stabilizes a radical at an adjacent carbon atom. The following resonance contributors are written to explain the effect.

In this case, the electron-withdrawing carbonyl group attached to the oxygen atom reduces the electron density at the oxygen atom, making the corresponding resonance contributor unfavorable. The product, with two good leaving groups on the same allylic carbon atom, might also be too unstable to isolate.

19.5 $CH_3CH_2CH_2CH_2CH \!=\! CHCH_3 \xrightarrow{\text{NBS}} CH_3CH_2CH_2\underset{\underset{Br}{|}}{C}HCH \!=\! CHCH_3$

60%

4-bromo-2-heptene   2-bromo-3-heptene

1-bromo-2-heptene   3-bromo-1-heptene

Concept Map 19.3   Selective free radical halogenations.

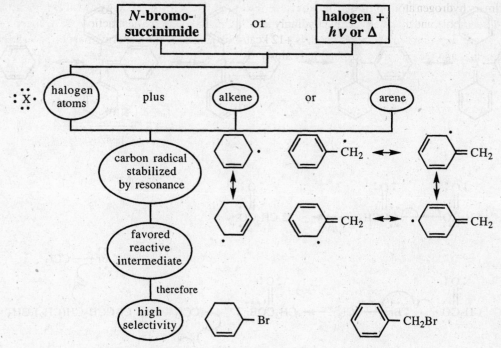

19.6   (a) PhCH$_2$CH$_3$ $\xrightarrow[\Delta]{\overset{Cl_2}{\underset{h\nu}{}}}$ PhCHCH$_3$

                                            |
                                           Cl
                                   major product

   (b) $\xrightarrow[\substack{benzoyl\ peroxide \\ carbon\ tetrachloride \\ \Delta}]{NBS}$

   (c) [naphthalene]—CH$_3$ $\xrightarrow[\substack{carbon \\ tetrachloride \\ \Delta}]{NBS}$ [naphthalene]—CH$_2$Br

   (d) PhCH$_2$CH$_3$ $\xrightarrow[h\nu]{Br_2}$ PhCHCH$_3$

                                            |
                                           Br
                                   major product

19.7   The unpaired electron can be delocalized over three rings.

19.7 (cont)

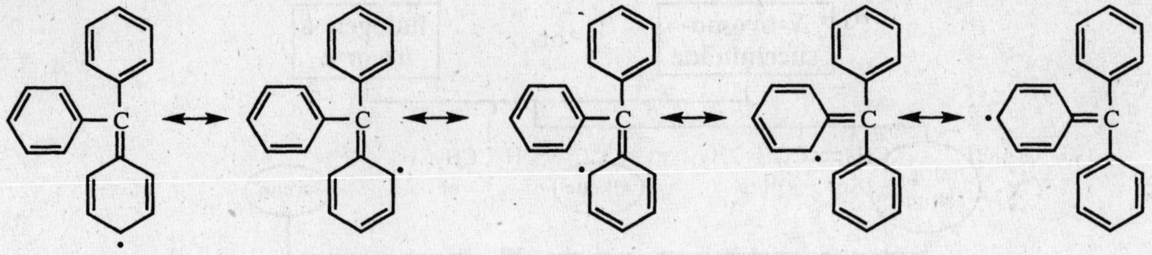

19.8 1.

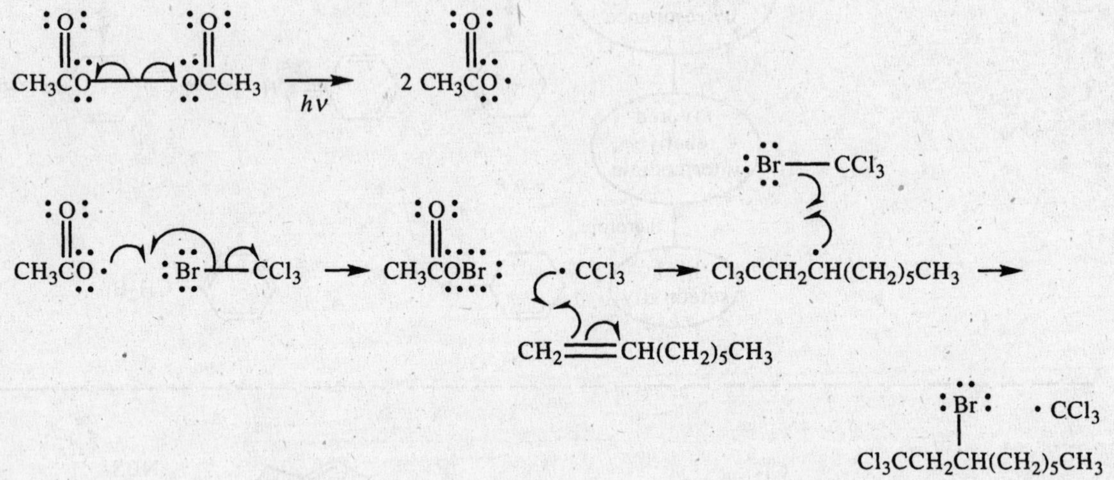
CH3CO
OCCH3
hν
2 CH3CO ·
CH3CO ·
Br
CCl3
CH3COBr
· CCl3
Cl3CCH2CH(CH2)5CH3
CH2═CH(CH2)5CH3
· CCl3
Cl3CCH2CH(CH2)5CH3

2.

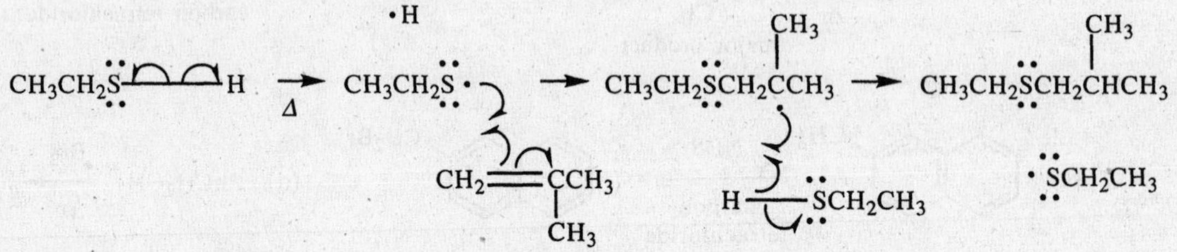
·H
CH3CH2S
H
Δ
CH3CH2S ·
CH2═CCH3
CH3
CH3
CH3CH2SCH2CCH3
H
SCH2CH3
CH3
CH3CH2SCH2CHCH3
· SCH2CH3

3.

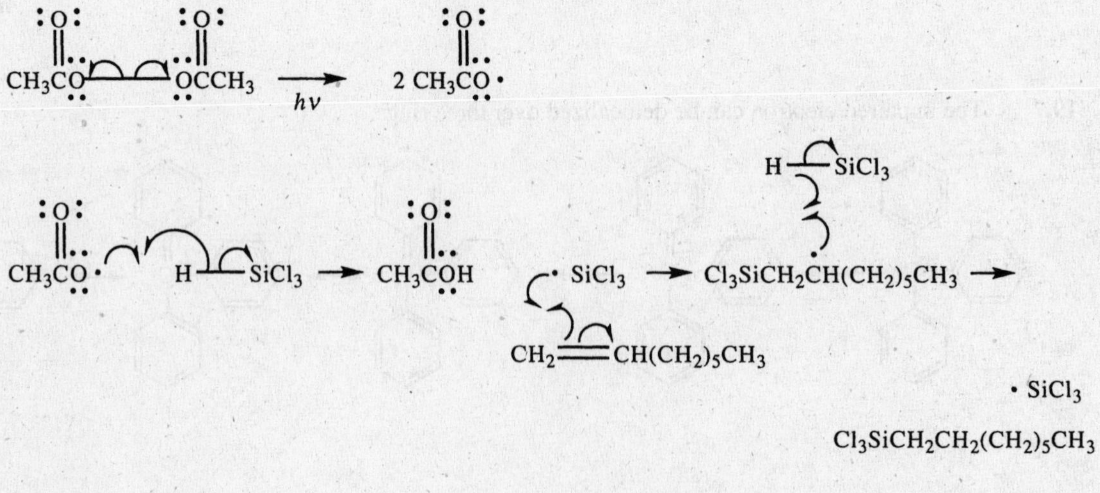
CH3CO
OCCH3
hν
2 CH3CO ·
CH3CO ·
H
SiCl3
CH3COH
· SiCl3
Cl3SiCH2CH(CH2)5CH3
CH2═CH(CH2)5CH3
H
SiCl3
· SiCl3
Cl3SiCH2CH2(CH2)5CH3

19.9   (a)   [cyclohexene] + CH₃SH $\xrightarrow[h\nu]{\text{acetone}}$ [cyclohexane-SCH₃]

(b) Cl_3CBr + $CH_2=\underset{\underset{CH_2CH_3}{|}}{C}CH_2CH_3$ $\xrightarrow{\Delta}$ $Cl_3C\underset{\underset{Br}{|}}{C}\overset{\overset{CH_2CH_3}{|}}{H_2}CCH_2CH_3$

(c) $CH_3(CH_2)_4SiH_3$ + $CH_3(CH_2)_5CH=CH_2$ $\xrightarrow{\text{peroxides}}$ $CH_3(CH_2)_4SiH_2(CH_2)_7CH_3$

- -

Concept Map 19.4   Free radical addition reactions of alkenes.

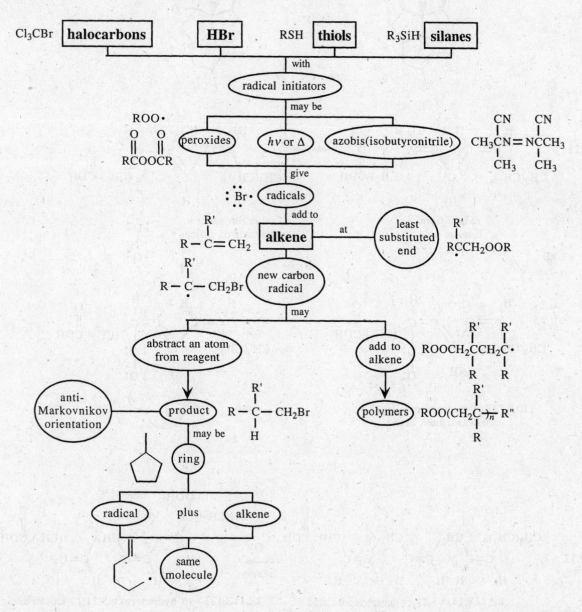

19.10

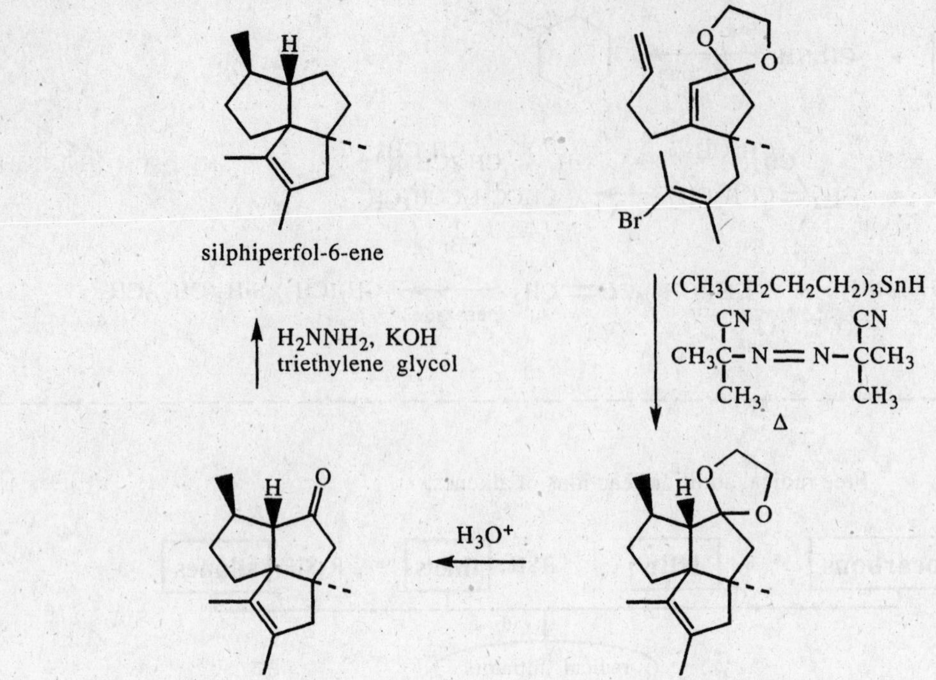

silphiperfol-6-ene

↑ H₂NNH₂, KOH
triethylene glycol

(CH₃CH₂CH₂CH₂)₃SnH

$$CH_3\overset{CN}{\underset{CH_3}{C}}-N=N-\overset{CN}{\underset{CH_3}{C}}CH_3$$

Δ

H₃O⁺ ←

19.11

radical from
linoleic acid

other resonance
contributor of radical

(10*E*,12*Z*)-9-hydroperoxy-10,12-
octadecadienoic acid

R·

R—H

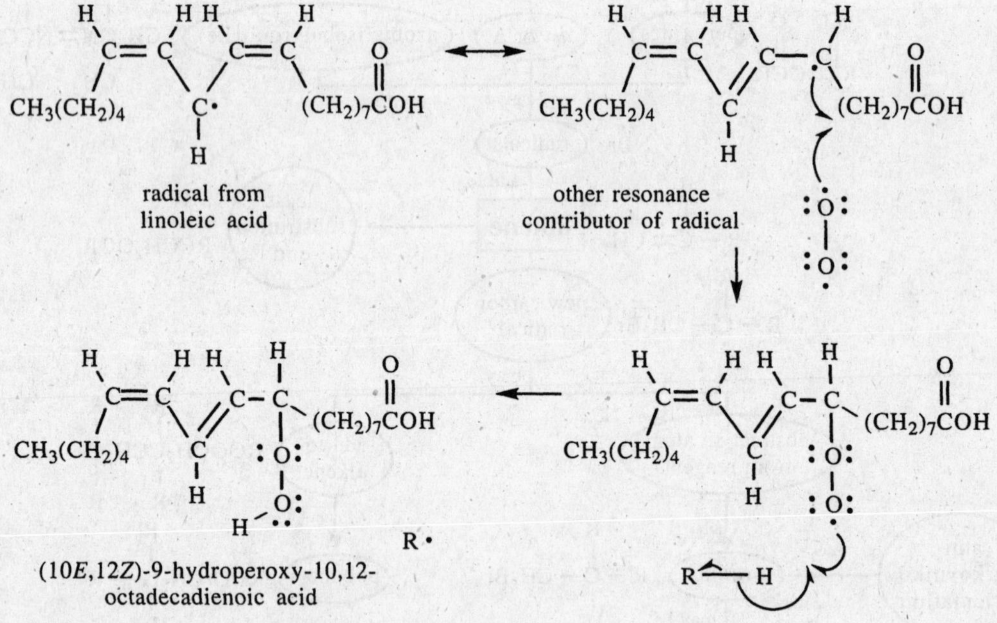

19.12

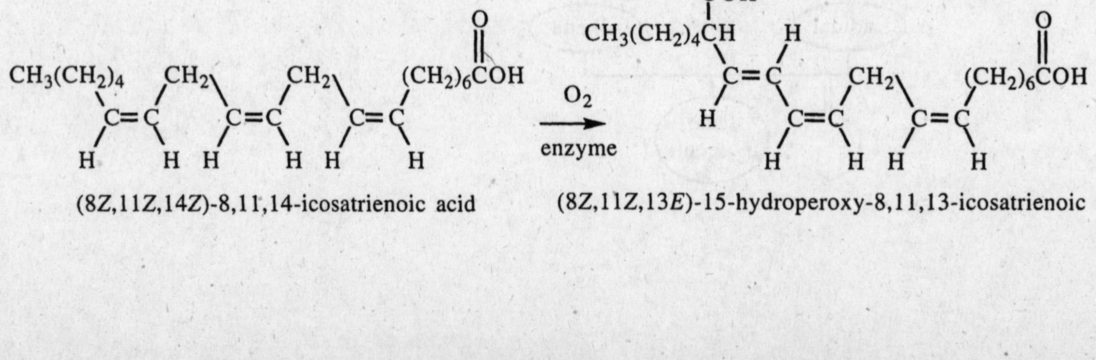

(8*Z*,11*Z*,14*Z*)-8,11,14-icosatrienoic acid

O₂
enzyme

(8*Z*,11*Z*,13*E*)-15-hydroperoxy-8,11,13-icosatrienoic acid

19.12 (cont)

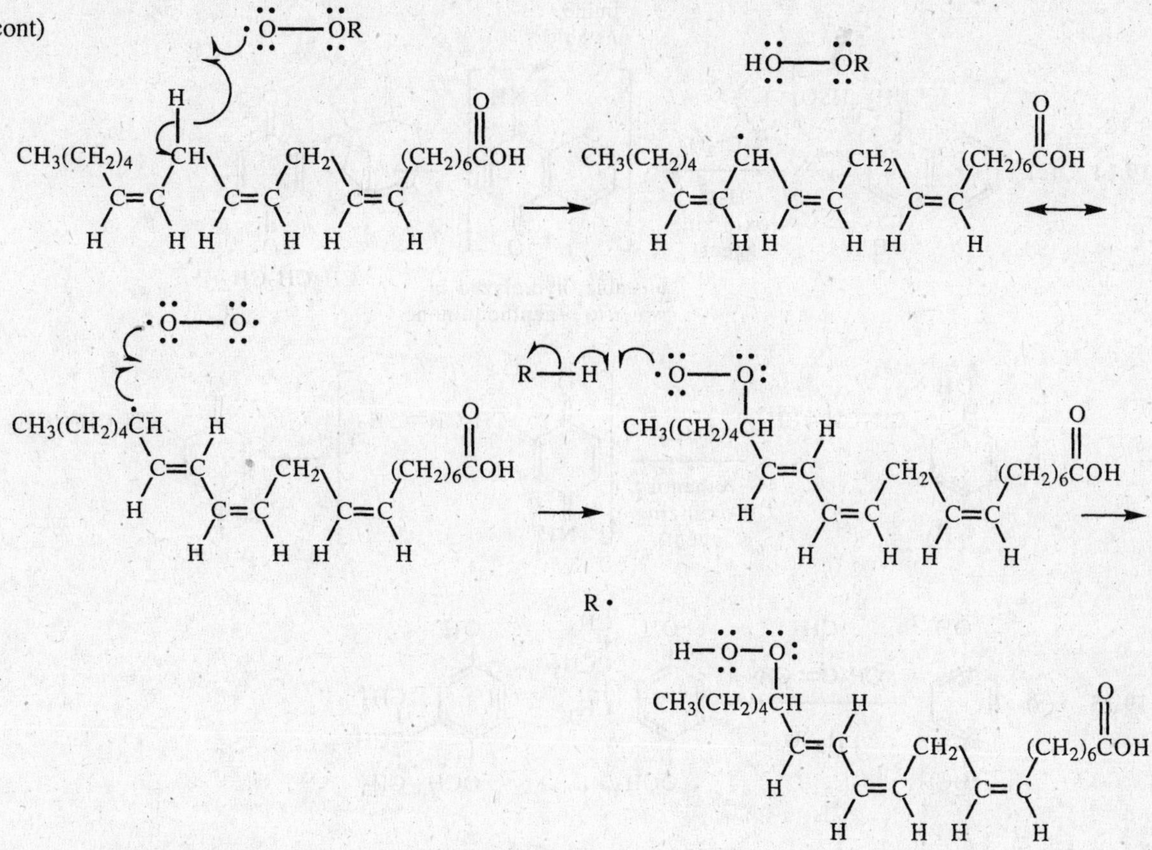

19.13

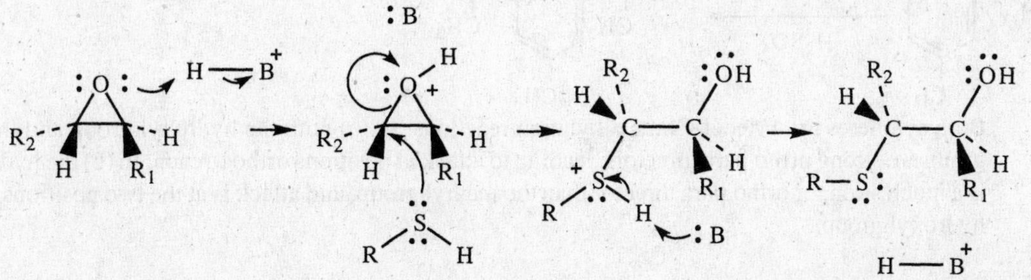

Note that while the opening of the ring in an unsymmetrical oxirane usually gives rise to a mixture of isomers, a reaction in biological systems gives rise to a single product. The mechanism shown above assumes that the reaction is taking place with a preferred orientation at the active site of an enzyme (see Section 23.8C for an example) and that, therefore, only one product is formed.

19.14 (a)

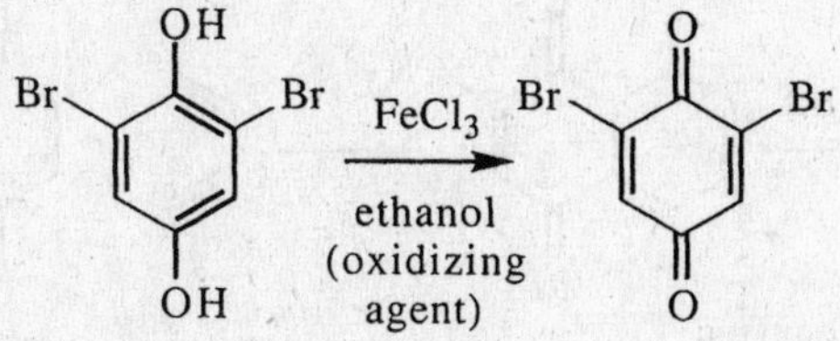

(b)

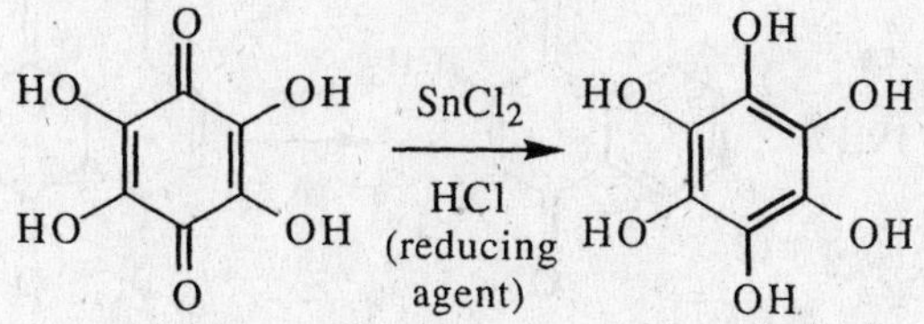

19.14 (c)

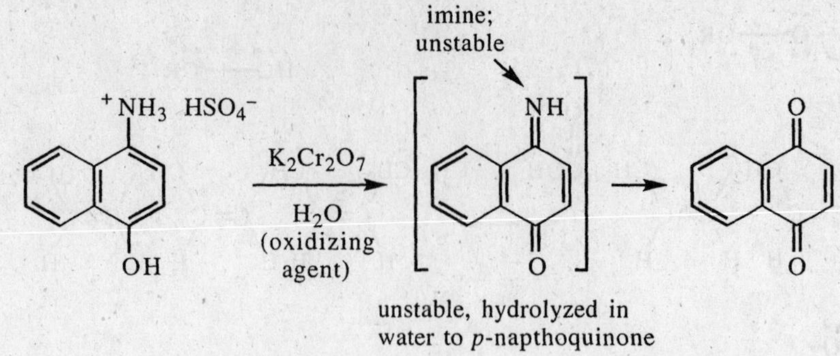

unstable, hydrolyzed in
water to *p*-napthoquinone

(d)

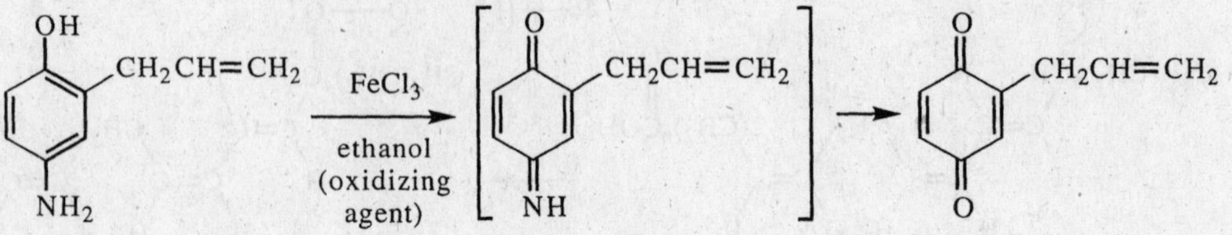

19.15 (a)

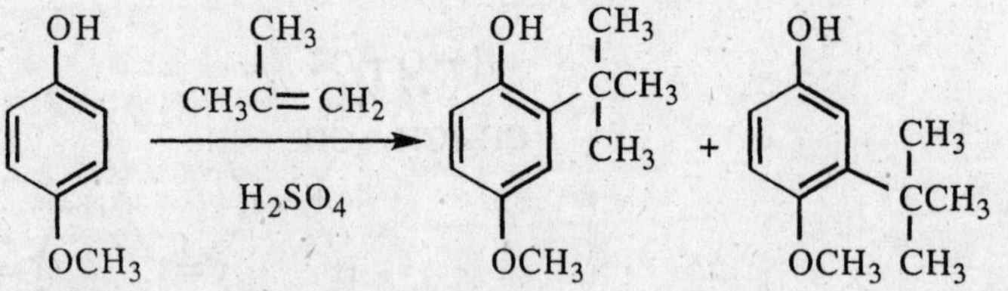

(b)

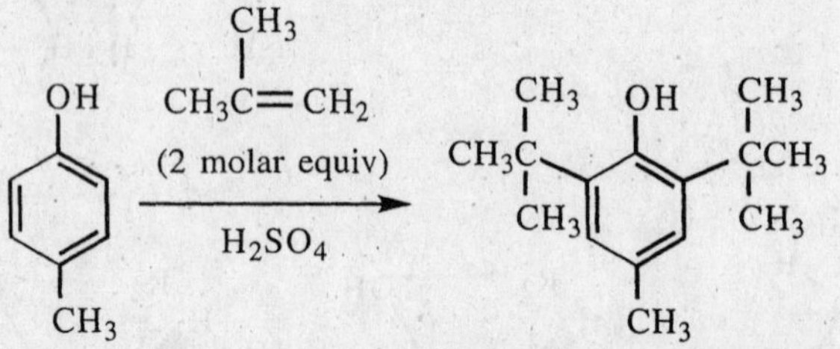

Both syntheses are Friedel-Crafts alkylation reactions. In (a) both the hydroxyl group and the methoxyl group are strong ortho,para directors, leading to attack at positions ortho to each. In (b) the hydroxyl group is a much stronger ortho,para director than the methyl group, and attack is at the two positions ortho to the hydroxyl group.

19.16

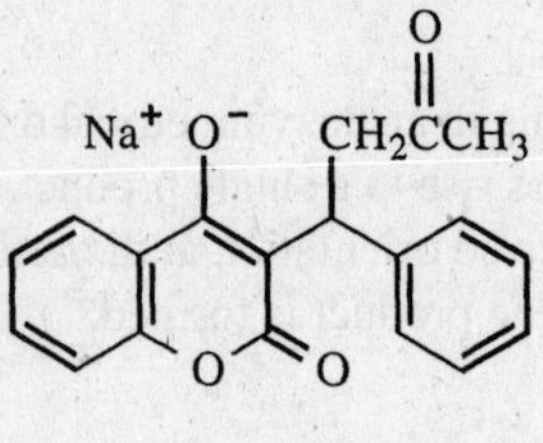

19.17

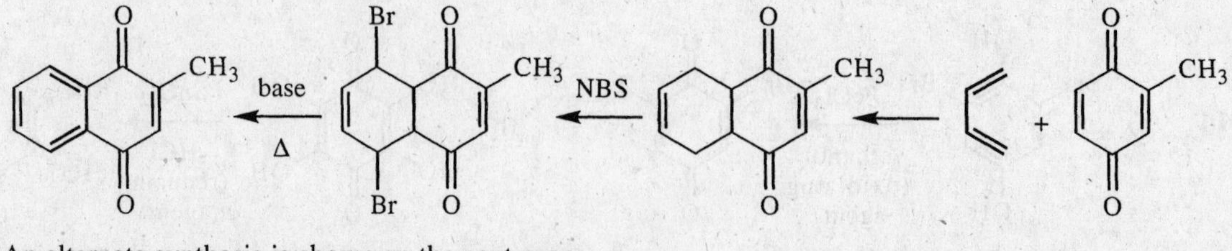

An alternate synthesis is shown on the next page:

19.17 (cont)

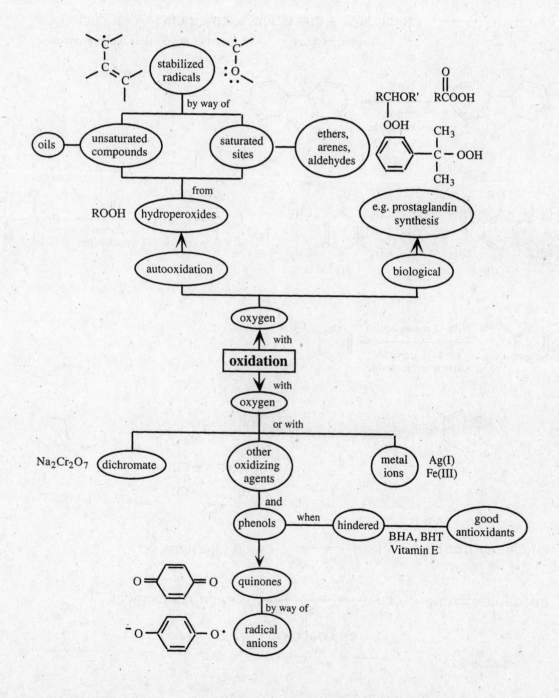

- -

Concept Map 19.5   Oxidation reactions as free radical reactions.

19.18

(a) $CH_3CH=CHCH=CHCH_3 \xrightarrow[\text{carbon tetrachloride}]{\substack{Br_2 \\ (1 \text{ molar equiv})}} CH_3CHBrCHBrCH=CHCH_3 + CH_3CHBrCH=CHCHBrCH_3$

(b) cyclopentene $\xrightarrow[\text{carbon tetrachloride}]{NBS}$ 3-bromocyclopentene (Br)

(c) cyclohexene $\xrightarrow{O_2}$ 2-cyclohexenyl hydroperoxide (OOH)

(d) $CH_3CH_2CH_3 \xrightarrow[h\nu]{Cl_2} CH_3CH_2CH_2Cl + CH_3CHClCH_3 + CH_3CH_2CHCl_2 + CH_3CHClCH_2Cl$

                                 major products   two of the possible minor products

(e) fluorene $\xrightarrow[\substack{\text{carbon} \\ \text{tetrachloride} \\ \Delta}]{NBS}$ 9-bromofluorene (Br)

(f) 2,3,5,6-tetramethylhydroquinone $\xrightarrow{Ag_2O}$ duroquinone (tetramethyl-1,4-benzoquinone)

(g) $\begin{array}{c}CH_3\;CH_3 \\ | \quad\; | \\ CH_3C\!-\!CHCH_3 \\ | \\ CH_3\end{array} \xrightarrow[\substack{\text{carbon} \\ \text{tetrachloride} \\ h\nu}]{Br_2} \begin{array}{c}CH_3\;CH_3 \\ | \quad\; | \\ CH_3C\!-\!CCH_3 \\ | \quad\; | \\ CH_3\;Br\end{array}$

(h) o-xylene (CH_3, CH_3) $\xrightarrow[\substack{\text{benzoyl peroxide} \\ \text{carbon tetrachloride} \\ \Delta}]{NBS\ (2\ \text{molar equiv})}$ o-bis(bromomethyl)benzene (CH_2Br, CH_2Br)

(i) $PhO\!-\!CH_2CH_2CH(Br)CH_2CH=CH_2$ + $(CH_3CH_2CH_2CH_2)_3SnH \xrightarrow[90\ ^\circ C]{\substack{CN\quad\quad CN \\ | \quad\quad\quad | \\ CH_3C\!-\!N\!=\!N\!-\!CCH_3 \\ | \quad\quad\quad | \\ CH_3\quad\quad CH_3}} PhO\!-\!CH_2CH_2\text{-cyclopentyl}$

(j) $CH_3CH_2CH_2CH=CH_2 + Cl_3SiH \xrightarrow{280\ ^\circ C} CH_3CH_2CH_2CH_2CH_2SiCl_3$

(k) $CH_3(CH_2)_5CH=CH_2 + CCl_4 \xrightarrow[h\nu]{\substack{O\quad\; O \\ \| \quad\; \| \\ CH_3COOCCH_3}} CH_3(CH_2)_5CHClCH_2CCl_3$

19.19

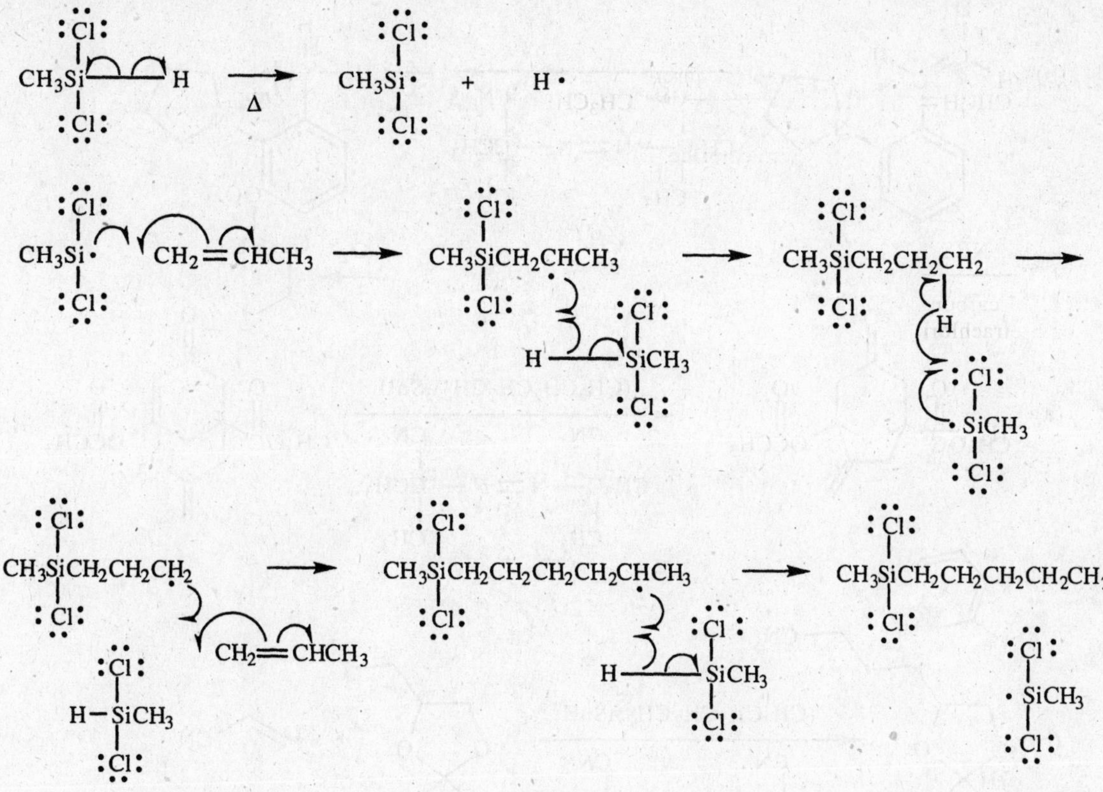

19.20 1. Conditions for the formation of free radicals, such as heat, light, or peroxides, would be recommended for these reactions.

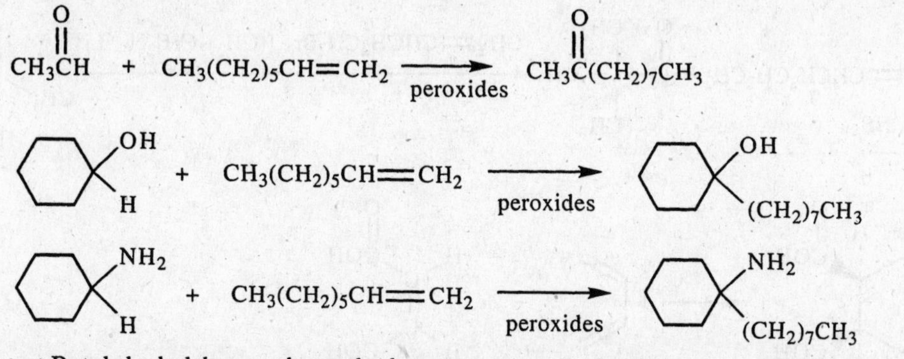

2. *tert*-Butyl alcohol does not have a hydrogen atom attached to the carbon atom bonded to the hydroxyl group and would, therefore, not react with 1-octene under free radical conditions.

19.21 (a)

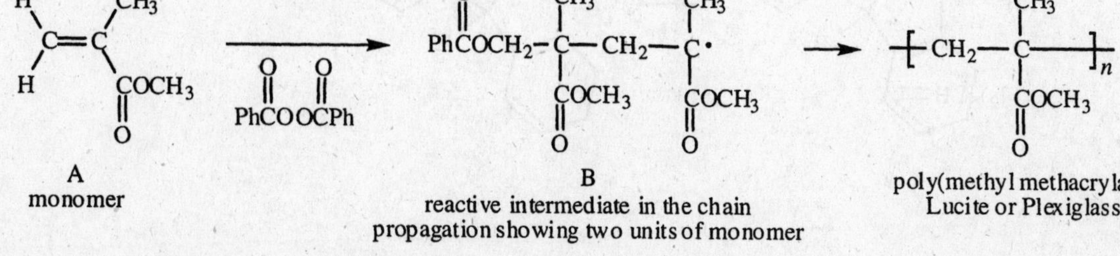

A monomer

B reactive intermediate in the chain propagation showing two units of monomer

poly(methyl methacrylate) Lucite or Plexiglass

19.21 (b)

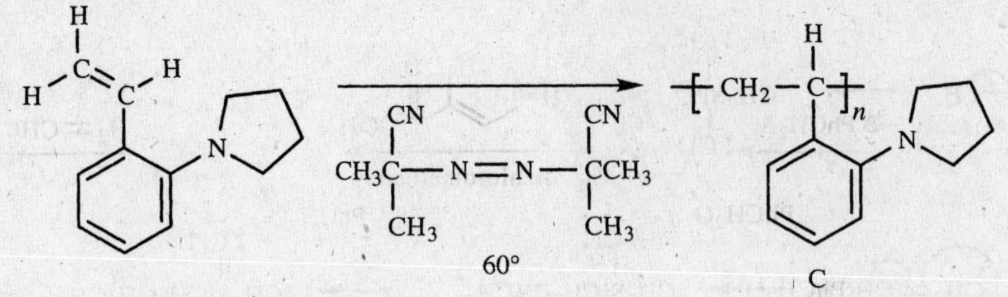

19.22 (a)

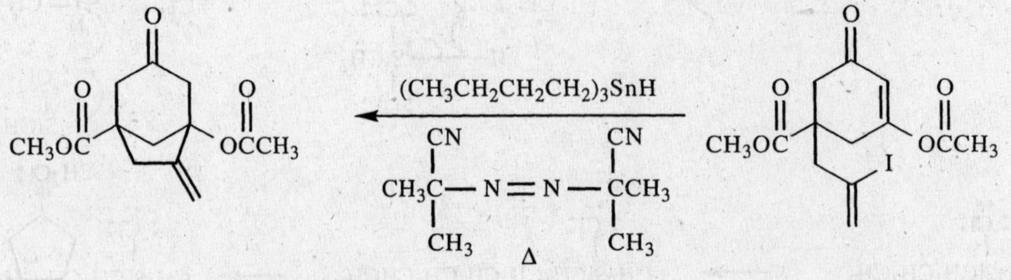

(b)

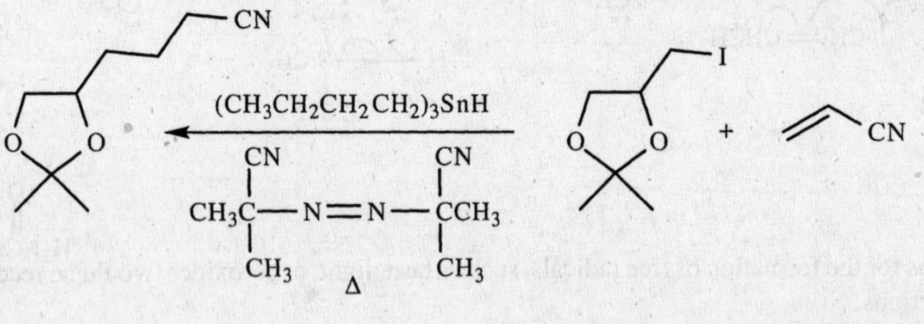

(c)

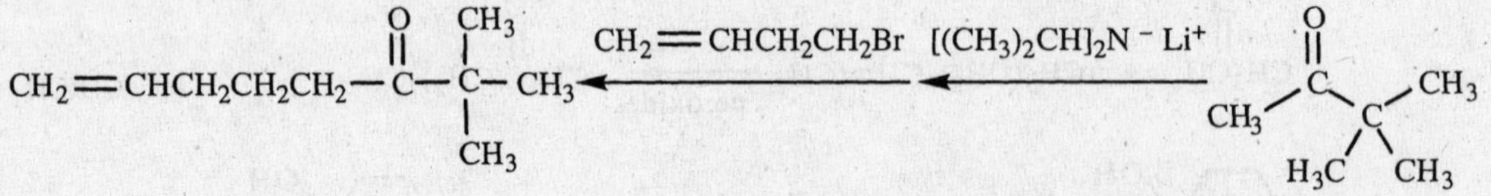

(d)

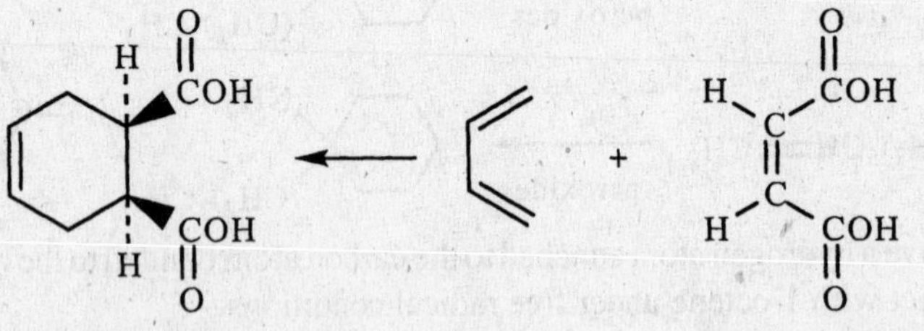

(e)

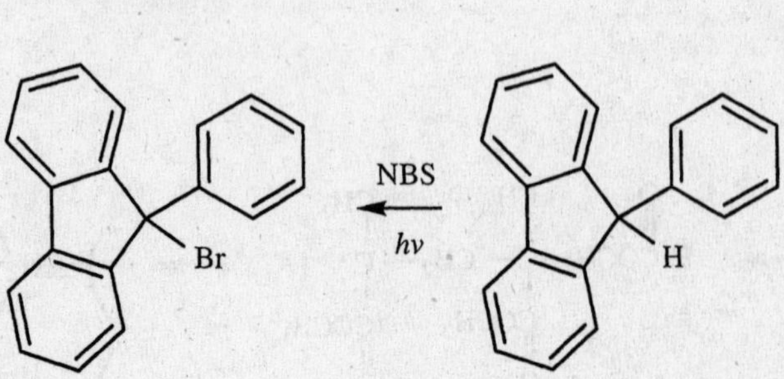

19.23

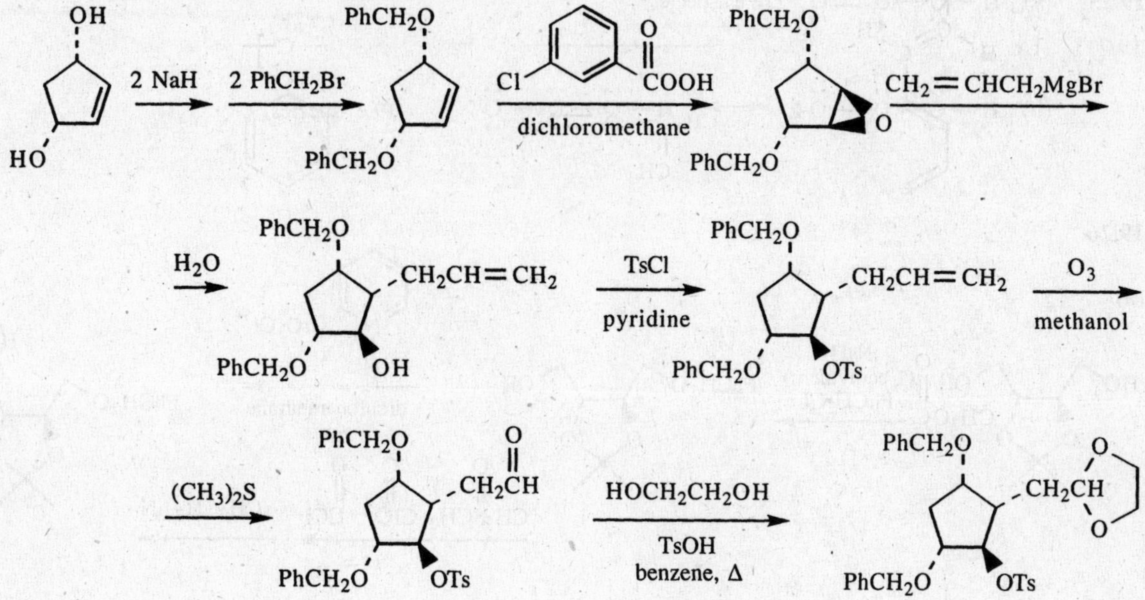

19.24

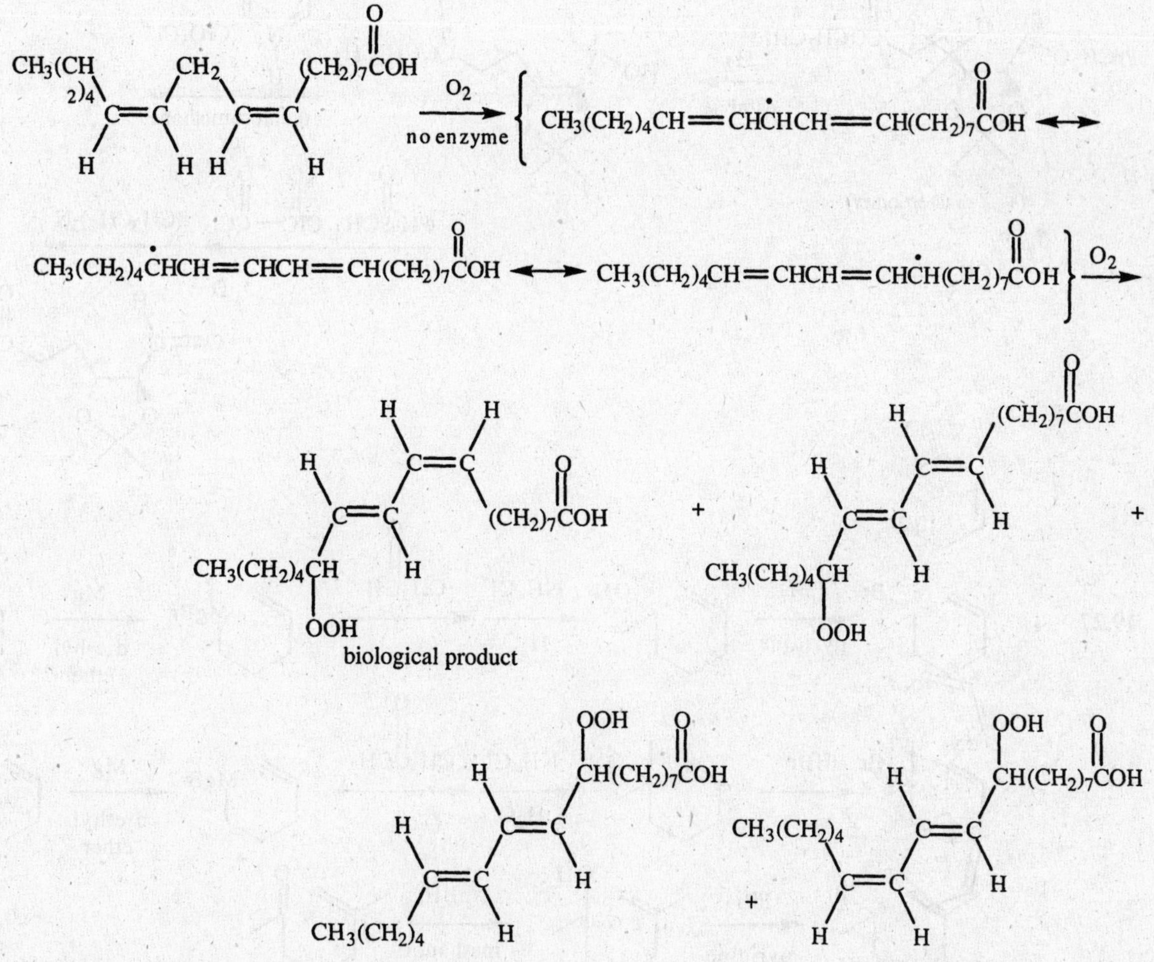

biological product

from Problem 19.11

19.25 (a) 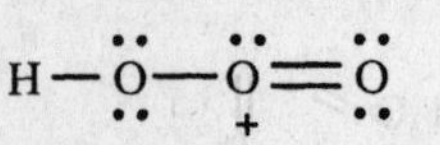   H—Ö—Ö=Ö   (18 e⁻ species)
$\underset{+}{}$

(b) 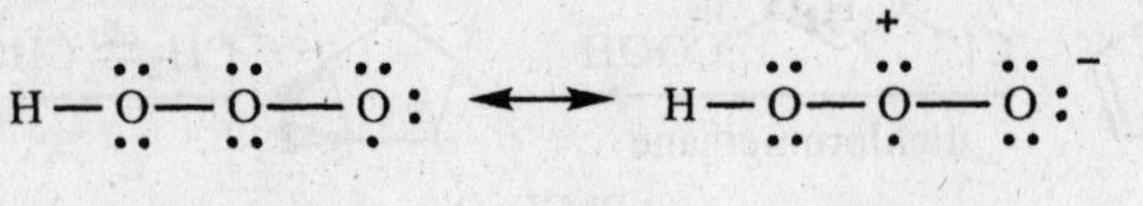   H—Ö—Ö—Ö: ⟷ H—Ö—Ö—Ö:⁻   (19 e⁻ species)

19.26

(a)

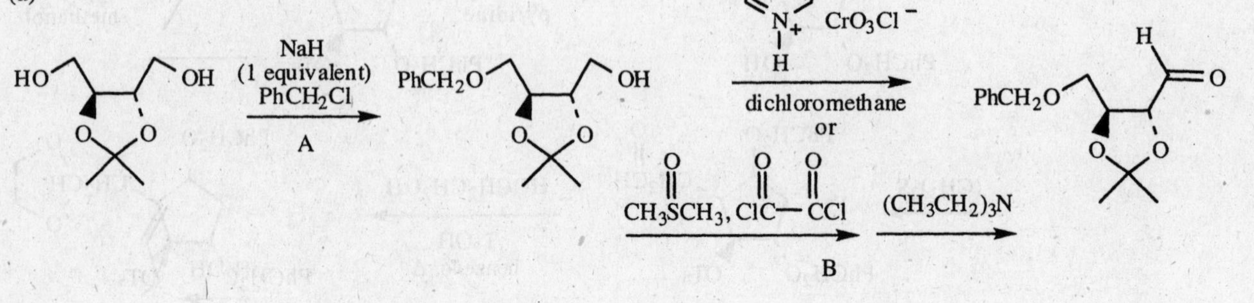

b)

(+ *Z* diastereomer)

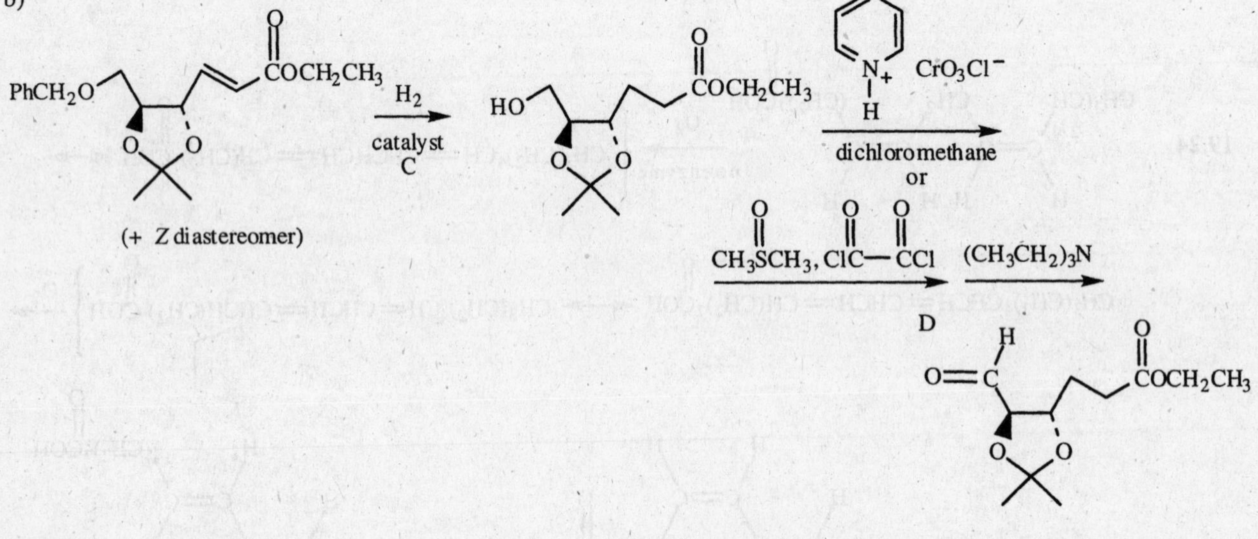

19.27   1.

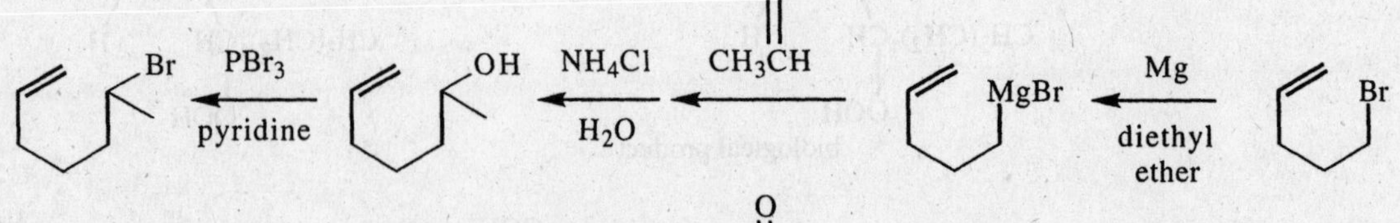

2.

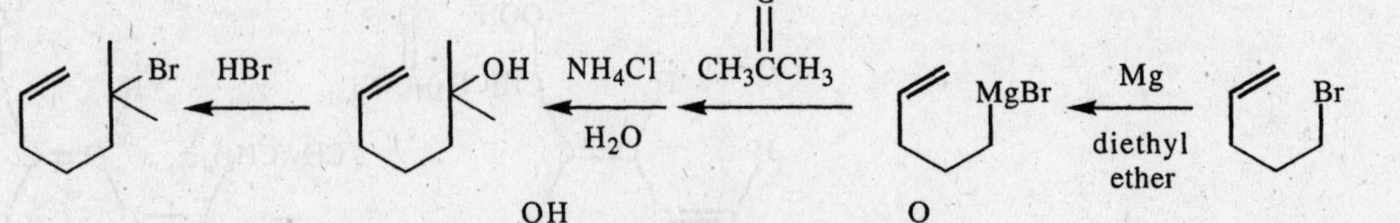

3.

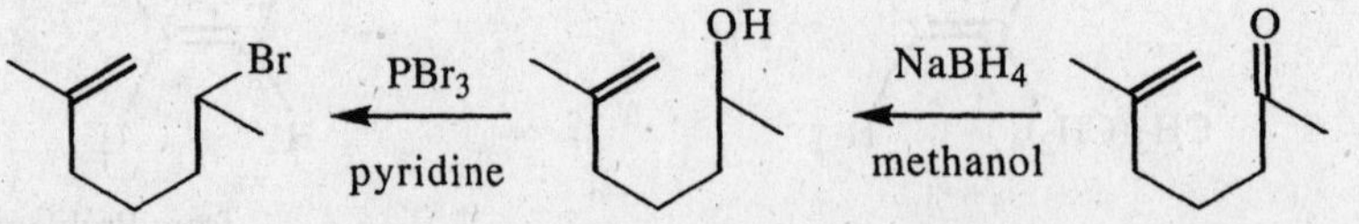

19.27   4.

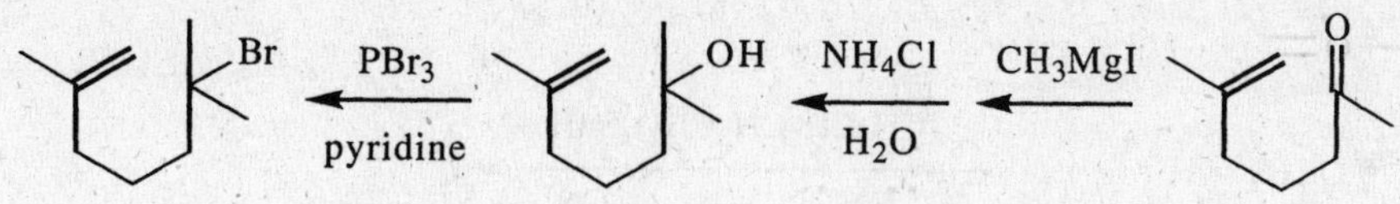

19.28

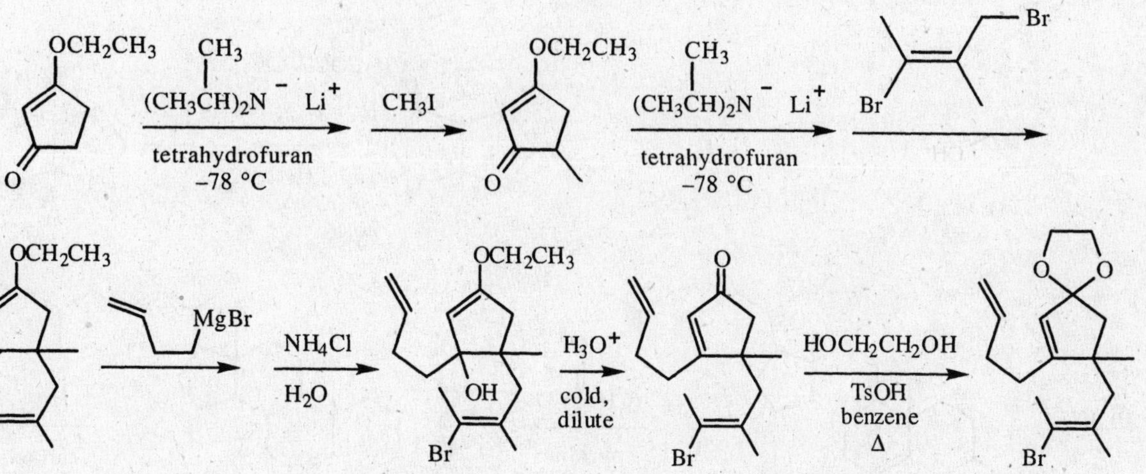

19.29   (a)

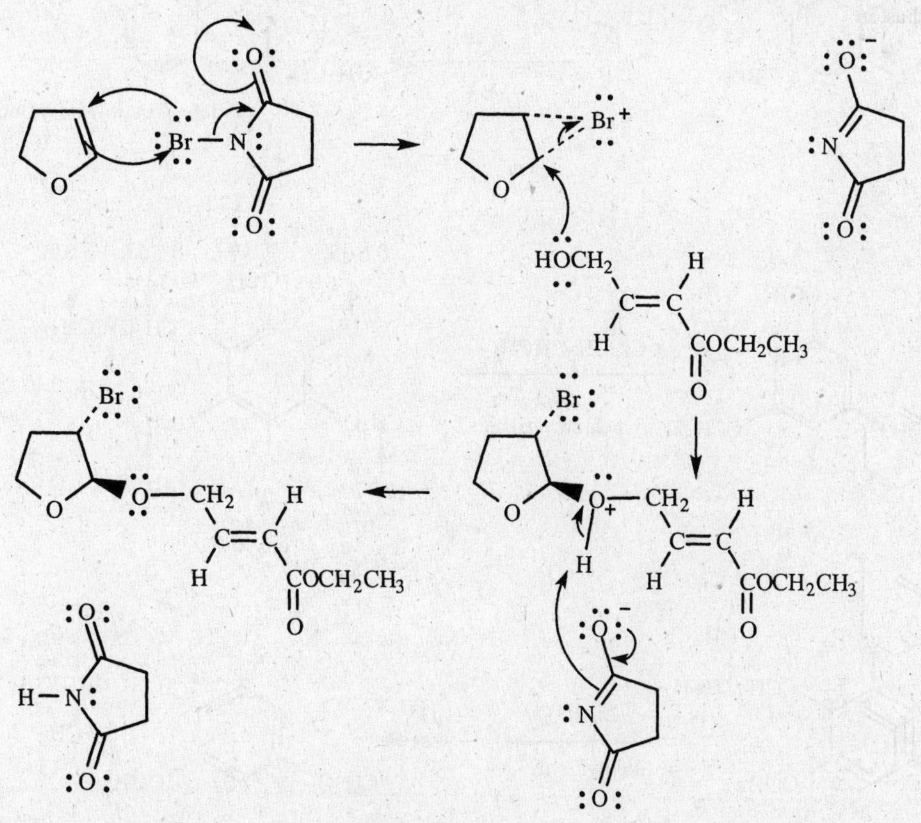

19.29 (b)

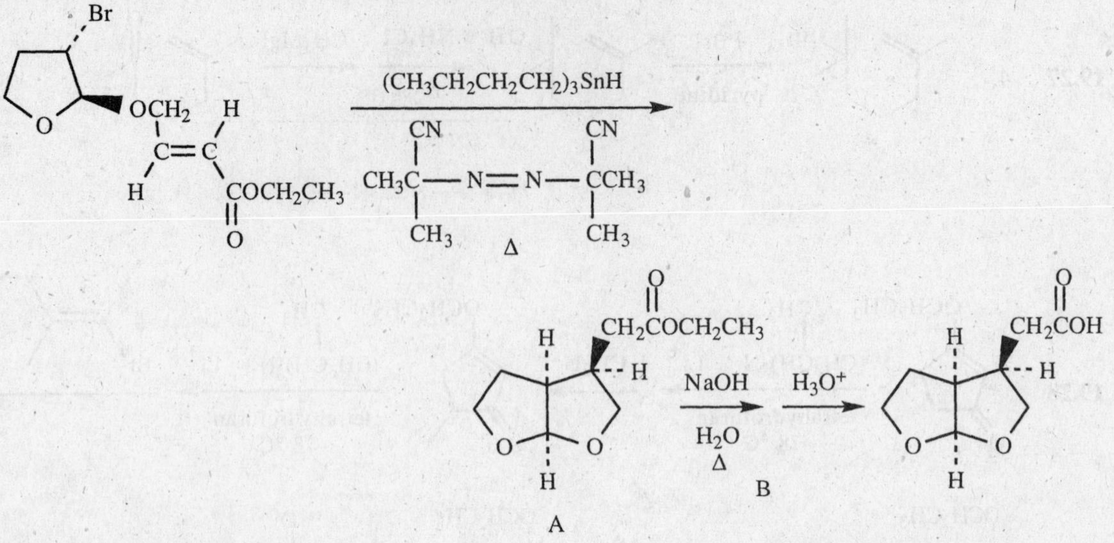

19.30

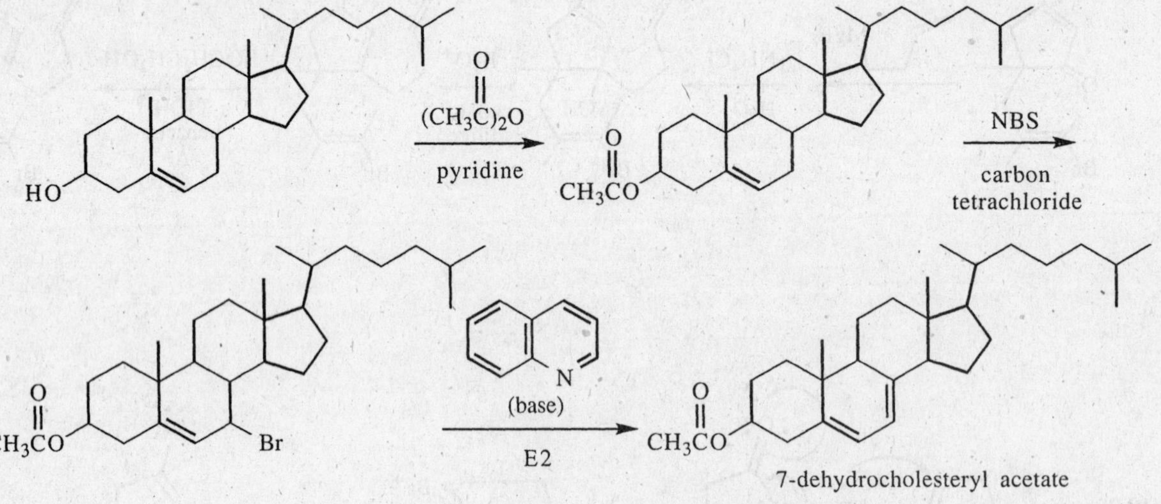

7-dehydrocholesteryl acetate

19.31 (a)

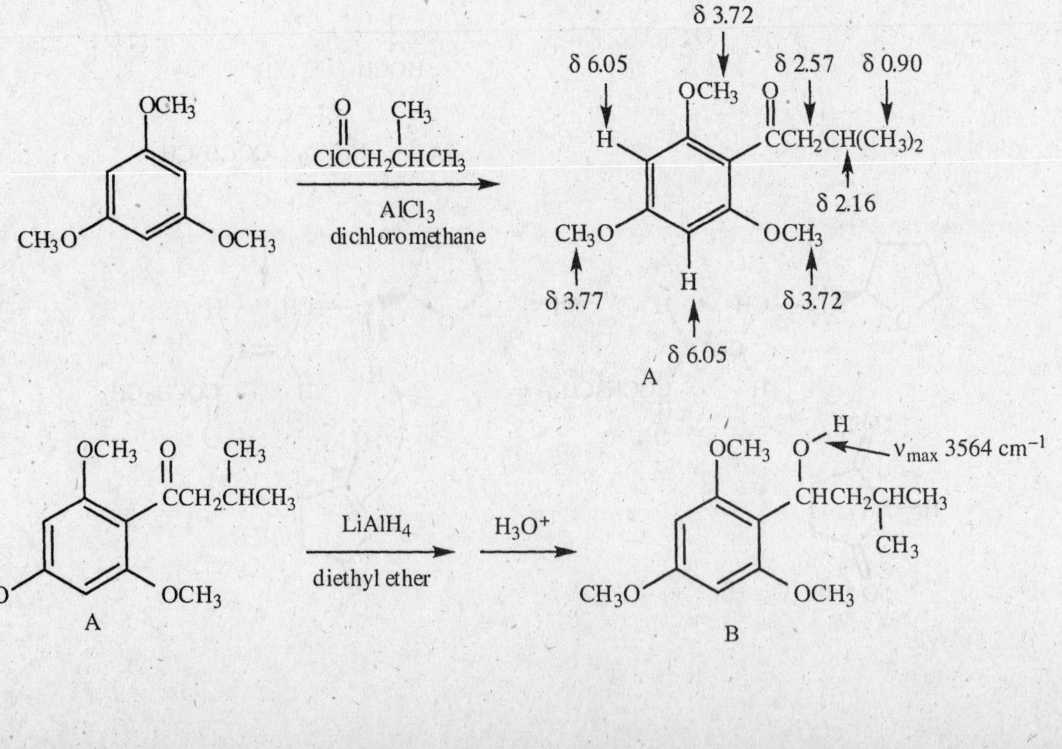

19.31 (b)

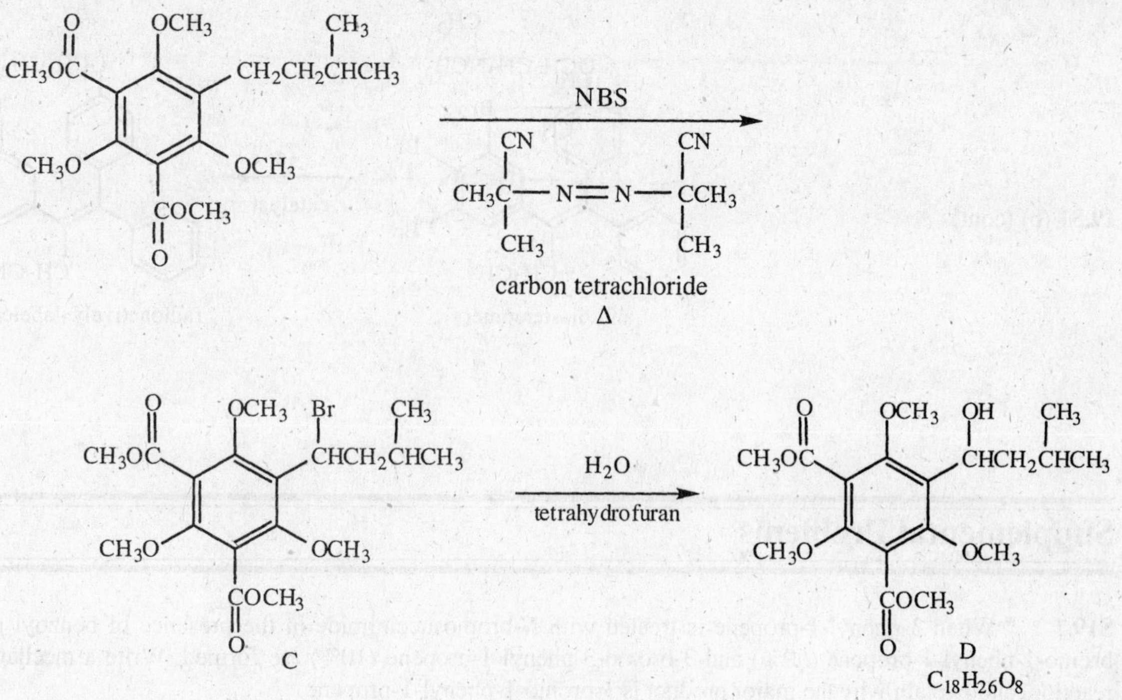

carbon tetrachloride
Δ

C

H$_2$O
tetrahydrofuran

D
C$_{18}$H$_{26}$O$_8$

19.32 (a)

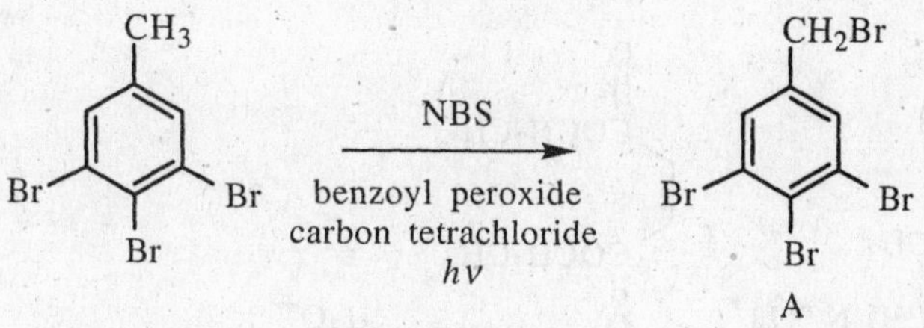

NBS
benzoyl peroxide
carbon tetrachloride
hv

A

(b)

NaH
B

CH$_3$CH$_2$I
C

tetrahydrofuran

HCl
methanol
Δ

NaOH

D

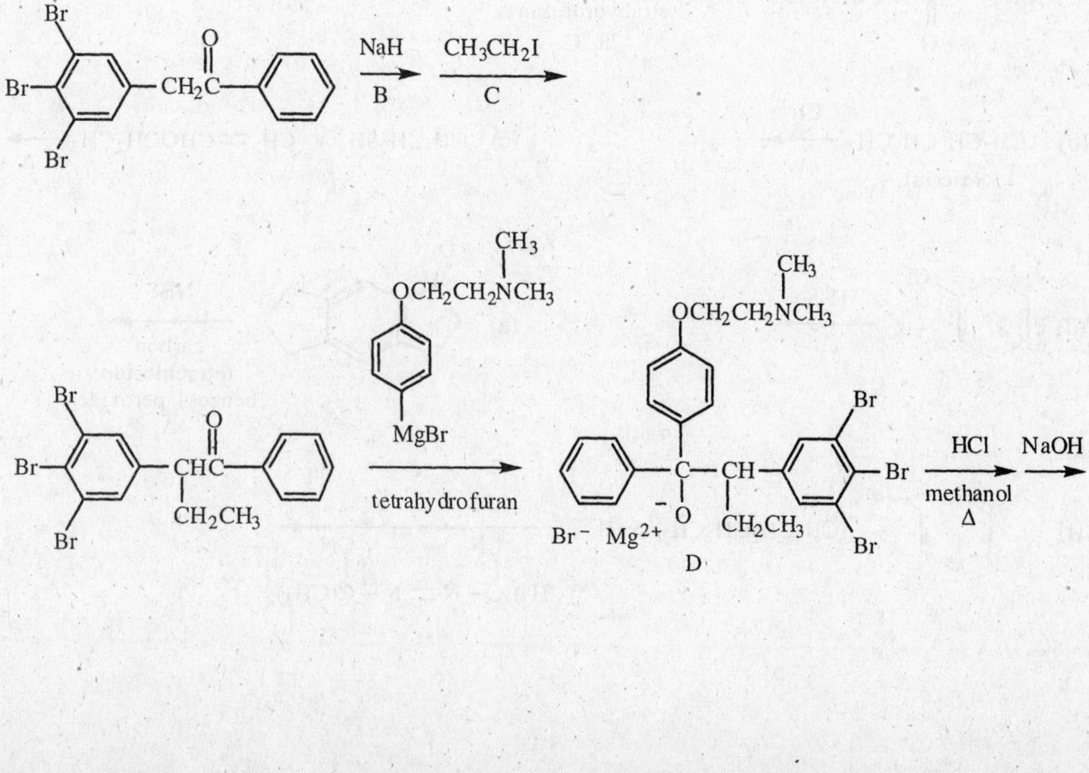

19.31 (b) (cont)

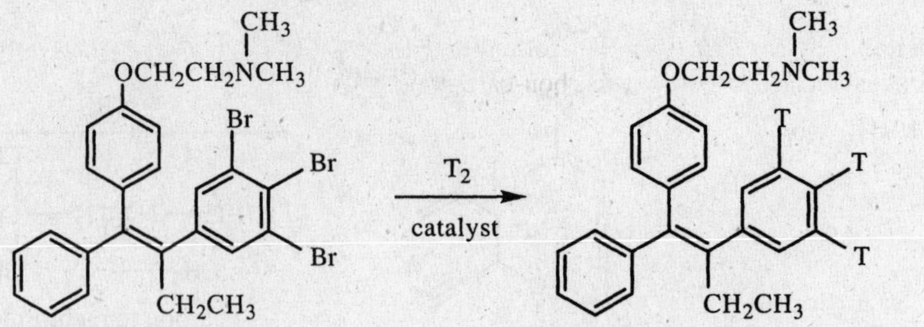

and diastereomer → (T₂, catalyst) → radioactively labeled tamoxifen

Supplemental Problems

S19.1 When 3-phenyl-1-propene is treated with *N*-bromosuccinimide in the presence of benzoyl peroxide, 3-bromo-1-phenyl-1-propene (50%) and 3-bromo-3-phenyl-1-propene (10%) are formed. Write a mechanism for the reaction, and explain why the major product is 3-bromo-1-phenyl-1-propene.

S19.2 Complete the following equations, showing the product(s) you expect. If the stereochemistry is known for a reaction, make sure your answer shows it.

(a)

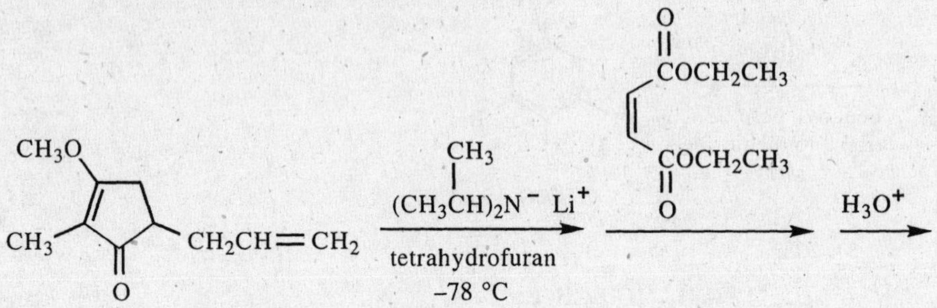

$(CH_3CH)_2N^- Li^+$, tetrahydrofuran, $-78\ °C$ → → H_3O^+ →

(b) $CH_3CH_2CH_2CH_3$ (excess) $\xrightarrow[h\nu]{Cl_2}$

(c) $CH_3CH_2SH + CH_2={}CHOCH_2CH_3 \xrightarrow{\Delta}$

(d)

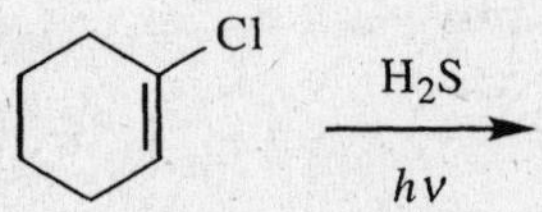

$\xrightarrow[h\nu]{H_2S}$

(e)

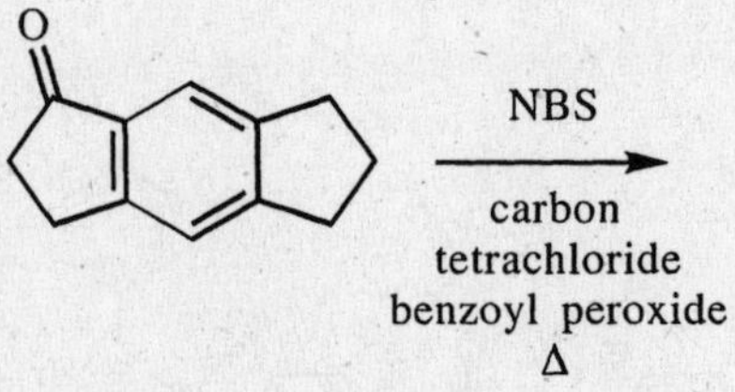

$\xrightarrow[\substack{\text{carbon} \\ \text{tetrachloride} \\ \text{benzoyl peroxide} \\ \Delta}]{NBS}$

(f)

$+ (CH_3CH_2CH_2CH_2)_3SnH \xrightarrow{\substack{(CH_3)_2\overset{CN}{\underset{}{C}}-N{=}N-\overset{CN}{\underset{}{C}}(CH_3)_2 \\ \Delta}}$

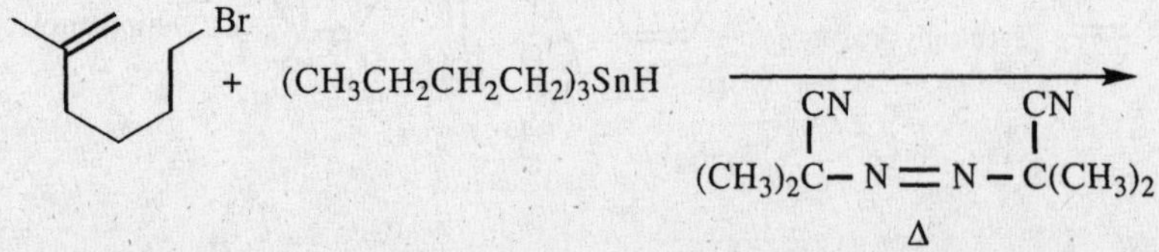

S19.3   The following transformation was observed on the way to the synthesis of compounds for the study of free-radical reactions that form carbon-carbon bonds.

The product is the result of four successive free-radical carbon-carbon bond formation steps. Construct the final product from the two starting materials. Which carbon-carbon bond-forming reaction will be the first? Remember that if an intramolecular reaction is possible, it will occur more easily than an intermolecular reaction.

20 The Chemistry of Amines

20.1 (a) 2-methyl-1-propanamine (isobutylamine)   (b) (1*R*, 2*S*)-2-ethyl-1-cyclopentanamine [(1*R*, 2*S*)-1-amino-2-ethylcyclopentane)]   (c) *N*,*N*-diethylbutanamine

(d) 3-cyclopropyl-1-propanamine (3-cyclopropyl-1-propylamine)   (e) 3-nitroaniline (*m*-nitroaniline)   (f) 2,4-dibromoaniline

20.2 (a)

(b)

(c)

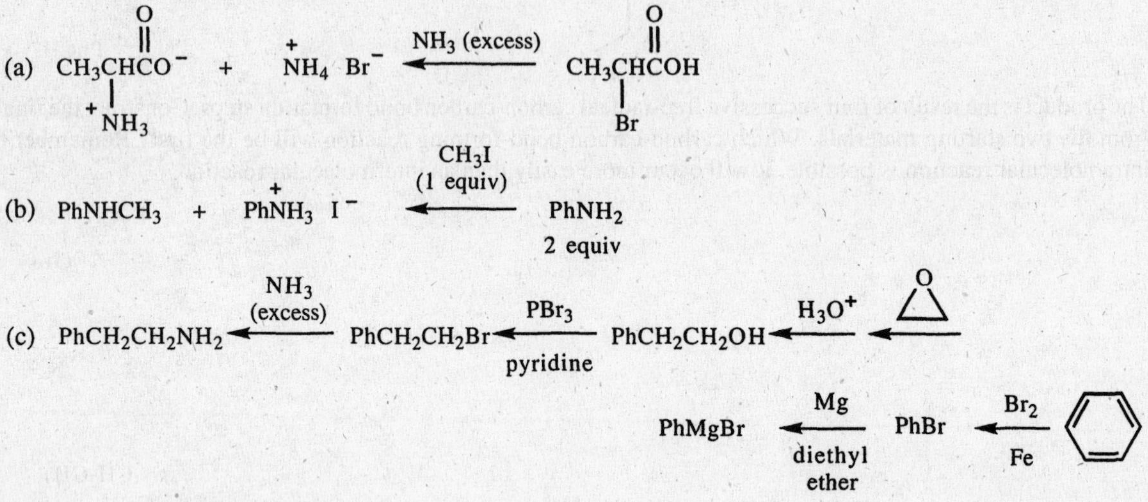

20.3

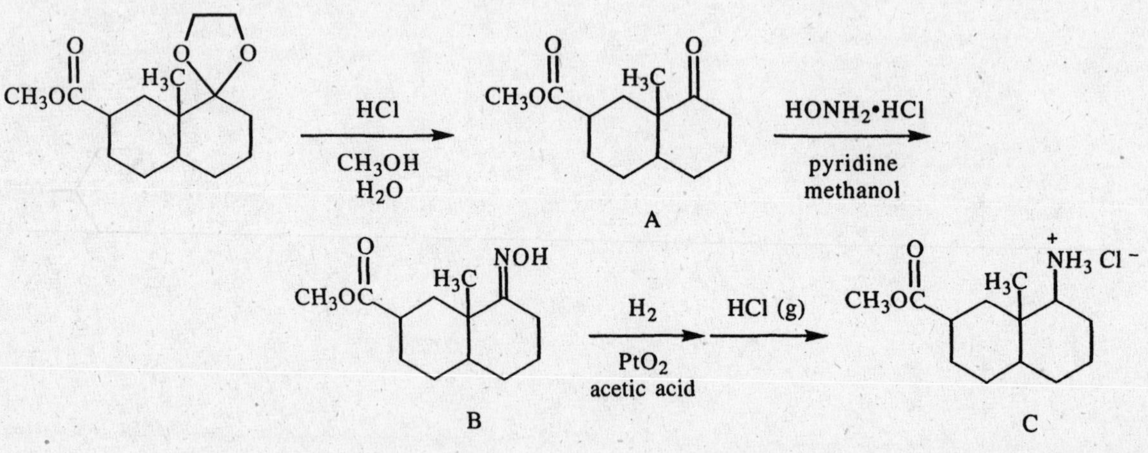

20.4

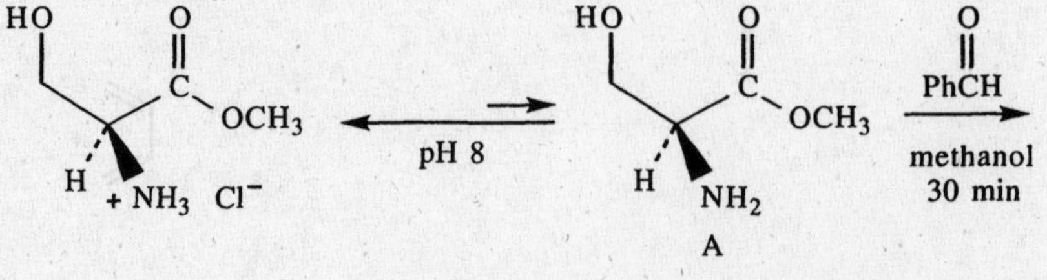

20.4 (cont)

$$\left[\begin{array}{c} \text{HO} \quad \overset{\text{O}}{\underset{\|}{\text{C}}} \text{OCH}_3 \\ \overset{|}{\underset{\text{N}=\text{CHPh}}{\text{H}}} \\ \text{B} \end{array} \right] \xrightarrow{\text{NaBH}_4 \quad \text{HCl (g)}} \begin{array}{c} \text{HO} \quad \overset{\text{O}}{\underset{\|}{\text{C}}} \text{OCH}_3 \\ \overset{|}{\underset{\overset{+}{\text{NH}_2\text{CH}_2\text{Ph}}}{\text{H}}} \\ \text{Cl}^- \qquad \text{C} \end{array} \xrightarrow[\text{D}]{\text{NH}_3} \begin{array}{c} \text{HO} \quad \overset{\text{O}}{\underset{\|}{\text{C}}} \text{NH}_2 \\ \overset{|}{\underset{\text{NHCH}_2\text{Ph}}{\text{H}}} \end{array}$$

20.5   (a) $\text{PhCH}_2\text{CH}_2\text{NH}_2 \xleftarrow[\text{PtO}_2]{\text{H}_2} \text{PhCH}=\text{CHNO}_2 \xleftarrow{\text{CH}_3\text{NO}_2,\ \text{NaOH}} \underset{\text{dichloromethane}}{\text{PhCH}} \xleftarrow{} \text{PhCH}_2\text{OH}$

(with pyridinium chlorochromate reagent $\overset{+}{\text{N}}\text{H}\ \text{CrO}_3\text{Cl}^-$ over the arrow, PhCH has =O)

(b) $\underset{\overset{|}{\text{NH}_2}}{\text{CH}_3\text{CH}_2\text{CH}_2\text{CHCH}_2\text{CH}_3} \xleftarrow[\text{Ni}]{\text{NH}_3,\ \text{H}_2} \underset{}{\text{CH}_3\text{CH}_2\text{CH}_2\overset{\overset{\text{O}}{\|}}{\text{C}}\text{CH}_2\text{CH}_3} \xleftarrow[\text{acetone}]{\text{CrO}_3} \underset{\overset{|}{\text{OH}}}{\text{CH}_3\text{CH}_2\text{CH}_2\text{CHCH}_2\text{CH}_3}$

$\text{NH}_4\text{Cl} \xleftarrow[\text{diethyl ether}]{\text{CH}_3\text{CH}_2\text{MgBr}} \underset{}{\text{CH}_3\text{CH}_2\text{CH}_2\overset{\overset{\text{O}}{\|}}{\text{C}}\text{H}} \xleftarrow[\text{dichloromethane}]{\overset{+}{\text{N}}\text{H}\ \text{CrO}_3\text{Cl}^-} \text{CH}_3\text{CH}_2\text{CH}_2\text{CH}_2\text{OH}$
H_2O

(c)

(d)

20.6   *enolization*   *condensation*   *protonation*

CH₃CH₂—OH

enolization   *protonation*

CH₃COH

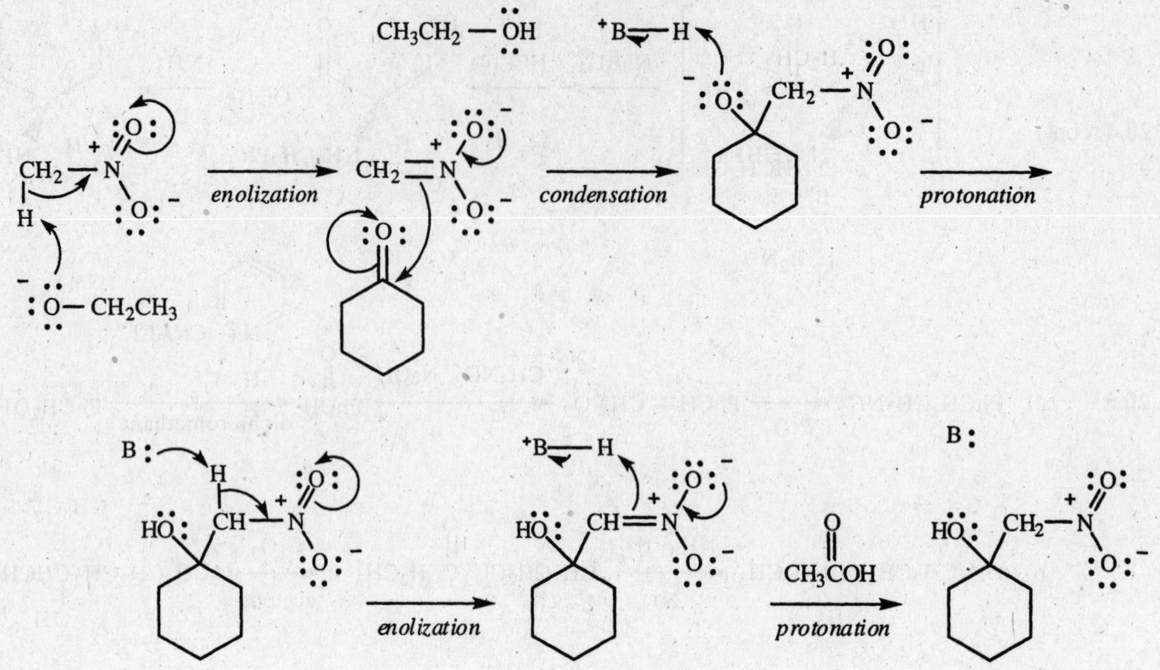

20.7   (a) PhCH + NH₃ $\xrightarrow[\text{Raney Ni}]{H_2}$ PhCH₂NH₂   (b) PhCH₂CH₂N₃ $\xrightarrow[\substack{\text{diethyl}\\ \text{ether}}]{LiAlH_4}$ $\xrightarrow{H_2O}$ PhCH₂CH₂NH₂

(c) 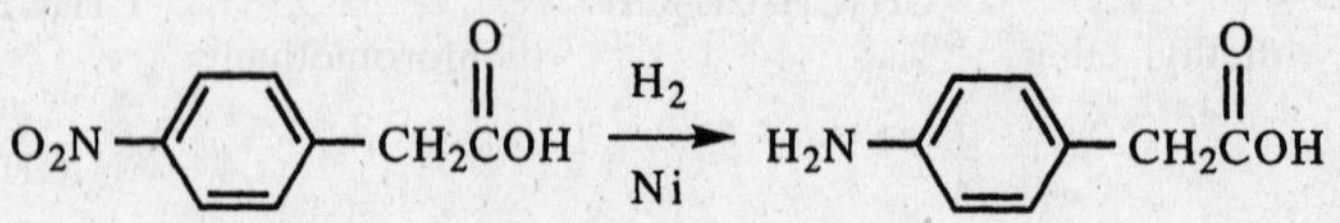 O₂N—⬡—CH₂COH $\xrightarrow[\text{Ni}]{H_2}$ H₂N—⬡—CH₂COH

(d) 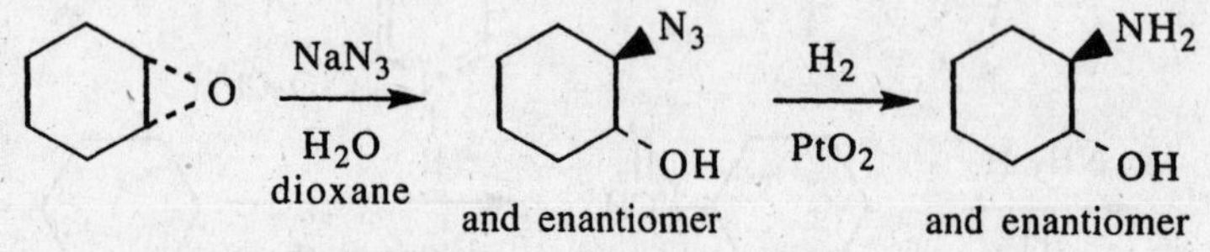 ⬡O $\xrightarrow[\substack{H_2O\\ \text{dioxane}}]{NaN_3}$ (N₃, OH) and enantiomer $\xrightarrow[\text{PtO}_2]{H_2}$ (NH₂, OH) and enantiomer

(e) 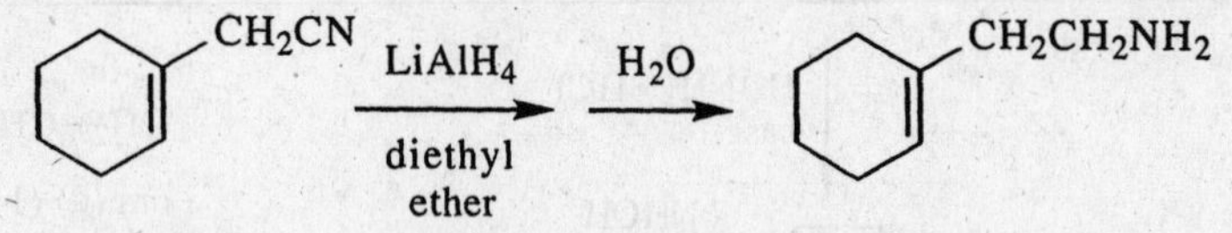 ⬡CH₂CN $\xrightarrow[\substack{\text{diethyl}\\ \text{ether}}]{LiAlH_4}$ $\xrightarrow{H_2O}$ ⬡CH₂CH₂NH₂

20.8   (a) 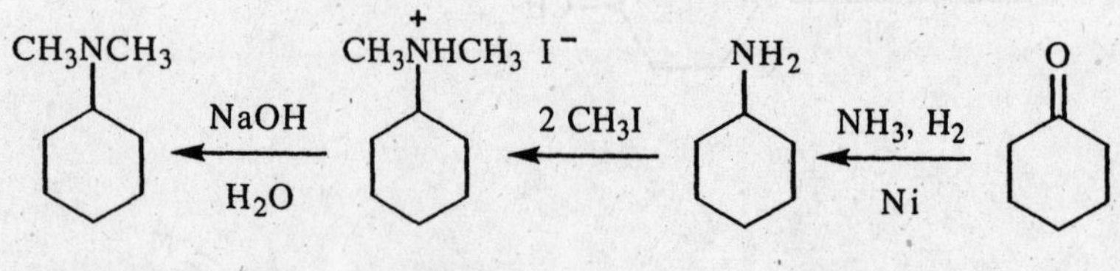   CH₃NCH₃ ⬡ $\xleftarrow[\substack{H_2O}]{NaOH}$ CH₃⁺NHCH₃ I⁻ ⬡ $\xleftarrow{2 CH_3I}$ NH₂ ⬡ $\xleftarrow[\text{Ni}]{NH_3, H_2}$ cyclohexanone

(b) 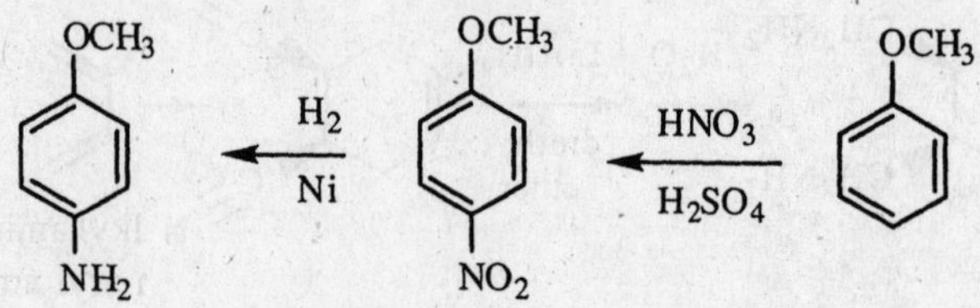   OCH₃ ⬡ NH₂ $\xleftarrow[\text{Ni}]{H_2}$ OCH₃ ⬡ NO₂ $\xleftarrow[\substack{H_2SO_4}]{HNO_3}$ OCH₃ ⬡

20.8   (c)   CH₂CH₂CH₂CH₃
          |
       PhNCH₂CH₂CH₂CH₃ ←—NaOH / H₂O—← CH₃CH₂CH₂CH₂Br

$$PhNHCH_2CH_2CH_2CH_3 \xleftarrow[H_2O]{NaBH_4} PhN=CHCH_2CH_2CH_3 \xleftarrow{PhNH_2} CH_3CH_2CH_2CH (O)$$

(d) (1) $CH_3(CH_2)_6CH_2NH_2 \xleftarrow{NH_3 \text{ (excess)}} CH_3(CH_2)_6CH_2Br \xleftarrow[\text{pyridine}]{PBr_3}$

$CH_3(CH_2)_6CH_2OH \xleftarrow{H_3O^+} \xleftarrow{\triangle O} CH_3(CH_2)_4CH_2MgBr \xleftarrow[\text{diethyl ether}]{Mg} CH_3(CH_2)_4CH_2Br \xleftarrow[\text{pyridine}]{PBr_3}$

$CH_3(CH_2)_4CH_2OH \xleftarrow{H_3O^+} \xleftarrow{\triangle O} CH_3CH_2CH_2CH_2MgBr \xleftarrow[\text{diethyl ether}]{Mg} CH_3CH_2CH_2CH_2Br$

(2) $CH_3(CH_2)_6CH_2NH_2 \xleftarrow[\substack{Pd \\ \text{acetic acid}}]{H_2} CH_3(CH_2)_5CH=CHNO_2 \xleftarrow{CH_3NO_2, NaOH}$

$CH_3(CH_2)_5CH (O) \xleftarrow[\substack{\text{dichloromethane} \\ \text{or}}]{\overset{+}{N}H \, CrO_3Cl^-} CH_3(CH_2)_5CH_2OH \xleftarrow{H_3O^+} \xleftarrow{\square O} CH_3CH_2CH_2CH_2MgBr$
[from (d) (1)]

$(CH_3CH_2)_3N \xleftarrow{} \xleftarrow{CH_3SCH_3, \, ClC-CCl \, (O)(O)}$

(3) $CH_3(CH_2)_6CH_2NH_2 \xleftarrow[Ni]{NH_3, H_2} CH_3(CH_2)_6CH (O) \xleftarrow[\text{dichloromethane}]{\overset{+}{N}H \, CrO_3Cl^-} CH_3(CH_2)_6CH_2OH$
[from (d) (1)]

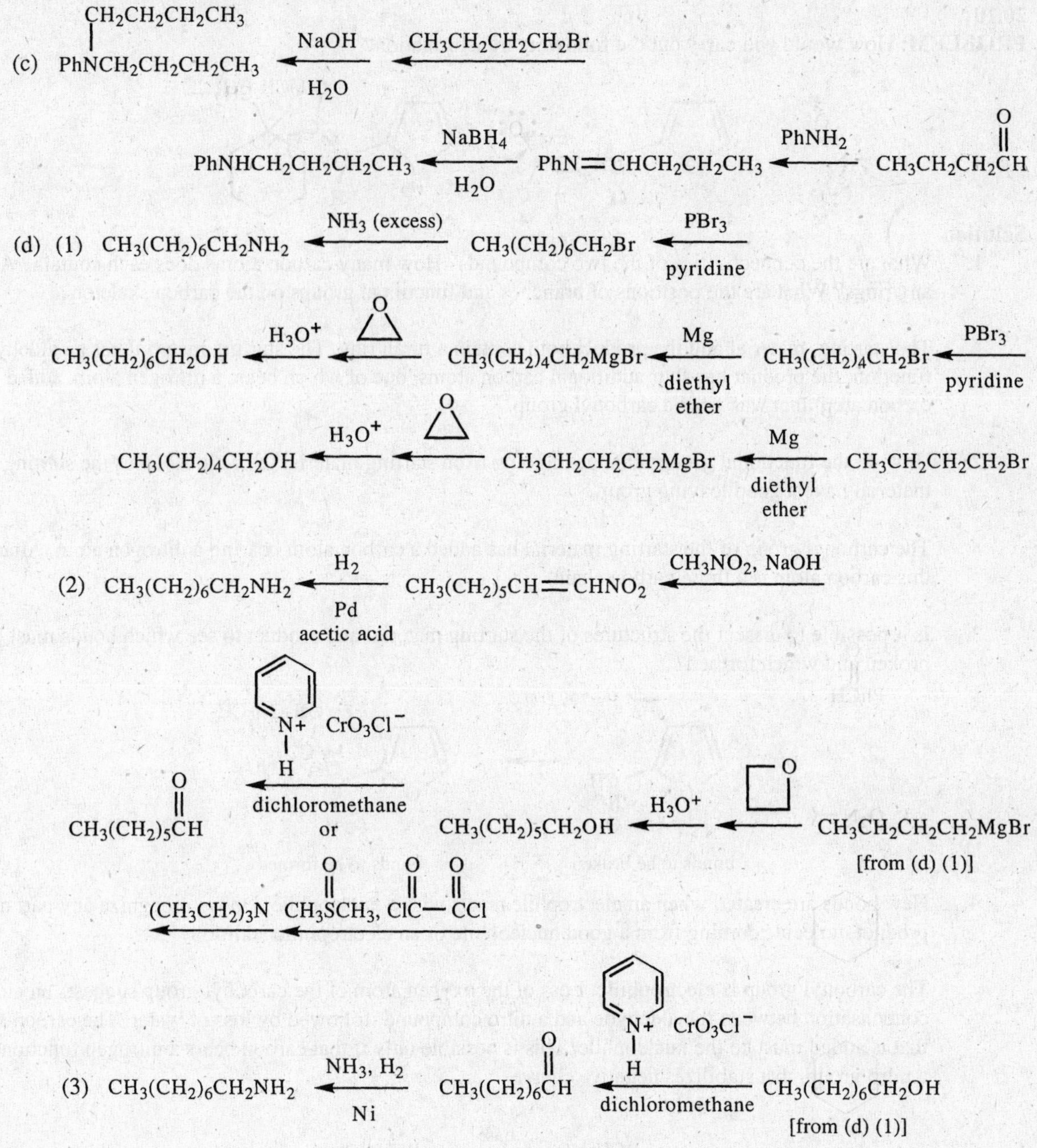

20.9

O   CH₃
‖ |
COCH₂CH₂NHCH₃ Cl⁻
              +

(benzene ring with NH₂)

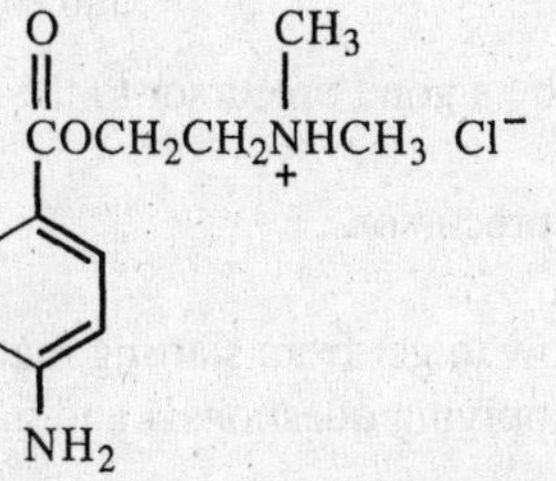

Procaine has two amine functions, an amino group on the aromatic ring and a tertiary alkyl amine function. The alkyl amine is much more basic (pK_a of a substituted ammonium ion, ~10) than an aryl amine (pK_a of the anilinium ion, 4.6). Therefore procaine hydrochloride has the structure shown above.

20.10
PROBLEM: How would you carry out the following transformation?

Solution

1. What are the connectivities of the two compounds? How many carbon atoms does each contain? Are there any rings? What are the positions of branches and functional groups on the carbon skeleton?

   The starting material and the product both contain a furan ring. The starting material has an aldehyde function; the product has four additional carbon atoms, one of which bears a nitrogen atom, added to the carbon atom that was once a carbonyl group.

2. How do the functional groups change in going from starting material to product? Does the starting material have a good leaving group?

   The carbonyl group of the starting material has added a carbon atom bearing a nitrogen atom. Attached to this carbon atom is a three-carbon chain.

3. Is it possible to dissect the structures of the starting material and product to see which bonds must be broken and which formed?

   bonds to be broken   bonds to be formed

4. New bonds are created when an electrophile reacts with a nucleophile. Do we recognize any part of the product molecule coming from a good nucleophile or an electrophilic addition?

   The carbonyl group is electrophilic. Loss of the oxygen atom of the carbonyl group suggests an aldol type condensation between the aldehyde and a nitro compound, followed by loss of water. The carbon atom that is added must be the nucleophile. This is possible only if that carbon bears a nitrogen function, such as a nitro group, that stabilizes negative charge.

$$RCH_2NO_2 \xrightarrow{\text{base}} R\overset{\cdot\cdot}{C}HNO_2^{-}$$

   enolate of an alkyl
   nitro compound

5. What type of compound would be a good precursor to the product?

   1-Nitrobutane would be a good precursor.

6. After this last step, do we see how to get from starting material to product? If not, we need to analyze the structure obtained in step 5 by applying questions 4 and 5 to it.

   In this case, easy dehydration to give a double bond conjugated with the aromatic furan ring and the nitro group takes place.

20.10 (6) (cont)

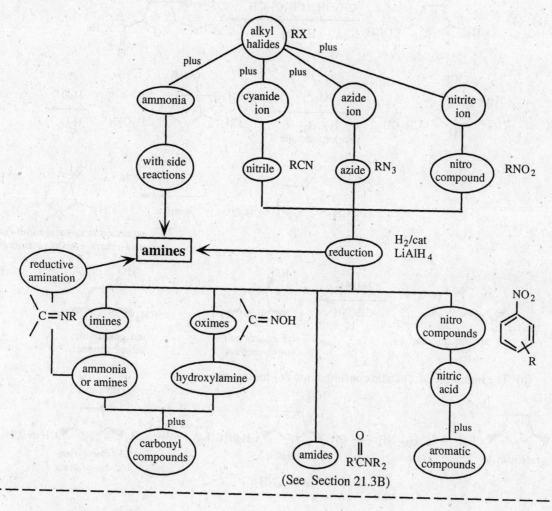

Concept Map 20.1   Preparation of amines.

20.11

1° alcohol site
reacts first

20.11 (cont)

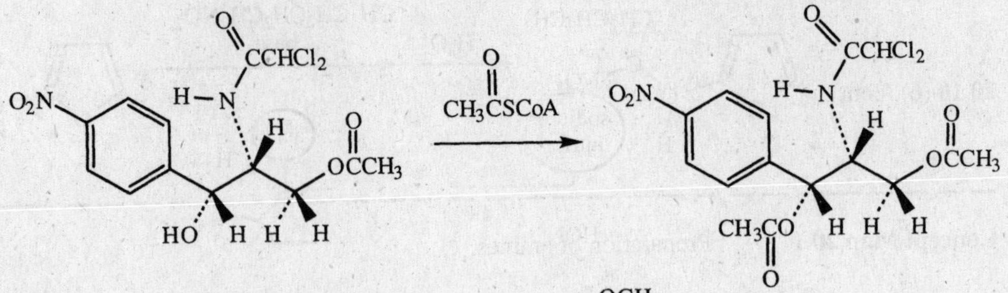
O₂N
H—N
CCHCl₂
OCCH₃
HO H H H
CH₃CSCoA
CH₃CO
H H H

20.12 (a) (1)

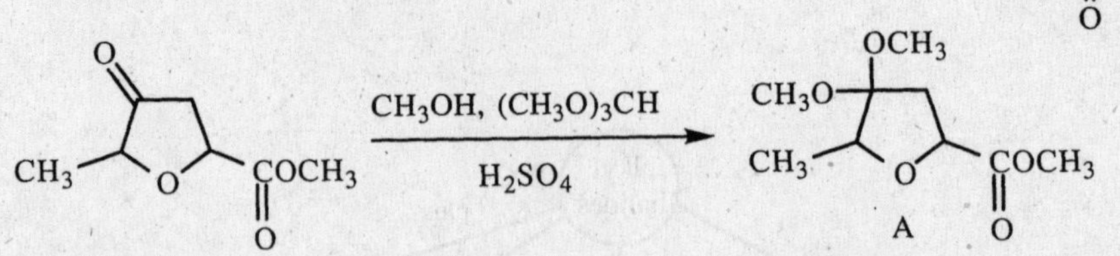
CH₃
COCH₃
CH₃OH, (CH₃O)₃CH
H₂SO₄
OCH₃
CH₃O
A

(2)

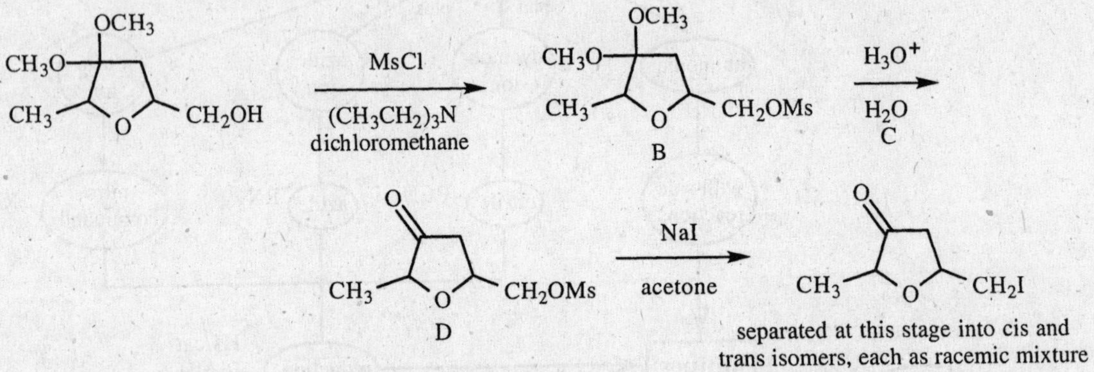
OCH₃
CH₃O
CH₃
CH₂OH
MsCl
(CH₃CH₂)₃N
dichloromethane
CH₂OMs
B
H₃O⁺
H₂O
C
D
NaI
acetone
CH₂I
separated at this stage into cis and trans isomers, each as racemic mixture

(3)

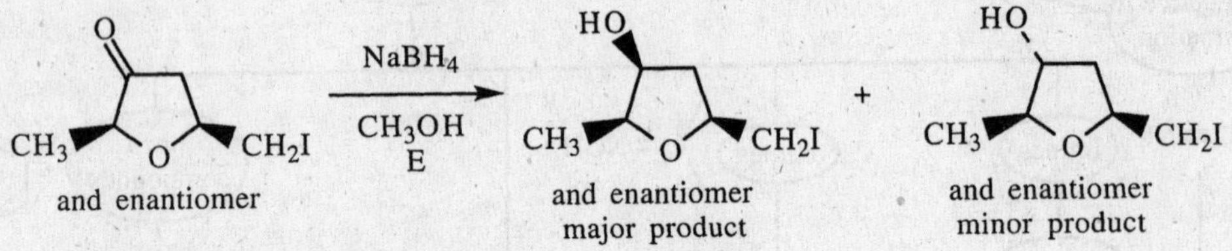
CH₃
CH₂I
and enantiomer
NaBH₄
CH₃OH
E
HO
and enantiomer
major product
+
and enantiomer
minor product

(b) The minor product can be converted into (+)-muscarine.

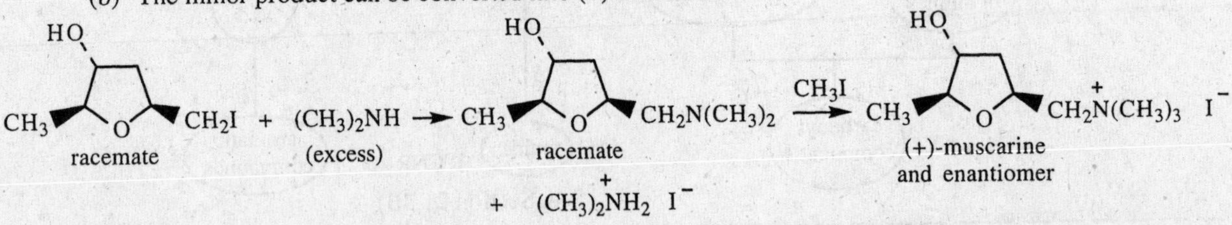
HO
CH₃
CH₂I
racemate
+ (CH₃)₂NH
(excess)
CH₂N(CH₃)₂
racemate
+ (CH₃)₂NH₂ I⁻
CH₃I
CH₂N(CH₃)₃ I⁻
(+)-muscarine
and enantiomer

20.13 $R_2NH + HCl \longrightarrow R_2\overset{+}{N}H_2\ Cl^-$

water soluble

$R_2NH + HNO_2 \xrightarrow[HCl]{} R_2\ddot{N}-\ddot{N}=\ddot{O} \longleftrightarrow R_2\overset{+}{N}=\ddot{N}-\ddot{O}:^-$

The nonbonding electrons on the amine nitrogen atom are delocalized to the nitroso group. The nitrosamine is therefore much less basic than the original amine and will not form a salt. A nitrosamine resembles an amide in basicity.

Concept Map 20.2   Nitrosation reactions.

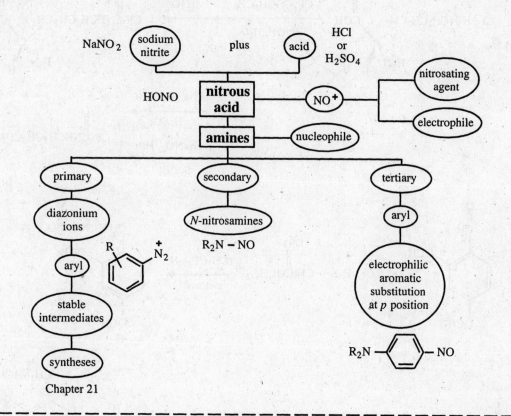

20.14   (a)

(b)

(c)

(d)

20.15   (a)

CF$_3$COCH$_2$CH$_3$, NaOH, H$_2$O; H$_3$O$^+$ → CF$_3$CNHCH$_2$CH$_2$CH$_2$—COH (A)

NaNO$_2$, KBr, H$_2$SO$_4$ or NaNO$_2$, HBr (B) → CF$_3$CNHCH$_2$CH$_2$CH$_2$—COH with Br

(b)

2,4-dimethoxybenzaldehyde + Cl$^-$ H$_3$N—CH(COCH$_3$)$_2$ → (CH$_3$CH$_2$)$_3$N, methanol → CH=NCH(COCH$_3$)$_2$ ... OCH$_3$ ... OCH$_3$ (C)

H$_2$, Pd/C (D) → CH$_2$NHCH(COCH$_3$)$_2$ ... OCH$_3$ ... OCH$_3$

- -

Concept Map 20.3   Reactions of diazonium ions.

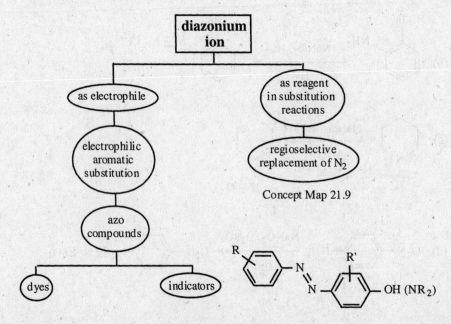

Concept Map 21.9

20.16

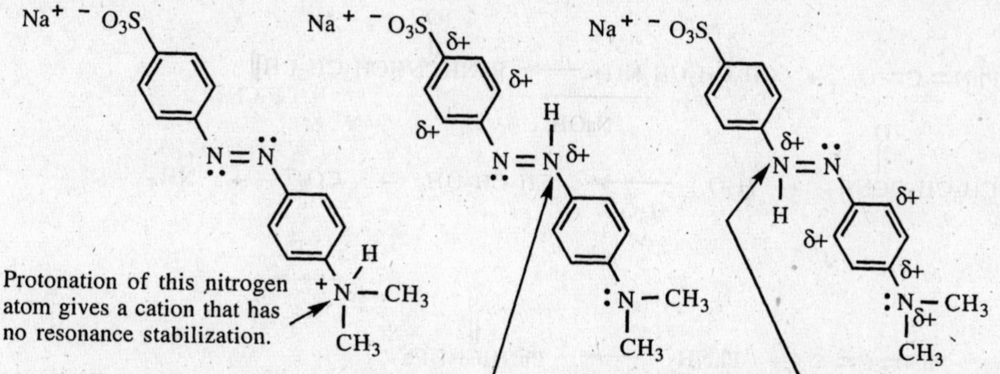

Protonation of this nitrogen atom gives a cation that has no resonance stabilization.

Protonation of this nitrogen atom gives a cation that is stabilized by resonance; the positive charge is delocalized to the carbon atoms of an aromatic ring.

Protonation of this nitrogen atom gives a cation that is stabilized by resonance; the positive charge is delocalized to the carbon atoms of an aromatic ring and to the nitrogen atom of the amino group. Therefore this is the most stable cation of the three.

20.17

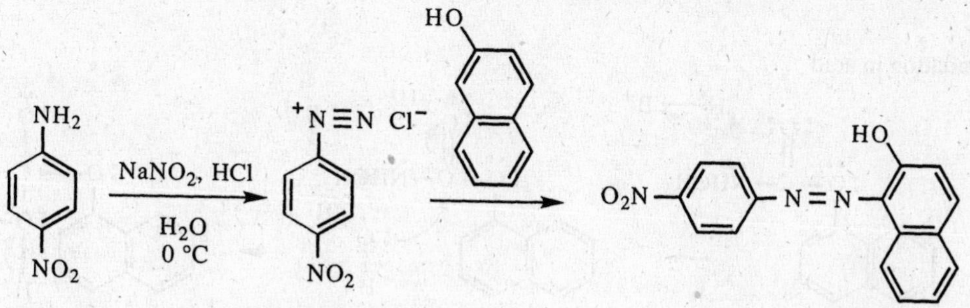

20.18

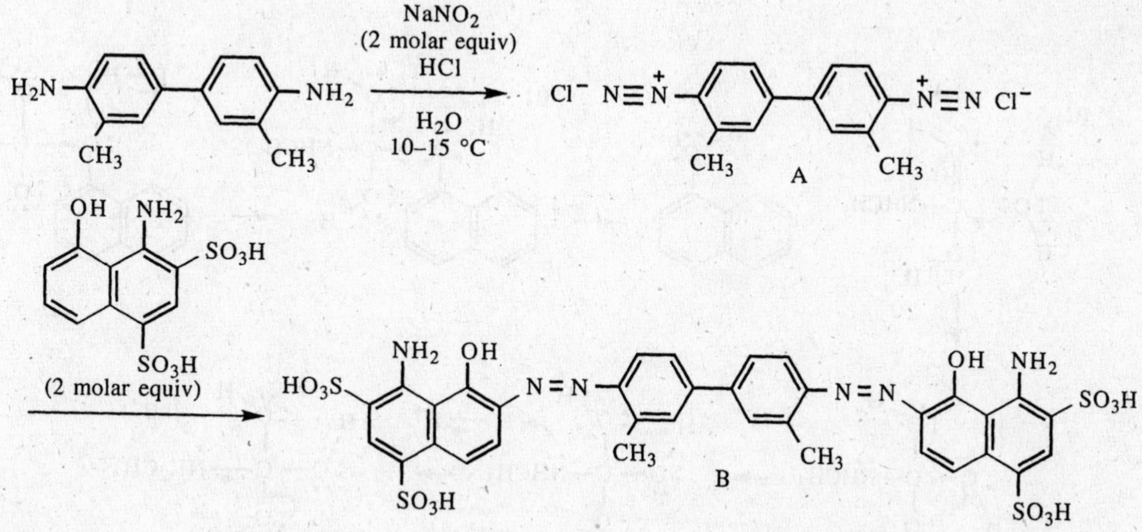

20.19 (a)

(b)

20.19 (c) $PhN = C = O$ + $CH_3CH_2CH_2NH_2$ $\longrightarrow$ $PhNHCNHCH_2CH_2CH_3$

$$O$$

(d) $CH_3CH_2OCNH_2$ + H_2O $\xrightarrow[\Delta]{H_3O^+}$ CH_3CH_2OH + $CO_2\uparrow$ + $\overset{+}{N}H_4$

(e) $PhN = C = S$ + $PhNH_2$ $\longrightarrow$ $PhNHCNHPh$
      phenyl isothiocyanate   a substituted thiourea

20.20   1. Degradation in acid

20.20 2. Degradation in base

20.21 (a) (1R, 2S)-2-ethyl-1-(N-methylamino)cyclobutane   (b) 2-methylpyridine   (c) triethylamine
(d) 2-amino-4-methylbenzoic acid   (e) 4-aminobutanoic acid   (f) 2,4,6-trichloroaniline
(g) 5-methyl-2-hexanamine (2-amino-5-methylhexane) (h) (1R, 2S)-2-aminocyclohexanol
(i) N,N-diethyl-4-nitroaniline (N,N-diethyl-p-nitroaniline)

20.22 (a)   (b)   (c)

(d) (e)   (f) $CH_3CH_2CH_2CHCH_2CH_2CHCH_3$

(g) $CH_3CH_2CH_2CHCH_2CH_3$ (h)

20.23 (a) PhCH=NPh $\xrightarrow[\text{diethyl ether}]{LiAlH_4}$ $\xrightarrow{H_2O}$ PhCH$_2$NHPh
                                                                                            A

(b) $\xrightarrow[H_2O]{NH_3}$

20.23 (c)

$$CH_3\overset{\underset{\displaystyle NO_2}{|}}{\underset{\displaystyle CH_3}{C}}CH_2CH_2\overset{\underset{\displaystyle O}{\|}}{C}OCH_3 \xrightarrow[\Delta]{\underset{\displaystyle Ni}{H_2}} \left[CH_3\overset{\underset{\displaystyle NH_2}{|}}{\underset{\displaystyle CH_3}{C}}CH_2CH_2\overset{\underset{\displaystyle O}{\|}}{C}OCH_3 \right] \longrightarrow$$

+ CH_3OH

(d)   H_2N— —SO_2NH_2 $\xrightarrow[Fe]{Br_2}$ H_2N— —SO_2NH_2

D

(e)

$$CH_3\overset{\underset{\displaystyle }{\|}}{\underset{\displaystyle O}{C}}(CH_2)_4CH_3 \xrightarrow[\underset{\displaystyle \Delta}{Ni}]{NH_3, H_2} CH_3\overset{\underset{\displaystyle NH_2}{|}}{C}H(CH_2)_4CH_3$$

E

(f)   CH_3CH_2CH_2NO_2 $\xrightarrow{\overset{\displaystyle O}{\overset{\displaystyle \|}{HCH}},\ NaOH}$ CH_3CH_2$\overset{\underset{\displaystyle NO_2}{|}}{C}$HCH_2OH $\xrightarrow[\Delta]{\underset{\displaystyle Ni}{H_2}}$ CH_3CH_2$\overset{\underset{\displaystyle NH_2}{|}}{C}$HCH_2OH

F   G

20.24 (a)

$\xrightarrow[\underset{\displaystyle \Delta}{pyridine}]{HO\overset{+}{N}H_3\ Cl^-}$ A $\xrightarrow[\underset{\displaystyle ether}{diethyl}]{LiAlH_4\quad H_3O^+}$ B $\xrightarrow[H_2O]{NaOH}$ C

(b) ^-O_3S— —$\overset{+}{N}H_3$ $\xrightarrow[H_2O]{Na_2CO_3}$ Na^+ $\ ^-O_3S$— —NH_2 $\xrightarrow[\underset{\displaystyle 0-5\ ^\circ C}{H_2O}]{NaNO_2,\ HCl}$ ^-O_3S— —$\overset{+}{N}\equiv N$

D   E

(c) $\xrightarrow[\underset{\displaystyle ether}{diethyl}]{LiAlH_4\quad H_2O}$

F

(d) $\xrightarrow[\underset{\displaystyle \Delta,\ pressure}{Ni}]{NH_3,\ H_2}$

G

(e) $CH_3\overset{\underset{\displaystyle }{\|}}{\underset{\displaystyle O}{C}}CH_3$ $\xrightarrow[HCl\ (catalyst)]{}$ H $\xrightarrow{NaOH}$ I

20.24 (f)

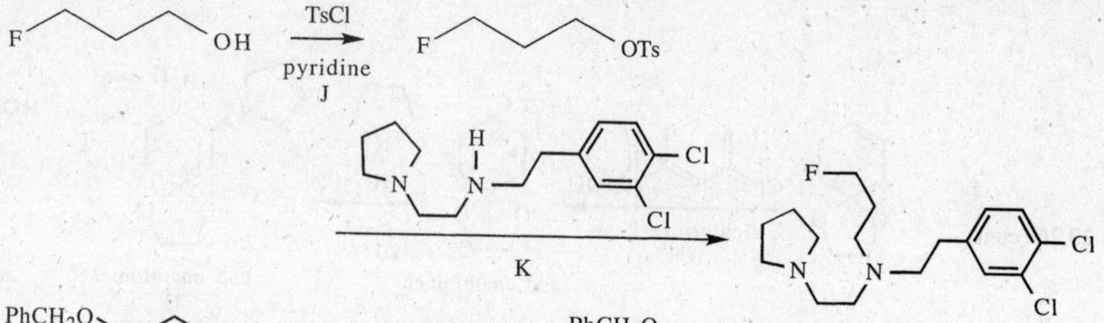
F
OH
TsCl
pyridine
J
OTs
H
N
Cl
K

(g)

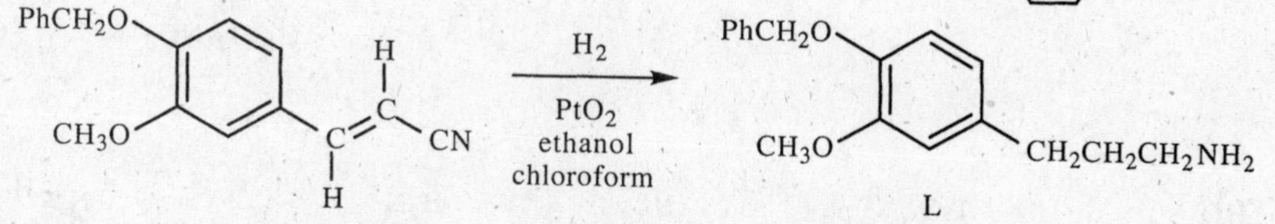
PhCH2O
CH3O
H
C
CN
H2
PtO2
ethanol
chloroform
CH2CH2CH2NH2
L

20.25

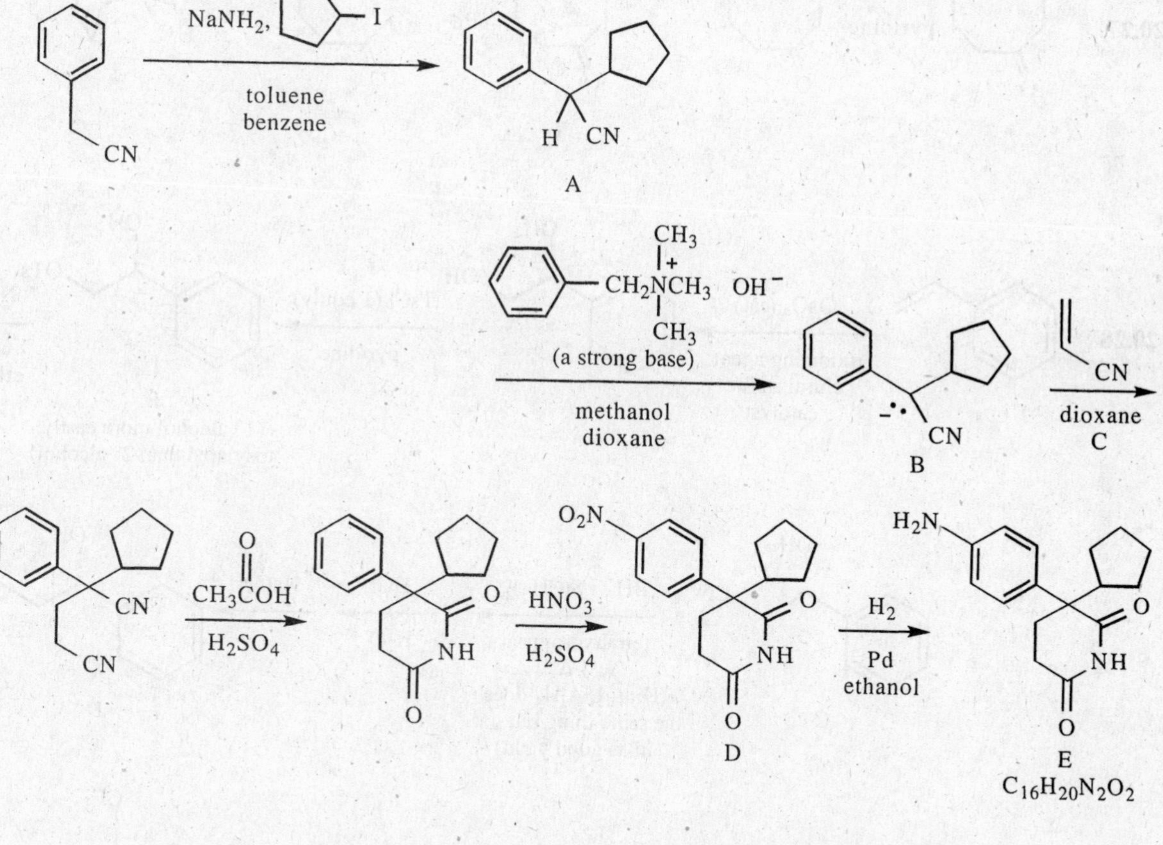
NaNH2,
I
toluene
benzene
CN
H
A
CH3
CH2NCH3 OH−
(a strong base)
methanol
dioxane
B
dioxane
C
CH3COH
H2SO4
O
NH
HNO3
O2N
D
H2
Pd
ethanol
H2N
E
C16H20N2O2

20.26

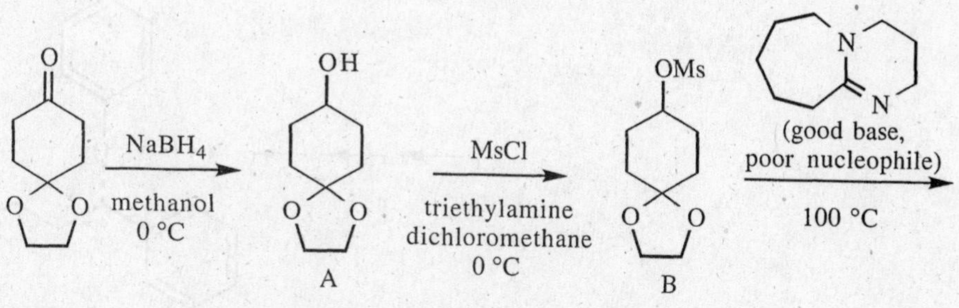
O
NaBH4
methanol
0 °C
OH
A
MsCl
triethylamine
dichloromethane
0 °C
OMs
B
N
(good base,
poor nucleophile)
100 °C

20.26 (cont)

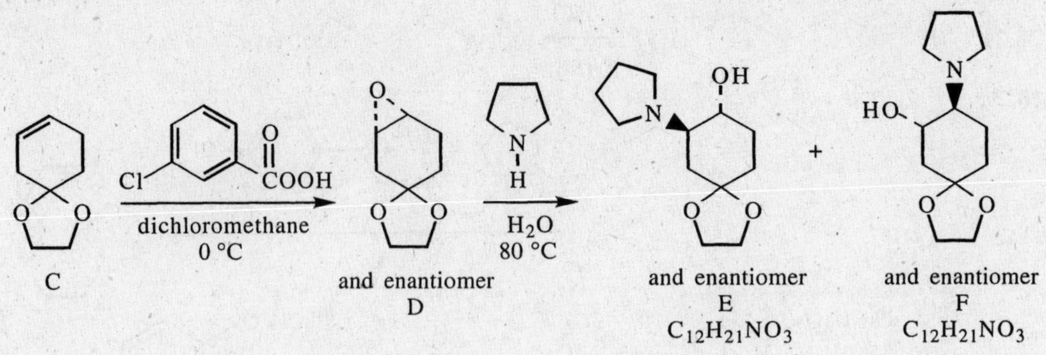
Cl
COOH
dichloromethane
0 °C
N
H
H_2O
80 °C
OH
N
+
HO
N
C
and enantiomer
D
and enantiomer
E
$C_{12}H_{21}NO_3$
and enantiomer
F
$C_{12}H_{21}NO_3$

20.27

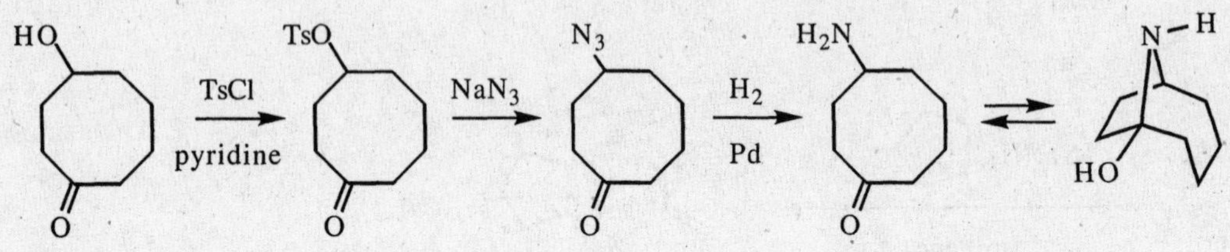
HO
TsCl
pyridine
TsO
NaN_3
N_3
H_2
Pd
H_2N
N–H
HO

20.28

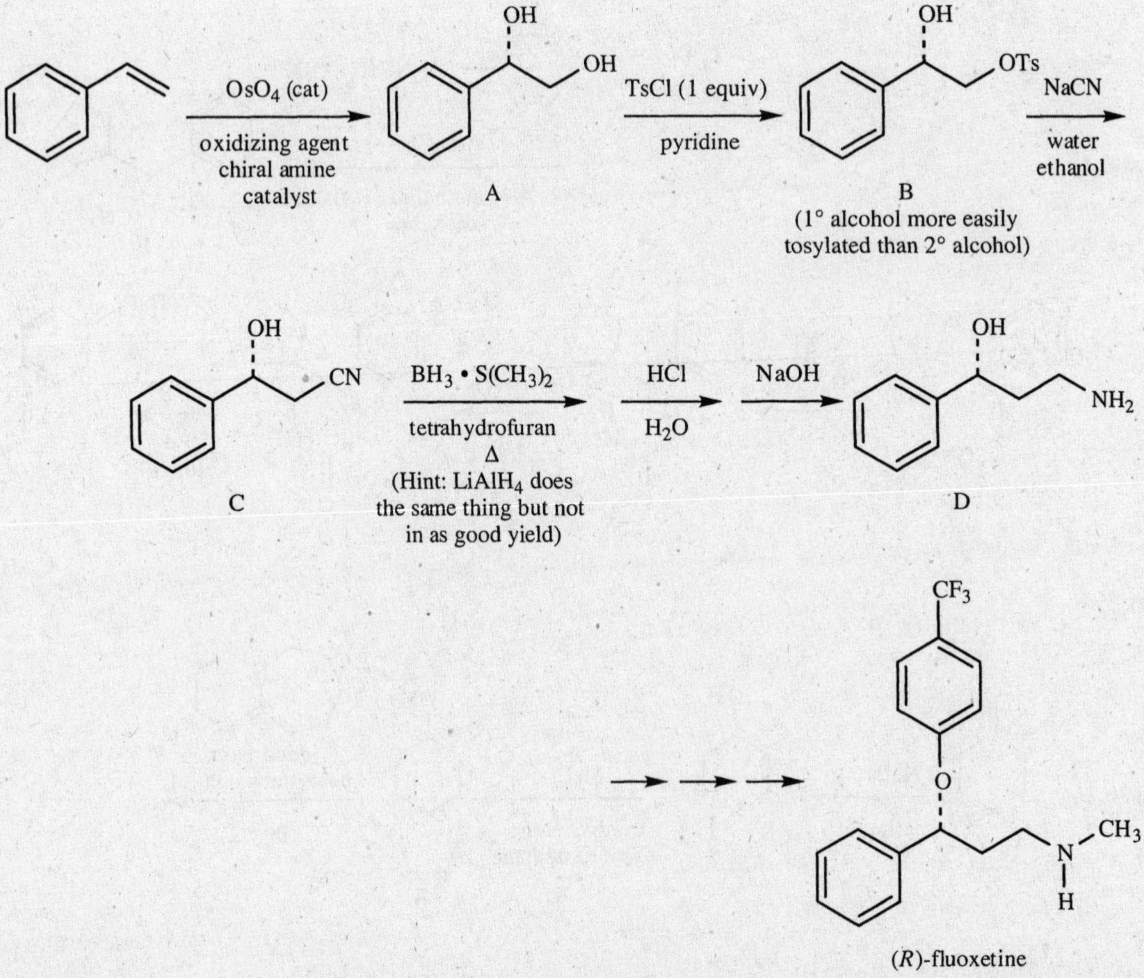
OsO_4 (cat)
oxidizing agent
chiral amine
catalyst
OH
OH
A
TsCl (1 equiv)
pyridine
OH
OTs
B
(1° alcohol more easily
tosylated than 2° alcohol)
NaCN
water
ethanol
OH
CN
C
$BH_3 \cdot S(CH_3)_2$
tetrahydrofuran
Δ
(Hint: $LiAlH_4$ does
the same thing but not
in as good yield)
HCl
H_2O
NaOH
OH
NH_2
D
CF_3
O
CH_3
N
H
(*R*)-fluoxetine

20.29

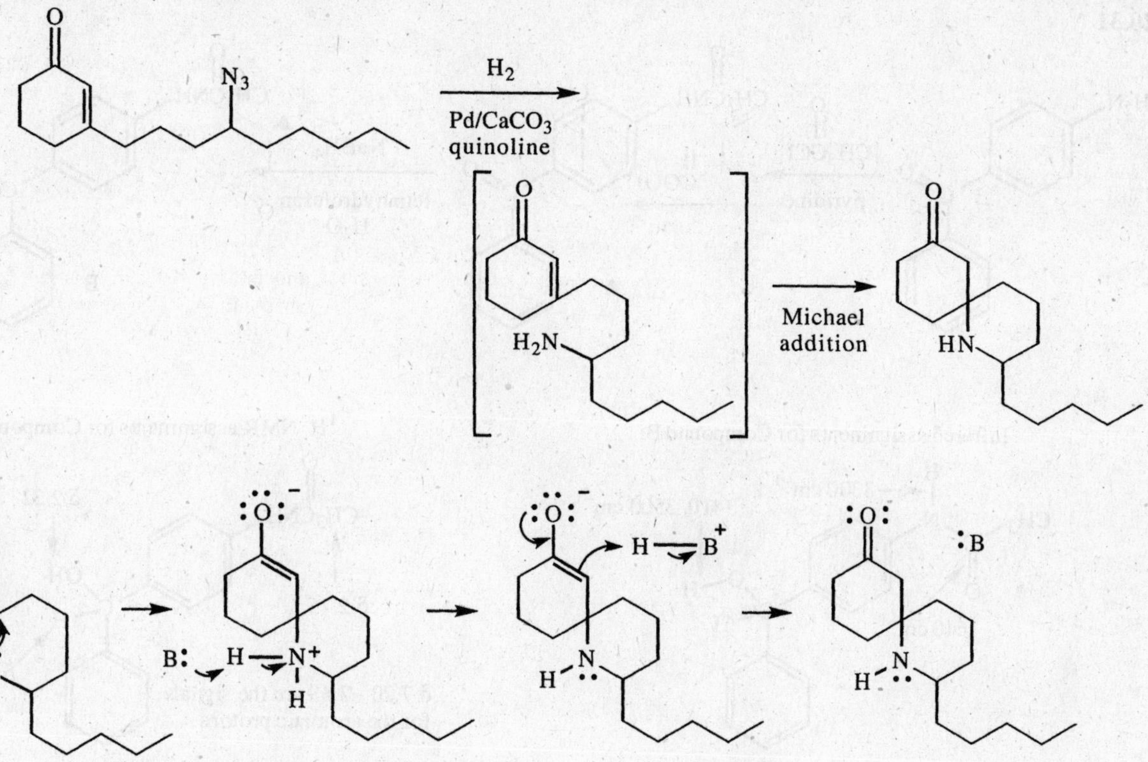
H_2
$Pd/CaCO_3$
quinoline
Michael addition

20.30

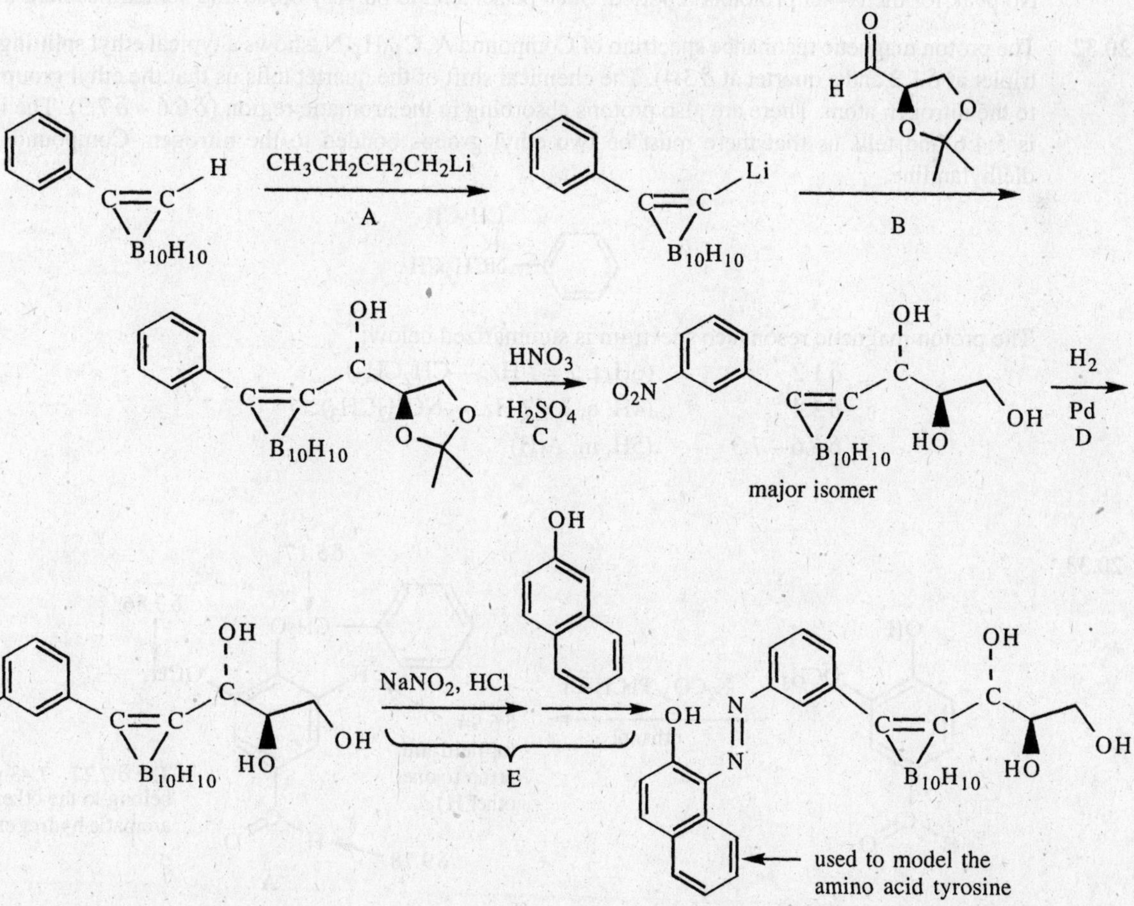
$CH_3CH_2CH_2CH_2Li$
A
B
HNO_3
H_2SO_4
C
major isomer
H_2
Pd
D
$NaNO_2$, HCl
E
used to model the amino acid tyrosine

20.31

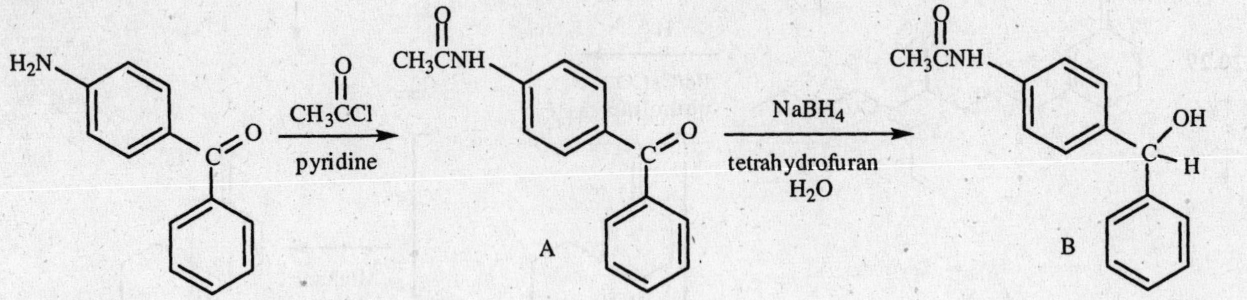

Infrared assignments for Compound B:

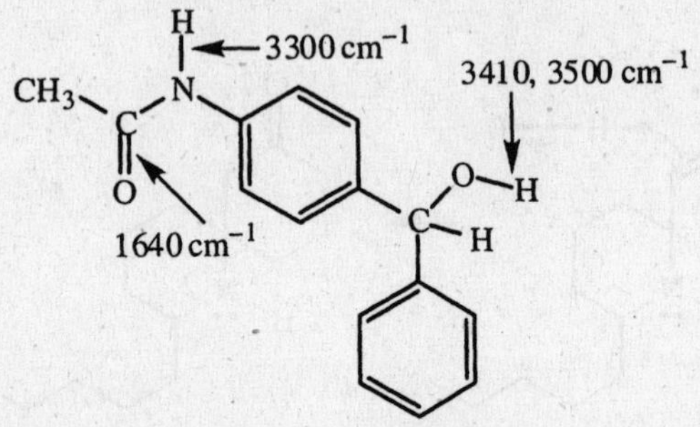

^{1}H NMR assignments for Compound B:

δ 7.20 - 7.49 are the signals for the aromatic protons.

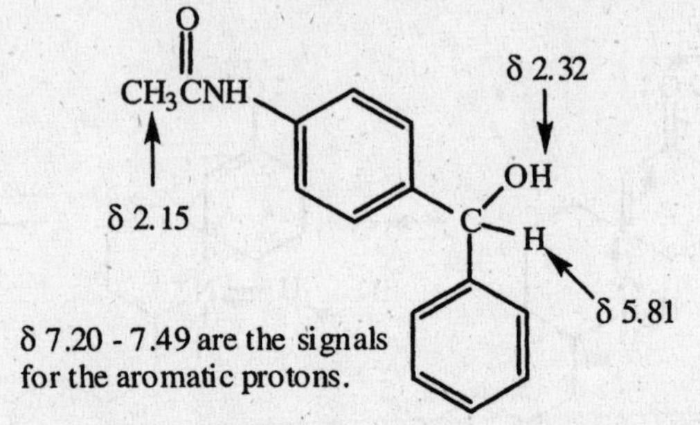

No peak for the N—H proton is reported. Such peaks tend to be very broad and sometimes hard to see.

20.32 The proton magnetic resonance spectrum of Compound A, $C_{10}H_{15}N$, shows a typical ethyl splitting pattern (a triplet at δ 1.2 and a quartet at δ 3.4). The chemical shift of the quartet tells us that the ethyl group is bonded to the nitrogen atom. There are also protons absorbing in the aromatic region (δ 6.6 – δ 7.3). The integration is 5:4:6 and tells us that there must be two ethyl groups bonded to the nitrogen. Compound A is *N,N*-diethylaniline.

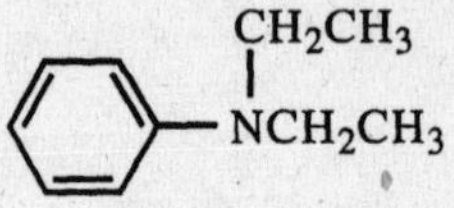

The proton magnetic resonance spectrum is summarized below:

   δ 1.2   (6H, t, J = 7 Hz, —CH$_2$C$\underline{H}_3$)
   δ 3.4   (4H, q, J = 7 Hz, —NC$\underline{H}_2$CH$_3$)
   δ 6.6 – 7.3   (5H, m, Ar$\underline{H}$)

20.33

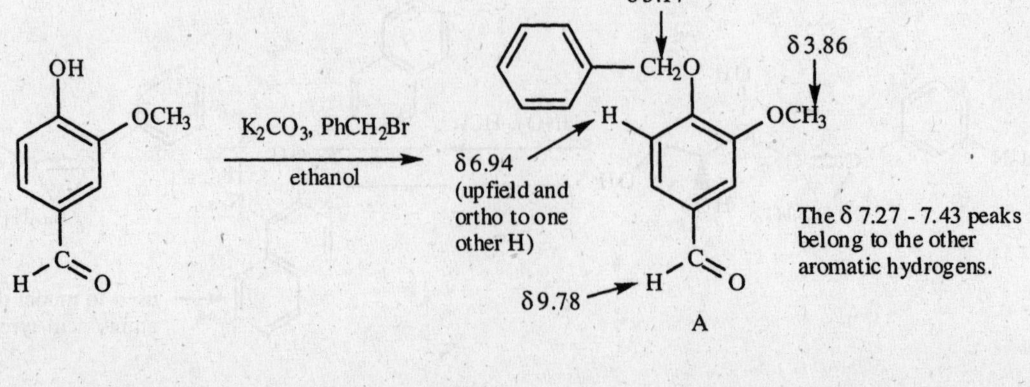

δ 5.17

δ 3.86

δ 6.94 (upfield and ortho to one other H)

The δ 7.27 - 7.43 peaks belong to the other aromatic hydrogens.

δ 9.78

20.33 (cont)

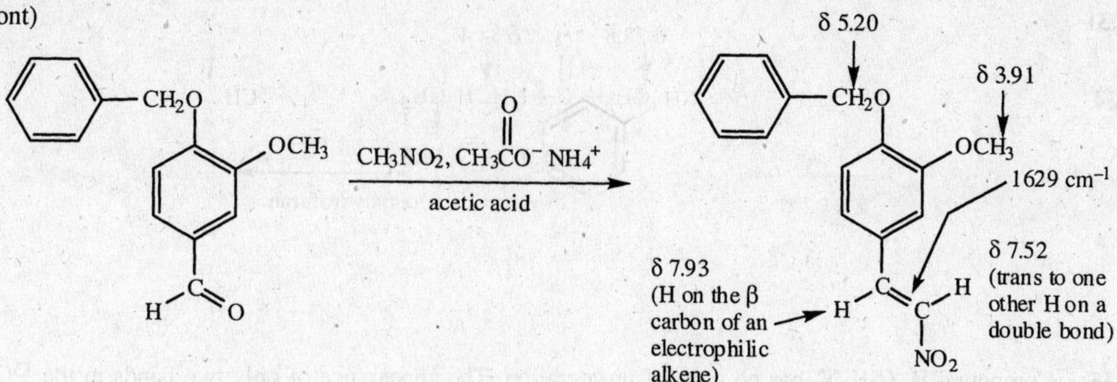

δ 5.20

δ 3.91

1629 cm⁻¹

δ 7.93
(H on the β
carbon of an
electrophilic
alkene)

δ 7.52
(trans to one
other H on a
double bond)

CH₃NO₂, CH₃CO⁻ NH₄⁺

acetic acid

B

The δ 7.32 - 7.42 and δ 6.89 - 7.09
peaks belong to the aromatic hydrogens.

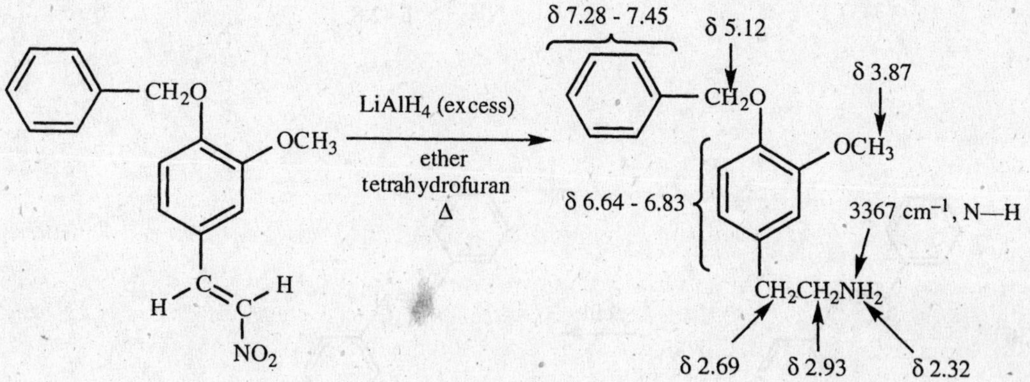

δ 7.28 - 7.45   δ 5.12

δ 3.87

LiAlH₄ (excess)

ether
tetrahydrofuran
Δ

δ 6.64 - 6.83

3367 cm⁻¹, N—H

δ 2.69   δ 2.93   δ 2.32

C

Note that LiAlH₄ used in excess in this reaction reduced the double bond in conjugation with the nitro
group, as well as the nitro group.

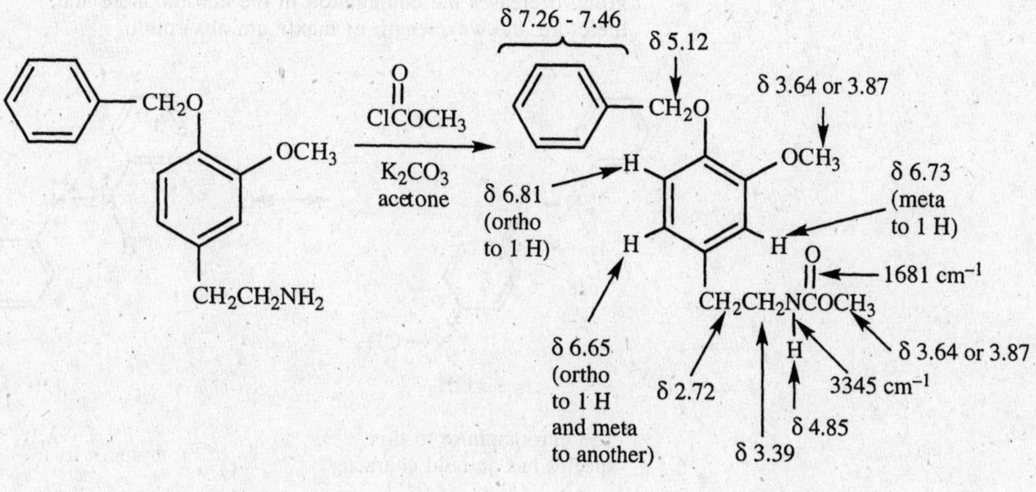

δ 7.26 - 7.46   δ 5.12

δ 3.64 or 3.87

O
||
ClCOCH₃

K₂CO₃
acetone

δ 6.81
(ortho
to 1 H)

δ 6.73
(meta
to 1 H)

1681 cm⁻¹

δ 6.65
(ortho
to 1 H
and meta
to another)

δ 2.72

δ 3.39

δ 4.85

3345 cm⁻¹

δ 3.64 or 3.87

CH₂CH₂NCOCH₃

D

20.34

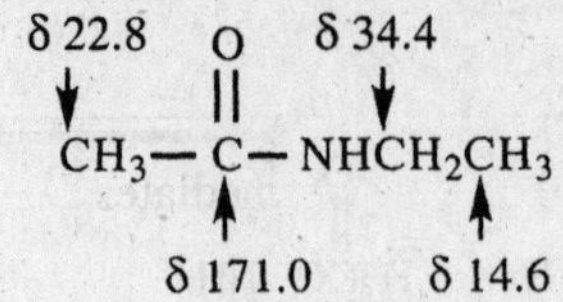

$$\underset{\underset{\delta\ 171.0}{\uparrow}}{\overset{\overset{\delta\ 22.8}{\downarrow}}{CH_3}} - \underset{}{\overset{\overset{O}{\parallel}}{C}} - \underset{\underset{\delta\ 14.6}{}}{\overset{\overset{\delta\ 34.4}{\downarrow}}{NHCH_2CH_3}}$$

20.35   Compound B, C_3H_9N, has no units of unsaturation. The appearance of only two bands in the ^{13}C nuclear magnetic resonance spectrum tells us that Compound B is symmetrical. Compound B is isopropylamine (2-aminopropane).

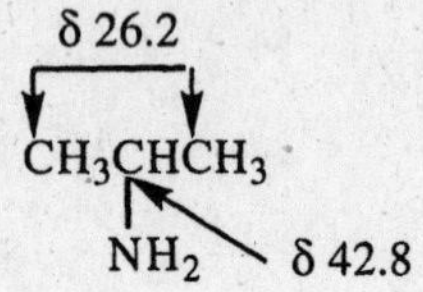

$$\overset{\delta\ 26.2}{\overbrace{\qquad\qquad}}$$
$$CH_3CHCH_3$$
$$\underset{NH_2}{\big|} \quad \delta\ 42.8$$

20.36

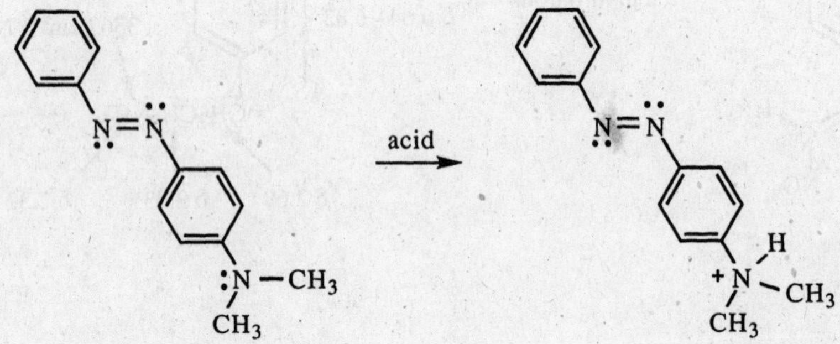

butter yellow
λ_{max} 408 nm

acid

λ_{max} 320 nm

Protonation of the nonbonding electrons on the amino group decreases the conjugation in the chromophore and, therefore, the wavelength of maximum absorption.

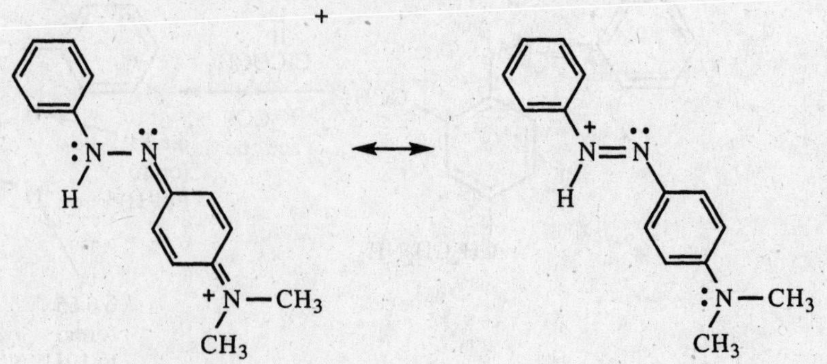

The chromophore in this species has quinoid character.

λ_{max} 510 nm

Supplemental Problems

S20.1   Write structural formulas for all products and intermediates designated by letters in the following equations.

(a)

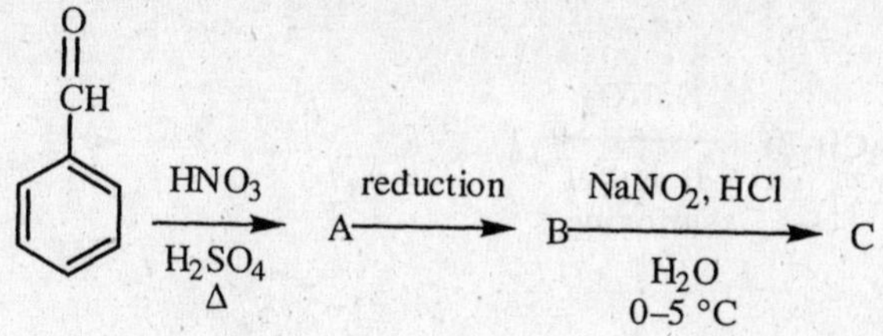

benzaldehyde $\xrightarrow[\substack{H_2SO_4 \\ \Delta}]{HNO_3}$ A $\xrightarrow{\text{reduction}}$ B $\xrightarrow[\substack{H_2O \\ 0-5\ °C}]{NaNO_2,\ HCl}$ C

(b)

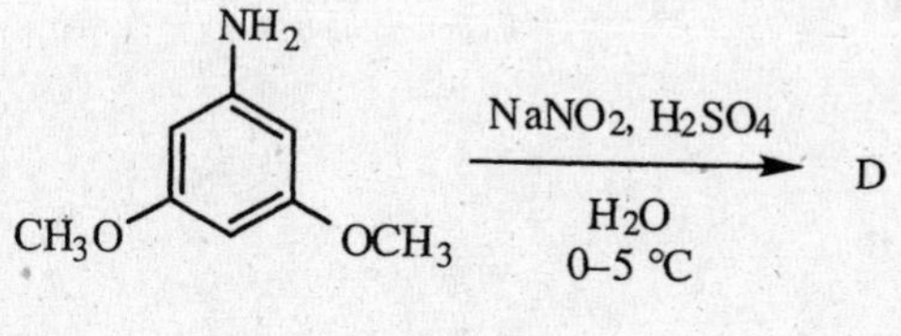

3,5-dimethoxyaniline $\xrightarrow[\substack{H_2O \\ 0-5\ °C}]{NaNO_2,\ H_2SO_4}$ D

(c)

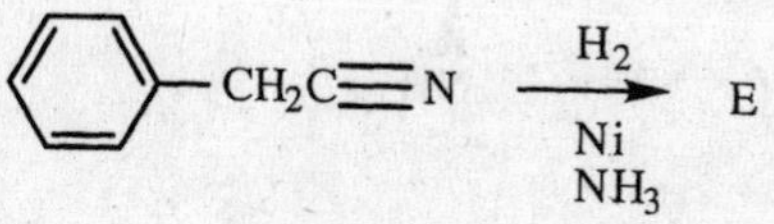

C_6H_5—$CH_2C{\equiv}N$ $\xrightarrow[\substack{Ni \\ NH_3}]{H_2}$ E

(d)

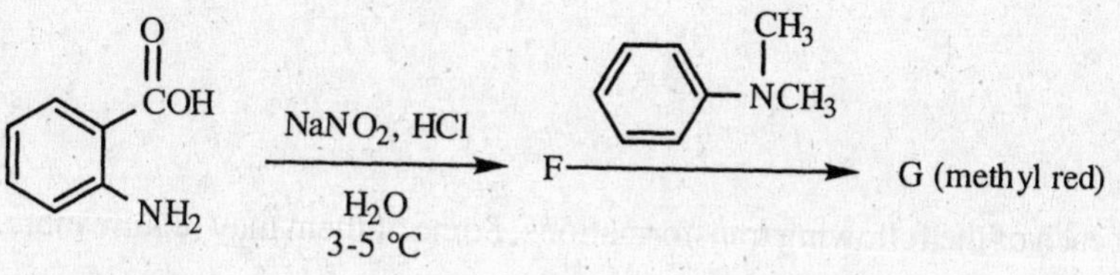

anthranilic acid (2-aminobenzoic acid) $\xrightarrow[\substack{H_2O \\ 3-5\ °C}]{NaNO_2,\ HCl}$ F $\xrightarrow{}$ G (methyl red)

(e)

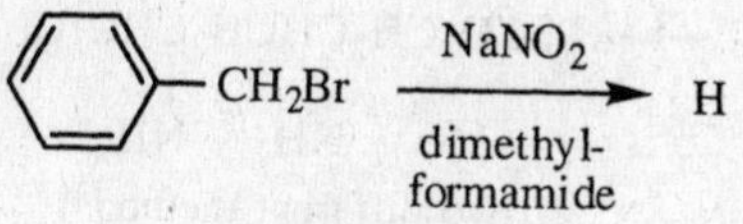

C_6H_5—CH_2Br $\xrightarrow[\substack{\text{dimethyl-} \\ \text{formamide}}]{NaNO_2}$ H

(f)

4-nitro-2-hydroxyphenyl ethyl ketone $\xrightarrow[\substack{Ni \\ \text{ethyl} \\ \text{acetate}}]{H_2}$ I

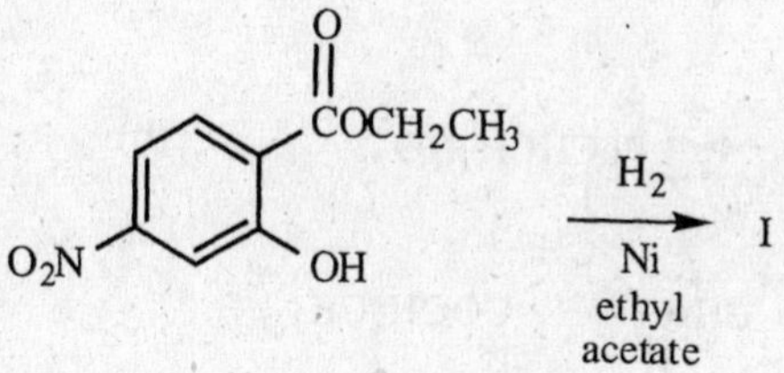

S20.2   For still more practice in recognizing reactions, write structural formulas for all compounds symbolized by letters in the following equations.

(a) 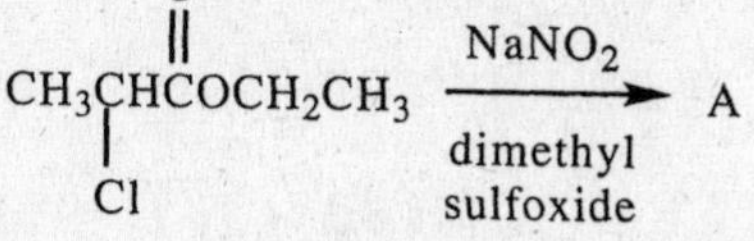 $CH_3\underset{\underset{Cl}{|}}{C}HCOCH_2CH_3$ $\xrightarrow[\substack{\text{dimethyl} \\ \text{sulfoxide}}]{NaNO_2}$ A

(b)   2-methylbenzonitrile $\xrightarrow[Ni]{H_2}$ B

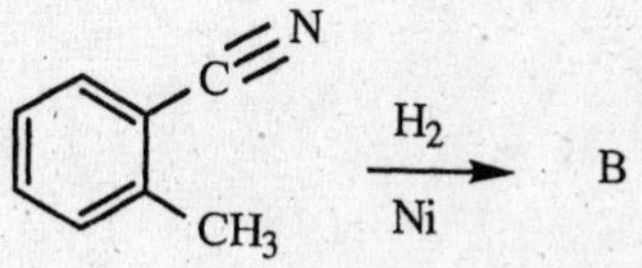

(c) 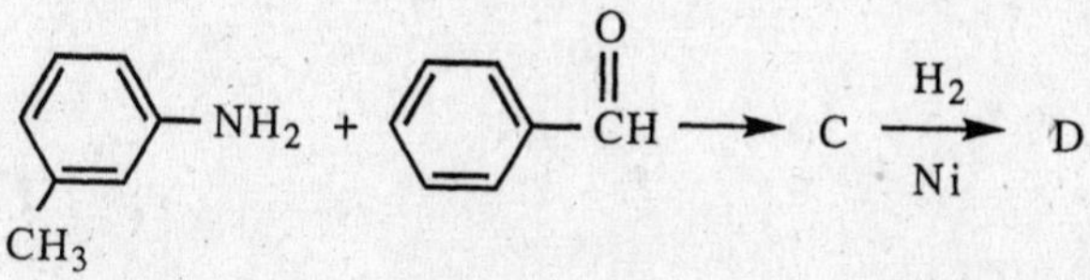   3-methylaniline + benzaldehyde $\longrightarrow$ C $\xrightarrow[Ni]{H_2}$ D

S20.2   (d) $CH_3CH_2CHCH_2CH_2CH_3$ $\xrightarrow[\text{pyridine}]{\text{TsCl}}$ E $\xrightarrow{CH_3NHCH_3}$ $\xrightarrow{\text{NaOH}}$ F $\xrightarrow[\text{acetonitrile}]{CH_3I}$ G

with OH substituent

(e) $PhNHCH_3$ $\xrightarrow{CH_3OCC \equiv CCOCH_3 \text{ (each with =O)}}$ H

(f) 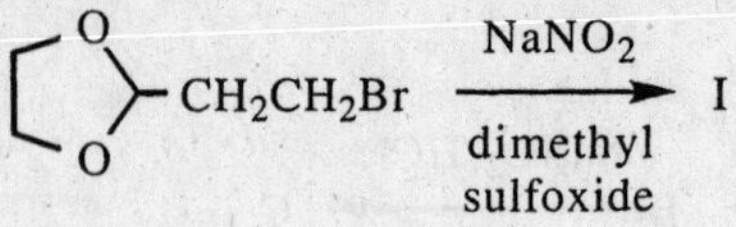   (1,3-dioxolane)$-CH_2CH_2Br$ $\xrightarrow[\text{dimethyl sulfoxide}]{\text{NaNO}_2}$ I

(g) 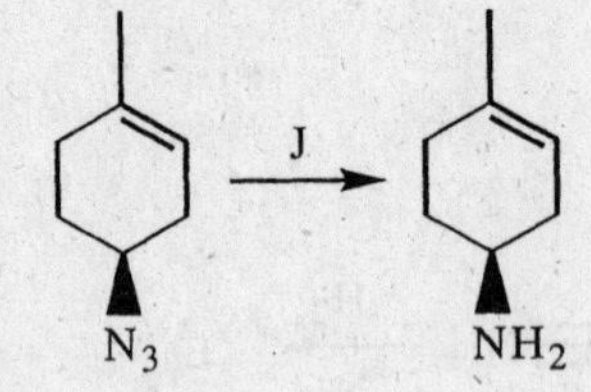   (methylcyclohexene with N_3) $\xrightarrow{J}$ (methylcyclohexene with NH_2)

S20.3   Show how you would carry out each of the following transformations. Some of them may require more than one step.

(a) $PhCH$ (with =O) $\longrightarrow$ $PhCHCH_2NO_2$ with CH bearing CH_3C(=O) and $COCH_2CH_3$(=O)

(b) $PhCH_2CH_2CCH_2CCH_3$ (two =O) $\longrightarrow$ $PhCH_2CH_2CHCH_2CHCH_3$ (with NH_2 and NH_2)

(two different methods)

(c) $HOCH_2CH_2CH_2OH$ $\longrightarrow$ $PhCH_2OCH_2CH_2CH_2NH_2$

(two different methods)

21 Synthesis

21.1 $CH_3CH_2CH_2Br \xrightarrow[\text{diethyl ether}]{Mg} CH_3CH_2CH_2MgBr$ + $H_2C \overset{O}{-} CH_2 \longrightarrow$

$CH_3CH_2CH_2CH_2CH_2O^- \ Mg^{2+} \ Br^- \xrightarrow[\text{H}_2\text{O}]{HCl} CH_3CH_2CH_2CH_2CH_2OH$

21.2 $CH_3CH_2CH_2CH_2CH_2OH \xleftarrow[\text{H}_2\text{O}]{\text{H}_2\text{O}_2,\ \text{NaOH}} \xleftarrow{BH_3} CH_3CH_2CH_2CH=CH_2 \xleftarrow[\substack{Pd,\ BaSO_4 \\ quinoline}]{H_2}$

$CH_3CH_2CH_2C\equiv CH \xleftarrow{CH_3CH_2CH_2Br} HC\equiv C:^-Li^+ \xleftarrow{CH_3CH_2CH_2CH_2Li} HC\equiv CH$

21.3
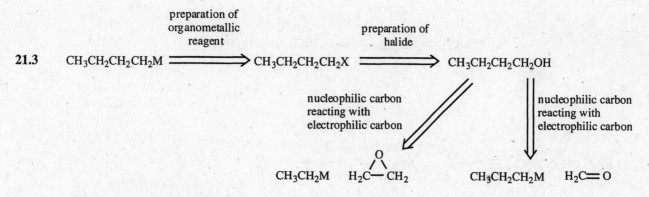

An actual synthesis might be:

$CH_3CH_2CH_2Br \xrightarrow{Li} CH_3CH_2CH_2Li \xrightarrow{H_2C=O} \xrightarrow{H_3O^+} CH_3CH_2CH_2CH_2OH$

$\downarrow PBr_3$

$CH_3CH_2CH_2CH_2CH_2OH \xleftarrow{H_3O^+} \xleftarrow{H_2C=O} CH_3CH_2CH_2CH_2MgBr \xleftarrow{Mg} CH_3CH_2CH_2CH_2Br$

21.4

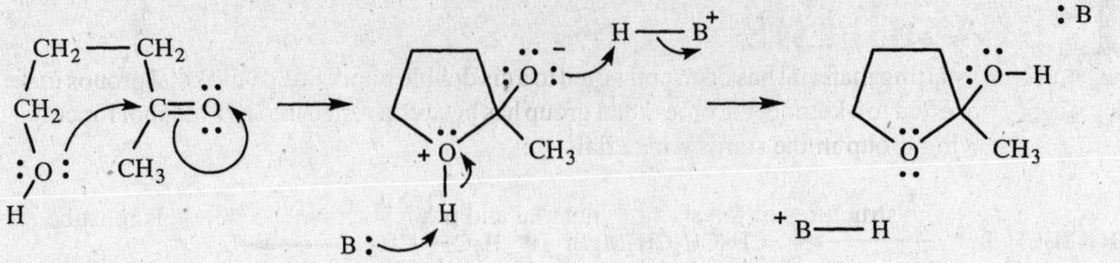

Note that the protonation and deprotonation step is most likely two separate steps.

21.5

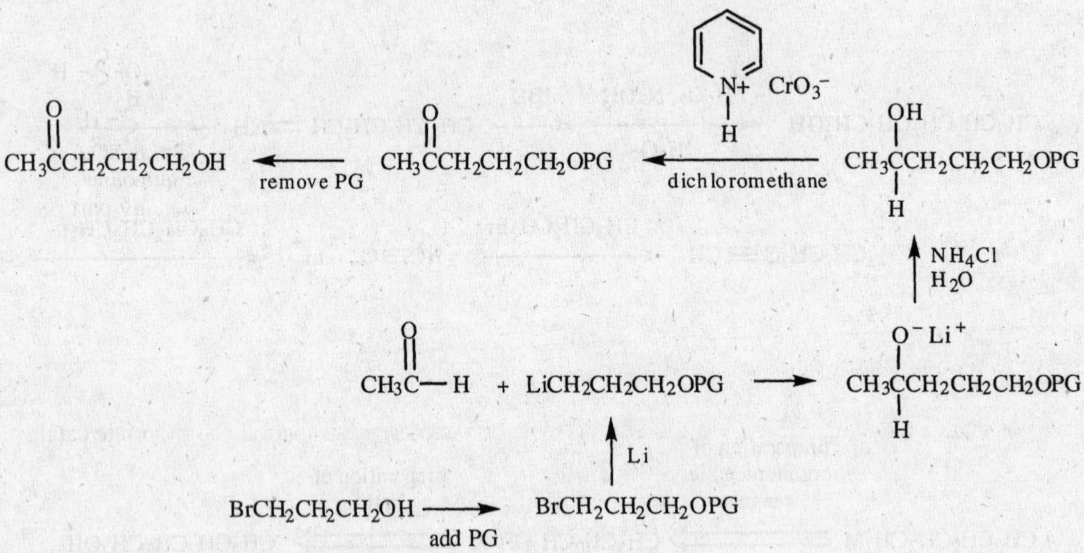

21.6

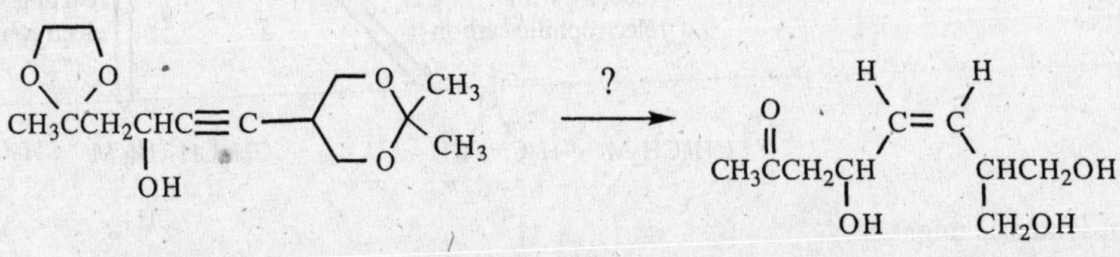

1. What are the connectivities of the two compounds? How many carbon atoms does each contain? Are there any rings? What are the positions of branches and functional groups on the carbon skeletons?

The atoms of the starting material and product molecules that are the same are shown below. The starting material has a triple bond, an alcohol, and two ketal functional groups; the product has a cis double bond, a ketone and three alcohol functional groups. The rings that are present are cyclic ketals.

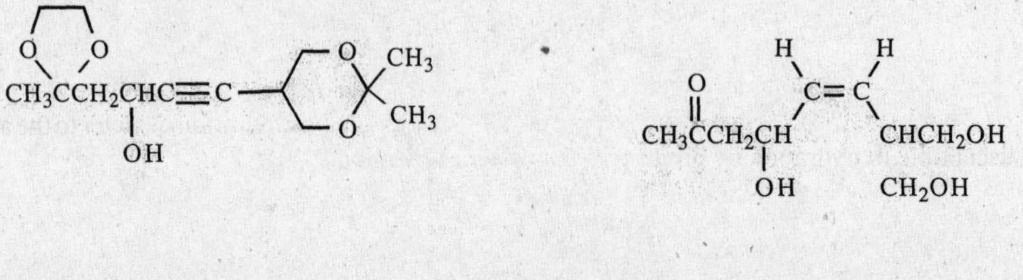

2. How do the functional groups change in going from starting material to product? Does the starting material have a good leaving group?

The triple bond in the starting material has been converted to a cis double bond. One of the ketal groups in the starting material has been converted to a ketone; the other ketal group has been converted into two alcohol functional groups. There is no good leaving group in the starting material.

3. Is it possible to dissect the structures of the starting material and product to see which bonds must be broken and which formed?

21.6 (cont)

bonds broken   bonds formed

4. New bonds are created when an electrophile reacts with a nucleophile. Do we recognize any part of the product molecule as coming from a good nucleophile or an electrophilic addition?

No.

5. What type of compound would be a good precursor to the product?

An alkyne that can be converted to a cis alkene by reduction with hydrogen and the appropriate catalyst.

The ketal groups in the starting material can be removed by acid hydrolysis.

6. After this last step, do we see how to get from starting material to product? If not, we need to analyze the structure obtained in step 5 by applying questions 4 and 5 to it.

The ketal groups are removed before reduction to prevent the acid-catalyzed addition of water to the alkene. Alkynes are less susceptible to hydration by dilute acid than alkenes (Section 9.4B).

Protection step:

21.7

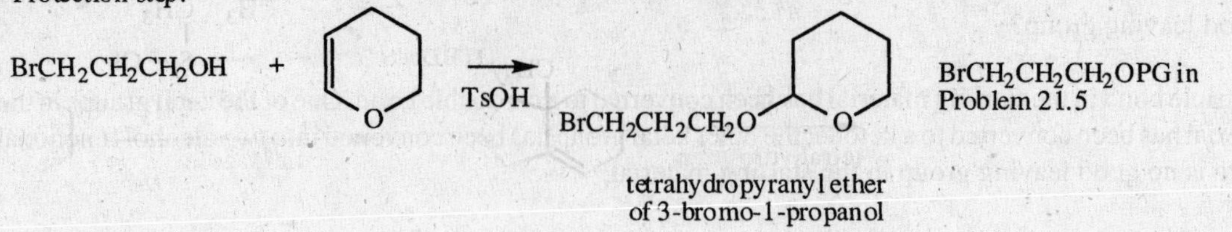

$BrCH_2CH_2CH_2OH$ +   [dihydropyran structure] $\xrightarrow{\text{TsOH}}$ $BrCH_2CH_2CH_2O$—[tetrahydropyranyl ether structure] $BrCH_2CH_2CH_2OPG$ in Problem 21.5

tetrahydropyranyl ether
of 3-bromo-1-propanol

Deprotection step:

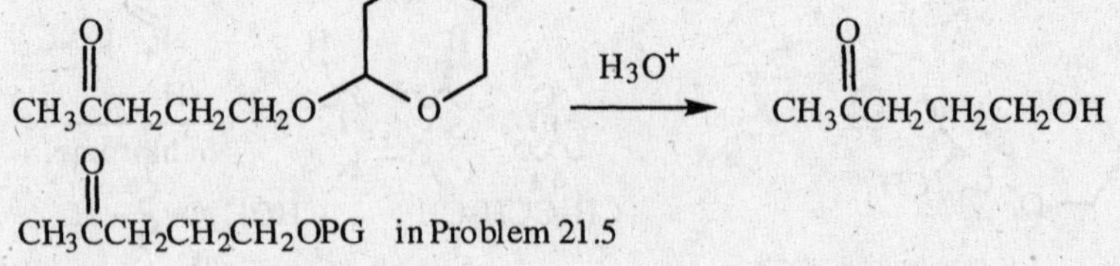

$CH_3\overset{O}{\overset{\|}{C}}CH_2CH_2CH_2O$—[tetrahydropyranyl ether structure] $\xrightarrow{H_3O^+}$ $CH_3\overset{O}{\overset{\|}{C}}CH_2CH_2CH_2OH$

$CH_3\overset{O}{\overset{\|}{C}}CH_2CH_2CH_2OPG$ in Problem 21.5

21.8 $HC\equiv CCH_2OH$ +   [dihydropyran structure] $\xrightarrow{HCl}$ $HC\equiv CCH_2O$—[THP structure] $\xrightarrow[\text{dimethyl sulfoxide}]{\text{NaH}}$ $Na^+ \, {}^-C\equiv CCH_2O$—[THP structure]

A   B

$\xrightarrow{CH_2=CH(CH_2)_8CH_2OTs}$ $CH_2=CH(CH_2)_9C\equiv CCH_2O$—[THP structure] $\xrightarrow[\substack{HCl \\ methanol}]{H_2O}$ $CH_2=CH(CH_2)_9C\equiv CCH_2OH$

C   D

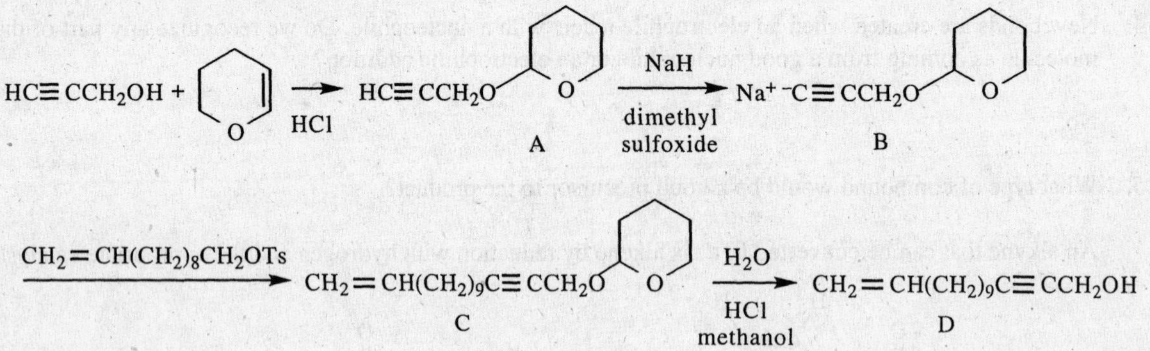

21.9   The product has a protecting group on the alcohol function and a ketone at one of the carbon atoms of the double bond of the starting material. A ketone can be prepared by oxidizing a secondary alcohol, which must be why the original alcohol group was protected.

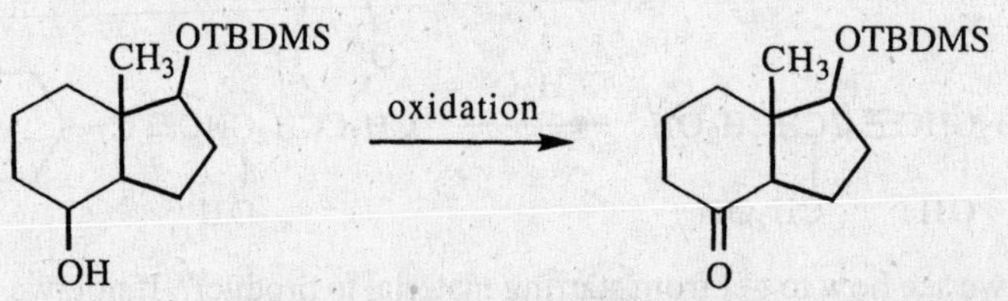

[bicyclic structure with CH₃, OTBDMS, and OH groups] $\xrightarrow{\text{oxidation}}$ [bicyclic structure with CH₃, OTBDMS, and O (ketone) groups]

The alcohol can be prepared by anti-Markovnikov addition of H_2O to the double bond. The synthesis must be:

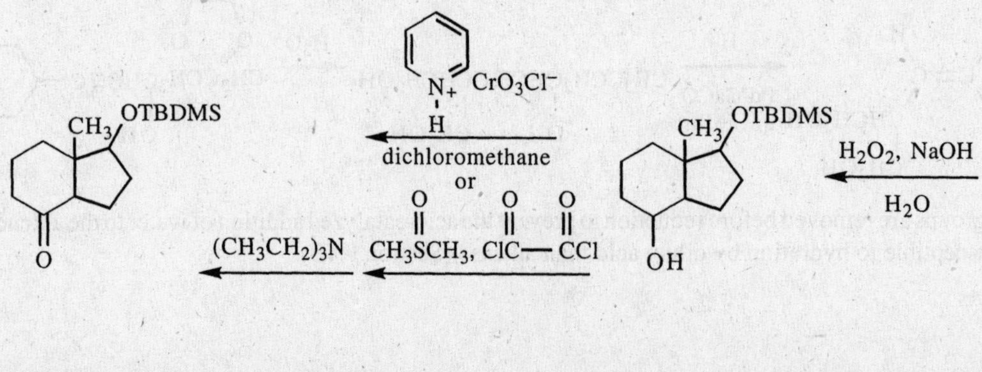

[bicyclic structure with CH₃, OTBDMS, and O (ketone)] $\xleftarrow[\substack{\text{dichloromethane} \\ \text{or} \\ (CH_3CH_2)_3N \quad CH_3SCH_3, \; ClC-CCl}]{\overset{\displaystyle \text{pyridinium } N^+\!-H \; CrO_3Cl^-}{}}$ [bicyclic structure with CH₃, OTBDMS, and OH] $\xleftarrow[H_2O]{H_2O_2, \, NaOH}$

21.9 (cont)

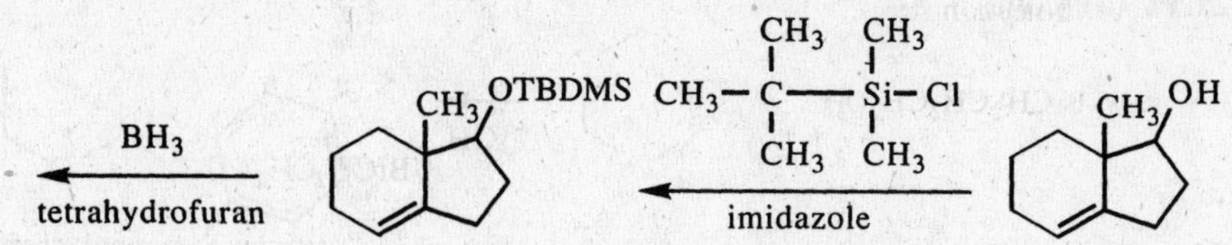

$$\xleftarrow{\underset{\text{tetrahydrofuran}}{BH_3}}$$

(structure: bicyclic compound with CH₃ and OTBDMS substituents)

$$\underset{\substack{CH_3 \\ | \\ CH_3-C-Si-Cl \\ | \quad\quad | \\ CH_3 \quad CH_3}}{} \xleftarrow{\text{imidazole}}$$

(structure: bicyclic compound with CH₃ and OH substituents)

21.10

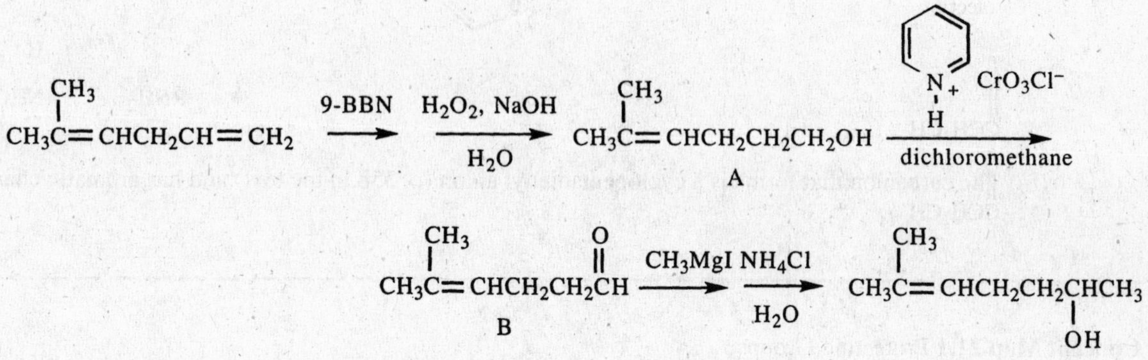

$$\underset{CH_3}{\overset{CH_3}{CH_3C}}\!=\!CHCH_2CH\!=\!CH_2 \xrightarrow{\text{9-BBN}} \xrightarrow[\substack{H_2O}]{H_2O_2,\ NaOH} \underset{A}{\overset{CH_3}{CH_3C}}\!=\!CHCH_2CH_2CH_2OH \xrightarrow[\text{dichloromethane}]{\overset{\displaystyle N^+\!-\!H}{}\ CrO_3Cl^-}$$

$$\underset{B}{\overset{CH_3}{CH_3C}}\!=\!CHCH_2CH_2\overset{\displaystyle O}{\overset{\|}{C}}H \xrightarrow[\substack{H_2O}]{CH_3MgI\ \ NH_4Cl} \underset{\substack{| \\ OH}}{\overset{CH_3}{CH_3C}}\!=\!CHCH_2CH_2CHCH_3$$

21.11

$$\underset{\substack{| \\ OTHP}}{CH_3CH_2CH}\!=\!CHCH\!=\!CHCHC\!\equiv\!CH \xleftarrow[\substack{25\ ^\circ C}]{TsOH} \text{(dihydropyran structure with O)}$$

$$\underset{\substack{| \\ OH}}{CH_3CH_2CH}\!=\!CHCH\!=\!CHCHC\!\equiv\!CH \xleftarrow[\substack{H_2O}]{NH_4Cl\ \ \ Na^+\ {}^-C\equiv CH} CH_3CH_2CH\!=\!CHCH\!=\!CHC\overset{\displaystyle O}{\overset{\|}{}}H$$

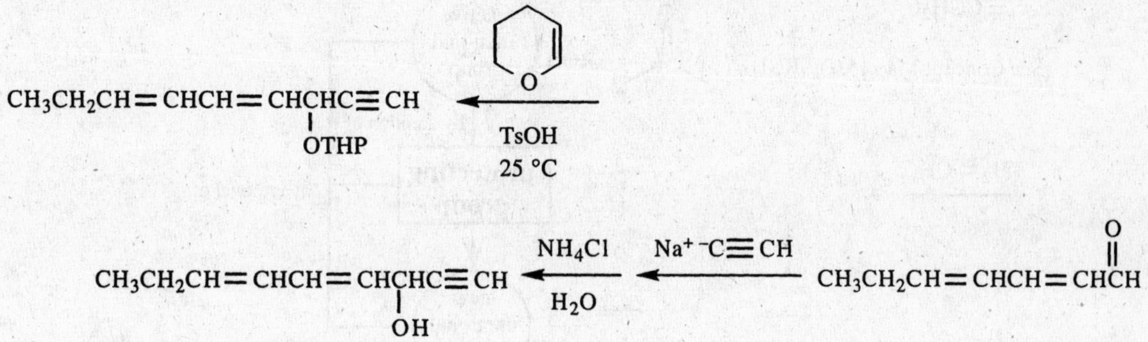

21.12   (a)

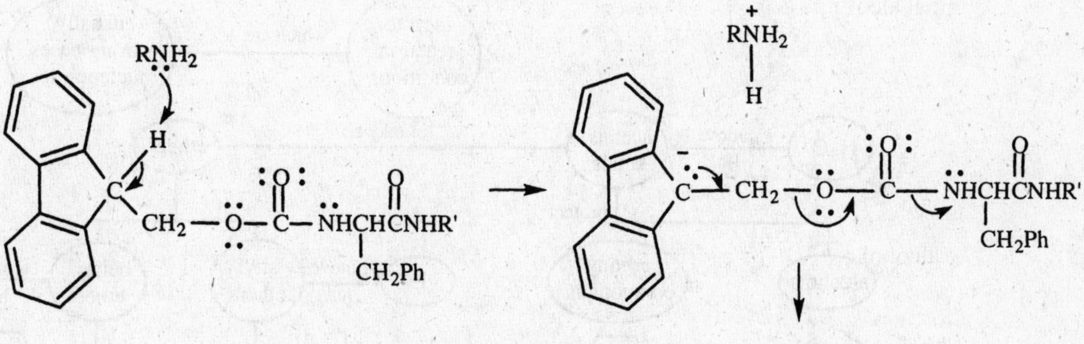

21.12 (a) (cont)

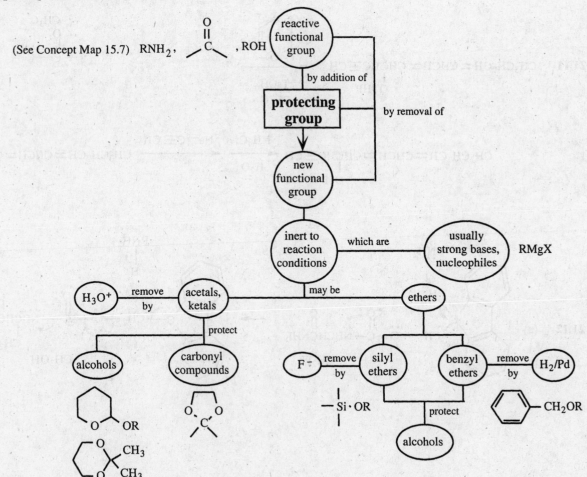

(b) The carbanion that forms is a cyclopentadienyl anion (p. 358 in the text) and has aromatic character.

- -

Concept Map 21.1 Protecting Groups

(See Concept Map 15.7) RNH_2, $\underset{C}{\overset{O}{\parallel}}$, ROH

21.13

$$CF_3\overset{O}{\underset{\parallel}{C}}NH(CH_2)_4\overset{CH_2Ph}{\underset{\mid}{N}}(CH_2)_3NHBoc \xrightarrow[\text{Pd}]{H_2} CF_3\overset{O}{\underset{\parallel}{C}}NH(CH_2)_4NH(CH_2)_3NHBoc \xrightarrow[\text{base}]{Ph-\overset{O}{\underset{\parallel}{C}}Cl}$$

$$CF_3\overset{O}{\underset{\parallel}{C}}NH(CH_2)_4\overset{\overset{O}{\underset{\parallel}{\overset{C}{\mid}}}{\underset{\mid}{N}}}{\underset{Ph}{}}(CH_2)_3NHBoc \xrightarrow[\text{H}_2\text{O}]{K_2CO_3} H_2N(CH_2)_4\overset{\overset{O}{\underset{\parallel}{\overset{C}{\mid}}}{\underset{\mid}{N}}}{\underset{Ph}{}}(CH_2)_3NHBoc \xrightarrow[\text{base}]{CH_3\overset{O}{\underset{\parallel}{C}}Cl}$$

$$CH_3\overset{O}{\underset{\parallel}{C}}NH(CH_2)_4\overset{\overset{O}{\underset{\parallel}{\overset{C}{\mid}}}{\underset{\mid}{N}}}{\underset{Ph}{}}(CH_2)_3NHBoc \xrightarrow{CF_3COH} \xrightarrow{\text{base}} CH_3\overset{O}{\underset{\parallel}{C}}NH(CH_2)_4\overset{\overset{O}{\underset{\parallel}{\overset{C}{\mid}}}{\underset{\mid}{N}}}{\underset{Ph}{}}(CH_2)_3NH_2 \xrightarrow[\text{base}]{}$$

(with 2,3-dimethoxybenzoyl chloride)

$$CH_3\overset{O}{\underset{\parallel}{C}}NH(CH_2)_4\overset{\overset{O}{\underset{\parallel}{\overset{C}{\mid}}}{\underset{\mid}{N}}}{\underset{Ph}{}}(CH_2)_3NH-\overset{O}{\underset{\parallel}{C}}\text{(2,3-dimethoxyphenyl)}$$

21.14   (a)

(b)

(c)

$$CH_3CH=CHCH_2CH_2\overset{O}{\underset{\parallel}{C}}OCH_3 \xrightarrow[\substack{\text{diethyl}\\\text{ether}}]{LiAlH_4} \xrightarrow{H_3O^+} CH_3CH=CHCH_2CH_2CH_2OH + CH_3OH$$

(d)

21.15

$$\xrightarrow[\text{diethyl ether}]{\text{LiAlH}_4 \quad \text{H}_3\text{O}^+}$$

A

+   CH$_3$CH$_2$OH

$$\xrightarrow[\text{dichloromethane}]{}$$

B

21.16

(a)   PhCOH $\xrightarrow[\substack{\text{benzene} \\ 45\,°\text{C, 8 h}}]{\substack{\overset{\text{CH}_3}{\underset{|}{(\text{CH}_3\text{CHCH}_2)_2\text{AlH}}} \\ (3 \text{ equiv})}}$ $\xrightarrow{\text{CH}_3\text{OH, H}_2\text{O}}$ PhCH$_2$OH

(one equivalent of the hydride is used
up in deprotonating the carboxylic acid)

(b)   CH$_3$CH$_2$OCCH$_2$CH$_2$COCH$_2$CH$_3$ $\xrightarrow[\text{tetrahydrofuran}]{\text{BH}_3\text{•S(CH}_3)_2}$ $\xrightarrow[\text{H}_2\text{O}]{\text{K}_2\text{CO}_3}$ HOCH$_2$CH$_2$CH$_2$CH$_2$OH + 2 CH$_3$CH$_2$OH

(c) $\underset{\overset{|}{\text{CH}_3}}{\overset{\overset{\text{CH}_3}{|}}{\text{CH}_3\text{C}}}$—$\overset{\overset{\text{O}}{\|}}{\text{CN(CH}_3)_2}$ $\xrightarrow[\text{tetrahydrofuran}]{\text{BH}_3\text{•S(CH}_3)_2}$ $\xrightarrow[\substack{\text{H}_2\text{O} \\ \Delta}]{\text{HCl}}$ $\xrightarrow[\text{H}_2\text{O}]{\text{NaOH}}$ $\underset{\overset{|}{\text{CH}_3}}{\overset{\overset{\text{CH}_3}{|}}{\text{CH}_3\text{C}}}$—CH$_2$N(CH$_3$)$_2$

(d)   NCCH$_2$CH$_2$CH$_2$CH$_2$CN $\xrightarrow[\text{tetrahydrofuran}]{\text{BH}_3\text{•S(CH}_3)_2}$ $\xrightarrow[\substack{\text{H}_2\text{O} \\ \Delta}]{\text{HCl}}$ $\xrightarrow[\text{H}_2\text{O}]{\text{NaOH}}$ H$_2$NCH$_2$CH$_2$CH$_2$CH$_2$CH$_2$CH$_2$NH$_2$

21.17   (1)

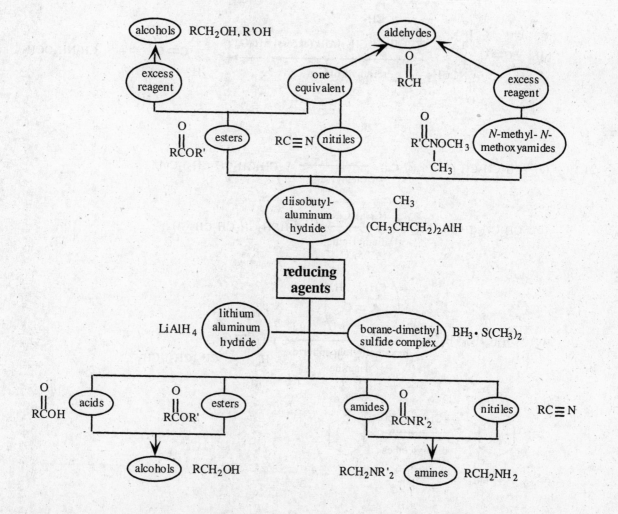

21.17 (1) cyclohexanol + PBr₃/pyridine → cyclohexyl bromide + NaCN →

cyclohexyl–CN, (CH₃CHCH₂)₂AlH₂ (1 eq), hexane −70 °C, then H₃O⁺ → cyclohexanecarbaldehyde

(2) cyclohexanol + PBr₃/pyridine → cyclohexyl bromide, Mg/diethyl ether → cyclohexyl–MgBr, CO₂, H₃O⁺ →

cyclohexanecarboxylic acid + CH₃OH/H₂SO₄/Δ → methyl ester, (CH₃CHCH₂)₂AlH₂ (1 eq), hexane −70 °C, H₂O → cyclohexanecarbaldehyde + CH₃OH

Concept Map 21.2   Reduction of acids and acid derivatives.

21.18   (a) $CH_3CH_2OCH_2(CH_2)_5COCH_2CH_3$ $\xrightarrow[\substack{\text{hexane} \\ -70\,°C}]{(CH_3CHCH_2)_2AlH \ (1\ eq)}$ $\xrightarrow{H_2O}$

(with CH_3 group on the $(CH_3CHCH_2)_2AlH$ reagent)

$CH_3CH_2OCH_2(CH_2)_5CH\!=\!\!O$ + CH_3CH_2OH

(b) $\xrightarrow[\text{diethyl ether}]{Li\,AlH_4}$ $\xrightarrow[\text{protonation}]{H_3O^+}$

BocNH–C(–CH₂=CHCH₂)(–H)–CONH–OCH₃ → BocNH–C(–CH₂=CHCH₂)(–H)–CHO

(c) $\xrightarrow[\substack{\text{benzene} \\ 25\,°C}]{(CH_3CHCH_2)_2AlH \ (1\ eq) \quad H_3O^+}$

(d) $\xrightarrow[\text{tetrahydrofuran}]{(CH_3CHCH_2)_2AlH \ (excess) \quad H_3O^+}$ + $CH_3NH_2OCH_3^+$

21.19   (a) $CH_3CH_2CH_2CH_2Li$ + CuI $\xrightarrow[0\,°C]{\text{diethyl ether}}$ $(CH_3CH_2CH_2CH_2)_2CuLi$

(b) $CH_3CH_2CH_2CH_2CH_2I$ $\xrightarrow[\substack{\text{diethyl ether} \\ 25\,°C}]{(CH_3)_2CuLi}$ $CH_3CH_2CH_2CH_2CH_2CH_3$

(c) $CH_3(CH_2)_6CH_2Cl$ $\xrightarrow[\substack{\text{hexamethylphosphoric} \\ \text{triamide} \\ 25\,°C}]{\left(\substack{CH_3 \quad\ H \\ \diagdown C=C \diagup \\ H \qquad}\right)_2 CuLi}$ $\substack{CH_3 \qquad\quad H \\ \diagdown C=C \diagup \\ H \qquad CH_2(CH_2)_6CH_3}$

(d) $\xrightarrow[\substack{\text{diethyl ether} \\ 0\,°C}]{(CH_3)_2CuLi}$

21.20

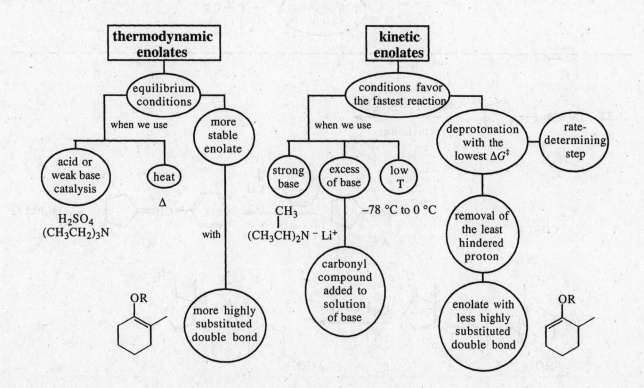

(a), (b), (c) reaction schemes

- -

Concept Map 21.3   Thermodynamic and kinetic enolates.

21.21   (a)   CH₃CH₂CCH₃ $\xrightarrow[\substack{\text{dimethylformamide} \\ \Delta}]{\text{(CH}_3)_3\text{SiCl, (CH}_3\text{CH}_2)_3\text{N}}$ CH₃CH=CCH₃

with carbonyl O on the ketone and OSi(CH₃)₃ on the product.

(b)   CH₃CH₂CCH₃ $\xrightarrow[\text{1,2-dimethoxyethane}]{\substack{\text{CH}_3 \\ \text{(CH}_3\text{CH)}_2\text{N}^-\text{Li}^+}}$ $\xrightarrow{\text{(CH}_3)_3\text{SiCl, (CH}_3\text{CH}_2)_3\text{N}}$ CH₃CH₂C=CH₂

with OSi(CH₃)₃ on the product.

Concept Map 21.4   Ylides.

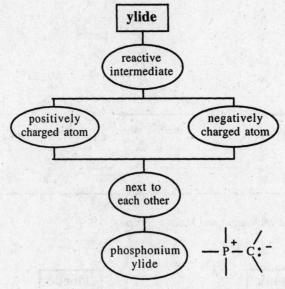

21.22   (a)   Ph₃P⁺—⟨cyclopentyl⟩   I⁻ $\xrightarrow[\text{diethyl ether}]{\text{CH}_3\text{CH}_2\text{CH}_2\text{CH}_2\text{Li}}$

Ph₃P=⟨cyclopentylidene⟩ $\xrightarrow{\text{C}_6\text{H}_5\text{CHO}}$ ⟨C}_6\text{H}_5—CH=cyclopentylidene⟩   +   Ph₃P=O

A   B   C

(b)   [steroid with ketone, HO—] $\xrightarrow{\text{Ph}_3\text{P}=\text{CH}_2}$ [steroid with =CH₂, HO—]

D

21.22   (c)   Br⁻   Ph₃PCH₂COCH₂CH₃ $\xrightarrow[\text{ethanol}]{\text{CH}_3\text{CH}_2\text{ONa}}$

Ph₃P=CHCOCH₂CH₃ $\xrightarrow{\text{CH}_3\text{CH}=\text{CHCH}}$ CH₃CH=CHCH=CHCOCH₂CH₃   +   Ph₃P=O

   E   F   G

(d)   Br⁻
      Ph₃PCH₂ $\xrightarrow[\text{ethanol}]{\text{CH}_3\text{CH}_2\text{ONa}}$ Ph₃P=

      H

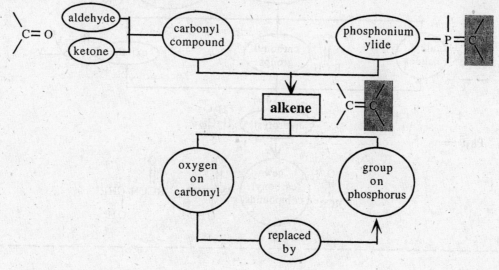

                                                                                        I   J

- -

Concept Map 21.5   The Wittig reaction.

21.23

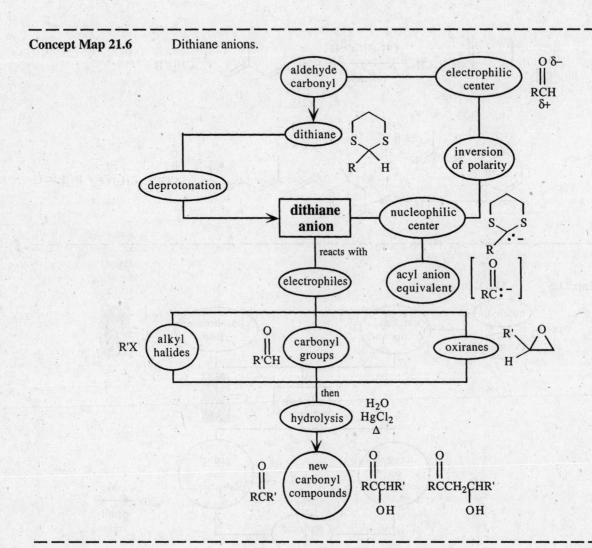

___

Concept Map 21.6   Dithiane anions.

___

21.24   (a)

21.24 (b)

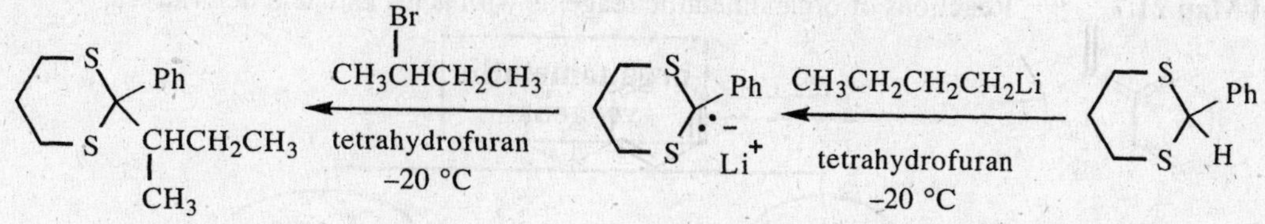

(c)

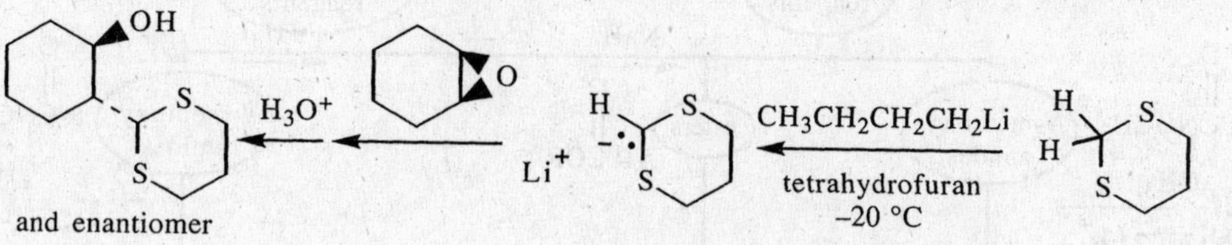

and enantiomer

(d)

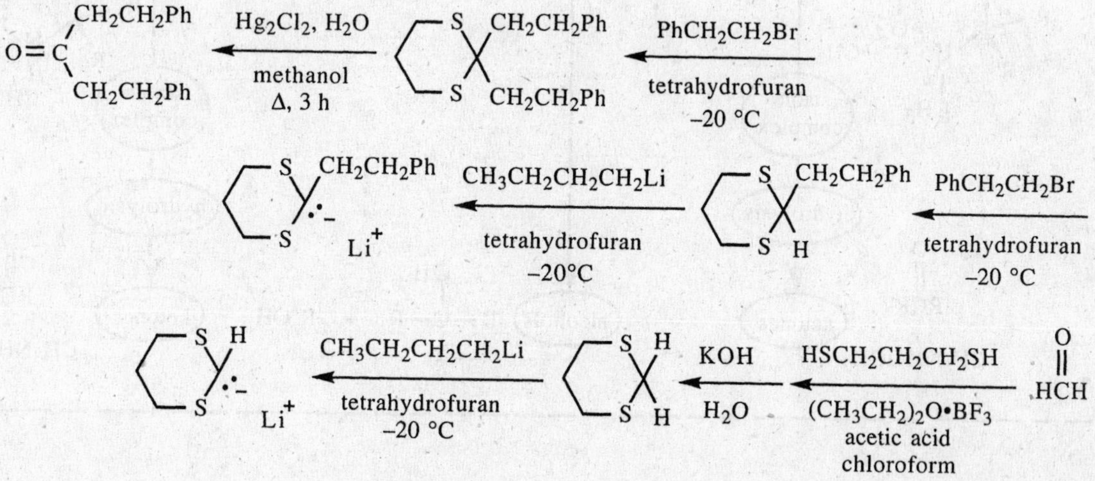

21.25

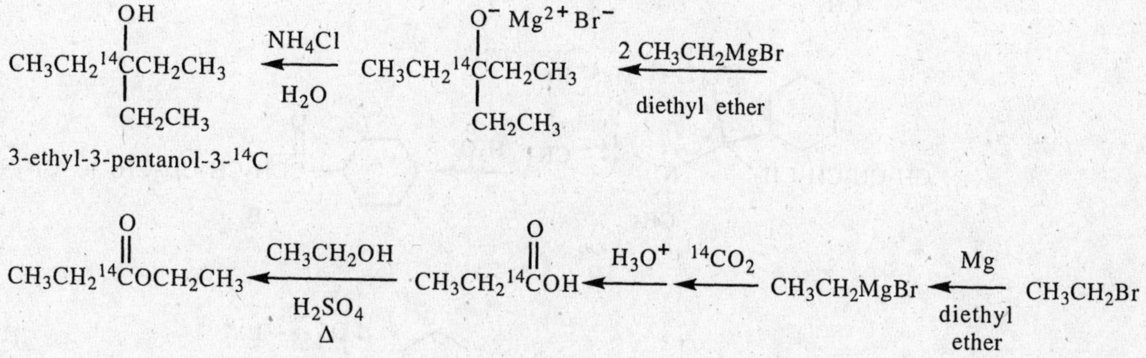

$$CH_3CH_2\overset{OH}{\underset{CH_2CH_3}{\overset{|}{\underset{|}{^{14}C}}}CH_2CH_3} \xleftarrow[H_2O]{NH_4Cl} CH_3CH_2\overset{O^- \; Mg^{2+} \; Br^-}{\underset{CH_2CH_3}{\overset{|}{\underset{|}{^{14}C}}}CH_2CH_3} \xleftarrow[\text{diethyl ether}]{2 \; CH_3CH_2MgBr}$$

3-ethyl-3-pentanol-3-^{14}C

$$CH_3CH_2\overset{O}{\overset{||}{^{14}C}}OCH_2CH_3 \xleftarrow[\substack{H_2SO_4 \\ \Delta}]{CH_3CH_2OH} CH_3CH_2\overset{O}{\overset{||}{^{14}C}}OH \xleftarrow[]{H_3O^+ \quad ^{14}CO_2} CH_3CH_2MgBr \xleftarrow[\substack{\text{diethyl} \\ \text{ether}}]{Mg} CH_3CH_2Br$$

21.26 $PhBr \xrightarrow[\substack{\text{diethyl} \\ \text{ether}}]{Mg} PhMgBr \xrightarrow[]{\overset{O}{\overset{||}{PhCPh}}} Ph_3CO^- \; Mg^{2+} \; Br^- \xrightarrow{H_3O^+} Ph_3COH$

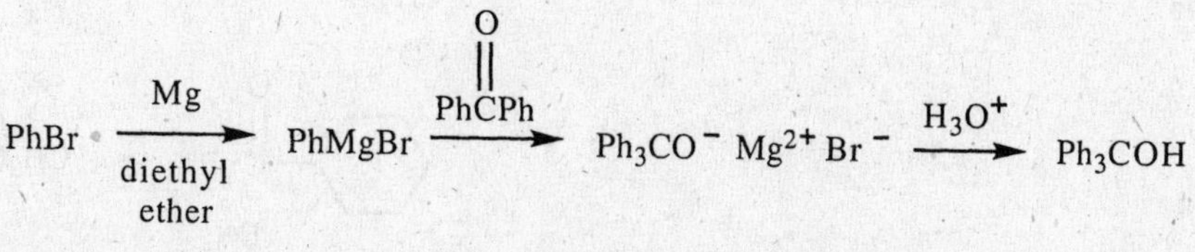

Concept Map 21.7   Reactions of organometallic reagents with acids and acid derivatives.

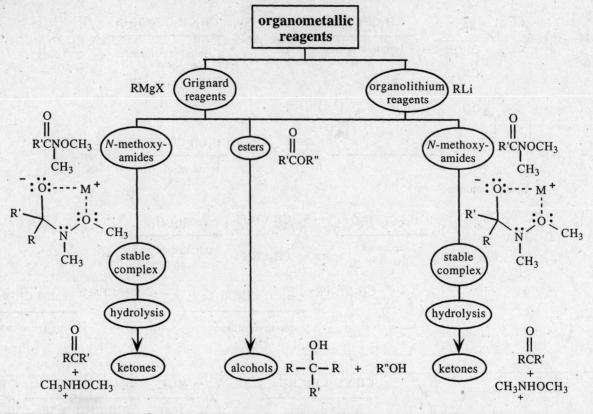

- -

21.27 (a)

21.27 (c) PhBr $\xrightarrow[\text{diethyl ether}]{\text{Mg}}$ PhMgBr $\xrightarrow{\text{CH}_3\text{COCH}_2\text{CH}_3}$ Ph—$\underset{\underset{\text{G}}{\overset{|}{\text{O}^-\text{Mg}^{2+}\text{Br}^-}}}{\overset{\overset{\text{CH}_3}{|}}{\text{C}}}$—Ph $\xrightarrow[\text{H}_2\text{O}]{\text{NH}_4\text{Cl}}$ Ph—$\underset{\underset{\text{I}}{\overset{|}{\text{OH}}}}{\overset{\overset{\text{CH}_3}{|}}{\text{C}}}$—Ph

21.28 $\text{CH}_3\text{CCH}_2\text{COCH}_2\text{CH}_3$ $\xrightarrow[\text{tetrahydrofuran}]{\text{NaH}}$ $\underset{\underset{\text{A}}{\text{Na}^+}}{\text{CH}_3\text{C}-\overset{-}{\text{CH}}-\text{COCH}_2\text{CH}_3}$ $\xrightarrow[\text{hexane}]{\text{CH}_3\text{CH}_2\text{CH}_2\text{CH}_2\text{Li}}$

$\text{Li}^+ \; \underset{\underset{\text{B}}{\text{Na}^+}}{{}^-\text{CH}_2\text{C}-\overset{-}{\text{CH}}-\text{COCH}_2\text{CH}_3}$ $\xrightarrow[\text{tetrahydrofuran}]{\text{PhOCH}_2\text{CH}_2\text{Br}}$ $\xrightarrow{\text{H}_3\text{O}^+}$ $\underset{\text{C}}{\text{PhOCH}_2\text{CH}_2 \, \text{CH}_2\text{CCH}_2\text{COCH}_2\text{CH}_3}$ $\xrightarrow{\text{LiAlH}_4}$

$\xrightarrow{\text{H}_3\text{O}^+}$ $\underset{\text{D}}{\text{PhOCH}_2\text{CH}_2 \, \text{CH}_2\overset{\overset{\text{OH}}{|}}{\text{CH}}\text{CH}_2\text{CH}_2\text{OH}}$ + $\underset{\text{E}}{\text{CH}_3\text{CH}_2\text{OH}}$

21.29 $\text{CH}_3\text{OCCH}_2\text{COCH}_3$ $\xrightarrow[\text{tetrahydrofuran}]{\text{NaH}}$ $\underset{\underset{\text{A}}{\text{Na}^+}}{\text{CH}_3\text{OC}-\overset{-}{\text{CH}}-\text{COCH}_3}$ $\xrightarrow[\text{tetrahydrofuran}]{\overset{\overset{\text{CH}_3}{|}}{\text{CH}_3\text{CHI}}}$

$\underset{\underset{\text{B}}{\overset{|}{\text{CH}_3\text{CHCH}_3}}}{\text{CH}_3\text{OCCHCOCH}_3}$ $\xrightarrow[\text{tetrahydrofuran}]{\text{LiAlH}_4}$ $\xrightarrow{\text{H}_2\text{O}}$ $\underset{\underset{\text{C}}{\overset{|}{\text{CH}_3\text{CHCH}_3}}}{\text{HOCH}_2\text{CHCH}_2\text{OH}}$ $\xrightarrow{\text{(0.3 equivalents)} \quad \text{(a mild acid catalyst)}}$

$\underset{\underset{\text{D}}{\overset{|}{\text{CH}_3\text{CHCH}_3}}}{\text{THPOCH}_2\text{CHCH}_2\text{OH}}$ $\xrightarrow{\text{CH}_3\text{SCH}_3, \; \text{ClC}-\text{CCl}}$ $\xrightarrow{\text{(CH}_3\text{CH}_2)_3\text{N}}$ $\underset{\underset{\text{E}}{\overset{|}{\text{CH}_3\text{CHCH}_3}}}{\text{THPOCH}_2\text{CHCH}}$

21.30

$$CH_3\overset{O}{\overset{\|}{C}}CH_2\overset{O}{\overset{\|}{C}}OCH_2CH_3 \xrightarrow[\text{ethanol}]{CH_3CH_2ONa} CH_3\overset{O}{\overset{\|}{C}}-\overset{-}{C}H-\overset{O}{\overset{\|}{C}}OCH_2CH_3 \xrightarrow{CH_3(CH_2)_4Br} CH_3\overset{O}{\overset{\|}{C}}CHC\overset{O}{\overset{\|}{C}}OCH_2CH_3$$

$$Na^+$$

A

B
(the most acidic hydrogen
is the one between the
two carbonyl groups)

$(CH_2)_4CH_3$

$$\xrightarrow[\text{tetrahydro-furan}]{NaH} CH_3\overset{O}{\overset{\|}{C}}-\overset{Na^+}{\overset{-}{C}}-\overset{O}{\overset{\|}{C}}OCH_2CH_3 \xrightarrow[\text{hexane}]{CH_3CH_2CH_2CH_2Li} \overset{Li^+}{\overset{-}{C}H_2}\overset{O}{\overset{\|}{C}}-\overset{Na^+}{\overset{-}{C}}-\overset{O}{\overset{\|}{C}}OCH_2CH_3 \xrightarrow{\triangle-CH_3}$$

$(CH_2)_4CH_3$

C
(the most acidic hydrogens are the
ones α to the carbonyl group)

$(CH_2)_4CH_3$

D
(the most nucleophilic
carbon is the one α to
only one carbonyl group)

$$CH_3\underset{O^-Li^+}{\overset{}{CH}}CH_2CH_2\overset{O}{\overset{\|}{C}}-\overset{Na^+}{\overset{-}{C}}-\overset{O}{\overset{\|}{C}}OCH_2CH_3 \xrightarrow[\Delta]{NaOH \\ H_2O} CH_3\underset{O^-Li^+}{\overset{}{CH}}CH_2CH_2\overset{O}{\overset{\|}{C}}-\overset{Na^+}{\overset{-}{C}}-\overset{O}{\overset{\|}{C}}O^-\ Na^+ \xrightarrow{H_2SO_4}$$

$(CH_2)_4CH_3$

E

$(CH_2)_4CH_3$

F

$$CO_2 + CH_3\underset{OH}{\overset{}{CH}}CH_2CH_2\overset{O}{\overset{\|}{C}}CH_2(CH_2)_4CH_3 \xrightarrow{CrO_3} CH_3\overset{O}{\overset{\|}{C}}CH_2CH_2\overset{O}{\overset{\|}{C}}CH_2(CH_2)_4CH_3 \xrightarrow[\text{ethanol}]{NaOH \\ H_2O}$$

G

H

dihydrojasmone

$$CH_3\overset{O}{\overset{\|}{C}}CH_2CH_2\overset{O}{\overset{\|}{C}}CH_2(CH_2)_4CH_3 \xrightarrow[\text{ethanol}]{NaOH \\ H_2O} CH_3\overset{O}{\overset{\|}{C}}CH_2CH_2\overset{O^-Na^+}{\overset{}{C}}=CH(CH_2)_4CH_3 \longrightarrow$$

more stable enolate anion,
containing the more highly
substituted double bond

dihydrojasmone

$$CH_2=\overset{O^-Na^+}{\overset{}{C}}CH_2CH_2\overset{O}{\overset{\|}{C}}CH_2(CH_2)_4CH_3 \longrightarrow$$

less stable enolate anion,
with the less highly
substituted double bond

this product does not form

21.31   (a)

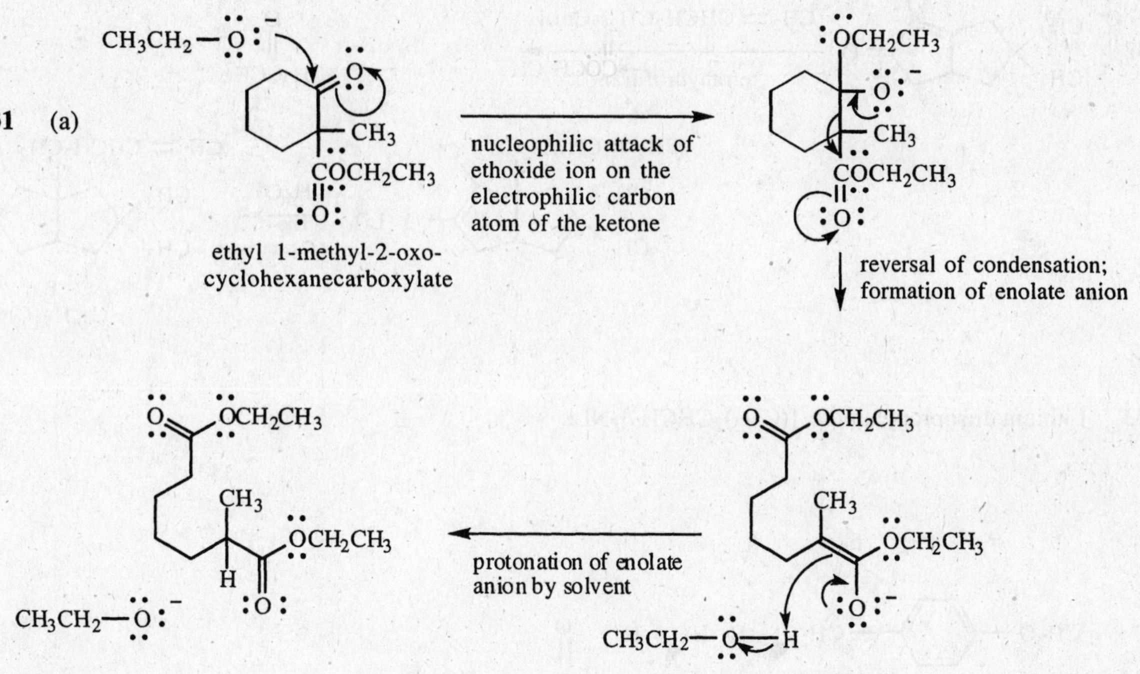

ethyl 1-methyl-2-oxo-
cyclohexanecarboxylate

nucleophilic attack of
ethoxide ion on the
electrophilic carbon
atom of the ketone

reversal of condensation;
formation of enolate anion

protonation of enolate
anion by solvent

(b)

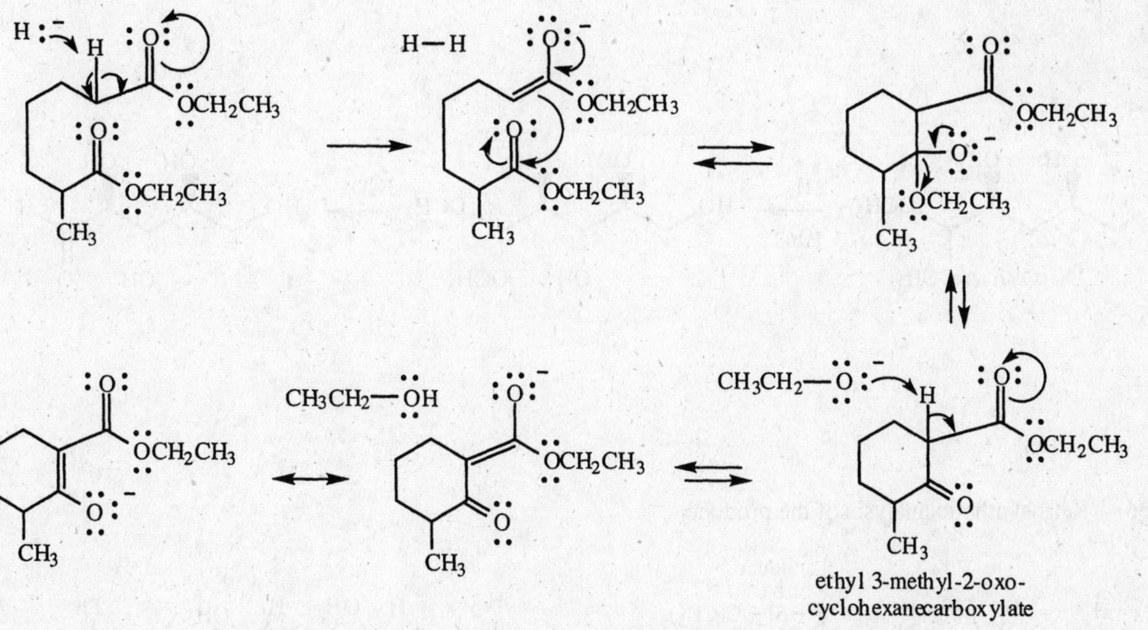

ethyl 3-methyl-2-oxo-
cyclohexanecarboxylate

The product β-ketoesters in Claisen condensations are stabilized by deprotonation to the corresponding enolates. Ethyl 1-methyl-2-oxocyclohexanecarboxylate does not have a hydrogen atom on the carbon atom between the two carbonyl groups, and, therefore, cannot be stabilized this way. It undergoes reversal of the cyclization reaction. The product resulting from the other cyclization reaction, ethyl 3-methyl-2-oxocyclohexanecarboxylate, does have an active hydrogen. Under the conditions for the reaction, it is the thermodynamically more stable product.

21.32

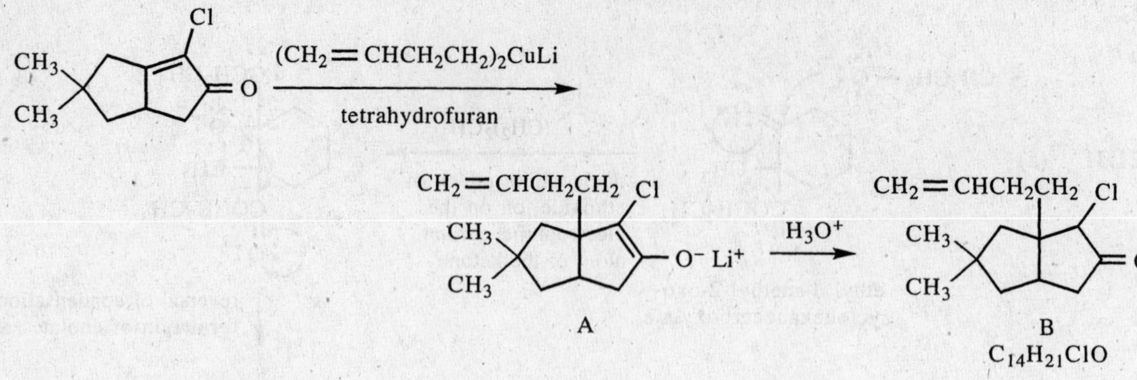

21.33   Lithium diisopropylamide, [(CH$_3$)$_2$CHCH$_2$]$_2$NLi

21.34

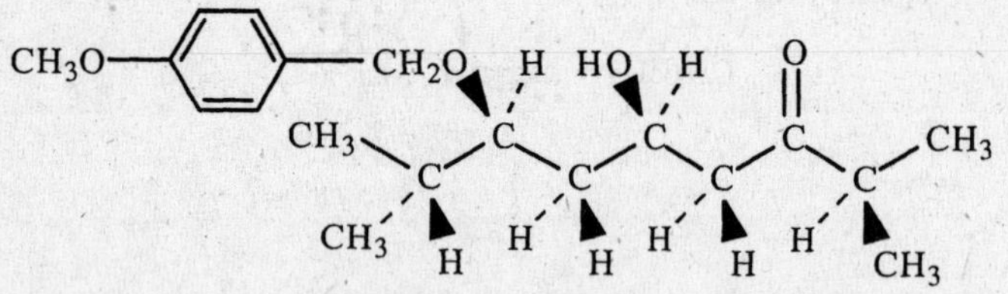

21.35

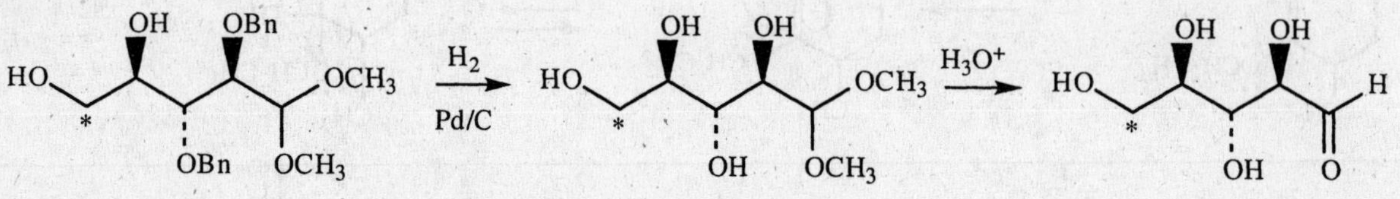

21.36   Retrosynthetic analysis of the product:

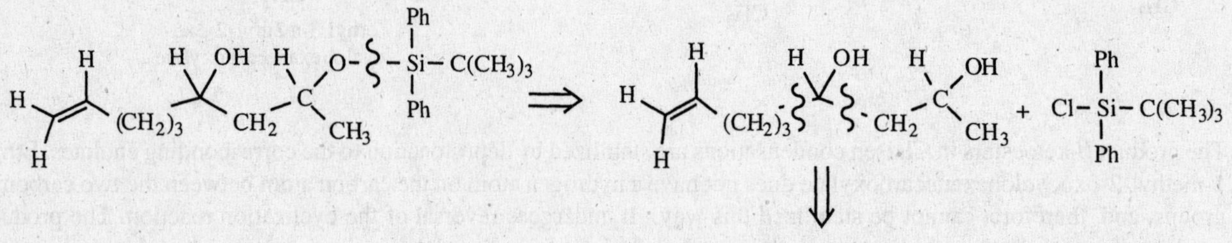

21.36 (cont)

electrophile   must be nucleophile   electrophile

Concept Map 21.8   The Diels-Alder reaction.

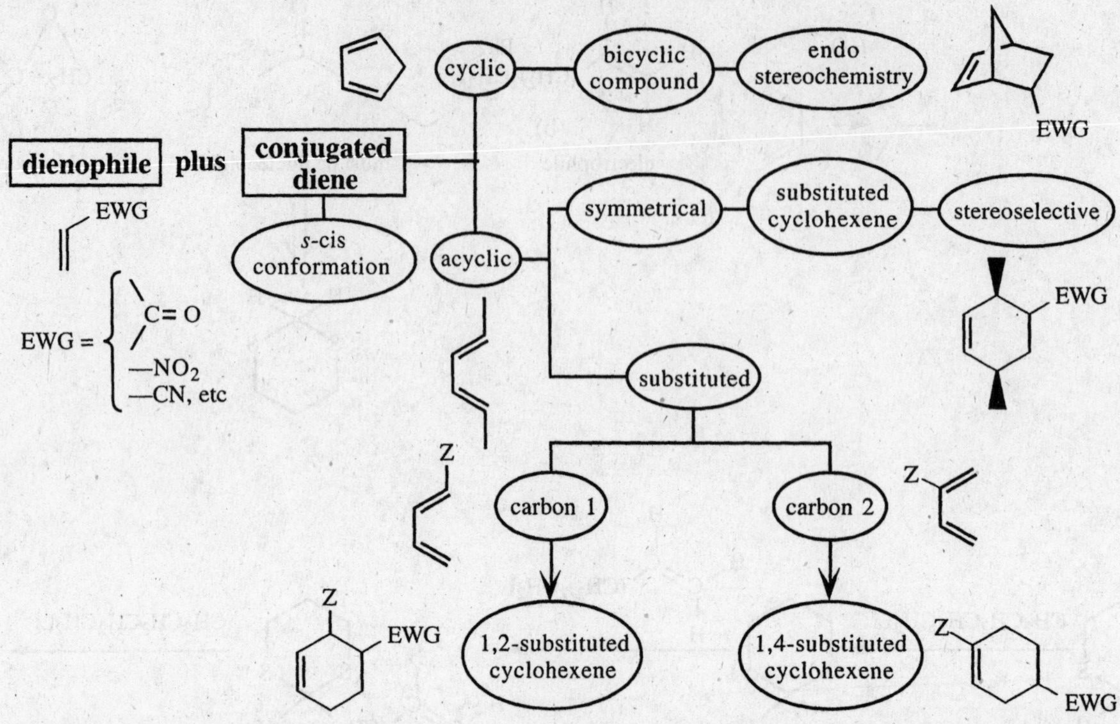

21.37

(a)

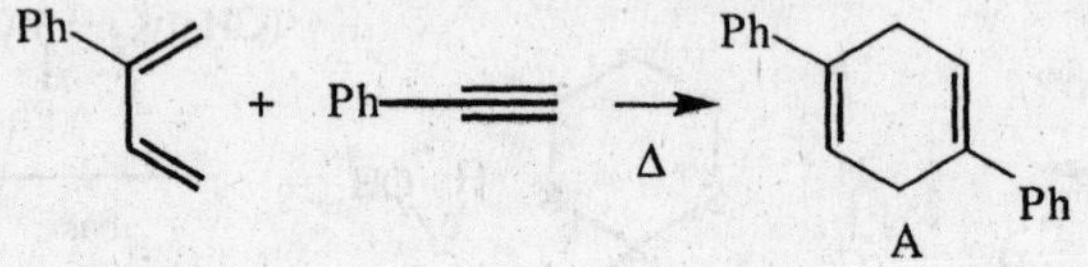

(b)

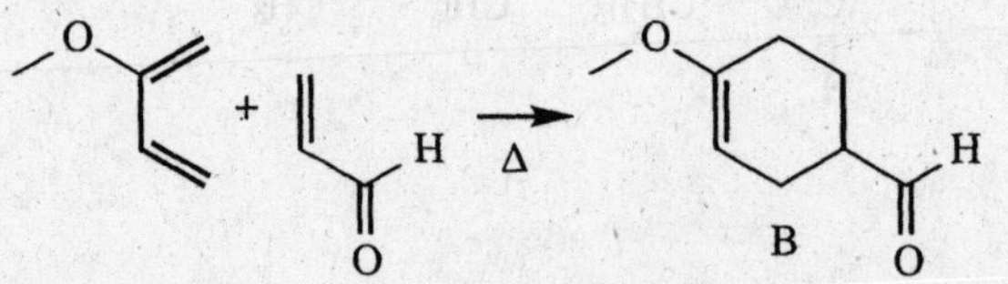

(c) 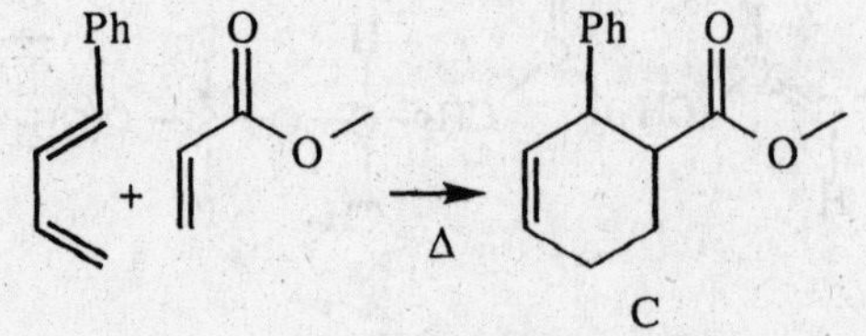   (d)

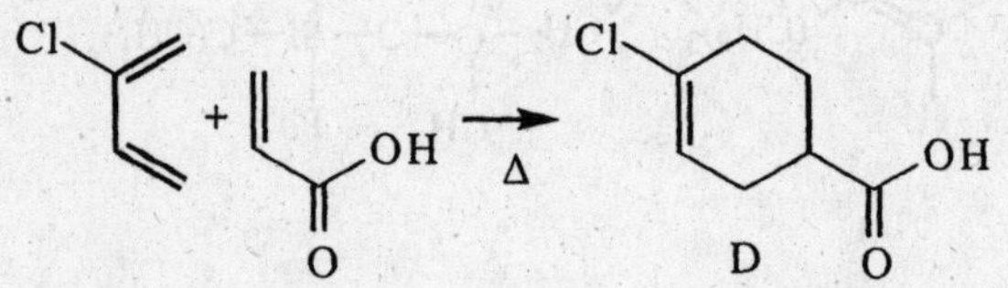

(e)

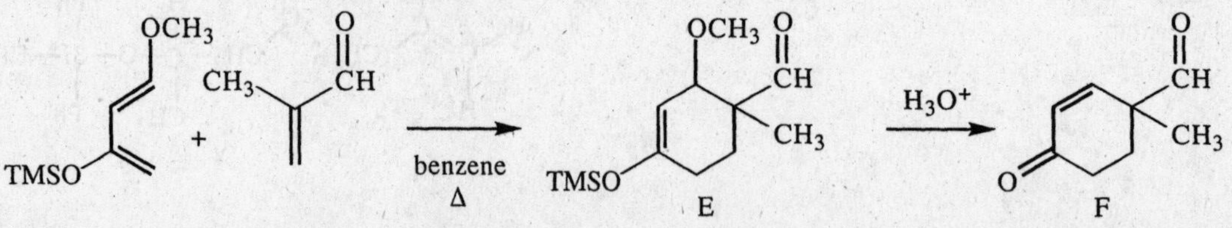

21.38

(a)

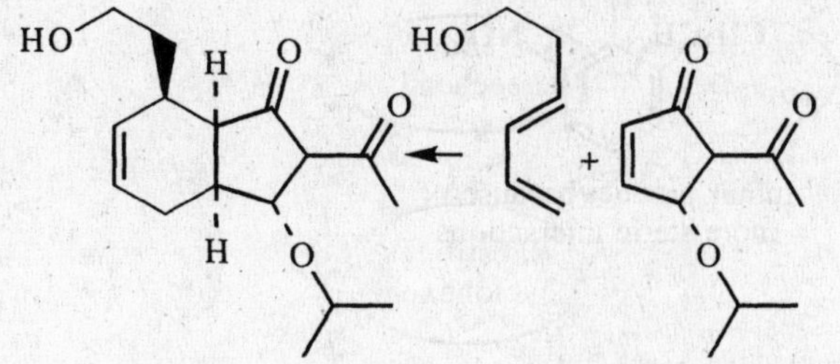

(b)

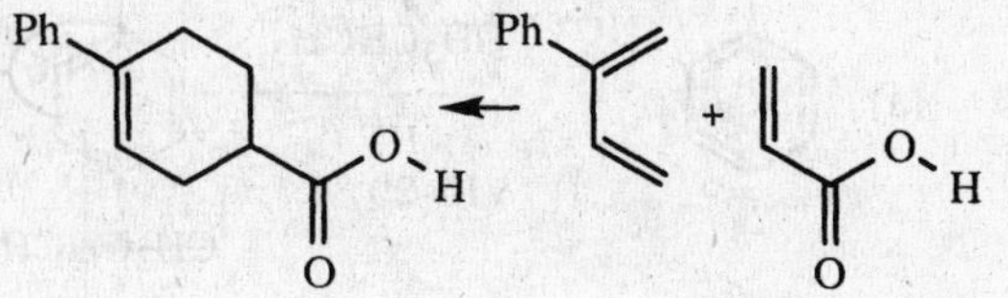

(c)

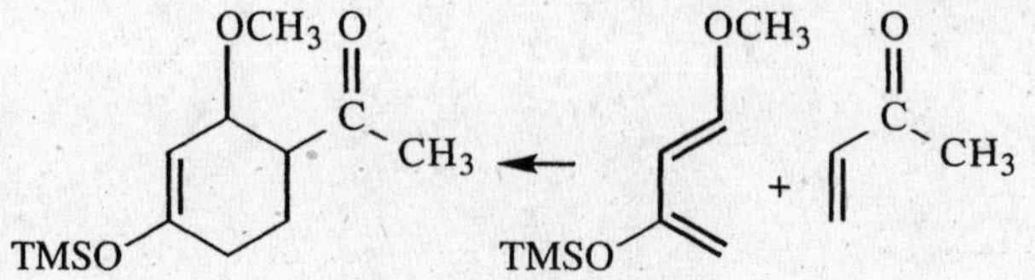

(d)

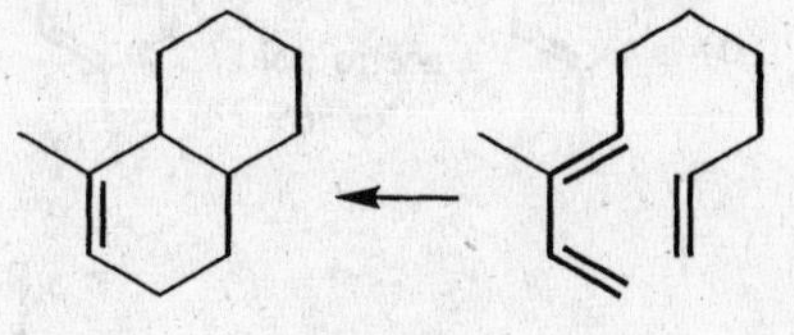

21.39

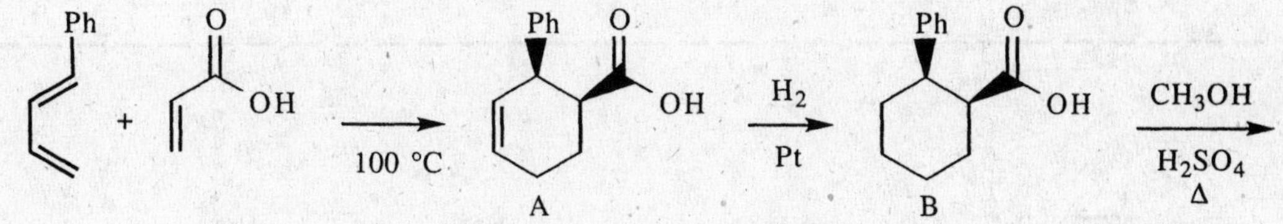

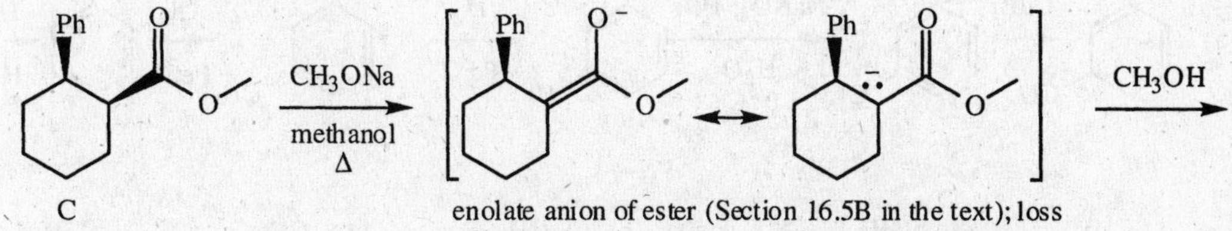

enolate anion of ester (Section 16.5B in the text); loss
of stereochemistry at the α-carbon atom of the ester

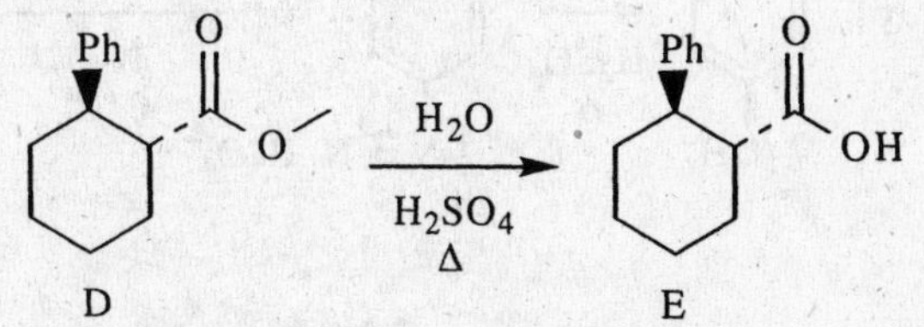

Generation of the carbanion from the methyl ester followed by reprotonation is an equilibrium reaction and
gives the thermodynamically more stable trans stereoisomer. Compound D is different from Compound C, and
Compound E is different from A. Therefore, the initial Diels-Alder product, Compound A, must have been the
cis stereoisomer.

21.40   (a)

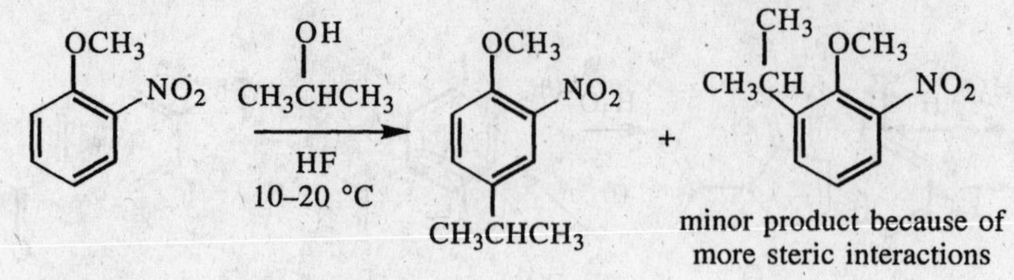

minor product because of
more steric interactions

(b)

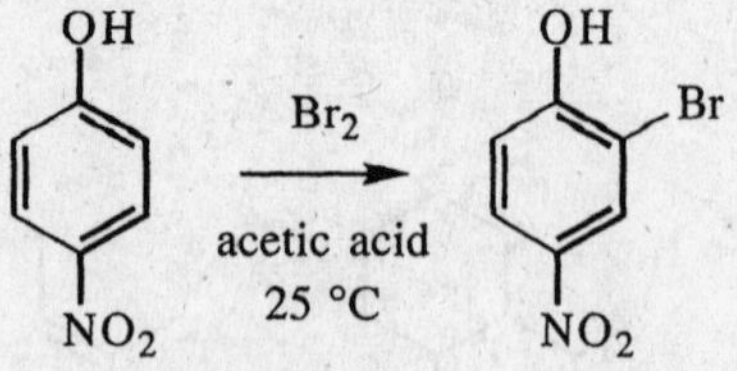

(c)

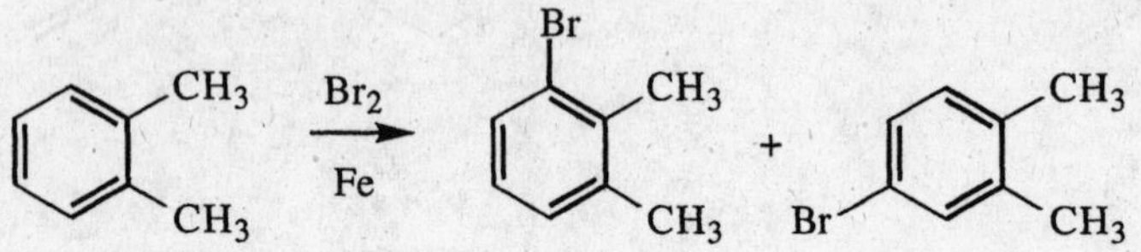

Concept Map 21.9 (see page 514)

21.41

(a)

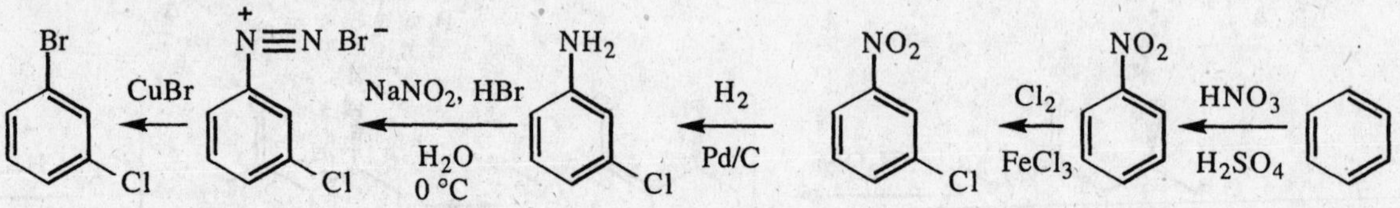

(b)

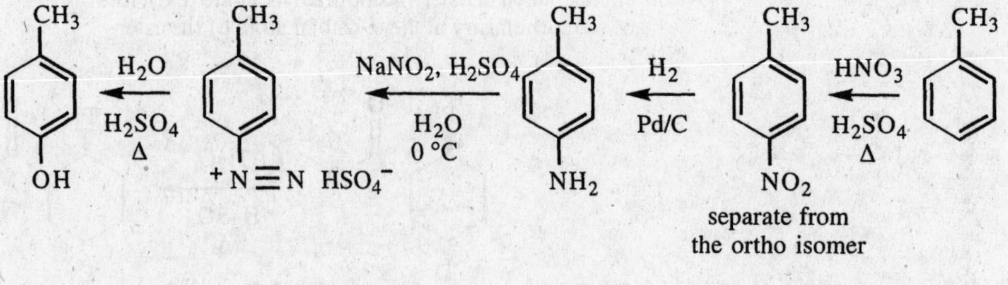

separate from
the ortho isomer

21.41 (c)

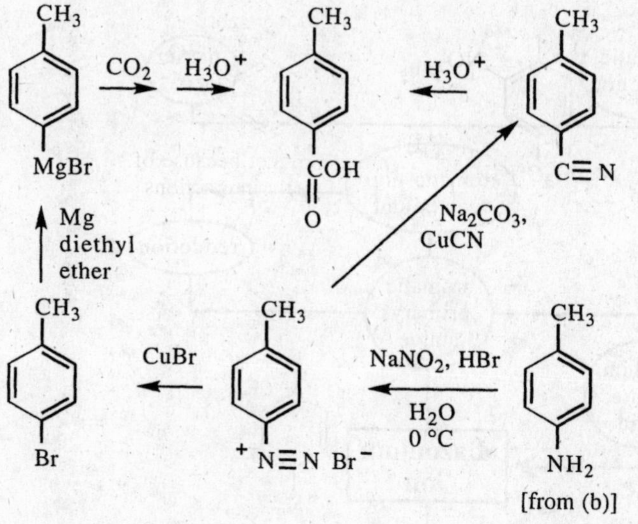

p-Bromotoluene can also be synthesized by treating toluene with bromine in the presence of ferric bromide, but it will be contaminated with *o*-bromotoluene. Separation of *o*-bromotoluene from *p*-bromotoluene is difficult because the two compounds differ by only 3° in boiling point. *p*-Nitrotoluene will also have the ortho isomer as a contaminant, but the two nitrotoluenes are easy to separate because the ortho isomer is a liquid at room temperature and the para isomer is a solid.

(d)

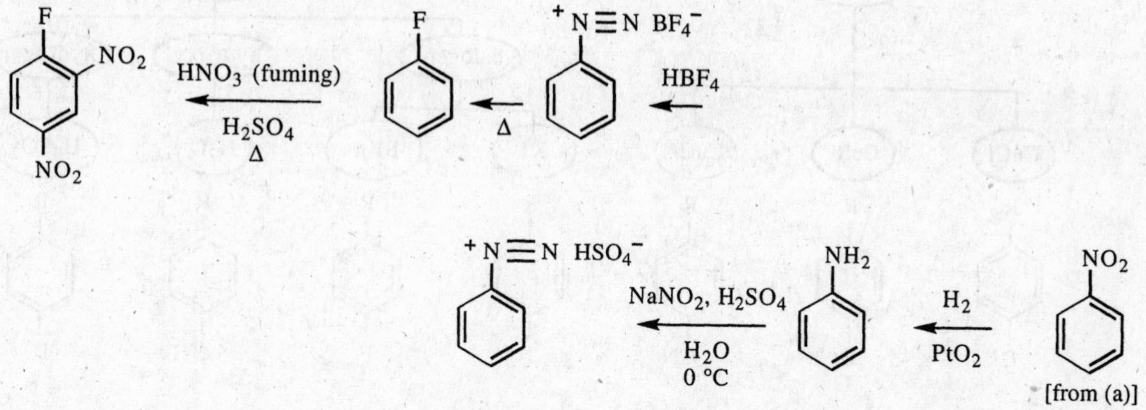

(e)

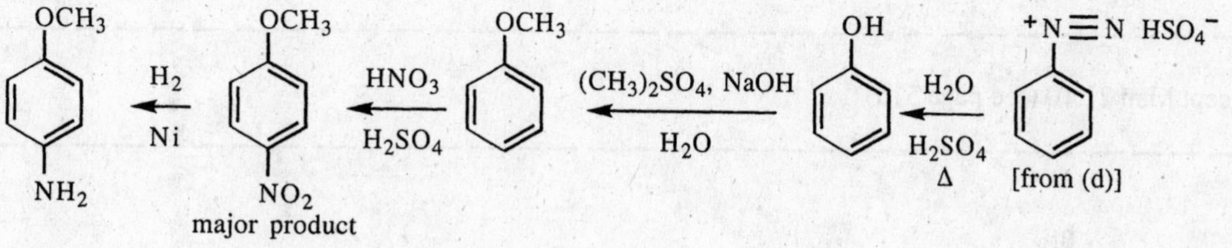

(f)

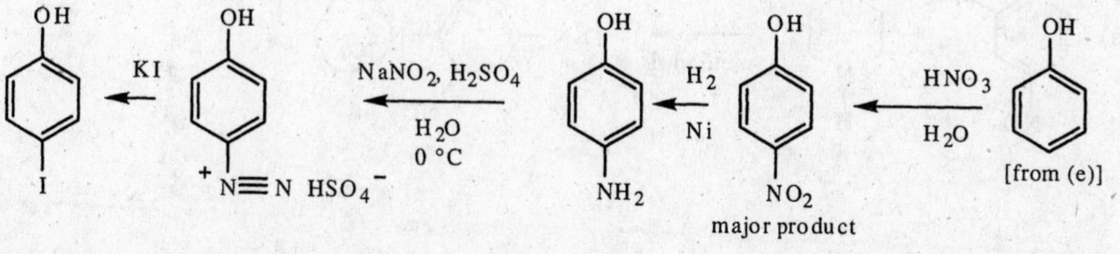

Concept Map 21.9   Diazonium ions in synthesis.

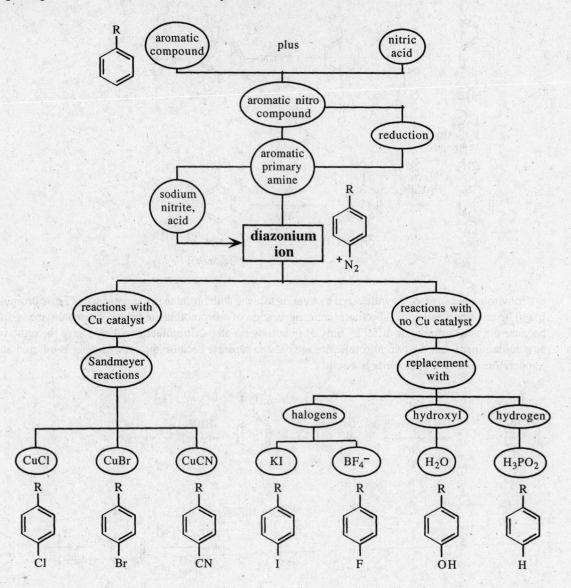

- -

Concept Map 21.10 (see page 515)

- -

21.42   (a)

21.42

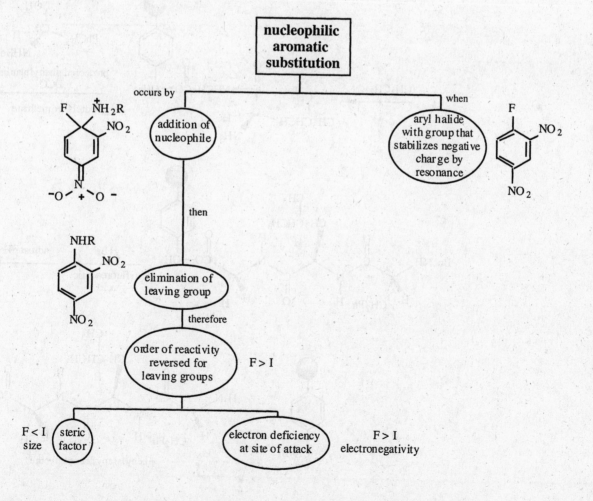

- -

Concept Map 21.10   Nucleophilic aromatic substitution.

21.43

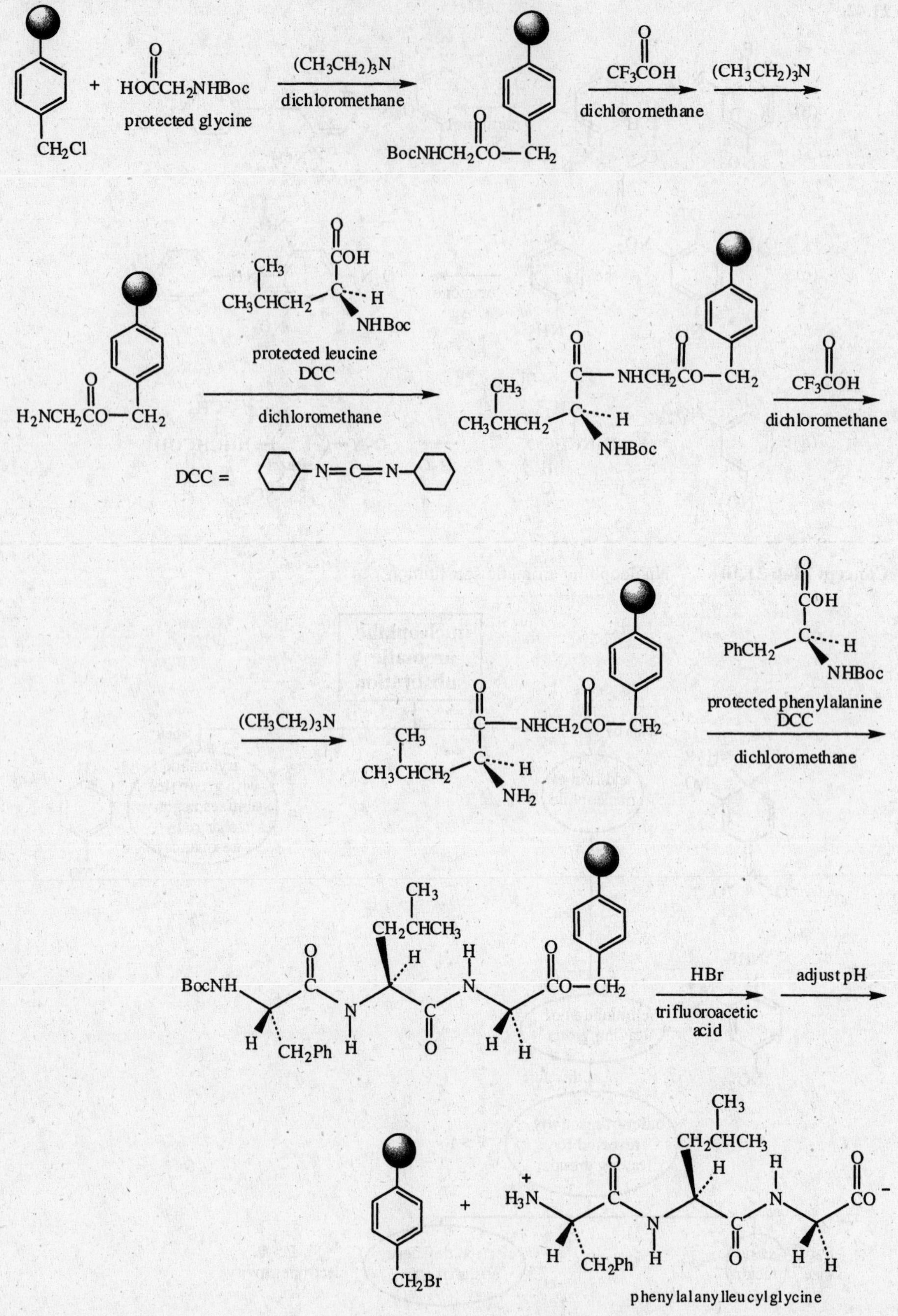
HOCCH2NHBoc
protected glycine
(CH3CH2)3N
dichloromethane
BocNHCH2CO—CH2
CF3COH
(CH3CH2)3N
H2NCH2CO—CH2
protected leucine
DCC
DCC =
NHCH2CO—CH2
NHBoc
NH2
protected phenylalanine
HBr
trifluoroacetic acid
adjust pH
CH2Br
phenylalanylleucylglycine

21.44

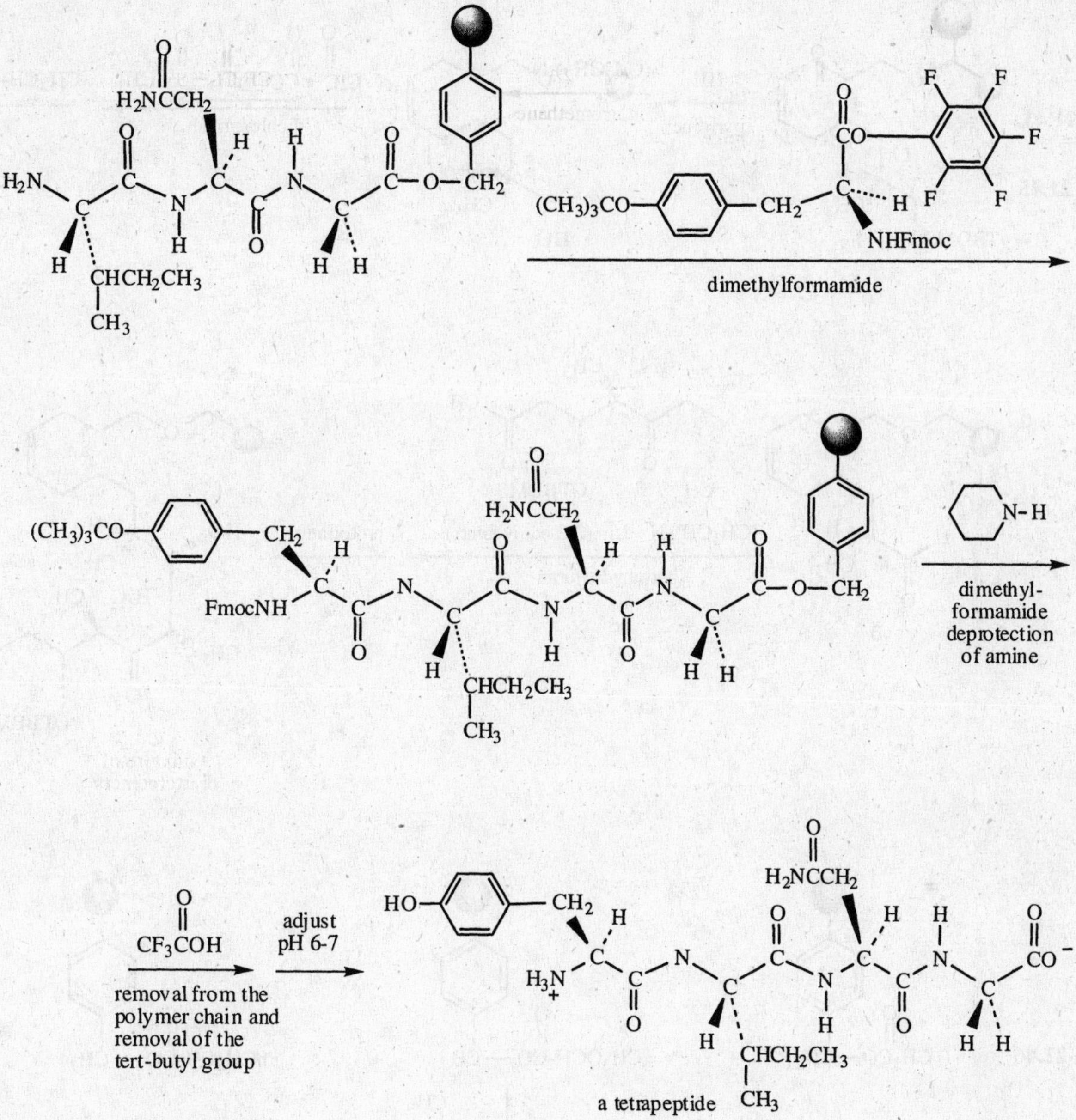

dimethylformamide

piperidine N–H

dimethyl-
formamide
deprotection
of amine

CF₃COH

removal from the
polymer chain and
removal of the
tert-butyl group

adjust
pH 6-7

a tetrapeptide

21.45

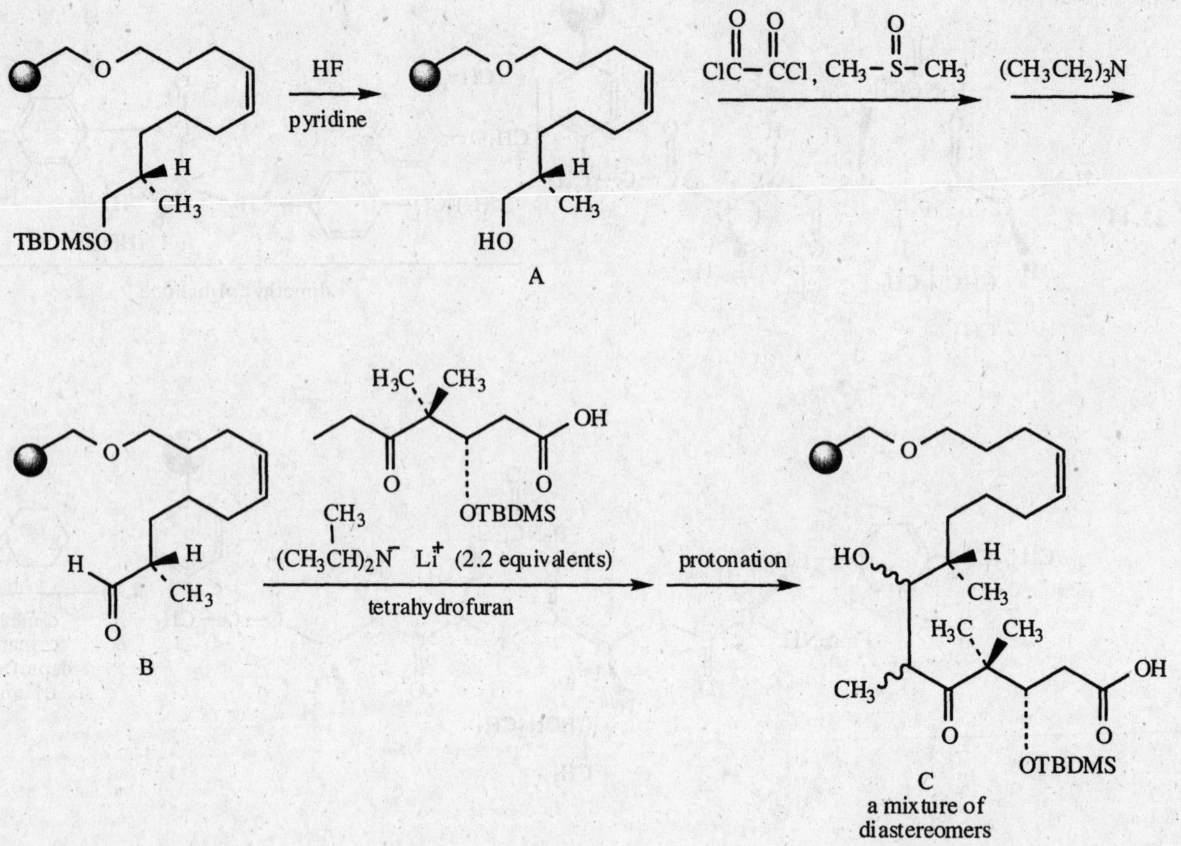

HF / pyridine

$ClC-CCl$, CH_3-S-CH_3 $(CH_3CH_2)_3N$

A

protonation

C
a mixture of
diastereomers

B

$(CH_3CH)_2N^-$ Li^+ (2.2 equivalents)
tetrahydrofuran

21.46

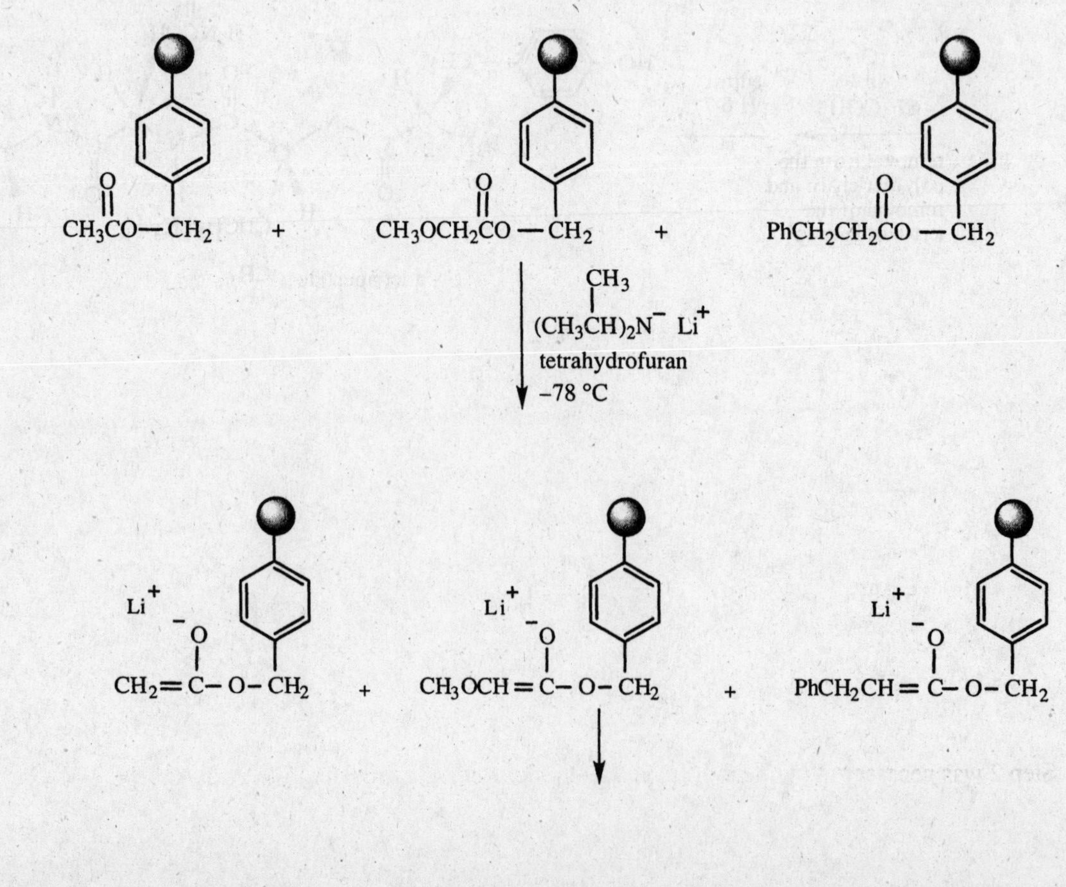

CH_3CO-CH_2 + $CH_3OCH_2CO-CH_2$ + $PhCH_2CH_2CO-CH_2$

$(CH_3CH)_2N^-$ Li^+
tetrahydrofuran
$-78\,°C$

$CH_2=C-O-CH_2$ + $CH_3OCH=C-O-CH_2$ + $PhCH_2CH=C-O-CH_2$

Li^+ ^-O Li^+ ^-O Li^+ ^-O

21.46 (cont)

Step 2 was necessary to protect the aldehyde group from oxidation during ozonolysis.

21.48 $BrCH_2CH_2CH_2CH_2OH$ $\xrightarrow[\text{dichloromethane}]{\overset{\underset{\displaystyle N^+}{\displaystyle |}}{H}\ CrO_3Cl^-}$ $\underset{A}{BrCH_2CH_2CH_2\overset{\displaystyle O}{\overset{\|}{C}}H}$ $\xrightarrow[\substack{\text{TsOH} \\ \text{benzene, } \Delta}]{HOCH_2CH_2OH}$ $\underset{B}{BrCH_2CH_2CH_2\overset{\overset{\displaystyle O \diagdown \diagup O}{\diagup \ \diagdown}}{C}H}$ $\xrightarrow[\substack{\text{tetrahydro-} \\ \text{furan}}]{Mg}$

$\underset{C}{BrMgCH_2CH_2CH_2\overset{\overset{\displaystyle O \diagdown \diagup O}{\diagup \ \diagdown}}{C}H}$ $\longrightarrow$ $\underset{D}{\underset{\displaystyle O^-Mg^{2+}Br^-}{\overset{\displaystyle CH_3 \quad\quad H}{\overset{\displaystyle \diagdown \ C=C \diagup}{\underset{H}{\diagup}}}}CHCH_2CH_2CH_2\overset{\overset{\displaystyle O \diagdown \diagup O}{\diagup \ \diagdown}}{C}H}$ $\xrightarrow{H_3O^+}$

$\left[\begin{array}{c} \overset{\displaystyle CH_3 \quad H}{C=C} \\ \overset{OH}{CH} \\ CH_2 \\ CH_2 \end{array} \right]$ $\longrightarrow$ $\underset{\substack{E \\ C_8H_{14}O_2 \\ \text{a stable, cyclic hemiacetal}}}{\text{(structure)}}$ $\xrightarrow[\text{dichloromethane}]{\overset{\underset{\displaystyle N^+}{\displaystyle |}}{H}\ CrO_3Cl^-}$ $\underset{\substack{F \\ C_8H_{12}O_2 \\ \text{three units of unsaturation} \\ \text{(2 double bonds and a ring)}}}{\text{(structure)}}$

21.49

(a) $\underset{\substack{\displaystyle OH \\ \text{7-methyl-6-octen-3-ol}}}{\overset{\displaystyle CH_3}{CH_3C=CHCH_2CH_2CHCH_2CH_3}}$ $\xleftarrow{H_3O^+}$ $\overset{\displaystyle O}{\overset{\|}{CH_3CH_2CH}}$ $\xleftarrow{}$ $\overset{\displaystyle CH_3}{CH_3C=CHCH_2CH_2MgBr}$ $\xleftarrow{Mg}$ $\overset{\displaystyle CH_3}{CH_3C=CHCH_2CH_2Br}$

(b) $\underset{OH}{\overset{\displaystyle CH_3}{CH_3C=CHCH_2CH_2CHCH_2CH_3}}$ $\xrightarrow[\substack{\text{dimethylformamide}}]{\substack{\displaystyle CH_3 \quad CH_3 \\ CH_3C-SiCl \\ \displaystyle CH_3 \quad CH_3}}$ $\underset{OTBDMS}{\overset{\displaystyle CH_3}{CH_3C=CHCH_2CH_2CHCH_2CH_3}}$

21.49

(c) $CH_3\overset{\underset{\displaystyle CH_3}{|}}{C}=CHCH_2CH_2\overset{\underset{\displaystyle OTBDMS}{|}}{C}HCH_2CH_3$ $\xrightarrow[\text{dichloromethane}]{\text{Cl}\text{—C}_6\text{H}_4\text{—COOH}}$

$\xrightarrow[\substack{\text{tetrahydrofuran}\\ 0\ °C}]{(CH_3CH_2CH_2CH_2)_4N^+F^-}$

The alcohol functional group is a nucleophile that can open the oxirane ring under acid catalysis (see Section 13.5A in the text).

21.50 $HOCH_2C\equiv CH$ $\xrightarrow[\text{TsOH}]{}$ $THPOCH_2C\equiv CH$ $\xrightarrow{CH_3CH_2CH_2CH_2Li}$ $THPOCH_2C\equiv C^-Li^+$
                                                            A   B

$\xrightarrow{BrCH_2CH_2CH_2Cl}$ $THPOCH_2C\equiv CCH_2CH_2CH_2Cl$ $\xrightarrow[\substack{\text{dimethyl}\\ \text{sulfoxide}}]{NaCN}$ $THPOCH_2C\equiv CCH_2CH_2CH_2CN$ $\xrightarrow[\substack{\text{ethanol}\\ \Delta}]{NaOH, H_2O}$
                                                C   D
                                  (Bromide ion is a better leaving
                                     group than choride ion.)

$\xrightarrow[\substack{0\ °C\\ 5\ min}]{H_3O^+}$ $THPOCH_2C\equiv CCH_2CH_2CH_2\overset{\displaystyle O}{\overset{\|}{C}}OH$ $\xrightarrow[\substack{\text{dimethyl-}\\ \text{formamide}}]{K_2CO_3}$ $THPOCH_2C\equiv CCH_2CH_2CH_2\overset{\displaystyle O}{\overset{\|}{C}}O^-\ K^+$ $\xrightarrow{CH_3I}$
                                                      E   F
                                  (The tetrahydropyranyl group is not
                                   hydrolyzed under these conditions.)

$THPOCH_2C\equiv CCH_2CH_2CH_2\overset{\displaystyle O}{\overset{\|}{C}}OCH_3$ $\xrightarrow[\substack{\overset{O}{\overset{\|}{PhSOH}}\\ \overset{\|}{O}\\ 3\ h}]{CH_3OH}$ $HOCH_2C\equiv CCH_2CH_2CH_2\overset{\displaystyle O}{\overset{\|}{C}}OCH_3$ $\xrightarrow[\substack{Pd/CaCO_3\\ \text{quinoline}}]{H_2}$
                              G   H

$+$

$\xrightarrow[\substack{\text{pyridine}\\ \text{tetrahydro-}\\ \text{furan}}]{CH_3\overset{\displaystyle O}{\overset{\|}{C}}O\overset{\displaystyle O}{\overset{\|}{C}}CH_3}$

$C_{10}H_{16}O_4$

21.50 (cont) The infrared spectrum of Compound H shows the presence of an alcohol (3450 cm^{-1} for the O—H stretch) and an ester (1720 cm^{-1} for the C=O stretch). Analysis of the proton magnetic resonance spectrum of Compound H is shown below:

δ 1.6-2.0   (2H, *m*, —CH$_2$CH$_2$—)

δ 2.2-2.6   (5H, *m*, —(≡CCH$_2$CH$_2$CH$_2$C=O and —OH)

δ 3.68   (3H, *s*, —OCH$_3$)

δ 4.2-4.3   (2H, *m*, —OCH$_2$C≡)

Compound H is methyl 7-hydroxy-5-heptynoate

21.51

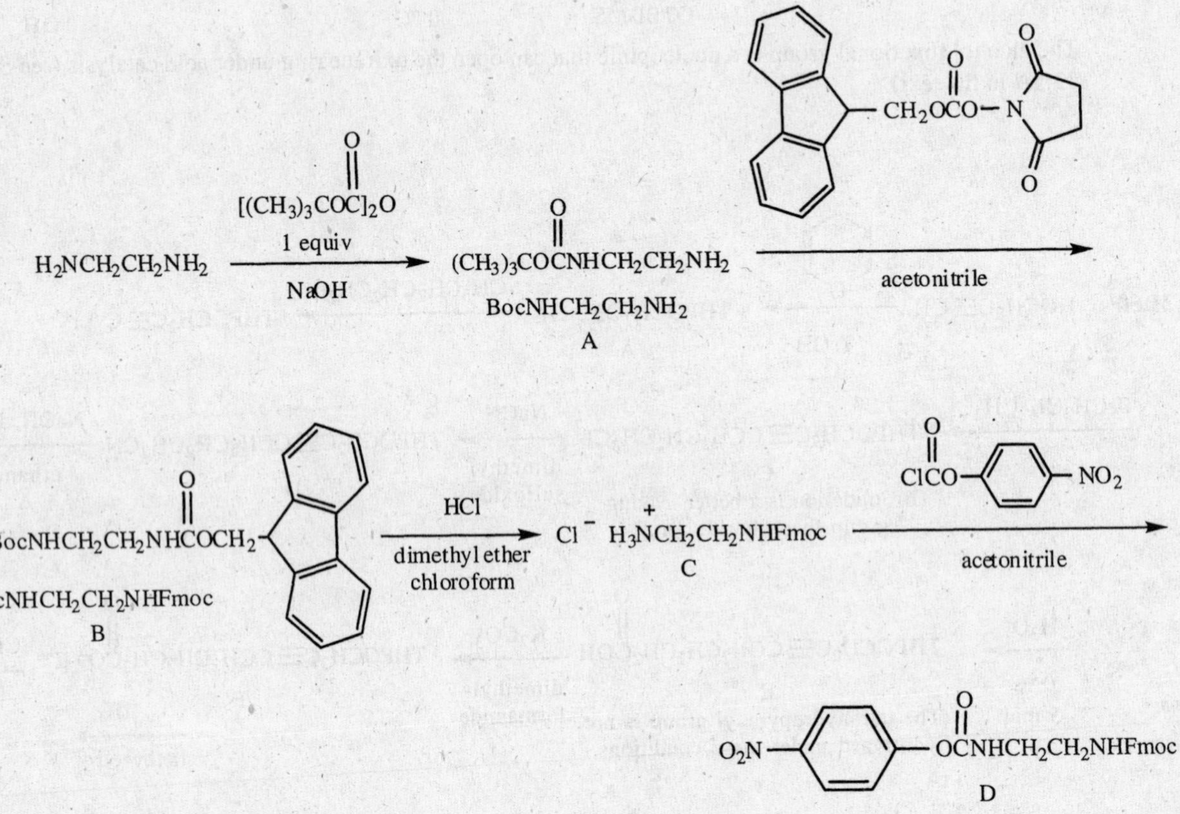

H$_2$NCH$_2$CH$_2$NH$_2$ $\xrightarrow[\text{NaOH}]{\substack{[(CH_3)_3COC]_2O \\ \text{1 equiv}}}$ (CH$_3$)$_3$COCNHCH$_2$CH$_2$NH$_2$

BocNHCH$_2$CH$_2$NH$_2$

A

$\xrightarrow{\text{acetonitrile}}$

BocNHCH$_2$CH$_2$NHCOCH$_2$

BocNHCH$_2$CH$_2$NHFmoc

B

$\xrightarrow[\substack{\text{dimethyl ether} \\ \text{chloroform}}]{\text{HCl}}$ Cl$^-$ $\overset{+}{\text{H}_3}$NCH$_2$CH$_2$NHFmoc

C

$\xrightarrow{\text{acetonitrile}}$ ClCO—NO$_2$

O$_2$N—OCNHCH$_2$CH$_2$NHFmoc

D

21.52 (a)

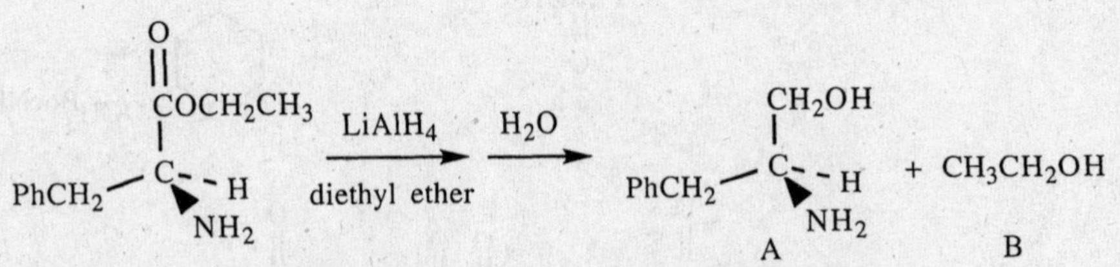

$\overset{\overset{O}{\|}}{\underset{}{\text{COCH}_2\text{CH}_3}}$

PhCH$_2$—C—H $\xrightarrow[\text{diethyl ether}]{\text{LiAlH}_4 \quad \text{H}_2\text{O}}$ PhCH$_2$—C—H + CH$_3$CH$_2$OH

NH$_2$ CH$_2$OH / NH$_2$

A   B

(b)

$\xrightarrow[\substack{\text{diethyl} \\ \text{ether}}]{\text{LiAlH}_4 \qquad \text{H}_2\text{O}}$

C

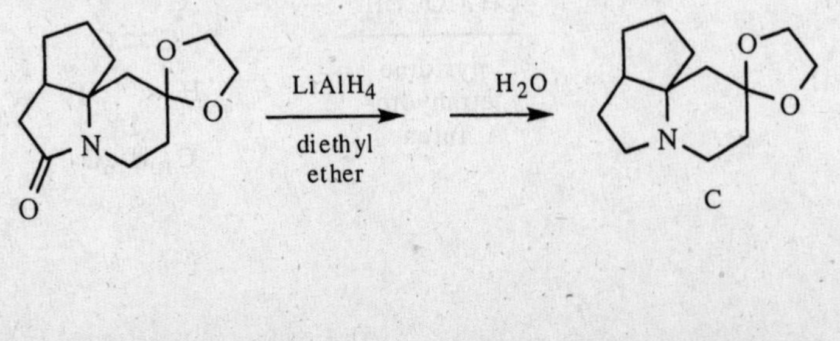

21.52 (c)

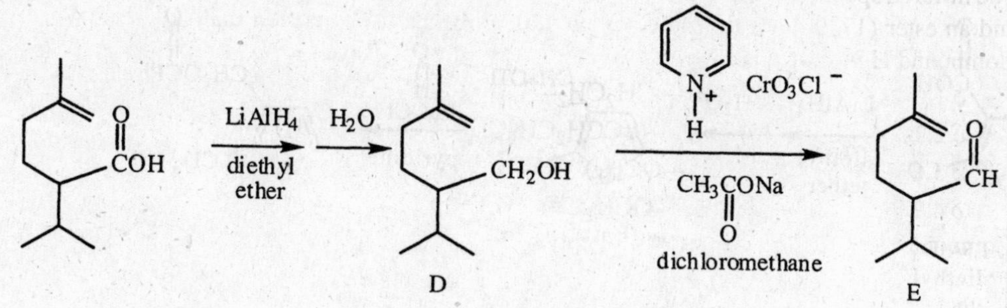
COH
LiAlH4
diethyl ether
H2O
CH2OH
D
CrO3Cl −
CH3CONa
dichloromethane
CH
E

(d)

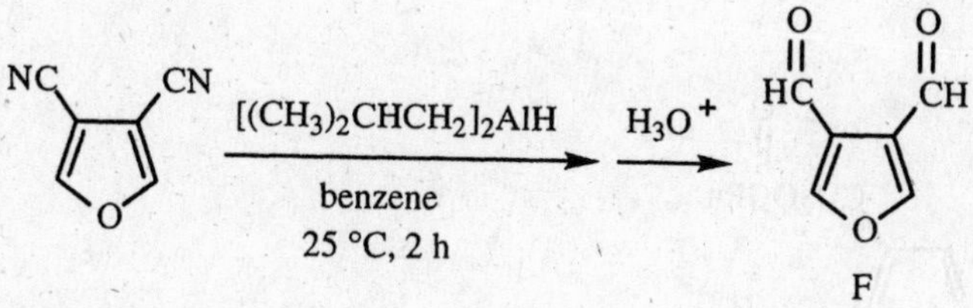
NC
CN
[(CH3)2CHCH2]2AlH
benzene
25 °C, 2 h
H3O+
HC
CH
F

(e)

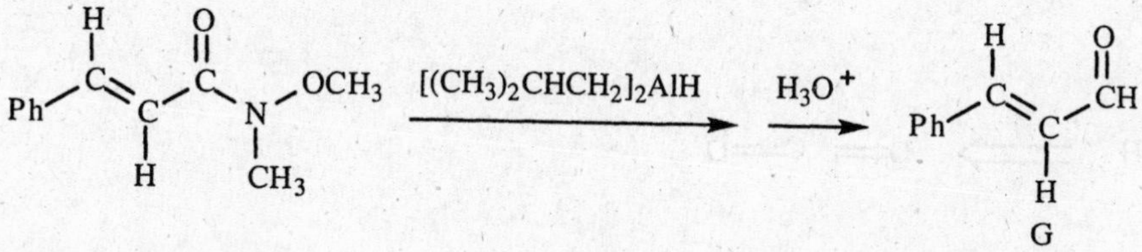
Ph
OCH3
CH3
[(CH3)2CHCH2]2AlH
H3O+
CH
G

(f)

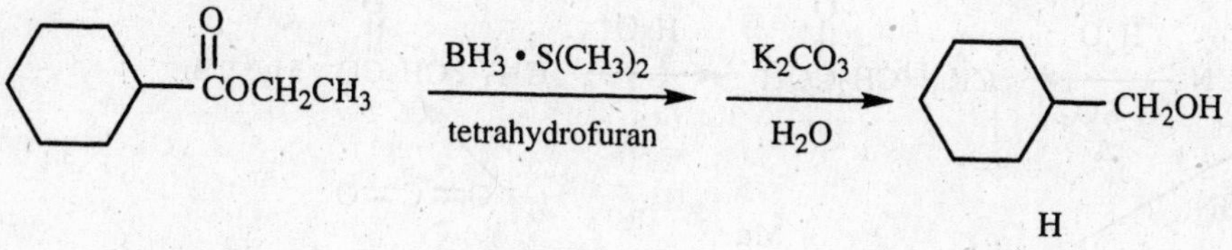
COCH2CH3
BH3 • S(CH3)2
tetrahydrofuran
K2CO3
H2O
CH2OH
H

(g)

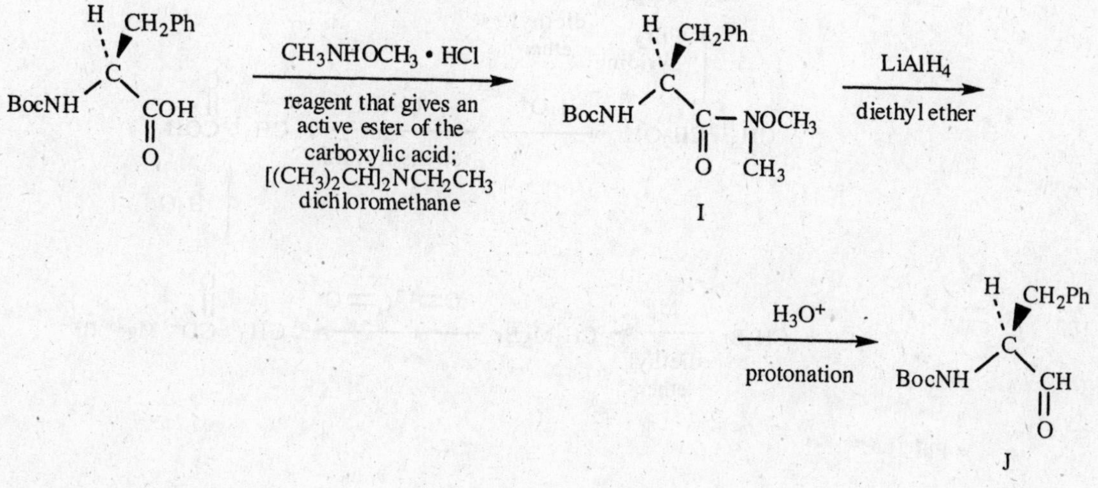
CH2Ph
BocNH
COH
CH3NHOCH3 • HCl
reagent that gives an active ester of the carboxylic acid; [(CH3)2CH]2NCH2CH3
dichloromethane
NOCH3
CH3
I
LiAlH4
diethyl ether
H3O+
protonation
CH
J

21.53

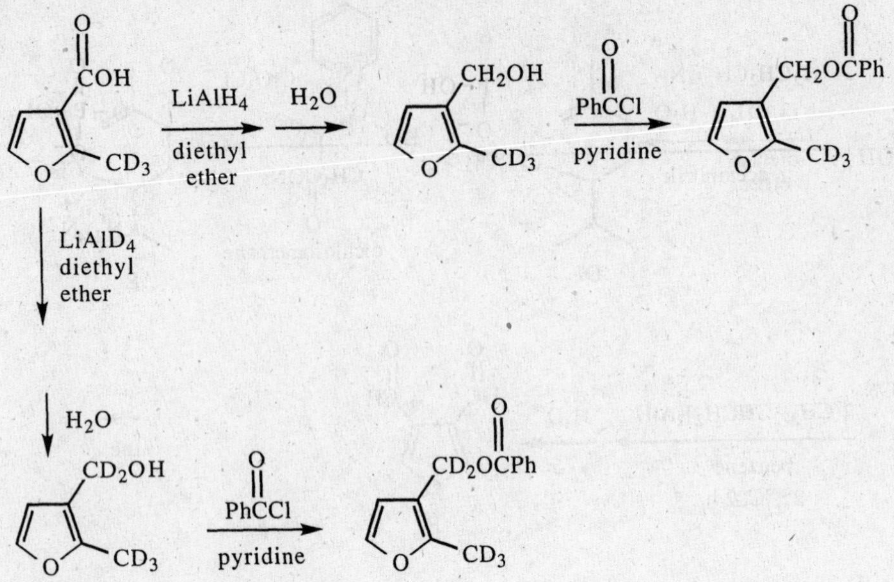

21.54

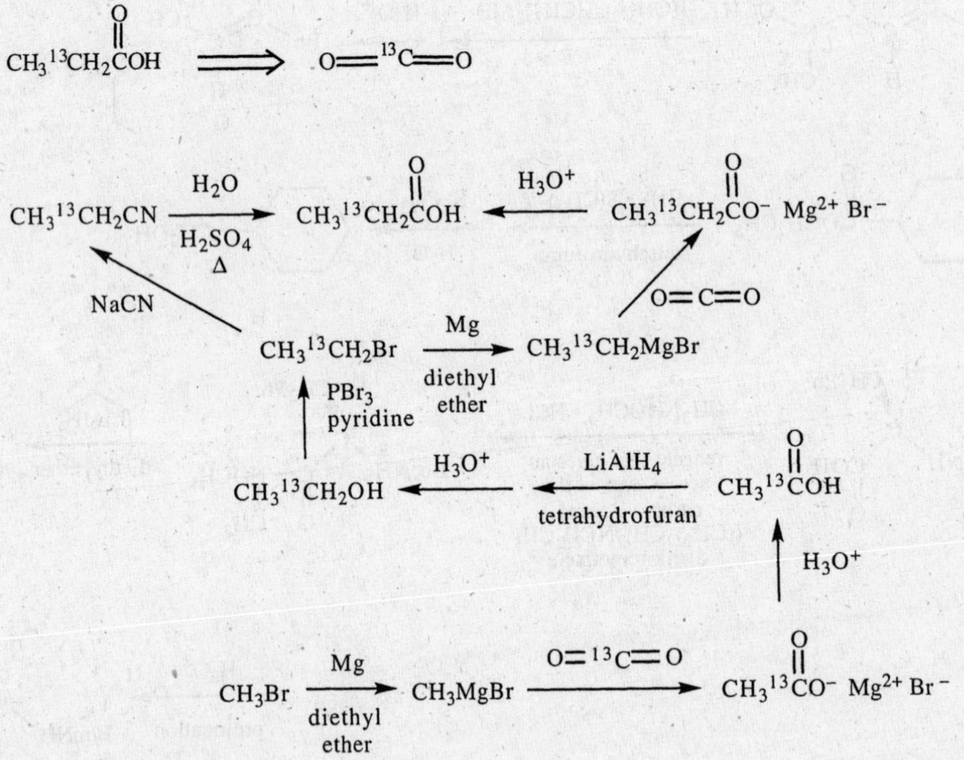

$$CH_3{}^{13}CH_2COH \Longrightarrow O={}^{13}C=O$$

21.55

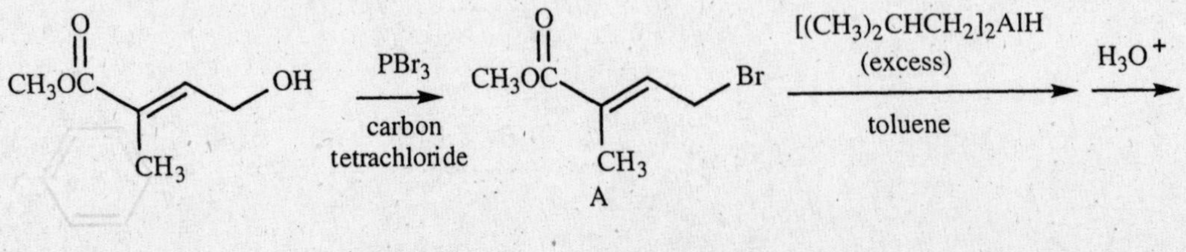

21.55 (cont)

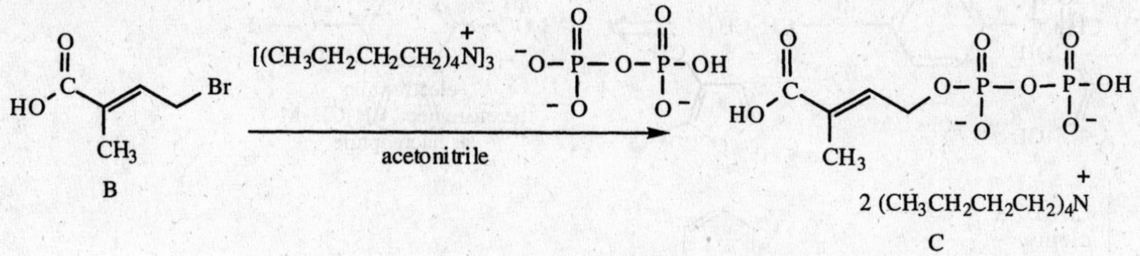

21.56

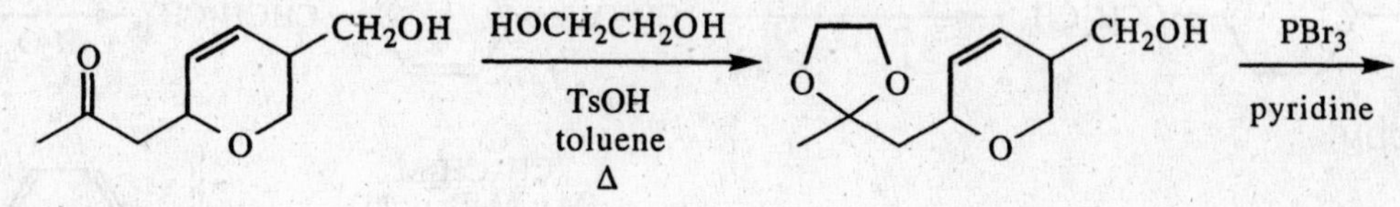

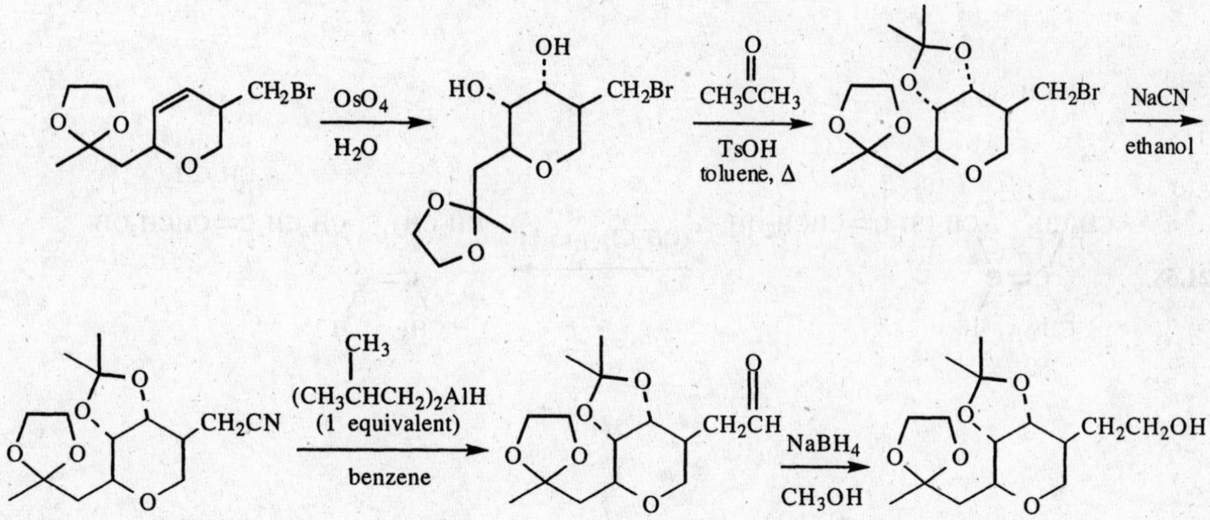

21.57 (1)

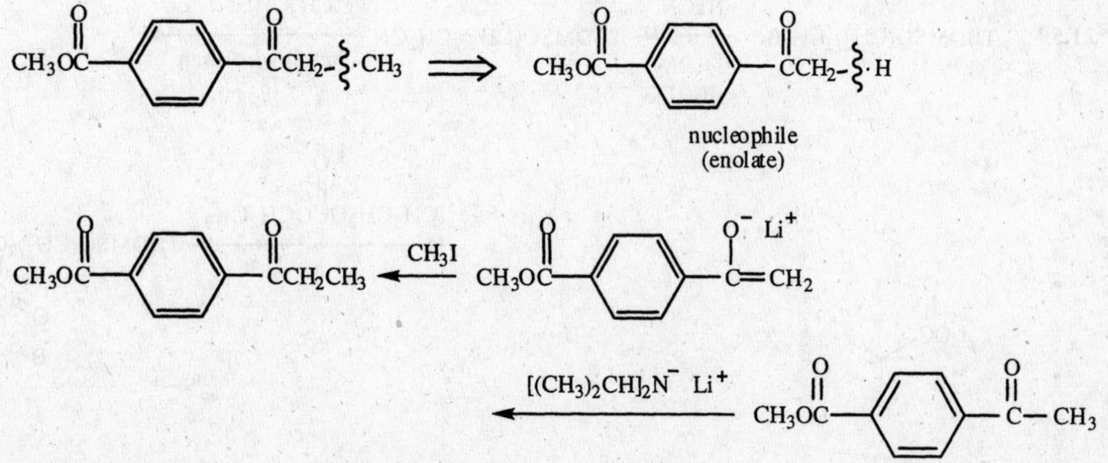

21.57 (2) $CH_3OC(=O)$—[benzene]—$C(=O)$—CH_2CH_3 $\Longrightarrow$ $CH_3OC(=O)$—[benzene]—$C(=O)$—H

electrophile
therefore need CH_3CH_2M
as nucleophile

$CH_3OC(=O)$—[benzene]—$C(=O)CH_2CH_3$ $\xleftarrow{\text{pyridinium } CrO_3\bar{Cl}}$ $CH_3OC(=O)$—[benzene]—$CH(OH)CH_2CH_3$ $\xleftarrow[\text{H}_2\text{O}]{\text{NH}_4\text{Cl}}$

$\xleftarrow[\text{(M = MgBr or Li)}]{CH_3CH_2M}$ $CH_3OC(=O)$—[benzene]—$C(=O)H$

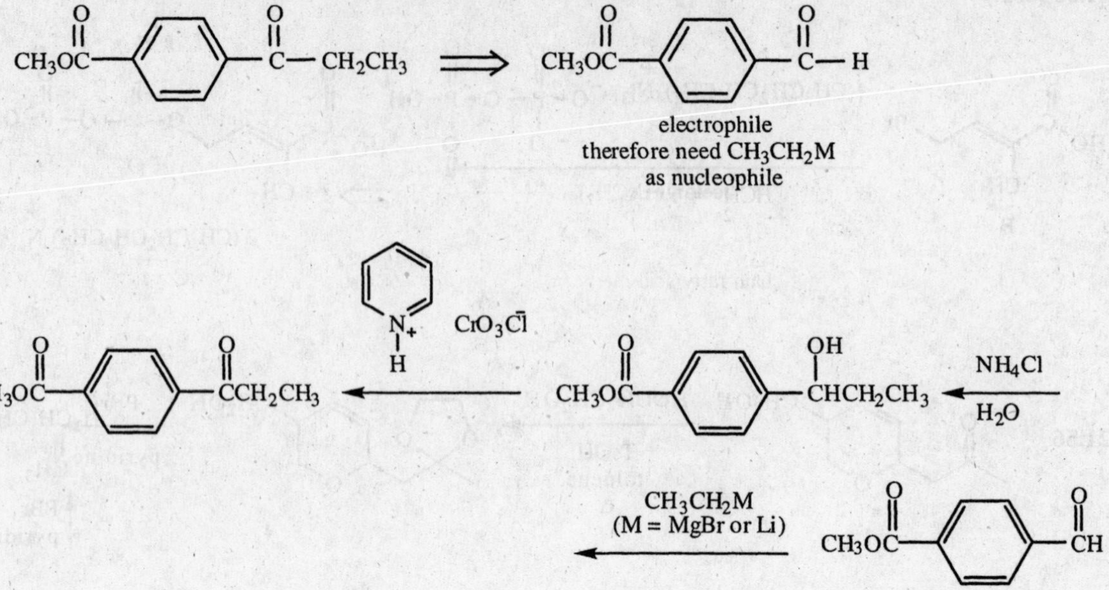

21.58

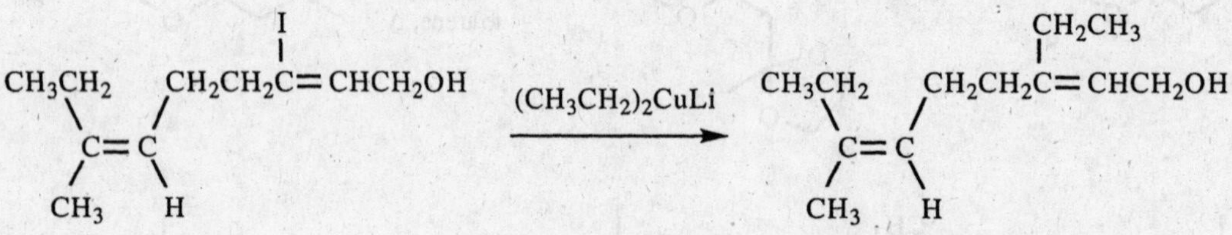

CH_3CH_2, CH_3 >C=C< H, with $CH_2CH_2C(I)=CHCH_2OH$ $\xrightarrow{(CH_3CH_2)_2CuLi}$ CH_3CH_2, CH_3 >C=C< H, with $CH_2CH_2C(CH_2CH_3)=CHCH_2OH$

21.59 $TBDMSO(CH_2)_2CH_2Br$ $\xrightarrow[\text{dimethyl sulfoxide}]{NaCN}$ $TBDMSO(CH_2)_2CH_2CN$ $\xrightarrow[\substack{\text{tetrahydrofuran} \\ -78\,°C}]{(CH_3CH)_2N^-\,Li^+ \ (CH_3)}$

A

$\xrightarrow{CH_3CH_2OC(=O)COCH_2CH_3 \ (O)}$ $TBDMSO(CH_2)_2CHCN$ with $C(=O)OCH_2CH_3$

B

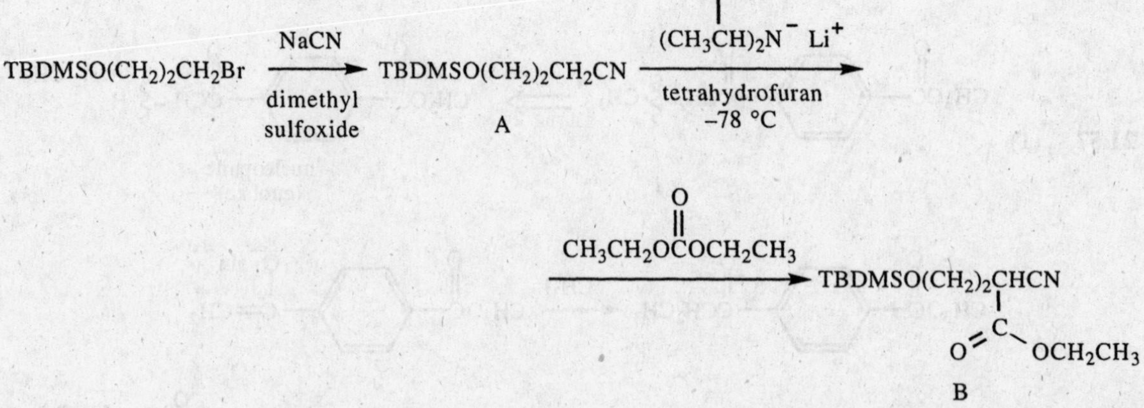

21.60

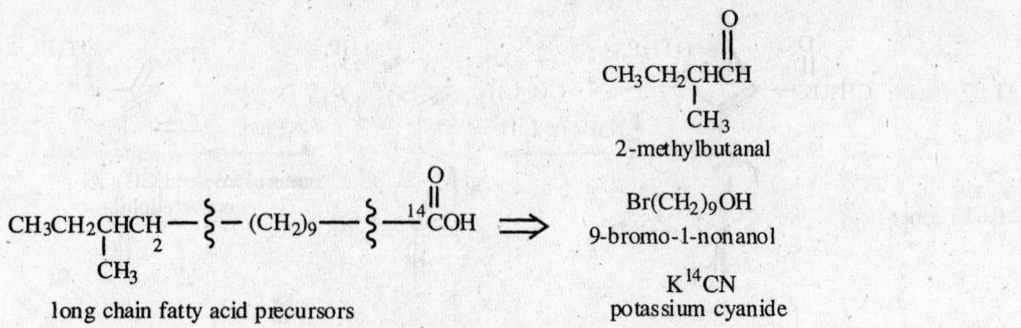

$$CH_3CH_2CHCH_2\text{---}\{\text{---}(CH_2)_9\text{---}\}\text{---}^{14}COH \Rightarrow$$
with CH_3 branch

long chain fatty acid precursors

O
‖
CH₃CH₂CHCH
  |
  CH₃
2-methylbutanal

Br(CH₂)₉OH
9-bromo-1-nonanol

K¹⁴CN
potassium cyanide

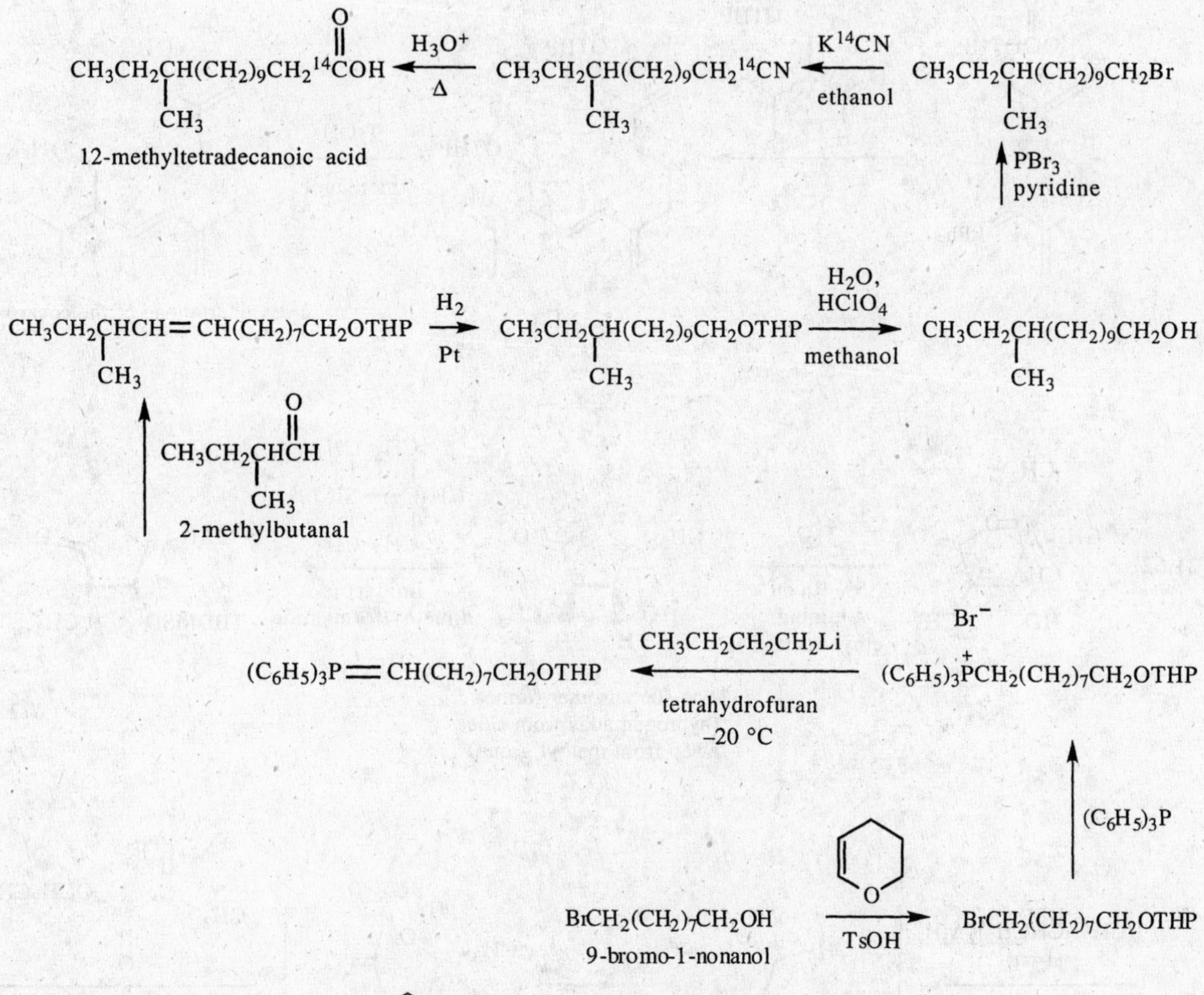

$$CH_3CH_2CH(CH_2)_9CH_2{}^{14}COH \xleftarrow[\Delta]{H_3O^+} CH_3CH_2CH(CH_2)_9CH_2{}^{14}CN \xleftarrow[\text{ethanol}]{K^{14}CN} CH_3CH_2CH(CH_2)_9CH_2Br$$

12-methyltetradecanoic acid

PBr₃
pyridine

$$CH_3CH_2CHCH\!\!=\!\!CH(CH_2)_7CH_2OTHP \xrightarrow[\text{Pt}]{H_2} CH_3CH_2CH(CH_2)_9CH_2OTHP \xrightarrow[\text{methanol}]{\substack{H_2O,\\ HClO_4}} CH_3CH_2CH(CH_2)_9CH_2OH$$
with CH₃ branches

O
‖
CH₃CH₂CHCH
  |
  CH₃
2-methylbutanal

$$(C_6H_5)_3P\!\!=\!\!CH(CH_2)_7CH_2OTHP \xleftarrow[\substack{\text{tetrahydrofuran}\\ -20\,^\circ C}]{CH_3CH_2CH_2CH_2Li} (C_6H_5)_3\overset{+}{P}CH_2(CH_2)_7CH_2OTHP \quad Br^-$$

(C₆H₅)₃P

$$BrCH_2(CH_2)_7CH_2OH \xrightarrow{\text{TsOH}} BrCH_2(CH_2)_7CH_2OTHP$$
9-bromo-1-nonanol

21.61

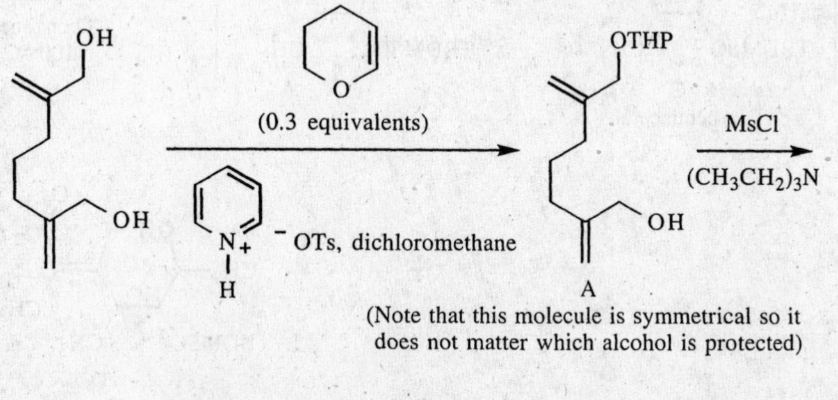

(0.3 equivalents)

MsCl
(CH₃CH₂)₃N

pyridinium OTs, dichloromethane

A
(Note that this molecule is symmetrical so it
does not matter which alcohol is protected)

21.61 (cont)

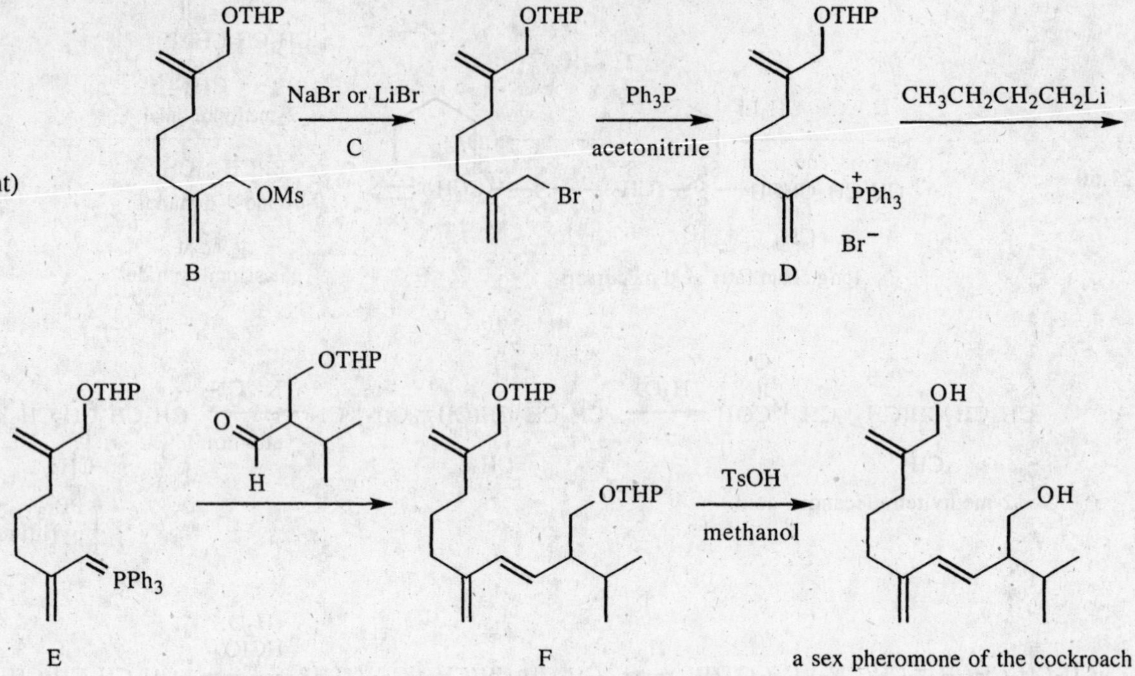
OTHP
OMs
B
NaBr or LiBr
C
Br
Ph3P
acetonitrile
PPh3
Br−
D
CH3CH2CH2CH2Li
OTHP
PPh3
E
O
H
OTHP
F
TsOH
methanol
OH
a sex pheromone of the cockroach

21.62

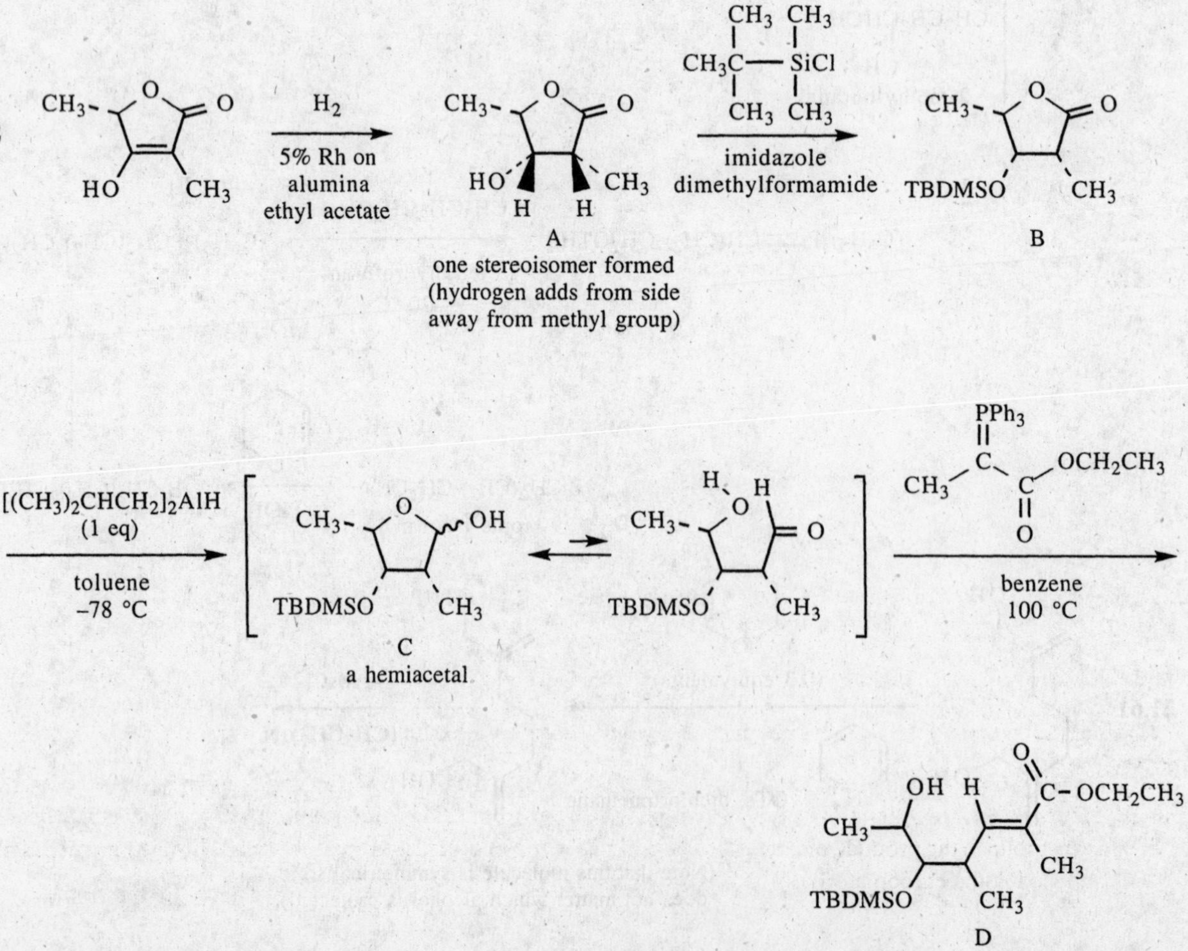
CH3
O
HO
CH3
H2
5% Rh on
alumina
ethyl acetate
A
one stereoisomer formed
(hydrogen adds from side
away from methyl group)
CH3 CH3
CH3C—SiCl
CH3 CH3
imidazole
dimethylformamide
TBDMSO
B
[(CH3)2CHCH2]2AlH
(1 eq)
toluene
−78 °C
OH
C
a hemiacetal
H
PPh3
C
OCH2CH3
benzene
100 °C
C-OCH2CH3
D

21.63

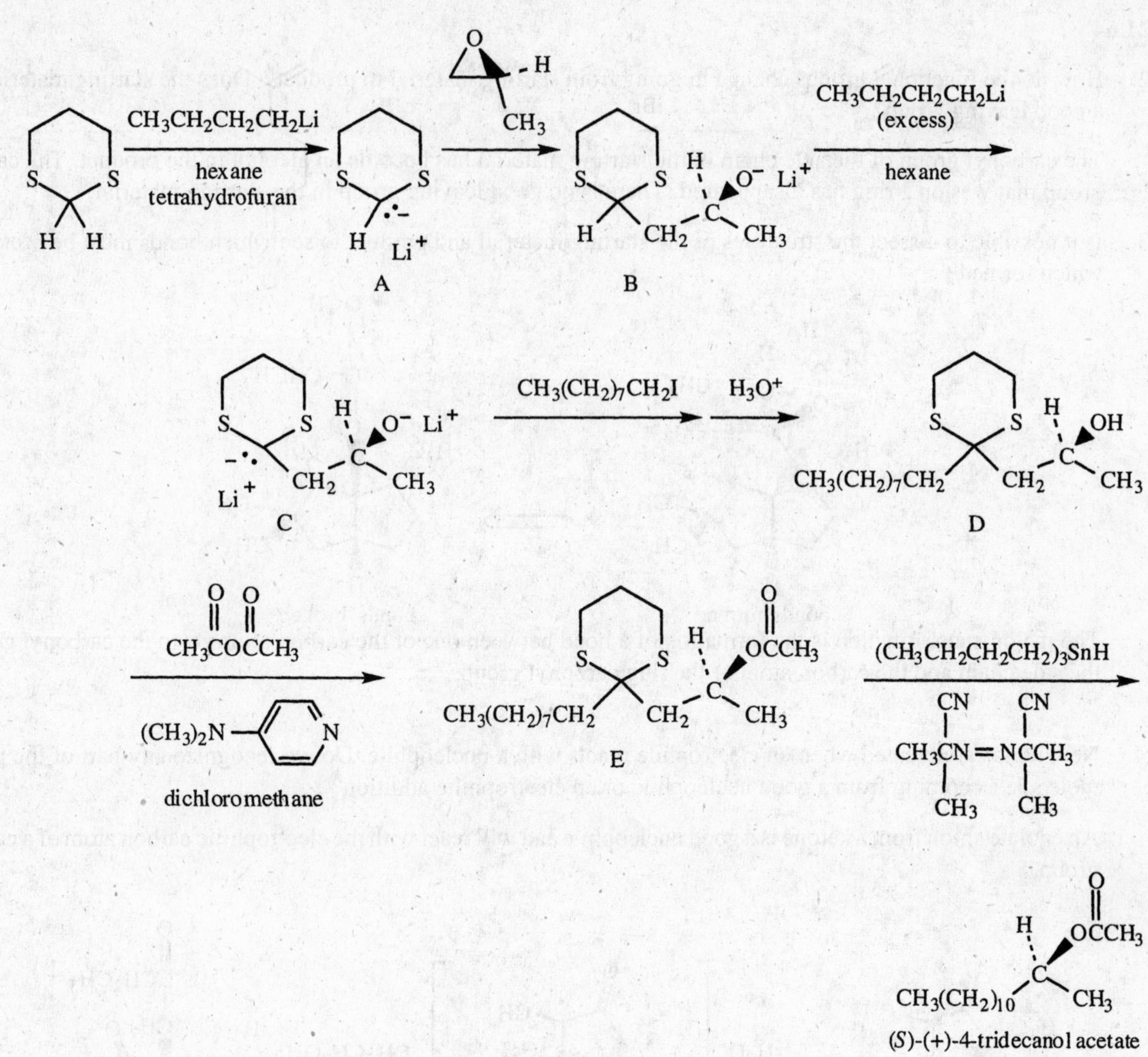

Reaction scheme:

1,3-dithiane (S—CH₂—S ring, with H H at C2) $\xrightarrow[\text{tetrahydrofuran}]{\text{hexane}}^{CH_3CH_2CH_2CH_2Li}$ **A** (dithiane anion, Li⁺)

$\xrightarrow{\text{epoxide (CH}_3\text{ substituted, with H)}}$ **B** (dithiane with CH_2—$\overset{H}{\underset{CH_3}{C}}$—O⁻ Li⁺)

$\xrightarrow[\text{hexane}]{CH_3CH_2CH_2CH_2Li \text{ (excess)}}$

C (dithiane dianion with Li⁺ and CH_2—$\overset{H}{\underset{CH_3}{C}}$—O⁻ Li⁺)

$\xrightarrow[]{CH_3(CH_2)_7CH_2I \quad H_3O^+}$ **D** $CH_3(CH_2)_7CH_2$—(dithiane)—CH_2—$\overset{H}{\underset{CH_3}{C}}$—OH

$\xrightarrow[\underset{\text{dichloromethane}}{(CH_3)_2N-\text{(pyridine)}}]{CH_3\overset{O}{C}\overset{O}{C}CH_3}$ **E** $CH_3(CH_2)_7CH_2$—(dithiane)—CH_2—$\overset{H}{\underset{CH_3}{C}}$—$O\overset{O}{C}CH_3$

$\xrightarrow[\underset{CH_3}{\overset{CN}{\underset{CH_3}{C}}}CH_3CN=NCCH_3\overset{CN}{\underset{CH_3}{}}]{(CH_3CH_2CH_2CH_2)_3SnH}$

$CH_3(CH_2)_{10}$—$\overset{H}{\underset{CH_3}{C}}$—$O\overset{O}{C}CH_3$

(*S*)-(+)-4-tridecanol acetate

21.64

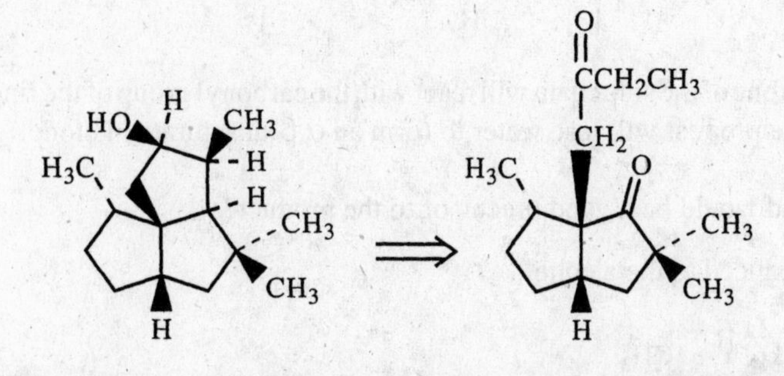

structure (left): bicyclic system with HO, H, CH₃, H, CH₃, CH₃, H₃C, H substituents $\Longrightarrow$ structure (right): bicyclic ketone with $\overset{O}{\overset{\|}{C}}CH_2CH_3$ side chain, CH₂, O, H₃C, CH₃, CH₃, H substituents

1.   What are the connectivities of the two compounds? How many carbon atoms does each contain? Are there any rings? What are the positions of branches and functional groups on the carbon skeletons?

The product has the same number of carbon atoms as the starting material. There are two five-membered rings in the reactant and three in the product. The carbonyl group of the side chain in the starting material has become an alcohol group in the product, and the side chain has been incorporated into the third ring. The new ring in the product is attached at the carbon atom of the carbonyl group in a ring of the starting material.

21.64

2. How do the functional groups change in going from starting material to product? Does the starting material have a good leaving group?

The carbonyl group of the side chain in the starting material has become an alcohol in the product. The carbonyl group that was on a ring has disappeared. There is no good leaving group in the starting material.

3. Is it possible to dissect the structures of the starting material and product to see which bonds must be broken and which formed?

bonds formed   bonds broken

The major transformation is the formation of a bond between one of the carbon atoms α to the carbonyl group of the side chain and the carbon atom of the ring carbonyl group.

4. New bonds are created when an electrophile reacts with a nucleophile. Do we recognize any part of the product molecule as coming from a good nucleophile or an electrophilic addition?

An enolate anion from a ketone is a good nucleophile and will react with the electrophilic carbon atom of a carbonyl group.

The enolate from the ketone of the side chain will react with the carbonyl group of the ring ketone in an intramolecular aldol condensation. The product will lose water to form an α,β-unsaturated ketone.

5. What type of compound would be a good precursor to the product?

A ketone that can be reduced to the alcohol.

The ketone that we need for step 5 is formed by the reduction of the α,β-unsaturated ketone we got in step 4.

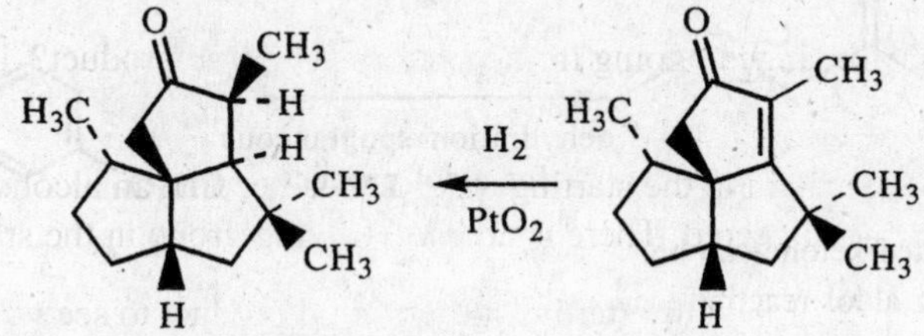

6.   After this last step, do we see how to get from starting material to product? If not, we need to analyze the structure obtained in step 5 by applying questions 4 and 5 to it.

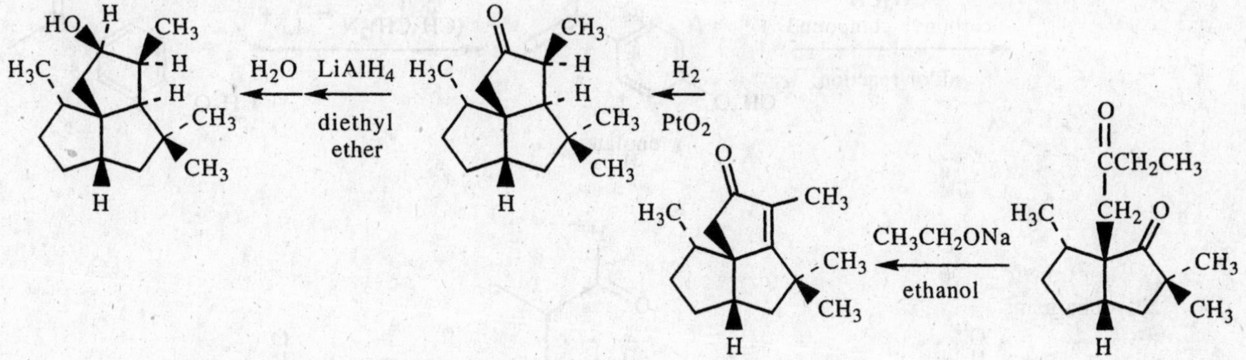

21.65

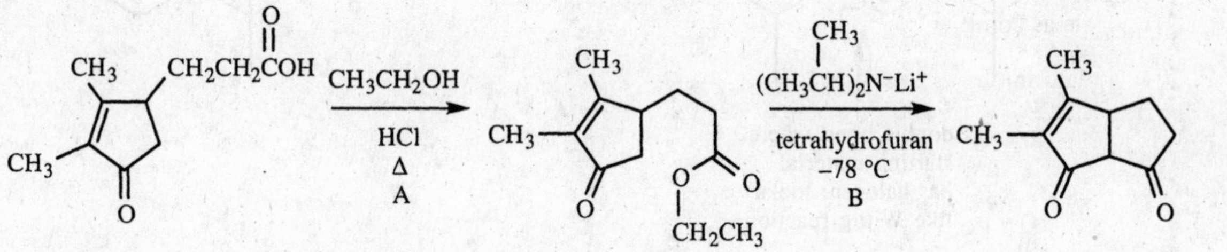

21.66   (a)

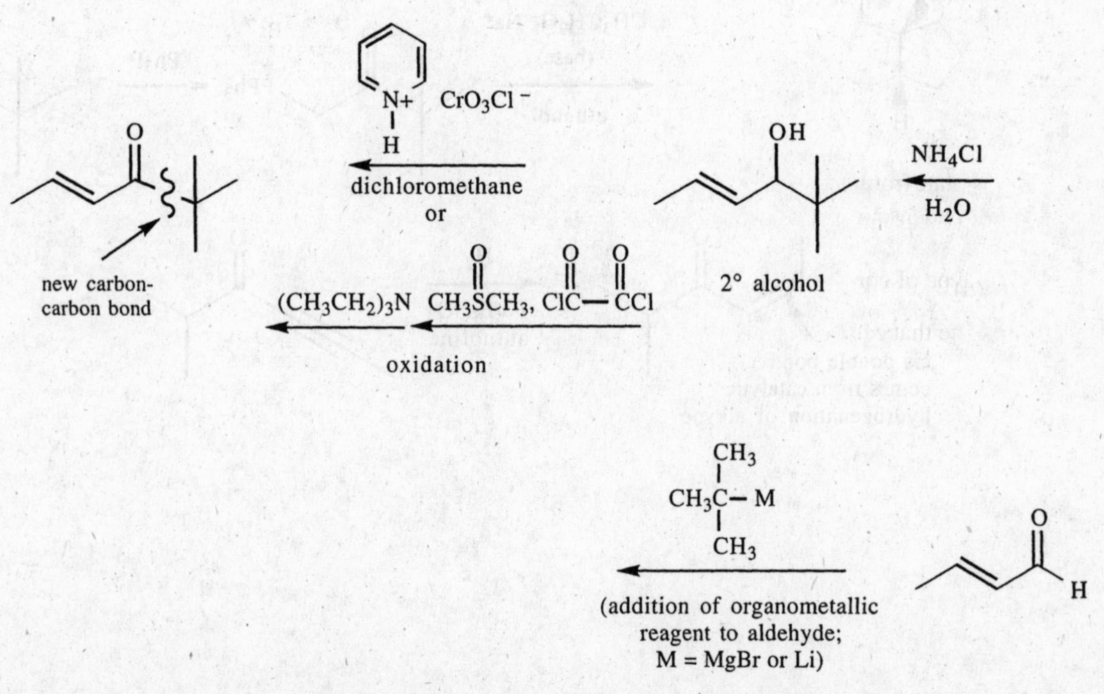

21.66   (b)

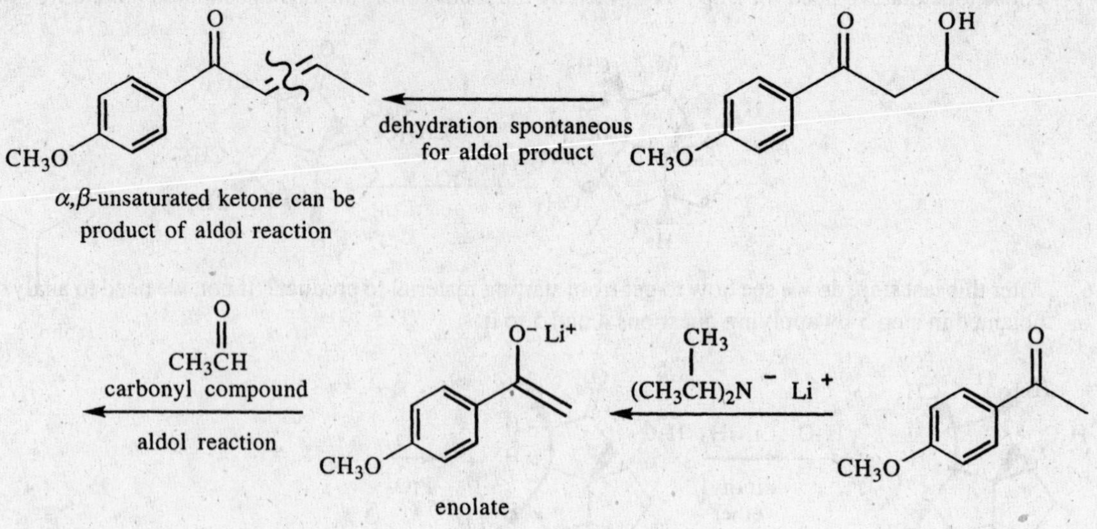

α,β-unsaturated ketone can be
product of aldol reaction

dehydration spontaneous
for aldol product

enolate

aldol reaction

carbonyl compound

(c)

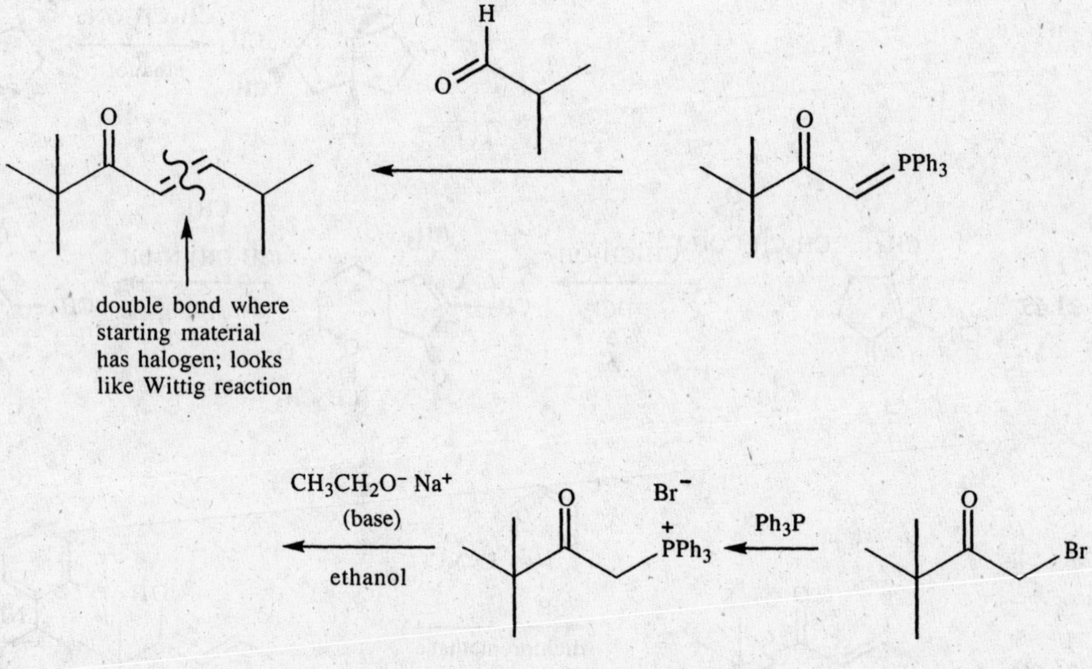

double bond where
starting material
has halogen; looks
like Wittig reaction

CH₃CH₂O⁻ Na⁺
(base)

ethanol

Ph₃P

(d)

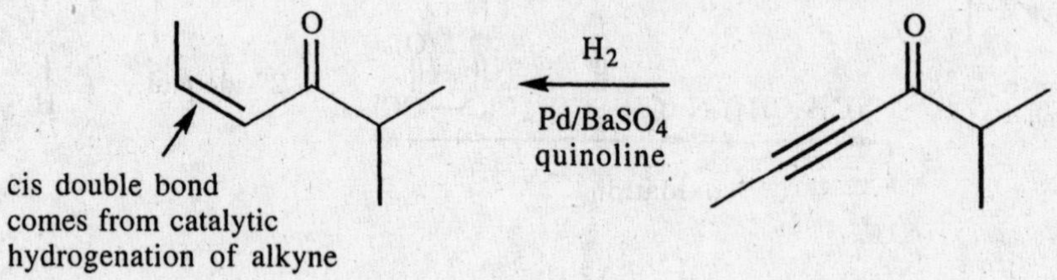

cis double bond
comes from catalytic
hydrogenation of alkyne

H₂

Pd/BaSO₄
quinoline

21.66 (d) (cont)

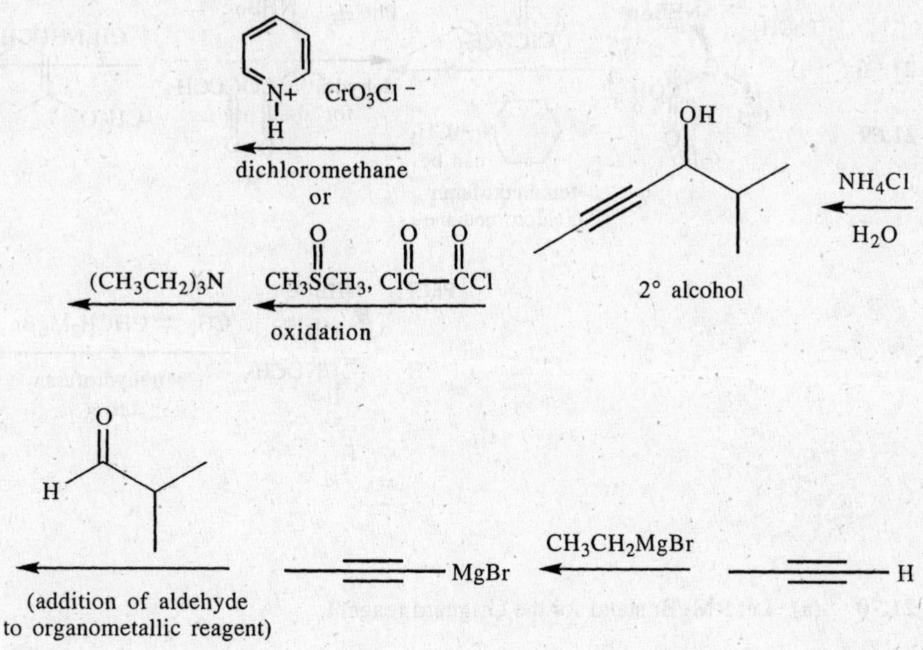

21.67

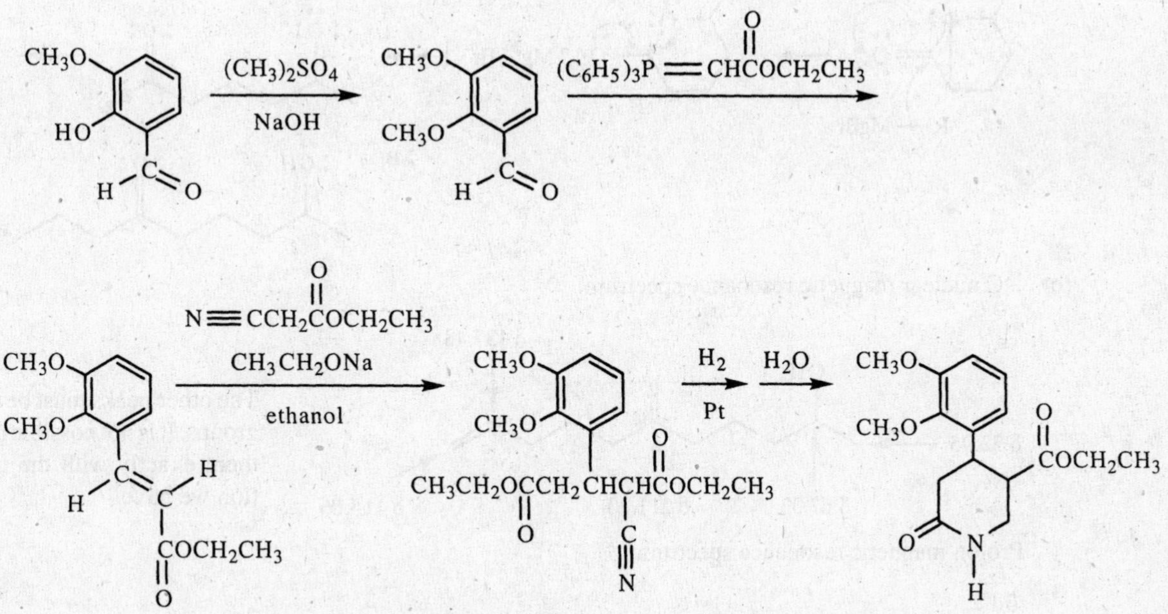

21.68

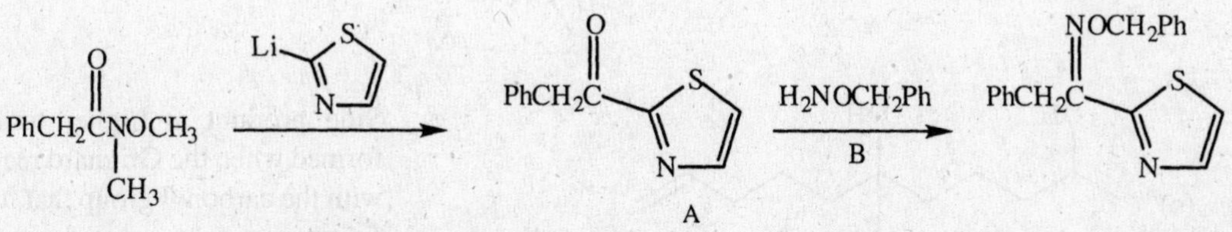

21.69

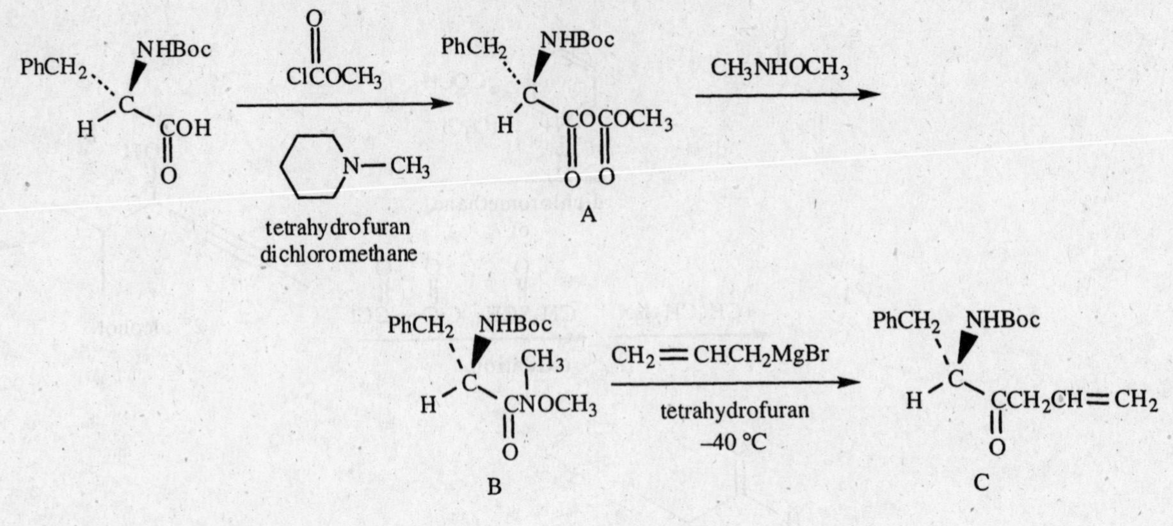

21.70   (a)   Let RMgBr stand for the Grignard reagent.

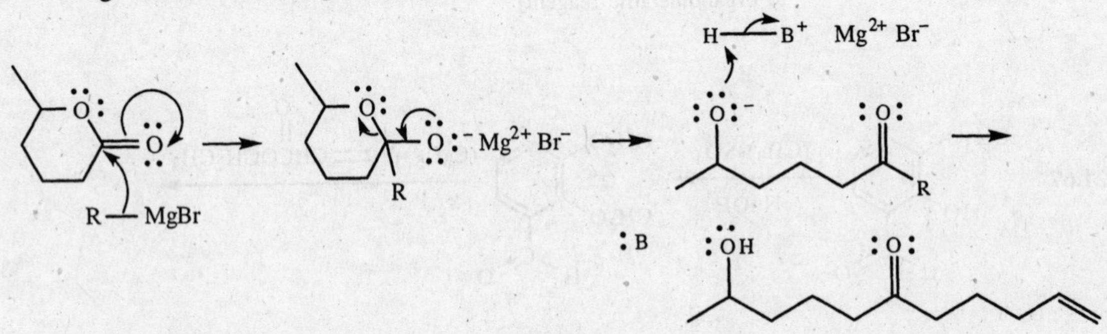

(b) ^{13}C nuclear magnetic resonance spectrum:

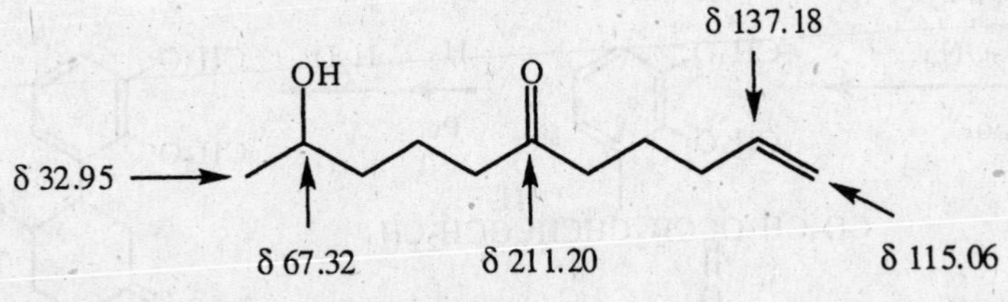

The other peaks must be the CH_2 groups. It is not possible to assign them exactly with the information we have.

Proton magnetic resonance spectrum:

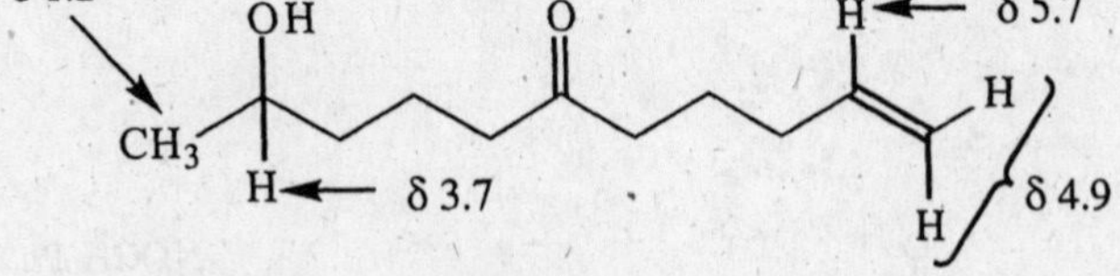

All the other protons absorb between δ 1.0 – 2.4.

(c)

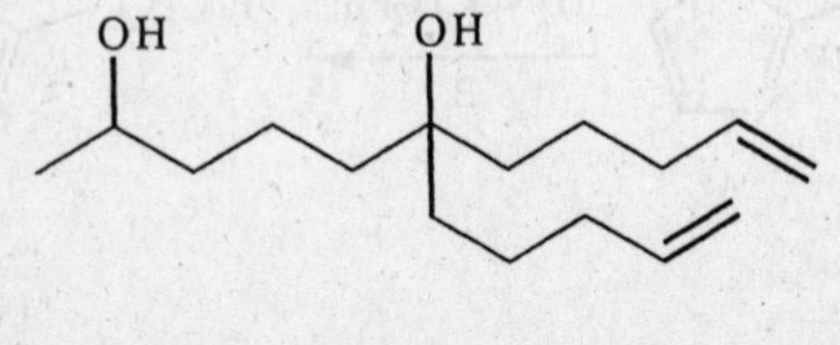

Side product at higher temperatures, formed when the Grignard reagent reacts with the carbonyl group that forms in the first part of the reaction [part (a)].

21.71   (a)

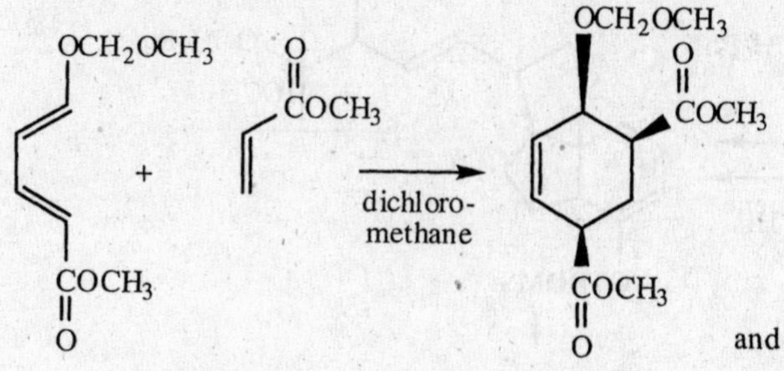

and enantiomer
A

(b)

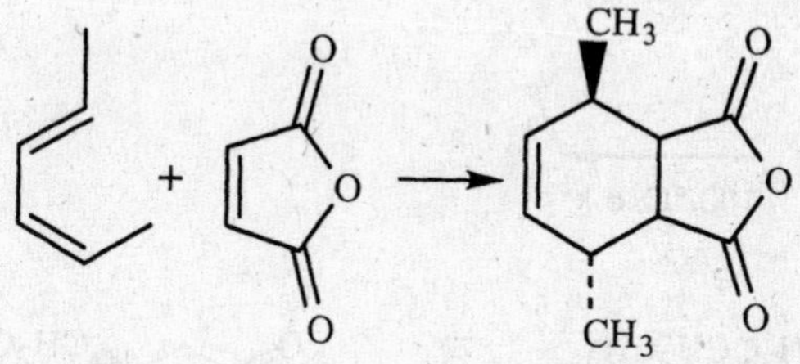

C   D   and enantiomer

(c)

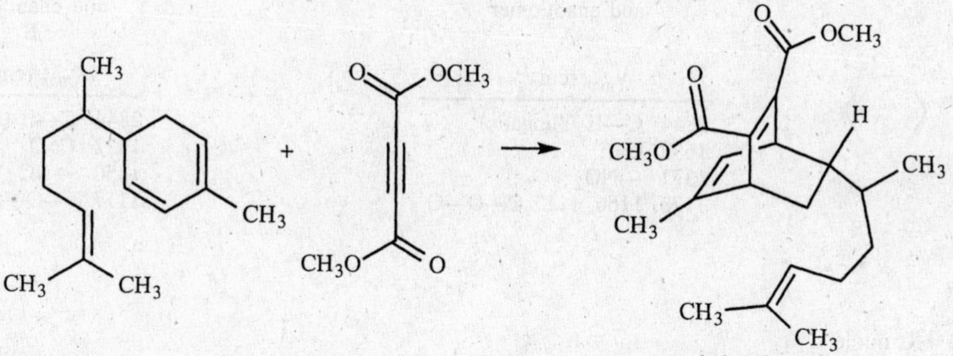

and enantiomer
E

(d)

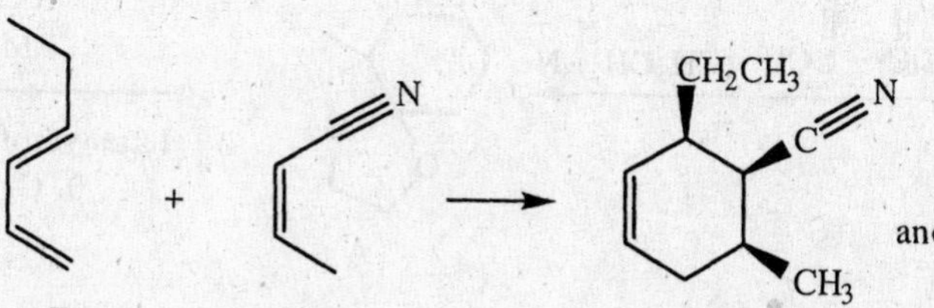

F   G   and enantiomer

(e)

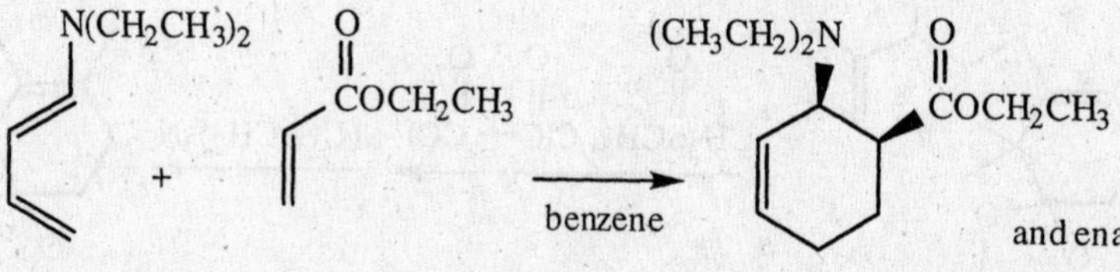

and enantiomer
H

21.71   (f)

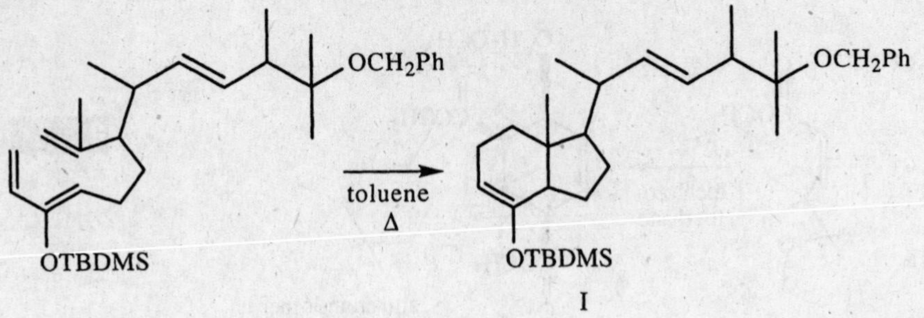

toluene
Δ

21.72

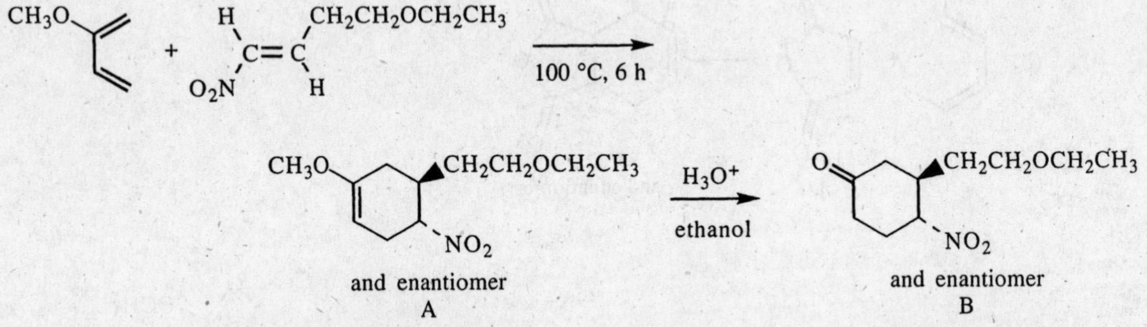

100 °C, 6 h

and enantiomer
A

ν_{max} (cm^{-1})

2844 C—H (aliphatic)
1658 C=C
1377 —NO$_2$
1178, 1166, 1117 C—O—C

and enantiomer
B

ν_{max} (cm^{-1})

2844 C—H (aliphatic)
1722 C=O
1350 —NO$_2$
1117 C—O—C

21.73

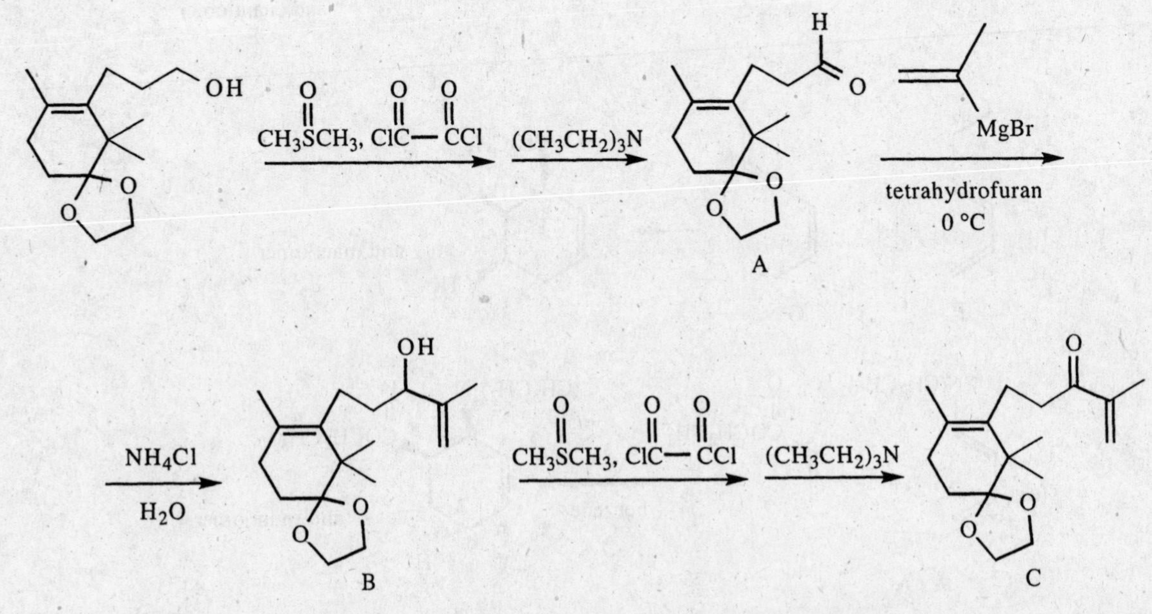

CH_3SCH_3, $ClC\!-\!CCl$ → $(CH_3CH_2)_3N$ →

A

MgBr

tetrahydrofuran
0 °C

NH_4Cl
H_2O

B

CH_3SCH_3, $ClC\!-\!CCl$ → $(CH_3CH_2)_3N$ →

C

21.73 (cont)

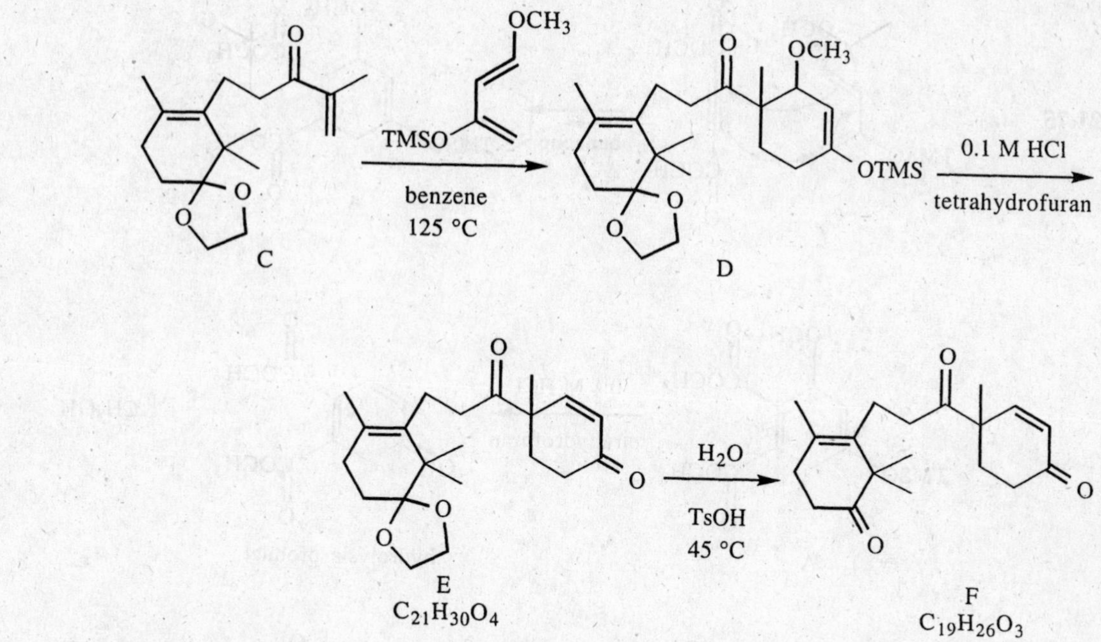

O
OCH3
TMSO
benzene
125 °C
OTMS
0.1 M HCl
tetrahydrofuran
C
D
H2O
TsOH
45 °C
E
C21H30O4
F
C19H26O3

21.74 (a)

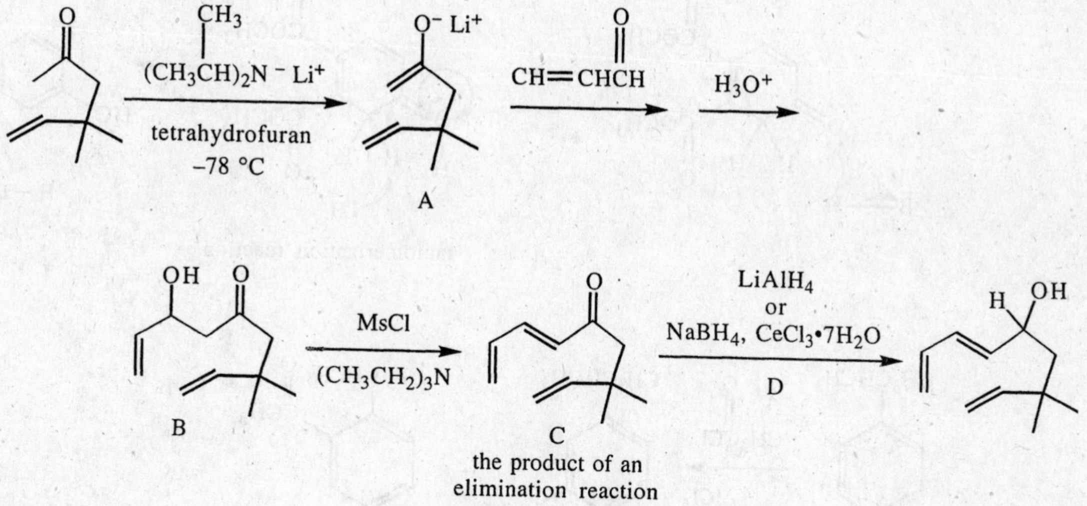

CH3
(CH3CH)2N − Li+
tetrahydrofuran
−78 °C
O− Li+
A
CH═CHCH
H3O+
OH
B
MsCl
(CH3CH2)3N
C
the product of an
elimination reaction
LiAlH4
or
NaBH4, CeCl3•7H2O
D
H
OH

(b)

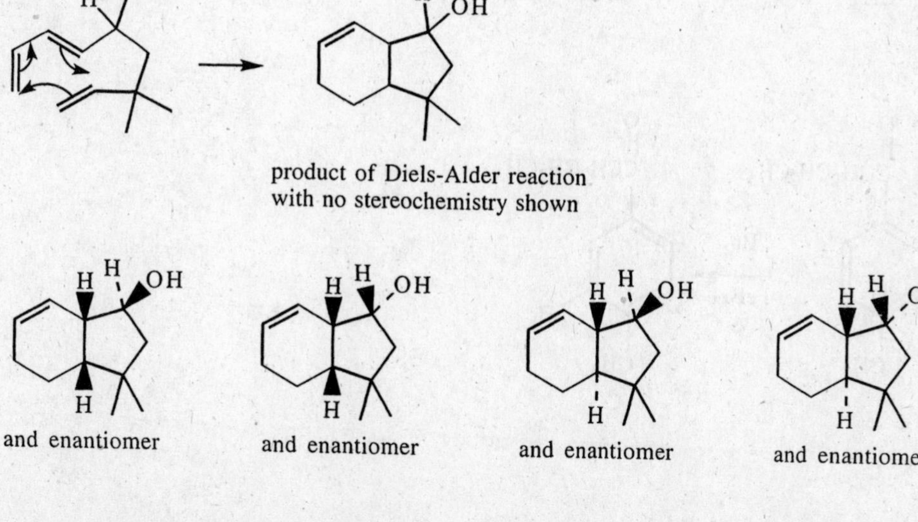

H
OH
product of Diels-Alder reaction
with no stereochemistry shown
and enantiomer
and enantiomer
and enantiomer
and enantiomer

21.75 (a)

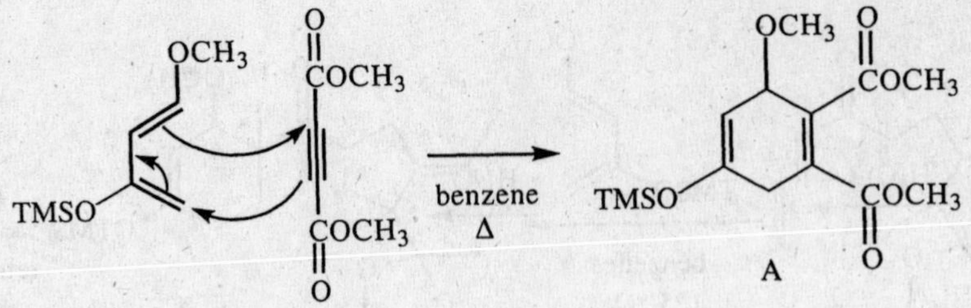

(b)

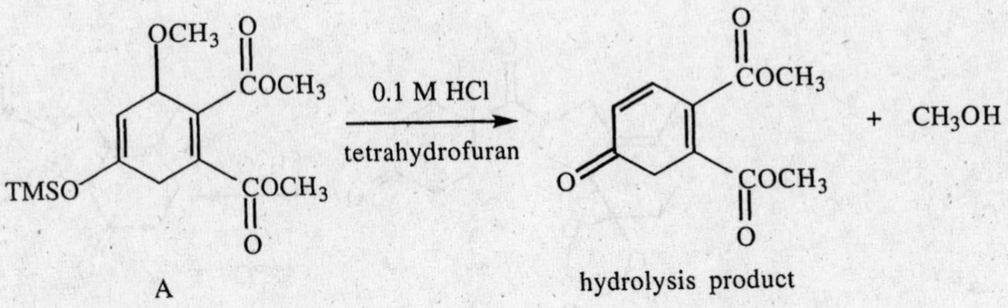

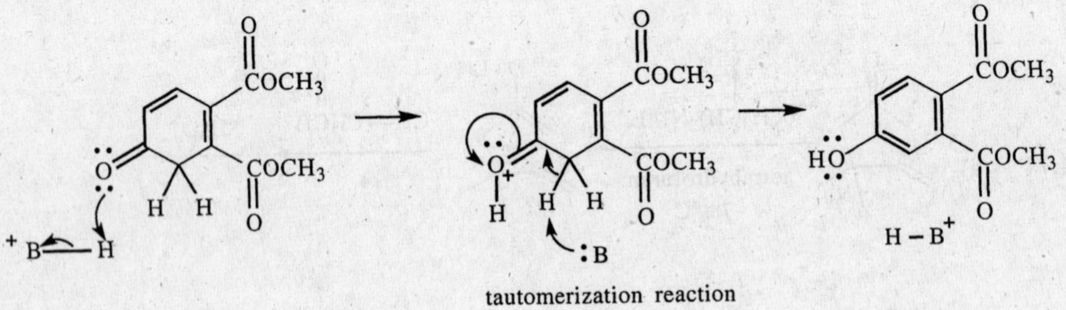

A   hydrolysis product

tautomerization reaction

21.76 (a)

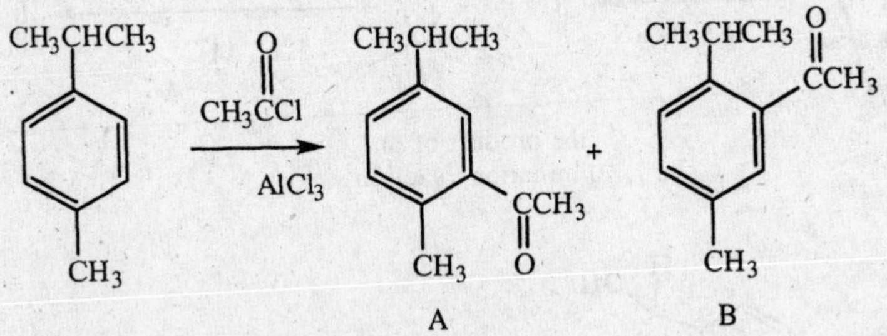

A   B

(b)

C

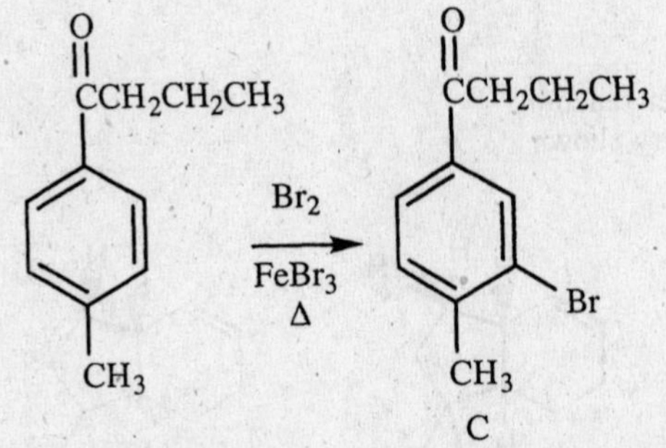

21.76 (c)

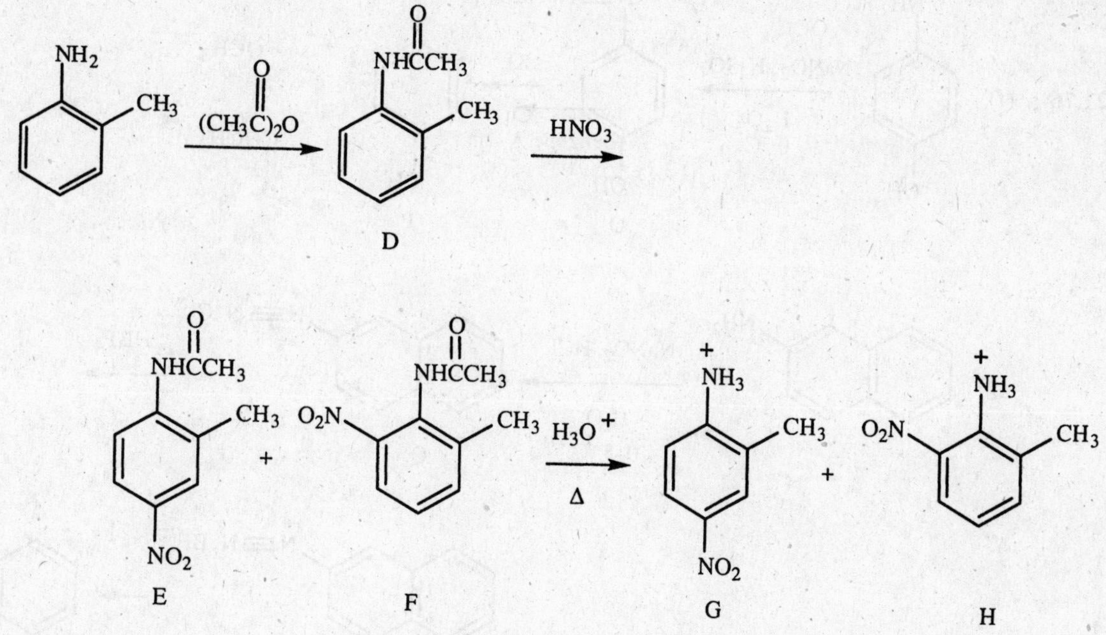
NH2
CH3
(CH3C)2O
NHCCH3
CH3
HNO3
D
NHCCH3
CH3
O2N
NHCCH3
CH3
H3O+
Δ
NH3
CH3
O2N
NH3
CH3
NO2
NO2
E
F
G
H

(d)

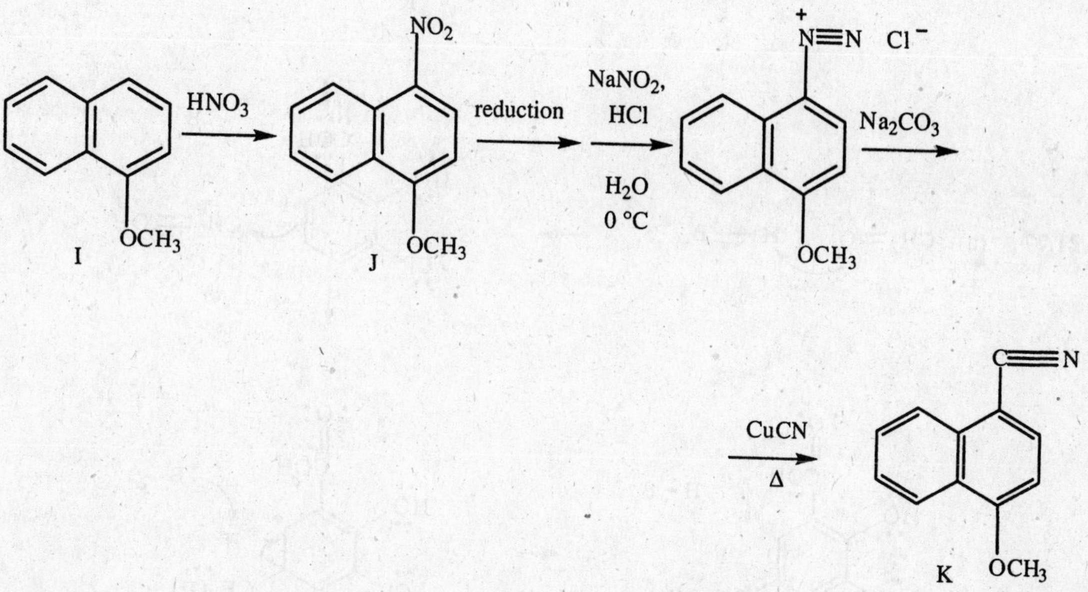
NO2
HNO3
reduction
NaNO2,
HCl
H2O
0 °C
N≡N Cl−
Na2CO3
OCH3
OCH3
OCH3
I
J
CuCN
Δ
C≡N
K
OCH3

(e)

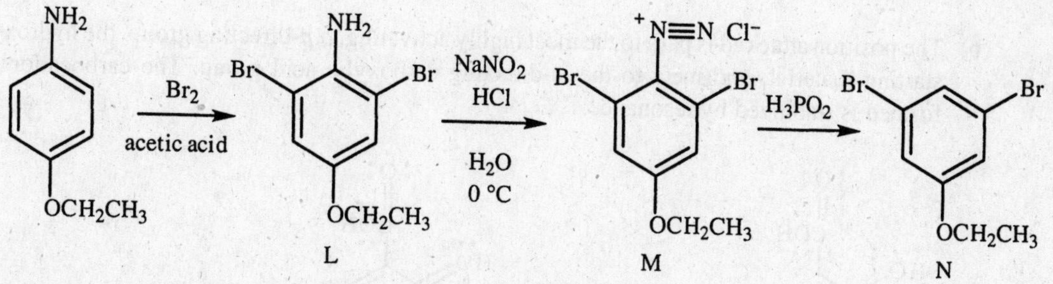
NH2
Br2
acetic acid
NH2
Br
Br
NaNO2
HCl
H2O
0 °C
N≡N Cl−
Br
Br
H3PO2
Br
Br
OCH2CH3
OCH2CH3
OCH2CH3
OCH2CH3
L
M
N

21.76 (f)

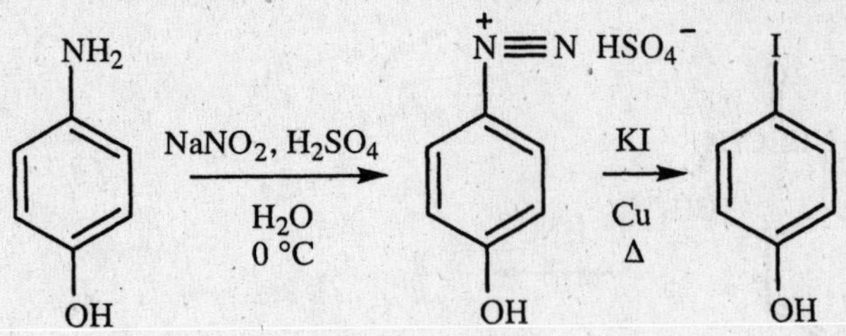

(g)

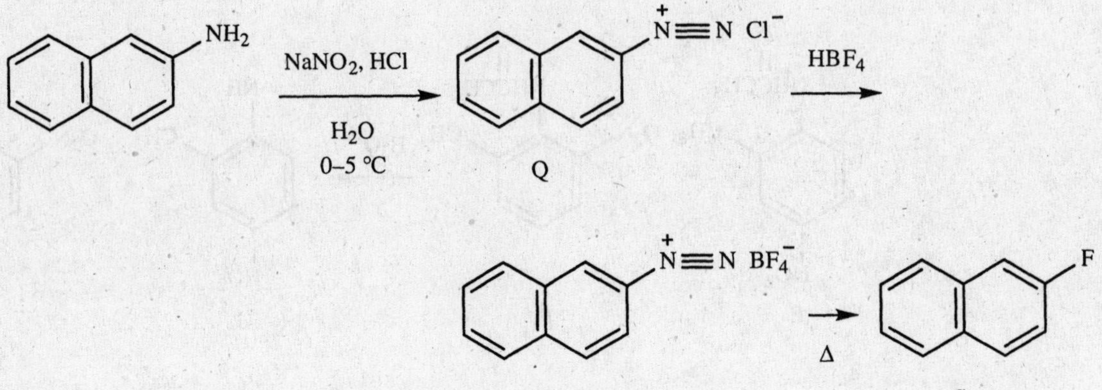

21.77 (a)

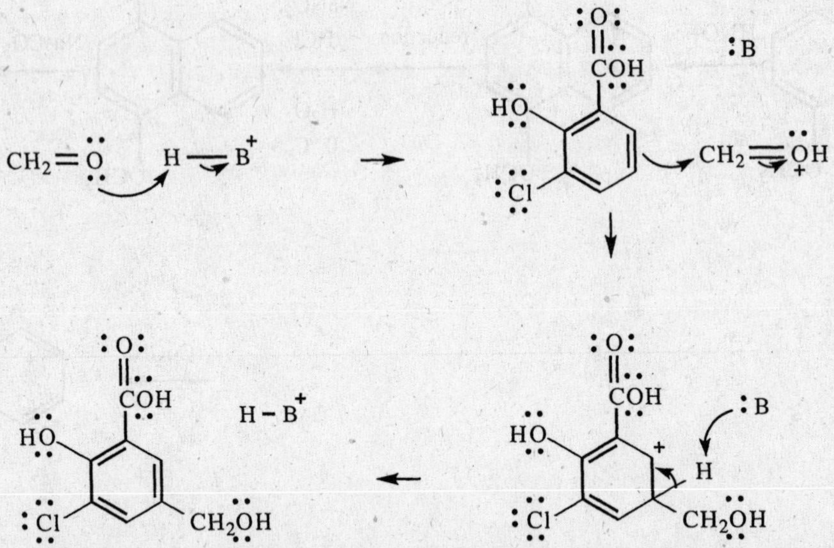

(b) The position attacked is para to the most highly activating, *o,p*-directing group, the hydroxyl group on the starting material, and meta to the *m*-directing carboxylic acid group. The carbocationic intermediate formed is stabilized by resonance.

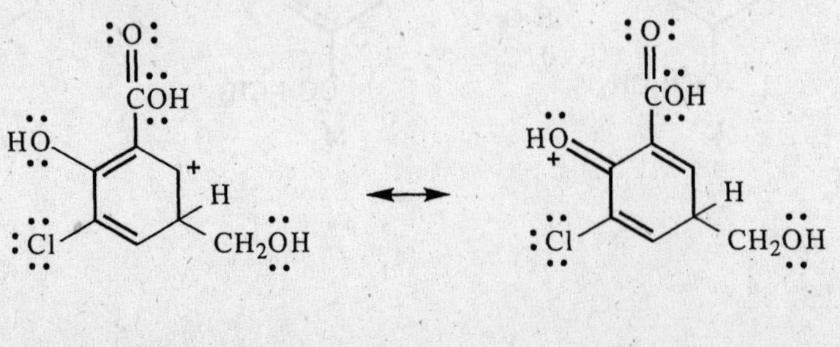

21.77   (c)

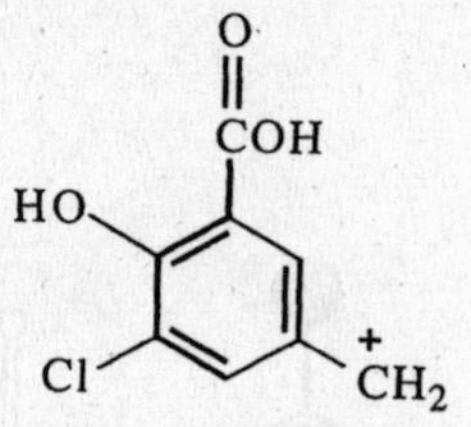

benzylic carbocation

21.78

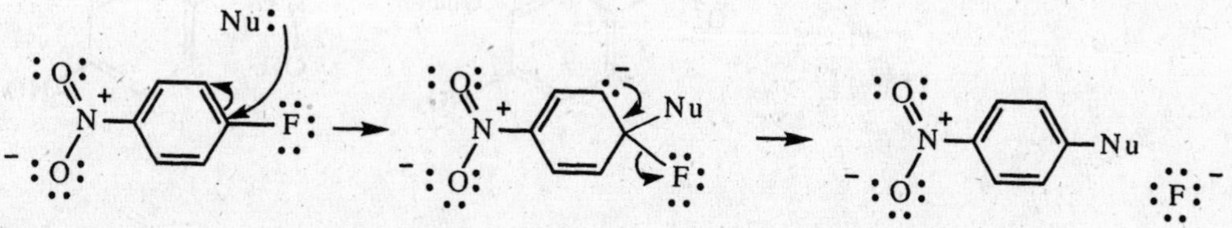

The reaction is a nucleophilic aromatic substitution reaction. The relative rates of reaction for the different nucleophiles reflect the relative nucleophilicities of the different species.

Methoxide and thiophenolate ions are the best nucleophiles. Methoxide ion is a good base, meaning that it is willing to share its electrons. The thiophenolate ion is a good nucleophile because of the polarizability of the sulfur atom. Aniline is the weakest nucleophile. It bears no negative charge, and its nonbonding electrons are delocalized to the aromatic ring by resonance and therefore not easily available for bond formation.

21.79   (a)

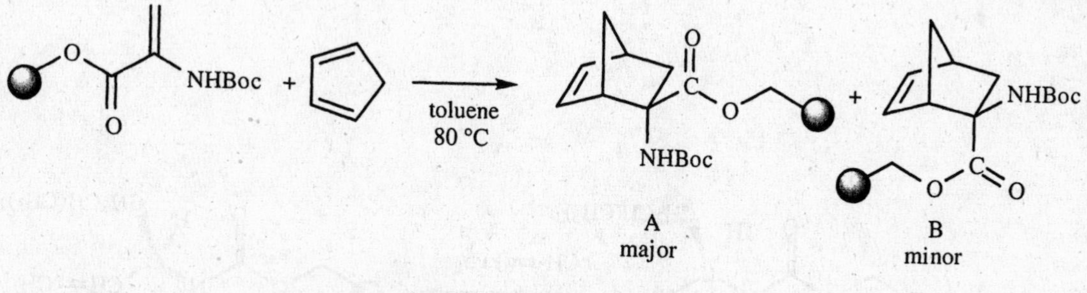

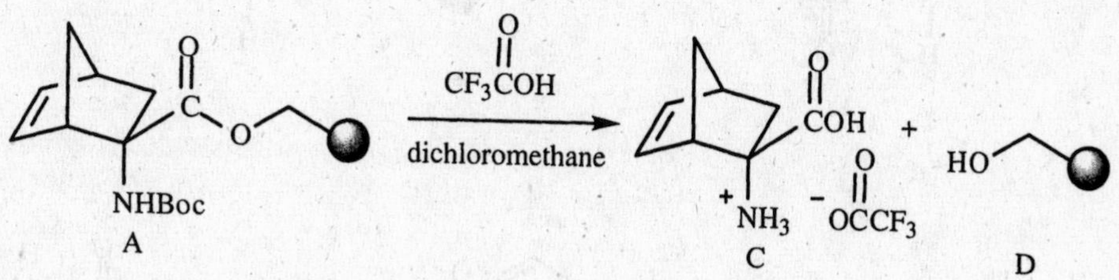

A
major

B
minor

21.79

(b)

(c)

21.79 (d) 1.

2.

(e)

21.80

Supplemental Problems

S21.1   Give structural formulas for the reagents, intermediates, and products symbolized by letters in the following equations.

(a)

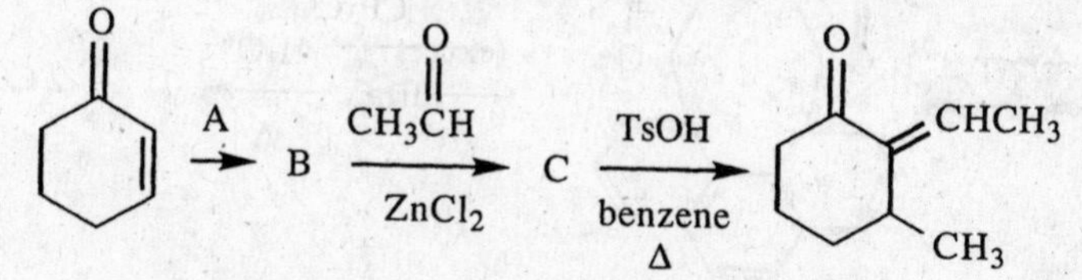

(b)

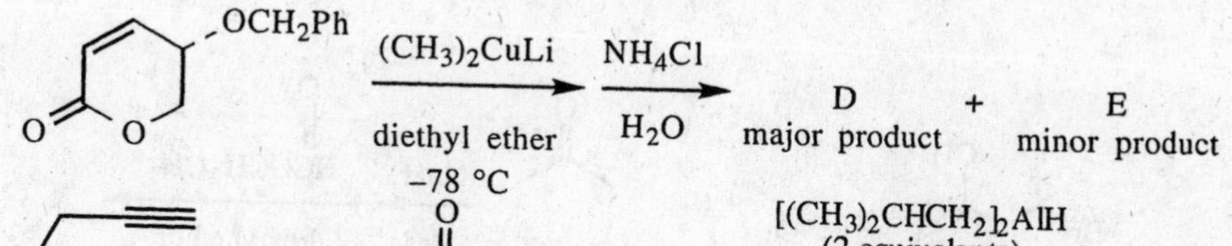

(c)

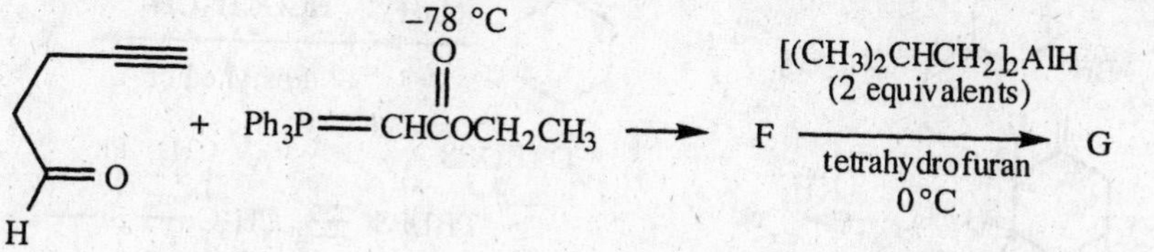

(d)

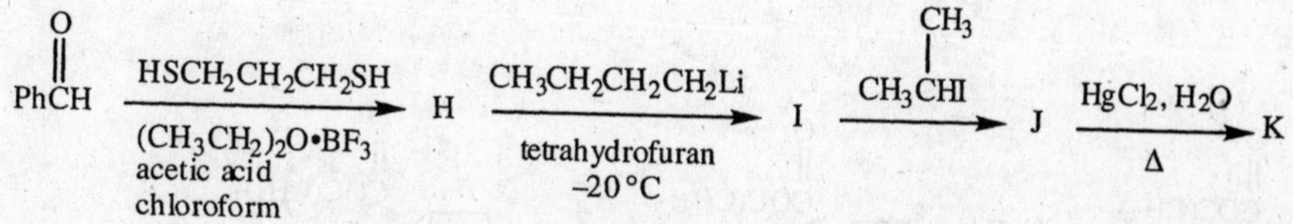

(e)

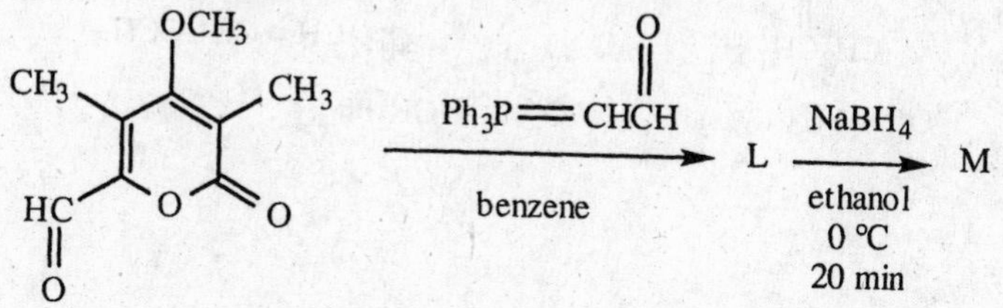

(f)

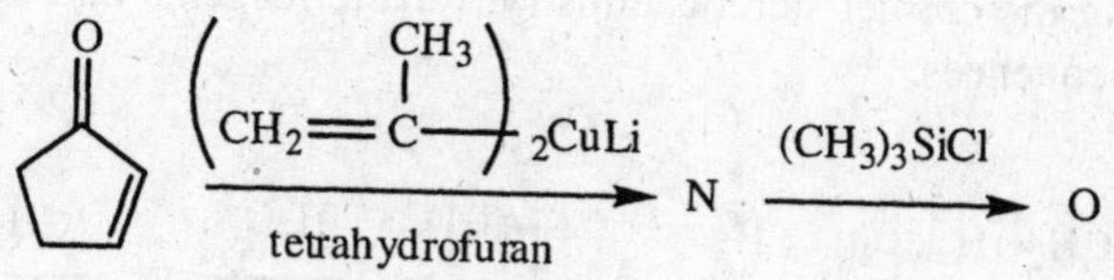

(g)

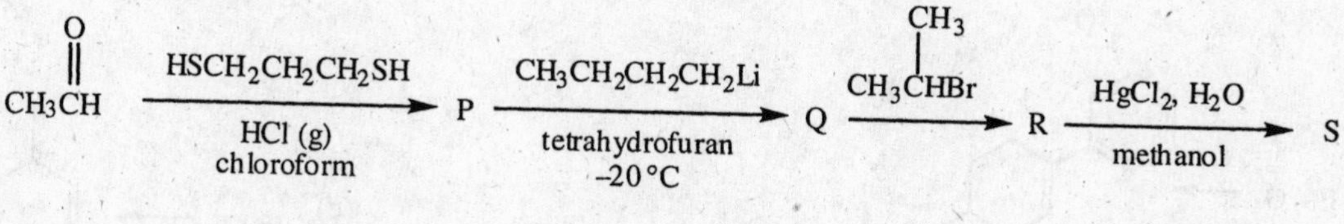

S21.1 (cont)

(h)

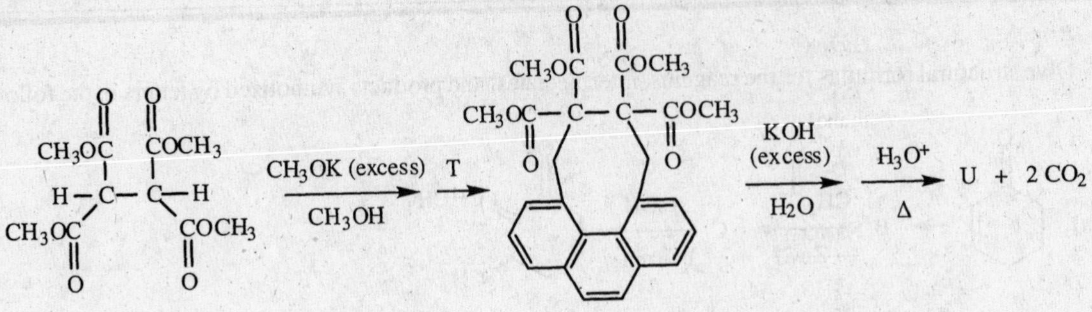

(i)

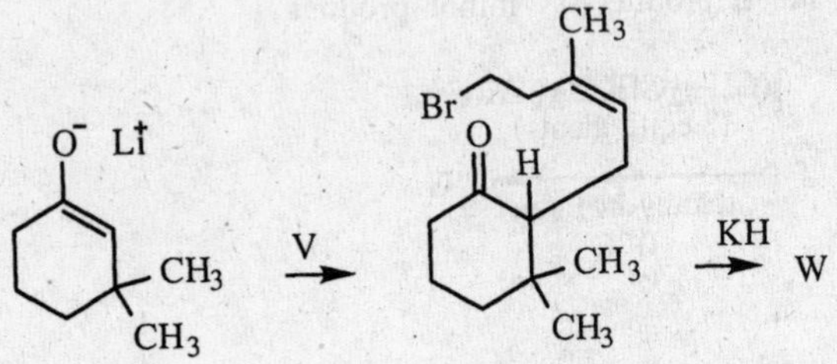

(j)

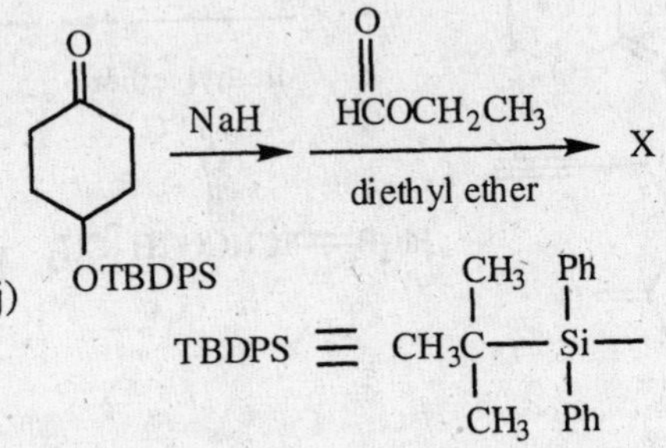

(k)

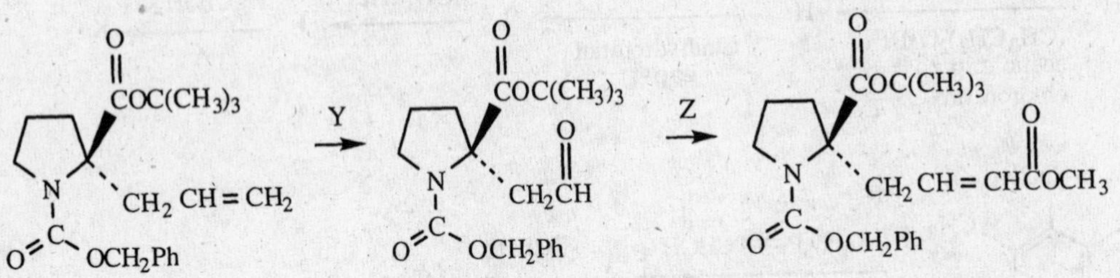

S21.2   Give the structural formulas, including the correct stereochemistry, for the reagents and major products represented by letters in the following synthetic sequences.

(a)

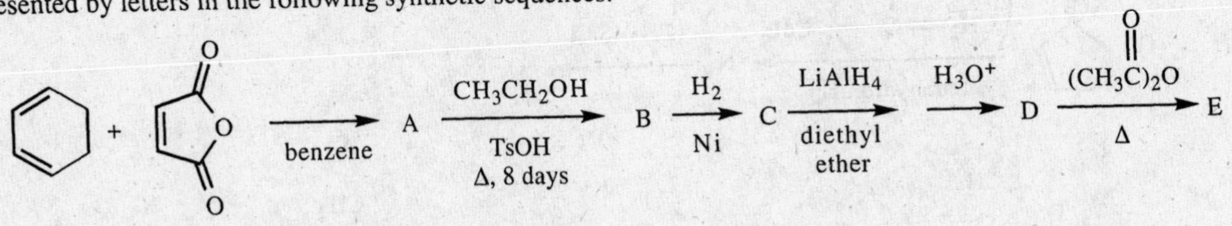

(b)

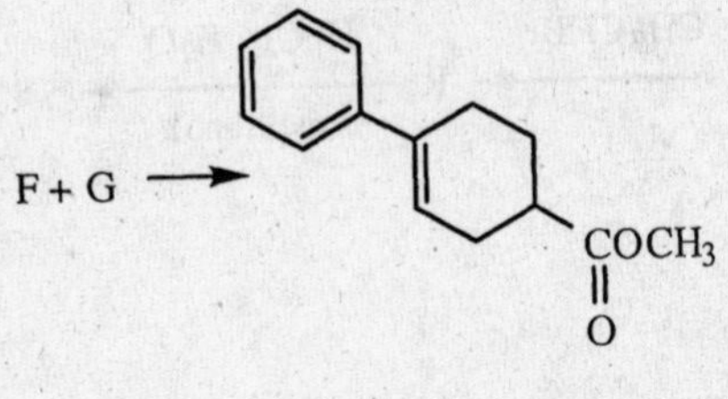

(c)

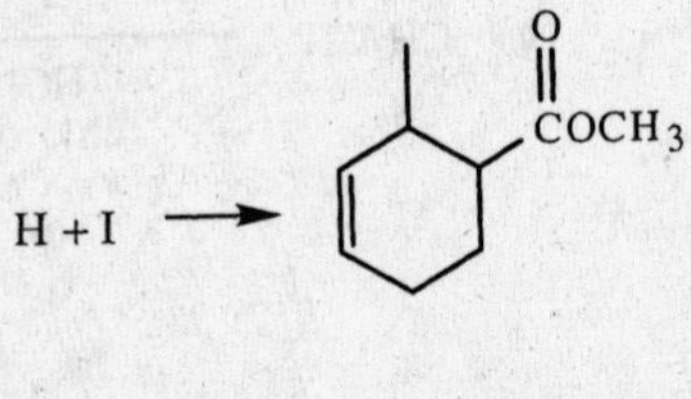

S21.2 (cont)

(d)

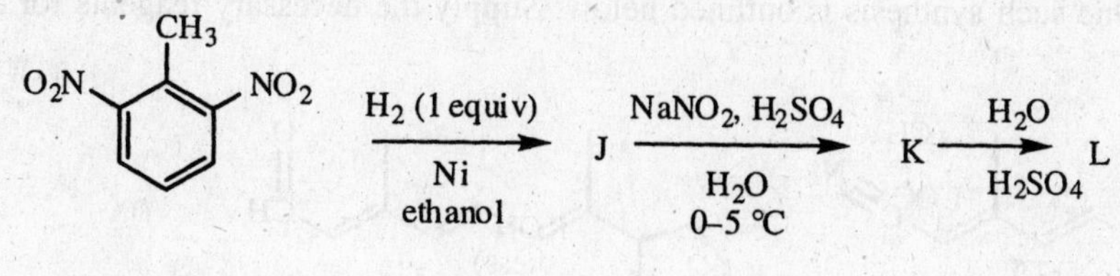

$$O_2N \quad CH_3 \quad NO_2 \xrightarrow[\substack{Ni \\ ethanol}]{H_2 \ (1 \ equiv)} J \xrightarrow[\substack{H_2O \\ 0-5 \ °C}]{NaNO_2, H_2SO_4} K \xrightarrow[\substack{H_2SO_4}]{H_2O} L$$

(e)

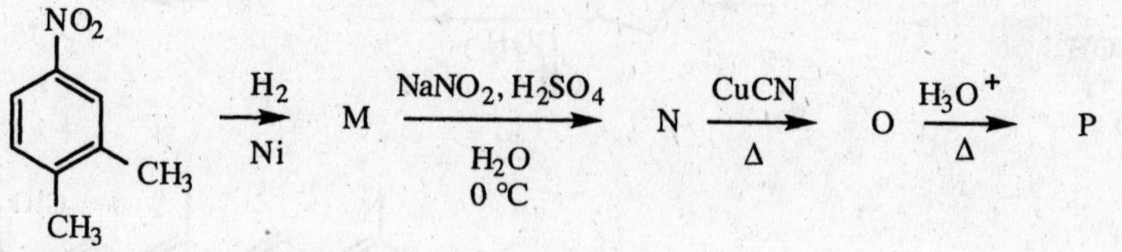

$$\xrightarrow[Ni]{H_2} M \xrightarrow[\substack{H_2O \\ 0 \ °C}]{NaNO_2, H_2SO_4} N \xrightarrow[\Delta]{CuCN} O \xrightarrow[\Delta]{H_3O^+} P$$

(f)

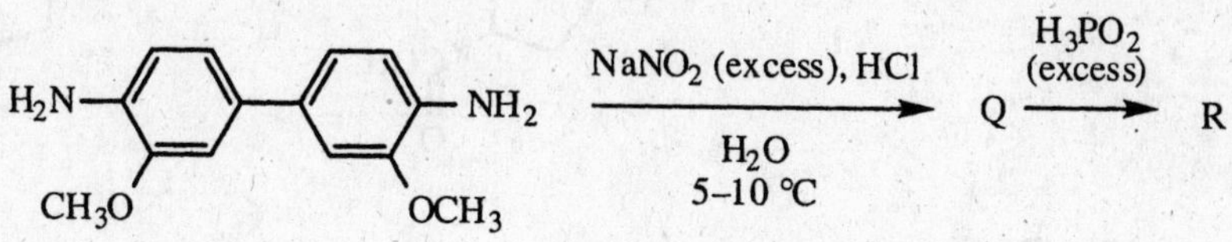

$$H_2N \cdots NH_2, \ CH_3O \quad OCH_3 \xrightarrow[\substack{H_2O \\ 5-10 \ °C}]{NaNO_2 \ (excess), \ HCl} Q \xrightarrow[]{\substack{H_3PO_2 \\ (excess)}} R$$

(g)

$$O_2N \quad NO_2 \quad Cl \quad NO_2 + H_2N-N \xrightarrow[chloroform]{} S$$

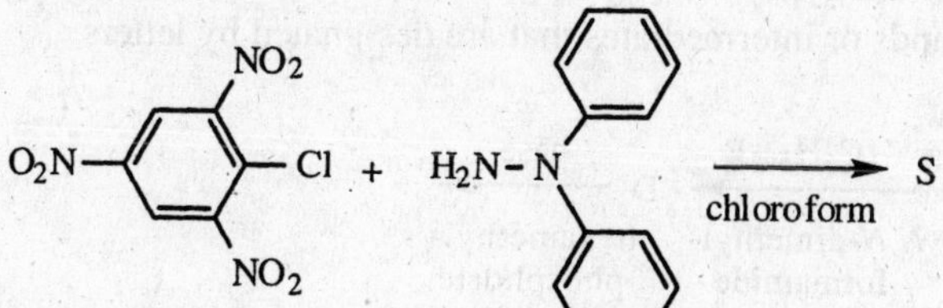

S21.3   Tell what reagents could be used to carry out the following transformations. More than one step may be required.

(a) 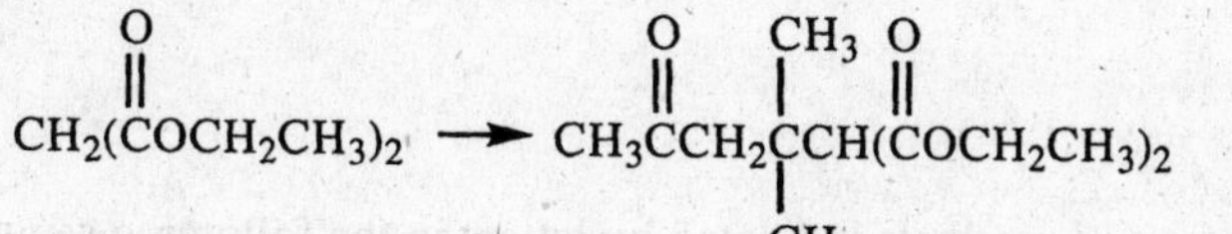 $CH_2(COCH_2CH_3)_2 \longrightarrow CH_3CCH_2CCH(COCH_2CH_3)_2$
         with CH_3 groups

(b) 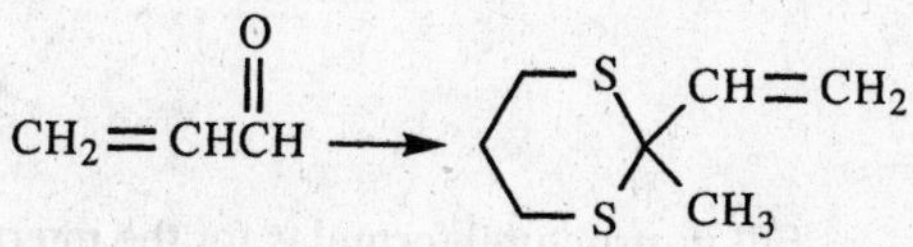 $CH_2=CHCH \longrightarrow$

(c)

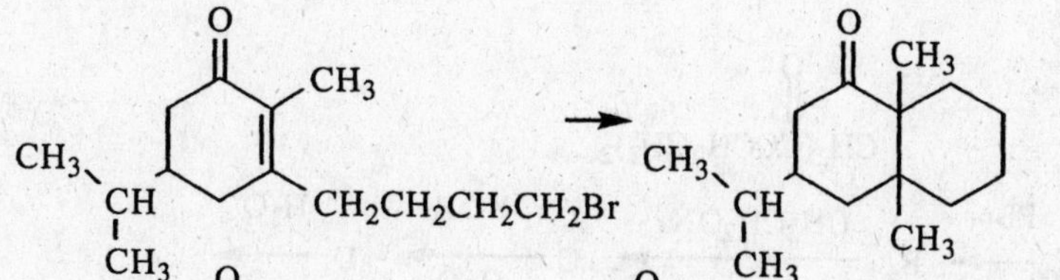

(d)

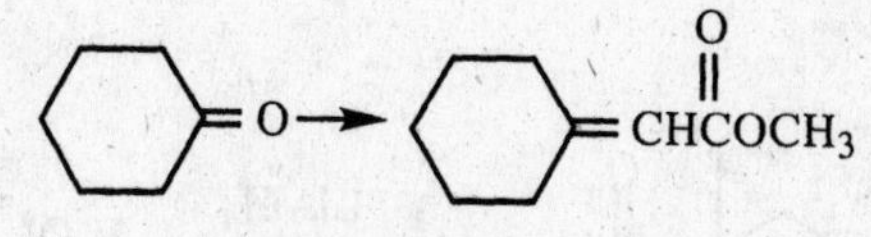

(e) 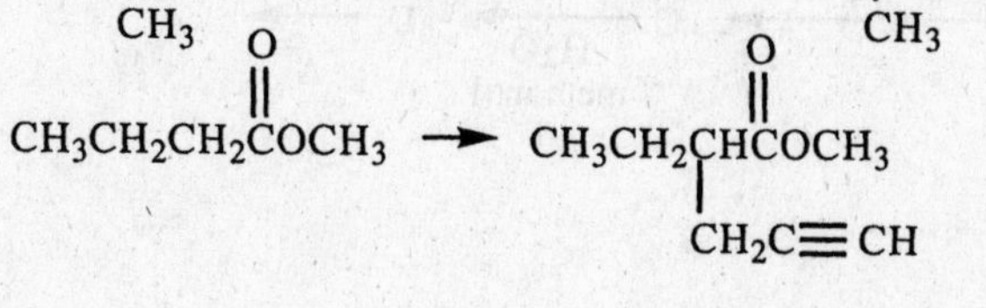 $CH_3CH_2CH_2COCH_3 \longrightarrow CH_3CH_2CHCOCH_3$
                                          $CH_2C\equiv CH$

(f)

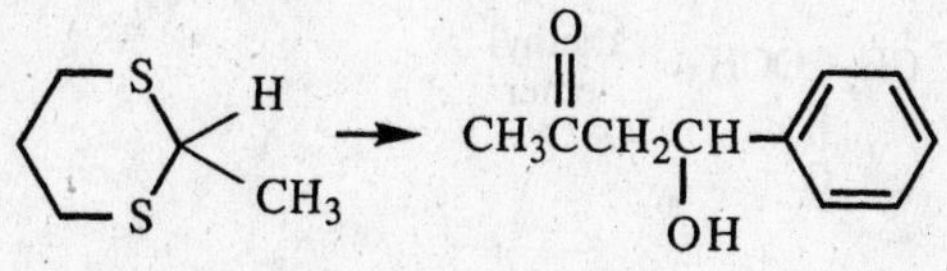

S21.4   Much current research is focused on the chemistry of vision. Compounds similar to rhodopsin (p. 758 in the text) have been synthesized for study. One such synthesis is outlined below. Supply the necessary reagents for the conversions shown.

S21.5   A synthesis of disparlure, the sex pheromone of the gypsy moth (p. 748 in the text) was carried out in the following way. Provide structural formulas for all compounds or intermediates that are designated by letters.

The Wittig reaction in hexamethylphosphoric triamide (p. 254 in the text) gives mostly the Z alkene.

S21.6   Fill in structural formulas for the intermediates and products indicated by letters in the following sequence of reactions.

S21.7   Write mechanisms that rationalize the following experimental observations.

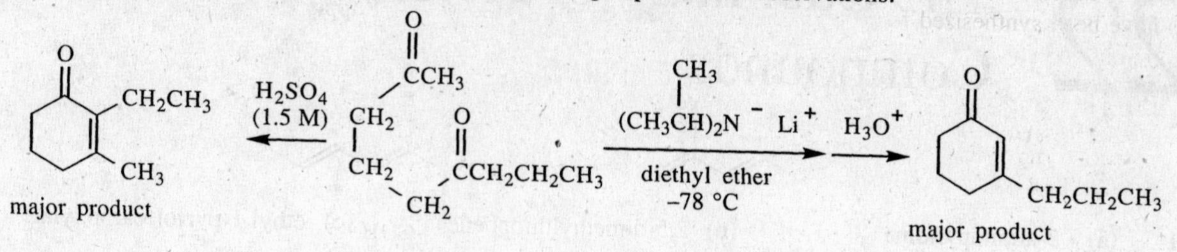

S21.8   Write a complete mechanism for the following reaction.

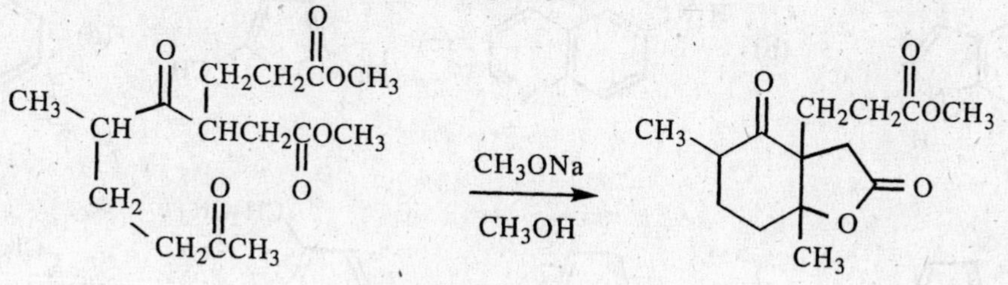

22 The Chemistry of Heterocyclic Compounds

22.1 (a) 3-aminopyridine   (b) 2,5-dimethylthiophene   (c) ethyl 1-pyrrolecarboxylate

    (d) 4-ethylisoquinoline

    (e) 3-methyl-2-isoxazoline (Note that the number 2 defines the position of the double bond.)

    (f) 4,4-dimethylazetidinone   (g) 3-indolecarboxylic acid   (h) 3-methylfuran

    (i) 8-hydroxyquinoline   (j) 1,4-dimethylpyrazole   (k) 2,3-diphenyloxazole

22.2 (a) ... (l)

Concept Map 22.1   Classification of cyclic compounds.

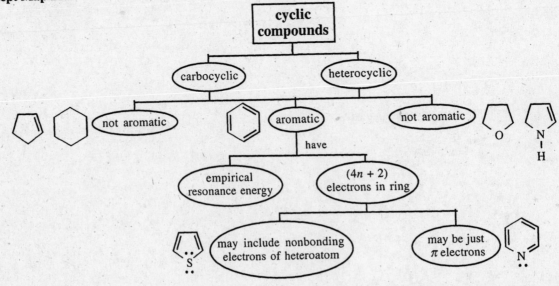

550

22.3

(resonance structures of furan)

major

(resonance structures of thiophene)

major

(resonance structures of pyrrole)

major

22.4

no resonance stabilization;
localized charge.

delocalization of charge by resonance

22.5 (a) The basicity of an amine is determined by the availability of the nonbonding electrons on the nitrogen atom. The pK_a values for their conjugate acids indicate that 2-methylpyridine is a slightly stronger base than pyridine, and 3-nitropyridine is a much weaker base than pyridine. The methyl group is weakly electron-donating; the nitro group is strongly electron-withdrawing.

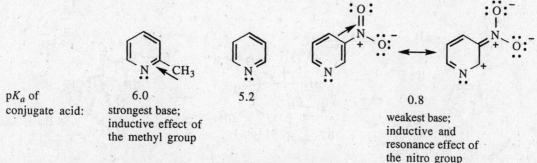

pK_a of conjugate acid:	6.0	5.2	0.8
	strongest base; inductive effect of the methyl group		weakest base; inductive and resonance effect of the nitro group

(b) The arguments used in part (a) of this problem can be applied to the compounds in this series also.

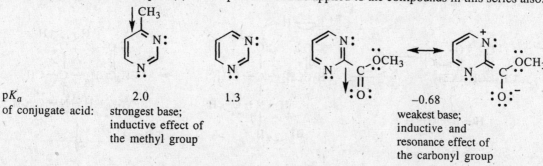

pK_a of conjugate acid:	2.0	1.3	−0.68
	strongest base; inductive effect of the methyl group		weakest base; inductive and resonance effect of the carbonyl group

22.6

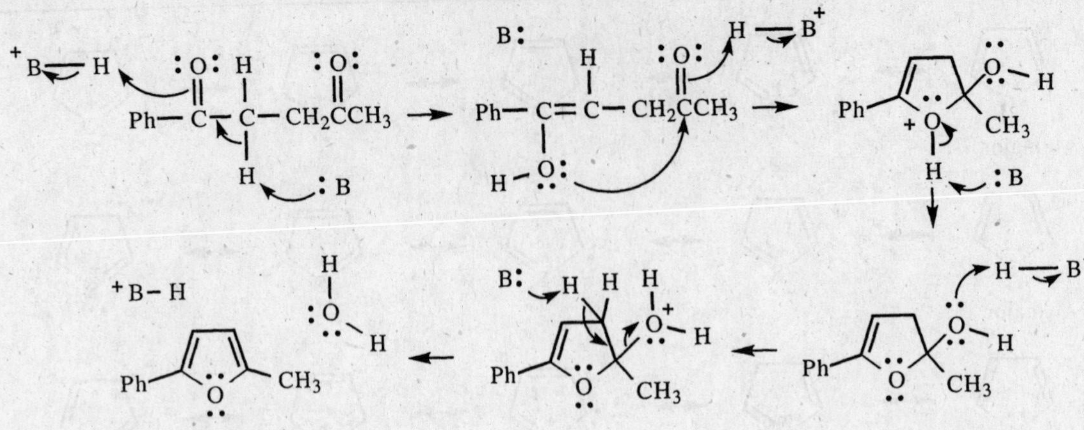

22.7 (a)

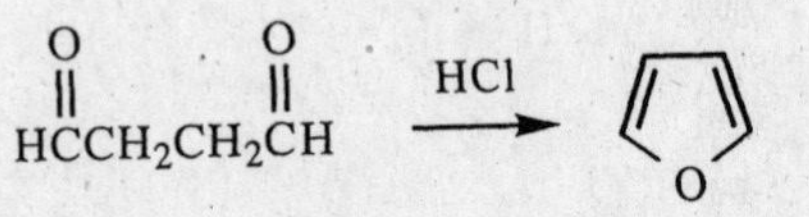

(b)

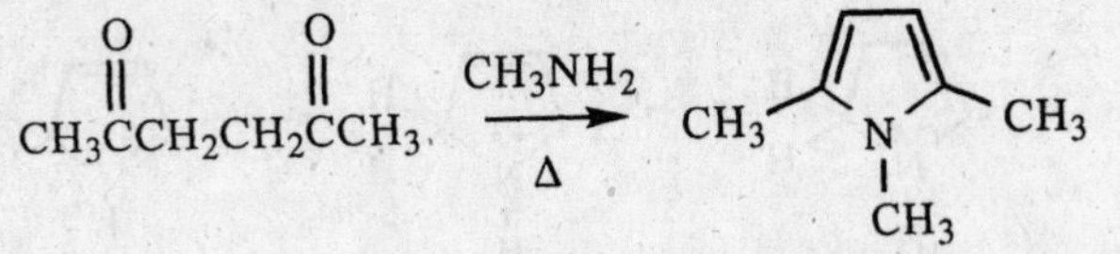

22.8

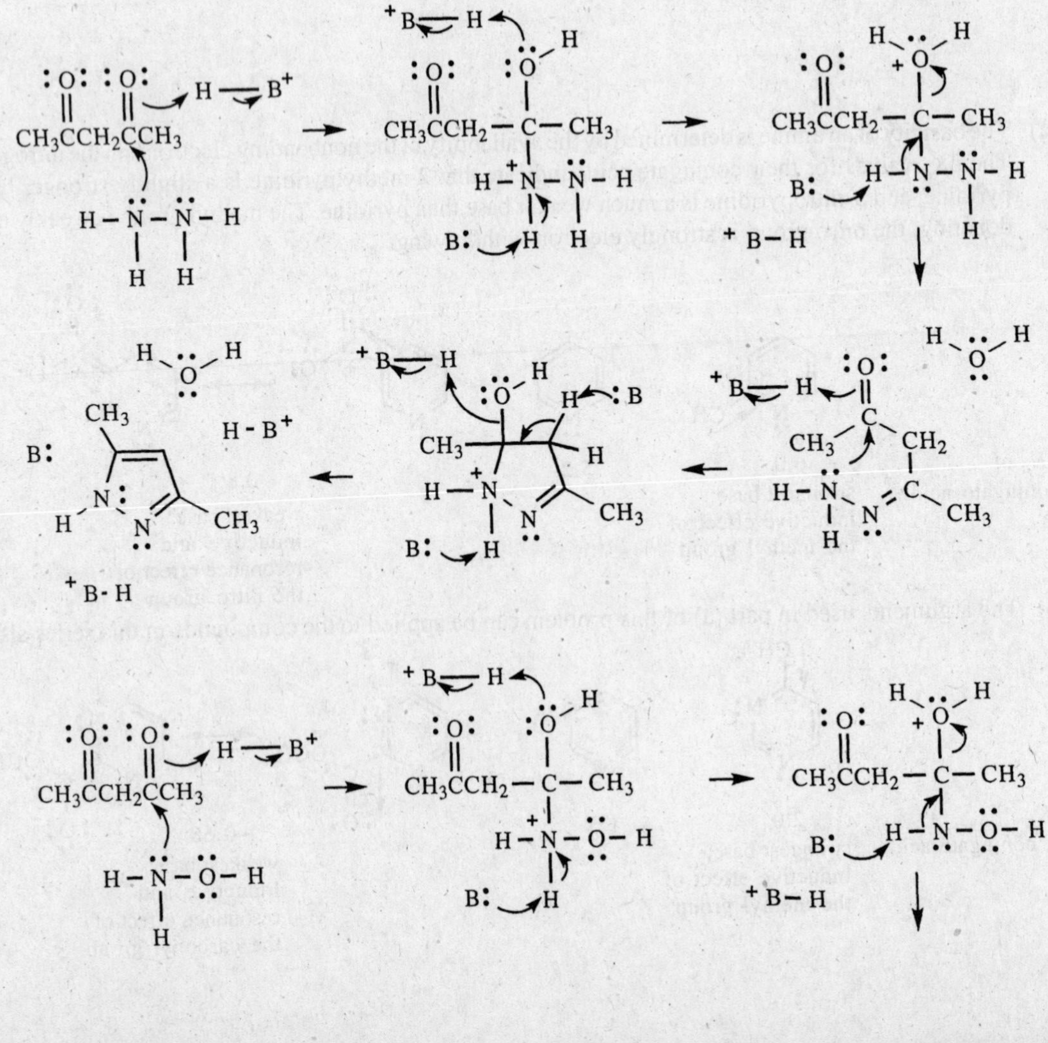

22.8 (cont)

22.9   (a) $PhCCH_2CPh \xrightarrow[\Delta]{HO\overset{+}{N}H_3\,Cl^-,\ NaOH}$ (5-Ph, 3-Ph isoxazole)

(b) $PhC-CPh$ + CH_3CHCH $\xrightarrow[\Delta]{CH_3CO^-\ \overset{+}{N}H_4,\ acetic\ acid}$

(c) $HCCH_2Br$ + $PhCNH_2$ $\xrightarrow[\substack{ethanol\\\Delta}]{\overset{piperidine}{NH}}$ (2-Ph thiazole)

22.10   (a) $CH_3CCH_2COCH_2CH_3$ + H_2NCNH_2 $\xrightarrow[ethanol]{HCl}$

(b) $CH_3CCH_2CCH_3$ + $CH_3ONHCNH_2$ $\xrightarrow[ethanol]{HCl}$

(c) $PhCCH_2CCH_3$ + H_2NCNH_2 $\xrightarrow[ethanol]{HCl}$

Concept Map 22.2   Synthesis of heterocycles from carbonyl compounds.

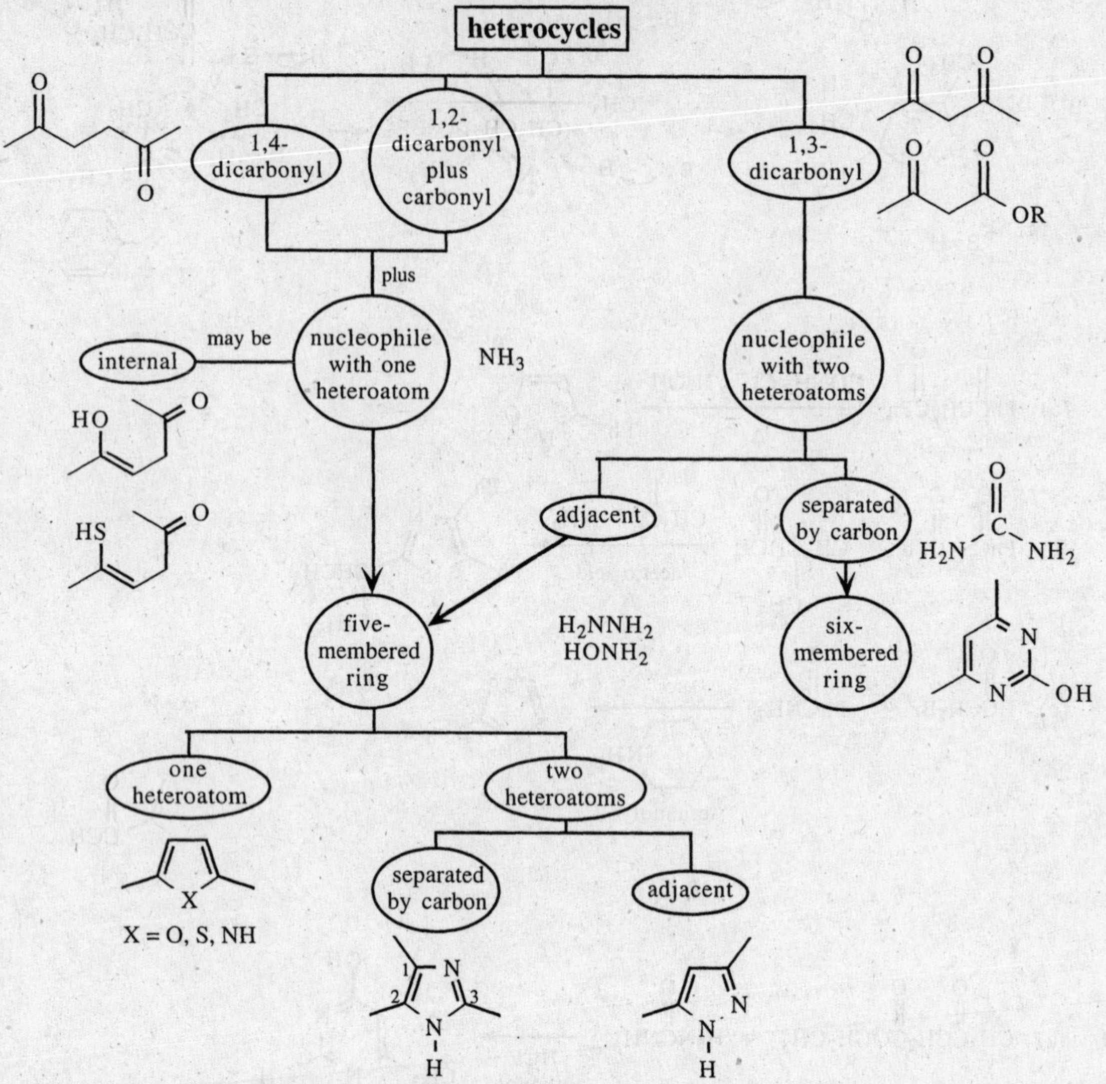

- -

22.11   (a)

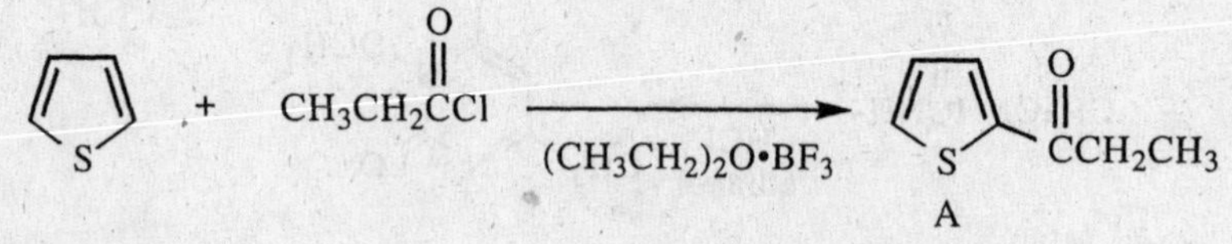

(b)

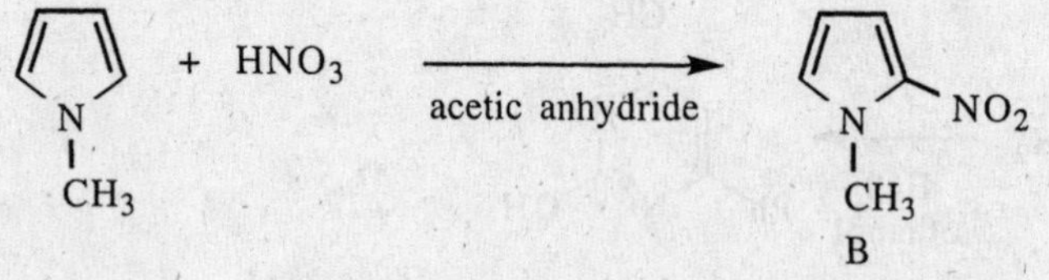

(c)

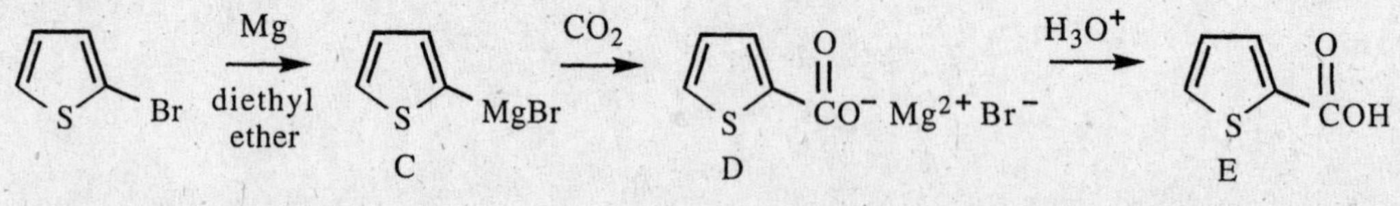

22.11 (d)

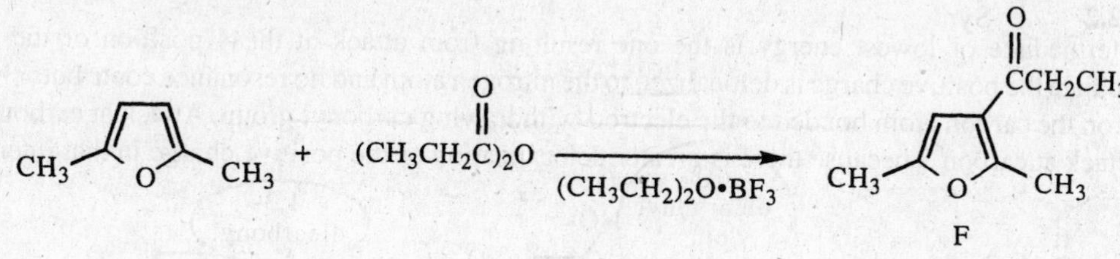

(e)

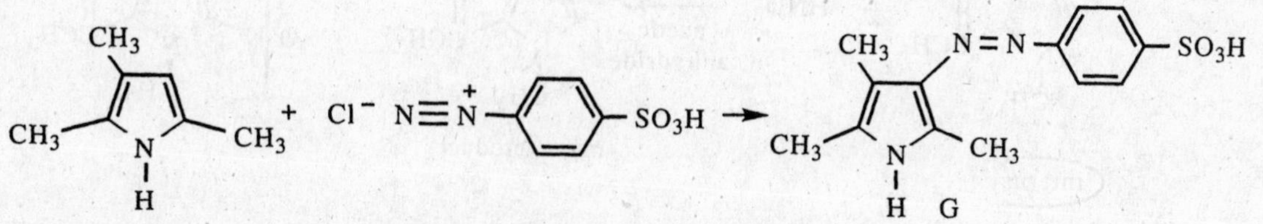

(f)

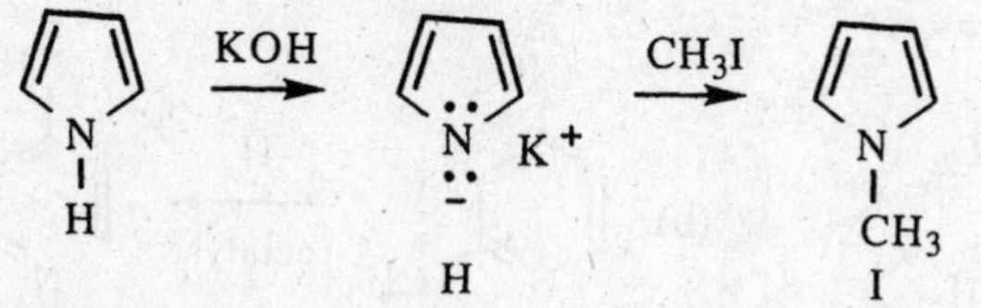

(g)

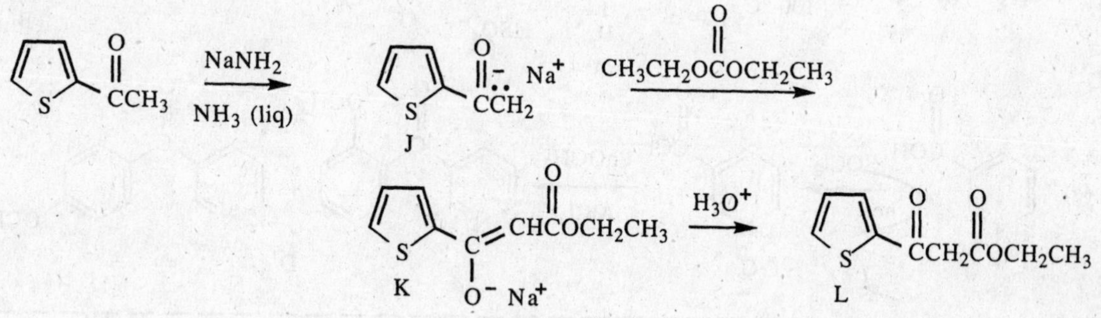

22.12 Three possible intermediates can form:

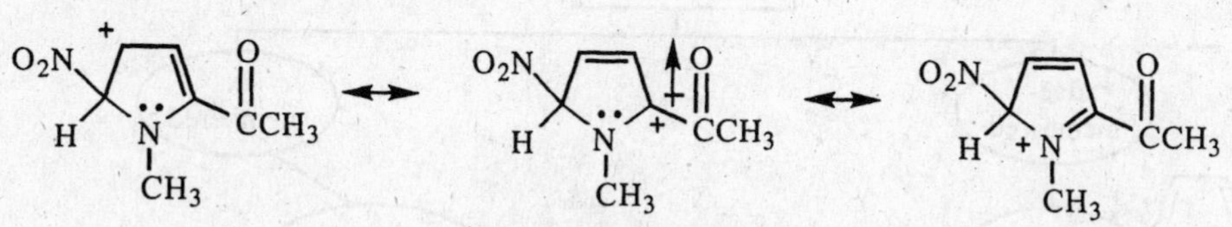

from attack at carbon 5

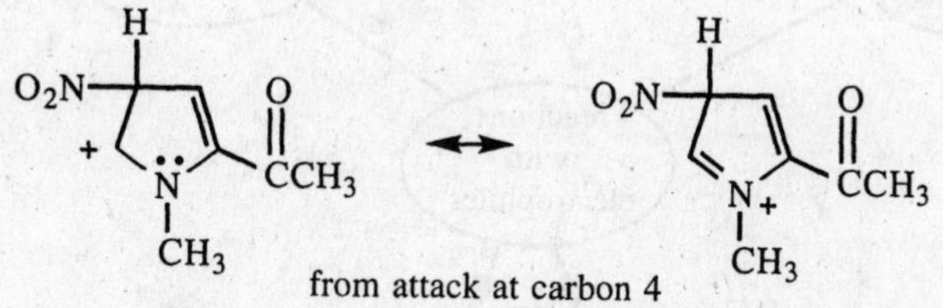

from attack at carbon 4

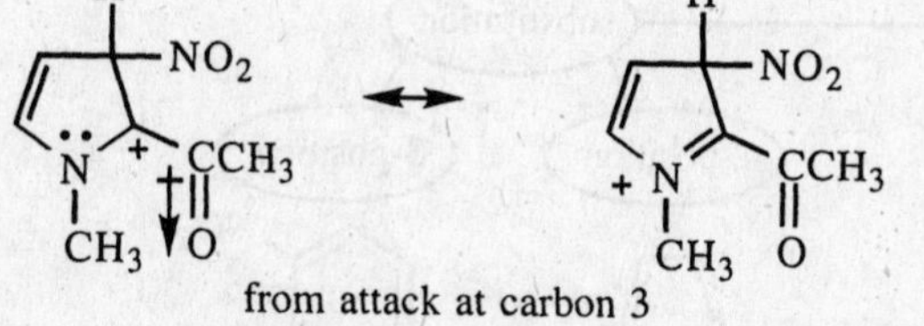

from attack at carbon 3

22.12 (cont)

The intermediate of lowest energy is the one resulting from attack at the 4 position of the ring. In this intermediate the positive charge is delocalized to the nitrogen atom and no resonance contributor has a positive charge on the carbon atom bonded to the electron-withdrawing carbonyl group. Attack at carbon 5 is favored over attack at carbon 3 because there is greater delocalization of the positive charge in that intermediate.

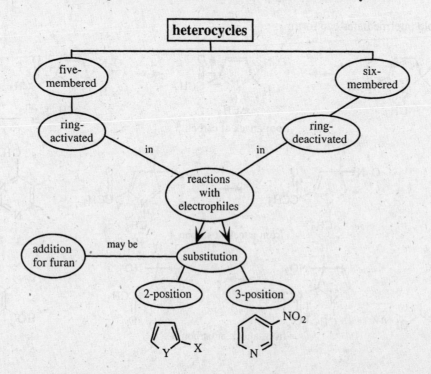

major product

22.13

(a)

(b)

(c)

- -

Concept Map 22.3   Electrophilic aromatic substitution reactions of heterocycles.

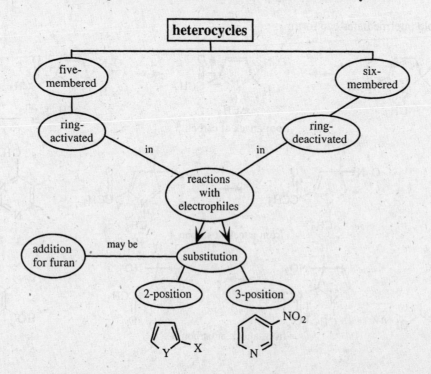

22.14

(a)

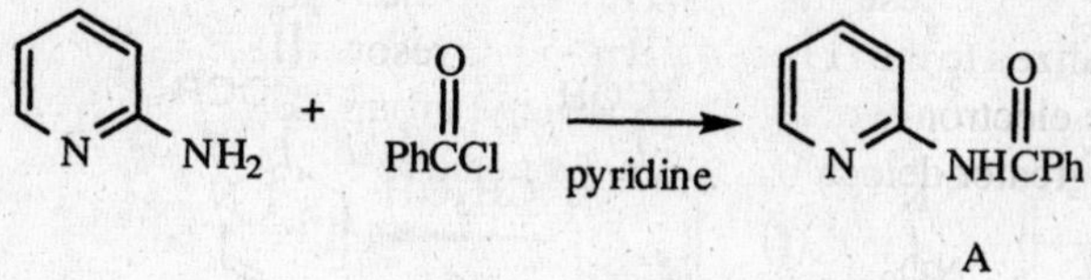

(b)

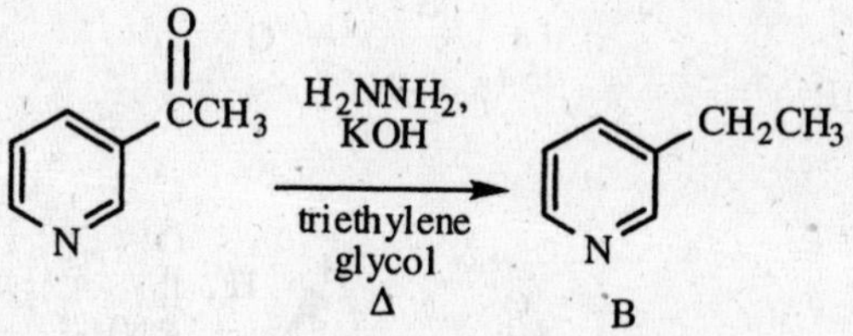

(c)

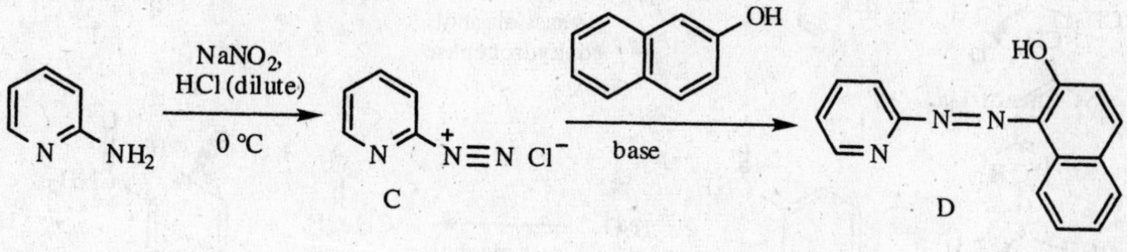

22.15

(a)

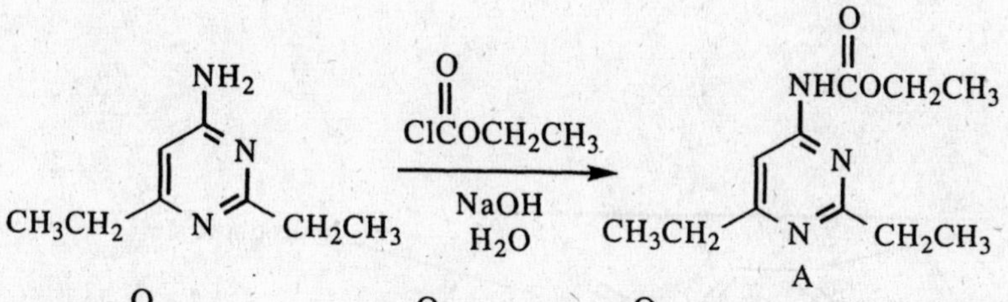

(b)

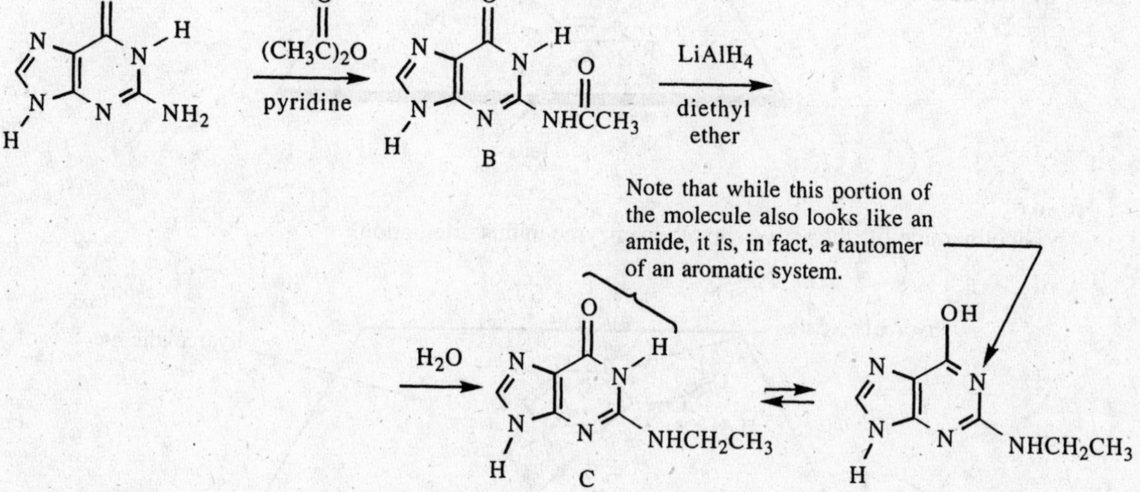

Note that while this portion of the molecule also looks like an amide, it is, in fact, a tautomer of an aromatic system.

(c)

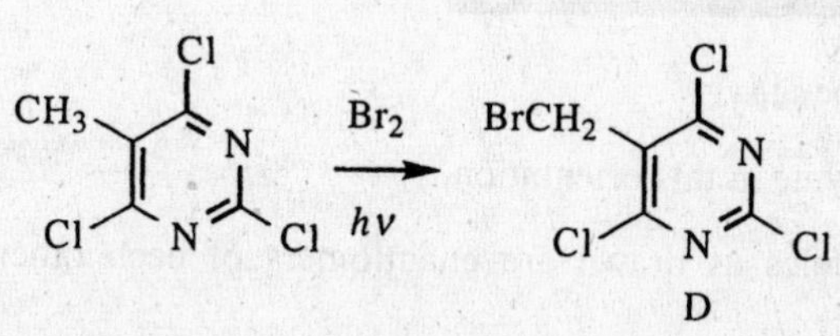

(d)

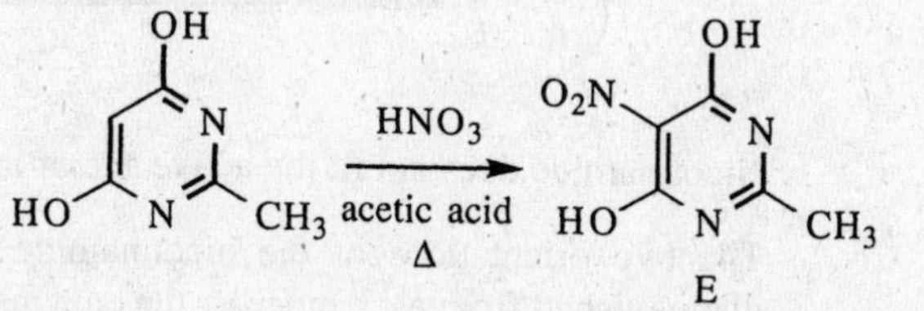

22.15 (e)

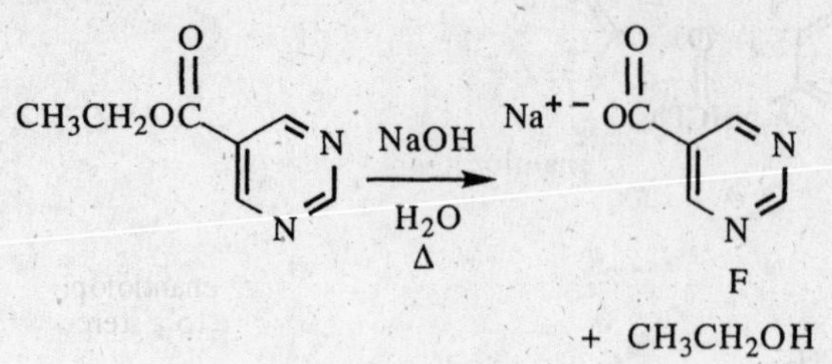

(f)

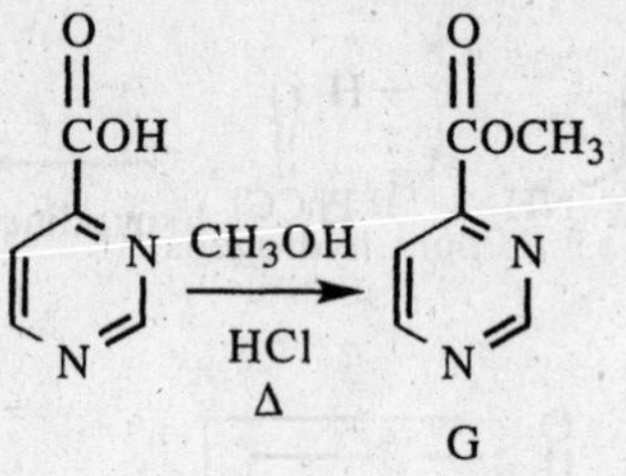

22.16

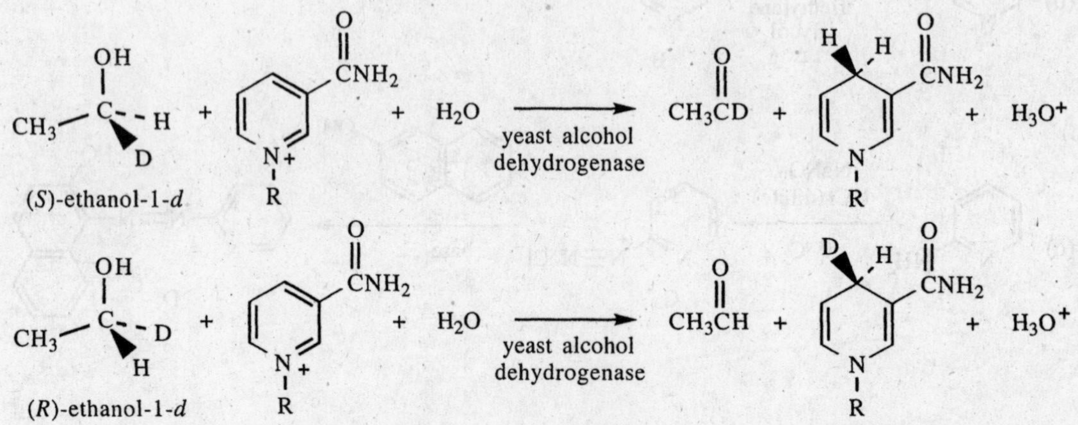

(S)-ethanol-1-*d*

(R)-ethanol-1-*d*

22.17

hydrophobic site   hydrophilic site

ionic site

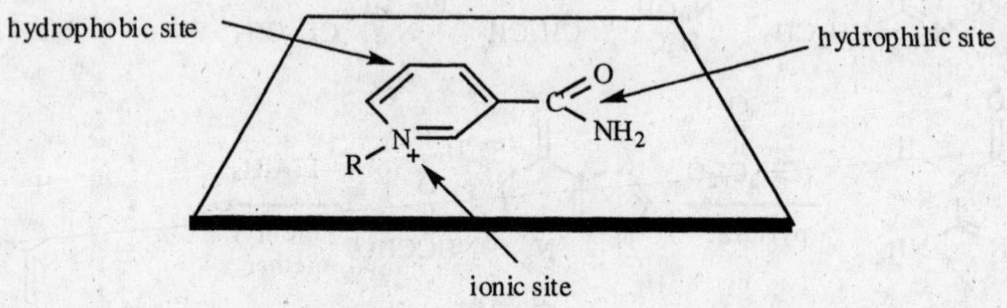

Nicotinamide fits the active site of the enzyme in this orientation.

hydrophobic site   hydrophilic site

ionic site

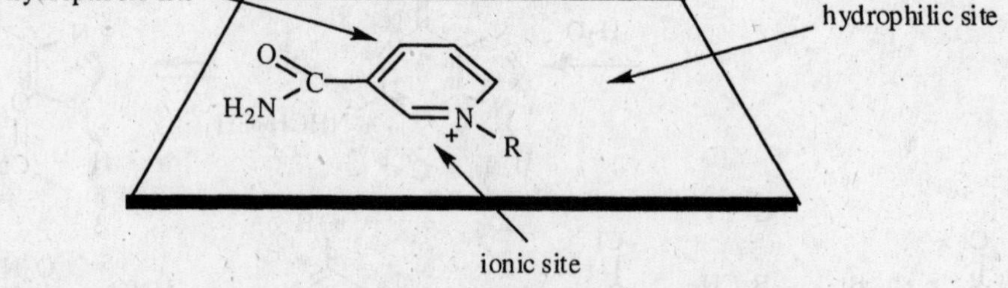

Nicotinamide does not fit the active site of the enzyme in this orientation.

The two orientations of the nicotinamide molecules as drawn are enantiomers of each other and are distinguished from each other by the enzyme.

22.18 (a) 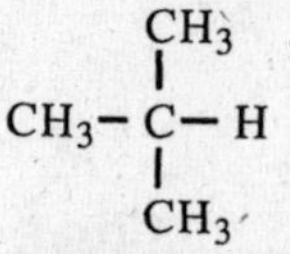

no enantiotopic hydrogens

(b) 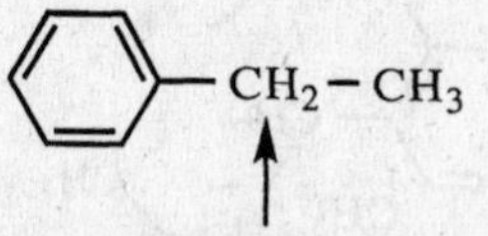

enantiotopic hydrogens

(c) 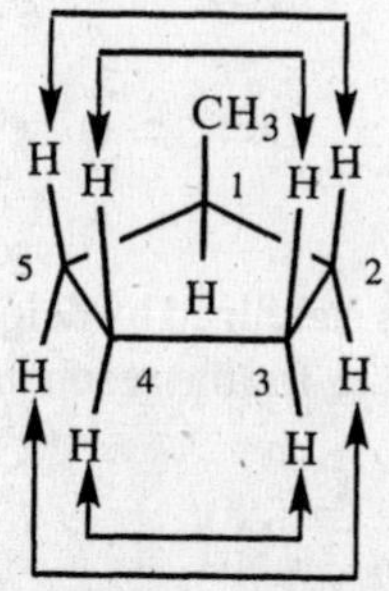

four sets of enantiotopic hydrogens; the two sets of cis hydrogens on carbons 2 and 5 and the two sets of cis hydrogens on carbons 3 and 4

these two methyl groups are enantiotopic; replacement of one gives rise to a stereocenter on the adjacent carbon

(d) 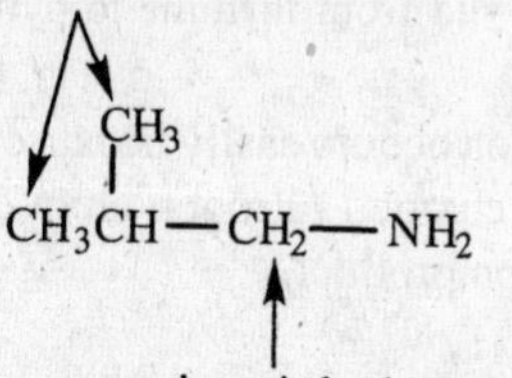

enantiotopic hydrogens

22.19 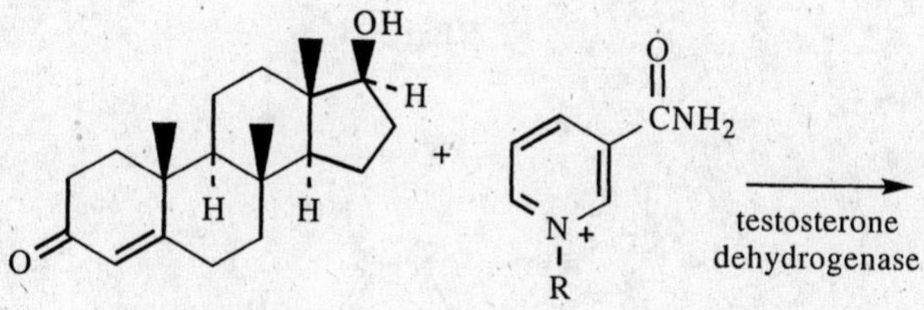

testosterone

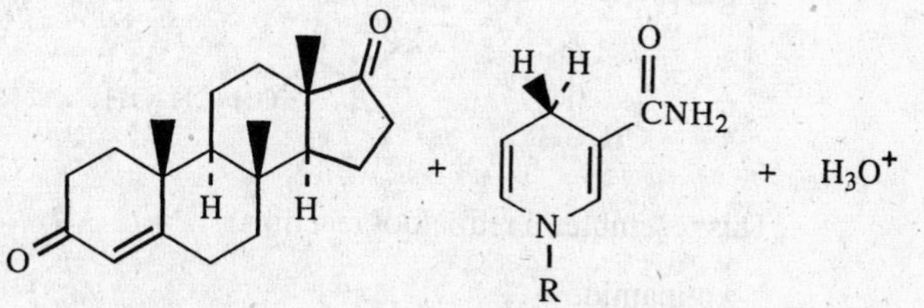

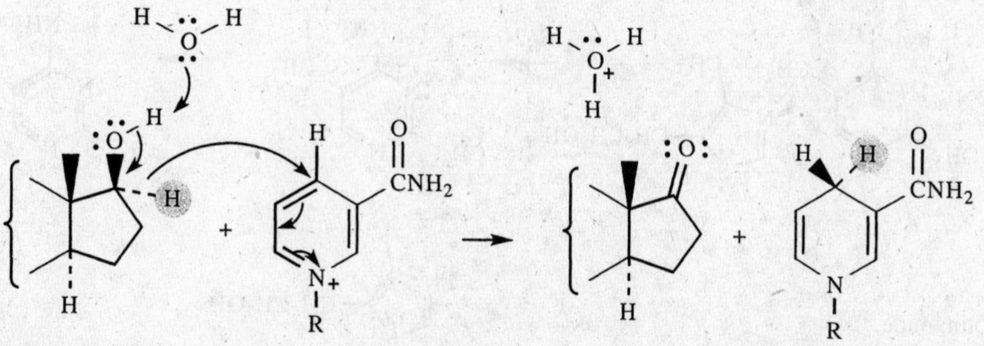

One hydrogen atom from testosterone goes to the solvent, the other, to the back face of the nicotinamide group.

22.20

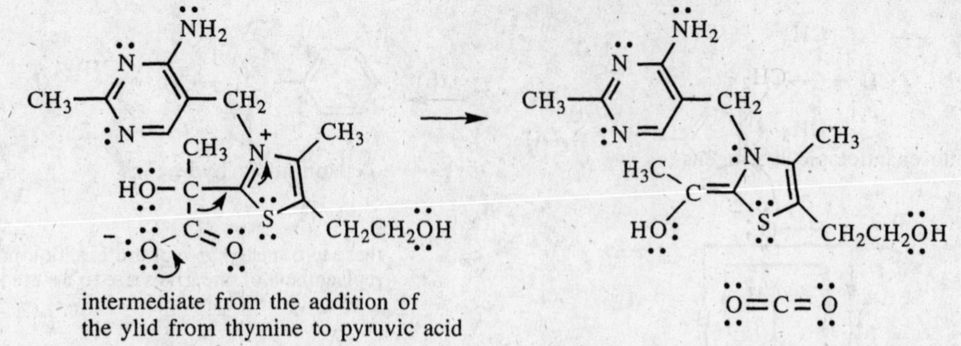

intermediate from the addition of
the ylid from thymine to pyruvic acid

Decarboxylation occurs easily because the carbanion that is formed is stabilized by delocalization of charge to the positively charged nitrogen atom of the ylide. (See Problem 22.21 for the resonance contributors of the decarboxylation product.)

22.21

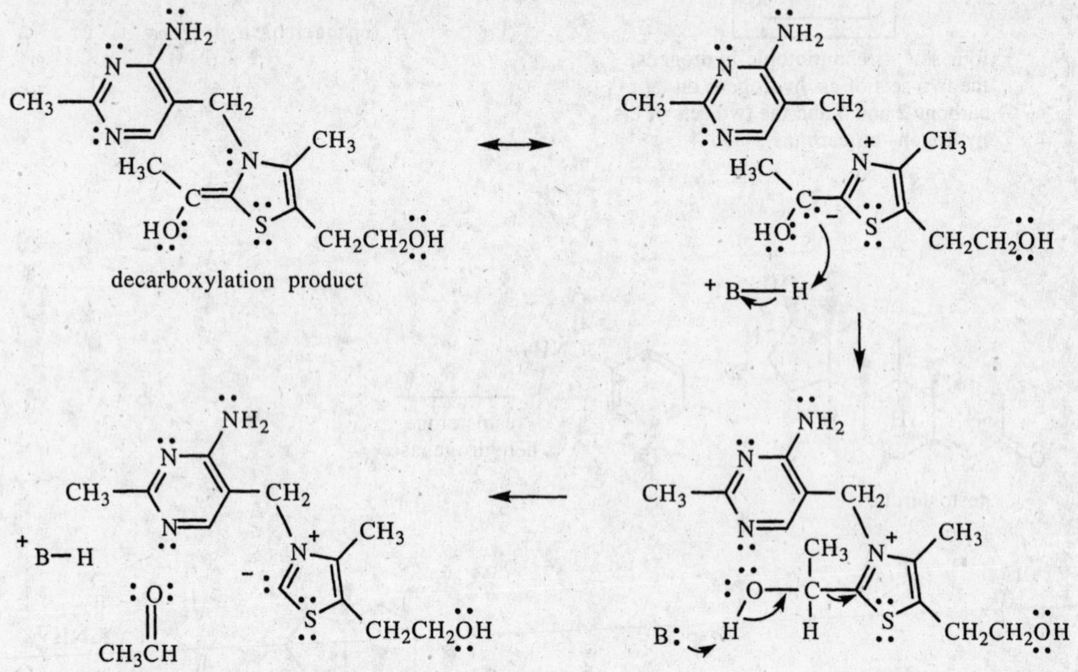

decarboxylation product

This resembles a retroaldol reaction.

22.22 (a)

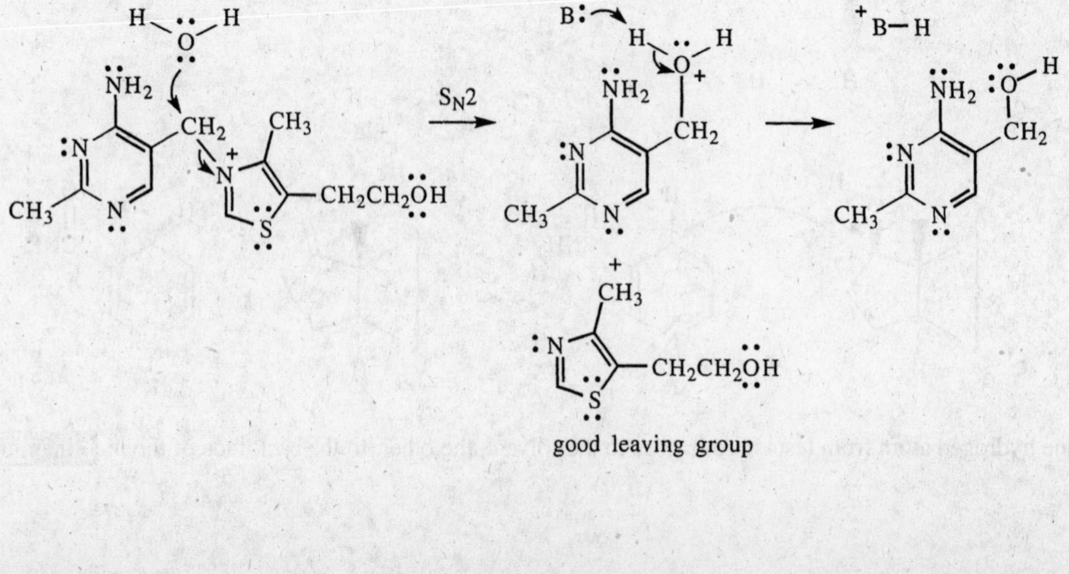

good leaving group

22.22   (b)

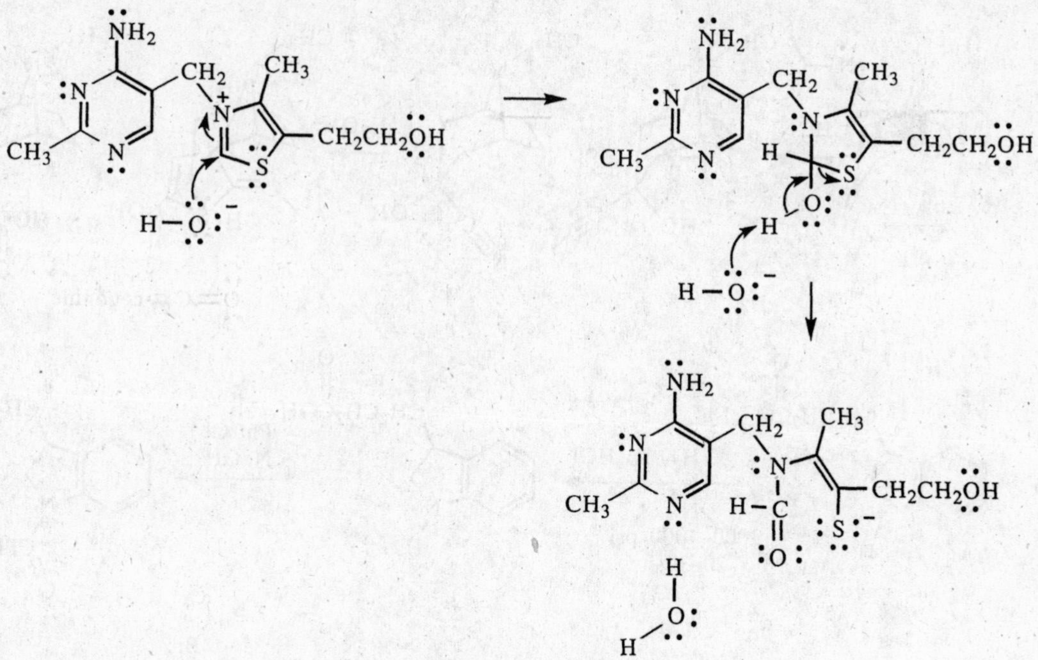

22.23

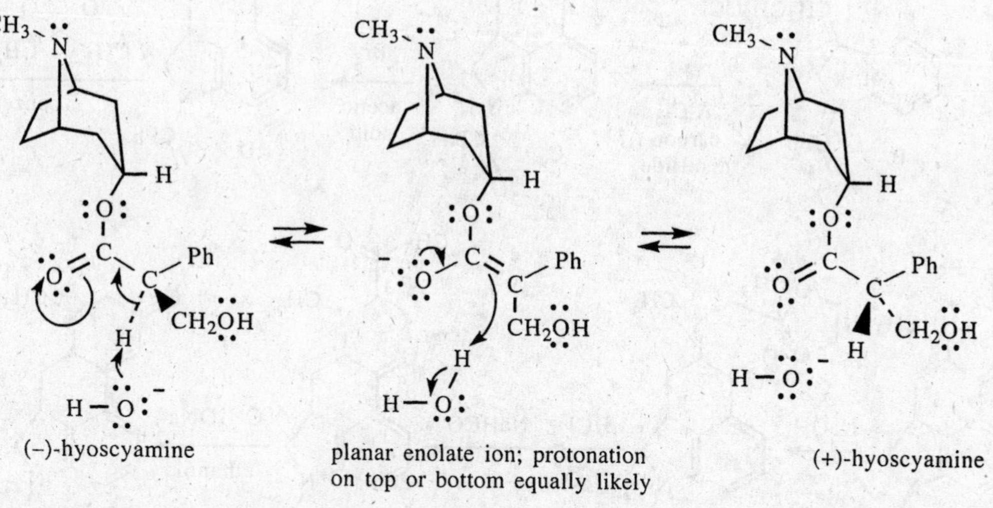

(–)-hyoscyamine   planar enolate ion; protonation   (+)-hyoscyamine
                         on top or bottom equally likely

Atropine is a racemic hyoscyamine or the equilibrium mixture of the two enantiomers.

22.24

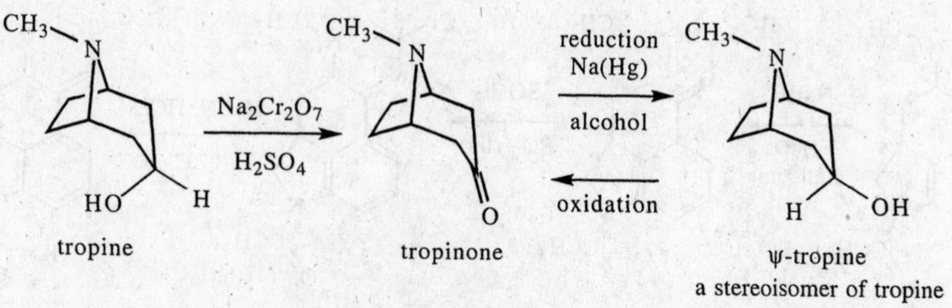

tropine   tropinone   ψ-tropine
                                                                       a stereoisomer of tropine

22.25

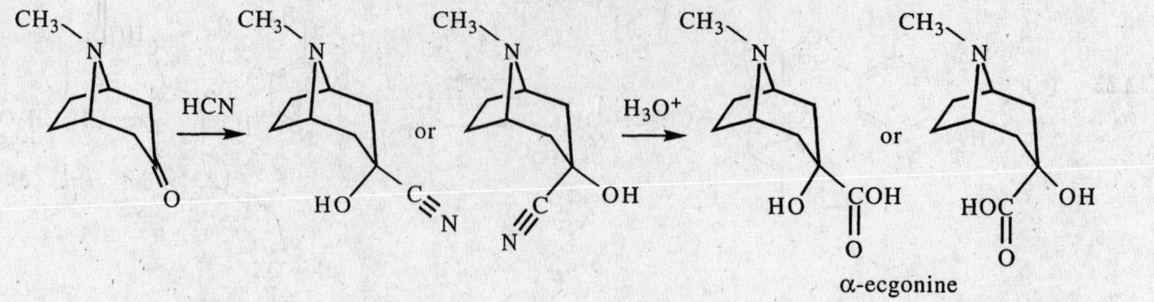
HCN
or
H3O+
or
α-ecgonine

22.26

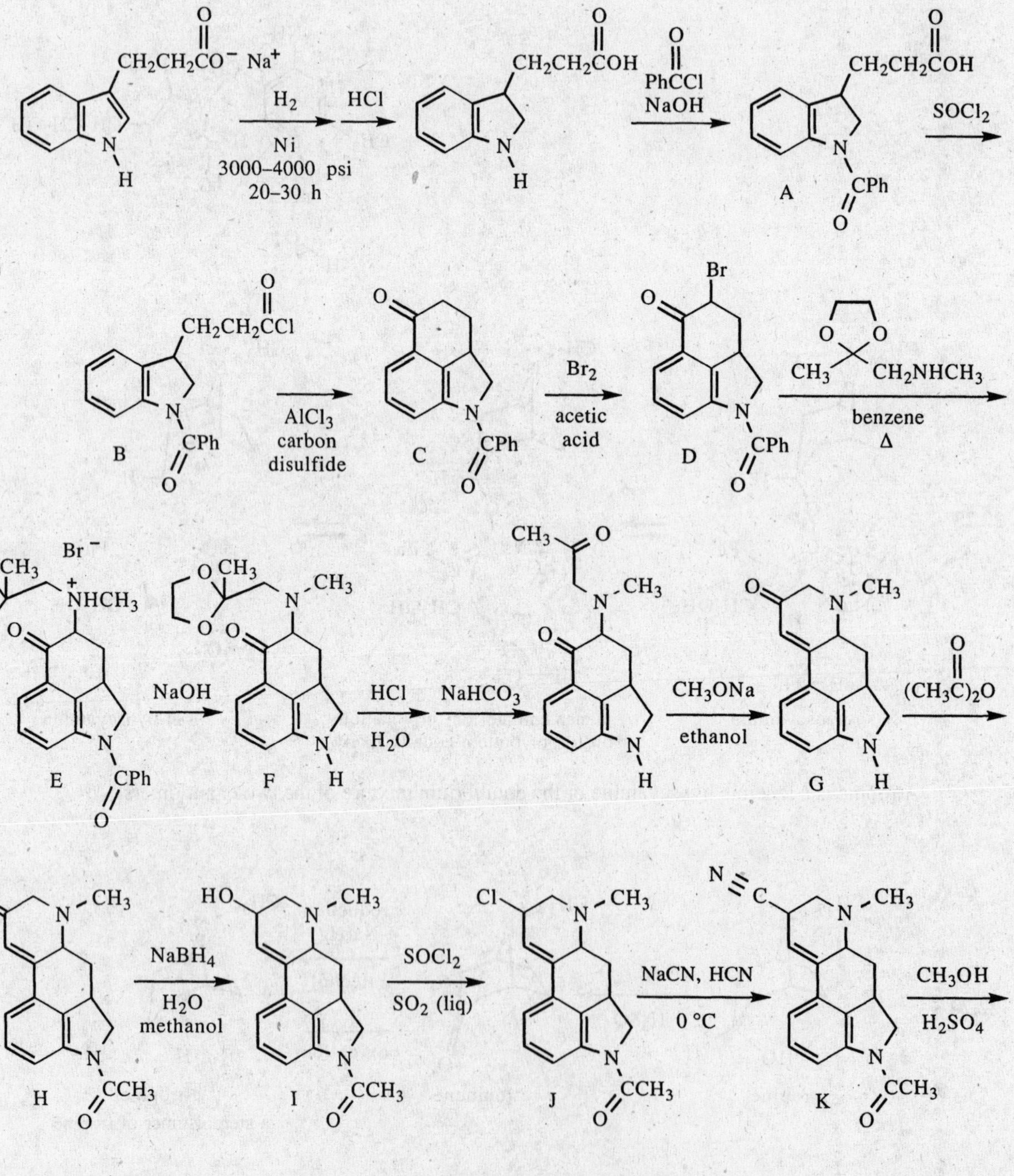
H2
Ni
3000–4000 psi
20–30 h
HCl
PhCCl
NaOH
A
SOCl2
B
AlCl3
carbon
disulfide
C
Br2
acetic
acid
D
CH3 CH2NHCH3
benzene
Δ
E
NaOH
F
HCl
NaHCO3
H2O
CH3ONa
ethanol
G
(CH3C)2O
H
NaBH4
H2O
methanol
I
SOCl2
SO2 (liq)
J
NaCN, HCN
0 °C
K
CH3OH
H2SO4

22.26 (cont)

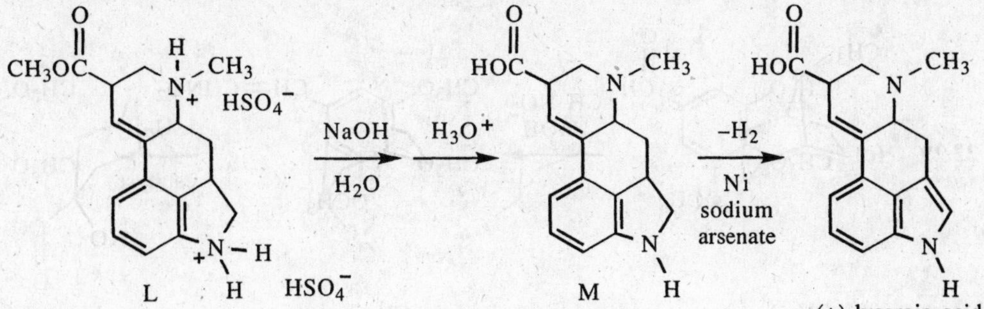

(±)-lysergic acid

22.27

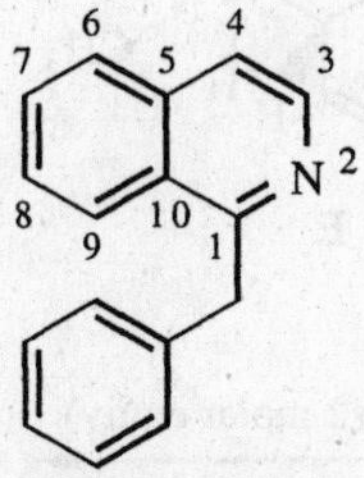

numbered around the isoquinoline ring (*not* the numbering used in naming the compound); benzyl group on carbon 1

benzyl unit on carbon 1

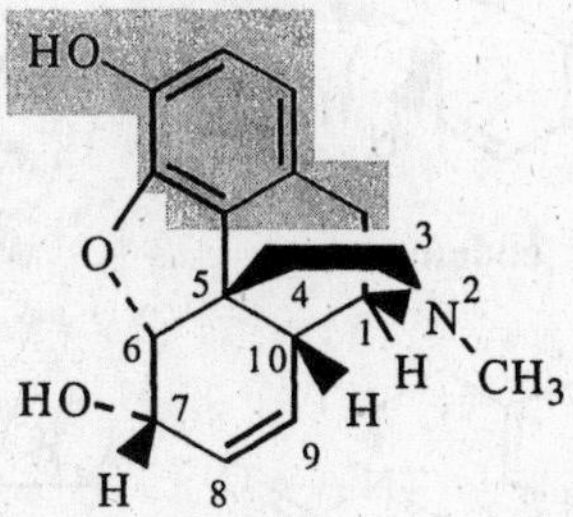

22.28

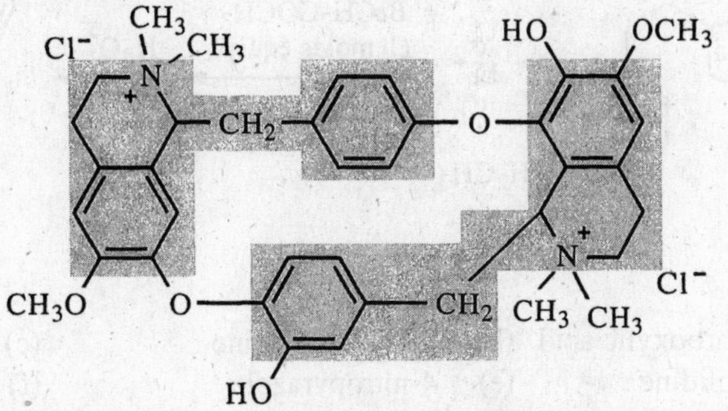

22.29 (a)

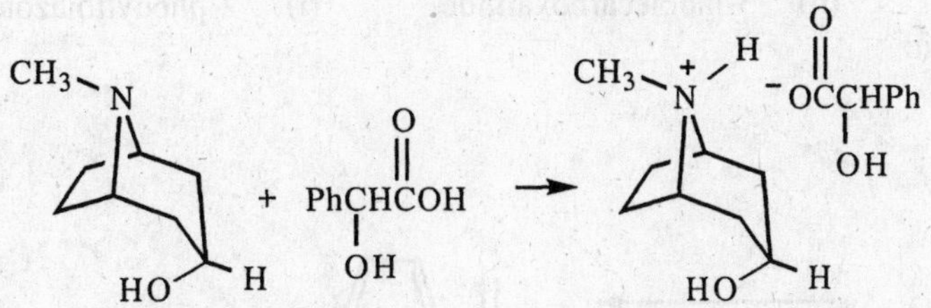

A

(b)

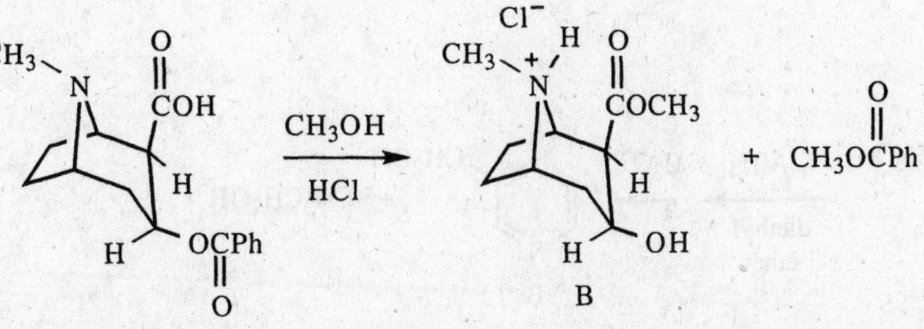

B

22.29 (c)

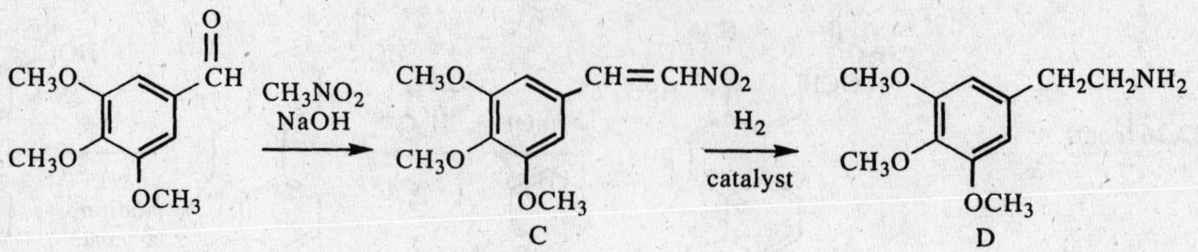
CH₃O
CH₃O
OCH₃
CH
O
CH₃NO₂
NaOH
CH=CHNO₂
C
H₂
catalyst
CH₂CH₂NH₂
D

(d)

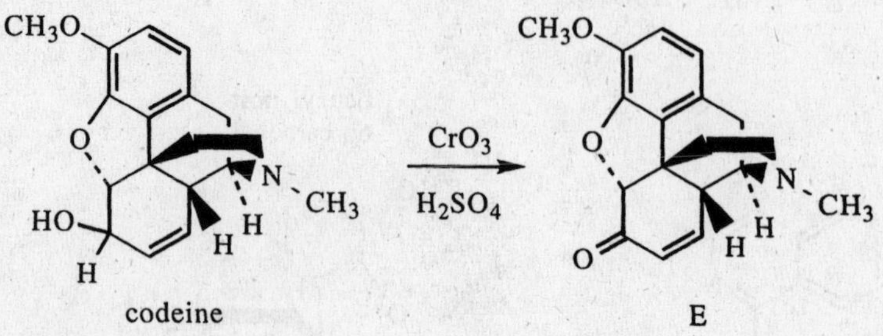
CH₃O
O
HO
H
N
CH₃
CrO₃
H₂SO₄
codeine
E

(e)

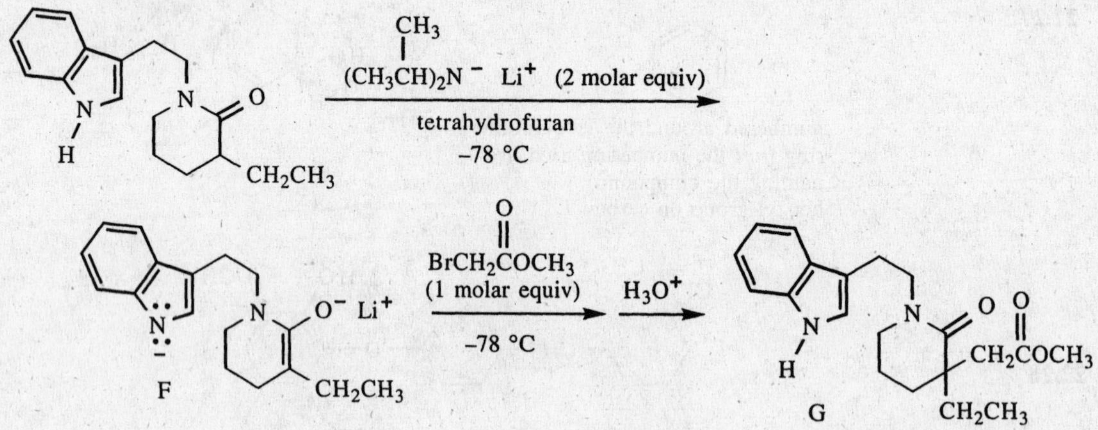
N
H
O
CH₂CH₃
CH₃
(CH₃CH)₂N⁻ Li⁺ (2 molar equiv)
tetrahydrofuran
−78 °C
O⁻ Li⁺
F
BrCH₂COCH₃
(1 molar equiv)
−78 °C
H₃O⁺
CH₂COCH₃
G

22.30 (a) 3-isoquinolinecarboxylic acid (b) 3-chloropyridine (c) 2-aminopyrimidine
(d) 3-phenylisoxazolidine (e) 4-nitropyrazole (f) methyl 3-furancarboxylate
(g) 1,4-dimethylimidazole (h) 3-indolecarboxamide (i) 2-phenylthiazole
(j) 2-thiophenecarboxylic acid

22.31 (a)

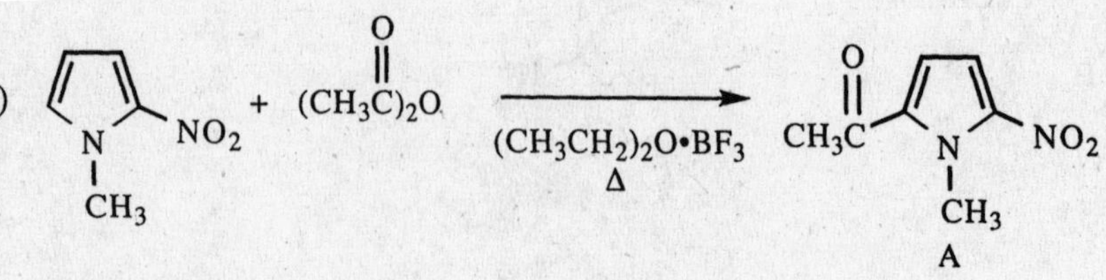
N
CH₃
NO₂
(CH₃C)₂O
(CH₃CH₂)₂O•BF₃
Δ
CH₃C
A

(b)

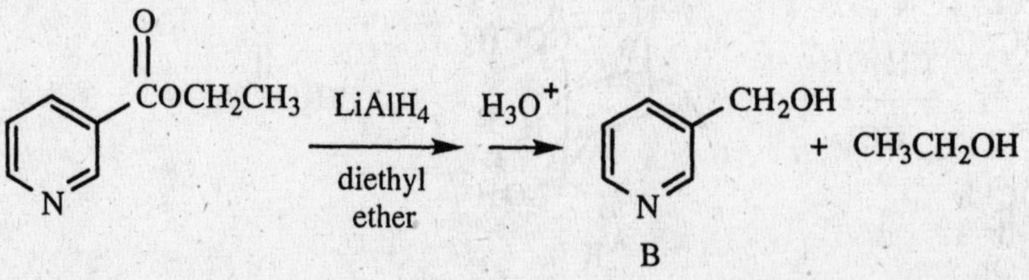
COCH₂CH₃
N
LiAlH₄
diethyl
ether
H₃O⁺
CH₂OH
+ CH₃CH₂OH
B

22.31 (c)

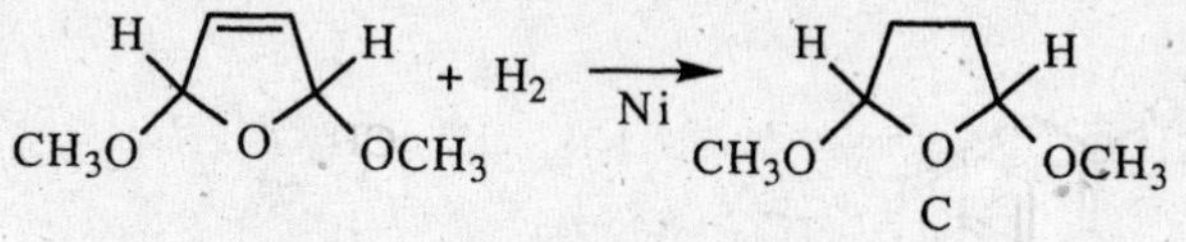

(d)

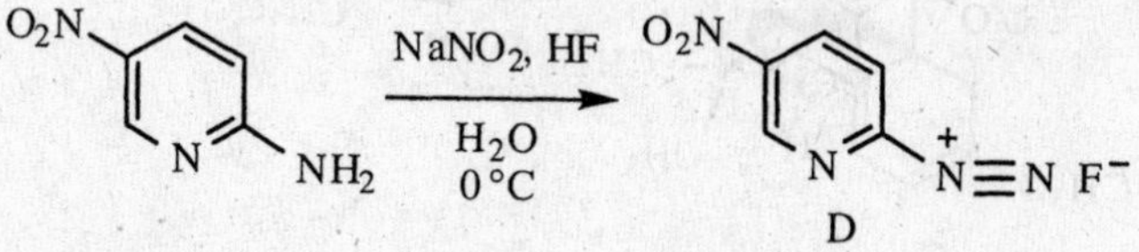

(e)

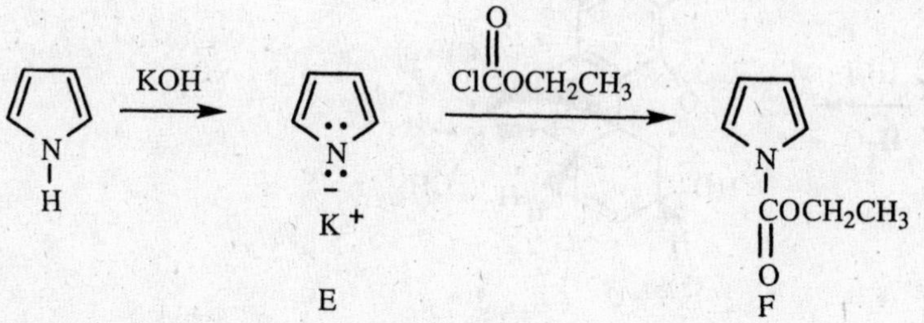

(f)

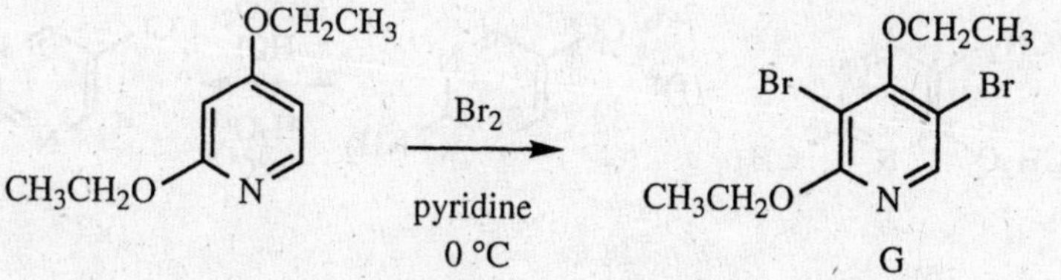

(g)

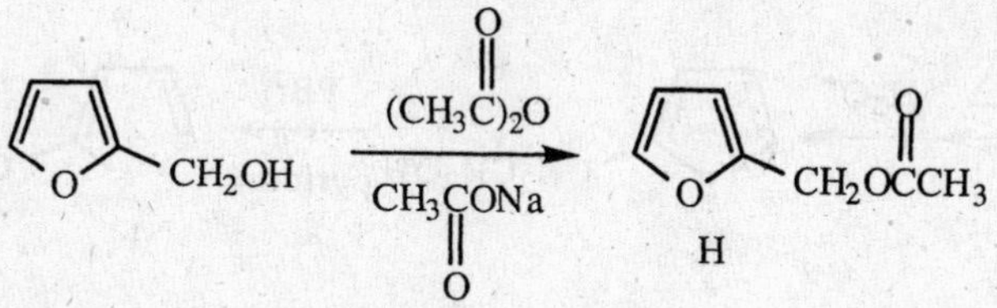

(h)

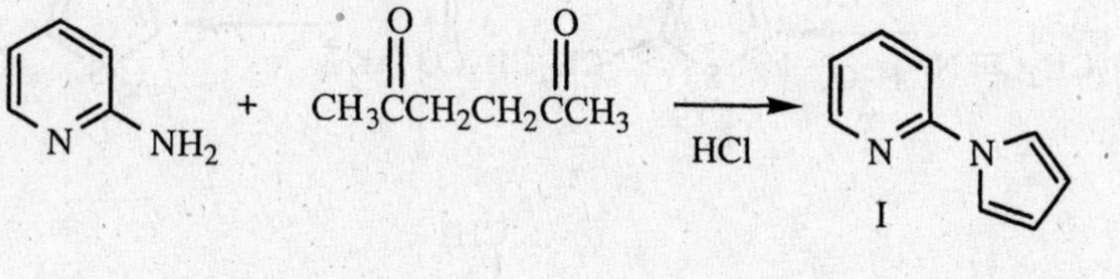

22.32

(a)

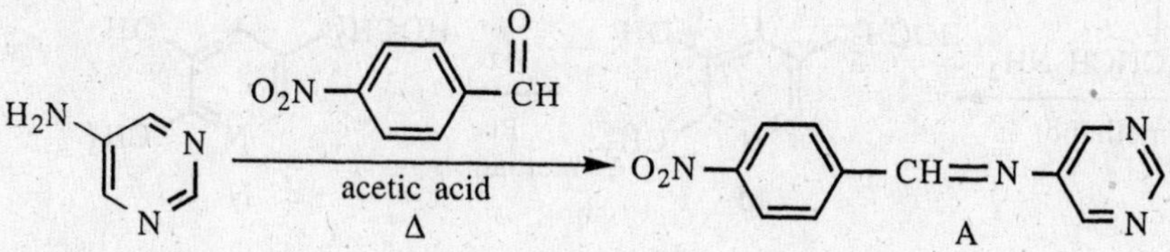

(b)

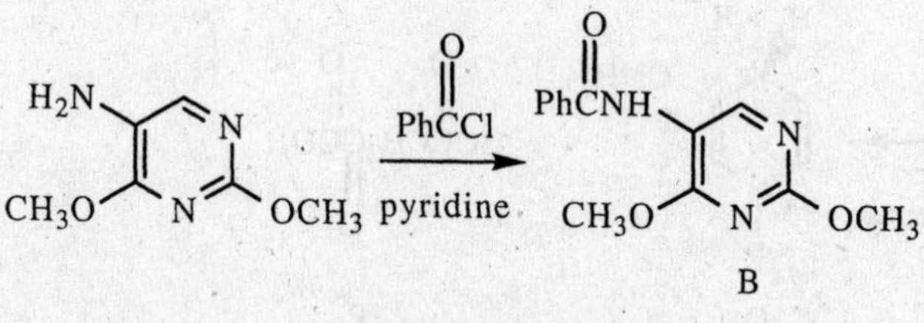

22.32

(c)

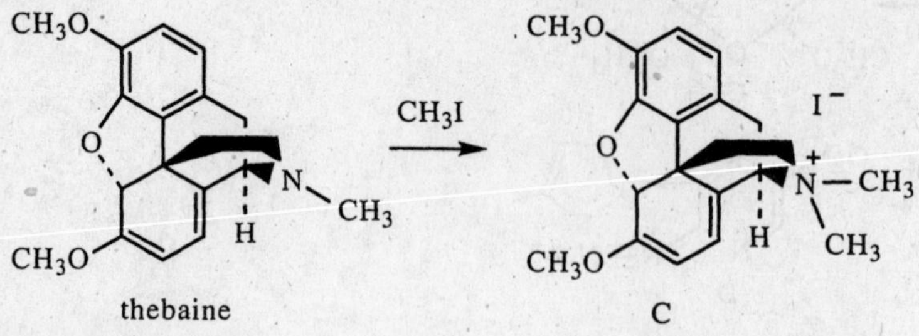

thebaine   C

(d)

morphine   D

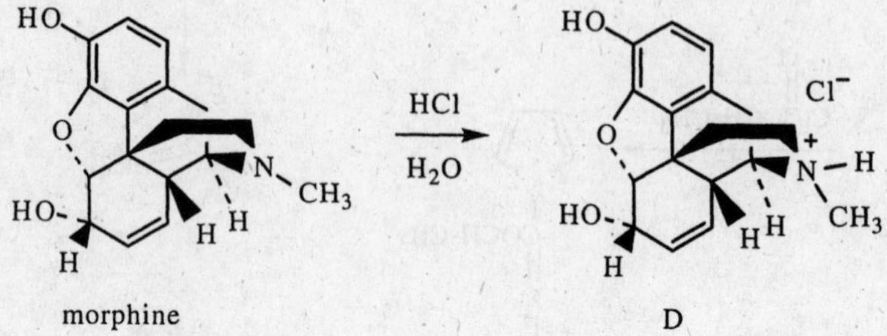

(e)   E

(f)   F

(g)   G 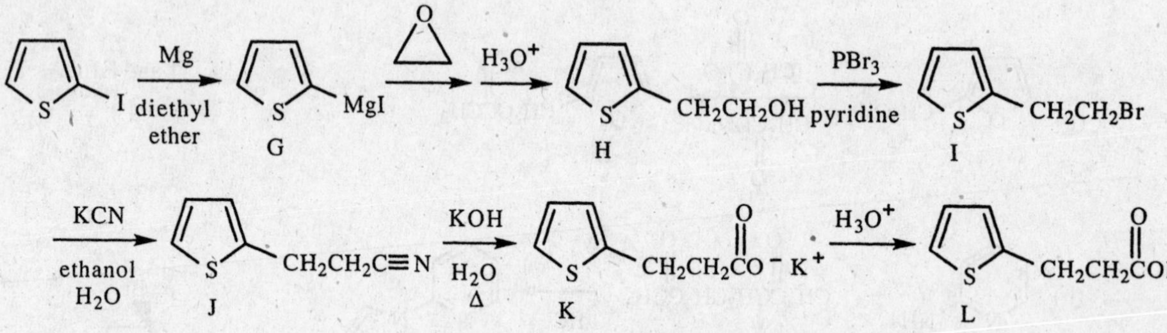   H 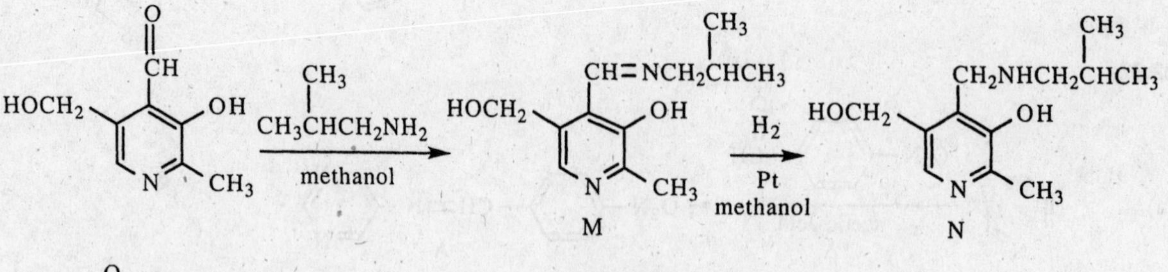   I

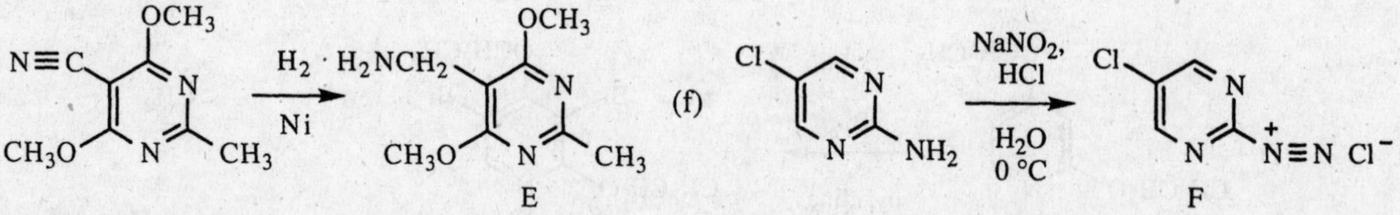

J   K   L

(h)   M   N

(i)   O   P

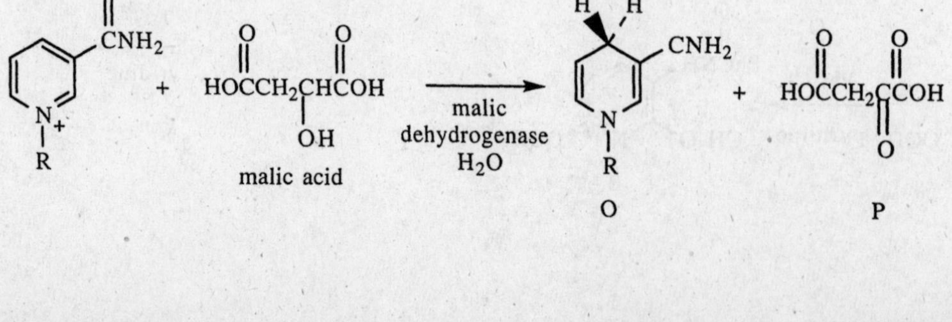

22.33

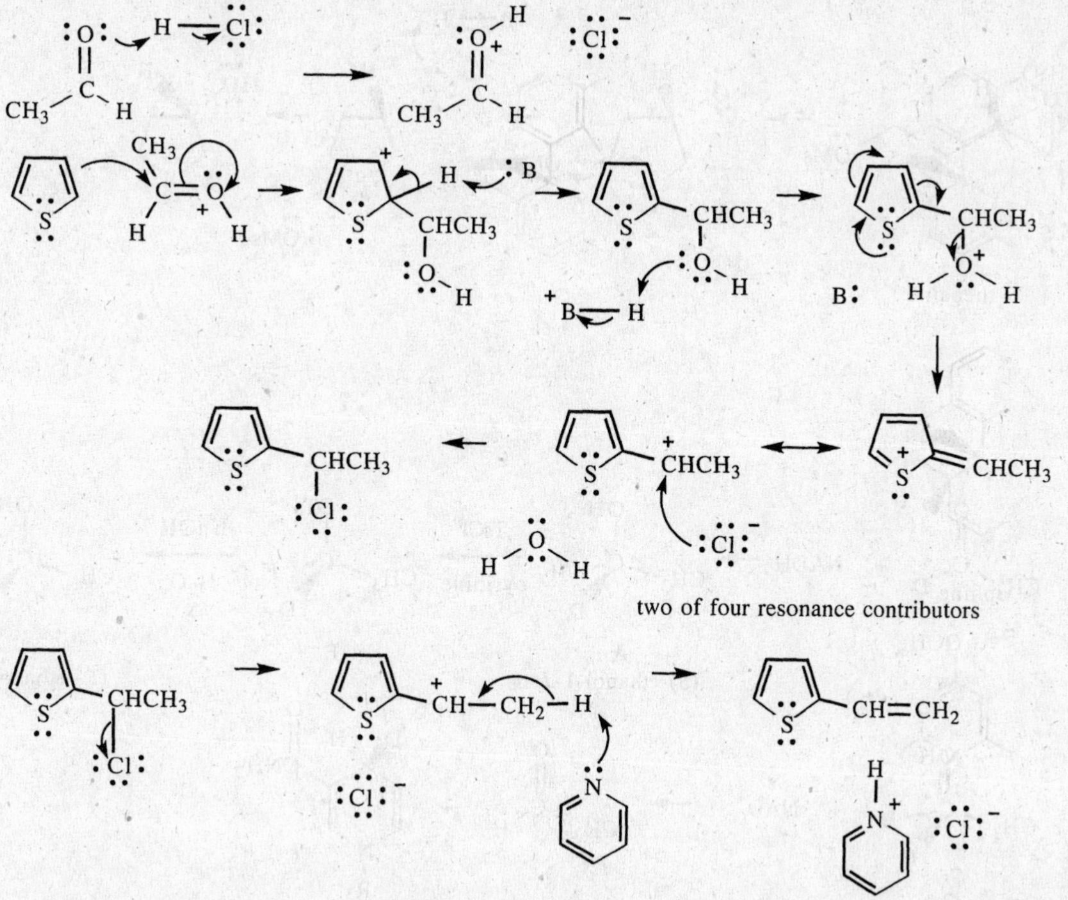

A carbocation adjacent to the thiophene ring is like a benzylic cation in stability, so S_N1 and E1 reactions have been shown. S_N2 and E2 reactions are also a possibility.

two of four resonance contributors

22.34

(a)

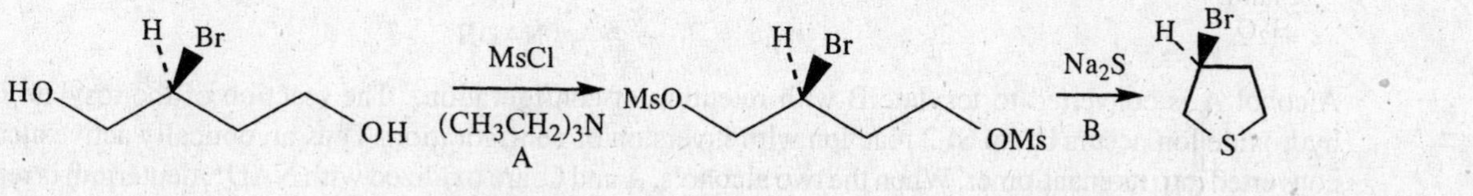

(b)

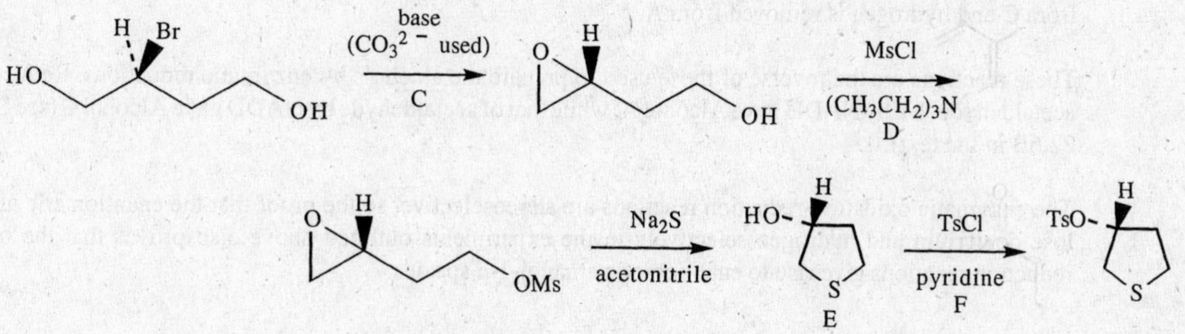

22.34

(c)

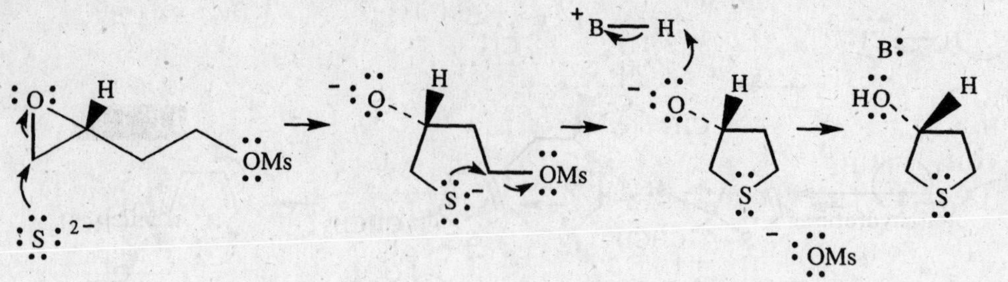

22.35

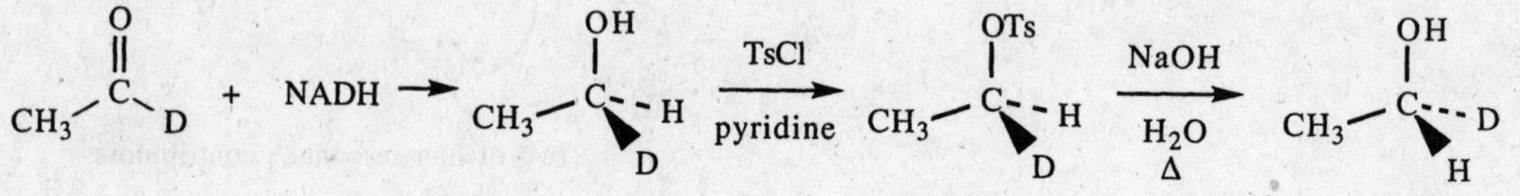

A
(S)-ethanol-1-d

B

C
(R)-ethanol-1-d

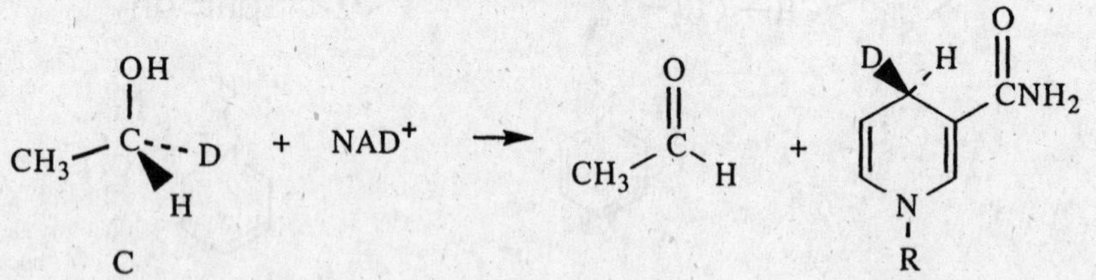

C

NADD

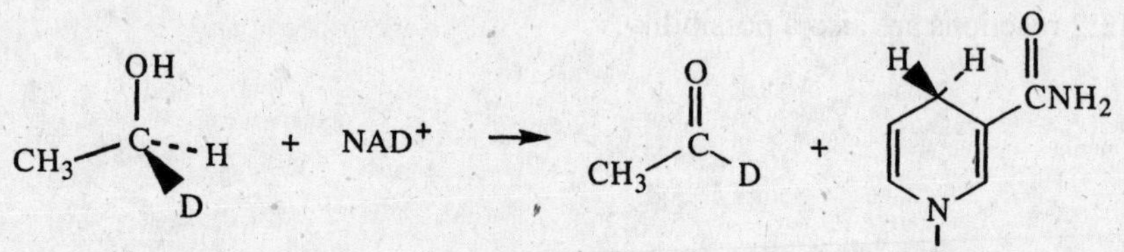

A

NADH

Alcohol A is converted to tosylate B with retention of configuration. The reaction of the tosylate B with hydroxide ion occurs by an S_N2 reaction with inversion of configuration. Thus an optically active alcohol is converted into its enantiomer. When the two alcohols, A and C, are oxidized with NAD^+, deuterium is removed from C and hydrogen is removed from A.

These reactions are the reverse of those used to prepare the alcohols by enzymatic reductions. Reduction of acetaldehyde-1-d by NADH gave Alcohol A, while that of acetaldehyde by NADD gave Alcohol C (see Section 22.6B in the text).

The enzymatic oxidation-reduction reactions are stereoselective, so the proof that the enantiomeric alcohols lose deuterium and hydrogen selectively in the experiments outlined above also proves that the original reduction reactions give rise to enantiomeric ethanol-1-d species.

22.36

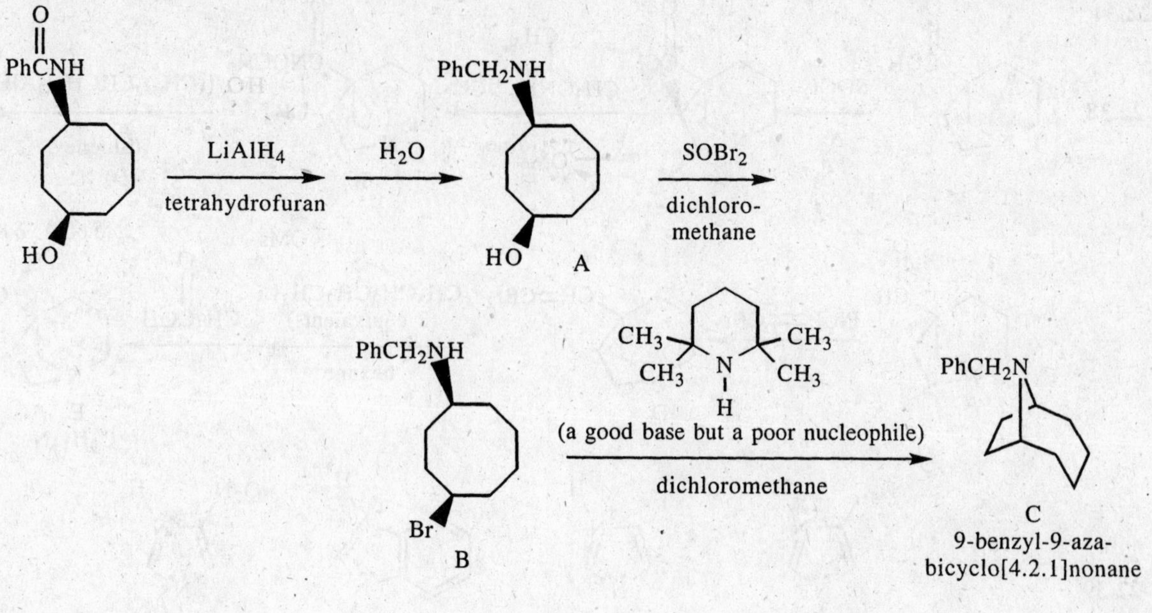
PhCNH
HO
LiAlH4
tetrahydrofuran
H2O
PhCH2NH
HO
A
SOBr2
dichloro-
methane
PhCH2NH
Br
B
CH3
CH3
N
H
(a good base but a poor nucleophile)
dichloromethane
PhCH2N
C
9-benzyl-9-aza-
bicyclo[4.2.1]nonane

22.37

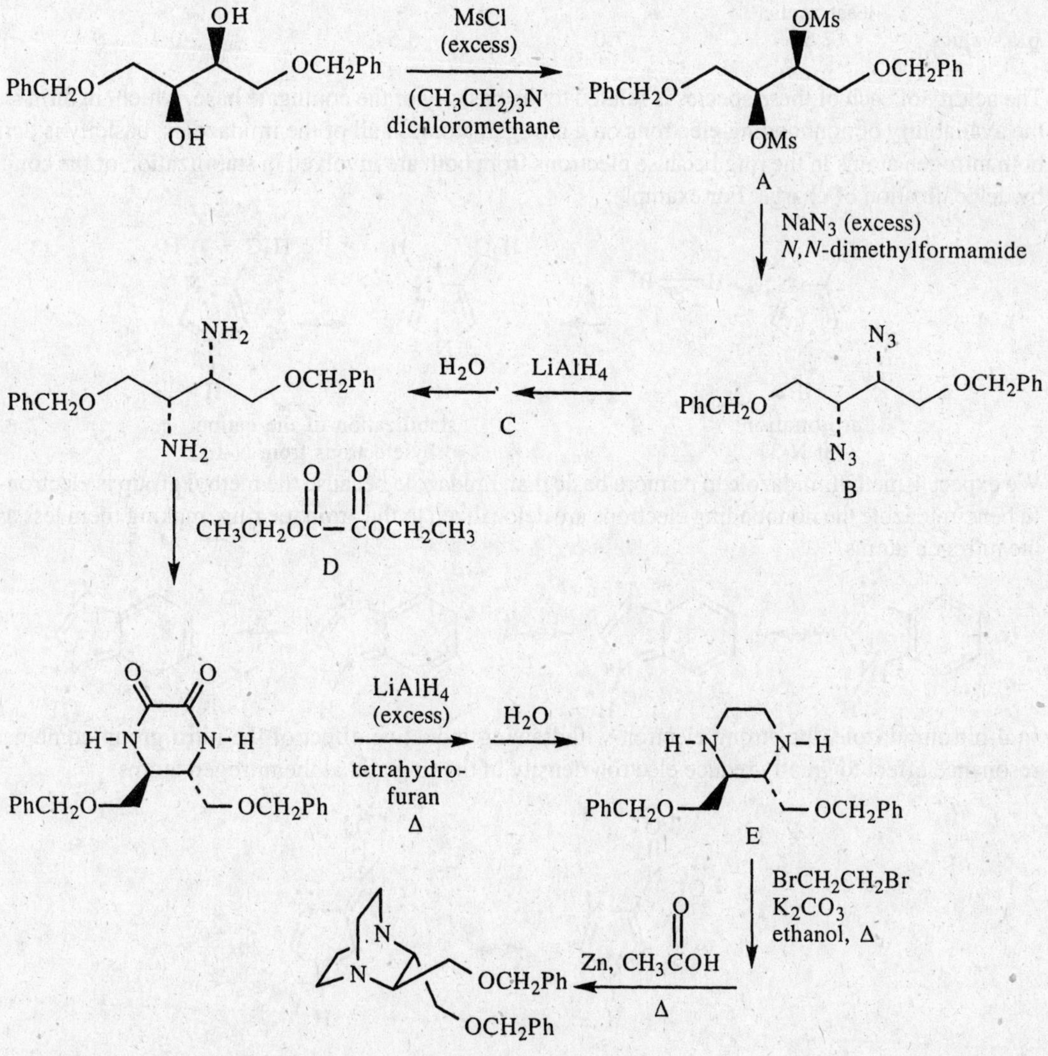
OH
PhCH2O
OCH2Ph
MsCl
(excess)
(CH3CH2)3N
dichloromethane
OMs
A
NaN3 (excess)
N,N-dimethylformamide
N3
B
LiAlH4
H2O
C
NH2
CH3CH2OC— COCH2CH3
D
H-N
N-H
LiAlH4
(excess)
tetrahydro-
furan
Δ
E
BrCH2CH2Br
K2CO3
ethanol, Δ
Zn, CH3COH
N
OCH2Ph

22.38

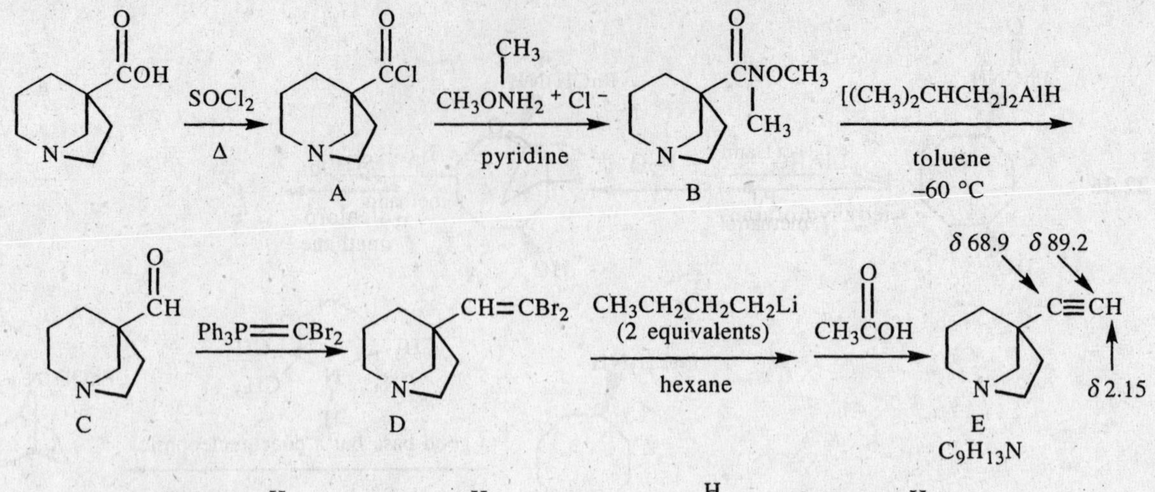

22.39

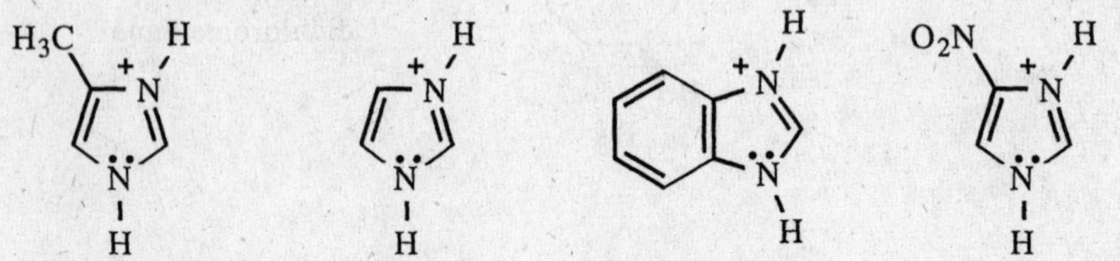

	4-methylimidazole	imidazole	benzimidazole	4-nitroimidazole
	least acidic			
pK_a values	7.5	7.0	5.5	−0.1

The acidity of each of these species is related to the basicity of the conjugate base, which, in turn, is related to the availability of nonbonding electrons on a nitrogen atom. In all of the imidazoles, basicity is derived from both nitrogen atoms in the ring because electrons from both are involved in stabilization of the conjugate acid by delocalization of charge. For example:

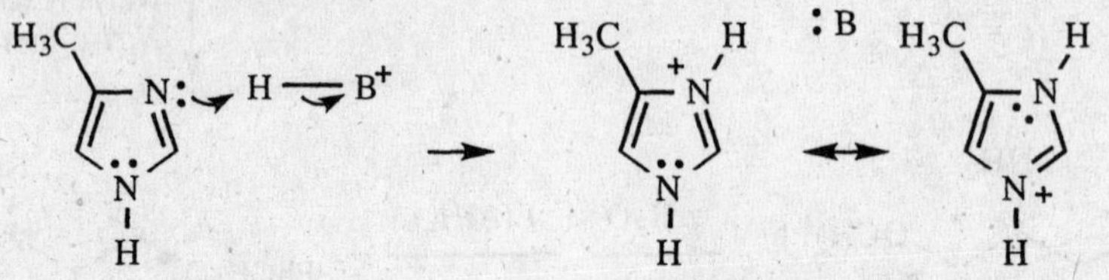

protonation   stabilization of the cation
at N-3   by electrons from N-1

We expect 4-methylimidazole to be more basic than imidazole because the methyl group is electron-donating. In benzimidazole the nonbonding electrons are delocalized to the aromatic ring, making them less available at the nitrogen atoms.

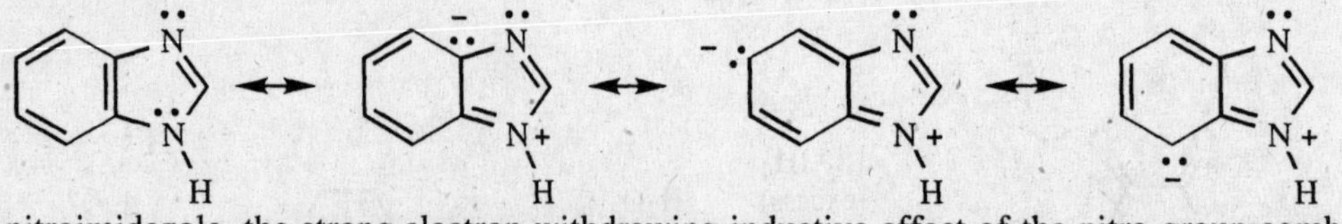

In 4-nitroimidazole, the strong electron-withdrawing inductive effect of the nitro group combines with the resonance effect to greatly reduce electron density in the ring and at the nitrogen atoms.

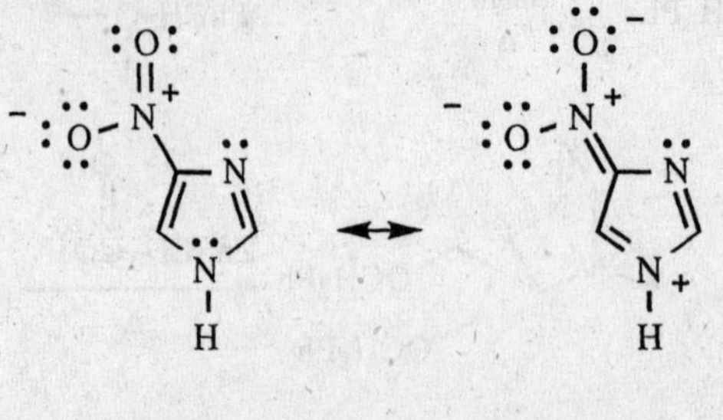

22.40

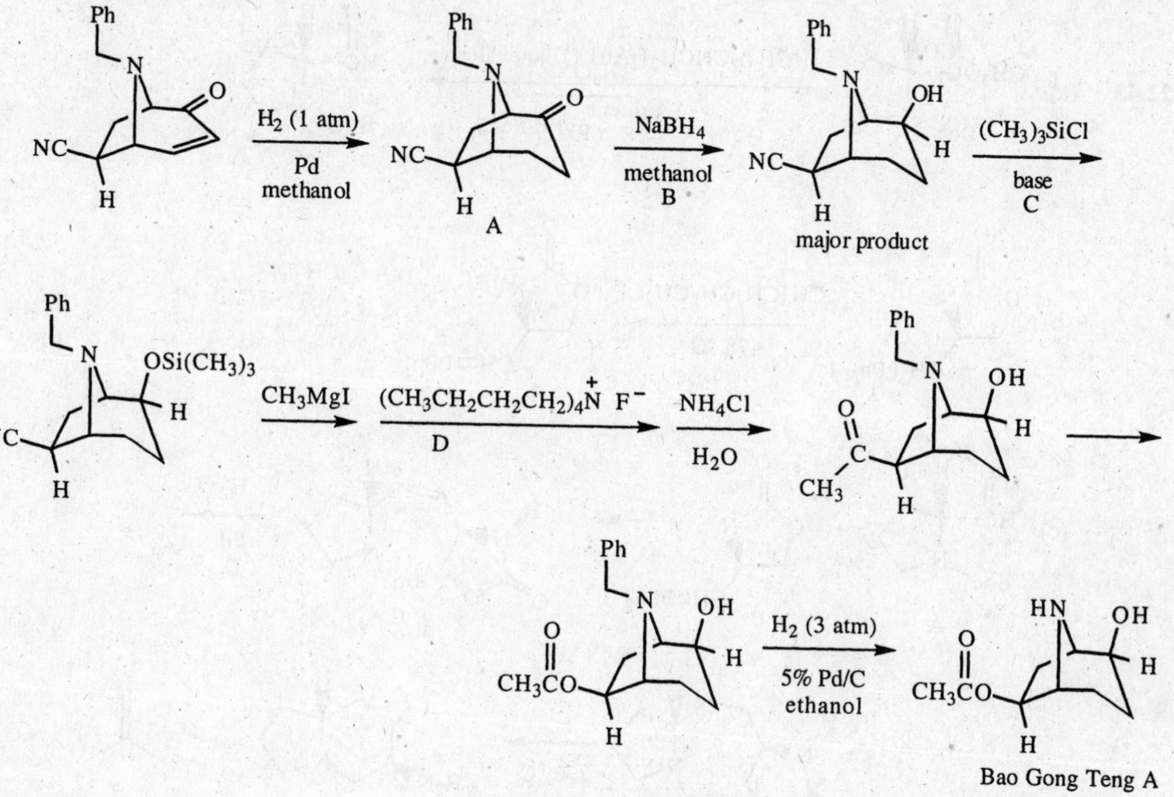

Bao Gong Teng A

22.41 $PhNHNH_2$ + $CH_3CH=CHCOOH$ →

product from nucleophilic addition to α,β-unsaturated compound

$C_{10}H_{12}N_2O$

$PhNHNH_2$ + $CH_3CCH_2COCH_2CH_3$ →

product from nucleophilic addition to carbonyl of ketone followed by loss of water to give a phenylhydrazone

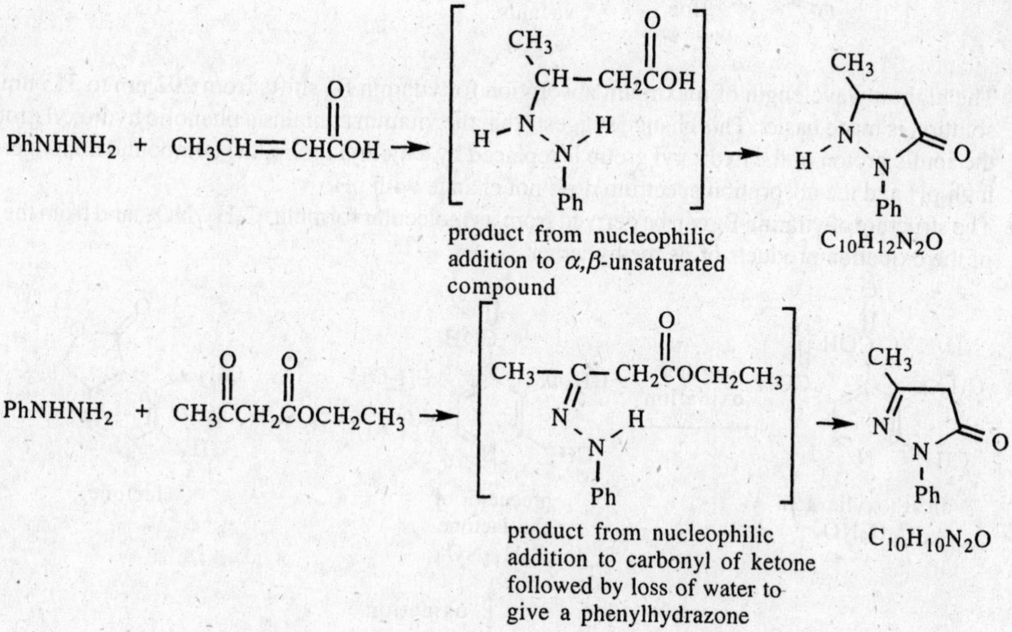

$C_{10}H_{10}N_2O$

22.42   Protein—NH_2 + $CH_3CCH_2CH_2CCH_3$ → Protein—N + 2 H_2O

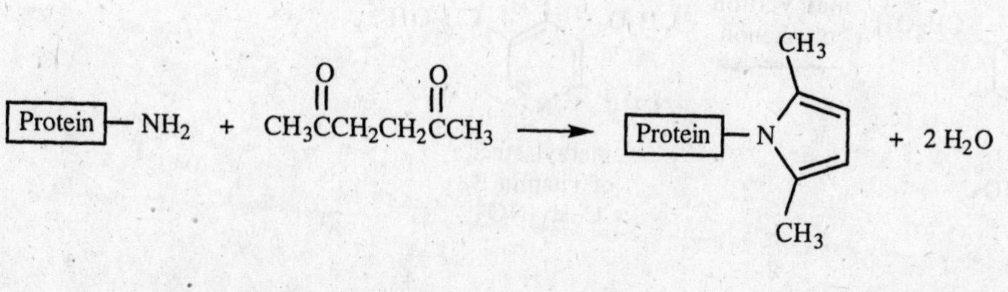

22.43   (a)

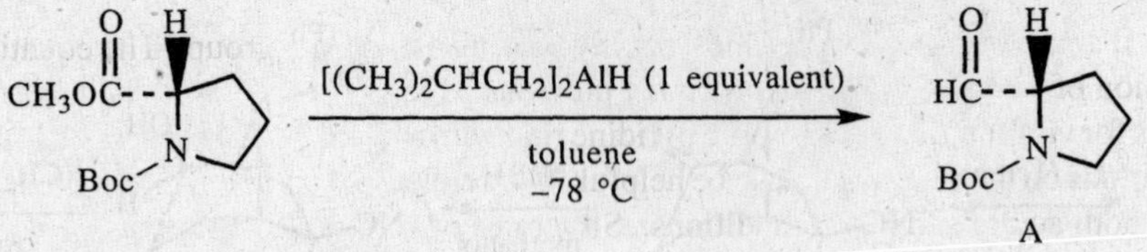

(b)

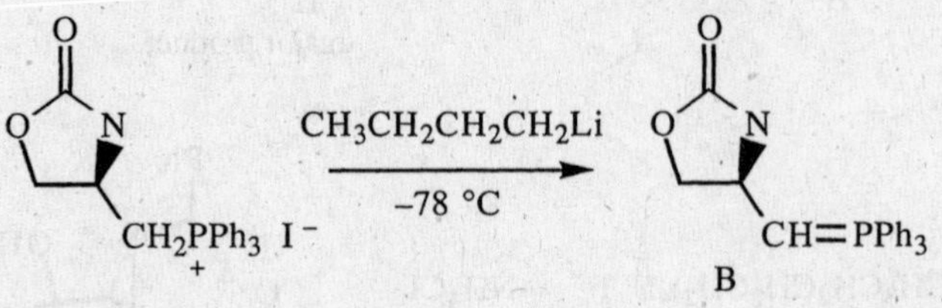

(c)

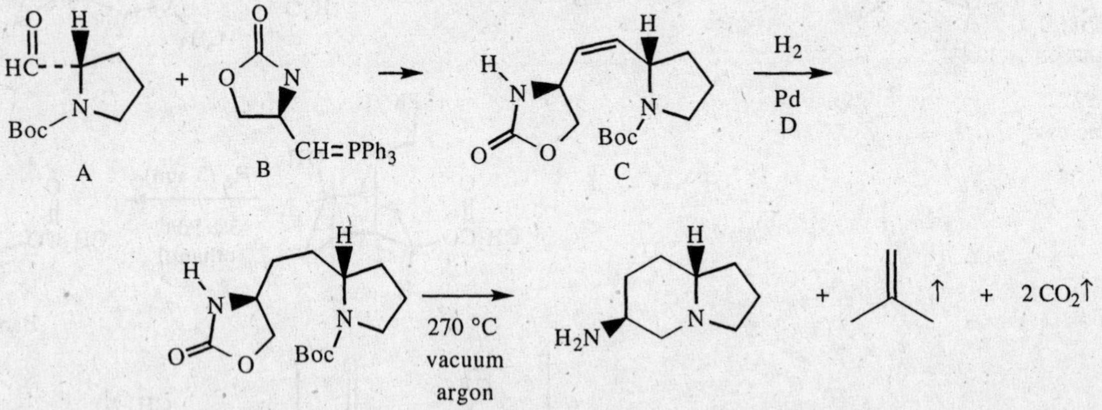

22.44   (a)   The highest wavelength of maximum absorption for vitamin B_6 shifts from 292 nm to 315 nm when the solution is made basic. This change suggests that the vitamin contains a phenolic hydroxyl group. When the acidic proton of the hydroxyl group is replaced by a methyl group, the compound no longer ionizes at high pH and the absorption spectrum does not change with pH.

  (b)   The structure of vitamin B_6 can be derived from its molecular formula, $C_8H_{11}NO_3$, and from the structures of the oxidation products of its methyl ether.

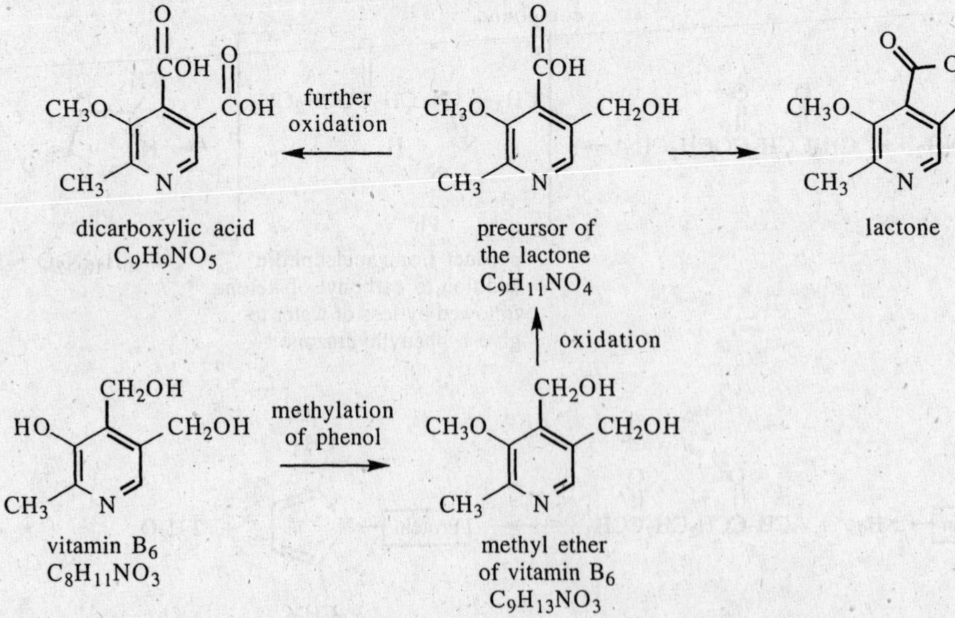

dicarboxylic acid
$C_9H_9NO_5$

precursor of
the lactone
$C_9H_{11}NO_4$

lactone

vitamin B_6
$C_8H_{11}NO_3$

methyl ether
of vitamin B_6
$C_9H_{13}NO_3$

22.44 (b) (cont)

The side chains that are oxidized are the ones already bearing a hydroxyl group. The equations outlining the transformation of vitamin B_6 are given on the previous page.

(c) Possible models for vitamin B_6 should have a pyridine ring with a hydroxyl substituent. Alkyl substituents at different positions on the ring would also be helpful. Ultraviolet spectra of the model compounds should be taken under both acidic and basic conditions. Structures of some possible models are:

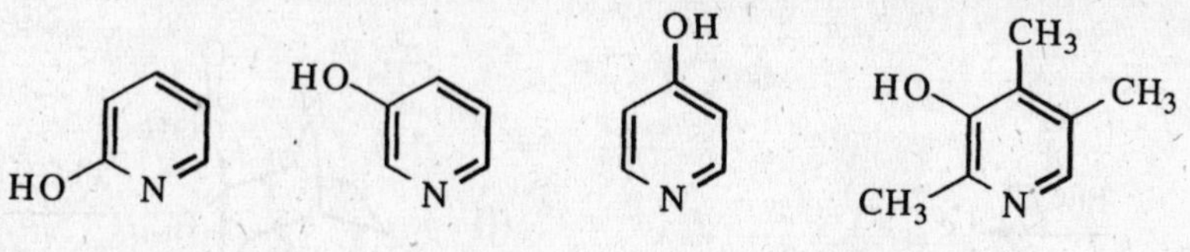

Supplemental Problems

S22.1   For the following set of bases, discuss the trend observed for the pK_a values given, which are for the conjugate acids of the compounds shown.

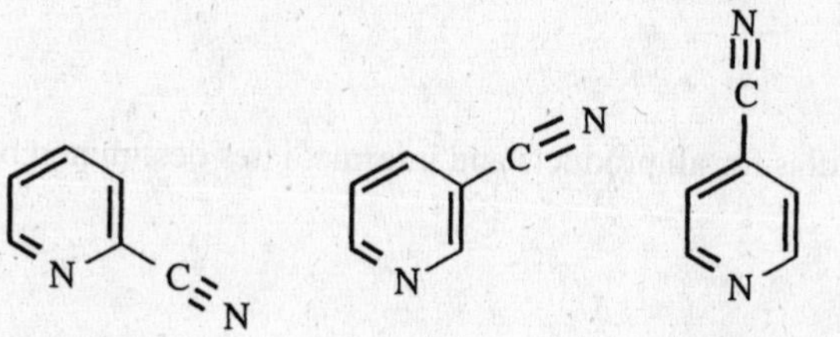

pK_a -0.3 pK_a 1.5 pK_a 1.9

S22.2   Write structural formulas for all intermediates and products designated by letters in the following equations.

(a) $\underset{O}{\overset{O}{\underset{\|}{\overset{\|}{PhCCH_2CCH_3}}}}$ + PhNHNH$_2$ $\xrightarrow{\Delta}$ A

(Hint: Which carbonyl group is more likely to undergo nucleophilic attack first?)

(b) $\underset{O}{\overset{O}{\underset{\|}{\overset{\|}{PhCCH_2CCH_3}}}}$ $\xrightarrow[\Delta]{\overset{+}{H_2NNH_3}\ HSO_4^-,\ NaOH}$ B

(c) 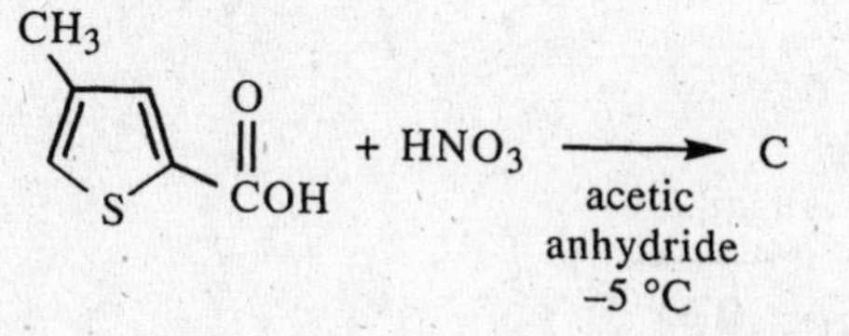   + HNO$_3$ $\xrightarrow[\substack{acetic \\ anhydride \\ -5\ ^\circ C}]{}$ C

(d)   + (CH$_3$C)$_2$O $\xrightarrow[\substack{(CH_3CH_2)_2O \cdot BF_3 \\ 100\ ^\circ C}]{}$ D

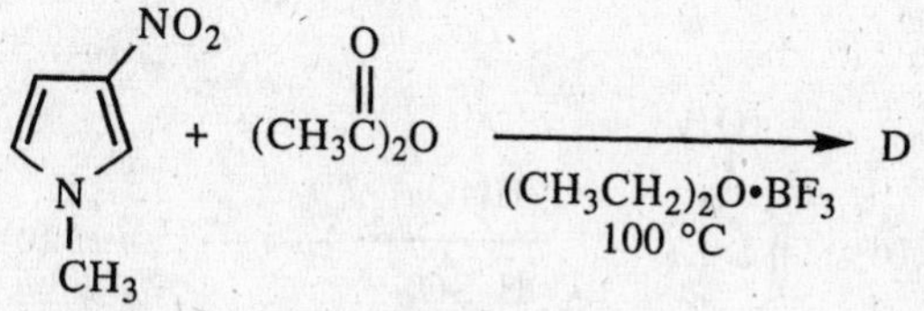

(e) 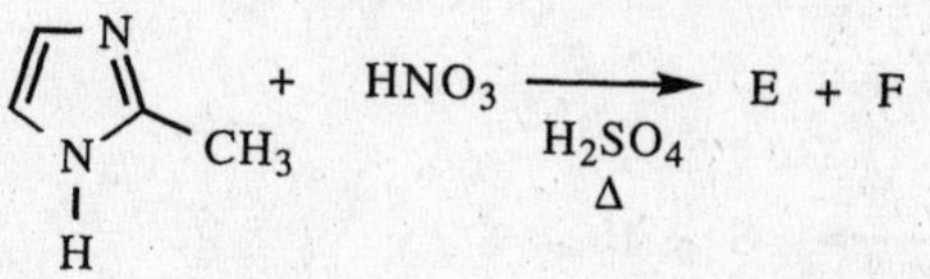   + HNO$_3$ $\xrightarrow[\substack{H_2SO_4 \\ \Delta}]{}$ E + F

(f) $\xrightarrow[\substack{H_2O \\ 0\ ^\circ C}]{NaNO_2,\ HI}$ G

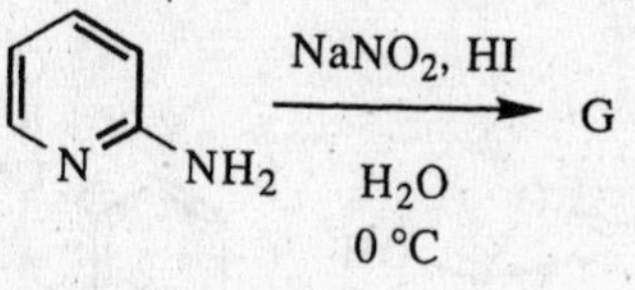

S22.2 (g)

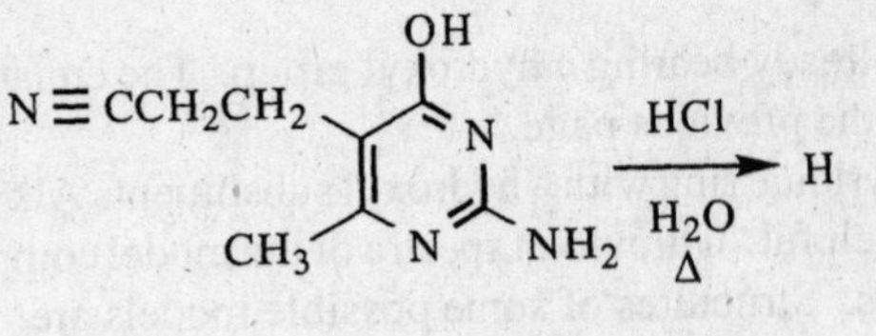

(h)

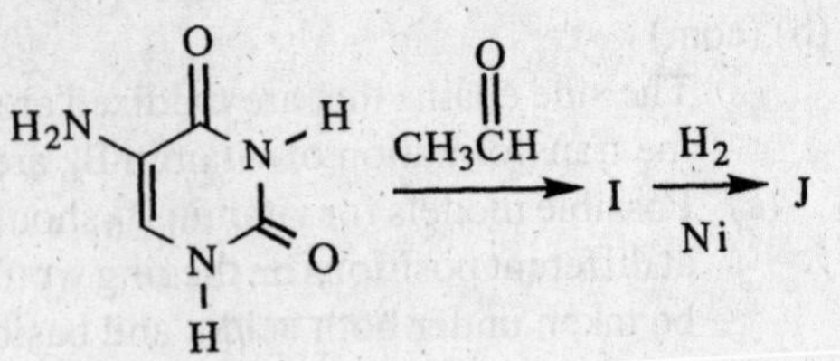

(i)

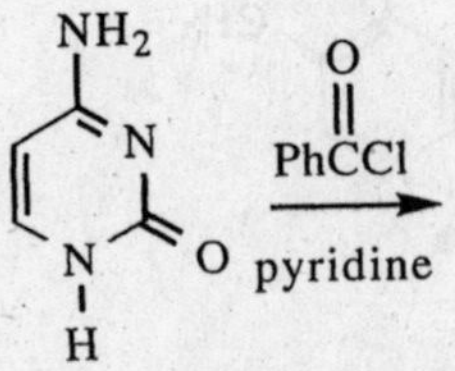

(j)

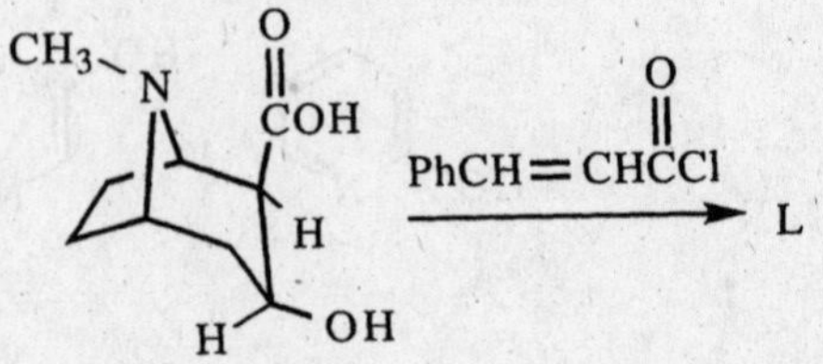

(k)

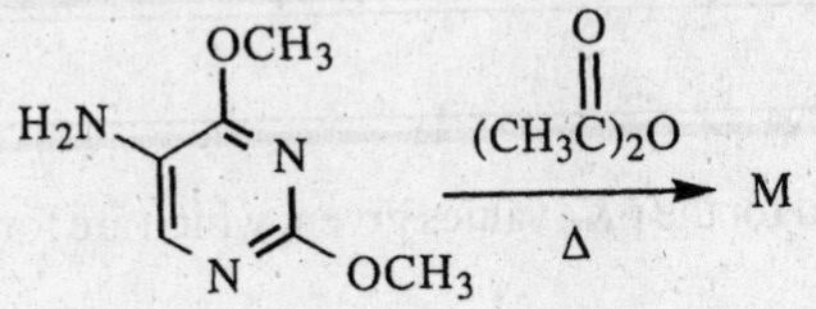

S22.3 Give structural formulas for all products and intermediates designated by letters in the following equations.

(a)

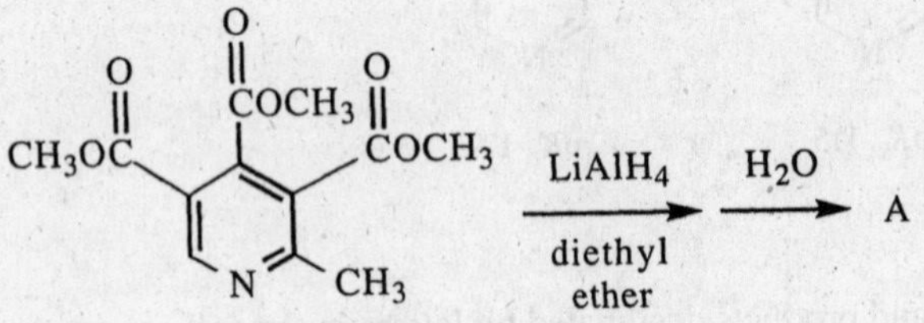

(b)

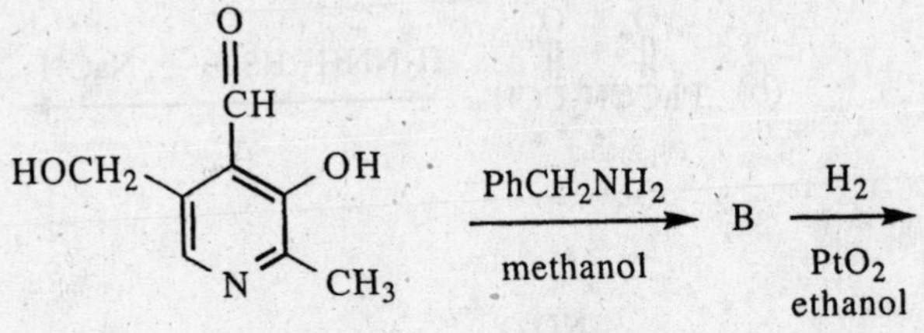

(c)

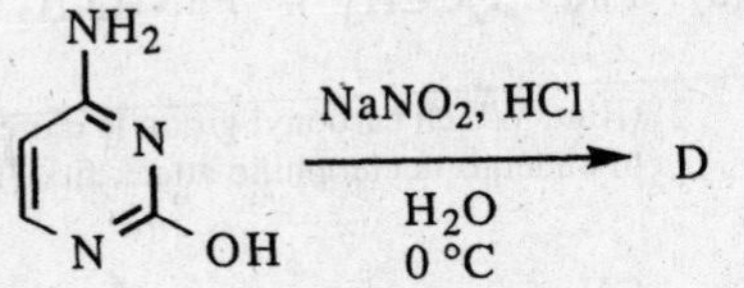

(d)

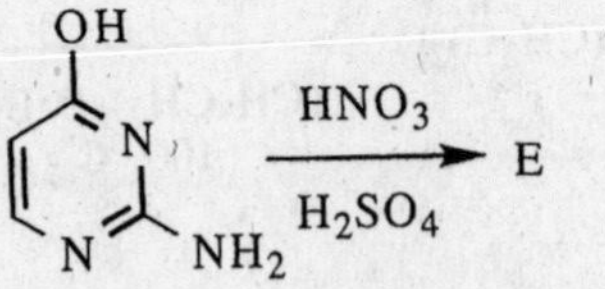

(e)

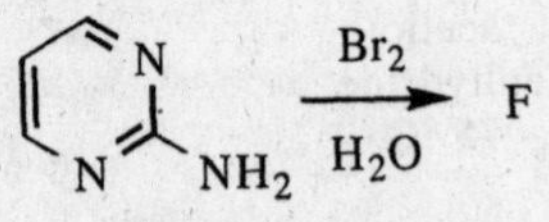

(f)

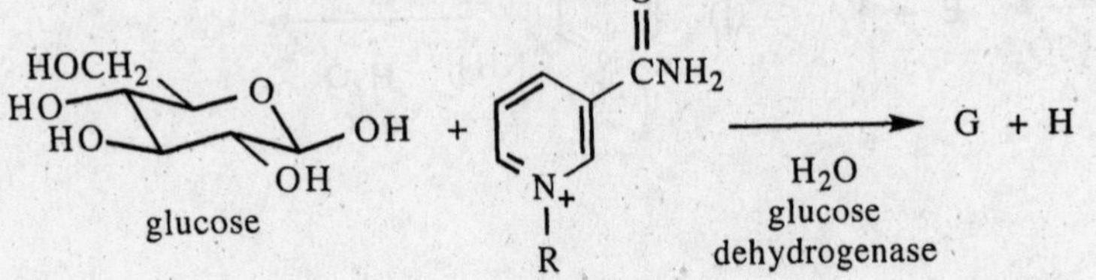

S22.3

(g)

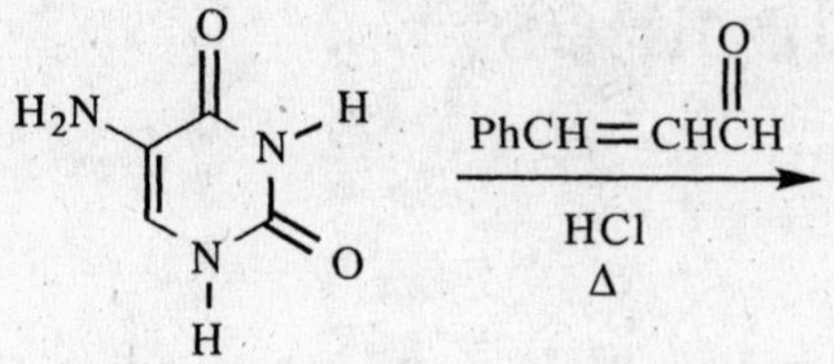

$$\text{PhCH}=\text{CHCH} \xrightarrow[\Delta]{\text{HCl}} \text{I}$$

(h)

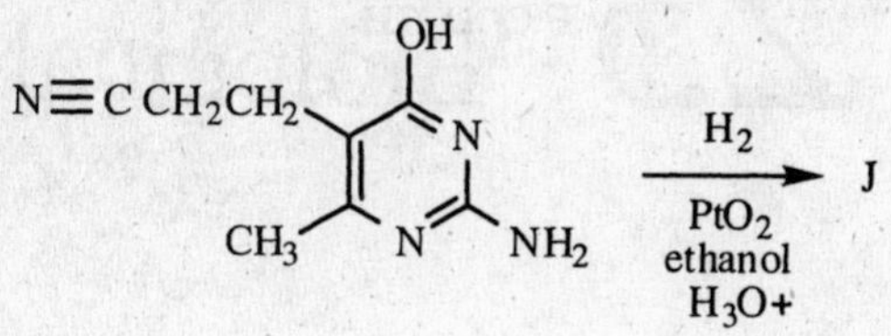

$$\xrightarrow[\substack{\text{PtO}_2 \\ \text{ethanol} \\ \text{H}_3\text{O+}}]{\text{H}_2} \text{J}$$

(i)

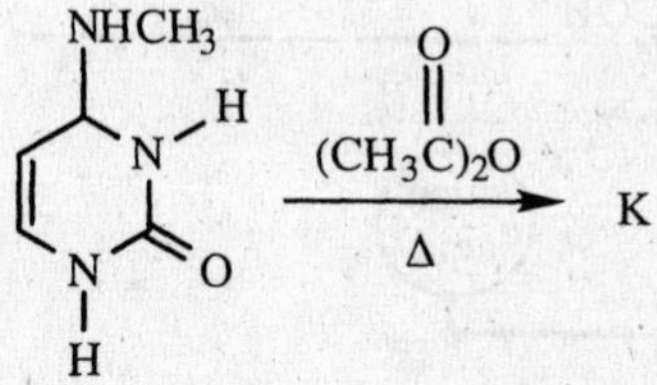

$$\xrightarrow[\Delta]{(\text{CH}_3\text{C})_2\text{O}} \text{K}$$

(j)

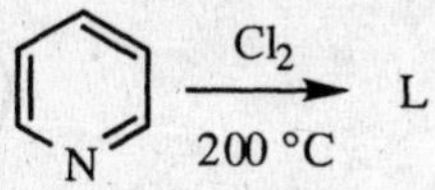

$$\xrightarrow[200\ °\text{C}]{\text{Cl}_2} \text{L}$$

(k)

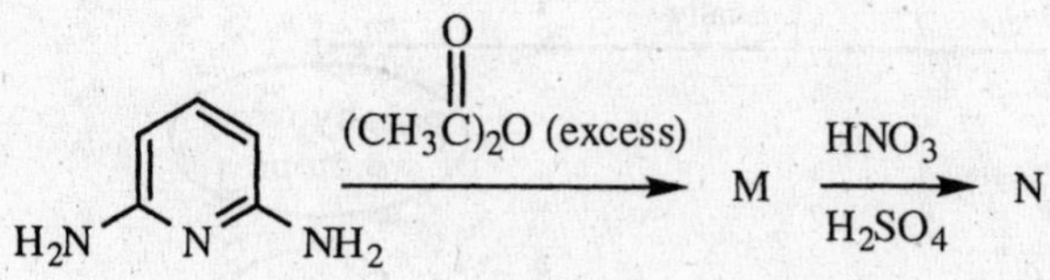

$$\xrightarrow{(\text{CH}_3\text{C})_2\text{O (excess)}} \text{M} \xrightarrow[\text{H}_2\text{SO}_4]{\text{HNO}_3} \text{N}$$

(l)

$$\xrightarrow[\substack{\text{acetic acid} \\ h\nu}]{\text{NBS}} \text{O}$$

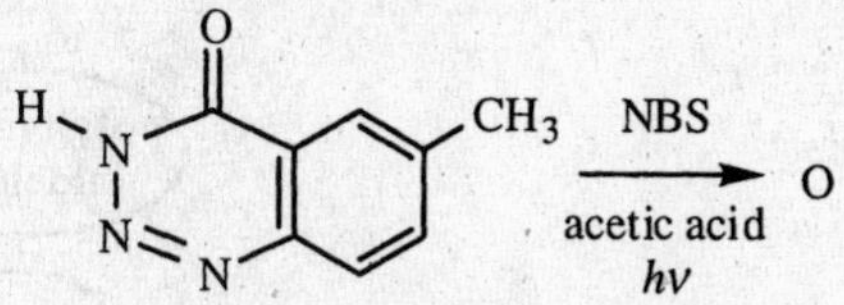

S22.4   Part of a synthesis of an analog of folic acid (p. 814 in the text) involves the following transformation. How would you carry it out?

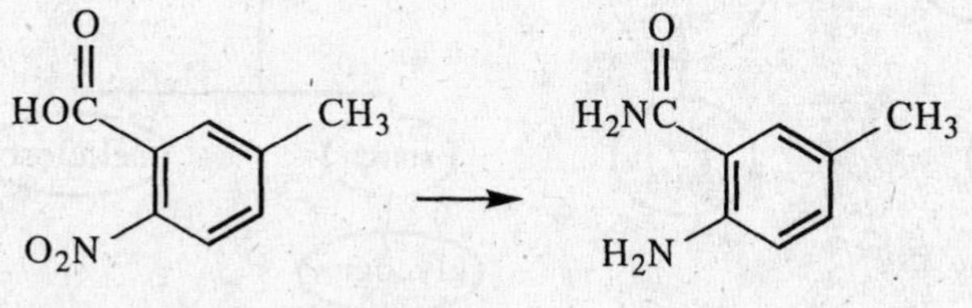

S22.5   A synthesis of an analog of folic acid was carried forward by treatment of the amine synthesized in Supplemental Problem S22.4 with nitrous acid. The product of this reaction is the heterocycle shown below. Show by writing key intermediates and a mechanism how the heterocyclic ring came into being.

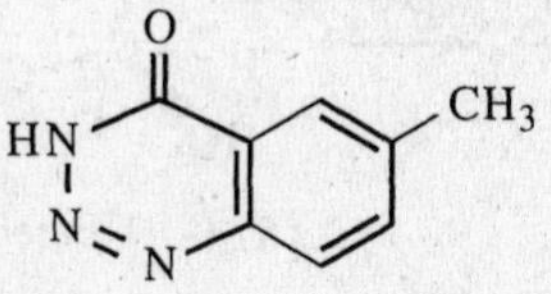

23.1   A is a ketotriose   B is an aldotriose   C is an aldotetraose   D is a ketopentose

Concept Map 23.1   Classification of carbohydrates.

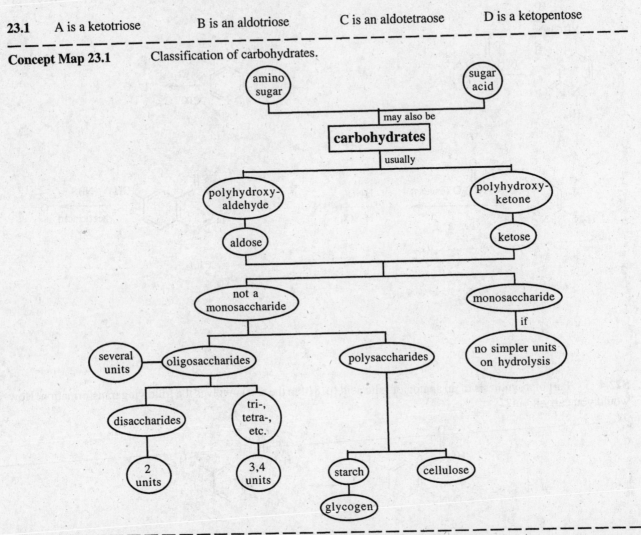

23.2   From the known configuration of (*R*)-(–)-lactic acid, we can assign the *R* configuration to (–)-1-buten-3-ol and (–)-2-butanol, because the reactions used for the interconversions do not break bonds to the stereocenter.

23.3   (a)

$$\begin{array}{c} CH_3 \\ | \\ H\!-\!\!\!-\!\!\!-\!Br \\ | \\ CH_2CH_3 \end{array}$$

(*S*)-2-bromobutane

$$\begin{array}{c} Br \\ | \\ H\!-\!\!\!-\!\!\!-\!CH_3 \\ | \\ CH_2CH_3 \end{array}$$

(*R*)-2-bromobutane

enantiomers

(b)

$$\begin{array}{c} \overset{O}{\overset{\|}{C}}OH \\ | \\ H\!-\!\!\!-\!\!\!-\!OH \\ | \\ CH_3 \end{array}$$

(*R*)-lactic acid

$$\begin{array}{c} \overset{\quad O}{H\quad \overset{\|}{C}OH} \\ | \\ HO\!-\!\!\!-\!\!\!-\! \\ | \\ CH_3 \end{array}$$

same

(c)

$$\begin{array}{c} CH_3 \\ | \\ H\!-\!\!\!-\!\!\!-\!Br \\ | \\ H\!-\!\!\!-\!\!\!-\!Br \\ | \\ CH_3 \end{array}$$

(2*S*, 3*R*)-2,3-
dibromobutane

$$\begin{array}{c} CH_3 \\ | \\ H\!-\!\!\!-\!\!\!-\!Br \\ | \\ H\!-\!\!\!-\!\!\!-\!CH_3 \\ | \\ Br \end{array}$$

(2*S*, 3*S*)-2,3-
dibromobutane

diastereomers

(d)

$$\begin{array}{c} \overset{O}{\overset{\|}{C}OH} \\ | \\ H\!-\!\!\!-\!\!\!-\!OH \\ | \\ HO\!-\!\!\!-\!\!\!-\!H \\ | \\ \underset{O}{\overset{}{C}OH} \end{array}$$

(2*R*, 3*R*)-2,3-
dihydroxy
butanedioic acid

$$\begin{array}{c} \overset{\quad O}{H\quad \overset{\|}{C}OH} \\ | \\ HO\!-\!\!\!-\!\!\!-\!COH \\ | \\ HO\!-\!\!\!-\!\!\!-\!COH \\ | \\ H\quad \overset{}{O} \end{array}$$

(2*R*, 3*S*)-2,3-
dihydroxy
butanedioic acid

diastereomers

23.4   D-(+)-glyceraldehyde ≡ (*R*)-(+)-glyceraldehyde
D-(−)-lactic acid ≡ (*R*)-(−)-lactic acid
D-(−)-tartaric acid ≡ (2*S*, 3*S*)-(−)-2,3-dihydroxybutanedioic acid
D-(−)-ribose ≡ (2*R*, 3*R*, 4*R*)-(−)-2,3,4,5-tetrahydroxypentanal
D-(+)-glucose ≡ (2*R*, 3*S*, 4*R*, 5*R*)-(+)-2,3,4,5-pentahydroxyhexanal
D-(−)-fructose ≡ (3*S*, 4*R*, 5*R*)-(−)-1,3,4,5,6-pentahydroxy-2-hexanone
L-(−)-glyceraldehyde ≡ (*S*)-(−)-glyceraldehyde
L-(+)-lactic acid ≡ (*S*)-(+)-lactic acid
L-(+)-tartaric acid ≡ (2*R*, 3*R*)-(+)-2,3-dihydroxybutanedioic acid
L-(+)-alanine ≡ (*S*)-(+)-alanine
L-(+)-arabinose ≡ (2*R*, 3*S*, 4*S*)-(+)-2,3,4,5-tetrahydroxypentanal

23.5

$$\begin{array}{c} \overset{O}{\overset{\|}{C}H} \\ | \\ HO\!-\!\!\!-\!\!\!-\!H \\ | \\ H\!-\!\!\!-\!\!\!-\!OH \\ | \\ HO\!-\!\!\!-\!\!\!-\!H \\ | \\ HO\!-\!\!\!-\!\!\!-\!H \\ | \\ CH_2OH \end{array}$$

L-(−)-glucose

23.6 (a)

```
        O
        ‖
        CH
   H ——— OH
  HO ——— H
  HO ——— H
   H ——— OH
   H ——— OH
      CH2OH
```

(b)

```
        O
        ‖
        CH
  HO ——— H
   H ——— H
   H ——— OH
   H ——— OH
       CH3
```

(c)

```
        O
        ‖
        CH
   H ——— OH
  HO ——— H        O
                  ‖
   H ——— NHCCH3
   H ——— OH
       CH3
```

Concept Map 23.2   Amino acids, polypeptides, proteins.

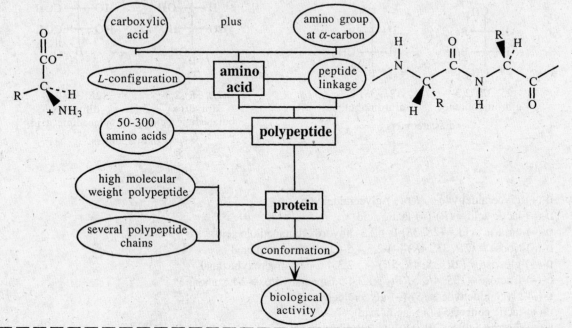

Concept Map 23.3   Acid-base properties of amino acids.

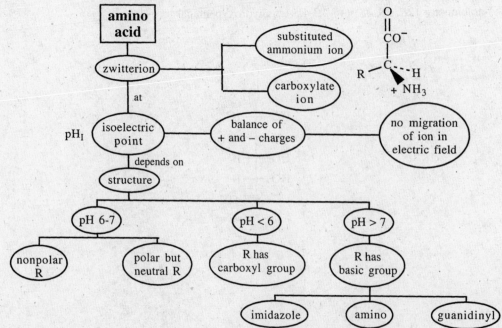

23.7 (a)

$$HOCCH_2CHCO^- \underset{H_3O^+}{\overset{OH^-}{\rightleftharpoons}} {}^-OCCH_2CHCO^-$$

aspartic acid

$$HOCCH_2CH_2CHCO^- \underset{H_3O^+}{\overset{OH^-}{\rightleftharpoons}} {}^-OCCH_2CH_2CHCO^-$$

glutamic acid

The carboxylic acid group on carbon 3 of aspartic acid is more acidic because it is closer to the electron-withdrawing effect of the ammonium ion on carbon 2 than is the acid group on carbon 4 of glutamic acid.

(b) The carboxylic acid group on carbon 3 in aspartic acid is more acidic than the carboxylic acid group on carbon 4 in glutamic acid, therefore the carboxylic acid group in aspartic acid has a lower pK_a. The pH required to keep that group protonated and the molecule in the balanced zwitterionic state would also be lower than the pH required for glutamic acid.

(c)

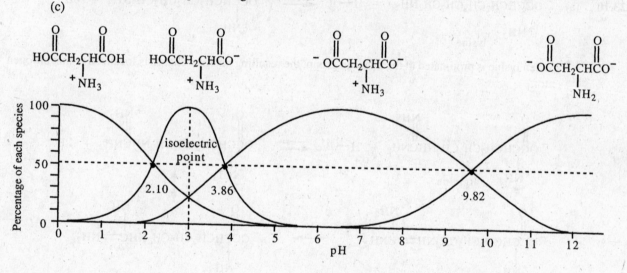

23.8 All of them have the *S* configuration except for cysteine, which has the *R* configuration.

In all other amino acids the carboxylate group is of higher priority than the rest of the chain. The presence of sulfur in cysteine makes that group on the stereocenter of higher priority than the carboxylate group.

23.9   Isoleucine, threonine, and hydroxyproline have more than one stereocenter.

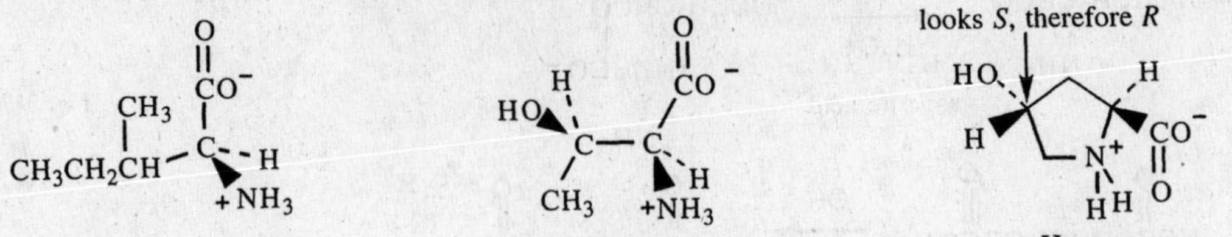

Ile	Thr	Hypro
configuration not specified at carbon 3	(2S, 3R)-2-amino-3-hydroxybutanoic acid	(2S, 4R)-4-hydroxyproline

23.10   (a)

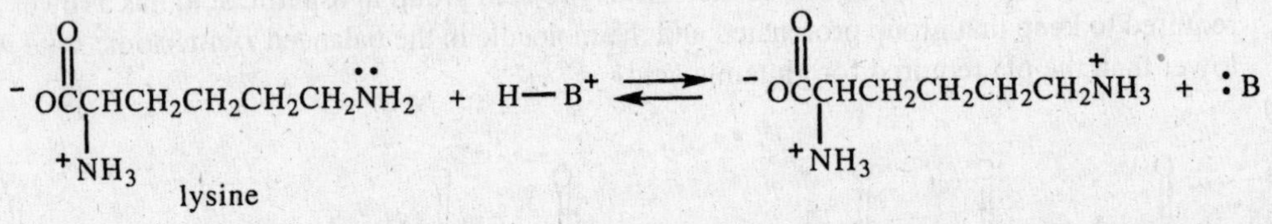

$$^-OCCHCH_2CH_2CH_2CH_2NH_2 \;+\; H-B^+ \;\rightleftharpoons\; {}^-OCCHCH_2CH_2CH_2CH_2\overset{+}{N}H_3 \;+\; :B$$

lysine

When lysine is protonated at the ε-amino group, the resulting positive charge is localized at that nitrogen atom.

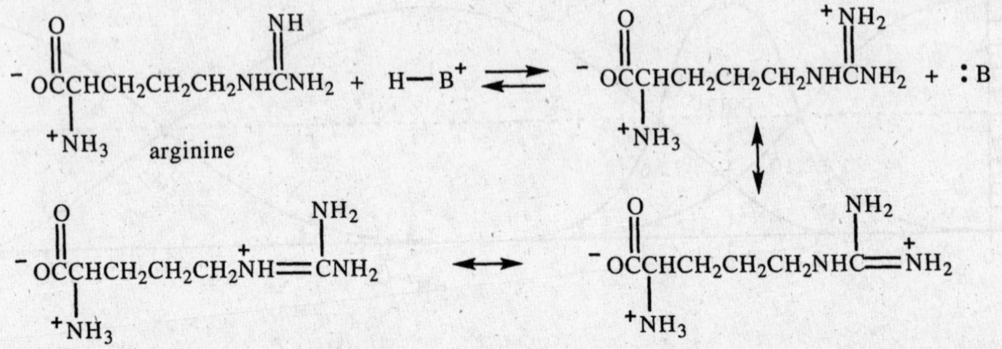

When the guanidinyl group in arginine is protonated, the resulting cation is stabilized by delocalization of the charge to three nitrogen atoms. The guanidinyl group is, therefore, more basic, more easily protonated, than an amino group. The conjugate acid of guanidine, with a pK_a of 12.5, is a weaker acid than the conjugate acid of a primary amine, $pK_a \sim 10.5$.

23.10   (b)

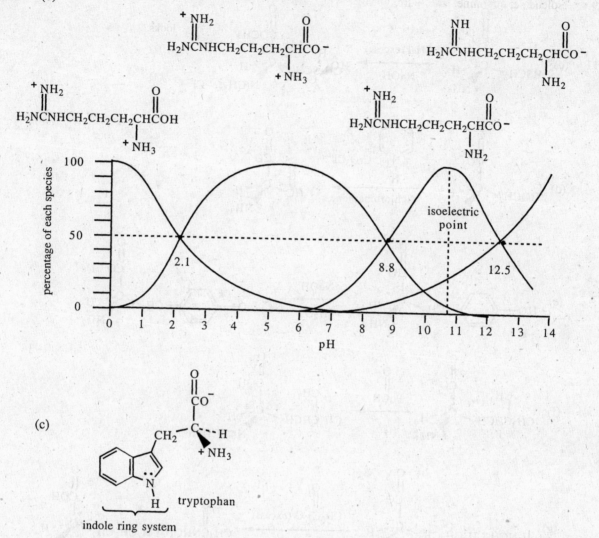

(c)

tryptophan

indole ring system

The nonbonding electrons on the nitrogen atom in the indole ring are part of an aromatic sextet and are therefore not available for protonation. Protonation of that nitrogen atom would lead to loss of aromaticity, so the difference in energy between indole and its conjugate acid would be very large.

23.11   (a)

23.11 (c)

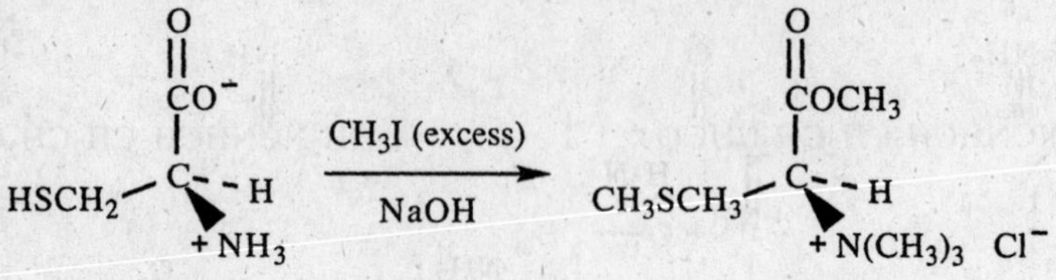

(d)

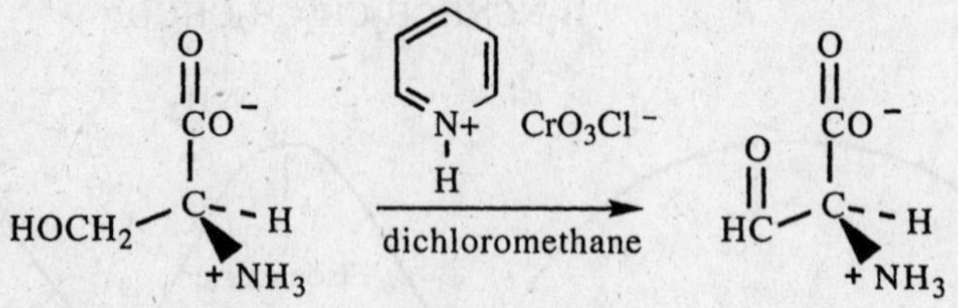

(e)

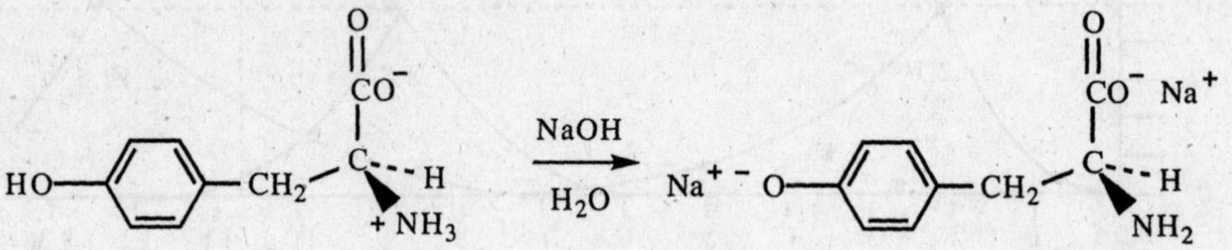

(f)

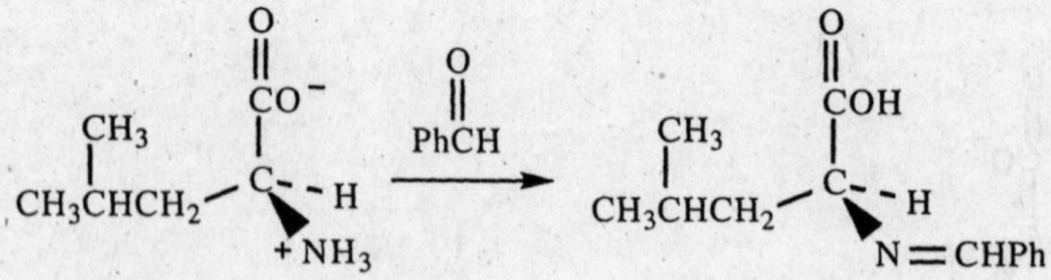

(g)

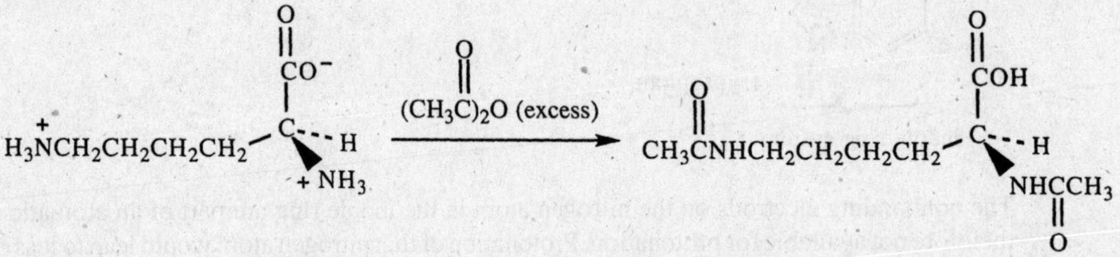

(h)

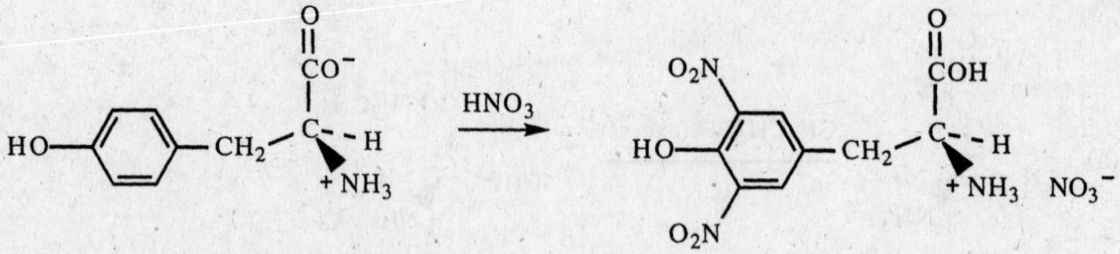

23.12 (a)

(b)

(c)

(d)

(e)

(f)

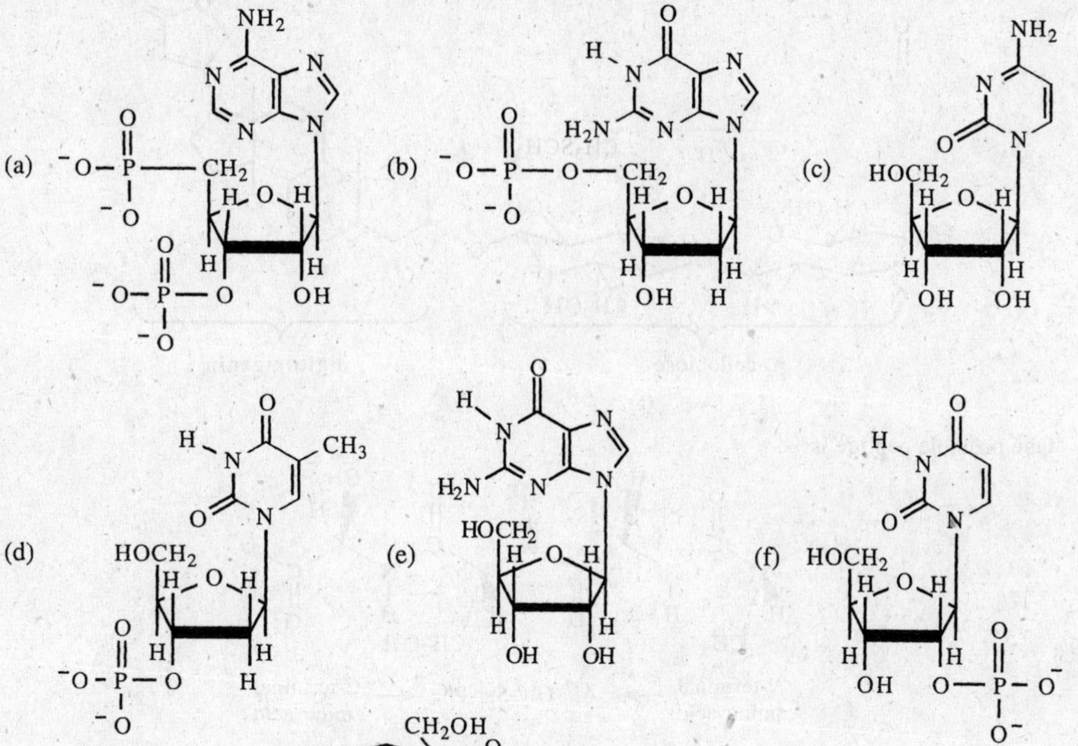

23.13

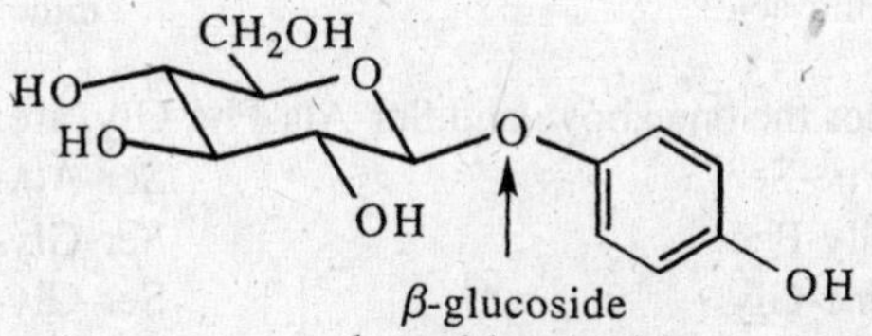

β-glucoside
cleaved by emulsin
arbutin

23.14 Yes, lactose has a hemiacetal function at carbon 1 of its glucose unit, which means that it is in equilibrium with its open chain aldehyde form and can undergo easy oxidation to a carboxylic acid.

23.15

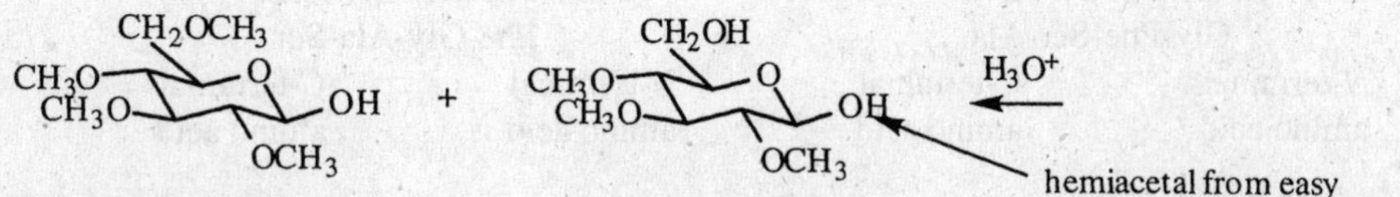

2,3,4,6-tetra-*O*-methyl-
D-glucopyranose
(therefore hydroxyl group
at C-1 in glycosidic linkage)

+

2,3,4-tri-*O*-methyl-
D-glucopyranose
(therefore hydroxyl group
at C-6 in glycosidic linkage)

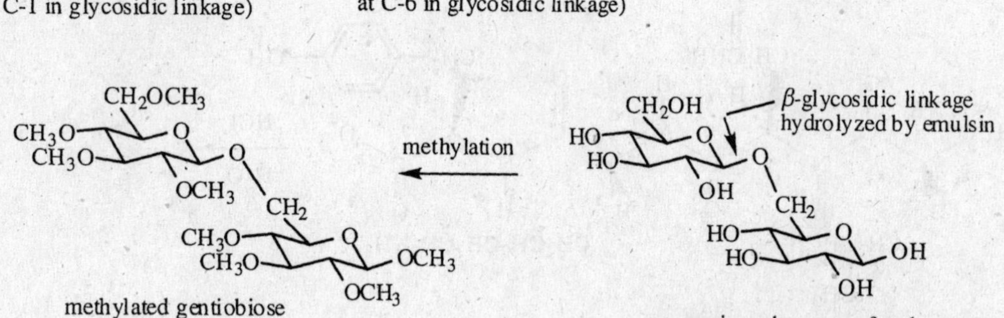

H_3O^+

hemiacetal from easy
hydrolysis in acid of full
methyl acetal

methylated gentiobiose

methylation

β-glycosidic linkage
hydrolyzed by emulsin

anomeric carbon atom free because
gentiobiose is a reducing sugar
O-β-D-glucopyranosyl-(1 → 6)-D-glucopyranose

23.16

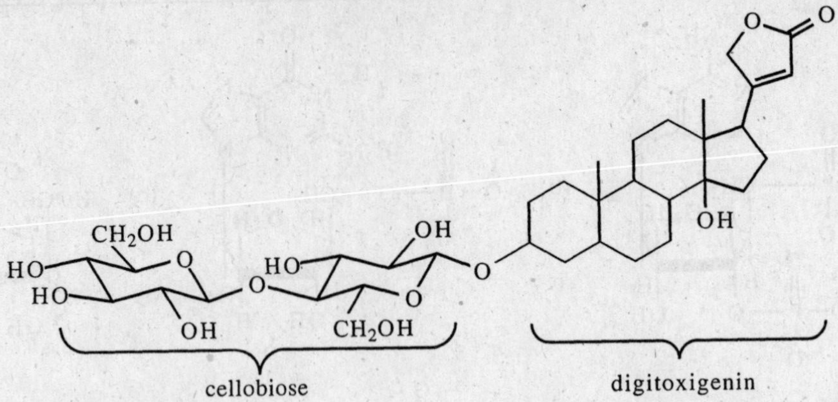

cellobiose   digitoxigenin

23.17   One possible peptide is:

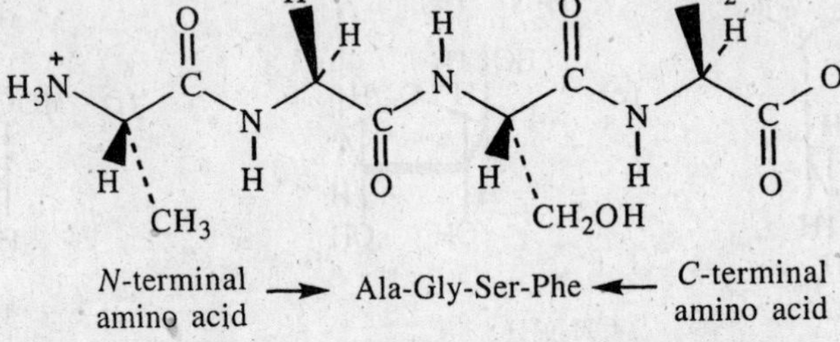

N-terminal ⟶ Ala-Gly-Ser-Phe ⟵ *C*-terminal
amino acid   amino acid

The other twenty-two, besides the one above and Ser-Ala-Phe-Gly, are:

Ala-Gly-Phe-Ser	Ser-Ala-Gly-Phe
Ala-Ser-Gly-Phe	Ser-Gly-Ala-Phe
Ala-Ser-Phe-Gly	Ser-Gly-Phe-Ala
Ala-Phe-Ser-Gly	Ser-Phe-Ala-Gly
Ala-Phe-Gly-Ser	Ser-Phe-Gly-Ala
Gly-Ala-Ser-Phe	Phe-Ala-Ser-Gly
Gly-Ala-Phe-Ser	Phe-Ala-Gly-Ser
Gly-Ser-Ala-Phe	Phe-Ser-Gly-Ala
Gly-Ser-Phe-Ala	Phe-Ser-Ala-Gly
Gly-Phe-Ala-Ser	Phe-Gly-Ser-Ala
Gly-Phe-Ser-Ala	Phe-Gly-Ala-Ser

N-terminal   *C*-terminal   *N*-terminal   *C*-terminal
amino acid   amino acid   amino acid   amino acid

23.18

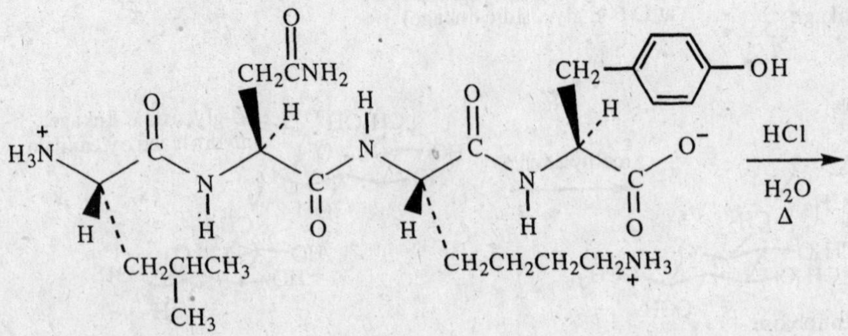

23.18 (cont)

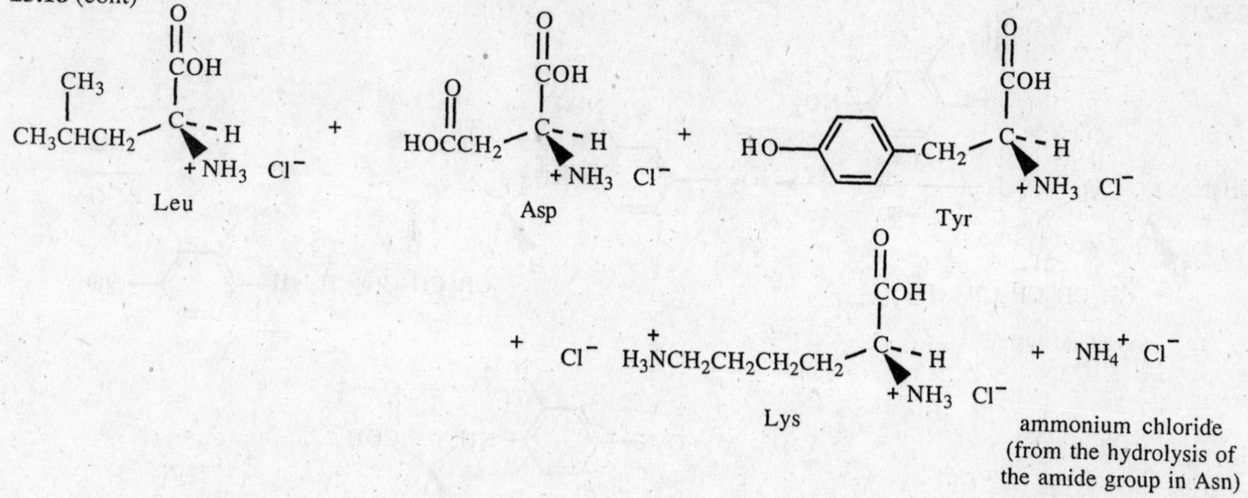

Leu   Asp   Tyr

Lys   ammonium chloride
(from the hydrolysis of
the amide group in Asn)

The hydrolysis was done in strong acid. Therefore, all amino acids are shown in their fully protonated forms.

23.19   (a) Gly-Phe-Thr-Lys will have $pH_I > 6$; it will move to the negative pole in an electric field at pH 6.
   (b) Tyr-Ala-Val-Asn will have $pH_I \sim 6$; it will not move in an electric field at pH 6.
   (c) Trp-Glu-Leu will have $pH_I < 6$; it will move to the positive pole in an electric field at pH 6.
   (d) Pro-Hypro-Gly will have $pH_I \sim 6$; it will not move in an electric field at pH 6.

23.20

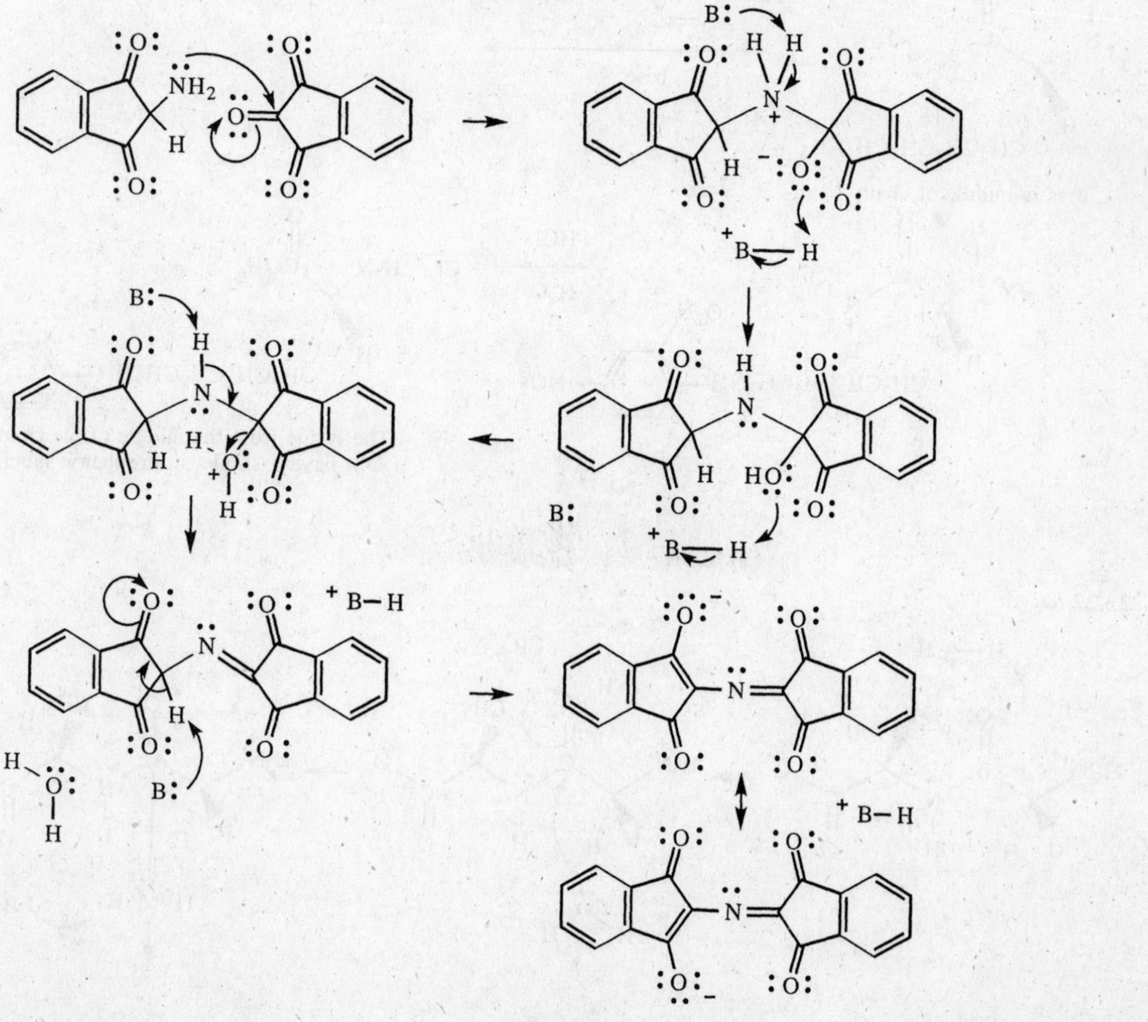

23.21

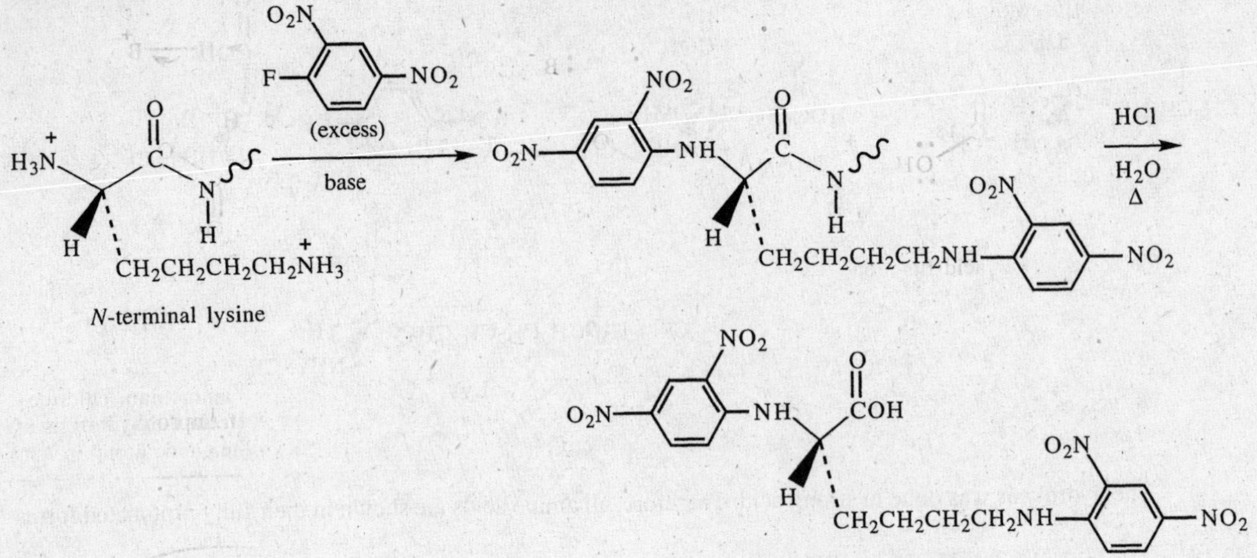

The *N*-terminal lysine
will be doubly labeled.

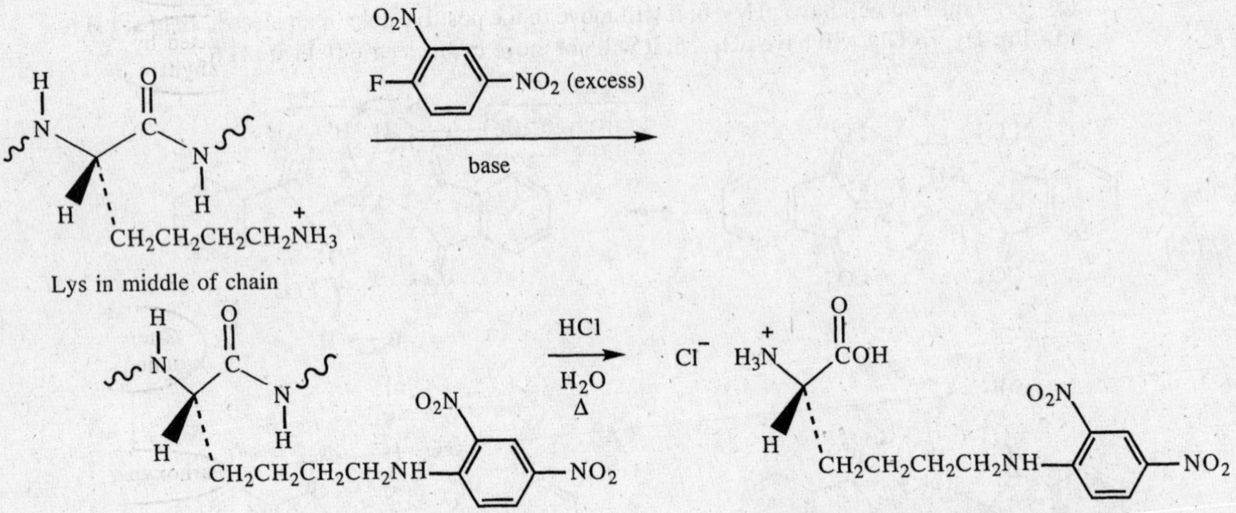

Lys in middle of chain

The lysine from the middle of the chain
will have a single dinitrophenyl label.

23.22

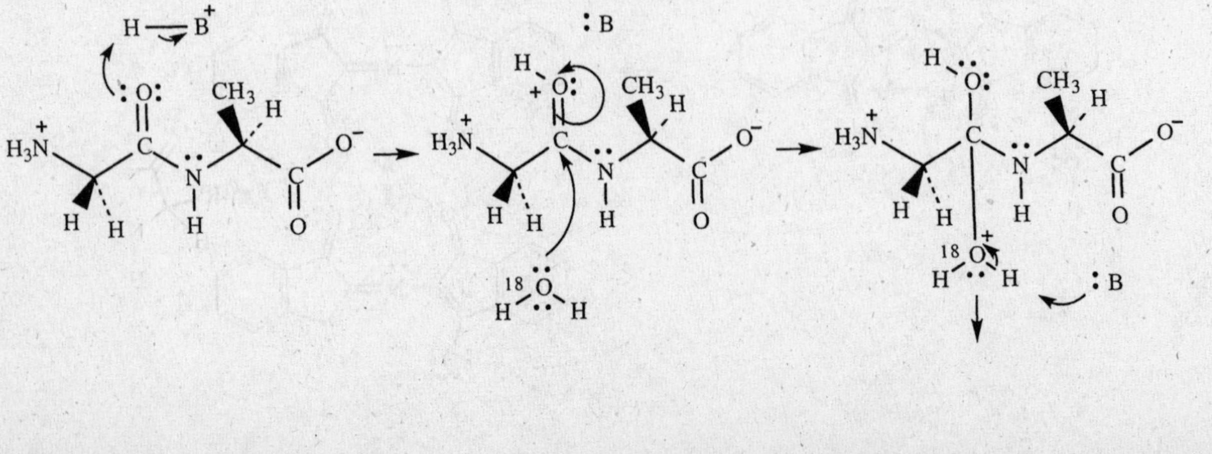

23.22 (cont)

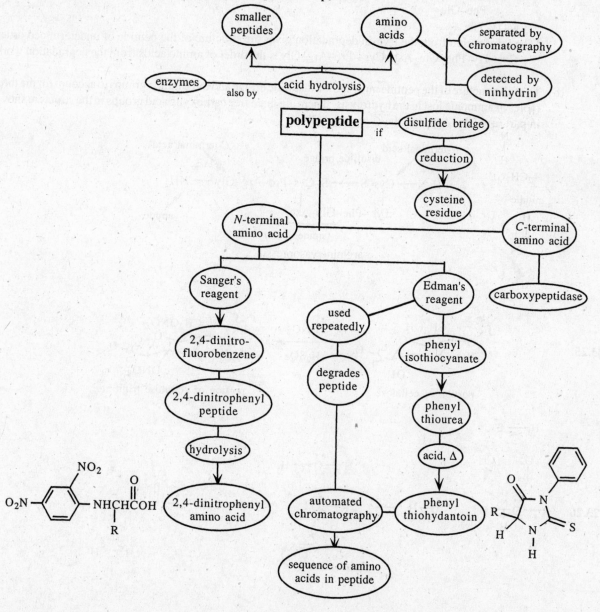

The carboxylic acid function of glycine, which was part of a peptide linkage, is labeled with ^{18}O.

No ^{18}O turns up in the carboxylic acid group of alanine, which was the C-terminal acid.

Note that glycine is shown with a protonated carboxylic acid function, and alanine has an unprotonated amino function. The final step in the reaction would be a proton transfer from glycine to alanine.

- -

Concept Map 23.4   Proof of structure of peptides and proteins.

23.23   (a) The *N*-terminal amino acid is arginine.

(b) The fragments are shown lined up below:

Arg–Gly–Pro

        Pro–Pro

               Pro–Phe–Ile–Val

               Pro–Phe–Ile

Therefore, the structure of the peptide is: Arg–Gly–Pro–Pro–Phe–Ile–Val.

23.24   (a) Cysteine is the *N*-terminal amino acid.

Glycinamide is the *C*-terminal amino acid.

Arginine is the second amino acid from the *C*-terminal end.

The peptide fragments are shown lined up below:

                     Cys–Pro–Arg–Gly (peptide of undetermined order*)

          Asp–Cys

       Glu–Asp–Cys

     Phe–Glu–Asp

     Phe–Glu

 Cys–Try–Phe

*The information from the trypsin degradation gives the structure of the peptide of undetermined order.

     Cys–Try–Phe–Glu–Asp–Cys–Pro–Arg–Gly is the order of amino acids from the degradation work.

(b) The full structure of the peptide must show a disulfide bridge and three amide groups to account for the three moles of ammonia lost in the hydrolysis. There are three free carboxylic acid groups in the structure shown in part (a): Glu, Asp, and the *C*-terminal Gly.

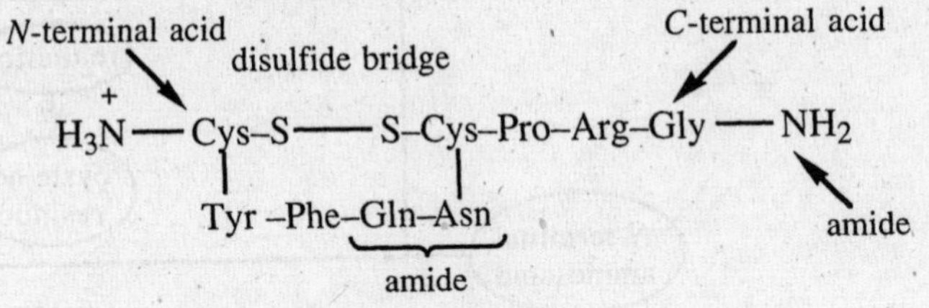

arginine vasopressin

23.25

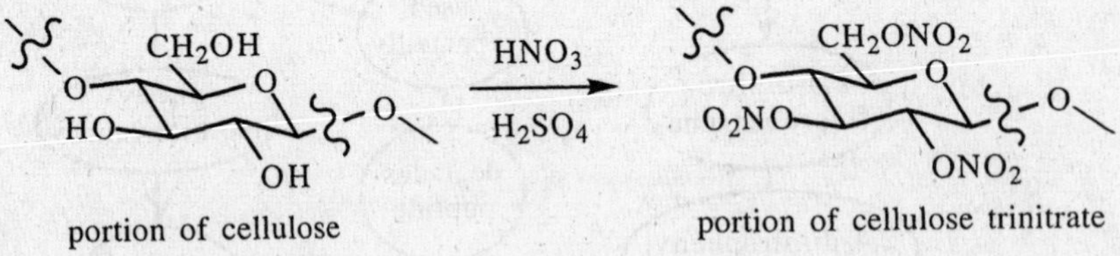

portion of cellulose   portion of cellulose trinitrate

23.26

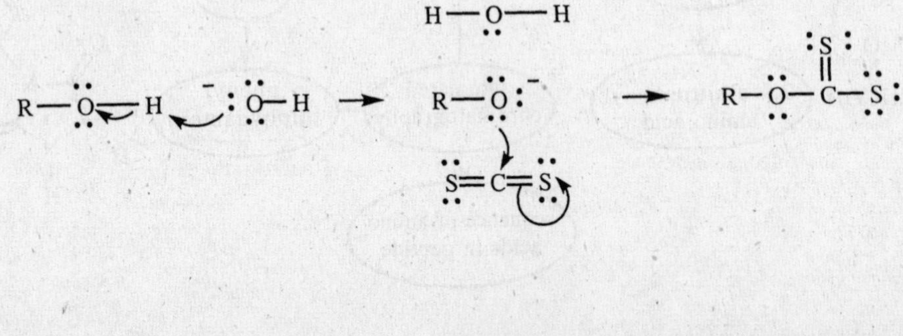

23.27

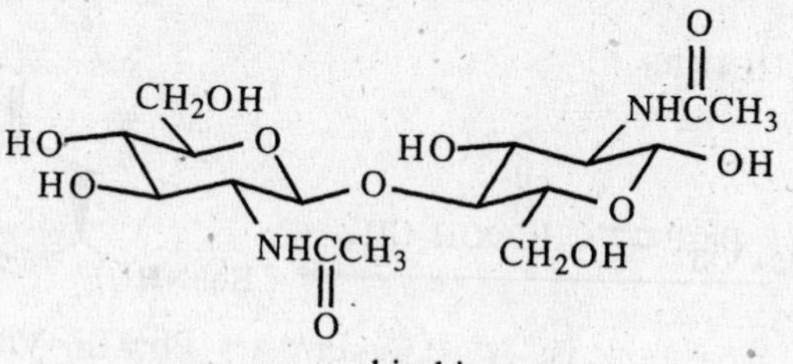

chitobiose

Concept Map 23.5   Conformation and structure in proteins.

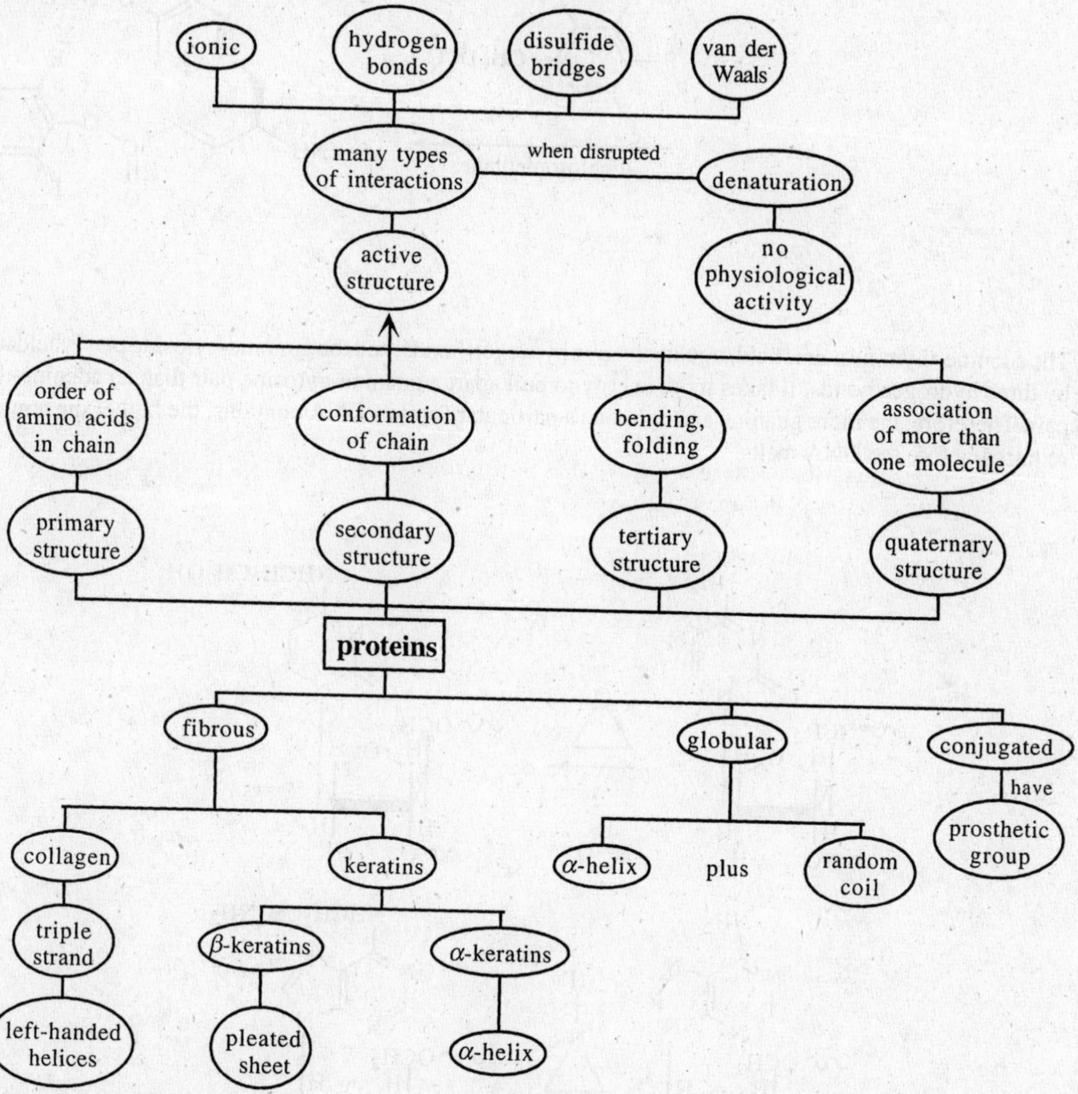

23.28

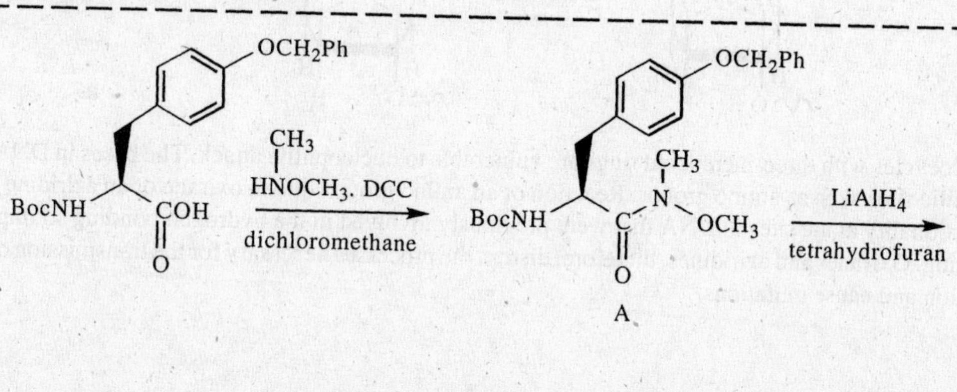

23.28 (cont)

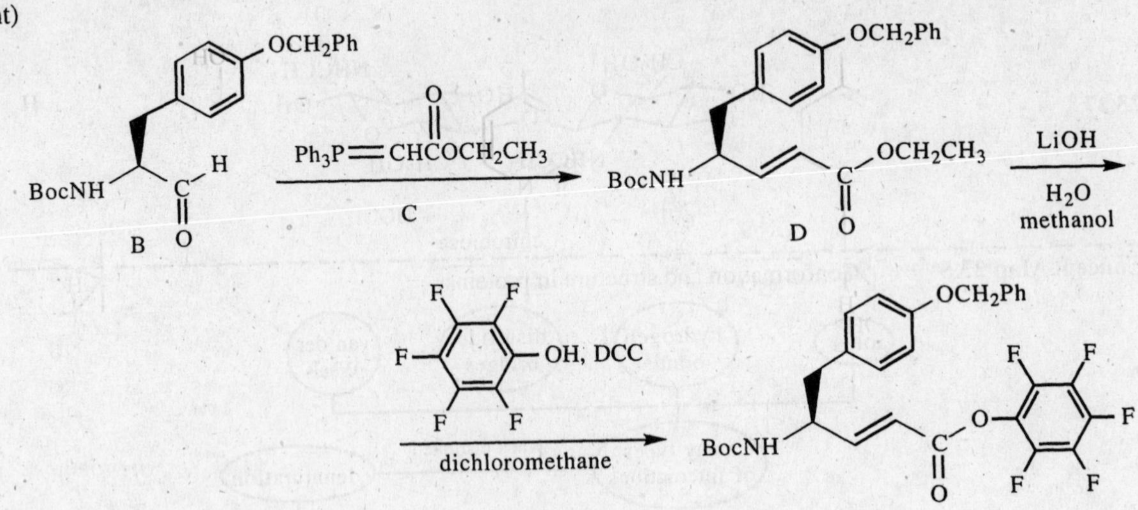

23.29 The adenine-thymine pair is held together by two hydrogen bonds, and the guanine-cytosine pair is held together by three hydrogen bonds. It takes more energy to pull apart a guanine-cytosine pair than an adenine-thymine pair. Therefore, the more guanine and cytosine a particular form of DNA contains, the higher the temperature required before the DNA melts.

23.30

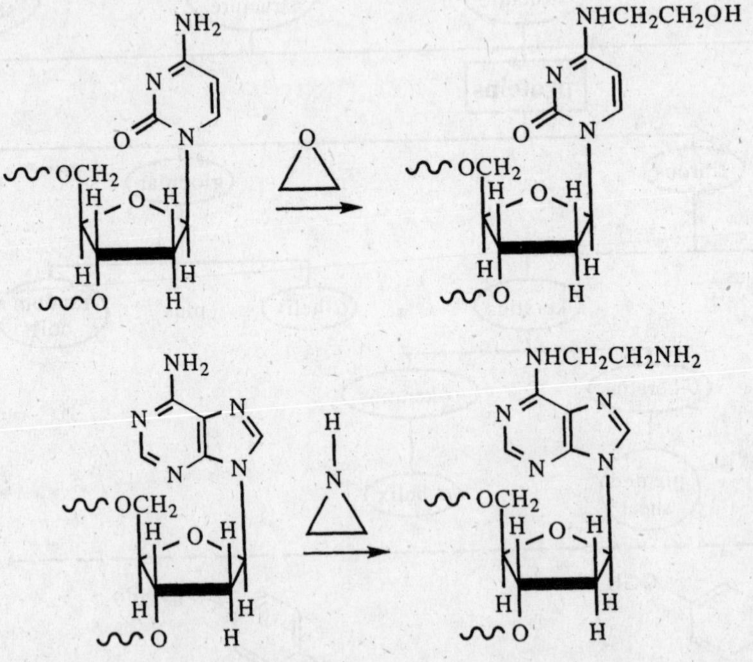

The heterocycles with three-membered rings are vulnerable to nucleophilic attack. The bases in DNA contain nucleophilic sites such as amino groups. Reaction of an amino group with an oxirane or an aziridine creates a new functionality at the sites in DNA that were previously involved in the hydrogen bonding so important to base pairing. Oxiranes and aziridines, therefore, disrupt the processes necessary for the transmission of genetic information and cause mutations.

23.31

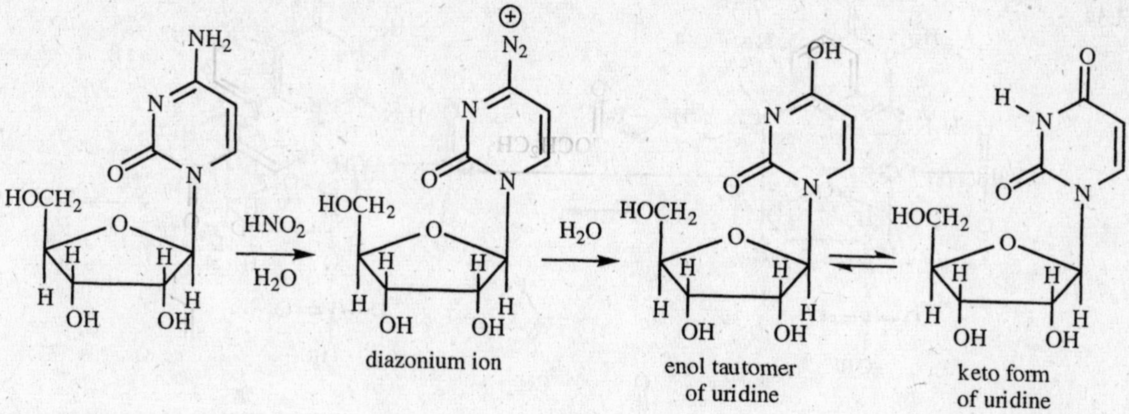
HNO2
H2O
H2O
diazonium ion
enol tautomer of uridine
keto form of uridine

A change from C to U in RNA would change the three letter code for an amino acid in protein synthesis.

23.32

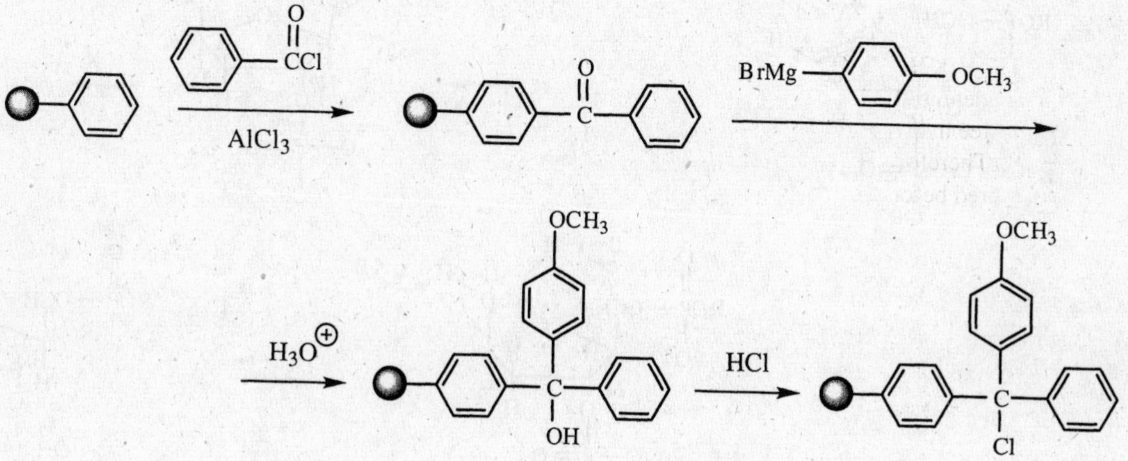
AlCl3
BrMg—C6H4—OCH3
H3O+
HCl

23.33

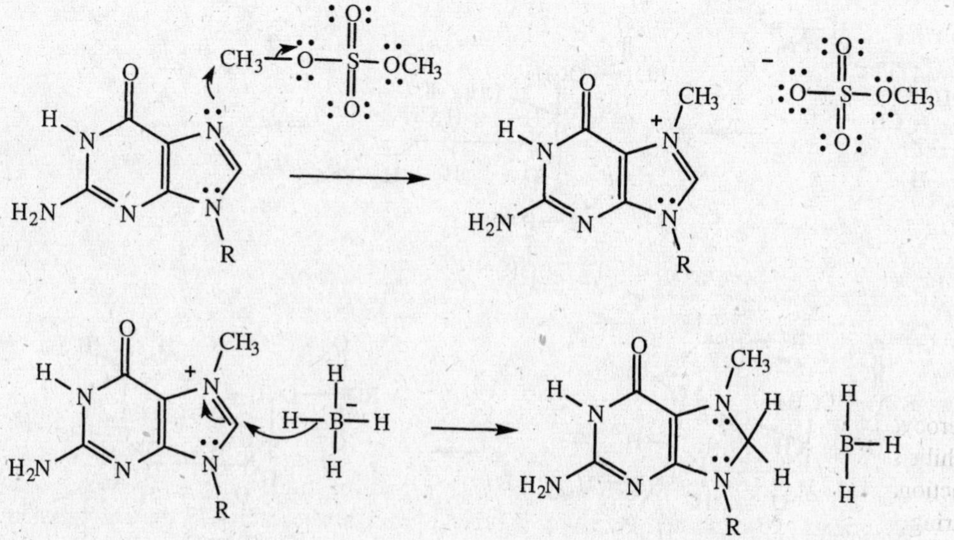

23.34

23.34 (cont)

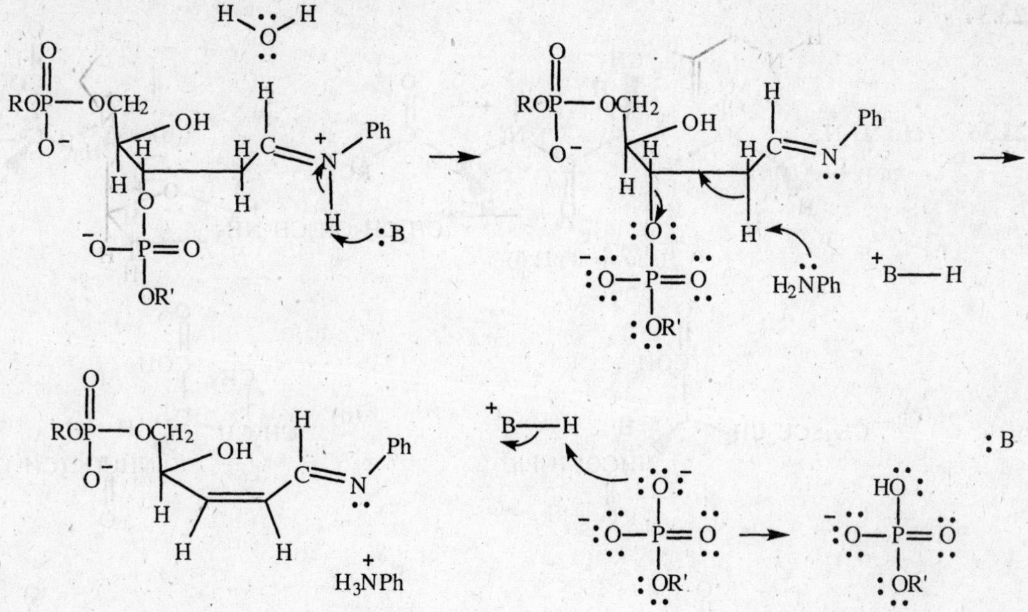

23.35

(a)

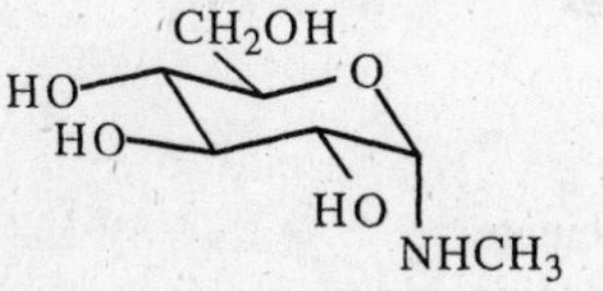

(b)

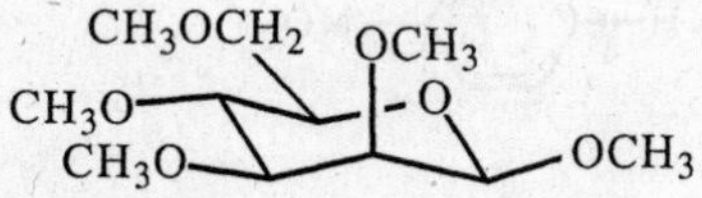

(c)

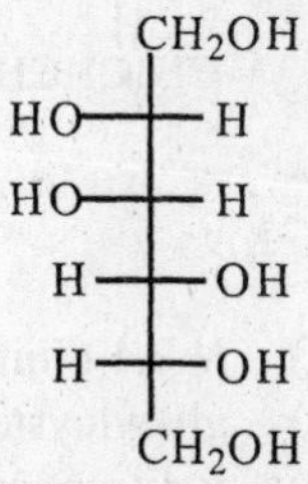

(d)

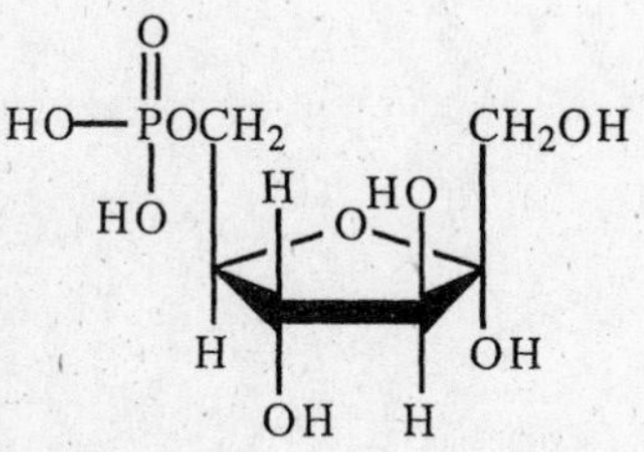

(e)

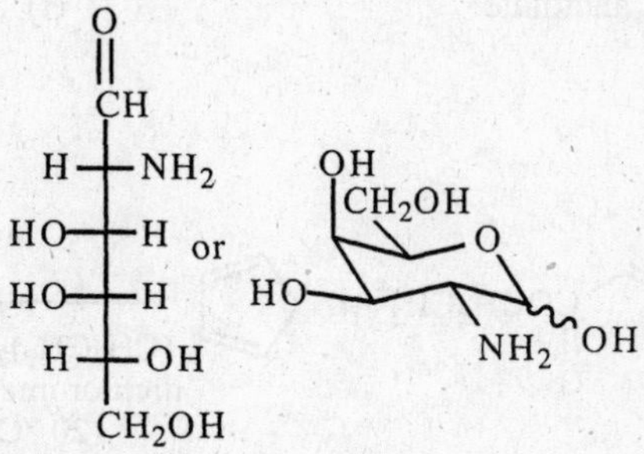

(f)

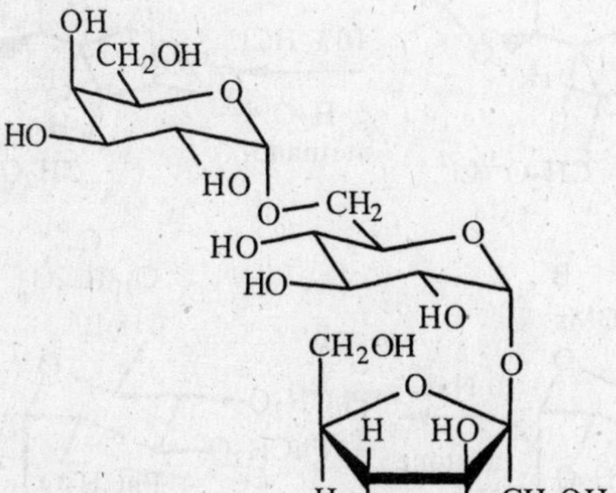

g)

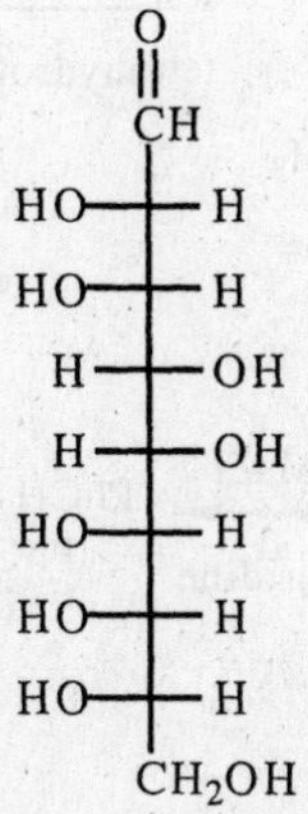

23.36 (a)

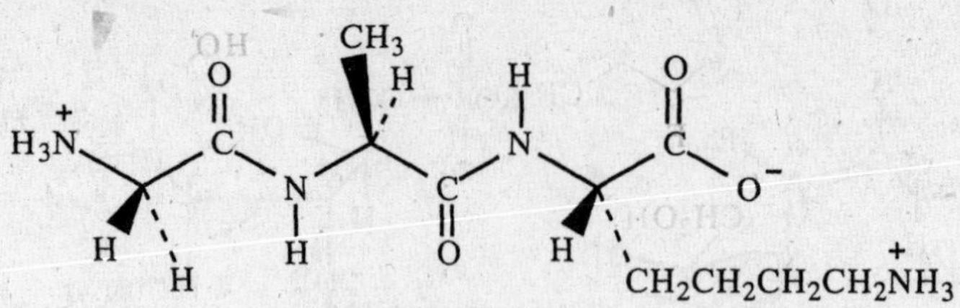

(shown at pH 6)

(b)

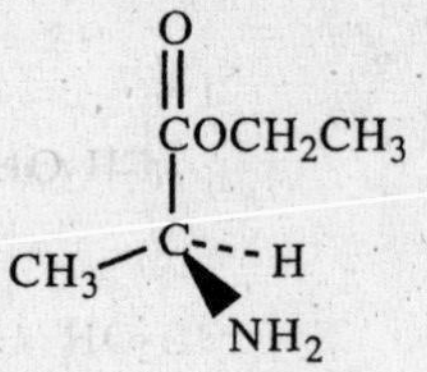

(c)

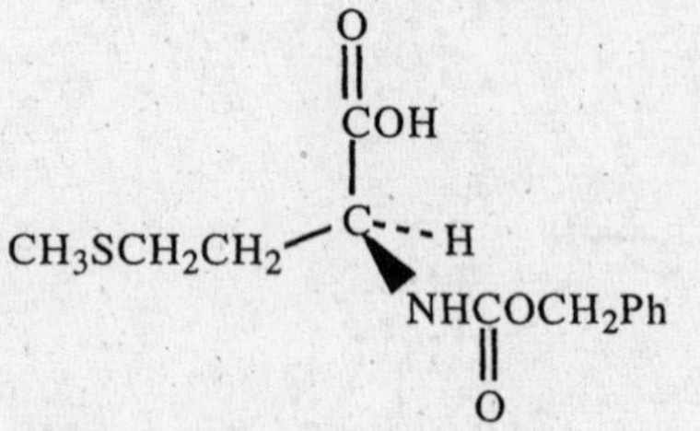

(d)

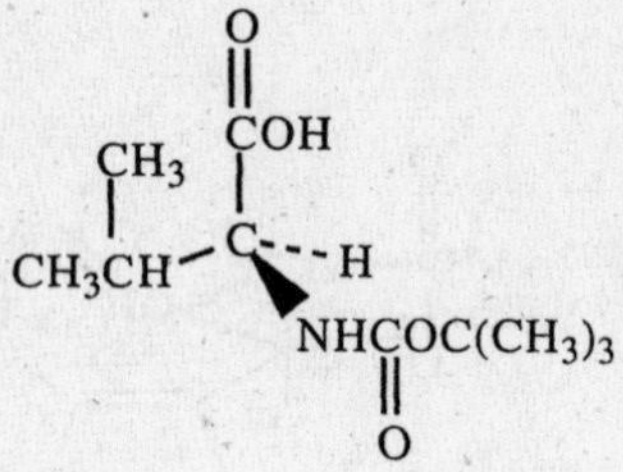

(e)

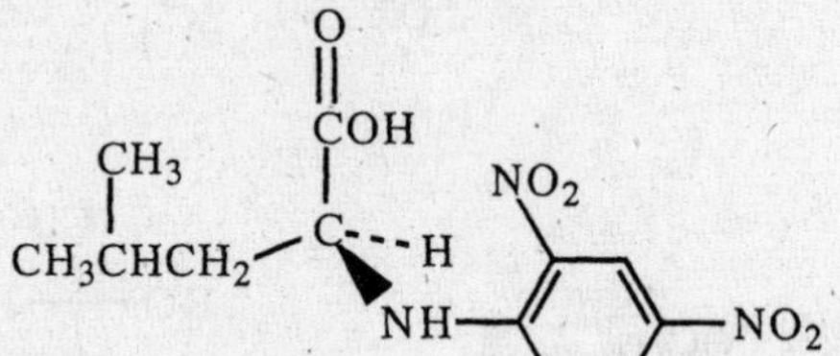

(f)

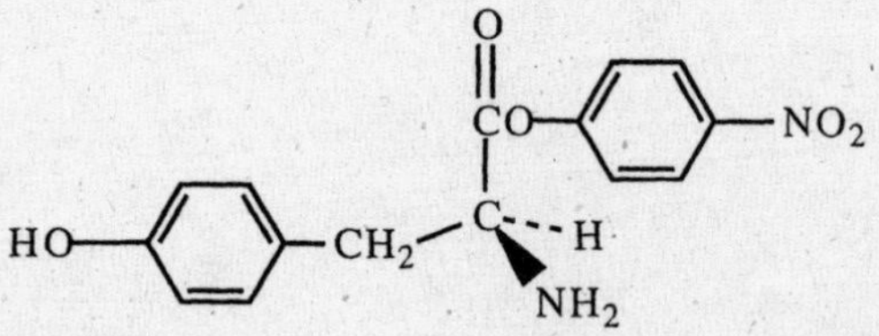

23.37
(a) *N*-2,4-dinitrophenylphenylalanine
(b) methyl cysteinate
(c) glycylcysteinylhistidine
(d) *N*-benzoylglycine; hippuric acid
(e) *p*-nitrophenyl alaninate
(f) *N*-*tert*-butoxycarbonylleucine

23.38 (a)

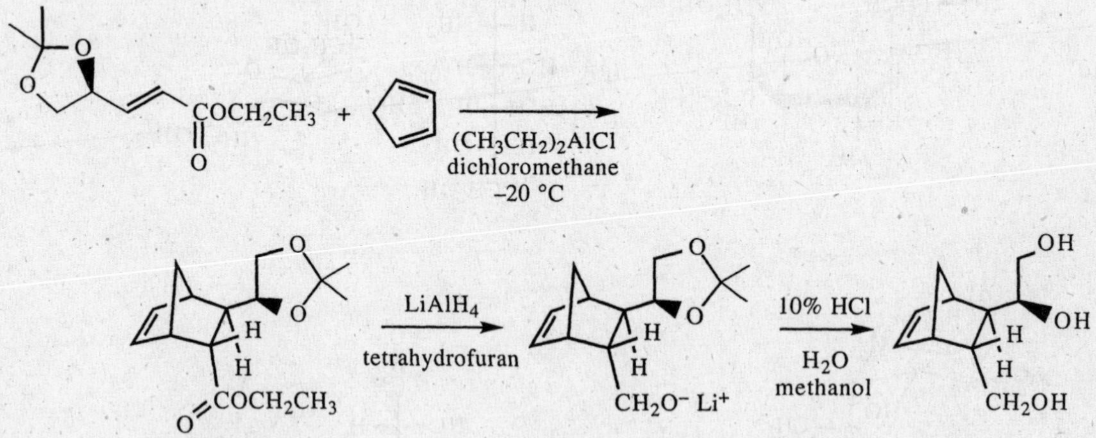

and enantiomer
A
B
C
$C_{10}H_{16}O_3$

(b)

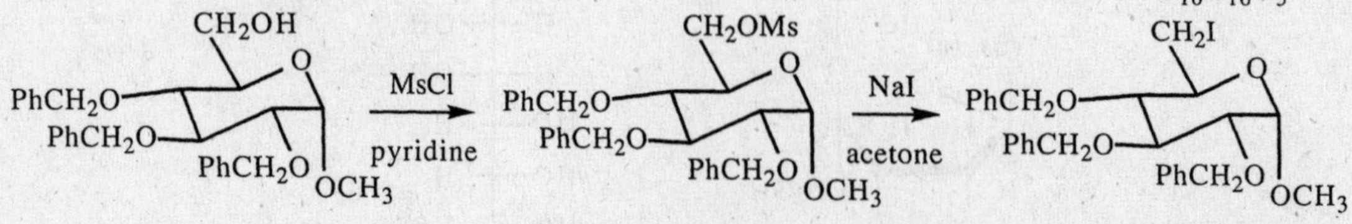

D
E

23.38 (c)

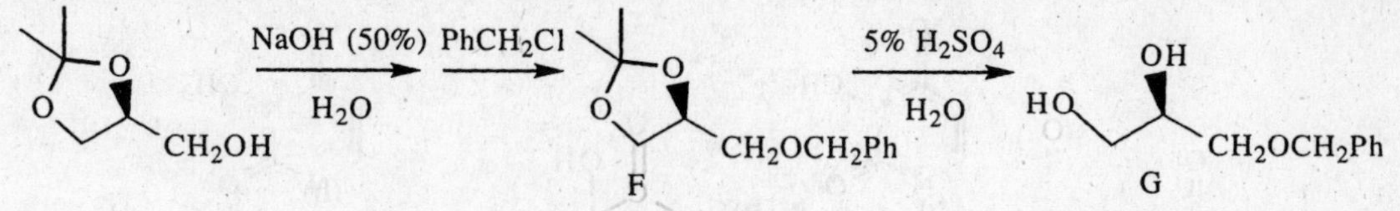

(d)

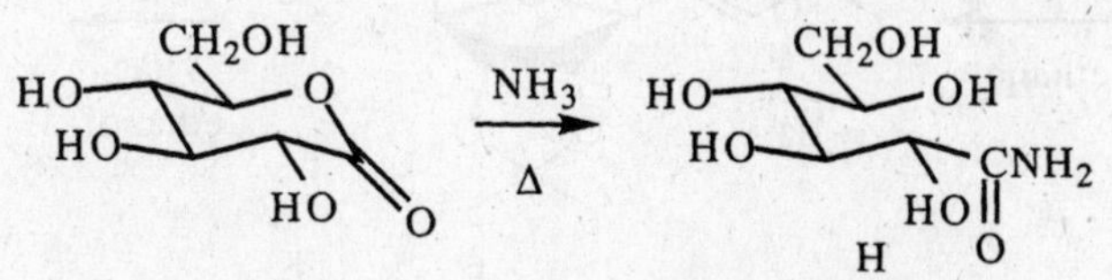

(e)

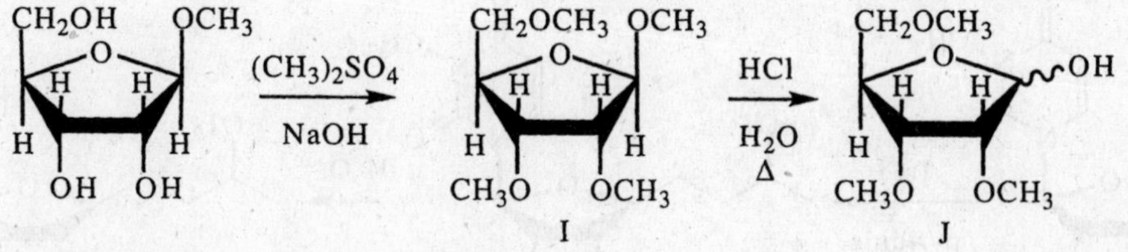

(f)

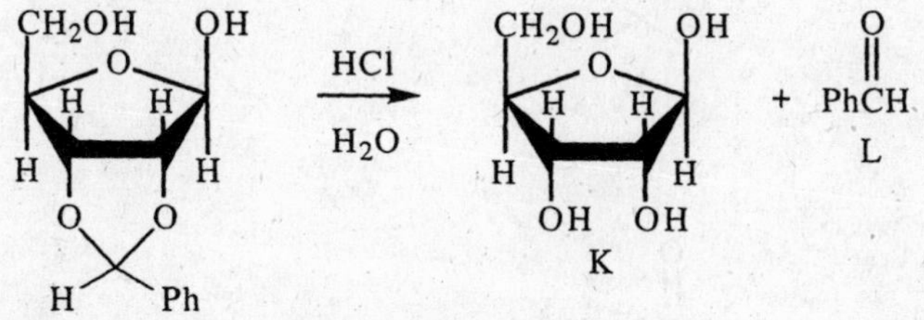

23.39 (a)

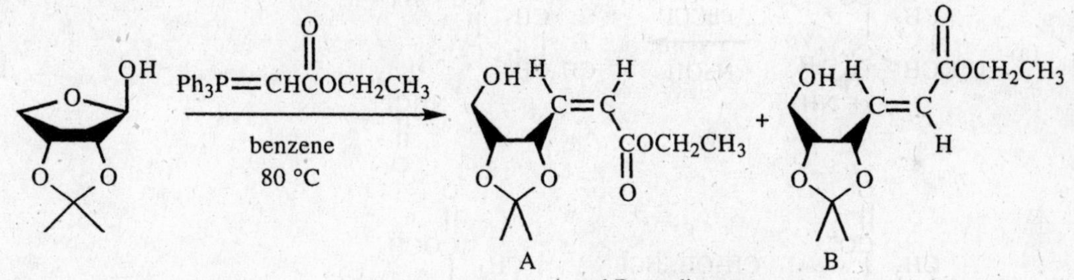

A and B are diastereomers

(b)

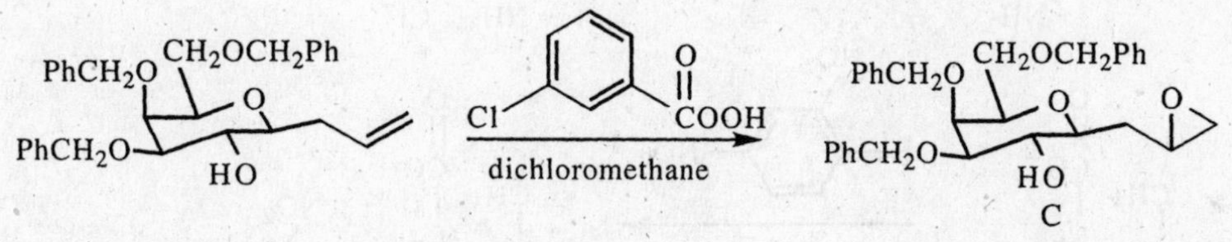

(c)

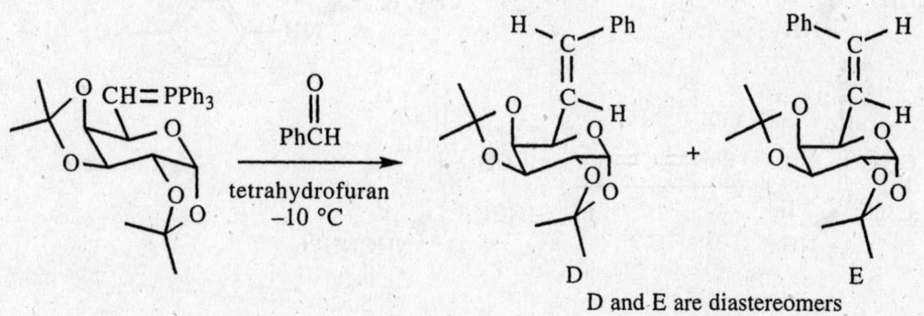

D and E are diastereomers

23.39 (d)

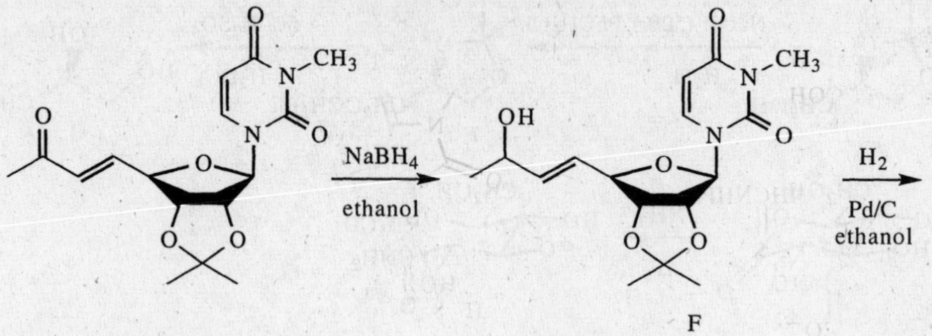

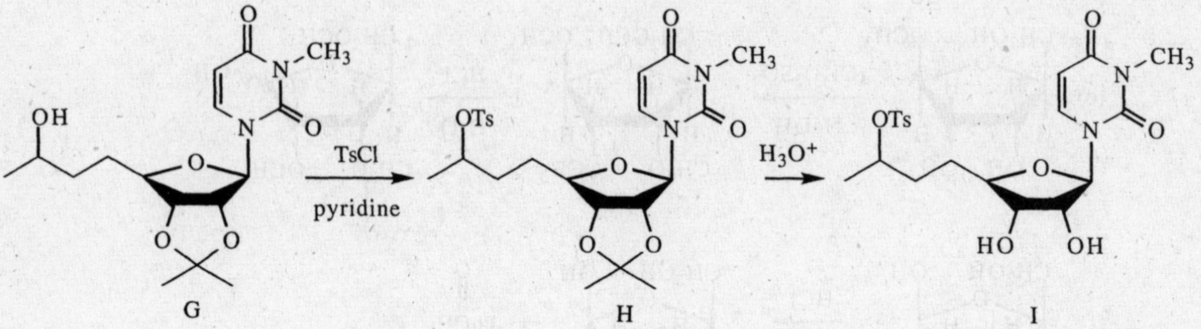

23.40 (a)

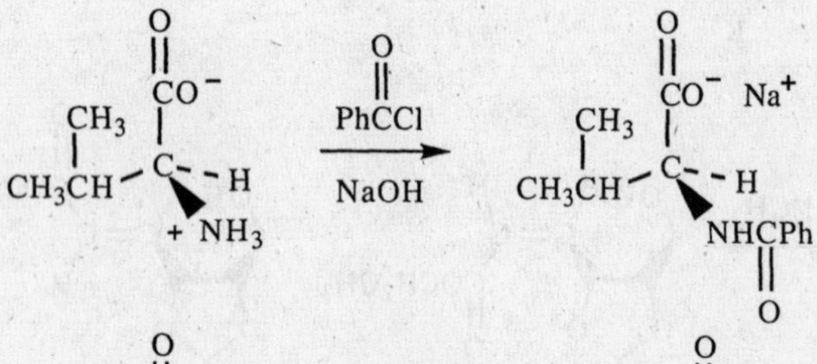

(b)

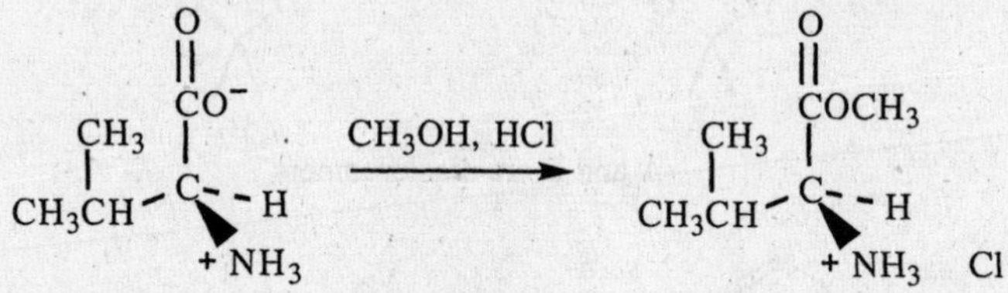

(c)

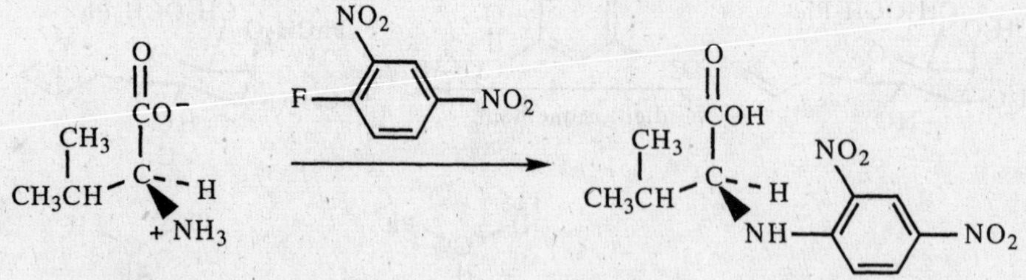

(d)

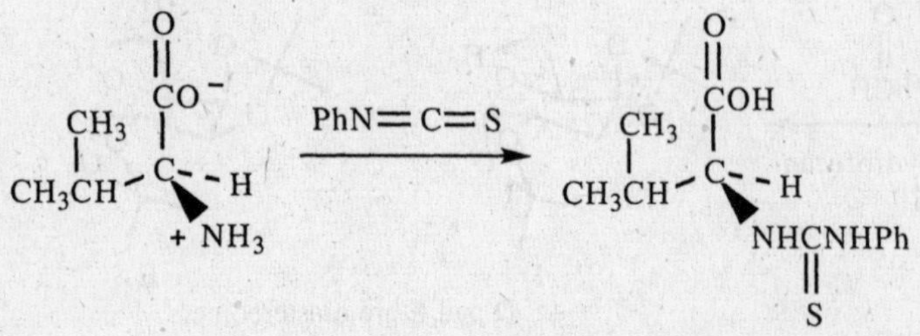

23.40 (e)

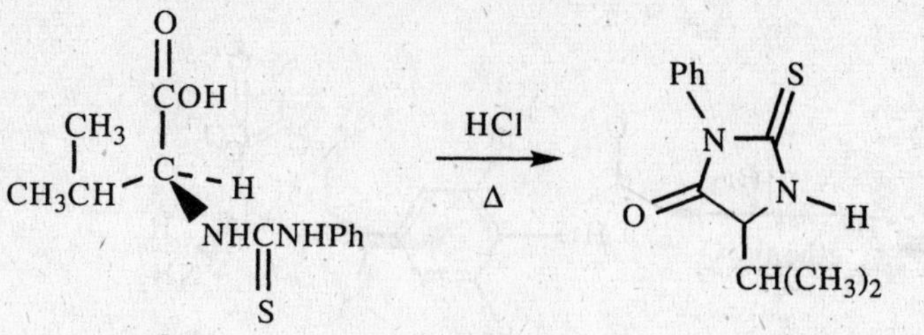

O
COH
CH3
CH3CH
C
H
NHCNHPh
S
HCl
Δ
Ph
S
N
N
H
O
CH(CH3)2

(f)

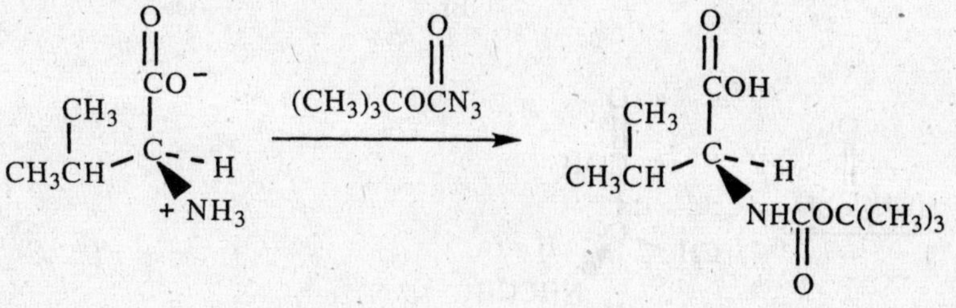

O
CO⁻
CH3
CH3CH
C
H
+ NH3
O
(CH3)3COCN3
O
COH
CH3
CH3CH
C
H
NHCOC(CH3)3
O

(g)

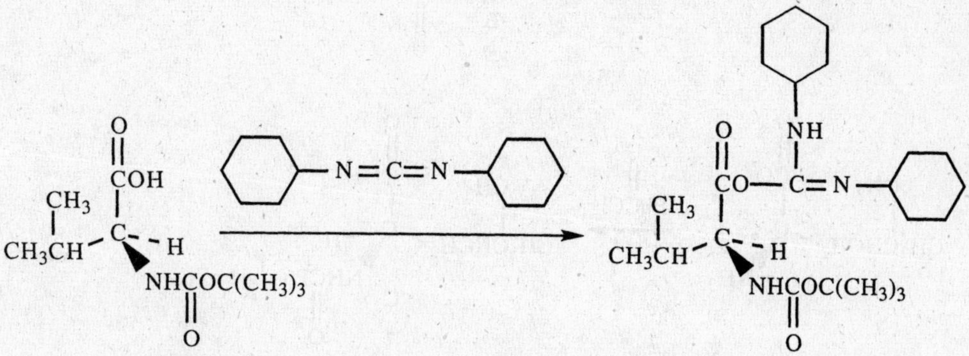

O
COH
CH3
CH3CH
C
H
NHCOC(CH3)3
O
N=C=N
O
NH
CO
C=N
CH3
CH3CH
C
H
NHCOC(CH3)3
O

(h)

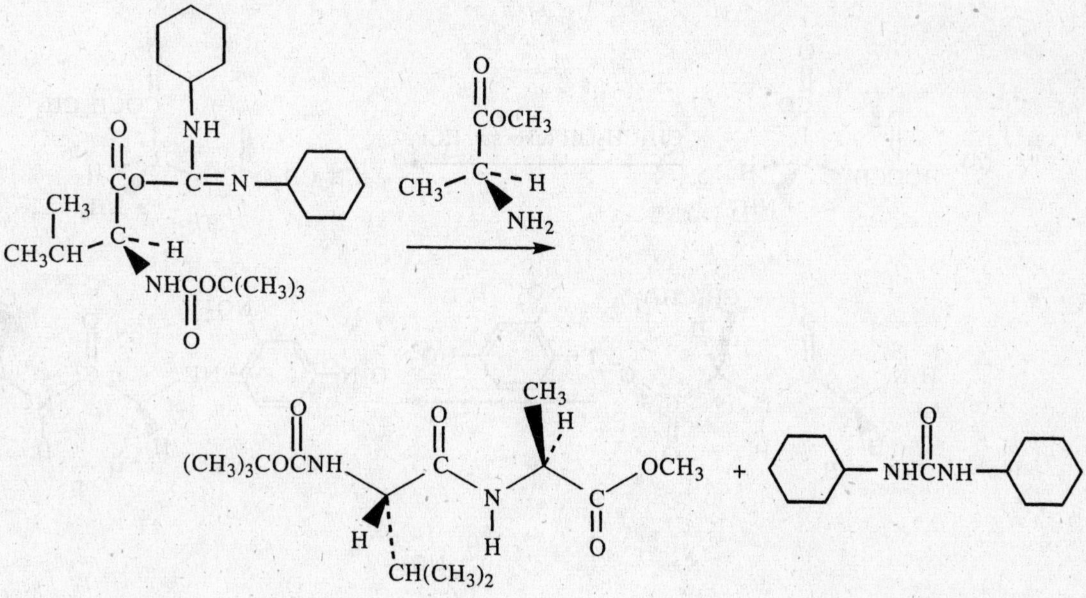

O
NH
CO
C=N
CH3
CH3CH
C
H
NHCOC(CH3)3
O
O
COCH3
CH3
C
H
NH2
CH3
O
H
O
(CH3)3COCNH
C
N
OCH3
H
H
C
O
CH(CH3)2
+
O
NHCNH

(i)

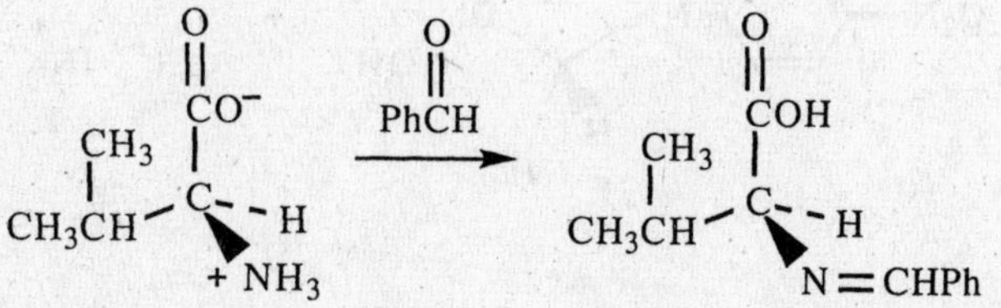

O
CO⁻
CH3
CH3CH
C
H
+ NH3
O
PhCH
O
COH
CH3
CH3CH
C
H
N=CHPh

23.41 (a)

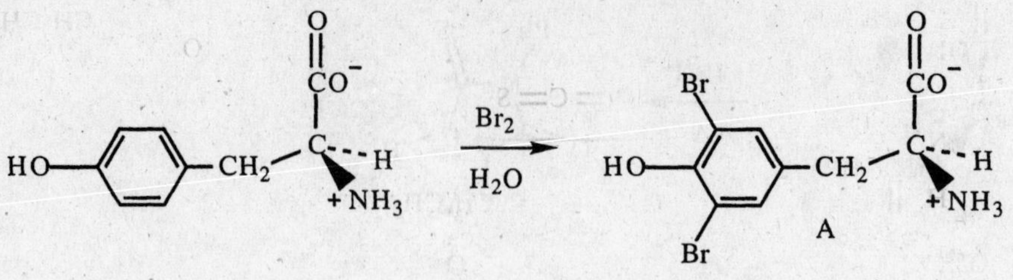

(b)

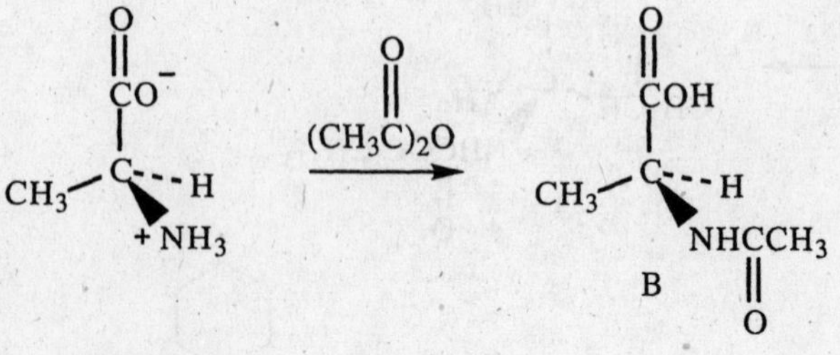

(c)

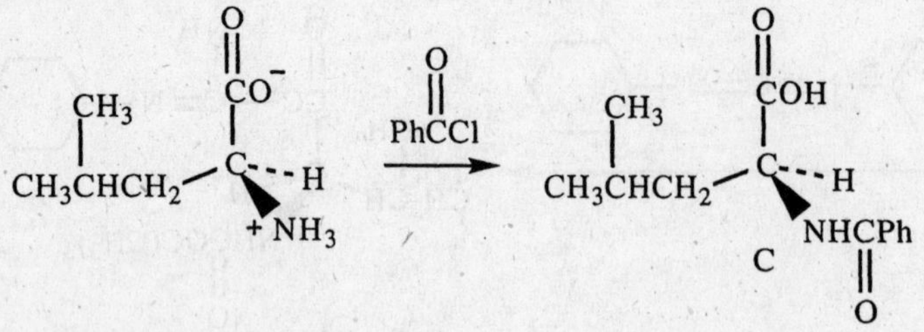

(d)

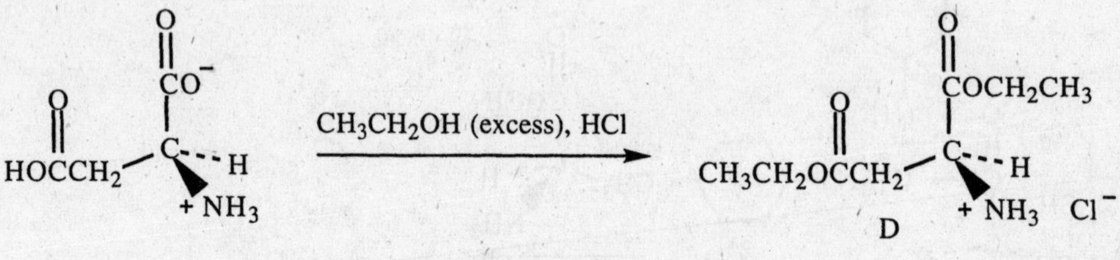

(e)

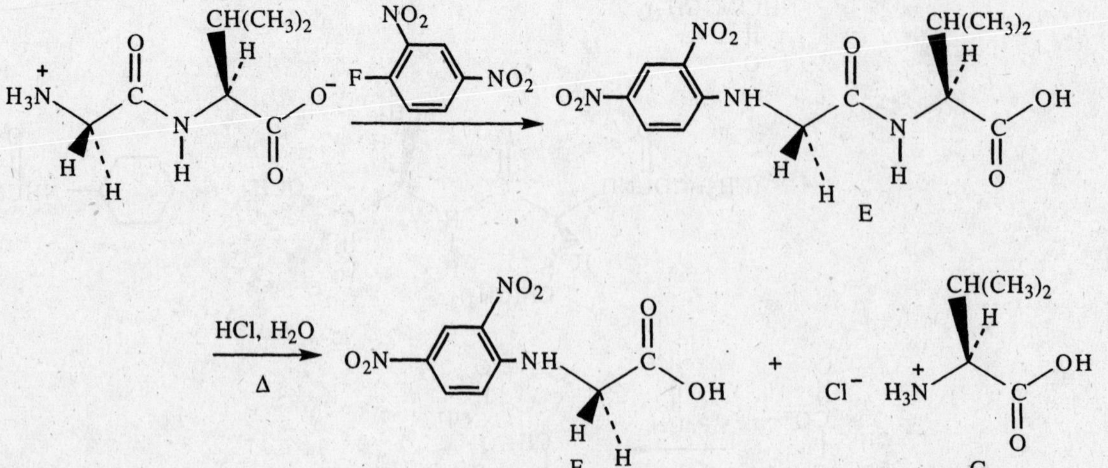

23.42

(a)

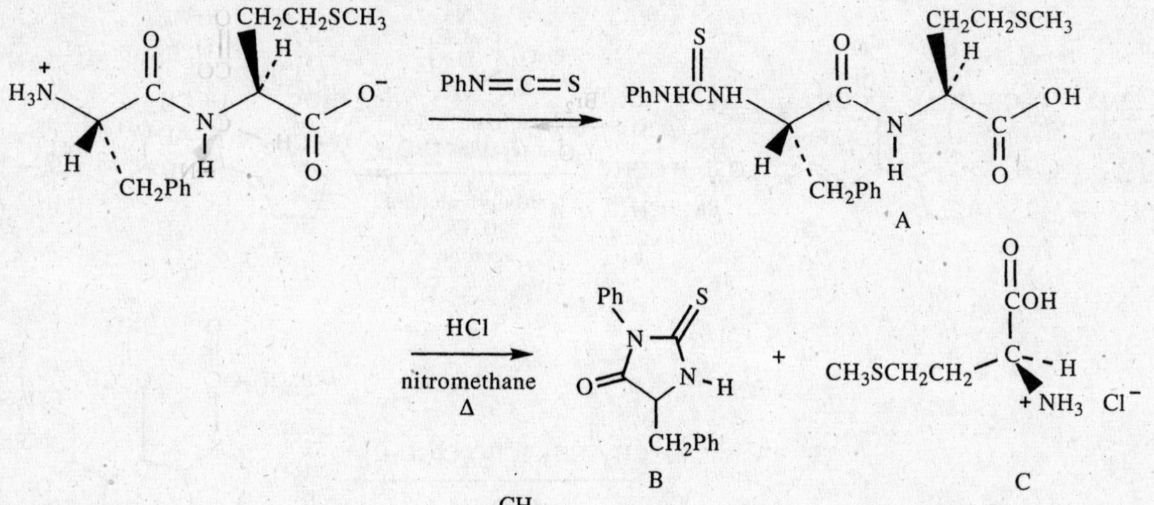

$$PhN=C=S \longrightarrow A$$

$$\xrightarrow[\substack{\text{nitromethane} \\ \Delta}]{HCl} B + C$$

(b)

$$\xrightarrow[\substack{(CH_3CH_2)_3N \\ \text{dichloromethane}}]{Ph-N=C=N-Ph} D$$

(c)

$$\xrightarrow[\substack{Pd/C \\ \text{methanol} \\ HCl}]{H_2} E + CO_2 + CH_3-C_6H_5$$

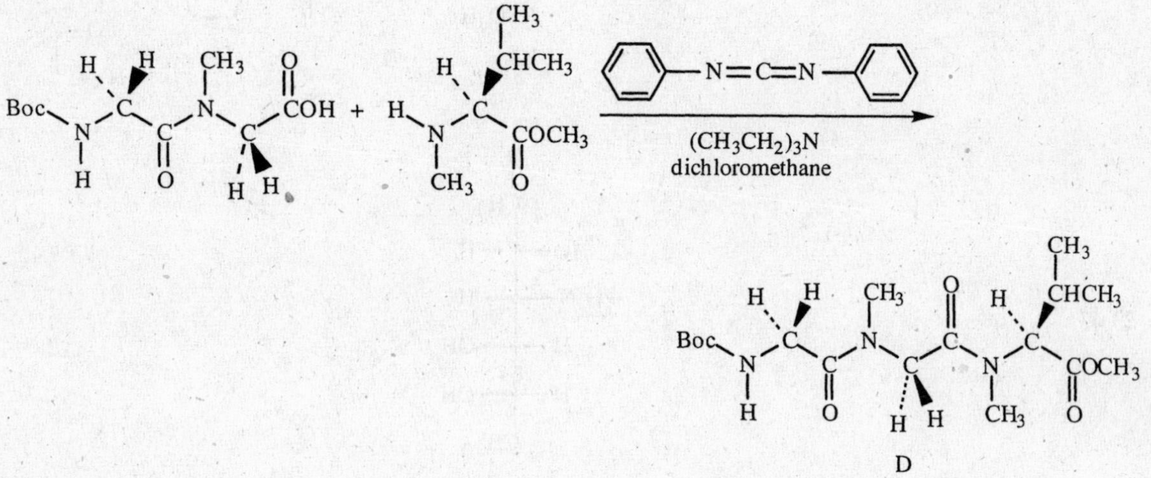

(d)

$$\xrightarrow[]{HO-N(succinimide), DCC} F$$

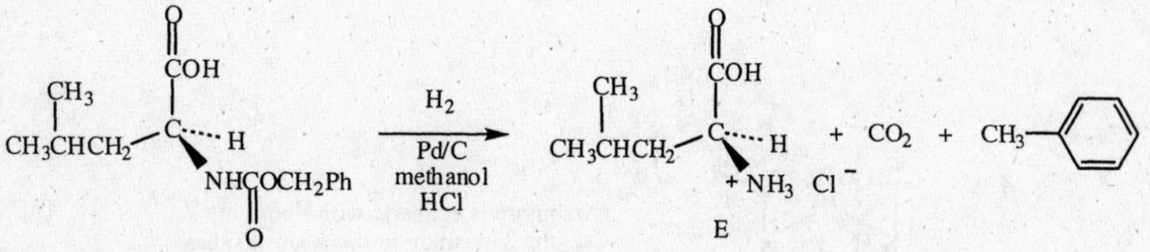

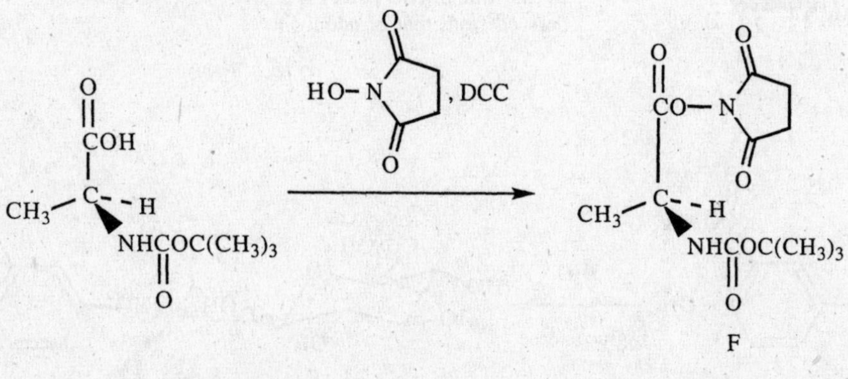

23.42 (e)

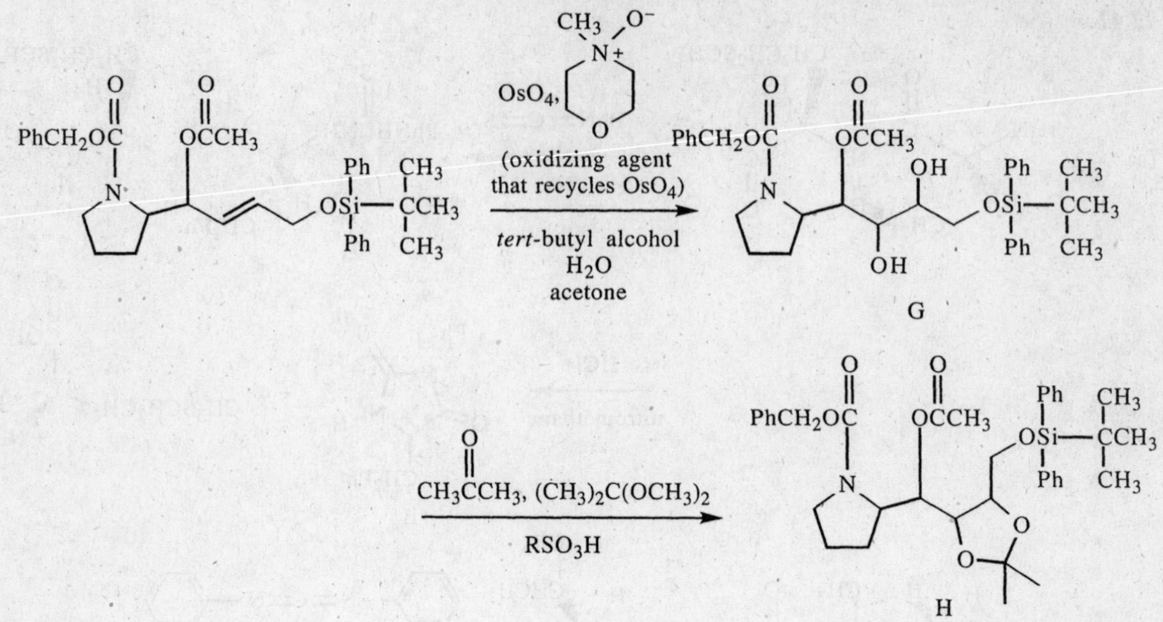

23.43

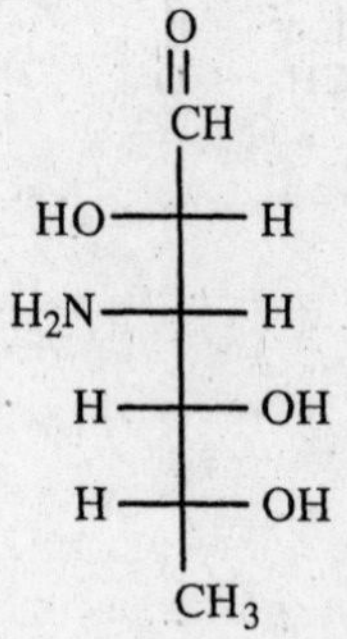

23.44

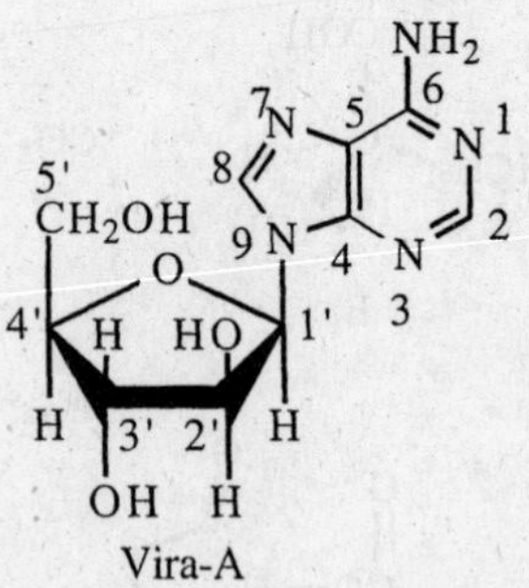

Vira-A

(Arabinose is epimeric with ribose at C-2, the 2' position in the sugar residue, so the structure of Vira-A is easily derived from that of adenosine.)

23.45

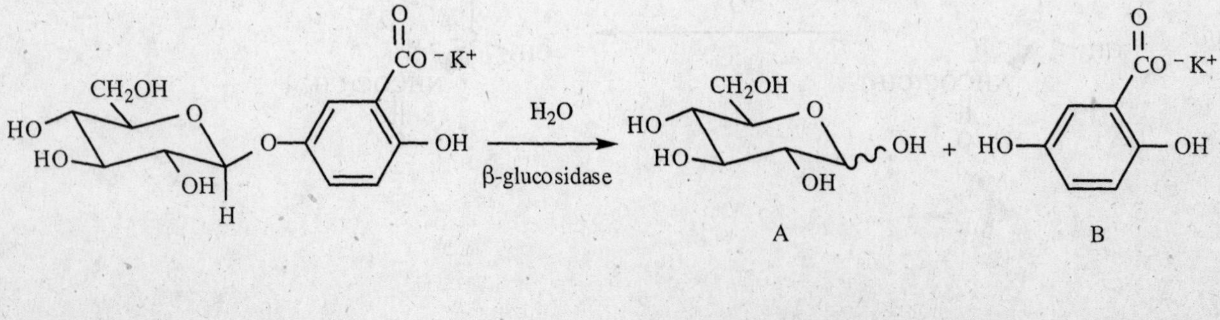

A B

23.46

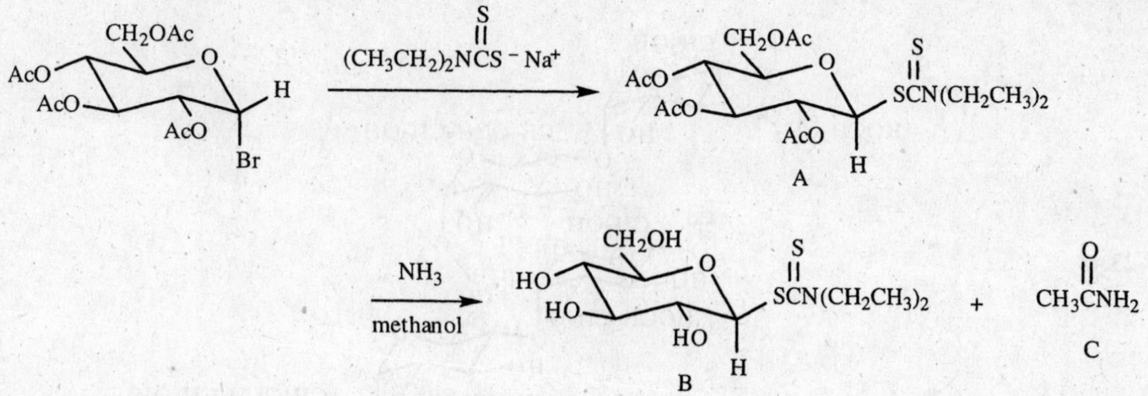

23.47 (a)

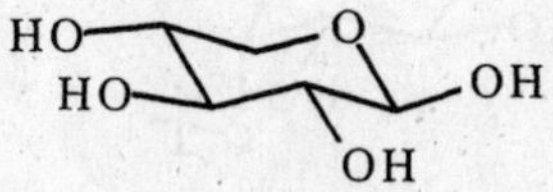

xylose
β-D-xylopyranose

(b) The glucose unit has the free hemiacetal function.

(c)

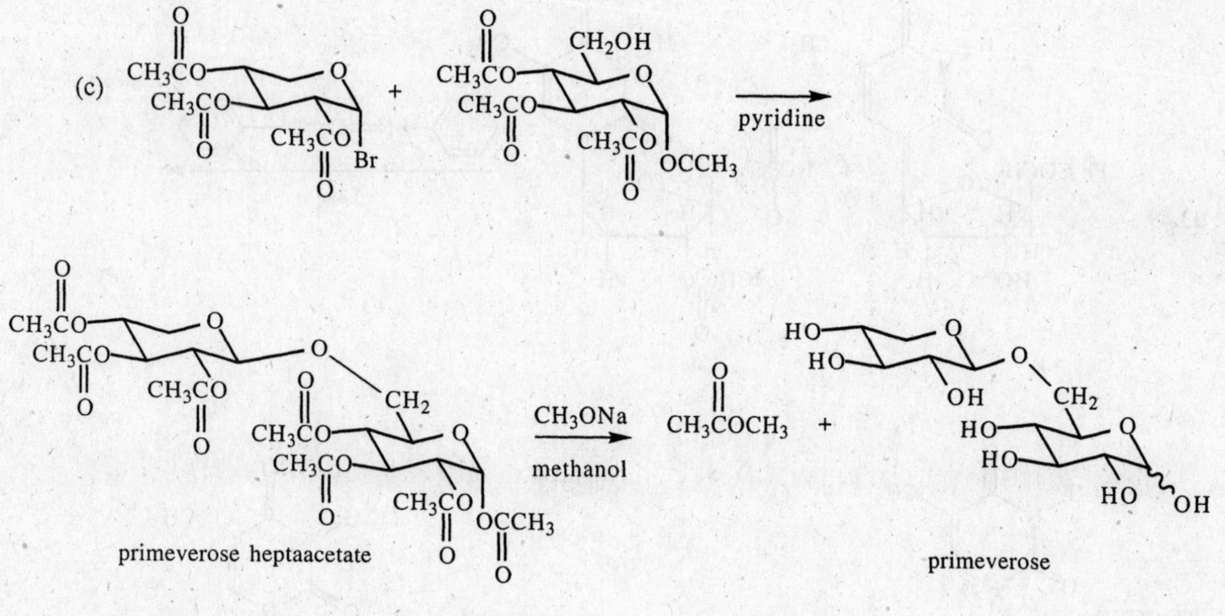

primeverose heptaacetate   primeverose

(d)

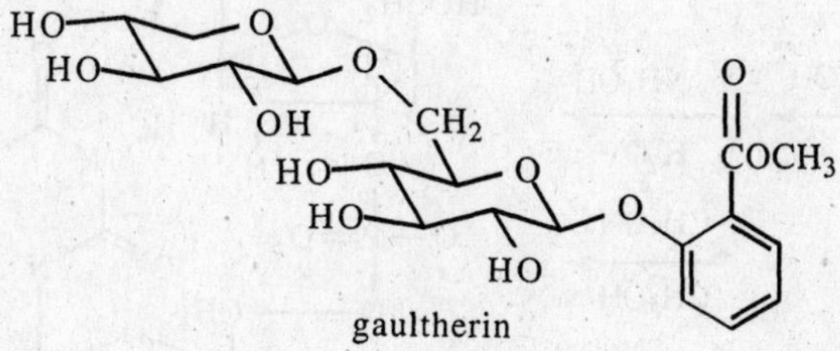

gaultherin

23.48

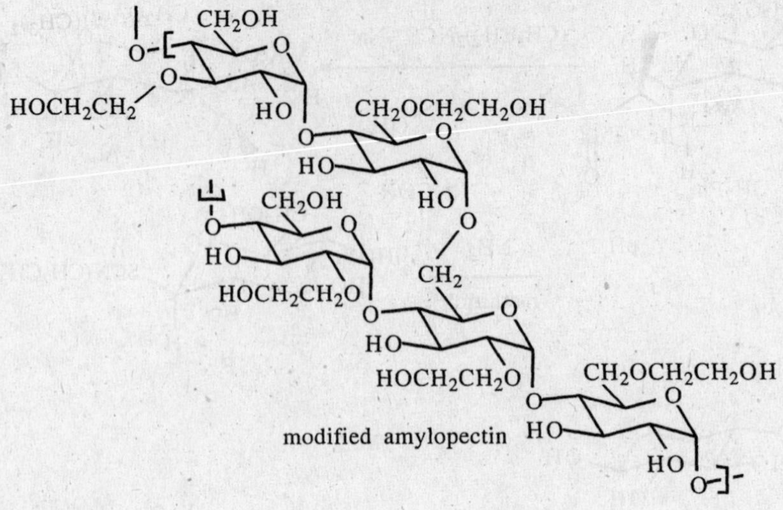
CH2OH
HOCH2CH2
HO
CH2OCH2CH2OH
HO
CH2OH
HO
HOCH2CH2O
CH2
HOCH2CH2O
CH2OCH2CH2OH
modified amylopectin

23.49

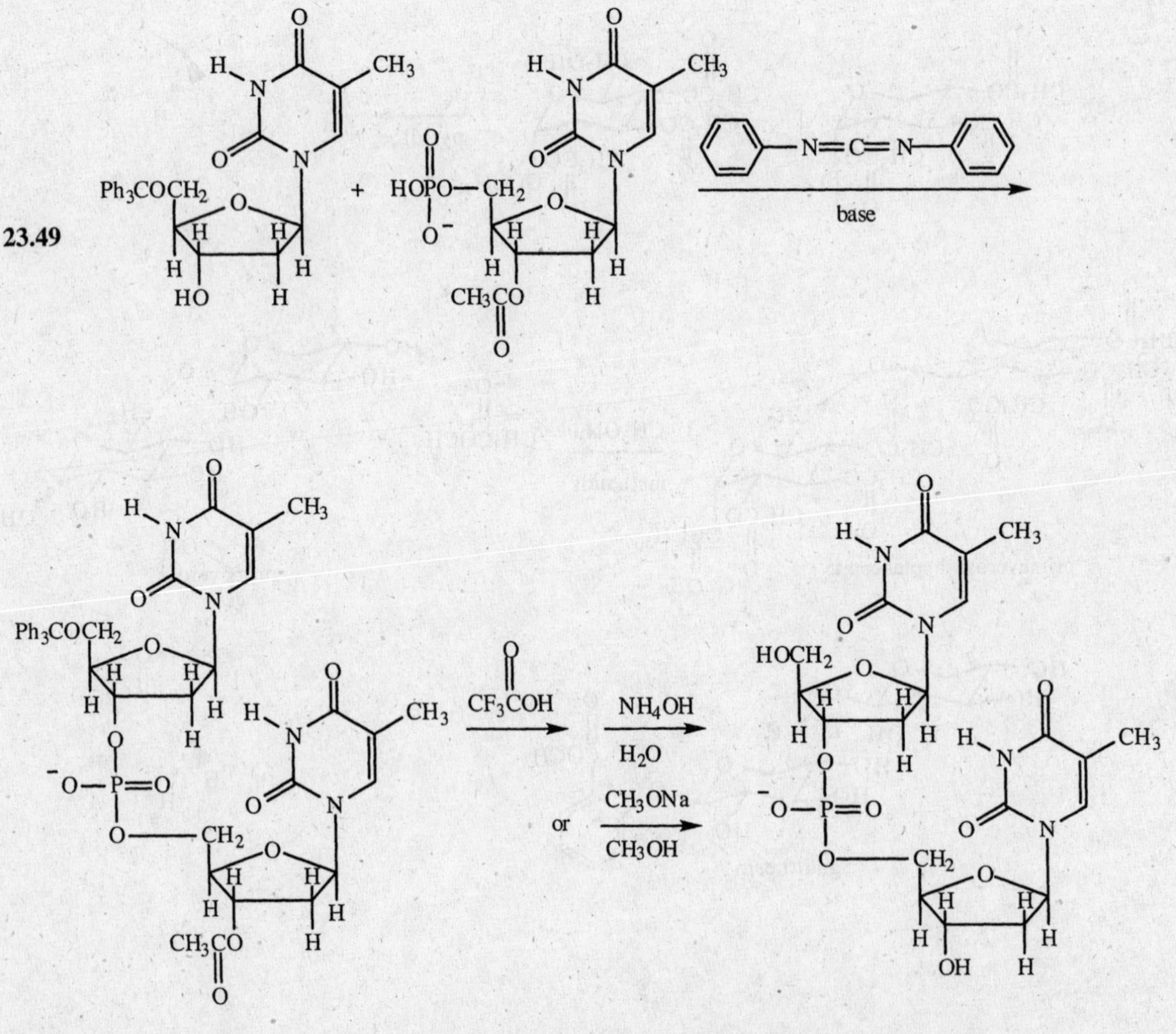
CH3
Ph3COCH2
HO
HOPO—CH2
CH3CO
N=C=N
base
CF3COH
NH4OH
H2O
or
CH3ONa
CH3OH
HOCH2
OH

23.50 (a)

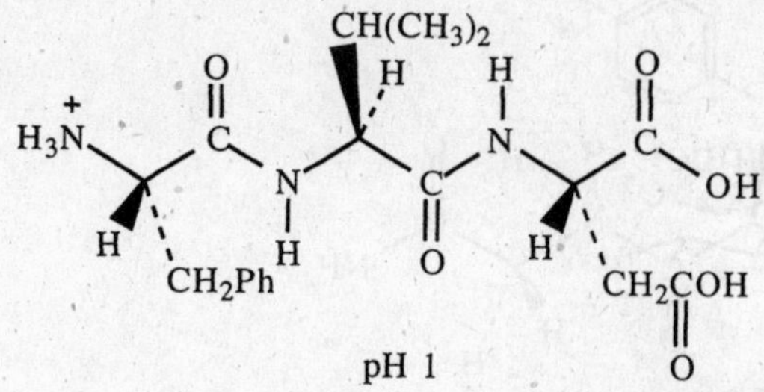

pH 1

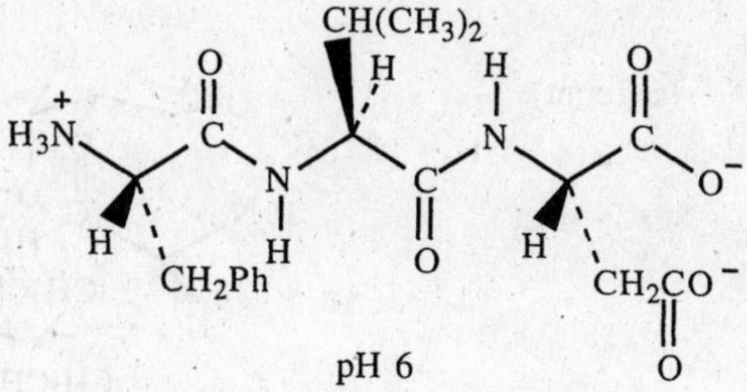

pH 6

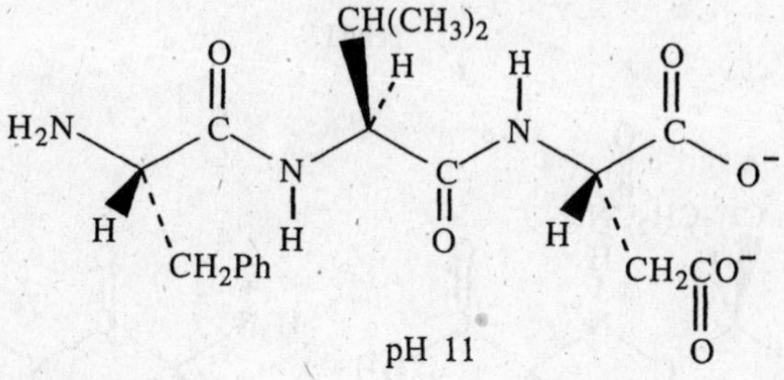

pH 11

(b)

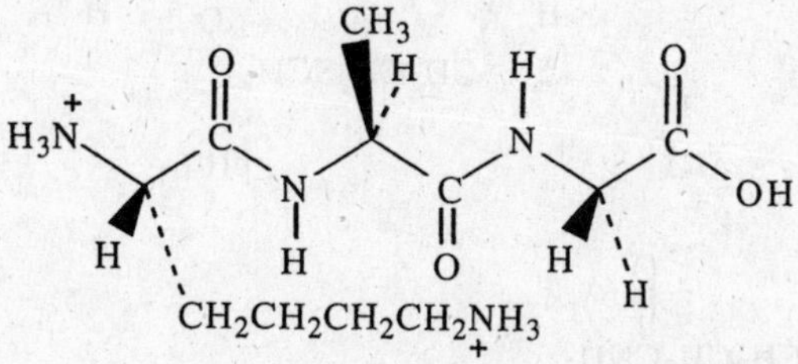

pH 1

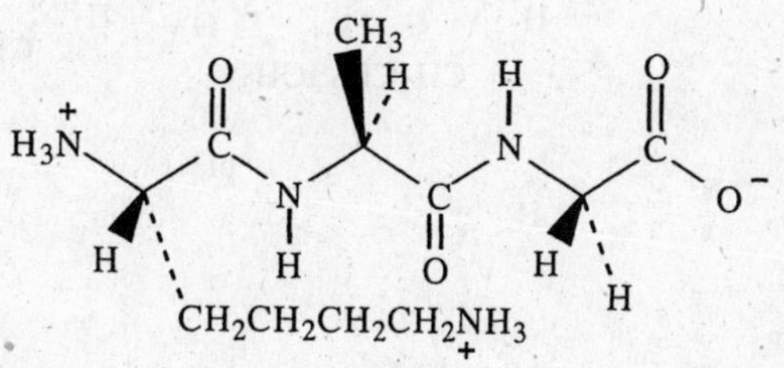

pH 6

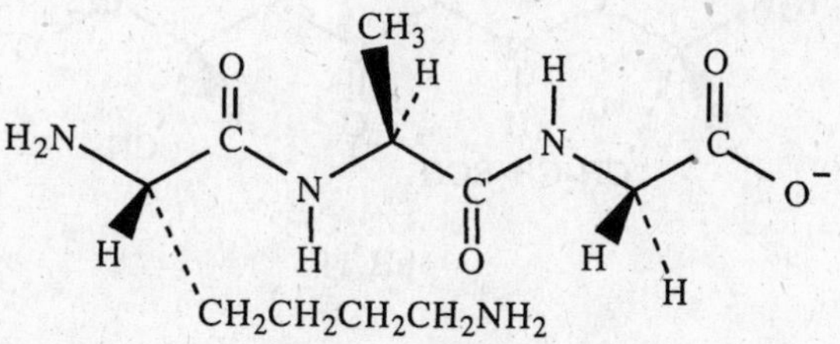

pH 11

23.50 (c)

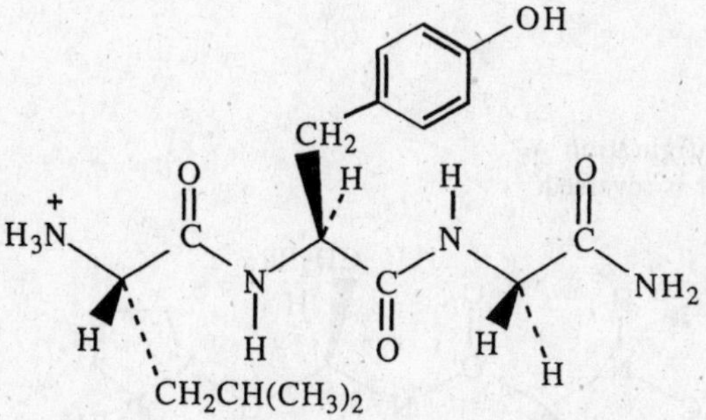

pH 1

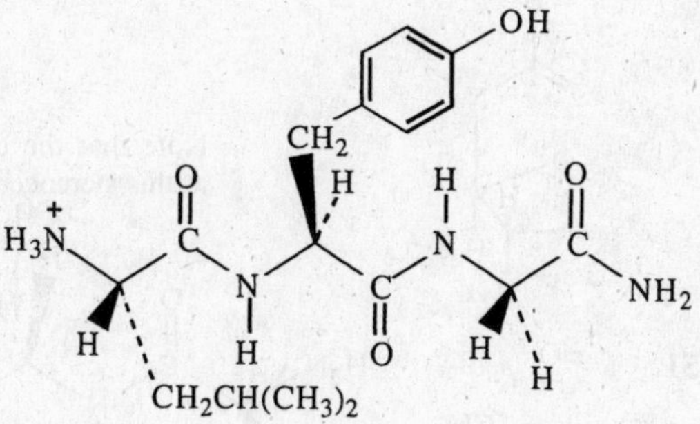

pH 6

23.50   (c) (cont)

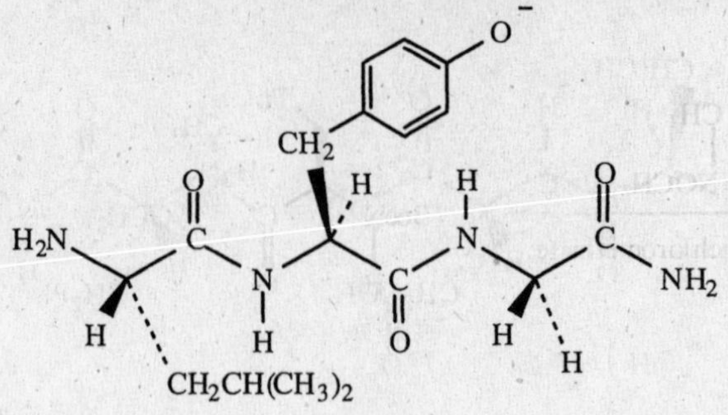

pH 11

(d)

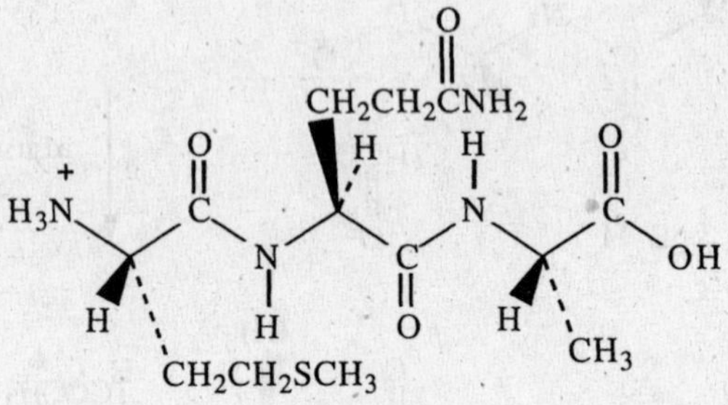

pH 1

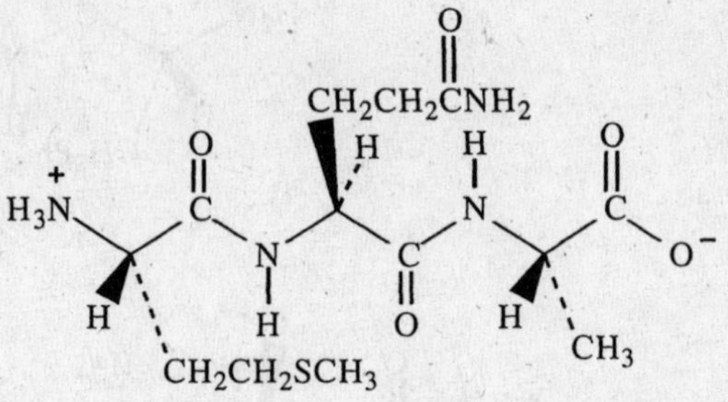

pH 6

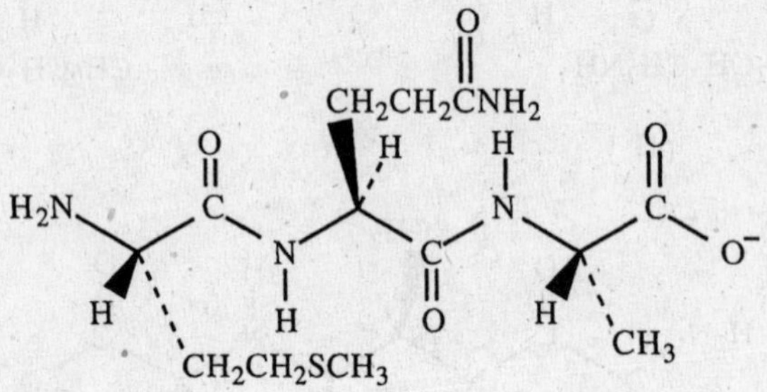

pH 11

23.51

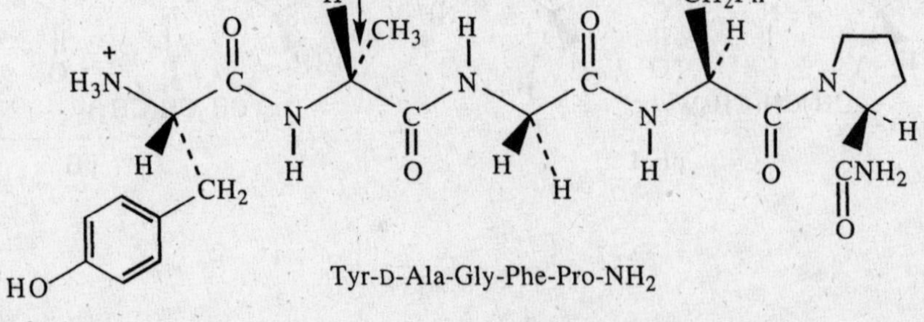

Note that the configuration
at this stereocenter is reversed.

Tyr-D-Ala-Gly-Phe-Pro-NH$_2$

23.52

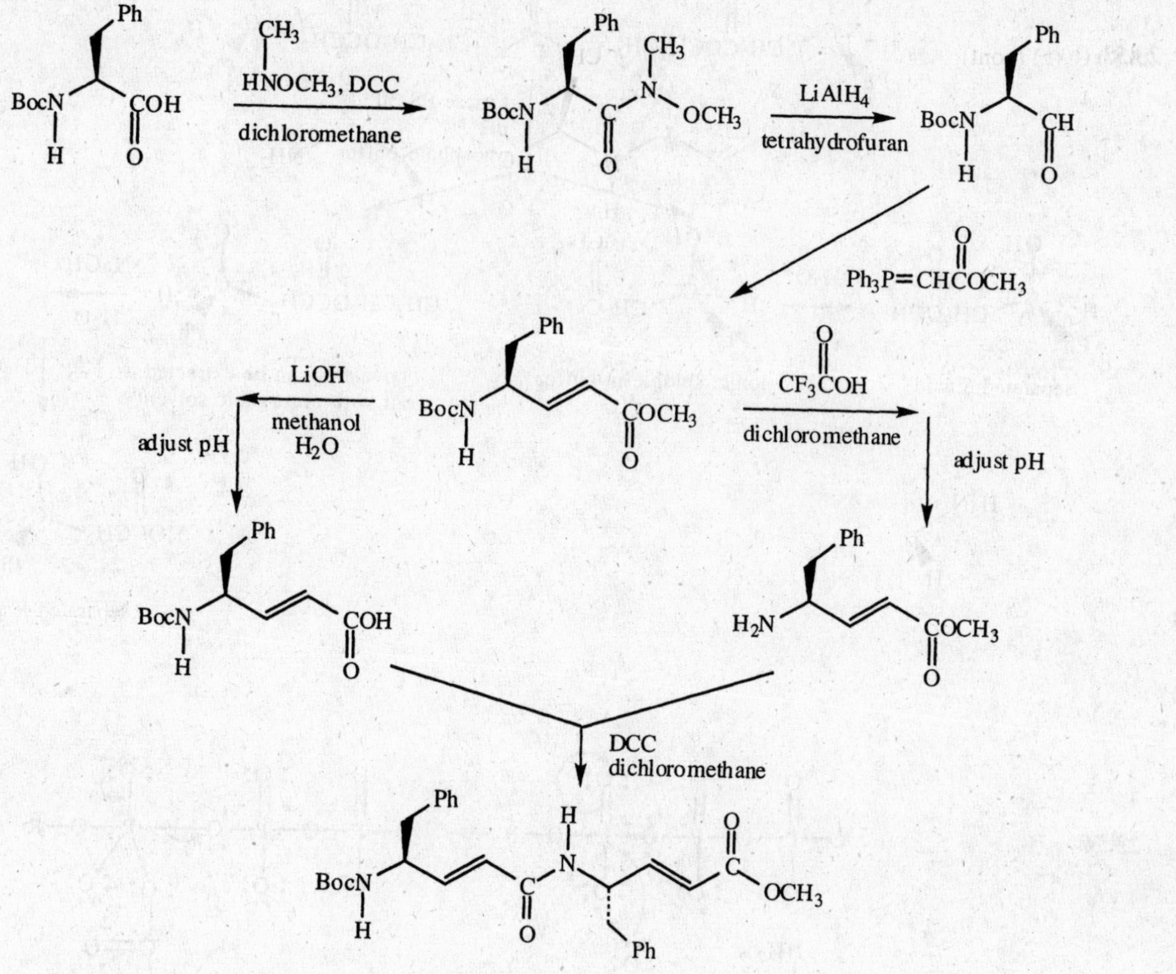

Ph
BocN
H
COH
O
CH3
HNOCH3, DCC
dichloromethane
N
OCH3
LiAlH4
tetrahydrofuran
CH
Ph3P=CHCOCH3
LiOH
methanol
H2O
adjust pH
CF3COH
dichloromethane
COCH3
H2N
DCC
dichloromethane

23.53 (a)

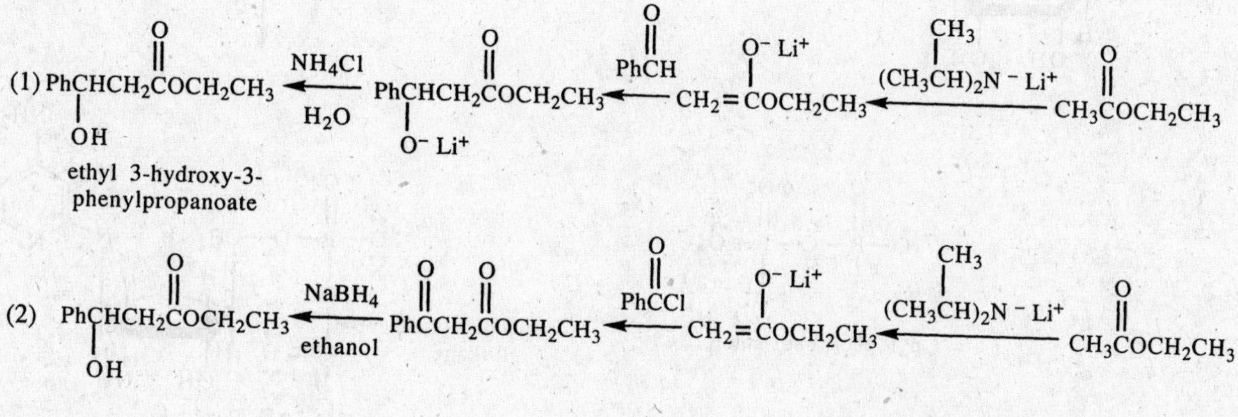

(1) PhCHCH2COCH2CH3
OH
ethyl 3-hydroxy-3-phenylpropanoate
NH4Cl
H2O
PhCHCH2COCH2CH3
O⁻ Li⁺
PhCH
CH2=COCH2CH3
(CH3CH)2N⁻ Li⁺
CH3COCH2CH3
(2) PhCHCH2COCH2CH3
NaBH4
ethanol
PhCCH2COCH2CH3
PhCCl

23.53 (b)

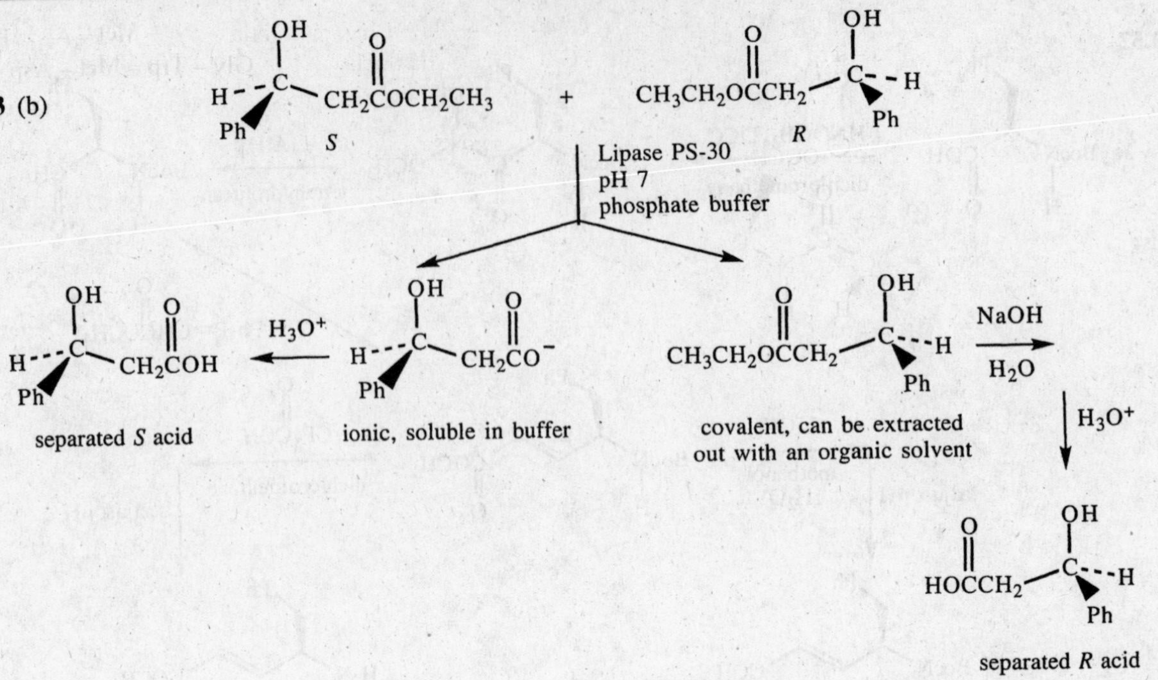
S
R
Lipase PS-30
pH 7
phosphate buffer
H_3O^+
NaOH
H_2O
H_3O^+
separated S acid
ionic, soluble in buffer
covalent, can be extracted out with an organic solvent
separated R acid

23.54

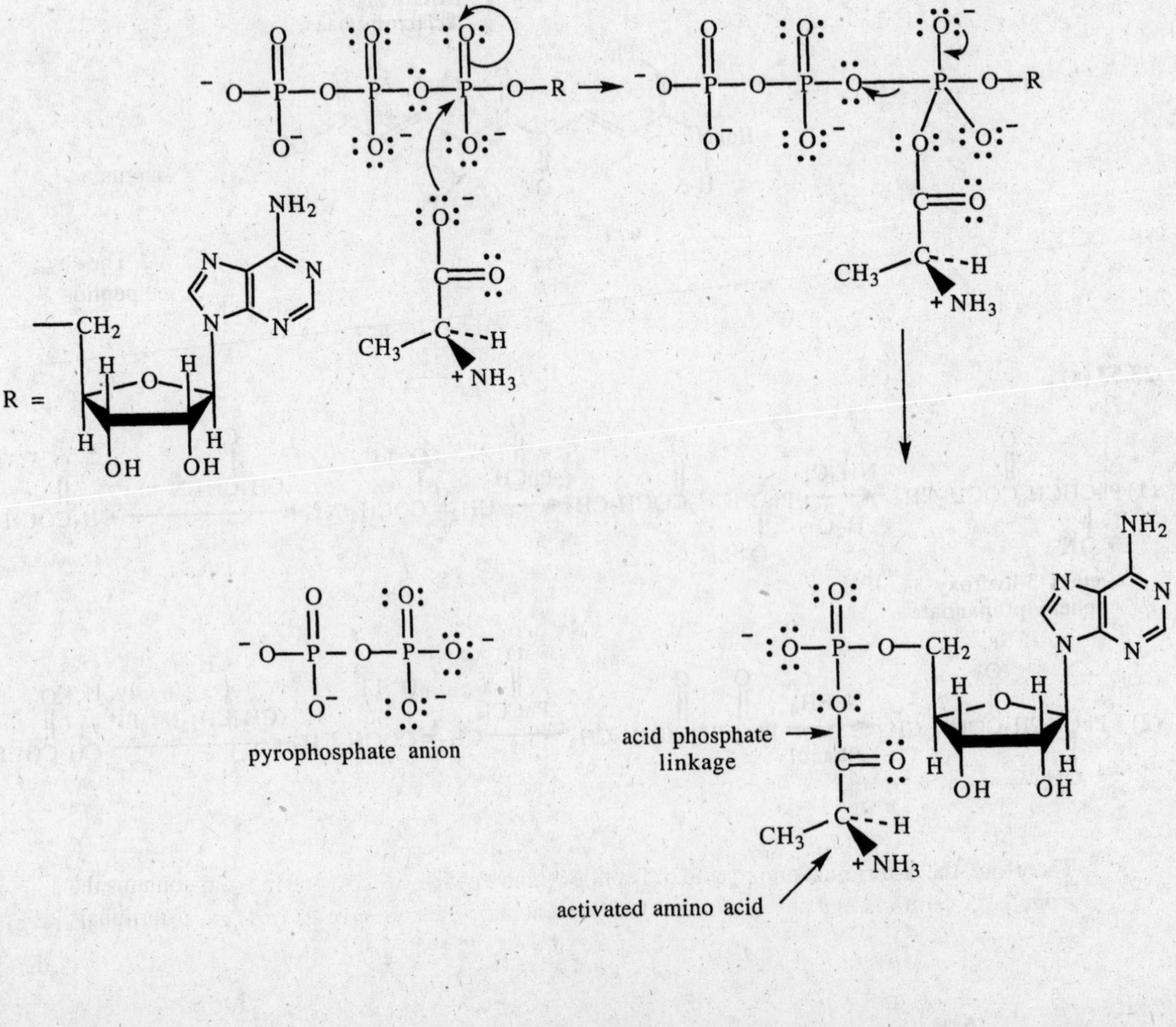
R =
pyrophosphate anion
acid phosphate linkage
activated amino acid

23.55 (a) Glu – Gly – Pro – Trp – Gly – Trp – Leu – Glu – Glu – Glu – Glu – Ala – Ala – Tyr – Met – Asp - Phe or
Glu – Gly – Pro – Trp – Leu – Glu – Glu – Glu – Glu – Ala – Ala – Tyr – Gly – Trp – Met – Asp - Phe

(b)

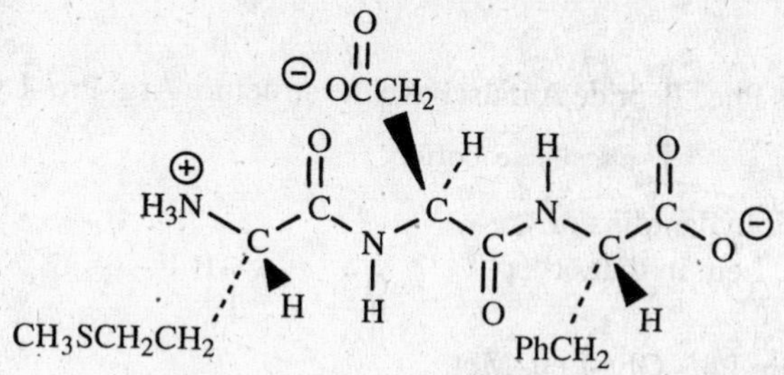

(c)

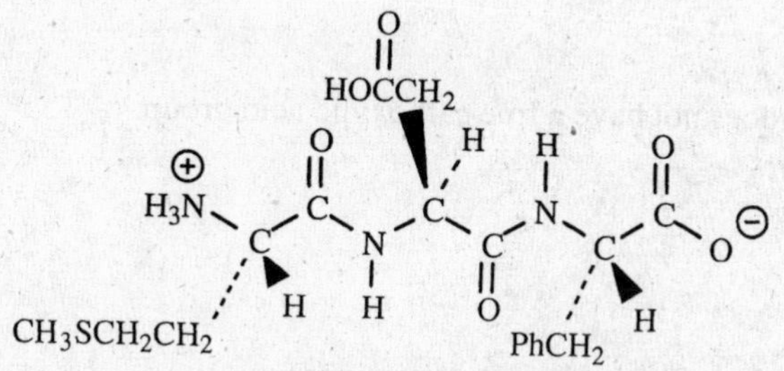

23.56 Combining the fragments from trypsin and chymotrypsin digestion gives the following three segments:

A peptide containing 17 amino acids from the *N*-terminal end of viscotoxin A$_2$:
Lys–Ser–Cys–Cys–Pro–Asn–Thr–Thr–Gly–Arg–Asn–Ile–Tyr–Asn–Thr–Cys–Arg

A peptide containing 17 amino acids from the *C*-terminal end:
Ser–Gly–Cys–Lys–Ile–Ile–Ser–Ala–Ser–Thr–Cys–Pro–Ser–Tyr–Pro–Asp–Lys

A 12 amino acid peptide; no overlap with either of the *N*- or *C*-terminal fragments:
Phe–Gly–Gly–Gly–Ser–Arg–Glu–Val–Cys–Ala–Ser–Leu

The three peptides add up to 46 amino acids, the total number contained in viscotoxin A$_2$. The complete order of amino acids in viscotoxin A$_2$ is thus the order that exists in the sequence of the three peptide fragments.

Lys–Ser–Cys–Cys–Pro–Asn–Thr–Thr–Gly–Arg–Asn–Ile–Tyr–Asn–Thr
                                               |

Ser–Leu–Ser–Ala–Cys–Val–Glu–Arg–Ser–Gly–Gly–Gly–Phe–Arg–Cys
|

Gly–Cys–Lys–Ile–Ile–Ser–Ala–Ser–Thr–Cys–Pro–Ser–Tyr–Pro–Asp–Lys

23.57 Peptide P contains 11 amino acids:

Arg, Gln (2), Gly, Leu, Lys, Met, Phe (2), Pro (2)

Peptide A contains Arg, Gln (2), Lys, Phe, Pro (2). Therefore, Peptide B must contain Gly, Leu, Met, Phe.

Peptide P (and Peptide A) $\xrightarrow{\text{Edman's reagent}}$ phenylthiohydantoins of
Arg, Pro, Lys, Pro, in that order

Therefore, the *N*-terminal amino acid of Peptide P (and Peptide A) is Arg. Peptide A contains the amino acids from the *N*-terminal end of P, and Peptide B must contain the amino acids from the *C*-terminal end of P.

23.57 (cont)

Peptide A $\xrightarrow[\text{carboxypeptidase}]{}$ first Phe and then Gln

Therefore, the *C*-terminal amino acid of A is Phe. Peptide A must have the structure Arg–Pro–Lys–Pro–Gln–Gln–Phe.

Peptide B $\xrightarrow{\text{Edman's reagent}}$ phenylthiohydantoins of Phe, Gly, Leu, in that order

Therefore, Peptide B must have the structure Phe–Gly–Leu–Met.

Peptide P $\xrightarrow[\text{carboxypeptidase}]{}$ no amino acid

Therefore, the *C*-terminal amino acid of P does not have a free carboxylic acid group.

Peptide P $\xrightarrow[\substack{\text{(hydrolysis of} \\ \text{amide bonds)}}]{\text{HCl (0.03 M)}}$ Peptide C

Peptide C $\xrightarrow[\text{carboxypeptidase}]{}$ Gly, Met, Leu, Phe (order not determined)

Therefore, Peptide C has a *C*-terminal amino acid with a free carboxylic acid group. The same amino acids as those present in Peptides B and P are released to the solution, so Peptide C has the same sequence of amino acids as Peptide P. Both have Met as the *C*-terminal amino acid. In Peptide P, the *C*-terminal acid exists as the amide, Met—NH$_2$.

These conclusions can be summarized as answers to the subsections of the problem.

(a) The *N*-terminal amino acid of Peptide P is Arg.
(b) The *C*-terminal amino acid of Peptide P is Met.
(c) The sequence of amino acids in Peptide A is Arg–Pro–Lys–Pro–Gln–Gln–Phe.
(d) The sequence of amino acids in Peptide B is Phe–Gly–Leu–Met.
(e) The complete structure of Peptide B is Phe–Gly–Leu–Met—NH$_2$.
(f) The sequence of amino acids in Peptide C is Arg–Pro–Lys–Pro–Gln–Gln–Phe–Phe–Gly–Leu–Met.
(g) The complete structure of Peptide P is Arg–Pro–Lys–Pro–Gln–Gln–Phe–Phe–Gly–Leu–Met—NH$_2$.

23.58   The solid state synthesis of Peptide P:

(a)

23.58

(b) Structure:

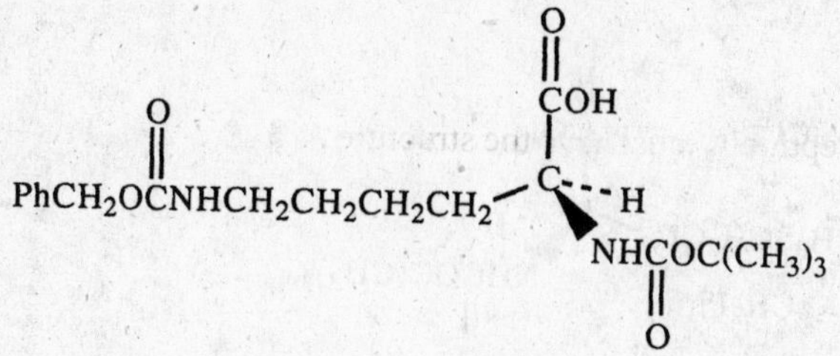

PhCH₂OCNHCH₂CH₂CH₂CH₂ — C(COOH)(H) — NHCOC(CH₃)₃

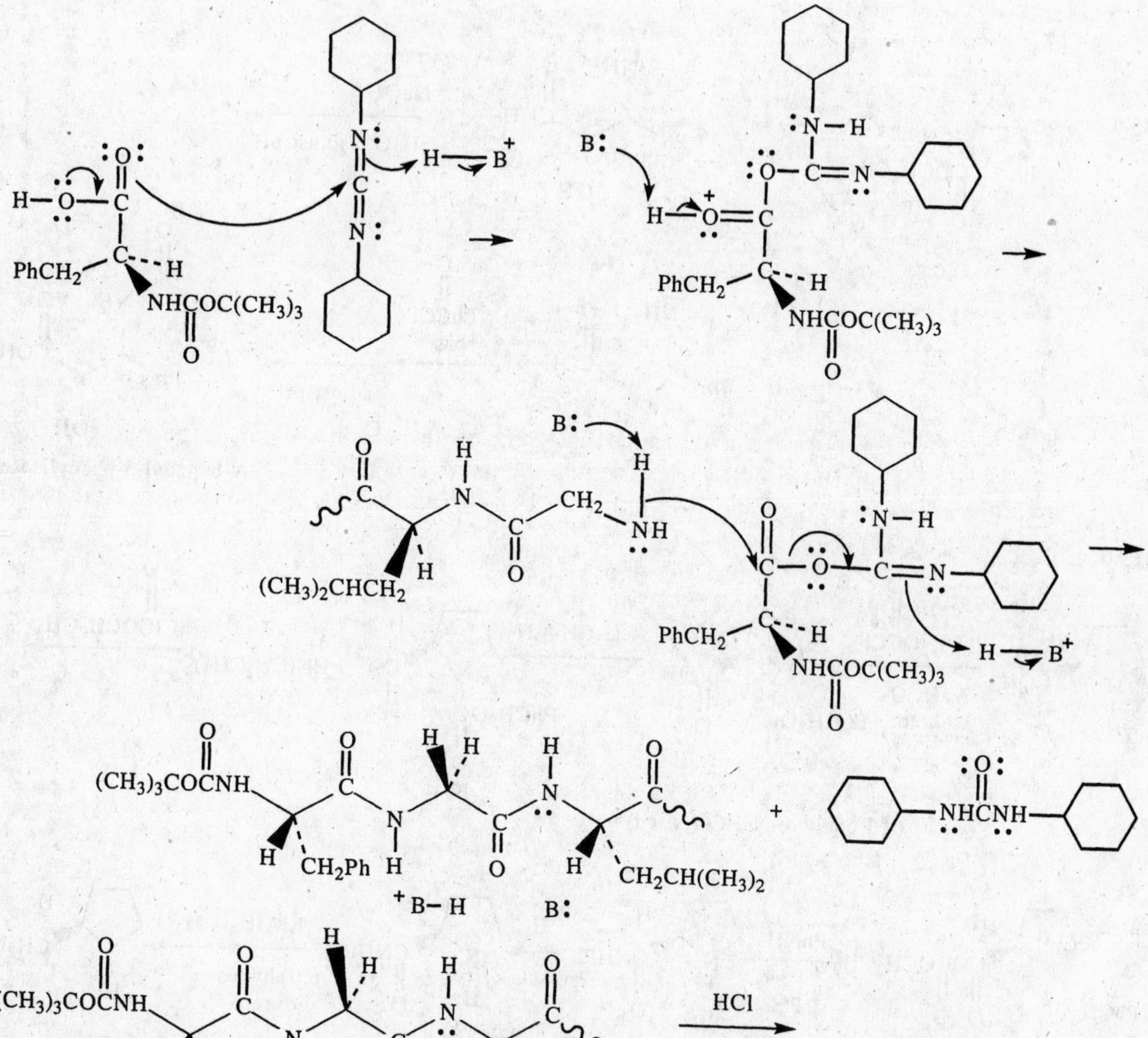

(c) Mechanism diagrams

(d) Reaction with HCl / dioxane

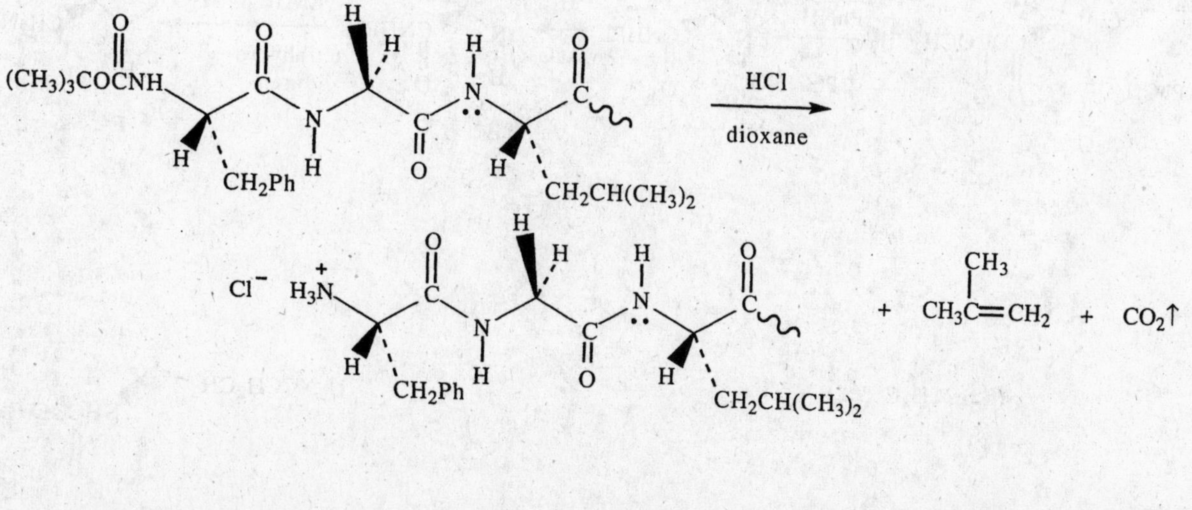

$$(CH_3)_3COCNH\cdots \xrightarrow[\text{dioxane}]{HCl} Cl^-\ H_3\overset{+}{N}\cdots + CH_3CH{=}CH_2 + CO_2\uparrow$$

23.58

(e)

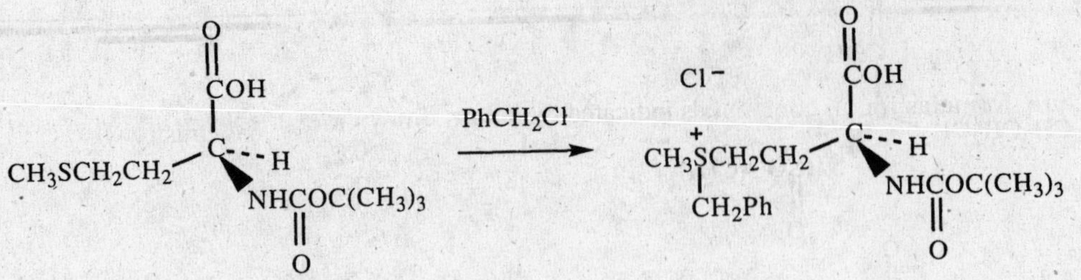

$$CH_3SCH_2CH_2-\underset{\underset{NHCOC(CH_3)_3}{|}}{\overset{\overset{COH}{\overset{\|}{O}}}{C}}-H \quad \xrightarrow{PhCH_2Cl} \quad CH_3\overset{+}{\underset{CH_2Ph}{S}}CH_2CH_2-\underset{\underset{\underset{O}{\|}}{NHCOC(CH_3)_3}}{\overset{\overset{COH}{\overset{\|}{O}}}{C}}-H \quad Cl^-$$

23.59

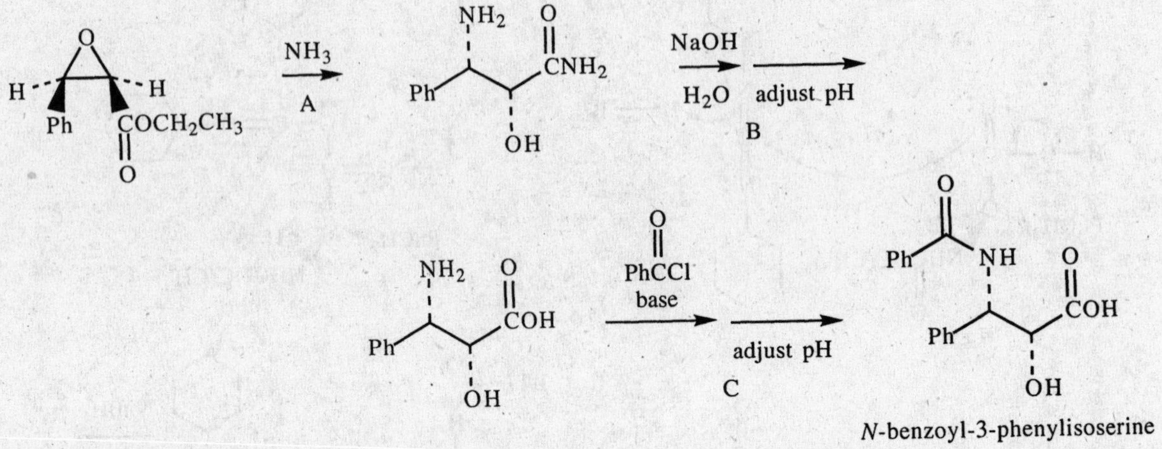

N-benzoyl-3-phenylisoserine

23.60

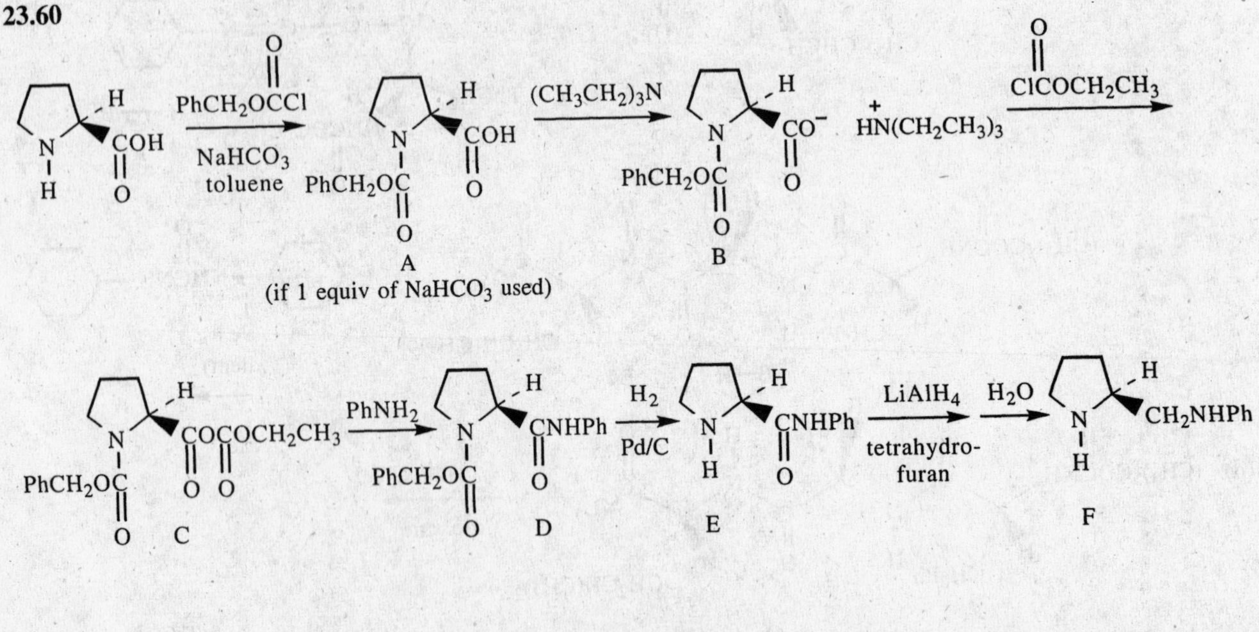

(if 1 equiv of NaHCO₃ used)

Supplemental Problems

S23.1 Write structural formulas for all compounds indicated by letters. Show the stereochemistry whenever it is known.

(a)

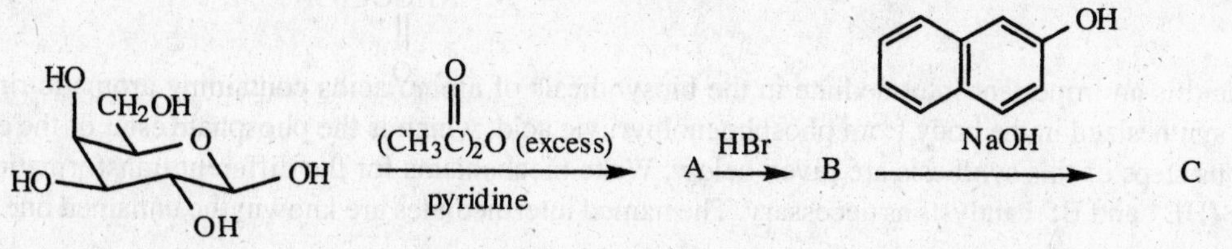

$$\text{glucose structure} \xrightarrow[\text{pyridine}]{(CH_3C)_2O \text{ (excess)}} A \xrightarrow{HBr} B \xrightarrow{NaOH} C$$

(b)

$$\text{ribose structure} \xrightarrow[\Delta]{PhNH_2} D \xrightarrow[NaOH]{(CH_3)_2SO_4} E$$

(c)

$$\text{glucoside structure} \xrightarrow[\text{NaOH}]{\substack{PhCH_2Cl \\ (excess)}} F$$

(d)

$$\text{lactone structure} \xrightarrow[H_2O]{NaBH_4} G$$

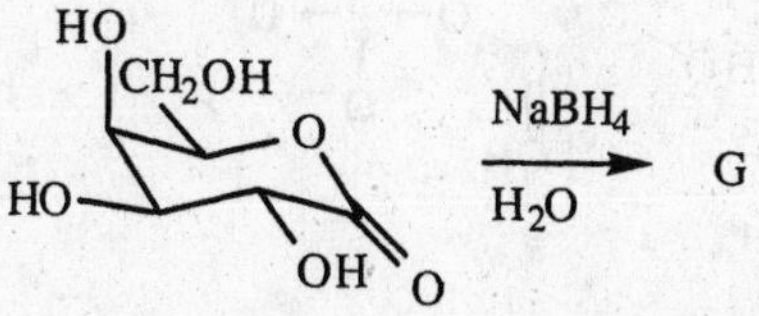

S23.2 Complete the following equations:

(a)

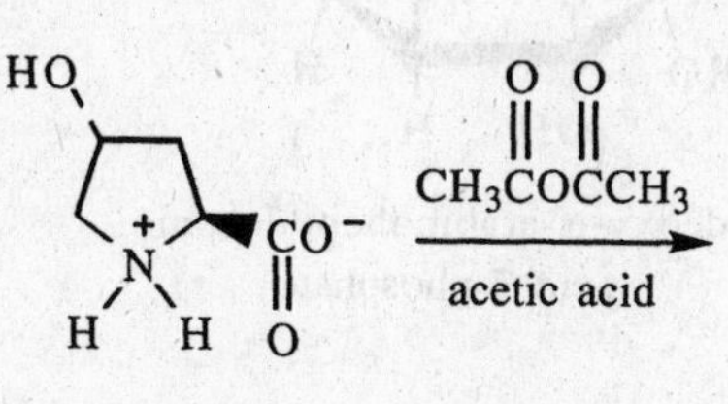

$$\text{hydroxyproline} \xrightarrow[\text{acetic acid}]{\underset{CH_3COCCH_3}{O\ \ O}}$$

(b)

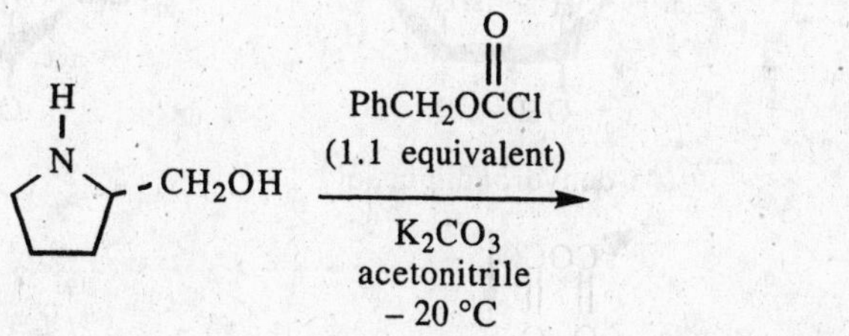

$$\text{prolinol} \xrightarrow[\substack{K_2CO_3 \\ acetonitrile \\ -20\ °C}]{\substack{O \\ PhCH_2OCCl \\ (1.1\ equivalent)}}$$

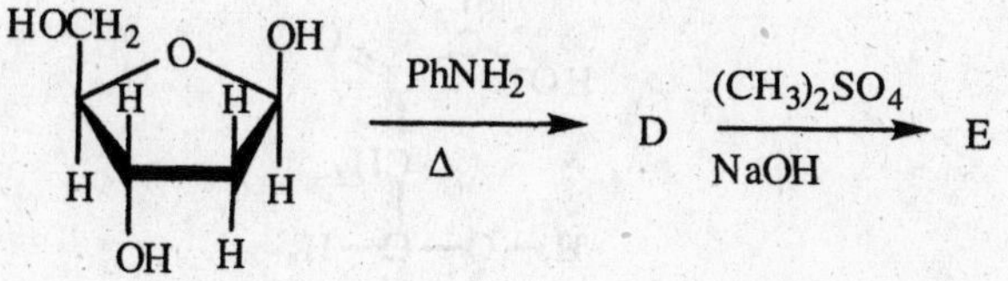

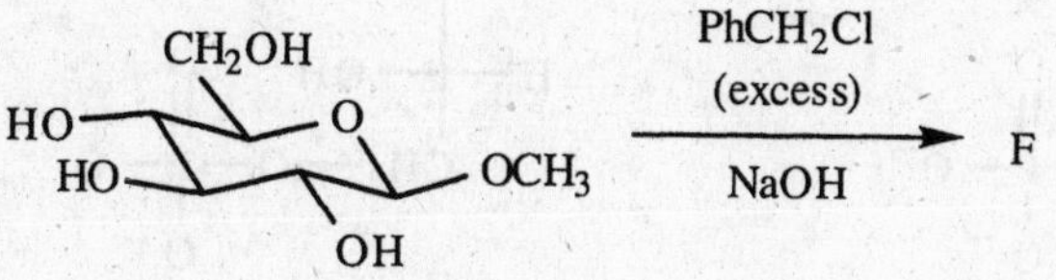

(c)

$$\text{N-Cbz-prolinol} \xrightarrow{CH_3SCH_3,\ ClC-CCl} \xrightarrow{(CH_3CH_2)_3N}$$

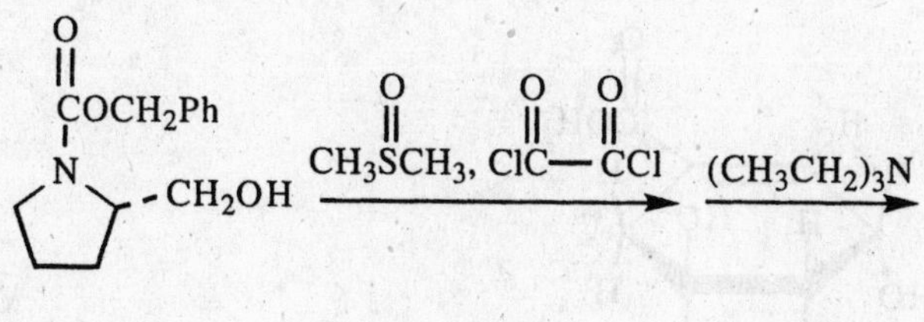

(d)

$$\text{acetoxy-proline derivative} \xrightarrow[\text{pyridine}]{TsCl}$$

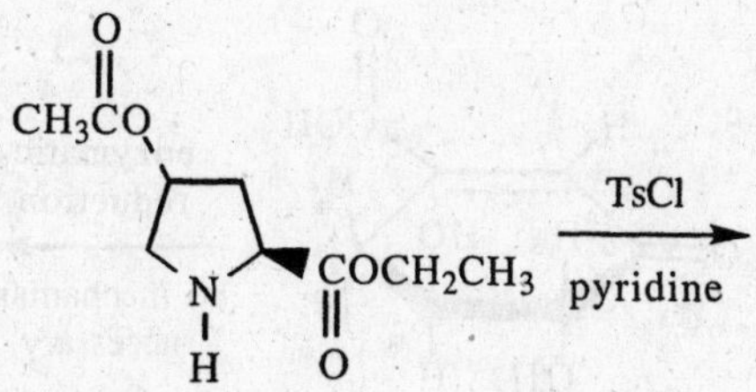

S23.3   α-D-Glucose reacts in its furanose form with 2 equivalents of acetone to give cyclic ketals with the hydroxyl groups at carbons 1 and 2 and carbons 5 and 6. Such cyclic ketals are known as isopropylidene derivatives of sugars. Draw the structural formula for the 1,2,5,6-di-*O*-isopropylidene-D-glucofuranose, and propose a mechanism for its formation.

S23.4   Shikimic acid is an important intermediate in the biosynthesis of amino acids containing aromatic rings. Shikimic acid itself is synthesized in the body from phosphoenolpyruvic acid, which is the phosphate ester of the enol of pyruvic acid. Various steps of this synthesis are given below. Write mechanisms for the different transformations, supplying acid or base (HB$^+$ and B:) catalysts as necessary. The named intermediates are known; the unnamed one has not been proven.

phosphoenolpyruvic acid   erythrose 4-phosphate   III

5-dehydroquinic acid   3-deoxy-α-arabinoheptulosonic acid 7-phosphate

5-dehydroshikimic acid   enzymatic reduction   no mechanism necessary   shikimic acid

S23.5   Synthetic, racemic leucine is resolved using an enzyme, hog renal acylase, which is isolated from hog kidneys. Hog renal acylase catalyzes the hydrolysis of the amides of L-amino acids only. The reactions are

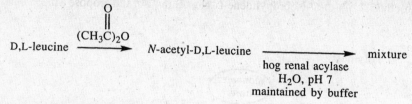

Some useful information is that acetic acid has pK_a 4.8 and leucine has pI 6.0. Also, leucine and *N*-acetylleucine are crystalline solids when bearing no net charge.

(a) Write equations for the reactions described, being sure to show the correct stereochemistry for the species present at each step. What are the components of the mixture formed at the end of the enzymatic hydrolysis?
(b) How would you separate the components of the mixture that results from the enzymatic treatment? Write equations showing what you would have to do to recover the individual compounds. (Hint: What ionic species are present in the solution when the reaction is over?)
(c) Write equations to describe what you would have to do to recover pure D-leucine.

S23.6   The octapeptide xenopsin from the frog *Xenopus laevis* has a powerful contractile effect on muscles. The *N*-terminal amino acid of xenopsin is glutamic acid, which exists in the natural peptide as the cyclic amide pyroglutamic acid (p. 998 in the text). Carboxypeptidase releases leucine first. Trypsin digestion yields ⌐Glu·Gly·Lys·Arg and ⌐Glu·Gly·Lys·Arg·Pro·Trp . Chymotrypsin gives the same hexapeptide as is obtained with trypsin and Ile-Leu. What is the structure of xenopsin?

S23.7   The human growth hormone isolated from the pituitary gland has 188 amino acid residues and has been synthesized by a solid-state technique similar to that of Merrifield (p. 901 in the text). The first five amino acids in the chain, starting from the *C*-terminal end, are phenylalanine, glycine, cysteine, serine, and glycine, in that order. Devise a solid-state synthesis of this portion of the chain. Assume that the amino acids cysteine and serine are available with benzyl groups protecting the thiol and the hydroxyl groups on their side chains. The benzyl group can be removed by reduction using sodium in ammonia.

24 Macromolecular Chemistry

Concept Map 24.1   Polymers.

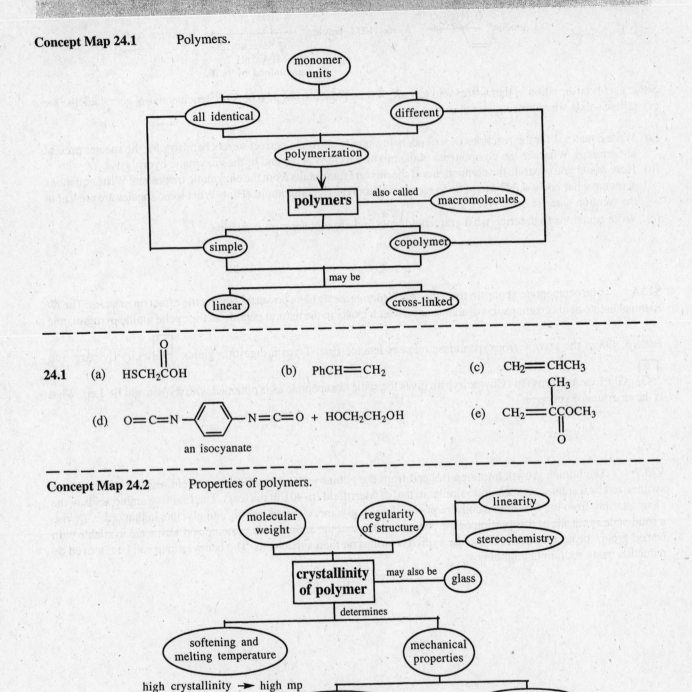

24.1 (a) HSCH$_2$COH (with =O on the carbonyl)

(b) PhCH=CH$_2$

(c) CH$_2$=CHCH$_3$

(d) O=C=N—⟨benzene ring⟩—N=C=O + HOCH$_2$CH$_2$OH

an isocyanate

(e) CH$_2$=CCOCH$_3$ (with CH$_3$ substituent and =O)

Concept Map 24.2   Properties of polymers.

24.2

(a) $CH_2=\overset{\overset{\displaystyle Ph}{|}}{C}CH_3$ and $N\equiv C-CH=CH-C\equiv N$; a copolymer   (b) $CH_2=CCl_2$; a simple polymer

(c) $PhCH=CHPh$ and 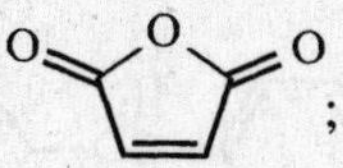 ; a copolymer   (d) $CH_2=\overset{\overset{\displaystyle CH_3}{|}}{C}Ph$ and $CH_2=\overset{\overset{\displaystyle CH_3}{|}}{C}-C\equiv N$; a copolymer

24.3

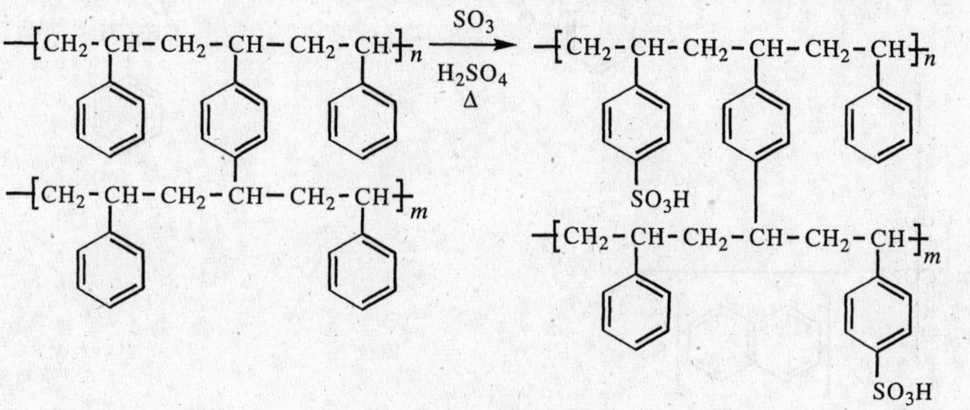

24.4   (a)

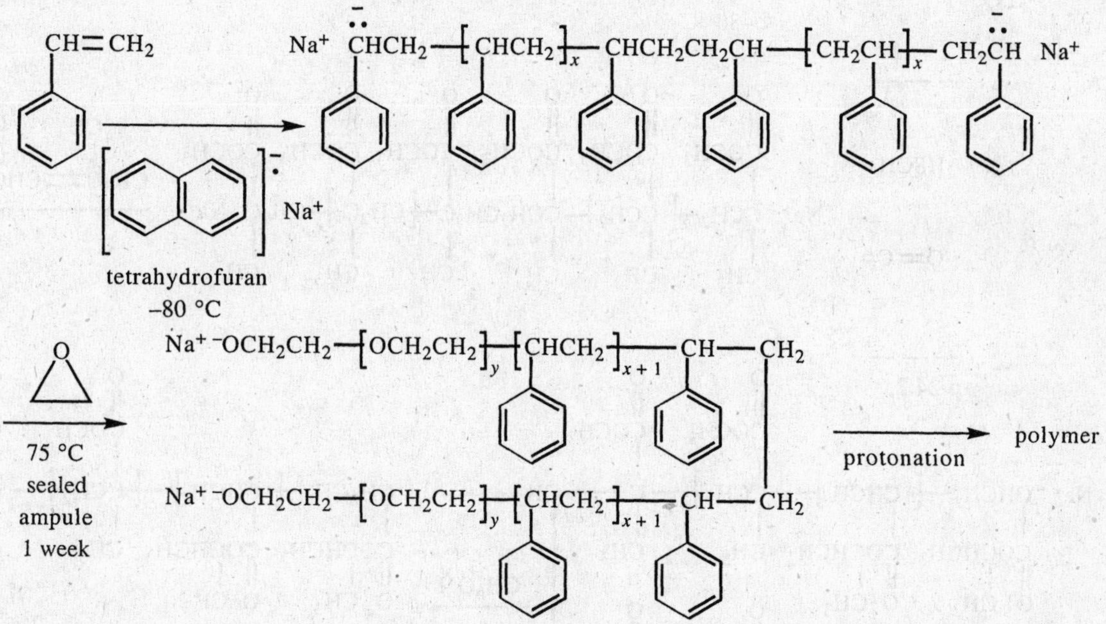

(b) $CH_2=CHCl + CH_2=CHCOCH_3 \xrightarrow[\substack{\text{acetone} \\ 50\,°C}]{\text{benzoyl peroxide}}$

or $\longrightarrow$ polymer

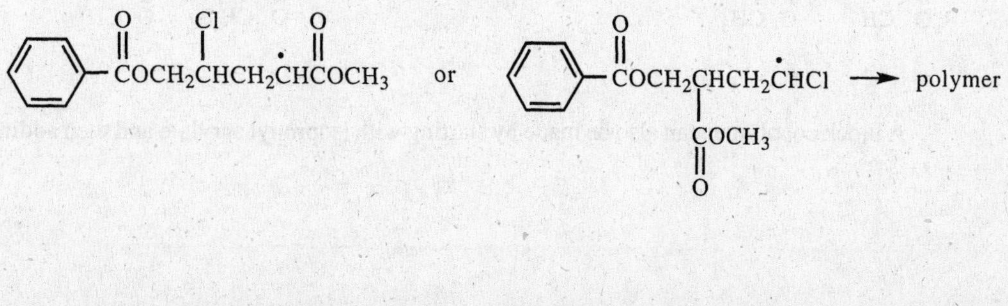

24.4   (c)

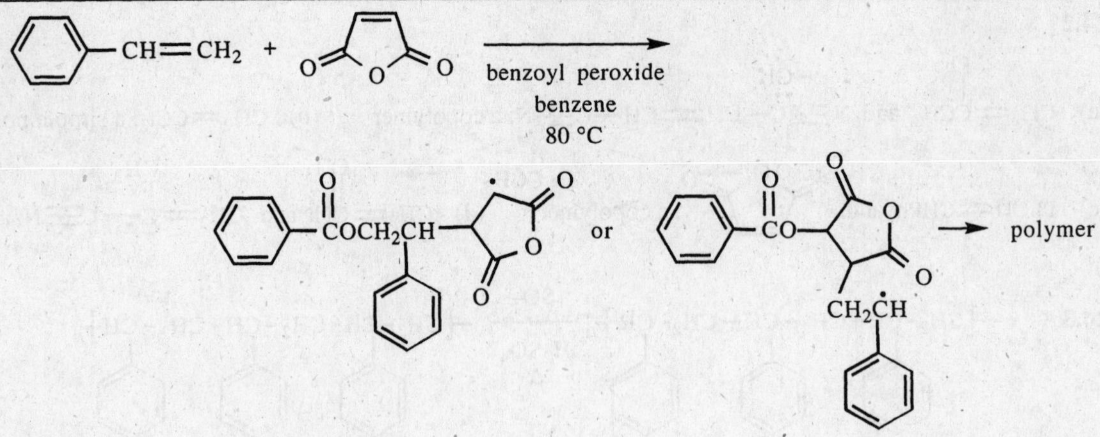

24.5

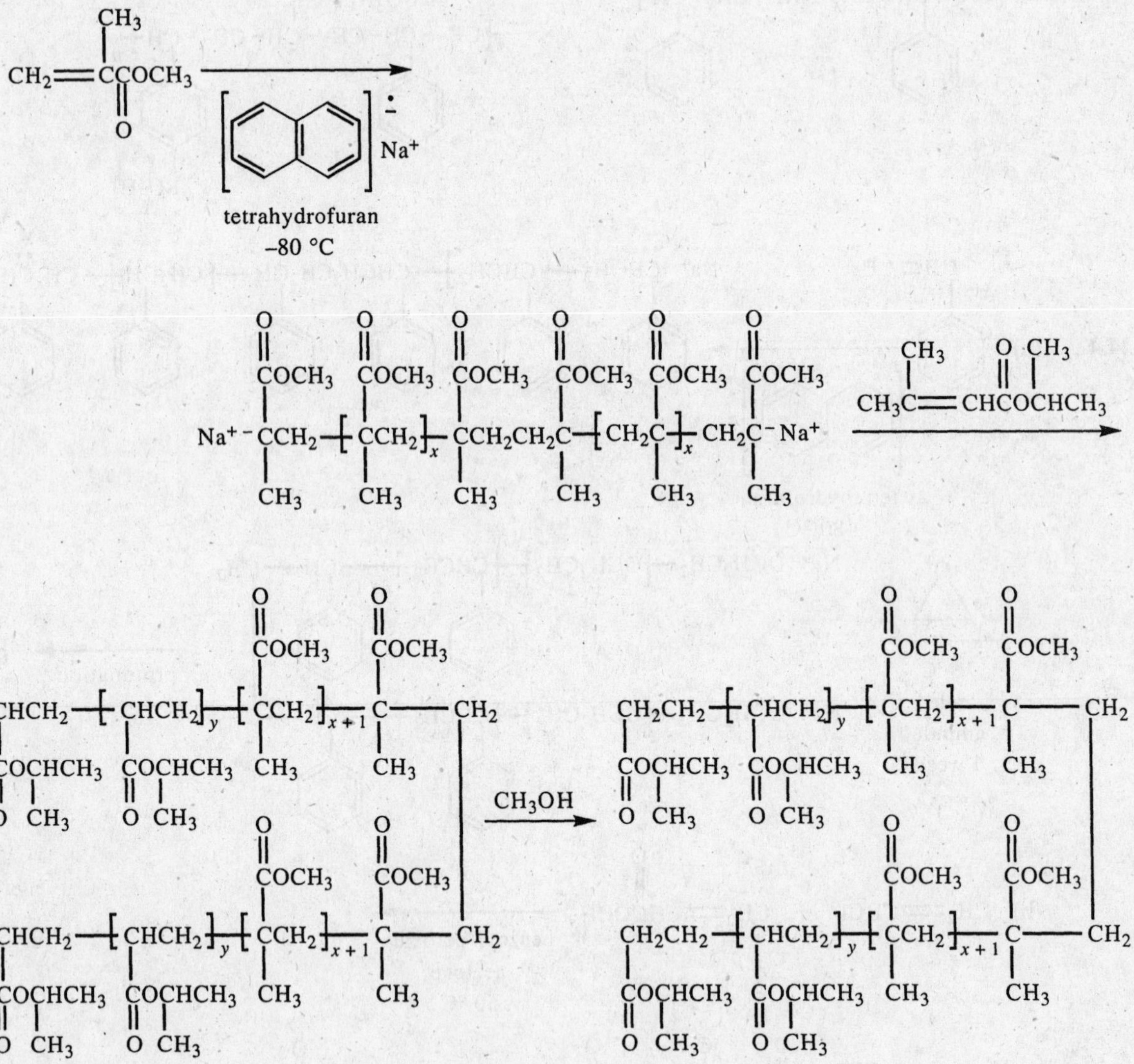

A block copolymer can also be made by starting with isopropyl acrylate and then adding methyl methacrylate.

24.6 $AlCl_3$ + $CH_3\overset{:O:}{\overset{\|}{C}}-\overset{..}{\overset{..}{Cl}}:$ $\longrightarrow$ $CH_3C\equiv\overset{+}{O}:$ + $^-AlCl_4$

initiation

propagation

polymer $\longleftarrow$

24.7   (a) $\overset{S}{\triangle}$ $\xrightarrow{BF_3}$ $\overset{+}{\triangle}S-\overset{-}{B}F_3$ $\longrightarrow$ $\overset{+}{\triangle}S-CH_2CH_2\overset{-}{S}BF_3$ $\longrightarrow$ polymer

(b) $\overset{H}{\underset{}{\triangle N}}$ $\xrightarrow[\Delta,\ 2\ h]{0.1\ M\ HCl}$ $\overset{H\ \ H}{\underset{}{\triangle \overset{+}{N}}}$ Cl^- $\longrightarrow$ $\triangle\overset{H}{\overset{+}{N}}-CH_2CH_2NH_2$ Cl^- $\longrightarrow$ polymer

(c) $\xrightarrow{acid}$ $HO(CH_2)_4\overset{O}{\overset{\|}{C}}-Nu$ $\longrightarrow$ $HO(CH_2)_4\overset{O}{\overset{\|}{C}}O(CH_2)_4\overset{O}{\overset{\|}{C}}-Nu$ $\longrightarrow$ polymer

(Nu represents the conjugate base of the acid initiator.)

(d) $\xrightarrow{base}$ $^-O(CH_2)_4\overset{O}{\overset{\|}{C}}-B$ $\longrightarrow$ $^-O(CH_2)_4\overset{O}{\overset{\|}{C}}O(CH_2)_4\overset{O}{\overset{\|}{C}}-B$ $\longrightarrow$ polymer

(B represents the basic initiator.)

(e) $\xrightarrow[\substack{solid\ state\\(trace\ of\ acid\\usually\ present)}]{}$ $HOCH_2CH_2O\overset{O}{\overset{\|}{C}}-\overset{O}{\overset{\|}{C}}OH$ $\longrightarrow$

$HOCH_2CH_2O\overset{O}{\overset{\|}{C}}-\overset{O}{\overset{\|}{C}}OCH_2CH_2O\overset{O}{\overset{\|}{C}}-\overset{O}{\overset{\|}{C}}OH$ $\longrightarrow$ $\longrightarrow$ polymer

(f) $\overset{CH_3}{\underset{}{CH_3C}}=CH_2$ $\xrightarrow[H_2O\ (trace)]{BF_3}$ $CH_3\overset{CH_3}{\underset{+}{C}}CH_3$ $\xrightarrow{CH_3C\overset{CH_3}{=}CH_2}$ $CH_3\overset{CH_3}{\underset{}{C}}CH_2\overset{CH_3}{\underset{+\ CH_3}{C}}CH_3$ $\longrightarrow$ polymer

24.8 $HO-\langle C_6H_4\rangle-CH_2CH_2-\langle C_6H_4\rangle-OH$ + $BrCH_2(CH_2)_6CH_2Br$ $\xrightarrow{\text{NaOH}}$

1,2-bis(4-hydroxyphenyl)ethane   1,8-dibromooctane

$\left[O-\langle C_6H_4\rangle-CH_2CH_2-\langle C_6H_4\rangle-OCH_2(CH_2)_6CH_2\right]_n$

24.9

isotactic polystyrene   syndiotactic polystyrene   atactic polystyrene

fragment of the chain, showing only one
of several possible orientations of the
phenyl groups with respect to each other

24.10   In order for a polymer to be highly crystalline, it has to be highly ordered so it can pack easily into a crystal structure. Isotactic and syndiotactic polymers, with their very regular primary structures, have strong interactions between chains and form crystals. Atactic polymers, with their much less regular structures, do not have as good interactions between chains and are more amorphous.

24.11

atactic poly(methyl methacrylate)
fragment of chain showing one of several possible
arrangements of the ester groups on the chain

24.12   (a) $CH_3CH_2\underset{\underset{CH_3}{|}}{C}HCH=CH_2$ $\xrightarrow[\text{(CH}_3\text{CH}_2)_2\text{AlCl}]{\text{TiCl}_4}$ $\xrightarrow{CH_3CH_2OH}$

(b) $CH_2=CHOCH_3$ $\xrightarrow[\substack{\text{CH}_3 \\ | \\ (\text{CH}_3\text{CHCH}_2)_3\text{Al} \\ \text{heptane}}]{\text{VCl}_3}$ $\xrightarrow{CH_3CH_2OH}$

24.12 (c) $CH_2{=}CH_2$ + [structure of 2-butene] $\xrightarrow{\underset{(CH_3CHCH_2)_3Al}{\underset{CH_3}{VCl_3}}}$ $\xrightarrow{CH_3CH_2OH}$

heptane
−30 °C

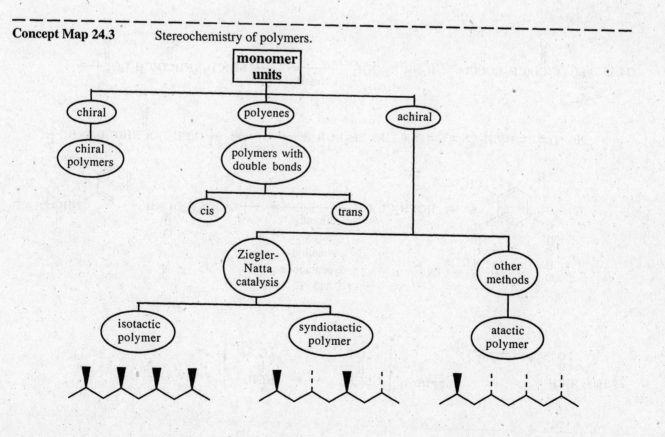

(d) $CH_2{=}\underset{\underset{O}{\overset{\overset{CH_3}{|}}{C}}}{C}COCH_3$ $\xrightarrow{\underset{(CH_3CHCH_2)_3Al}{\underset{CH_3}{VCl_3}}}$ $\xrightarrow{CH_3CH_2OH}$ $\left[CH_2\underset{\underset{\underset{O}{||}}{COCH_3}}{\overset{\overset{CH_3}{|}}{C}} \right]_n$

heptane

(e) $CH_3CH_2CH{=}CH_2$ $\xrightarrow{\underset{(CH_3CH_2)_2AlCl}{FeCl_3}}$ $\xrightarrow{CH_3CH_2OH}$ $\left[CH_2\underset{\overset{|}{CH}}{\overset{\overset{CH_2CH_3}{|}}{CH}} \right]_n$

Concept Map 24.3   Stereochemistry of polymers.

24.13

HOC(=O)—(benzene ring)—C(=O)OH, with NH$_2$ substituent + PhCH(=O) $\xrightarrow[\Delta]{N,N\text{-dimethyl-formamide}}$ HOC(=O)—(benzene ring)—C(=O)OH, with N=CHPh substituent **A** $\xrightarrow[\text{pyridine}]{SOCl_2 \text{ (excess)}}$

ClC(=O)—(benzene ring)—C(=O)Cl, with N=CHPh substituent **B** + H$_2$N—(C$_6$H$_4$)—CH$_2$—(C$_6$H$_4$)—NH$_2$ $\longrightarrow$ $\left[\!\!\!\begin{array}{c} \text{C(=O)—(benzene ring, N=CHPh)—C(=O)NH—(C}_6\text{H}_4\text{)—CH}_2\text{—(C}_6\text{H}_4\text{)—NH} \end{array}\!\!\!\right]_n$ **polymer**

24.14

HO—(C$_6$H$_4$)—C(CH$_3$)$_2$—(C$_6$H$_4$)—OH + CH$_3$OCOCH$_3$ $\xrightarrow{\text{high temperature}}$ $\left[\text{O—(C}_6\text{H}_4\text{)—C(CH}_3\text{)}_2\text{—(C}_6\text{H}_4\text{)—OC(=O)}\right]_n$ + 2n CH$_3$OH

CH$_3$OC(=O)—(C$_6$H$_4$)—C(=O)OCH$_3$ + HO(CH$_2$)$_4$OH $\xrightarrow{\text{high temperature}}$ $\left[\text{O(CH}_2\text{)}_4\text{OC(=O)—(C}_6\text{H}_4\text{)—C(=O)}\right]_n$ + 2n CH$_3$OH

24.15   (a) ClC(=O)(CH$_2$)$_4$OC(=O)Cl + H$_2$N(CH$_2$)$_6$NH$_2$ $\xrightarrow[\text{H}_2\text{O}]{\text{NaOH}}$ $\left[\text{NH(CH}_2\text{)}_6\text{NHC(=O)(CH}_2\text{)}_4\text{OC(=O)}\right]_n$

(b) O=C=N(CH$_2$)$_6$N=C=O + HO(CH$_2$)$_4$OH $\xrightarrow{185-195\,°C}$ $\left[\text{O(CH}_2\text{)}_4\text{OCNH(CH}_2\text{)}_6\text{NHC}\right]_n$

(c) (benzene ring with CH$_3$ and two N=C=O groups) + HOCH$_2$CH$_2$OH $\xrightarrow[\substack{\text{4-methyl-}\\ \text{2-pentanone}\\ 115\,°C}]{\text{dimethyl sulfoxide}}$ $\left[\text{OCH}_2\text{CH}_2\text{OCNH—(benzene ring, CH}_3\text{)—NHC}\right]_n$

24.16   :O:—HCH $\rightarrow$ H$\rightleftharpoons$B$^+$ $\longrightarrow$ $\left[\;\overset{+}{:}\overset{H}{O}\text{—HCH}\;\longleftrightarrow\;\ddot{O}\text{—H, }\overset{+}{H}\text{CH}\;\right]$:B

24.16 (cont)

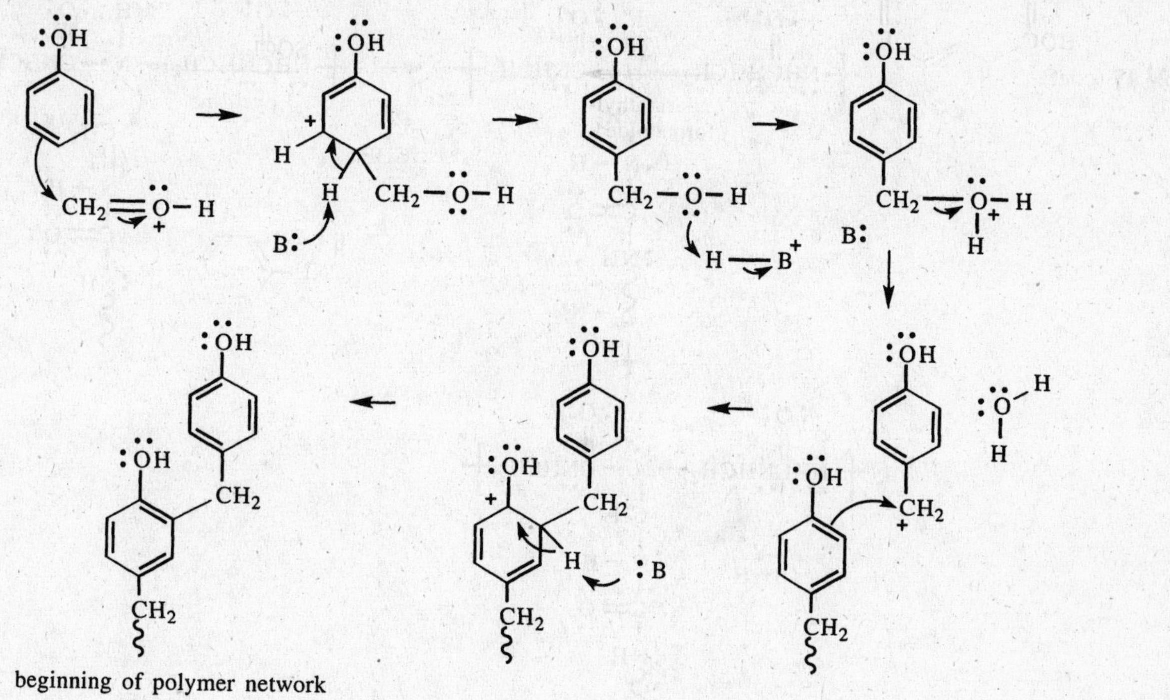

beginning of polymer network

24.17

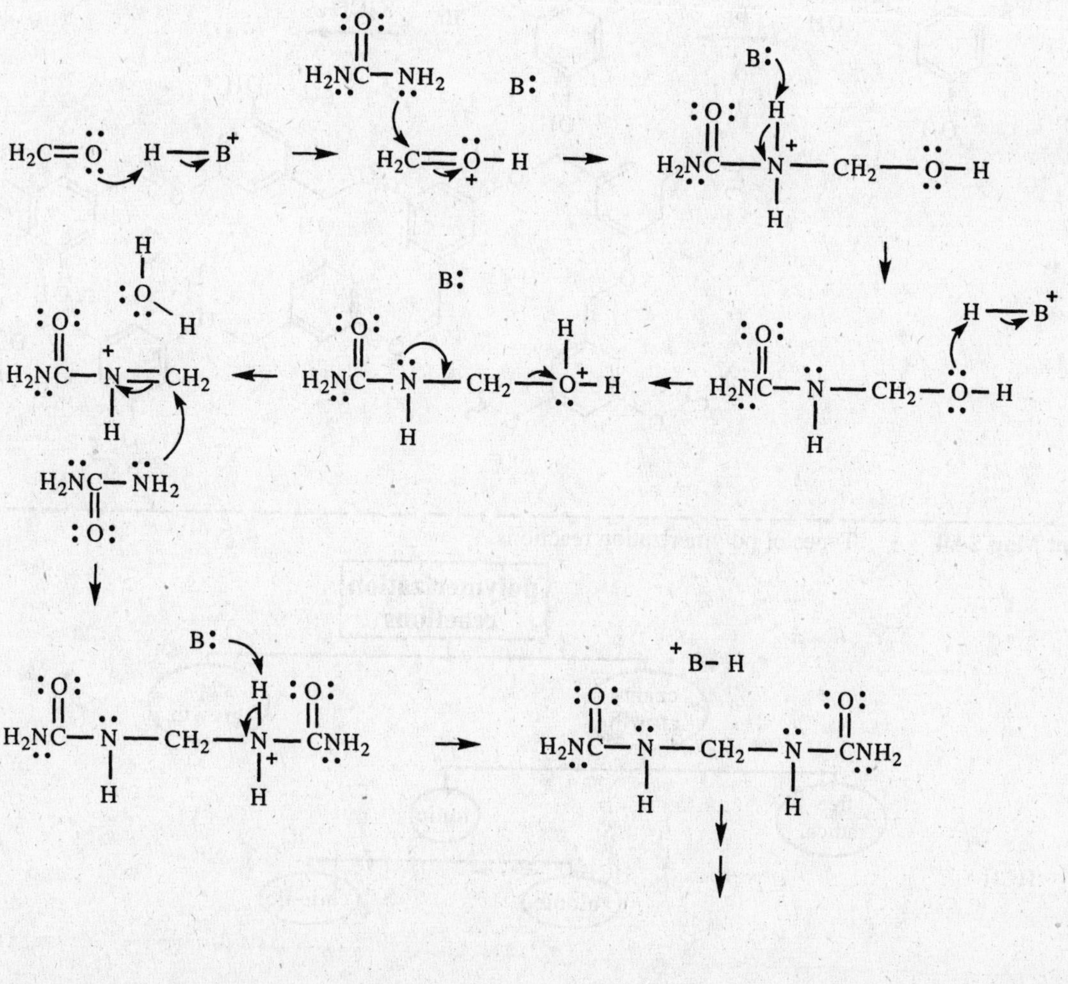

24.17 (cont)

24.18

$\xrightarrow{\text{PBr}_3}$ A

$\xrightarrow{\text{K}_2\text{CO}_3}$

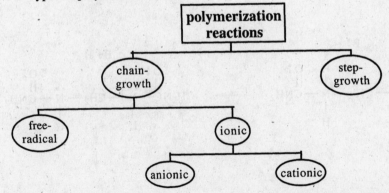

Concept Map 24.4   Types of polymerization reactions.

polymerization reactions
- chain-growth
  - free-radical
  - ionic
    - anionic
    - cationic
- step-growth

24.19

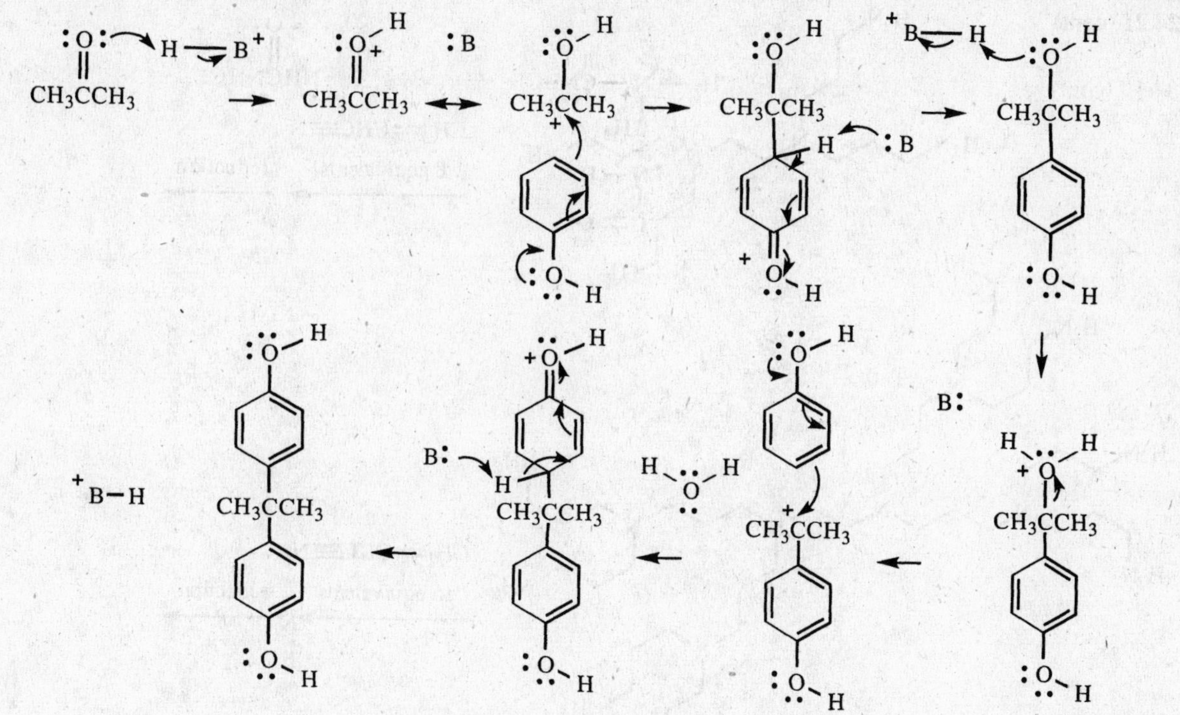

24.20

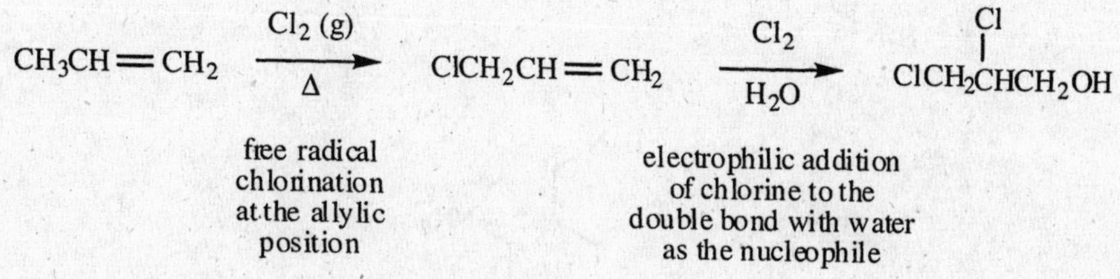

$$CH_3CH=CH_2 \xrightarrow[\Delta]{Cl_2\ (g)} ClCH_2CH=CH_2 \xrightarrow[H_2O]{Cl_2} ClCH_2CHCH_2OH$$

free radical
chlorination
at the allylic
position

electrophilic addition
of chlorine to the
double bond with water
as the nucleophile

24.21

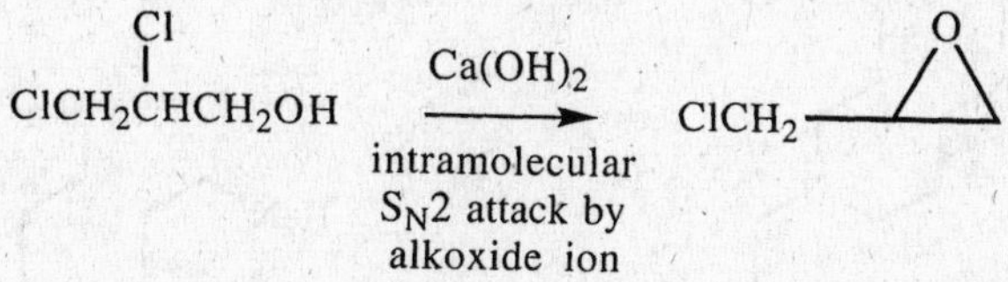

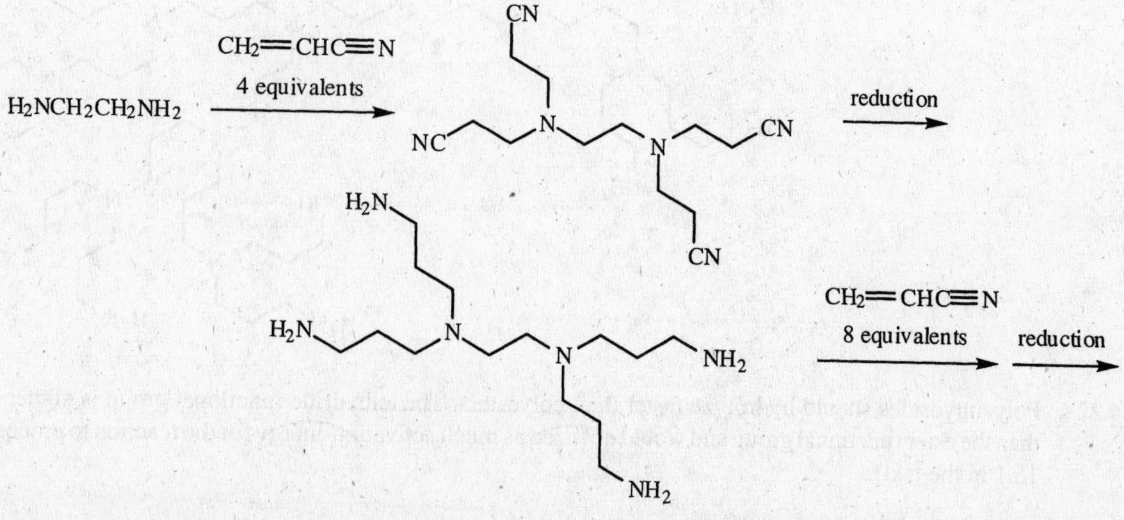

$$ClCH_2CHCH_2OH \xrightarrow[\substack{\text{intramolecular} \\ S_N2 \text{ attack by} \\ \text{alkoxide ion}}]{Ca(OH)_2} ClCH_2\!-\!\!\triangle\!\!-O$$

24.21 (cont)

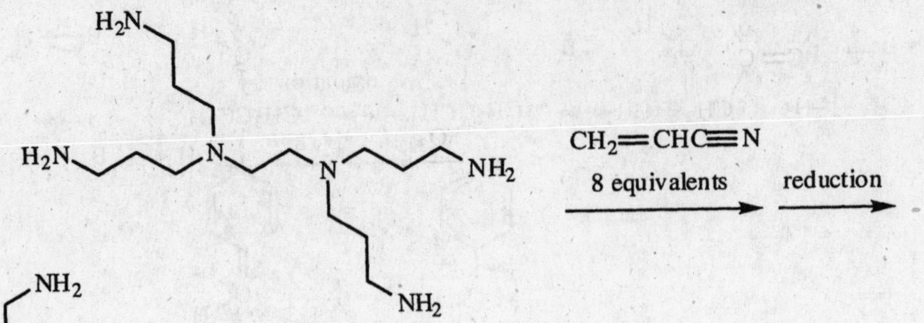

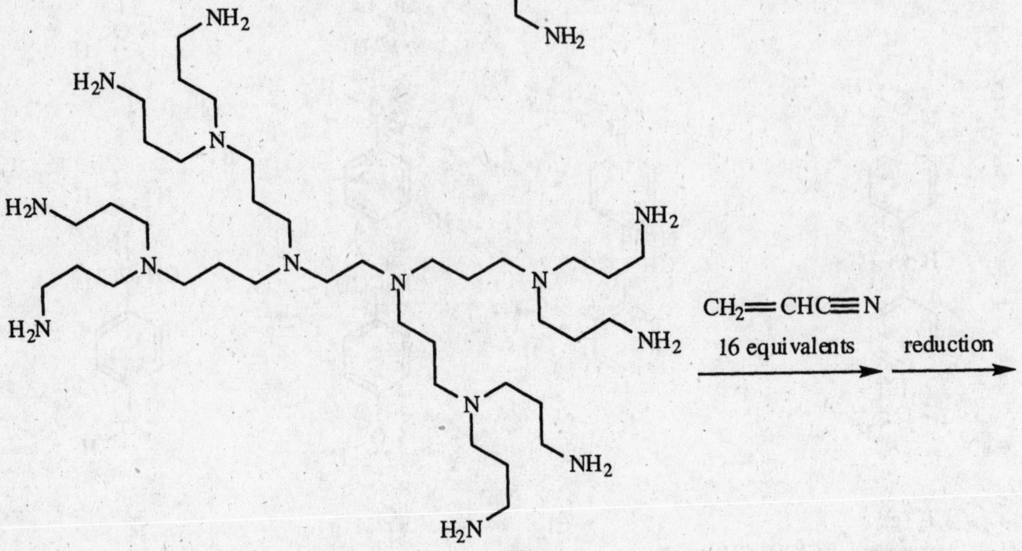

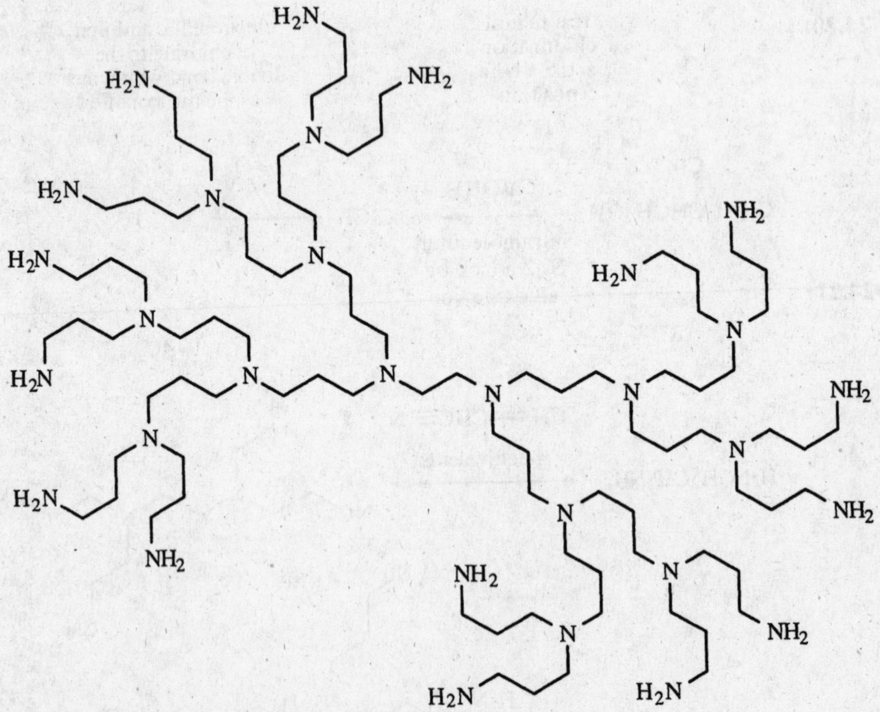

CH₂=CHC≡N
8 equivalents → reduction →

CH₂=CHC≡N
16 equivalents → reduction →

24.22 Polyanhydrides should hydrolyze faster than polyesters. The anhydride functional group is higher in energy than the ester functional group and would not need as much activation energy for the reaction to proceed (Figure 15.1 in the text).

24.23

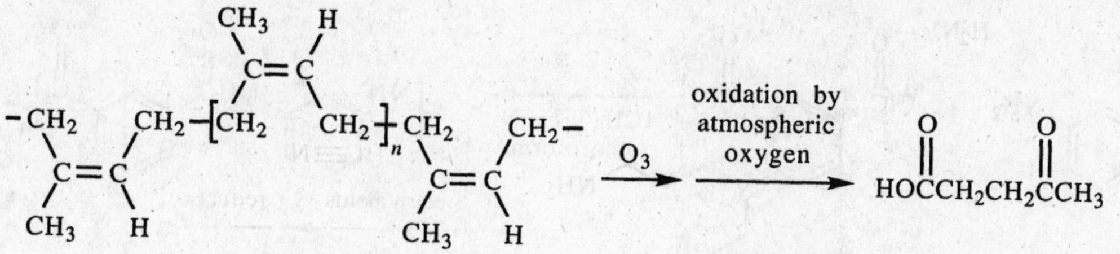

$$\xrightarrow{O_3} \xrightarrow{\substack{\text{oxidation by}\\ \text{atmospheric}\\ \text{oxygen}}} HOCCH_2CH_2CCH_3$$

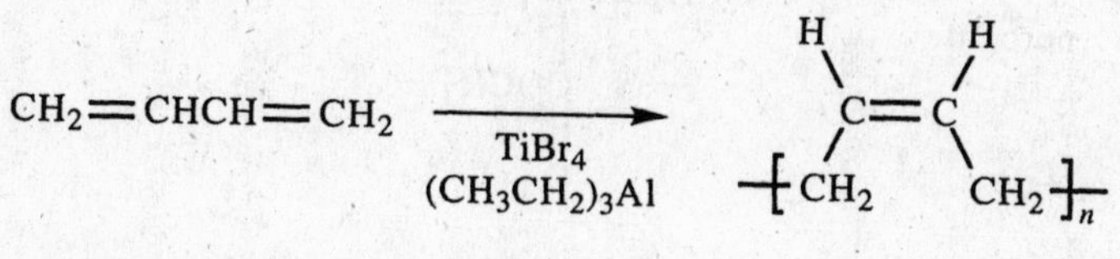

24.24 $CH_2=CHCH=CH_2 \xrightarrow[\substack{TiBr_4\\ (CH_3CH_2)_3Al}]{}$

24.25 $CH_3OCH=CHCH=CH_2 \xrightarrow[\substack{\text{free radical}\\ \text{initiator}}]{}$

amorphous polymer;
cis double bonds

$CH_3OCH=CHCH=CH_2 \xrightarrow[\substack{VCl_3\\ CH_3\\ (CH_3CHCH_2)_3Al}]{}$

crystalline polymer;
trans double bonds

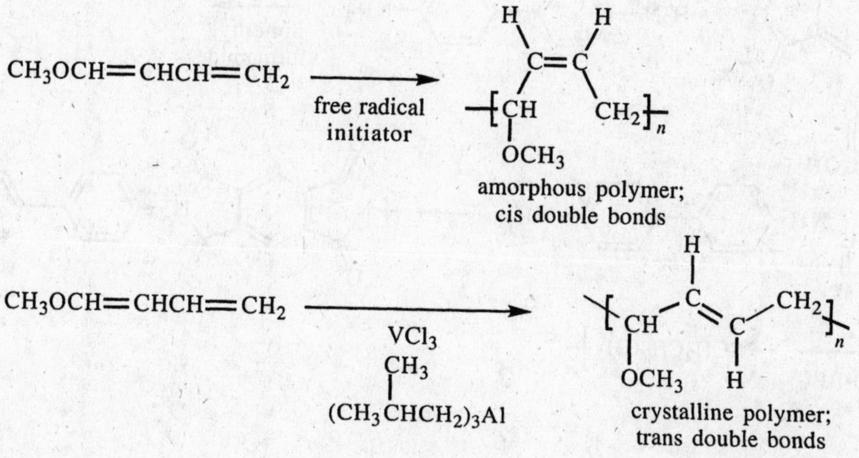

24.26

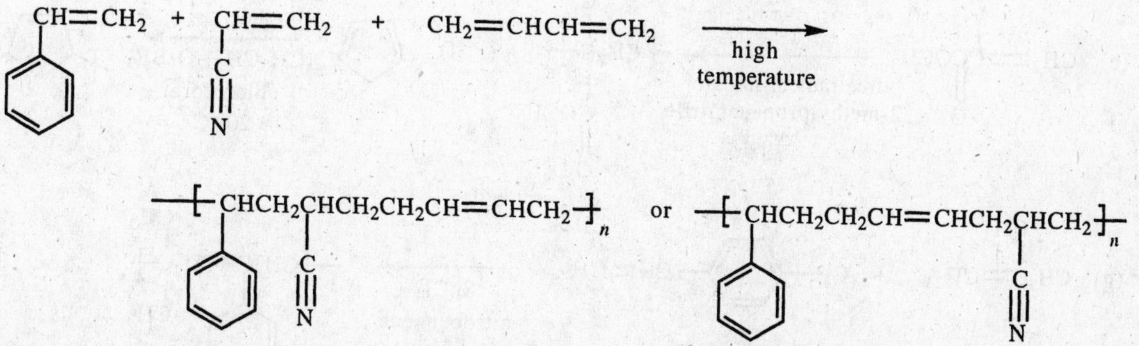

There are many different copolymers that can be written for these three monomers, depending on the order in which they are arranged and on whether 1,2- or 1,4-addition to 1,3-butadiene takes place. Two of the possibilities are shown.

24.27

(a)

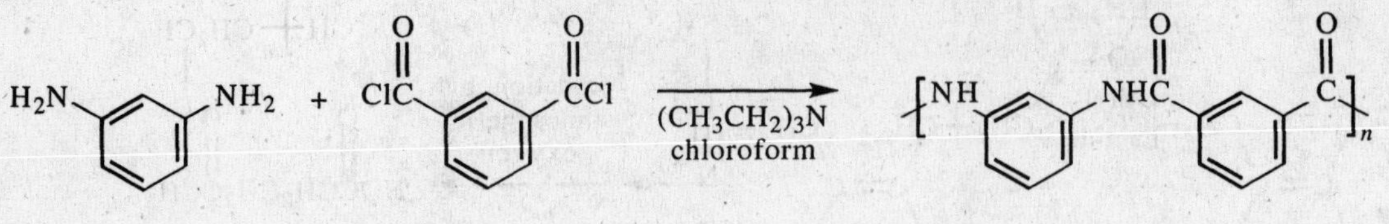

(b)

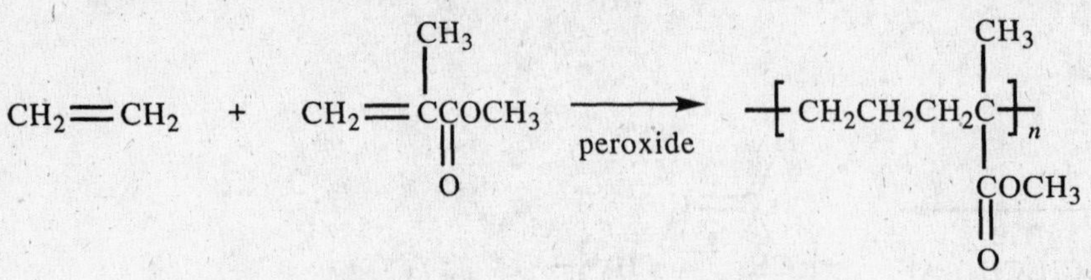

24.27 (c)

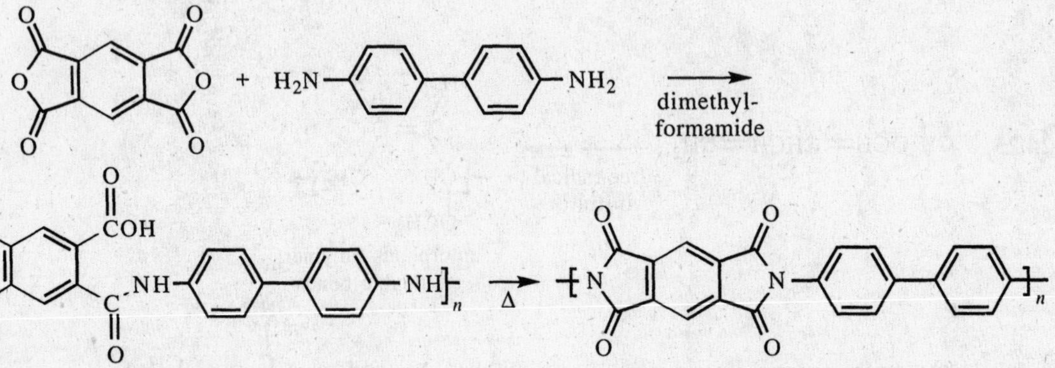

(d)

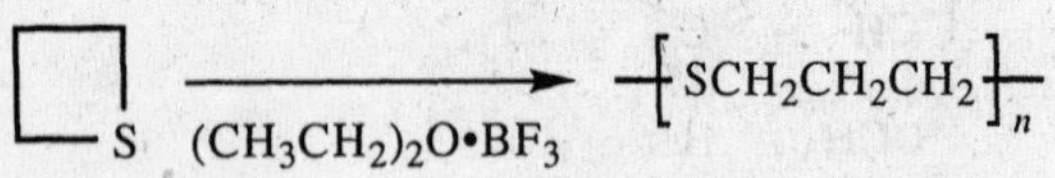

(e) 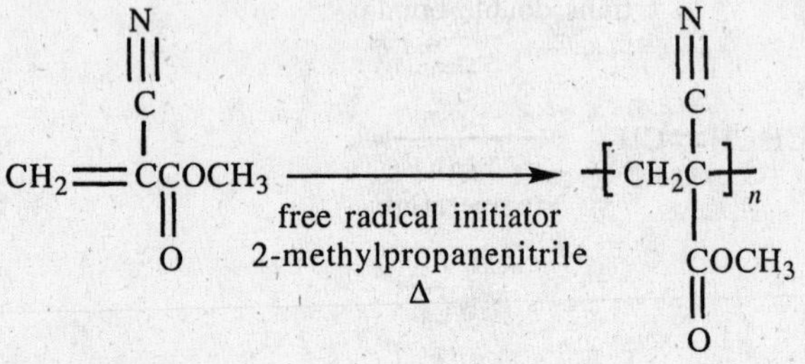   (f)

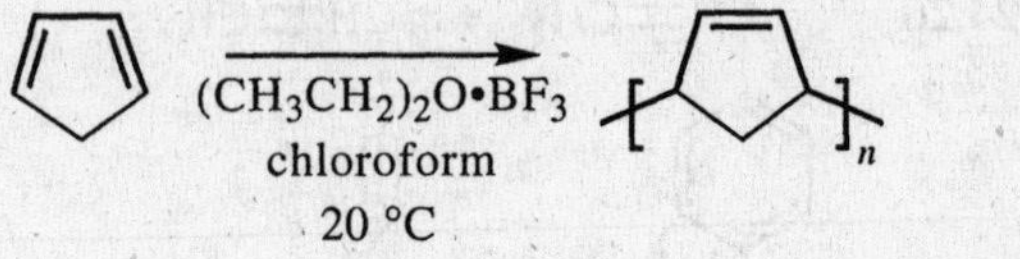

(g)

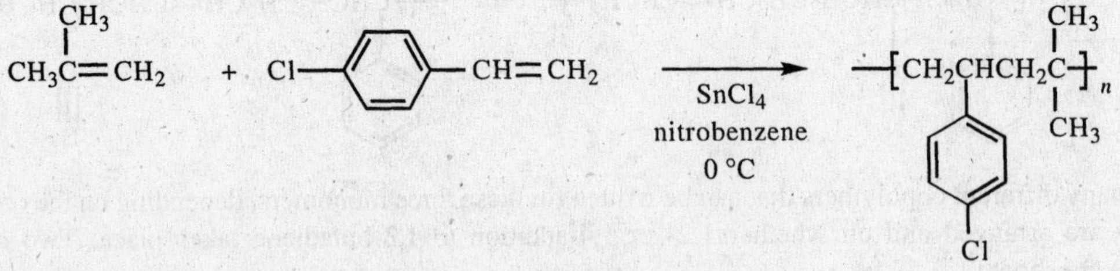

(h)

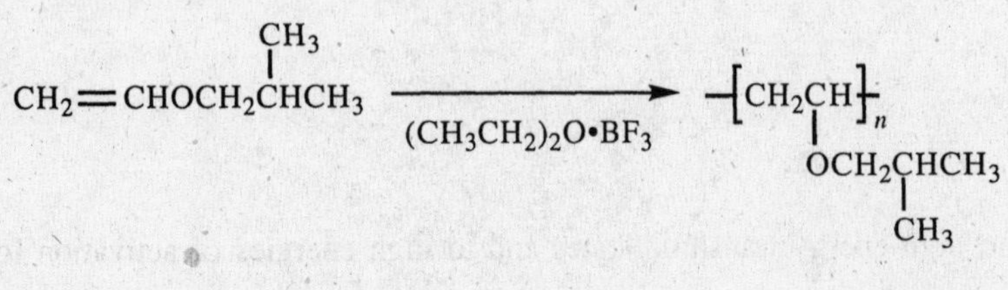

(i) $CH_2=CHC\equiv N$ $\xrightarrow[\substack{CaO \\ dimethyl- \\ formamide \\ 20\ °C}]{}$ $^-O\left[CH_2CH\right]_n CH_2\ddot{C}H$ $\xrightarrow{HCl}$ $HO\left[CH_2CH\right]_n CH_2CH_2$

with cyano groups ($C{\equiv}N$) on the CH and CH₂ carbons as drawn.

(j) $CH_2=CHOCHCH_3$ (with CH_3 substituent) $\xrightarrow[\substack{VCl_3 \\ CH_3 \\ (CH_3CHCH_2)_3Al \\ heptane}]{}$ $\xrightarrow{CH_3CH_2OH}$ polymer chain with $OCHCH_3$ (CH_3) side groups and H substituents

24.28   Cyano and ester groups are electron-withdrawing and stabilize a carbanionic intermediate by their inductive effects and also by resonance.

$$R'\!-\!CH_2\ddot{\overset{-}{C}}HC\equiv N\colon \longleftrightarrow R'\!-\!CH_2CH=C=\ddot{\overset{-}{N}}\colon$$

$$R'\!-\!CH_2\overset{:\overset{\displaystyle :O:}{\|}}{\underset{}{\ddot{C}HC}}-\ddot{O}R \longleftrightarrow R'\!-\!CH_2CH=\overset{\displaystyle :\ddot{O}:^-}{\underset{}{C}}-\ddot{O}R$$

where R' is the initiator.

Radicals are also stabilized by groups such as the cyano group or carbonyl group.

$$R'\!-\!CH_2\overset{\displaystyle \bullet}{C}HC\equiv N\colon \longleftrightarrow R'\!-\!CH_2CH=C=\overset{\displaystyle \bullet}{\ddot{N}}\colon$$

$$R'\!-\!CH_2\overset{:\overset{\displaystyle :O:}{\|}}{\underset{}{C}HC}-\ddot{O}R \longleftrightarrow R'\!-\!CH_2CH=\overset{\displaystyle :\overset{\bullet}{O}:}{\underset{}{C}}-\ddot{O}R$$

Reactions of these monomers that go through anionic or radical intermediates have low energies of activation and proceed easily.

Cationic intermediates are destabilized by the electron-withdrawing cyano and ester groups. Any resonance contributors that are written for such species delocalize the positive charge from carbon to the more electronegative nitrogen or oxygen atom.

$$R'\!-\!CH_2\overset{\displaystyle +}{C}HC\equiv N\colon \longleftrightarrow R'\!-\!CH_2CH=C=\overset{\displaystyle +}{\ddot{N}}\colon$$

$$R'\!-\!CH_2\overset{\overset{\displaystyle :O:}{\|}}{\underset{}{\overset{+}{C}HC}}-\ddot{O}R \longleftrightarrow R'\!-\!CH_2CH=\overset{\displaystyle :\overset{+}{O}:}{\underset{}{C}}-\ddot{O}R$$

Such intermediates correspond to high-energy transition states and to high energies of activation for the reactions.

24.29   The rate of a reaction depends on the free energy of activation of the rate-determining step. The free energy of activation is the difference in energy between the reagents and the transition state. In rationalizing most relative rates, it is necessary only to look at the relative stabilization of the transition states. In general, the more stabilized the transition state is, the lower the free energy of activation and the faster the reaction. However, if the energy of a reagent is lowered even more by stabilization, the free energy of activation will be larger, and the rate correspondingly smaller.

In free radical polymerization the rate-determining step is the addition of an unstable radical to the alkene monomer. Substituents play an important role in stabilizing this radical, in which one species bears full radical character, and a less important role in stabilizing the transition state, in which radical character is distributed between two species. The consequences of this are illustrated graphically below.

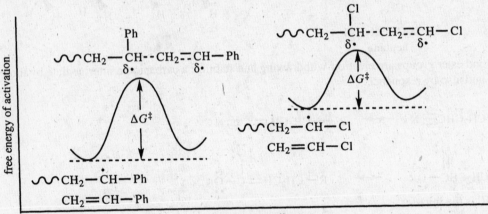

The phenyl group stabilizes the reagent radical more strongly than does the chlorine atom. Even though the transition state for the reaction of the vinyl chloride may be higher in energy in absolute terms than the transition state for the reaction of styrene, the energy of activation for the reaction of vinyl chloride is smaller. The other three monomers fall in between vinyl chloride and styrene in terms of the relative stabilities of their free radicals. The relative rates of the polymerization reactions thus reflect the relative stabilization of radicals by the substituents on the double bond.

24.30   The first method would give a copolymer in which the amide bonds would be formed more or less randomly by the two different amines reacting with the acid chloride. There would be no pattern to the polymer. If we designate the siloxanediamine as H_2N—(Si)—NH_2 and the aryldiamine as H_2N—(Ar)—NH_2, the first method would give polymers such as

a random copolymer

24.30 (cont)

The second method would give a block copolymer. It would be formed from the polymerization of oligomer A and oligomer B, each of which is a simple polymer. Oligomer A is an amine because a deficiency of acid chloride molecules is used in making it. Oligomer B is an acid chloride.

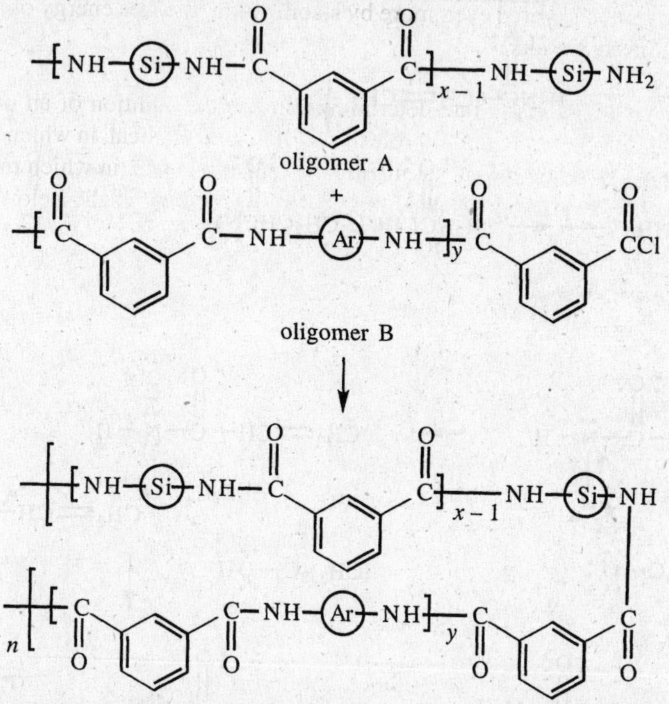

oligomer A

+

oligomer B

In the block copolymer, stretches of the polymer chain have structural regularity. The amine is either the siloxanediamine or the aryldiamine for many repeating units before a change occurs.

24.31

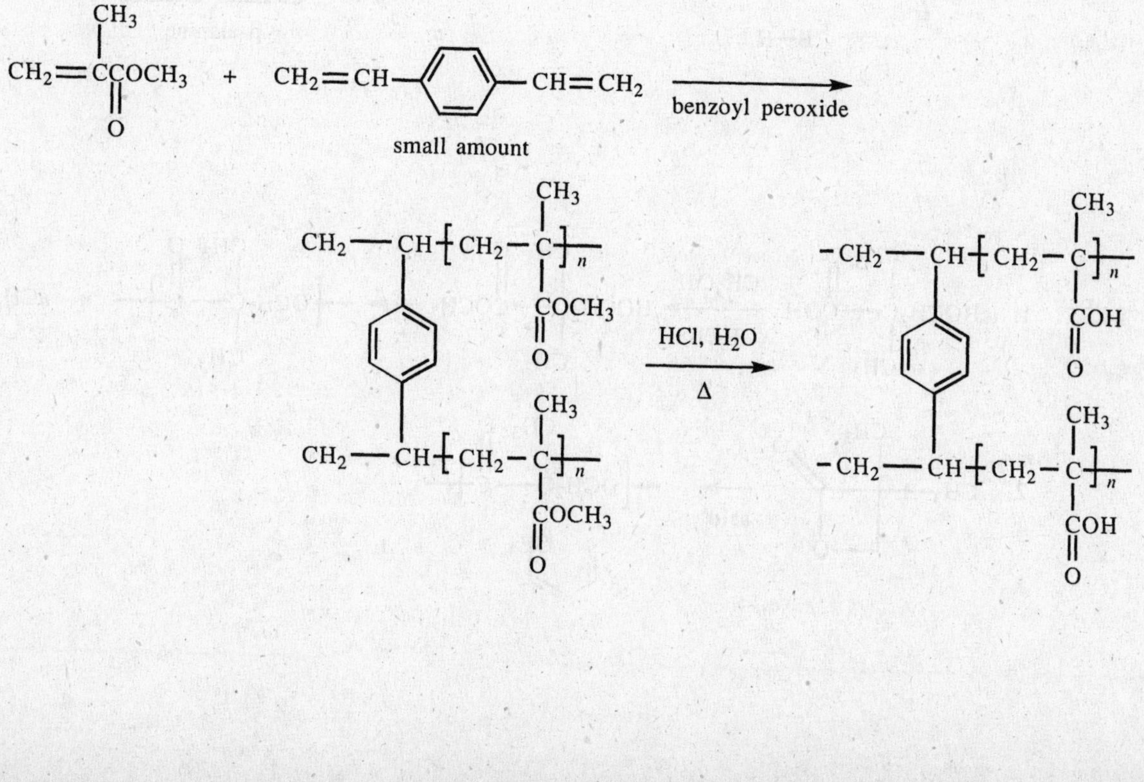

benzoyl peroxide

small amount

HCl, H_2O
Δ

24.32 1.

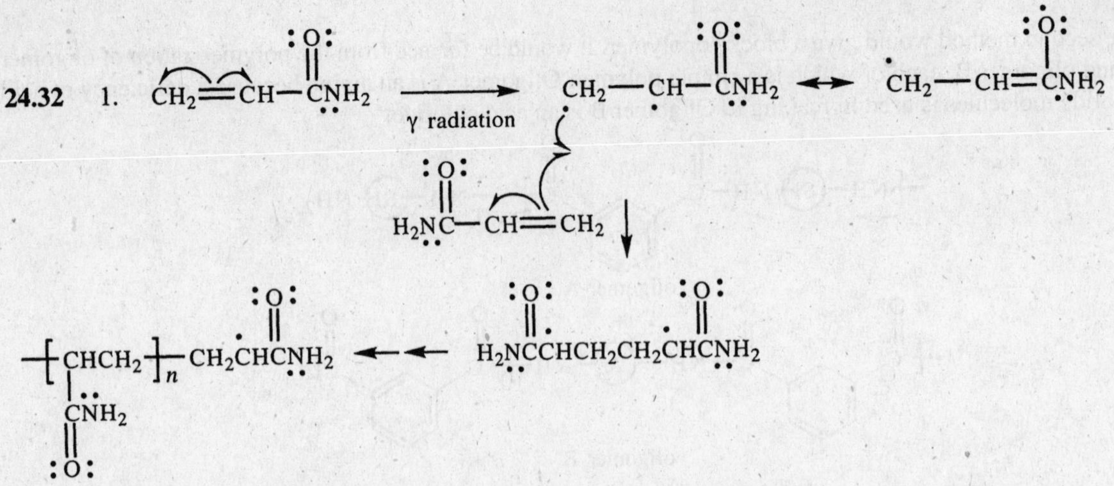

2.

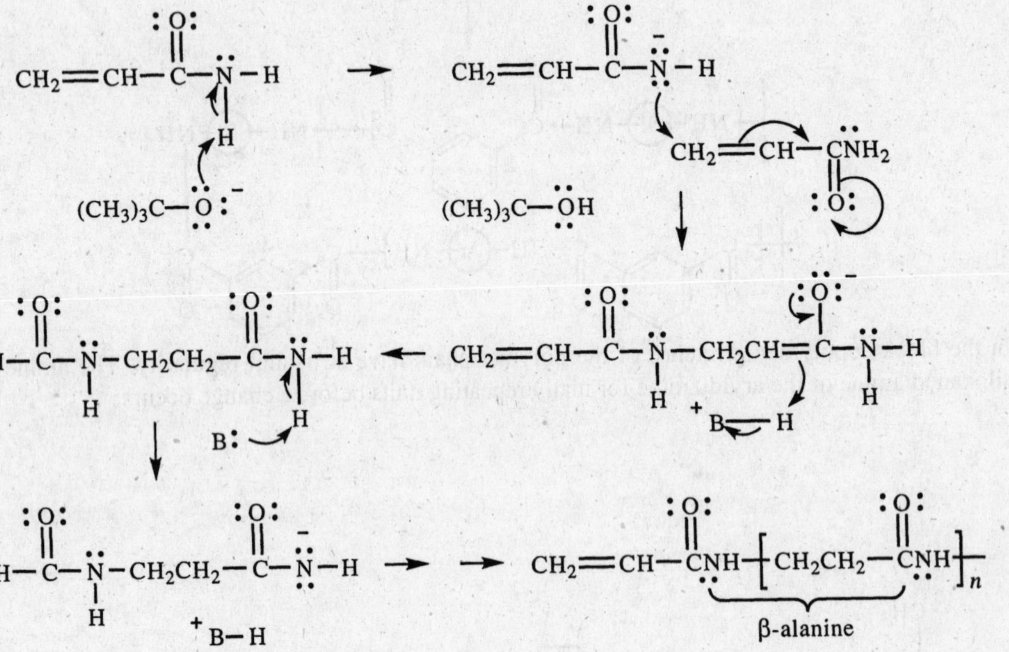

24.33 1.

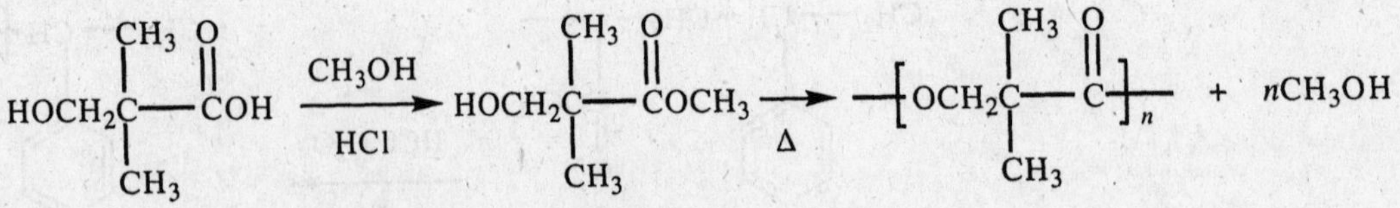

2.

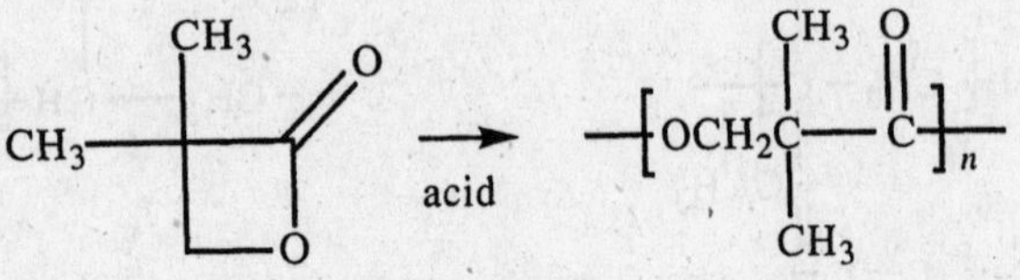

24.34   1.

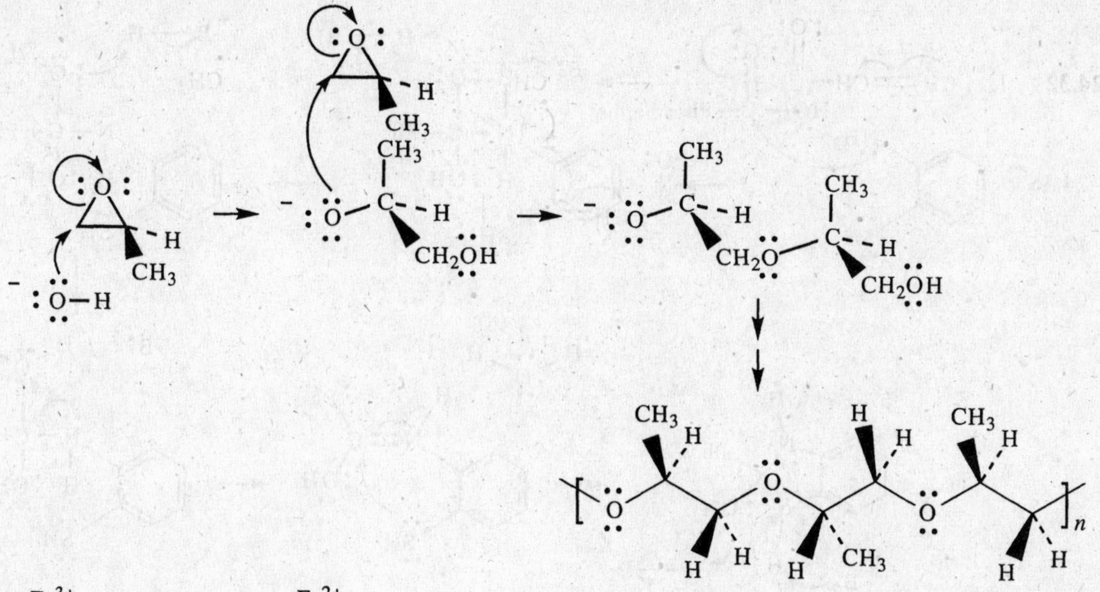

2.

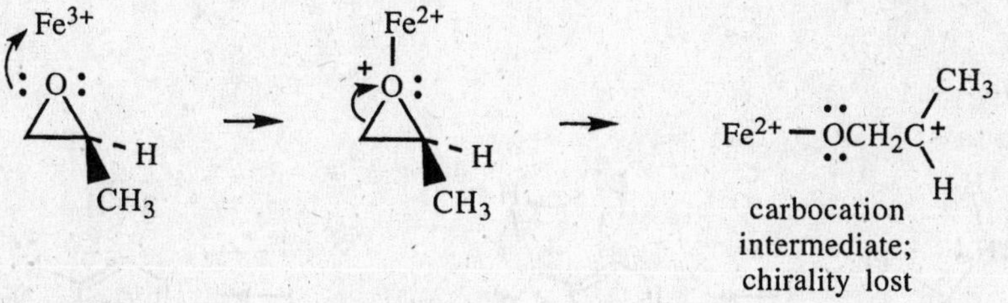

carbocation
intermediate;
chirality lost

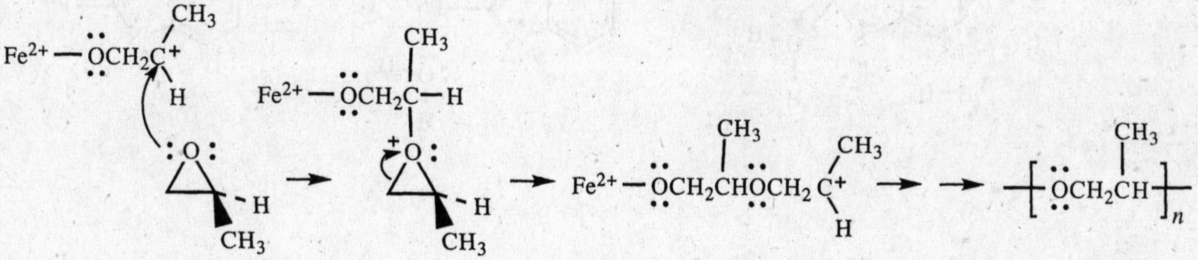

24.35   1.

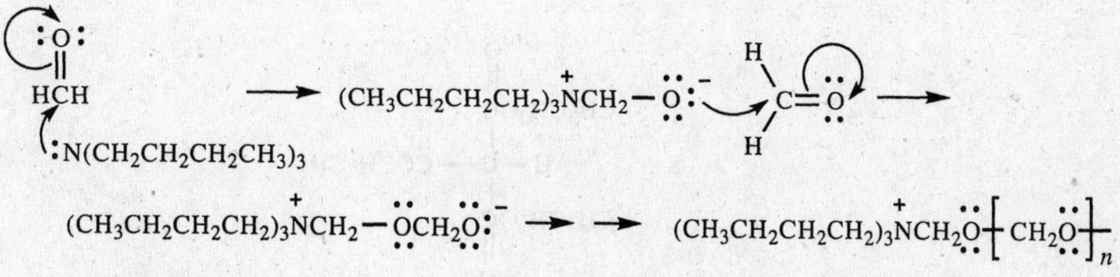

2.

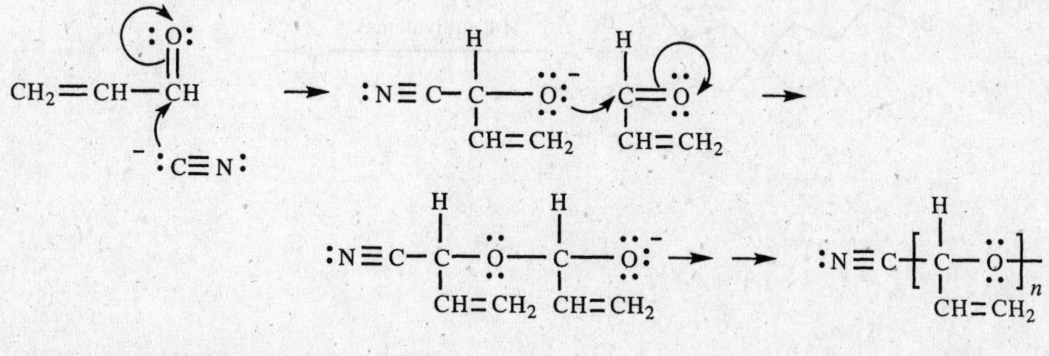

24.36

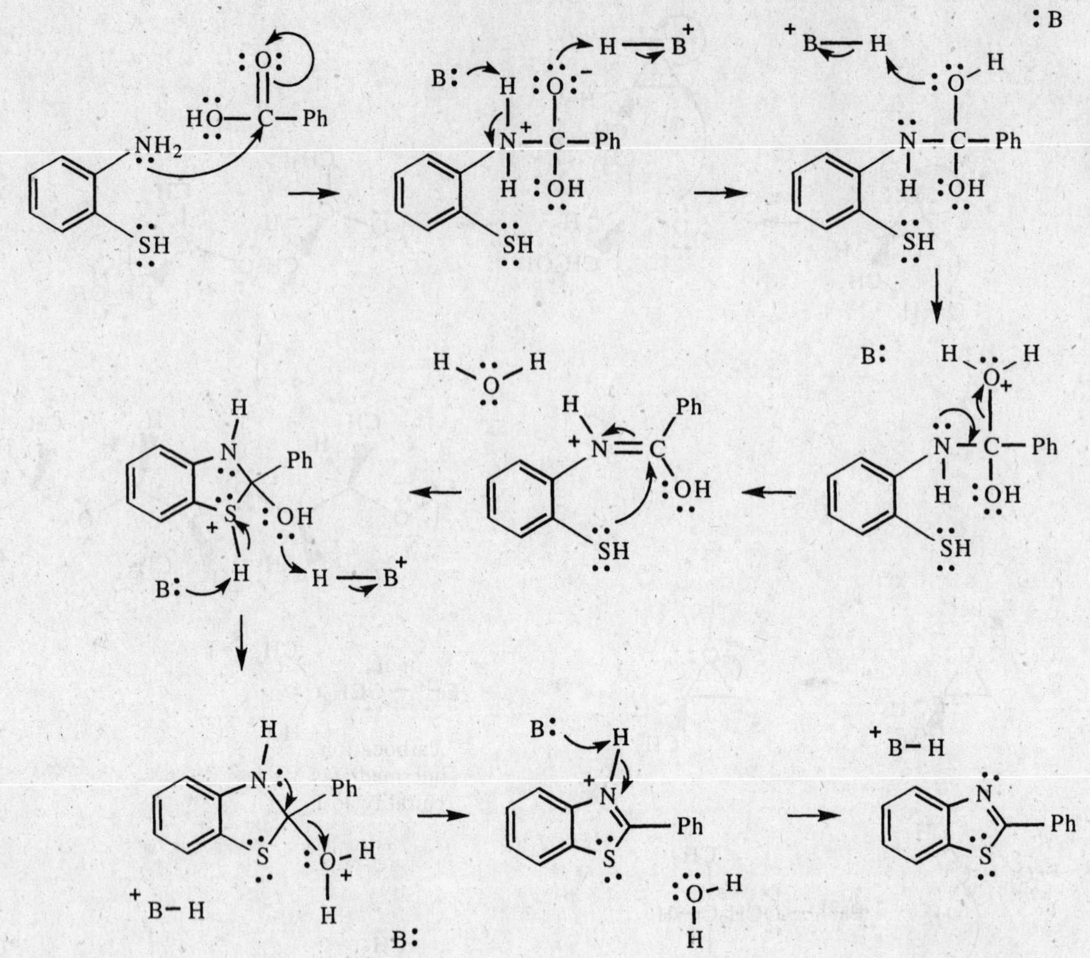
B:
H
N
Ph
S
OH
C
O
SH
NH2
HO
Ph
+B–H
B+

24.37

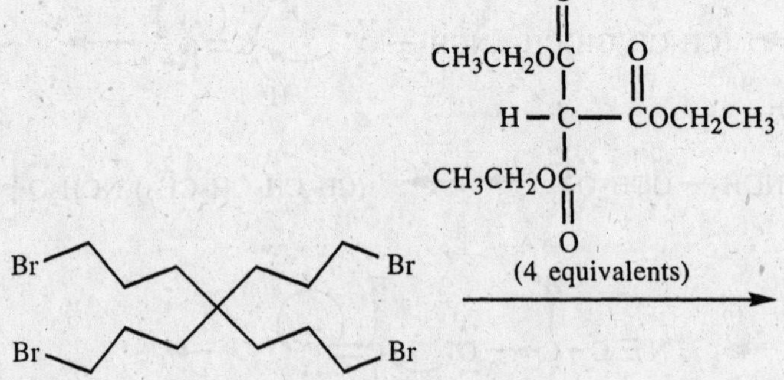
CH3CH2OC
H
COCH2CH3
CH3CH2OC
Br
Br
Br
Br
(4 equivalents)

24.37 (cont)

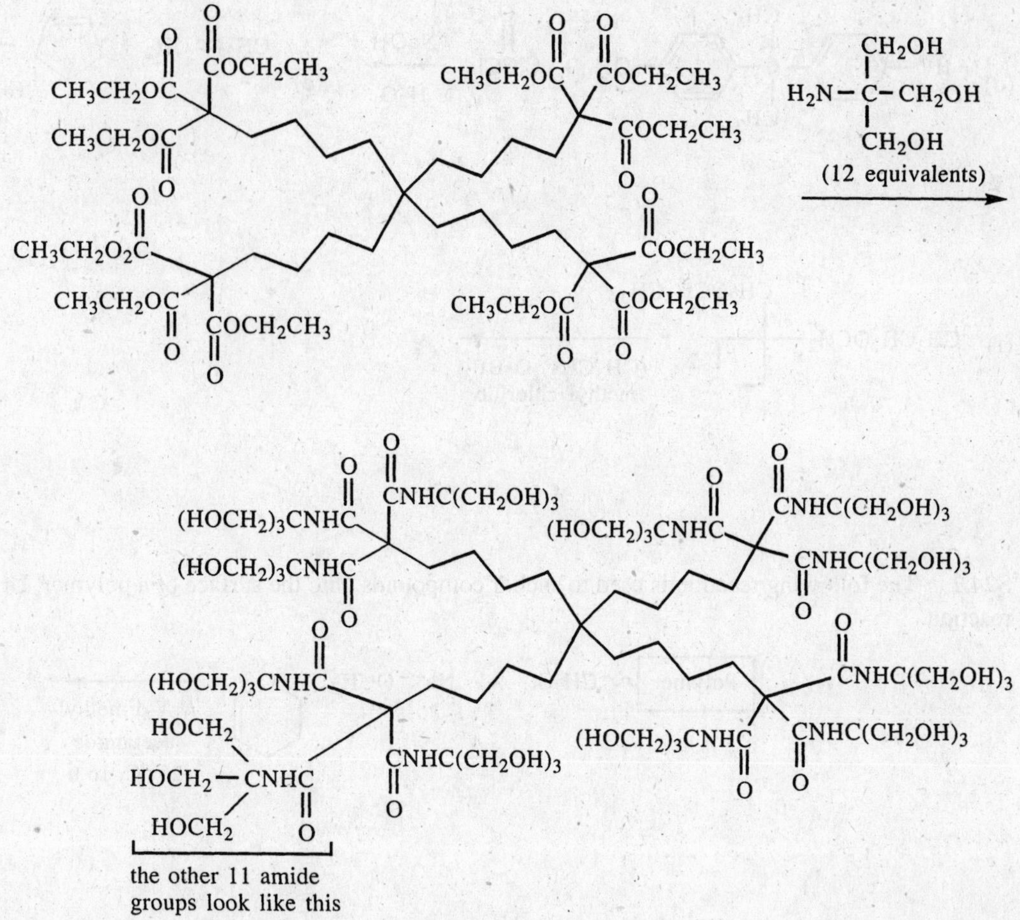

(HOCH₂)₃CNHC...

the other 11 amide
groups look like this

Supplemental Problems

S24.1   Predict the structures of the polymers that will result from the following reactions.

(a) 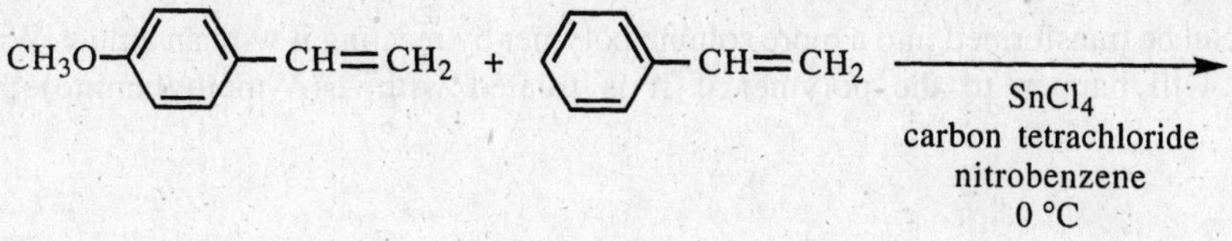 CH_3O—⬡—$CH{=}CH_2$ + ⬡—$CH{=}CH_2$ $\xrightarrow[\substack{\text{carbon tetrachloride}\\ \text{nitrobenzene}\\ 0\,°C}]{SnCl_4}$

(b) $CH_2{=}CHCl$ $\xrightarrow{\text{benzoyl peroxide}}$

(c) 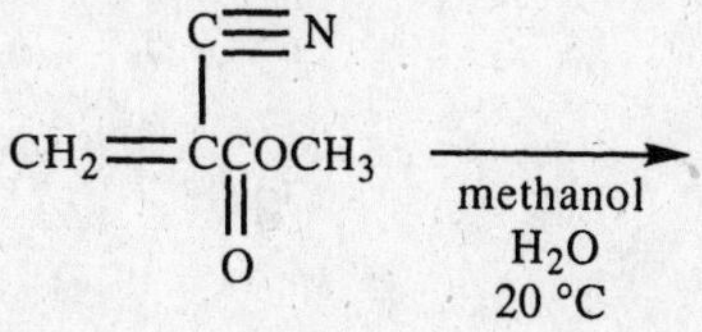 $CH_2{=}\underset{\underset{O}{\|}}{C}\overset{C{\equiv}N}{C}OCH_3$ $\xrightarrow[\substack{\text{H}_2\text{O}\\ 20\,°C}]{\text{methanol}}$

S24.1

(d)

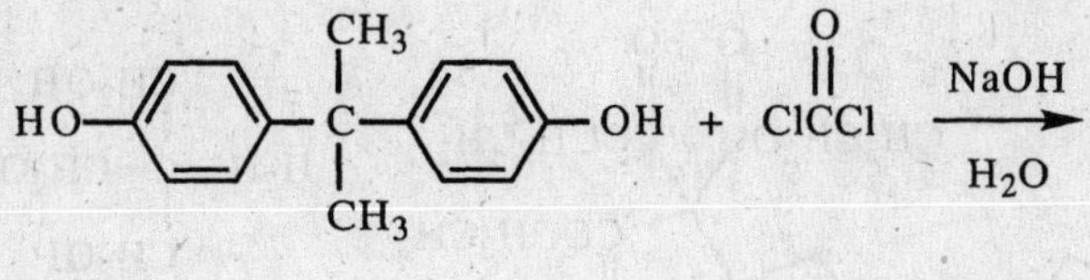

(e)

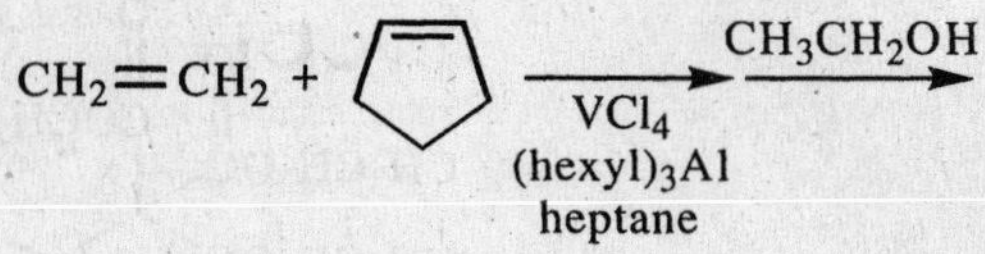

(f)

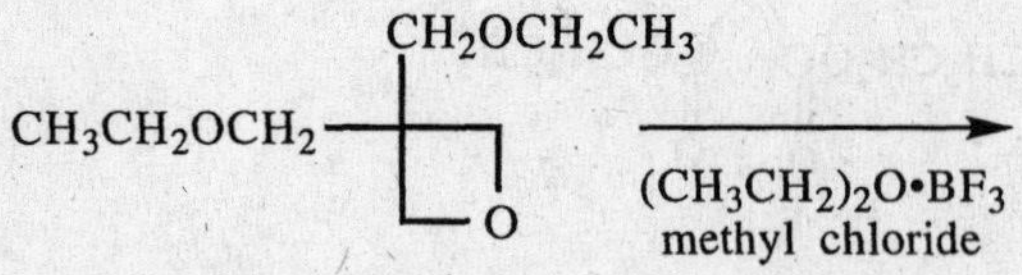

S24.2   The following reaction is used to anchor compounds onto the surface of a polymer. Draw the product of this reaction.

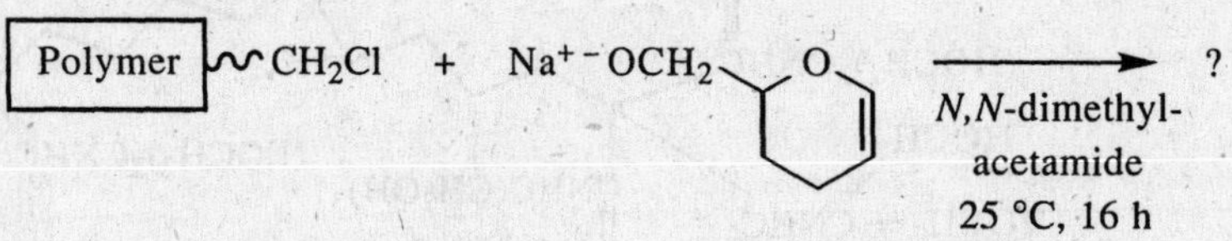

S24.3

(a) The following polyimide has been used in a plastic film that retains its usefulness over a very wide range of temperatures. The repeating unit of the polymer is shown below. What monomeric units would you need to synthesize this polymer?

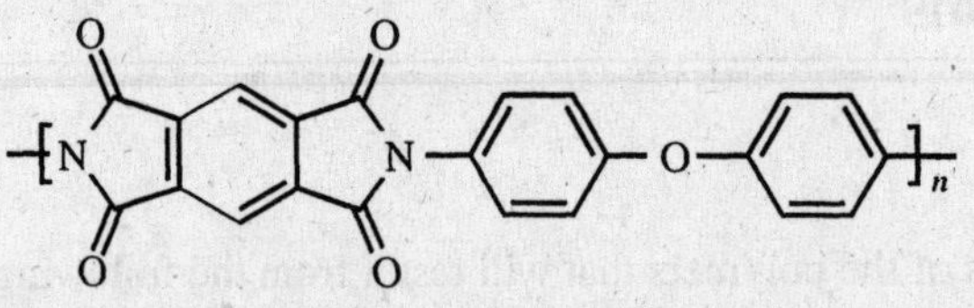

(b) The polymer shown above can be transformed into a more soluble polymer by reacting it with an amine. Write an equation predicting what will happen to the polymer if it is treated with 2-(*N*-methylamino)ethanol, $CH_3NHCH_2CH_2OH$.

25 Concerted Reactions

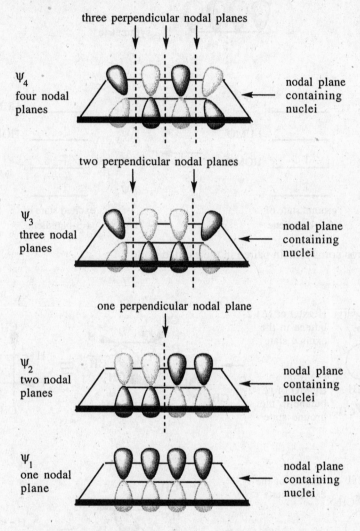

25.1

Ψ_4
four nodal
planes

three perpendicular nodal planes

nodal plane
containing
nuclei

two perpendicular nodal planes

Ψ_3
three nodal
planes

nodal plane
containing
nuclei

one perpendicular nodal plane

Ψ_2
two nodal
planes

nodal plane
containing
nuclei

Ψ_1
one nodal
plane

nodal plane
containing
nuclei

25.2 The symmetry of a molecular orbital can be determined by looking at the symmetry of the atomic *p*-orbitals from which it is constructed since the molecular orbital retains that symmetry.

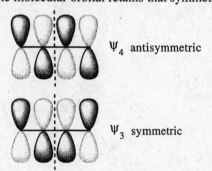

Ψ_4 antisymmetric

Ψ_3 symmetric

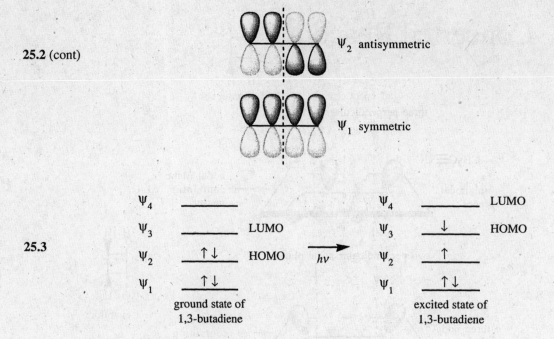

25.2 (cont) Ψ_2 antisymmetric

Ψ_1 symmetric

25.3

	ground state of 1,3-butadiene			excited state of 1,3-butadiene	
Ψ_4	_____		Ψ_4	_____	LUMO
Ψ_3	_____	LUMO	Ψ_3	$\downarrow$	HOMO
Ψ_2	$\uparrow\downarrow$	HOMO	Ψ_2	$\uparrow$	
Ψ_1	$\uparrow\downarrow$		Ψ_1	$\uparrow\downarrow$	

$\xrightarrow{h\nu}$

Note that electron spins remain paired in the excited state.

25.4   HOMO of (E)-2-butene in the excited state

LUMO of (E)-2-butene in the ground state

25.5   HOMO of the excited state

LUMO of the ground state

HOMO of the excited state

LUMO of the ground state

25.6 (a)

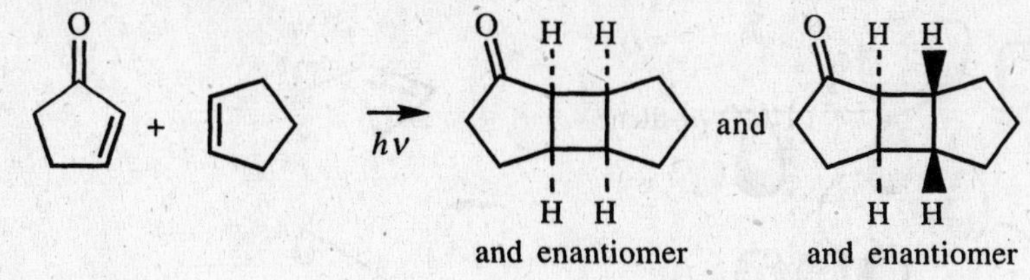
hν
and
H H
and enantiomer
and enantiomer

(b)

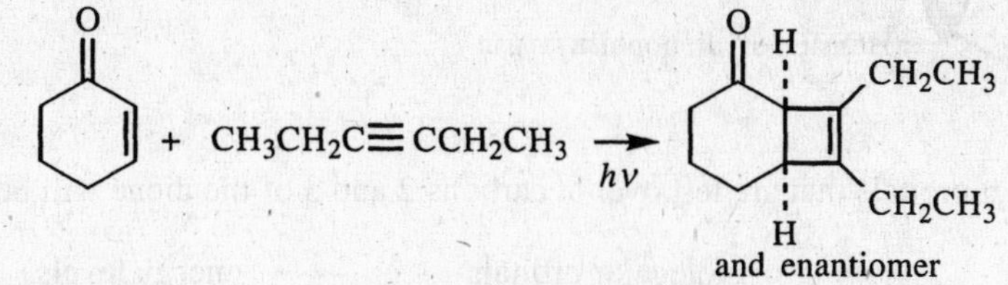
+ $CH_3CH_2C{\equiv}CCH_2CH_3$
hν
CH_2CH_3
CH_2CH_3
and enantiomer

(c)

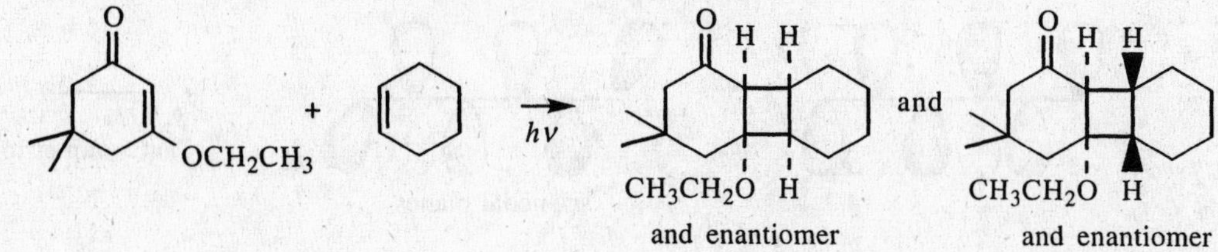
OCH_2CH_3
hν
and
CH_3CH_2O H
and enantiomer
and enantiomer

25.7 (a)

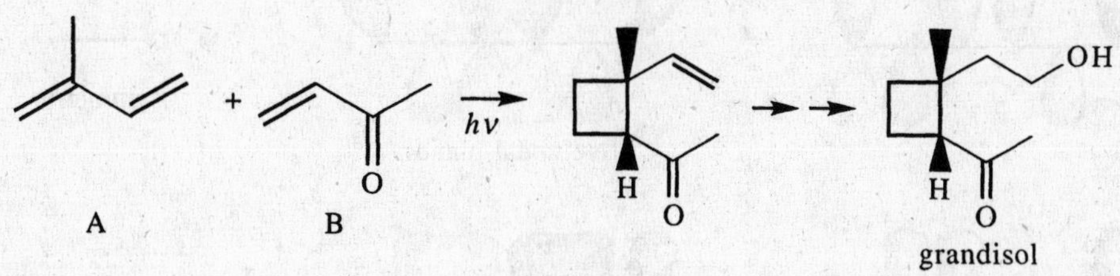
hν
A
B
OH
grandisol

(b)

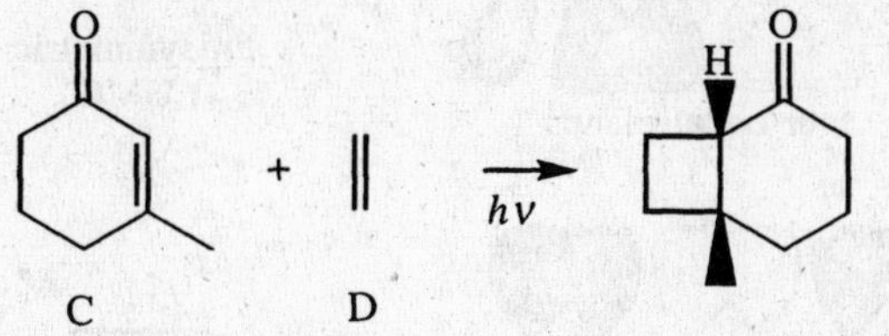
hν
C
D

(c)

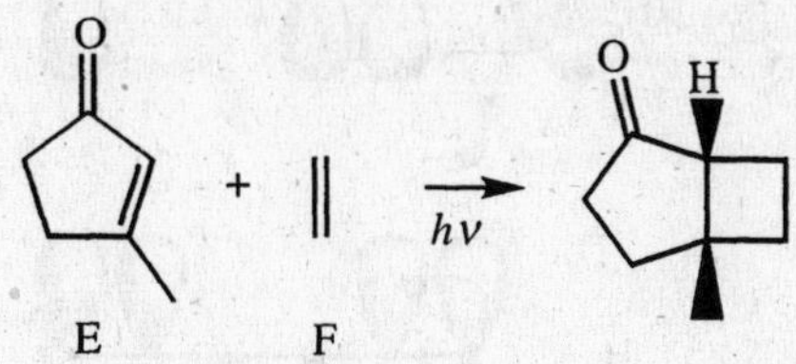
hν
E
F

(d)

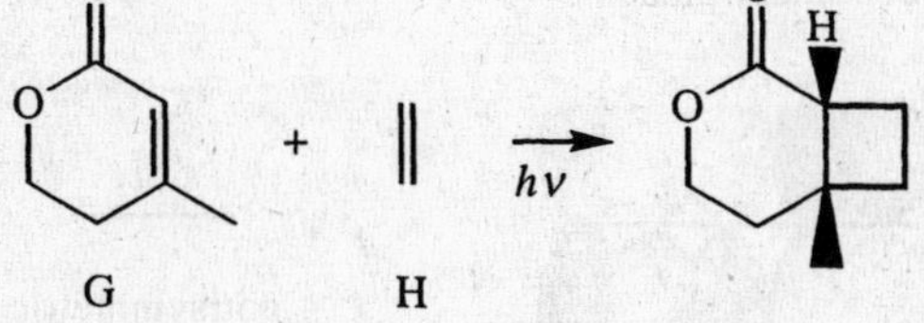
hν
G
H

25.8

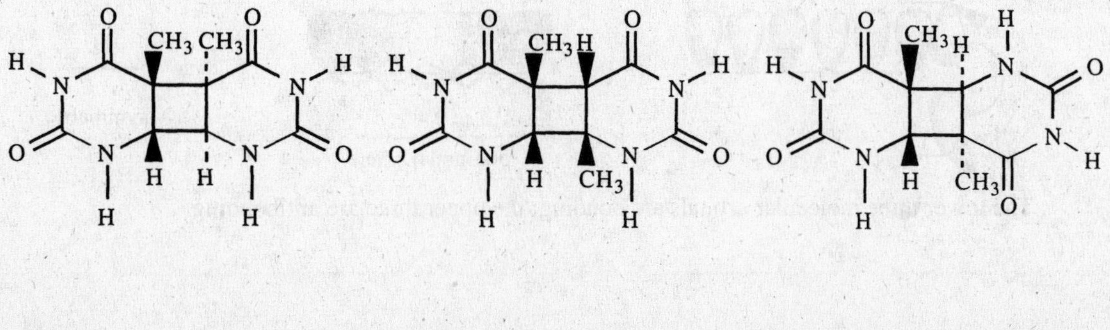
CH_3 CH_3
CH_3 H
CH_3 H

25.9

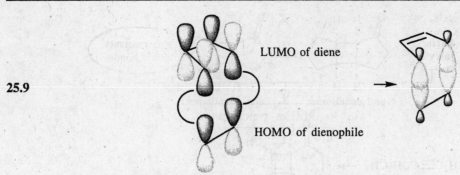

LUMO of diene

HOMO of dienophile

In this picture the lobes of the *p*-orbitals that are left over at carbons 2 and 3 of the diene will be bonding.

25.10

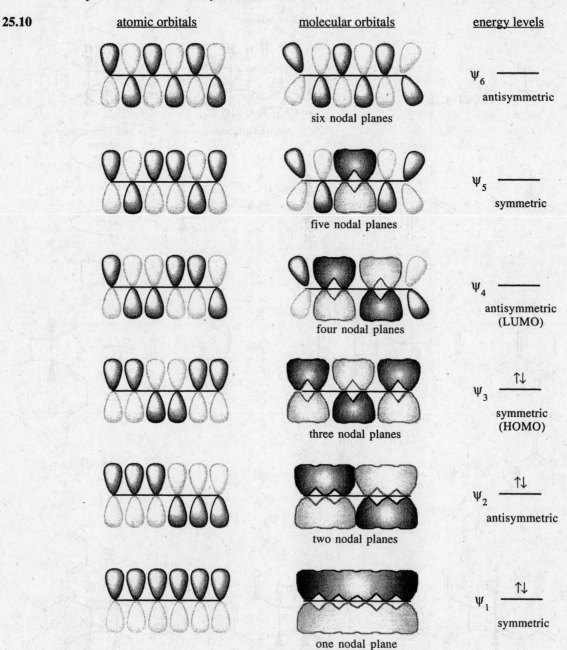

| atomic orbitals | molecular orbitals | energy levels |

ψ_6 ——— antisymmetric

six nodal planes

ψ_5 ——— symmetric

five nodal planes

ψ_4 ——— antisymmetric (LUMO)

four nodal planes

ψ_3 ⇅ symmetric (HOMO)

three nodal planes

ψ_2 ⇅ antisymmetric

two nodal planes

ψ_1 ⇅ symmetric

one nodal plane

The lower three molecular orbitals are bonding; the upper three are antibonding.

Concept Map 25.1   Cycloaddition reactions.

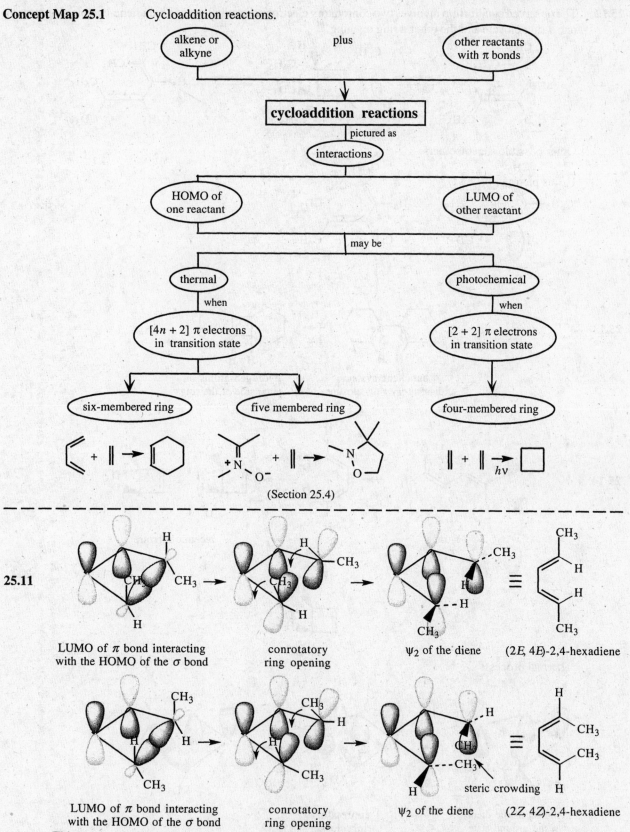

(Section 25.4)

LUMO of π bond interacting
with the HOMO of the σ bond

conrotatory
ring opening

ψ_2 of the diene

(2E, 4E)-2,4-hexadiene

LUMO of π bond interacting
with the HOMO of the σ bond

conrotatory
ring opening

ψ_2 of the diene

steric crowding

(2Z, 4Z)-2,4-hexadiene

This product is also possible from a conrotatory ring opening of trans-3,4-dimethylcyclobutene, but it is not formed because of steric crowding in the transition state.

25.12   The observed equilibrium involves two conrotatory electrocyclic reactions of a butadiene-cyclobutene system, one a ring closure and the other a ring opening.

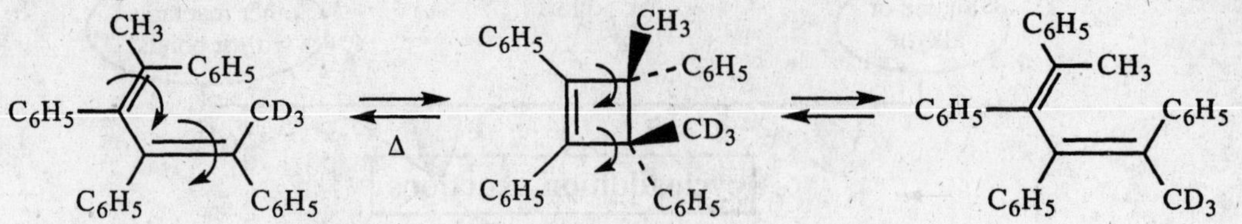

Other possible stereoisomers:

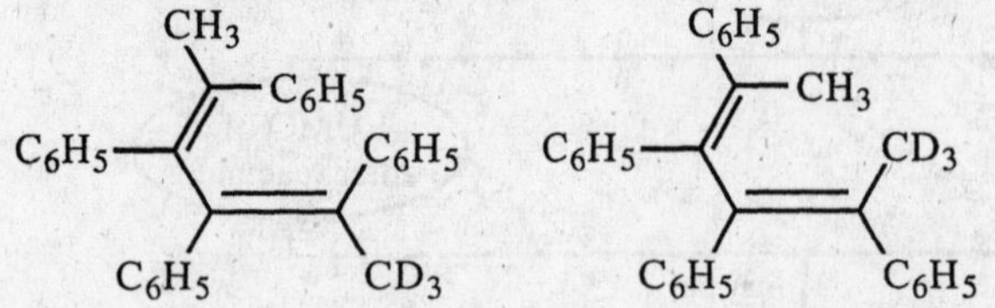

25.13

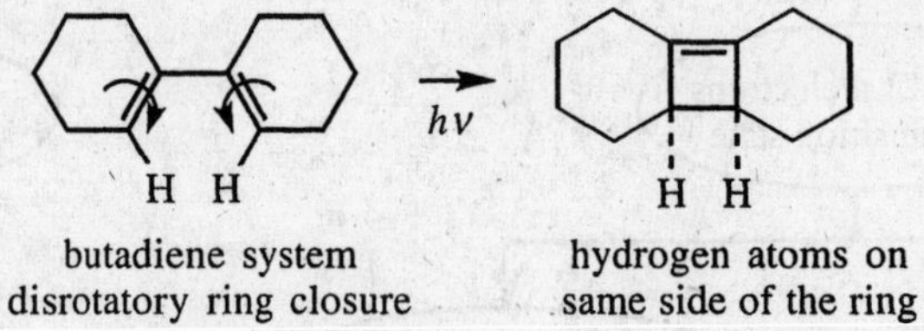

butadiene system
disrotatory ring closure

hydrogen atoms on
same side of the ring

25.14

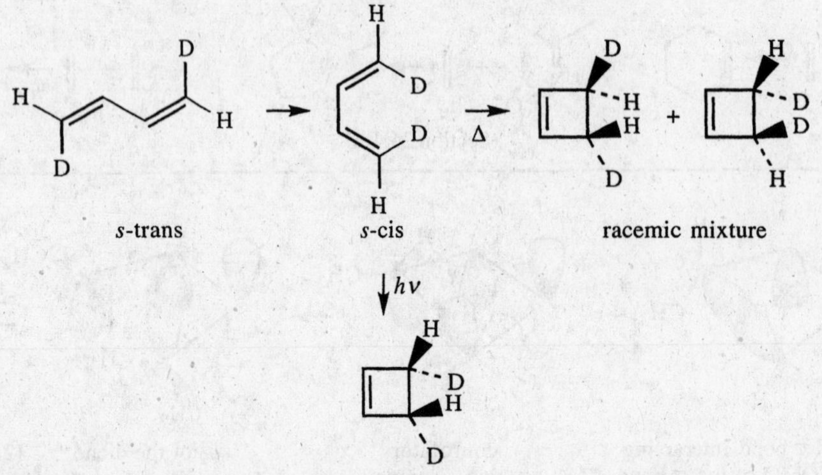

s-trans   *s*-cis   racemic mixture

Thermal process:

HOMO of the diene
in the ground state

conrotatory
ring closing

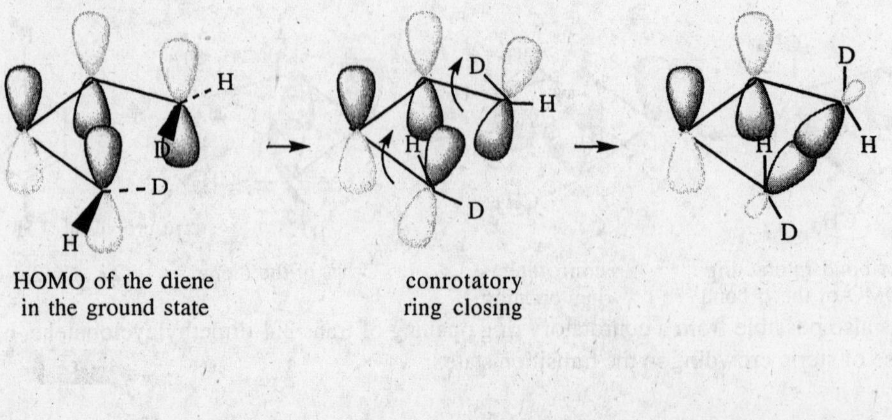

25.14 (cont)

Photochemical process:

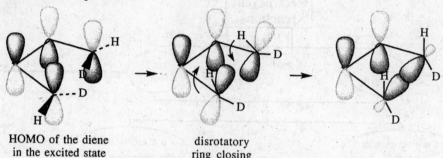

HOMO of the diene
in the excited state

disrotatory
ring closing

25.15

HOMO of the triene
in the ground state

disrotatory →

(See Problem 25.10 for the molecular orbitals of the triene.)

25.16

HOMO of the triene
in the ground state;
thermal ring closure

disrotatory →

same molecule as that
shown in Figure 25.11

HOMO of the triene
in the excited state;
photochemical ring closure

conrotatory
hv →

enantiomer of the molecule
shown in Figure 25.12

Concept Map 25.2   Electrocyclic reactions.

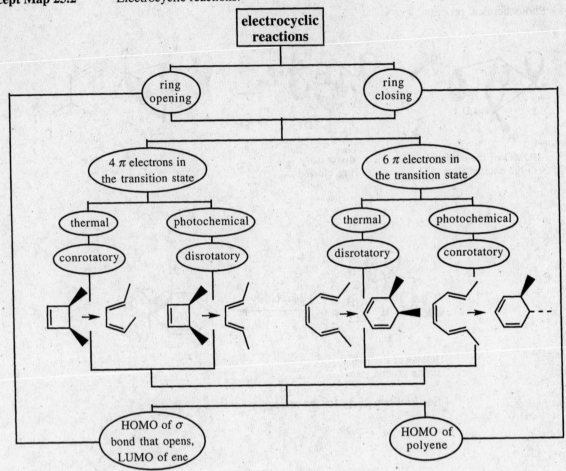

25.17   HOMO of triene in the ground state:

25.17 (cont)

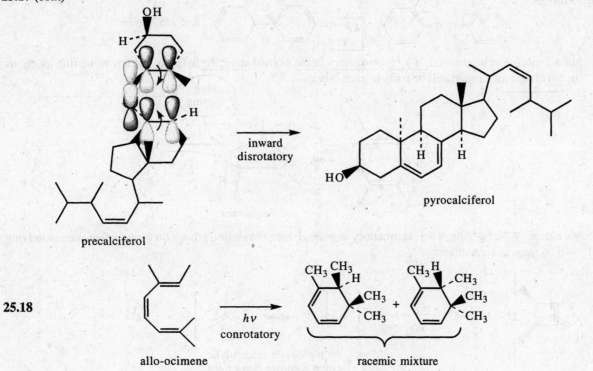

precalciferol → inward disrotatory → pyrocalciferol

25.18

allo-ocimene $\xrightarrow[\text{conrotatory}]{h\nu}$ racemic mixture

---

Concept Map 25.3   Woodward-Hoffmann rules.

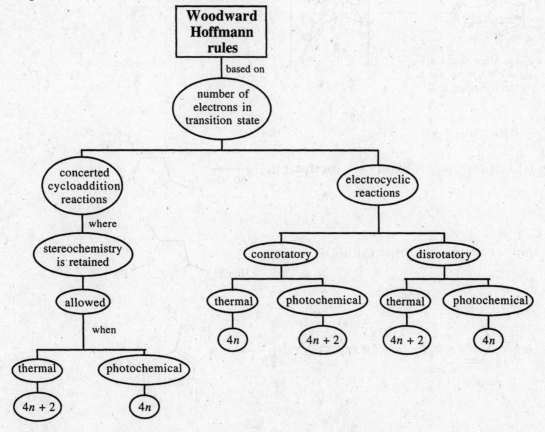

25.19

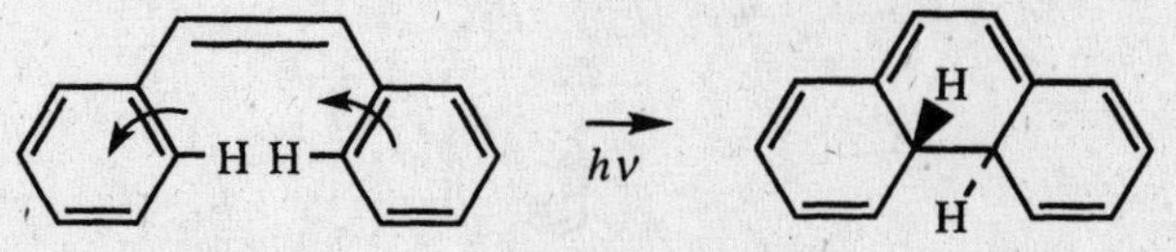

and enantiomer

$4n + 2$ electrocyclic reaction; $n = 1$; conrotatory in the excited state; the hydrogen atoms at the ring junctions in dihydrophenanthrene will be trans to each other.

25.20

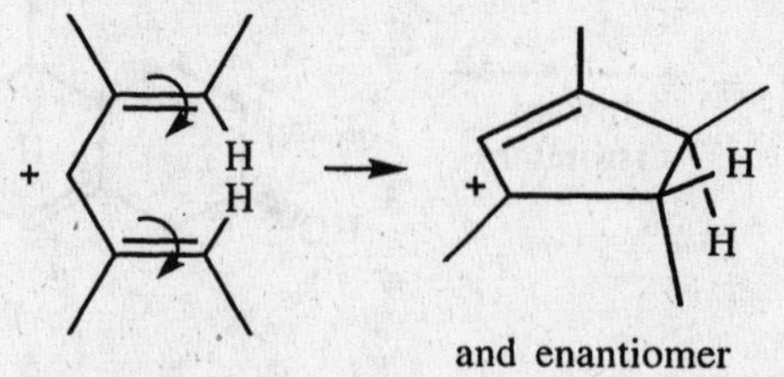

and enantiomer

$4n$ electrocyclic reaction, $n = 1$; conrotatory in ground state; the methyl groups on the new five-membered ring will be trans to each other.

25.21

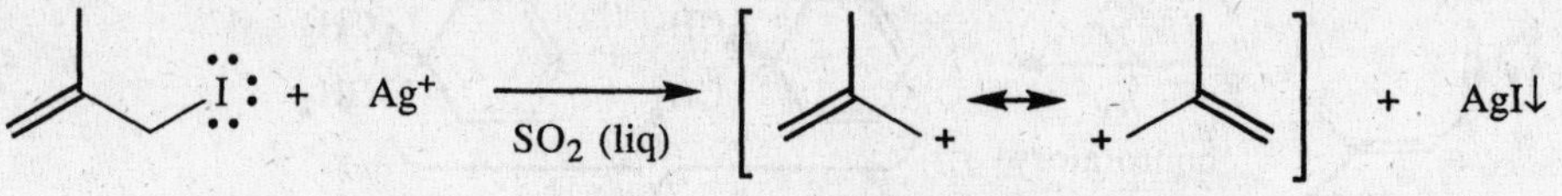

the dienophile is an allylic cation with two π electrons

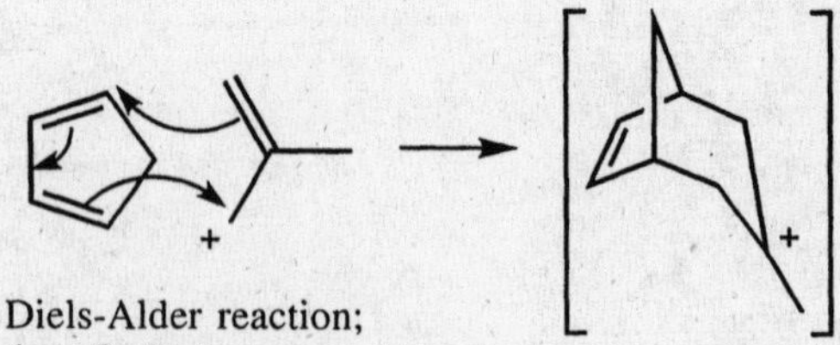

Diels-Alder reaction;
$4n + 2$ electrons, $n = 1$,
in the transition state

25.22 (a)

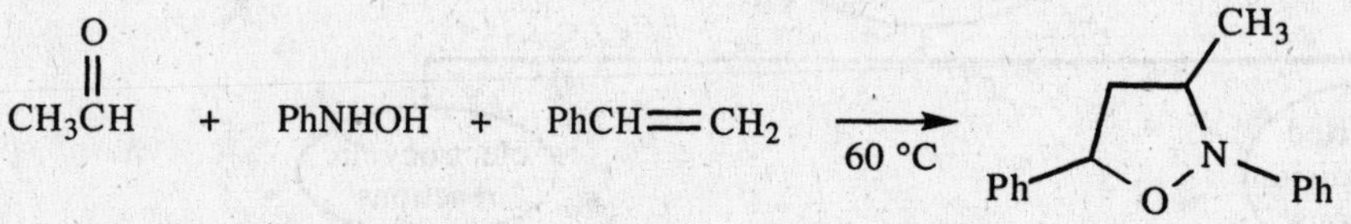

(b)

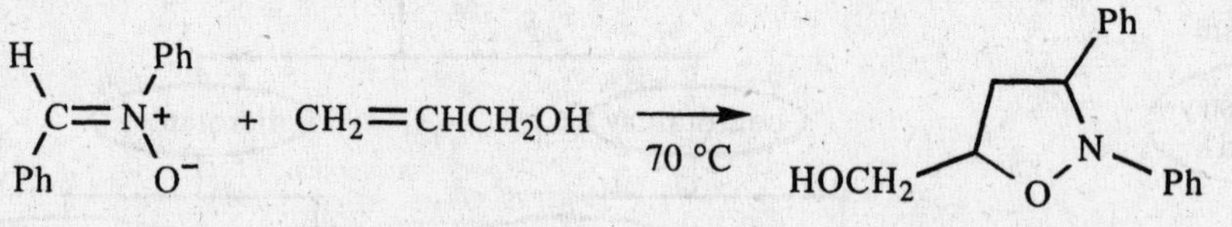

(c)

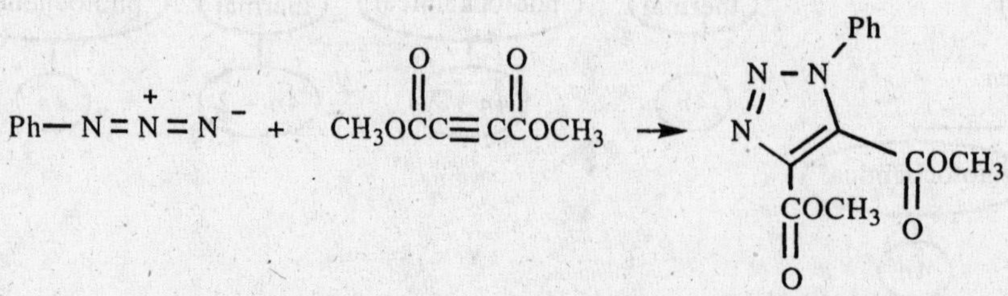

25.22 (d) [structure: 1,4-naphthoquinone] + Ph—N=N⁺=N⁻ ⟶ [tricyclic triazoline product with H, N, N, Ph, and two C=O groups]

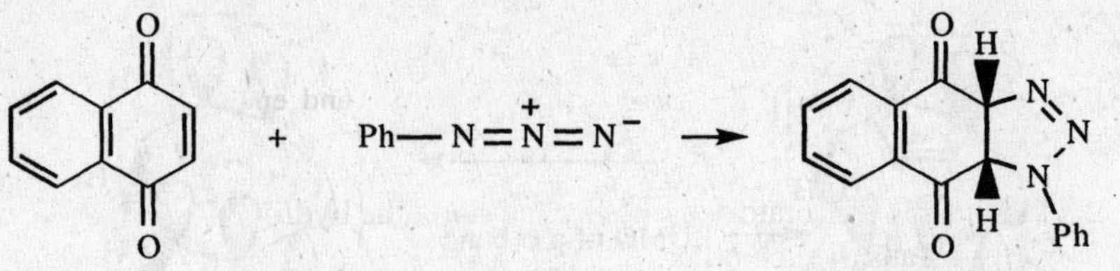

and enantiomer

(e) [nitrone: H, Ph, C=N⁺, Ph, O⁻] + CH₂=C(OCH₂CH₃)(OCH₂CH₃) $\xrightarrow{\text{toluene}\ 100\ °C}$ [isoxazolidine: CH₃CH₂O, CH₃CH₂O, Ph, N, Ph, O product]

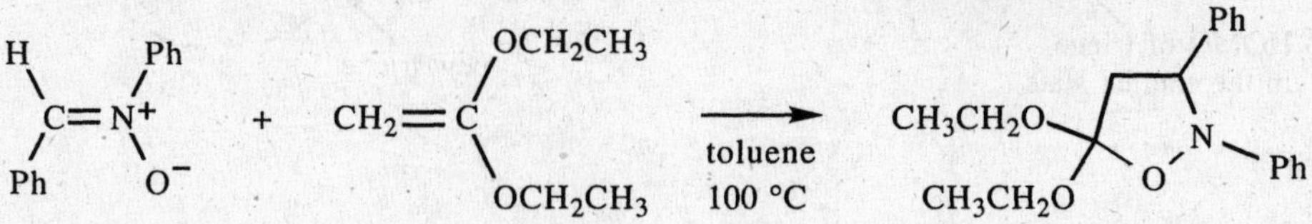

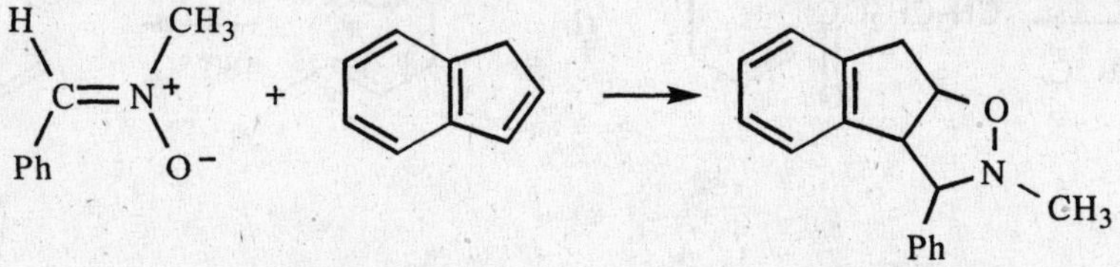

25.23 (a) Ph—N=N⁺=N⁻ + CH₃CH₂CH₂CH₂OCH=CH₂ ⟶ [triazoline product with Ph—N, N, N, CH₃CH₂CH₂CH₂O]

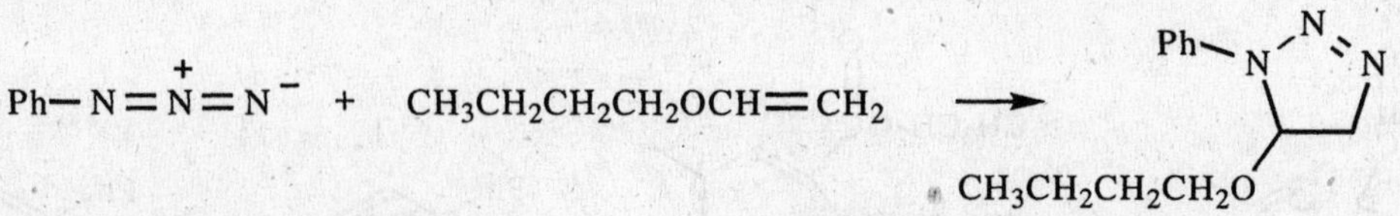

(b) [nitrone: H, CH₃, C=N⁺, Ph, O⁻] + [indene] ⟶ [fused tricyclic product with O, N—CH₃, Ph]

(c) [cyclic nitrone: H, C=N⁺, O⁻] + CH₂=CHCH₃ ⟶ [bicyclic product N—O, CH₃]

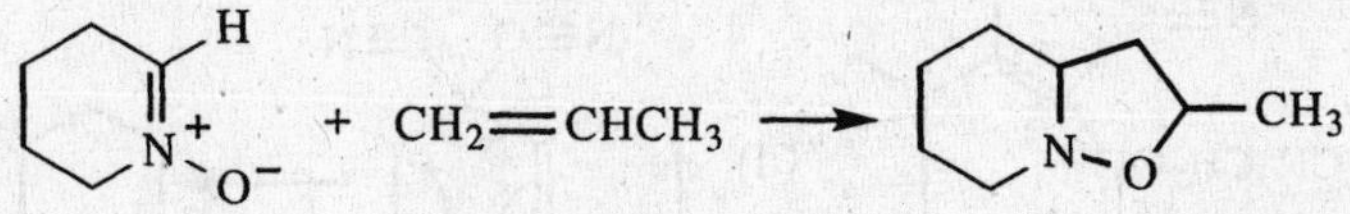

(d) [nitrone: H, CH₃, C=N⁺, Ph, O⁻] + CH₂=CHCOCH₂CH₃ ⟶ [isoxazolidine with COCH₂CH₃, O, Ph, N—CH₃]

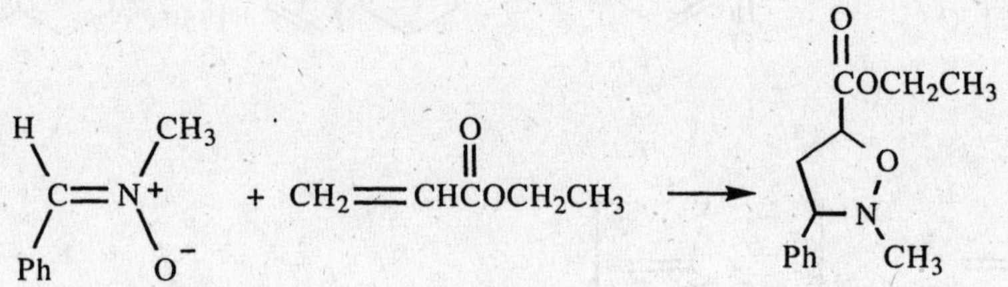

(e) [nitrone: H, Ph, C=N⁺, Ph, O⁻] + CH₂=CHCOCH₂CH₃ ⟶ [isoxazolidine with Ph, Ph, N—O, COCH₂CH₃]

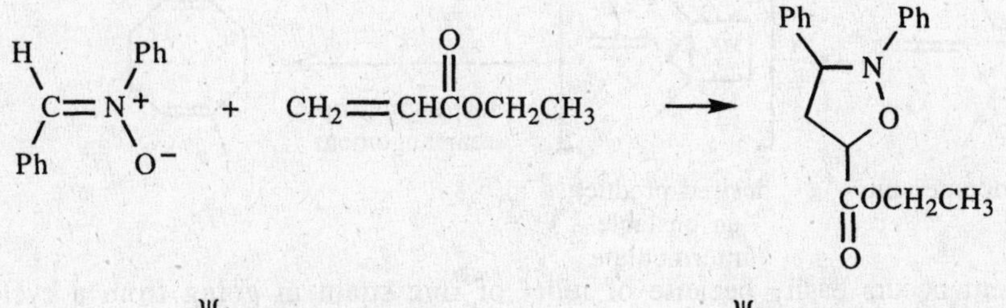

25.24

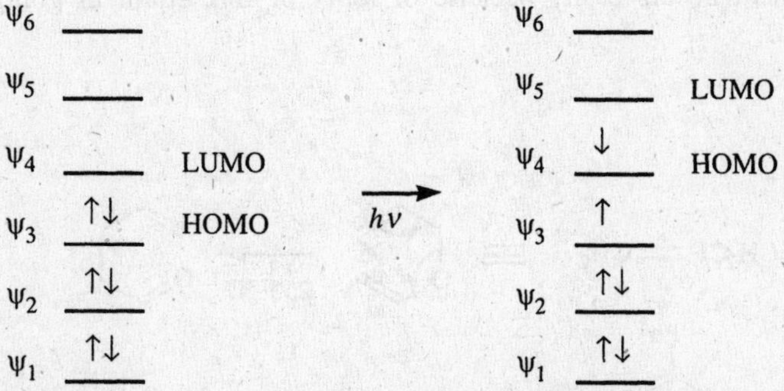

ψ_6 _____ ψ_6 _____

ψ_5 _____ ψ_5 _____ LUMO

ψ_4 _____ LUMO $\xrightarrow{h\nu}$ ψ_4 ↓ HOMO

ψ_3 ↑↓ HOMO ψ_3 ↑

ψ_2 ↑↓ ψ_2 ↑↓

ψ_1 ↑↓ ψ_1 ↑↓

triene in the   triene in the
ground state   excited state

(See Problem 25.10 for pictures of the orbitals.)

25.24 (cont)

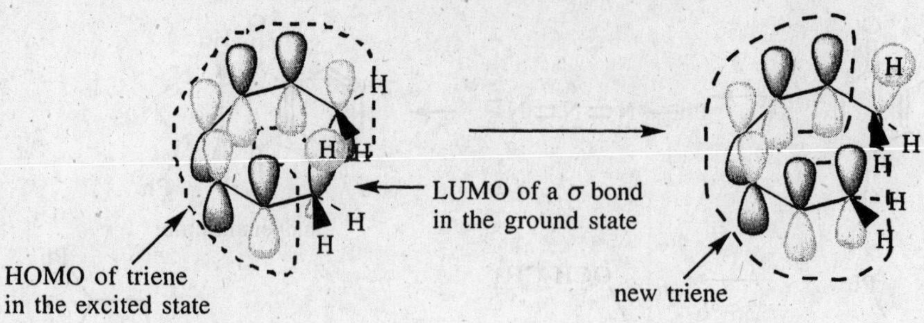

HOMO of triene
in the excited state

← LUMO of a σ bond
in the ground state

new triene

25.25

(a) 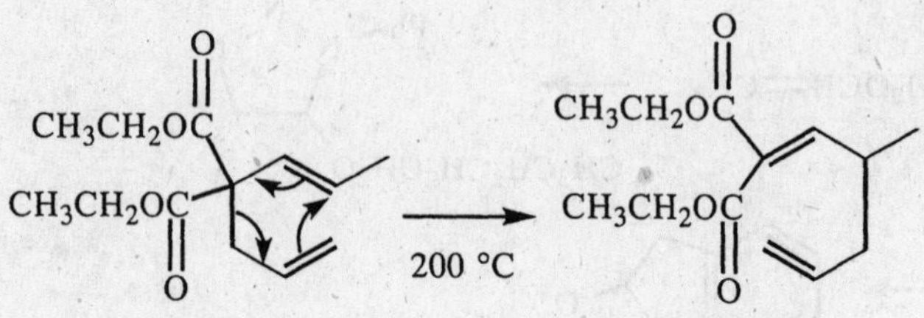 200 °C

(b) 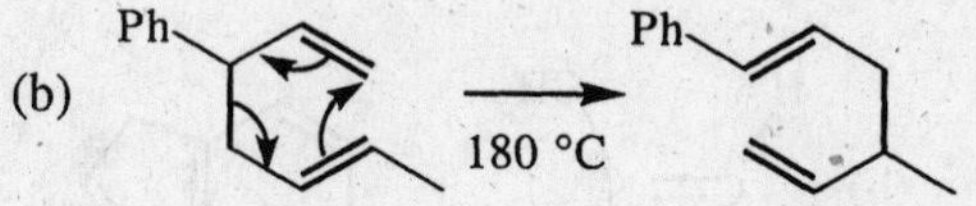 180 °C

(c) 160 °C

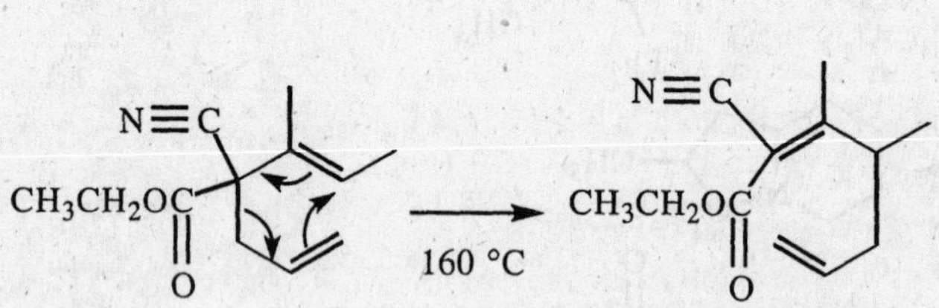

(d) 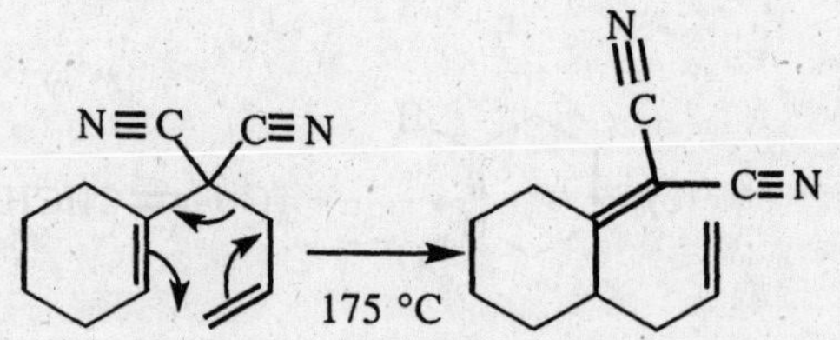 175 °C

25.26

two isoprene units   desired product;
                            an unstable
                            intermediate

[3,3]-sigmatropic
rearrangement

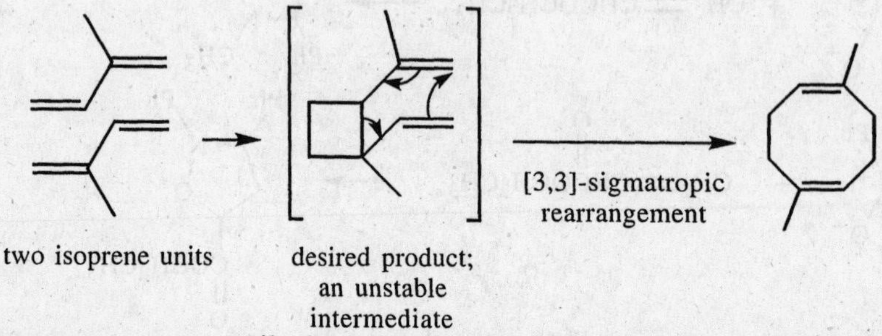

The Cope rearrangement occurs easily because of relief of ring strain in going from a cyclobutane to a cyclooctadiene system.

25.27 (a) $CH_2{=}CHOCH_2CH{=}CH_2$ ≡ 255 °C

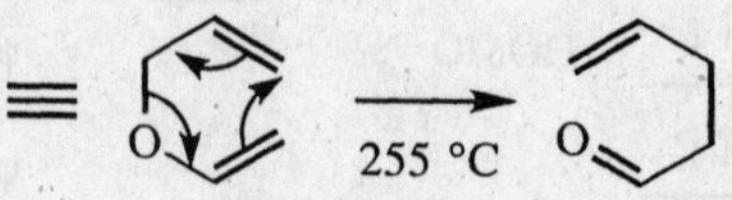

(b) 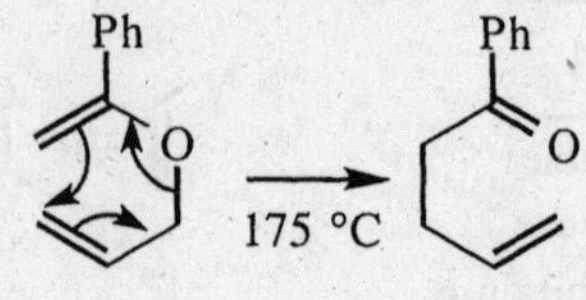 175 °C

25.27 (c)

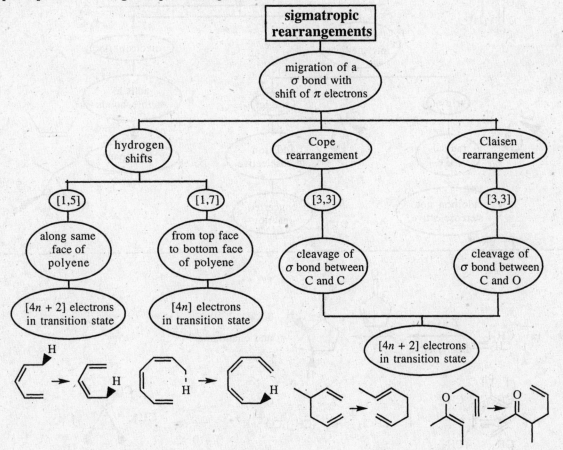

Concept Map 25.4   Sigmatropic rearrangements.

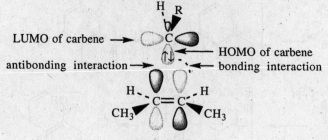

25.28

LUMO of carbene →

← HOMO of carbene

antibonding interaction →

← bonding interaction

LUMO of alkene

- -

Concept Map 25.5   Carbenes.

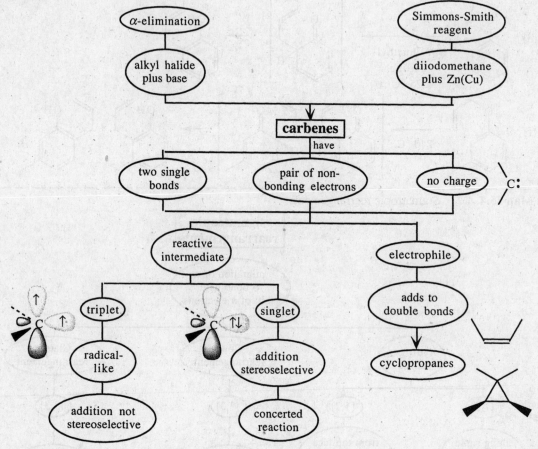

- -

25.29 (a) CH_2I_2 $\xrightarrow[\text{diethyl ether}]{Zn(Cu)}$ and enantiomer

(b) CH_3 CH_3 C=C CH_3 CH_3 + $CHCl_3$ + $(CH_3)_3CO^- K^+$ $\xrightarrow{-10\ ^\circ C}$

25.29 (c)

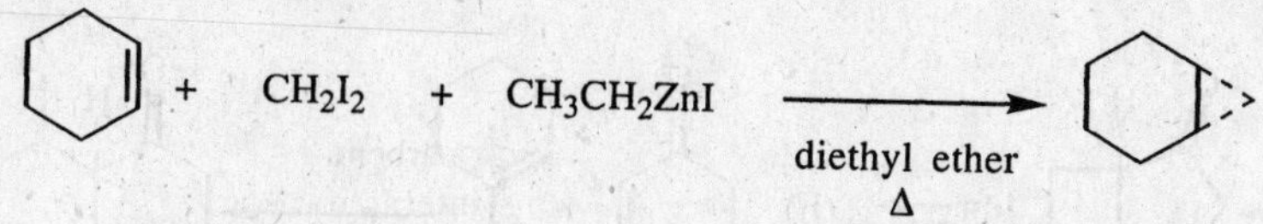

(d) $CH_2=CHOCH_2CH_3$ + $CHCl_3$ + $(CH_3)_3CO^-$ K^+ $\xrightarrow{-10\ °C}$

and enantiomer

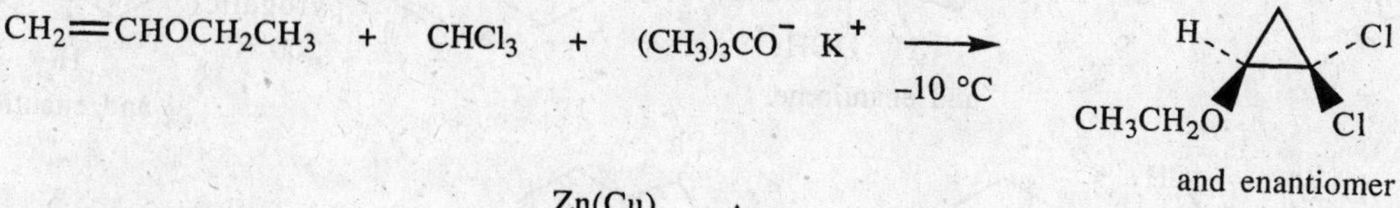

(e) $CH_3(CH_2)_5CH=CH_2$ + CH_2I_2 $\xrightarrow[\substack{\text{diethyl}\\ \text{ether}\\ \Delta}]{Zn(Cu)}$

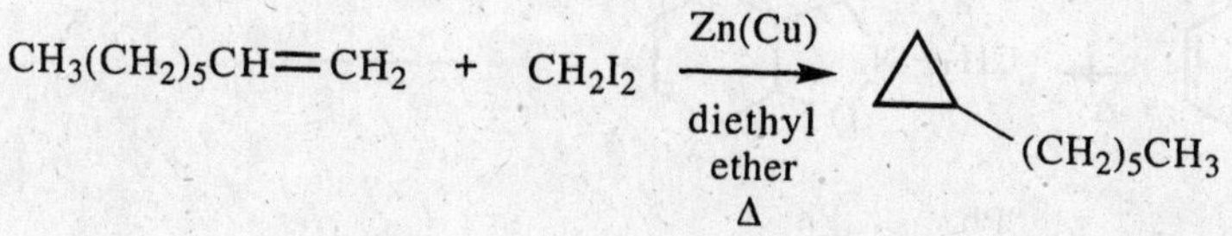

(f) + CH_2I_2 $\xrightarrow[\substack{\text{diethyl ether}\\ \Delta}]{Zn(Cu)}$ and enantiomer

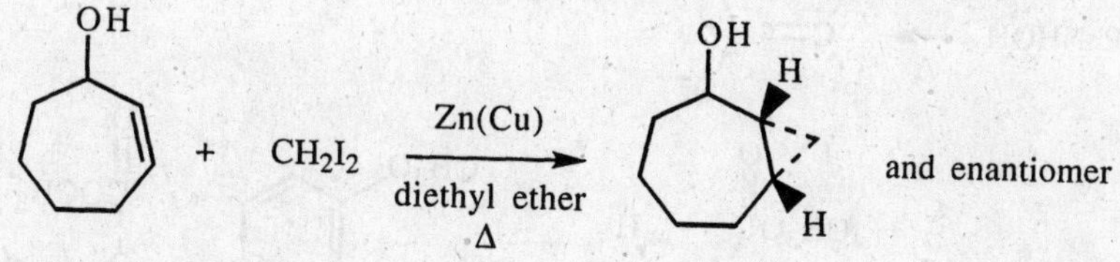

(g) + CH_2I_2 $\xrightarrow[\substack{\text{diethyl ether}\\ \Delta}]{Zn(Cu)}$

and enantiomer

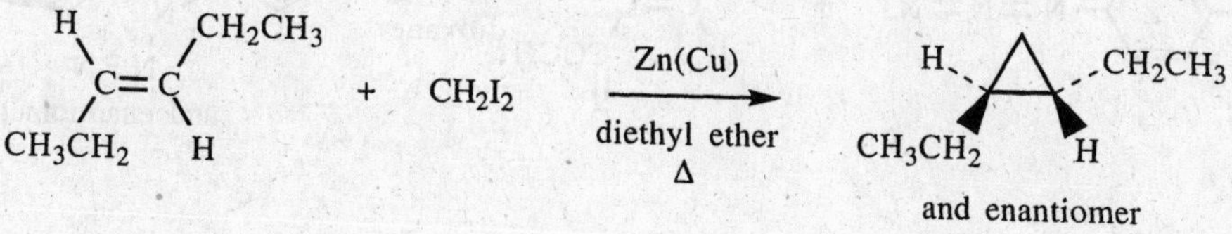

25.30

(a) + $CH_3C\equiv CCH_3$ $\xrightarrow{h\nu}$

and enantiomer

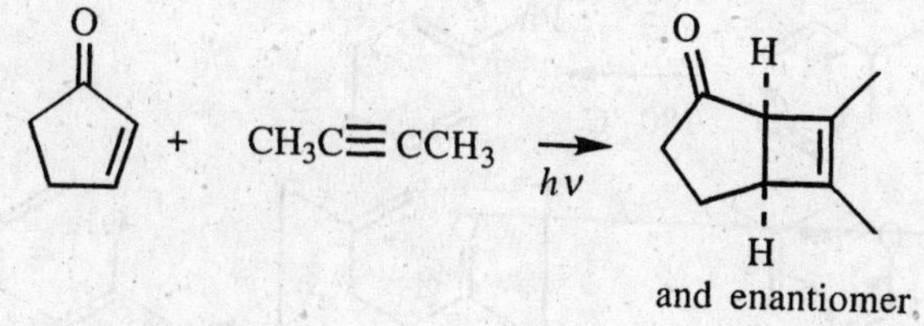

(b) + $CH_2=CHCH_2C\equiv N$ $\xrightarrow{\Delta}$

and enantiomer

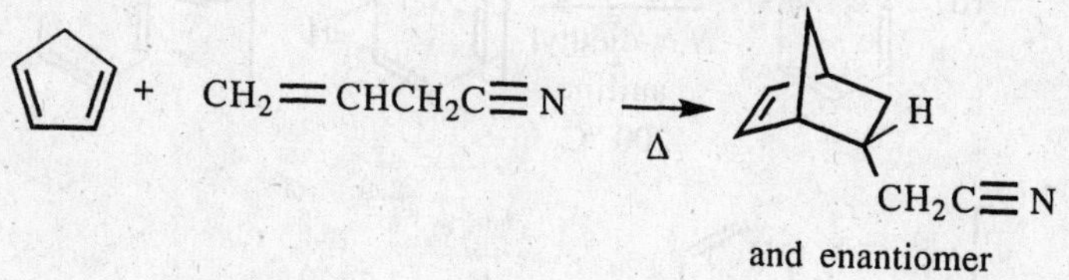

(c) + $HC\equiv CCCH_3$ (with O) $\xrightarrow{130\ °C}$

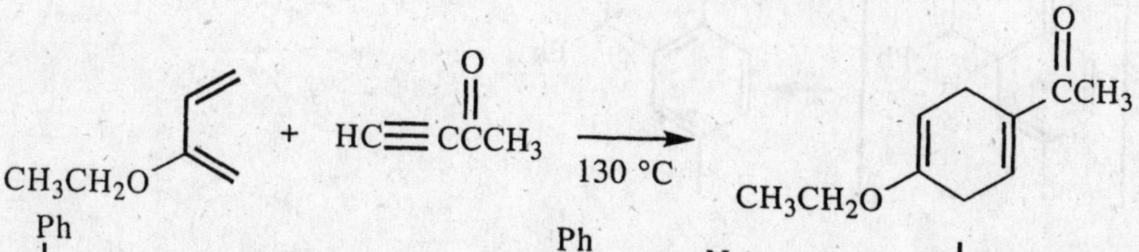

(d) + $CH_2=CHC\equiv N$ $\xrightarrow{\Delta}$

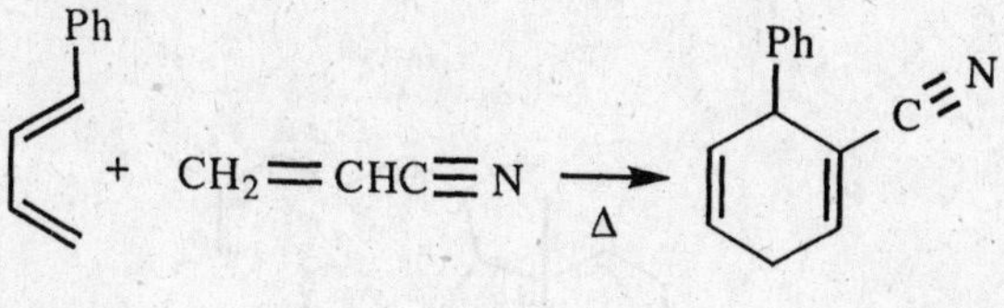

(e) + $\xrightarrow{100\ °C}$

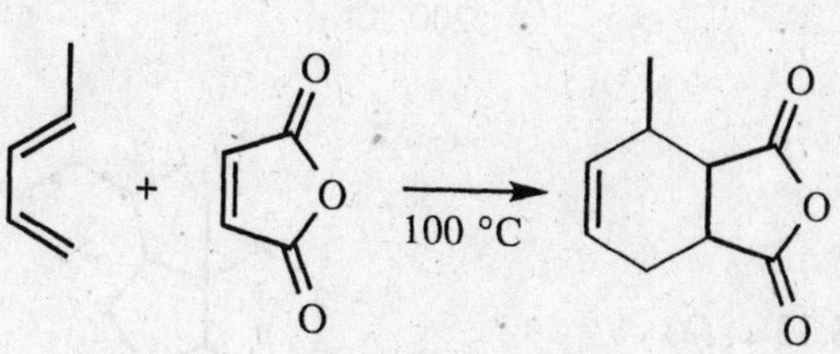

(f) + $CH_2=CHCCH_3$ (with O) $\xrightarrow[\Delta]{\text{benzene}}$

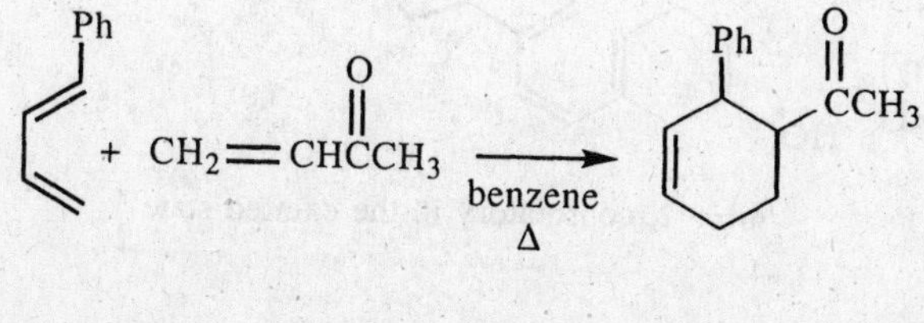

25.30

(g)

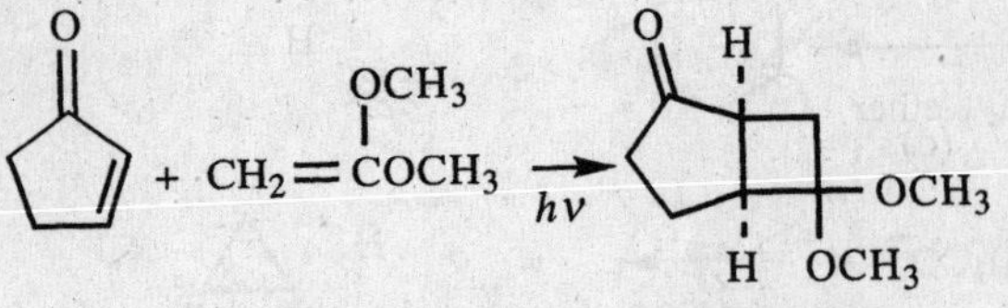

and enantiomer

(h)

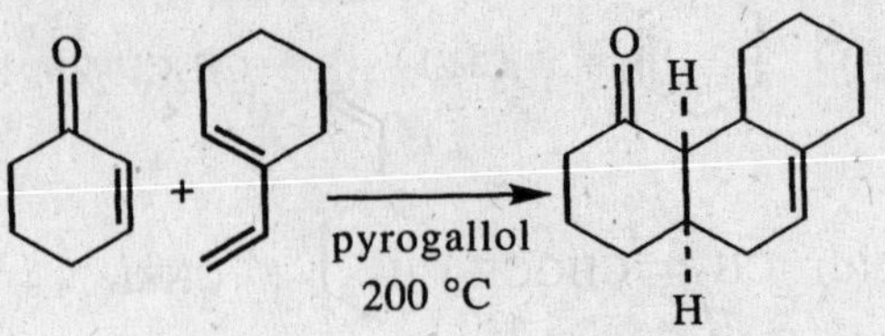

and enantiomer

25.31 (a)

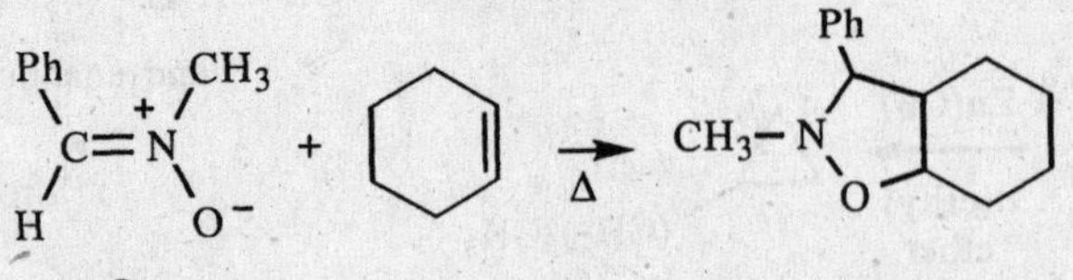

(b)

(c)

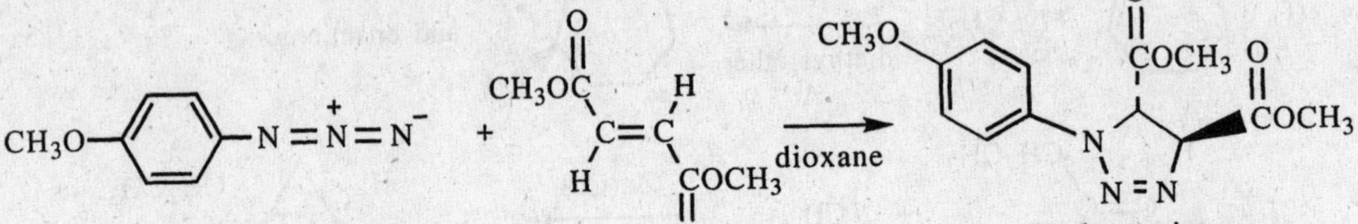

and enantiomer

25.32

(a)

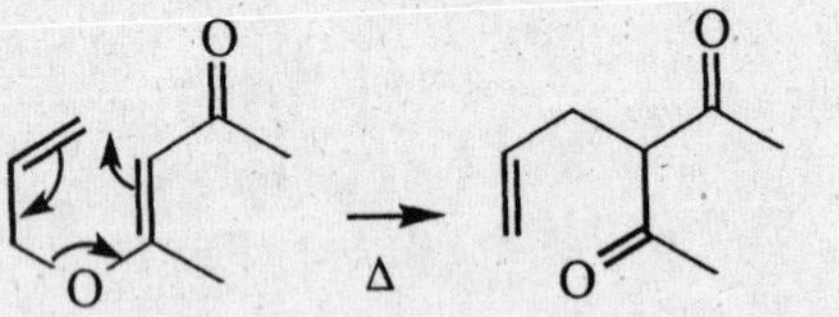

(b)

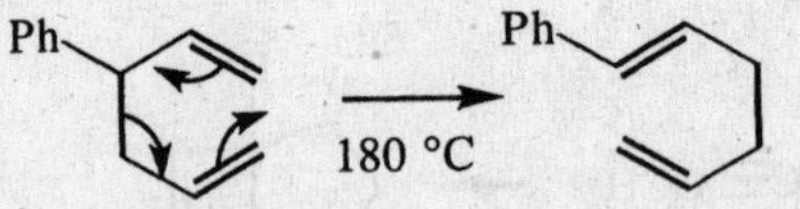

(c)

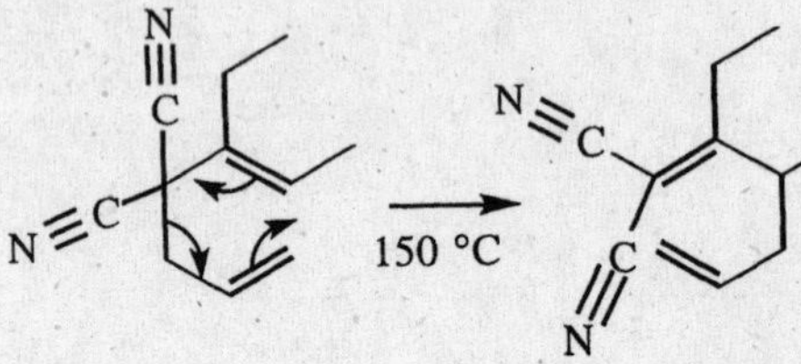

(d)

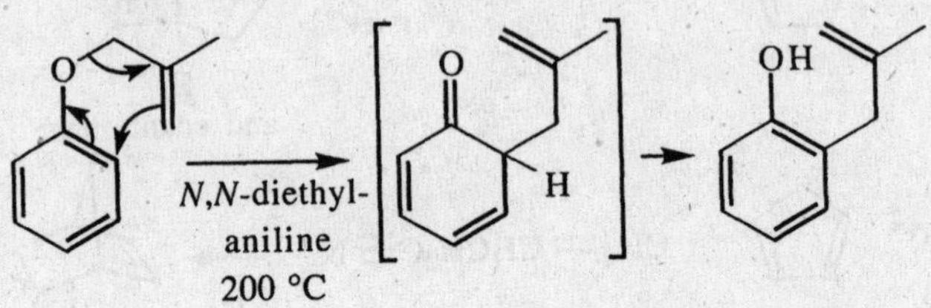

(e)

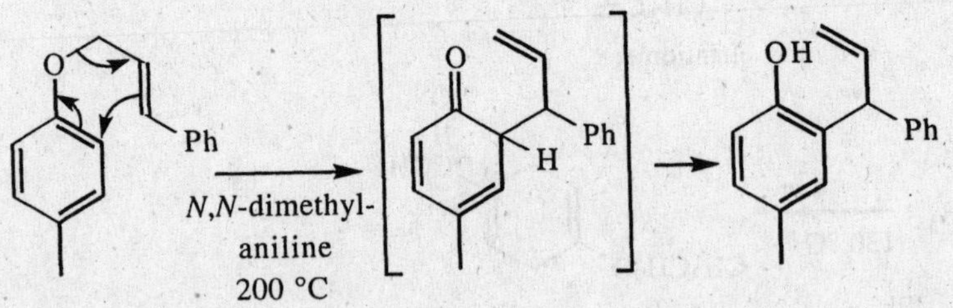

25.33 (a)

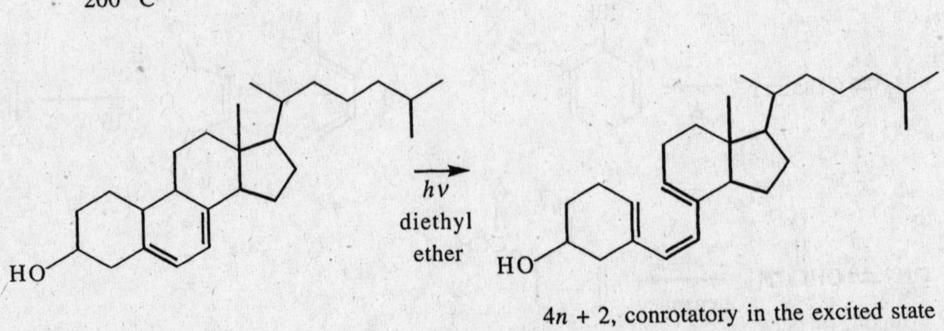

$4n + 2$, conrotatory in the excited state

25.33   (b)

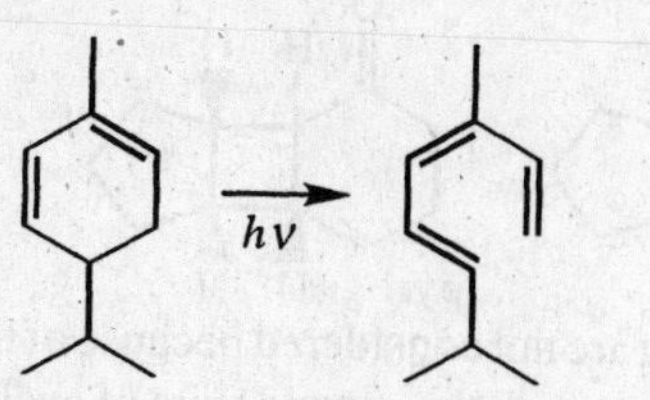

4n + 2,
conrotatory in
the excited state

(c)

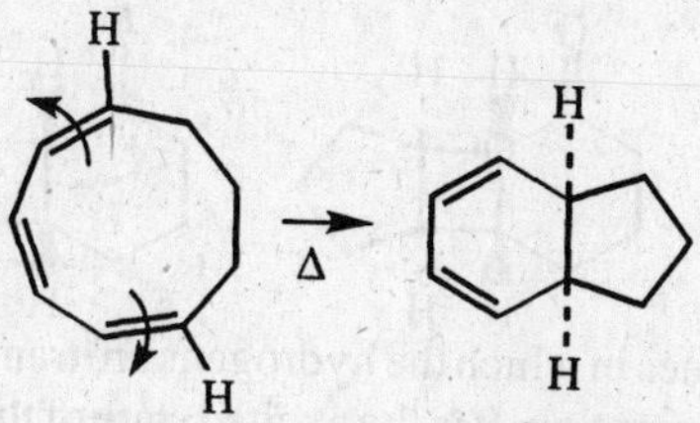

4n + 2,
disrotatory in
the ground state

(d)

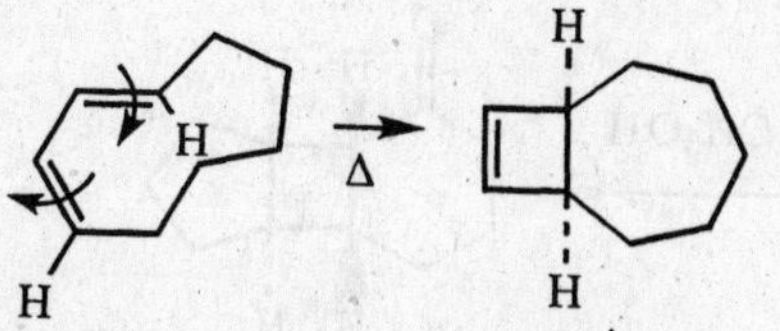

4n,
conrotatory in
the ground state

(e)

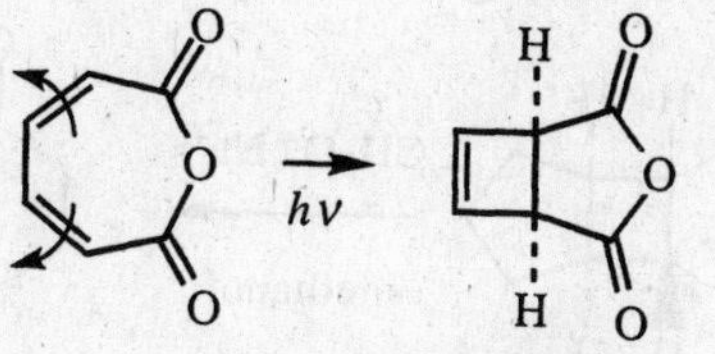

4n,
disrotatory in
the ground state

(f)

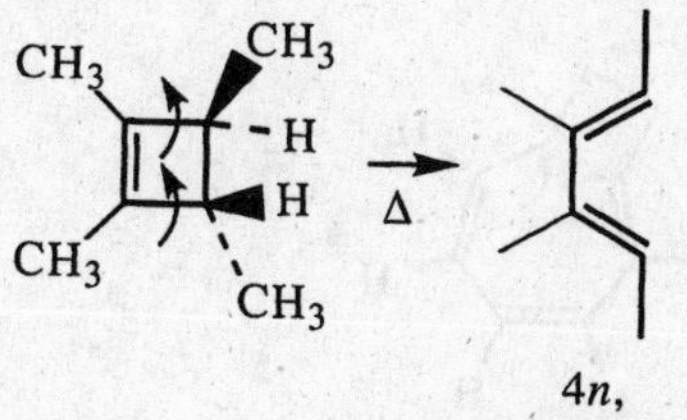

4n,
conrotatory in
the ground state

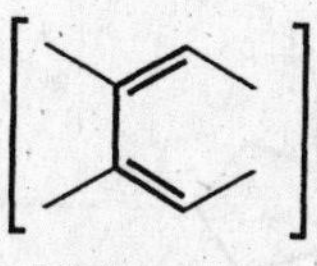

too hindered to form

25.34   (a)

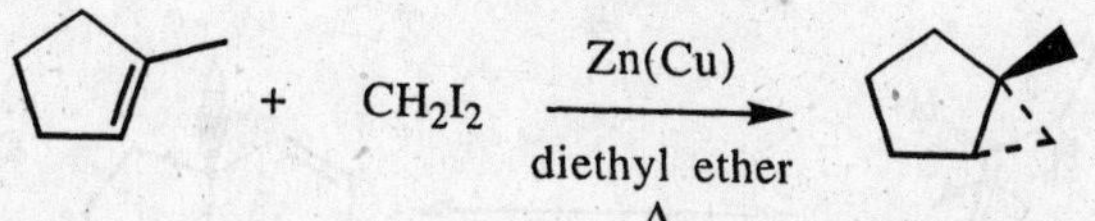

and enantiomer

(b)

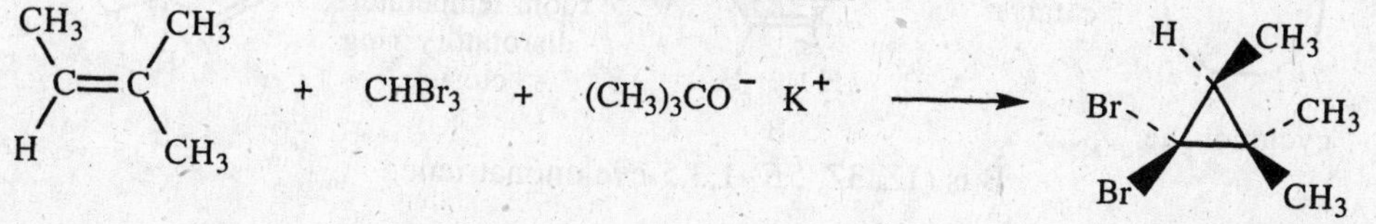

$CH_2=C$ ... $+$ CHBr$_3$ $+$ (CH$_3$)$_3$CO$^-$ K$^+$ $\longrightarrow$

and enantiomer

(c) $CH_2=CHCCH_3$ $+$ CH_2I_2

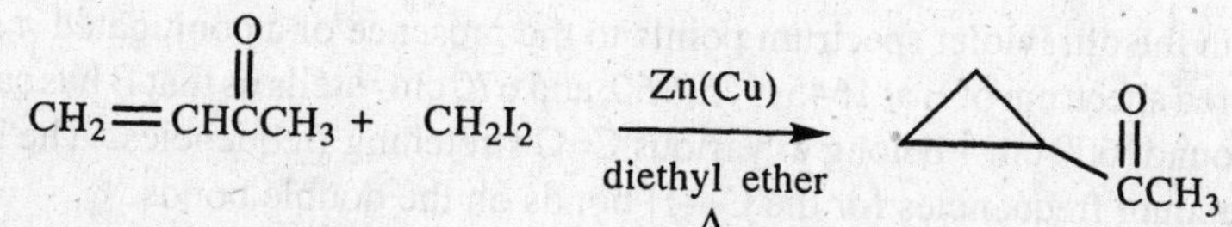

(d)

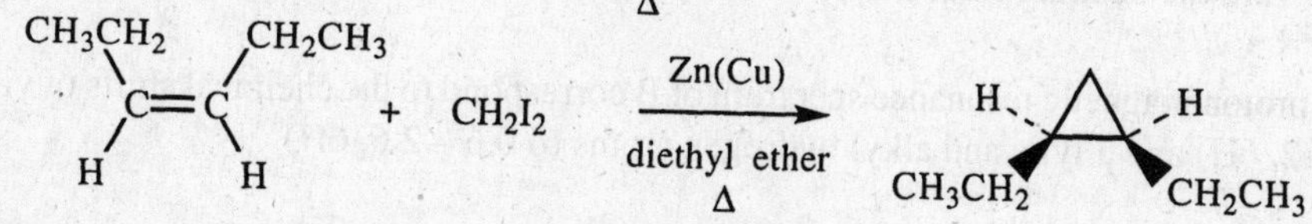

25.35

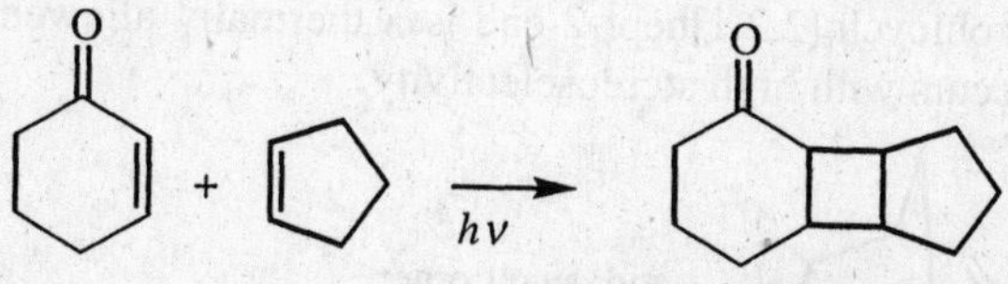

the only possible isomerism
is stereoisomerism

The probable structures are shown on the next page. Each structure also has an enantiomer.

25.35 (cont)

The ones in which the hydrogens are trans on the cyclopentane ring are not considered because of the mechanism of the reaction. It is the excited state of the cyclohexenone that reacts with the ground state of cyclopentene. The double bond in cyclohexenone may become twisted in the excited state, giving rise to a trans fusion of the six- and four-membered rings. The trans-fused compounds will isomerize in base through enolization and reprotonation to give the more stable cis-fused ring system. For example,

This is the compound that is formed in the largest amount in the initial photochemical reaction.

25.36

$4n + 2$ system
$n = 1$
A

hv
methanol
conrotatory
ring opening

B

cyclononane

H_2
catalyst

B

room temperature;
disrotatory ring
closure

C

B is (1Z, 3Z, 5E)-1,3,5-cyclononatriene.

A λ_{max} of 290 nm (ε 2050) in the ultraviolet spectrum points to the presence of a conjugated π system. The presence of bands in the infrared spectrum of B at 1645, 975, 960, and 670 cm^{-1} tells us that B has carbon-carbon double bonds. The bands around 1600 cm^{-1} belong to various C=C stretching frequencies. The bands at 975, 960 and 670 cm^{-1} are the bending frequencies for the C—H bonds on the double bonds.

The peaks in the proton magnetic resonance spectrum of B correspond to the chemical shifts of vinyl hydrogen atoms (δ 4.7 – 6.2, 6H) and allylic and alkyl hydrogen atoms (δ 0.6 – 2.6, 6H).

25.37 The formation of *trans*-5,6-dichloro-5,6-difluorobicyclo[2.2.1]hept-2-ene is a thermally allowed [$4n + 2$] cycloaddition (a Diels-Alder) reaction, which occurs with high stereoselectivity.

and enantiomer

25.37 (cont)

The concerted thermal cycloaddition reaction for a system with 4*n* electrons is forbidden. The formation of 6,7-dichloro-6,7-difluorobicyclo[3.2.0]hept-2-ene goes by a radical mechanism and gives mixtures of stereoisomers.

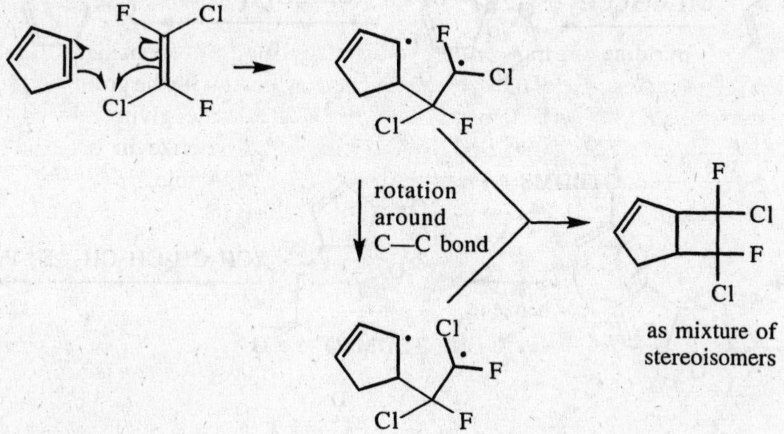
rotation
around
C—C bond
as mixture of
stereoisomers

25.38

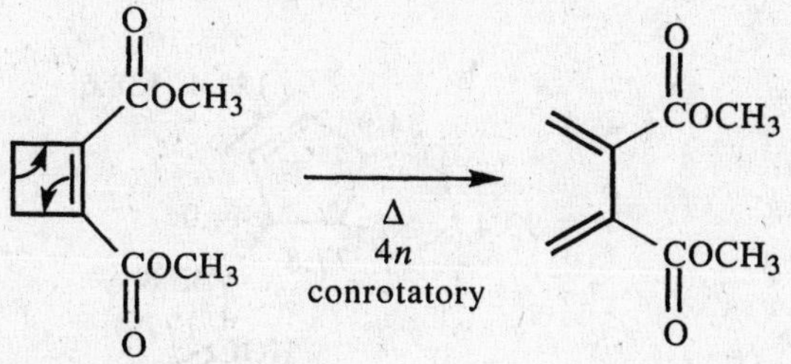
Δ
4n
conrotatory

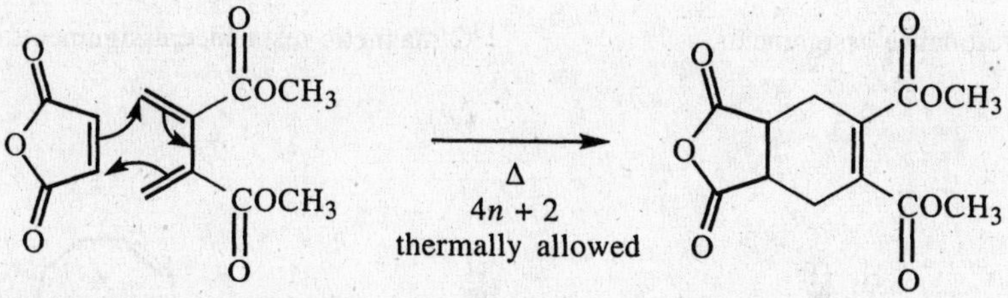
Δ
4n + 2
thermally allowed

Diels-Alder reaction

25.39

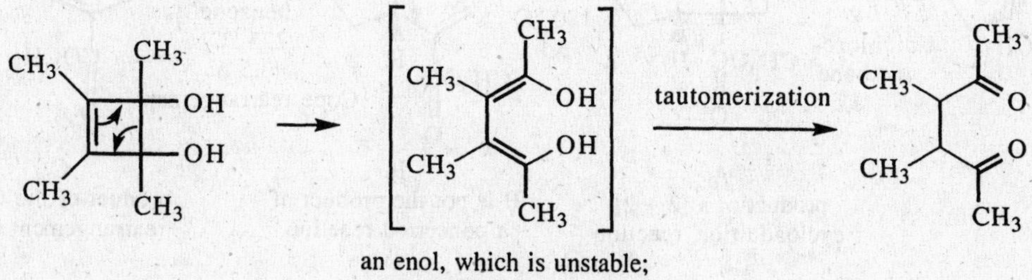
tautomerization
an enol, which is unstable;
no stereochemistry implied

25.40

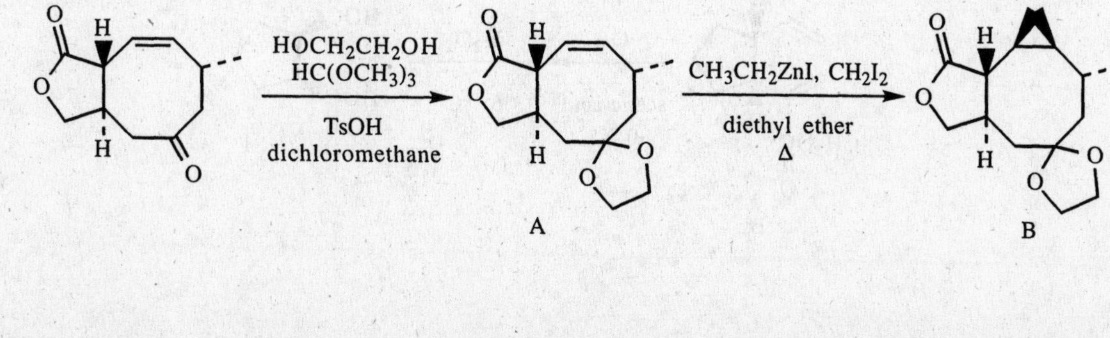
HOCH2CH2OH
HC(OCH3)3
TsOH
dichloromethane
CH3CH2ZnI, CH2I2
diethyl ether
Δ
A
B

25.41 (a)

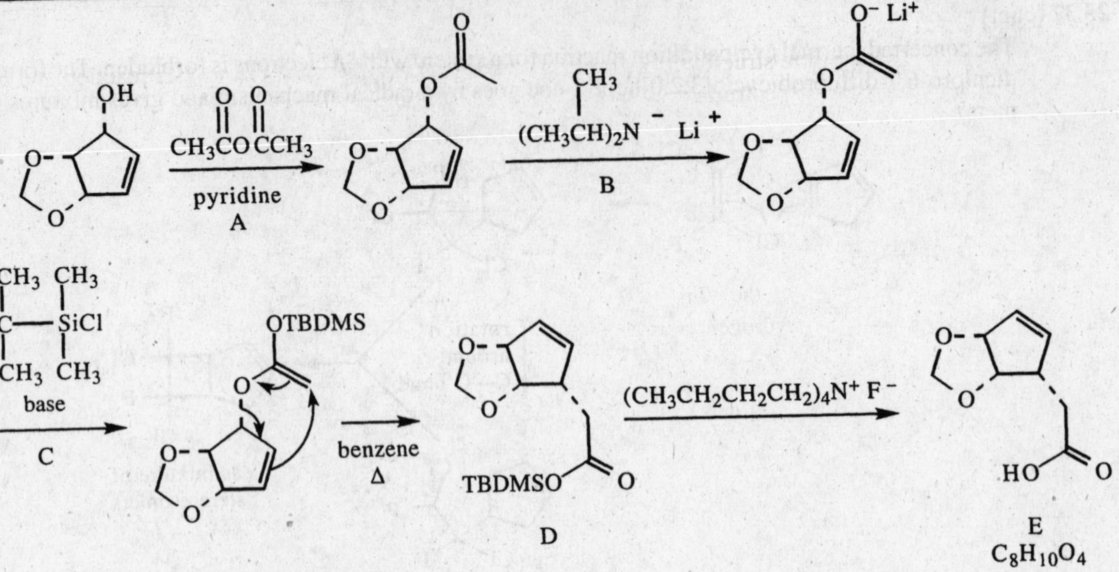

(b)

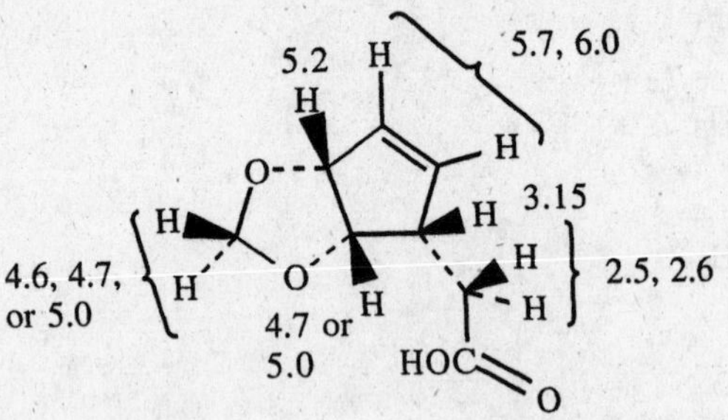

proton magnetic resonance assignments

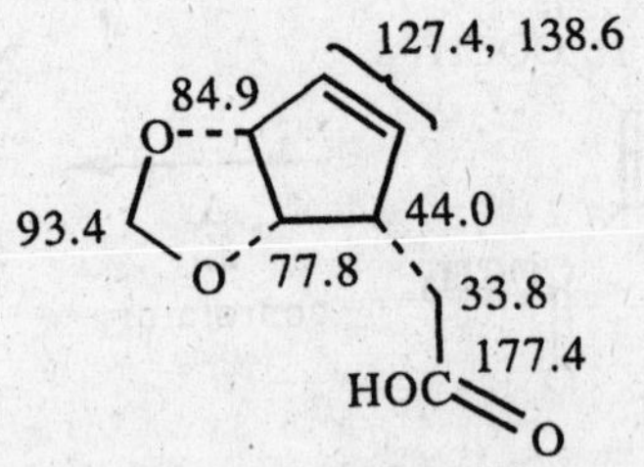

^{13}C magnetic resonance assignments

25.42

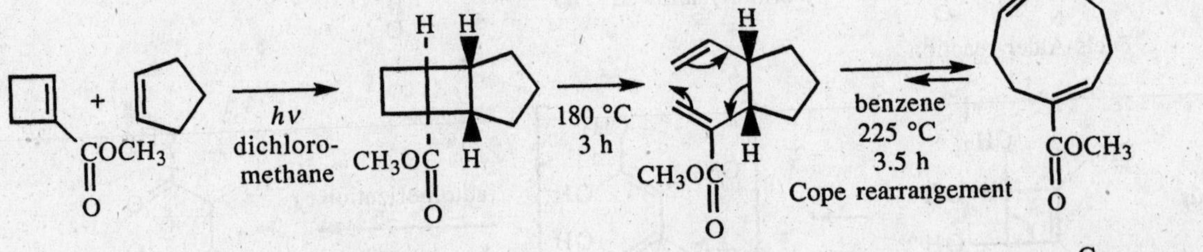

A
product of a [2 + 2]
cycloaddition reaction

B is not the product of
a concerted reaction

C
product of the Cope
rearrangement of B

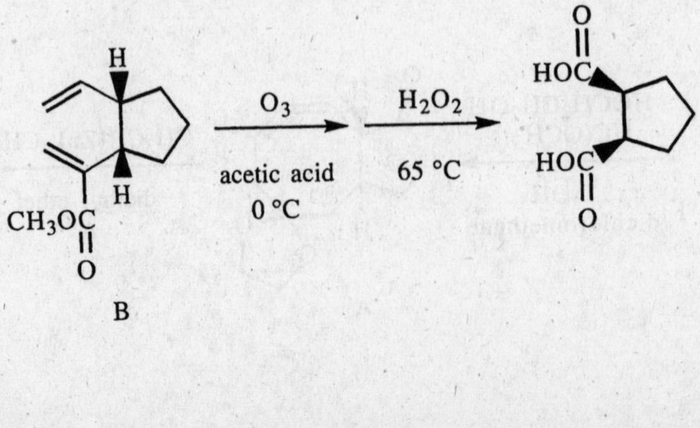

B

25.42 (cont)

The assignment of the structure of B is supported by the results of the ozonolysis reaction and by the proton magnetic resonance spectrum.

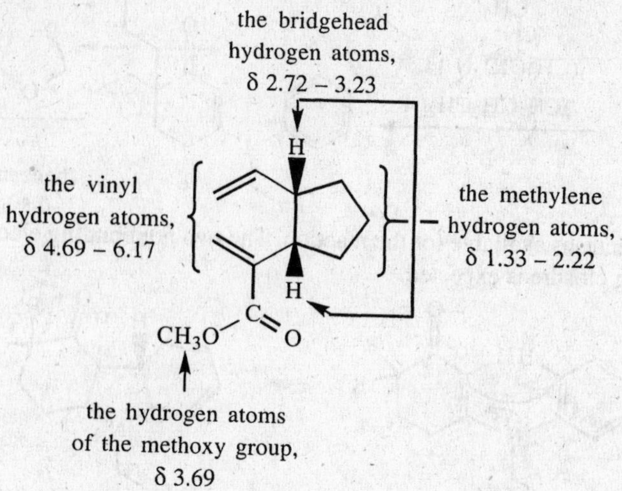
the bridgehead
hydrogen atoms,
δ 2.72 – 3.23
the vinyl
hydrogen atoms,
δ 4.69 – 6.17
the methylene
hydrogen atoms,
δ 1.33 – 2.22
the hydrogen atoms
of the methoxy group,
δ 3.69

25.43 (a)

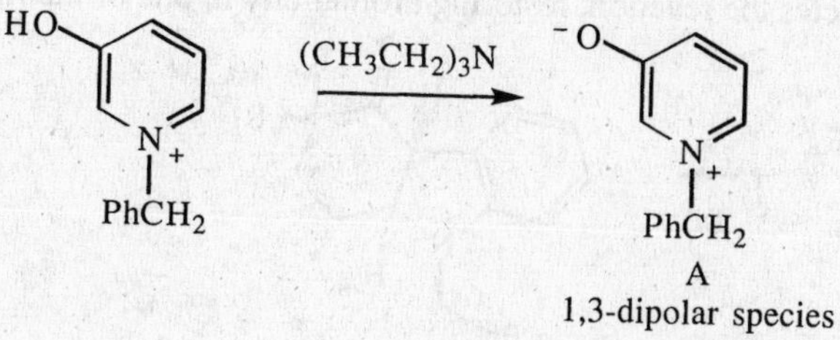
(CH3CH2)3N
PhCH2
A
1,3-dipolar species

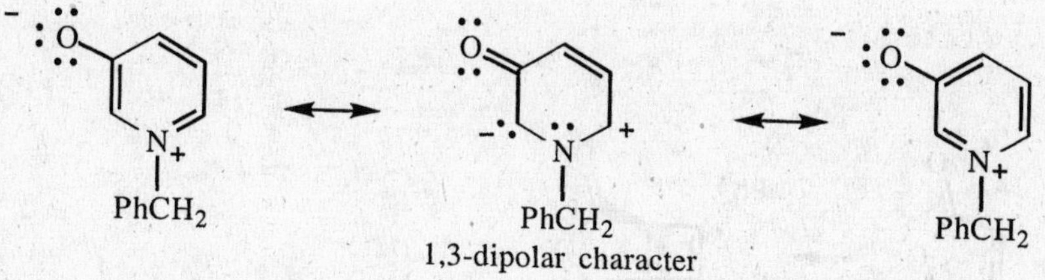
PhCH2
1,3-dipolar character

(b)

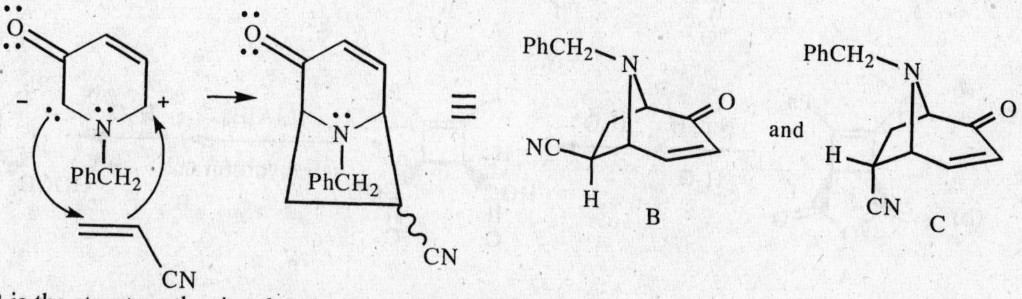
PhCH2
CN
B
and
C

B is the structure that is related to Bao Gong Teng A.

25.44 (a)

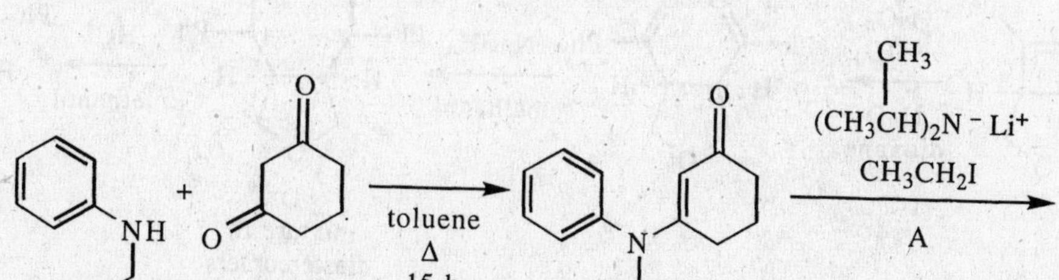
toluene
Δ
15 h
(CH3CH)2N⁻ Li+
CH3CH2I
A

25.44 (a) (cont)

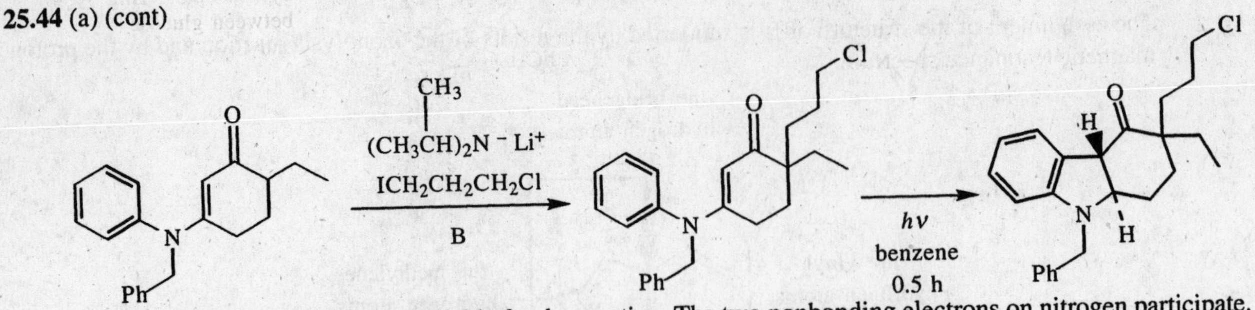

(b) There are six electrons available for the reaction. The two nonbonding electrons on nitrogen participate. Conrotatory ring closure is expected.

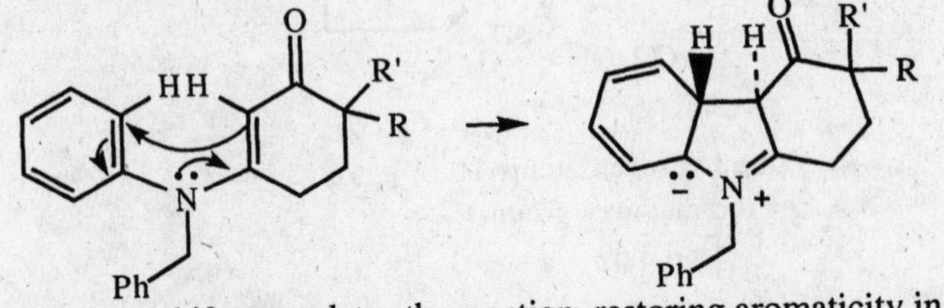

A pair of 1,2-hydrogen shifts completes the reaction, restoring aromaticity in one of the rings.

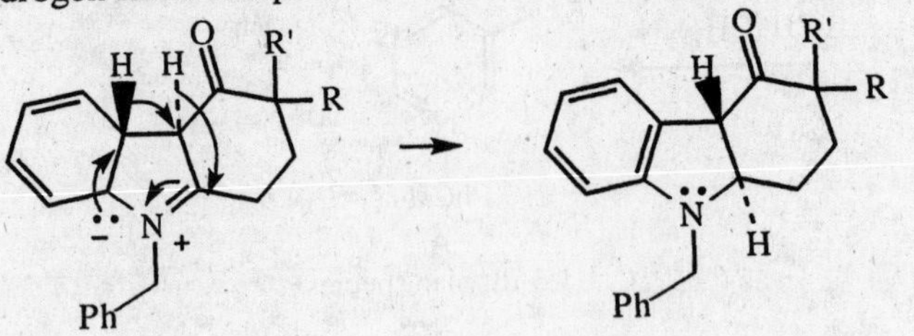

25.45   (a)   PhC≡CPh +

(b)

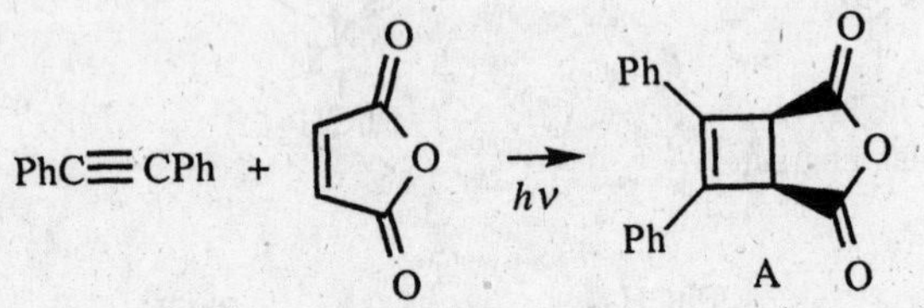

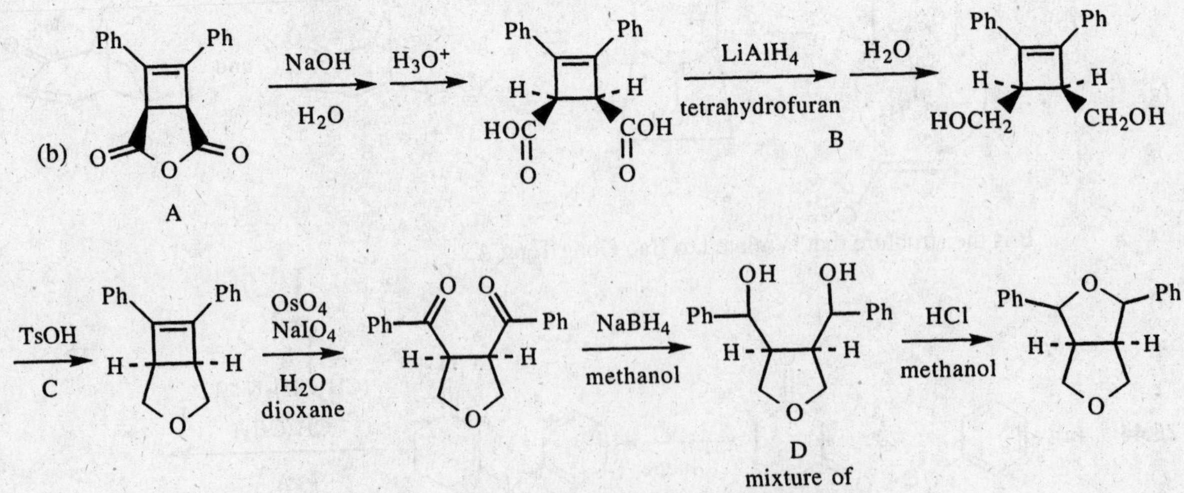

25.46

Supplemental Problems

S25.1   Complete the following equations. Be sure to show stereochemistry if it can be predicted.

S25.1

(l)

(m) 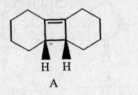 + CH_2=$CHOCH_2CH_2CH_2CH_3$ $\xrightarrow{\Delta}$

S25.2 When (1Z, 3E)-1,3-cyclooctadiene is heated to 80 °C, it converts to bicyclo[4.2.0]oct-7-ene (shown below). (1Z, 3Z)-1,3-Cyclooctadiene undergoes a photochemical ring closure to give the same compound, with the same stereochemistry, but it does not give a cyclobutene product when heated. How do you explain these observations?

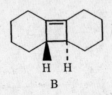

bicyclo[4.2.0]oct-7-ene

When bicyclo[4.2.0]oct-7-ene is heated above 300 °C, the cyclobutene ring opens and (1Z, 3Z)-1,3-cyclooctadiene is formed. Is this opening of the ring a concerted reaction?

S25.3 Of the two compounds shown below, one undergoes opening of the cyclobutene ring easily, whereas the other requires high temperatures. Predict which one reacts with ease, and explain why.

A   B